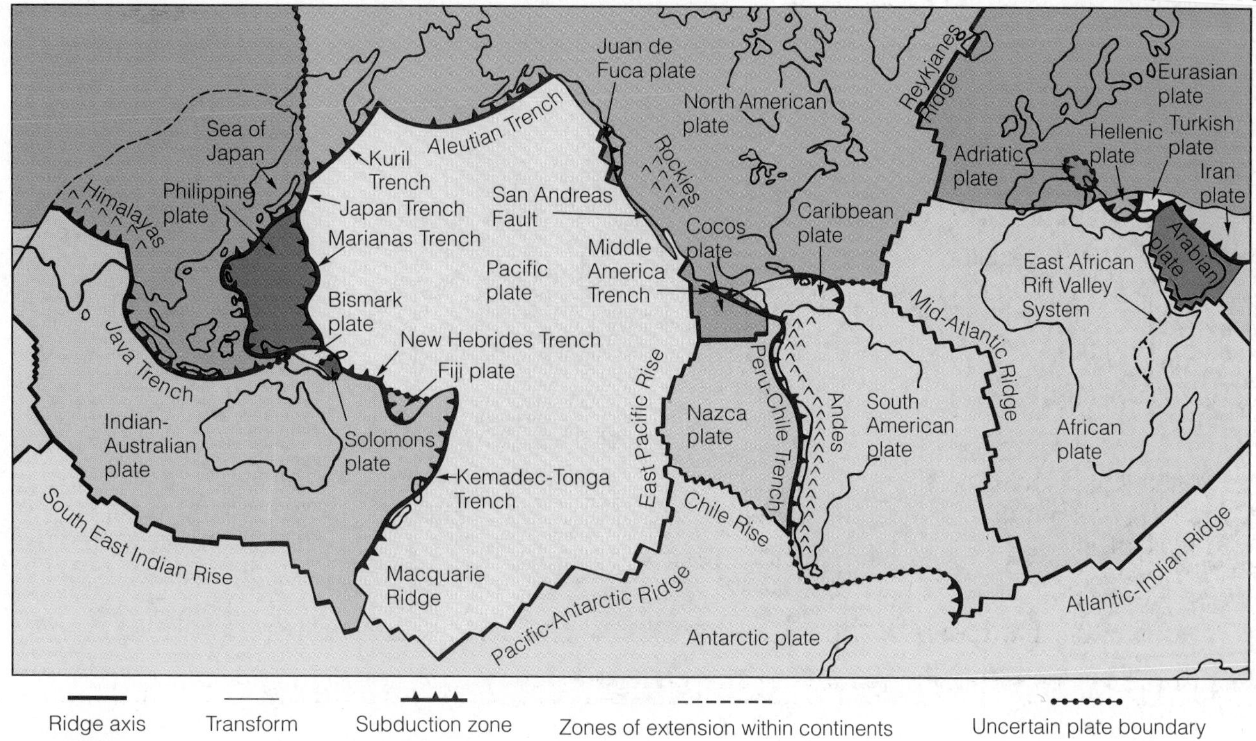

▶ The Earth's Plates (Figure 1-14)

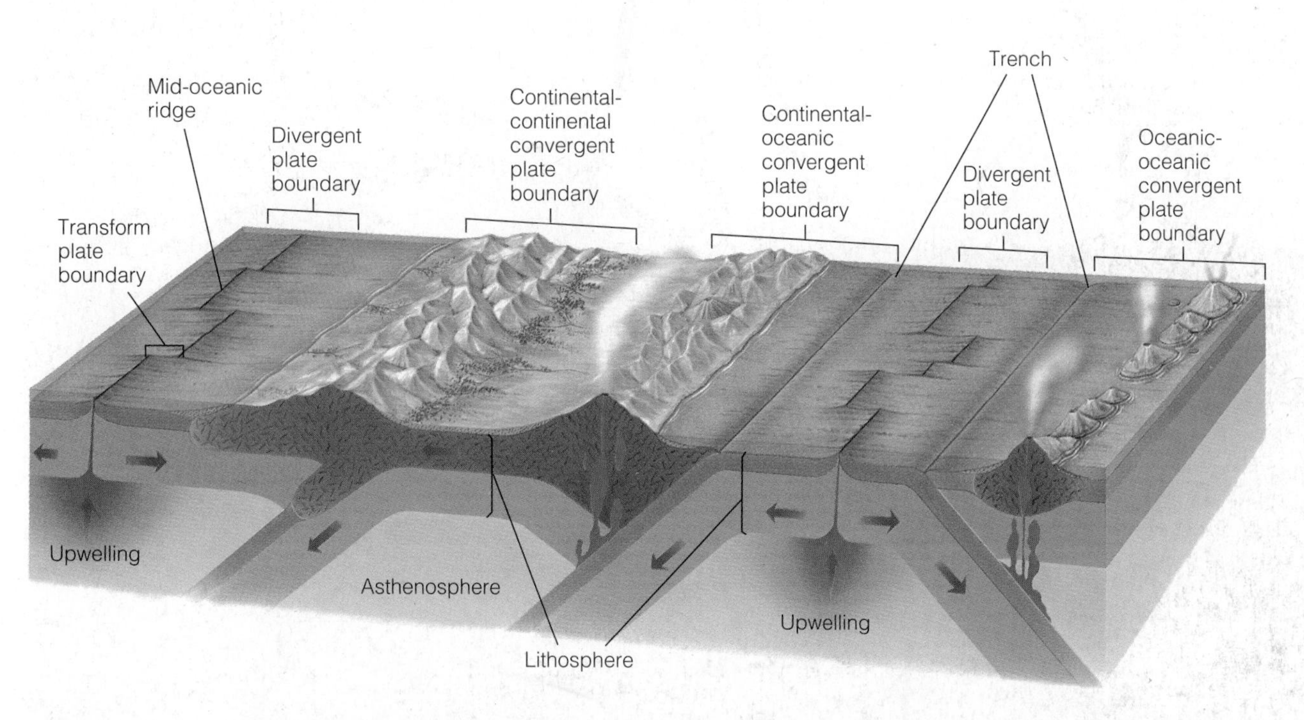

▶ Three Principal Types of Plate Boundaries (Figure 1-15)

S ECOND EDITION

PHYSICAL GEOLOGY

Exploring the Earth

James S. Monroe
Reed Wicander
Central Michigan University

WEST PUBLISHING COMPANY
Minneapolis/St. Paul New York Los Angeles San Francisco

PRODUCTION CREDITS

Copyediting and Indexing Patricia Lewis
Interior Design, Cover Design, and Page Layout Diane Beasley
Artwork Darwen and Vally Hennings, Carlyn Iverson, Precision Graphics, and Victor Royer. Individual credits follow index.
Composition Carlisle Communications, Ltd.
Cover Image The cover painting, *The Falls of Kaaterskill,* is by Thomas Cole (dated 1826), and represents the Hudson River School of Painting. Cole, the most noted of this group of artists, was known for his inspiring views of the natural world. A number of geologic phenomena are represented in this painting, as well as in the other paintings in this book. Courtesy of The Warner Collection, Gulf States Paper Corporation.

WEST'S COMMITMENT TO THE ENVIRONMENT

In 1906, West Publishing Company began recycling materials left over from the production of books. This began a tradition of efficient and responsible use of resources. Today, up to 95 percent of our legal books and 70 percent of our college and school texts are printed on recycled, acid-free stock. West also recycles nearly 22 million pounds of scrap paper annually-the equivalent of 181,717 trees. Since the 1960s, West has devised ways to capture and recycle waste inks, solvents, oils, and vapors created in the printing process. We also recycle plastics of all kinds, wood, glass, corrugated cardboard, and batteries, and have eliminated the use of Styrofoam book packaging. We at West are proud of the longevity and the scope of our commitment to the environment.

Production, Prepress, Printing and Binding by West Publishing Company.

610 Opperman Drive
P.O. Box 64526
St. Paul, MN 55164–0526

Printed in the United States of America

02 01 00 99 98 97 96 95 8 7 6 5 4 3 2 1

Library of Congress Cataloging-in-Publication Data

Monroe, J. S. (James S.)
Physical geology : exploring the earth / James S. Monroe, Reed Wicander. -- 2nd ed.
p. cm.
ISBN 0-314-04273-3
1. Physical geology. I. Wicander, Reed, 1946- . II. Title.
QE28.2.M655 1995 94-26808
550--dc20 CIP

British Library Cataloguing-in-Publication Data. A catalogue record for this book is available from the British Library.

About the Authors

James S. Monroe

James S. Monroe is a professor of geology at Central Michigan University where he teaches physical geology, historical geology, prehistoric life, and stratigraphy and sedimentology. He has co-authored several textbooks with Reed Wicander for West Publishing Company, and has interests in Cenozoic geology and geologic education.

Reed Wicander

Reed Wicander is a geology professor at Central Michigan University where he teaches physical geology, historical geology, prehistoric life, and invertebrate paleontology. He has co-authored several geology textbooks with James S. Monroe. His main research interests involve various aspects of Paleozoic palynology, specifically the study of acritarchs, on which he has published many papers. He is currently the President of the American Association of Stratigraphic Palynologists.

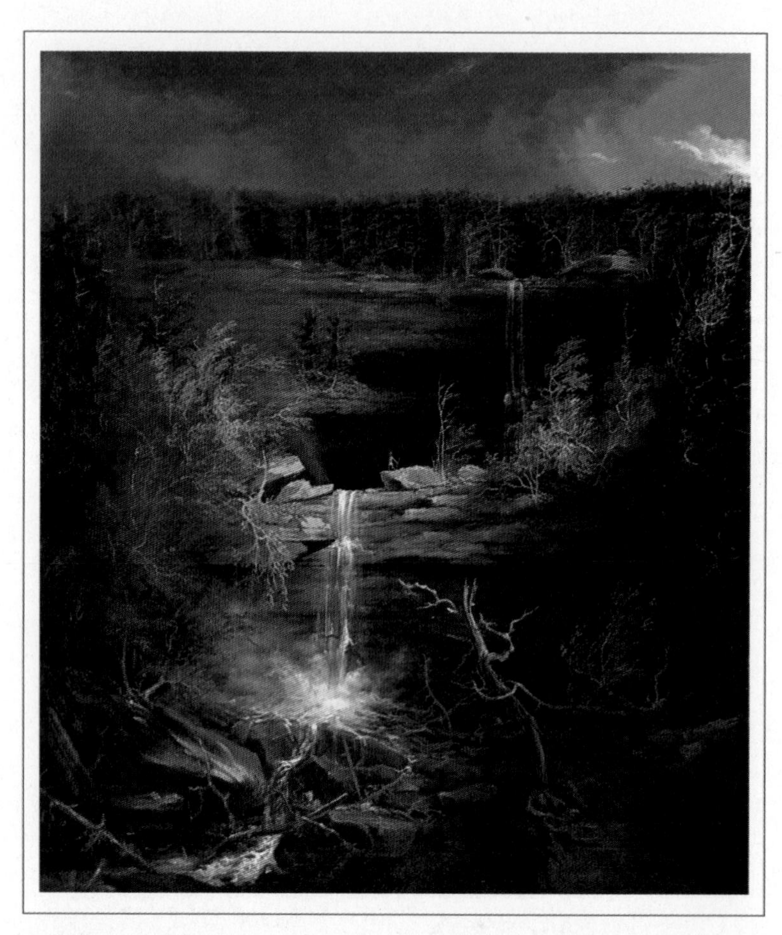

Falls of Kaaterskill, Thomas Cole, Oil on Canvas, 43 × 36 inches. The Warner Collection of Gulf States Paper Corporation, Tuscaloosa, Alabama, dated 1826.

BRIEF CONTENTS

CONTENTS

Chapter 1

UNDERSTANDING THE EARTH: AN INTRODUCTION TO PHYSICAL GEOLOGY

Chapter 2

MINERALS

Chapter 3
IGNEOUS ROCKS AND INTRUSIVE IGNEOUS ACTIVITY

Chapter 4
VOLCANISM

Chapter 5

WEATHERING, EROSION, AND SOIL

Chapter 6

SEDIMENT AND SEDIMENTARY ROCKS

 *Chapter 7*

METAMORPHISM AND METAMORPHIC ROCKS

 *Chapter 8*

GEOLOGIC TIME

Chapter 9

EARTHQUAKES

Chapter 10

THE INTERIOR OF THE EARTH

Chapter 11

THE SEA FLOOR

Chapter 12

PLATE TECTONICS: A UNIFYING THEORY

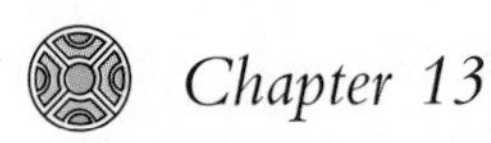

Chapter 13

DEFORMATION, MOUNTAIN BUILDING, AND THE EVOLUTION OF CONTINENTS

Chapter 14

MASS WASTING

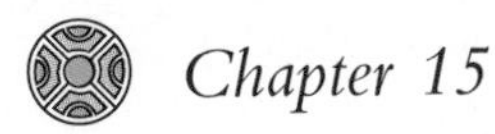 *Chapter 15*

RUNNING WATER

 *Chapter 16*

GROUNDWATER

 *Chapter 17*

GLACIERS AND GLACIATION

 *Chapter 18*

THE WORK OF WIND AND DESERTS

 *Chapter 19*

SHORELINES AND SHORELINE PROCESSES

 *Chapter 20*

A HISTORY OF THE UNIVERSE, SOLOR SYSTEM, AND PLANETS

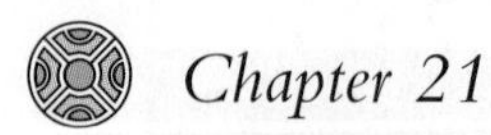

Chapter 21

GEOLOGY, THE ENVIRONMENT, AND NATURAL RESOURCES

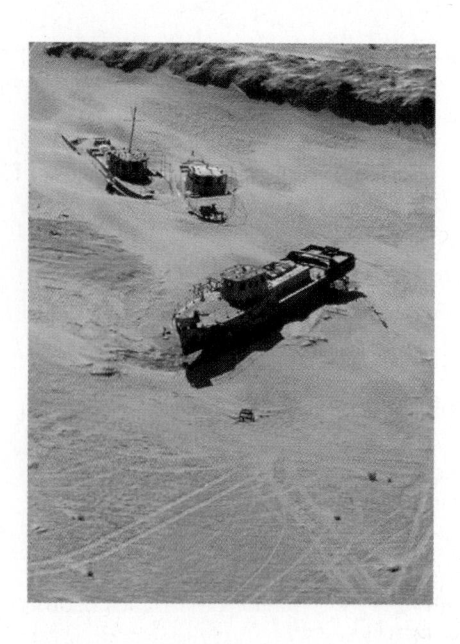

PREFACE

The Earth is a dynamic planet that has changed continuously during its 4.6 billion years of existence. The size, shape, and geographic distribution of the continents and ocean basins have changed through time, as have the atmosphere and biota. Over the past 20 years, bold new theories and discoveries concerning the Earth's origin and how it works have sparked a renewed interest in geology. We have become increasingly aware of how fragile our planet is and, more importantly, how interdependent all of its various systems are. We have learned that we cannot continually pollute our environment and that our natural resources are limited and, in most cases, nonrenewable. Furthermore, we are coming to realize how central geology is to our everyday lives. For these and other reasons, geology is one of the most important college or university courses a student can take.

Physical Geology: Exploring the Earth was designed for a one-semester introductory course in geology that serves both majors and nonmajors in geology and the Earth sciences. One of the problems with any introductory science course is that students are overwhelmed by the amount of material that must be learned. Furthermore, most of the material does not seem to be linked by any unifying theme and does not always appear to be relevant to their lives.

One of the goals of this book is to provide students with a basic understanding of geology and its processes and, more importantly, with an understanding of how geology relates to the human experience; that is, how geology affects not only individuals, but society in general. With this goal in mind, we introduce the major themes of the book in the first chapter to provide students with an overview of the subject and enable them to see how the various systems of the Earth are interrelated. We also discuss the economic and environmental aspects of geology throughout the book rather than treating these topics in separate chapters. In this way students can see, through relevant and interesting examples, how geology impacts our lives.

 NEW FEATURES IN THE SECOND EDITION

This second edition has undergone considerable rewriting and updating to produce a book that is easier to read, maintains a high level of currency, and has an improved organization. Drawing on the comments and suggestions of reviewers, we have incorporated many new features into this edition. For example, Chapters 3 and 4 on igneous rocks and volcanism, respectively, have been reorganized and updated so that a discussion of igneous processes and rocks precedes the discussion on volcanism. Chapter 9 on earthquakes has also been reorganized and updated to include information on the recent earthquakes in India and Northridge, California. Large parts of Chapter 13 (Deformation, Mountain Building, and the Evolution of Continents) have been rewritten to more clearly define and explain the nature and genesis of geologic structures, mountains, and the evolution of continents.

The chapter on planetology, which was the second chapter in the first edition, is now placed near the end of the book (Chapter 20). At the same time, a short section on the origin of the solar system and Earth was added to the introductory chapter (Chapter 1), so that it precedes the section on Earth materials. In many chapters more emphasis is placed on the relationship between geology and its impact on our lives. The chapters on surface processes (Chapters 14–19) are still largely descriptive, but the dynamic nature of these processes is more fully emphasized. Finally, a new capstone chapter (Chapter 21) has been added to bring together the various economic and environmental aspects of geology discussed throughout the book.

Updated information on resources and environmental issues has been added to many chapters. The perspectives on radon and asbestos contain new data, and the discussions of mineral and energy resources in several chapters have more current estimates of reserves and quantities extracted and used. In Chapter 5 (Weathering, Erosion, and Soil), current information on soil degradation and deforestation has been included.

Other important changes include a number of new Prologues, such as the Northridge, California earthquake (Chapter 9) and the Flood of '93 (Chapter 15). New Perspectives also appear, including one on the practical uses of geologic maps (Chapter 13) and one about the origin of Cape Cod, Massachusetts (Chapter 19).

Many photos in the first edition have been replaced, including most of the chapter-opening photographs, and a number of photographs within chapters have been enlarged to enhance their visual impact. Many illustrations have been modified or replaced so that various geologic processes and features are depicted more clearly. Roughly 75% of the photos and art are new or improved for this edition.

We feel these changes are a marked improvement over the first edition and make the book easier to read and comprehend, as well as a more effective teaching tool. Additionally, improvements have been made in the ancillary package that accompanies the book. In particular, a new CD-ROM product, linked to the textbook, provides a wonderful teaching/learning vehicle for you and your students.

 TEXT ORGANIZATION

Plate tectonic theory is the unifying theme of geology and this book. This theory has revolutionized geology because it provides a global perspective of the Earth and allows geologists to treat many seemingly unrelated geologic phenomena as part of a total planetary system. Because plate tectonic theory is so important, it is introduced in Chapter 1 and is discussed in most subsequent chapters in terms of the subject matter of that chapter.

We have organized *Physical Geology: Exploring the Earth* into several informal categories. Chapter 1 is an introduction to geology, its relevance to the human experience, plate tectonic theory, the rock cycle, geologic time and uniformitarianism, and the origin of the solar system and Earth. Chapters 2–7 examine the Earth's materials (minerals and igneous, sedimentary, and metamorphic rocks) and the geologic processes associated with them including the role of plate tectonics in their origin and distribution. Chapter 8 discusses geologic time, introduces several dating methods, and explains how geologists correlate rocks. Chapters 9–13 deal with the related topics of the Earth's interior, the sea floor, earthquakes, deformation and mountain building, and plate tectonics. Chapters 14–19 cover the Earth's surface processes. Chapter 20 discusses the origin of the universe and solar system and the Earth's place in the evolution of these larger systems. In Chapter 21, we bring everything together to show the importance of geology in environmental issues and the search for natural resources and to explain how the study of the Earth can provide a perspective for discussion of such topics as global warming.

We have found, that presenting the material in this order works well for most students. We know, however, that many instructors prefer an entirely different order of topics depending on the emphasis in their course. We have therefore written this book so that instructors can present the chapters in any order that suits the needs of their course.

 CHAPTER ORGANIZATION

All chapters have the same organizational format. Each chapter opens with a photograph relating to the chapter material, a detailed outline, and a prologue, which is designed to stimulate interest in the chapter material by discussing some aspect of the material.

The text is written in a clear informal style, making it easy for students to comprehend. Numerous color diagrams and photographs complement the text, providing a visual representation of the concepts and information presented. Each chapter contains at least one Perspective that presents a brief discussion of an interesting aspect of geology or geological research. Mineral and energy resources are discussed in the final sections of a number of chapters to provide interesting, relevant information in the context of the chapter topics.

The end-of-chapter materials begin with a concise review of important concepts and ideas in the Chapter Summary. The Important Terms, which are printed in boldface type in the chapter text, are listed at the end of each chapter for easy review, and a full glossary of important terms appears at the end of the text. The Review Questions are another important feature of this book; they include multiple-choice questions with answers as well as short answer and essay questions, and thought-provoking and quantitative questions under the Points to Ponder heading. Each chapter concludes with a list of Additional Readings, many of which are written at a level appropriate for beginning students interested in pursuing a particular topic.

 SPECIAL FEATURES

This book contains a number of special features that set it apart from other physical geology textbooks. Among them are a critical thinking and study skills section, the chapter Prologues, guest essays by people who chose geology or geologically related fields for their careers, the integration of economic and environmental geologic issues throughout the book, and a set of multiple-choice questions with answers for each chapter. A separate section titled "Points to Ponder," which contains thought-provoking and quantitative questions, has been added to each chapter.

Study Skills

Immediately following the Preface is a section devoted to developing critical thinking and study skills. This section contains hints to help students improve their study habits, prepare for exams, and generally get the most out of every course they take. While these tips can be helpful in any course, many of them are particularly relevant to geology. Whether you are just beginning college or about to graduate, take a few minutes to read over this section as these suggestions can help you in your studies and later in life.

Prologues

Many of the introductory Prologues focus on the human aspects of geology such as the eruption of Krakatau (Chapter 1), the eruption of Mount Pinatubo (Chapter 4), the Northridge earthquake (Chapter 9), and the Flood of '93 (Chapter 15).

Economic and Environmental Geology

The topics of environmental and economic geology are discussed throughout the text. Integrating economic and environmental geology with the chapter material helps students see the importance and relevance of geology to their lives. In addition, many chapters close with a section on resources, further emphasizing the importance of geology in today's world.

Figures

Many of the illustrations depicting geologic processes or events are block diagrams rather than cross sections so that students can more easily visualize the salient features of these processes and events. Some illustrations have also been updated and revised to make them easier to understand. Great care has been taken to ensure that the art and captions provide an attractive, informative, and accurate illustration program.

Figure and Table Reference System

A color cue (⯈) will be found in the text next to the first reference for each figure, and a (◉) appears next to the first reference for each table. This system is designed to help students quickly return to their place in the text when they interrupt their reading to examine an illustration or table.

Perspectives

The chapter Perspectives focus on aspects of environmental, economic, or planetary geology such as asbestos and graphite (Chapter 7), radioactive waste disposal (Chapter 16), and wind activity on Mars (Chapter 18). The topics for the Perspectives were chosen to provide students with an overview of the many fascinating aspects of geology. The perspectives can be assigned as part of the chapter reading, used as the basis for lecture or discussion topics, or even used as the starting point for student papers.

Guest Essays

Eighteen guest essays, nine of which are new, are interspersed throughout the book. These essays focus on three themes—how and why the individuals became interested in geology as a potential career, their current areas of research, and the possible sociopolitical ramifications of their specific field. The essayists include Bruce Babbitt (U.S. Secretary of the Interior and a geology major as an undergraduate), Randolph H. Bromery (University of Massachusetts at Amherst and former president of the Geological Society of America), Susan M. Landon (a consulting geologist), Kate Hutton (Caltech geologist often interviewed on television), Robert Ballard (noted oceanographer), and Steve Stow (head of nuclear waste disposal at Oak Ridge National Laboratories).

Capstone Chapter

A capstone chapter brings together the various economic and environmental aspects of geology discussed throughout the book and shows how geology plays an important role in society today. In addition, such topics as acid rain and global warming are covered, and we discuss how geology applies to these problems.

Ancillary Materials

To assist you in teaching this course and supplying your students with the best in teaching aids, West Publishing Company has prepared a complete supplemental package available to all adopters:

- The comprehensive instructor's manual and test bank includes teaching ideas, learning objectives, lecture outlines in the form of a point-by-point summary, discussions of common student misconceptions, a list of media sources, Consider This lecture questions, enrichment topics, lists of acetates and slides that accompany the text, and a test bank, The test bank has been revised and contains approximately two thousand multiple-choice, true/false, fill-in-the-blank, matching, and short-answer questions.
- The entire test bank is provided on diskette along with WESTEST, a computerized testing package. Using WESTEST 3.0, it is possible to generate examinations using questions selected by the instructor or those that are randomly generated by the computer. The WESTEST 3.0 edit function makes it possible to modify these questions, add new questions, or delete existing questions. Additionally, West's Classroom Management Software allows student data to be recorded, stored, and used for various reports.
- The new CD-ROM disk, *In-TERRA-Active*, developed through West by Phil Brown (University of Wisconsin-Madison) and Jeremy Dunning (University of Indiana-Bloomington), provides instructors and students with meaningful new ways to enhance the textbook. Tied to the text, the illustrations, animations, photos, and video make this a teaching/learning tool of great value. Additionally, the interactive modules provide students with a new mode for mastering the material.
- Three slide sets are available to adopters. The first set includes approximately 150 of the most important and attractive illustrations and photographs from the second edition, and the second set contains over 450 slides illustrating important geologic features. A majority of these photographs are from North America, but the set also includes examples from around the world and the solar system. The third set of slides contain images from the *In-TERRA-Active* CD, plus twenty slides from the January, 1994 Northridge earthquake.
- A new set of 200 full-color transparency acetates provides clear and effective illustrations of important artwork and maps from the text.
- A new ancillary containing all acetates printed and bound with perforated, three-hole punched pages allows students to take notes as the acetates are shown in lecture. It also contains Study Skills for Science Students by Daniel Chiras, which is described below.

- *Current Issues in Geology: Selected Readings*, prepared by Michael L. McKinney and Robert L. Tolliver of the University of Tennessee—Knoxville, is a collection of 65 very current articles to supplement material that students will encounter in their coursework. The articles are collected from a number of general interest and science magazines. West can make this supplement available with the text as a set, or it can be purchased separately.
- Although the text has many built-in study mechanisms, a full study guide is also available to help students master the material.
- West's Geology Videodisc allows instructors to display photographic images, illustrations, video segments, and computer animations in lecture or lab settings. Developed specifically for *Physical Geology: Exploring the Earth*, the disk contains more than 1,500 still photographs of geologic features organized by region. These can be used to show students examples of the formations discussed that are local, from other regions of the country, or from around the world. The videodisc also includes all illustrations and diagrams from the text.
- West's Geology Videotape Library includes the entire Planet Earth film series as well as additional programs that discuss earthquakes, mineral resources, and environmental geology topics. For a complete, up-to-date listing of the titles available, please contact your local West representative.
- *The Changing Earth Update*; West's biannual geology newsletter, is provided to adopters twice each year to update the book with recent and relevant research news. This will ensure that your students have the most current information available.
- A copy of *Great Ideas for Teaching Geology* is available free to all adopters of the textbook. This 100-page book is a collection of lecture suggestions, demonstrations, analogies, and other ideas contributed by geology teachers from across the country. These ideas are intended to provide instructors with a variety of approaches to teaching some of the difficult concepts of geology.
- West Publishing also has available tutorial software for students' review and software for lecture presentations. For example, GeoTutor by Vicki Harder is a hypercard stack designed as a review program in the format of a question and answer game. Interactive Geoscience Tutorials take topics and figures in geology and illustrate them using full-color animation. The twelve modules include the Rock Cycle, Igneous Rocks, Sedimentary Rocks, Plate Tectonics, Weathering and Erosion, Earthquakes, Minerals, and Mass Wasting, among others. Each tutorial begins with an introduction file that illustrates a concept related to the title for the tutorial. Three software programs are available from Micro-Innovations, Inc. *Quake* helps students understand the distribution of earthquakes. *Groundwater* allows students to manipulate several hydrologic variables simultaneously, then rapidly solve the modified program, and display the results graphically, and *Coastal* is an instructional program to simulate the effects of wave action on beach shape. For lecture, a presentation software package, ASTOUND, is available to qualified adopters. Among other things, it contains images from the text in electronic form.
- *Perspectives in Canadian Geology*, prepared by I. Peter Martini and Ward Chesworth of the University of Guelph, is a collection of essays that expands upon topics in the text by highlighting the geological features of Canada.
- Lastly, West has available *Study Skills for Science Students* by Daniel Chiras. This supplement emphasizes critical thinking and developing a positive lifestyle and provides students simple ways to improve memory, learn more quickly, get the most out of lectures, prepare for tests, produce top-notch term papers, and improve critical-thinking skills. West can make this supplement available with the text as a set, or it can be purchased separately.

Acknowledgments

As the authors, we are, of course, responsible for the organization, style, and accuracy of the text, and any mistakes, omissions, or errors are our responsibility. The finished product is the culmination of several years of work during which we received numerous comments and advice from many geologists who reviewed parts of the text. We wish to express our sincere appreciation to the following reviewers whose contributions were invaluable:

Theodore G. Benitt
Nassau Community College

Bruce A. Blackerby
California State University-Fresno

Gerald F. Brem
California State University-Fullerton

Ronald E. Davenport
Louisiana Tech University

Isabella M. Drew
Ramapo College of New Jersey

Jeremy Dunning
Indiana University

Norman K. Grant
Miami University

Norris W. Jones
University of Wisconsin-Oshkosh

Patricia M. Kenyon
University of Alabama

Peter L. Kresan
University of Arizona

Dave B. Loope
University of Nebraska-Lincoln

David N. Lumsden
Memphis State University

Steven K. Reid
Morehead State University

M.J. Richardson
Texas A&M University

Charles J. Ritter
University of Dayton

Gary D. Rosenberg
Indiana University-Purdue University-Indianapolis

J. Alexander Speer
North Carolina State University

James C. Walters
University of Northern Iowa

We would also like to provide thanks to the individuals who reviewed the first edition of this book in manuscript form. Their comments helped make the book a success for students and instructors alike.

Gary C. Allen
University of New Orleans

R. Scott Babcock
Western Washington University

Kennard Bork
Denison University

Thomas W. Broadhead
University of Tennessee at Knoxville

Anna Buising
California State University at Hayward

F. Howard Campbell III
James Madison University

Larry E. Davis
Washington State University

Noel Eberz
California State University at San Jose

Allan A. Ekdale
University of Utah

Stewart S. Farrar
Eastern Kentucky University

Richard H. Fluegeman, Jr.
Ball State University

William J. Fritz
Georgia State University

Kazuya Fujita
Michigan State University

Norman Gray
University of Connecticut

Jack Green
California State University at Long Beach

David R. Hickey
Lansing Community College

R. W. Hodder
University of Western Ontario

Cornelis Klein
University of New Mexico

Lawrence W. Knight
William Rainey Harper College

Martin B. Lagoe
University of Texas at Austin

Richard H. Lefevre
Grand Valley State University

I.P. Martini
University of Guelph, Ontario

Michael McKinney
University of Tennessee at Knoxville

Robert Merrill
California State University at Fresno

Carleton Moore
Arizona State University

Alan P. Morris
University of Texas at San Antonio

Harold Pelton
Seattle Central Community College

James F. Petersen
Southwest Texas State University

Katherine H. Price
DePauw University

William D. Romey
St. Lawrence University

Gary Rosenberg
Indiana University, Purdue University at Indianapolis

David B. Slavsky
Loyola University of Chicago

Edward F. Stoddard
North Carolina State University

Charles P. Thornton
Pennsylvania State University

Samuel B. Upchurch
University of South Florida

John R. Wagner
Clemson University

We also wish to thank Professor Emeritus Richard V. Dietrich of Central Michigan University for reading various drafts of the book, providing us with several photographs, and discussing various aspects of the text with us on numerous occasions. In addition, we are grateful to the other members of the Geology Department of Central Michigan University for reading various drafts and providing us with photographs. They are Eric L. Johnson, David J. Matty, Jane M. Matty, Wayne E. Moore, and Stephen D. Stahl. We also thank Martha Brian of the Geology Department, whose word processing skills and general efficiency were invaluable during the preparation of the manuscript, and Bruce M. C. Pape of the Geography Department for providing photographs. We also thank Pam Iacco, Dawn Anderson, and Kathleen Butzier for their assistance in preparing the revised manuscript. We are also grateful for the generosity of the various agencies and individuals from many countries who provided photographs.

Special thanks must go to Jerry Westby, college editorial manager for West Publishing Company, who made many valuable suggestions and patiently guided us through both editions. His continued encouragement provided constant inspiration and helped us produce the best possible book. Thanks to West Publishing developmental editors Dean DeChambeau, who has overseen the extensive ancillary package, and Betsy Friedman, who has coordinated the Guest Essays in this edition. We are equally indebted to our production editor, Matt Thurber, whose attention to detail and consistency is greatly appreciated, as are his unflagging efforts and diligence in securing many of the photographs and paintings used in the book. Matt was especially helpful in responding to our last-minute concerns as he guided the book through final production. We would also like to thank Patricia Lewis for her excellent copyediting and indexing skills. We appreciate her help in improving our manuscript. Because geology is such a visual science, we extend special thanks to Carlyn Iverson who rendered the reflective art and to the artists at Precision Graphics who were responsible for much of the rest of the art program. They did an excellent job, and we enjoyed working with them. We would also like to acknowledge our promotion manager, Stephanie Buss, for her help in the development of the promotional material for this edition, Maureen Rosener, media editor, who developed the excellent videodisc that accompanies this book, and Lucinda Gatch and Robyn Thorson for their hard work developing the CD-ROM project.

Our families were patient and encouraging when most of our spare time and energy were devoted to this book. We thank them for their support and understanding.

DEVELOPING CRITICAL THINKING AND STUDY SKILLS

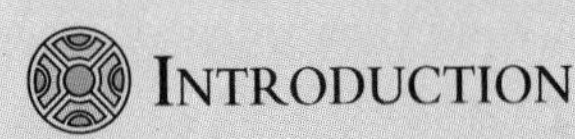

INTRODUCTION

College is a demanding and important time, a time when your values will be challenged, and you will try out new ideas and philosophies. You will make personal and career decisions that will affect your entire life. One of the most important lessons you can learn in college is how to balance your time among work, study, and recreation. If you develop good time management and study skills early in your college career, you will find that your college years will be successful and rewarding.

This section offers some suggestions to help you maximize your study time and develop critical thinking and study skills that will benefit you, not only in college, but throughout your life. While mastering the content of a course is obviously important, learning how to study and to think critically is, in many ways, far more important. Like most things in life, learning to think critically and study efficiently will initially require additional time and effort, but once mastered, these skills will save you time in the long run.

You may already be familiar with many of the suggestions and may find that others do not directly apply to you. Nevertheless, if you take the time to read this section and apply the appropriate suggestions to your own situation, we are confident that you will become a better and more efficient student, find your classes more rewarding, have more time for yourself, and get better grades. We have found that the better students are usually also the busiest. Because these students are busy with work or extracurricular activities, they have had to learn to study efficiently and manage their time effectively.

One of the keys to success in college is avoiding procrastination. While procrastination provides temporary satisfaction because you have avoided doing something you did not want to do, in the long run it leads to stress. While a small amount of stress can be beneficial, waiting until the last minute usually leads to mistakes and a subpar performance. By setting clear, specific goals and working toward them on a regular basis, you can greatly reduce the temptation to procrastinate. It is better to work efficiently for short periods of time than to put in long, unproductive hours on a task, which is usually what happens when you procrastinate.

Another key to success in college is staying physically fit. It is easy to fall into the habit of eating junk food and never exercising. To be mentally alert, you must be physically fit. Try to develop a program of regular exercise. You will find that you have more energy, feel better, and study more efficiently.

GENERAL STUDY SKILLS

Most courses, and geology in particular, build upon previous material, so it is extremely important to keep up with the coursework and set aside regular time for study in each of your courses. Try to follow these hints, and you will find you do better in school and have more time for yourself:

- Develop the habit of studying on a daily basis.
- Set aside a specific time each day to study. Some people are day people, and others are night people. Determine when you are most alert and use that time for study.
- Have an area dedicated for study. It should include a well-lighted space with a desk and the study materials you need, such as a dictionary, thesaurus, paper, pens and pencils, and a computer if you have one.
- Study for short periods and take frequent breaks, usually after an hour of study. Get up and move around and do something completely different. This will help you stay alert, and you'll return to your studies with renewed vigor.
- Try to review each subject every day or at least the day of the class. Develop the habit of reviewing lecture material from a class the same day.
- Become familiar with the vocabulary of the course. Look up any unfamiliar words in the glossary of your textbook or in a dictionary. Learning the language of the discipline will help you learn the material.

GETTING THE MOST FROM YOUR NOTES

If you are to get the most out of a course and do well on exams, you must learn to take good notes. This does not mean you should try to take down every word your professor says. Part of being a good note taker is knowing what is important and what you can safely leave out.

Early in the semester, try to determine whether the lecture will follow the textbook or be predominantly new material. If much of the material is covered in the textbook, your notes do not have to be as extensive or detailed as when the material is new. In any case, the following suggestions should make you a better note taker and enable you to derive the maximum amount of information from a lecture:

- Regardless of whether the lecture discusses the same material as the textbook or supplements the reading assignment, read or scan the chapter the lecture will cover before class. This way you will be somewhat familiar with the concepts and can listen critically to what is being said rather than trying to write down everything. Later a few key words or phrases will jog your memory as to what was said.
- Before each lecture, briefly review your notes from the previous lecture. Doing this will refresh your memory and provide a context for the new material.
- Develop your own style of note taking. Do not try to write down every word. These are notes you're taking, not a transcript. Learn to abbreviate and develop your own set of abbreviations and symbols for common words and phrases: for example, w/o (without), w (with), = (equals), $\wedge$ (above or increases), $\vee$ (below or decreases), < (less than), > (greater than), & (and), u (you).
- Geology lends itself to many abbreviations that can increase your note-taking capability: for example, pt (plate tectonics), ig (igneous), meta (metamorphic), sed (sedimentary), rx (rock or rocks), ss (sandstone), my (million years), and gts (geologic time scale).
- Rewrite your notes soon after the lecture. Rewriting your notes helps reinforce what you heard and gives you an opportunity to determine whether you understand the material.
- By learning the vocabulary of the discipline before the lecture, you can cut down on the amount you have to write—you won't have to write down a definition if you already know the word.
- Learn the mannerisms of the professor. If he or she says something is important or repeats a point, be sure to write it down and highlight it in some way. Students have told me (RW) that when I stated something twice during a lecture, they knew it was important and probably would appear on a test. (They were usually right!)
- Check any unclear points in your notes with a classmate or look them up in your textbook. Pay particular attention to the professor's examples. These usually elucidate and clarify an important point and are easier to remember than an abstract concept.
- Go to class regularly, and sit near the front of the class if possible. It is easier to hear and see what is written on the board or projected onto the screen, and there are fewer distractions.
- If the professor allows it, tape record the lecture, but don't use the recording as a substitute for notes. Listen carefully to the lecture and write down the important points; then fill in any gaps when you replay the tape.
- If your school allows it, and they are available, buy class lecture notes. These are usually taken by a graduate student who is familiar with the material; typically they are quite comprehensive. Again use these notes to supplement your own.
- Ask questions. If you don't understand something, ask the professor. Many students are reluctant to do this, especially in a large lecture hall, but if you don't understand a point, other people are probably confused as well. If you can't ask questions during a lecture, talk to the professor after the lecture or during office hours.

 GETTING THE MOST OUT OF WHAT YOU READ

The old adage that "you get out of something what you put into it" is very true when it comes to reading textbooks. By carefully reading your text and following these suggestions, you can greatly increase your understanding of the subject:

- Look over the chapter outline to see what the material is about and how it flows from topic to topic. If you have time, skim through the chapter before you start to read in depth.
- Pay particular attention to the tables, charts, and figures. They contain a wealth of information in abbreviated form and illustrate important concepts and ideas. Geology, in particular, is a visual science, and the figures and photographs will help you visualize what is being discussed in the text and provide actual examples of features such as faults or unconformities.
- As you read your textbook, highlight or underline key concepts or sentences, but make sure you don't highlight everything. Make notes in the margins. If you don't understand a term or concept, look it up in the glossary.
- Read the chapter summary carefully. Be sure you understand all of the key terms, especially those in boldface or italic type. Because geology builds on previous material, it is imperative that you understand the terminology.
- Go over the end-of-chapter questions. Write out your answers as if you were taking a test. Only when you see your answer in writing will you know if you really understood the material.

 DEVELOPING CRITICAL THINKING SKILLS

Few things in life are black and white, and it is important to be able to examine an issue from all sides and come to a logical conclusion. One of the most important things you will learn in college is to think critically and not accept everything you read and hear at face value. Thinking critically is particularly important in learning new material and relating it to what you already know. Although you can't know everything, you can learn to question effectively and arrive at conclusions consistent with the facts. Thus, these suggestions for critical thinking can help you in all your courses:

- Whenever you encounter new facts, ideas, or concepts, be sure you understand and can define all of the terms used in the discussion.
- Determine how the facts or information was derived. If the facts were derived from experiments, were the experiments well executed and free of bias? Can they be

repeated? The controversy over cold fusion is an excellent example. Two scientists claim to have produced cold fusion reactions using simple experimental laboratory apparatus, yet other scientists have as yet been unable to achieve the same reaction by repeating the experiments.

- Do not accept any statement at face value. What is the source of the information? How reliable is the source?
- Consider whether the conclusions follow from the facts. If the facts do not appear to support the conclusions, ask questions and try to determine why they don't. Is the argument logical or is it somehow flawed?
- Be open to new ideas. After all, the underlying principles of plate tectonic theory were known early in this century, yet were not accepted until the 1970s in spite of overwhelming evidence.
- Look at the big picture to determine how various elements are related. For example, how will constructing a dam across a river that flows to the sea affect the stream's profile? What will be the consequences to the beaches that will be deprived of sediment from the river? One of the most important lessons you can learn from your geology course is how interrelated the various systems of the Earth are. When you alter one feature, you affect numerous other features as well.

Improving Your Memory

Why do you remember some things and not others? The reason is that the brain stores information in different ways and forms, making it easy to remember some things and difficult to remember others. Because college requires that you learn a vast amount of information, any suggestions that can help you retain more material will help you in your studies:

- Pay attention to what you read or hear. Focus on the task at hand, and avoid daydreaming. Repetition of any sort will help you remember material. Review the previous lecture before going to class, or look over the last chapter before beginning the next. Ask yourself questions as you read.
- Use mnemonic devices to help you learn unfamiliar material. For example, the order of the Paleozoic periods (Cambrian, Ordovician, Silurian, Devonian, Mississippian, Pennsylvanian, and Permian) of the geologic time scale can be remembered by the phrase, *C*ampbell's *O*nion *S*oup *D*oes *M*ake *P*eter *P*ale, or the order of the Cenozoic epochs (Paleocene, Eocene, Oligocene, Miocene, Pliocene, and Pleistocene) can be remembered by the phrase, *P*ut *E*ggs *O*n *M*y *P*late *P*lease. Using rhymes can also be helpful.
- Look up the roots of important terms. If you understand where a word comes from, its meaning will be easier to remember. For example, *pyroclastic* comes from *pyro* meaning fire and *clastic* meaning broken pieces. Hence a pyroclastic rock is one formed by volcanism and composed of pieces of other rocks. We have provided the roots of many important terms throughout this text to help you remember their definitions.
- Outline the material you are studying. This will help you see how the various components are interrelated. Learning a body of related material is much easier than learning unconnected and discrete facts. Looking for relationships is particularly helpful in geology because so many things are interrelated. For example, plate tectonics explains how mountain building, volcanism, and earthquakes are all related. The rock cycle relates the three major groups of rocks to each other and to subsurface and surface processes (Chapter 1).
- Use deductive reasoning to tie concepts together. Remember that geology builds on what you learned previously. Use that material as your foundation and see how the new material relates to it.
- Draw a picture. If you can draw a picture and label its parts, you probably understand the material. Geology lends itself very well to this type of memory device because so much is visual. For example, instead of memorizing a long list of glacial terms, draw a picture of a glacier and label its parts and the type of topography it forms.
- Focus on what is important. You can't remember everything, so focus on the important points of the lecture or the chapter. Try to visualize the big picture, and use the facts to fill in the details.

Preparing For Exams

For most students, tests are the critical part of a course. To do well on an exam, you must be prepared. These suggestions will help you focus on preparing for examinations:

- The most important advice is to study regularly rather than try to cram everything into one massive study session. Get plenty of rest the night before an exam, and stay physically fit to avoid becoming susceptible to minor illnesses that sap your strength and lessen your ability to concentrate on the subject at hand.
- Set up a schedule so that you cover small parts of the material on a regular basis. Learning some concrete examples will help you understand and remember the material.
- Review the chapter summaries. Construct an outline to make sure you understand how everything fits together. Drawing diagrams will help you remember key points. Make up flash cards to help you remember terms and concepts.
- Form a study group, but make sure your group focuses on the task at hand, not on socializing. Quiz each other and compare notes to be sure you have covered all the material. We have found that students dramatically improved their grades after forming or joining a study group.
- Write out the answers to all of the Review Questions. Before doing so, however, become thoroughly familiar with the subject matter by reviewing your lecture notes

and reading the chapter. Otherwise, you will spend an inordinate amount of time looking up answers.

- If you have any questions, visit the professor or teaching assistant. If review sessions are offered, be sure to attend. If you are having problems with the material, ask for help as soon as you have difficulty. Don't wait until the end of the semester.
- If old exams are available, look at them to see what is emphasized and what type of questions are asked. Find out whether the exam will be all objective or all essay or a combination. If you have trouble with a particular type of question (such as multiple choice or essay), practice answering questions of that type—your study group or a classmate may be able to help.

Taking Exams

The most important thing to remember when taking an exam is not to panic. This, of course, is easier said than done. Almost everyone suffers from test anxiety to some degree. Usually, it passes as soon as the exam begins, but in some cases, it is so debilitating that an individual does not perform as well as he or she could. If you are one of those people, get help as soon as possible. Most colleges and universities have a program to help students overcome test anxiety or at least keep it in check. Don't be afraid to seek help if you suffer test anxiety. Your success in college depends to a large extent on how well you perform on exams, so by not seeking help, you are only hurting yourself. In addition, the following suggestions may be helpful:

- First of all, relax. Then look over the exam briefly to see its format and determine which questions are worth the most points. If it helps, quickly jot down any information you are afraid you might forget or particularly want to remember for a question.
- Answer the questions that you know the best first. Make sure, however, that you don't spend too much time on any one question or on one that is worth only a few points.
- If the exam is a combination of multiple choice and essay, answer the multiple-choice questions first. If you are not sure of an answer, go on to the next one. Sometimes the answer to one question can be found in another question. Furthermore, the multiple-choice questions may contain many of the facts needed to answer some of the essay questions.
- Read the question carefully and answer only what it asks. Save time by not repeating the question as your opening sentence to the answer. Get right to the point. Jot down a quick outline for longer essay questions to make sure you cover everything.
- If you don't understand a question, ask the examiner. Don't assume anything. After all, it is your grade that will suffer if you misinterpret the question.
- If you have time, review your exam to make sure you covered all the important points and answered all the questions.
- If you have followed our suggestions, by the time you finish the exam, you should feel confident that you did well and will have cause for celebration.

Concluding Comments

We hope that the suggestions we have offered will be of benefit to you not only in this course, but throughout your college career. Though it is difficult to break old habits and change a familiar routine, we are confident that following these suggestions will make you a better student. Furthermore, many of the suggestions will help you work more efficiently, not only in college, but also throughout your career. Learning is a lifelong process that does not end when you graduate. The critical thinking skills that you learn now will be invaluable throughout your life, both in your career and as an informed citizen.

PHYSICAL GEOLOGY

Exploring the Earth

Chapter 1

UNDERSTANDING THE EARTH: AN INTRODUCTION TO PHYSICAL GEOLOGY

OUTLINE

As a result of numerous eruptions like the one shown here, Anak Krakatau emerged above sea level in 1928 from the 275 meter deep caldera formed by the 1883 eruption of Krakatau.

PROLOGUE

On August 26, 1883, Krakatau, a small, uninhabited volcanic island in the Sunda Straits between Java and Sumatra, exploded (▶ Figure 1-1). In less than one day, 18 cubic kilometers (km^3) of rock were erupted in an ash cloud 80 kilometers (km) high. The explosion was heard as far away as Australia and Rodriguez Island, 4,653 km to the west in the Indian Ocean. Where the 450 meter (m) high peak of Danan once stood, the water was now 275 m deep, and only one-third of the 5 × 9 km island remained above sea level (▶ Figure 1-2). The explosions and the collapse of the magma (molten rock) chamber beneath the volcano produced giant sea waves, some as high as 40 m. On nearby islands, at least 36,000 people were killed and 165 coastal villages destroyed by the sea waves. An eyewitness in Anjer in Java described the sea waves as follows:

> At first sight, it seemed like a low range of hills rising out of the water, but I knew there was nothing of the kind in that part of the Sunda Strait. A second glance—and a very hurried one it was—convinced me that it was a lofty ridge of water many feet high. . . . There was no time to give any warning, and so I turned and ran for my life. My running days have long gone by, but you may be sure that I did my best. In a few minutes, I heard the water with a loud roar break upon the shore. Everything was engulfed. Another glance around showed the houses being swept away, and the trees thrown down on every side. Breathless and exhausted I still pressed on. . . . A few yards more brought me to some rising ground, and here the torrent of water overtook me. I gave up all for lost. . . . I was soon taken off my feet and borne inland by the force of the resistless mass. I remember nothing more until a violent blow revived me. . . . I found myself clinging to a coconut palm. Most of the trees near the town were uprooted and thrown down for miles, but this one fortunately had escaped and myself with it. . . . As I clung to the palm-tree, wet and exhausted, there floated past the dead bodies of many a friend and neighbour. Only a mere handful of the population escaped.*

So much ash was blown into the stratosphere that the Sunda Straits were completely dark from 10 A.M., August 27, until dawn the next day. Ash was reported falling on ships as far away as 6,076 km. In 13 days volcanic dust, ash, and aerosols had encircled the Earth. For the next three years, vivid red sunsets were common around the world due to these airborne products (▶ Figure 1-3). The volcanic dust in the stratosphere not only created spectacular sunsets, it also reflected incoming solar radiation back into space; the average global temperature dropped as much as 1/2°C during the following year and did not return to normal until 1888.

Of course, all animal life was destroyed on Krakatau. The remaining portion of the original island was blanketed by tens of meters of volcanic ash and pumice; two months later, the ash and pumice were still so hot that walking on it was difficult! A year after the eruption, a few shoots of grass appeared, and three years later 26 species of plants had colonized the island, thus providing a suitable habitat for the animals that soon followed. The first creatures to reach Krakatau probably flew or were lofted in by the wind; later, others either swam or were rafted to the island on driftwood or other flotsam.

Why have we chosen the eruption of Krakatau as an introduction to physical geology? The eruption was dramatic and interesting in its own right, but it also illustrates several of the aspects of geology that we will be examining, including the way the Earth's interior, surface, and atmosphere are all interrelated.

Sumatra, Java, Krakatau, and the Lesser Sunda Islands are part of a 3,000 km long chain of volcanic islands comprising the nation of Indonesia. Their location is a result of

▶ **FIGURE 1-1** Krakatau's climactic explosion in August 1883 was preceded by several smaller eruptions. This photograph was taken on May 27, 1883, one week after Krakatau's initial eruption. It shows ash and steam erupting from a vent on the south side of the island.

*C. Officer and J. Page, *Tales of the Earth* (New York: Oxford University Press, 1993), p. 17–18.

➢ FIGURE 1-2 (*a*) Krakatau, part of the island nation of Indonesia, is located in the Sunda Straits between Java and Sumatra. (*b*) Krakatau before and after the 1883 eruption. After the eruption, only one-third of the island remained above sea level.

a collision between two pieces of the Earth's outer layer, generally called the lithosphere. The theory that the Earth's lithosphere is divided into rigid plates that move over a plastic zone is known as *plate tectonics* (see Chapter 12). This unifying theory explains and ties together such apparently unrelated geologic phenomena as volcanic eruptions, earthquakes, and the origin of mountain ranges.

In tropical areas such as Indonesia, physical and chemical processes rapidly break down ash falls and lava flows, converting them into rich soils that are agriculturally productive and can support large populations (see Chapter 5). In spite of the dangers of living in a region of active volcanism, a strong correlation exists between volcanic activity and population density. Indonesia has experienced 972 eruptions during historic time, 83 of which have caused fatalities. Yet these same eruptions are also ultimately responsible for the high food production that can support large numbers of people.

Volcanic eruptions also affect weather patterns; recall that the eruption of Krakatau caused a global cooling of 1/2°C. More recently, the 1991 eruption of Mount Pinatubo in the Philippines resulted in lower global temperatures and abnormal weather patterns the following summer (see Chapter 4). A study issued by the Earth System Science Lab of the University of Alabama in Huntsville showed that summer temperatures were at a 10-year low in the Northern Hemisphere and at a 15-year low in the Southern Hemisphere.

➢ FIGURE 1-3 Airborne volcanic ash and dust particles from the eruption of Krakatau soon encircled the globe, producing exceptionally long, beautiful sunsets. This sunset was sketched by William Ascroft in London, England, at 4:40 P.M. on November 26, 1883, three months after Krakatau erupted.

As you read this book, keep in mind that the different topics you are studying are parts of dynamic interrelated systems, and not isolated pieces of information. For example, volcanic eruptions such as Krakatau are the result of complex interactions involving the Earth's interior and surface. These eruptions not only have an immediate effect on the surrounding area, but also contribute to climatic changes that affect the entire planet.

 INTRODUCTION

One major benefit of the space age is the ability to look back from space and view our planet in its entirety. Every astronaut has remarked in one way or another on how the Earth stands out as an inviting oasis in the otherwise black void of space (⯈ Figure 1-4).

The Earth is unique among the planets of our solar system in that it supports life and has oceans of water, a hospitable atmosphere, and a variety of climates. It is ideally suited for life as we know it because of a combination of factors, including its distance from the Sun and the evolution of its interior, crust, oceans, and atmosphere. Over time, changes in the Earth's atmosphere, oceans, and, to some extent, its crust have been influenced by life processes. In turn, these physical changes have affected the evolution of life.

⯈ FIGURE 1-4 *Apollo 17* view of the Earth. Almost the entire coastline of Africa is clearly shown in this view with Madagascar visible off its eastern coast. The Arabian Peninsula can be seen at the northeastern edge of Africa, while the Asian mainland is on the horizon toward the northeast. The present location of continents and ocean basins is the result of plate movement. The interaction of plates through time has affected the physical and biological history of the Earth.

The Earth is not a simple, unchanging planet. Rather, it is a complex dynamic body in which innumerable interactions are occurring among its many components. The continuous evolution of the Earth and its life makes geology an exciting and ever-changing science in which new discoveries are constantly being made.

 WHAT IS GEOLOGY?

Just what is geology and what is it that geologists do? **Geology**, from the Greek *geo* and *logos*, is defined as the study of the Earth. It is generally divided into two broad areas—physical geology and historical geology. *Physical geology* is the study of Earth materials, such as minerals and rocks, as well as the processes operating within the Earth and upon its surface. *Historical geology* examines the origin and evolution of the Earth, its continents, oceans, atmosphere, and life.

The discipline of geology is so broad that it is subdivided into many different fields or specialties. ◉ Table 1-1 shows many of the diverse fields of geology and their relationship to the sciences of astronomy, physics, chemistry, and biology.

Nearly every aspect of geology has some economic or environmental relevance. Many geologists are involved in exploration for mineral and energy resources, using their specialized knowledge to locate the natural resources on which our industrialized society is based. As the world demand for these nonrenewable resources increases, geologists are intensifying their search and applying the basic principles of geology in increasingly sophisticated ways (⯈ Figure 1-5).

Although locating mineral and energy resources is extremely important, geologists are also being asked to use their expertise to help solve many of our environmental problems. Some geologists are involved in finding groundwater for the ever-burgeoning needs of communities and industries or in monitoring surface and underground water pollution and suggesting ways to clean it up. Geological engineers help find safe locations for dams, waste disposal sites, and power plants and design earthquake-resistant buildings.

Geologists are also involved in making short- and long-range predictions about earthquakes and volcanic eruptions and the potential destruction that may result. In addition, they are working with civil defense planners to help draw up contingency plans should such natural disasters occur.

As this brief survey illustrates, geologists are employed in a wide variety of pursuits. As the world's population in-

TABLE 1-1 Specialties of Geology and Their Broad Relationship to the Other Sciences

Specialty	Area of Study	Related Science
Geochronology Planetary geology	Time and history of the Earth Geology of the planets	Astronomy
Paleontology	Fossils and life history	Biology
Economic geology Environmental geology Geochemistry Hydrogeology Mineralogy Petrology	Mineral and energy resources Environment Geology of chemical change Water resources Minerals Rocks	Chemistry
Geophysics Structural geology Seismology	Earth's interior Rock deformation Earthquakes	Physics
Geomorphology Oceanography Paleogeography Stratigraphy/sedimentology	Landforms Oceans Ancient geographic features and locations Layered rocks and sediments	

creases and greater demands are made on the Earth's limited resources, the need for geologists and their expertise will become even greater.

 GEOLOGY AND THE HUMAN EXPERIENCE

Many people are surprised at the extent to which we depend on geology in our everyday lives and also at the numerous references to geology in the arts, music, and literature. Rocks and landscapes are realistically represented in many sketches and paintings. Examples by famous artists include Leonardo da Vinci's *Virgin of the Rocks* and *Virgin and Child with Saint Anne*, Giovanni Bellini's *Saint Francis in Ecstasy* and *Saint Jerome*, and Asher Brown Durand's *Kindred Spirits* (▶ Figure 1-6).

In the field of music, Ferde Grofé's *Grand Canyon Suite* was, no doubt, inspired by the grandeur and timelessness of

(a)

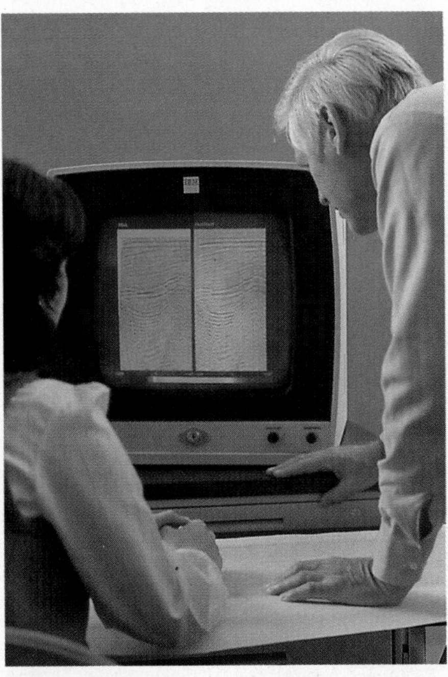

(b)

▶ FIGURE 1-5 (*a*) Geologists measuring the amount of erosion on a glacier in Alaska. (*b*) Geologists increasingly use computers in their search for petroleum and other natural resources.

⫸ **FIGURE 1-6** *Kindred Spirits* by Asher Brown Durand (1849) realistically depicts the layered rocks occurring along gorges in the Catskill Mountains of New York State. Durand was one of numerous artists of the nineteenth-century Hudson River School, who were known for their realistic landscapes. This painting was done to show Durand conversing with the recently deceased Thomas Cole, the original founding force of the Hudson River School. Cole's work is featured on the cover of this text. These artists clearly demonstrate the linkage between geology and art.

Arizona's Grand Canyon and its vast rock exposures. The rocks on the Island of Staffa in the Inner Hebrides provided the inspiration for Felix Mendelssohn's famous *Hebrides* Overture (⫸ Figure 1-7).

References to geology abound in *The German Legends of the Brothers Grimm*. Jules Verne's *Journey to the Center of the Earth* describes an expedition into the Earth's interior (see Chapter 10 Prologue). On one level, the poem "Ozymandias" by Percy

⫸ **FIGURE 1-7** Felix Mendelssohn was inspired by the rocks on the Island of Staffa in the Inner Hebrides, when he wrote the famous *Hebrides* (also known as *Fingal's Cave*) Overture. Mendelssohn wrote the opening bars of this overture while visiting Staffa.

B. Shelley deals with the fact that nothing lasts forever and even solid rock eventually disintegrates under the ravages of time and weathering. References to geology can even be found in comics, two of the best known being *B.C.* by Johnny Hart and *The Far Side* by Gary Larson (▶ Figure 1-8).

Geology has also played an important role in history. Wars have been fought for the control of such natural resources as oil, gas, gold, silver, diamonds, and other valuable minerals. Empires throughout history have risen and fallen on the distribution and exploitation of natural resources. The configuration of the Earth's surface, or its topography, which is shaped by geologic agents, plays a critical role in military tactics. Natural barriers such as mountain ranges and rivers have frequently served as political boundaries.

How Geology Affects Our Everyday Lives

Destructive volcanic eruptions, devastating earthquakes, disastrous landslides, large sea waves, floods, and droughts are headline-making events that affect many people (▶ Figure 1-9). Although we are unable to prevent most of these natural disasters, the more we know about them, the better we are able to predict, and possibly control, the severity of their impact. The environmental movement has forced everyone to take a closer look at our planet and the delicate balance among its various systems.

▶ FIGURE 1-8 References to geology are frequently found in comics as this *Far Side* cartoon by Gary Larson illustrates.

THE FAR SIDE By GARY LARSON

"You know, I used to like this hobby ... But shoot! Seems like *everybody's* got a rock collection."

▶ FIGURE 1-9 As these headlines from various newspapers indicate, geology affects our everyday lives.

Most readers of this book will not become professional geologists. Everyone, however, should have a basic understanding of the geological processes that ultimately affect all of us. Such an understanding of geology is important so that one can avoid, for example, building in an area prone to landslides or flooding. Just ask anyone who purchased a home in the Portuguese Bend area of southern California during the 1950s and saw it destroyed by landslides, or who built along a lakeshore and later saw the lake level rise and the beach and sometimes even the house disappear.

As society becomes increasingly complex and technologically oriented, we, as citizens, need an understanding of science so that we can make informed choices about those things that affect our lives. We are already aware of some of the negative aspects of an industrialized society, such as problems relating to solid waste disposal, contaminated groundwater, and acid rain. We are also learning the impact that humans, in increasing numbers, have on the environment and that we can no longer ignore the role that we play in the dynamics of the global ecosystem.

Most people are unaware of the extent to which geology affects their everyday lives. For many people, the connection between geology and such well-publicized problems as nonrenewable energy and mineral resources, let alone waste disposal and pollution, is simply too far removed or too complex to be fully appreciated. But consider for a moment just how dependent we are on geology in our daily routines.

Much of the electricity for our appliances comes from the burning of coal, oil, or natural gas or from uranium consumed in nuclear-generating plants. It is geologists who locate the coal, petroleum, and uranium. The copper or other metal wires through which electricity travels are manufactured from materials found as the result of mineral exploration. The buildings that we live and work in owe their very existence to geological resources. A few examples are the concrete foundation (concrete is a mixture of clay, sand, or gravel, and limestone), the drywall (made largely from the mineral gypsum), the windows (the mineral quartz is the principal ingredient in the manufacture of glass), and the metal or plastic plumbing fixtures inside the building (the metals are from ore deposits and the plastics are most likely manufactured from petroleum distillates of crude oil).

Furthermore, when we go to work, the car or public transportation we use is powered and lubricated by some type of petroleum by-product and is constructed of metal alloys and plastics. And the roads or rails we ride over come from geologic materials, such as gravel, asphalt, concrete, or steel. All of these items are the result of processing geologic resources.

It is quite apparent that as individuals and societies, the standard of living we enjoy is directly dependent on the consumption of geologic materials. Therefore, we need to be aware of geology and of how our use and misuse of geologic resources may affect the delicate balance of nature and irrevocably alter our culture as well as our environment.

The Origin of the Solar System and the Differentiation of the Early Earth

According to the currently accepted theory for the origin of the solar system (➤ Figure 1-10), interstellar material in a spiral arm of the Milky Way Galaxy condensed and began collapsing. As this cloud gradually collapsed under the influence of gravity, it flattened and began rotating counterclockwise, with about 90% of its mass concentrated in the central part of the cloud. As the rotation and concentration of material continued, an embryonic Sun formed, surrounded by a turbulent, rotating cloud of material called a *solar nebula*.

The turbulence in this solar nebula formed localized eddies where gas and solid particles condensed. As this condensation proceeded, gaseous, liquid, and solid particles began accreting into ever-larger masses called *planetesimals* that eventually became true planetary bodies. While the planets were accreting, material that had been pulled into the center of the nebula also condensed, collapsed, and was heated to several million degrees by gravitational compression. The result was the birth of a star, our Sun.

➤ **FIGURE 1-10** The currently accepted theory for the origin of our solar system involves (*a*) a huge nebula condensing under its own gravitational attraction, then (*b*) contracting, rotating, and (*c*) flattening into a disk, with the Sun forming in the center and eddies gathering up material to form planets. As the Sun contracted and began to visibly shine, (*d*) intense solar radiation blew away unaccreted gas and dust until finally, (*e*) the Sun began burning hydrogen and the planets completed their formation.

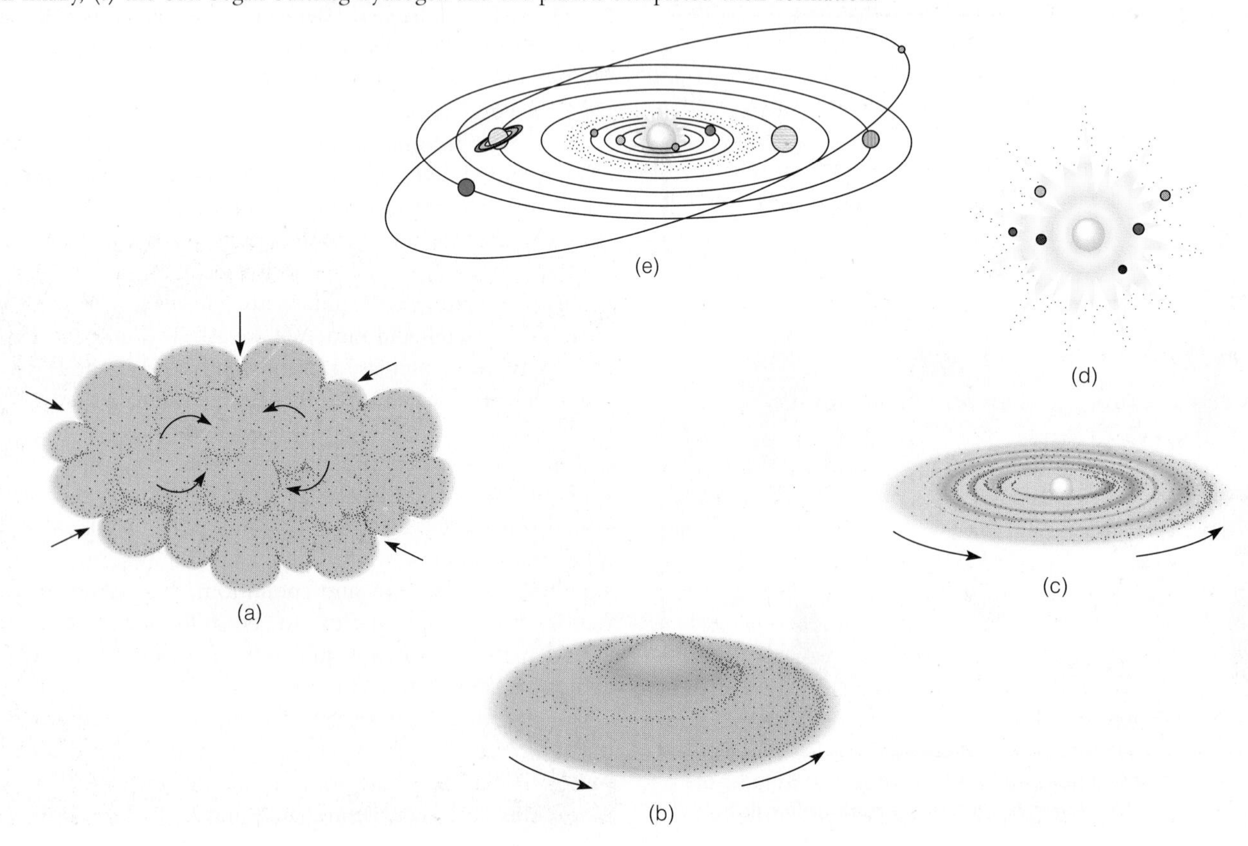

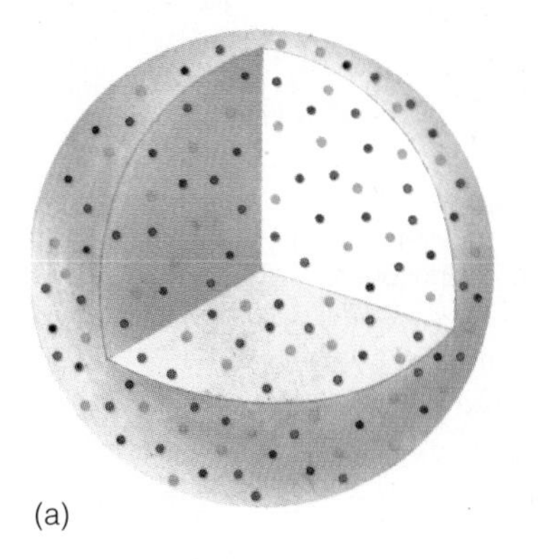

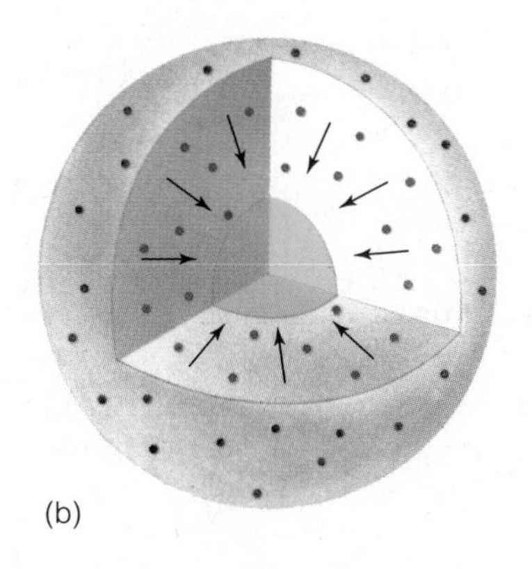

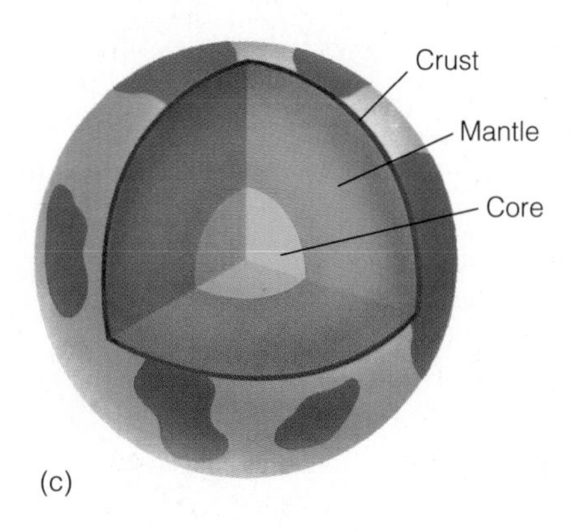

▶ **FIGURE 1-11** (*a*) The early Earth was probably of uniform composition and density throughout. (*b*) Heating of the early Earth reached the melting point of iron and nickel, which, being denser than silicate minerals, settled to the Earth's center. At the same time, the lighter silicates flowed upward to form the mantle and the crust. (*c*) In this way, a differentiated Earth formed, consisting of a dense iron-nickel core, an iron-rich silicate mantle, and a silicate crust with continents and ocean basins.

Some 4.6 billion years ago, enough material eventually gathered together in one of the turbulent eddies that swirled around the early Sun to form the planet Earth. Scientists think this early Earth was rather cool, so that the accreting elements and nebular rock fragments were solids rather than gases or liquids. This early Earth was also thought to be of generally uniform composition and density throughout (▶ Figure 1-11a). It was composed mostly of compounds of silicon, iron, magnesium, oxygen, aluminum, and smaller amounts of all the other chemical elements. Subsequently, when the Earth underwent heating, this homogeneous composition disappeared, and the result was a differentiated planet, consisting of a series of concentric layers of differing composition and density (Figure 1-11c). This differentiation into a layered planet is probably the most significant event in the history of the Earth. Not only did it lead to the formation of a crust and eventually to continents, but it was also probably responsible for the emission of gases from the interior that eventually led to the formation of the oceans and the atmosphere.

 THE EARTH AS A DYNAMIC PLANET

The Earth is a dynamic planet that has changed continuously during its 4.6-billion-year existence. The size, shape, and geographic distribution of continents and ocean basins have changed through time, the composition of the atmosphere has evolved, and life-forms existing today differ from those that lived during the past. We can easily visualize how mountains and hills are worn down by erosion and how landscapes are changed by the forces of wind, water, and ice. Volcanic eruptions and earthquakes reveal an active interior, and folded and broken rocks indicate the tremendous power of the Earth's internal forces.

The Earth consists of three concentric layers: the core, the mantle, and the crust (▶ Figure 1-12). This orderly division results from density differences between the layers as a function of variations in composition, temperature, and pressure.

The **core** has a calculated density of 10 to 13 grams per cubic centimeter (g/cm^3) and occupies about 16% of the Earth's total volume. Seismic (earthquake) data indicate that the core consists of a small, solid inner core and a larger, apparently liquid, outer core. Both are thought to consist largely of iron and a small amount of nickel.

▶ **FIGURE 1-12** A cross section of the Earth illustrating the core, mantle, and crust. The enlarged portion shows the relationship between the lithosphere, composed of the continental crust, oceanic crust, and upper mantle, and the underlying asthenosphere and lower mantle.

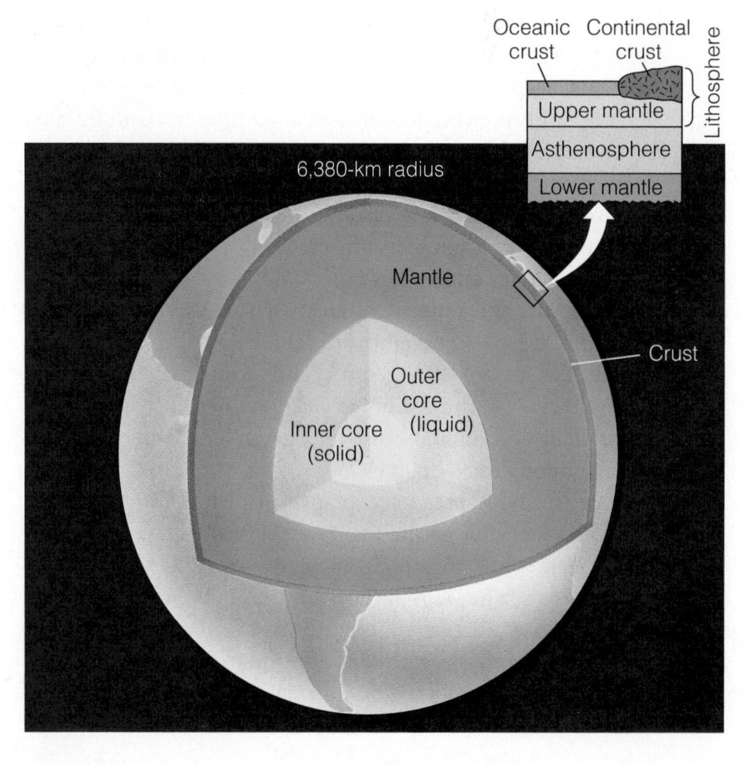

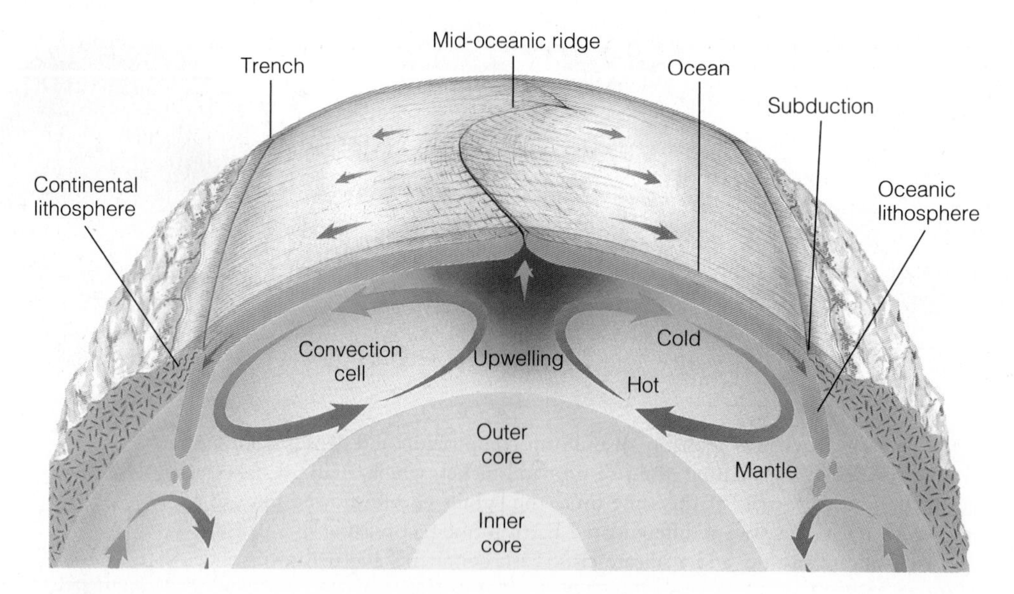

➢ **FIGURE 1-13** The Earth's plates are thought to move as a result of underlying mantle convection cells in which warm material from deep within the Earth rises toward the surface, cools, and then, upon losing heat, descends back downward into the interior.

The **mantle** surrounds the core and comprises about 83% of the Earth's volume. It is less dense than the core (3.3–5.7 g/cm^3) and is thought to be composed largely of *peridotite,* a dark, dense igneous rock containing abundant iron and magnesium. The mantle can be divided into three distinct zones based on physical characteristics. The lower mantle is solid and forms most of the volume of the Earth's interior. The **asthenosphere** surrounds the lower mantle. It has the same composition as the lower mantle but behaves plastically and slowly flows. Partial melting within the asthenosphere generates magma, some of which rises to the Earth's surface because it is less dense than the rock from which it was derived. The upper mantle surrounds the asthenosphere. The solid upper mantle and the overlying crust constitute the **lithosphere**, which is broken into numerous individual pieces called **plates** that move over the asthenosphere as a result of underlying *convection cells* (➢ Figure 1-13). Interactions of these plates are responsible for such phenomena as earthquakes, volcanic eruptions, and the formation of mountain ranges and ocean basins.

The **crust**, the outermost layer of the Earth, consists of two types. The *continental crust* is thick (20–90 km), has an average density of 2.7 g/cm^3, and contains considerable silicon and aluminum. The *oceanic crust* is thin (5–10 km), denser than continental crust (3.0 g/cm^3), and is composed of the igneous rock *basalt.*

Since the widespread acceptance of plate tectonic theory about 25 years ago, geologists have viewed the Earth from a global perspective in which all of its systems are interconnected. Thus, the distribution of mountain chains, major fault systems, volcanoes, and earthquakes, the origin of new ocean basins, the movement of continents, and several other geological processes and features are perceived to be interrelated.

 GEOLOGY AND THE FORMULATION OF THEORIES

The term **theory** has various meanings. In colloquial usage, it means a speculative or conjectural view of something—hence the widespread belief that scientific theories are little more than unsubstantiated wild guesses. In scientific usage, however, a theory is a coherent explanation for one or several related natural phenomena that is supported by a large body of objective evidence. From a theory are derived predictive statements that can be tested by observation and/or experiment so that their validity can be assessed. The law of universal gravitation is an example of a theory describing the attraction between masses (an apple and the Earth in the popularized account of Newton and his discovery).

Theories are formulated through the process known as the **scientific method**. This method is an orderly, logical approach that involves gathering and analyzing the facts or data about the problem under consideration. Tentative explanations or **hypotheses** are then formulated to explain the observed phenomena. Next, the hypotheses are tested to see if what they predicted actually occurs in a given situation. Finally, if one of the hypotheses is found, after repeated tests, to explain the phenomena, then that hypothesis is proposed as a theory. One should remember, however, that in science, even a theory is still subject to further testing and refinement as new data become available.

The fact that a scientific theory can be tested and is subject to such testing separates science from other forms of human inquiry. Because scientific theories can be tested, they have the potential of being supported or even proved wrong. Accordingly, science must proceed without any appeal to beliefs or supernatural explanations, not because such beliefs or explanations are necessarily untrue, but because we have no way to investigate them. For this reason, science makes no claim about the existence or nonexistence of a supernatural or spiritual realm.

Each scientific discipline has certain theories that are of particular importance for that discipline. In geology, the formulation of plate tectonic theory has changed the way geologists view the Earth. Geologists now view Earth history in terms of interrelated events that are part of a global pattern of change. Before plate tectonic theory was generally accepted by geologists, however, numerous interrelated hypotheses were proposed and tested. Thus, the evolution of this theory illustrates the scientific method at work (see Perspective 1-1).

Plate Tectonic Theory

The acceptance of **plate tectonic theory** is recognized as a major milestone in the geological sciences. It is comparable to the revolution caused by Darwin's theory of evolution in biology. Plate tectonics has provided a framework for interpreting the composition, structure, and internal processes of the Earth on a global scale. It has led to the realization that the continents and ocean basins are part of a lithosphere-atmosphere-hydrosphere (water portion of the planet) system that evolved together with the Earth's interior.

According to plate tectonic theory, the lithosphere is divided into plates that move over the asthenosphere (▶ Figure 1-14). Most of the boundaries between plates are marked by zones of earthquake activity, volcanic activity, or both. Along these boundaries, plates diverge, converge, or slide sideways past each other.

At **divergent plate boundaries**, plates move apart as magma rises to the surface from the asthenosphere (▶ Figure 1-15). The magma solidifies to form rock, which attaches to the moving plates. The margins of divergent plate boundaries are marked by mid-oceanic ridges in oceanic crust and are recognized by linear rift valleys where newly forming divergent boundaries occur beneath continental crust. The separation of South America from Africa and the formation of the South Atlantic Ocean occurred along a divergent plate boundary, the Mid-Atlantic Ridge (Figure 1-14).

Plates move toward one another along **convergent plate boundaries** (Figure 1-15). When an oceanic plate collides with a continental plate, for example, the denser oceanic plate sinks beneath the continental plate along what is known as a **subduction zone.** As the subducting oceanic plate descends into the Earth, it becomes hotter and hotter, and its interaction with the mantle produces a magma. As this magma rises, it may erupt at the Earth's surface, forming a chain of volcanoes. The Andes Mountains on the west coast of South America are a good example of a volcanic mountain range formed as a result of subduction of the Nazca plate beneath the South American plate along a convergent plate boundary (Figure 1-14).

Crust is produced and consumed at divergent and convergent plate boundaries, respectively. In contrast, **transform**

▶ **FIGURE 1-14** The Earth's lithosphere is divided into rigid plates of various sizes that move over the asthenosphere.

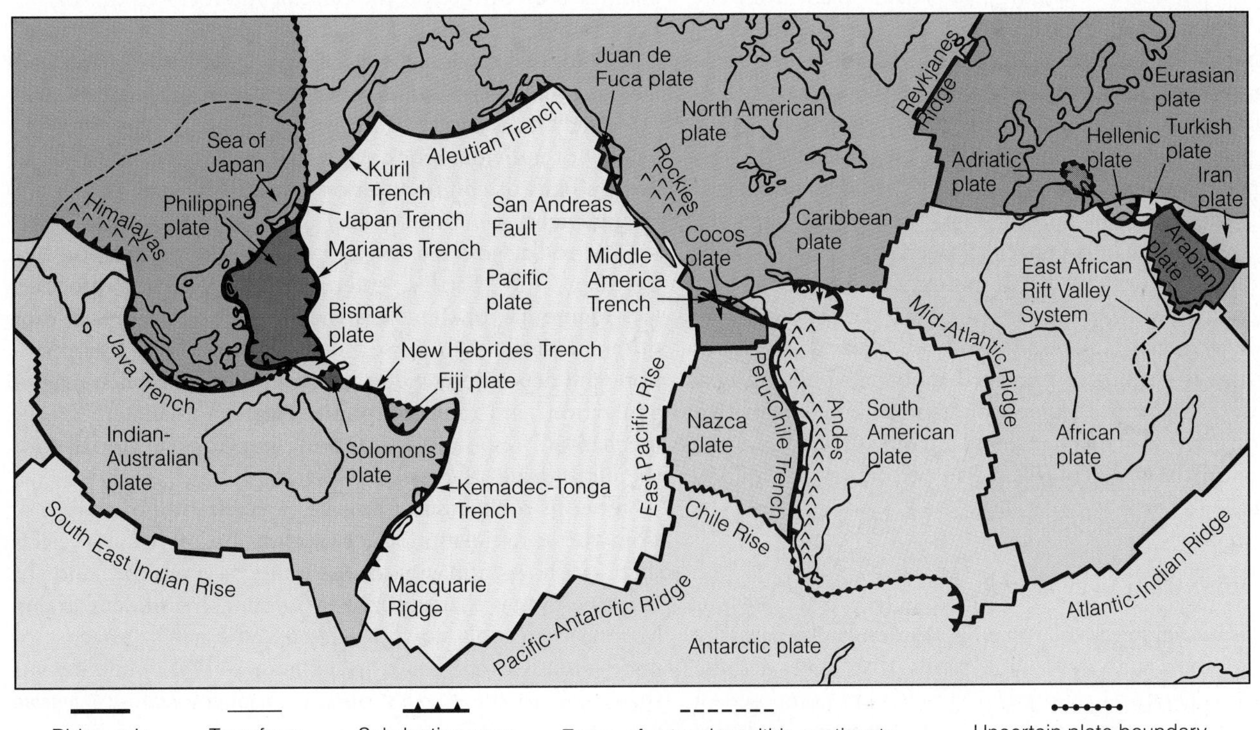

Perspective 1-1

The Formulation of Plate Tectonic Theory

The idea that continents moved during the past goes back to the time when people first noticed that the margins of eastern South America and western Africa looked as if they fit together. Geologists also noticed that similar or identical fossils occur on widely separated continents, that the same types of rocks from the same time period are found on different continents, and that ancient rocks and features indicating former glacial conditions occur in today's tropical areas. As more and more facts were gathered, hypotheses were proposed to explain them. In 1912, Alfred Wegener, a German meteorologist, amassed a tremendous amount of geological, paleontological, and climatological data that indicated continents moved through time; he proposed the hypothesis of *continental drift* to explain and synthesize this myriad of facts.

Wegener stated that at one time all of the continents were united into one single supercontinent that he named *Pangaea.* Pangaea later broke apart, and the individual continents drifted to their current locations. The continental drift hypothesis explained why the shorelines of different continents fit together, how different mountain ranges were once part of a larger continuous mountain range, why the same fossil animals and plants are found on different continents, and why rocks indicating glacial conditions are now found on continents located in the tropics.

Wegener's hypothesis could be tested by asking what type of rocks or fossils would one expect to find at a given location on a continent if that continent was in the tropics 180 million years ago. To test the hypothesis of continental drift, all researchers had to do was to go into the field and examine the rocks and fossils for a particular time period on any continent to see if they indicated what the hypothesis predicted for the proposed location of that continent. In almost all cases, the data fit the hypothesis. However, there was one problem with Wegener's hypothesis: it did not explain how continents moved over oceanic crust and what the mechanism of continental movement was.

During the late 1950s and early 1960s, new data about the sea floor emerged that enabled geologists to propose the

---

plate boundaries are sites where plates slide sideways past each other, and crust is neither produced nor consumed (Figure 1-15). The San Andreas fault in California is a transform plate boundary separating the Pacific plate from the North American plate (Figure 1-14). The earthquake activity along the San Andreas fault results from the Pacific plate moving northward relative to the North American plate.

A revolutionary theory when it was proposed in the 1960s, plate tectonic theory has had significant and far-reaching consequences in all fields of geology because it provides the basis for relating many seemingly unrelated geologic phenomena. Its impact has been particularly notable in the interpretation of Earth history. For example, the Appalachian Mountains in eastern North America and the mountain ranges of Greenland, Scotland, Norway, and Sweden are not the result of unrelated mountain-building episodes, but are part of a larger mountain-building event that involved the closing of an ancient "Atlantic Ocean" and the formation of the supercontinent Pangaea about 245 million years ago.

 The Rock Cycle

A **rock** is an aggregate of minerals. **Minerals** are naturally occurring, inorganic, crystalline solids that have definite physical and chemical properties. Minerals are composed of elements such as oxygen, silicon, and aluminum, and elements are made up of atoms, the smallest particles of matter that still retain the characteristics of an element. More than 3,500 minerals have been identified and described, but only about a dozen make up the bulk of the rocks in the Earth's crust (see Table 2-6).

Geologists recognize three major groups of rocks—*igneous*, *sedimentary*, and *metamorphic*—each of which is characterized by its mode of formation. Each group contains a variety of individual rock types that differ from one another on the basis of composition or texture (the size, shape, and arrangement of mineral grains).

The **rock cycle** is a way of viewing the interrelationships between the Earth's internal and external processes (▶ Figure 1-16). It relates the three rock groups to each other; to surficial processes such as weathering, transportation, and deposition; and to internal processes such as magma generation and metamorphism. Plate movement is the mechanism responsible for recycling rock materials and therefore drives the rock cycle.

Igneous rocks result from the crystallization of magma or from the accumulation and consolidation of volcanic ejecta such as ash. As a magma cools, minerals crystallize, and the resulting rock is characterized by interlocking mineral grains. Magma that cools slowly beneath the Earth's surface produces *intrusive igneous rocks* (▶ Figure 1-17a), while magma that cools at the Earth's surface produces *extrusive igneous rocks* (Figure 1-17b).

hypothesis of *sea-floor spreading*. This hypothesis suggested that the continents and segments of oceanic crust move together as single units, and that some type of thermal convection cell system operating within the Earth was the mechanism responsible for plate movements.

Sea-floor spreading and continental drift were then combined into a single hypothesis. According to this hypothesis, moving rigid plates are composed of continental and/or oceanic crust and the underlying upper mantle. These plates are bounded by mid-oceanic ridges, oceanic trenches, faults, and mountain belts. In this hypothesis, plates move away from mid-oceanic ridges and toward oceanic trenches. Furthermore, new crust is added along the mid-oceanic ridges and consumed or destroyed along oceanic trenches, and mountain chains are formed adjacent to the oceanic trenches.

According to this later hypothesis, Europe and North America should be steadily moving away from each other at a rate of up to several centimeters per year. Precise measurements of continental positions by satellites have verified this, thus confirming the validity of the plate movement hypothesis.

Furthermore, if plates are moving away from mid-oceanic ridges as predicted by the plate tectonic hypothesis, then rocks of the oceanic crust should become progressively older with increasing distance from the mid-oceanic ridges. To test this prediction, deep-sea sediment and oceanic crust were drilled as part of a massive scientific study of the ocean basins called the *Deep Sea Drilling Project*. Analysis of the oceanic crust and the layer of sediment immediately above it showed that the age of the oceanic crust does indeed increase with distance from the mid-oceanic ridges, and that the oldest oceanic crust is adjacent to the continental margins.

With the confirmation of these and other predictions of the plate tectonic hypothesis, most geologists accept that the hypothesis is correct and therefore call it the *plate tectonic theory*. Its acceptance has been so widespread not only because of the overwhelming evidence supporting it but also because it appears to explain the relationships among many seemingly unrelated geologic features and events.

▶ **FIGURE 1-15** An idealized cross section illustrating the relationship between the lithosphere and the underlying asthenosphere and the three principal types of plate boundaries: divergent, convergent, and transform.

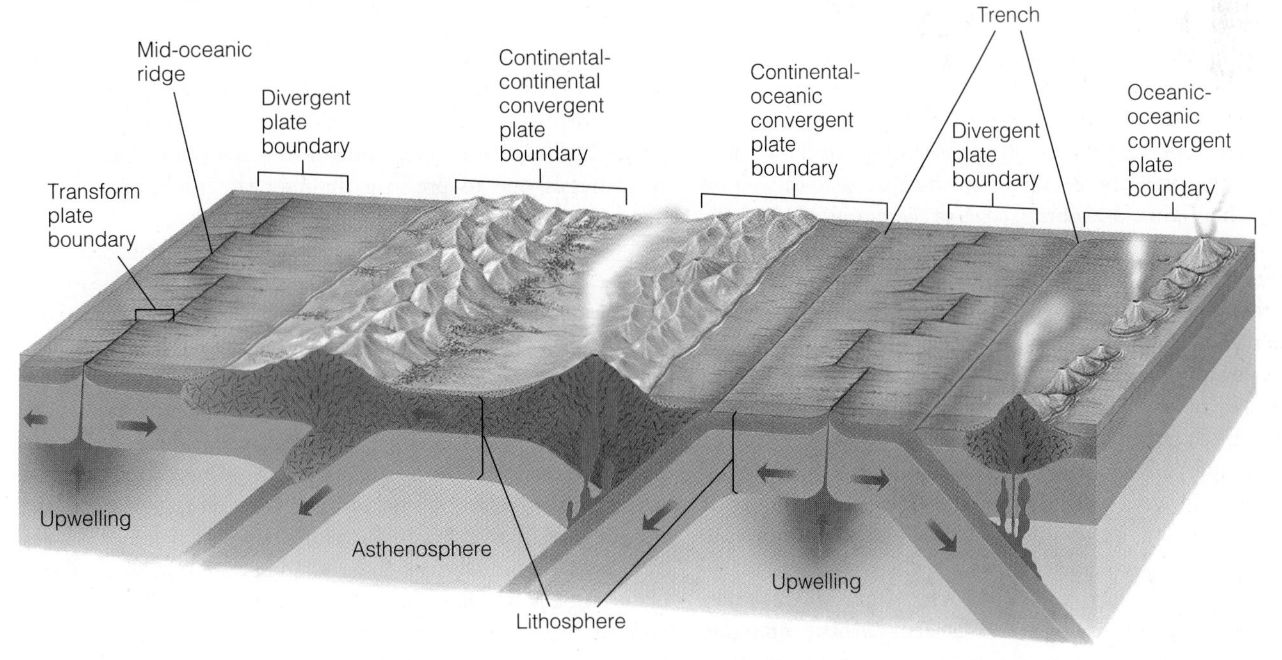

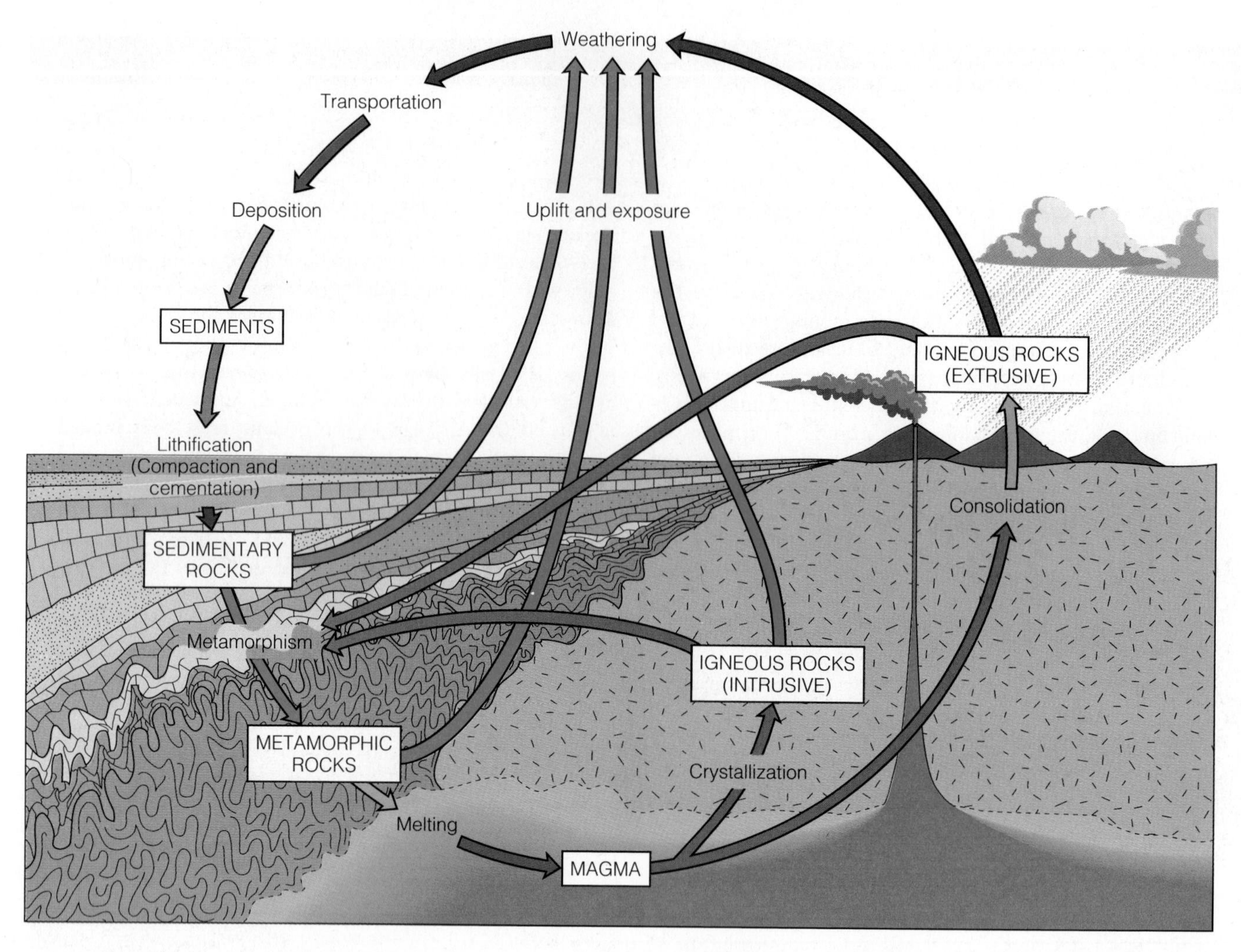

▶ **FIGURE 1-16** The rock cycle showing the interrelationships between the Earth's internal and external processes and how each of the three major rock groups is related to the others.

Rocks exposed at the Earth's surface are broken into particles and dissolved by various weathering processes. The particles and dissolved material may be transported by wind, water, or ice and eventually deposited as *sediment*. This sediment may then be compacted or cemented into sedimentary rock.

Sedimentary rocks originate by consolidation of rock fragments, precipitation of mineral matter from solution, or compaction of plant or animal remains (Figure 1-17c and d). Because sedimentary rocks form at or near the Earth's surface, geologists can make inferences about the environment in which they were deposited, the type of transporting agent, and perhaps even something about the source from which the sediments were derived (see Chapter 6). Accordingly, sedimentary rocks are very useful for interpreting Earth history.

Metamorphic rocks result from the alteration of other rocks, usually beneath the Earth's surface, by heat, pressure, and the chemical activity of fluids. For example, marble, a rock preferred by many sculptors and builders, is a metamorphic rock produced when the agents of metamorphism are applied to the sedimentary rock limestone or dolostone. Metamorphic rocks are either *foliated* (Figure 1-17e) or *nonfoliated* (Figure 1-17f). Foliation is the parallel alignment of minerals due to pressure. It gives the rock a layered or banded appearance.

The Rock Cycle and Plate Tectonics

Interactions among plates determine, to a certain extent, which of the three rock groups will form (▶ Figure 1-18). For example, weathering and erosion produce sediments that are transported by agents such as running water from the continents to the oceans, where they are deposited and accumulate. These sediments, some of which may be lithified and become sedimentary rock, become part of a moving plate along with the underlying oceanic crust. When plates converge, heat and pressure generated along the plate boundary may lead to igneous activity and metamorphism within the descending oceanic plate, thus producing various igneous and metamorphic rocks.

▶ **FIGURE 1-17** Hand specimens of common igneous (*a, b*), sedimentary (*c, d*), and metamorphic (*e, f*) rocks. (*a*) Granite, an intrusive igneous rock. (*b*) Basalt, an extrusive igneous rock. (*c*) Conglomerate, a sedimentary rock formed by the consolidation of rock fragments. (*d*) Marine limestone, a sedimentary rock formed by the extraction of mineral matter from seawater by organisms or by the inorganic precipitation of the mineral calcite from seawater. (*e*) Gneiss, a foliated metamorphic rock. (*f*) Quartzite, a nonfoliated metamorphic rock. (Photos courtesy of Sue Monroe.)

Some of the sediment and sedimentary rock is subducted and melts, while other sediments and sedimentary rocks along the boundary of the nonsubducted plate are metamorphosed by the heat and pressure generated along the converging plate boundary. Later, the mountain range or chain of volcanic islands formed along the convergent plate boundary will once again be weathered and eroded, and the new sediments will be transported to the ocean to begin yet another cycle.

Guest Essay Bruce Babbitt

SCIENCE: PRESERVING OUR HERITAGE

The following essay is based on Secretary Babbitt's statement before the Committee on Natural Resources of the U.S. House of Representatives in February 1993 and his speech to the National Press Club in April 1993.

I grew up in Flagstaff, Arizona, a small town in the West. I was always looking at the land and collecting fossils, always wanting to get to the next thing on the horizon, whether it was a mountaintop or another trail on the Grand Canyon. In those days Flagstaff was the center of a small regional economy dependent on the management of natural resources—mining, forestry, reclamation, and grazing—and on decisions made about those resources at the Department of the Interior. Our horizons, then as now, were dominated by the Grand Canyon National Park and the Navajo Indian Reservation. Our opportunities and our problems, then as now, were inextricably intertwined with the management of the federal and Indian lands of the West.

I learned early on that the development of the West was guided, for good or ill, by the policies set by Congress and the federal government and administered by the Department of the Interior. The role of the department as land manager, natural resources steward, wildlife conservator, parks curator, and trustee were not abstract notions. Each decision made by the department echoes through the economy, politics, and quality of life in the West.

But the Interior Department is not just the "Department of the West." The department's minerals management responsibilities extend from the outer continental shelf of Alaska to George's Bank; its Office of Surface Mining has nationwide regulatory responsibilities; and the National Park Service plays an increasingly important role in offering recreational opportunities to city dwellers in the East as well as to citizens across the rest of America.

I believe we have a particular responsibility toward the great parks of the nation. And when I say "great parks," I mean all parks because each in its own right inspires its visitors. We must care for the crown jewels of the park system, such as Yellowstone, Grand Canyon, Yosemite, Acadia, and the Great Smokies. At the same time, we must care for our urban parks, such as Gateway and Golden Gate.

To this end, we have proceeded with plans to protect the extraordinary wildlife and ecological resources of the California desert. This area is truly one of our national treasures. The abundance of wildlife and the amazing ecological diversity of the desert deserve to be protected for future generations of Americans. At the same time, these areas need to remain accessible and available as a place where citizens today—particulary those whose lives are mainly spent in the metropolitan regions of our country—can achieve the encounter with their natural heritage that is the fundamental purpose of our national parks and wilderness areas.

My goal is to improve the management of the nation's natural resources and to balance needed development with a renewed emphasis on stewardship and conservation, so that the United States can meet its needs in the twenty-first century. One way we will accomplish this is through the use of market principles in resource allocation. For one thing, this is an issue of fairness: it is simply unreasonable to say to the American people: everyone should pay their fair share—except miners, timber companies, ranchers, and reclamation water users. Secondly, the move to market has important environmental benefits—market pricing of resources encourages conservation and the efficient allocation of limited resources.

One priority will be reform of the Mining Law of 1872. The law has been tinkered with before, but many problems remain—such as disposal of valuable public resources for nominal fees, inadequate environmental regulation, and lack of secure tenure for mineral exploration. The mining industry, other users of the public lands, and above all the American people will benefit if we have a modern mining law—one that takes full account of the public interest in the lands and minerals owned by the American people.

The transition to market principles for resource allocation also has major implications for the development and use of water in the West. The search for more water from ever more distant sources with greater environmental destruction is a time-honored western tradition that must now give way to a simple reality: there is plenty of water developed and available in the West if only we will allow market principles to replace heavy-handed bureaucratic allocations. By pricing water at its true cost, and thereby encouraging its conservation and wise use, there will be plenty of water for everyone.

Above all, then, we are set on creating a new American land ethic. We will bring old and true economic values to a new and urgent cause: the imperative to live more lightly and productively on the land—our land.

BRUCE BABBITT graduated from the University of Notre Dame where he majored in geology and later received a master's degree in geophysics from the University of Newcastle in England. He had planned on a career in mining until a trip to Bolivia in 1962. As he says, the poverty he saw there "crystallized my social awareness." Instead of mining, he became a lawyer, graduating from Harvard Law School. He returned to his home state of Arizona where he became the state's attorney general in 1974. He subsequently served two terms as governor. In 1993 President Bill Clinton named Babbitt to be secretary of the interior.

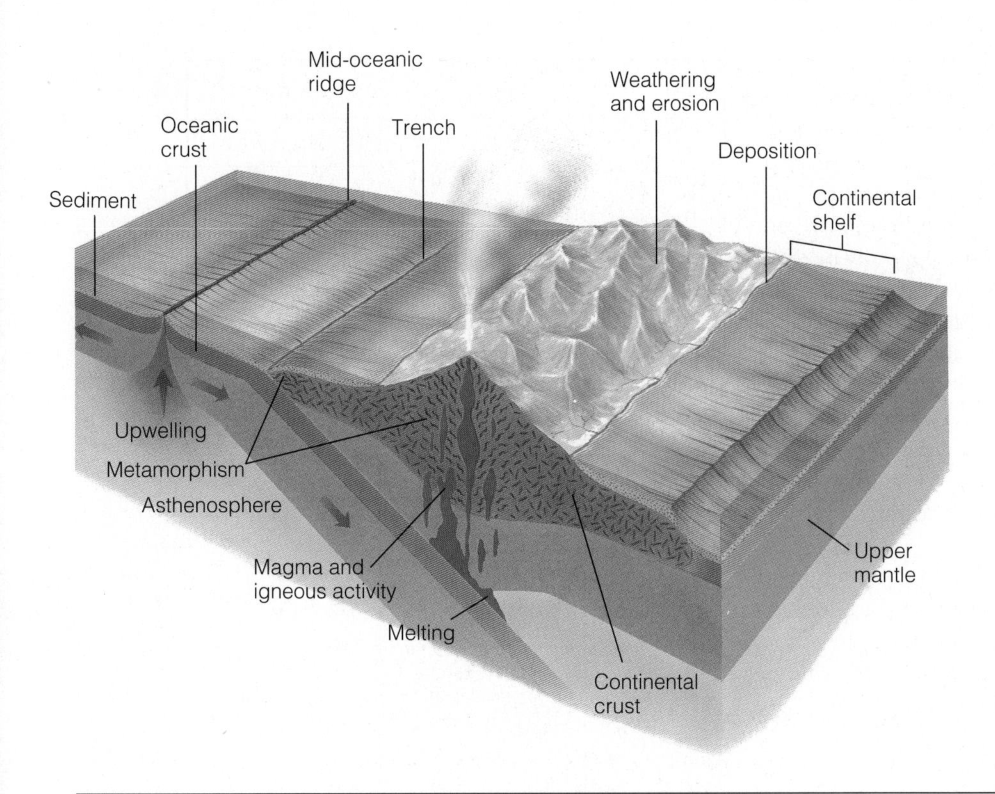

FIGURE 1-18 Plate tectonics and the rock cycle. The cross section shows how the three major rock groups—igneous, metamorphic, and sedimentary—are recycled through both the continental and oceanic regions.

 GEOLOGIC TIME AND UNIFORMITARIANISM

An appreciation of the immensity of geologic time is central to understanding the evolution of the Earth and its biota. Indeed, time is one of the main aspects that sets geology apart from the other sciences. Most people have difficulty comprehending geologic time because they tend to think in terms of the human perspective—seconds, hours, days, and years. Ancient history is what occurred hundreds or even thousands of years ago. When geologists talk of ancient geologic history, however, they are referring to events that happened hundreds of millions or even billions of years ago. To a geologist, recent geologic events are those that occurred within the last million years or so.

One popular analogy geologists use to convey the immensity of geologic time is to compare the history of the Earth to a calendar year (◉ Table 1-2). The time when the Earth formed 4.6 billion years ago corresponds to 12:00:00 A.M., January 1. On this calendar, we see that the oldest fossils, simple, microscopic bacteria, which first appeared

TABLE 1-2 Geologic Time and Significant Events in Earth History Condensed into One Calendar Year

Years before Present	Event	Days since January 1	Date and Time
10,000	Ice Age ends	364.9	December 31, 11:58:51 P.M.
1.6 million	Ice Age begins	364.9	December 31, 8:57:11 P.M.
4 million	First humans	364.6	December 31, 4:11:29 P.M.
53 million	First horses	360.8	December 27, 7:04:10 P.M.
66 million	Dinosaurs become extinct	359.8	December 26, 6:18:47 P.M.
115 million	First flowering plants	355.9	December 22, 9:00:00 A.M.
145 million	First birds	353.5	December 20, 11:52:10 A.M.
222 million	First mammals	347.4	December 14, 9:14:05 A.M.
230 million	First dinosaurs	344.8	December 11, 7:08:52 P.M.
360 million	First amphibians	336.4	December 3, 10:26:02 A.M.
430 million	First land plants	330.9	November 27, 9:07:50 P.M.
510 million	First fish	324.5	November 21, 12:46:57 P.M.
700 million	First multicellular animals	309.5	November 4, 10:57:23 A.M.
3.6 billion	Oldest fossils	71.0	March 13, 1:08:52 A.M.
4.6 billion	Earth formed	0	January 1, 12:00:00 A.M.

about 3.6 billion years ago, are in mid-March; dinosaurs, which existed between 230 million and 66 million years ago, are between December 11 and December 26; and all of recorded human history occurs during the last few seconds of December 31. Furthermore, all of the scientific and technological discoveries that have brought us to our present level of knowledge take place in the final second of the year!

The **geologic time scale** resulted from the work of many nineteenth-century geologists who pieced together information from numerous rock exposures and constructed a sequential chronology based on changes in the Earth's biota through time. However, with the discovery of radioactivity in 1895, and the development of various radiometric dating techniques, geologists have since been able to assign absolute age dates in years to the subdivisions of the geologic time scale (▶ Figure 1-19).

One of the cornerstones of geology is the **principle of uniformitarianism**. It is based on the premise that present-day processes have operated throughout geologic time. Therefore, in order to understand and interpret the geologic record, we must first understand present-day processes and their results.

Uniformitarianism is a powerful principle that allows us to use present-day processes as the basis for interpreting the past and for predicting potential future events. We should keep in mind that uniformitarianism does not exclude such sudden or catastrophic events as volcanic eruptions, earthquakes, landslides, or flooding. These are processes that shape our modern world, and, in fact, some geologists view the history of the Earth as a series of such short-term or punctuated events. Such a view is certainly in keeping with the modern principle of uniformitarianism.

Furthermore, uniformitarianism does not require that the rates and intensities of geological processes be constant through time. We know that volcanic activity was more intense in North America 5 to 10 million years ago than it is today, and that glaciation has been more prevalent during the last several million years than in the previous 300 million years.

What uniformitarianism means is that even though the rates and intensities of geological processes have varied during the past, the physical and chemical laws of nature have remained the same. Although the Earth is in a dynamic state of change and has been ever since it was formed, the processes that have shaped it during the past are the same ones in operation today.

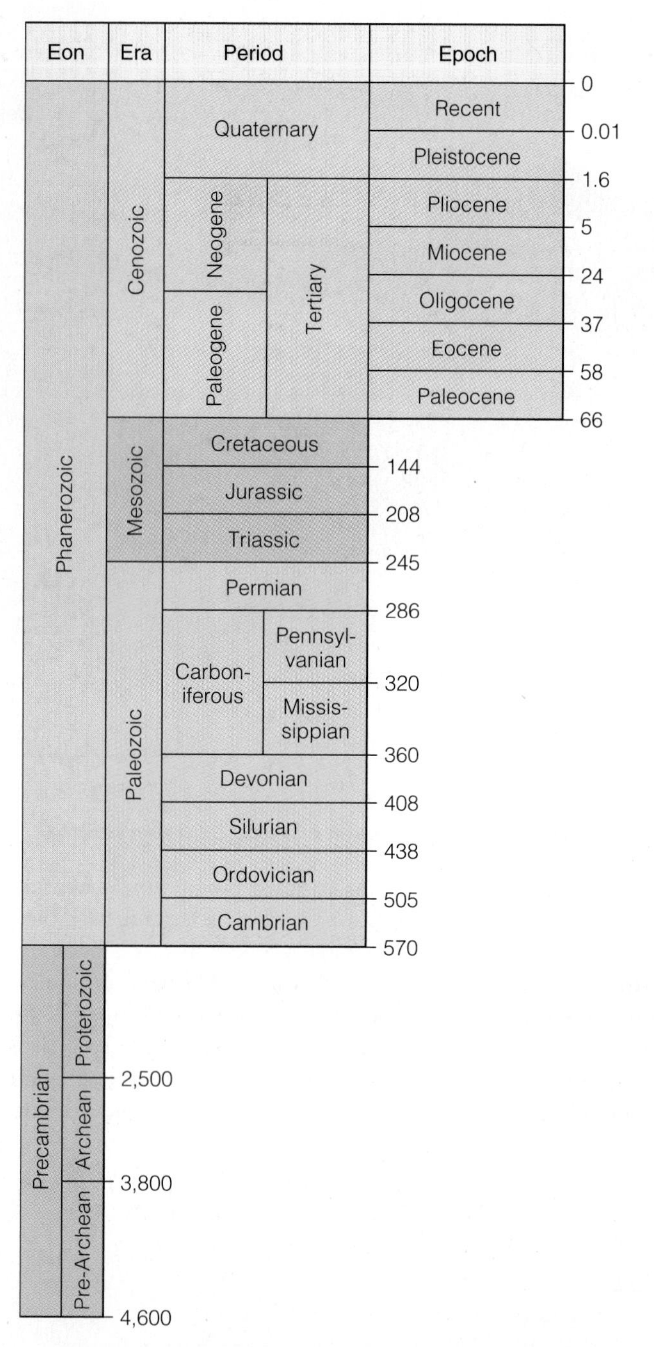

▶ FIGURE 1-19 The geologic time scale. Numbers to the right of the columns are ages in millons of years before the present.

CHAPTER SUMMARY

1. Geology is the study of the Earth. It is divided into two broad areas: physical geology is the study of the composition of Earth materials as well as the processes that operate within the Earth and upon its surface; historical geology examines the origin and evolution of the Earth, its continents, oceans, atmosphere, and life.

2. Geology is part of the human experience. We can find examples of it in the arts, music, and literature. A basic understanding of geology is also important for dealing with the many environmental problems and issues facing society.
3. Geologists engage in a variety of occupations, the main one being exploration for mineral and energy resources. They are also becoming increasingly involved in environmental issues and making short- and long-range predictions of the potential dangers from such natural disasters as volcanic eruptions and earthquakes.
4. About 4.6 billion years ago, the solar system formed from a rotating cloud of interstellar matter. Eventually, as this cloud condensed, it collapsed under the influence of gravity and flattened into a rotating disk. Within this rotating disk, the Sun, planets, and moons formed from the turbulent eddies of nebular gases and solids.
5. The Earth is differentiated into layers. The outermost layer is the crust, which is divided into continental and oceanic portions. Below the crust is the upper mantle. The crust and upper mantle, or lithosphere, overlie the asthenosphere, a zone that behaves plastically. The asthenosphere is underlain by the solid lower mantle. The Earth's core consists of an outer liquid portion and an inner solid portion.
6. The lithosphere is broken into a series of plates that diverge, converge, and slide sideways past one other.
7. The scientific method is an orderly, logical approach that involves gathering and analyzing facts about a particular phenomenon, formulating hypotheses to explain the phenomenon, testing the hypotheses, and finally proposing a theory. A theory is a testable explanation for some natural phenomenon that has a large body of supporting evidence.
8. Plate tectonic theory provides a unifying explanation for many geological features and events. The interaction between plates is responsible for volcanic eruptions, earthquakes, the formation of mountain ranges and ocean basins, and the recycling of rock material.
9. Igneous, sedimentary, and metamorphic rocks comprise the three major groups of rocks. Igneous rocks result from the crystallization of magma or the consolidation of volcanic ejecta. Sedimentary rocks are formed by the consolidation of rock fragments, precipitation of mineral matter from solution, or compaction of plant or animal remains. Metamorphic rocks are produced from other rocks, generally beneath the Earth's surface, by heat, pressure, and chemically active fluids.
10. The rock cycle illustrates the interactions between internal and external processes and shows how the three rock groups of the Earth are interrelated.
11. Time sets geology apart from the other sciences, except astronomy. The geologic time scale is the calendar geologists use to date past events; it is divided into eras, periods, and epochs.
12. The principle of uniformitarianism is basic to the interpretation of Earth history. This principle holds that the laws of nature have been constant through time and that the same processes operating today have operated in the past, albeit at different rates.

Important Terms

asthenosphere
convergent plate boundary
core
crust
divergent plate boundary
geologic time scale
geology
hypothesis
igneous rock
lithosphere
mantle
metamorphic rock
mineral
plate
plate tectonic theory
principle of uniformitarianism
rock
rock cycle
scientific method
sedimentary rock
subduction zone
theory
transform plate boundary

Review Questions

1. Krakatau is in:
   a. ____ Italy;
   b. ____ the United States;
   c. ____ Indonesia;
   d. ____ Japan;
   e. ____ Australia.
2. The study of Earth materials is:
   a. ____ paleontology;
   b. ____ stratigraphy;
   c. ____ physical geology;
   d. ____ historical geology;
   e. ____ environmental geology.
3. The eruption of Krakatau:
   a. ____ killed thousands of people;
   b. ____ created giant sea waves;
   c. ____ produced spectacular sunsets around the world;
   d. ____ caused a global cooling of about 1/2°C;
   e. ____ all of these.
4. Which of the following is not a subdivision of geology?
   a. ____ paleontology;

b. ____ astronomy;
c. ____ mineralogy;
d. ____ petrology;
e. ____ stratigraphy.

5. Into how many major concentric layers is the Earth divided?
a. ____ 1;
b. ____ 2;
c. ____ 3;
d. ____ 4;
e. ____ 5.

6. The Earth's core is inferred to be:
a. ____ hollow;
b. ____ composed of rock with a high silica content;
c. ____ completely molten;
d. ____ composed mostly of iron;
e. ____ completely solid.

7. The asthenosphere:
a. ____ lies beneath the lithosphere;
b. ____ is composed primarily of peridotite;
c. ____ behaves plastically and flows slowly;
d. ____ is the zone over which plates move;
e. ____ all of these.

8. The layer between the core and the crust is the:
a. ____ mantle;
b. ____ lithosphere;
c. ____ hydrosphere;
d. ____ biosphere;
e. ____ asthenosphere.

9. What fundamental process is thought to be responsible for plate motion?
a. ____ hot spot activity;
b. ____ subduction;
c. ____ spreading ridges;
d. ____ convection cells;
e. ____ density differences.

10. Which of the following statements about a scientific theory is not true?
a. ____ it is an explanation for some natural phenomenon;
b. ____ it has a large body of supporting evidence;
c. ____ it is a conjecture or guess;
d. ____ it is testable;
e. ____ none of these.

11. The man who proposed the hypothesis of continental drift was:
a. ____ Hutton;
b. ____ Wegener;
c. ____ Hess;
d. ____ Lyell;
e. ____ Lovelock.

12. Mid-oceanic ridges are examples of what type of boundary?
a. ____ divergent;
b. ____ convergent;
c. ____ transform;
d. ____ subduction;
e. ____ answers (b) and (d).

13. The San Andreas fault separating the Pacific plate from the North American plate is an example of what type of boundary?
a. ____ divergent;
b. ____ convergent;
c. ____ transform;
d. ____ subduction;
e. ____ answers (b) and (d).

14. A plate is composed of the:
a. ____ core and lower mantle;
b. ____ lower mantle and asthenosphere;
c. ____ asthenosphere and upper mantle;
d. ____ upper mantle and crust;
e. ____ continental and oceanic crust.

15. Which of the following is not a major rock group?
a. ____ meteorites;
b. ____ igneous;
c. ____ metamorphic;
d. ____ sedimentary;
e. ____ none of these.

16. Which rock group forms from the cooling of a magma?
a. ____ igneous;
b. ____ metamorphic;
c. ____ sedimentary;
d. ____ all of these;
e. ____ none of these.

17. The premise that present-day processes have operated throughout geologic time is known as the principle of:
a. ____ plate tectonics;
b. ____ sea-floor spreading;
c. ____ continental drift;
d. ____ volcanism;
e. ____ uniformitarianism.

18. The rock cycle implies that:
a. ____ metamorphic rocks are derived from magma;
b. ____ any rock type can be derived from any other rock type;
c. ____ igneous rocks only form beneath the Earth's surface;
d. ____ sedimentary rocks only form from the weathering of igneous rocks;
e. ____ all of these.

19. Why is it important for people to have a basic understanding of geology?

20. Describe some of the ways in which geology affects our everyday lives.

21. Explain both the difference between physical and historical geology and how they are related.

22. Name the major layers of the Earth, and describe their general composition.

23. Describe the scientific method, and explain how it may lead to a scientific theory.

24. Briefly discuss the origin of the Earth and its differentiation into three concentric layers.

25. Briefly describe the plate tectonic theory, and explain why it is a unifying theory of geology.

26. Describe the rock cycle, and explain how it may be related to plate tectonics.

27. Does the principle of uniformitarianism allow for catastrophic events? Explain.

Points to Ponder

1. Propose a pre–plate tectonic hypothesis explaining the formation and distribution of mountain ranges.
2. Why is an accurate geologic time scale particularly important for geologists in their search for mineral and energy resources and for reconstructing the history of the Earth?
3. Provide several examples of how a knowledge of geology would be useful in planning a military campaign against another country.

Additional Readings

Albritton, C. C., Jr. 1980. *The abyss of time.* San Francisco, Calif.: Freeman, Cooper & Co.

Dietrich, R. V. 1989. Rock music. *Earth Science* 42, no. 2: 24–25.

______. 1990. Rocks depicted in painting and sculpture. *Rocks & Minerals* 65, no. 3: 224–36.

______. 1991. How can I get others interested in rocks? *Rocks & Minerals* 67, no. 2: 124–28.

Dietrich, R. V., and B. J. Skinner. 1990. *Gems, granites, and gravels.* New York: Cambridge University Press.

Ernst, W. G. 1990. *The dynamic planet.* Irvington, N.Y.: Columbia University Press.

Francis, P., and S. Self. 1983. The eruption of Krakatau. *Scientific American* 249, no. 5: 172–87.

Hively, W. 1988. How much science does the public understand? *American Scientist* 76, no. 5: 439–44.

Mirsky, A. 1989. Geology in our everyday lives. *Journal of Geological Education* 37, no. 1: 9–12.

Officer, C., and J. Page. 1993. *Tales of the Earth.* New York: Oxford University Press.

Pestrong, R. 1994. Geoscience and the arts. *Journal of Geological Education* 42, no. 3: 249–57.

Rhodes, F. H. T., and R. O. Stone. 1981. *Language of the Earth.* Elmsford, N.Y.: Pergamon Press.

Siever, R. 1983. The dynamic Earth. *Scientific American* 249, no. 3: 46–55.

Chapter 2

MINERALS

OUTLINE

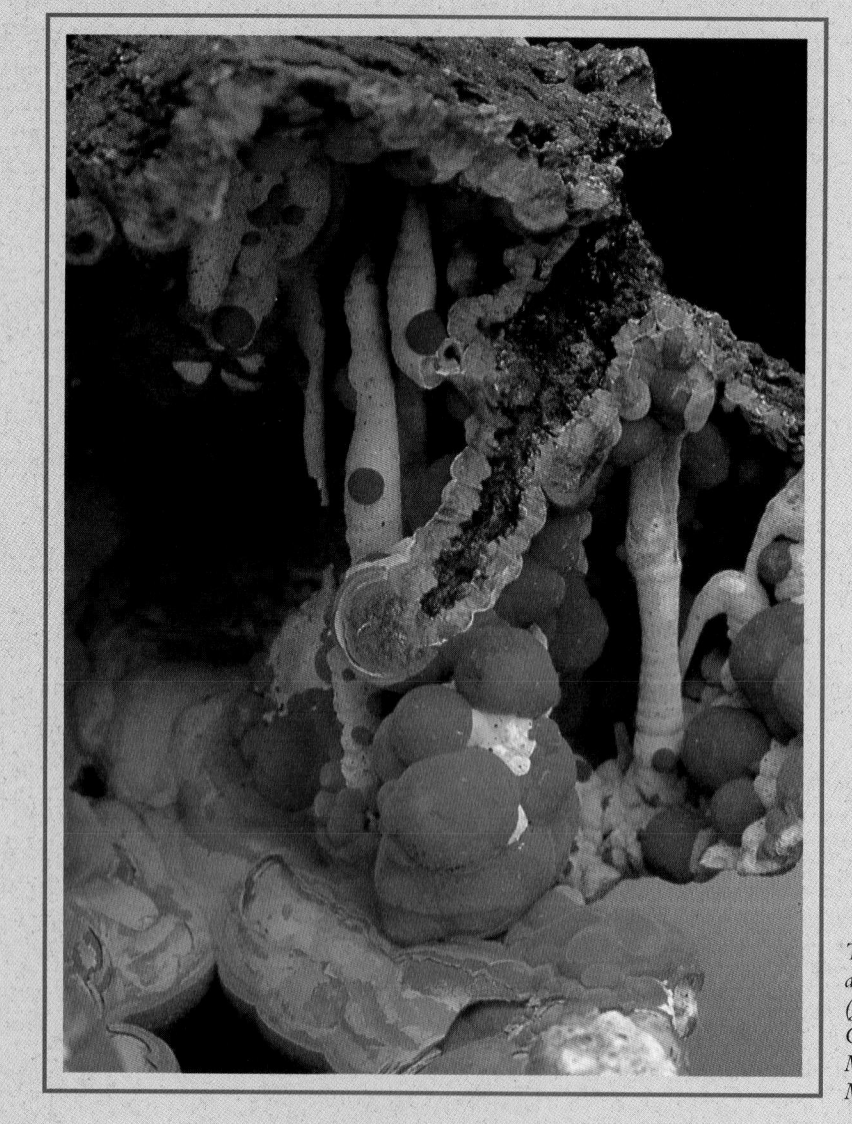

Two copper minerals, azurite (blue) and malachite (green), in the Flagg Collection, Arizona Mining and Mineral Museum.

PROLOGUE

Among the hundreds of minerals used by humans none is so highly prized and eagerly sought as gold (⊳ Figure 2-1). This deep yellow mineral has been the cause of feuds and wars and was one of the incentives for the exploration of the Americas. Gold has been mined for at least 6,000 years, and archaeological evidence indicates that people in Spain possessed small quantities of gold 40,000 years ago. Probably no other substance has caused so much misery, but at the same time provided so many benefits for those who possessed it.

Why is gold so highly prized? Certainly not for use in tools or weapons, for it is too soft and pliable to hold a cutting edge. Furthermore, it is too heavy to be practical for most utilitarian purposes (it weighs about twice as much as lead). During most of historic time, gold has been used for jewelry, ornaments, and ritual objects and has served as a symbol of wealth and as a monetary standard. Gold is so desired for several reasons: (1) its pleasing appearance, (2) the ease with which it can be worked, (3) its durability, and (4) its scarcity (it is much rarer than silver).

Central and South American natives used gold extensively long before the arrival of Europeans. In fact, the Europeans' lust for gold was responsible for the ruthless conquest of the natives in those areas. In the United States, gold was first profitably mined in North Carolina in 1801 and in Georgia in 1829, but the truly spectacular finds occurred in California in 1848. This latter discovery culminated in the great gold rush of 1849 when tens of thousands of people flocked to California to find riches. Unfortunately, only a few found what they sought. Nevertheless, during the five years from 1848 to 1853, which constituted the gold rush proper, more than $200 million in gold was recovered.

Another gold rush occurred in 1876 following the report by Lieutenant Colonel George Armstrong Custer that "gold in satisfactory quantities can be obtained in the Black Hills [South Dakota]." The flood of miners into the Black Hills, the Holy Wilderness of the Sioux Indians, resulted in the Indian War during which Custer and some 260 of his men were annihilated at the Battle of the Little Bighorn in Montana in June 1876. Despite this stunning victory, the Sioux could not sustain a war against the U.S. Army, and in September 1876, they were forced to relinquish the Black Hills.

Canada, too, has had its gold rushes. The first discovery came in 1850 in the Queen Charlotte Islands on the Pacific coast, and by 1858 about 10,000 people were panning for gold there. The greatest Canadian gold rush occurred between 1897 and 1899 when as many as 35,000 men and women traveled to the remote, hostile Klondike region of the Yukon Territory. In fact, Dawson City grew so rapidly that hundreds of people had to be evacuated during the autumn and winter of 1897 because of food shortages. As in other gold rushes, local merchants made out better than the miners, most of whom barely eked out a living.

For 50 years following the California gold rush, the United States led the world in gold production, and it still produces a considerable amount, mostly from mines in Nevada, California, and South Dakota. Currently the leading producer is South Africa with the United States a distant second, followed by Russia, Australia, and Canada where gold valued at $2.1 billion was mined in 1992. Although much gold still goes into jewelry, its uses have expanded those of earlier times; now gold is used in the chemical industry and for gold plating, electrical circuitry, and glass making. Consequently, the quest for gold has not ceased or even abated. In many industrialized nations, including the United States, domestic production cannot meet the demand, and much of the gold used must be imported.

⊳ FIGURE 2-1 Specimen of gold from Grass Valley, California—National Museum of Natural History (NMNH) specimen #R121297. (Photo by D. Penland, courtesy of Smithsonian Institution.)

Introduction

The term "mineral" commonly brings to mind dietary substances essential for good nutrition such as calcium, iron, potassium, and magnesium. These substances are actually chemical elements, not minerals in the geologic sense. Mineral is also sometimes used to refer to any substance that is neither animal nor vegetable. Such usage implies that minerals are inorganic substances, which is correct, but not all inorganic substances are minerals. Water, for example, is not a mineral even though it is inorganic and is composed of the same chemical elements as ice, which is a mineral. Ice, of course, is a solid whereas water is a liquid; minerals are solids rather than liquids or gases. In fact, geologists have a very specific definition of the term **mineral**: a naturally occurring, inorganic, crystalline solid. Crystalline means it has a regular internal structure. Furthermore, a mineral has a narrowly defined chemical composition and characteristic physical properties such as density, color, and hardness. Most rocks are solid aggregates of one or more minerals, and thus minerals are the building blocks of rocks.

Obviously, minerals are important to geologists as the constituents of rocks, but they are important for other reasons as well. Gemstones such as diamond and topaz are minerals, and rubies are simply red-colored varieties of the mineral corundum. The sand used in the manufacture of glass is composed of the mineral quartz, and ore deposits are natural concentrations of economically valuable minerals. Indeed, industrialized societies depend directly upon finding and using mineral resources such as iron, copper, gold, and many others.

Matter and Its Composition

Anything that has mass and occupies space is *matter*. The atmosphere, water, plants and animals, and minerals and rocks are all composed of matter. Matter occurs in one of three states or phases, all of which are important in geology: *solids, liquids,* and *gases* (◉ Table 2-1). Atmospheric gases and liquids such as surface water and groundwater will be discussed later in this book, but here we are concerned chiefly with solids because all minerals are solids.

Elements and Atoms

All matter is made up of chemical **elements**, each of which is composed of incredibly small particles called **atoms**. Atoms are the smallest units of matter that retain the characteristics of an element. Ninety-two naturally occurring elements have been discovered, some of which are listed in ◉ Table 2-2, and more than a dozen additional elements have been made in laboratories. Each naturally occurring element and most artificially produced ones have a name and a chemical symbol (Table 2-2).

Atoms consist of a compact **nucleus** composed of one or more **protons**, which are particles with a positive electrical charge, and **neutrons**, which are electrically neutral (▶ Figure 2-2). The nucleus of an atom makes up most of its mass. Encircling the nucleus are negatively charged particles called **electrons**. Electrons orbit rapidly around the

▶ **FIGURE 2-2** The structure of an atom. The dense nucleus consisting of protons and neutrons is surrounded by a cloud of orbiting electrons.

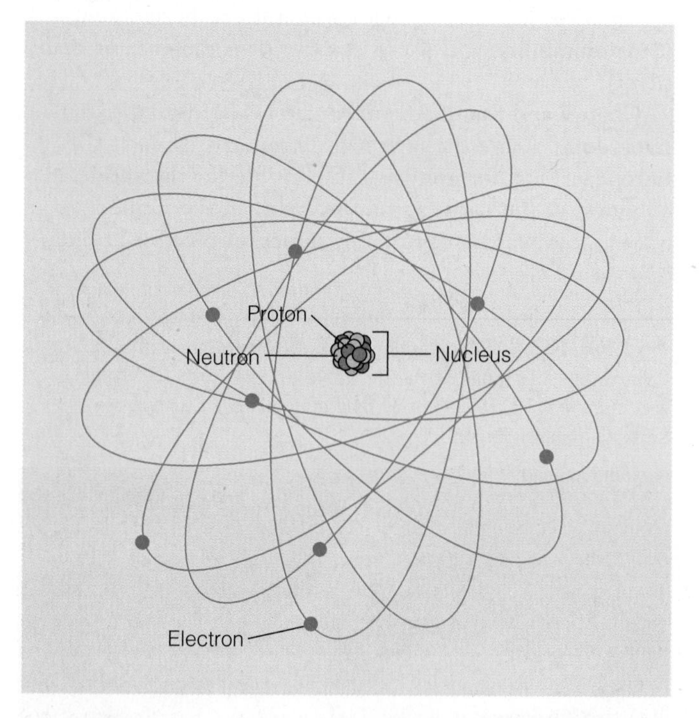

TABLE 2-1 Phases or States of Matter

Phase	Characteristics	Examples
Solid	Rigid substance that retains its shape unless distorted by a force	Minerals, rocks, iron, wood
Liquid	Flows easily and conforms to the shape of the containing vessel; has a well-defined upper surface and greater density than a gas	Water, lava, wine, blood, gasoline
Gas	Flows easily and expands to fill all parts of a containing vessel; lacks a well-defined upper surface; is compressible	Helium, nitrogen, air, water vapor

 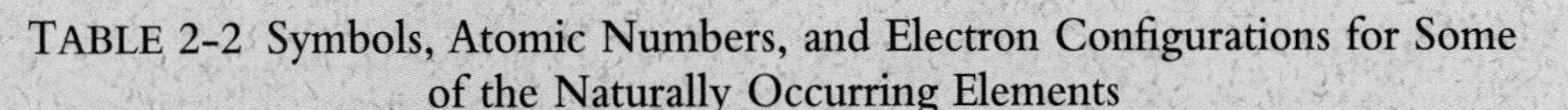 

TABLE 2-2 Symbols, Atomic Numbers, and Electron Configurations for Some of the Naturally Occurring Elements

Element	Symbol	Atomic Number	Number of Electrons in Each Shell			
			1	*2*	*3*	*4*
Hydrogen	H	1	1			
Helium	He	2	2			
Lithium	Li	3	2	1		
Beryllium	Be	4	2	2		
Boron	B	5	2	3		
Carbon	C	6	2	4		
Nitrogen	N	7	2	5		
Oxygen	O	8	2	6		
Fluorine	F	9	2	7		
Neon	Ne	10	2	8		
Sodium	Na	11	2	8	1	
Magnesium	Mg	12	2	8	2	
Aluminum	Al	13	2	8	3	
Silicon	Si	14	2	8	4	
Phosphorus	P	15	2	8	5	
Sulfur	S	16	2	8	6	
Chlorine	Cl	17	2	8	7	
Argon	Ar	18	2	8	8	
Potassium	K	19	2	8	8	1
Calcium	Ca	20	2	8	8	2

nucleus at specific distances in one or more **electron shells** (Figure 2-2). The number of protons in the nucleus of an atom determines what the element is and determines the **atomic number** for that element. For example, each atom of the element hydrogen (H) has one proton in its nucleus and thus has an atomic number of l. Helium (He) possesses 2 protons, carbon (C) has 6, and uranium (U) has 92, so their atomic numbers are 2, 6, and 92, respectively.

Atoms are also characterized by their **atomic mass number**, which is determined by adding together the number of protons and neutrons in the nucleus (electrons contribute negligible mass to an atom) (Figure 2-2). However, not all atoms of the same element have the same number of neutrons in their nuclei. In other words, atoms of the same element may have different atomic mass numbers. For example, different carbon (C) atoms have atomic mass numbers of 12, 13, and 14. All of these atoms possess 6 protons—otherwise they would not be carbon—but the number of neutrons varies. Forms of the same element with different atomic mass numbers are **isotopes** (▶ Figure 2-3).

A number of elements have a single isotope but many, such as uranium and carbon, have several (Figure 2-3). Some isotopes are unstable and spontaneously change to a stable form. This process, called *radioactive decay,* occurs because the forces that bind the nucleus together are not strong enough. Such decay occurs at known rates and is the basis for several techniques for determining age that will be discussed in Chapter 8. All isotopes of an element behave the same chemically. For example, both carbon 12 and carbon 14 are present in carbon dioxide (CO_2).

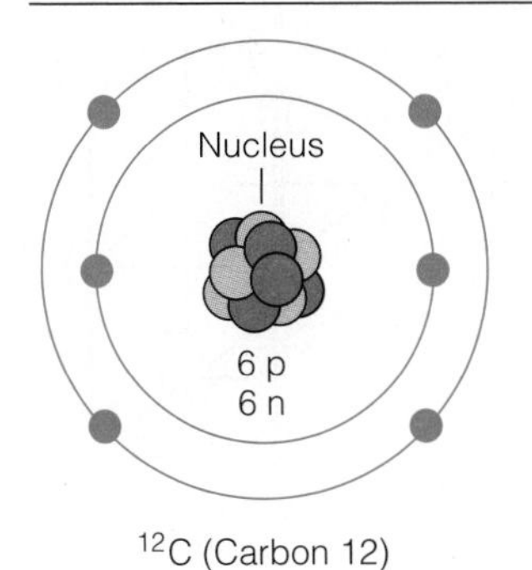

▶ **FIGURE 2-3** Schematic representation of isotopes of carbon. A carbon atom has an atomic number of 6 and an atomic mass number of 12, 13, or 14 depending on the number of neutrons in its nucleus.

Bonding and Compounds

The process whereby atoms are joined to other atoms is called **bonding**. When atoms of two or more different elements are bonded, the resulting substance is a **compound**. A chemical substance such as gaseous oxygen, which consists entirely of oxygen atoms, is an element, whereas ice, which consists of hydrogen and oxygen, is a compound. Most minerals are compounds although there are several important exceptions, such as gold and silver.

To understand bonding, it is necessary to delve deeper into the structure of atoms. Recall that negatively charged electrons in electron shells orbit the nuclei of atoms. With the exception of hydrogen, which has only one proton and one electron, the innermost electron shell of an atom contains only two electrons. The other shells contain various numbers of electrons, but the outermost shell never contains more than eight (Table 2-2). The electrons in the outermost shell are those that are usually involved in chemical bonding.

Two types of chemical bonds, *ionic* and *covalent,* are particularly important in minerals, and many minerals contain both types of bonds. Two other types of chemical bonds, *metallic* and *van der Waals,* are much less common, but are extremely important in determining the properties of some very useful minerals.

Ionic Bonding. Notice in Table 2-2 that most atoms have fewer than eight electrons in their outermost electron shell. Some elements, however, including neon and argon, have complete outer shells containing eight electrons; they are known as the *noble gases.* The noble gases do not react readily with other elements to form compounds because of this electron configuration. Interactions among atoms tend to produce electron configurations similar to those of the noble gases. That is, atoms interact such that their outermost electron shell is filled with eight electrons, unless the first shell (with two electrons) is also the outermost electron shell as in helium.

One way that the noble gas configuration can be attained is by the transfer of one or more electrons from one atom to another. Common salt, for example, is composed of the elements sodium (Na) and chlorine (Cl), each of which is poisonous, but when combined chemically, they form the compound sodium chloride (NaCl), the mineral halite or common salt. Notice in ➢ Figure 2-4a that sodium has 11 protons and 11 electrons; thus, the positive electrical charges of the protons are exactly balanced by the negative charges of the electrons, and the atom is electrically neutral. Likewise, chlorine with 17 protons and 17 electrons is electrically neutral (Figure 2-4a). However, neither sodium nor chlorine has eight electrons in its outermost electron shell; sodium has only one whereas chlorine has seven. In order to attain a stable configuration, sodium loses the electron in its outermost electron shell, leaving its next shell with eight electrons as the outermost one (Figure 2-4a). Sodium now has one fewer electron (negative charge) than it has protons (positive charge) so it is an electrically charged particle.

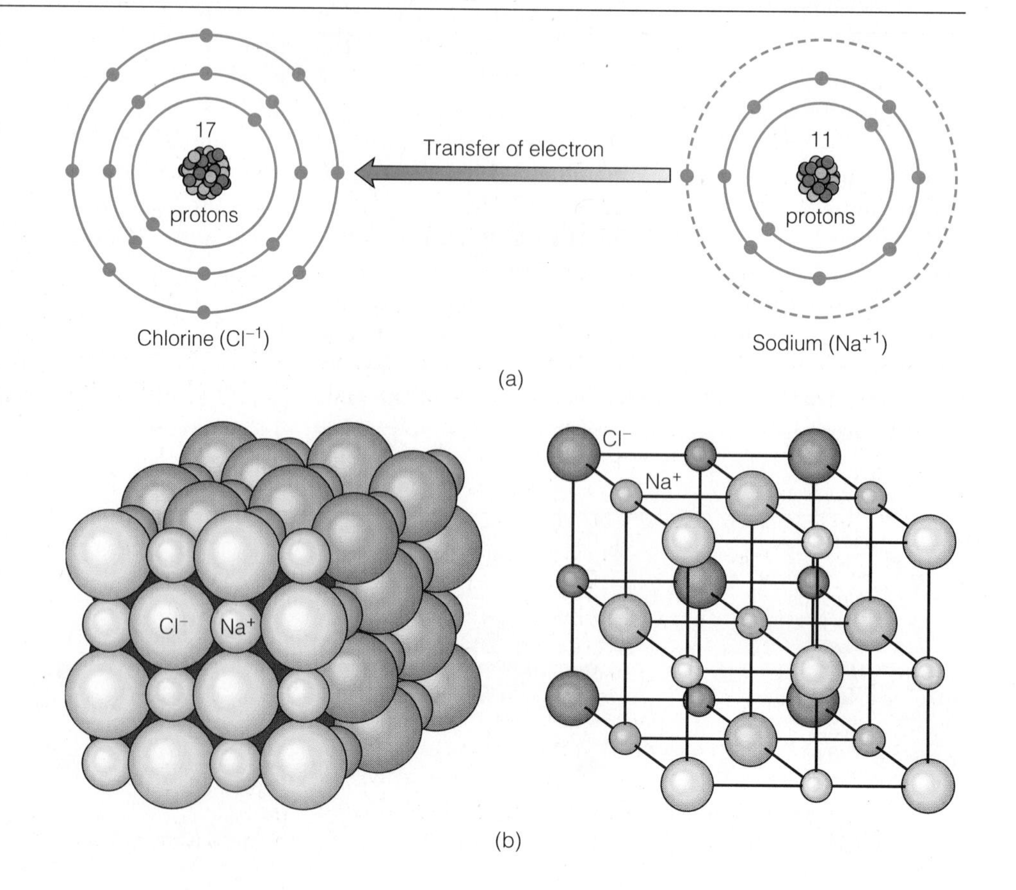

➢ FIGURE 2-4 (*a*) Ionic bonding. The electron in the outermost shell of sodium is transferred to the outermost electron shell of chlorine. Once the transfer has occurred, sodium and chlorine are positively and negatively charged ions, respectively. (*b*) The crystal structure of sodium chloride, the mineral halite. The diagram on the left shows the relative sizes of sodium and chlorine ions, and the diagram on the right shows the locations of the ions in the crystal structure.

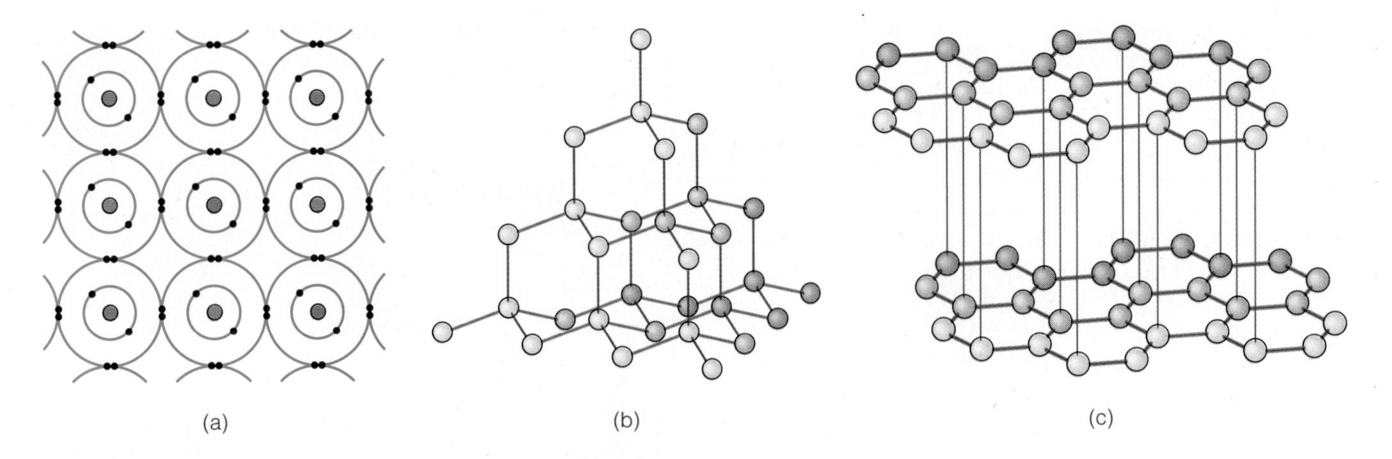

▶ FIGURE 2-5 (*a*) Covalent bonds formed by adjacent atoms sharing electrons in diamond. (*b*) The three-dimensional framework of carbon atoms in diamond. (*c*) Covalent bonding also occurs in graphite, but here the carbon atoms are bonded together to form sheets that are held to one another by van der Waals bonds. The sheets themselves are strong, but the bonds between sheets are weak.

Such a particle is an **ion** and, in the case of sodium, is symbolized Na^{+1}.

The electron lost by sodium is transferred to the outermost electron shell of chlorine, which had seven electrons to begin with. The addition of one more electron gives chlorine an outermost electron shell of eight electrons, the configuration of a noble gas. But its total number of electrons, is now 18, which exceeds by one the number of protons. Accordingly, chlorine also becomes an ion, but it is negatively charged (Cl^{-1}). An **ionic bond** forms between sodium and chlorine because of the attractive force between the positively charged sodium ion and the negatively charged chlorine ion (Figure 2-4a).

In ionic compounds, such as sodium chloride (the mineral halite), the ions are arranged in a three-dimensional framework that results in overall electrical neutrality. In halite, sodium ions are bonded to chlorine ions on all sides, and chlorine ions are surrounded by sodium ions (Figure 2-4b).

Covalent Bonding. **Covalent bonds** form between atoms when their electron shells overlap and electrons are shared. For example, atoms of the same element, such as oxygen in oxygen gas, cannot bond by transferring electrons from one atom to another. Carbon (C), which forms the minerals graphite and diamond, has four electrons in its outermost electron shell (▶ Figure 2-5a). If these four electrons were transferred to another carbon atom, the atom receiving the electrons would have the noble gas configuration of eight electrons in its outermost electron shell, but the atom contributing the electrons would not.

In such situations, adjacent atoms share electrons by overlapping their electron shells. A carbon atom in diamond, for instance, shares all four of its outermost electrons with a neighbor to produce a stable noble gas configuration (Figure 2-5a).

Covalent bonds are not restricted to substances composed of atoms of a single kind. Among the most common minerals, the silicates (discussed later in this chapter), the element silicon forms partly covalent and partly ionic bonds with oxygen.

Metallic and van der Waals Bonds. Metallic bonding results from an extreme type of electron sharing. The electrons of the outermost electron shell of such metals as gold, silver, and copper readily move about from one atom to another. This electron mobility accounts for the fact that metals have a metallic luster (their appearance in reflected light), provide good electrical and thermal conductivity, and can be easily reshaped. Only a few minerals possess metallic bonds, but those that do are very useful; copper, for example, is used for electrical wiring because of its high electrical conductivity.

Some electrically neutral atoms and molecules* have no electrons available for ionic, covalent, or metallic bonding. They nevertheless have a weak attractive force between them when in proximity. This weak attractive force is a *van der Waals* or *residual bond.* The carbon atoms in the mineral graphite are covalently bonded to form sheets, but the sheets are weakly held together by van der Waals bonds (Figure 2-5c).

MINERALS

Most minerals are compounds of two or more elements. Mineral composition is generally shown by a chemical formula, which is a shorthand way of indicating the numbers of atoms of different elements composing a mineral. The mineral quartz consists of one silicon (Si) atom for every two oxygen (O) atoms, and thus has the formula SiO_2; the

*A molecule is the smallest unit of a substance having the properties of that substance. A water molecule (H_2O), for example, possesses two hydrogen atoms and one oxygen atom.

Mineralogy: A Career with Diverse Pursuits

My pathway toward becoming a professional geoscientist was somewhat circuitous. As an undergraduate, I first majored in forestry and then zoology, fully intending to become a marine biologist. After graduation I obtained a temporary position with the U.S. Geological Survey where, fortuitously, I learned to operate the electron microscope and then applied my newfound skills to the study of clay minerals—tiny mineral particles that generally can be seen only by utilizing the very high magnifications obtainable with this instrument. A whole new world opened up to me as I photographed mineral particles magnified as much as half a million times. This experience convinced me to redirect my interests once again, and I went on to graduate school to study the science of mineralogy.

My first serious scientific studies were concerned with the elucidation of the crystal symmetries of fine-grained vanadium-bearing minerals, utilizing the techniques of electron microscopy and diffraction. This work was followed by studies of the crystal structures of several uranium-bearing minerals as part of the U.S. Atomic Energy Program and, then, by an extended study of the crystal structures and physical chemistry of several of the silicon-bearing minerals that make up a major portion of the Earth's crust—the micas, pyroxenes, and amphiboles.

In 1968 I submitted a proposal to the National Aeronautics and Space Administration to study the important silicate minerals composing the surface of the Moon. Many (but not all) geoscientists then thought that the lunar surface was composed of rocks and minerals similar to those found on the Earth. One influential cosmologist suggested that the lunar surface consisted of substances that have no counterpart on the Earth. Those were heady times. The first lunar samples were returned to Earth by the *Apollo 11* crew on July 24, 1969. Within a few weeks, samples of lunar rocks and soils were sent to earth scientists all over the world, and after four months of intensive study, over 500 of these scientists converged on Houston, Texas, on January 5, 1970, to report on their investigations. Indeed, it turned out that the rocks and minerals of the Moon and Earth are similar in many respects; the Moon was not made of green cheese or some other exotic substance. There were, however, some striking differences noted between lunar and terrestrial rocks. The lunar rocks are very old (3.7 to 4.3 billion years); the absence of younger rocks implies that geochemical processes stopped very early in the Moon's history in contrast to Earth processes that still go on today. New minerals were found on the Moon that had not been observed before on the Earth, and a complete absence of water-bearing minerals was also noted.

The research on the terrestrial and lunar minerals led to a new understanding of the temperature and pressure under which various silicate minerals form and of the nature of their transformation during the long slow cooling from temperatures as high as 1,200° C. During this cooling process, the high-temperature silicate minerals precipitate other minerals, the original crystals becoming a virtual zoo of secondary minerals as revealed by X-ray diffraction and electron microscopy.

In the late 1970s, I became involved in the "asbestos and health" issue due to the increasing national concern over the effects of exposure to asbestos dust. Four types of asbestos minerals have been used in international commerce. The most utilized type is chrysotile asbestos, which accounts for 95% of the total world production. The three other forms of asbestos, anthophyllite, amosite, and crocidolite, account for the rest of the world production. Until recently, few medical investigators had paid any attention to the relationship between the type and degree of asbestos disease (lung cancer, mesothelioma, and asbestosis) and the type of asbestos to which the individual was exposed. By comparing the medical studies of miners and millers who were exposed to only one form of asbestos, I found that the common form of asbestos, chrysotile, is not a hazard at low to moderate exposures and offers no danger to children attending school in buildings that contain this mineral. Since most buildings contain only chrysotile, removal of asbestos from such buildings is unnecessary and even counterproductive.

More recently, I completed a study of the effects of "acid rain" on limestone and marble building materials. I found that air pollutants generated within large cities, rather than the sulfur dioxide originating from midwestern power plants, were primarily responsible for stone deterioration. This observation has particular importance as we look for the most effective and economical ways of mitigating the effects of acid rain in urban areas. After devoting the last few years to a mineral resource investigation in the Hot Springs area of Arkansas, I have now returned to the subject of the health effects of mineral dusts—a subject that I believe is causing undo concern among the citizens of the United States.

Perhaps this essay can give the reader some idea of how interesting and diversified geoscience can be. It combines laboratory work with field studies in fascinating localities and encompasses both basic research and research directly related to humans' benefit.

MALCOLM ROSS earned his Ph.D. degree in geology from Harvard University in 1962. He has been employed by the U.S. Geological Survey since that time, specializing in studies related to the occurrence, chemistry, structure, and health effects of a wide variety of minerals.

subscript number indicates the number of atoms. Orthoclase is composed of one potassium, one aluminum, three silicon, and eight oxygen atoms, so its formula is $KAlSi_3O_8$. A few minerals are composed of a single element. Known as **native elements**, they include such minerals as gold (Au), silver (Ag), platinum (Pt), and graphite and diamond, both of which are composed of carbon (C).

Before we discuss minerals in more detail, let us recall our formal definition: a mineral is a naturally occurring, inorganic, crystalline solid, with a narrowly defined chemical composition and characteristic physical properties. The next sections will examine each part of this definition.

Naturally Occurring, Inorganic Substances

"Naturally occurring" excludes from minerals all substances that are manufactured by humans. Accordingly, synthetic diamonds and rubies and a number of other artificially synthesized substances are not regarded as minerals by most geologists. This criterion is particularly important to those who buy and sell gemstones, most of which are minerals, because some human-made substances are very difficult to distinguish from natural gem minerals.

Some geologists think the term "inorganic" in the mineral definition is superfluous. It does, however, remind us that animal matter and vegetable matter are not minerals. Nevertheless, some organisms including corals, clams, and a number of other animals construct their shells of the compound calcium carbonate ($CaCO_3$), which is either aragonite or calcite, both of which are minerals.

The Nature of Crystals

By definition minerals are **crystalline solids**, in which the constituent atoms are arranged in a regular, three-dimensional framework, as in the mineral halite (Figure 2-4b). Under ideal conditions, such as in a cavity, mineral crystals can grow and form perfect crystals that possess planar surfaces (crystal faces), sharp corners, and straight edges (▶ Figure 2-6). In other words, the regular geometric shape of a well-formed mineral crystal is the exterior manifestation of an ordered internal atomic arrangement. Not all rigid substances are crystalline solids, however; natural and manufactured glass lack the ordered arrangement of atoms and are said to be *amorphous,* meaning without form.

As early as 1669, a well-known Danish scientist, Nicholas Steno, determined that the angles of intersection of equivalent crystal faces on different specimens of quartz are identical. Since then the *constancy of interfacial angles* has been demonstrated for many other minerals, regardless of their size, shape, or geographic occurrence (▶ Figure 2-7). Steno postulated that mineral crystals are composed of very small, identical building blocks and that the arrangement of these blocks determines the external form of the crystals. Such regularity of the external form of minerals must surely mean that external crystal form is controlled by internal structure.

In many cases numerous minerals grow in proximity, as in a cooling lava flow, and do not have an opportunity to develop well-formed crystals. Even though well-formed mineral crystals are rare, all minerals of a given species have the same internal atomic structure.

Crystalline structure can be demonstrated even in minerals lacking obvious crystals. For example, many minerals possess a property known as *cleavage,* meaning that they break or split along closely spaced, smooth planes. The fact that these minerals can be split along such smooth planar surfaces indicates that the mineral's internal structure controls such breakage. The behavior of light and X-ray beams transmitted through minerals also provides compelling evidence for an orderly arrangement of atoms within minerals.

Chemical Composition

The definition of a mineral contains the phrase "a narrowly defined chemical composition," because some minerals actually have a range of compositions. For many minerals the chemical composition is constant: quartz is always composed of silicon and oxygen (SiO_2), and halite contains only sodium and chlorine (NaCl). Other minerals have a range of

▶ **FIGURE 2-6** Mineral crystals occur in a variety of shapes, several of which are shown here. (*a*) Cubic crystals typically develop in the minerals halite, galena, and pyrite. (*b*) Dodecahedron crystals such as those of garnet have 12 sides. (*c*) Diamond has octahedral or 8-sided crystals. (*d*) A prism terminated by a pyramid is found in quartz.

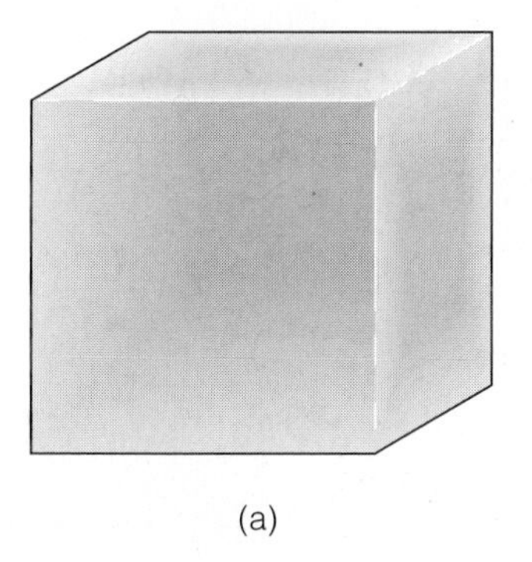
(a)

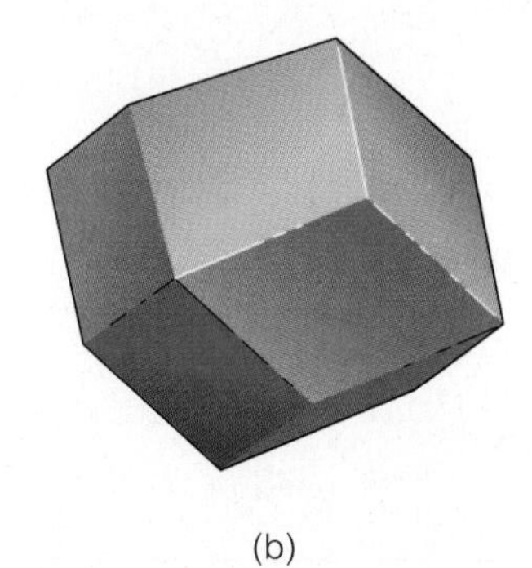
(b)

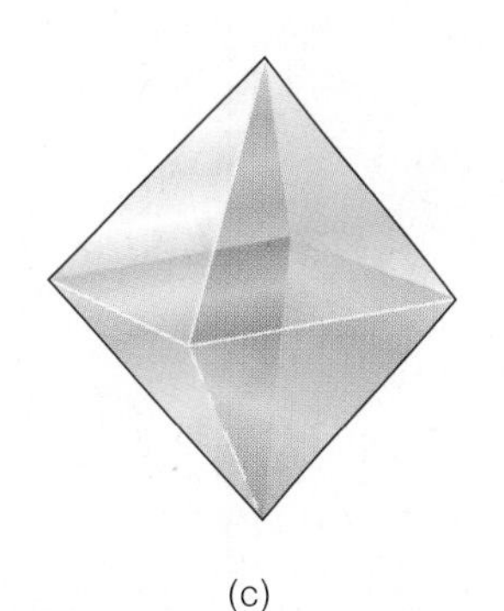
(c)

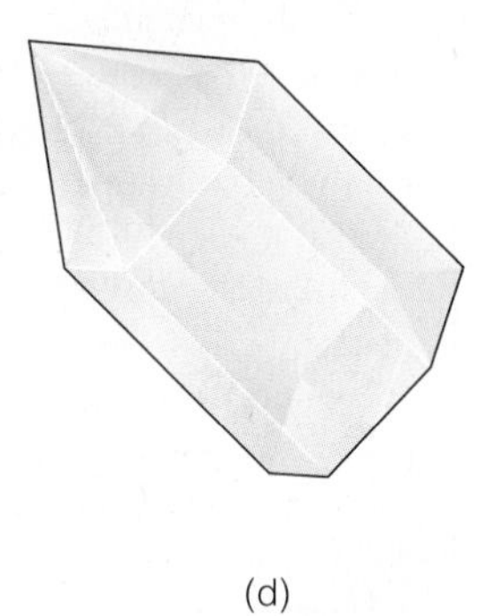
(d)

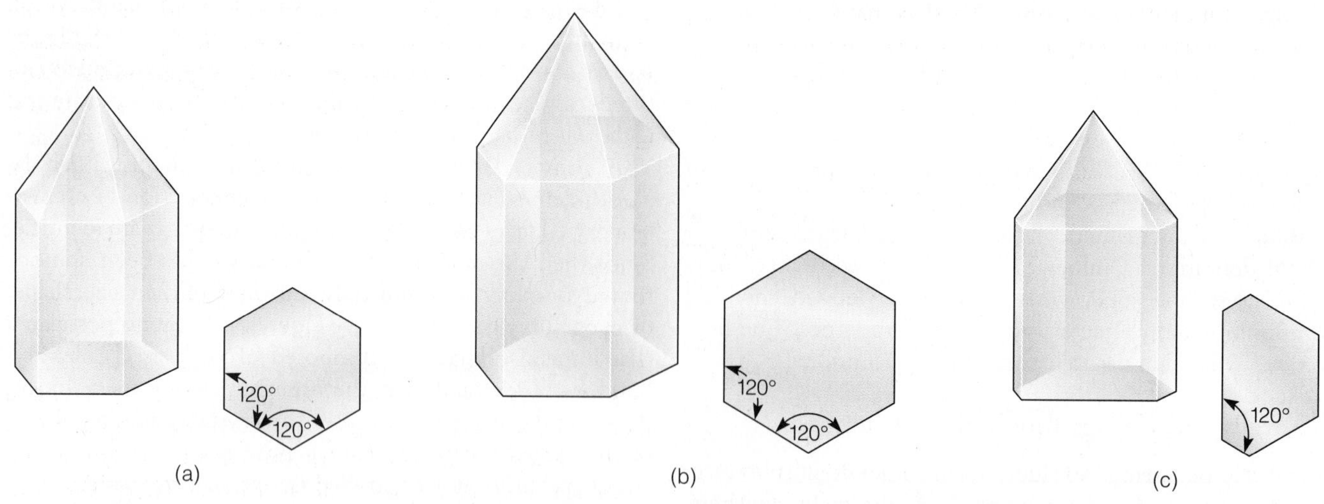

➢ **FIGURE 2-7** Side views and cross sections of three quartz crystals showing the constancy of interfacial angles: (*a*) a well-shaped crystal; (*b*) a larger crystal; and (*c*) a poorly shaped crystal. The angles formed between equivalent crystal faces on different specimens of the same mineral are the same regardless of the size or shape of the specimens.

compositions because one element may substitute for another if the atoms of two or more elements are nearly the same size and the same charge. Notice in ➢ Figure 2-8 that iron and magnesium atoms are about the same size; therefore they can substitute for one another. The chemical formula for the mineral olivine is $(Mg,Fe)_2SiO_4$, meaning that, in addition to silicon and oxygen, it may contain only magnesium, only iron, or a combination of both. A number of other minerals also have ranges of compositions, so these are actually mineral groups with several members.

Physical Properties

The last criterion in our definition of a mineral, "characteristic physical properties," refers to such properties as hardness, color, and crystal form. These properties are controlled

➢ **FIGURE 2-8** Electrical charges and relative sizes of ions common in minerals. The numbers within the ions are the radii shown in Ångstrom units.

Negatively charged ions		Positively charged ions			
2^-	1^-	1^+	2^+	3^+	4^+
Oxygen 1.40	Fluorine 1.36	Sodium 0.99	Calcium 1.00	Aluminum 0.39	Silicon 0.26
Sulfur 1.84	Chlorine 1.81	Potassium 1.37	$Iron^{2+}$ 0.63	$Iron^{3+}$ 0.49	Carbon 0.15
			Magnesium 0.72		

1 Ångstrom = 10^{-8} cm

by composition and structure. We shall have more to say about physical properties of minerals later in this chapter.

 MINERAL DIVERSITY

More than 3,500 minerals have been identified and described, but only a very few—perhaps two dozen—are particularly common. Considering that 92 naturally occurring elements have been discovered, one might think that an extremely large number of minerals could be formed, but several factors limit the number of possible minerals. For one thing, many combinations of elements are chemically impossible; no compounds are composed of only potassium and sodium or of silicon and iron, for example. Another important factor restricting the number of common minerals is that only eight chemical elements make up the bulk of the Earth's crust (◉ Table 2-3). As Table 2-3 shows, oxygen and silicon constitute more than 74% (by weight) of the Earth's crust and nearly 84% of the atoms available to form compounds. By far the most common minerals in the Earth's crust consist of silicon and oxygen, combined with one or more of the other elements listed in Table 2-3.

 MINERAL GROUPS

Objects such as organisms, rocks, and minerals can be classified in many ways, such as by color, size, shape, density, composition, structure, or a combination of such features. In any case, all classification schemes share two characteristics: (1) they systematically categorize similar objects, and (2) their purpose is to convey information. Geologists recognize mineral classes or groups, each of which contains members that share the same negatively charged ion or ion group; several of these mineral groups are listed in ◉ Table 2-4.

Silicate Minerals

Because silicon and oxygen are the two most abundant elements in the Earth's crust (Table 2-3), it is not surprising that many minerals contain these elements. A combination of silicon and oxygen is called **silica**, and the minerals containing silica are **silicates**. Quartz (SiO_2) is composed entirely of silicon and oxygen so it is pure silica. Most silicates have one or more additional elements, as in orthoclase ($KAlSi_3O_8$) and olivine [$(Mg,Fe)_2SiO_4$]. Silicate minerals include about one-third of all known minerals, but their abundance is even more impressive when one considers that they make up perhaps as much as 95% of the Earth's crust.

The basic building block of all silicate minerals is the **silica tetrahedron**, which consists of one silicon atom and four oxygen atoms (▶ Figure 2-9). These atoms are arranged so that the four oxygen atoms surround a silicon atom, which occupies the space between the oxygen atoms; thus, a four-faced pyramidal structure is formed. The silicon atom has a positive charge of 4, while each of the four oxygen atoms has a negative charge of 2, resulting in an ion group with a total negative charge of 4 $(SiO_4)^{-4}$.

Because the silica tetrahedron has a negative charge, it does not exist in nature as an isolated ion group; rather it combines with positively charged ions or shares its oxygen atoms with other silica tetrahedra. In the simplest silicate minerals, the silica tetrahedra exist as single units bonded to positively charged ions. In minerals containing isolated tetrahedra, the

▶ FIGURE 2-9 The silica tetrahedron. (*a*) Expanded view showing oxygen atoms at the corners of a tetrahedron and a small silicon atom at the center. (*b*) View of the silica tetrahedron as it really exists with the oxygen atoms touching. (*c*) The silica tetrahedron represented diagramatically; the oxygen atoms are at the four points of the tetrahedron.

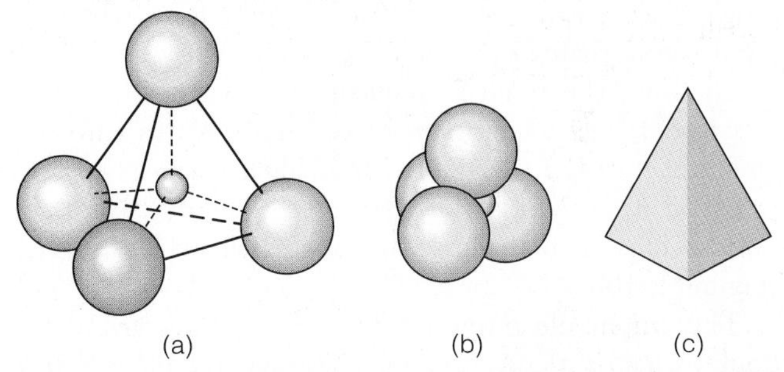

TABLE 2-3 Common Elements in the Earth's Crust

Element	Symbol	Percentage of Crust (by Weight)	Percentage of Crust (by Atoms)
Oxygen	O	46.6%	62.6%
Silicon	Si	27.7	21.2
Aluminum	Al	8.1	6.5
Iron	Fe	5.0	1.9
Calcium	Ca	3.6	1.9
Sodium	Na	2.8	2.6
Potassium	K	2.6	1.4
Magnesium	Mg	2.1	1.8
All others		1.5	0.1

TABLE 2-4 Some of the Mineral Groups Recognized by Geologists

Mineral Group	Negatively Charged Ion or Ion Group	Examples	Composition
Carbonate	$(CO_3)^{-2}$	Calcite	$CaCO_3$
		Dolomite	$CaMg(CO_3)_2$
Halide	Cl^{-1}, F^{-1}	Halite	NaCl
		Fluorite	CaF_2
Native element	—	Gold	Au
		Silver	Ag
		Diamond	C
		Graphite	C
Oxide	O^{-2}	Hematite	Fe_2O_3
		Magnetite	Fe_3O_4
Silicate	$(SiO_4)^{-4}$	Quartz	SiO_2
		Potassium feldspar	$KAlSi_3O_8$
		Olivine	$(Mg,Fe)_2SiO_4$
Sulfate	$(SO_4)^{-2}$	Anhydrite	$CaSO_4$
		Gypsum	$CaSO_4 \cdot 2H_2O$
Sulfide	S^{-2}	Galena	PbS
		Pyrite	FeS_2

silicon to oxygen ratio is 1:4, and the negative charge of the silica ion is balanced by positive ions (▶ Figure 2-10a). Olivine [$(Mg,Fe)_2SiO_4$], for example, has either two magnesium (Mg^{+2}) ions, two iron (Fe^{+2}) ions, or one of each to offset the −4 charge of the silica ion.

Silica tetrahedra may also be arranged so that they join together to form chains of indefinite length (Figure 2-10b). Single chains, as in the pyroxene minerals, form when each tetrahedron shares two of its oxygens with an adjacent tetrahedron; the result is a silicon to oxygen ratio of 1:3. Enstatite, a pyroxene group mineral, reflects this ratio in its chemical formula, $MgSiO_3$. Individual chains, however, possess a net −2 electrical charge, so they are balanced by positive ions, such as Mg^{+2}, that link parallel chains together (Figure 2-10b).

The amphibole group of minerals is characterized by a double-chain structure in which alternate tetrahedra in two parallel rows are cross-linked (Figure 2-10b). The formation of double chains results in a silicon to oxygen ratio of 4:11, so that each double chain possesses a −6 electrical charge. Mg^{+2}, Fe^{+2}, and Al^{+2} are usually involved in linking the double chains together.

In sheet structure silicates, three oxygens of each tetrahedron are shared by adjacent tetrahedra (Figure 2-10c). Such structures result in continuous sheets of silica tetrahedra with silicon to oxygen ratios of 2:5. Continuous sheets also possess a negative electrical charge that is satisfied by positive ions located between the sheets. This particular structure is what accounts for the characteristic sheet structure of the *micas,* such as biotite and muscovite, and the *clay minerals.*

Three-dimensional frameworks of silica tetrahedra form when all four oxygens of the silica tetrahedron are shared by adjacent tetrahedra (Figure 2-10d). Such sharing of oxygen atoms results in a silicon to oxygen ratio of 1:2, which is electrically neutral. Quartz is a common framework silicate.

Ferromagnesian Silicates. The **ferromagnesian silicates** are silicate minerals containing iron (Fe), magnesium (Mg), or both. Such minerals are commonly dark colored and more dense than nonferromagnesian silicates. Some of the common ferromagnesian silicate minerals are olivine, the pyroxenes, the amphiboles, and biotite (▶ Figure 2-11). Olivine, an olive green mineral, is common in some igneous rocks, but uncommon in most other rock types. The pyroxenes and amphiboles are actually mineral groups, but the varieties augite and hornblende are the most common. Biotite mica is a common, dark-colored ferromagnesian silicate with a distinctive sheet structure.

Nonferromagnesian Silicates. The **nonferromagnesian silicates**, as their name implies, lack iron and magnesium, are generally light colored, and are less dense than ferromagnesian silicates (▶ Figure 2-12). The most common minerals in the Earth's crust are nonferromagnesian silicates known as *feldspars.* Feldspar is a general name, however, and two distinct groups are recognized, each of which includes several species. The *potassium feldspars,* represented by microcline and orthoclase ($KAlSi_3O_8$), are common in igneous, metamorphic, and some sedimentary rocks. Like all feldspars, microcline and orthoclase have two internal planes of weakness along which they break or cleave.

The second group of feldspars, the *plagioclase feldspars,* range from calcium-rich ($CaAl_2Si_2O_8$) to sodium-rich ($NaAlSi_3O_8$) varieties. They possess the characteristic feldspar cleavage and typically are white or cream to

			Formula of negatively charged ion group	Silicon to oxygen ratio	Example
(a)	Isolated tetrahedra		$(SiO_4)^{-4}$	1:4	Olivine
(b)	Continuous chains of tetrahedra	Single chain	$(SiO_3)^{-2}$	1:3	Pyroxene group
		Double chain	$(Si_4O_{11})^{-6}$	4:11	Amphibole group
(c)	Continuous sheets		$(Si_4O_{10})^{-4}$	2:5	Micas
(d)	Three-dimensional networks	Too complex to be shown by a simple two-dimensional drawing	$(SiO_2)^0$	1:2	Quartz

➢ FIGURE 2-10 Structures of some of the common silicate minerals shown by various arrangements of silica tetrahedra: (*a*) isolated tetrahedra; (*b*) continuous chains; (*c*) continuous sheets; and (*d*) networks. The arrows adjacent to single-chain, double-chain, and sheet silicates indicate that these structures continue indefinitely in the directions shown.

medium gray. Plagioclase cleavage surfaces commonly show numerous distinctive, closely spaced, parallel lines called striations.

Quartz (SiO_2), another common nonferromagnesian silicate (see Perspective 2-1), is common in the three major rock groups, especially in such rocks as granite, gneiss, and

➢ FIGURE 2-11 Common ferromagnesian silicates: (*a*) olivine; (*b*) augite, a pyroxene group mineral; (*c*) hornblende, an amphibole group mineral; and (*d*) biotite mica. (Photo courtesy of Sue Monroe.)

➢ FIGURE 2-12 Common nonferromagnesian silicates: (*a*) quartz; (*b*) the potassium feldspar orthoclase; (*c*) plagioclase feldspar; and (*d*) muscovite mica. (Photo courtesy of Sue Monroe.)

Perspective 2-1

Quartz—A Common Useful Mineral

During the Middle Ages, quartz crystals were thought to be ice, frozen so solidly that they would not melt (▶ Figure 1). In fact, the term "crystal" is derived from a Greek word meaning ice. Even today, crystal refers not only to transparent quartz, but also to clear, colorless glass of high quality, such as crystal ware, crystal chandeliers, or the transparent glass or plastic cover of a watch or clock dial.

Quartz is a common mineral in the Earth's crust. Most of the sand on beaches, in sand dunes, and in stream channels is quartz. Sand deposits composed mostly of quartz are called silica sands and are used in the manufacture of glass. Quartz is also used in optical equipment, for abrasives such as sandpaper, in the manufacture of steel alloys, and for a variety of other industrial applications.

Quartz occurs in several color varieties. Milky white quartz is common and frequently occurs as well-formed crystals. A milky white quartz crystal weighing 11.8 metric tons and measuring 3.5 m long and 1.7 m in diameter was discovered in Siberia. Color varieties of quartz include amethyst (purple), smoky (smoky brown to black), citrine (yellow to yellowish brown), and rose (pale pink to deep rose) (Figure 1). Agate is a very finely crystalline variety of quartz commonly used as a decorative stone (Figure 1d).

Colorless quartz in particular has been used as a semiprecious stone for jewelry. The term "rhinestone" originally referred to transparent quartz crystals used for jewelry made in Germany. Herkimer "diamonds" are simply colorless quartz crystals from Herkimer County, New York. During the past, large, transparent quartz crystals were shaped into spheres for the fortune teller's crystal ball.

The property of piezoelectricity (which literally means "pressure" electricity) is what enables quartz to be such an accurate timekeeper. When pressure is applied to a quartz crystal, an electric current is generated. If an electric current is applied to a quartz crystal, the crystal expands and compresses extremely rapidly and regularly (about 100,000 times per second). In a quartz movement watch, an electrical current supplied by the watch's battery causes a thin wafer of a quartz crystal to vibrate.

The first clock driven by a quartz crystal was developed in 1928. Today quartz clocks and watches are commonplace, and even inexpensive quartz timepieces are extremely accurate. Precision-manufactured quartz clocks used in astronomical observatories do not gain or lose more than one second every 10 years.

An interesting historical note is that during World War II the United States had difficulty obtaining quartz crystals from Brazil needed for making radios. This shortage prompted the development of artificially synthesized quartz, and now most of the quartz used in watches and clocks is synthetic.

sandstone. It is a framework silicate that can usually be recognized by its glassy appearance and hardness (Figure 2-12a).

Another fairly common nonferromagnesian silicate is muscovite, which is a mica. Like biotite it is a sheet silicate, but muscovite is typically nearly colorless whereas biotite is dark colored (Figure 2-12d). Various clay minerals also possess the sheet structure typical of the micas, but their crystals are so small that they can be seen only with extremely high magnification. These clay minerals are important constituents of several types of rocks and are essential components of soils.

Carbonate Minerals

Carbonate minerals are those that contain the negatively charged carbonate ion $(CO_3)^{-2}$. An example is calcium carbonate ($CaCO_3$), the mineral *calcite* (Table 2-4). Calcite is the main constituent of the sedimentary rock *limestone*. A number of other carbonate minerals are known, but only one of these need concern us: *dolomite* [$CaMg(CO_3)_2$] is formed by the chemical alteration of calcite by the addition of magnesium. Sedimentary rock composed of the mineral dolomite is *dolostone* (see Chapter 6).

Other Mineral Groups

In addition to silicates and carbonates, several other mineral groups are recognized (Table 2-4). In the oxides, an element is combined with oxygen as in *hematite* (Fe_2O_3). Hematite and another iron oxide called *magnetite* (Fe_3O_4) are both commonly present in small quantities in a variety of rocks. Rocks containing high concentrations of hematite and magnetite, such as those in the Lake Superior region of Canada and the United States, are important sources of iron ores for the manufacture of steel.

The sulfides have a positively charged ion combined with sulfur (S^{-2}), such as in the mineral galena (PbS), which contains lead (Pb) and sulfur (▶ Figure 2-13a). Sulfates contain an element combined with the complex sulfate ion $(SO_4)^{-2}$; gypsum ($CaSO_4 \cdot 2H_2O$) is a good example (Figure 2-13b). The halides contain halogen elements

(a) (b) (c) (d) (e)

▶ **FIGURE 1** Varieties of quartz. (*a*) Colorless crystals from the Jeffrey Stone Quarry, Arkansas—National Museum of Natural History specimen #R12804. (Photo by Chip Clark, courtesy of Smithsonian Institution.) (*b*) Smoky quartz. (*c*) Amethyst. (*d*) Agate, a variety of very finely crystalline quartz. (*e*) Rose quartz. (Photos, b, c, d, and e courtesy of Sue Monroe).

such as chlorine (Cl^{-1}) and fluorine (F^{-1}); examples include the minerals halite (NaCl) and fluorite (CaF_2) (Figure 2-13c).

 Physical Properties of Minerals

All minerals possess characteristic physical properties that are determined by their internal structure and chemical composition. Many physical properties are remarkably constant for a given mineral species, but some, especially color, may vary. Though a professional geologist may use sophisticated techniques in studying and identifying minerals, most common minerals can be identified by using the following physical properties.

Color and Luster

For many minerals, color varies because of minute amounts of impurities (▶ Figure 1, Perspective 2-1). The fact that some minerals occur in a variety of colors is distressing to beginning students because the most obvious mineral property is not particularly useful for identification. However, some generalizations about color can be made. Ferromagnesian silicates are typically black, brown, or dark green, although olivine is olive green (Figure 2-11a). Nonferromagnesian silicates, on the other hand, can vary considerably in color, but are only rarely dark (Figure 2-12). Furthermore, for those minerals that have the appearance of metals, color is rather consistent.

Luster (not to be confused with color) is the appearance of a mineral in reflected light. Two basic types of luster are recognized: *metallic* and *nonmetallic* (▶ Figure 2-14). They are distinguished by observing the quality of light reflected from a mineral and determining if it has the appearance of a metal or a nonmetal. Several types of nonmetallic luster are recognized including glassy or vitreous, greasy, waxy, brilliant (as in diamond), and dull or earthy.

Crystal Form

As previously noted, mineral crystals are rare, so many mineral specimens you encounter will not show the perfect

(a)

(b)

(c)

▶ FIGURE 2-13 Representative examples of minerals from (*a*) the sulfides (galena—PbS); (*b*) the sulfates (gypsum—$CaSO_4 \cdot 2H_2O$); and (*c*) the halides (halite—NaCl).

crystal form typical of that mineral species. Keep in mind, however, that even though crystals may not be apparent, minerals nevertheless possess the atomic structure that would have yielded well-formed crystals if they had developed under ideal conditions.

▶ FIGURE 2-14 Luster is the appearance of a mineral in reflected light. Galena (*left*), the ore of lead, has the appearance of a metal and is said to have a metallic luster, whereas orthoclase has a nonmetallic luster.

Some minerals do typically occur as crystals. For example, 12-sided crystals of garnet are common, as are 6- and 12-sided crystals of pyrite (▶ Figure 2-15). Minerals that grow in cavities or are precipitated from circulating hot water (hydrothermal solutions) in cracks and crevices in rocks also commonly occur as crystals.

Crystal form can be a very useful characteristic for mineral identification, but a number of minerals have the same crystal form. For example, pyrite (FeS_2), galena (PbS), and halite (NaCl) all occur as cubic crystals. Nevertheless, such minerals can usually be easily identified by other properties such as color, luster, hardness, and density.

Cleavage and Fracture

Cleavage is a property of individual mineral crystals. Not all minerals possess cleavage, but those that do tend to break, or split, along a smooth plane or planes of weakness determined by the strength of the bonds within the mineral crystal. Cleavage can be characterized in terms of quality (perfect, good, poor), direction, and angles of intersection of cleavage planes. Biotite, a common ferromagnesian silicate, has perfect cleavage in one direction (▶ Figure 2-16a). The fact that biotite preferentially cleaves along a number of closely spaced, parallel planes is related to its structure; it is a sheet silicate with the sheets of silica tetrahedra weakly bonded to one another by iron and magnesium ions (Figure 2-10c).

Feldspars possess two directions of cleavage that intersect at right angles, and the mineral halite has three directions of cleavage, all of which intersect at right angles (Figure 2-16c). Calcite also possesses three directions of cleavage, but none of the intersection angles is a right angle, so cleavage fragments of calcite are rhombohedrons (Figure 2-16d). Minerals with four directions of cleavage include fluorite and diamond. Ironically, diamond, the hardest mineral, can be easily cleaved (see Perspective 2-2). A few minerals such as sphalerite, an ore of zinc, have six directions of cleavage (Figure 2-16f).

▶ **FIGURE 2-15** (*a*) Crystals of pyrite from Spain—NMNH specimen #R18657. (Photo by D. Penland, courtesy of Smithsonian Institution.) (*b*) Garnet crystals from Alaska.

▶ **FIGURE 2-16** Several types of mineral cleavage: (*a*) one direction; (*b*) two directions at right angles; (*c*) three directions at right angles; (*d*) three directions, not at right angles; (*e*) four directions; and (*f*) six directions.

(a) Cleavage in one direction — Cleavage plane — Micas—biotite and muscovite

(b) Cleavage in two directions at right angles — Potassium feldspars, plagioclase feldspars

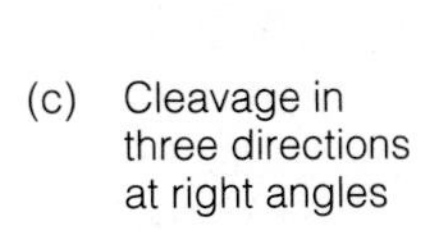

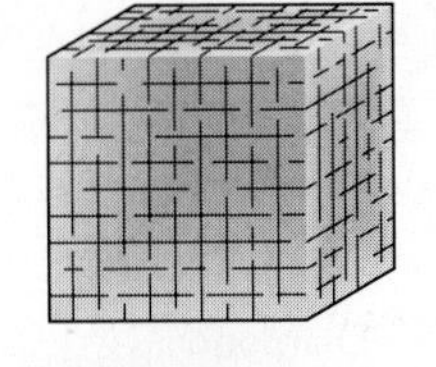

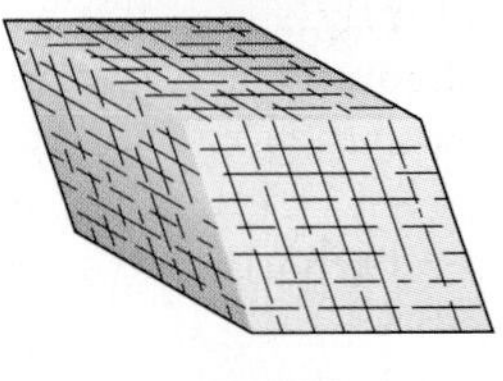

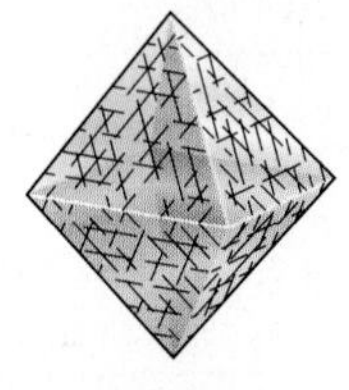

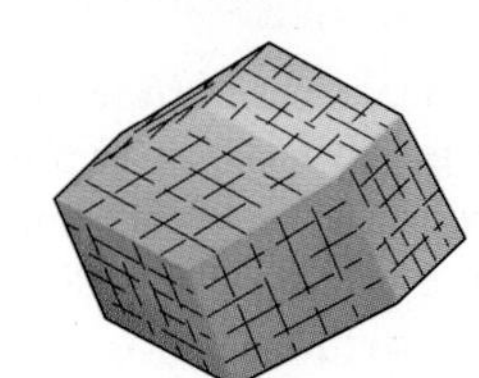

Cleavage is a very important diagnostic property of minerals, and its recognition is essential in distinguishing between some minerals. The pyroxene mineral augite and the amphibole mineral hornblende, for example, look much alike: both are generally dark green to black, have the same hardness, and possess two directions of cleavage. But the cleavage planes of augite intersect at about 90°, whereas the cleavage planes of hornblende intersect at angles of 56° and 124° (▶ Figure 2-17).

▶ **FIGURE 2-17** Cleavage in augite and hornblende. (*a*) Augite crystal and cross section of crystal showing cleavage. (*b*) Hornblende crystal and cross section of crystal showing cleavage.

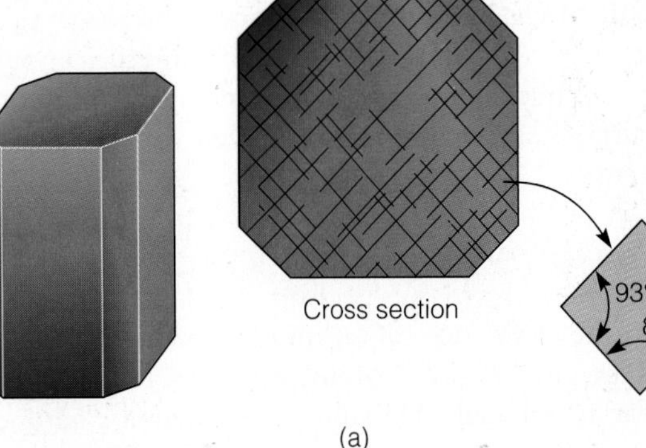

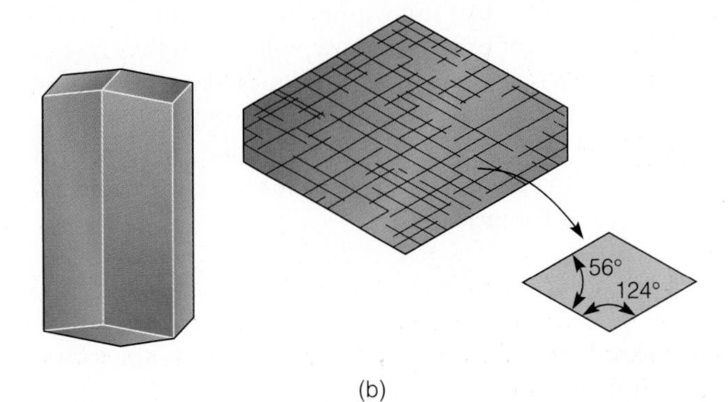

Perspective 2-2

DIAMONDS AND PENCIL LEADS

You may be surprised to learn that diamonds and pencil "lead" (graphite) are composed of the same substance, carbon. Both diamonds and graphite are crystalline solids and are therefore minerals; because they each contain only a single element, they are also native elements. Other than composition, however, diamond and graphite have little in common: diamond is the hardest mineral, whereas graphite is so soft that it can be scratched by a fingernail; diamond may be colorless, red, yellow, blue, gray, or black, while graphite is invariably steel gray (▶ Figure 1). Obviously, the same chemical substance occurs in vastly different forms, but what could possibly control such differences?

Diamond and graphite differ mostly because of their internal structure—both are crystalline but the atoms within crystals of diamond and graphite are arranged quite differently. Such minerals sharing the same composition but differing in structure are *polymorphs* (poly = many; morph = shape or form). Notice in Figure 2-5 that in a diamond crystal the carbon atoms are arranged such that they all are bonded to one another. In graphite the carbon atoms are bonded together to form sheets, but the sheets are weakly held together by van der Waals bonds (Figure 2-5c).

Graphite can be used for pencil leads because it has good cleavage in one direction. When a pencil lead is moved across a piece of paper, small pieces of graphite flake off along the planes held together by van der Waals bonds and adhere to the paper.

Most of the diamonds mined are not of gem quality and are used in such industrial applications as diamond drill bits, diamond-tipped cutting blades, or abrasives. Most gem-quality diamonds are mined in South Africa, although in terms of total diamond production South Africa is in fifth place, with Australia being the largest producer.

How does one "cut" a diamond, the hardest substance known? Diamond cutting is actually done by several processes, one of which is cleaving. Diamond possesses four directions of cleavage, and if a diamond is cleaved such that all four cleavage planes are perfectly developed, the resulting "stone" will be shaped like two pyramids placed base to base. Diamonds are cleaved by placing a knife parallel with a cleavage plane and then tapping the knife with a mallet. Large diamonds are commonly preshaped by cleaving them into smaller pieces that are then further shaped by sawing and grinding with diamond dust.

In contrast to cleavage, *fracture* is mineral breakage along irregular surfaces. Any mineral can be fractured if enough force is applied, but the fracture surfaces will not be smooth. Fracture surfaces are commonly uneven or conchoidal (smoothly curved).

Hardness

Hardness is the resistance of a mineral to abrasion. An Austrian geologist, Friedrich Mohs, devised a relative hardness scale for 10 minerals. He arbitrarily assigned a hardness value of 10 to diamond, the hardest mineral known, and lesser values to the other minerals. Relative hardness can be determined easily by using the Mohs hardness scale (◉ Table 2-5). For example, quartz will scratch fluorite but cannot be scratched by fluorite, gypsum can be scratched by a fingernail, and so on. Hardness is controlled mostly by internal structure. Both graphite and diamond are composed of carbon, but the former has a hardness of 1 to 2 whereas the latter has a hardness of 10.

TABLE 2-5 Mohs Hardness Scale

Hardness	Mineral	Hardness of Some Common Objects
10	Diamond	
9	Corundum	
8	Topaz	
7	Quartz	
		Steel file (6½)
6	Orthoclase	
		Glass (5½–6)
5	Apatite	
4	Fluorite	
3	Calcite	Copper penny (3)
		Fingernail (2½)
2	Gypsum	
1	Talc	

Specific Gravity

The *specific gravity* of a mineral is the ratio of its weight to the weight of an equal volume of water. A mineral with a specific gravity of 3.0 is three times as heavy as water. Like all ratios, specific gravity is not expressed in units such as grams per cubic centimeter—it is a dimensionless number.

Specific gravity varies in minerals depending upon their composition and structure. Among the common silicates,

(a)

(b)

▶ **FIGURE 1** Two minerals composed of carbon. (*a*) Graphite. (Photo courtesy of Sue Monroe.) (*b*) The Oppenheimer diamond—NMNH specimen #117538. (Photo by D. Penland, courtesy of Smithsonian Institution.)

the ferromagnesian silicates have specific gravities ranging from 2.7 to 4.3, whereas the nonferromagnesian silicates vary from 2.6 to 2.9. Obviously, the ranges of values overlap somewhat, but for the most part ferromagnesian silicates have greater specific gravities than nonferromagnesian silicates. In general, the metallic minerals, such as galena (7.58) and hematite (5.26), are heavier than nonmetals. Structure as a control of specific gravity is illustrated by the native element carbon (C): the specific gravity of graphite varies from 2.09 to 2.33; that of diamond is 3.5.

Other Properties

A number of other physical properties characterize some minerals. Talc has a distinctive soapy feel, graphite writes on paper, halite tastes salty, and magnetite is magnetic (▶ Figure 2-18). Calcite possesses the property of *double refraction,* meaning that an object when viewed through a transparent piece of calcite will have a double image (Figure 2-18c). Some minerals are plastic and, when bent into a new shape, will retain that shape, whereas others are flexible and, if bent, will return to their original position when the forces that bent them are removed.

A simple chemical test to identify the minerals calcite and dolomite involves applying a drop of dilute hydrochloric acid to the mineral specimen. If the mineral is calcite, it will react vigorously with the acid and release carbon dioxide, which causes the acid to bubble or effervesce. Dolomite, on the other hand, will not react with hydrochloric acid unless it is powdered.

 GEMSTONES

Gemstones of one kind or another have been sought for thousands of years. Archaeological evidence indicates that 75,000 years ago people in Spain and France were carving objects from bone, ivory, horn, and various stones. The ancient Egyptians mined turquoise more than 5,000 years ago, and by 3400 B.C. they were using rock crystal (colorless quartz), amethyst (purple quartz), lapis lazuli (a rock composed of a variety of minerals), and several other stones to make ornaments. Turquoise remains a popular gemstone in many cultures, including those of the Native Americans of the southwestern United States (▶ Figure 2-19).

Most *gemstones* are minerals, more rarely rocks, that are cut and polished for jewelry. To qualify as a gemstone, a mineral or rock must be appealing for some reason. Many gemstones are desired because of one or several features such as their brilliance, beauty, durability, and scarcity. Furthermore, they become even more desirable when some kind of

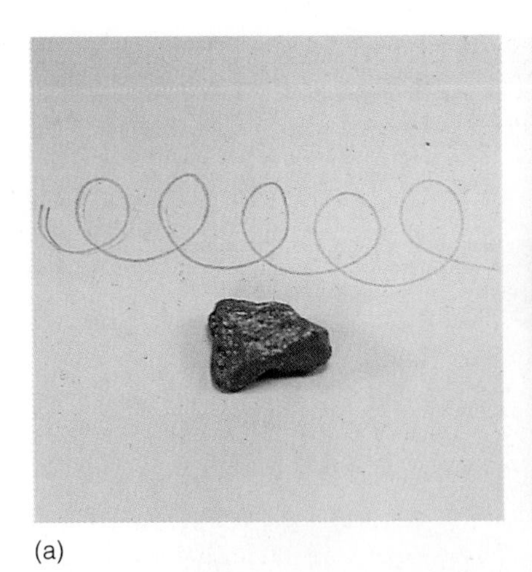
(a)

(b)

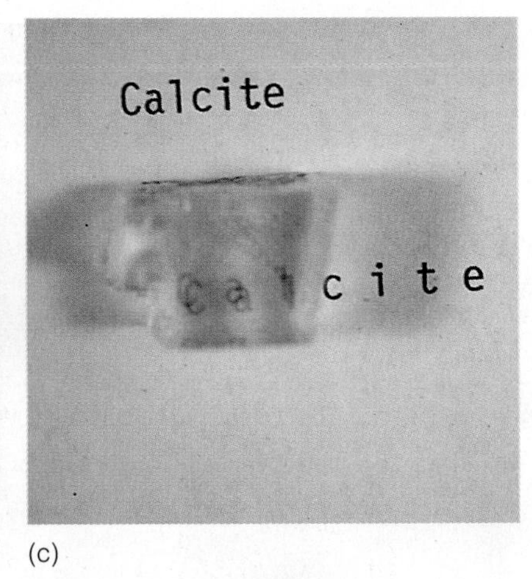

(c)

▶ **FIGURE 2-18** Various properties of minerals. (*a*) Graphite, the mineral from which pencil leads are made, writes on paper. (*b*) Magnetite is magnetic. (*c*) Calcite shows double refraction.

lore is associated with them. Relating gemstones to birth month gives them an added appeal to many people.

Many minerals and rocks are attractive but fail to qualify as gemstones. Cut and polished fluorite is beautiful, but it is fairly common and too soft to be durable. Diamond, on the other hand, meets most of the criteria for a gemstone. In fact, only about two dozen gemstones are used widely. Some are recognized as *precious,* and others are considered to be *semiprecious.*

The precious gemstones include diamond; ruby and sapphire, which are red and blue varieties of the mineral corundum, respectively; emerald, a bright green variety of the mineral beryl; and precious opal (▶ Figure 2-20). Some of the semiprecious gemstones are garnet, jade, tourmaline, topaz, peridot (olivine), aquamarine (light bluish green beryl), turquoise, and several varieties of quartz such as amethyst, agate, and tiger's eye. In addition, amber, a fossil resin from coniferous trees that often contains insects, is included among the semiprecious gemstones. Recall that insects preserved in amber played a pivotal role in the novel and movie *Jurassic Park.*

Transparent gemstones are most often cut and faceted in various ways to enhance the quality of reflected or refracted light. For example, the brilliance of diamonds is maximized by faceting the back of the gemstone so that as much light as possible is reflected. Most opaque and translucent gemstones are not faceted. Rather they are cut and polished into dome-shaped cabochons to emphasize their most interesting features, or they are simply polished by tumbling.

Throughout history gemstones have been used for many purposes including personal adornment and as symbols of

▶ **FIGURE 2-19** Turquoise, a sky blue, blue-green, or light green hydrated copper aluminum phosphate, is a semiprecious gemstone used for jewelry and as a decorative stone.

▶ **FIGURE 2-20** Gem-quality diamonds. Their hardness, brilliance, beauty, durability, and scarcity make them the most sought-after gemstones.

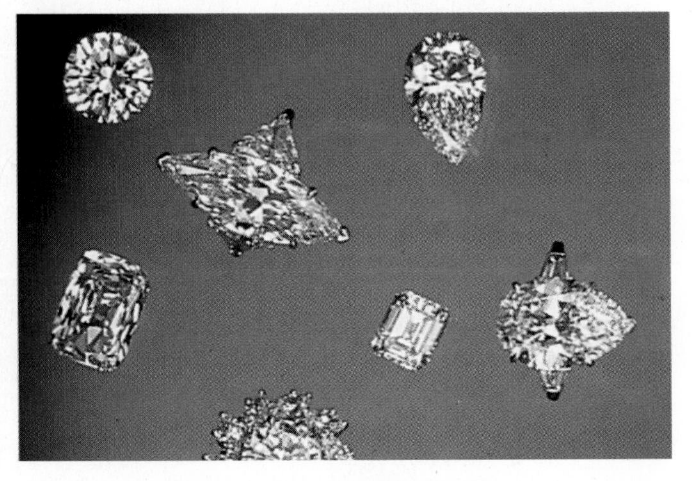

wealth and power. Gemstones and other minerals, rocks, and fossils have also served as religious symbols and talismans, or they have been worn or carried for their presumed mystical powers. Many people own small gemstones, but most of the truely magnificent ones are in museums or collections of crown jewels.

IMPORTANT ROCK-FORMING MINERALS

Rocks are generally defined as solid aggregates of grains of one or more minerals. Two important exceptions to this definition are natural glass such as obsidian (see Chapters 3 and 4) and the sedimentary rock coal (see Chapter 6). Although many minerals occur in various kinds of rocks, only a few varieties are common enough to be designated as **rock-forming minerals**. Most of the others occur in such small amounts that they can be disregarded in the identification and classification of rocks; these are generally called *accessory minerals*. Granite, an igneous rock consisting largely of potassium feldspar and quartz, commonly contains such accessory minerals as sodium plagioclase, biotite, hornblende, muscovite, and, rarely, pyroxene (➢ Figure 2-21).

We have already emphasized that the Earth's crust is composed largely of silicate minerals. This being the case, one would suspect that most rocks are also composed of silicate minerals, and this is correct. Only a few of the hundreds of known silicates are common in rocks, however, although many occur as accessories. The common rock-forming silicates are summarized in ◉ Table 2-6. Several varieties of clay minerals, all of which are sheet silicates, are also common rock-forming minerals. These clay minerals form mostly by the chemical alteration of other silicate minerals, such as feldspars, and are particularly common in some sedimentary and metamorphic rocks, as well as in soils.

The most common nonsilicate rock-forming minerals are the two carbonates, calcite ($CaCO_3$) and dolomite [$CaMg(CO_3)_2$], the primary constituents of the sedimentary rocks limestone and dolostone, respectively. Among the sulfates and halides, gypsum ($CaSO_4 \cdot 2H_2O$) and halite ($NaCl$), respectively, are the only common rock-forming minerals.

MINERAL RESOURCES AND RESERVES

Geologists of the U.S. Geological Survey and the U.S. Bureau of Mines define a **resource** as follows:

> A concentration of naturally occurring solid, liquid, or gaseous material in or on the Earth's crust in such form and amount that economic extraction of a commodity from the concentration is currently or potentially feasible.

Accordingly, resources include such substances as metals (*metallic resources*), sand, gravel, crushed stone, and sulfur (*nonmetallic resources*) and uranium, coal, oil, and natural gas (*energy resources*). An important distinction must be made between a resource, which is the total amount of a commodity whether discovered or undiscovered, and a **reserve**, which is that part of the resource base that can be extracted economically.

The amount of mineral resources used has steadily increased since Europeans settled this continent; during 1988, each resident of North America used about 14 metric tons, a large part of which was bulk items such as sand, gravel, and crushed stone. It is no exaggeration to say that industrialized societies are totally dependent on mineral resources. Unfortunately, they are being used at rates far faster than they form. Thus, mineral resources are nonrenewable, meaning that once the resources from a deposit

(a)

➢ **FIGURE 2-21** The igneous rock granite is composed largely of potassium feldspar and quartz, lesser amounts of plagioclase feldspar, and accessory minerals such as biotite mica. (*a*) Hand specimen of granite. (*b*) Photomicrograph showing the various minerals.

(b)

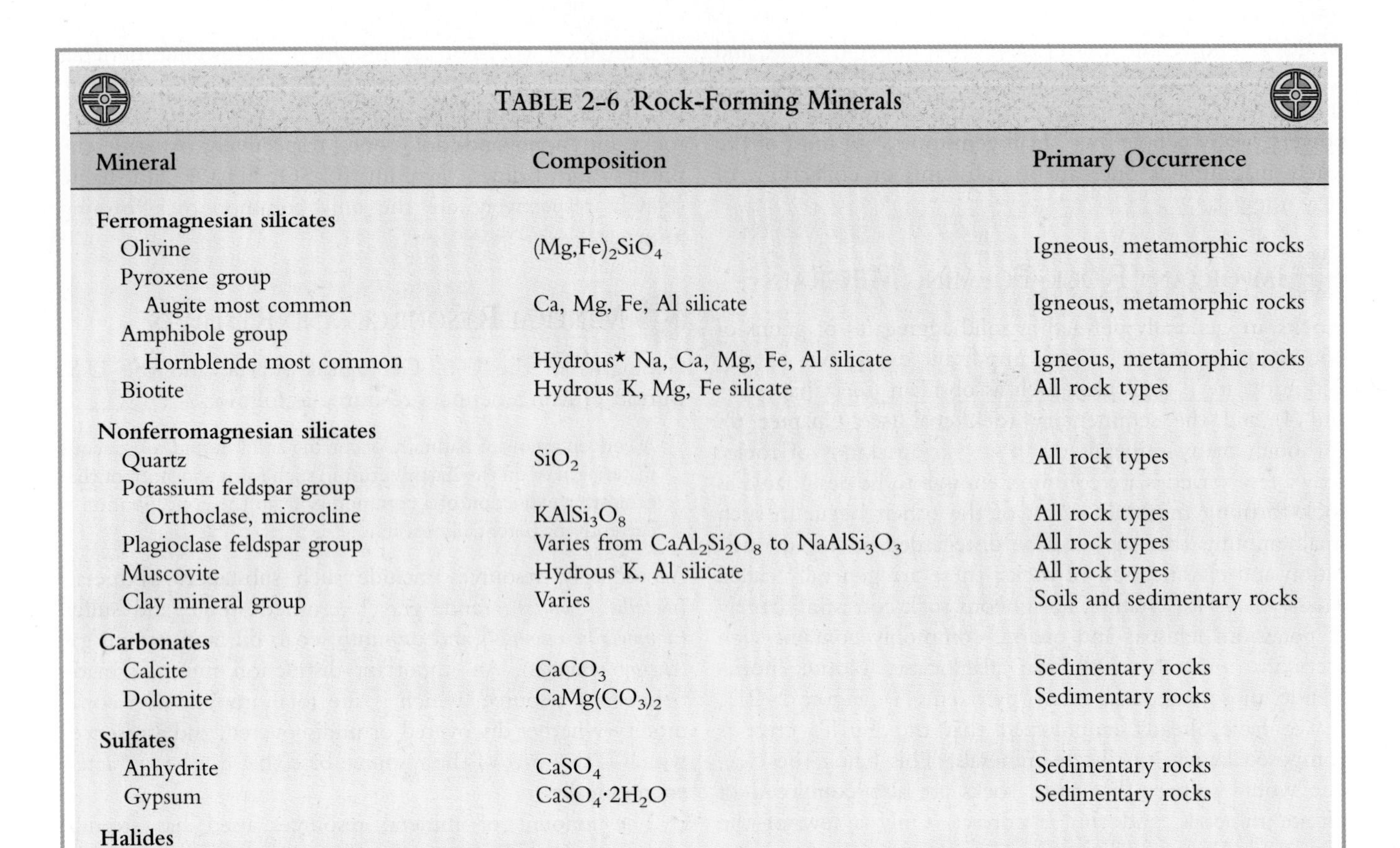

TABLE 2-6 Rock-Forming Minerals

Mineral	Composition	Primary Occurrence
Ferromagnesian silicates		
Olivine	$(Mg,Fe)_2SiO_4$	Igneous, metamorphic rocks
Pyroxene group		
Augite most common	Ca, Mg, Fe, Al silicate	Igneous, metamorphic rocks
Amphibole group		
Hornblende most common	Hydrous* Na, Ca, Mg, Fe, Al silicate	Igneous, metamorphic rocks
Biotite	Hydrous K, Mg, Fe silicate	All rock types
Nonferromagnesian silicates		
Quartz	SiO_2	All rock types
Potassium feldspar group		
Orthoclase, microcline	$KAlSi_3O_8$	All rock types
Plagioclase feldspar group	Varies from $CaAl_2Si_2O_8$ to $NaAlSi_3O_3$	All rock types
Muscovite	Hydrous K, Al silicate	All rock types
Clay mineral group	Varies	Soils and sedimentary rocks
Carbonates		
Calcite	$CaCO_3$	Sedimentary rocks
Dolomite	$CaMg(CO_3)_2$	Sedimentary rocks
Sulfates		
Anhydrite	$CaSO_4$	Sedimentary rocks
Gypsum	$CaSO_4 \cdot 2H_2O$	Sedimentary rocks
Halides		
Halite	NaCl	Sedimentary rocks

*Contains elements of water in some kind of union.

have been exhausted, new supplies or suitable substitutes must be found.

Adequate supplies of some mineral resources are available for indefinite periods (sand and gravel, for example), but for others, supplies are limited or must be imported from other parts of the world (◉ Table 2-7). The United States is almost totally dependent on imports of manganese, an essential element in the manufacture of steel. Even though the United States is a leading producer of gold, it still depends on imports for more than half of its gold needs. More than half of the crude oil used in the United States is imported, much of it from the Middle East where more than 50% of the Earth's proven reserves exist. A poignant reminder of our dependence on the availability of resources was the response to the takeover of Kuwait by Iraq during August 1990.

In terms of mineral and energy resources, Canada is more self-reliant than the United States. Canada meets most of its domestic needs for mineral resources, although it must import phosphate, chromium, manganese, and bauxite, the ore of aluminum. Canada also produces more crude oil and natural gas than it uses, and it is the world leader in the production and export of uranium.

What constitutes a resource as opposed to a reserve depends on several factors. For example, iron-bearing minerals occur in many rocks, but in quantities or ways that make their recovery uneconomical. As a matter of fact, most minerals that are concentrated in economic quantities are mined in only a few areas; 75% of all the metals mined in the world come from about 150 locations. Geographic location is also an important consideration. A mineral resource in a remote region may not be mined because transportation costs are too high, and what may be considered a resource in the United States or Canada may be mined in a developing country where labor costs are low. The market price of a commodity is, of course, important in evaluating a potential resource. From 1935 to 1968, the U.S. government maintained the price of gold at $35 per troy ounce (= 31.1 g). When this restriction was removed and the price of gold became subject to supply and demand, the price rose (it reached an all-time high of $843 per troy ounce during January 1980). As a consequence, many marginal deposits became reserves and many abandoned mines were reopened.

Technological developments can also change the status of a resource. The rich iron ores of the Great Lakes region of the United States and Canada had been depleted by World War II. However, the development of a method of separating the iron from previously unusable rocks and shaping it into pellets that are ideal for use in blast furnaces made it feasible to mine poorer grade ores.

Most of the largest and richest mineral deposits have probably already been discovered and, in some cases, depleted. In order to ensure continued supplies of essential

TABLE 2-7 World, U.S., and Canadian Reserves of Various Metals (in Thousands of Metric Tons)

Mineral Resource	Uses	World Reserves	U.S. Reserves	Canadian Reserves	Major Producing Countries
Bauxite	Ore of aluminum	21,559,000	38,000	0	Australia, Guinea, Jamaica
Chromium	Alloys, electroplating	418,900	0	0	South Africa, CIS,* India, Turkey
Copper	Alloys, electric wires	321,000	55,000	12,000	Chile, USA, Canada, CIS
Gold	Jewelry; circuitry in computers, communications equipment; dentistry	42	5	1.8	South Africa, USA, CIS, Australia, Canada
Iron ore	Iron and steel	64,648,000	3,800,000	4,600,000	CIS, Brazil, Australia, China
Lead	Storage batteries; solder; pipes	70,440	11,000	7,000	CIS, USA, Mexico
Manganese	Iron and steel production	812,800	0	0	CIS, South Africa, Gabon, Australia, Brazil
Nickel	Stainless steel	48,660	30	8,130	CIS, Canada, New Caledonia
Silver	Jewelry; photography; dentistry	780	190	26.8	Mexico, USA, Peru, CIS, Canada
Tin	Coating on metals; tin cans; alloys; solder	59,300	20	60	China, Brazil, Indonesia, Malaysia
Titanium	Alloys; white pigment in paint, paper, and plastics	288,600	8,100	27,000	Australia, Norway, CIS
Zinc	Iron and steel alloys; rubber products, medicines	143,910	20,000	21,000	Canada, Australia, CIS, China, Peru

*Commonwealth of Independent States (includes much of the former Soviet Union).

SOURCES: *World resources 1992–1993: A guide to the global environment* (New York: Oxford University Press). *Minerals yearbook 1990,* vols. 1 and 3 (Metals and Minerals, U.S. Department of the Interior, Bureau of Mines); *Canada Year Book 1994.*

minerals, geologists are using increasingly sophisticated geophysical and geochemical mineral exploration techniques. The U.S. Geological Survey and the U.S. Bureau of Mines continually assess the status of resources in view of changing economic and political conditions and developments in science and technology. In the following chapters, we will discuss the origin and distribution of various mineral resources and reserves.

CHAPTER SUMMARY

1. All matter is composed of chemical elements, each of which consists of atoms. Individual atoms consist of a nucleus, containing protons and neutrons, and electrons that circle the nucleus in electron shells.
2. Atoms are characterized by their atomic number (the number of protons in the nucleus) and their atomic mass number (the number of protons plus the number of neutrons in the nucleus).
3. Bonding is the process whereby atoms are joined to other atoms. If atoms of different elements are bonded, they form a compound. Ionic and covalent bonds are most common in minerals, but metallic and van der Waals bonds also occur in a few.
4. Most minerals are compounds, but a few, including gold and silver, are composed of a single element and are called native elements.
5. All minerals are crystalline solids, meaning that they possess an orderly internal arrangement of atoms.
6. Some minerals vary in chemical composition because atoms of different elements can substitute for one another

provided that the electrical charge is balanced and the atoms are of about the same size.

7. Of the more than 3,500 known minerals, most are silicates. Ferromagnesian silicates contain iron (Fe) and magnesium (Mg), and nonferromagnesian silicates lack these elements.
8. In addition to silicates, several other mineral groups are recognized, including carbonates, oxides, sulfides, sulfates, and halides.
9. The physical properties of minerals such as color, hardness, cleavage, and crystal form are controlled by composition and structure.
10. Some minerals and a few rocks are known as gemstones. They are valued because of their appearance and scarity.
11. A few minerals are common enough constituents of rocks to be designated rock-forming minerals. Most rock-forming minerals are silicates, but the carbonates calcite and dolomite are also common.
12. Many resources are concentrations of minerals of economic importance.
13. Reserves are that part of the resource base that can be extracted economically.
14. The United States is dependent on imports for many mineral resources, whereas Canada is more self-reliant.

Important Terms

atom
atomic mass number
atomic number
bonding
carbonate mineral
cleavage
compound
covalent bond
crystalline solid
electron
electron shell
element
ferromagnesian silicate
ion
ionic bond
isotope
mineral
native element
neutron
nonferromagnesian silicate
nucleus
proton
reserve
resource
rock-forming mineral
silica
silica tetrahedron
silicate

Review Questions

1. The atomic number of an element is determined by the:
   a. ____ number of electrons in its outermost shell;
   b. ____ number of protons in its nucleus;
   c. ____ diameter of its most common isotope;
   d. ____ number of neutrons plus electrons in its nucleus;
   e. ____ total number of neutrons orbiting the nucleus.
2. To which of the following groups do most minerals in the Earth's crust belong?
   a. ____ oxides;
   b. ____ carbonates;
   c. ____ sulfates;
   d. ____ halides;
   e. ____ silicates.
3. When an atom loses or gains electrons, it becomes a(n):
   a. ____ isotope;
   b. ____ proton;
   c. ____ ion;
   d. ____ neutron;
   e. ____ native element.
4. The two most abundant elements in the Earth's crust are:
   a. ____ iron and magnesium;
   b. ____ carbon and potassium;
   c. ____ sodium and nitrogen;
   d. ____ silicon and oxygen;
   e. ____ sand and clay.
5. The sharing of electrons by adjacent atoms is a type of bonding called:
   a. ____ van der Waals;
   b. ____ covalent;
   c. ____ silicate;
   d. ____ tetrahedral;
   e. ____ ionic.
6. Many minerals break along closely spaced planes and are said to possess:
   a. ____ specific gravity;
   b. ____ cleavage;
   c. ____ covalent bonds;
   d. ____ fracture;
   e. ____ double refraction.
7. The chemical formula for olivine is $(Mg,Fe)_2SiO_4$, which means that in addition to silica:
   a. ____ magnesium and iron can substitute for one another;
   b. ____ magnesium is more common than iron;
   c. ____ magnesium is heavier than iron;
   d. ____ all olivine contains both magnesium and iron;
   e. ____ more magnesium than iron occurs in the Earth's crust.
8. The basic building block of all silicate minerals is the:
   a. ____ silicon sheet;
   b. ____ oxygen-silicon cube;
   c. ____ silica tetrahedron;
   d. ____ silicate double chain;
   e. ____ silica framework.
9. An example of a common nonferromagnesian silicate mineral is:
   a. ____ calcite;
   b. ____ quartz;
   c. ____ biotite;
   d. ____ hematite;
   e. ____ halite.
10. The ratio of a mineral's weight to the weight of an equal volume of water is its:
    a. ____ specific gravity;
    b. ____ luster;

c. ____ hardness;
d. ____ atomic mass number;
e. ____ cleavage.

11. Those chemical elements having eight electrons in their outermost electron shell are the:
a. ____ noble gases;
b. ____ native elements;
c. ____ carbonates;
d. ____ halides;
e. ____ isotopes.

12. Minerals are solids possessing an orderly internal arrangement of atoms, meaning that they are:
a. ____ amorphous substances;
b. ____ crystalline;
c. ____ composed of at least three different elements;
d. ____ composed of a single element;
e. ____ ionic compounds.

13. The silicon ion has a positive charge of 4, and oxygen has a negative charge of 2. Accordingly, the ion group (SiO_4) has a:
a. ____ positive charge of 2;
b. ____ negative charge of 2;
c. ____ negative charge of 1;
d. ____ positive charge of 4;
e. ____ negative charge of 4.

14. Calcite and dolomite are:
a. ____ oxide minerals of great value;
b. ____ ferromagnesian silicates possessing a distinctive sheet structure;
c. ____ common rock-forming carbonate minerals;
d. ____ minerals used in the manufacture of pencil leads;
e. ____ important energy resources.

15. How does a crystalline solid differ from a liquid and a gas?

16. An atom of the element magnesium is shown below.

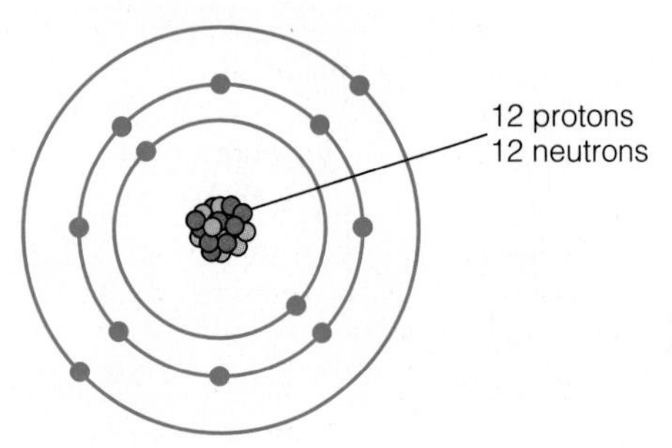

If the two electrons in its outer electron shell are lost, what is the electrical charge of the magnesium ion?

17. What is the atomic mass number of the magnesium atom shown above?
18. Compare ionic and covalent bonding.
19. How do compounds and native elements differ.
20. What accounts for the fact that some minerals have a range of chemical compositions?
21. Why are the angles between the same crystal faces on all specimens of a mineral species always the same?
22. What is a silicate mineral? How do the two subgroups of silicate minerals differ from one another?
23. In sheet silicates, individual sheets composed of silica tetrahedra possess a negative electrical charge. How is this negative charge satisfied?
24. What do all carbonate minerals have in common?
25. Describe the mineral property of cleavage, and explain what controls cleavage.
26. What are rock-forming minerals and accessory minerals?

Points to Ponder

1. How would the color and density of a rock composed mostly of nonferromagnesian silicate minerals differ from one made up primarily of ferromagnesian silicates?
2. Why must the United States, a natural resource–rich nation, import a large part of the resources it needs? What are some of the problems created by dependence on imports?
3. Where would be some good places to look for mineral crystals?
4. Explain how the composition and structure of minerals control such mineral properties as hardness, cleavage, color, and specific gravity.

Additional Readings

Berry, L. G., B. Mason, and R. V. Dietrich. 1983. *Mineralogy.* 2d ed. San Francisco, Calif.: W. H. Freeman and Co.

Blackburn, W. H., and W. H. Dennen. 1988. *Principles of mineralogy.* Dubuque, Iowa: William C. Brown Publishing Co.

Cepeda, J. C. 1994. *Introduction to minerals and rocks.* New York: Macmillan College Publishing Co.

Dietrich, R. V., and B. J. Skinner. 1979. *Rocks and rock minerals.* New York: John Wiley & Sons.

______. 1990. *Gems, granites, and gravels: Knowing and using rocks and minerals.* New York: Cambridge University Press.

Frye, K. 1993. *Mineral science: An introductory survey.* New York: Macmillan Publishing Co.

Klein, C., and C. S. Hurlbut Jr. 1985. *Manual of mineralogy* (after James D. Dana). 20th ed. New York: John Wiley & Sons.

Pough, F. H. 1987. *A field guide to rocks and minerals.* 4th ed. Boston, Mass.: Houghton Mifflin.

Vanders, I., and P. F. Kerr. 1967. *Mineral recognition.* New York: John Wiley & Sons.

Chapter 3

IGNEOUS ROCKS AND INTRUSIVE IGNEOUS ACTIVITY

OUTLINE

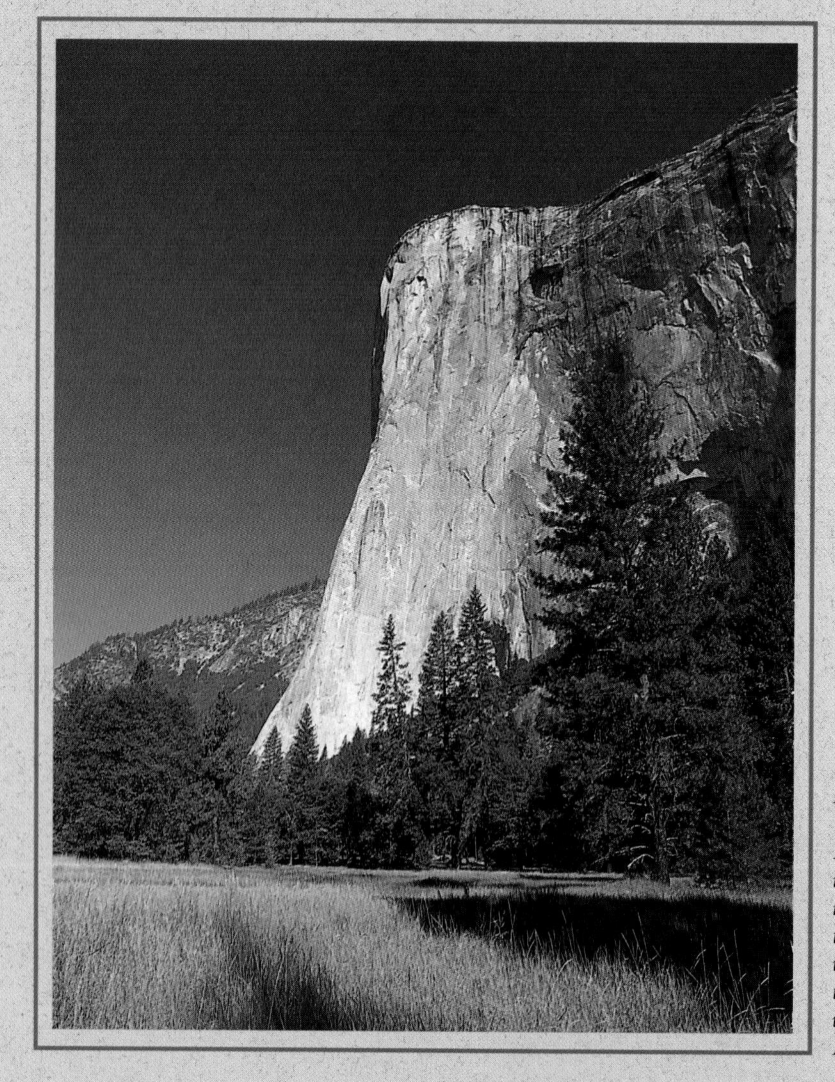

El Capitan, meaning "The Chief," in Yosemite National Park, California, is composed of intrusive igneous rock. It rises more than 900 m above the valley floor and is the tallest unbroken cliff in the world. (Photo courtesy of Richard L. Chambers.)

PROLOGUE

About 45 to 50 million years ago, several small masses of molten rock were intruded into the Earth's crust in what is now northeastern Wyoming. The molten rock cooled and solidified, forming igneous rock bodies. The best known of these, Devil's Tower, was established as our first national monument by President Theodore Roosevelt in 1906. Devil's Tower is a remarkable landform; rising nearly 260 m above its base (▶ Figure 3-1) and visible from 48 km away, it served as a landmark for early travelers in the area.

Devil's Tower and similar, nearby bodies such as Bear Butte in South Dakota play an important role in the legends of the Cheyenne and Lakota Sioux Indians. These Native Americans call Devil's Tower *Mateo Tepee*, which means "Grizzly Bear Lodge." It was also called the "Bad God's Tower," and reportedly, "Devil's Tower" is a translation of this phrase. According to one legend, the tower formed when the Great Spirit caused it to rise up from the ground, carrying with it several children who were trying to escape from a gigantic grizzly bear. Another legend tells of six brothers and a woman who were also being pursued by a grizzly bear. The youngest brother carried a small rock, and when he sang a song, the rock grew to the present size of Devil's Tower. In both legends, the bear's attempts to reach the Indians left deep scratch marks in the tower's rocks (▶ Figure 3-2).

Geologists have a less dramatic explanation for the tower's origin. The near-vertical striations (the bear's scratch marks) are simply the lines formed by the intersections of columnar joints. Columnar joints form in response to cooling and contraction in some igneous bodies and in some lava flows. Many of the columns are six-sided, but columns with four, five, and seven sides occur as well. The larger columns measure about 2.5 m across. The pile of rubble at the tower's base has accumulated as columns have fallen from the tower.

Geologists agree that Devil's Tower originated as a small body intruded into the Earth's crust and that subsequent erosion exposed it in its present form. The type of igneous body and the extent of its modification by erosion are debatable. Some geologists think that Devil's Tower is the eroded remnant of a more extensive body of intrusive rock, whereas others think it is simply the remnant of the magma that solidified in a pipelike conduit of a volcano and that it has been little modified by erosion.

▶ FIGURE 3-1 Devil's Tower in Wyoming exhibits well-developed columnar jointing. (Photo courtesy of R. V. Dietrich.)

▶ FIGURE 3-2 An artist's rendition of a Cheyenne legend about the origin of Devil's Tower.

Introduction

Rocks resulting from volcanic eruptions are widespread, but they probably represent only a small portion of the total rocks formed by the cooling and crystallization of molten rock material, which is called magma. Most magma cools below the Earth's surface and forms bodies of rock known as *plutons*. Although volcanism and plutonic, or intrusive igneous, activity are discussed in separate chapters, they are related phenomena. The same types of magmas are involved in both volcanism and plutonism, but some magmas are more mobile than others and therefore more commonly reach the surface. Plutons typically underlie areas of extensive volcanism and are the sources of the overlying lavas and fragmental materials ejected during explosive eruptions. Furthermore, like volcanism, most plutonism occurs at or near plate margins. In this chapter we are concerned primarily with the textures, composition, and classification of igneous rocks and with plutonic activity.

Magma and Lava

Magma is molten rock material below the Earth's surface, and **lava** is magma at the Earth's surface. Magma is less dense than the solid rock from which it was derived, so it tends to move upward toward the surface. Some magma is erupted onto the surface as **lava flows,** and some is forcefully ejected into the atmosphere as particles called **pyroclastic materials** (from the Greek *pyro* = fire and *klastos* = broken).

Igneous rocks (from the Latin *ignis* = fire) form when magma cools and crystallizes, or when pyroclastic materials such as volcanic ash become consolidated. Magma extruded onto the Earth's surface as lava and pyroclastic materials forms **volcanic** or **extrusive igneous rocks,** whereas magma that crystallizes within the Earth's crust forms **plutonic** or **intrusive igneous rocks.**

Composition

Recall from Chapter 2 that the most abundant minerals in the Earth's crust are silicates, composed of silicon and oxygen and the other elements listed in Table 2-3. Accordingly, when crustal rocks melt and form magma, the magma is typically silica rich and also contains considerable aluminum, calcium, sodium, iron, magnesium and potassium as well as many other elements in lesser quantities. Not all magmas originate by melting of crustal rocks, however; some are derived from upper mantle rocks that are composed largely of ferromagnesian silicates. A magma from this source contains comparatively less silicia and more iron and magnesium.

Although silica is the primary constituent of nearly all magmas, silica content varies and serves to distinguish **felsic, intermediate,** and **mafic** magmas (◉ Table 3-1). A felsic magma contains more than 65% silica and considerable

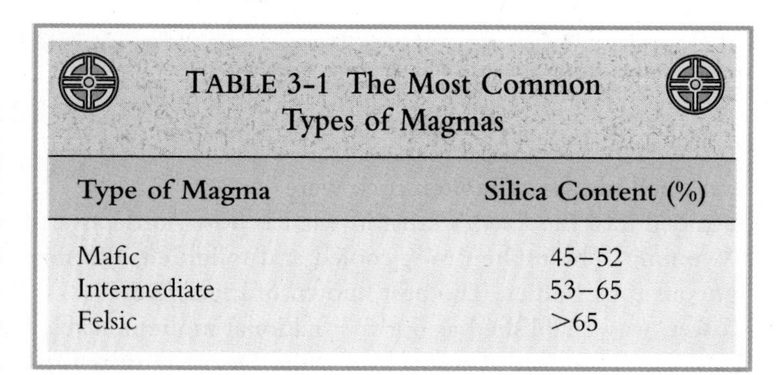

Table 3-1 The Most Common Types of Magmas

Type of Magma	Silica Content (%)
Mafic	45–52
Intermediate	53–65
Felsic	>65

sodium, potassium, and aluminum, but little calcium, iron, and magnesium.

In contrast to felsic magmas, mafic magmas are silica poor and contain proportionately more calcium, iron, and magnesium. As one would expect, intermediate magmas have compositions that are intermediate between those of mafic and felsic magmas (Table 3-1).

Temperature

No direct measurements of temperatures of magma below the Earth's surface have been made. Erupting lavas generally have temperatures in the range of 1,000° to 1,200°C, although temperatures of 1,350°C have been recorded above Hawaiian lava lakes where volcanic gases reacted with the atmosphere.

Most direct temperature measurements have been taken at volcanoes characterized by little or no explosive activity where geologists can safely approach the lava. Therefore, little is known of the temperatures of felsic lavas, because eruptions of such lavas are rare, and when they do occur, they tend to be explosive. The temperatures of some lava domes, most of which are bulbous masses of felsic magma, have been measured at a distance by using an instrument called an optical pyrometer. The surfaces of these domes have temperatures up to 900°C, but the exterior of a dome is probably much cooler than its interior.

When Mount St. Helens erupted in 1980, it ejected felsic magma as particulate matter in pyroclastic flows. Two weeks later, these flows still had temperatures between 300° and 420°C.

Viscosity

Magma is also characterized by its **viscosity,** or resistance to flow. The viscosity of some liquids, such as water, is very low; they are highly fluid and flow readily. The viscosity of some other liquids is so high, however, that they flow much more slowly. Motor oil and syrup flow readily when they are hot, but become stiff and flow very slowly when they are cold. Thus, one might expect that temperature controls the viscosity of magma, and such an inference is partly correct. We can generalize and say that hot lava flows more readily than cooler lava. However, temperature is not the only control of viscosity.

Magma viscosity is strongly controlled by silica content. In a felsic lava, numerous networks of silica tetrahedra retard flow, because the strong bonds of the networks must be ruptured for flow to occur. Mafic lavas, on the other hand, contain fewer silica tetrahedra networks and consequently flow more readily. Felsic lavas form thick, slow-moving flows, whereas mafic lavas tend to form thinner flows that move rather rapidly over great distances. One such flow in Iceland in 1783 flowed about 80 km, and some ancient flows in the state of Washington can be traced for more than 500 km.

Igneous Rocks

All intrusive and many extrusive igneous rocks form when minerals crystallize from magma. The process of crystallization involves the formation of crystal nuclei and their subsequent growth. The atoms in a magma are in constant motion, but when cooling begins, some atoms bond to form small groups, or nuclei, whose arrangement of atoms corresponds to the arrangement in mineral crystals. As other atoms in the liquid chemically bond to these nuclei, they do so in an ordered geometric arrangement, and the nuclei grow into crystalline *mineral grains*, the individual particles that comprise a rock. During rapid cooling, the rate of nuclei formation exceeds the rate of growth, and an aggregate of many small grains results (▶ Figure 3-3a). With slow cooling, the rate of growth exceeds the rate of nucleation, so relatively large grains form (Figure 3-3b).

▶ **FIGURE 3-3** The effect of the cooling rate of a magma on nucleation and growth of crystals: (*a*) Rapid cooling results in many small grains and a fine-grained or aphanitic texture. (*b*) Slow cooling results in a coarse-grained or phaneritic texture.

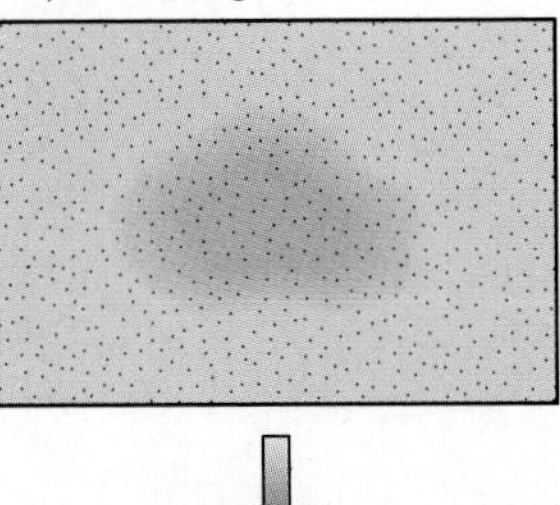

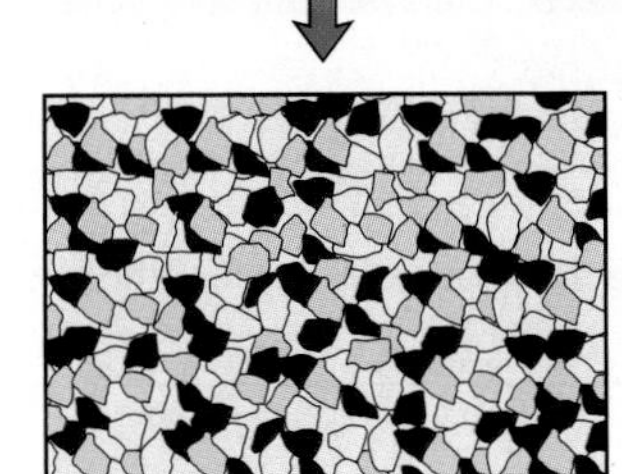

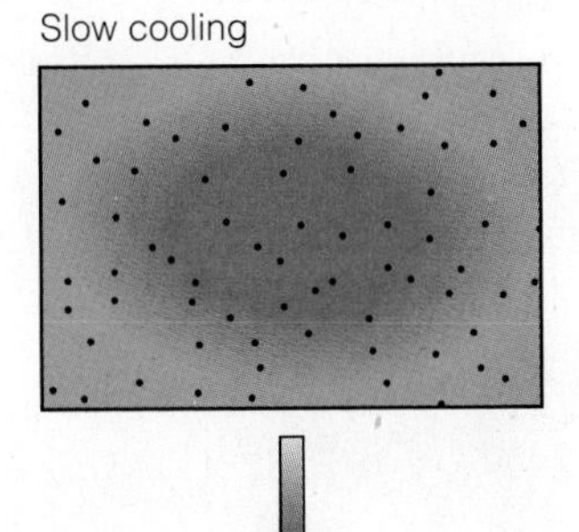

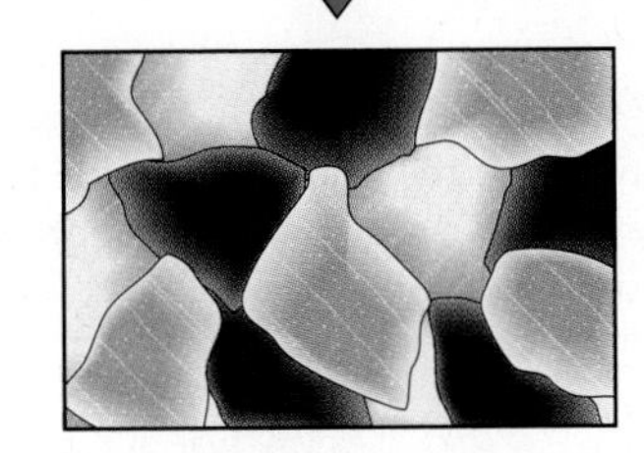

Textures

Several textures of igneous rocks are related to the cooling history of a magma or lava. Rapid cooling, as occurs in lava flows or some near-surface intrusions, results in a fine-grained texture termed **aphanitic**. In an aphanitic texture, individual mineral grains are too small to be observed without magnification (▶ Figure 3-4a). In contrast, igneous rocks with a coarse-grained or **phaneritic** texture have mineral grains that are easily visible without magnification (Figure 3-4b). Such large mineral grains indicate slow cooling and generally an intrusive origin, but a phaneritic texture can also develop in the interiors of some thick lava flows.

Rocks with **porphyritic** textures have a somewhat more complex cooling history. Such rocks have a combination of mineral grains of markedly different sizes. The larger grains are *phenocrysts*, and the smaller ones are referred to as *groundmass* (Figure 3-4c). Suppose that a magma begins cooling slowly as an intrusive body, and that some mineral-crystal nuclei form and begin to grow. Suppose further that before the magma has completely crystallized, the remaining liquid phase and solid mineral grains within it are extruded onto the Earth's surface where it cools rapidly, forming an aphanitic texture. The resulting igneous rock would have large mineral grains (phenocrysts) suspended in a finely crystalline groundmass, and the rock would be characterized as a *porphyry*.

A lava may cool so rapidly that its constituent atoms do not have time to become arranged in the ordered, three-dimensional frameworks typical of minerals. As a consequence of such rapid cooling, a *natural glass* such as *obsidian* forms (▶ Figure 3-5a). Even though obsidian is not composed of minerals, it is still considered to be igneous rock.

Some magmas contain large amounts of water vapor and other gases. These gases may be trapped in cooling lava where they form numerous small holes or cavities called **vesicles**; rocks possessing numerous vesicles are termed *vesicular*, as in vesicular basalt (Figure 3-5b). The igneous rock known as *scoria* contains more cavities than solid rock (Figure 3-5c).

A **pyroclastic** or **fragmental texture** characterizes igneous rocks formed by explosive volcanic activity. For example, ash may be discharged high into the atmosphere and eventually settle to the surface where it accumulates; if it is turned into solid rock, it is considered to be a pyroclastic igneous rock.

Composition

Magmas are characterized as mafic (45–52% silica), intermediate (53–65% silica), or felsic (>65% silica) (see Table 3-1).

➢ **FIGURE 3-4** Textures of igneous rocks. (*a*) Aphanitic or fine-grained texture in which individual minerals are too small to be seen without magnification. (*b*) Phaneritic or coarse-grained texture in which minerals are easily discerned without magnification. (*c*) Porphyritic texture consisting of minerals of markedly different sizes. (Photos courtesy of Sue Monroe.)

➢ **FIGURE 3-5** (*a*) The natural glass obsidian forms when lava cools too quickly for mineral crystals to form. (*b*) Vesicular texture. (*c*) Scoria contains more vesicles than solid rock. (Photos courtesy of Sue Monroe.)

The parent magma plays a significant role in determining the mineral composition of igneous rocks, yet it is possible for the same magma to yield different igneous rocks because its composition can change as a consequence of crystal settling, assimilation, magma mixing, and the sequence in which minerals crystallize.

Bowen's Reaction Series. During the early part of this century, N. L. Bowen hypothesized that mafic, intermediate, and felsic magmas could all derive from a parent mafic magma. He knew that minerals do not all crystallize simultaneously from a cooling magma, but rather crystallize in a predictable sequence. Based on his observations and laboratory experiments, Bowen proposed a mechanism, now called **Bowen's reaction series**, to account for the derivation of intermediate and felsic magmas from a basaltic (mafic) magma (▶ Figure 3-6). Bowen's reaction series consists of two branches: a *discontinuous branch* and a *continuous branch* (Figure 3-6). Crystallization of minerals occurs along both branches simultaneously, but for convenience we will discuss them separately.

In the discontinuous branch, which contains only ferromagnesian minerals, one mineral changes to another over specific temperature ranges (Figure 3-6). As the temperature decreases, a temperature range is reached in which a given mineral begins to crystallize. A previously formed mineral reacts with the remaining liquid magma (the melt) such that it forms the next mineral in the sequence. For example, olivine $[(Mg,Fe)_2SiO_3]$ is the first ferromagnesian mineral to crystallize. As the magma continues to cool, it reaches the temperature range at which pyroxene is stable; a reaction occurs between the olivine and the remaining melt, and pyroxene forms.

A similar reaction takes place between pyroxene and the melt as further cooling occurs, and the pyroxene structure is rearranged to form amphibole. Further cooling causes a reaction between the amphibole and the melt, and its structure is rearranged so that the sheet structure typical of biotite mica forms. Although the reactions just described tend to convert one mineral to the next in the series, the reactions are not always complete. Olivine, for example, might have a rim of pyroxene, indicating an incomplete reaction. If a magma cools rapidly enough, the early-formed minerals do not have time to react with the melt, and thus all the ferromagnesian minerals in the discontinuous branch can be in one rock. In any case, by the time biotite has

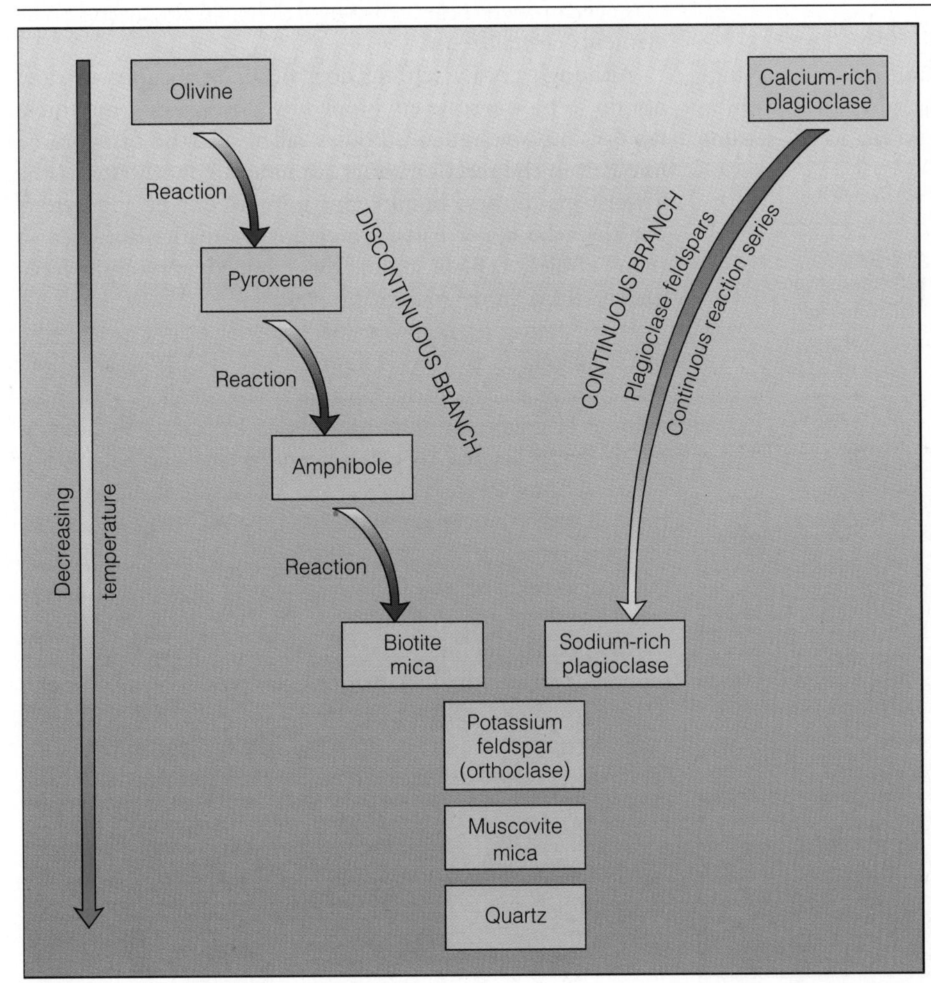

▶ **FIGURE 3-6** Bowen's reaction series. Note that it consists of a discontinuous branch and a continuous branch.

crystallized, essentially all magnesium and iron present in the original magma have been used up.

Plagioclase feldspars are the only minerals in the continuous branch of Bowen's reaction series (Figure 3-6). Calcium-rich plagioclase crystallizes first. As cooling of the magma proceeds, calcium-rich plagioclase reacts with the melt, and plagioclase containing proportionately more sodium crystallizes until all of the calcium and sodium are used up. In many cases cooling is too rapid for a complete transformation from calcium-rich to sodium-rich plagioclase to occur. Plagioclase forming under these conditions is *zoned*, meaning that it has a calcium-rich core surrounded by zones progressively richer in sodium (▶ Figure 3-7).

Magnesium and iron on the one hand and calcium and sodium on the other are used up as crystallization occurs along the two branches in Bowen's reaction series. Accordingly, any magma left over is enriched in potassium, aluminum, and silicon. These elements combine to form potassium feldspar orthoclase ($KAlSi_3O_8$), and if the water pressure is high, the sheet silicate muscovite mica will form. Any remaining magma is predominantly silicon and oxygen (silica) and forms the mineral quartz (SiO_2). The crystallization of potassium feldspar and quartz is not a true reaction series, however, because they form independently rather than from a reaction of the orthoclase with the remaining melt.

Crystal Settling. A magma's composition may also change by **crystal settling**, which involves the physical separation of minerals by crystallization and gravitational settling (▶ Figure 3-8). Olivine, the first ferromagnesian mineral to form in the discontinuous branch of Bowen's reaction series, has a specific gravity greater than that of the remaining magma and tends to sink downward in the melt. Accordingly, the remaining melt becomes relatively rich in silica, sodium, and potassium, because much of the iron and magnesium were removed when minerals containing these elements crystallized.

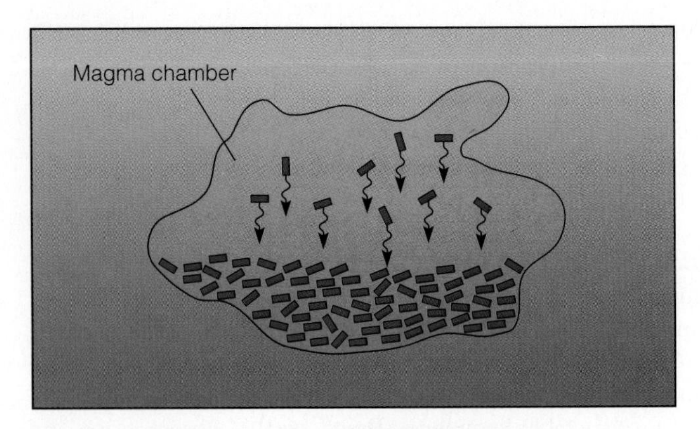

▶ FIGURE 3-8 Differentiation by crystal settling. Early-formed ferromagnesian minerals have a specific gravity greater than that of the magma so they settle and accumulate in the lower part of the magma chamber.

▶ FIGURE 3-7 Photomicrograph of zoned plagioclase crystals. The magma in which these crystals formed cooled too quickly for a complete transformation from calcium-rich to sodium-rich plagioclase to occur. They contain cores rich in calcium surrounded by zones progressively richer in sodium. (Photo courtesy of R. V. Dietrich.)

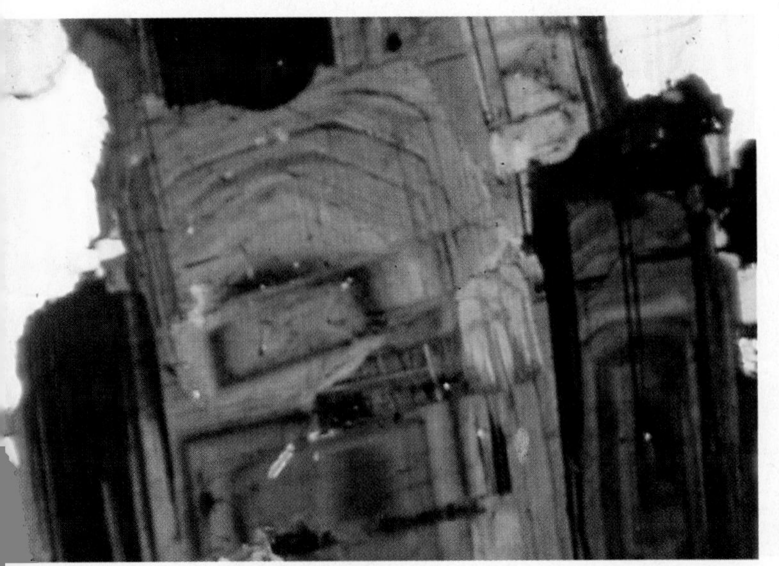

Although crystal settling does occur in magmas, it does not do so on the scale envisioned by Bowen. In some thick, tabular, intrusive igneous bodies called sills, the first-formed minerals in the reaction series are indeed concentrated. The lower parts of these bodies contain more olivine and pyroxene than the upper parts, which are less mafic. But even in these bodies, crystal settling has yielded very little felsic magma from an original mafic magma.

If felsic magma could be derived on a large scale from mafic magma as Bowen thought, there should be far more mafic magma than felsic magma. To yield a particular volume of granite (a felsic igneous rock), about 10 times as much mafic magma would have to be present initially for crystal settling to yield the volume of granite in question. If this were so, then mafic intrusive igneous rocks should be much more common than felsic ones. However, just the opposite is the case, so it appears that mechanisms other than crystal settling must account for the large volume of felsic magma. Partial melting of mafic oceanic crust and silica-rich sediments of continental margins during subduction yields magma richer in silica than the source rock. Furthermore, magma rising through the continental crust can absorb some felsic materials by *assimilation* and become more enriched in silica.

Assimilation. The composition of a magma can be changed by **assimilation**, a process whereby a magma reacts with preexisting rock, called **country rock**, with which it comes in contact (▶ Figure 3-9). The walls of a volcanic conduit

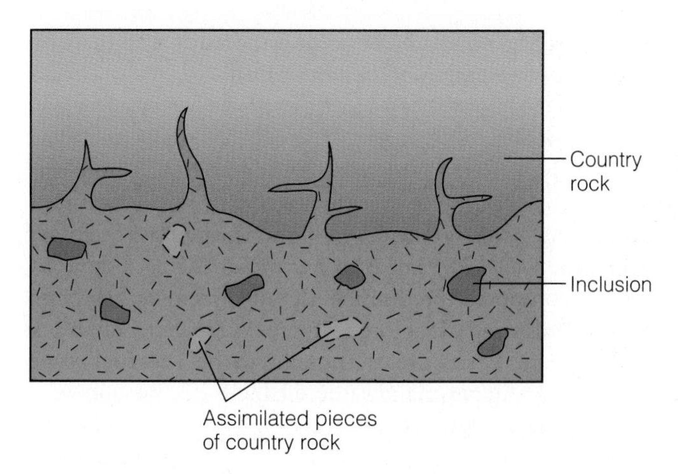

⧐ **FIGURE 3-9** As magma moves upward, fragments of country rock are dislodged and settle into the magma. If they have a lower melting temperature than the magma, they may be incorporated into the magma by assimilation. Incompletely assimilated pieces of country rock are inclusions.

or magma chamber are, of course, heated by the adjacent magma, which may reach temperatures of 1,300°C. Some of these rocks can be partly or completely melted, provided their melting temperature is less than that of the magma. Since the assimilated rocks seldom have the same composition as the magma, the composition of the magma is changed.

The fact that assimilation occurs can be demonstrated by *inclusions*, incompletely melted pieces of rock that are fairly common within igneous rocks. Many inclusions were simply wedged loose from the country rock as the magma forced its way into preexisting fractures (Figures 3-9 and ⧐ 3-10).

There is no doubt that assimilation occurs, but its effect on the bulk composition of most magmas must be slight. The reason is that the heat for melting must come from the magma itself, and this would have the effect of cooling the magma. Only a limited amount of rock can be assimilated by a magma, and that amount is usually insufficient to bring about a major compositional change.

Neither crystal settling nor assimilation can produce a significant amount of felsic magma from a mafic one. But both processes, if operating concurrently, can change the composition of a mafic magma much more than either process acting alone. Some geologists think that this is one way that many intermediate magmas form where oceanic lithosphere is subducted beneath continental lithosphere.

Magma Mixing. The fact that a single volcano can erupt lavas of different composition indicates that magmas of differing composition must be present. It seems likely that some of these magmas would come into contact and mix with one another. If this is the case, we would expect that the composition of the magma resulting from **magma mixing** would be a modified version of the parent magmas. Suppose that a rising mafic magma mixes with a felsic magma of about the same volume (⧐ Figure 3-11). The resulting "new" magma would have a more intermediate composition.

Classification

Most igneous rocks are classified on the basis of textural features and composition (⧐ Figure 3-12). Notice in Figure 3-12 that all of the rocks, except peridotite, constitute pairs; the members of a pair have the same composition but

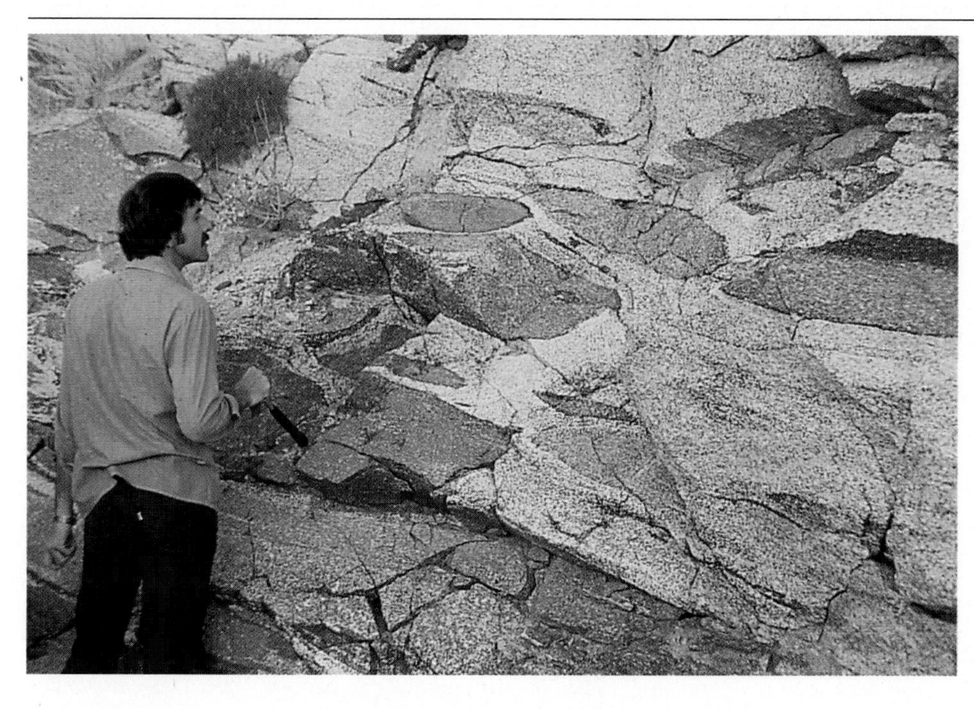

⧐ **FIGURE 3-10** Dark-colored inclusions in granitic rock in California. (Photo courtesy of David J. Matty.)

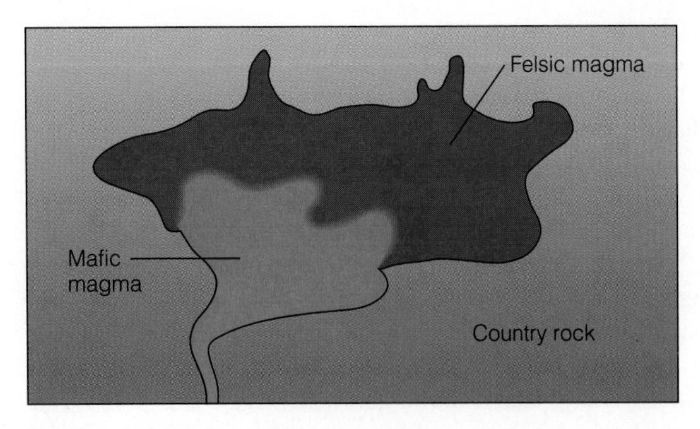

➢ **FIGURE 3-11** Magma mixing. Two rising magmas mix and produce a magma with a composition different from either of the parent magmas.

different textures. Basalt and gabbro, andesite and diorite, and rhyolite and granite are compositional (mineralogical) equivalents, but basalt, andesite, and rhyolite are aphanitic and most commonly extrusive, whereas gabbro, diorite, and granite have phaneritic textures that generally indicate an intrusive origin. All of these pairs exist in a textural continuum. The extrusive and intrusive members of each pair can usually be differentiated by texture, but many shallow intrusive rocks have textures that cannot be readily distinguished from those of extrusive igneous rocks.

The igneous rocks shown in Figure 3-12 are also differentiated by composition. Reading across the chart from rhyolite to andesite to basalt, for example, the relative proportions of nonferromagnesian and ferromagnesian minerals differ. The differences in composition are gradual, however, so that a compositional continuum exists. In other words, there are rocks whose compositions are intermediate between rhyolite and andesite, and so on.

Ultramafic Rocks. Ultramafic rocks (<45% silica) are composed largely of ferromagnesian silicate minerals. The ultramafic rock *peridotite* contains mostly olivine, lesser amounts of pyroxene, and generally a little plagioclase feldspar (➢ Figures 3-12, 3-13a). Another ultramafic rock (pyroxenite) is composed predominantly of pyroxene. Because these minerals are dark colored, the rocks are generally black or dark green. Peridotite is thought to be the rock type composing the upper mantle (see Chapter 10), but ultramafic rocks are rare at the Earth's surface. In fact, ultramafic lava flows are rare in rocks younger than 2.5 billion years (see Perspective 3-1). Ultramafic rocks are generally thought to have originated by concentration of the early-formed ferromagnesian minerals that separated from mafic magmas.

Basalt-Gabbro. *Basalt* and *gabbro* (45–52% silica) are the fine-grained and coarse-grained rocks, respectively, that crystallize from mafic magmas (Figure 3-13). Thus, both have the same composition—mostly calcium-rich plagioclase and pyroxene, with smaller amounts of olivine and amphibole (Figure 3-12). Because they contain a large proportion of ferromagnesian minerals, basalt and gabbro are dark colored; those that are porphyritic typically contain calcium plagioclase or olivine phenocrysts.

Basalt is generally considered to be the most common extrusive igneous rock. Extensive basalt lava flows were erupted in vast areas in Washington, Oregon, Idaho, and northern California (see Chapter 4). Oceanic islands such as Iceland, the Galapagos, the Azores, and the Hawaiian Islands are composed mostly of basalt. Furthermore, the

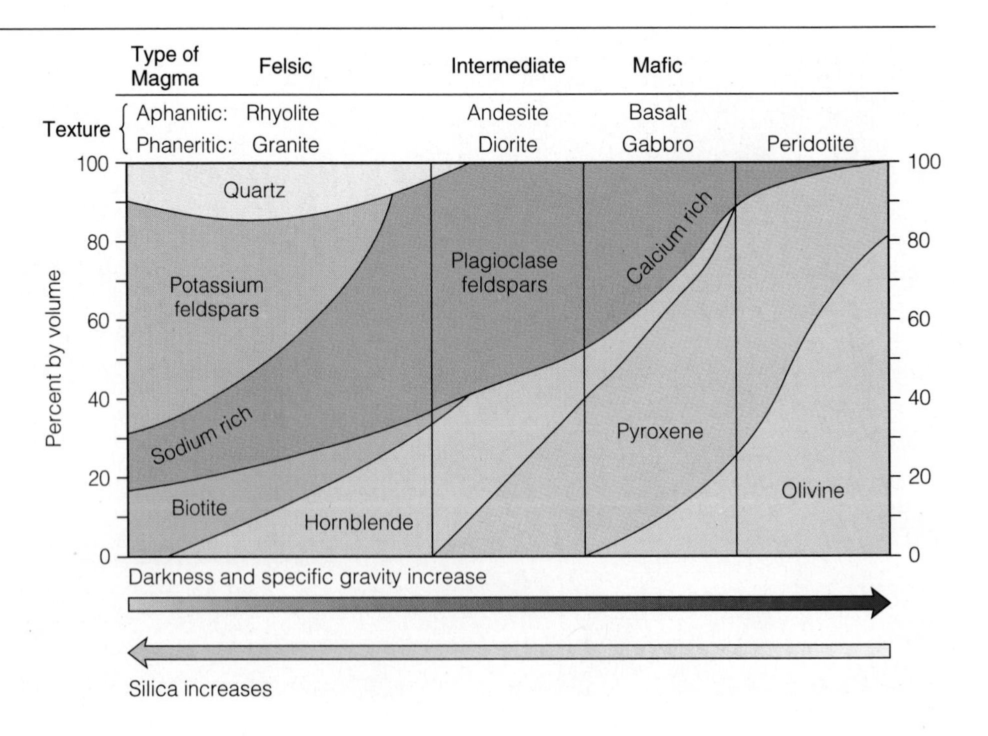

➢ **FIGURE 3-12** Classification of igneous rocks. The diagram illustrates the relative proportions of the chief mineral components of common igneous rocks.

(a)

(b)

(a)

(b)

(c)

▶ FIGURE 3-13 (*a*) The ultramafic rock peridotite. Mafic igneous rocks: (*b*) basalt; and (*c*) gabbro. (Photos courtesy of Sue Monroe.)

▶ FIGURE 3-14 Intermediate igneous rocks: (*a*) andesite and (*b*) diorite. (Photos courtesy of Sue Monroe.)

upper part of the oceanic crust is composed almost entirely of basalt.

Gabbro is much less common than basalt, at least in the continental crust or where it can be easily observed. Small intrusive bodies of gabbro do occur in the continental crust, but intermediate to felsic intrusive rocks such as diorite and granite are much more common. The lower part of the oceanic crust is composed of gabbro, however.

Andesite-Diorite. Magmas intermediate in composition (53–65% silica) crystallize to form *andesite* and *diorite*, which are compositionally equivalent fine- and coarse-grained igneous rocks (▶ Figure 3-14). Andesite and diorite are composed predominantly of plagioclase feldspar, with the typical ferromagnesian component being amphibole or biotite (Figure 3-12). Andesite is generally medium to dark gray, but diorite has a salt and pepper appearance because of its white to light gray plagioclase and dark ferromagnesian minerals (Figure 3-14).

Perspective 3-1

ULTRAMAFIC LAVA FLOWS

Geologists refer to the interval of geologic time from 3.8 to 2.5 billion years ago as the Archean Eon. Some of the most interesting rocks that formed during the Archean Eon are ultramafic lava flows, which are rare in younger rocks and are not forming at present. Archean ultramafic lava flows are generally parts of large, complex associations of rocks known as greenstone belts. An idealized greenstone belt consists of three major rock units: the lower and middle units are dominated by volcanic rocks, and the upper unit is composed mostly of sedimentary rocks (➢ Figure 1). The lower volcanic units of some Archean greenstone belts contain ultramafic lava flows.

Why did ultramafic lava flows occur during early Earth history, but only rarely later? The answer is related to the heat produced within the Earth. When it first formed, the Earth possessed a considerable amount of residual heat inherited from the formative processes. Rock is a poor conductor of heat, so this primordial heat was slowly lost.

Another source of heat within the Earth is related to the phenomenon of radioactive decay. Recall from Chapter 2 that as the isotopes of some elements decay to a more stable state, they generate heat. Accordingly, the Earth has a mechanism whereby heat can be generated internally. As the radioactive isotopes decay to a stable state, however, they become less abundant (except carbon 14, which is produced in the atmosphere), so the heat generated within the Earth has decreased.

In order to erupt, an ultramafic magma requires near-surface temperatures of more than 1,600°C; the surface temperatures of present-day basalt lava flows are generally between 1,000° and 1,200°C. During early Earth history, there was considerably more residual heat and more heat generated by radioactive decay; the mantle was hotter than it is today—perhaps 300°C hotter—and ultramafic magmas could be erupted onto the surface. Because the amount of heat has decreased through time, the Earth has cooled, and the eruption of ultramafic lava flows has largely ceased.

Andesite is a common extrusive igneous rock formed from lavas erupted in volcanic chains at convergent plate margins. The volcanoes of the Andes Mountains of South America and the Cascade Range in western North America are composed in part of andesite. Intrusive bodies composed of diorite are fairly common in the continental crust. However, diorite is not nearly as abundant as granitic rocks and is uncommon outside the areas where andesites occur.

Rhyolite-Granite. *Rhyolite* and *granite* (>65% silica) crystallize from felsic magmas and are therefore silica-rich rocks (➢ Figure 3-15). Rhyolite and granite consist largely of potassium feldspar, sodium-rich plagioclase, and quartz, with

➢ **FIGURE 3-15** Felsic igneous rocks: (*a*) granite and (*b*) rhyolite. (Photos courtesy of Sue Monroe.)

(a)

(b)

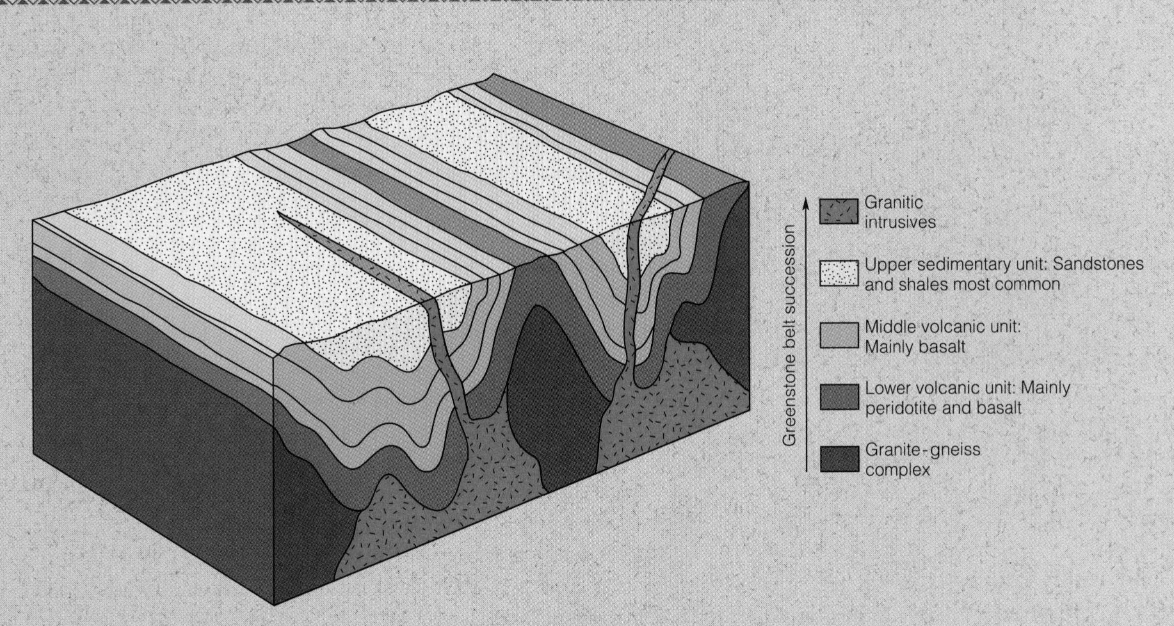

▶ **FIGURE 1** Two adjacent greenstone belts showing their structure and sequence of rock types. The lower volcanic units of some greenstone belts contain ultramafic lava flows.

perhaps some biotite and rarely amphibole (Figure 3-12). Because nonferromagnesian minerals predominate, rhyolite and granite are generally light colored. Rhyolite is fine grained, although most often it contains phenocrysts of potassium feldspar or quartz, and granite is coarse-grained. Granite porphyry is also fairly common.

Rhyolite lava flows are much less common than andesite and basalt flows. Recall that the greatest control of viscosity in a magma is the silica content. Thus, if a felsic magma rises to the surface, it begins to cool, the pressure on it decreases, and gases are released explosively, usually yielding rhyolitic pyroclastic particles. The rhyolitic lava flows that do occur are thick and highly viscous and move only short distances.

Among geologists, granite has come to mean any coarsely crystalline igneous rock with a composition corresponding to that of the field shown in Figure 3-12. Strictly speaking, not all rocks in this field are granites. For example, a rock with a composition close to the line separating granite and diorite is usually called *granodiorite.* To avoid the confusion that might result from introducing more rock names, we will follow the practice of referring to rocks to the left of the granite-diorite line in Figure 3-12 as *granitic.*

Granitic rocks are by far the most common intrusive igneous rocks, although they are restricted to the continents. Most granitic rocks were intruded at or near convergent plate margins during episodes of mountain building. When these mountainous regions are uplifted and eroded, the vast bodies of granitic rocks forming their cores are exposed. The granitic rocks of the Sierra Nevada of California form a composite body measuring about 640 km long and 110 km wide (see chapter-opening photo), and the granitic rocks of the Coast Ranges of British Columbia, Canada, are even more voluminous.

Pegmatite. *Pegmatite* is a very coarsely crystalline igneous rock. Such rocks contain minerals measuring at least 1 cm across, and many crystals are much larger (▶ Figure 3-16). The name pegmatite refers to texture rather than a specific composition, but most pegmatites are composed largely of quartz, potassium feldspar, and sodium-rich plagioclase and correspond to the composition of granite. Some are mafic or intermediate in composition, and are appropriately called gabbro and diorite pegmatites, respectively. Many pegmatites are associated with large granite intrusive bodies and appear to represent the minerals that formed from the fluid and vapor phases that remained after most of the granite crystallized.

The water-rich vapor phase that exists after most of a magma has crystallized as granite has properties that differ from the magma from which it separated. It has a lower density and viscosity, and commonly invades the country rock where it crystallizes. The water-rich vapor phase ordinarily contains a number of elements that rarely enter

⧐ FIGURE 3-16 Pegmatite is a textural term for very coarse-grained igneous rock, which generally has a composition close to that of granite. The mineral grains in this specimen measure 2 to 3 cm.

into the common minerals that form granite. Pegmatites crystallizing to form very coarsely crystalline granite are simple pegmatites, whereas those with minerals containing elements such as lithium, beryllium, cesium, tin, and several others are complex pegmatites. Some complex pegmatites contain 300 different mineral species, a few of which are important economically. In addition, several gem minerals such as emerald and aquamarine, both of which are varieties of the silicate mineral beryl, and tourmaline are found in some pegmatites (⧐ Figure 3-17). Many rare minerals of lesser value and well-formed crystals of common minerals, such as quartz, are also mined and sold to collectors and museums.

The formation and growth of mineral-crystal nuclei in pegmatites are similar to those processes in magma, but with one critical difference: the vapor phase from which pegmatites crystallize inhibits the formation of nuclei. The silica tetrahedra are inhibited from forming the ordered configuration of minerals. However, some nuclei do form, and because the appropriate ions in the liquid can move easily and attach themselves to a growing crystal, individual mineral grains have the opportunity to grow to very large sizes.

Other Igneous Rocks. Some igneous rocks, including tuff, volcanic breccia, obsidian, and pumice, are identified solely by their textures. Much of the fragmental material erupted by volcanoes is *ash*, a designation for pyroclastic materials less than 2.0 mm in diameter; much ash consists of broken pieces or shards of volcanic glass. The consolidation of ash forms the pyroclastic rock *tuff* (⧐ Figure 3-18). Most tuff is silica rich and light colored and is appropriately called *rhyolite tuff*. Some ash flows are so hot that as they come to rest, the ash particles fuse together and form a *welded tuff*. Consolidated deposits of larger pyroclasts, such as cinders, blocks, and bombs, are *volcanic breccia*.

Both *obsidian* and *pumice* are varieties of volcanic glass (⧐ Figure 3-19). Obsidian may be black, dark gray, red, or brown, with the color depending on the presence of tiny particles of iron minerals. Obsidian breaks with the conchoidal fracture that is typical of glass (Figure 3-19a). Analyses of numerous samples indicate that most obsidian is compositionally similar to rhyolite and thus has a high silica content.

Pumice is a variety of volcanic glass containing numerous bubble-shaped vesicles that develop when gas escapes through lava and forms a froth (Figure 3-19b). Some

⧐ FIGURE 3-17 Tourmaline from the Dunton pegmatite mine in Maine.

⊳ **FIGURE 3-18** Exposure of tuff in Colorado. (Photo courtesy of David J. Matty.)

⊳ **FIGURE 3-19** (*a*) Obsidian and (*b*) pumice. (Photos courtesy of Sue Monroe.)

(a)

(b)

pumice forms as crusts on lava flows, and some forms as particles erupted from explosive volcanoes. If pumice falls into water, it can be carried great distances because it is so porous and light that it floats.

 INTRUSIVE IGNEOUS BODIES: PLUTONS

Unlike volcanism and the origin of extrusive or volcanic igneous rocks, which can be observed, intrusive igneous activity can be studied only indirectly. Intrusive igneous bodies called **plutons** form when magma cools and crystallizes within the Earth's crust (▶ Figure 3-20). Although plutons can be observed after erosion has exposed them at the surface, we cannot duplicate the conditions that exist deep in the crust, except in small-scale laboratory experiments. Geologists face a greater challenge in interpreting the mechanisms whereby plutons originate than in studying the origins of extrusive igneous rocks.

Several types of plutons are recognized, all of which are defined by their geometry (three-dimensional shape) and their relationship to the country rock (Figure 3-20). Geometrically, plutons may be characterized as massive or irregular, tabular, cylindrical, or mushroom shaped. Plutons are also described as concordant or discordant. A **concordant** pluton, such as a sill, has boundaries that are parallel to the layering in the country rock. A **discordant** pluton, such as a dike, has boundaries that cut across the layering of the country rock (Figure 3-20).

Dikes and Sills

Both **dikes** and **sills** are tabular or sheetlike plutons, but dikes are discordant whereas sills are concordant (Figure 3-20). Dikes are common intrusive features (▶ Figure 3-21). Many are small bodies measuring 1 or 2 m across, but they range from a few centimeters to more than 100 m thick. Dikes are emplaced within preexisting zones of weakness where fractures already exist or where the fluid pressure is great enough for them to form their own fractures during emplacement.

Erosion of the Hawaiian volcanoes exposes dikes in rift zones, the large fractures that cut across these volcanoes. The Columbia River basalts (discussed in Chapter 4) issued from long fissures, and the magma that cooled in the fissures formed dikes. Some of the large historic fissure eruptions are underlain by dikes; for example, dikes underlie both the Laki

▶ FIGURE 3-20 Block diagram showing the various types of plutons. Notice that some of these plutons cut across the layering in the country rock and are thus discordant, whereas others parallel the layering and are concordant.

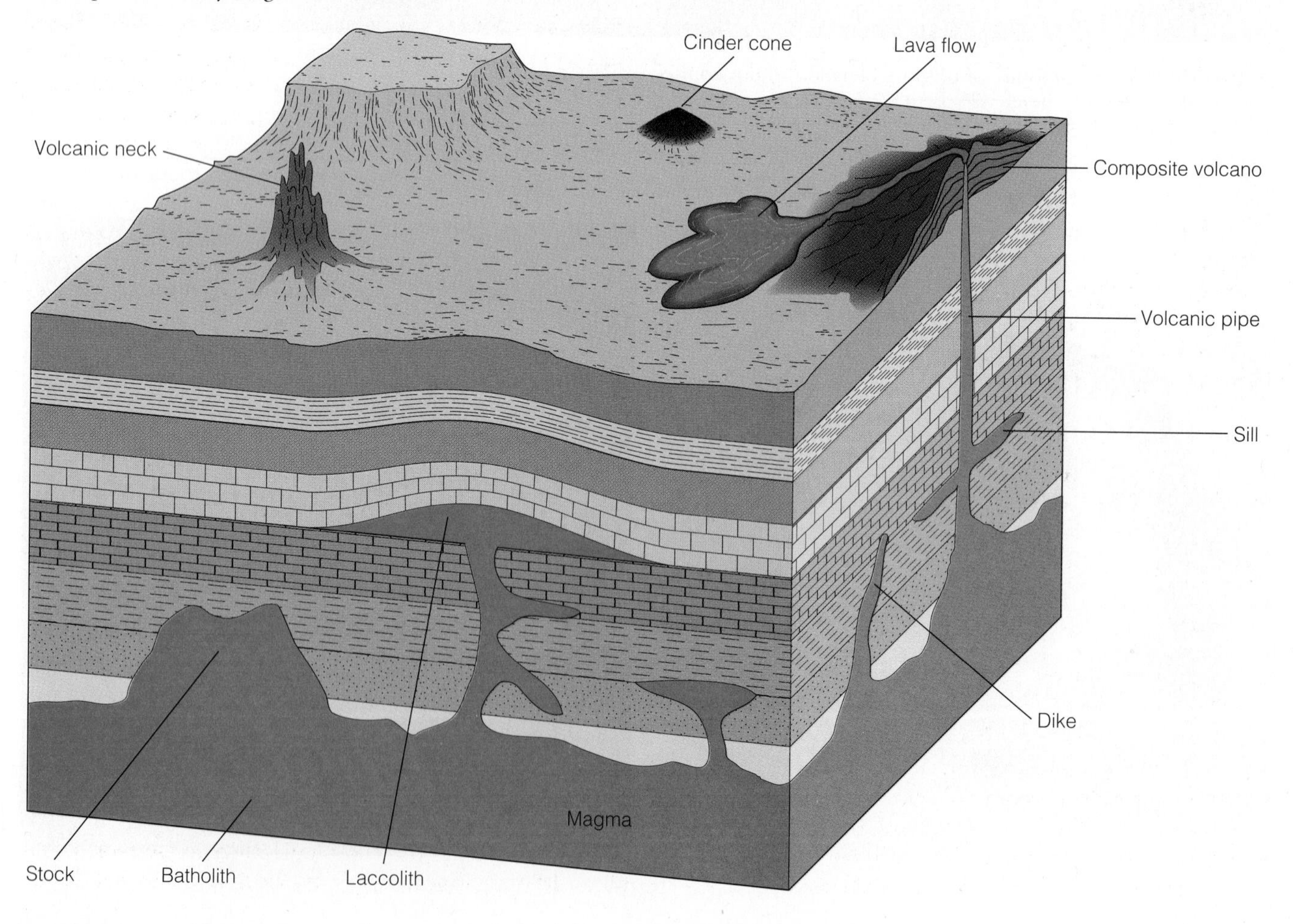

⋗ **FIGURE 3-21** The dark layer cutting diagonally across the rock layers is a dike. The other dark layer is a sill because it parallels the layering.

fissure eruption of 1783 in Iceland and the Eldgja fissure, also in Iceland, where eruptions occurred in A.D. 950 from a fissure 300 km long.

Sills are concordant plutons, many of which are a meter or less thick, although some are much thicker (Figures 3-20 and 3-21). A well-known sill in the United States is the Palisades sill that forms the Palisades along the west side of the Hudson River in New York and New Jersey. It is exposed for 60 km along the river and is up to 300 m thick. Most sills have been intruded into sedimentary rocks, but eroded volcanoes also reveal that sills are commonly injected into piles of volcanic rocks. In fact, some of the inflation of volcanoes preceding eruptions may be caused by the injection of sills (see Perspective 4-2).

In contrast to dikes, which follow zones of weakness, sills are emplaced when the fluid pressure is so great that the intruding magma actually lifts the overlying rocks. Because emplacement requires fluid pressure exceeding the force exerted by the weight of the overlying rocks, sills are typically shallow intrusive bodies.

Laccoliths

Laccoliths are similar to sills in that they are concordant, but instead of being tabular, they have a mushroomlike geometry (Figure 3-20). They tend to have a flat floor and are domed up in their central part. Like sills, laccoliths are rather shallow intrusive bodies that actually lift up the overlying strata. In this case, however, the strata are arched upward over the pluton (Figure 3-20). Most laccoliths are rather small bodies. The best-known laccoliths in the United States are in the Henry Mountains of southeastern Utah.

Volcanic Pipes and Necks

The conduit connecting the crater of a volcano with an underlying magma chamber is a **volcanic pipe** (Figure 3-20). In other words, it is the structure through which magma rises to the surface. When a volcano ceases to erupt, it is eroded as it is attacked by water, gases, and acids. The volcanic mountain eventually erodes away, but the magma that solidified in the pipe is commonly more resistant to weathering and erosion and is often left as an erosional remnant, a **volcanic neck** (Figure 3-20). A number of volcanic necks are present in the southwestern United States, especially in Arizona and New Mexico (see Perspective 3-2), and others are recognized elsewhere.

Batholiths and Stocks

Batholiths are the largest intrusive bodies. By definition they must have at least 100 km^2 of surface area, and most are much larger than this (Figure 3-20). **Stocks** have the same general features as batholiths but are smaller, although some stocks are simply the exposed parts of much larger intrusions that, once revealed by erosion, are batholiths (⋗ Figure 3-22). Batholiths are generally discordant, and most consist of multiple intrusions. In other words, a batholith is a large composite body produced by repeated, voluminous intrusions of magma in the same area. The coastal batholith of Peru, for example, was emplaced over a period of 60 to 70 million years and consists of perhaps as many as 800 individual plutons.

The igneous rocks composing batholiths are mostly granitic, although diorite may also occur. Most batholiths are

Perspective 3-2

SHIPROCK, NEW MEXICO

According to Navajo legend, a young man named Nayenezgani asked his grandmother where the mythical birdlike creatures known as Tse'na'hale lived. She replied, "They dwell at Tsae-bidahi," which means Winged Rock or Rock with Wings. We know Winged Rock as Shiprock, a volcanic neck rising nearly 550 m above the surrounding plain. Radiating outward from this conical volcanic neck are three dikes (▶ Figure 1). Navajo legend holds that Winged Rock represents a giant bird that brought the Navajo people from the north, and the dikes are snakes that have turned to stone.

Shiprock is the most impressive of many volcanic necks exposed in the Four Corners region of the southwestern United States. (Four Corners is a designation for the point where the boundaries of Colorado, Utah, Arizona, and New Mexico converge.) Shiprock is visible from 160 km and was a favorite with rock climbers for many years until the Navajos put a stop to all climbing on the reservation. The country rock penetrated by this volcanic neck includes ancient metamorphic and igneous rocks and about 1,000 m of overlying sedimentary rocks. The rock unit exposed at the surface is the Mancos Shale, a sedimentary rock unit composed mostly of mud that was deposited in an arm of the sea that existed in North America during the Cretaceous Period. Absolute dating of one of the dikes indicates that the magma that solidified to form Shiprock was emplaced about 27 million years ago.

Shiprock is one of several volcanic necks in the Navajo volcanic field that formed as a result of explosive volcanic eruptions. During these eruptions, volcanic materials along with large pieces of country rock torn from the vent walls were hurled high into the air and fell randomly around the area. The material composing Shiprock itself is characterized as a tuff-breccia consisting of fragmental volcanic debris along with inclusions of various sedimentary rocks and some granite and metamorphic rocks. Because Shiprock now stands about 550 m above the surrounding plain, at least that much erosion must have occurred to expose it in its present form. We can only speculate as to how much higher and larger it was when it was part of an active volcano.

The dikes radiating from Shiprock (Figure 1b) formed when magma ascended rather quietly and was emplaced in the country rocks. However, the fractures along which this magma rose may have formed as a result of the explosive emplacement of the tuff-breccia that filled the volcanic vent. The dike on the northeast side of Shiprock extends more than 2,900 m outward from the vent and averages 2.3 m thick. Because the dike rock, like the material composing the volcanic neck, is more resistant to erosion than the adjacent Mancos Shale, the dikes stand as near-vertical walls above the surrounding plain.

▶ **FIGURE 3-22** Some stocks are small intrusive bodies, but others are simply exposed parts of larger plutons. In this example, erosion to the dashed line would expose a batholith.

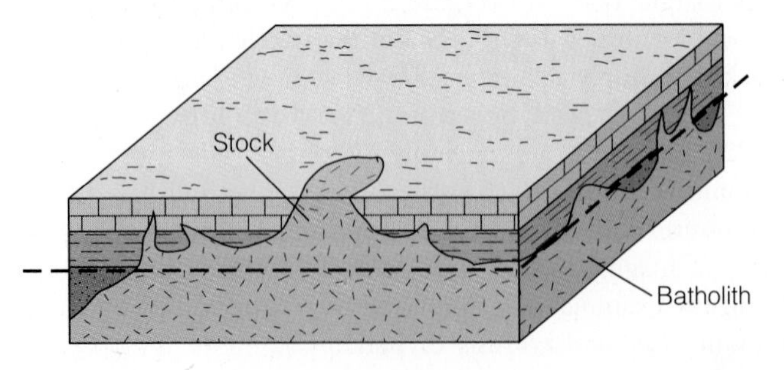

emplaced near continental margins during episodes of mountain building. For example, the Sierra Nevada batholith of California (▶ Figure 3-23) was emplaced over a period of millions of years during a mountain-building episode known as the Nevadan orogeny. Later uplift and erosion exposed this huge composite pluton at the Earth's surface. Other large batholiths in North America include the Idaho batholith and the Coast Range batholith in British Columbia, Canada.

A number of mineral resources occur in rocks of batholiths and stocks and in the country rocks adjacent to them. For example, silica-rich igneous rocks, such as granite, are the primary source of gold, which forms from mineral-rich solutions moving through cracks and fractures of the igneous body. The copper deposits at Butte, Montana, are in rocks near the margins of the granitic rocks of the Boulder

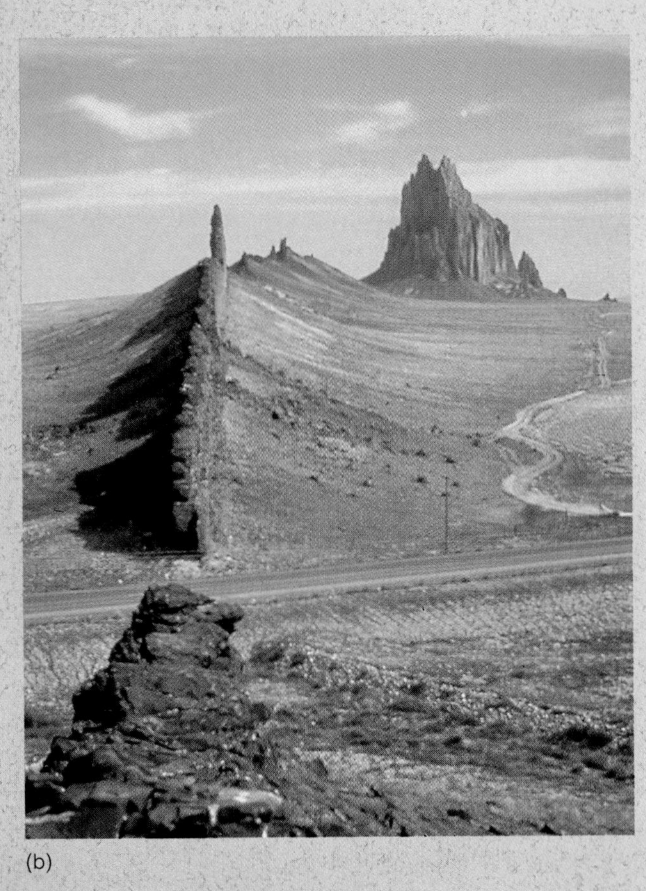

(a) (b)

▶ **FIGURE 1** (*a*) Shiprock, a volcanic neck in northwestern New Mexico, rises nearly 550 m above the surrounding plain. (Photo courtesy of Sue Monroe.) (*b*) View of one of the dikes radiating from Shiprock. (Photo courtesy of Frank Hanna.)

batholith (▶ Figure 3-24). Near Salt Lake City, Utah, copper is mined from the mineralized rocks adjacent to the Bingham stock, a composite pluton composed of granite and granite porphyry.

As noted above, batholiths appear to be emplaced in the cores of mountain ranges that resulted from plate collisions. However, large exposures of granitic rocks also occur within the interiors of continents where mountains are absent. A large area in Canada is underlain by extensive granitic rocks as well as by other rock types. These granites were apparently emplaced during mountain-building episodes that occurred during the early history of the Earth, and the mountains that were once present have long since been eroded away. The remaining rocks represent the eroded "roots" of these ancient mountains.

 MECHANICS OF BATHOLITH EMPLACEMENT

Geologists realized long ago that the emplacement of batholiths posed a space problem; that is, what happened to the rock that formerly occupied the space now occupied by a granite batholith? One proposed answer was that no displacement had occurred, but rather that the granite had been formed in place by alteration of the country rock through a process called *granitization*. According to this view, granite did not originate as a magma, but rather from hot, ion-rich solutions that simply altered the country rock and transformed it into granite. Granitization is a solid-state phenomenon so it is essentially an extreme type of metamorphism (see Chapter 7).

▶ FIGURE 3-23 View of granitic rocks in part of the Sierra Nevada batholith in Yosemite National Park, California.

▶ FIGURE 3-24 A copper mine in Butte, Montana. The copper deposits occur in rocks at the margin of the Boulder batholith.

MONITORING VOLCANIC ACTIVITY

My interest in geology stems from a long-standing fascination with the outdoors and nature. This fascination was strongly influenced by my early childhood in Yellowstone National Park and later visits with my father (who worked for the National Park Service) to Yosemite and other western parks. I began college studying biology with the vague idea of becoming a naturalist with the National Park Service. In response to a growing interest in the more analytical aspects of the physical sciences, however, I switched to geology at the end of my sophomore year and went on to study geophysics and seismology in graduate school.

I began my research with the U.S. Geological Survey using earthquake waves and small variations in the Earth's gravity field to study the structure of the Earth's crust beneath the western United States. Scientifically, the thickness and physical properties of the crust and upper mantle are key elements for understanding the geologic processes that form the outer layers of the Earth. They are also keys to understanding seismic wave propagation in the Earth. Support for this work derived from a national interest in using seismology to discriminate underground nuclear explosions from earthquakes and, ultimately, to provide a technical means for verifying a comprehensive test ban treaty.

More recently, my research has turned toward the study of earthquakes and the clues they provide on deformation of the Earth's crust (seismotectonics) and volcanic processes. Since 1983, I have been in charge of the U.S. Geological Survey's efforts to monitor and better understand the recurring episodes of earthquake swarms and ground uplift in Long Valley caldera in eastern California. Long Valley caldera is a large oval depression (15 by 30 km) at the base of the eastern escarpment of the Sierra Nevada that was formed by a massive volcanic eruption 730,000 years ago. Volcanic activity has continued in the area with the most recent eruptions occurring just 500 years ago. The current earthquake activity and ground deformation are symptomatic of the movement of magma in the upper 5–10 km of the crust and are typical of the sort of geological unrest that often precedes volcanic eruptions.

This activity raises a host of fascinating scientific questions, but it also raises important issues regarding the interaction of science and society. The resort town of Mammoth Lakes lies at the southwestern margin of Long Valley caldera, and the initial news in 1982 that the activity might be related to volcanic activity was met with disbelief and anger by the residents of Mammoth Lakes and Mono County. Responding to the continuing activity in the caldera over subsequent years has proved to be an educational process for both local residents and scientists studying and monitoring the activity. The local residents have come to better appreciate the geologic processes that have sculpted the spectacular setting in which they live, and the scientists have come to appreciate the challenge involved in effectively communicating the results of their research to the public in a useful way. The latter is a challenge that faces scientists in general as taxpayers and politicians increasingly demand to know what they are getting for their money; it is a particularly acute challenge for those of us pursuing research on geologic hazards (earthquakes, volcanoes, landslides, and the like) because the results of our research and the manner in which we present them can have an immediate impact on the economy and social well-being of the communities involved as well as on public safety.

Over the last couple of decades, earth scientists have learned a great deal about the nature of volcanic unrest from detailed studies of volcanoes around the world. A number of these volcanoes have erupted, and in several cases, the eruptions were successfully predicted. Notable examples of successful predictions include five eruptions of Mount Redbout, Alaska, from December 14, 1989, through April 15, 1990; eruptions of Mount Spurr, Alaska, on June 27 and September 17, 1992; and the catastrophic eruption of Mount Pinatubo in the Philippines on June 15, 1991. The successful prediction of the climactic Mount Pinatubo eruption (the second largest eruption in the world this century after the 1915 Mount Katmi eruption in Alaska) provided time for the evacuation of over 60,000 people and the removal of several hundred million dollars worth of aircraft equipment from nearby Clark Air Force Base. The high-stakes drama between scientists monitoring the volcano and local military and civilian authorities in the days and hours before and during the eruption was documented on the PBS network's "Nova" program on Mount Pinatubo.

The Earth sciences are central to our understanding of the risks posed by geologic hazards and a growing number of environmental issues that modern society faces. They will continue to offer a wide range of scientifically exciting and socially significant opportunities in the future.

DAVID P. HILL graduated from San Jose State University and earned an M.S. in geophysics from the Colorado School of Mines and a Ph.D. from the California Institute of Technology. He has been a geophysicist with the U.S. Geological Survey since 1961. His work in the Mammoth Lakes area has been widely cited in newspapers and magazines, including *Science News*, as a good example of how scientific research can have an impact on the general public.

Many granites show clear evidence of an intrusive origin. For example, the contact between these granites and the adjacent country rock is sharp rather than gradational, and elongated mineral crystals are commonly oriented parallel with the contact. Some granitic rocks lack sharp contacts, and gradually change in character until they resemble the adjacent country rocks. Some of these have likely been altered by granitization. Most geologists think that only small quantities of granite are formed by this process, and that it cannot account for the huge granite batholiths of the world. These geologists think an igneous origin for granite is clear, but then they must deal with the space problem. One solution, proposed by some geologists, is that these large igneous bodies melted their way into the crust. In other words, they simply assimilated the country rock as they moved upward (Figure 3-9). The presence of inclusions, especially near the tops of such intrusive bodies, indicates that assimilation does occur. Nevertheless, as we noted previously, assimilation is a limited process because magma is cooled as country rock is assimilated; calculations indicate that far too little heat is available in a magma to assimilate the huge quantities of country rock necessary to make room for a batholith.

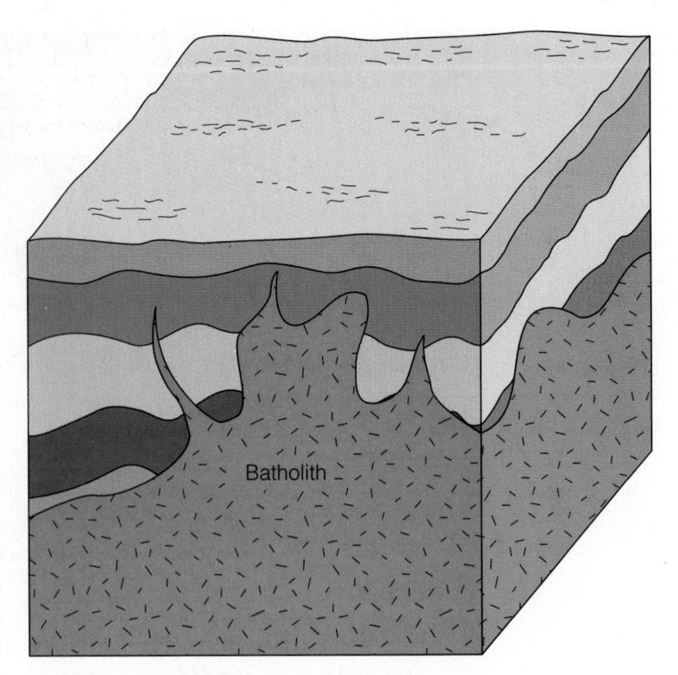

▶ FIGURE 3-25 Emplacement of a hypothetical batholith. As the magma rises, it shoulders aside and deforms the country rock.

Most geologists now agree that batholiths were emplaced by *forceful injection* as magma moved upward toward the surface. Recall that granite is derived from viscous felsic magma and, therefore, rises slowly. It appears that the magma deforms and shoulders aside the country rock, and as it rises further, some of the country rock fills the space beneath the magma (▶ Figure 3-25). A somewhat analogous situation was discovered in which large masses of sedimentary rock called rock salt rise through the overlying rocks to form salt domes.

Salt domes are recognized in several areas of the world including the Gulf Coast of the United States. Layers of rock salt exist at some depth, but salt is less dense than most other types of rock materials. When under pressure, it rises toward the surface even though it remains solid, and as it moves upward, it pushes aside and deforms the country rock (▶ Figure 3-26). Natural examples of rock salt flowage are known, and it can easily be demonstrated experimentally. In the arid Middle East, for example, salt moving upward in the manner described actually flows out at the surface.

Some batholiths do indeed show evidence of having been emplaced forcefully by shouldering aside and deforming the country rock. This mechanism probably occurs in the deeper parts of the crust where temperature and pressure are high and the country rocks are easily deformed in the manner described. At shallower depths, the crust is more rigid and tends to deform by fracturing. In this environment, batholiths may be emplaced by **stoping,** a process in which rising magma detaches and engulfs pieces of country rock (▶ Figure 3-27). According to this concept, magma moves upward along fractures and the planes separating layers of country rock. Eventually, pieces of country rock are detached and settle into the magma. No new room is created during stoping; the magma simply fills the space formerly occupied by country rock (Figure 3-27).

▶ FIGURE 3-26 Three stages in the origin of a salt dome. Rock salt is a low-density sedimentary rock that (*a*) when deeply buried (*b*) tends to rise toward the surface, (*c*) pushing aside and deforming the country rock and forming a dome. Salt domes are thought to rise in much the same manner as batholiths are intruded into the Earth's crust.

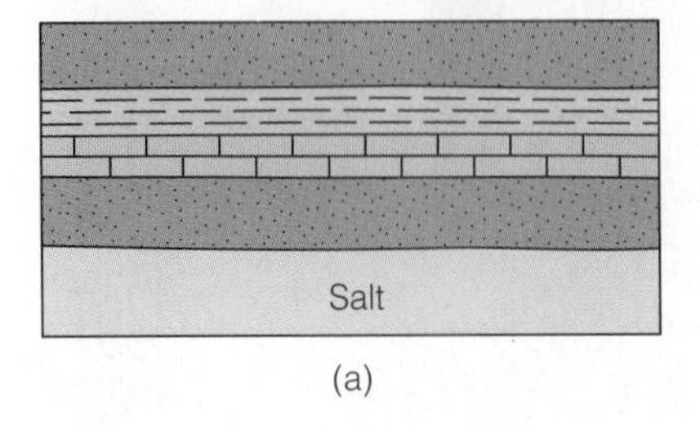

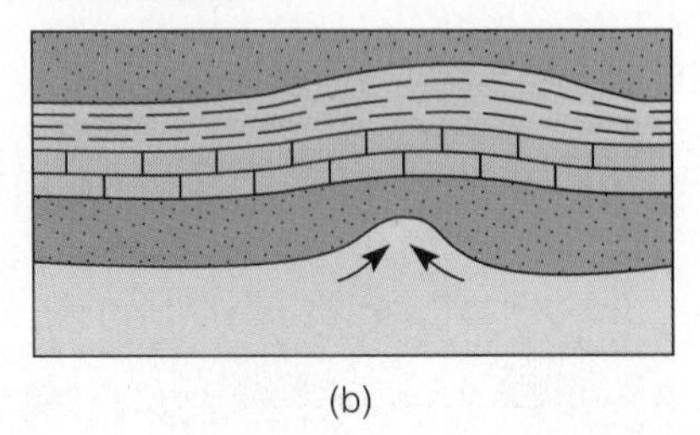

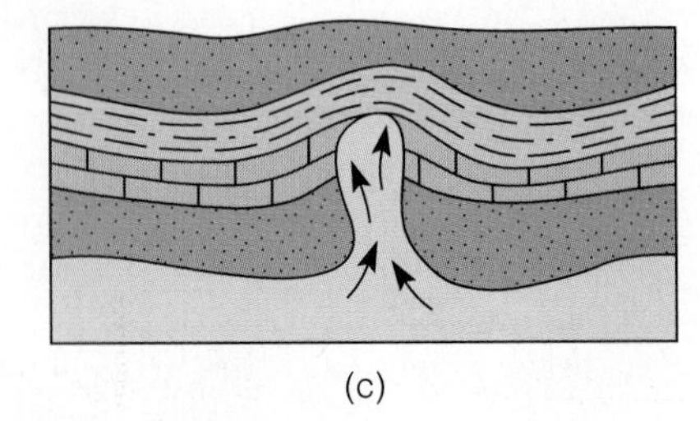

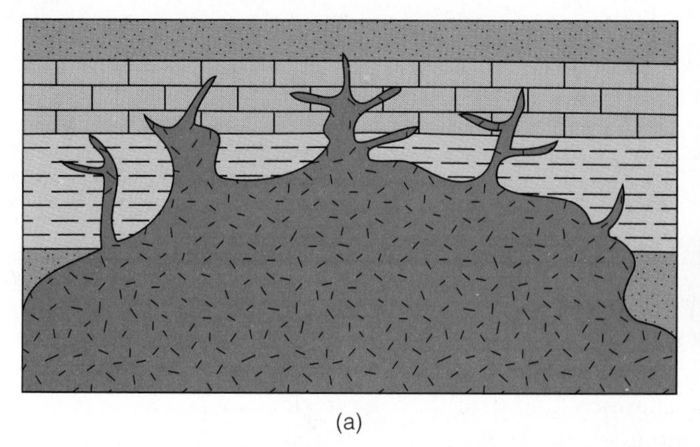
(a)

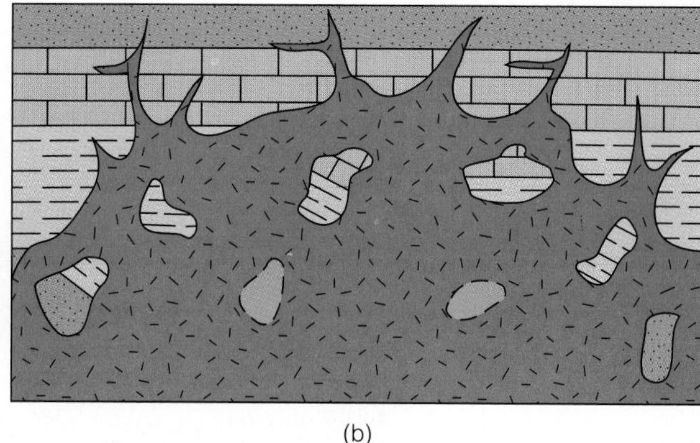
(b)

➢ **FIGURE 3-27** Emplacement of a batholith by stoping. (*a*) The magma is injected into the country rock along fractures and planes between layers. (*b*) Blocks of country rock are detached and engulfed in the magma by stoping. Some of these blocks may be assimilated.

CHAPTER SUMMARY

1. Magma is molten rock material below the Earth's surface, whereas lava is magma that reaches the surface. The silica content of magmas varies and serves to differentiate felsic, intermediate, and mafic magmas.
2. The viscosity of lava flows depends mostly on their temperature and composition. Silica-rich (felsic) lava is more viscous than silica-poor (mafic) lava.
3. Minerals crystallize from magma and lava when small crystal nuclei form and grow.
4. Volcanic rocks generally have aphanitic textures because of their rapid cooling, whereas slow cooling and phaneritic textures characterize plutonic rocks. Igneous rocks with a porphyritic texture have mineral crystals of markedly different sizes. Other igneous rock textures include vesicular and pyroclastic.
5. The composition of igneous rocks is determined largely by the composition of the parent magma. It is possible, however, for an individual magma to yield igneous rocks of differing compositions.
6. Under ideal cooling conditions, a mafic magma yields a sequence of different minerals that are stable within specific temperature ranges. This sequence, called Bowen's reaction series, consists of a discontinuous branch and a continuous branch.
   a. The discontinuous branch contains only ferromagnesian minerals, each of which reacts with the melt to form the next mineral in the sequence.
   b. The continuous branch involves changes only in plagioclase feldspar as sodium replaces calcium in the crystal structure.
7. The ferromagnesian minerals that form first in Bowen's reaction series can settle and become concentrated near the base of a magma chamber or intrusive body. Such settling of iron- and magnesium-rich minerals causes a chemical change in the remaining melt.
8. A magma can be changed compositionally when it assimilates country rock, but this process usually has only a limited effect. Magma mixing may also bring about compositional changes in magmas.
9. Most igneous rocks are classified on the basis of their textures and composition. Two fundamental groups of igneous rocks are recognized: volcanic or extrusive rocks, most of which are aphanitic, and plutonic or intrusive rocks, most of which are phaneritic.
   a. Common volcanic rocks include rhyolite, andesite, basalt, and tuff.
   b. Common plutonic rocks include granite, diorite, and gabbro.
10. Pegmatites are very coarse-grained igneous rocks, most of which have an overall composition similar to that of granite. Crystallization from a vapor-rich phase left over after the crystallization of granite accounts for the very large mineral crystals in pegmatites.
11. Plutons are igneous bodies that formed in place or were intruded into the Earth's crust. Various types of plutons are classified by their geometry and whether they are concordant or discordant.
12. Common plutons include dikes (tabular geometry, discordant); sills (tabular geometry, concordant); volcanic necks (cylindrical geometry, discordant); laccoliths (mushroom shaped, concordant); and batholiths and stocks (irregular geometry, discordant).
13. By definition batholiths must have at least 100 km^2 of surface area; stocks are similar to batholiths but smaller. Many batholiths are large composite bodies consisting of many plutons emplaced over a long period of time.

14. Most batholiths appear to have formed in the cores of mountain ranges during episodes of mountain building.
15. Some geologists think that granite batholiths are emplaced when felsic magma moves upward and shoulders aside and deforms the country rock. The upward movement of rock salt and the formation of salt domes provide a somewhat analogous situation.

Important Terms

aphanitic
assimilation
batholith
Bowen's reaction series
concordant
country rock
crystal settling
dike
discordant
felsic magma
igneous rock
intermediate magma
laccolith
lava
lava flow
mafic magma
magma
magma mixing
phaneritic
pluton
plutonic (intrusive igneous) rock
porphyritic
pyroclastic materials
pyroclastic (fragmental) texture
sill
stock
stoping
vesicle
viscosity
volcanic neck
volcanic pipe
volcanic (extrusive igneous) rock

Review Questions

1. The first minerals to crystallize from a mafic magma are:
   a. ____ quartz and potassium feldspar;
   b. ____ calcium-rich plagioclase and olivine;
   c. ____ biotite and muscovite;
   d. ____ amphibole and pyroxene;
   e. ____ andesite and basalt.
2. The most common aphanitic igneous rock is:
   a. ____ basalt;
   b. ____ granite;
   c. ____ pumice;
   d. ____ obsidian;
   e. ____ rhyolite.
3. Volcanic rocks can usually be distinguished from plutonic rocks by:
   a. ____ color;
   b. ____ composition;
   c. ____ iron-magnesium content;
   d. ____ the size of their mineral grains;
   e. ____ specific gravity.
4. An example of a concordant pluton having a tabular geometry is a:
   a. ____ sill;
   b. ____ batholith:
   c. ____ volcanic neck;
   d. ____ lava flow;
   e. ____ dike.
5. Most pegmatites are essentially:
   a. ____ light-colored gabbro;
   b. ____ thick accumulations of pyroclastic materials;
   c. ____ very coarse-grained granite;
   d. ____ rhyolite porphyry;
   e. ____ cylindrical plutons.
6. An igneous rock possessing a combination of mineral grains with markedly different sizes is:
   a. ____ a natural glass;
   b. ____ the product of very rapid cooling;
   c. ____ formed by explosive volcanism;
   d. ____ a porphyry;
   e. ____ a tuff.
7. Which of the following minerals is likely to be separated from a mafic magma by crystal settling?
   a. ____ sodium-rich plagioclase;
   b. ____ muscovite;
   c. ____ quartz;
   d. ____ olivine;
   e. ____ potassium feldspar.
8. The process whereby a magma reacts with and incorporates preexisting rock is:
   a. ____ crystal differentiation;
   b. ____ granitization;
   c. ____ plutonism;
   d. ____ magma mixing;
   e. ____ assimilation.
9. Igneous rocks composed largely of ferromagnesian minerals are characterized as:
   a. ____ pyroclastic;
   b. ____ ultramafic;
   c. ____ intermediate;
   d. ____ felsic;
   e. ____ mafic.
10. Which of the following pairs of igneous rocks have the same mineral composition?
    a. ____ granite-tuff;
    b. ____ andesite-rhyolite;
    c. ____ pumice-diorite;
    d. ____ basalt-gabbro;
    e. ____ peridotite-andesite.
11. Which of the following is a concordant pluton?
    a. ____ sill;
    b. ____ stock;
    c. ____ volcanic neck;
    d. ____ dike;
    e. ____ batholith.
12. Batholiths are composed mostly of what type of rock?
    a. ____ granitic;
    b. ____ gabbro;
    c. ____ basalt;

d. ___ andesite;
e. ___ peridotite.

13. An igneous rock possessing mineral grains large enough to be seen without magnification is said to have a ____________ texture.
    a. ___ porphyritic;
    b. ___ aphanitic;
    c. ___ fragmental;
    d. ___ phaneritic;
    e. ___ vesicular.
14. What are the two major kinds of igneous rocks? How do they differ?
15. Describe the process whereby mineral crystals form and grow. Why are volcanic rocks generally aphanitic?
16. What is a natural glass, and how does it form?
17. In terms of composition, how are granite and diorite similar and dissimilar?
18. Compare the continuous and discontinuous branches of Bowen's reaction series.
19. Describe how the composition of a magma can be changed by crystal settling; by assimilation. Cite evidence indicating that both of these processes occur.
20. What is a welded tuff?
21. How do dikes and sills differ? How is each emplaced?
22. Describe the sequence of events in the formation of a volcanic neck.
23. Briefly explain where and how batholiths form.
24. What are pegmatites? Explain why some pegmatites contain very large mineral crystals.
25. Are extrusive and intrusive igneous activity related, or are these completely separate phenomena?
26. Why are felsic lava flows so much more viscous than mafic lava flows?

Points to Ponder

1. In the discontinuous branch of Bowen's reaction series, olivine forms in a specific temperature range, but as the magma continues to cool, it reacts with the remaining melt and changes to pyroxene. Pyroxene in turn changes to amphibole, and amphibole changes to biotite with continued cooling. How is it possible to have any of these minerals other than biotite in an igneous rock?
2. What kinds of evidence would indicate that a batholith was emplaced as magma rather than by granitization?
3. How can color and the size of mineral grains in igneous rock give clues about the composition and cooling history of a magma?
4. Two rock specimens have the following compositions:
   Specimen 1: 15% biotite, 15% sodium-rich plagioclase, 60% potassium feldspar, and 10% quartz.
   Specimen 2: 10% olivine, 55% pyroxene, 5% hornblende, and 30% calcium-rich plagioclase.
   How would these two rocks differ in color and specific gravity? Also, what was the viscosity of the magmas from which these rocks crystallized?

Additional Readings

Baker, D. S. 1983. *Igneous rocks.* Englewood Cliffs, N.J.: Prentice-Hall.

Best, M. G. 1982. *Igneous and metamorphic petrology.* San Francisco, Calif.: W. H. Freeman and Co.

Dietrich, R. V., and B. J. Skinner. 1979. *Rocks and rock minerals.* New York: John Wiley & Sons.

Dietrich, R. V. and R. Wicander. 1983. *Minerals, rocks, and fossils.* New York: John Wiley & Sons.

Ernst, W. G. 1969. *Earth materials.* Englewood Cliffs, N.J.: Prentice-Hall.

Hall, A. 1987. *Igneous petrology.* Essex, England: Longman Scientific and Technical.

Hess, P. C. 1989. *Origins of igneous rocks.* Cambridge, Mass.: Harvard University Press.

McBirney, A. R. 1984. *Igneous petrology.* San Francisco, Calif.: Freeman, Cooper and Co.

MacKenzie, W. S., C. H. Donaldson, and C. Guilford. 1982. *Atlas of igneous rocks and their textures.* New York: Halsted Press.

Middlemost, E. A. K. 1985. *Magma and magmatic rocks.* London: Longman Group.

Chapter 4

VOLCANISM

OUTLINE

Parícutin volcano in Mexico soon after it formed. On February 20, 1943, a farmer noticed fumes emanating from a crack in his cornfield, and a few minutes later ash and cinders were erupted. Within a month a cone 300 m high had formed. Lava flows broke through the flanks and base of the volcano and covered two nearby towns. Activity ceased in 1952.

PROLOGUE

On June 15, 1991, about 90 km northwest of Manila, the capital of the Philippines, 1,800-m-high Mount Pinatubo erupted violently, discharging huge quantities of ash and gases into the atmosphere. An estimated 3 to 5 km^3 of particulate matter, mostly volcanic ash, was discharged, making this the largest volcanic eruption in more than half a century. By contrast, the eruption of Mount St. Helens in 1980 ejected only 0.9 km^3 of ash. During the eruption, much of Mount Pinatubo's summit was destroyed as it collapsed following the ejection of material from below the mountain. The summit was replaced by a caldera, a large oval depression measuring 2 km in diameter.

This eruption and continuing activity during June resulted in the evacuation of thousands of U.S. military personnel and their families from Clark Air Force Base (➤ Figure 4-1). Casualty reports were sketchy at first, but by October 1991, the death toll had risen to 722; 281 were killed during the initial eruption, 83 were killed by later mudflows, and 358 died of illness. An additional 200,000 people were evacuated from areas around the volcano, and at least 42,000 homes were destroyed.

➤ FIGURE 4-1 Mount Pinatubo erupting on June 12, 1991. Within 17 minutes of the eruption, this huge cloud of ash and steam had risen about 18 km.

During Mount Pinatubo's eruptions, about 20 million tons of sulfur dioxide (SO_2) were ejected into the atmosphere and reacted with water vapor to form an aerosol of sulfuric acid. Within 21 days of the initial eruption, this aerosol cloud had encircled the Earth. Volcanic ash settles rather quickly and has little long-lasting atmospheric effect, but sulfur gases may remain suspended for several years. In fact, a weather study issued by the Earth System Science Lab at the University of Alabama in Huntsville indicates that the cool summer of 1992 can be attributed to these eruptions.

Mount Pinatubo is but one of approximately 550 known active volcanoes, at least several of which are erupting at any given time. Before the 1991 eruptions, Mount Pinatubo had been quiet for 600 years, unlike some of the other dozen volcanoes on the island of Luzon. For example, Mayon volcano erupted for the twelfth time this century on February 2, 1993. It erupted a dense cloud of ash and gases known as a pyroclastic flow that killed at least 80 people and prompted officials to evacuate 60,000. Both Mount Pinatubo and Mayon volcano erupt along a convergent plate boundary where a plate is subducted at the Manila Trench (➤ Figure 4-2). In fact, all of the Philippine Islands are part of a volcanic island arc.

The first signs of renewed activity at Mount Pinatubo came when farmers noticed small steam explosions from a line of vents on the volcano on April 2, 1991. Geologists of the Philippine Institute of Volcanology and Seismology

➤ FIGURE 4-2 Map showing the location of Mount Pinatubo on the island of Luzon and its proximity to the Manila Trench.

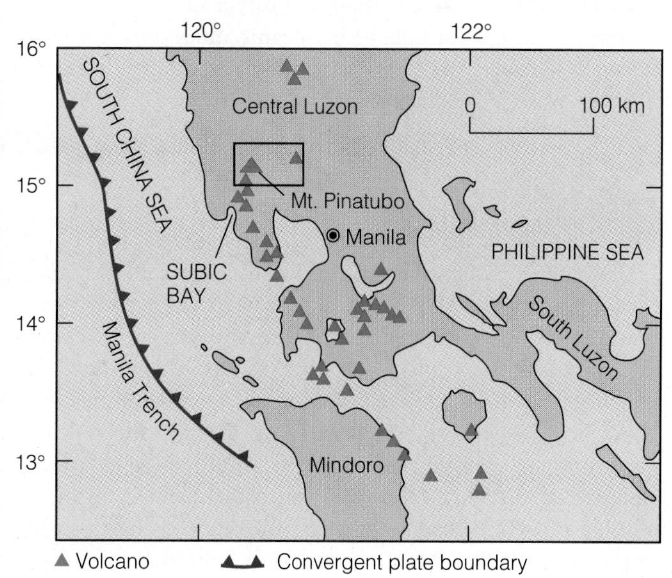

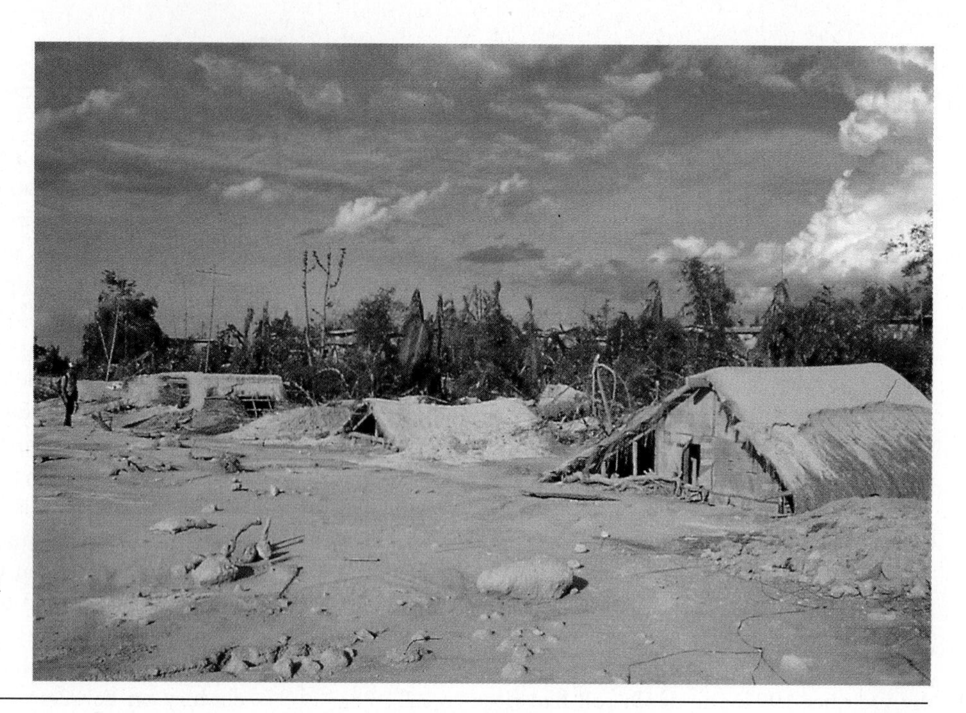

▶ FIGURE 4-3 Homes partly buried by a volcanic mudflow (lahar) on June 15, 1991. Note that the roof at the far right is still partly covered by pyroclastic materials.

and the U.S. Geological Survey quickly installed seismometers to monitor earthquake activity and tiltmeters to see if the volcano's slope changed as a result of bulging prior to an eruption. By late May, molten rock injected into the mountain formed a lava dome, and emissions of sulfur dioxide increased markedly, indicating an impending eruption.

Without such careful monitoring, the death toll would very likely have been much higher. Just four days before the major eruption, emissions of ash and gases increased even more, and the lava dome grew rapidly. Fearing an imminent explosion, officials ordered the evacuation of everyone within 24 km of the volcano. When the eruption occurred on June 15, most people within the zone of immediate danger had been evacuated. Nevertheless, an estimated 364 people perished as a direct result of the eruption. Most were killed by volcanic mudflows, formed from rain-soaked ash, that destroyed everything in their paths (▶ Figure 4-3) or by pyroclastic flows. Measuring 220 m thick in some places, these pyroclastic flows traveled 12 to 18 km from the mountain's summit at speeds of up to 130 km/hr.

Mount Pinatubo remained active throughout the rest of 1991 and well into 1992. On April 4, 1992, a volcanic mudflow on the northeast flank of the mountain disrupted a 1991 pyroclastic flow, resulting in a steam explosion that produced an ash cloud that rose 1,200 m. Even in June 1992, a year after the eruption, more mudflows roared down the mountain's slopes as heavy rains saturated older ash deposits.

In addition to the direct effects of the eruptions and the secondary effects of climate cooling, Mount Pinatubo's renewed activity has prompted the Philippine government to curtail exploration and assessment of the Pinatubo geothermal field, which is close to the large population center around Manila.

INTRODUCTION

Erupting volcanoes are the most impressive manifestations of the dynamic processes operating within the Earth. During many eruptions, molten rock rises to the surface and flows as incandescent streams or is ejected into the atmosphere in fiery displays that are particularly impressive at night (▶ Figure 4-4). In some parts of the world, volcanic eruptions are commonplace events. The residents of the Philippines, Iceland, Hawaii, and Japan are fully cognizant of volcanoes and their effects. In the United States, other than Alaska and Hawaii, volcanic eruptions are not particularly common and are currently localized in the Cascade Range of western North America.

Ironically, eruptions of volcanoes are constructive processes when considered in the context of Earth history. The Hawaiian Islands and Iceland, for example, owe their existence to volcanism. The oceanic crust is continually produced by volcanism at spreading ridges, and volcanic eruptions during the early history of the Earth released gases that are thought to have formed the atmosphere and surface waters.

➢ FIGURE 4-4 Lava fountains such as this one in Hawaii are particularly impressive at night.

VOLCANISM

Volcanism refers to the processes whereby magma and its associated gases rise through the Earth's crust and are extruded onto the surface or into the atmosphere. Currently, about 550 volcanoes are *active*—that is, they have erupted during historic time. Well-known examples of active volcanoes include Mauna Loa and Kilauea on the island of Hawaii, Mount Etna on Sicily, Fujiyama in Japan, and Mount St. Helens in Washington. Only two other bodies in the solar system possess active volcanoes: Io, a moon of Jupiter, and perhaps Triton, one of Neptune's moons (see the Prologue to Chapter 20).

In addition to active volcanoes, numerous *dormant volcanoes* exist that have not erupted recently but may do so again. Mount Vesuvius in Italy had not erupted in human memory until A.D. 79 when it erupted and destroyed the cities of Herculaneum and Pompeii (◉ Table 4-1). Some volcanoes have not erupted during recorded history and show no evidence of doing so again; thousands of these *extinct* or *inactive* volcanoes are known.

Volcanic Gases

Samples of gases taken from present-day volcanoes indicate that 50 to 80% of all volcanic gases are water vapor. Lesser amounts of carbon dioxide, nitrogen, sulfur gases, especially sulfur dioxide and hydrogen sulfide, and very small amounts of carbon monoxide, hydrogen, and chlorine are also commonly emitted. In areas of recent volcanism, such as Lassen Volcanic National Park in California, gases continue to be emitted (➢ Figure 4-5). One cannot help but notice the rotten-egg odor of hydrogen sulfide gas in such areas.

When magma rises toward the surface, the pressure is reduced and the contained gases begin to expand. However, in felsic magmas, which are highly viscous, expansion is inhibited and gas pressure increases. Eventually, the pressure may become great enough to cause an explosion and produce pyroclastic materials such as ash. In contrast, low-viscosity mafic magmas allow gases to expand and escape easily. Accordingly, mafic magmas generally erupt rather quietly.

➢ FIGURE 4-5 Gases being emitted at the Sulfur Works in Lassen Volcanic National Park, California.

TABLE 4-1 Some Notable Volcanic Eruptions

Date	Volcano	Deaths
Aug. 24, 79	Mt. Vesuvius, Italy	3,360 killed in Pompeii and Herculaneum.
1586	Kelut, Java	Mudflows kill 10,000.
Dec. 16, 1631	Mt. Vesuvius, Italy	3,500 killed.
Aug. 4, 1672	Merapi, Java	3,000 killed by mudflows and pyroclastic flows.
Dec. 10, 1711	Awu, Indonesia	3,000 killed by pyroclastic flows.
Sept. 22, 1760	Makian, Indonesia	Eruption kills 2,000; island evacuated for seven years.
June 8, 1783	Lakagigar, Iceland	Largest historic lava flows: 12 km^3; 9,350 die.
July 26, 1783	Asama, Japan	Pyroclastic flows and floods kill 1,200+.
May 21, 1782	Unzen, Japan	14,500 die in debris avalanche and tsunami.
Apr. 10, 1815	Tambora, Indonesia	92,000 killed; another 80,000 reported to have died from famine and disease.
Oct. 8, 1822	Galunggung, Java	4,011 die in pyroclastic flows and mudflows.
Mar. 2, 1856	Awu, Indonesia	Pyroclastic flows kill 2,806.
Aug. 27, 1883	Krakatau, Indonesia	36,417 die; most killed by tsunami.
June 7, 1892	Awu, Indonesia	1,532 die in pyroclastic flows.
May 8, 1902	Mt. Pelée, Martinique	St. Pierre destroyed by pyroclastic flow; 28,000 killed.
Oct. 24, 1902	Santa María, Guatemala	5,000 killed.
June 6, 1912	Novarupta, Alaska	Largest 20th-century eruption: about 33 km^3 of pyroclastic materials erupted; no fatalities.
May 19, 1919	Kelut, Java	Mudflows kill 5,110, devastate 104 villages.
Jan. 21, 1951	Lamington, New Guinea	2,942 killed by pyroclastic flows.
Mar. 17, 1963	Agung, Indonesia	1,148 killed.
Aug. 12, 1976	Soufrière, Guadeloupe	74,000 residents evacuated.
May 18, 1980	Mt. St. Helens, Washington	63 killed; 600 km^2 of forest devastated.
Mar. 28, 1982	El Chichón, Mexico	Pyroclastic flows kill 1,877.
Nov. 13, 1985	Nevado del Ruiz, Colombia	Mudflows kill 23,000.
Aug. 21, 1986	Oku volcanic field, Cameroon	1,746 asphyxiated by cloud of CO_2 released from Lake Nyos.
June 1991	Unzen, Japan	43 killed; at least 8,500 fled.
June 1991	Mt. Pinatubo, Philippines	~281 killed during initial eruption; 83 killed by later mudflows; 358 died of illness; 84,000 evacuated.
Feb. 2, 1993	Mayon, Philippines	At least 70 killed; 60,000 evacuated.

SOURCE: American Geological Institute Data Sheets, except for last two entries.

The amount of gases contained in magmas varies, but it is rarely more than a few percent by weight. Even though volcanic gases constitute a small proportion of a magma, they can be dangerous and, in some cases, have had far-reaching climatic effects (see Perspective 4-1).

Lava Flows and Pyroclastic Materials

Lava flows are frequently portrayed in movies and on television as fiery streams of incandescent rock material posing a great danger to humans. Actually, lava flows are the least dangerous manifestation of volcanism, although they may destroy buildings and cover agricultural land. Most lava flows do not move particularly fast, and because they are fluid, they follow existing low areas. Thus, once a flow erupts from a volcano, determining the path it will take is fairly easy, and anyone in areas likely to be affected can be evacuated.

The geometry of lava flows differs considerably, depending on their viscosity and the preexisting topography. Unless they are confined to a valley, comparatively fluid flows are thin and widespread, whereas more viscous flows tend to be lobate and to have distinct margins (▶ Figure 4-6).

Two types of lava flows, both of which were named for Hawaiian flows, are generally recognized. A **pahoehoe** (pronounced pah-hoy-hoy) flow has a ropy surface almost like taffy (▶ Figure 4-7a). The surface of an **aa** (pronounced ah-ah) flow is characterized by rough, jagged angular blocks and fragments (Figure 4-7b). Some flows solidify as pahoehoe or aa throughout, but some pahoehoe flows change to aa in the downflow direction; an aa flow will not change to pahoehoe in a downflow direction. Pahoehoe flows are less viscous than aa flows; indeed, the latter are viscous enough to break up into blocks and move forward as a wall of rubble.

The surfaces of lava flows may be marked by such features as pressure ridges and spatter cones. **Pressure ridges** are

(a)

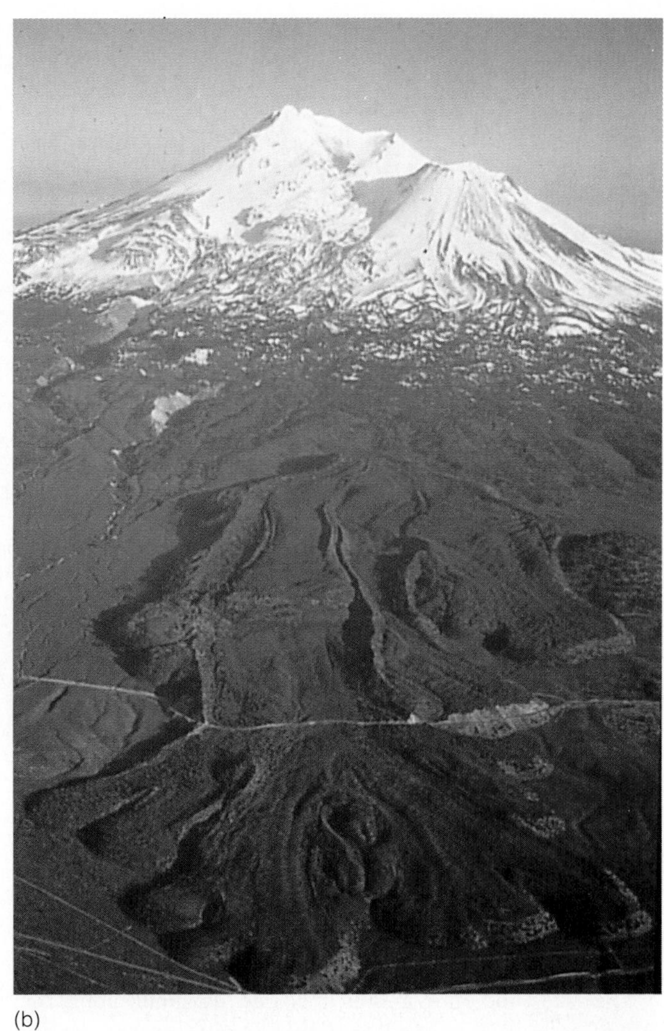

(b)

▶ **FIGURE 4-6** (*a*) A fluid lava flow erupted during the 1969–1971 Mauna Ulu eruption of Kilauea volcano, Hawaii. (*b*) A viscous lava flow (foreground) at Mount Shasta in California showing distinct margins. The peak on the right is Shastina, a parasitic cone that developed on the flank of Mount Shasta.

▶ **FIGURE 4-7** (*a*) A pahoehoe flow in the east rift zone of Kilauea volcano, Hawaii, in 1972. (*b*) An aa flow in the east rift zone of Kilauea volcano in 1983. The flow front is about 2.5 m high.

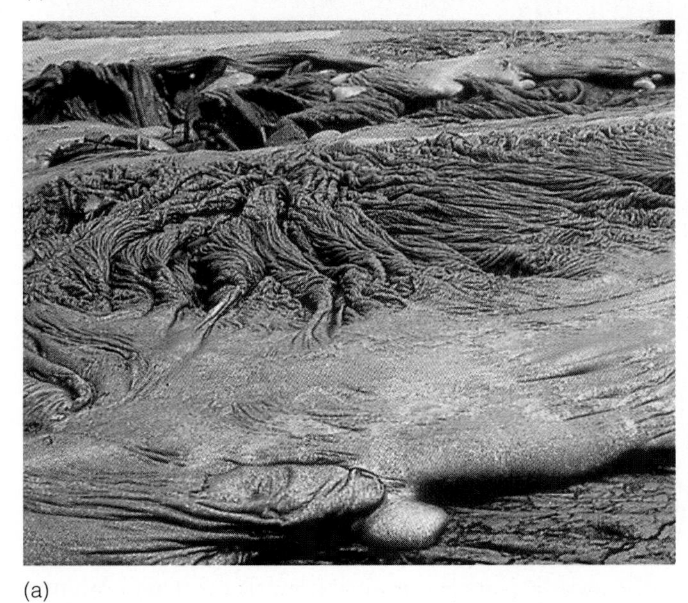

(a)

(b)

Perspective 4-1

Volcanic Gases and Climate

Most volcanic gases quickly dissipate in the atmosphere and pose little danger to humans, but on several occasions volcanic gases have caused numerous fatalities. In 1783, toxic gases, probably sulfur dioxide, erupted from Laki fissure in Iceland had devastating effects. About 75% of the nation's livestock died, and the haze resulting from the gas caused lower temperatures and crop failures; about 24% of Iceland's population died as a result of the ensuing Blue Haze Famine.

Obviously, large volcanic eruptions can devastate local areas, but they can also affect climate over much larger regions—in some cases worldwide. The 1783 Laki fissure eruption produced what Benjamin Franklin called a "dry fog" that was responsible for dimming the intensity of sunlight in Europe. The severe winter of 1783–1784 in Europe and eastern North America is attributed to the presence of this "dry fog" in the upper atmosphere. In Iceland, the winter temperature was 4.8°C below the long-term average; the country suffered its coldest winter in 225 years.

More recently, in 1986, in the African nation of Cameroon 1,746 people died when a cloud of carbon dioxide engulfed them. The gas accumulated in the waters of Lake Nyos, which occupies a volcanic crater. No agreement exists on what caused the gas to suddenly burst forth from the lake, but once it did, it flowed downhill along the surface because it was denser than air. In fact, the density and velocity of the gas cloud were great enough to flatten vegetation, including trees, a few kilometers from the lake. Unfortunately, thousands of animals and many people, some as far as 23 km from the lake, were asphyxiated.

Volcanic ash erupted into the upper atmosphere has some effect on climate, but all particles except the smallest settle quickly and produce no long-lasting effect. Sulfur gases emitted during large eruptions have more important effects; small gas molecules remain in the upper atmosphere for years, absorbing incoming solar radiation and reflecting it back into space. In 1816, a persistent "dry fog" caused unusually cold spring and summer weather in Europe, the eastern United States, and eastern Canada. In North America, 1816 was called "The Year without a Summer" or "Eighteen Hundred and Froze to Death." Killing frosts occurred throughout the summer in New England, resulting in crop failures and food shortages.

The particularly cold spring and summer of 1816 are attributed to the 1815 eruption of Tambora in Indonesia, the largest and most deadly eruption during historic time. The eruption of Mayon volcano in the Philippines during the previous year may have contributed to the cool spring and summer of 1816 as well. Another large historic eruption that had widespread climatic effects was the eruption of Krakatau in 1883 (see the Prologue to Chapter 1).

In comparison with Tambora and Krakatau, the 1980 Mount St. Helens eruption was small. Furthermore, it did not emit much sulfur gas, and its explosion was directed laterally so that most of the particulate matter did not enter the upper atmosphere. In fact, the much smaller 1982 eruption of El Chichón in Mexico had a greater effect on the climate, because so much sulfur gas, other gases, and ash were ejected vertically that a large amount entered the upper atmosphere.

buckled areas on the surface of a lava flow (▶ Figure 4-8a) that form because of pressure on the partly solid crust of a moving flow. **Spatter cones** form when gases escaping from a flow hurl globs of molten lava into the air. These globs fall back to the surface and adhere to one another, forming small, steep-sided cones (Figure 4-8b).

Columnar joints are common in many lava flows, especially mafic flows, but they also occur in other kinds of flows and in some intrusive igneous rocks (▶ Figure 4-9). A lava flow contracts as it cools and produces forces that cause fractures called *joints* to open up. On the surface of a flow, these joints commonly form polygonal (often six-sided) cracks. These cracks also extend downward into the flow, forming parallel columns with their long axes perpendicular to the principal cooling surface. Excellent examples of columnar joints can be seen at Devil's Postpile National Monument in California (Figure 4-9), Devil's Tower National Monument in Wyoming (see the Prologue to Chapter 3), the Giant's Causeway in Ireland, and many other areas.

Much of the igneous rock in the upper part of the oceanic crust is of a distinctive type; it consists of bulbous masses of basalt resembling pillows, hence the name **pillow lava**. It was long recognized that pillow lava forms when lava is rapidly chilled beneath water, but its formation was not observed until 1971. Divers near Hawaii saw pillows form when a blob of lava broke through the crust of an underwater lava flow and cooled almost instantly, forming a glassy exterior. Remaining fluid inside then broke through the crust of the pillow, resulting in an accumulation of interconnected pillows (▶ Figure 4-10).

Much pyroclastic material is erupted as **ash**, a designation for pyroclastic particles measuring less than 2.0 mm. Ash may

(a) (b)

➢ **FIGURE 4-8** (*a*) A pressure ridge on a 1982 lava flow in Hawaii. (*b*) A row of spatter cones formed on February 25, 1983, on a flow at Kilauea volcano, Hawaii.

be erupted in two ways: an ash fall or an ash flow (➢ Figure 4-11a). During an ash fall, ash is ejected into the atmosphere and settles to the surface over a wide area. In 1947, ash that erupted from Mount Hekla in Iceland fell 3,800 km away on Helsinki, Finland. About 10 million years ago, in what is now northeastern Nebraska, numerous rhinoceroses, horses, camels, and other mammals were buried by volcanic ash that was apparently erupted in New Mexico, more than 1,000 km away. Ash is also erupted in ash flows, which are coherent clouds of ash and gas that commonly flow along or close to the land surface. Such flows can move at more than 100 km per hour, and some of them cover vast areas.

Pyroclastic materials larger than ash are also erupted by explosive volcanoes. Particles measuring from 2 to 64 mm are known as *lapilli,* and any particle larger than 64 mm is called a *bomb* or *block* depending on its shape. Bombs have twisted, streamlined shapes that indicate they were erupted as globs of fluid that cooled and solidified during their flight through the air (Figure 4-11b). Blocks are angular pieces of rock ripped from a volcanic conduit or pieces of a solidified crust of a magma. Because of their large size, volcanic bomb and block accumulations are not nearly as widespread as ash deposits; instead, they are confined to the immediate area of eruption.

Volcanoes

Conical mountains formed around a vent where lava, gases, and pyroclastic materials are erupted are **volcanoes**. Volca-

➢ **FIGURE 4-9** (*a*) Columnar joints in a lava flow at Devil's Postpile National Monument, California. (*b*) Surface view of the same columnar joints showing their polygonal pattern. The straight lines and polish resulted from glacial ice moving over this surface.

(a) (b)

➢ **FIGURE 4-10** These bulbous masses of pillow lava form when magma is erupted under water.

noes, which are named for *Vulcan*, the Roman deity of fire, come in many shapes and sizes, but geologists recognize several major categories, each of which has a distinctive eruptive style. One must realize, however, that each volcano is unique in terms of its overall history of eruptions and development. The frequency of eruptions, for example, varies considerably; the Hawaiian volcanoes have erupted repeatedly during historic time, whereas others, such as Mount St. Helens, have erupted periodically after long periods of inactivity. One of the duties of the U.S. Geological Survey is monitoring active volcanoes and developing methods of forecasting eruptions (see Perspective 4-2).

Most volcanoes have a circular depression or **crater** at their summit. Craters form as a result of the extrusion of gases and lava from a volcano and are connected via a conduit to a magma chamber below the surface. It is not unusual, though, for magma to erupt from vents on the flanks of large volcanoes where smaller, parasitic cones develop. For example, Shastina is a large parasitic cone on the flank of Mount Shasta in California (Figure 4-6b), and Mount Etna on Sicily has some 200 smaller vents on its flanks.

Some volcanoes are characterized by a **caldera** rather than a crater. Craters are generally less than 1 km in diameter, whereas calderas exceed this dimension and have steep sides. One of the best-known calderas in the United States is the misnamed Crater Lake in Oregon—Crater Lake is actually a caldera (➢ Figure 4-12). It formed about 6,600 years ago after voluminous eruptions partially drained the magma chamber. This drainage left the summit of the

(a)

(b)

➢ **FIGURE 4-11** Pyroclastic materials. (*a*) Volcanic ash being erupted from Mount Ngauruhoe, New Zealand, during January 1974. Ash rising into the atmosphere will form an ash fall, whereas the cloud rushing down the volcano's slope is an ash flow. (*b*) Volcanic bombs collected in Hawaii. Their streamlined shape indicates they were erupted as globs of magma that cooled and solidified as they descended.

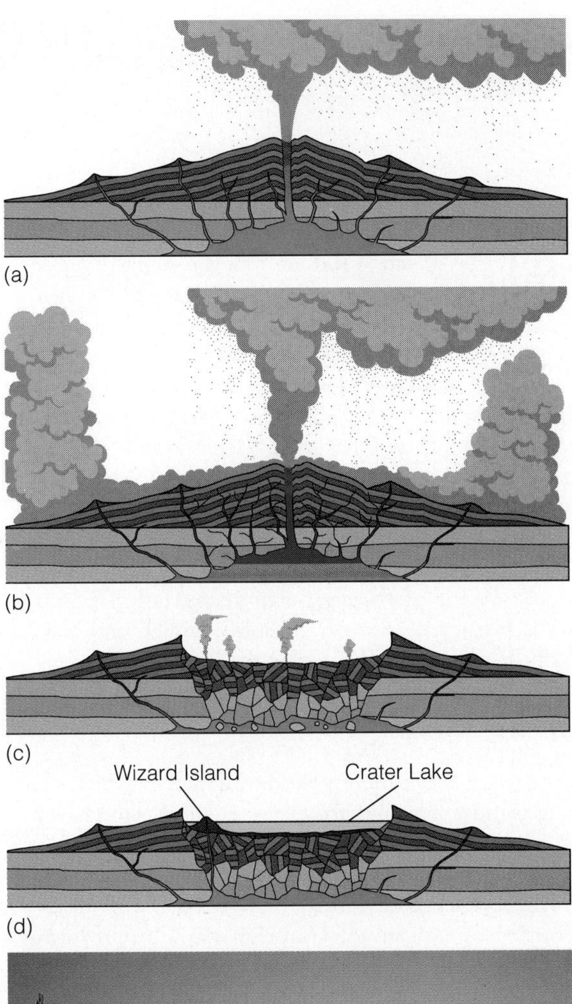

▶ FIGURE 4-12 The sequence of events leading to the origin of Crater Lake, Oregon. (*a–b*) Ash clouds and ash flows partly drain the magma chamber beneath Mount Mazama. (*c*) The collapse of the summit and formation of the caldera. (*d*) Post-caldera eruptions partly cover the caldera floor, and the small volcano known as Wizard Island forms. (*e*) View from the rim of Crater Lake showing Wizard Island.

mountain, Mount Mazama, unsupported, and it collapsed into the magma chamber, forming a caldera more than 1,200 m deep and measuring 9.7 by 6.5 km. Most calderas probably formed when a summit collapsed during particularly large, explosive eruptions as in the case of Crater Lake, but a few small ones apparently formed when the top of the original volcano was blasted away.

Shield Volcanoes. **Shield volcanoes** resemble the outer surface of a shield lying on the ground with the convex side up (▶ Figure 4-13a). They have low, rounded profiles with gentle slopes ranging from about 2 to 10 degrees. Their low slopes reflect the fact that they are composed mostly of mafic lava flows that had low viscosity, so the flows spread out and formed thin layers. Eruptions from shield volcanoes, sometimes called *Hawaiian-type volcanoes*, are quiet compared to those of volcanoes such as Mount St. Helens; lavas most commonly rise to the surface with little explosive activity, so they usually pose little danger to humans. Lava fountains, some up to 400 m high, contribute some pyroclastic materials to shield volcanoes (Figure 4-4), but otherwise these volcanoes are composed largely of basalt lava flows; flows comprise more than 99% of the Hawaiian volcanoes above sea level.

Although eruptions of shield volcanoes tend to be rather quiet, some of the Hawaiian volcanoes have, on occasion, produced sizable explosions. Such explosions occur when magma comes in contact with groundwater, causing it to vaporize instantly. One such explosion occurred in 1790 while Chief Keoua was leading about 250 warriors across the summit of Kilauea volcano to engage a rival chief in battle. About 80 of Keoua's warriors were killed by a cloud of hot volcanic gases.

Shield volcanoes are most common in oceanic areas, such as the Hawaiian Islands and Iceland, but some are also present on the continents—for example, in east Africa. The island of Hawaii consists of five huge shield volcanoes, two of which, Kilauea and Mauna Loa, are active much of the time. These Hawaiian volcanoes are the largest volcanoes in the world (Figure 4-13b). Mauna Loa is nearly 100 km across at the base and stands more than 9.5 km above the surrounding sea floor. Its volume is estimated at about 50,000 km^3. By contrast, the largest volcano in the continental United States, Mount Shasta in northern California (Figure 4-6b), has a volume of only about 205 km^3.

Shield volcanoes have a summit crater or caldera and a number of smaller cones on their flanks through which lava is erupted (Figure 4-13a). A vent opened on the flank of Kilauea and grew to more than 250 m high between June 1983 and September 1986.

Cinder Cones. Volcanic peaks composed of pyroclastic materials that resemble cinders are known as **cinder cones** (▶ Figure 4-14 and the chapter-opening photo). They form when pyroclastic materials are ejected into the atmosphere and fall back to the surface to accumulate around the

Perspective 4-2

Monitoring Volcanoes and Forecasting Eruptions

Two facilities in the United States staffed by geologists of the U.S. Geological Survey (USGS) are devoted to volcano monitoring; Hawaiian Volcano Observatory on the rim of the crater of Kilauea volcano and the David A. Johnston Cascades Volcano Observatory in Vancouver, Washington. The latter was established in 1981 and named in memory of a USGS geologist killed during the 1980 Mount St. Helens eruption. This facility is responsible for monitoring the various Cascade Range volcanoes.

Numerous volcanoes on the margins of the Earth's tectonic plates have erupted explosively during historic time and have the potential to do so again. As a matter of fact, volcanic eruptions are not as unusual as one might think; 376 separate outbursts occurred between 1975 and 1985. Fortunately, none of these compared to the 1815 eruption of Tambora; nevertheless, fatalities occurred in several instances, the worst being in 1985 in Colombia where about 23,000 perished in mudflows generated by an eruption (Table 4-1). Only a few of these potentially dangerous volcanoes are monitored, including some in Italy, Japan, New Zealand, Russia, and the Cascade Range.

Many of the methods for monitoring active volcanoes were developed at the Hawaiian Volcano Observatory. These methods involve recording and analyzing various changes in both the physical and chemical attributes of volcanoes. Tiltmeters are used to detect changes in the slopes of a volcano when it inflates as magma is injected into it, while a geodimeter uses a laser beam to measure horizontal distances, which also change when a volcano inflates (▶ Figure 1). Geologists also monitor gas emissions and changes in the local magnetic and electrical fields of volcanoes.

Of critical importance in volcano monitoring and eruption forecasting are a sudden increase in earthquake activity and the detection of *harmonic tremor*. Harmonic tremor is continuous ground motion as opposed to the sudden jolts produced by earthquakes. It precedes all eruptions of Hawaiian volcanoes and also preceded the eruption of Mount St. Helens. Such activity indicates that magma is moving below the surface.

The analysis of data gathered during monitoring is not by itself sufficient to forecast eruptions; the past history of a particular volcano must also be known. To determine the eruptive history of a volcano, the record of previous eruptions as preserved in rocks must be studied and analyzed. Indeed, prior to 1980, Mount St. Helens was considered one of the most likely Cascade volcanoes to erupt because detailed studies indicated that it has had a record of explosive activity for the past 4,500 years.

For the better monitored volcanoes, such as those in Hawaii, it is now possible to make accurate short-term forecasts of eruptions. In 1960 the warning signs of an eruption of Kilauea were recognized soon enough to evacuate the residents of a small village that was subsequently buried by lava flows. Unfortunately, current forecasting is limited to just a few months in the future.

For some volcanoes, little or no information is available for making predictions. For example, on January 14, 1993, Colombia's Galeras volcano erupted without warning, killing 6 of 10 volcanologists on a field trip and 3 Colombian tourists. Ironically, the volcanologists were attending a conference on improving methods for predicting volcanic eruptions.

vent, forming small, steep-sided cones. The slope angle may be as much as 33 degrees, depending on the angle that can be maintained by the irregularly shaped pyroclastic materials. Cinder cones are rarely more than 400 m high (Figure 4-13b), and many have a large, bowl-shaped crater. Many cinder cones are very nearly symmetric in shape; that is, the pyroclastic materials accumulate uniformly around the vent, forming a symmetrical cone. The symmetry may be less than perfect, however, when prevailing winds cause the pyroclastic materials to build up higher in the downwind direction.

Many cinder cones form on the flanks or within the calderas of larger volcanic mountains and appear to represent the final stages of activity, particularly in areas formerly characterized by basalt lava flows. Wizard Island in Crater Lake, Oregon, is a small cinder cone that formed after the

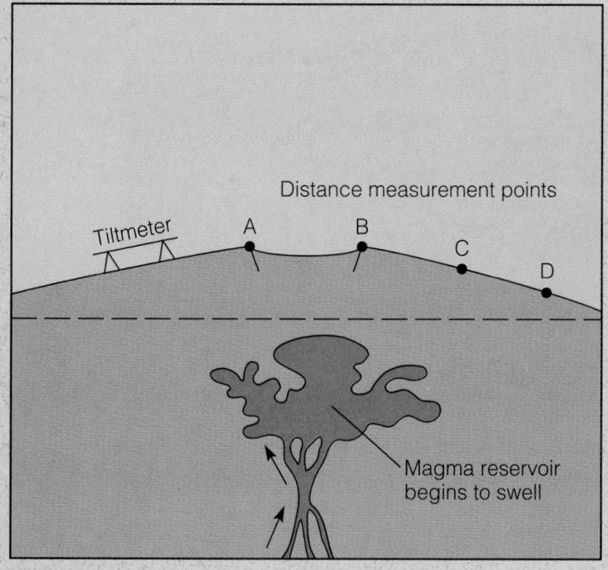

(a) Stage 1

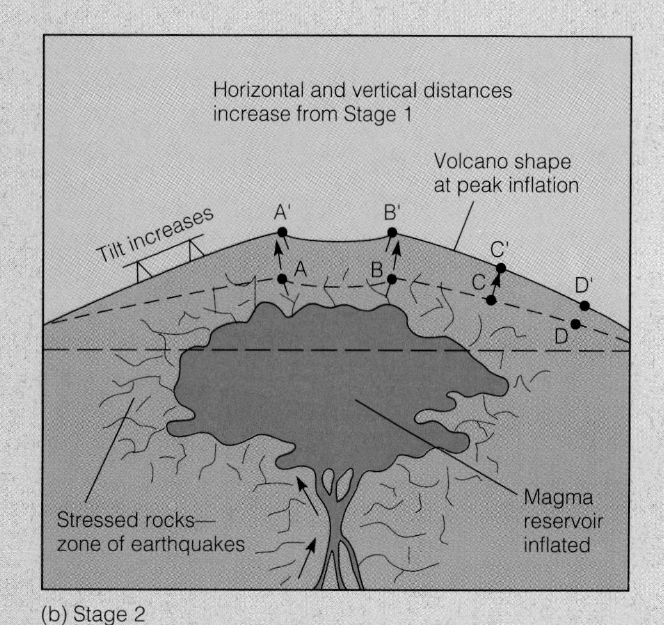

(b) Stage 2

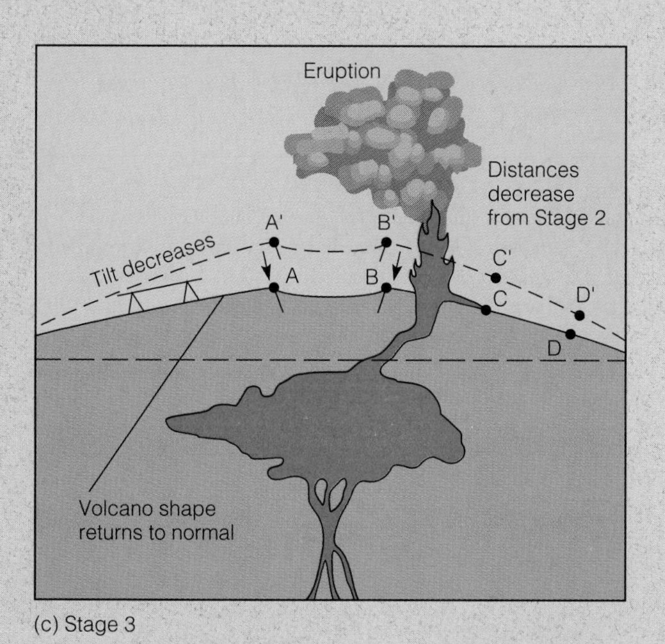

(c) Stage 3

▶ **FIGURE 1** Volcano monitoring. These diagrams show three stages in a typical eruption of a Hawaiian volcano: (*a*) The volcano begins to inflate; (*b*) inflation reaches its peak; (*c*) the volcano erupts and then deflates, returning to its normal shape.

summit of Mount Mazama collapsed to form a caldera (Figure 4-12). Cinder cones are common in the southern Rocky Mountain states, particularly New Mexico and Arizona, and many others occur in northern California, Oregon, and Washington.

In 1973, on the Icelandic island of Heimaey, the town of Vestmannaeyjar was threatened by a new cinder cone. The initial eruption began on January 23, and within two days a cinder cone, later named Eldfell, rose to about 100 m above the surrounding area (Figure 4-14b). Pyroclastic materials from the volcano buried parts of the town, and by February a massive aa lava flow was advancing toward the town. The flow's leading edge ranged from 10 to 20 m thick, and its central part was as much as 100 m thick. By spraying the leading edge of the flow with sea water, which caused it to cool and solidify, the residents of Vestmannaeyjar

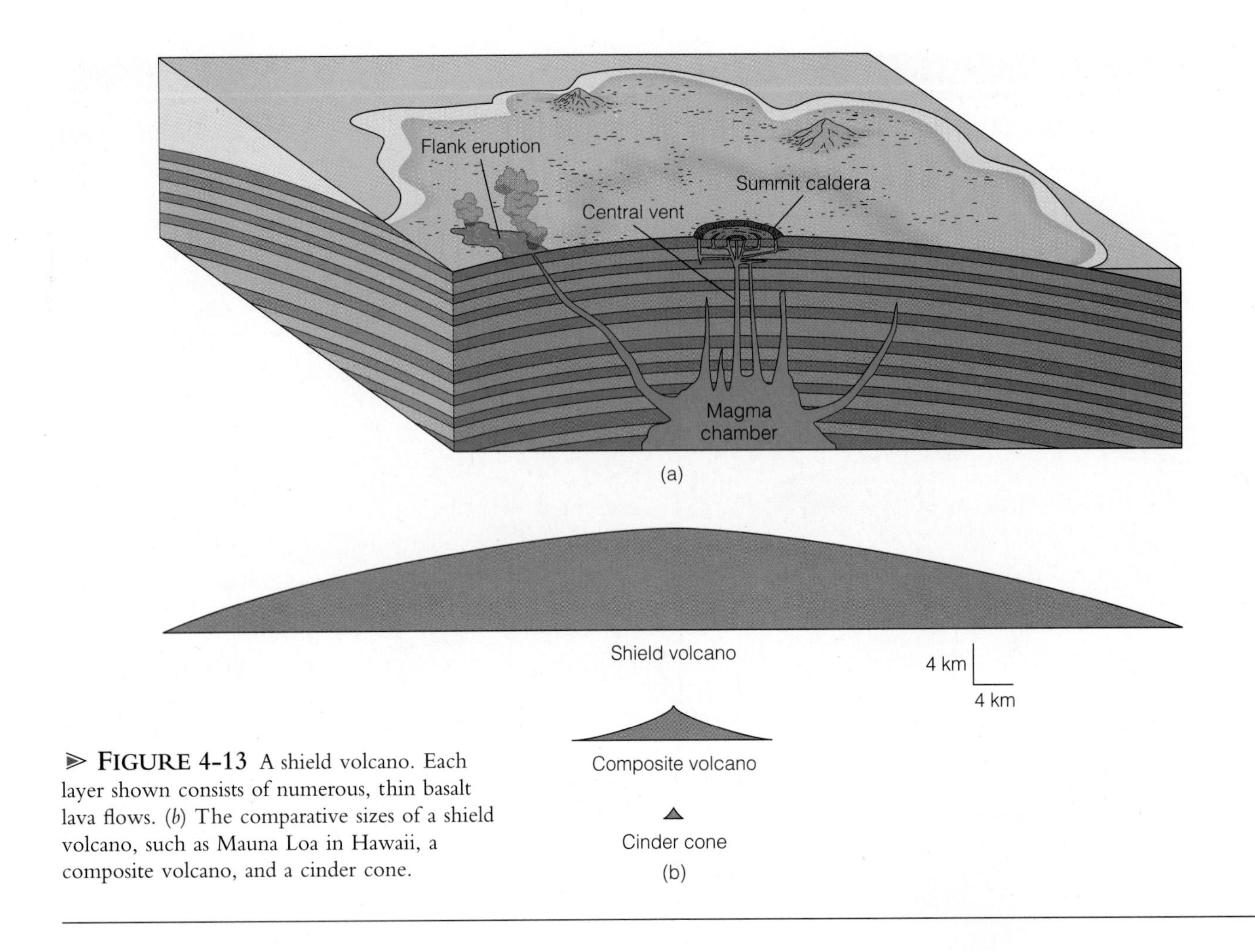

▶ **FIGURE 4-13** A shield volcano. Each layer shown consists of numerous, thin basalt lava flows. (*b*) The comparative sizes of a shield volcano, such as Mauna Loa in Hawaii, a composite volcano, and a cinder cone.

▶ **FIGURE 4-14** (*a*) Cinder cones are composed of layers of angular pyroclastic materials. (*b*) The town of Vestmannaeyjar in Iceland was threatened by lava flows from Eldfell, a cinder cone that formed in 1973. Within two days of the initial eruption on January 23, the new volcano had grown to about 100 m high. Another cinder cone called Helgafell is also visible.

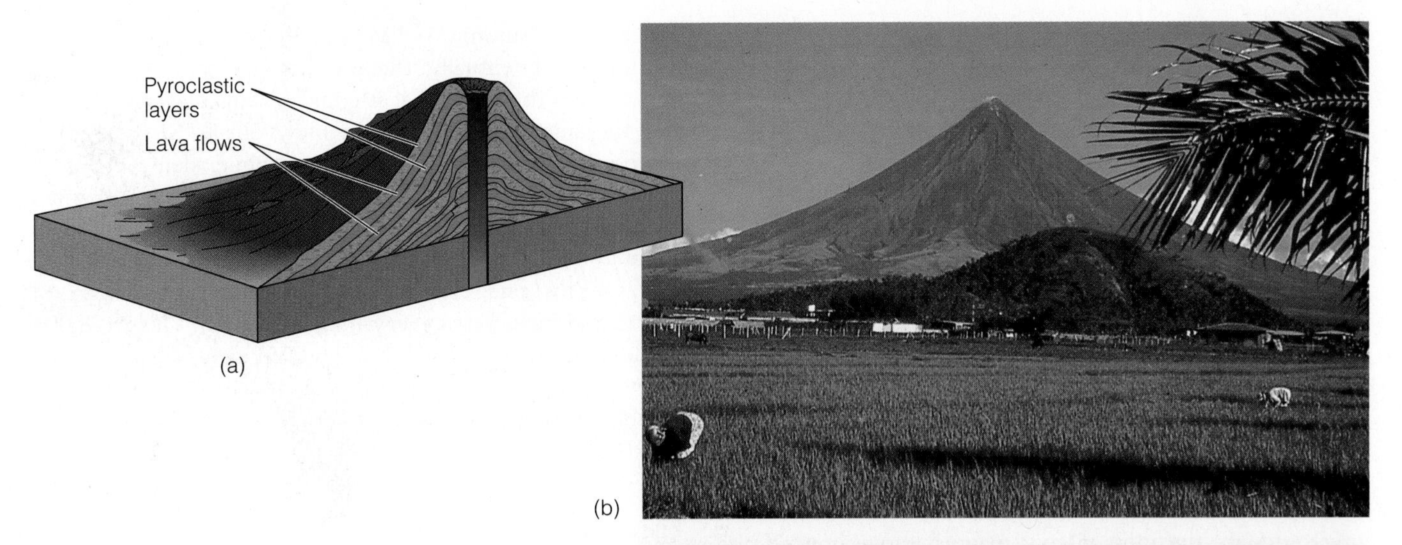

▶ FIGURE 4-15 (*a*) Composite volcanoes are the typical large volcanic mountains on continents. They are composed of lava flows, pyroclastic layers, and volcanic mudflows. Note that the slope is steep in the summit area but decreases toward the base. (*b*) Mayon volcano in the Philippines is one of the world's most nearly symmetrical composite volcanoes. It erupted in 1993 for the twelfth time in this century. (*c*) View of Mount St. Helens, Washington from the southwest in 1978.

successfully diverted the flow before it did much damage to the town.

Composite Volcanoes. **Composite volcanoes**, also called *stratovolcanoes*, are composed of both pyroclastic layers and lava flows (▶ Figure 4-15). Typically, both materials have an intermediate composition, and the flows cool to form andesite. Recall that lava of intermediate composition is more viscous than mafic lava. In addition to lava flows and pyroclastic layers, a significant proportion of a composite volcano is made up of **lahars** (volcanic mudflows). Some lahars form when rain falls on layers of loose pyroclastic materials and creates a muddy slurry that moves downslope. On November 13, 1985, mudflows resulting from a rather minor eruption of Nevado del Ruiz in Colombia killed about 23,000 people (Table 4-1).

Composite volcanoes are steep-sided near their summits, perhaps as much as 30 degrees, but the slope decreases toward the base where it is generally less than 5 degrees (Figure 4-15a). Mayon volcano in the Philippines is one of the most perfectly symmetrical composite volcanoes on Earth. Its concave slopes rise ever steeper to the summit with its central vent through which lava and pyroclastic materials are periodically erupted (Figure 4-15b).

Composite volcanoes are the typical large volcanoes of the continents and island arcs. Familiar examples include Fujiyama in Japan and Mount Vesuvius in Italy as well as many of the volcanic peaks such as Mount St. Helens (Figure 4-15c) in the Cascade Range of western North America (see Perspective 4-3).

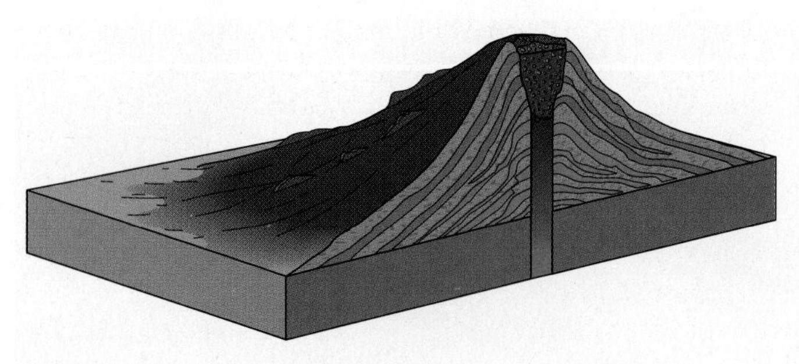

⊳ **FIGURE 4-16** A cross section showing the internal structure of a lava dome. Lava domes form when a viscous mass of magma, generally of felsic composition, is forced up through a volcanic conduit.

Lava Domes. If the upward pressure in a volcanic conduit is great enough, the most viscous magmas move upward and form bulbous, steep-sided **lava domes** (⊳ Figure 4-16). Lava domes are generally composed of felsic lavas although some are of intermediate composition. Because the magma is so viscous, it moves upward very slowly; the lava dome that formed in Santa María volcano in Guatemala in 1922 took two years to grow to 500 m high and 1,200 m across. Lava domes contribute significantly to many composite volcanoes. Beginning in 1980, a number of lava domes were emplaced in the crater of Mount St. Helens; most of these were destroyed during subsequent eruptions. Since 1983, Mount St. Helens has been characterized by sporadic dome growth. In June 1991, a dome in Japan's Unzen volcano collapsed causing a flow of debris and hot ash that killed 43 people in a nearby town (Table 4-1).

Lava domes are often responsible for extremely explosive eruptions. In 1902, viscous magma accumulated beneath the summit of Mount Pelée on the island of Martinique. Eventually, the pressure within the mountain increased to the point that it could no longer be contained, and the side of the mountain blew out in a tremendous explosion. When this occurred, a mobile, dense cloud of pyroclastic materials and gases called a **nuée ardente** (French for "glowing cloud") was ejected and raced downhill at about 100 km/hr, engulfing the city of St. Pierre (⊳ Figure 4-17). This nuée ardente had internal temperatures of 700°C and incinerated everything in its path. Of the 28,000 residents of St. Pierre, only 2 survived, a prisoner in a cell below the ground surface and a man on the surface who was terribly burned by the nuée ardente.

Fissure Eruptions

During the Miocene and Pliocene epochs (between about 17 million and 5 million years ago), some 164,000 km^2 of eastern Washington and parts of Oregon and Idaho were covered by overlapping basalt lava flows. These Columbia River basalts, as they are called, are now well exposed in the walls of the canyons eroded by the Snake and Columbia rivers (⊳ Figure 4-18). These lavas, which were erupted from long fissures, were so fluid that volcanic cones failed to develop. Such **fissure eruptions** yield flows that spread out over large areas and form **basalt plateaus** (⊳ Figure 4-19). The Columbia River basalt flows have an aggregate thickness of about 1,000 m, and some individual flows cover huge areas—for example, the Roza flow, which is 30 m thick, advanced along a front about 100 km wide and covered 40,000 km^2.

Fissure eruptions and basalt plateaus are not common, although several large areas with these features are known.

⊳ **FIGURE 4-17** St. Pierre, Martinique after it was destroyed by a nuée ardente erupted from Mount Pelée in 1902. Only 2 of the city's 28,000 inhabitants survived.

⊳ **FIGURE 4-18** The Columbia River basalts.

Currently, such activity is occurring only in Iceland. A number of volcanic mountains are present in Iceland, but the bulk of the island is composed of basalt flows erupted from fissures. Two large fissure eruptions, one in A.D. 930 and the other in 1783, account for about half of the magma erupted in Iceland during historic time. The 1783 eruption occurred along the Laki fissure, which is more than 30 km long; about 12.5 km^3 of lava flowed several tens of kilometers from the fissure; the lava covered an area of more than 560 km^2 and in one place filled a valley to a depth of about 200 m.

Pyroclastic Sheet Deposits

More than 100 years ago, geologists were aware of vast areas covered by felsic volcanic rocks a few meters to hundreds of meters thick. It seemed improbable that these could have formed as vast lava flows, but it also seemed equally unlikely that they were ash fall deposits. Based on observations of historic pyroclastic flows, such as the nuée ardente erupted by Mount Pelée in 1902, it now seems probable that these ancient rocks originated as pyroclastic flows, hence the name **pyroclastic sheet deposits**. They cover far greater areas than any observed during historic time, and apparently erupted from long fissures rather than from a central vent. The pyroclastic materials of many of these flows were so hot that they fused together to form *welded tuff*.

It now appears that major pyroclastic flows issue from fissures formed during the origin of calderas. The Yellowstone Tuff, for example, was erupted during the formation of a large caldera in the area of present-day Yellowstone National Park in Wyoming (⊳ Figure 4-20). Similarly, the Bishop Tuff of eastern California appears to have been erupted shortly before the formation of the Long Valley caldera. Interestingly, earthquake activity in the Long Valley caldera and nearby areas beginning in 1978 may indicate that magma is moving upward beneath part of the caldera. Thus, the possibility of future eruptions in that area cannot be discounted.

⊳ **FIGURE 4-19** A block diagram showing fissure eruptions and the origin of a basalt plateau.

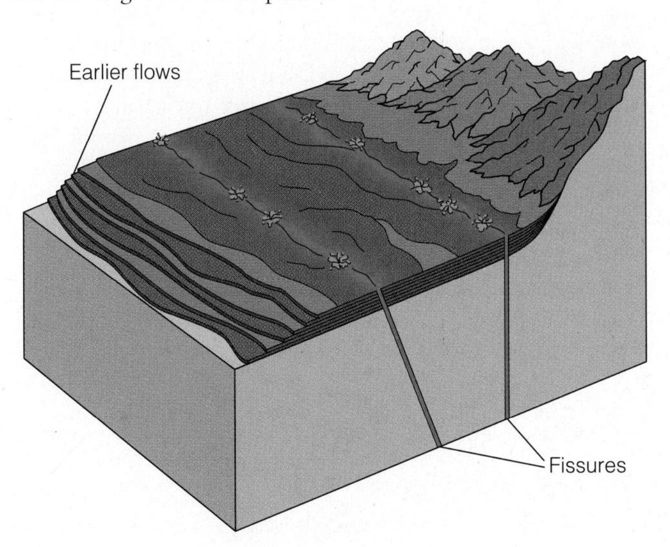

⊳ **FIGURE 4-20** The Yellowstone Tuff in the walls of the Grand Canyon of the Yellowstone, Yellowstone National Park, Wyoming. Tuff is a volcanic rock composed of consolidated ash. (Photo courtesy of Jerry Westby.)

Perspective 4-3

Eruptions of Cascade Range Volcanoes

During the summer of 1914, Mount Lassen in northern California began erupting without warning and culminated with the "Great Hot Blast," a huge steam explosion on May 22, 1915 (▶ Figure 1). Fortunately, the area was sparsely settled, and little property damage and no deaths resulted, even though a large area of forest on the volcano's eastern and northeastern flanks was devastated. Activity largely ceased by 1921, but hot springs, boiling mud pots, and gas vents known as *fumaroles* remind us that a source of heat still exists beneath the surface (Figure 4-5).

Mount Lassen is one of 15 large volcanoes in the Cascade Range of northern California, Oregon, Washington, and southern British Columbia, Canada (Figures 4-15c and ▶ 2). After its 1914–1921 eruptions, the Cascade volcanoes remained quiet for 63 years. Then, on March 16, 1980, following an inactive period of 123 years, Mount St. Helens in southern Washington showed signs of renewed activity, and on May 18 it erupted violently, causing the worst volcanic disaster in U.S. history (▶ Figure 3).

▶ **FIGURE 1** Mount Lassen in northern California erupted numerous times between 1914 and 1921. This eruption occurred in 1915.

The awakening of Mount St. Helens came as no surprise to geologists of the U.S. Geological Survey (USGS) who warned in 1978 that Mount St. Helens is ". . . an especially dangerous volcano because [of] its past behavior and [its] relatively high frequency of eruptions during the last 4,500 years . . . ."* Although no one could predict precisely when Mount St. Helens would erupt, the USGS report included maps showing areas in which damage from an eruption could be expected. Forewarned with such data, local officials were better prepared to formulate policies when the eruption did occur.

On March 27, 1980, Mount St. Helens began erupting steam and ash and continued to do so during the rest of March and most of April. By late March, a visible bulge had developed on its north face as molten rock was injected into the mountain, and the bulge continued to expand at about 1.5 m per day. On May 18, an earthquake shook the area, the unstable bulge collapsed, and the pent-up volcanic gases below expanded rapidly, creating a tremendous northward-directed lateral blast that blew out the north side of the mountain (Figure 3). The lateral blast accelerated from 350 to 1,080 km/hr, obliterating virtually everything in its path. Some 600 km^2 of forest were completely destroyed; trees were snapped off at their bases and strewn about the countryside, and trees as far as 30 km from the bulge were seared by the intense heat. Tens of thousands of animals were killed; roads, bridges, and buildings were destroyed; and 63 people perished.

Shortly after the lateral blast, volcanic ash and steam erupted, forming a cloud above the volcano 19 km high (▶ Figure 4). The ash cloud drifted east-northeast, and the resulting ash fall at Yakima, Washington, 130 km to the east, caused almost total darkness at midday. Detectable amounts of ash were deposited over a huge area. Flows of hot gases and volcanic ash raced down the north flank of the mountain, causing steam explosions when they encountered bodies of water or moist ground. Steam explosions continued for weeks, and at least one occurred a year later.

Snow and glacial ice on the upper slopes of Mount St. Helens melted and mixed with ash and other surface debris to form thick, pasty volcanic mudflows. The largest and most destructive mudflow surged down the valley of the North Fork of the Toutle River. Ash and mudflows displaced water in lakes and streams and flooded downstream areas. Ash and other particles carried by the flood waters were deposited in stream channels; many kilometers

* D. R. Crandell and D. R. Mullineaux, "Potential Hazards from Future Eruptions of Mt. St. Helens Volcano, Washington," *United States Geological Survey Bulletin 1383-C* (1978): C1.

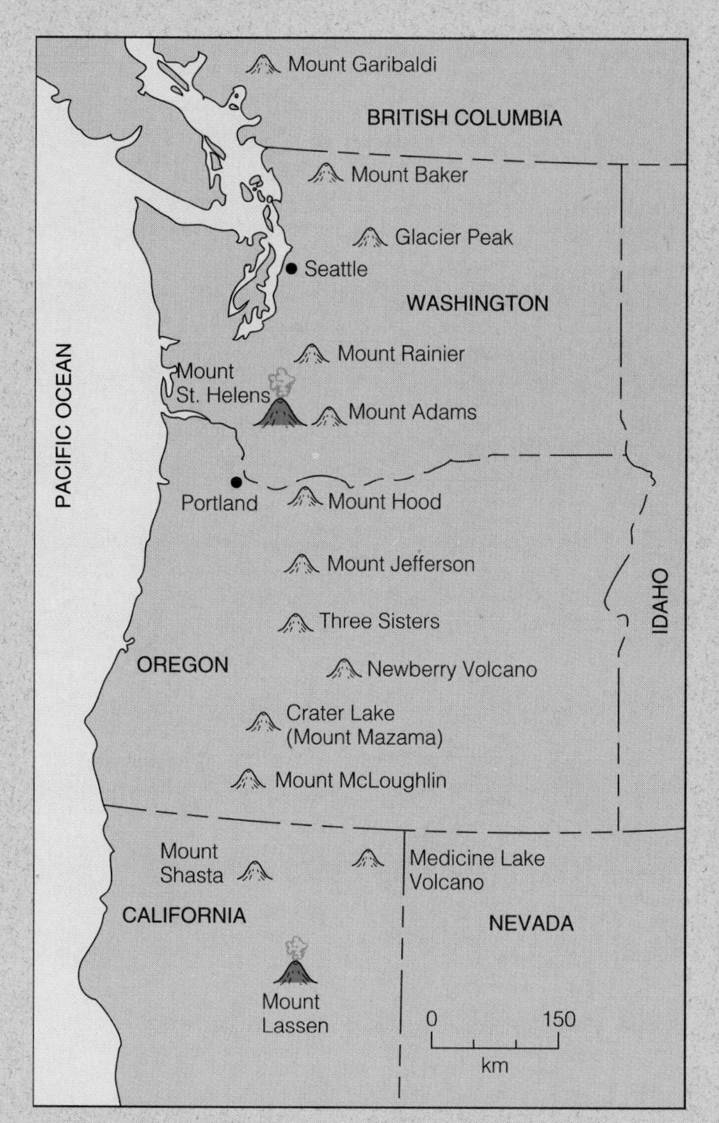

▶ **FIGURE 2** The major volcanoes of the Cascade Range. Mount Lassen and Mount St. Helens have erupted during this century, and several others have been active during the last two hundred years.

▶ **FIGURE 3** The eruption of Mount St. Helens on May 18, 1980. The lateral blast occurred when a bulge on the north face of the mountain collapsed and reduced the pressure on the molten rock within the mountain.

▶ **FIGURE 4** Shortly after the lateral blast of May 18, 1980, Mount St. Helens erupted a steam and ash cloud that rose about 19 km.

from Mount St. Helens, the navigation channel of the Columbia River was reduced from 12 m to less than 4 m as a result of such deposition.

Although the damage resulting from the eruption of Mount St. Helens was significant and the deaths were tragic, it was not a particularly large or deadly eruption compared with some historic eruptions (see Table 4-1). The eruption of Tambora in 1815 is the greatest volcanic eruption in recorded history in terms of both casualties and the amount of material erupted; it produced as least 80 times more ash than the 0.9 km^3 that spewed forth from Mount St. Helens.

Guest Essay Stanley N. Williams

Volcanology: The Challenge of Volcanoes

I remember very clearly what made me a volcanologist. During my second undergraduate course in geology, the professor told us that the unusual sample he held was a product of the 1902 eruption of Mount Pelée in Martinique. Some of the sample's internal structures looked as if the rock had been formed by a lava flow, while others seemed to be related to ash falling from the sky. The professor explained that when Pelée erupted, clouds of ash and hot gas swept down the slopes of the volcano at incredible speed with seemingly no friction between the fragments of ash. At rest, the deposits were still so hot that they were partially glued together. Hence the confusing appearance of such rocks and my note to myself "this is what I want to study."

That was quite a few years ago, and I have since been on the scene of many volcanoes showing threatening activity and at several disasters—the aftermath of eruptions where thousands of people died. Each example has produced very important progress in our ability to understand how volcanoes erupt.

Mount St. Helens has been recognized by two U.S. Geological Survey geologists as the volcano of the Cascade Mountains most likely to erupt in the twentieth century. Just about one year after they published a paper calling attention to the potential for eruptions, an earthquake produced a crack in the volcano's summit and caused a small explosion. My mentor, Dick Stoiber, and I were there within 18 hours to measure the gases emitted by the eruptions. We used a remote sensor to quantify the amount of sulfur dioxide (SO_2) released. We were amazed that the eruptions were releasing the smallest amounts of SO_2 we had ever measured at an erupting volcano. We had learned something important—goundwater can enter cracks on active volcanoes and become so hot that it flashes to steam and produces an eruption. Two months later came the famous lateral blast and subsequent partial collapse of Mount St. Helens.

As months and years pass, volcanologists are recognizing that many volcanoes have had very similar eruptions. Thus, we need direct observations to understand that a thin layer of ash we might find over a large area is the record of powerful lateral blasts, and the irregular landscape of large blocks (some the size of buildings) and small depressions represents collapse deposits. Just to keep the eruptions a mystery, many volcanoes have completely filled the hole (caldera) and show no sign of the past on the edifice (St. Helens is doing that).

Volcanologists learn a great deal from each tragedy but avoiding the next one is the really serious challenge. There are about 500–600 active volcanoes around the world, and many millions of people live very close to them. To understand what is happening when a volcano becomes active, we need much information: What has this volcano done historically and prehistorically? What areas are most likely to be impacted by a future eruption? How can the public be trained to mitigate the hazards? How does what is happening now compare to the "normal" activity for this volcano? Unfortunately, very little is usually known about the recent eruptions, and almost nothing is known about the "normal" activity. Furthermore, we volcanologists are usually not able to forecase with any certainty whether a volcano may erupt, what kind of eruption it may be, or how large it may be. Nevertheless, we must talk with the authorities about what we do know and how we interpret it. In doing so, we need to avoid jargon and try to help them understand the natural processes. We also must not fall into the trap of being *the* authorities on what should be done. Many different people—medical doctors, engineers, architects, civil defense personnel, economists, political leaders, and others—must be involved. It is usually not realistic to evacuate cities, and we cannot tell the authorities when to do it anyway. Together, we try to make good plans to inform people, so that they are not terrified by volcanoes and are able to help themselves if there is an eruption.

As I write this, I am struggling with information about one of the biggest volcanoes in the world (Popocatépetl), which stands close to the world's largest city (Mexico City). It seems to have been releasing much more gas than before for about a year, but the number and energy of earthquakes do not seem to be much different. What are we going to do that might help minimize the chance of another crisis or disaster? The search for answers is what makes volcanology so exciting and challenging.

STANLEY N. WILLIAMS was a geology major at Beloit College and then went to Dartmouth College for his M.S. study with Dick Stoiber on the 1902 eruption of Santa María volcano in Guatemala, which was one of the largest of the twentieth century. He stayed and worked on his Ph.D. in Nicaragua where the volcano Masaya has frequently erupted during historic times. Studying active volcanoes has meant working on more than 100 volcanoes in 20 countries. It has also meant being the survivor in 1993 when six colleagues died, when they tried to understand the danger of a volcano in Colombia. He is returning to field studies of active volcanoes because he is caught by curiosity. He has been a professor at Arizona State University since 1991.

Distribution of Volcanoes

Rather than being distributed randomly around the Earth, volcanoes occur in well-defined zones or belts. More than 60% of all active volcanoes are in the **circum-Pacific belt** that nearly encircles the margins of the Pacific Ocean basin (▶ Figure 4-21). This belt includes the volcanoes along the west coast of South America, those in Central America, Mexico, and the Cascade Range, and the Alaskan volcanoes in the Aleutian Island arc. The belt continues on the western side of the Pacific Ocean basin where it extends through Japan, the Philippines, Indonesia, and New Zealand. The circum-Pacific belt also includes the southernmost active volcano, Mount Erebus in Antarctica, and a large caldera at Deception Island that erupted during 1970 (Figure 4-21).

About 20% of all active volcanoes are in the **Mediterranean belt** (Figure 4-21). Included in this belt are the famous Italian volcanoes such as Mount Etna, Stromboli, and Mount Vesuvius.

The large volcanoes in the circum-Pacific and Mediterranean belts are mostly composite volcanoes, but a number of them have had lava domes emplaced in their craters or calderas, making them especially dangerous. These composite volcanoes consist largely of lava flows and pyroclastic layers of intermediate and felsic composition. Accordingly, most of the rocks in these volcanoes are andesite and tuff.

Most of the rest of the active volcanoes are at or near mid-oceanic ridges (Figure 4-21). The longest of these ridges is the Mid-Atlantic Ridge, which is near the middle of the Atlantic Ocean basin and curves around the southern tip of Africa where it continues as the Indian Ridge. Branches of the Indian Ridge extend into the Red Sea and East Africa where a number of volcanoes are located including Mount Kilimanjaro (Figure 4-21). Much of the volcanism along the mid-oceanic ridges is submarine and goes largely undetected, but in some places, such as Iceland, it occurs above sea level. The volcanoes that form at or near these ridges are mostly shield volcanoes composed of mafic lavas that cool to form the volcanic rock basalt.

Volcanism is occurring in a few other areas at present, most notably on and near the island of Hawaii (Figure 4-21). Only two volcanoes are currently active on the island, Mauna Loa and Kilauea, although a submarine volcano named Loihi exists about 32 km to the south; Loihi rises more than 3,000 m above the sea floor, but its summit is still about 940 m below sea level. The Hawaiian volcanoes above sea level are also composed mostly of basalt that cooled from mafic lava flows.

▶ **FIGURE 4-21** Most volcanoes are at or near plate boundaries. Two major volcano belts are recognized: the circum-Pacific belt contains about 60% of all active volcanoes, about 20% are in the Mediterranean belt, and most of the rest are located along mid-oceanic ridges.

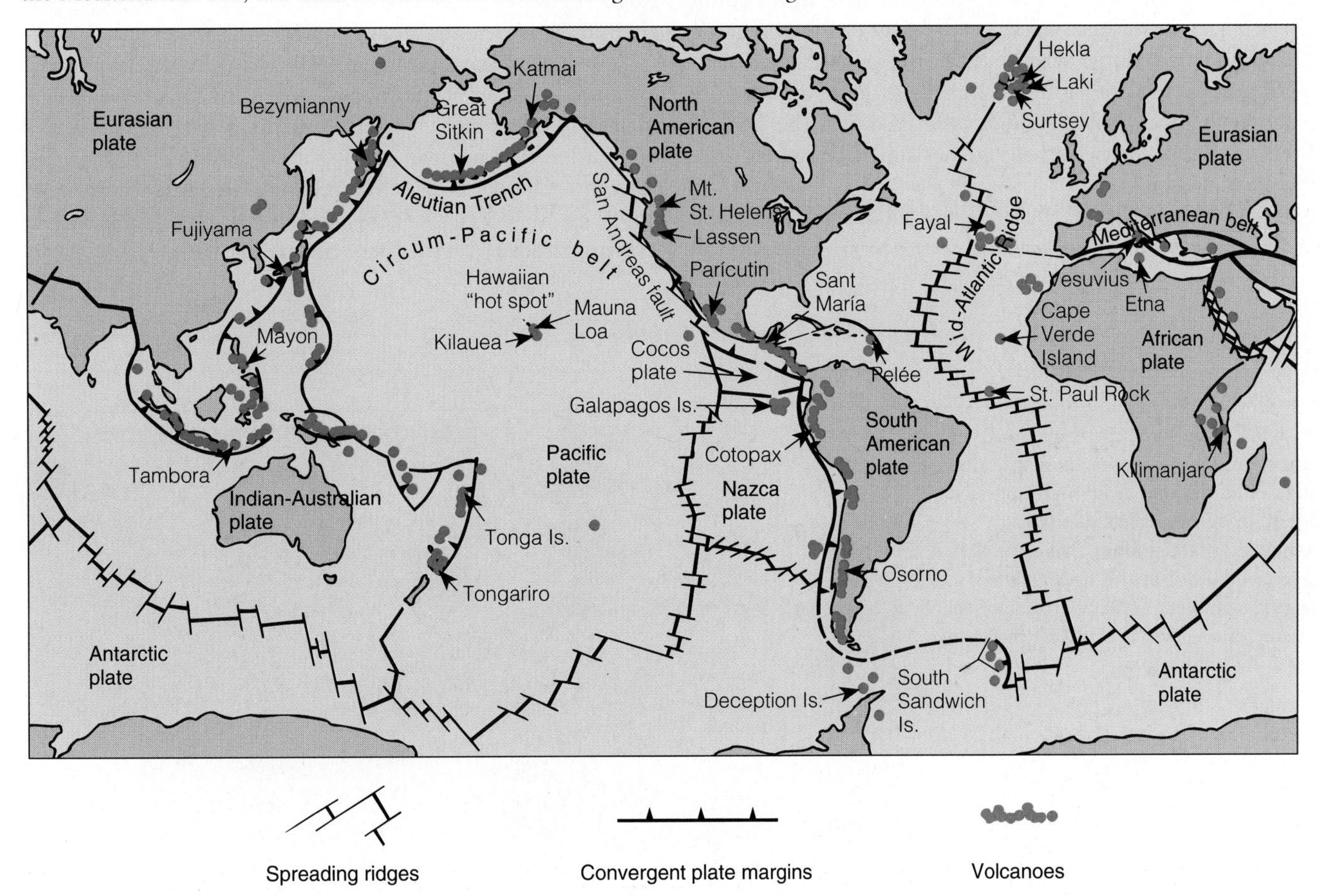

Plate Tectonics and Igneous Activity

At this point, two questions might be asked regarding volcanoes: (1) What accounts for the alignment of volcanoes in belts? (2) Why do magmas erupted within ocean basins and magmas erupted at or near continental margins have different compositions? In addition, plutons emplaced within the ocean basins are invariably mafic, mostly gabbro, whereas the vast batholiths emplaced at continental margins are composed largely of felsic and intermediate rocks such as granite and diorite. Recall from Chapter 1 that the outer part of the Earth is divided into large plates, which are sections of the lithosphere. Lithosphere consists of upper mantle and oceanic crust or upper mantle and continental crust, called oceanic and continental lithosphere, respectively. Most igneous activity occurs at spreading ridges where plates diverge or along subduction zones where plates converge.

Igneous Activity at Spreading Ridges

Spreading ridges are areas where new oceanic lithosphere is produced by igneous activity as plates diverge and move away from one another (see Figure 1-15). Most spreading ridges are in the ocean basins, but some extend into continents as in East Africa (Figure 4-21). According to plate tectonic theory, the Atlantic Ocean basin began developing when rifting of a large plate and subsequent plate divergence resulted in the breakup of the supercontinent Pangaea about 250 million years ago. The Mid-Atlantic Ridge is the spreading ridge along which plate divergence began and continues at present. Similar spreading is also currently occurring at divergent margins in the Red Sea, the Gulf of Aden, and East Africa and along the Indian Ridge and the East Pacific Rise.

Recall that most of the magma generated within the Earth forms plutons rather than being erupted onto the surface. For example, mafic magma originates beneath spreading ridges, and some is erupted at the surface as lava flows and/or pyroclastic materials; however, much of it is simply emplaced at depth as vertical dikes and gabbro plutons (➤ Figure 4-22). In fact, the oceanic crust is composed largely of such mafic rock.

Some of the volcanism at spreading ridges is apparent because it occurs above sea level, but, as previously noted, much of it is submarine and goes undetected. Research submarines have descended into the rifts at the crests of spreading ridges where scientists have observed pillow lavas that formed during submarine eruptions (Figure 4-10).

The fact that volcanism occurs at spreading ridges is undisputed, but how magma originates beneath the ridges is not fully understood. One explanation is related to the manner in which the Earth's temperature increases with depth. We know from deep mines and deep drill holes that a temperature increase, called the *geothermal gradient*, does occur and that, on average, the gradient is about 25°C/km. Accordingly, rocks at depth are hot, but remain solid because their melting temperature rises with increasing pressure.

Beneath spreading ridges the temperature locally exceeds the melting temperature, at least in part, because pressure decreases. That is, rifting probably causes a decrease in pressure on the hot rocks at depth, thus initiating melting (➤ Figure 4-23a). Furthermore, the presence of water can also decrease the melting temperature beneath spreading ridges because water aids thermal energy in breaking the chemical bonds in minerals (Figure 4-23b).

Another explanation for spreading ridge volcanism is that localized, cylindrical plumes of hot mantle material, called **mantle plumes** rise beneath spreading ridges and spread outward in all directions (➤ Figure 4-24). Perhaps localized concentrations of radioactive minerals within the crust and upper mantle decay and generate the heat responsible for the melting associated with these hot mantle plumes.

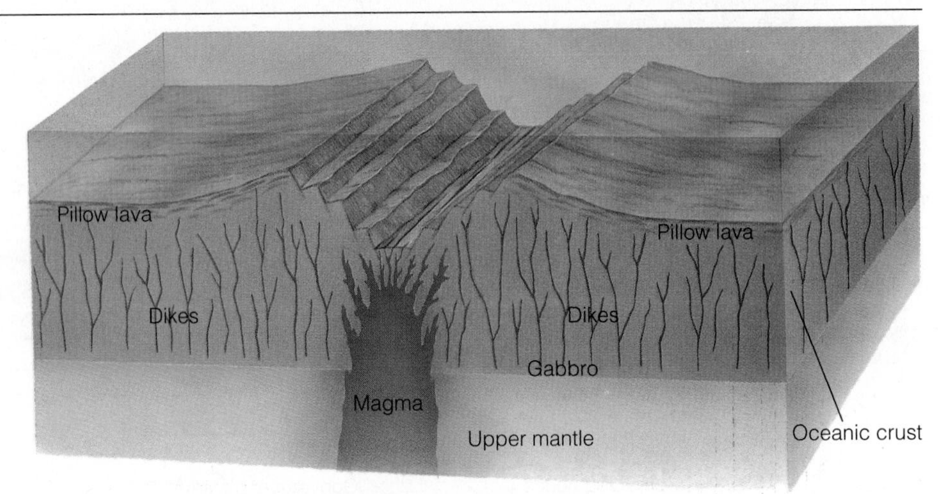

➤ **FIGURE 4-22** Intrusive and extrusive igneous activity at a spreading ridge. The oceanic crust is composed largely of vertical dikes of basaltic composition and gabbro that appears to have crystallized in the upper part of a magma chamber. The upper part of the oceanic crust consists of submarine lavas, especially pillow lavas.

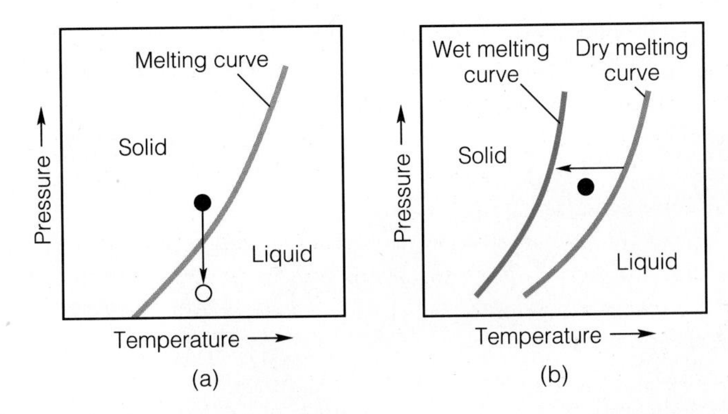

▶ **FIGURE 4-23** (*a*) Melting temperature increases with increasing temperature, so a decrease in pressure on already hot rocks can initiate melting. (*b*) When water is present, the melting curve shifts to the left because of the additional agent to break chemical bonds.

The lavas erupted at spreading ridges are invariably mafic and cool to form basalt. But the upper mantle, from which these lavas are derived, is composed of ultramafic rock, probably peridotite, which consists largely of ferromagnesian silicates and lesser amounts of nonferromagnesian silicates. To explain how mafic magma (45–52% silica) originates from ultramafic rock (≤45% silica), geologists propose that the magma is formed from source rock that only partially melts. This phenomenon of *partial melting* occurs because various minerals have different melting temperatures. Recall the sequence of minerals in Bowen's reaction series (see Figure 3-6). The order in which these minerals melt is the opposite of their order of crystallization. Accordingly, quartz, potassium feldspar, and sodium-rich plagioclase melt before most of the ferromagnesian silicates and the calcic varieties of plagioclase. So, when ultramafic rock begins to melt, the minerals richest in silica melt first followed by those containing less silica. If melting is not complete, a mafic magma containing proportionately more silica than the source rock results. Once this mafic magma is formed, some of it rises to the surface where it is erupted, cools, and crystallizes to form basalt.

Igneous Activity at Subduction Zones

Three types of convergence are recognized: convergence between two oceanic plates, convergence between an oceanic plate and a continental plate, and convergence between two continental plates (see Figure 1-15). In the first two types of convergence, an oceanic plate is subducted beneath another plate, and volcanism occurs near the leading margin of the overriding plate. Continental lithosphere is not dense enough to be subducted, so little volcanism occurs where two such plates converge.

To illustrate how igneous activity is related to subduction, let us consider what occurs along a continental–oceanic plate boundary (▶ Figure 4-25). As the oceanic and continental lithosphere converge, the denser oceanic plate is subducted

▶ **FIGURE 4-24** Some of the "hot spots" in the Earth's crust that are thought to overlie rising mantle plumes.

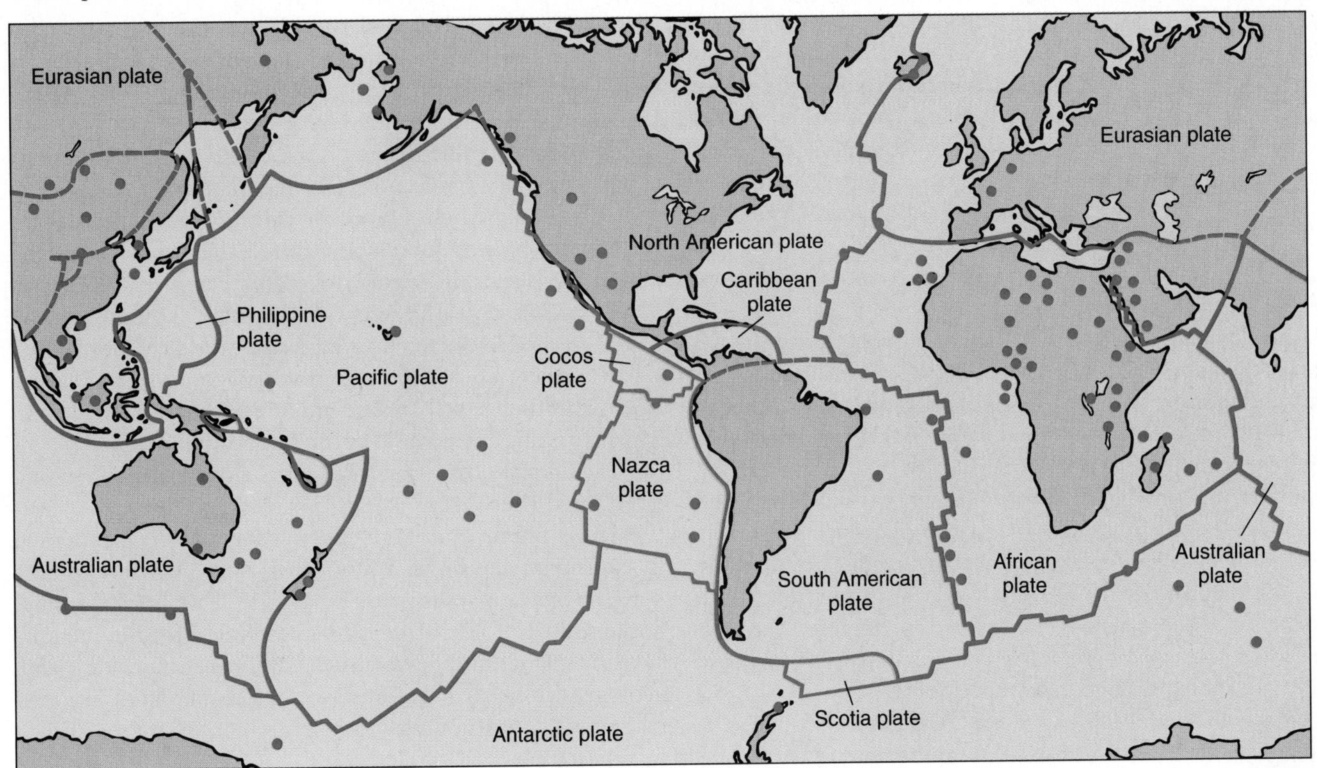

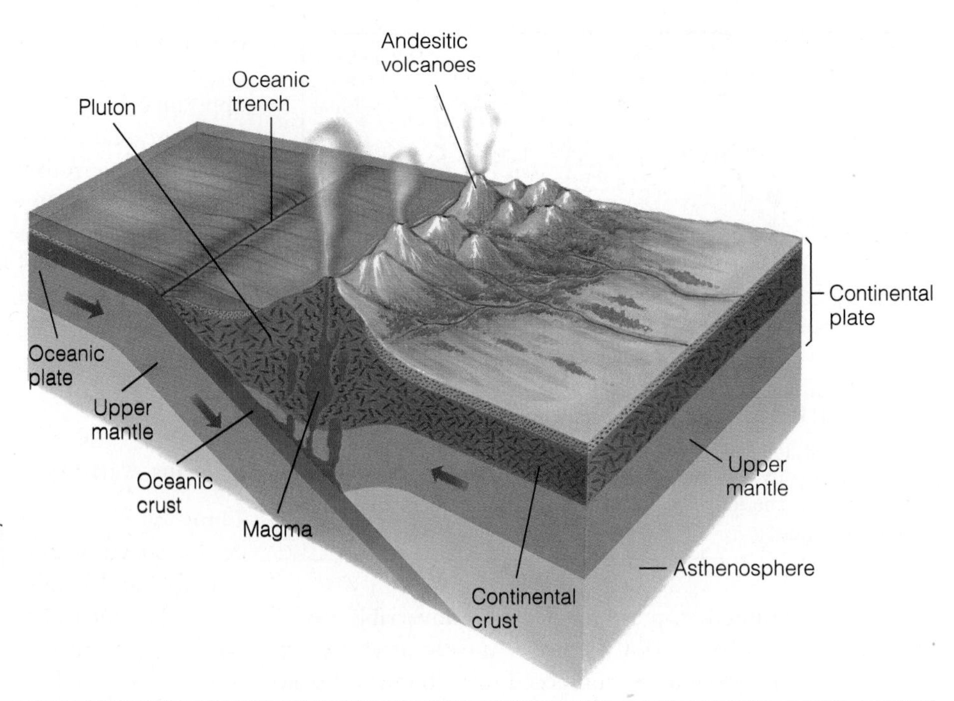

▶ FIGURE 4-25 The subduction of an oceanic plate beneath a continental plate produces magma. Some of the magma forms plutons, especially batholiths, and some is erupted to form andesitic volcanoes.

beneath the continental plate (Figure 4-25). As the subducted plate descends toward the asthenosphere, it is heated by the Earth's geothermal gradient. When the descending plate reaches a depth where the temperature is high enough, partial melting occurs and magma is generated (Figure 4-25). Additionally, the wet oceanic crust descends to a depth at which dewatering occurs. As the water rises into the overlying mantle, it enhances melting, and a magma may be generated. A belt of large composite volcanoes and granitic plutons near the leading margin of the continental plate form from the magma created by partial melting of the subducted plate.

Igneous activity also occurs where two oceanic plates converge and one is subducted beneath the other. Partial melting of the subducted plate generates magma that rises through the overriding plate and forms a curved chain of volcanic islands called a volcanic island arc. Excellent examples of volcanic island arcs are the islands of Japan, the Philippines, and the Aleutian Islands of Alaska (Figure 4-21).

Partial melting is one phenomenon accounting for the fact that magmas generated at subduction zones are intermediate and felsic in composition. Recall that partial melting of ultramafic rock of the upper mantle yields mafic magma. Likewise, partial melting of oceanic crust, which has a mafic composition, may yield magma richer in silica than the source rock. Additionally, some of the silica-rich sediments and sedimentary rocks of continental margins are probably carried downward with the subducted plate and contribute their silica to the magma. Also, mafic magma rising through the lower continental crust may be contaminated with felsic materials, which change its composition.

Intermediate and felsic magmas are typically produced at convergent plate margins where subduction occurs. The intermediate magma that is erupted is more viscous than mafic magma and tends to form composite volcanoes. Much felsic magma is intruded into the continental crust where it forms various plutons, especially granitic batholiths, but some is erupted as pyroclastic materials or emplaced as lava domes, thus accounting for the explosive eruptions that characterize convergent plate margins.

Intraplate Volcanism

Mauna Loa and Kilauea on the island of Hawaii and Loihi just to the south are within the interior of a rigid plate far from any spreading ridge or subduction zone (Figure 4-21). It is postulated that a mantle plume creates a local "hot spot" beneath Hawaii. The magma is mafic and relatively fluid, so it builds up shield volcanoes (▶ Figure 4-26a).

Even though these Hawaiian volcanoes are unrelated to spreading ridges or subduction zones, the evolution of the Hawaiian Islands is related to plate tectonics. Notice in Figure 4-26b that the ages of the rocks composing the islands in the Hawaiian chain increase toward the northwest; Kauai formed 3.8 to 5.6 million years ago, whereas Hawaii began forming less than one million years ago, and Loihi began forming even more recently. Continuous motion of the Pacific plate over the "hot spot," now beneath Hawaii, has created the various islands in succession.

Mantle plumes and "hot spots" have also been proposed to explain volcanism in a few other areas. A mantle plume may be beneath Yellowstone National Park in Wyoming. Some source of heat at depth is responsible for the present-day hot springs and geysers such as Old Faithful, but many geologists think that the source of heat is a body of intruded magma that has not yet completely cooled rather than a mantle plume.

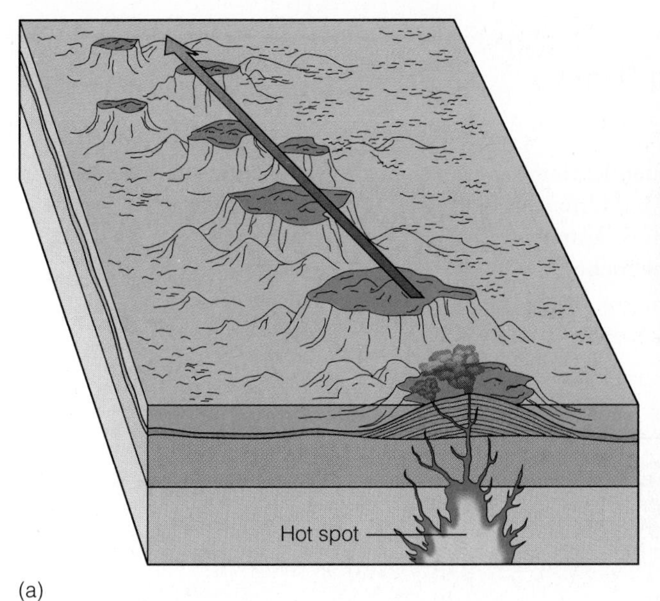

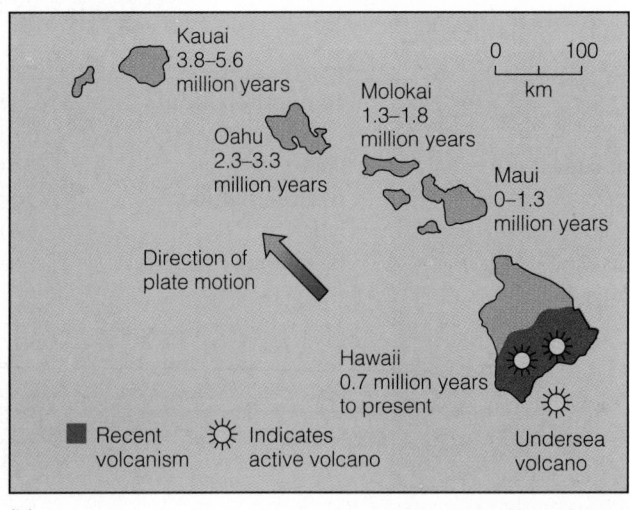

▶ **FIGURE 4-26** (*a*) Generalized diagram showing the origin of the Hawaiian Islands. As the lithospheric plate moves over a hot spot, a succession of volcanoes forms. Present-day volcanism occurs only on Hawaii and beneath the sea just to the south. (*b*) Map showing the age of the islands in the Hawaiian chain.

Chapter Summary

1. Volcanism is the process whereby magma and its associated gases erupt at the surface. Some magma erupts as lava flows, and some is ejected explosively as pyroclastic materials.
2. Only a few percent by weight of a magma consists of gases, most of which is water vapor. Sulfur gases emitted during large eruptions can have far-reaching climatic effects.
3. The surfaces of aa lava flows consist of rough, jagged, angular blocks, whereas pahoehoe flows have smoothly wrinkled surfaces.
4. Many lava flows are characterized by pressure ridges and spatter cones. Columnar joints form in some lava flows when they cool. Pillow lavas are erupted under water and consist of interconnected bulbous masses.
5. Volcanoes are conical mountains built up around a vent where lava flows and/or pyroclastic materials are erupted.
6. Shield volcanoes have low, rounded profiles and are composed mostly of mafic flows that have cooled and formed basalt. Cinder cones form where pyroclastic materials that resemble cinders are erupted and accumulate as small, steep-sided cones. Composite volcanoes are composed of lava flows of intermediate composition, layers of pyroclastic materials, and volcanic mudflows.
7. Viscous masses of lava, generally of felsic composition, are forced up through the conduits of some volcanoes and form bulbous, steep-sided lava domes. Volcanoes with lava domes are dangerous because they erupt explosively and frequently eject nuée ardentes.
8. The summits of volcanoes are characterized by a circular or oval crater or a much larger caldera. Many calderas form by summit collapse when an underlying magma chamber is partly drained.
9. Fluid mafic lava erupted from long fissures (fissure eruptions) spreads over large areas to form basalt plateaus.
10. Pyroclastic flows erupted from fissures formed during the origin of calderas cover vast areas. Such eruptions of pyroclastic materials form sheetlike deposits.
11. Most active volcanoes are distributed in linear belts. The circum-Pacific belt and Mediterranean belt contain more than 80% of all active volcanoes.
12. Volcanism in the circum-Pacific and Mediterranean belts is at convergent plate margins where subduction occurs. Partial melting of the subducted plate generates intermediate and felsic magmas, most of which form plutons.
13. Magma derived by partial melting of the upper mantle beneath spreading ridges accounts for the mafic plutons and lavas of ocean basins. Melting in these areas may be caused by reduction in pressure and/or hot mantle plumes.
14. The two active volcanoes on the island of Hawaii and one just to the south are thought to lie above a hot mantle plume. The Hawaiian Islands developed as a series of volcanoes formed on the Pacific plate as it moved over the mantle plume.

Important Terms

aa
ash
basalt plateau
caldera
cinder cone
circum-Pacific belt
columnar joint
composite volcano (stratovolcano)
crater
fissure eruption
lahar
lava dome
mantle plume
Mediterranean belt
nuée ardent
pahoehoe
pillow lava
pressure ridge
pyroclastic sheet deposit
shield volcano
spatter cone
volcanism
volcano

Review Questions

1. Which of the following is most dangerous to humans?
   a. ____ nuée ardente;
   b. ____ lava flows;
   c. ____ volcanic bombs;
   d. ____ pahoehoe;
   e. ____ pillow lava.
2. A lava flow with a surface of jagged blocks is termed:
   a. ____ lapilli;
   b. ____ vesicular;
   c. ____ aa;
   d. ____ obsidian;
   e. ____ pyroclastic sheet deposit.
3. Most calderas form by:
   a. ____ summit collapse;
   b. ____ explosions;
   c. ____ fissure eruptions;
   d. ____ forceful injection;
   e. ____ erosion of lava domes.
4. Basalt plateaus form as a result of:
   a. ____ repeated eruptions of cinder cones;
   b. ____ widespread ash falls;
   c. ____ accumulation of thick layers of pyroclastic materials;
   d. ____ the origin of lahars on composite volcanoes;
   e. ____ eruptions of fluid lava from long fissures.
5. One other Cascade Range volcano besides Mount St. Helens has erupted during this century. It is:
   a. ____ Mount Hood, Oregon;
   b. ____ Mount Lassen, California;
   c. ____ Mount Garibaldi, British Columbia;
   d. ____ Mount Adams, Washington;
   e. ____ Mount Mazama, Oregon.
6. Volcanic or extrusive igneous rocks form by the cooling and crystallization of lava flows and the:
   a. ____ crystallization of magma beneath the surface;
   b. ____ consolidation of pyroclastic materials;
   c. ____ reaction of volcanic gases with the atmosphere;
   d. ____ heating of sedimentary rocks beneath lava flows;
   e. ____ all of these.
7. The most commonly emitted volcanic gas is:
   a. ____ carbon dioxide;
   b. ____ hydrogen sulfide;
   c. ____ nitrogen;
   d. ____ chlorine;
   e. ____ water vapor.
8. Small, steep-sided cones that form on the surfaces of lava flows where gases escape are:
   a. ____ lava tubes;
   b. ____ spatter cones;
   c. ____ columnar joints;
   d. ____ pahoehoe;
   e. ____ volcanic bombs.
9. Much of the upper part of the oceanic crust is composed of interconnected bulbous masses of igneous rock called:
   a. ____ pillow lava;
   b. ____ lapilli;
   c. ____ pyroclastic material;
   d. ____ parasitic cones;
   e. ____ blocks.
10. Shield volcanoes have low slopes because they are composed of:
    a. ____ mostly pyroclastic layers;
    b. ____ lahars and viscous lava flows;
    c. ____ fluid mafic lava flows;
    d. ____ felsic magma;
    e. ____ pillow lavas.
11. Crater Lake in Oregon is an excellent example of a:
    a. ____ caldera;
    b. ____ cinder cone;
    c. ____ shield volcano;
    d. ____ basalt plateau;
    e. ____ lava dome.
12. The volcanic conduit of a lava dome is most commonly plugged by:
    a. ____ mafic magma;
    b. ____ columnar joints;
    c. ____ viscous, felsic magma;
    d. ____ volcanic mudflows;
    e. ____ spatter cones.
13. Most active volcanoes are in:
    a. ____ the Mediterranean belt;
    b. ____ the Hawaiian Islands;
    c. ____ Iceland;
    d. ____ the circum-Pacific belt;
    e. ____ the oceanic ridge belt.
14. The magma generated beneath spreading ridges is mostly:
    a. ____ mafic;
    b. ____ felsic;
    c. ____ intermediate;
    d. ____ all of these;
    e. ____ answers (a) and (b) only.
15. The volcanoes of ________ are unrelated to either a divergent or a convergent plate margin.
    a. ____ East Africa;
    b. ____ the mid-oceanic ridges;
    c. ____ the Cascade Range;

d. ____ the Hawaiian Islands;
e. ____ Iceland.

16. The only area where fissure eruptions are currently occurring is:
a. ____ the Red Sea;
b. ____ western South America;
c. ____ the Pacific Northwest;
d. ____ Iceland;
e. ____ Japan.
17. Explain how pyroclastic materials and volcanic gases can affect climate.
18. How do spatter cones and columnar joints form?
19. Explain how most calderas form.
20. What kinds of warning signs enable geologists to forecast eruptions?
21. Why do shield volcanoes have such low slopes?
22. How do pahoehoe and aa lava flows differ?
23. Draw a cross section of a composite volcano. Indicate its constituent materials, and show how and where a flank eruption might occur.
24. Why do composite volcanoes occur in belts near convergent plate margins? Are such volcanoes present at all convergent plate margins?
25. Why are lava domes so dangerous?
26. Give a brief summary of the origin and development of the Hawaiian Islands.

Points to Ponder

1. During this century, two Cascade Range volcanoes have erupted. What kinds of evidence would indicate that some of the other volcanoes in this range might erupt in the future?
2. Suppose that you found rocks on land consisting of layers of pillow lava overlain by deep-sea sedimentary rocks. Where did the pillow lava layers originally form, and what type of rock would you expect to lie beneath them?
3. If several eruptions the size of Mount Pinatubo's occurred in one year, what types of widespread effects could be expected?
4. What geologic events would have to occur in order for a chain of volcanoes to form along the east coasts of Canada and the United States?

Additional Readings

Aylesworth, T. G., and V. Aylesworth. 1983. *The Mount St. Helens disaster: What we've learned.* New York: Franklin Watts.

Bullard, F. M. 1984. *Volcanoes of the Earth.* 2d ed. Austin, Tex.: University of Texas Press.

Coffin, M. F., and O. Eldholm. 1993. Large igneous provinces. *Scientific American* 269, no. 4: 42–49.

Decker, R. W., and Decker, B. B. 1991. *Mountains of fire: The nature of volcanoes.* New York: Cambridge University Press.

Erickson, J. 1988. *Volcanoes & earthquakes.* Blue Ridge Summit, Pa.: Tab Books.

Harris, S. L. 1976. *Fire and ice: The Cascade volcanoes.* Seattle, Wash.: The Mountaineers.

Lipman, P. W., and D. R. Mullineaux, eds. 1981. The 1980 eruptions of Mount St. Helens, Washington. *United States Geological Survey Professional Paper 1250.*

McClelland, L., T. Simkin, M. Summers, E. Nielsen, and T. C. Stein, eds. 1989. *Global volcanism 1975–1985.* Englewood Cliffs, N.J.: Prentice-Hall.

Rampino, M. R., S. Self, and R. B. Strothers. 1988. Volcanic winters. *Annual Review of Earth and Planetary Sciences* 16: 73–99.

Simkin, T., et al. 1981. *Volcanoes of the world: A regional gazetteer, and chronology of volcanism during the last 10,000 years.* Stroudsburg, Pa.: Hutchison Ross Publishing Co.

Stager, C. 1987. Silent death from Cameroon's killer lake. *National Geographic* v. 172, no. 3: 404-420.

Tilling, R. I. 1987 *Eruptions of Mount St. Helens: Past, present, and future.* U.S. Geological Survey.

Tilling, R. I., C. Heliker, and T. L. Wright. 1987. *Eruptions of Hawaiian volcanoes: Past, present, and future.* U. S. Geological Survey.

Volcanoes and the Earth's interior. 1982. Readings from Scientific American. San Francisco, Calif.: W. H. Freeman and Co.

Wenkam, R. 1987. *The edge of fire: Volcano and earthquake country in western North America and Hawaii.* San Francisco, Calif.: Sierra Club Books.

Wolfe, G. W. 1992. The 1991 eruptions of Mount Pinatubo, Philippines. *Earthquakes and Volcanoes* 23, no. 1: 5–37.

Wright, T. L., and T. C. Pierson. 1992. Living with volcanoes: The U.S. Geological Survey's volcano hazards program. *U.S. Geological Survey Circular 1073.*

Chapter 5

WEATHERING, EROSION, AND SOIL

OUTLINE

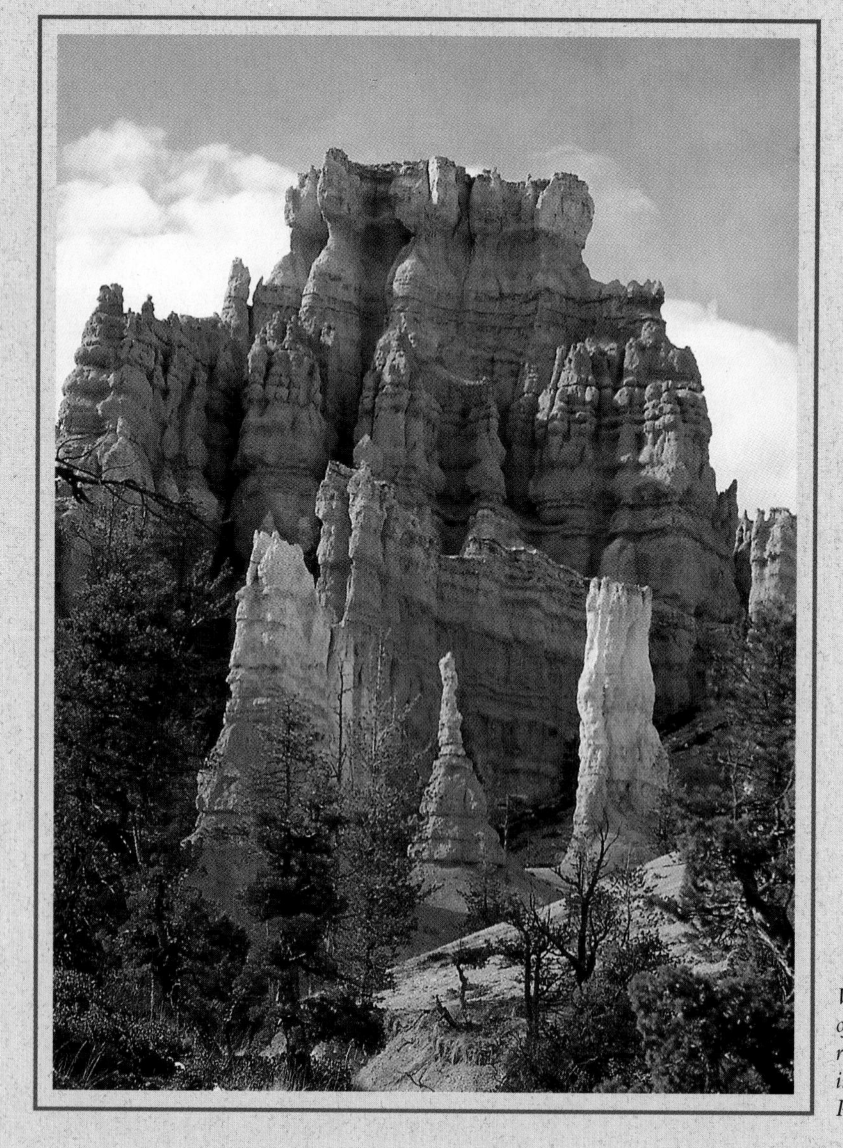

Weathering and erosion of sedimentary rocks are responsible for the scenery in Bryce Canyon National Park, Utah.

PROLOGUE

Most of the middle-latitude forests of Europe and North America were destroyed during the past several centuries as a result of human activities. Today, the rainforests of the tropics are disappearing much more rapidly and over a much larger area. Indeed, since 1950 about half of the original 8.8 million km^2 of tropical rainforest has been cleared! Currently, more than 52 million acres are being cleared annually, an area equivalent to the state of Utah. Scientists find these figures disturbing because clearing tropical rainforests has several far-reaching implications.

Prior to the 1950s, tropical rainforests were relatively undisturbed because they were in remote areas, there was little demand for lumber from these forests, the climate was considered unhealthy for humans, and the soils were of low fertility. Accordingly, the rainforests were largely ignored until the 1950s and 1960s when economic conditions changed.

Persistent population pressure has contributed significantly to the destruction of the rainforests. In many developing nations in the tropics, population increases far surpass the number of jobs created in the depressed economies, and with all fertile soils already devoted to agriculture, poor farmers are increasingly spreading into areas of poor soils, namely, the rainforests. There they clear the land for crops and use the wood for lumber and fuel.

In addition, despite a number of difficulties, lumbering in tropical rainforests has become increasingly profitable. Many species of tropical trees that were unwanted a decade or two ago are now in demand. As a result, roads have been built into formerly inaccessible areas, heavy machinery has been brought in, and trees are being cut using modern lumbering practices (▶ Figure 5-1). If present trends continue, by the year 2000 only two large areas of tropical rainforest will remain, one in the western Amazon Basin in South America and one in the central Congo Basin of Africa.

Deforestation has several adverse effects. One is the removal of the vegetation itself. Even plants that are not cut during logging operations are often damaged. If left undisturbed, cleared areas may eventually return to their original condition, but such areas are commonly used for crops or grazing, thus effectively eliminating the possibility of regrowth. Furthermore, trees in rainforests are particularly effective at absorbing soil nutrients into their roots before the nutrients can be leached out of the soil by the abundant rainfall. In fact, most of the nutrients in rainforests are in the trees and are slowly released when they die and decompose.

Soil erosion rates generally increase when tropical rainforest is cleared because the surface is left unprotected (▶ Figure 5-2). More rainwater runs off at the surface, gullying becomes more common, and flooding is more frequent because the trees with their huge water-holding capacity are no longer present. Furthermore, the soil becomes more compacted and less absorbent, inhibiting infiltration and increasing runoff.

▶ **FIGURE 5-1** Logging operation and large-scale deforestation in the rain forests of Brazil.

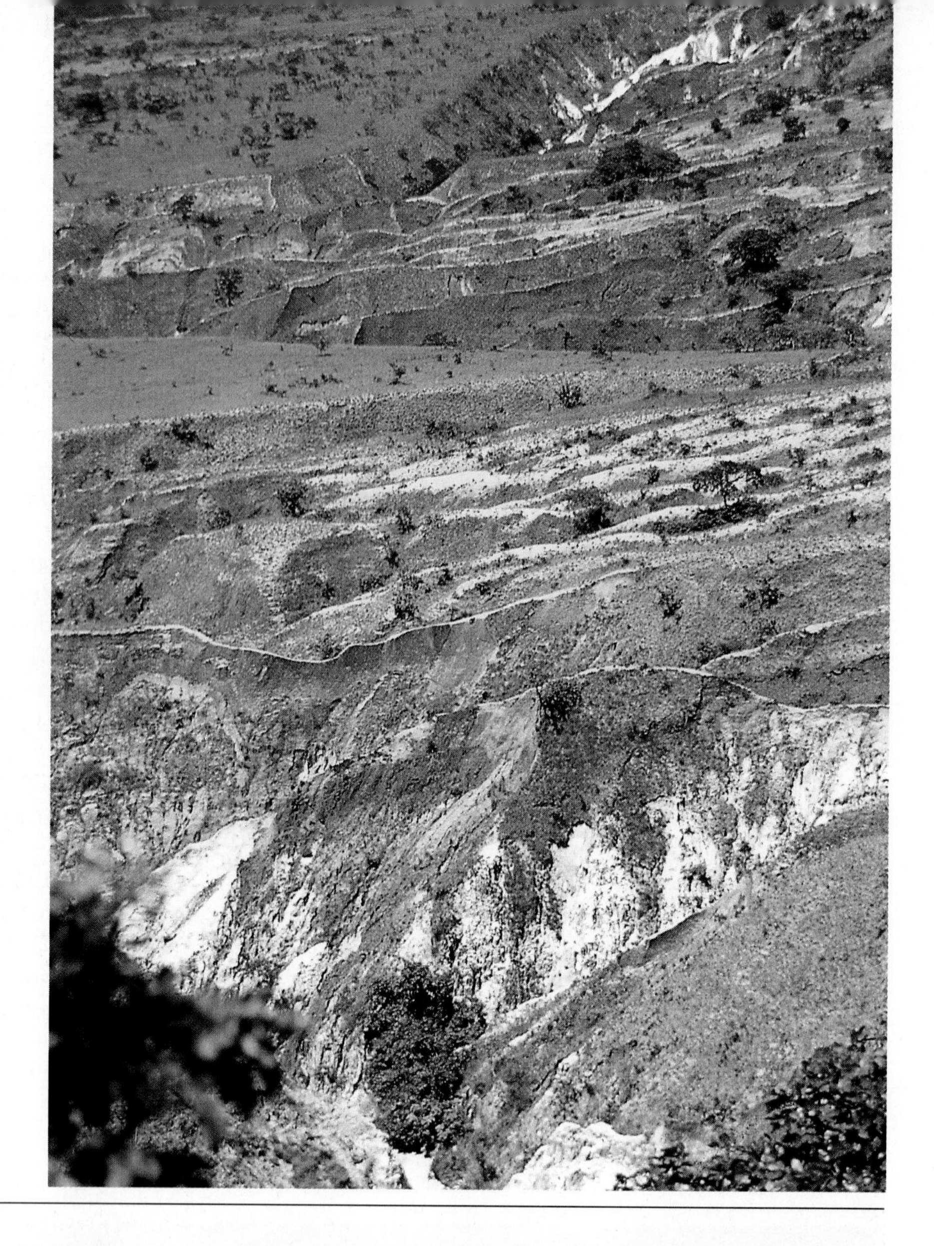

▶ **FIGURE 5-2** Accelerated soil erosion on a bare surface in Madagascar that was once covered by lush forest.

Other negative effects of deforestation include a decrease in biodiversity and possibly climatic changes. The rainforests are the home of millions of species of plants and animals; indeed, more species are found in the tropics than anywhere else on Earth. Many of these species have not been studied or named, yet some will become extinct because of habitat destruction.

The relationship between deforestation and climate is somewhat speculative, but some have suggested that deforestation exacerbates the greenhouse effect and could contribute to global warming. More carbon dioxide (CO_2), an important greenhouse gas, is present in the vegetation of tropical rainforests than in all other vegetation on Earth. Without vegetation to hold this huge reservoir of carbon dioxide, more will remain in the atmosphere and perhaps contribute to global warming.

 INTRODUCTION

Weathering is the physical breakdown (disintegration) and chemical alteration (decomposition) of rocks and minerals at or near the Earth's surface. It is the process whereby rocks and minerals are physically and chemically altered such that they are more nearly in equilibrium with a new set of environmental conditions. For example, many rocks form within the Earth's crust where little or no water or oxygen is present and where temperatures, pressures, or both are high. At or near the surface, however, the rocks are exposed to low temperatures and pressures and are attacked by atmospheric gases, water, acids, and organisms.

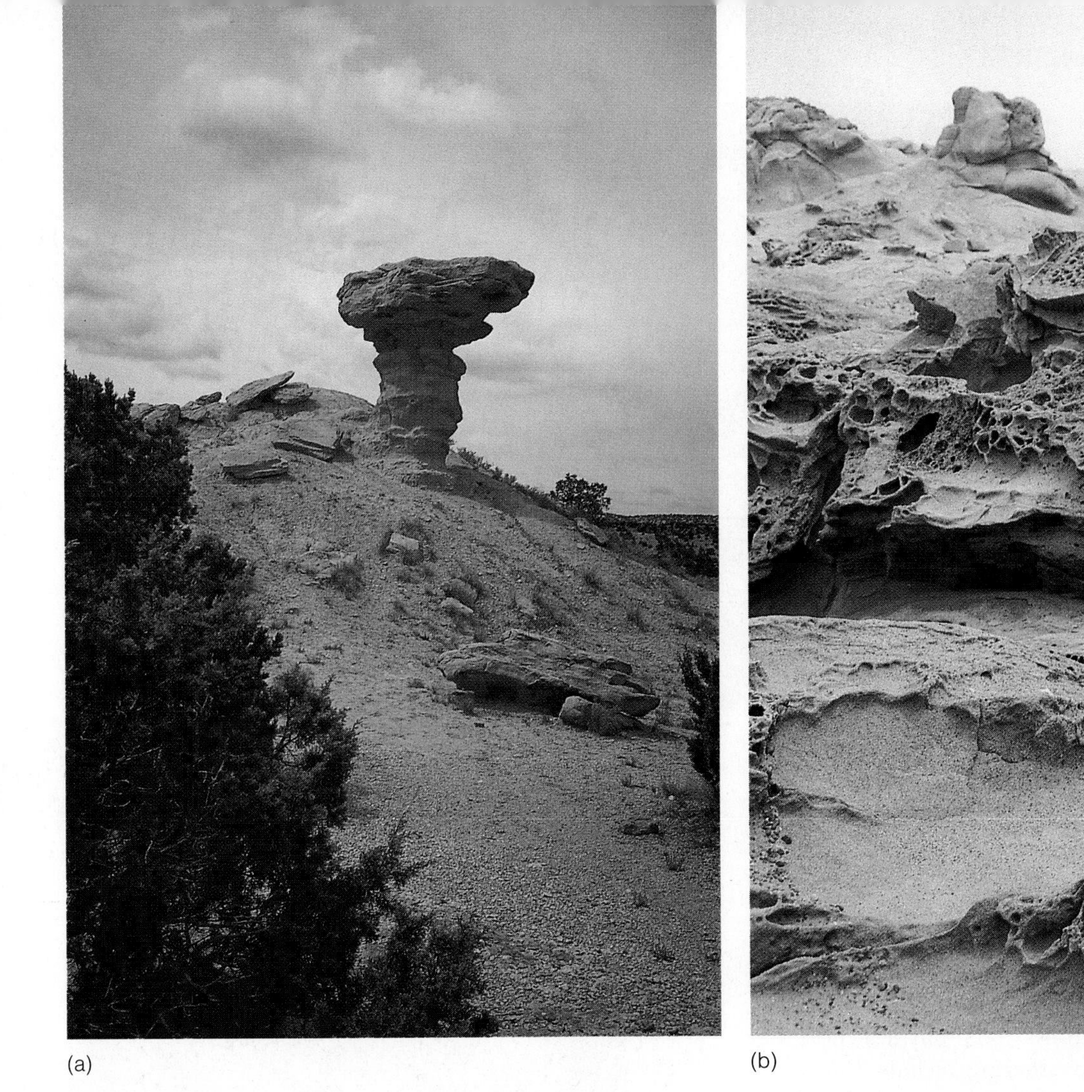
(a) (b)

▶ **FIGURE 5-3** Differential weathering and erosion. (*a*) Camel Rock near Santa Fe, New Mexico. (*b*) An intricate, uneven weathering surface known as honeycomb weathering at Pebble Beach, California. (Photos courtesy of Sue Monroe.)

Geologists are interested in the phenomenon of weathering because it is an essential part of the rock cycle (see Figure 1-16). The **parent material,** or rock being weathered, is broken down into smaller pieces, and some of its constituent minerals are dissolved or altered and removed from the weathering site. The removal of the weathered materials is known as **erosion.** Running water, wind, or glaciers commonly **transport** the weathered materials elsewhere, where they are deposited as sediment, which may become sedimentary rock (see Figure 1-16). Whether they are eroded or not, weathered rock materials can also be further modified to form a soil. Thus, weathering provides the raw materials for both sedimentary rocks and soils. Weathering is also important in the origin of some mineral resources such as aluminum ores, and it is responsible for the enrichment of other deposits of economic importance.

Weathering is such a pervasive phenomenon that many people take it for granted or completely overlook it. Nevertheless, it occurs continuously although its rate and impact vary from area to area or even within the same area. Because rocks vary in composition and structure, they do not all weather at the same rate. Even in a single rock layer, the response to weathering is variable because of slight differences in composition and structure. Weathering is more intense on fractures than on adjacent unfractured areas of rocks. As a consequence of these variations, **differential weathering** occurs which means that rocks weather at different rates producing uneven surfaces. In Bryce Canyon National Park, Utah, differential weathering and erosion of sedimentary rocks cut by intersecting fractures have produced oddly shaped rock formations (see the chapter-opening photo). Peculiar surfaces and shapes formed by differential weathering are also found in many other areas (▶ Figure 5-3). Like rocks in natural exposures, the rock-like materials of roadways, sidewalks, and foundations also disintegrate and decompose due to weathering.

Two types of weathering are recognized, *mechanical* and *chemical.* Both types occur simultaneously at the weathering site, during erosion and transport, and even in the environments where weathered materials are deposited.

▶ **FIGURE 5-4** Mechanically weathered granite. The sandy material consists of small pieces of granite (rock fragments) and minerals such as quartz and feldspars liberated from the parent material.

 MECHANICAL WEATHERING

Mechanical weathering occurs when physical forces break rock materials into smaller pieces that retain the chemical composition of the parent material. Granite, for example, may be mechanically weathered to yield smaller pieces of granite, or disintegration may liberate individual mineral grains from it (▶ Figure 5-4). The physical processes responsible for mechanical weathering include *frost action, pressure release, thermal expansion and contraction, salt crystal growth*, and the *activities of organisms.*

Frost Action

Frost action involves the repeated freezing and thawing of water in cracks and crevices in rocks. When water seeps into a crack and freezes, it expands by about 9% and exerts great force on the walls of the crack, thereby widening and extending it by **frost wedging.** As a consequence of repeated freezing and thawing, pieces of rock are eventually detached from the parent material (▶ Figure 5-5). Frost wedging is particularly effective if the crack is convoluted. If the crack is a simple wedge-shaped opening, much of the force of expansion is released upward toward the surface.

▶ **FIGURE 5-5** Frost wedging occurs when water seeps into cracks and expands as it freezes. Angular pieces of rock are pried loose by repeated freezing and thawing.

Frost action is most effective in areas where temperatures commonly fluctuate above and below freezing. In the high mountains of the western United States and Canada, frost action is very effective even during summer months. In the tropics and in areas where water is permanently frozen, frost action is of little or no importance.

The debris produced by frost wedging in mountains commonly accumulates as large cones of **talus** lying at the bases of slopes (▶ Figure 5-6). The materials that form the talus are simply angular pieces of rock from a larger body that has been mechanically weathered. Most rocks have a system of fractures called *joints* along which frost action is particularly effective. Water seeps along the joint surfaces and eventually wedges pieces of rock loose; these then fall downslope to accumulate with other loosened rocks.

In the phenomenon known as **frost heaving,** a mass of sediment or soil undergoes freezing, expansion, and actual lifting, followed by thawing, contraction, and lowering of the mass. Frost heaving is particularly evident where water freezes beneath roadways and sidewalks.

▶ **FIGURE 5-6** Talus in the Canadian Rocky Mountains.

Pressure Release

Pressure release is a mechanical weathering process that is especially evident in rocks that formed as deeply buried intrusive bodies such as batholiths, but it occurs in other types of rocks as well. When a batholith forms, the magma crystallizes under tremendous pressure (the weight of the overlying rock) and is stable under these pressure conditions. When the batholith is uplifted and the overlying rock is stripped away by erosion, the pressure is reduced. But the rock contains energy that is released by expansion and the formation of **sheet joints,** large fractures that more or less parallel the rock surface (▶ Figure 5-7). Slabs of rock bounded by sheet joints may slip, slide, or spall (break) off of the host rock—a process called **exfoliation**—and accumulate as talus. The large rounded domes of rock resulting from this process are **exfoliation domes;** examples are found in Yosemite National Park in California and Stone Mountain in Georgia (▶ Figure 5-8). Sheet-jointing and exfoliation constitute an engineering problem in many areas (see Perspective 5-1).

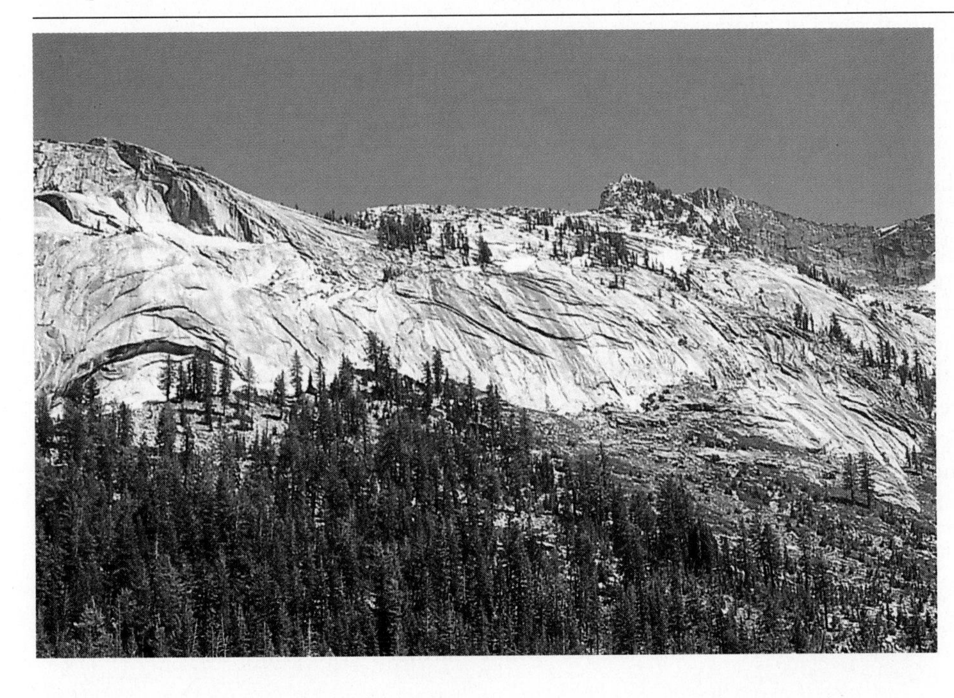

▶ **FIGURE 5-7** Sheet joints in granite in the Sierra Nevada of California.

➢ **FIGURE 5-8** Exfoliation domes in Yosemite National Park, California.

Thermal Expansion and Contraction

During **thermal expansion and contraction,** the volume of solids, such as rocks, changes in response to heating and cooling. In a desert, where the temperature may vary as much as 30°C in one day, rocks expand when heated and contract as they cool. Expansion and contraction, however, do not occur uniformly throughout rocks. For one thing, a rock is a poor conductor of heat, so its outside heats up more than its inside. Consequently, the surface expands more than the interior, producing stresses that may cause fracturing. Furthermore, dark minerals absorb heat faster than light-colored minerals, so differential expansion occurs even between the mineral grains of some rocks.

Repeated thermal expansion and contraction is a common phenomenon, but are the forces generated sufficient to overcome the internal strength of a rock? Experiments in which rocks are heated and cooled repeatedly to simulate years of such activity indicate that thermal expansion and contraction is not an important agent of mechanical weathering.* Despite these experimental results, some rocks in deserts do indeed appear to show the effects of this process.

Daily temperature variation is the most common cause of alternate expansion and contraction, but these changes occur over periods of hours. In contrast, fire can cause very rapid expansion. During a forest fire, rocks may heat very rapidly, especially near the surface because they conduct heat so poorly. The heated surface layer expands more rapidly than the interior, and thin sheets paralleling the rock surface become detached.

*Thermal expansion and contraction may be a significant mechanical weathering process on the Moon where extreme temperature changes occur quickly.

Salt Crystal Growth

Under some circumstances, salt crystals forming from solution can cause disaggregation of rocks. Growing crystals exert enough force to widen cracks and crevices or dislodge particles in porous, granular rocks such as sandstones. Even in crystalline rocks such as granite, salt crystal growth may pry loose individual mineral grains. To the extent that salt crystal growth produces forces that expand openings in rocks, it is similar to frost wedging. Most **salt crystal growth** occurs in hot arid areas, although it probably affects rocks in some coastal regions as well.

Activities of Organisms

Animals, plants, and bacteria all participate in the mechanical and chemical alteration of rocks. Burrowing animals, such as worms, reptiles, rodents, and many others, constantly mix soil and sediment particles and bring material from depth to the surface where further weathering may occur. Even materials ingested by worms are further reduced in size, and animal burrows allow gases and water to have easier access to greater depths. The roots of plants, especially large bushes and trees, wedge themselves into cracks in rocks and further widen them (▶ Figure 5-9). Tree roots that grow under or through sidewalks and foundations may do considerable damage.

Chemical Weathering

Chemical weathering is the process whereby rock materials are decomposed by chemical alteration of the parent material. A number of clay minerals, for example, form as the chemically altered products of other minerals. Some minerals are completely decomposed during chemical weathering, but others, which are more resistant, are simply liberated from the parent material. Such weathering is accomplished by the action of atmospheric gases, especially oxygen, and water and acids. Organisms also play an important role in chemical weathering. Rocks that have lichens (composite organisms consisting of fungi and algae) growing on their surfaces undergo more extensive chemical alteration than lichen-free rocks (▶ Figure 5-10). Plants remove ions from soil water and reduce the chemical stability of soil minerals, and their roots release organic acids.

Solution

During **solution** the ions of a substance become dissociated from one another in a liquid, and the solid substance dissolves. Water is a remarkable solvent because its molecules have an asymmetric shape; they consist of one oxygen atom

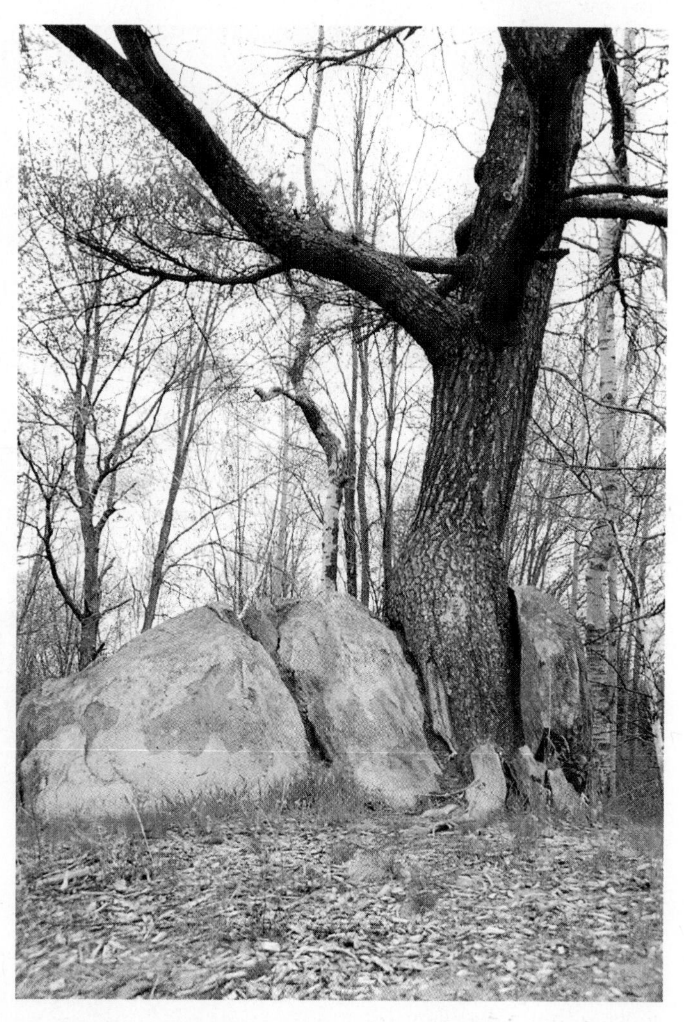

▶ **FIGURE 5-9** The contribution of organisms to mechanical weathering. Tree roots enlarge cracks in rocks.

▶ **FIGURE 5-10** The greenish, irregular masses on these rocks are lichens, composite organisms consisting of fungi and algae. Lichen-covered rocks are chemically weathered much more rapidly than lichen-free rocks.

Perspective 5-1

Bursting Rocks and Sheet Joints

The fact that solid rock can expand and produce fractures is a well-known phenomenon. In deep mines, masses of rock suddenly detach from the sides of the excavation, often with explosive violence. Such *rock bursts* generally occur below depths of about 600 m; spectacular rock bursts have been recorded in deep gold mines in South Africa and Canada and in zinc mines in Idaho. Obviously, rock bursts and related phenomena, such as less violent *popping*, pose a danger to mine workers. In South Africa, about 20 miners are killed by rock bursts every year.

In some quarrying operations,* the removal of surface materials to a depth of only 7 or 8 m has led to the formation of sheet joints in the underlying rock (▶ Figure 1). At quarries in Vermont and Tennessee, the excavation of marble exposed rocks that were formerly buried and under great pressure. When the overlying rock was removed, the marble expanded and sheet

*A quarry is a surface excavation, generally for the extraction of building stone.

▶ FIGURE 1 Sheet joints formed by expansion in the Mount Airy Granite in North Carolina. (Photo courtesy of W. D. Lowry.)

with two hydrogen atoms arranged such that the angle between the two hydrogens is about 104 degrees (▶ Figure 5-11). Because of this asymmetry, the oxygen end of the molecule retains a slight negative electrical charge, whereas the hydrogen end retains a slight positive charge. When a soluble substance such as the mineral halite (NaCl) comes in contact with a water molecule, the positively charged sodium ions are attracted to the negative end of the water molecule, and the negatively charged chloride ions are attracted to the positively charged end of the water molecule (Figure 5-11b). Thus, ions are liberated from the crystal structure, and the solid dissolves.

Most minerals are not very soluble in pure water because the attractive forces of water molecules are not sufficient to overcome the forces between particles in minerals. The mineral calcite ($CaCO_3$), the major constituent of the sedimentary rock limestone and the metamorphic rock marble, is practically insoluble in pure water, but rapidly dissolves if a small amount of acid is present. An easy way to make water acidic is by dissociating the ions of carbonic acid as follows:

$$\underset{\text{water}}{H_2O} + \underset{\text{carbon dioxide}}{CO_2} \rightleftharpoons \underset{\text{carbonic acid}}{H_2CO_3} \rightleftharpoons \underset{\text{hydrogen ion}}{H^+} + \underset{\text{bicarbonate ion}}{HCO_3^-}$$

According to this chemical equation, water and carbon dioxide combine to form *carbonic acid*, a small amount of which dissociates to yield hydrogen and bicarbonate ions. The concentration of hydrogen ions determines the acidity of a solution; the more hydrogen ions present, the stronger the acid.

There are several sources of carbon dioxide that may combine with water and react to form acid solutions. The

joints formed. Some slabs of rock that were bounded by sheet joints burst so violently that quarrying machines weighing more than a ton were thrown from their tracks, and some quarries had to be abandoned because fracturing rendered the stone useless.

Sheet joints paralleling the walls of the Vaiont River valley in Italy contributed to the worst reservoir disaster in history. On October 9, 1963, more than 240 million m^3 of rock slid into the Vaiont Reservoir. Although several factors contributed to this slide, it moved partly along a system of sheet joints. The slide displaced water in the reservoir, causing a large wave to overtop the dam and flood the downstream area where nearly 3,000 people drowned. (See Perspective 14-2 for a more complete discussion of the Vaiont Reservoir disaster.)

The Sierra Nevada of California are composed of granitic rocks, many of which contain numerous sets of sheet joints parallel to the canyon walls. Large slabs of granite bounded by sheet joints lie on steeply inclined surfaces above highways and railroad tracks where they pose a danger to the road or track below (▶ Figure 2). Occasionally, a mass of this unsupported rock slides or falls, blocking highways and railroad tracks.

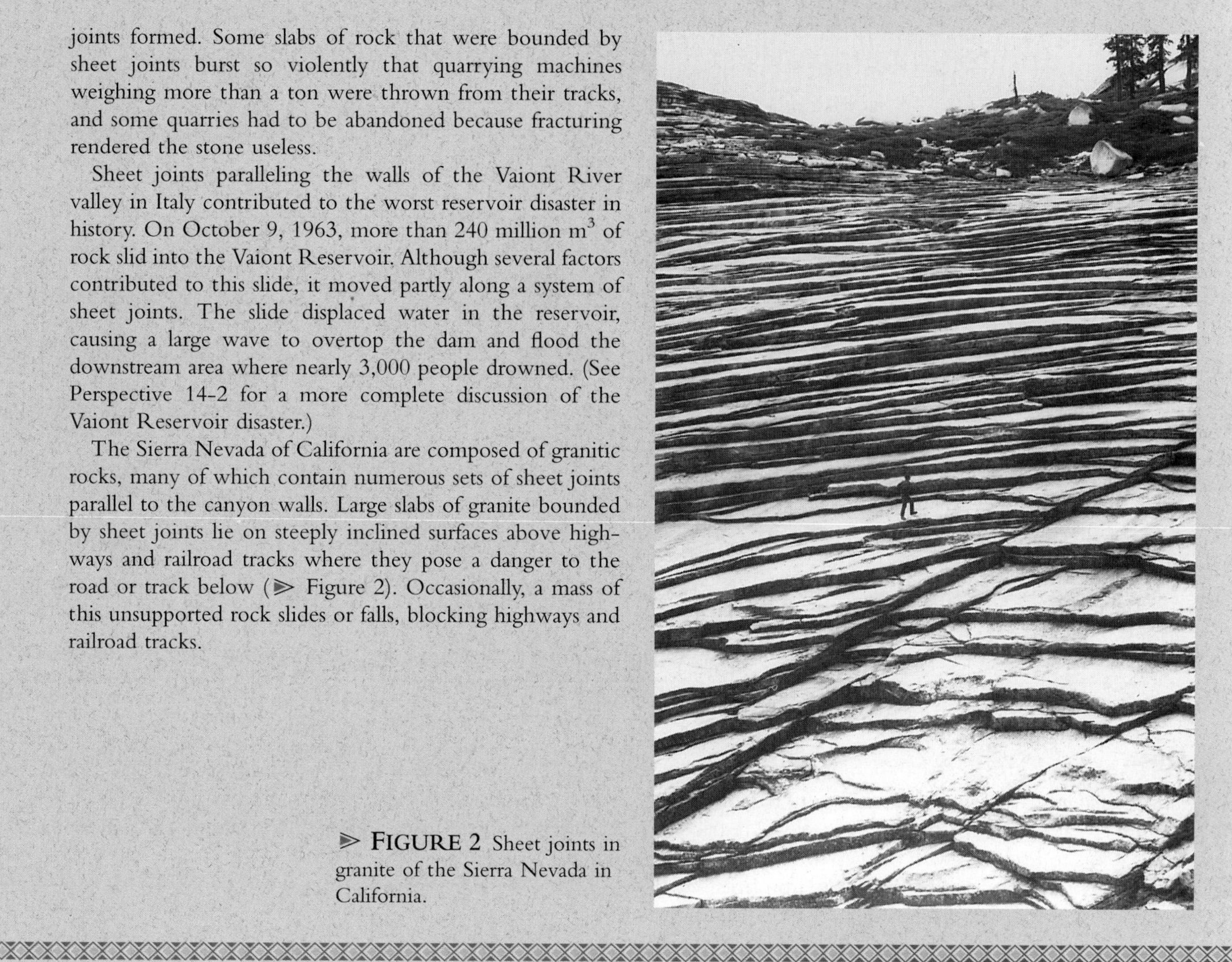

▶ FIGURE 2 Sheet joints in granite of the Sierra Nevada in California.

atmosphere is mostly nitrogen and oxygen, but about 0.03% is carbon dioxide, causing rain to be slightly acidic. Carbon dioxide is also produced in soil by the decay of organic matter and the respiration of organisms, so groundwater is also generally slightly acidic. Arid regions have sparse vegetation and few soil organisms, so groundwater has a limited supply of carbon dioxide and tends to be alkaline rather than acidic, that is, it has a low concentration of hydrogen ions.

Whatever the source of carbon dioxide, once an acidic solution is present, calcite rapidly dissolves according to the following reaction:

$$\underset{\text{calcite}}{CaCO_3} + \underset{\text{water}}{H_2O} + \underset{\text{carbon dioxide}}{CO_2} \rightleftharpoons \underset{\text{calcium ion}}{Ca^{++}} + \underset{\text{bicarbonate ion}}{2HCO_3^-}$$

Because of the dissociation of the ions in carbonic acid, this reaction may also be written as:

$$\underset{\text{calcite}}{CaCO_3} + \underset{\text{hydrogen ion}}{H^+} + \underset{\text{bicarbonate ion}}{HCO_3^-} \rightleftharpoons \underset{\text{calcium ion}}{Ca^{++}} + \underset{\text{bicarbonate ion}}{2HCO_3^-}$$

The dissolution of the calcite in limestone and marble has had dramatic effects in many places ranging from small cavities to large caverns such as Mammoth Cave in Kentucky and Carlsbad Caverns in New Mexico (see Chapter 16).

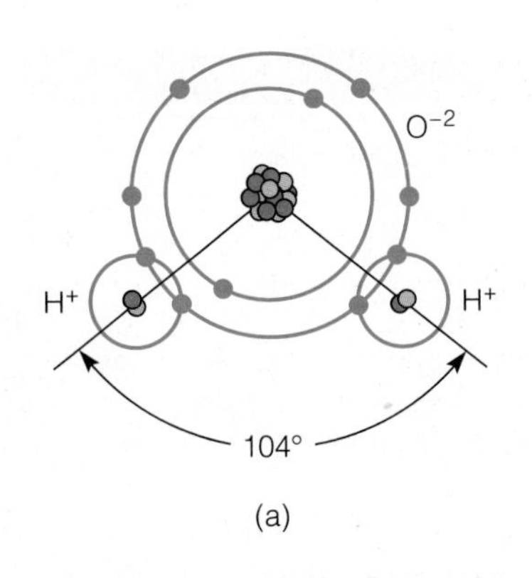

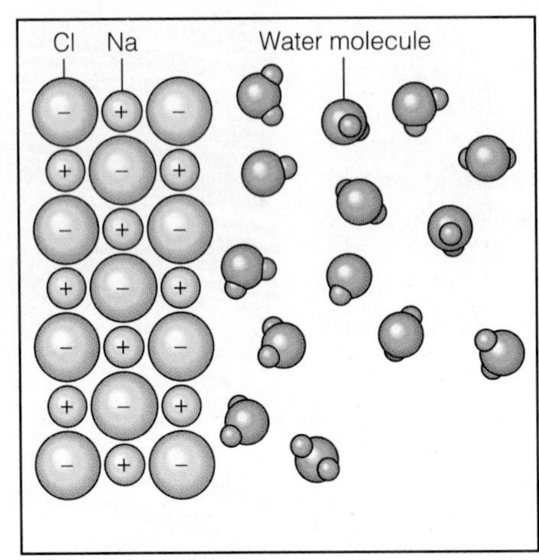

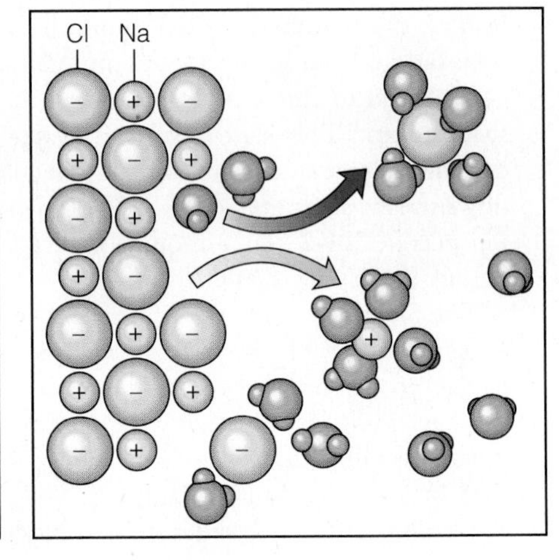

▶ **FIGURE 5-11** (*a*) The structure of a water molecule. The asymmetric arrangement of the hydrogen atoms causes the molecule to have a slight positive electrical charge at its hydrogen end and a slight negative charge at its oxygen end. (*b*) The dissolution of sodium chloride (NaCl) in water.

Oxidation

The term *oxidation* has a variety of meanings to chemists, but in chemical weathering its meaning is more restricted. **Oxidation** refers to reactions with oxygen to form oxides or, if water is present, hydroxides. For example, iron rusts when it combines with oxygen to form the iron oxide hematite:

$$\underset{\text{iron}}{4Fe} + \underset{\text{oxygen}}{3O_2} \rightarrow \underset{\text{iron oxide (hematite)}}{2Fe_2O_3}$$

Of course, atmospheric oxygen is abundantly available for oxidation reactions, but oxidation is generally a slow process unless water is present. Most oxidation is carried out by oxygen dissolved in water.

Oxidation is very important in the alteration of ferromagnesian minerals such as olivine, pyroxenes, amphiboles, and biotite. Iron in these minerals combines with oxygen to form the reddish iron oxide hematite (Fe_2O_3) or the yellowish or brown hydroxide limonite ($FeO(OH) \cdot nH_2O$). The yellow, brown, and red colors of many soils and sedimentary rocks are caused by the presence of small amounts of hematite or limonite.

An oxidation reaction of particular concern in some areas is the oxidation of iron sulfides such as the mineral pyrite (FeS_2). Pyrite is commonly associated with coal, so in mine tailings* pyrite oxidizes to form sulfuric acid (H_2SO_4) and iron oxide. Acid soils and waters in coal-mining areas are produced in this manner and present a serious environmental hazard (▶ Figure 5-12).

Hydrolysis

Hydrolysis is the chemical reaction between the hydrogen (H^+) ions and hydroxyl (OH^-) ions of water and a mineral's ions. In hydrolysis hydrogen ions actually replace positive ions in minerals. Such replacement changes the composition of minerals by liberating soluble compounds and iron that then may be oxidized.

As an illustration of hydrolysis, consider the chemical alteration of feldspars. Potassium feldspars such as orthoclase ($KAlSi_3O_8$) are common in many rock types, as are the plagioclase feldspars (which vary in composition from $CaAl_2Si_2O_8$ to $NaAlSi_3O_8$). All feldspars are framework silicates, but when altered, they yield materials in solution and clay minerals, such as kaolinite, which are sheet silicates.

The chemical weathering of potassium feldspar by hydrolysis occurs as follows:

*Tailings are the rock debris of mining; they are considered too poor for further processing and are left as heaps on the surface.

▶ **FIGURE 5-12** The oxidation of pyrite in mine tailings forms acid water as in this small stream. More than 11,000 km of U.S. streams, mostly in the Appalachian region, are contaminated by abandoned coal mines that leak sulfuric acid.

$$\underset{\text{orthoclase}}{2KAlSi_3O_8} + \underset{\text{hydrogen ion}}{2H^+} + \underset{\text{bicarbonate ion}}{2HCO_3^-} + \underset{\text{water}}{H_2O} \rightarrow$$

$$\underset{\text{clay (kaolinite)}}{Al_2Si_2O_5(OH)_4} + \underset{\text{potassium ion}}{2K^+} + \underset{\text{bicarbonate ion}}{2HCO_3^-} + \underset{\text{silica}}{4SiO_2}$$

In this reaction hydrogen ions attack the ions in the orthoclase structure, and some liberated ions are incorporated in a developing clay mineral, and others simply dissolve. On the right side of the equation is excess silica that would not fit into the crystal structure of the clay mineral. Such dissolved silica is an important source of cement in sedimentary rocks (see Chapter 6).

Plagioclase feldspars are altered by hydrolysis in the same way as orthoclase. The only difference is that soluble calcium and sodium compounds are formed rather than potassium compounds. In fact, these dissolved compounds are what make hard water hard. Calcium in water is a problem because it inhibits the reaction of detergents with dirt and precipitates as scaly mineral matter in water pipes and water heaters.

FACTORS CONTROLLING THE RATE OF CHEMICAL WEATHERING

Chemical weathering processes operate on the surfaces of particles; that is, chemically weathered rocks or minerals are altered from the outside inward. Several factors including particle size, climate, and parent material control the rate of chemical weathering.

Particle Size

Because chemical weathering affects particle surfaces, the greater the surface area, the more effective the weathering. It is important to realize that small particles have larger surface areas compared to their volume than do large particles. Notice in ▶ Figure 5-13 that a block measuring 1 m on a side has a total surface area of 6 m^2, but when the block is broken into particles measuring 0.5 m on a side, the total surface area increases to 12 m^2. And if these particles are all reduced to 0.25 m on a side, the total surface area increases to 24 m^2. Note that while the surface area in this example increases, the total volume remains the same at 1 m^3.

We can make two important statements regarding the block in Figure 5-13. First, as it is split into a number of smaller blocks, its total surface area increases. Second, the smaller any single block is, the more surface area it has

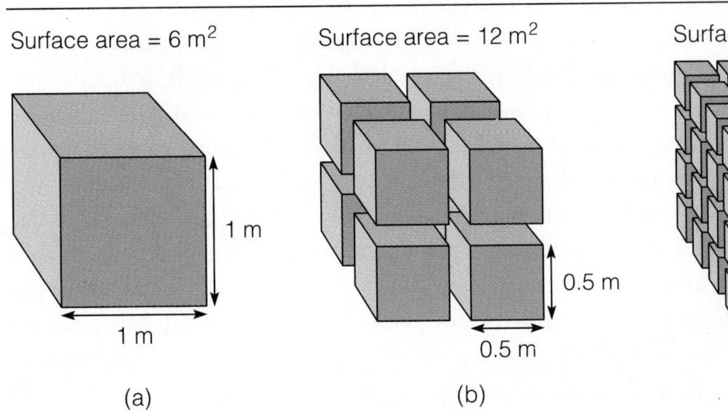

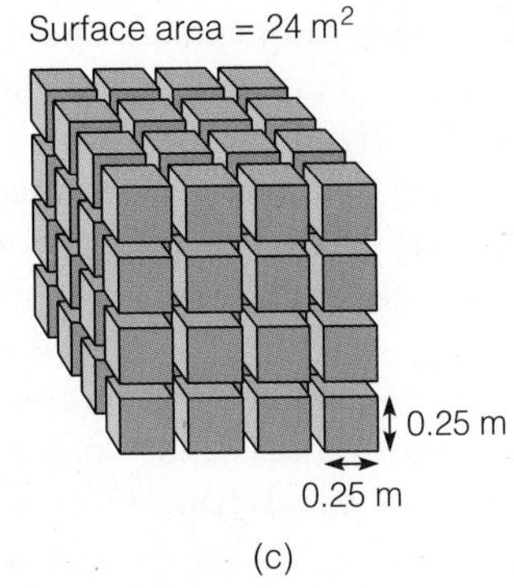

▶ **FIGURE 5-13** Particle size and chemical weathering. As a rock is reduced into smaller and smaller particles, its surface area increases but its volume remains the same. In (*a*) the surface area is 6 m^2, in (*b*) it is 12 m^2, and in (*c*) 24 m^2, but the volume remains the same at 1 m^3. Small particles have more surface area in proportion to their volume than do large particles.

compared to its volume. Because chemical weathering is a surface process, the fact that small objects have proportionately more surface area compared to volume than do large objects has profound implications. We can conclude that mechanical weathering, which reduces the size of particles, contributes to chemical weathering by exposing more surface area.

Climate

Most chemical processes occur more rapidly at high temperatures and in the presence of liquids. Accordingly, it is not surprising that chemical weathering is more effective in the tropics than in arid and arctic regions because temperatures and rainfall are high and evaporation rates are low (➢ Figure 5-14). In addition, vegetation and animal life are much more abundant in the tropics than in arid or cold regions. Consequently, the effects of weathering extend to depths of several tens of meters in the tropics, but commonly extend only centimeters to a few meters deep in arid and arctic regions. One should realize, though, that chemical weathering goes on everywhere, except perhaps where Earth materials are permanently frozen.

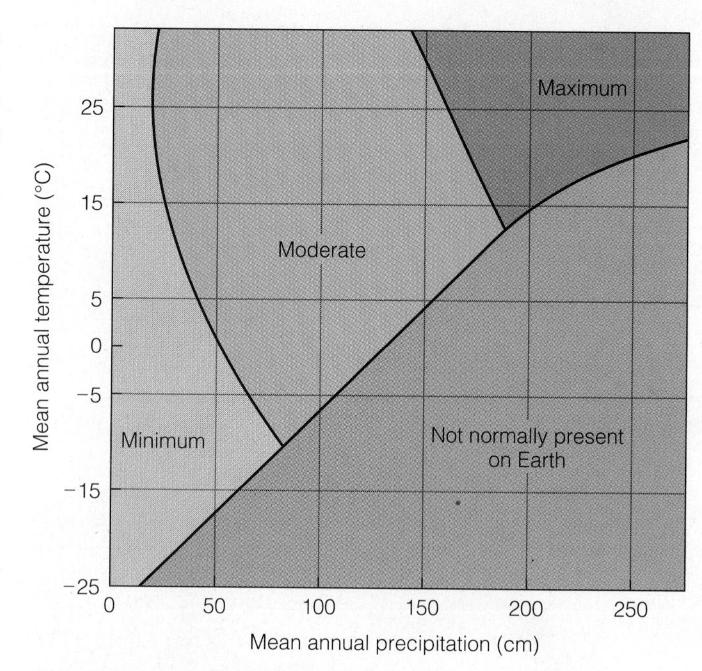

➢ **FIGURE 5-14** Relationships of chemical weathering rates and climate. Chemical weathering is at a maximum where temperature and rainfall are high. It is at a minimum in arid environments whether they are hot or cold.

Parent Material

Some rocks are chemically more stable than others and are not altered as rapidly by chemical processes. The metamorphic rock quartzite, composed of quartz, is an extremely stable substance that alters very slowly compared to most other rock types. In contrast, rocks such as granite, which contain large amounts of feldspar minerals, decompose rapidly because feldspars are chemically unstable. Ferromagnesian minerals are also chemically unstable and, when chemically weathered, yield clays, iron oxides, and ions in solution. In fact, the stability of common minerals is just the opposite of their order of crystallization in Bowen's reaction series (➢ Figure 5-15): the minerals that form last in this series are chemically stable, whereas those that form early are easily altered by chemical processes because they are most out of equilibrium with their conditions of formation.

One manifestation of chemical weathering is **spheroidal weathering** (➢ Figure 5-16). In spheroidal weathering, a stone, even one that is rectangular to begin with, weathers to form a spheroidal shape because that is the most stable shape it can assume. The reason is that on a rectangular stone the corners are attacked by weathering processes from three sides, and the edges are attacked from two sides, but the flat surfaces are weathered more or less uniformly (Figure 5-16). Consequently, the corners and edges are altered more rapidly, the material sloughs off them, and a more spherical shape develops. Once a spherical shape is present, all surfaces are weathered at the same rate.

Spheroidal weathering is often observed in granitic rock bodies containing joints. Fluids follow the joint surfaces and reduce rectangular joint-bounded blocks to a spherical shape (➢ Figure 5-17).

SOIL

In most places the land surface is covered by a layer of unconsolidated rock and mineral fragments called **regolith.** Regolith may consist of volcanic ash, sediment deposited by wind, streams, or glaciers, or weathered rock material formed in place as a residue. Some regolith that consists of weathered material, water, air, and organic matter and can support plant growth is recognized as **soil.** Soil is an essential link between the parent material below and the life above. Most land-dwelling organisms are dependent on soil for their existence. Plants derive their nutrients and most of their water from soils, and land-dwelling animals depend directly or indirectly on plants for nutrients.

About 45% of a good, fertile soil for gardening or farming is weathered rock material including sand, silt, and clay, but an essential constituent of such soils is **humus.** Humus, which gives many soils their dark color, is derived by bacterial decay of organic matter. It contains more carbon and less nitrogen than the original material and is nearly resistant to further bacterial decay. Although a fertile soil may contain only a small amount of humus, it is an essential source of plant nutrients and enhances moisture retention.

Some weathered materials in soils are simply sand- and silt-sized mineral grains, especially quartz, but other weathered materials may be present as well. Such solid particles are important because they hold soil particles apart, allowing oxygen and water to circulate more freely. Clay minerals are also important constituents of soils and aid in the retention of water as well as supplying nutrients to plants. Soils with

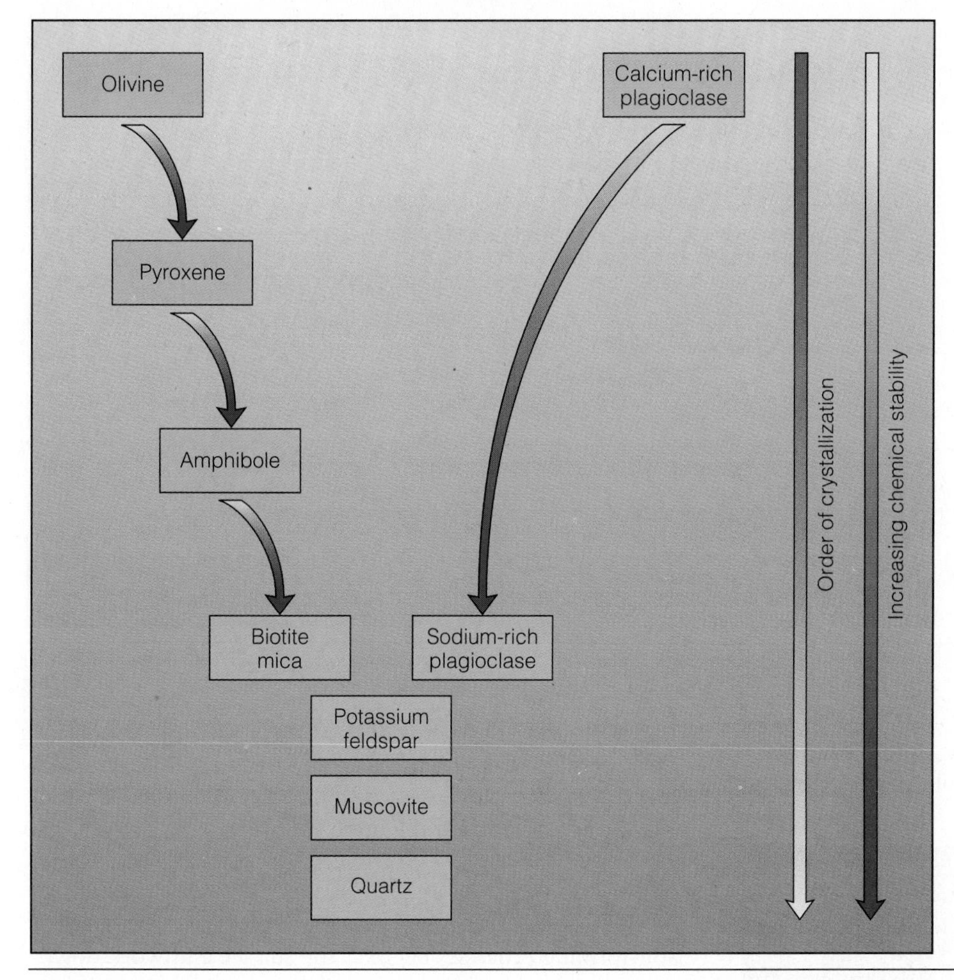

➢ FIGURE 5-15 Bowen's reaction series and chemical stability. The minerals forming first in this series are most out of equilibrium with their conditions of formation and are most chemically unstable.

excess clay minerals, however, drain poorly and are sticky when wet and hard when dry.

Soils are commonly characterized as residual or transported. If a body of rock weathers and the weathering residue accumulates over it, the soil so formed is residual, meaning that it formed in place (➢ Figure 5-18a). In contrast, transported soil develops on weathered material that has been eroded and transported from the weathering site

➢ FIGURE 5-16 Spheroidal weathering. (*a*) The rectangular blocks outlined by joints are attacked by chemical weathering processes, (*b*) but the corners and edges are weathered most rapidly. (*c*) When a block has been weathered so that it is spherical, its entire surface is weathered evenly, and no further change in shape occurs.

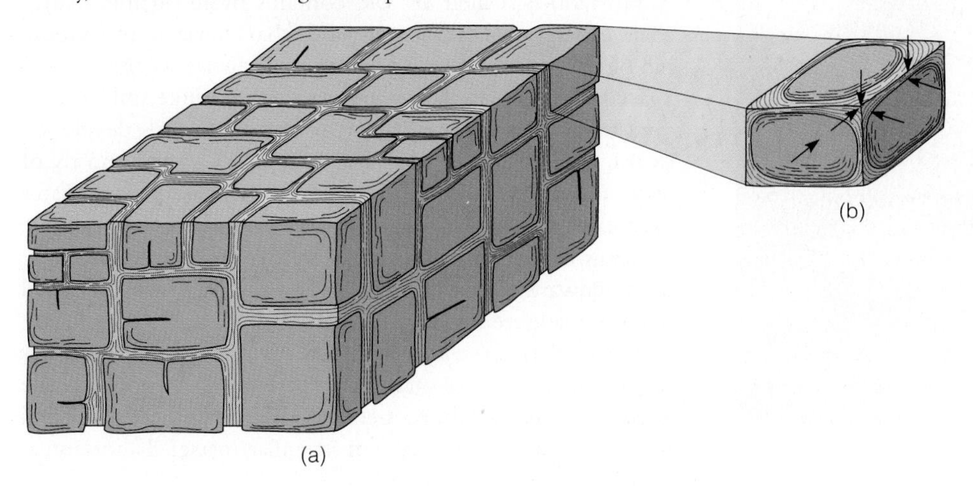

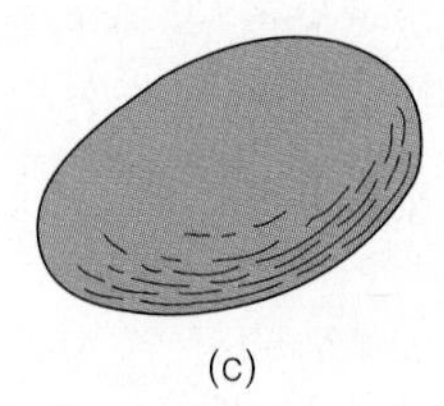

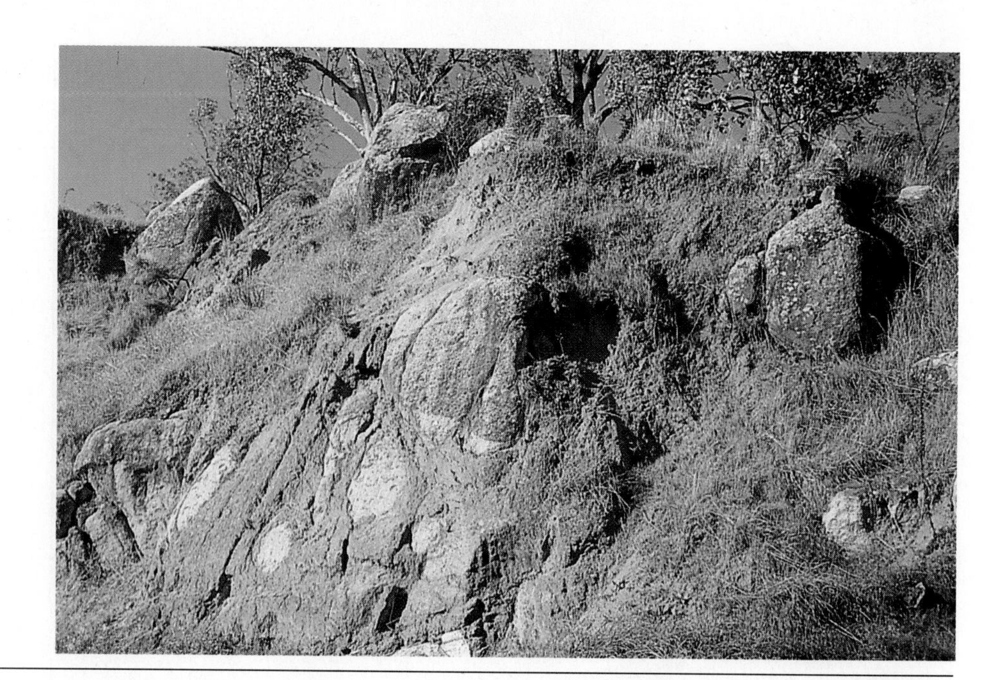

▶ FIGURE 5-17 Spheroidal weathering of granite in Australia.

▶ FIGURE 5-18 (*a*) Residual soil developed on bedrock near Denver, Colorado. (*b*) Transported soil developed on a windblown dust deposit.

(a)

(b)

and deposited elsewhere, such as on a stream's floodplain. Many fertile transported soils of the Mississippi River valley and the Pacific Northwest developed on deposits of windblown dust called *loess* (Figure 5-18b).

The Soil Profile

Soil-forming processes begin at the surface and work downward, so the upper layer of soil is more altered from the parent material than the layers below. Observed in vertical cross section, a soil consists of distinct layers or **soil horizons** that differ from one another in texture, structure, composition, and color (▶ Figure 5-19). Starting from the top, the horizons typical of soils are designated O, A, B, and C, but the boundaries between horizons are transitional rather than sharp.

The O horizon, which is generally only a few centimeters thick, consists of organic matter. The remains of plant materials are clearly recognizable in the upper part of the O horizon, but its lower part consists of humus.

Horizon A, called *top soil,* contains more organic matter than the layers below. It is also characterized by intense biological activity because plant roots, bacteria, fungi, and animals such as worms are abundant. Threadlike soil bacteria give freshly plowed soil its earthy aroma. In soils developed over a long period of time, the A horizon consists mostly of clays and chemically stable minerals such as quartz. Water percolating down through horizon A dissolves the soluble minerals that were originally present and carries them away or downward to lower levels in the soil by a process called **leaching** (Figure 5-19).

Horizon B, or *subsoil,* contains fewer organisms and less organic matter than horizon A. Horizon B is known as the **zone of accumulation,** because soluble minerals leached from horizon A accumulate as irregular masses. If horizon A

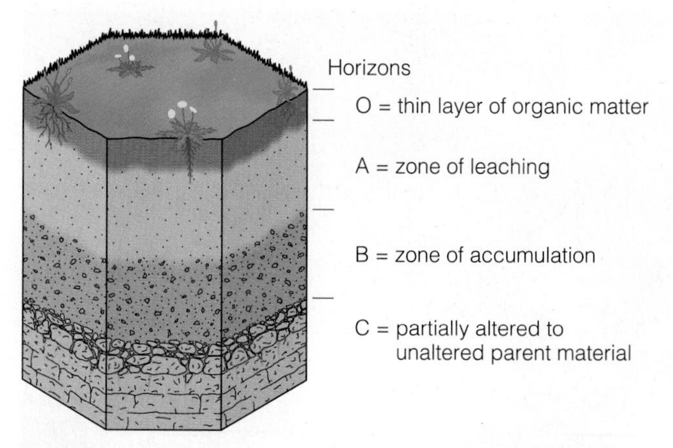

▶ **FIGURE 5-19** The soil horizons in a fully developed or mature soil.

is stripped away by erosion leaving horizon B exposed, plants do not grow as well, and if horizon B is clayey, it is harder when dry and stickier when wet than other soil horizons.

Horizon C, the lowest soil layer, consists of partially altered to unaltered parent material (Figure 5-19). In horizons A and B, the composition and texture of the parent material have been so thoroughly altered that the parent material is no longer recognizable. In contrast, rock fragments and mineral grains of the parent material retain their identity in horizon C. Horizon C contains little organic matter.

 FACTORS CONTROLLING SOIL FORMATION

Climate, parent material, organic activity, relief and slope, and time are the critical factors controlling soil formation (▶ Figure 5-20). Complex interactions among these factors are responsible for soil type, thickness, and fertility.

Climate

It has long been acknowledged that climate is the single most important factor influencing soil type and depth. Intense chemical weathering in the tropics generally yields deep soils from which most of the soluble mineral matter has been removed by leaching. In arctic and desert climates, on the other hand, soils tend to be thin, contain significant quantities of soluble minerals, and are composed mostly of materials derived by mechanical weathering (Figure 5-20).

A very general classification recognizes three major soil types characteristic of different climatic settings. Soils that develop in humid regions such as the eastern United States and much of Canada are **pedalfers,** a name derived from the Greek word *pedon,* meaning soil, and from the chemical

▶ **FIGURE 5-20** Schematic representation showing soil formation as a function of the relationships between climate and vegetation, which alter parent material over time. Soil-forming processes operate most vigorously where precipitation and temperatures are high.

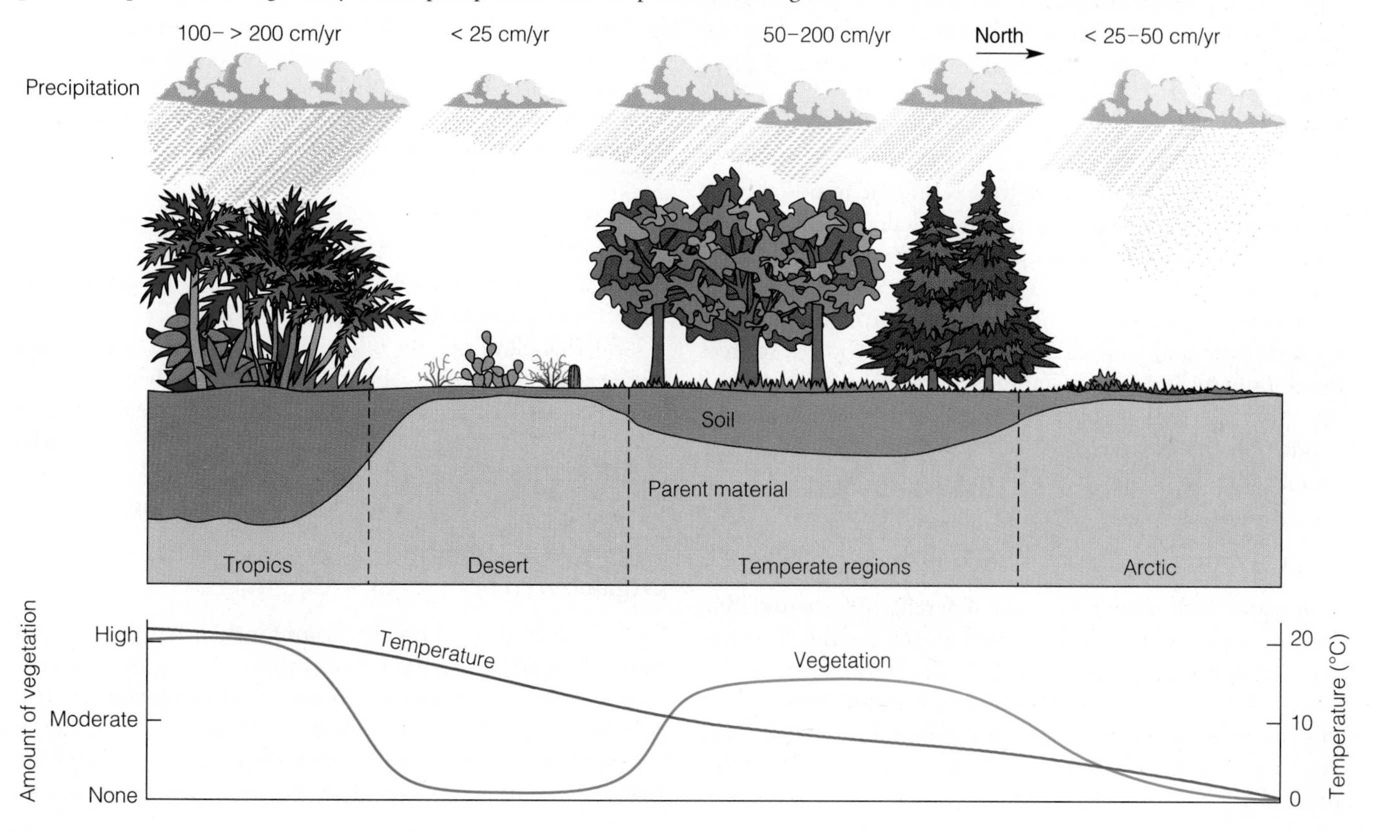

symbols for aluminum (Al) and iron (Fe). Because these soils form where abundant moisture is present, most of the soluble minerals have been leached from horizon A. Although it may be gray, horizon A is generally dark colored because of abundant organic matter, and aluminum-rich clays and iron oxides tend to accumulate in horizon B.

Pedocals are soils characteristic of arid and semiarid regions and are found in much of the western United States, especially the southwest. Pedocal derives its name in part from the first three letters of calcite. Such soils contain less organic matter than pedalfers, so horizon A is generally lighter colored and contains more unstable minerals because of less intense chemical weathering. As soil water evaporates, calcium carbonate leached from above commonly precipitates in horizon B where it forms irregular masses of *caliche*. Precipitation of sodium salts in some desert areas where soil water evaporation is intense yields *alkali soils* that are so alkaline that they cannot support plants.

Laterite is a soil formed in the tropics where chemical weathering is intense and leaching of soluble minerals is complete. Such soils are red, commonly extend to depths of several tens of meters, and are composed largely of aluminum hydroxides, iron oxides, and clay minerals; even quartz, a chemically stable mineral, is generally leached out (▶ Figure 5-21a).

Although laterites support lush vegetation, as in the tropical rainforests, they are not very fertile. The native vegetation is sustained by nutrients derived mostly from the surface layer of organic matter. When these soils are cleared of their native vegetation, the existing surface accumulation of organic matter is rapidly oxidized, and there is little to replace it. Consequently, when societies practicing slash-and-burn agriculture clear these soils, they can raise crops for only a few years at best. Then the soil is depleted of plant nutrients, the clay-rich laterite bakes brick hard in the tropical sun, and the farmers move on to another area where the process is repeated.

One aspect of laterites is of great economic importance. If the parent material is rich in aluminum, aluminum hydroxides may accumulate in horizon B as *bauxite*, the ore of aluminum (Figure 5-21b). Because such intense chemical weathering currently does not occur in North America, the United States and Canada are dependent on foreign sources for aluminum ores. Some aluminum ores do exist in Arkansas, Alabama, and Georgia, which had a tropical climate about 50 million years ago, but currently it is cheaper to import aluminum ore than to mine these deposits.

(a)

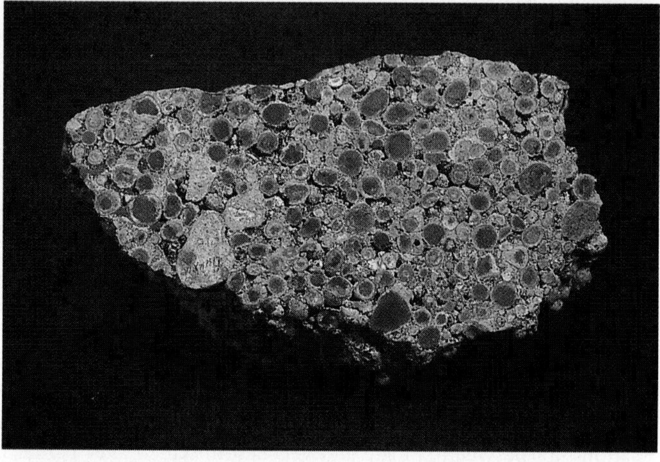

(b)

▶ **FIGURE 5-21** (*a*) Laterite, shown here in Madagascar, is a deep, red soil that forms in response to intense chemical weathering in the tropics. (*b*) Bauxite, the ore of aluminum, forms in horizon B of laterites derived from aluminum-rich parent materials. (Photo courtesy of Sue Monroe.)

Parent Material

The same rock type can yield different soils in different climatic regimes, and in the same climatic regime the same soils can develop on different rock types. Thus, it seems that climate is more important than parent material in determining the type of soil that develops. Nevertheless, rock type does exert some control. For example, the metamorphic rock quartzite will have a thin soil over it because it is chemically stable, whereas an adjacent body of granite will have a much deeper soil (▶ Figure 5-22a).

Soil that develops on basalt will be rich in iron oxides because basalt contains abundant ferromagnesian minerals, but rocks lacking such minerals will not yield an iron oxide–rich soil no matter how thoroughly they are weathered. Also, weathering of a pure quartz sandstone will yield no clay, whereas weathering of clay will yield no sand.

Organic Activity

Soils not only depend on organisms for their fertility, but also provide a suitable habitat for organisms ranging from microscopic, single-celled bacteria to large burrowing animals such as ground squirrels and gophers. Earthworms—as many as one million per acre—ants, sowbugs, termites, centipedes, millipedes, and nematodes, along with various types of

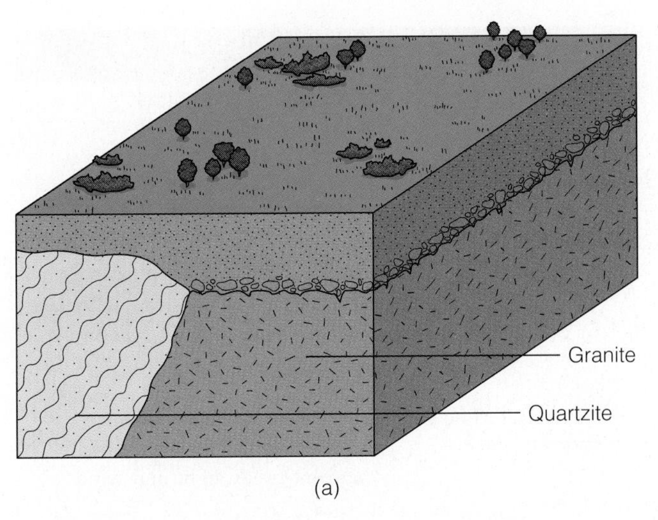

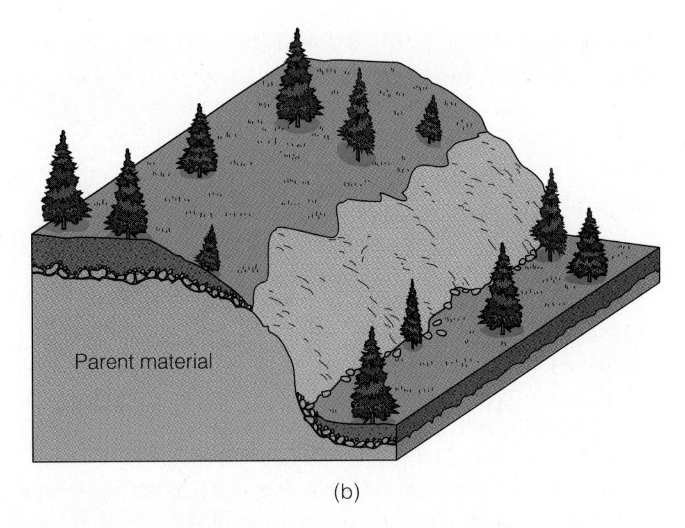

➢ FIGURE 5-22 (*a*) The influence of parent material on soil development. Quartzite is resistant to chemical weathering, whereas granite is altered quickly. (*b*) The effect of slope on soil formation. Where slopes are steep, erosion occurs faster than soil can form.

fungi, algae, and single-celled animals, make their homes in the soil. All of these contribute to the formation of soils and provide humus when they die and are decomposed by bacterial action.

Much humus in soils is provided by grasses or leaf litter that microorganisms decompose to obtain food. In so doing, they break down organic compounds within plants and release nutrients back into the soil. Additionally, organic acids produced by decaying soil organisms are important in further weathering of parent materials and soil particles.

Burrowing animals constantly churn and mix soils, and their burrows provide avenues for gases and water. Soil organisms, especially some types of bacteria, are extremely important in changing atmospheric nitrogen into a form of soil nitrogen suitable for use by plants.

Relief and Slope

Relief is the difference in elevation between high and low points in a region. Because climate changes with elevation, relief affects soil-forming processes largely through elevation. Slope affects soils in two ways. One is simply *slope angle*: the steeper the slope, the less opportunity for soil development because weathered material is eroded faster than soil-forming processes can work (Figure 5-22b). The other slope control is the direction the slope faces. In the Northern Hemisphere, north-facing slopes receive less sunlight than south-facing slopes. If a north-facing slope is steep, it may receive no sunlight at all. Consequently, north-facing slopes have soils with cooler internal temperatures, may support different vegetation, and, if in a cold climate, remain frozen longer.

Time

The degree of alteration of parent material in horizon A is complete because it has been undergoing change for the longest time. The properties of a soil are determined by the factors of climate and organisms altering parent material through time; the longer these processes have operated, the more fully developed the soil will be. If a soil is weathered for extended periods of time, however, its fertility decreases as plant nutrients are leached out, unless new materials are delivered. Agricultural lands adjacent to major streams such as the Nile River in Egypt have their soils replenished during yearly floods. In areas of active tectonism, uplift and erosion provide fresh materials that are transported to adjacent areas where they contribute to soils.

How much time is needed to develop a centimeter of soil or a fully developed soil a meter or so deep? No definitive answer can be given because weathering proceeds at vastly different rates depending on climate and parent material, but an overall average might be about 2.5 cm per century. However, a lava flow a few centuries old in Hawaii may have a well-developed soil on it, whereas a flow the same age in Iceland will have considerably less soil. Given the same climatic conditions, soil will develop faster on unconsolidated sediment than it will on solid bedrock.*

Under optimum conditions, the soil-forming process occurs at a rapid rate in the context of geologic time. From the human perspective, however, soil formation is a slow process; consequently, soil is regarded as a nonrenewable resource.

 SOIL DEGRADATION

Any decrease in soil productivity or loss of soil to erosion is referred to as **soil degradation.** According to recent studies by the World Resources Institute, between 1945 and 1990

*Bedrock is a general term for the rock underlying soil or unconsolidated sediment.

Perspective 5-2

THE DUST BOWL

The stock market crash of 1929 ushered in the Great Depression, a time when millions of people were unemployed and many had no means to provide food and shelter. Urban areas were affected most severely by the depression, but rural areas suffered as well, especially during the great drought of the 1930s. Prior to the 1930s, farmers had enjoyed a degree of success unparalleled in U.S. history. During World War I, the price of wheat soared, and after the war when Europe was recovering, the government subsidized wheat prices. High prices and mechanized farming practices resulted in more and more land being tilled. Even the weather cooperated, and land in the western United States that would otherwise have been marginally productive was plowed. Deep-rooted prairie grasses that held the soil in place were replaced by shallow-rooted wheat.

Beginning in about 1930, drought conditions prevailed throughout the country; only two states—Maine and Vermont—were not drought-stricken. Drought conditions varied from moderate to severe, and the consequences of the drought were particularly severe in the southern Great Plains. Some rain fell, but in amounts insufficient to maintain agricultural production. And since the land, even marginal land, had been tilled, the native vegetation was no longer available to keep the soil from blowing away. And blow away it did—in huge quantities. Nothing stopped the wind from removing large quantities of top soil.

A large region in the southern Great Plains that was particularly hard hit by the drought, dust storms, and soil erosion came to be known as the Dust Bowl. Although its boundaries were not well defined, it included parts of Kansas, Colorado, and New Mexico as well as the panhandles of Oklahoma and Texas (▶ Figure 1); together the Dust Bowl and its less affected fringe area covered more than 400,000 km^2!

Dust storms became common during the 1930s, and some reached phenomenal sizes (▶ Figure 2). One of the largest storms occurred in 1934 and covered more than 3.5 million km^2. It lifted dust nearly 5 km into the air, obscured the sky over large parts of six states, and blew hundreds of millions of tons of soil eastward where it settled on New York City, Washington, D.C., and other eastern cities as well as on ships as far as 480 km out in the Atlantic Ocean. The Soil Conservation

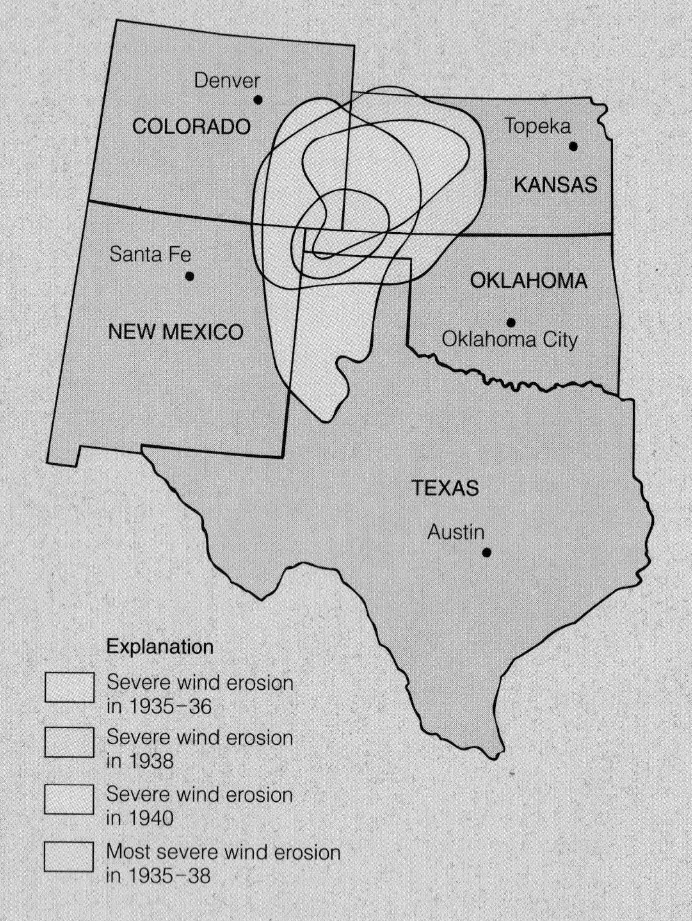

▶ **FIGURE 1** The Dust Bowl of the 1930s. Drought conditions extended far beyond the boundaries shown here, but this area was particularly hard hit by drought and wind erosion.

the soils of 17% of the world's vegetated land were degraded to some extent as a result of human activities (▶ Figure 5-23). In North America, 5.3% of the soil is estimated to be degraded, whereas the figure is much higher for all other continents.

Three types of soil degradation are generally recognized: erosion, chemical deterioration, and physical deterioration. Most soil erosion occurs by the action of wind and water. When the natural vegetation is removed and a soil is pulverized by plowing, the fine particles are easily blown away. The Dust Bowl of the 1930s in the western United States is a poignant reminder of just how effective wind erosion can be (see Perspective 5-2). Falling rain also disrupts soil particles and carries soil with it when it runs off

▶ **FIGURE 2** The huge dust storm of April 14, 1935, also known as Black Sunday, photographed at Hugoton, Kansas.

Service reported that dust storms of regional extent occurred on 140 occasions during 1936 and 1937. Dust was everywhere. It seeped into houses, suffocated wild animals and livestock, and adversely affected human health.

The dust was, of course, the material derived from the tilled lands; in other words, much of the soil in many regions was simply blown away. Blowing dust was not the only problem; sand piled up along fences, drifted against houses and farm machinery, and covered what otherwise might have been productive soils. Agricultural production fell precipitously in the Dust Bowl, farmers could not meet their mortgage payments, and by 1935 tens of thousands were leaving. Many of these people went west to California and became the migrant farm workers immortalized in John Steinbeck's novel *The Grapes of Wrath.*

The Dust Bowl was an economic disaster of great magnitude. Droughts had stricken the southern Great Plains before, and have done so since, but the drought of the 1930s was especially severe. Political and economic factors also contributed to the disaster. Due in part to the artificially inflated wheat prices, many farmers were deeply in debt—mostly because they had purchased farm machinery in order to produce more and benefit from the high prices. Feeling economic pressure because of their huge debts, they tilled marginal land and employed few, if any, soil conservation measures.

If the Dust Bowl has a bright side, it is that the government, farmers, and the public in general no longer take soil for granted or regard it as a substance that needs no nurturing. In addition, a number of soil conservation methods developed then have now become standard practices.

at the surface. This is particularly devastating on steep slopes from which the vegetative cover has been removed by overgrazing, deforestation, or construction. Two types of erosion by water are recognized: sheet erosion and rill erosion.

Sheet erosion is more or less evenly distributed over the surface and removes thin layers of soil. **Rill erosion,** on the other hand, occurs when running water scours small channels. If these channels can be eliminated by plowing, they are called *rills,* but if they are too deep (about 30 cm) to be plowed over, they are gullies (▶ Figure 5-24). Where gullying becomes extensive, croplands can no longer be tilled and must be abandoned.

Wind erosion causes about 28% of all soil degradation, and water erosion causes 56%. Both types of erosion most commonly result from agricultural practices, overgrazing, deforestation, and overexploitation by collecting vegetation for fuel.

Guest Essay *Stephen H. Stow*

ENVIRONMENTAL GEOLOGY: SUSTAINING THE EARTH

We can think of Earth as a spaceship, upon which all humans live. It has limited resources and a limited ability to respond to abuse. Our existence depends on our learning about our home, the Earth, about its behavior, its limits, and about how we, as passengers on this spaceship, can most efficiently live with our environment. Earth science is nothing more—and nothing less—than the study of the Earth and the way humans can exist with it. Geology is a specific field in earth science—and a very important one. Geology involves the study of the solid Earth whereas other earth sciences deal with water (hydrology), the oceans (oceanography), the atmosphere (climatology), and other aspects of the Earth. All these fields are interrelated, however, and cannot really be separated from each other.

Earth science touches almost every aspect of our lives. It encompasses natural disasters—volcanoes, earthquakes, tropical storms, and floods. Such natural events make us acutely aware of the dynamics of the Earth and the need to understand its processes. Of equal importance is our dependence on the Earth's resources; virtually everything we use owes its existence to water, mineral, energy, and soil resources. But most resources are limited, and they are not distributed evenly throughout the world, so shortages often arise, sometimes leading to confrontations between nations. The study of resources, from exploration for valuable minerals and energy sources to management of agricultural lands and conservation of groundwater, is an essential part of the earth sciences.

The impact of the human race on the Earth's environment is another important aspect of the earth sciences. We have all become aware of the ozone hole in the atmosphere and global warming with a resultant increase in sea level that could inundate coastal cities. These situations, though not yet fully understood, may be the result of humans' release of materials into the fragile atmosphere, altering the delicate heat balance that evolved over millions of years. Another problem is groundwater contamination due to unrestricted disposal of waste products over the last several decades. The challenges and costs of assessing and correcting these situations are immense, but must be undertaken. Everyone, not just professional earth scientists, should be aware of the fragility of our planet and the impact that we can have on it.

All earth scientists, including geologists, are much in demand. Historically, the job market for geologists was driven by the petroleum industry, but today there is an unprecedented need for earth scientists to undertake environmental studies. For instance, hydrology studies dealing with waste disposal issues are needed as are studies of how water resources respond to changes in global climates. Deciphering the rock record to identify past fluctuations in climate may help us predict future fluctuations. As populations grow, the proper use of precious land and resources has become an increasingly important issue, and earth scientists are becoming intimately involved in the decision process. For instance, we must understand volcanic and seismic hazards to avert large-scale disasters in urban areas. Energy resources are another crucial area. The demand for petroleum will continue, at least through our lifetimes, as will the need for geologists in that industry.

Although my first job after graduate school involved exploration and geochemistry for a major oil company, my current work deals almost entirely with the application of the earth sciences to environmental studies. My present position involves the study of the cleanup of historical waste disposal sites where the Department of Energy (and its predecessor agencies) disposed of nuclear and chemical wastes from nuclear energy and weapons manufacturing. This cleanup is a massive effort being undertaken throughout the entire United States; it requires sophisticated understanding of Earth processes, such as groundwater flow and contaminant transport, structural and stratigraphic aspects of disposal sites, and computer modeling of data obtained from field and laboratory studies. To function effectively in this area, earth scientists must not only have a sound base in their discipline, but must also be familiar with other sciences, mathematics, and legislation that guides many of the environmental studies. The job is challenging because it requires working with professionals in many scientific and engineering fields and addressing real-world problems.

My interest in the Earth goes back to my childhood days; in high school, I decided to major in geology in college. My interest was aroused by mineral-hunting field trips with the geology club as well as by an excellent chemistry teacher, who encouraged my interests in science. My professional interests are no longer the same as those initial enthusiasms, but that is to be expected because the profession has changed, too. It is gratifying to be applying fundamental knowledge to the solution of issues that confront us daily—issues that absolutely must be solved if our future existence is to be ensured.

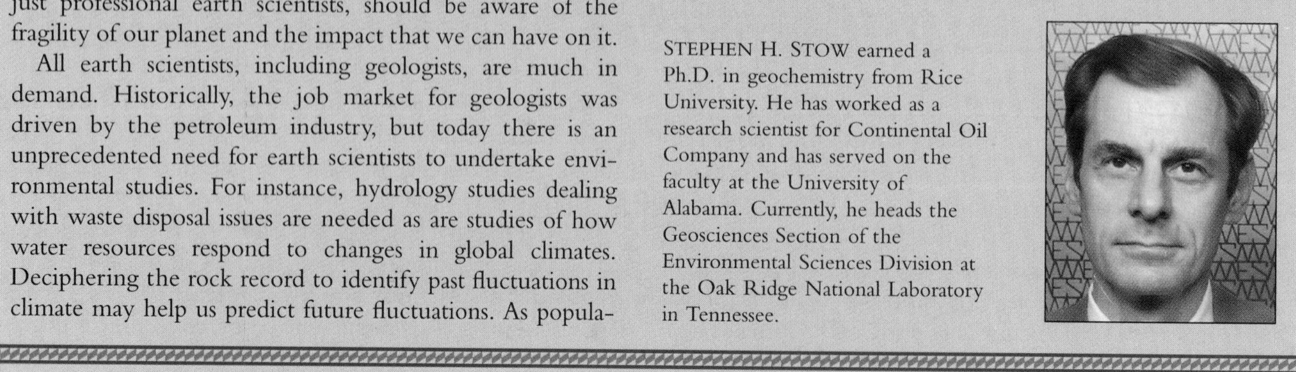

STEPHEN H. STOW earned a Ph.D. in geochemistry from Rice University. He has worked as a research scientist for Continental Oil Company and has served on the faculty at the University of Alabama. Currently, he heads the Geosciences Section of the Environmental Sciences Division at the Oak Ridge National Laboratory in Tennessee.

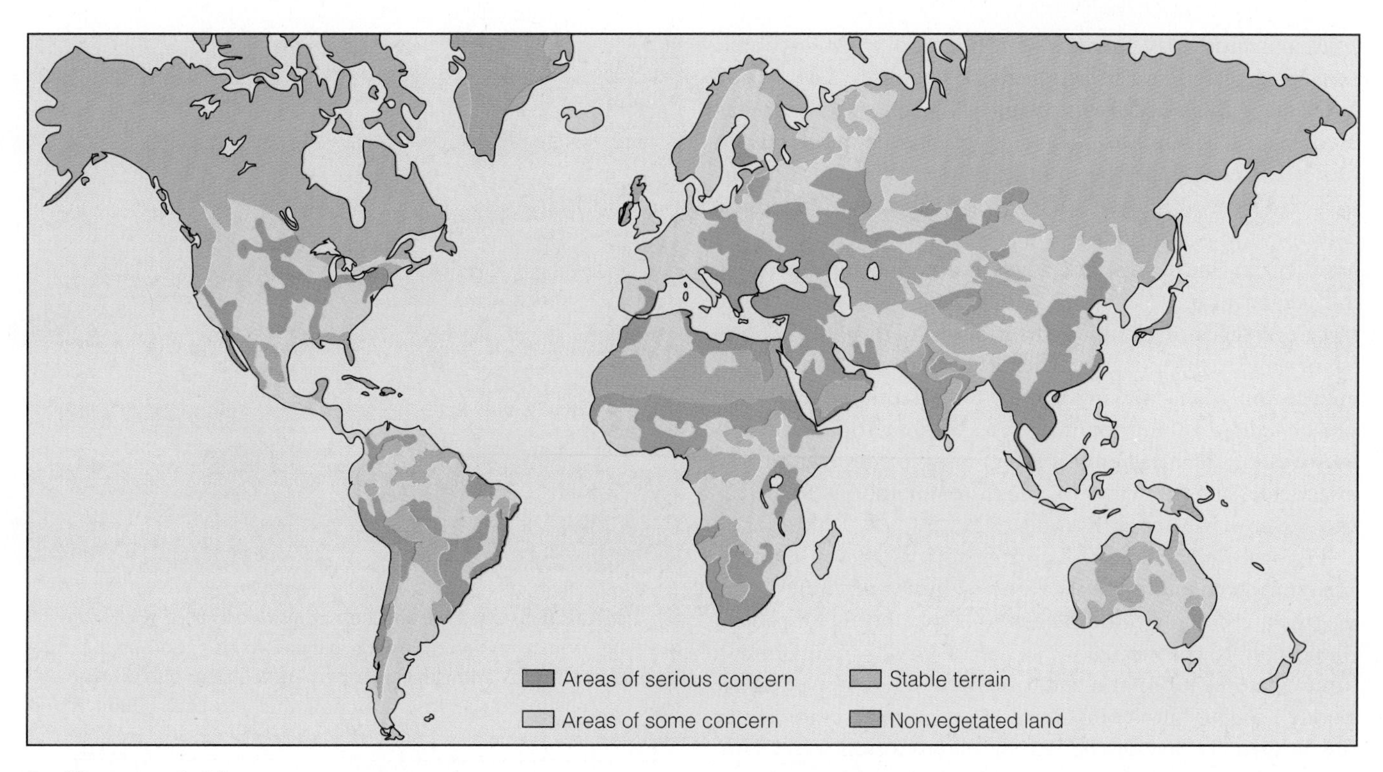

▶ **FIGURE 5-23** Areas of soil degradation.

▶ **FIGURE 5-24** (*a*) Rill erosion in a field during a rainstorm. This rill was later plowed over. (*b*) This small gully is too deep to be plowed.

(a)

(b)

A soil undergoes chemical deterioration when its nutrients are depleted and its productivity decreases. Loss of soil nutrients is most notable in countries where soils are overused in an attempt to maintain agricultural productivity. Other causes include the insufficient use of chemical fertilizers and clearing soils of their natural vegetation. Chemical deterioration of soils occurs on all continents, but it is most prevalent in South America where it accounts for 29% of all soil degradation.

Other types of chemical deterioration are pollution and *salinization,* which occurs when the concentration of salts increases in a soil, making it unfit for agriculture. Pollution can be caused by improper disposal of domestic and industrial wastes, oil and chemical spills, and the concentration of insecticides and pesticides in soils. Soil pollution is a particularly severe problem in Europe.

Physical deterioration of soils results when soil particles are compacted under the weight of heavy machinery and livestock, especially cattle. When soils have been compacted, they are more costly to plow, and plants have a more difficult time emerging from the soil. Furthermore, water does not readily infiltrate into compacted soils, so more runoff occurs; this in turn accelerates the rate of water erosion.

If soil losses to erosion are minimal, soil-forming processes can keep pace, and the soil remains productive. Should the loss rate exceed the formation rate, however, the most productive upper layer of soil, horizon A, is removed, exposing horizon B, which is much less productive. The Soil Conservation Service of the U.S. Department of Agriculture estimates that 25% of the cropland in the United States is eroding faster than soil-forming processes can replace it. Such losses are problems, of course, but there are additional consequences. For one thing, the eroded soil is transported elsewhere, perhaps onto neighboring cropland, onto roads, or into channels. Sediment accumulates in canals and irrigation ditches, and agricultural fertilizers and insecticides are carried into streams and lakes.

In North America, the rich prairie soils of the midwestern United States and the Great Plains of the United States and Canada are suffering significant soil degradation (Figure 5-23). Nevertheless, this degradation, which is characterized as moderate, is less serious than in many other parts of the world where it is considered severe or extreme. These soils remain productive, although their overall productivity has decreased somewhat during the last several decades and more fertilizers are needed to maintain productivity. Other areas of concern are the central valleys of California, an area in Washington State, and some parts of Mississippi and Missouri where water erosion rates are high (Figure 5-23).

Problems experienced during the past, particularly during the 1930s, have stimulated the development of methods to minimize soil erosion on agricultural lands. Crop rotation, contour plowing, and the construction of terraces have all proved helpful (▶ Figure 5-25). So has no-till planting in which the residue from the harvested crop is left on the ground to protect the surface from the ravages of wind and water.

▶ **FIGURE 5-25** One soil conservation practice is contour plowing, which involves plowing parallel to the contours of the land. The furrows and ridges are perpendicular to the direction that water would otherwise flow downhill and thus inhibit erosion.

Weathering and Mineral Resources

In a preceding section, we discussed intense chemical weathering in the tropics and the origin of *bauxite,* the chief ore of aluminum. Such an accumulation of valuable minerals formed by the selective removal of soluble substances is a *residual concentration.* It represents an insoluble residue of chemical weathering. In addition to bauxite, a number of other residual concentrations are economically important; for example, ore deposits of iron, manganese, clays, nickel, phosphate, tin, diamonds, and gold.

Some limestones contain small amounts of iron carbonate minerals. When the limestone is dissolved during chemical weathering, a residual concentration of insoluble iron oxides accumulates. Some of the sedimentary iron deposits (see Chapter 6) of the Lake Superior region were enriched by chemical weathering when the soluble constituents that were originally present were carried away. Residual concentrations of insoluble manganese oxides form in a similar fashion from manganese-rich source rocks.

Most commercial clay deposits were formed by hydrothermal alteration of granitic rocks or by sedimentary processes. However, some have formed in place as residual concentrations. A number of kaolinite deposits in the southern United States were formed by the chemical weathering of feldspars in pegmatites and of clay-bearing limestones and dolostones. Kaolinite is a type of clay mineral used in the manufacture of paper and ceramics.

Gossans, oxidized ores, and *supergene enrichment* of ores are interrelated, and all result from chemical weathering (▶ Figure 5-26). A gossan is a yellow to reddish deposit composed largely of hydrated iron oxides that formed by the

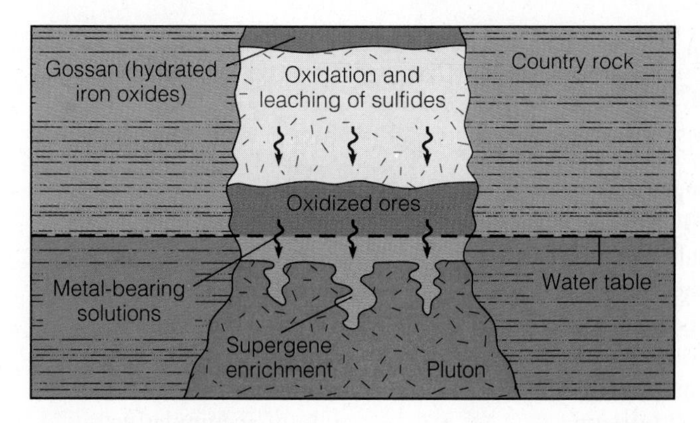

▶ **FIGURE 5-26** A cross section showing a gossan and the origin of oxidized ores and the supergene enrichment of ores.

oxidation and leaching of sulfide minerals such as pyrite (FeS_2). The dissolution of such sulfide minerals forms sulfuric acid, which causes other metallic minerals to dissolve, and these tend to be carried downward toward the groundwater table (Figure 5-26). Oxidized ores form just above the groundwater table as a result of chemical reactions with these descending solutions. Some of the minerals formed in this zone contain copper, zinc, and lead.

Supergene enrichment of ores occurs where metal-bearing solutions penetrate below the water table (Figure 5-26). Such deposits are characterized by the replacement of sulfide minerals of the primary deposit with sulfide minerals introduced by the descending solutions. The iron in iron sulfides may be replaced by other metals such as lead, zinc, nickel, and copper that have a greater affinity for sulfur. Indeed, supergene chalcocite (Cu_2S), an important copper ore, forms as a replacement of primary pyrite (FeS_2) and chalcopyrite ($CuFeS_2$). Notice that both chalcocite and chalcopyrite are copper-bearing minerals, but the former is a richer source of copper than the latter.

Gossans have been used occasionally as sources of iron, but they are far more important as indicators of underlying ore deposits. One of the oldest known underground mines exploited such ores about 3,400 years ago in what is now southern Israel. Supergene enriched ore bodies are generally small but extremely rich sources of various metals. The largest copper mine in the world, at Bingham, Utah, was originally mined for supergene ores, but currently only primary ores are being mined (see Figure 12-27).

CHAPTER SUMMARY

1. Mechanical and chemical weathering are processes whereby parent material is disintegrated and decomposed so that it is more nearly in equilibrium with new physical and chemical conditions. The products of weathering include solid particles, soluble compounds, and ions in solution.
2. The residue of weathering can be further modified to form soil, or it can be deposited as sediment, which might become sedimentary rock.
3. Mechanical weathering includes such processes as frost action, pressure release, thermal expansion and contraction, salt crystal growth, and the activities of organisms. Particles liberated by mechanical weathering retain the chemical composition of the parent material.
4. Solution, oxidation, and hydrolysis are chemical weathering processes; they result in a chemical change of the weathered products. Clay minerals, various ions in solution, and soluble compounds are formed during chemical weathering.
5. Chemical weathering proceeds most rapidly in hot, wet environments, but it occurs in all areas, except perhaps where water is permanently frozen.
6. Mechanical weathering aids chemical weathering by breaking parent material into smaller pieces, thereby exposing more surface area.
7. Mechanical and chemical weathering produce regolith, some of which is soil if it consists of solids, air, water, and humus and supports plant growth.
8. Soils are characterized by horizons that are designated, in descending order, as O, A, B, and C; soil horizons differ from one another in texture, structure, composition, and color.
9. The factors controlling soil formation include climate, parent material, organic activity, relief and slope, and time.
10. Soils called pedalfers develop in humid regions such as the eastern United States and much of Canada. Arid and semiarid regions soils are pedocals, many of which contain irregular masses of caliche in horizon B.
11. Laterite is a soil resulting from intense chemical weathering as in the tropics. Such soils are deep and red and are sources of aluminum ores if derived from aluminum-rich parent material.
12. Soil degradation caused by erosion (mostly sheet and rill erosion), chemical deterioration, and physical deterioration of the soil is a problem in some areas. Human activities such as construction, agriculture, and deforestation can accelerate soil degradation.
13. Intense chemical weathering is responsible for the origin of residual concentrations, many of which contain valuable minerals such as iron, lead, copper, and clay.
14. Gossans, oxidized ores, and supergene enrichment of ores all result from chemical weathering.

IMPORTANT TERMS

chemical weathering
differential weathering
erosion
exfoliation
exfoliation dome
frost action
frost heaving
frost wedging
humus
hydrolysis
laterite
leaching
mechanical weathering
oxidation
parent material
pedalfer
pedocal
pressure release
regolith
rill erosion
salt crystal growth
sheet erosion
sheet joint
soil
soil degradation
soil horizon
solution
spheroidal weathering
talus
thermal expansion and contraction
transport
weathering
zone of accumulation

REVIEW QUESTIONS

1. The type of soil typical of arid and semiarid regions is:
   a. ____ laterite;
   b. ____ pedocal;
   c. ____ gossan;
   d. ____ bauxite;
   e. ____ pedalfer.
2. Which mechanical weathering process forms exfoliation domes?
   a. ____ heating and cooling;
   b. ____ expansion and contraction;
   c. ____ the activities of organisms;
   d. ____ oxidation and reduction;
   e. ____ pressure release.
3. Limestone, which is composed of the mineral calcite ($CaCO_3$), is nearly insoluble in pure water but dissolves rapidly if ________ is present.
   a. ____ carbonic acid;
   b. ____ silicon dioxide;
   c. ____ calcium sulfate;
   d. ____ residual manganese;
   e. ____ clay.
4. Talus is an accumulation of:
   a. ____ calcium carbonate in horizon B of pedocals;
   b. ____ angular rock fragments at the base of a slope;
   c. ____ valuable minerals formed by selective removal of soluble substances;
   d. ____ debris produced mostly by the activities of organisms;
   e. ____ soil produced by intense weathering in the tropics.
5. When the ions in a substance become dissociated, the substance has been:
   a. ____ weathered mechanically;
   b. ____ altered to clay:
   c. ____ dissolved;
   d. ____ oxidized;
   e. ____ converted to soil.
6. The process whereby hydrogen and hydroxyl ions of water replace ions in minerals is:
   a. ____ supergene enrichment;
   b. ____ oxidation;
   c. ____ laterization;
   d. ____ hydrolysis;
   e. ____ carbonization.
7. Which of the minerals in Bowen's reaction series is most stable chemically?
   a. ____ calcium plagioclase;
   b. ____ quartz;
   c. ____ pyroxene;
   d. ____ biotite;
   e. ____ olivine.
8. Granite weathers more rapidly than quartzite because it contains abundant:
   a. ____ feldspars;
   b. ____ quartz;
   c. ____ ferromagnesian minerals;
   d. ____ carbonate minerals;
   e. ____ caliche.
9. Most of the Earth's land surface is covered by soil and unconsolidated rock material called:
   a. ____ regolith;
   b. ____ laterite;
   c. ____ humus;
   d. ____ parent material;
   e. ____ talus.
10. Horizon B of a soil is also known as the:
    a. ____ top soil;
    b. ____ humus layer;
    c. ____ alkali zone;
    d. ____ zone of accumulation;
    e. ____ organic-rich layer.
11. The chief ore of aluminum is:
    a. ____ caliche;
    b. ____ pedalfer;
    c. ____ subsoil;
    d. ____ gossan;
    e. ____ bauxite.
12. The removal of thin layers of soil by water over a more or less continuous surface is:
    a. ____ gullying;
    b. ____ sheet erosion;
    c. ____ weathering;
    d. ____ leaching;
    e. ____ exfoliation.
13. Oxidation and leaching of sulfide minerals yield a yellow to red deposit of hydrated iron oxides known as a(n):
    a. ____ residual deposit;
    b. ____ exfoliation dome;
    c. ____ gossan;
    d. ____ sheet joint;
    e. ____ clay deposit.
14. How does mechanical weathering differ from and contribute to chemical weathering?
15. What is differential weathering, and why does it occur?
16. Explain how sheet joints and exfoliation domes originate.
17. Describe the process whereby soluble minerals such as halite (NaCl) are dissolved.
18. Why are most minerals not very soluble in pure water?

19. What is an acid solution, and why are acid solutions important in chemical weathering?
20. What role do hydrogen ions play in the hydrolysis process?
21. Explain why particle size is an important factor in chemical weathering.
22. Draw a soil profile and list the characteristics of each soil horizon.
23. What is the significance of climate and parent material in the development of soil?
24. How do organisms contribute to soil formation?
25. Compare pedalfer, pedocal, and laterite.
26. What are residual concentrations, and how do they form?

Points to Ponder

1. Consider the following: A soil is 1.5 m thick, new soil forms at the rate of 2.5 cm per century, and the erosion rate is 4 mm per year. How much soil will be left after 100 years?
2. What kinds of weathering would you expect to occur on the Moon? On Venus? Explain your answer.
3. What would be the effects of chemical weathering on basalt, granite, quartzite, and limestone in a humid climate? How would these same rocks respond to chemical weathering in an arid climate?
4. How do human practices contribute to soil degradation? What can be done to minimize the impact of such practices on soils?

Additional Readings

Bear, F. E. 1986. *Earth: The stuff of life.* 2d revised ed. Norman, Okla.: University of Oklahoma Press.

Birkeland, P. W. 1984. *Soils and geomorphology.* New York: Oxford University Press.

Buol, S. W., F. D. Hole, and R. J. McCracken. 1980. *Soil genesis and classification.* Ames, Iowa: Iowa State University Press.

Carroll, D. 1970. *Rock weathering.* New York: Plenum Press.

Coughlin, R. C. 1984. *State and local regulations for reducing agricultural erosion.* American Planning Association, Planning Advisory Service Report No. 386.

Courtney, F. M., and S. T. Trudgill. 1984. *The soil: An introduction to soil study.* 2d ed. London: Arnold.

Daniels, R. B. and R. D. Hammer. 1992. *Soil Geomorphology.* New York: John Wiley & Sons, Inc.

Gibbons, B. 1984. Do we treat our soil like dirt? *National Geographic* 166, no. 3:350–89.

Loughnan, F. C. 1969. *Chemical weathering of the silicate minerals.* New York: Elsevier.

Ollier, C. 1969. *Weathering.* New York: Elsevier.

Parfit, M. 1989. The dust bowl. *Smithsonian* 20, no. 3:44–54, 56–57.

World Resources 1992–93. 1992. A Report by the World Resources Institute. New York: Oxford University Press.

Chapter 6

SEDIMENT AND SEDIMENTARY ROCKS

OUTLINE

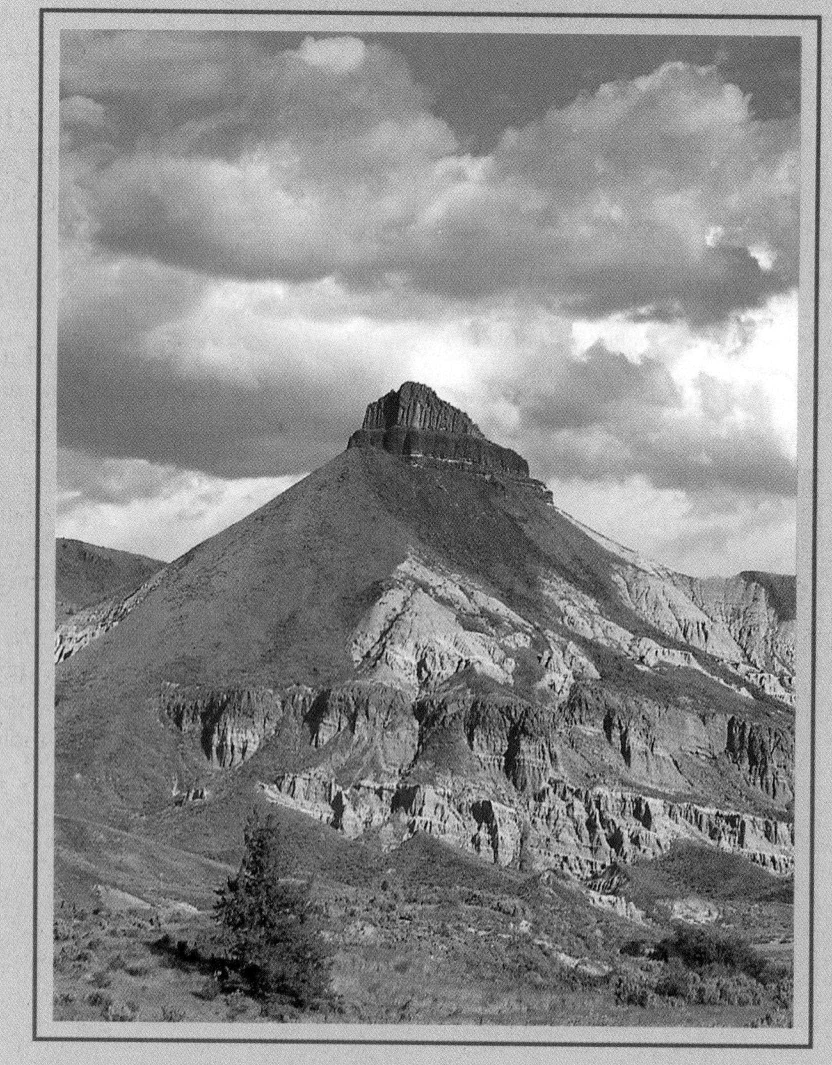

Sedimentary rocks exposed in the Sheep Rock area of John Day Fossil Beds National Monument, Oregon. This small hill is capped by the remnants of a lava flow.

Prologue

About 50 million years ago, two large lakes existed in what are now parts of Wyoming, Utah, and Colorado. Sand, mud, and dissolved minerals were carried into these lakes where they accumulated as layers of sediment that were subsequently converted into sedimentary rock. These sedimentary rocks, called the Green River Formation, contain the fossilized remains of millions of fish, plants, and insects and are a potential source of large quantities of oil, combustible gases, and other substances.

Thousands of fossilized fish skeletons are found on single surfaces within the Green River Formation, indicating that mass mortality must have occurred repeatedly (▶ Figure 6-1). The cause of these events is not known with certainty, but some geologists have suggested that blooms of blue-green algae produced toxic substances that killed the fish. Others propose that rapidly changing water temperatures or excessive salinity at times of increased evaporation was responsible. Whatever the cause, the fish died by the thousands and settled to the lake bottom where their decomposition was inhibited because the water contained little or no oxygen. One area of the formation in Wyoming where fossil plants are particularly abundant has been designated as Fossil Butte National Monument.

The Green River Formation is also well known for its huge deposits of oil shale (▶ Figure 6-2). Oil shale consists of small clay particles and an organic substance known as *kerogen.* When the appropriate processes are used, liquid oil and combustible gases can be extracted from the kerogen of oil shale. To be considered a true oil shale, however, the rock must yield a minimum of 10 gallons of oil per ton of rock. The use of oil shale as a source of fuel is not new, nor is oil shale restricted to the Green River Formation. During the Middle Ages, people in Europe used oil shale as solid fuel for domestic purposes, and during the 1850s, small oil shale industries existed in the eastern United States; the latter were discontinued when drilling and pumping of oil began in 1859. Oil shales occur on all continents, but the Green River Formation contains the most extensive deposits and has the potential to yield huge quantities of oil.

To produce oil from oil shale, the rock is heated to nearly 500°C in the absence of oxygen, and hydrocarbons are driven off as gases and recovered by condensation. During this process, 25 to 75% of the organic matter of oil shale can be converted to oil and combustible gases. The Green River Formation oil shales yield from 10 to 140 gallons of oil per ton of rock processed, and the total amount of oil recoverable with present processes is estimated at 80 billion barrels. Currently no oil is produced from oil shale in the United States because the process is more expensive than conventional drilling and pumping. Nevertheless, the Green River oil shale constitutes one of the largest untapped sources of oil in the world. If more effective processes are developed, it could eventually yield even more than the currently estimated 80 billion barrels.

One should realize, however, that at the current and expected consumption rates of oil in the United States, oil production from oil shale will not solve all U.S. energy needs. Furthermore, the large-scale mining that would be necessary would have considerable environmental impact. What would be done with the billions of tons of processed rock? Can such large-scale mining be conducted with

▶ **FIGURE 6-1** Fossil fish from the Green River Formation of Wyoming. (Photo courtesy of Sue Monroe.)

▶ **FIGURE 6-2** Layers of oil shale of the Green River Formation are exposed along these hillsides.

minimal disruption of wildlife habitats and groundwater systems? Where will the huge volumes of water necessary for processing come from—especially in an area where water is already in short supply?

These and other questions are currently being considered by scientists and industry. Perhaps at some future time, the Green River Formation will provide some of the energy used in the United States.

 INTRODUCTION

Mechanical and chemical weathering disintegrates and decomposes rocks, yielding the raw materials for both soils and **sedimentary rocks.** All **sediment** is derived from preexisting rocks and may be *detrital*, consisting of rock fragments and mineral grains liberated during mechanical weathering, or it may be *chemical*, consisting of minerals formed of the substances that were dissolved during chemical weathering. Once sediment has been derived from parent material, it is commonly eroded and transported to another location where it is deposited as an aggregate of loose slides, such as sand on a beach (▶ Figure 6-3).

Most sedimentary rocks formed from sediment that was transformed into solid rock, but a few sedimentary rocks skipped the unconsolidated sediment stage. Coral reefs, for example, form as solids when the reef organisms extract dissolved mineral matter from seawater for their skeletons. However, if a reef is broken apart during a storm, the solid pieces of reef material deposited on the sea floor are sediment.

One important criterion for classifying sedimentary particles is their size (◉ Table 6-1). *Gravel* refers to any sedimentary particle measuring more than 2 mm, whereas *sand* is any particle, regardless of composition, that measures 1/16 to 2 mm. Gravel- and sand-sized particles are large enough to be observed with the unaided eye or with low-power magnification, but silt- and clay-sized particles are too small to be observed except with very high magnification. Gravel generally consists of rock fragments, whereas sand, silt, and clay particles are mostly individual mineral grains. We should note, however, that *clay* has two meanings: in textural terms, clay refers to sedimentary grains less than 1/256 mm in size, and in compositional terms, clay refers to certain types of sheet silicate minerals (see Figure 2-10). Most clay-sized particles in sedimentary rocks are, in fact, clay minerals.

 SEDIMENT TRANSPORT AND DEPOSITION

Detrital sediment can be transported by any geologic agent possessing enough energy to move particles of a given size. Glaciers are very effective agents of transport and can move particles of any size. Wind, on the other hand, can transport only sand-sized and smaller sediment. Waves and marine currents also transport sediment, but by far the most effective

(a)

(b)

▶ FIGURE 6-3 Examples of sediment. (*a*) Sand and gravel on a Lake Michigan beach. (*b*) Broken shells of marine organisms on a Florida beach.

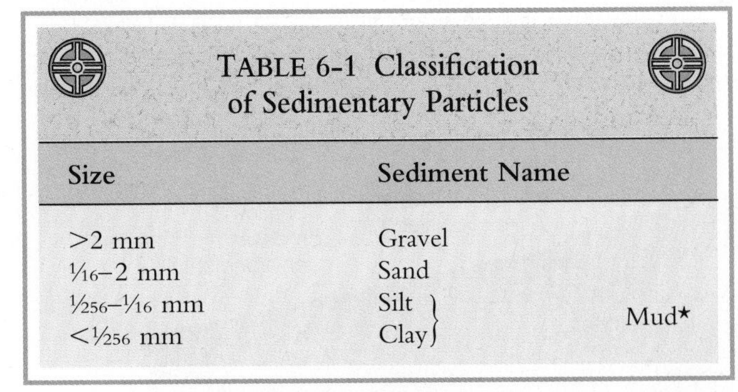

TABLE 6-1 Classification of Sedimentary Particles

Size	Sediment Name	
>2 mm	Gravel	
$1/16$–2 mm	Sand	
$1/256$–$1/16$ mm	Silt	Mud*
<$1/256$ mm	Clay	

*Mixtures of silt and clay are generally referred to as mud.

way to erode sediment from the weathering site and transport it elsewhere is by running water.

During sediment transport, *abrasion* reduces the size of particles, and the sharp corners and edges are worn smooth as gravel and sand particles collide with one another and become **rounded** (⊳ Figure 6-4a). Transport also results in **sorting,** which refers to the size distribution in an aggregate of sediment. If all the particles are about the same size, the sediment is characterized as well-sorted, but if a wide range of grain sizes occur, it is poorly sorted (Figure 6-4b). Both rounding and sorting are important properties used to determine the origin of sedimentary rocks; they are discussed more fully in a later section.

Sediment may be transported a considerable distance from its source area, but eventually it is deposited. Some of the sand and mud being deposited at the mouth of the Mississippi River at the present time came from such distant places as Ohio, Minnesota, and Wyoming. Any geographic area in which sediment is deposited is a **depositional environment.**

Although no completely satisfactory classification of depositional environments exists, geologists generally recognize three major depositional settings: continental, transitional, and marine (⊳ Figure 6-5). Major continental depositional environments include stream systems, deserts, lakes, and areas affected by glaciation. Environments, such as deltas and beaches, are transitional from continental to marine. Delta deposits, for example, accumulate where a river enters the sea, but the deposits are modified by marine processes such as waves and tides. Marine environments, such as the continental shelf, lie seaward of transitional environments (Figure 6-5).

Lithification: Sediment to Sedimentary Rock

At present, calcium carbonate mud (chemical sediment) is accumulating in the shallow waters of Florida Bay, and detrital sand is being deposited in river channels, on beaches, and in sand dunes. Such deposits of sediment may be compacted and/or cemented and thereby converted into sedimentary rock; the process of transforming sediment into sedimentary rock is **lithification.**

(a)

(b)

⊳ FIGURE 6-4 Rounding and sorting of sedimentary particles. (*a*) A deposit consisting of well-sorted and well-rounded gravel. (*b*) Poorly sorted, angular gravel. (Photos courtesy of R. V. Dietrich.)

When sediment is deposited, it consists of solid particles and *pore spaces,* which are the voids between particles. The amount of pore space varies depending on the depositional process, the size of the sediment grains, and sorting. When sediment is buried, **compaction,** resulting from the pressure exerted by the weight of overlying sediments, reduces the amount of pore space, and thus the volume of the deposit (⊳ Figure 6-6). When deposits of mud, which can have as

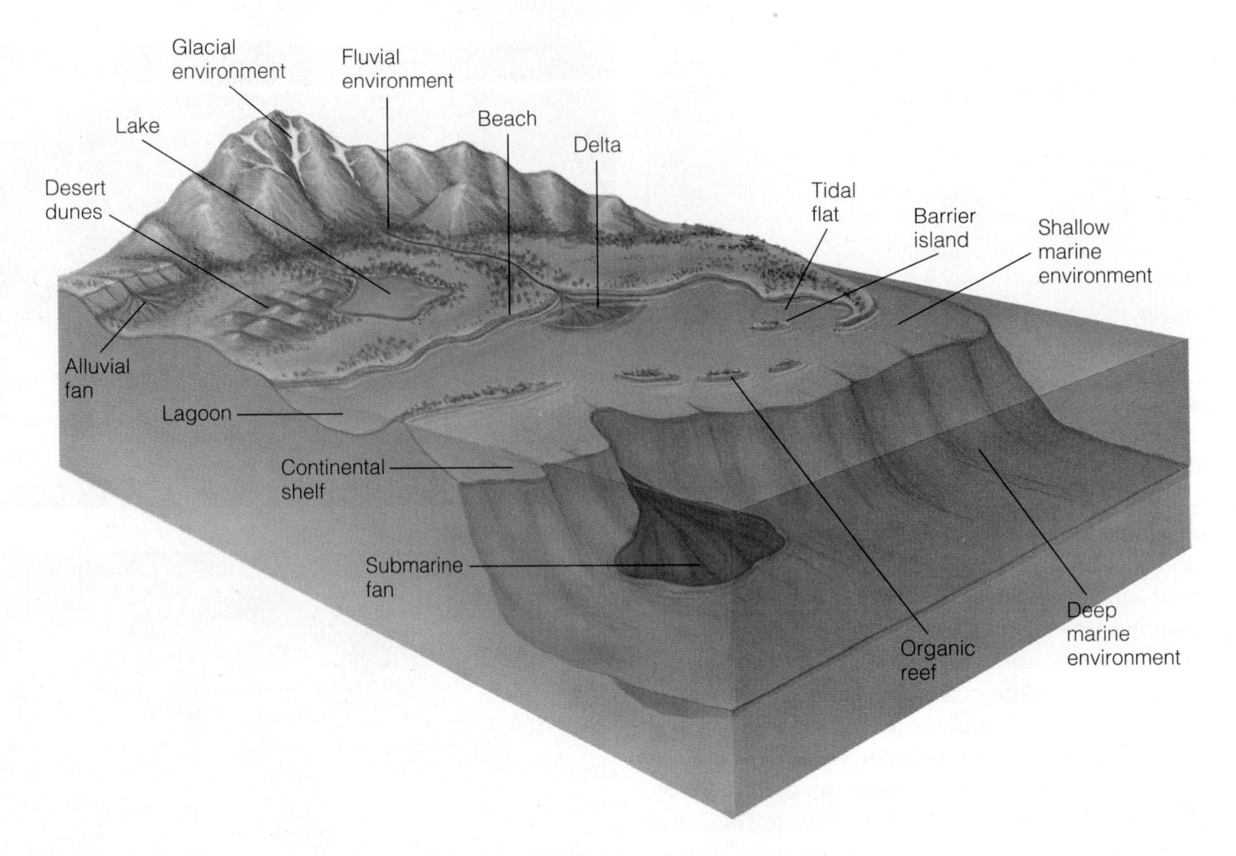

▶ **FIGURE 6-5** Major depositional environments. The environments along the shoreline are transitional from continental to marine. The shallow marine environment corresponds to the continental shelf and can be the site of either detrital or carbonate deposition.

▶ **FIGURE 6-6** Lithification of detrital sediments by compaction and cementation to form sedimentary rocks.

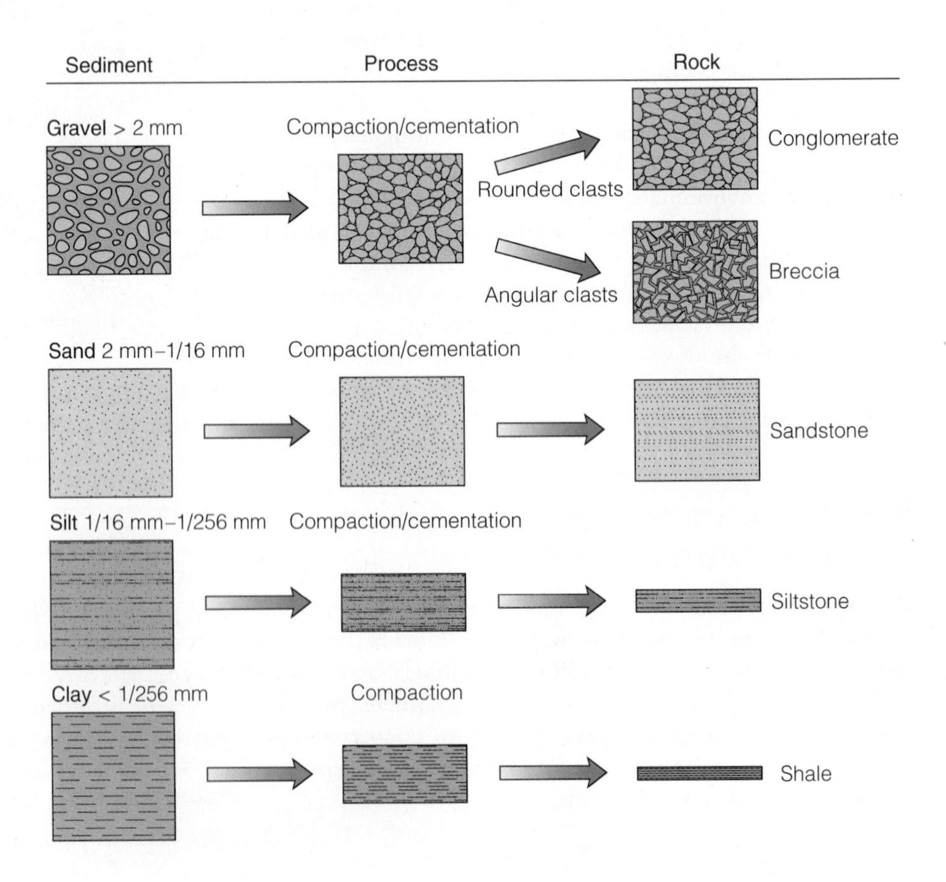

much as 80% water-filled pore space, are buried and compacted, water is squeezed out, and the volume can be reduced by up to 40%. Sand sometimes has up to 50% pore space, although it is generally somewhat less, and it, too, can be compacted so that the sand grains fit more tightly together.

Compaction alone is generally sufficient for lithification of mud, but for sand and gravel deposits **cementation** is necessary to convert the sediment into sedimentary rock (Figure 6-6). Recall from Chapter 5 that calcium carbonate ($CaCO_3$) readily dissolves in water containing a small amount of carbonic acid, and that chemical weathering of feldspars and other silicate minerals yields silica (SiO_2) in solution. These dissolved compounds may be precipitated in the pore spaces of sediments, where they act as a cement that effectively binds the sediment together (Figure 6-6).

Calcium carbonate and silica are the most common cements in sedimentary rocks, but iron oxides and hydroxides, such as hematite (Fe_2O_3) and limonite [FeO(OH)], respectively, also form a chemical cement in some rocks. Much of the iron oxide cement is derived from the oxidation of iron in ferromagnesian minerals present in the original deposit, although some is carried in by circulating groundwater. The yellow, brown, and red sedimentary rocks exposed in the southwestern United States are colored by small amounts of iron oxide or hydroxide cement (⯈ Figure 6-7).

⯈ **FIGURE 6-7** These sedimentary rocks at Jemez Pueblo, New Mexico are red because they contain iron oxide cement. (Photo courtesy of Sue Monroe.)

 SEDIMENTARY ROCKS

Even though about 95% of the Earth's crust is composed of igneous and metamorphic rocks, sedimentary rocks are the most common at or near the surface. About 75% of the surface exposures on continents consist of sediments or sedimentary rocks, and they cover most of the sea floor. Sedimentary rocks are generally classified as *detrital* or *chemical* (◉ Table 6-2).

Detrital Sedimentary Rocks

Detrital sedimentary rocks consist of *detritus,* the solid particles of preexisting rocks. Such rocks have a *clastic texture,* meaning that they are composed of fragments or particles also known as *clasts* (⯈ Figure 6-8a). Several varieties of detrital sedimentary rocks are recognized, each of which is characterized by the size of its constituent particles (Table 6-2).

Conglomerate and Sedimentary Breccia. Both *conglomerate* and *sedimentary breccia* consist of gravel-sized particles (Table 6-2; ⯈ Figure 6-9). Usually, the particles measure a few millimeters to a few centimeters, but boulders several meters in diameter are sometimes present. The only difference between conglomerate and sedimentary breccia is the shape of the gravel particles; conglomerate consists of rounded gravel, whereas sedimentary breccia is composed of angular gravel called *rubble.*

Conglomerate is a fairly common rock type, but sedimentary breccia is rather rare. The reason is that gravel-sized particles become rounded very quickly during transport. So if a sedimentary breccia is encountered, one can conclude that the rubble that composes it was not transported very far. High-energy transport agents such as rapidly flowing streams and waves are needed to transport gravel, so gravel tends to be deposited in high-energy environments such as stream channels and beaches (Figure 6-3a).

Sandstone. The term *sand* is simply a size designation, so *sandstone* may be composed of grains of any type of mineral or rock fragment. Most sandstones, however, consist primarily of the mineral quartz (⯈ Figure 6-10a) with small amounts of a number of other minerals. Geologists recognize several types of sandstones, each characterized by its composition. *Quartz sandstone* is the most common and, as its name suggests, is composed mostly of quartz. *Arkose,* which contains more than 25% feldspars, is also a fairly common variety of sandstone (Table 6-2).

It may seem odd that quartz is so common in sandstones since feldspars are so much more abundant in

TABLE 6-2 Classification of Sedimentary Rocks

Detrital Sedimentary Rocks			
Sediment Name and Size	*Description*	*Rock Name*	
Gravel (>2 mm)	Rounded gravel particles Angular gravel particles	Conglomerate Sedimentary breccia	
Sand (1/16-2mm)	Mostly quartz sand Quartz with >25% feldspar	Quartz sandstone Arkose	
Mud (<1/16 mm)	Mostly silt Silt and clay Mostly clay	Siltstone Mudstone* Claystone*	Mudrocks
Chemical Sedimentary Rocks			
Texture	*Composition*	*Rock Name*	
Crystalline Crystalline	Calcite ($CaCO_3$) Dolomite [$CaMg(CO_3)_2$]	Limestone Dolostone	Carbonates
Crystalline Crystalline	Gypsum ($CaSO_4 \cdot 2H_2O$) Halite (NaCl)	Rock gypsum Rock salt	Evaporites
Biochemical Sedimentary Rocks			
Texture	*Composition*	*Rock Name*	
Clastic	Calcium carbonate ($CaCO_3$) shells	Limestone (various types such as chalk and coquina)	
Usually crystalline	Altered microscopic shells of silicon dioxide (SiO_2)	Chert	
--------	Mostly carbon from altered plant remains	Coal	

*Mudrocks possessing the property of fissility, meaning they break along closely spaced planes, are commonly called *shale*.

➤ **FIGURE 6-8** (*a*) Photomicrograph of a sandstone showing a clastic texture consisting of fragments of minerals, mostly quartz averaging about 0.5 mm in diameter. (*b*) Photomicrograph of the crystalline texture of a limestone showing a mosaic of calcite crystals, which measure about 1 mm across.

(a)

(b)

(a)

(b)

➢ **FIGURE 6-9** Detrital sedimentary rocks: (*a*) conglomerate; and (*b*) sedimentary breccia. (Photos courtesy of Sue Monroe.)

source rocks.* However, the chance that any specific type of mineral will end up in a sedimentary rock depends on its availability, mechanical durability, and chemical stability. Both quartz and feldspars are abundant, but quartz is hard, lacks cleavage, and is chemically stable, whereas feldspars have two directions of cleavage and are readily chemically weathered to clays, soluble compounds, and ions in solution (see Chapter 5).

The only other particles of much consequence in sandstones are the micas (muscovite and biotite) and fragments of chert, a rock composed of microscopic crystals of quartz. Except for biotite, the ferromagnesian silicates are uncommon in sandstones because they are chemically unstable. One of the most common accessory minerals is the iron oxide magnetite.

Mudrocks. The *mudrocks* include all detrital sedimentary rocks composed of silt- and clay-sized particles (Table 6-2; Figure 6-10b). Among the mudrocks we can differentiate between *siltstone, mudstone,* and *claystone.* Siltstone, as the name implies, is composed of silt-sized particles; mudstone contains a mixture of silt- and clay-sized particles; and claystone is composed mostly of clay (Table 6-2). Some mudstones and claystones are designated as *shale* if they are fissile, which means they break along closely spaced parallel planes (Figure 6-10b).

The mudrocks comprise about 40% of all detrital sedimentary rocks, making them the most common of

*The Earth's crust contains an estimated 39% plagioclase feldspar, 12% potassium feldspar, and 12% quartz.

➢ **FIGURE 6-10** Detrital sedimentary rocks: (*a*) sandstone; and (*b*) the mudrock shale. (Photos courtesy of Sue Monroe.)

(a)

(b)

Guest Essay *Susan M. Landon*

EXPLORING FOR OIL AND NATURAL GAS

I am an independent petroleum geologist. I specialize in applying geological principles to frontier areas—places where little or no exploration has occurred and few or no hydrocarbons have been discovered. It is very much like solving a mystery. The earth provides a variety of clues—rock type, organic content, stratigraphic relationships, structure, and the like—that geologists must piece together to determine the potential for the presence of hydrocarbons.

An example of an exploration frontier is the Precambrian Midcontinent Rift located in the north central portion of the United States. Some rifts, like the Gulf of Suez and the North Sea, are characterized by significant hydrocarbon reserves, and the presence of an unexplored rift basin in the center of North America is intriguing. Rocks deposited in this rift basin are exposed along the shores of Lake Superior where they serve as the host for copper ores. One of the mines in the Upper Peninsula of Michigan, the White Pine Mine, has historically been plagued by oil bleeding out of fractures in the shale. For many years, this had been documented as academically interesting because the rocks are much older than those that typically have been associated with hydrocarbon production. Oil and natural gas are generated from organic material preserved in sediment that is subjected to increased temperature through time. However, the sediments associated with the one-billion-year-old rift had a very limited source of organisms (algae, fungi, and bacteria) to contribute to the organic content.

Field and laboratory work documented that the copper-bearing shale at the White Pine Mine contained adequate organic material to be the source of the oil. The thermal history of the basin was modeled to determine the timing of hydrocarbon generation. If hydrocarbons had been generated prior to deposition of an effective seal and formation of a trap, the hydrocarbons would have leaked out naturally into the atmosphere.

Further work identified sandstones with enough porosity to serve as reservoirs for hydrocarbons. Analogy with other hydrocarbon productive rifts gave the exploration team models for trap types. Seismic data were acquired and interpreted to identify specific traps. Coordination with geophysicists and engineers provided the final data necessary to generate a specific prospect. We then had to convince management that this prospect had high enough potential to contain hydrocarbon reserves to offset the significant risks and costs. An economic evaluation was conducted to determine the worth of the project given a probability of success. In this case, management agreed that the risk was offset by the potential for a very large accumulation of hydrocarbons, and a well was authorized. Amoco drilled a 5,441 m well in Iowa to test the prospect at a cost of nearly $5 million. The well was dry (economically unsuccessful), but the geologic information obtained as a result of drilling the well will be used to continue to define prospective drilling sites in the Midcontinent Rift.

My interest in geology began very early as a result of collecting rocks and growing up in an oil field in the Midwest. I completed my undergraduate work at a small liberal arts college and earned a master's degree from a larger state university. I believe that a well-rounded education provided me with a sound geological background and communication skills that have contributed to my successful career.

My career in the petroleum industry began with Amoco Production Company, and after 15 years, I made the decision to leave the company to work independently. For several years, I consulted for a variety of companies, assisting them in exploration projects. I am now involved in a partnership with several other exploration geologists and geophysicists, and we are actively exploring for oil and natural gas in the United States. We develop ideas, defining areas that we believe may be prospective, and attract other companies as partners to assist us in further exploration and drilling. Currently, I am working on projects located in Utah, southern Illinois, southern Michigan, Iowa, and Minnesota. I was Manager of Exploration Training when I left Amoco, and as a result, I currently teach a few courses (such as Petroleum Geology for Engineers), which allows me to travel to places like Cairo, Egypt, Quito, Ecuador, and Houston, Texas.

SUSAN M. LANDON began her career in 1974 with Amoco Production Company and, in 1989, opened her own consulting office in Denver, Colorado. She is currently a partner in the exploration group, Thomasson Partner Associates. In 1990, she was elected president of the American Institute of Professional Geologists and, in 1992, treasurer of the American Association of Petroleum Geologists.

these rocks. Turbulence in water keeps silt and clay suspended and must therefore be at a minimum if they are to settle. Consequently, deposition occurs in low-energy depositional environments where currents are weak such as in the quiet offshore waters of lakes and in lagoons.

Chemical and Biochemical Sedimentary Rocks

Chemical sedimentary rocks originate from the substances taken into solution during chemical weathering. These dissolved materials can be extracted from lake or ocean water to form minerals either by inorganic chemical processes or by the chemical activities of organisms. Some rocks formed by lithification of these minerals have a *crystalline texture,* meaning that they consist of a mosaic of interlocking mineral crystals (Figure 6-8b), whereas others have a clastic texture. Those chemical sedimentary rocks formed by the activities of organisms are generally referred to as **biochemical sedimentary rocks.**

Limestone-Dolostone. Calcite (the main component of limestone) and dolomite (the mineral comprising the rock dolostone) are both carbonate minerals; calcite is a calcium carbonate ($CaCO_3$), whereas dolomite [$CaMg(CO_3)_2$] is a calcium magnesium carbonate. Thus, limestone and dolostone are **carbonate rocks.** Recall from Chapter 5 that calcite readily dissolves in water containing a small amount of acid, but the chemical reaction leading to dissolution is reversible, so solid calcite can be precipitated from solution. Accordingly, some limestone, although probably not very much, results from inorganic chemical reactions.

Most limestones have a large component of calcite that was originally extracted from seawater by organisms. Corals, clams, algae, snails, and a number of other marine organisms construct their skeletons of aragonite, which is an unstable form of calcium carbonate that alters to calcite. When the organisms die, their skeletons may be broken up and accumulate as gravel-, sand-, silt-, and clay-sized sediment that becomes lithified and forms limestone (⪢ Figure 6-11). Because organisms play such a significant role in their origin, most limestones are conveniently classified as *biochemical sedimentary rocks* (Table 6-2). The limestone known as *coquina* consists entirely of broken shells cemented by calcium carbonate, and *chalk* is a soft variety of biochemical limestone composed largely of microscopic shells of organisms (Figure 6-11a and b).

One distinctive type of limestone contains small spherical grains called *ooids.* Ooids have a small nucleus, a sand grain or shell fragment perhaps, around which concentric layers of calcite precipitate; lithified deposits of ooids form *oolitic limestones* (Figure 6-11c).

The near-absence of recent dolostone and evidence from chemistry and studies of rocks indicate that most dolostone was originally limestone that has been changed to dolostone. Many geologists think most dolostones originated when magnesium replaced some of the calcium in calcite. ⪢ Figure 6-12 shows one way this can occur. Note that in a restricted environment, such as a lagoon, where evaporation rates are high, much of the calcium in solution is extracted as it is used to make the minerals calcite ($CaCO_3$) and gypsum ($CaSO_4 \cdot 2H_2O$). Magnesium (Mg) becomes concentrated in the water, which then becomes denser and permeates the preexisting limestone and converts it to dolostone by the addition of magnesium.

Evaporites. **Evaporites** include such rocks as *rock salt* and *rock gypsum,* which form by inorganic chemical precipitation of minerals from solution (Table 6-2; ⪢ Fig. 6-13a and b). Both are characterized by a crystalline texture. In Chapter 5 we noted that some minerals are dissolved during chemical weathering, but a solution can hold only a certain volume of dissolved mineral matter. If the volume of a solution is reduced by evaporation, the amount of dissolved mineral matter increases in proportion to the volume of the solution and eventually reaches the saturation limit, the point at which precipitation must occur. Perspective 6-1 discusses the formation of some notable evaporite deposits.

Rock salt, composed of the mineral halite (NaCl), is simply sodium chloride that was precipitated from seawater or, more rarely, lake water (Figure 6-13a). Rock gypsum, the most common evaporite rock, is composed of the mineral gypsum ($CaSO_4 \cdot H_2O$), which also precipitates from evaporating solutions (Figure 6-13b). A number of other evaporite rocks and minerals are known, but most of these are rare. Some are important, however, as sources of various chemical compounds; for example, sylvite, a potassium chloride (KCl), is used in the manufacture of fertilizers, dyes, and soaps.

Chert. *Chert* is a hard rock composed of microscopic crystals of quartz (SiO_2) (Table 6-2; Figure 6-13c). It is found in several varieties including *flint,* which is black because of inclusions of organic matter, and *jasper,* which is red or brown because of iron oxide inclusions. Because chert lacks cleavage and can be shaped to form sharp cutting edges, many cultures have used it for the manufacture of tools, spear points, and arrowheads.

Chert occurs as irregular masses or *nodules* in other rocks, especially limestones, and as distinct layers of rock called *bedded chert.* Most nodules in limestones are clearly secondary in origin; that is, they have replaced part of the host rock, apparently by being precipitated from solution.

Bedded chert can be precipitated directly from seawater, but because so little silica is dissolved in seawater, it seems unlikely that most bedded cherts are inorganic chemical precipitates. It appears that many bedded cherts are biochemical, resulting from accumulations of shells of silica-secreting, single-celled organisms such as radiolarians and diatoms (⪢ Figure 6-14). Unfortunately,

➢ FIGURE 6-11 Four types of limestone, two of which, (*a*) and (*b*), are biochemical sedimentary rocks. (*a*) Coquina is composed of the broken shells of organisms. (Photo courtesy of Sue Monroe.) (*b*) Chalk cliffs in Denmark. Chalk is made up of microscopic shells. (Photo courtesy of R. V. Dietrich.) (*c*) Present-day ooids from the Bahamas. (*d*) Limestone containing numerous fossil shells.

➢ FIGURE 6-12 One way that limestone can be converted to dolostone. In this example, the seawater in a lagoon becomes enriched in magnesium as evaporation occurs. This magnesium-rich water is denser than normal seawater so it permeates the older limestones and converts them to dolostone.

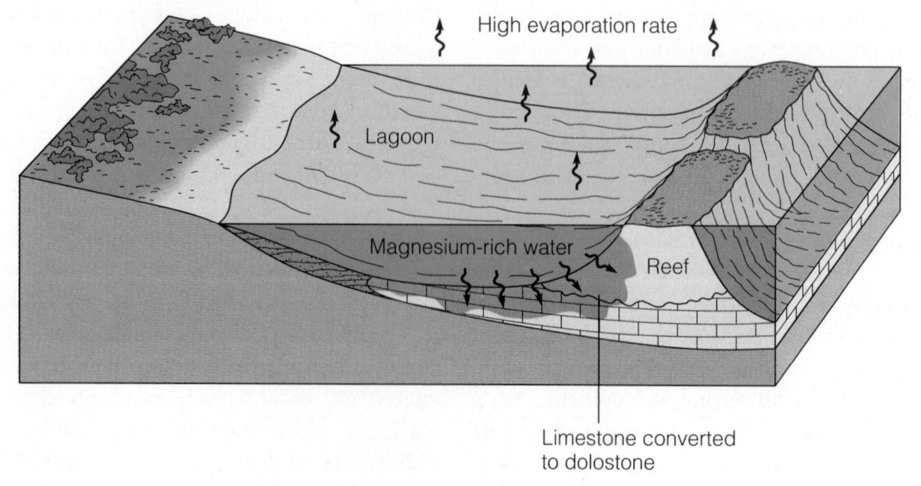

▶ FIGURE 6-13 (*a*) Core of rock salt from a well in Michigan. (*b*) Rock gypsum. (*c*) Chert. (Photos courtesy of Sue Monroe.)

▶ FIGURE 6-14 Bedded chert is composed of microscopic shells of silica-secreting, single-celled (*a*) radiolarians and (*b*) diatoms.

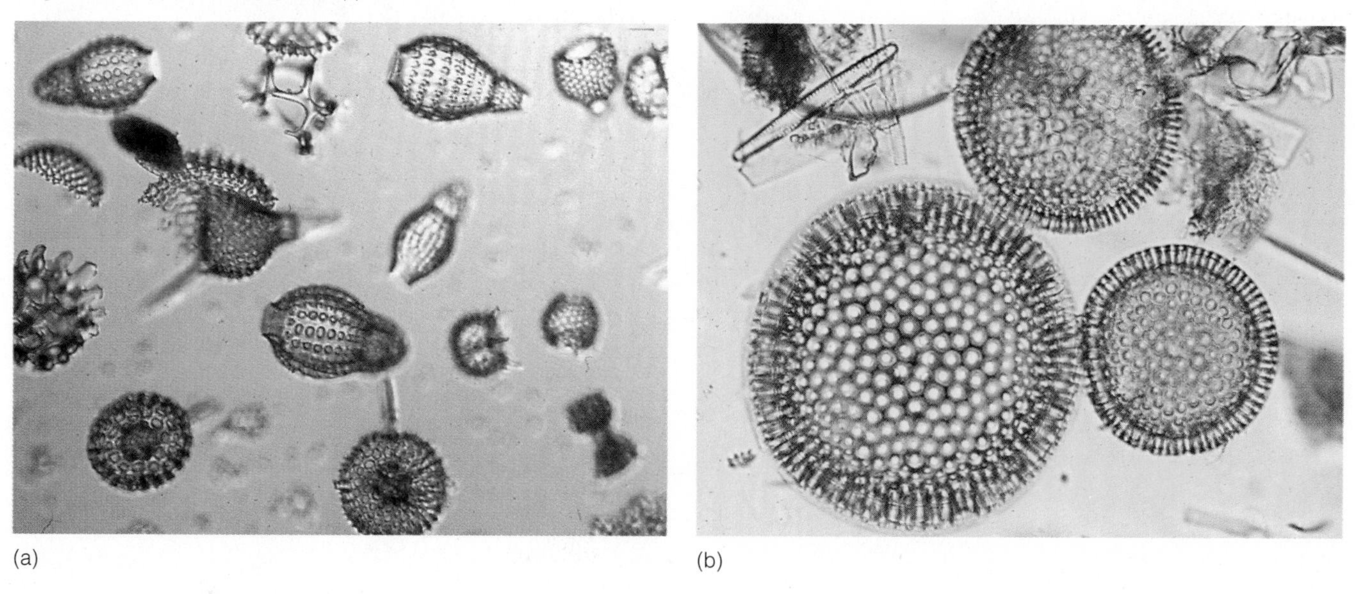

the shells of these organisms are easily altered, so the evidence for a biochemical origin of many bedded cherts is obscured.

Coal. *Coal* is a biochemical sedimentary rock composed of the compressed, altered remains of organisms, especially land plants (Table 6-2; ▶ Figure 6-15). It forms in swamps and bogs where the water is deficient in oxygen or where organic matter accumulates faster than it decomposes. The bacteria that decompose vegetation in swamps can exist without oxygen, but their wastes must be oxidized, and because no oxygen is present, the wastes accumulate and kill the bacteria. Thus, bacterial decay ceases and plant materials are not completely destroyed. These partly altered plant remains accumulate as layers of organic muck, which commonly smell of hydrogen sulfide (the rotten-egg odor of swamps). When buried, this organic muck becomes *peat,* which looks rather like coarse pipe tobacco. Where peat is abundant, as in Ireland and Scotland, it is burned as a fuel. Peat that is buried more deeply and compressed, especially if it is heated too, is altered to a type of dark brown coal called *lignite,* in which plant remains are still clearly visible. During the change from organic muck to coal, volatile elements of the vegetation such as oxygen, hydrogen, and nitrogen are partly vaporized and driven off, enriching the residue in carbon; lignite contains about 70% carbon as opposed to about 50% in peat.

▶ **FIGURE 6-15** Coal is a biochemical sedimentary rock composed of the altered remains of land plants. (Photo courtesy of Sue Monroe.)

Bituminous coal, which contains about 80% carbon, is a higher grade coal than lignite. It is dense and black and has been so thoroughly altered that plant remains can only rarely be seen. The highest grade coal is *anthracite,* which is a metamorphic type of coal (see Chapter 7). It contains up to 98% carbon and, when burned, yields more heat per unit volume than other types of coal.

Historically, most of the coal mined in the United States has been bituminous coal from the coal fields of the Appalachian coal basin (▶ Figure 6-16). These coal

▶ **FIGURE 6-16** Distribution of coal deposits in the United States.

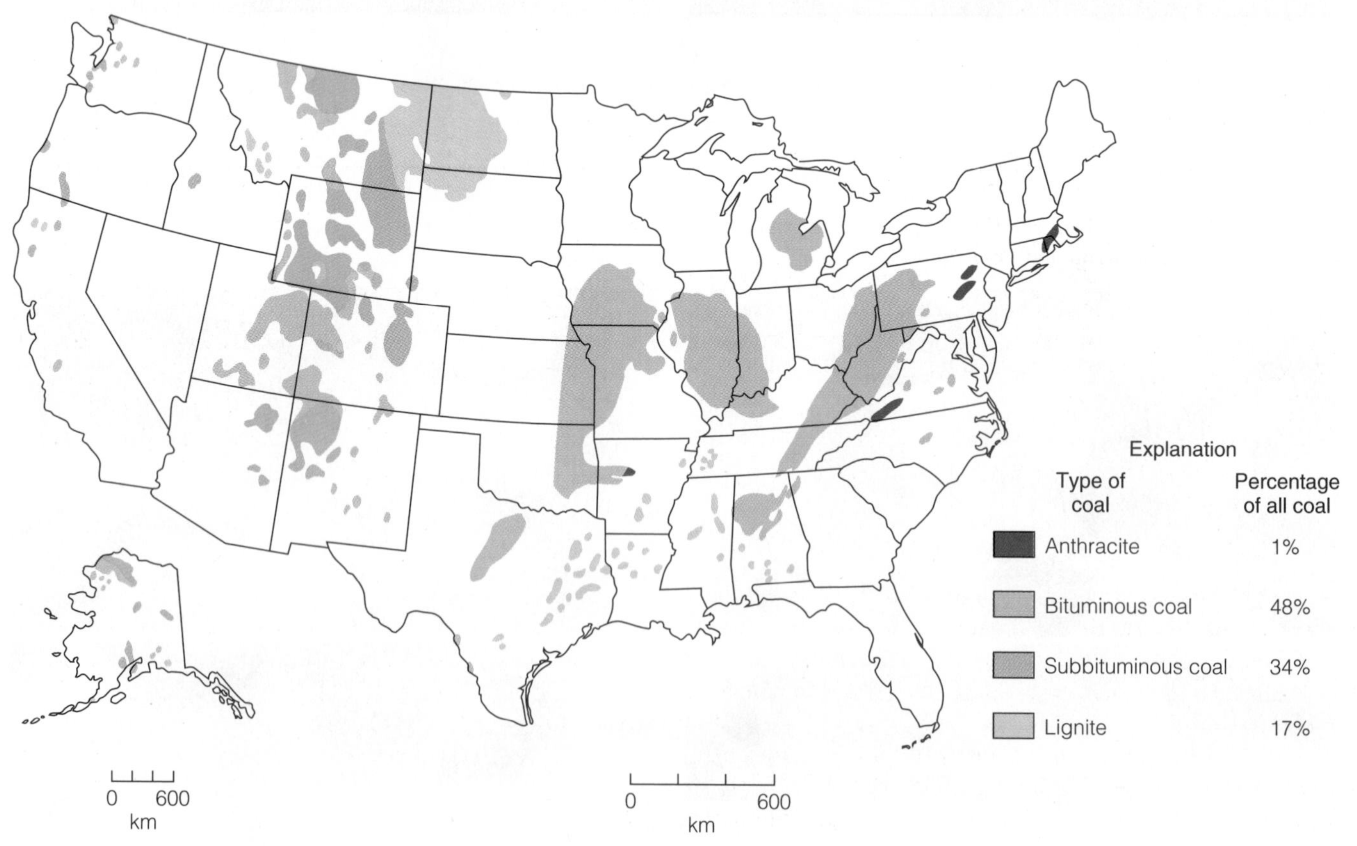

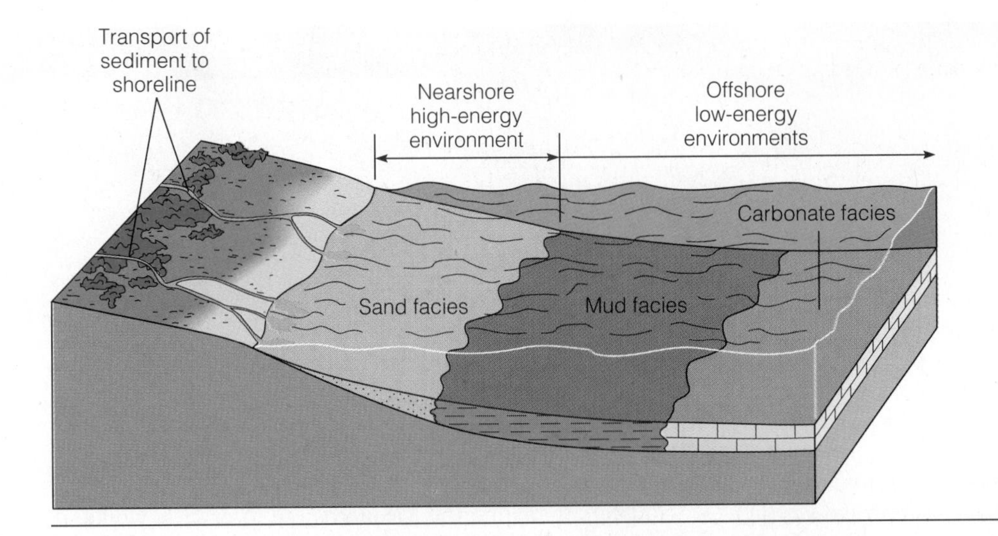

▶ **FIGURE 6-17** Deposition in adjacent environments yields distinct bodies of sediment, each of which is designated as a sedimentary facies.

deposits formed in coastal swamps during the Pennsylvanian Period between 286 and 320 million years ago. Huge lignite and subbituminous coal deposits also exist in the western United States, and these are becoming increasingly important resources (Figure 6-16).

 SEDIMENTARY FACIES

If a layer of sediment or sedimentary rock is traced laterally, it generally changes in composition, texture, or both. These lateral changes result from the simultaneous operation of different depositional processes in adjacent depositional environments. For example, sand may be deposited in a high-energy nearshore environment while mud and carbonate sediments accumulate simultaneously in the laterally adjacent low-energy offshore environments (▶ Figure 6-17). Deposition in each of these environments produces a body of sediment, each of which is characterized by a distinctive set of physical, chemical, and biological attributes. Such distinctive bodies of sediment, or sedimentary rock, are **sedimentary facies.**

Any aspect of sedimentary rocks that makes them recognizably different from adjacent rocks of the same age, or approximately the same age, can be used to establish a sedimentary facies. Figure 6-17 illustrates three sedimentary facies: a sand facies, a mud facies, and a carbonate facies. If lithified, these sediments become sandstone, mudstone (or shale), and limestone facies, respectively.

Marine Transgressions and Regressions

Many sedimentary rocks in the interiors of continents show clear evidence of having been deposited in marine environments. The rocks in ▶ Figure 6-18d, for example, consist of a sandstone facies that was deposited in a nearshore marine environment overlain by shale and limestone facies that were deposited in offshore environments. Such a vertical sequence of facies can be explained by deposition occurring during a time when sea level rose with respect to the continents.

▶ **FIGURE 6-18** (*a*), (*b*), and (*c*) Three stages of a marine transgression. (*d*) Diagrammatic representation of the vertical sequence of facies resulting from the transgression.

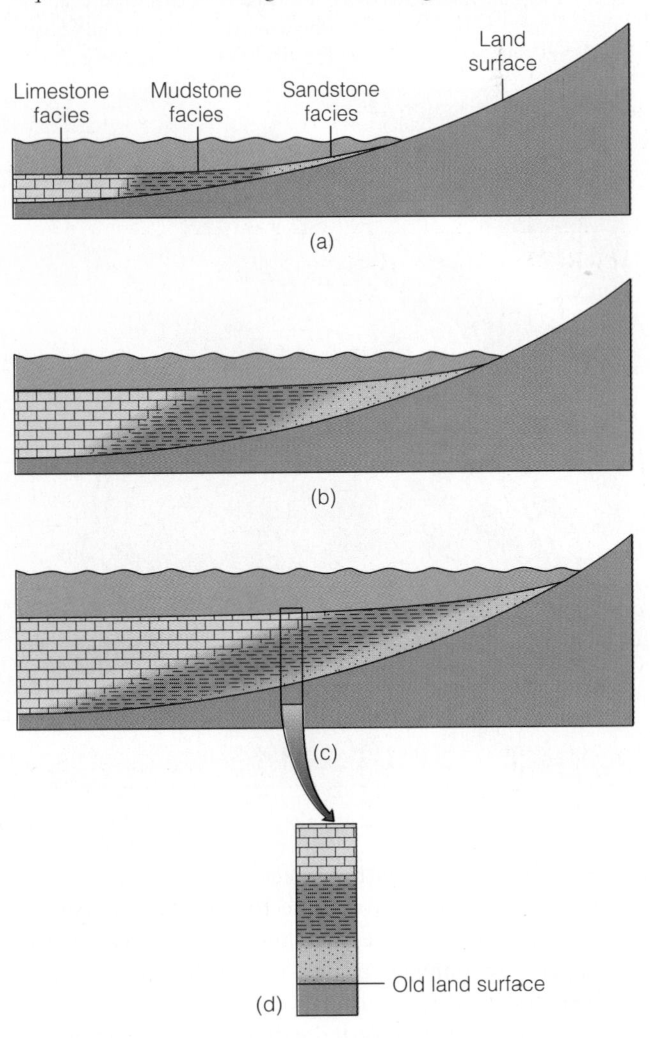

Perspective 6-1

THE MEDITERRANEAN DESERT

Vast, thick evaporite deposits are present in several areas, but one of the most notable is beneath the Mediterranean Sea. At the present time, the Mediterranean Sea is in an arid region where the rate of evaporation of seawater exceeds the rate at which water is added to the sea by rainfall runoff. If it were not for the connection with the Atlantic Ocean at the Strait of Gibraltar (▶ Figure 1), the Mediterranean would eventually dry up and form a vast desert basin far below sea level. Some geologists think that the Mediterranean did dry up during the Cenozoic Era, resulting in the deposition of rock gypsum and rock salt now present beneath the floor of the sea.

Studies of the Mediterranean evaporites indicate deposition in shallow water, rather than in a deep-ocean basin, which is what the Mediterranean is now. According to this hypothesis, the Mediterranean lost its connection with the Atlantic, evaporated to near dryness in as little as 1,000 years, and became a vast desert basin lying 3,000 m below sea level (▶ Figure 2).

Simply evaporating the Mediterranean to dryness would yield only 40 to 45 m of evaporites, not a layer 2 km thick.

▶ **FIGURE 1** View showing the western end of the Mediterranean Sea (upper right). The narrowly constricted area leading from the Mediterranean to the Atlantic Ocean (lower right) is the Strait of Gibraltar. About 6 million years ago, the Mediterranean was probably a vast desert lying 3,000 m below sea level, where evaporite minerals were deposited. At the end of the Miocene, an oceanic connection was reestablished at the Strait of Gibraltar, and the basin rapidly refilled. (Photo courtesy of NASA.)

When sea level rises with respect to a continent, the shoreline moves inland, giving rise to a **marine transgression** (Figure 6-18). As the shoreline advances inland, the depositional environments parallel to the shoreline do likewise. Remember that each laterally adjacent environment in Figure 6-18 is the depositional site of a different sedimentary facies. As a result of a marine transgression, the facies that formed in the offshore environments are superposed over the facies deposited in the nearshore environment, thus accounting for the vertical succession of sedimentary facies (Figure 6-18d).

Another important aspect of marine transgressions is that an individual facies can be deposited over a huge geographic area. Even though the nearshore environment is long and narrow at any particular time, deposition occurs continuously as the environment migrates landward during a marine

It appears that once the Mediterranean evaporated, it formed a low-lying basin in which vast, shallow saline lakes existed. Evaporites formed in these lakes and simply accumulated, forming thick deposits. Furthermore, apparently the connection with the Atlantic was periodically reestablished so that the Mediterranean refilled. During the times when an oceanic connection existed, sand and gravel were deposited near the margins of the sea, and deep-sea sediments, mostly clay, were deposited in the offshore areas. Subsequently, the Mediterranean was again isolated from the Atlantic, and the evaporation sequence was repeated, perhaps several times.

Supporting evidence for this view of Mediterranean history comes from southern Europe and North Africa. The present-day Mediterranean controls the level to which streams can erode downward: they can erode no lower than sea level. In Europe and Africa, however, there are canyons cut into solid bedrock that extend to depths far below present sea level. The contention is that during periods of lower sea level, when evaporites were deposited, streams such as the Nile River in Africa eroded downward in response to the lower sea level. The subsequent rise in sea level caused these streams to fill their valleys with sediment so that they now flow at higher levels.

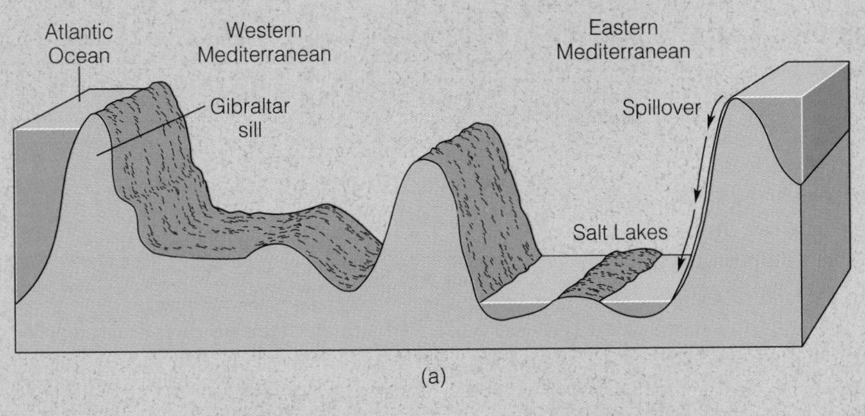

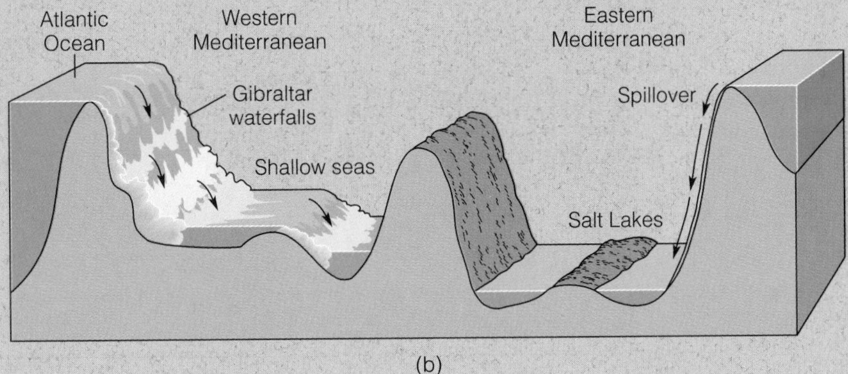

▶ **FIGURE 2** Proposed model to explain the deposition of thick evaporites in the Mediterranean Basin. (*a*) Isolation of the Mediterranean and evaporation to near dryness. (*b*) The refilling stage. To account for the thick evaporites beneath the basin, this cycle of isolation, evaporation, and refilling must have occurred many times. The vertical scale is greatly exaggerated here.

transgression. The sand deposited under these conditions may be tens to hundreds of meters thick, but have horizontal dimensions of length and width measured in hundreds of kilometers.

The opposite of a marine transgression is a **marine regression.** If sea level falls with respect to a continent, the shoreline and environments that parallel the shoreline move in a seaward direction. The vertical sequence produced by a marine regression has facies of the nearshore environment superposed over facies of offshore environments. Marine regressions can also account for the deposition of a facies over a large geographic area.

Environmental Analysis

When geologists investigate sedimentary rocks in the field, they are observing the products of events that occurred during the past. The only record of these events is preserved in the rocks, so geologists must evaluate those aspects of sedimentary rocks that allow inferences to be made about the original processes and the environment of deposition. Sedimentary textures such as sorting and rounding can give clues to the depositional process. Windblown dune sands, for example, tend to be well sorted and well rounded. The geometry or three-dimensional shape of rock bodies is another important criterion in environmental interpretation. Marine transgressions and regressions yield sediment bodies with a blanket or sheetlike geometry, whereas deposits in stream channels tend to be long and narrow and are therefore described as having a shoestring geometry (▶ Figure 6-19). Other aspects of sedimentary rocks that are important in environmental analysis include *sedimentary structures* and *fossils*.

Sedimentary Structures

When sediments are deposited, they contain a variety of features known as **sedimentary structures** that formed as a result of physical and biological processes operating in the depositional environment. One of the most common of these sedimentary structures is distinct layers known as **strata** or **beds** (▶ Figure 6-20). Beds vary in thickness from less than a millimeter up to many meters. Individual beds are separated from one another by *bedding planes* and are distinguished from one another by differences in composition, grain size, color, or a combination of features (Figure 6-20). Almost all sedimentary rocks show some kind of stratification or bedding; a few, such as limestones that formed as coral reefs, lack this feature.

In **graded bedding,** grain size decreases upward within a single bed (▶ Figure 6-21). Most graded bedding appears to have formed from turbidity current deposition, although some forms in stream channels during the waning stages of floods. *Turbidity currents* are underwater flows of sediment-

▶ **FIGURE 6-20** Like these sandstones in Montana, most sedimentary rocks show some kind of layering or bedding.

▶ **FIGURE 6-21** (*a*) Turbidity currents flow downslope along the sea floor (or lake bottom) because of their density. (*b*) Graded bedding formed by deposition from a turbidity current.

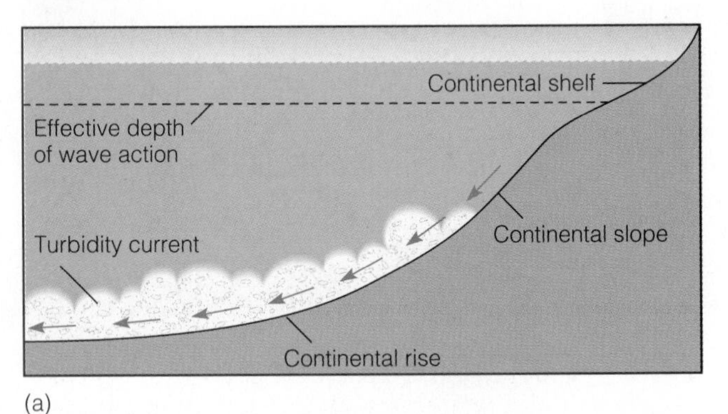

▶ **FIGURE 6-19** Two different geometries of sedimentary rock bodies. The shale, sandstone, and limestone all have blanket geometries. The elongate sandstones within the shale have a shoestring geometry.

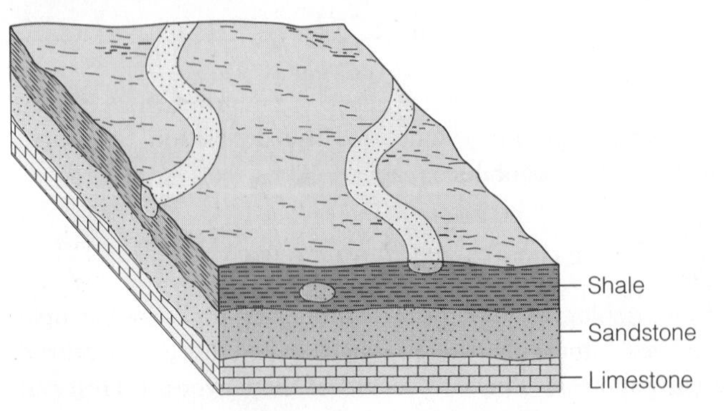

▶ **FIGURE 6-22** In cross-bedding, the beds are inclined with respect to the surface upon which they accumulate. Cross-beds indicate ancient current directions by their dip, to the right in this case.

water mixtures that are denser than sediment-free water. Such flows move downslope along the bottom of the sea or a lake until they reach the relatively level sea or lake floor. There, they rapidly slow down and begin depositing transported sediment, the coarsest first followed by progressively smaller particles.

Many sedimentary rocks are characterized by **cross-bedding;** cross-beds are arranged at an angle to the surface on which they are deposited (▶ Figure 6-22). Cross-bedding is common in desert dunes and in sediments in stream channels and shallow marine environments. Invariably, cross-beds result from transport by wind or water currents and deposition on the downcurrent side of dunelike structures (Figure 6-22). Cross beds are inclined downward, or dip, in the direction of flow. Because their orientation depends on the direction of flow, cross-beds are good indicators of ancient current directions or **paleocurrents** (Figure 6-22).

In sand deposits one can commonly observe small-scale, ridgelike **ripple marks** on bedding planes. Two common types of ripple marks are recognized. One type is asymmetrical in cross section with a gentle upstream slope and a steep downstream slope. Known as *current ripple marks* (▶ Figure 6-23a), they form as a result of currents that move in one direction as in a stream channel. Like cross-bedding, current ripple marks are good paleocurrent indicators. In contrast, the to-and-fro motion of waves produces ripples that tend to be symmetrical in cross section; they are known as *wave-formed ripple marks* which form mostly in the shallow, nearshore waters of oceans and lakes. (Figure 6-23b).

Mud cracks are found in clay-rich sediment that has dried out (▶ Figure 6-24). When the sediment dries, it shrinks and forms intersecting fractures (mud cracks). Such features in ancient sedimentary rocks indicate that the sediment was deposited where periodic drying was possible, as on a river floodplain, near a lake shore, or where muddy deposits are exposed on marine shorelines at low tide. Many other sedimentary structures are known, including some that form long after deposition (see Perspective 6-2).

Fossils

Fossils are the remains or traces of ancient organisms (▶ Figure 6-25). These remains are mostly the hard skeletal parts of organisms such as shells, bones, and teeth, but under exceptional conditions, even the soft-part anatomy may be preserved. For example, several frozen woolly mammoths have been discovered in Alaska and Siberia with hair, flesh, and internal organs preserved. The remains of organisms are known as *body fossils* to distinguish them from

(a)

(b)

➤ **FIGURE 6-23** (*a*) Current ripple marks on the bed of a stream in California. The shape of the ripples indicates that the current moved from right to left. (*b*) Wave-formed ripples on ancient rocks in Montana.

trace fossils such as tracks, trails, and burrows (Figure 6-25), which are indications of ancient organic activity.

For any potential fossil to be preserved, it must escape the ravages of such destructive processes as running water, waves, scavengers, exposure to the atmosphere, and bacterial decay. Obviously, the soft parts of organisms are devoured or decomposed most rapidly, but even the hard skeletal parts will be destroyed unless they are buried and protected in mud, sand, or volcanic ash. Even if buried, bones and shells may be dissolved by groundwater or destroyed by alteration of the host rock. Nevertheless, fossils are quite common. The remains of microscopic plants and animals are the most common, but these require specialized methods of recovery, preparation, and study and are not sought out by casual fossil collectors. Shells of marine animals are also very common and easily collected in many areas, and even the bones and teeth of dinosaurs are much more common than most people realize.

➤ **FIGURE 6-24** Mud cracks form in clay-rich sediments when they dry and shrink.

Some fossils retain their original composition and structure, and are preserved as unaltered remains, but many have been altered in some way. Dissolved mineral matter can be precipitated in the pores of bones, teeth, and shells or can fill the spaces within cells of wood. Wood may be preserved by silica replacing the woody tissues; it then is referred to as *petrified,* a term that means "to become stone" (➤ Figure 6-26). Silicon dioxide (SiO_2) or iron sulfide (FeS_2) can completely replace the calcium carbonate ($CaCO_3$) shells of marine animals (➤ Figure 6-27a). Insects and the leaves, stems, and roots of plants are commonly preserved as thin carbon films that show the details of the original organism (Figure 6-27b). Shells in sediment may be dissolved leaving a cavity called a *mold* that is shaped like the shell (Figure 6-27d). If a mold is filled in, it becomes a *cast* (Figure 6-27e).

If it were not for fossils, we would have no knowledge of such extinct animals as trilobites and dinosaurs. Thus, fossils constitute our only record of ancient life. They are not simply curiosities, but have several practical uses. In many geologic studies it is necessary to correlate or determine age equivalence of sedimentary rocks in different areas. Such

(a)

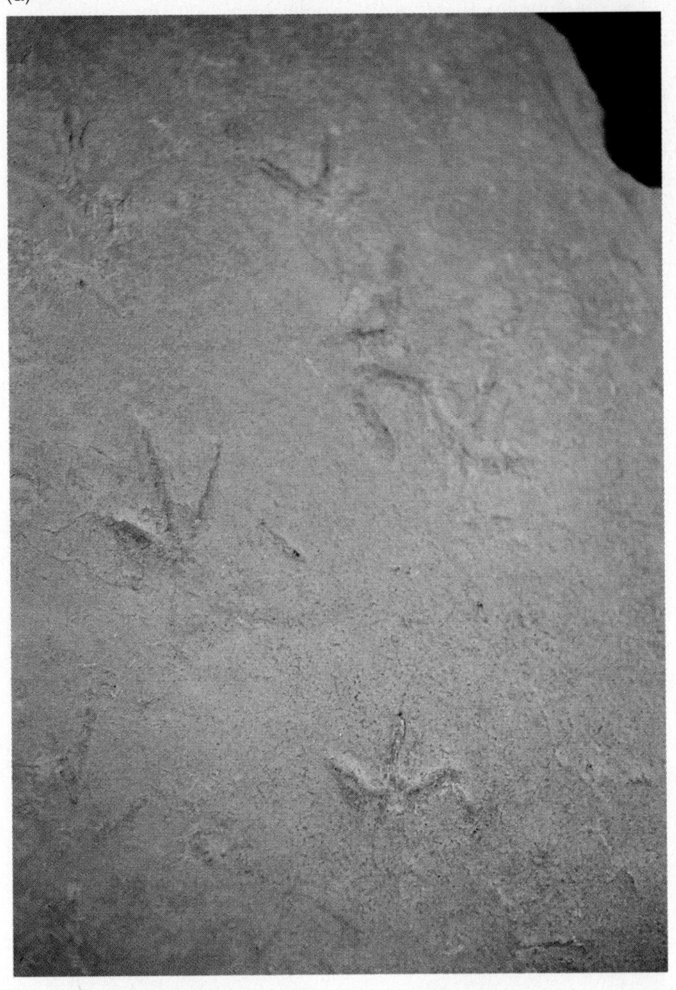

(b)

▶ FIGURE 6-25 (*a*) Body fossils consist of the actual remains of organisms. Fossil horse teeth are preserved in this rock. (*b*) Trace fossils are an indication of ancient organic activity. These bird tracks are preserved in mudrock of the Green River Formation of Wyoming.

correlations are most commonly demonstrated with fossils; we will discuss correlation more fully in Chapter 8. Fossils are also useful in determining environments of deposition.

▶ FIGURE 6-26 Specimen of petrified wood showing the original structure of the woody tissues. (Photo courtesy of Sue Monroe.)

Environment of Deposition

The sedimentary rocks in the geologic record acquired their various properties, in part, as a result of the physical, chemical, and biological processes that operated in the original depositional environment. One of geologists' major tasks is to determine the specific depositional environment of sedimentary rocks. Based on their knowledge of how sedimentary structures originate and present-day processes, such as sediment transport and deposition by streams, geologists can make inferences regarding the depositional environments of ancient sedimentary rocks.

While conducting field studies, geologists commonly make some preliminary interpretations. Some sedimentary particles, for example, such as ooids in limestones most commonly form in shallow marine environments where currents are vigorous. Large-scale cross-bedding is typical of, but not restricted to, desert dunes. Fossils of land plants and animals can be washed into transitional environments, but most of them are preserved in deposits of continental environments. Fossil shells of such marine-dwelling animals as corals obviously indicate marine depositional environments.

Much environmental interpretation is done in the laboratory where the data and rock samples collected during field work can be more fully analyzed. The analysis may include microscopic and chemical examination of rock samples, identification of fossils, and graphic representations showing the three-dimensional shapes of rock units and their relationships to other rock units. In addition, the features of sedimentary rocks are compared with those of sediments from present-day depositional environments; the contention is that features in ancient rocks, such as ripple marks, formed during the past in response to the same processes responsible for them now. Finally, when all data have been analyzed, an environmental interpretation is made.

The following examples illustrate how environmental interpretations are made. The Navajo Sandstone of the

CONCRETIONS AND GEODES

In addition to sedimentary structures such as ripple marks and cross-bedding, which are important for determining depositional environments, a variety of other features known as *secondary sedimentary structures* form in sediments long after deposition. Although these structures tell nothing of how a rock layer was deposited in the first place, they are interesting in their own right, and some are quite attractive. Many such structures are known, but we will discuss only two varieties, concretions and geodes.

Concretions consist of any mass of material that can easily be separated from the enclosing rock. Sandstone concretions, which are probably the most common type, form by more thorough cementation of sand deposits in local areas. These concretions occur in a variety of sizes ranging from 1 cm to 9 m in diameter, and although commonly spherical, their shapes can be quite irregular (▶ Figure 1). Because concretions are more thoroughly cemented and thus harder than the surrounding rock, they commonly accumulate at the surface during weathering (Figure 1a).

Imatra stones or *marlekor* are particularly interesting (Figure 1b). These concretions are typically disc-shaped, composed of calcium carbonate and silt, and occur in clay layers in glacial lake deposits. Most are small, perhaps 2 to 3 cm across; generally, they have a simple form, but some unusual shapes result when simple forms grow together or when outgrowths form (Figure 1b).

Perhaps the most attractive concretions are *septaria* or *septarian nodules*. These large spheroidal concretions are characterized by a series of cracks that widen toward the concretion's center and are crossed by smaller cracks more or less parallel with the margin (▶ Figure 2a). The cracks apparently formed when the original concretion shrank as a

(a)

(b)

▶ **FIGURE 1** Concretions. (*a*) Spherical concretions from the Pumpkin Patch in the Ocotillo Wells State Vehicular Recreation Area in the Anza Borrego Desert, California. (*b*) These disc-shaped concretions called *imatra stones* or *marlekor* are from glacial lake deposits in Connecticut. (Photo courtesy of Sue Monroe.)

⧐ **FIGURE 2** (*a*) Cross section showing the internal structure of a septarian nodule. (Photo courtesy of Sue Monroe.) (*b*) Surface view of septarian nodule resembling a turtle shell.

(a)

(b)

result of dehydration; later they were filled or nearly filled with mineral crystals, mostly calcite. Some septaria are released from the host rock and eroded so that the interior veins are visible, giving the concretions the appearance of a turtle shell (Figure 2b). Indeed, some fossil collectors have mistakenly labeled such specimens as "fossil turtles."

Geodes are attractive and commonly used as decorative stones. They form when minerals grow along the margins of a cavity, partially filling it (⧐ Figure 3). Most geodes are 30 cm or less in diameter, and they commonly occur in limestone beds although they occur in a variety of other rocks as well. The outermost layer of minerals is a thin, sometimes discontinuous layer of agate, a variety of color-banded, compact microcrystalline quartz (SiO_2). The banding in the agate conforms to the walls of the original cavity. Within the outermost layer, the cavity is partly filled with inward-pointing quartz or calcite ($CaCO_3$) crystals, or crystals of various sulfates such as barite ($BaSO_4$) and celestite ($SrSO_4$), all of which were precipitated from solution.

⧐ **FIGURE 3** A geode formed by the partial filling of a cavity by color-banded agate, which conforms to the walls of the cavity, and by inward-pointing quartz crystals. (Photo courtesy of Sue Monroe.)

(a)

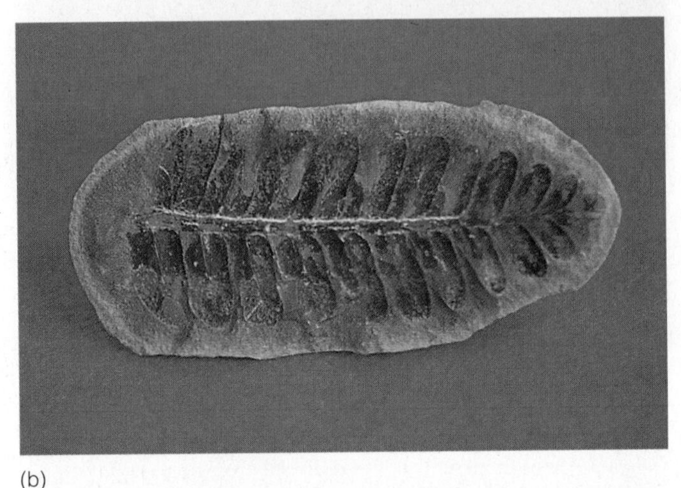

(b)

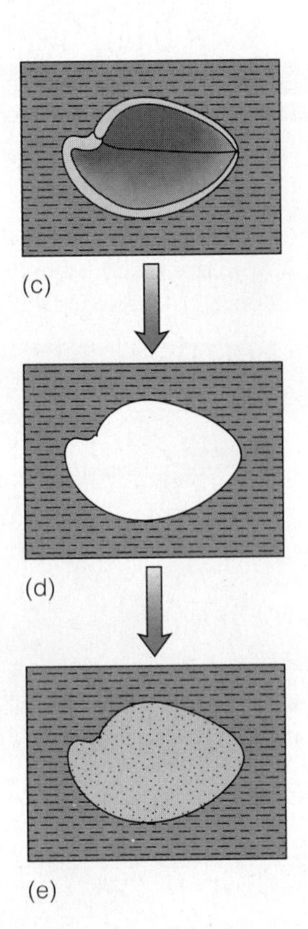

▶ FIGURE 6-27 Various types of fossilization. (*a*) Replacement by iron sulfide (FeS_2). (*b*) Carbonized leaf. (Photos a and b courtesy of Sue Monroe.) (c, d, and e) Origin of a mold and a cast. (*c*) Burial of a shell in sediment. (*d*) The shell is dissolved out leaving a cavity, or mold. (*e*) The mold is filled by sediment forming a cast.

southwestern United States covers a vast area, perhaps as much as 500,000 km^2. It has an irregular sheet geometry, reaches about 300 m thick in the area of Zion National Park, Utah, and consists mostly of well-sorted, well-rounded sand grains measuring about .2 to .5 mm in diameter. (▶ Figure 6-28). Some of the sandstone beds also possess tracks of dinosaurs and other land-dwelling animals, ruling out the possibility of a marine origin for the rock unit. These features, and the fact that the Navajo Sandstone has cross-beds up to 30 m high (Figure 6-28) and ripple marks (both of which appear to have formed in sand dunes), lead to the conclusion that the sandstone represents an ancient desert dune deposit. The cross-beds are inclined downward, or dip generally to the southwest, indicating that the wind blew mostly from the northeast.

In the Grand Canyon of Arizona, a number of formations are exposed, and many of these can be traced for great distances. Three of these, the Tapeats Sandstone, the Bright Angel Shale, and the Muav Limestone, occur in vertical sequence and contain features, including fossils, that clearly indicate that they were deposited in transitional and marine environments (▶ Figure 6-29). In fact, all three were forming simultaneously, but a marine transgression caused them to be superposed in the order now observed (Figure 6-18). Similar sequences of rocks of approximately the same age in Utah, Colorado, Wyoming, Montana, and South Dakota indicate that this marine transgression was widespread indeed.

 SEDIMENTS, SEDIMENTARY ROCKS, AND NATURAL RESOURCES

Sediments and sedimentary rocks or the materials they contain have a variety of uses. Sand and gravel are essential to the construction industry, pure clay deposits are used for ceramics, and limestone is used in the manufacture of cement and in blast furnaces where iron ore is refined to make steel. Evaporites are the source of common table salt as well as a number of chemical compounds, and rock gypsum is used to manufacture wallboard. Sand composed mostly of quartz, or what is called silica sand, has a variety of uses including the manufacture of glass, refractory bricks for blast furnaces, and molds for casting iron, aluminum, and copper alloys.

The tiny island nation of Nauru, with one of the highest per capita incomes in the world, has an economy based almost entirely on mining and exporting phosphate-bearing sedimentary rock that is used in fertilizers. More than half of Florida's mineral value in 1989 came from mining phosphate

⊳ **FIGURE 6-28** Outcrop of the Navajo Sandstone in Zion National Park, Utah. Notice the large cross-beds.

rock. In addition to fertilizers, phosphorous from phosphate rock is used in metallurgy, preserving foods, ceramics, and matches.

Dolostones in Missouri are the host rocks for ores of lead and zinc. Diatomite is a sedimentary rock composed of the microscopic shells of single-celled plants that have a skeleton of silica (SiO_2). This lightweight, porous rock is used in gas purification and to filter a number of fluids such as molasses, fruit juices, water, and sewage. The United States is the world leader in diatomite production, mostly from mines in California, Oregon, and Washington.

Anthracite coal is an especially desirable resource because it burns hot with a smokeless flame. Unfortunately, it is the least common type of coal, so most coal used for heating buildings and for generating electrical energy is bituminous (Figure 6-16). Bituminous coal is also used to make *coke,* a hard, gray substance consisting of the fused ash of bituminous coal; coke is prepared by heating the coal and driving off the volatile matter. Coke is used to fire blast furnaces during the production of steel (⊳ Figure 6-30). Synthetic oil and gas and a number of other products are also made from bituminous coal and lignite.

⊳ **FIGURE 6-29** View of the Tapeats Sandstone (lower cliff), Bright Angel Shale (forming the slope in the middle distance), and Muav Limestone (upper cliff) in the Grand Canyon in Arizona. These formations were deposited during a widespread marine transgression.

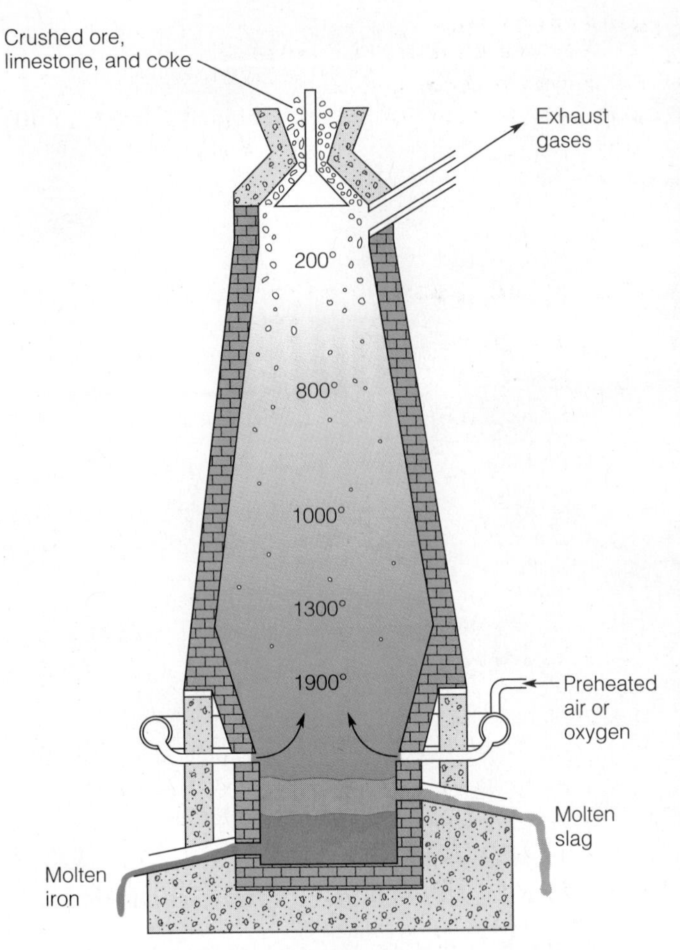

▶ **FIGURE 6-30** Diagrammatic representation of a blast furnace in which iron ore is refined. The raw materials needed are iron ore, coke made from bituminous coal, and limestone as a fluxing agent. The limestone combines with silica in the iron ore and forms a glassy slag that is drawn off near the bottom of the blast furnace. Much of the molten iron is further refined to produce steel.

Petroleum and Natural Gas

Both petroleum and natural gas are *hydrocarbons,* meaning that they are composed of hydrogen and carbon. Hydrocarbons form from the remains of microscopic organisms that exist in the seas and in some large lakes. When these organisms die, their remains settle to the sea or lake floor where little oxygen is available to decompose them. They are then buried under layers of sediment. As the depth of burial increases, they are heated and transformed into petroleum and natural gas. The rock in which the hydrocarbons formed is generally called the *source rock.*

For petroleum and natural gas to occur in economic quantities, they must migrate from the source rock into some kind of rock where they can be trapped. If there were no trapping mechanism, both would migrate upward and eventually seep out at the surface. Indeed, such seeps are known; one of the most famous is the La Brea Tar Pits in Los Angeles, California. The rock in which petroleum and natural gas accumulate is known as *reservoir rock* (▶ Figure 6-31). Effective reservoir rocks contain considerable pore space where appreciable quantities of hydrocarbons can accumulate. Furthermore, the reservoir rocks must possess high *permeability,* or the capacity to transmit fluids; otherwise hydrocarbons cannot be extracted in reasonable quantities. In addition, some kind of impermeable *cap rock* must be present over the reservoir rock to prevent upward migration of the hydrocarbons (Figure 6-31).

Many hydrocarbon reservoirs consist of nearshore marine sandstones in proximity to fine-grained, organic-rich source rocks. Such oil and gas traps are called *stratigraphic traps* because they owe their existence to variations in the rock layers or strata. Ancient coral reefs are also good stratigraphic traps. Indeed, some of the oil in the Persian Gulf region is trapped in ancient reefs. *Structural traps* result when rocks are deformed by folding, fracturing, or both. In areas where sedimentary rocks have been deformed into a series of folds, hydrocarbons migrate to the high parts of such structures (Figure 6-31b). Displacement of rocks along faults (fractures along which movement has occurred) also yields situations conducive to trapping hydrocarbons (Figure 6-31b).

In the Gulf Coast region, hydrocarbons are commonly found in structures adjacent to salt domes. A vast layer of rock salt was deposited in this region during the Jurassic Period as the ancestral Gulf of Mexico formed when North America separated from North Africa. Rock salt is a low-density sedimentary rock, and when deeply buried beneath more dense sediments such as sand and mud, it rises toward the surface in pillars known as *salt domes.* As the rock salt rises, it penetrates and deforms the overlying rock layers, forming structures along its margins that may trap petroleum and gas (▶ Figure 6-32).

Other sources of petroleum that will probably become increasingly important in the future include *oil shales* and *tar sands.* The United States has about two-thirds of all known oil shales, although large deposits also occur in South America, and all continents have some oil shale. The richest deposits in the United States are in the Green River Formation of Colorado, Utah, and Wyoming (see the Prologue).

Tar sand is a type of sandstone in which viscous, asphalt-like hydrocarbons fill the pore spaces. This substance is the sticky residue of once-liquid petroleum from which the volatile constituents have been lost. Liquid petroleum can be recovered from tar sand, but to do so, large quantities of rock must be mined and processed. Because the United States has few tar sand deposits, it cannot look to this source as a significant future energy resource. The Athabaska tar sands in Alberta, Canada, however, are one of the largest deposits of this type. These deposits are currently being mined, and it is estimated that they contain several hundred billion barrels of recoverable petroleum.

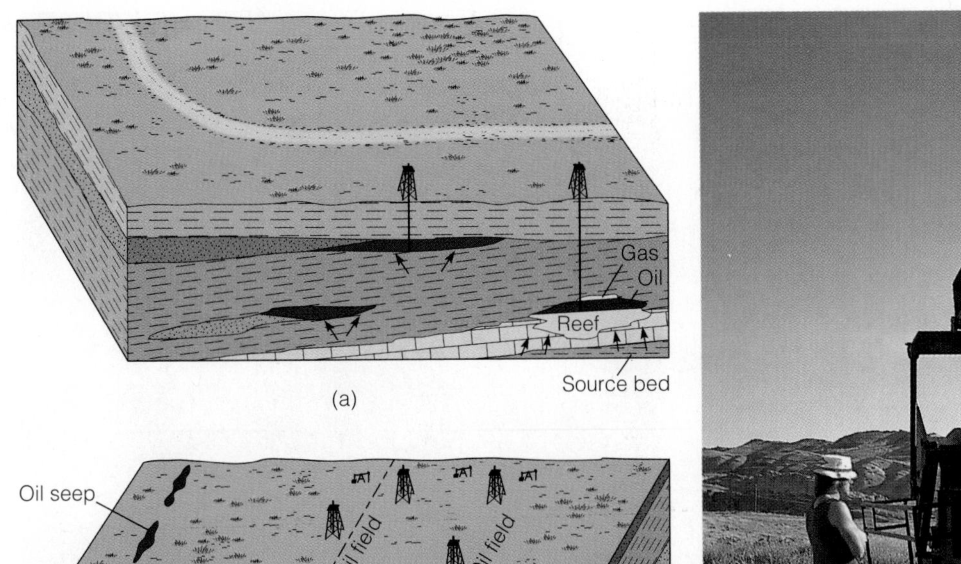

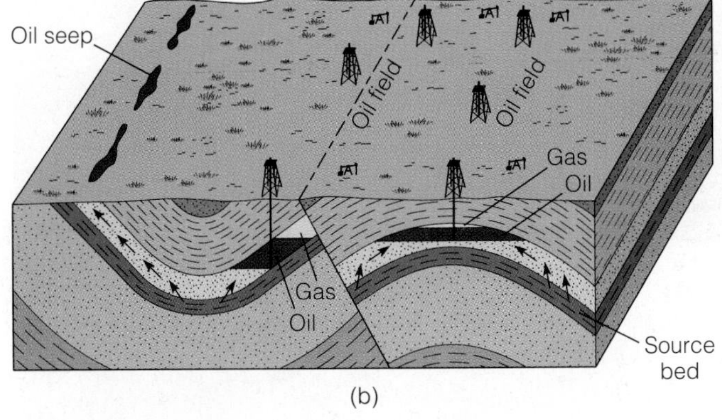

▶ **FIGURE 6-31** Oil and natural gas traps. The arrows in the diagrams indicate the migration of hydrocarbons. (*a*) Two examples of stratigraphic traps. (*b*) Two examples of structural traps, one formed by folding, the other by faulting. (*c*) Oil well in Wyoming on the crest of a buried fold like that on the right side of diagram (b).

Uranium

Most of the uranium used in nuclear reactors in North America comes from the complex potassium-, uranium-, vanadium-bearing mineral *carnotite* found in some sedimentary rocks. Some uranium is also derived from *uraninite* (UO_2), a uranium oxide that occurs in granitic rocks and hydrothermal veins. Uraninite is easily oxidized and dissolved in groundwater, transported elsewhere, and chemically reduced and precipitated in the presence of organic matter.

▶ **FIGURE 6-32** An example of structures adjacent to a salt dome in which oil and natural gas may be trapped.

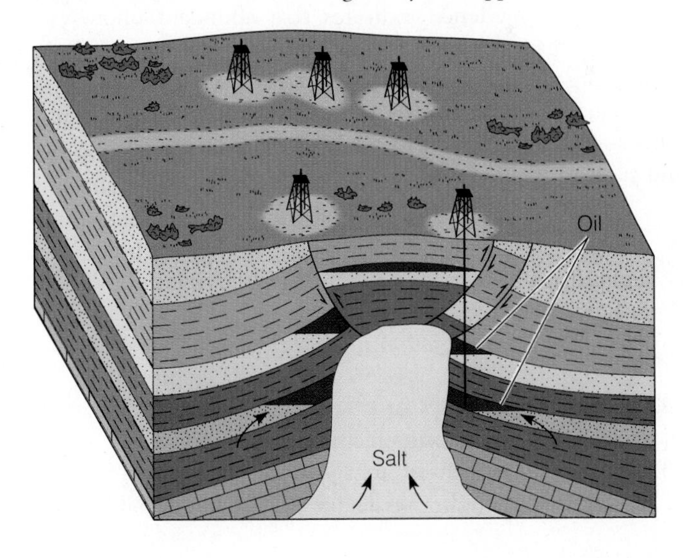

The richest uranium ores in the United States are widespread in the Colorado Plateau area of Colorado and adjoining parts of Wyoming, Utah, Arizona, and New Mexico. These ores, consisting of fairly pure masses and encrustations of carnotite, are associated with plant remains in sandstones that formed in ancient stream channels. Although most of these ores are associated with fragmentary plant remains, some petrified trees also contain large quantities of uranium.

Large reserves of low-grade uranium ore also occur in the Chattanooga Shale. The uranium is finely disseminated in this black, organic-rich mudrock that underlies large parts of several states including Illinois, Indiana, Ohio, Kentucky, and Tennessee.

Canada is the world's largest producer and exporter of uranium. In 1987, Canada exported an estimated 13,612 metric tons of uranium valued at more than $1 billion, much of it to the United States.

▶ **FIGURE 6-33** Outcrop of banded iron formation in northern Michigan.

Banded Iron Formation

Banded iron formation is a chemical sedimentary rock of great economic importance. Such rocks consist of alternating thin layers of chert and iron minerals, mostly the iron oxides hematite and magnetite (▶ Figure 6-33). Banded iron formations are present on all the continents and account for most of the iron ore mined in the world today. Vast banded iron formations are present in the Lake Superior region of the United States and Canada and in the Labrador trough of eastern Canada.

The origin of banded iron formations is not fully understood, and none are currently forming. Fully 92% of all banded iron formations were deposited in shallow seas during the Proterozoic Eon between 2.5 and 2.0 billion years ago. Iron is a highly reactive element that in the presence of oxygen combines to form rustlike oxides that are not readily soluble in water. During early Earth history, little oxygen was present in the atmosphere, so little was dissolved in seawater. However, soluble reduced iron (Fe^{+2}) and silica were present in seawater.

Geological evidence indicates that abundant photosynthesizing organisms were present about 2.5 billion years ago. These organisms, such as bacteria, release oxygen as a byproduct of respiration; thus, they released oxygen into seawater and caused large-scale precipitation of iron oxides and silica as banded iron formations.

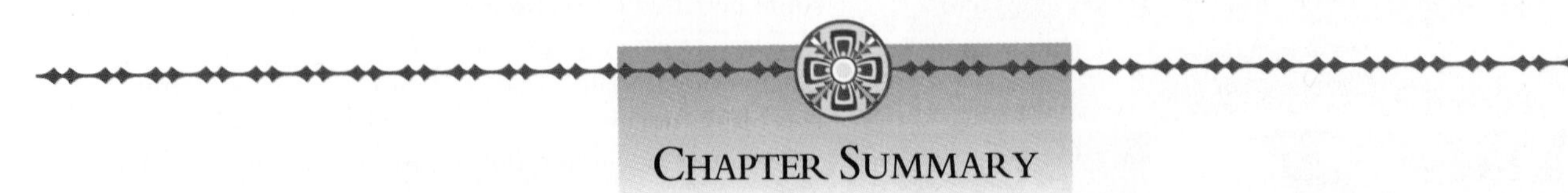

CHAPTER SUMMARY

1. Detrital sediment consists of mechanically weathered solid particles, whereas chemical sediment consists of minerals extracted from solution by inorganic chemical processes and the activities of organisms.
2. Sedimentary particles are designated in order of decreasing size as gravel, sand, silt, and clay.
3. Sedimentary particles are rounded and sorted during transport although the degree of rounding and sorting

depends on particle size, transport distance, and depositional process.

4. Any area where sediment is deposited is a depositional environment. Major depositional settings are continental, transitional, and marine, each of which includes several specific depositional environments.
5. Compaction and cementation are the processes of sediment lithification in which sediment is converted into sedimentary rock. Silica and calcium carbonate are the most common chemical cements, but iron oxide and iron hydroxide cements are important in some rocks.
6. Sedimentary rocks are generally classified as detrital or chemical:
   a. Detrital sedimentary rocks consist of solid particles derived from preexisting rocks.
   b. Chemical sedimentary rocks are derived from substances in solution by inorganic chemical processes or the biochemical activities of organisms. A subcategory called biochemical sedimentary rocks is recognized.
7. Carbonate rocks contain minerals with the carbonate ion $(CO_3)^{-2}$ as in limestone and dolostone. Dolostone probably forms when magnesium partly replaces the calcium in limestone.
8. Evaporites include rock salt and rock gypsum, both of which form by inorganic precipitation of minerals from evaporating water.
9. Coal is a type of biochemical sedimentary rock composed of the altered remains of land plants.
10. Sedimentary facies are bodies of sediment or sedimentary rock that are recognizably different from adjacent sediments or rocks.
11. Vertical sequences of facies with offshore facies overlying nearshore facies are formed when sea level rises with respect to the land, causing a marine transgression. A rise in the land relative to sea level causes a marine regression, which results in nearshore facies overlying offshore facies.
12. Sedimentary structures such as bedding, cross-bedding, and ripple marks commonly form in sediments when or shortly after they are deposited. Such features preserved in sedimentary rocks help geologists determine ancient current directions and depositional environments.
13. Sediments and sedimentary rocks are the host materials for most fossils. Fossils provide the only record of prehistoric life and are useful for correlation and environmental interpretations.
14. Depositional environments of ancient sedimentary rocks are determined by studying sedimentary textures and structures, examining fossils, and making comparisons with present-day depositional processes.
15. Many sediments and sedimentary rocks including sand, gravel, evaporites, coal, and banded iron formations are important natural resources. Most oil and natural gas are found in sedimentary rocks.

Important Terms

bed (bedding)
biochemical sedimentary rock
carbonate rock
cementation
chemical sedimentary rock
compaction
cross-bedding
depositional environment
detrital sedimentary rock
evaporite
fossil
graded bedding
lithification
marine regression
marine transgression
mud crack
paleocurrent
ripple mark
rounding
sediment
sedimentary facies
sedimentary rock
sedimentary structure
sorting
strata (straftification)

Review Questions

1. Which of the following is detrital sediment?
   a. ____ broken sea shells;
   b. ____ ions in solution;
   c. ____ quartz sand;
   d. ____ conglomerate;
   e. ____ graded bedding.
2. A clay-sized sedimentary particle measures:
   a. ____ greater than 2 m;
   b. ____ 2–10 cm;
   c. ____ ¼–½ mm;
   d. ____ less than 1⁄256 mm;
   e. ____ 5 mm.
3. If an aggregate of sediment consists of particles that are all about the same size, it is said to be:
   a. ____ well sorted;
   b. ____ poorly rounded;
   c. ____ completely abraded;
   d. ____ sandstone;
   e. ____ lithified.

4. The process whereby dissolved mineral matter precipitates in the pore spaces of sediment and binds it together is:
   a. ____ compaction;
   b. ____ rounding;
   c. ____ bedding;
   d. ____ weathering;
   e. ____ cementation.
5. Sedimentary breccia is a rare rock type because:
   a. ____ gravel is rounded quickly during transport;
   b. ____ clay is less abundant than other sedimentary particles;
   c. ____ feldspars are chemically unstable;
   d. ____ sand deposits are typically well sorted;
   e. ____ it forms only by evaporation of water.
6. The most abundant detrital sedimentary rocks are:
   a. ____ limestones;
   b. ____ sandstones;
   c. ____ evaporites;
   d. ____ mudrocks;
   e. ____ arkoses.
7. Most limestones have a large component of calcite that was originally extracted from seawater by:
   a. ____ inorganic chemical reactions;
   b. ____ organisms;
   c. ____ evaporation;
   d. ____ chemical weathering;
   e. ____ lithification.
8. Dolostone is formed by the addition of ________ to limestone.
   a. ____ calcium;
   b. ____ carbonate;
   c. ____ magnesium;
   d. ____ iron;
   e. ____ sodium.
9. The superposition of offshore facies over nearshore facies occurs when sea level rises and the shoreline migrates inland during a marine:
   a. ____ superposition;
   b. ____ regression;
   c. ____ facies;
   d. ____ invasion;
   e. ____ transgression.
10. Which of the following can be used to determine paleocurrent direction?
    a. ____ mud cracks;
    b. ____ graded bedding;
    c. ____ cross-bedding;
    d. ____ turbidity currents;
    e. ____ grain size.
11. Which of the following is a trace fossil?
    a. ____ dinosaur tooth;
    b. ____ frozen mammoth;
    c. ____ worm burrow;
    d. ____ bird bone;
    e. ____ clam shell.
12. Traps for petroleum and natural gas resulting from variations in the properties of sedimentary rocks are ________traps.
    a. ____ reservoir;
    b. ____ stratigraphic;
    c. ____ cap rock;
    d. ____ structural;
    e. ____ salt dome.
13. Most of the known oil shales are in:
    a. ____ Russia;
    b. ____ China;
    c. ____ Venezuela;
    d. ____ the United States;
    e. ____ Australia.
14. How does the gravel in sedimentary breccia differ from the gravel in conglomerate?
15. Explain why the sediment in wind-blown sand dunes is better sorted than that in glacial deposits.
16. What are the common chemical cements in sedimentary rocks, and how do they form?
17. Distinguish clastic and crystalline textures. Give an example of a sedimentary rock with each texture.
18. In what fundamental way do chemical sedimentary rocks differ from detrital sedimentary rocks?
19. What are the common evaporites, and how do they originate?
20. Briefly describe the origin of coal.
21. Name three sedimentary structures and explain how they form.
22. How can fossils be used to interpret ancient depositional environments?
23. What are marine transgressions and regressions? Explain how a marine transgression can account for beach sand being deposited over a vast region.
24. What kinds of data do geologists use to determine depositional environment?
25. What is oil shale, and how can liquid oil be extracted from it?

Points to Ponder

1. As a field geologist, you encounter a rock unit consisting of well-sorted, well-rounded sandstone. In addition, the unit has large cross-beds and contains reptile footprints. What can you infer about the depositional environment?
2. While visiting one of our national parks, you notice a vertical sequence of rocks consisting of sandstone at the base followed upward by shale and limestone. All of these rocks contain fossil sea lilies, corals, and clams. Explain how these rocks were deposited and how they came to be superposed in the vertical sequence observed.
3. The Earth's crust is estimated to contain 51% feldspars, 24% ferromagnesian silicates (biotite, pyroxenes, amphiboles), 12% quartz, and 13% other minerals. Considering these abundances, how can you explain the fact that quartz is by far the most common mineral in sandstones?
4. The United States has a total coal reserve of 243 billion metric tons and uses about 860 million metric tons of coal annually. Assuming that all of this coal can be mined, how long will it last at the current rate of consumption? Why is it improbable that all of the reserve can be mined?

ADDITIONAL READINGS

Blatt, H., G. Middleton, and R. Murray. 1980. *Origin of sedimentary rocks.* New York: W. H. Freeman.

Boggs, S., Jr. 1987. *Principles of sedimentology and stratigraphy.* Columbus, Ohio: Merrill Publishing Co.

Collinson, J. D., and D. B. Thompson. 1982. *Sedimentary structures.* London: Allen & Unwin.

Friedman, G. M., J. E. Sanders, and D. C. Kopaska-Merkel. 1992. *Principles of Sedimentary Deposits.* New York: Macmillan Publishing Co.

Fritz, W. J., and J. N. Moore. 1988. *Basics of physical stratigraphy and sedimentology.* New York: John Wiley & Sons.

LaPorte, L. F. 1979. *Ancient environments.* 2d ed. Englewood Cliffs, N.J.: Prentice-Hall.

Moody, R. 1986. *Fossils.* New York: Macmillan Publishing Co.

Selley, R. C. 1978. *Ancient sedimentary environments.* Ithaca, N.Y.: Cornell University Press.

______ . 1982. *An introduction to sedimentology.* 2d ed. New York: Academic Press.

Simpson, G. G. 1983. *Fossils and the history of life.* New York: Scientific American Books.

Chapter 7

METAMORPHISM AND METAMORPHIC ROCKS

OUTLINE

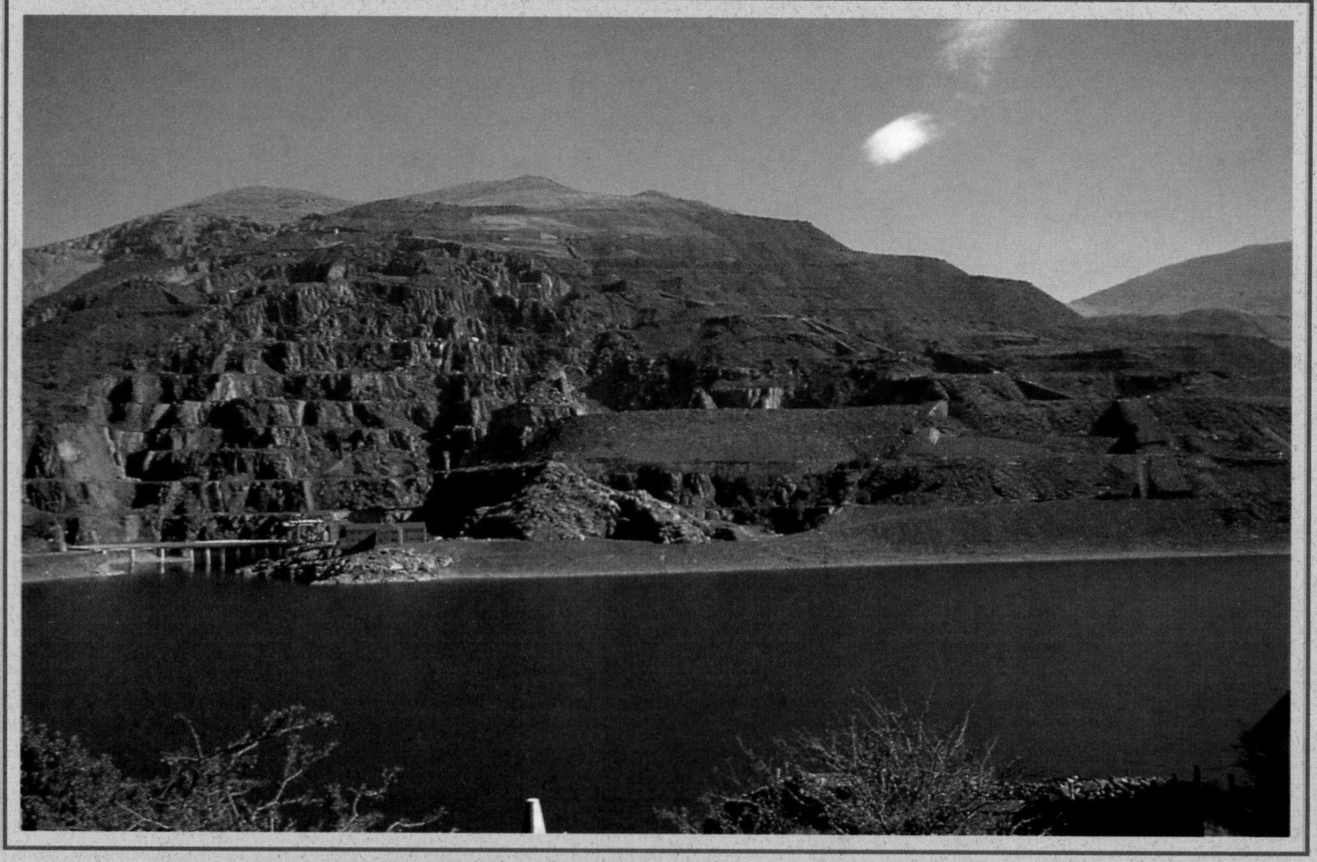

Slate quarry in Wales. Slate is the result of low-grade metamorphism of shale and has a variety of uses. These high-quality slates were formed by a mountain-building episode that occurred approximately 400 to 440 million years ago in the present-day countries of Iceland, Scotland, Wales, and Norway.

Prologue

Marble is a metamorphic rock that is formed from limestone or dolostone. Its homogeneity, softness, and textures have made marble a favorite rock of sculptors throughout history. As the value of authentic marble sculptures has increased through the years, the number of forgeries has also increased. With the price of some marble sculptures in the millions of dollars, private collectors and museums need some means of assuring the authenticity of the work they are buying. Aside from the monetary considerations, it is important that forgeries not become part of the historical and artistic legacy of human endeavor.

Experts have traditionally relied on the artistic style and weathering characteristics to determine whether a marble sculpture is authentic or a forgery. Because marble is not very resistant to weathering, forgers have resorted to a variety of methods to produce the weathered appearance of an authentic ancient work. Now, however, using new techniques, geologists can distinguish a naturally weathered marble surface from one that has been artificially altered.

Although marbles result when the agents of metamorphism (heat, pressure, and fluid activity) are applied to carbonate rocks, the type of marble formed depends, in part, on the original composition of the parent carbonate rock as well as the type and intensity of metamorphism. Therefore, one way to authenticate a marble sculpture is to determine the origin of the marble itself. During the Preclassical, Greek, and Roman periods, the islands of Naxos, Thasos, and Paros in the Aegean Sea as well as the Greek mainland, Turkey, and Italy were all sites of major marble quarries.

In order to identify the source of the marble in various sculptures, geologists have employed a wide variety of analytical techniques. These include hand specimen and thin-section analysis of the marble, trace element analysis by X-ray fluorescence, stable isotopic ratio analysis for carbon and oxygen, and other more esoteric techniques. Currently, however, carbon and oxygen isotopic analysis has proven to be the most powerful and reliable method for source area determination, because each quarry yields marble with a distinctive set of carbon and oxygen isotope values.

Recall that isotopes are forms of the same element with different atomic mass numbers. If the carbon and oxygen isotope ratios of a sculpture fall outside the typical range of the locality from which the marble supposedly comes, then it is probably a forgery. Using this technique, geologists showed that a marble head of Achilles owned by the J. Paul Getty Museum in Malibu, California, was a forgery. When the carbon and oxygen isotope ratios from the Getty Museum specimen were compared with those obtained from another marble head of known authenticity, they did not match, indicating that the two sculptures were carved from marbles that came from two different quarries.

Norman Herz of the Geology Department of the University of Georgia has sampled all of the major and many of the minor ancient marble quarries in the Aegean Sea region and assembled a large isotopic data base for these quarries. Using this data base for comparative purposes, Herz not only determined the source area of many marble pieces, but also demonstrated that some marble sculptures have been reassembled from marbles that came from different localities and therefore were not part of the original piece.

In one case, Herz determined that the five fragments composing the Antonia Minor portrait in the Fogg Museum at Harvard University (▶ Figure 7-1) are not all the same marble. The portrait was purchased by the earl of Pembroke in 1678 and its restoration was completed in 1758. Since that time, art historians have debated the portrait's authenticity and method of restoration.

The five fragments composing the portrait are the head, the end of the ponytail, the right shoulder and breast, the lower left shoulder, and the upper left shoulder and breast. Carbon and oxygen isotopic analysis of the five fragments revealed that three of the pieces were of Parian marble and

▶ FIGURE 7-1 Carbon and oxygen isotopic analysis of the 53-cm tall Antonia Minor portrait showed that the head is authentic, but unrelated to the other four pieces that compose it.

two were Carrara marble. It was concluded that the head is authentic, but unrelated to the other pieces, with the right shoulder and breast and the upper left shoulder and breast being comparatively recent additions.

Another interesting case involved the authentication of an ancient Greek kouros (a sculptured figure of a Greek youth) thought to have been carved around 530 B.C. (▶ Figure 7-2). The kouros was offered to the Getty Museum in 1984 for a reported price of $9 million. Some of its stylistic features, however, caused some experts to question its authenticity. Because of the value of the kouros, the museum had a variety of geochemical and mineralogical tests performed in an effort to determine its authenticity.

The kouros was carved from dolomitic marble and its surface is covered with a complex thin crust (0.01 to 0.05 mm thick) consisting mostly of whewellite, a calcium oxalate monohydrate mineral. In order to be sure that the crust is the result of long-term weathering, and not a modern forgery, dolomitic marble samples were subjected to a variety of forgery techniques to try and replicate the surface of the kouros. Samples were soaked or boiled in various mixtures for periods of time ranging from hours to months, and their surfaces treated and retreated to try and match the appearance of the weathered surface of the kouros. Such tests yielded only a few examples that appeared similar to the surface of kouros. Even those samples, however, were different when examined under high magnification or subjected to geochemical analysis. In fact, all of the samples clearly showed that they were the result of recent alteration and not long-term weathering processes.

While the scientific tests have been unable to unequivocably prove authenticity, they have shown that the weathered surface layer of the kouros bears more similarities to naturally occurring weathered surfaces than to known artifically produced surfaces. Furthermore, there is no evidence indicating that the surface alteration of the kouros is of modern origin.

In spite of intensive study by scientists, archeologists, and art historians, opinion is still divided as to the authenticity of the Getty kouros. Most scientists accept that the kouros was carved sometime around 530 B.C., but most art historians are doubtful. Pointing to inconsistencies in its style of sculpture for that period, they believe it is a modern forgery.

Regardless of the ultimate conclusion on the Getty kouros, geological testing to authenticate marble sculptures is now an important part of many museums' curatorial functions. In addition, a large body of data about the characteristics and origin of marble is being amassed as more sculptures and quarries are analyzed.

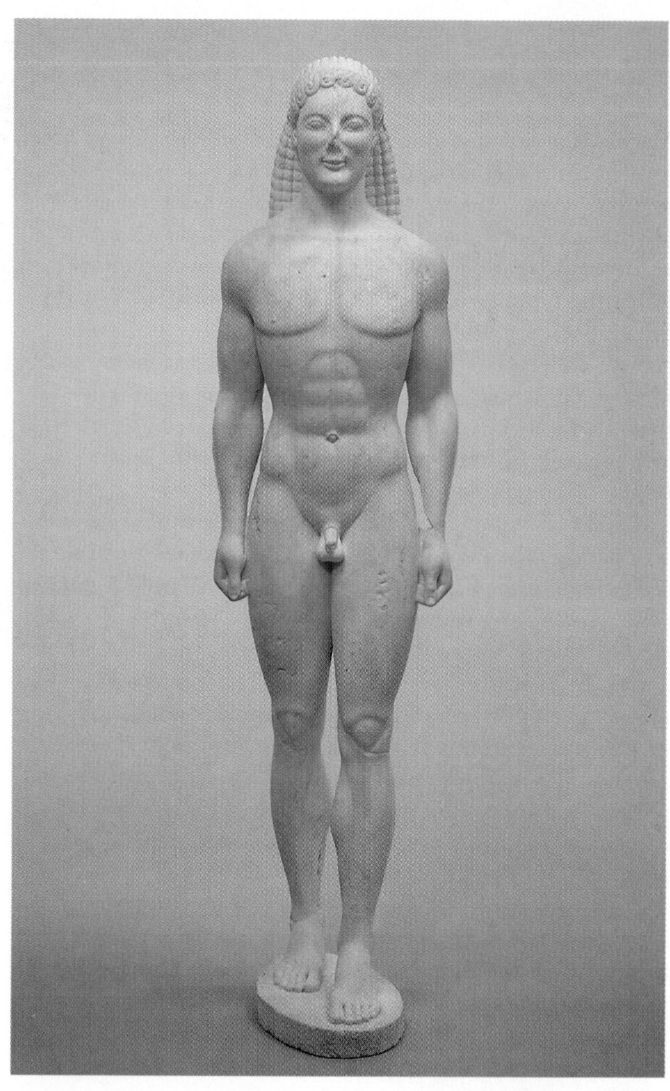

▶ FIGURE 7-2 This Greek kouros, which stands 206 cm tall, is the object of an intensive authentication study by the Getty Museum. Using several geological tests, it has been determined that the kouros was carved from dolomitic marble that probably came from the Cape Vathy quarries on Thasos.

INTRODUCTION

Metamorphic rocks (from the Greek *meta* meaning change and *morpho* meaning shape), are the third major group of rocks. They result from the transformation of other rocks by metamorphic processes that usually occur beneath the Earth's surface (see Figure 1-16). During metamorphism, rocks are subjected to sufficient heat, pressure, and fluid activity to change their mineral composition and/or texture,

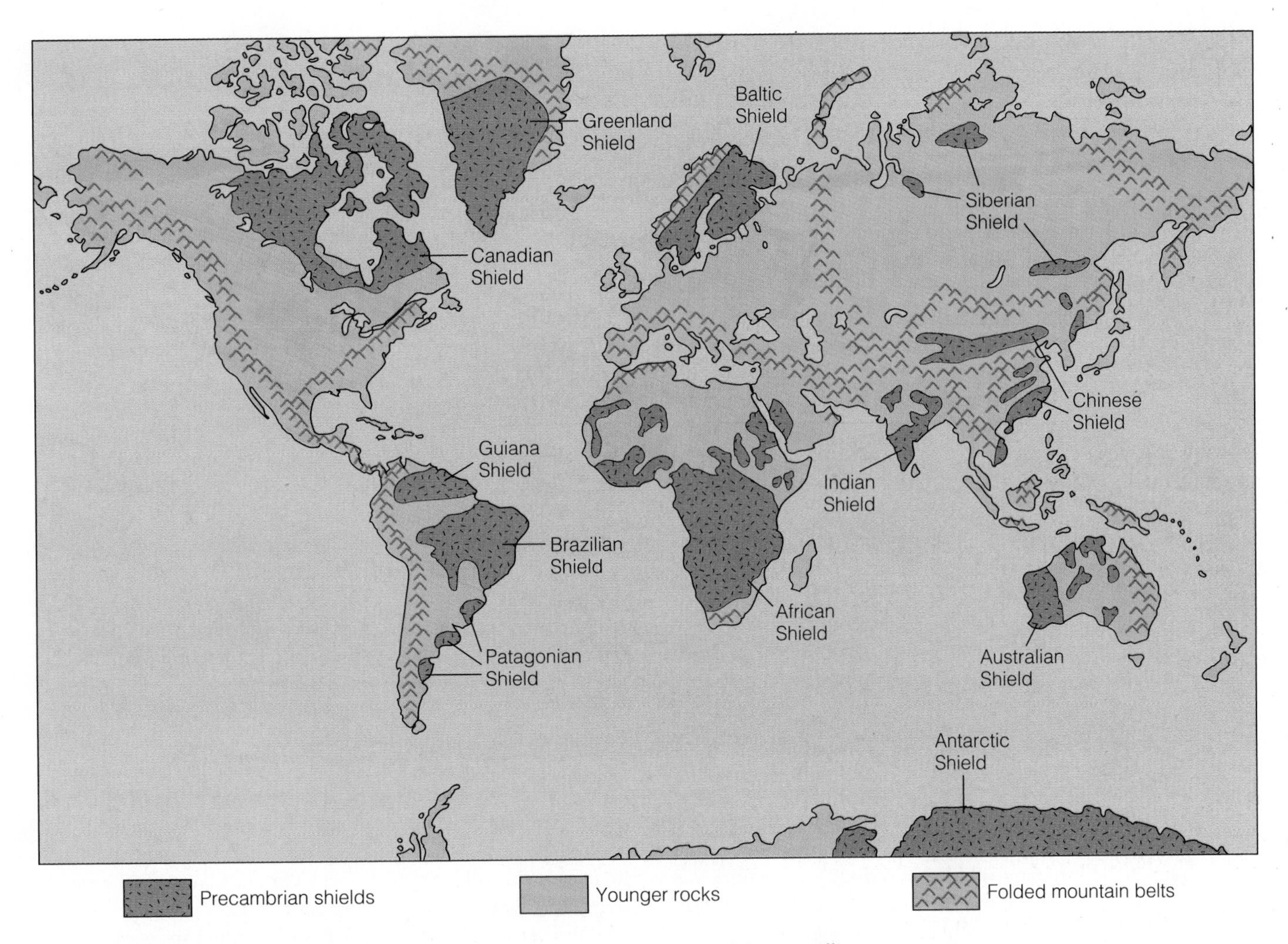

FIGURE 7-3 Shields of the world. Shields are the exposed portion of the crystalline basement rocks that underlie each continent; these areas have been very stable during the past 600 million years.

thus forming new rocks. These transformations take place in the solid state, and the type of metamorphic rock formed depends on the original composition and texture of the parent rock, the agents of metamorphism, and the amount of time the parent rock was subjected to the effects of metamorphism.

A large portion of the Earth's continental crust is composed of metamorphic and igneous rocks. Together, they form the crystalline basement rocks that underlie the sedimentary rocks of a continent's surface. This basement rock is widely exposed in regions of the continents known as *shields*; which have been very stable during the past 600 million years (Figure 7-3). Metamorphic rocks also constitute a sizable portion of the crystalline core of large mountain ranges. Some of the oldest known rocks, dated at 3.96 billion years from the Canadian Shield, are metamorphic, indicating they formed from even older rocks.

Why is it important to study metamorphic rocks? For one thing, they provide information about geological processes operating within the Earth and about the way these processes have varied through time. From the presence of certain minerals in metamorphic rocks, geologists can determine the approximate temperatures and pressures that parent rocks were subjected to during metamorphism and thus gain insights into the physical and chemical changes that occur at different depths within the Earth's crust. Furthermore, metamorphic rocks such as marble and slate are used as building materials, and certain metamorphic minerals are economically valuable. For example, garnets are used as gemstones or abrasives; talc is used in cosmetics, in the manufacture of paint, and as a lubricant; asbestos is used for insulation and fireproofing (see Perspective 7-1); and kyanite is used in the production of refractory materials such as sparkplugs.

The Agents of Metamorphism

The three agents of metamorphism are heat, pressure, and fluid activity. During metamorphism, the original rock undergoes change so as to achieve equilibrium with its new environment. The changes may result in the formation of new minerals and/or a change in the texture of the rock by

Perspective 7-1

ASBESTOS

Asbestos (from the Latin, meaning unquenchable) is a general term applied to any silicate mineral that easily separates into flexible fibers (▶ Figure 1). The combination of such features as noncombustibility and flexibility makes asbestos an important industrial material of considerable value. In fact, asbestos has more than 3,000 known uses. These include brake linings fireproof fabrics, heat insulators, and chemical equipment, to name a few.

The unique properties of asbestos were certainly known in the ancient world. The Romans used it to make lamp wicks that never burned out and also wove it into cremation clothes for the nobility.

Asbestos can be divided into two broad groups, serpentine and amphibole asbestos. *Chrysotile,* which is a hydrous magnesium silicate with the chemical formula $Mg_3Si_2O_5(OH)_4$, is the fibrous form of serpentine asbestos; it is the most valuable type and constitutes the bulk of all commercial asbestos. Chrysotile's strong, silky fibers are easily spun and can withstand temperatures up to 2,750°C.

The vast majority of chrysotile asbestos occurs in serpentine, a type of rock formed by the alteration of such ultramafic igneous rocks as peridotite under low- and medium-grade metamorphic conditions. Serpentine is thought to form from the alteration of olivine by hot, chemically active, residual fluids emanating from cooling magma. The chrysotile asbestos forms veinlets of fiber within the serpentine and may comprise up to 20% of the rock. Other chrysotile results when the metamorphism of magnesium limestone or dolostone produces discontinuous serpentine bands that develop within the carbonate beds.

At least five varieties of amphibole asbestos are known, but *crocidolite*, a sodium-iron amphibole with the chemical formula $Na_2(Fe^{+3})_2(Fe^{+2})_3Si_8O_{22}(OH)_2$, is the most common. Crocidolite, which is also known as blue asbestos, is a long, coarse, spinning fiber that is stronger but more brittle than chrysotile and also less resistant to heat. The other varieties of amphibole asbestos have little commercial value and are used chiefly for insulation.

Crocidolite is found in such metamorphic rocks as slates and schists. It is thought that crocidolite forms by the solid-state alteration of other minerals within the high-temperature and high-pressure environment that results from deep burial. Unlike chrysotile, crocidolite is rarely found associated with igneous intrusions.

In spite of its widespread use the federal Environmental Protection Agency (EPA) has instituted a gradual ban on all new asbestos products. The ban was imposed because some forms of asbestos can cause lung cancer and scarring of the lungs if its fibers are inhaled. Because the EPA apparently paid little attention to the issue of risks versus benefits when it enacted this rule, the U.S. Fifth Circuit Court of Appeals overturned the EPA ban on asbestos in 1991.

The threat of lung cancer has also resulted in legislation mandating the removal of asbestos already in place in all public buildings, including all public and private schools.

---

the reorientation of the original minerals. In some instances the change is minor, and features of the parent rock can still be recognized. In other cases the rock changes so much that the identity of the parent rock can be determined only with great difficulty, if at all.

In addition to heat, pressure, and fluid activity, time is also important to the metamorphic process. Chemical reactions proceed at different rates and thus require different amounts of time to complete. Reactions involving silicate compounds are particularly slow, and because most metamorphic rocks are composed of silicate minerals, it is thought that metamorphism is a slow geologic process. There is currently no way to effectively duplicate and extrapolate the time element in metamorphic processes in the laboratory. Nevertheless, based on laboratory experiments replicating the temperatures and pressures rocks are subjected to during metamorphism, it appears that the larger the minerals comprising a metamorphic rock, the longer the rock was subjected to metamorphic conditions. Accordingly, coarse-grained metamorphic rocks probably result from long periods of sustained metamorphism, while fine-grained metamorphic rocks reflect lower temperatures and pressures over a shorter length of time.

Heat

Heat is an important agent of metamorphism because it increases the rate of chemical reactions that may produce minerals different from those in the original rock. The heat may come from intrusive magmas or result from deep burial in the Earth's crust such as occurs during subduction along a convergent plate boundary.

When rocks are intruded by bodies of magma, they are subjected to intense heat that affects the surrounding rock; the most intense heating usually occurs adjacent to the

Recently, however, important questions have been raised concerning the threat posed by asbestos and the additional potential hazards that may arise from its improper removal.

The current policy (1993) of the EPA mandates that all forms of asbestos be treated as identical hazards. Yet studies indicate that only the amphibole forms constitute a known health hazard. Chrysotile, whose fibers tend to be curly, does not become lodged in the lungs. Furthermore, its fibers are generally soluble and disappear in tissue. In contrast, crocidolite has long, straight, thin fibers that penetrate the lungs and stay there. These fibers irritate the lung tissue and over a long period of time can lead to lung cancer. Thus, crocidolite, and not chrysotile, is overwhelmingly responsible for asbestos-related lung cancer. Because about 95% of the asbestos in place in the United States is chrysotile, many people are questioning whether the dangers from asbestos have been somewhat exaggerated.

Removing asbestos from buildings where it has been installed might cost as much as $100 billion, and some recent studies have indicated that the air in buildings containing asbestos has essentially the same amount of airborne asbestos fibers as the air outdoors. In fact, unless the material containing the asbestos is disturbed, asbestos does not shed fibers. Furthermore, improper removal of asbestos can lead to contamination. In most cases of improper removal, the concentration of airborne asbestos fibers is far higher than if the asbestos had been left in place.

The problem of asbestos contamination is a good example of how geology affects our lives and why a basic knowledge of science is important. As we have indicated, only the amphibole forms of asbestos have been shown to be a health hazard, and then only when its fibers are inhaled. Yet legislation has been passed requiring all asbestos to be removed from public buildings regardless of whether it is one of the hazardous amphibole varieties or benign chrysotile.

▶ **FIGURE 1** Hand specimen of chrysotile from Thetford, Quebec, Canada. Chrysotile is the fibrous form of serpentine asbestos.

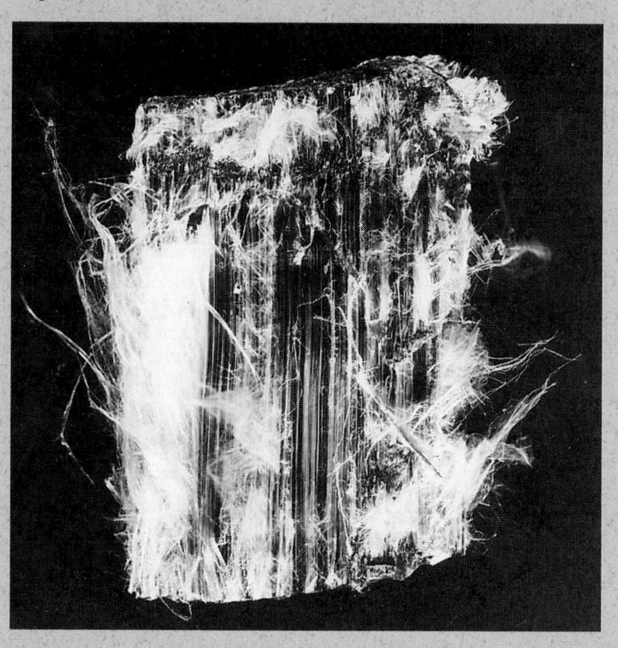

magma body and gradually decreases with distance from the intrusion. The zone of metamorphosed rocks that forms in the country rock adjacent to an intrusive igneous body is usually rather distinct and easy to recognize.

Recall that temperature increases with depth and that the Earth's geothermal gradient averages about 25°C/km. Rocks forming at the Earth's surface may be transported to great depths by subduction along a convergent plate boundary and subjected to increasing temperature and pressure. During subduction, some minerals may be transformed into other minerals that are more stable under the higher temperature and pressure conditions.

Pressure

When rocks are buried, they are subjected to increasingly greater **lithostatic pressure**; this pressure, which results from the weight of the overlying rocks, is applied equally in all directions (▶ Figure 7-4a). A similar situation occurs when an object is immersed in water. For example, the deeper a styrofoam cup is submerged in the ocean, the smaller it gets because pressure increases with depth and is exerted on the cup equally in all directions, thereby compressing the styrofoam (Figure 7-4b).

Just as in the styrofoam cup example, rocks are subjected to increasing lithostatic pressure with depth such that the mineral grains within a rock may become more closely packed. Under these conditions, the minerals may *recrystallize*; that is, they may form smaller and denser minerals.

In addition to the lithostatic pressure resulting from burial, rocks may also experience **differential pressures**. In this case, the pressures are not equal on all sides, and the rock is consequently distorted. Differential pressures typically occur during deformation associated with mountain building and can produce distinctive metamorphic textures and features (▶ Figure 7-5).

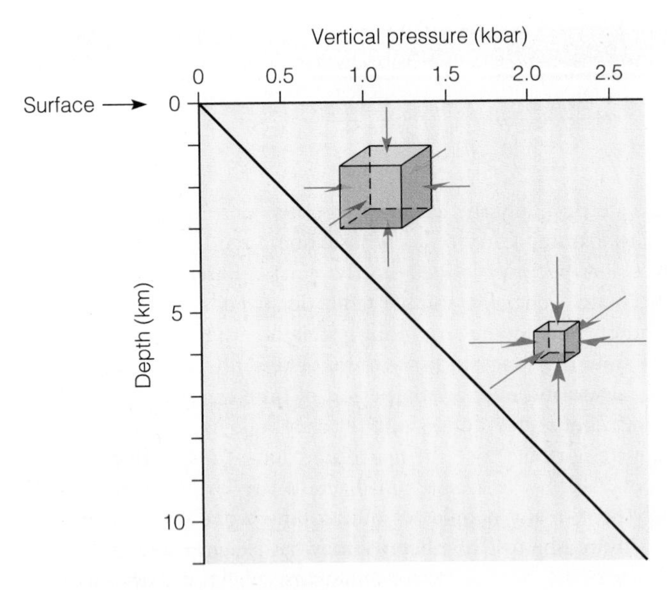

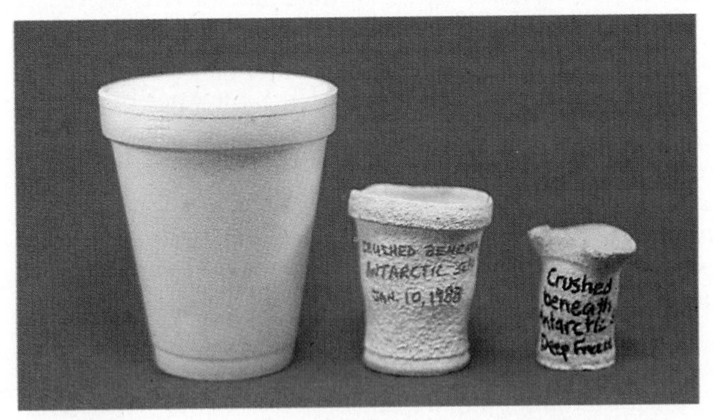

(b)

▶ **FIGURE 7-4** (*a*) Lithostatic pressure is applied equally in all directions in the Earth's crust due to the weight of the overlying rocks. Thus, pressure increases with depth. (*b*) A similar situation occurs when 200 ml styrofoam cups are lowered to ocean depths of approximately 750 m and 1,500 m. Increased pressure is exerted equally in all directions on the cups, and they consequently decrease in volume, while still maintaining their general shape. (Styrofoam cups courtesy of David J. Matty and Jane M. Matty. Photo courtesy of Sue Monroe.)

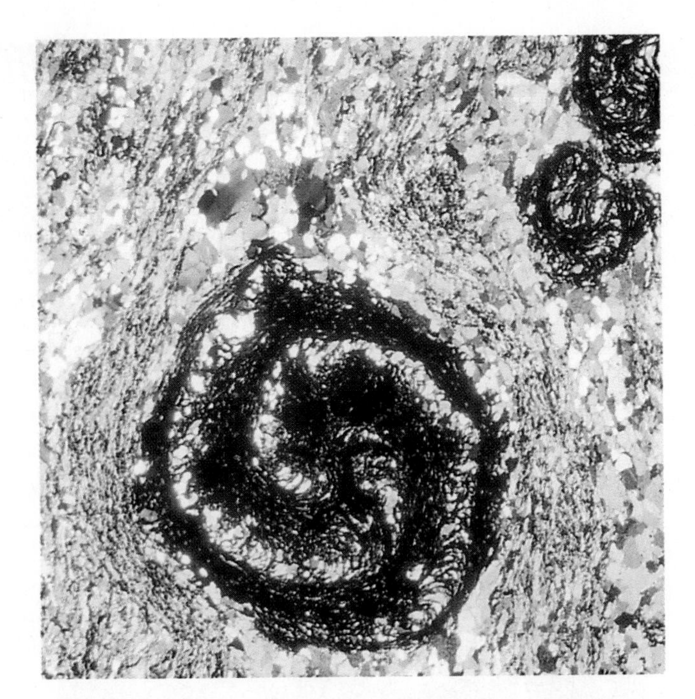

▶ **FIGURE 7-5** Differential pressure is pressure that is unequally applied to an object. Rotated garnets are a good example of the effects of differential pressure applied to a rock during metamorphism. These rotated garnets come from a calcareous schist of the Waits River Formation, north of Springfield, Vermont. (Photo courtesy of John L. Rosenfeld, University of California, Los Angeles.)

Fluid Activity

In almost every region where metamorphism occurs, water and carbon dioxide (CO_2) are present in varying amounts along mineral grain boundaries or in the pore spaces of rocks. These fluids, which may contain ions in solution, enhances metamorphism by increasing the rate of chemical reactions. Under dry conditions, most minerals react very slowly, but when even small amounts of fluid are introduced, reaction rates increase, mainly because ions can move readily through the fluid and thus enhance chemical reactions and the formation of new minerals.

The following reaction illustrates how new minerals can be formed by **fluid activity**. Seawater moving through hot basaltic rock transforms olivine into the metamorphic mineral serpentine:

$$\underset{\text{olivine}}{2Mg_2SiO_4} + \underset{\text{water}}{2H_2O} \rightarrow \underset{\text{serpentine}}{Mg_3Si_2O_5(OH)_4} + \underset{\text{carried away in solution}}{MgO}$$

The chemically active fluids that are part of the metamorphic process come primarily from three sources. The first is water trapped in the pore spaces of sedimentary rocks as they form; as these rocks are subjected to heat and pressure, the water is heated, thus accelerating the various chemical reaction rates. A second source is the volatile fluid within magma; as these hot fluids disperse through the surrounding rock, they frequently react with and alter the minerals of the country rock by adding or removing ions. The third source is the dehydration of water-bearing minerals such as gypsum ($CaSO_4 \cdot 2H_2O$) and some clays; when these minerals, which contain water as part of their crystal chemistry, are subjected to heat and pressure, the water may be driven off and enhance metamorphism.

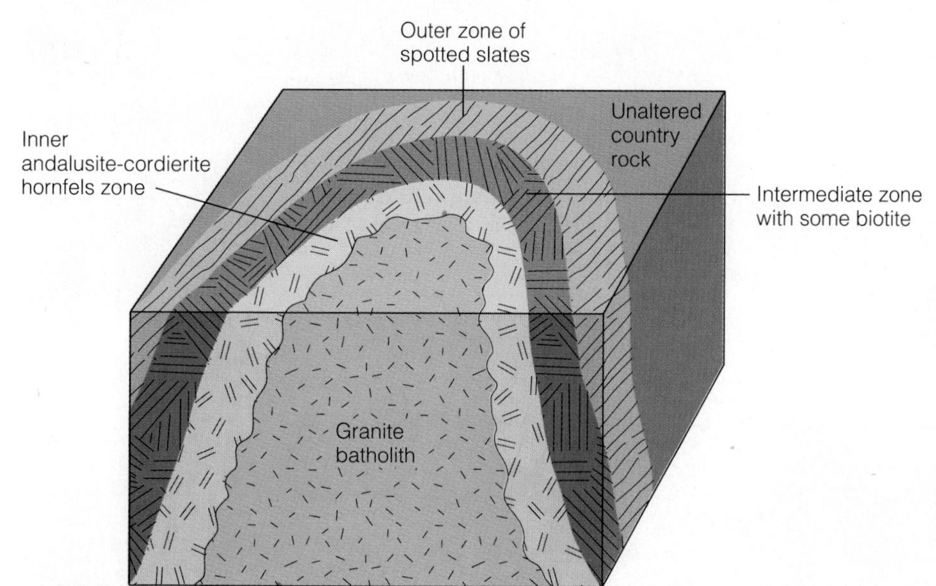

⊳ FIGURE 7-6 A metamorphic aureole typically surrounds igneous intrusions. The metamorphic aureole around this idealized granite batholith contains three zones of mineral assemblages reflecting the decreases in temperature with distance from the intrusion. An andalusite- cordierite hornfels forms the inner zone adjacent to the batholith. This is followed by an intermediate zone of extensive recrystallization in which some biotite develops, and farthest from the intrusion is the outer zone, which is characterized by spotted slates.

TYPES OF METAMORPHISM

Three major types of metamorphism are recognized: *contact metamorphism* in which magmatic heat and fluids act to produce change; *dynamic metamorphism*, which is principally the result of high differential pressures associated with intense deformation; and *regional metamorphism*, which occurs within a large area and is caused primarily by mountain-building forces. Even though we will discuss each type of metamorphism separately, the boundary between them is not always distinct and depends largely on which of the three metamorphic agents was dominant.

Contact Metamorphism

Contact metamorphism takes place when a body of magma alters the surrounding country rock. At shallow depths an intruding magma raises the temperature of the surrounding rock, causing thermal alteration. Furthermore, the release of hot fluids into the country rock by the cooling intrusion can also aid in the formation of new minerals.

Important factors in contact metamorphism are the initial temperature and size of the intrusion as well as the fluid content of the magma and/or country rock. The initial temperature of an intrusion is controlled, in part, by its composition: mafic magmas are hotter than felsic magmas (see Chapter 3) and hence have a greater thermal effect on the rocks directly surrounding them. The size of the intrusion is also important. In the case of small intrusions, such as dikes and sills, usually only those rocks in immediate contact with the intrusion are affected. Because large intrusions, such as batholiths, take a long time to cool, the increased temperature in the surrounding rock may last long enough for a larger area to be affected.

Fluids also play an important role in contact metamorphism. Many magmas are wet and contain hot, chemically active fluids that may emanate into the surrounding rock. These fluids can react with the rock and aid in the formation of new minerals. In addition, the country rock may contain pore fluids that, when heated by the magma, also increase reaction rates.

Temperatures can reach nearly 900°C adjacent to an intrusion, but they gradually decrease with distance. The effects of such heat and the resulting chemical reactions usually occur in concentric zones known as **aureoles** (⊳ Figure 7-6). The boundary between an intrusion and its aureole may be either sharp or transitional (⊳ Figure 7-7).

Metamorphic aureoles vary in width depending on the size, temperature, and composition of the intrusion as well as the compositon of the surrounding country rock. For example, small intrusive bodies such as sills and dikes may produce an aureole only a few centimeters wide, whereas large intrusive bodies such as batholiths may give rise to an aureole several kilometers wide. Typically, these large intrusive bodies have several metamorphic zones, each characterized by distinctive mineral assemblages indicating the decrease in temperature with distance from the intrusion (Figure 7-6). The zone closest to the intrusion, and hence subject to the highest temperatures, may contain high-temperature metamorphic minerals (that is, minerals in equilibrium with the higher temperature environment) such as sillimanite. The outer zones may be characterized by lower-temperature metamorphic minerals such as chlorite, talc, and epidote.

The formation of new minerals by contact metamorphism depends not only on proximity to the intrusion, but also on the composition of the country rock. Shales, mudstones, impure limestones, and impure dolostones are particularly susceptible to the formation of new minerals by contact metamorphism, whereas pure sandstones or pure limestones typically are not.

Two types of contact metamorphic rocks are generally recognized: those resulting from baking of country rock and

▶ **FIGURE 7-7** A sharp and clearly defined boundary occurs between the intruding light-colored igneous rock on the left and the dark-colored metamorphosed country rock on the right. The intrusion is part of the Peninsular Ranges Batholith, east of San Diego, California. (Photo courtesy of David J. Matty.)

those altered by hot solutions. Many of the rocks resulting from contact metamorphism have the texture of porcelain; that is, they are hard and fine grained. This is particularly true for rocks with a high clay content, such as shale. Such texture results because the clay minerals in the rock are baked, just as a clay pot is baked when fired in a kiln.

During the final stages of cooling, when the intruding magma begins to crystallize, large amounts of hot, watery solutions are often released. These solutions may react with the country rock and produce new metamorphic minerals. This process, which usually occurs near the Earth's surface, is called *hydrothermal alteration* and may result in valuable mineral deposits such as the Kuroko sulfide deposit in Japan. Geologists think that many of the world's ore deposits result from the migration of metallic ions in hydrothermal solutions. Examples include copper, gold, iron ores, tin, and zinc in various localities including Australia, Canada, China, Cyprus, Finland, Russia, and the western United States.

Dynamic Metamorphism

Most **dynamic metamorphism** is associated with fault (a fracture along which movement has occurred) zones where rocks are subjected to high differential pressures. The metamorphic rocks resulting from pure dynamic metamorphism are called *mylonites* and are typically restricted to narrow zones adjacent to faults. Mylonites are hard, dense, fine-grained rocks, many of which are characterized by thin laminations (▶ Figure 7-8). Mylonites are differentiated

▶ **FIGURE 7-8** Mylonite from the Adirondack Highlands, New York. (Photo courtesy of Eric Johnson.)

from fault breccias (rocks that are broken up by fault movement) by the intensity of the pressure applied to the rock. High shearing pressure completely pulverizes the country rock and essentially "smears" the fine particles together, producing a characteristic mylonite texture. Fault breccias, which are composed of broken particles, are not, strictly speaking, metamorphic rocks. Tectonic settings where mylonites occur include the Moine Thrust Zone in northwest Scotland, portions of the San Andreas fault in California, and the Swiss Alps.

Regional Metamorphism

Most metamorphic rocks result from **regional metamorphism**, which occurs over a large area and is usually caused by tremendous temperatures, pressures, and deformation within the deeper portions of the Earth's crust. Regional metamorphism is most obvious along convergent plate margins where rocks are intensely deformed and recrystallized during convergence and subduction. These metamorphic rocks usually exhibit a gradation of metamorphic intensity from areas that were subjected to the most intense pressures and/or highest temperatures to areas of lower pressures and temperatures. Such a gradation in metamorphism can be recognized by the metamorphic minerals that are present.

Regional metamorphism is not just confined to convergent margins. It also occurs in areas where plates diverge, though usually at much shallower depths because of the high geothermal gradient associated with these areas.

From field studies and laboratory experiments, certain minerals are known to form only within specific temperature and pressure ranges. Such minerals are known as **index minerals** because their presence allows geologists to recognize low-, intermediate-, and high-grade metamorphic zones (◉ Table 7-1). For example, in clay-rich rocks such as shale, the mineral chlorite is produced under relatively low temperatures of about 200°C, so its presence indicates low-grade metamorphism. At the high-grade end of the metamorphic spectrum for clay-rich rocks, sillimanite may form in environments where the temperature exceeds 500°C. A typical progression of index minerals from chlorite to sillimanite involves the sequential formation of the following minerals:

chlorite → biotite → garnet → staurolite → kyanite → sillimanite

Different rock compositions develop different index minerals. The sequence of index minerals just listed forms primarily in rocks that were originally clay rich. When sandy dolomites are metamorphosed, they produce an entirely different set of index minerals. Thus, a specific set of index minerals commonly forms in specific rock types as metamorphism progresses.

Although such common minerals as mica, quartz, and

TABLE 7-1 Metamorphic Zones and Their Mineral Assemblages for Different Country Rock Types

Metamorphic Grade	Metamorphic Zone for Clay-Rich Rocks	Mineral Assemblage Produced for Different Country Rocks		
		Mudrocks	*Limestones*	*Mafic Igneous Rocks*
Increasing				
Low	Chlorite	Chlorite,* quartz, muscovite, plagioclase	Chlorite,* calcite or dolomite, plagioclase	Chlorite,* plagioclase
	Biotite	Biotite,* quartz, plagioclase		
Medium	Garnet	Garnet,* mica, quartz, plagioclase	Garnet,* epidote, hornblende, calcite	Garnet,* chlorite, epidote, plagioclase
	Staurolite	Staurolite,* mica, garnet, quartz, plagioclase	Garnet, hornblende,* plagioclase	
High	Kyanite	Kyanite,* mica, garnet, quartz, plagioclase		
metamorphism				Hornblende,* plagioclase
	Sillimanite	Sillimanite,* garnet, mica, quartz, plagioclase	Garnet, augite,* plagioclase	

*Index mineral.

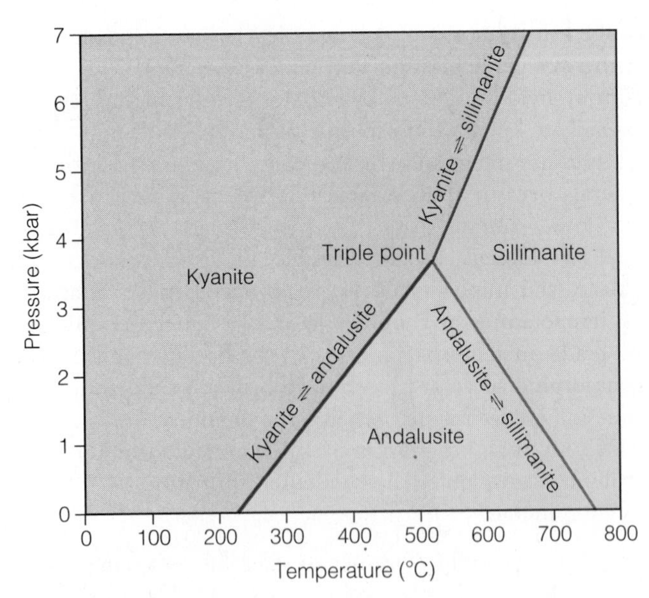

▶ **FIGURE 7-9** Andalusite, sillimanite, and kyanite all have the same chemical composition (Al_2SiO_5) but very different physical properties and crystal structures. This is because they form under different pressure-temperature conditions. The double arrows between the mineral combinations mean that the transformations are reversible; that is, they can go in either direction. The triple point where the three lines meet is the pressure-temperature value where, in theory, all three minerals can exist in equilibrium.

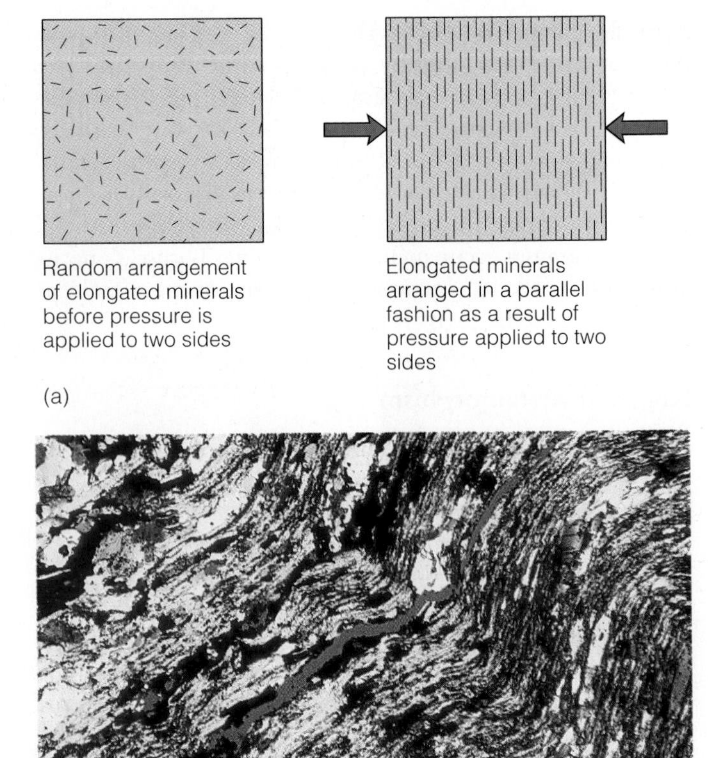

▶ **FIGURE 7-10** (*a*) When rocks are subjected to differential pressure, the mineral grains are typically arranged in a parallel fashion, producing a foliated texture. (*b*) Photomicrograph of a metamorphic rock with a foliated texture showing the parallel arrangement of mineral grains.

feldspar can occur in both igneous and metamorphic rocks, other minerals such as andalusite, sillimanite, and kyanite generally occur only in metamorphic rocks derived from clay-rich sediments. Although these three minerals all have the same chemical formula (Al_2SiO_5), they differ in crystal structure and other physical properties because each forms under a different range of pressures and temperatures (▶ Figure 7-9). Thus, they are sometimes used as index minerals for metamorphic rocks formed from clay-rich sediments.

 CLASSIFICATION OF METAMORPHIC ROCKS

For purposes of classification, metamorphic rocks are commonly divided into two groups: those exhibiting a *foliated texture* and those with a *nonfoliated texture* (◉ Table 7-2).

Foliated Metamorphic Rocks

Rocks subjected to heat and differential pressure during metamorphism typically have minerals arranged in a parallel fashion that gives them a **foliated texture** (▶ Figure 7-10). The size and shape of the mineral grains determine whether the foliation is fine or coarse. If the foliation is such that the individual grains cannot be recognized without magnification, the rock is said to be slate (▶ Figure 7-11a). A coarse foliation results when granular minerals such as quartz and feldspar are segregated into roughly parallel and streaky zones that differ in composition and color as in a gneiss. Foliated metamorphic rocks can be arranged in order of increasingly coarse grain size and perfection of foliation.

Slate is a very fine-grained metamorphic rock that commonly exhibits *slaty cleavage* (Figure 7-11b). Slate is the result of low-grade regional metamorphism of shale or, more rarely, volcanic ash. Because it can easily be split along cleavage planes into flat pieces, slate is an excellent rock for roofing and floor tiles, billiard and pool table tops, and blackboards. The different colors of most slates are caused by minute amounts of graphite (black), iron oxide (red and purple), and/or chlorite (green).

Phyllite is similar in composition to slate, but is coarser grained. However, the minerals are still too small to be identified without magnification. Phyllite can be distinguished from slate by its glossy or lustrous sheen. It represents an intermediate grain size between slate and schist.

Schist is most commonly produced by regional metamorphism. The type of schist formed depends on the intensity of metamorphism and the character of the parent rock (▶ Figure 7-12). Metamorphism of many rock types can

TABLE 7-2 Classification of Common Metamorphic Rocks

Texture	Metamorphic Rock	Typical Minerals	Metamorphic Grade	Characteristics of Rocks	Parent Rock
Foliated	Slate	Clays, micas, chlorite	Low	Fine-grained, splits easily into flat pieces	Mudrocks, claystones, volcanic ash
	Phyllite	Fine-grained quartz, micas, chlorite	Low to medium	Fine-grained, glossy or lustrous sheen	Mudrocks
	Schist	Micas, chlorite, quartz, talc, hornblende, garnet, staurolite, graphite	Low to high	Distinct foliation, minerals visible	Mudrocks, carbonates, mafic igneous rocks
	Gneiss	Quartz, feldspars, hornblende, micas	High	Segregated light and dark bands visible	Mudrocks, sandstones, felsic igneous rocks
	Amphibolite	Hornblende, plagioclase	Medium to high	Dark-colored, weakly foliated	Mafic igneous rocks
	Migmatite	Quartz, feldspars, hornblende, micas	High	Streaks or lenses of granite intermixed with gneiss	Felsic igneous rocks mixed with sedimentary rocks
Nonfoliated	Marble	Calcite, dolomite	Low to high	Interlocking grains of calcite or dolomite, reacts with HCI	Limestone or dolostone
	Quartzite	Quartz	Medium to high	Interlocking quartz grains, hard, dense	Quartz sandstone
	Greenstone	Chlorite, epidote, hornblende	Low to high	Fine-grained, green color	Mafic igneous rocks
	Hornfels	Micas, garnets, andalusite, cordierite, quartz	Low to medium	Fine-grained, equidimensional grains, hard, dense	Mudrocks
	Anthracite	Carbon	High	Black, lustrous, subconcoidal fracture	Coal

yield schist, but most schist appears to have formed from clay-rich sedimentary rocks.

All schists contain more than 50% platy and elongated minerals, all of which are large enough to be clearly visible. Their mineral composition imparts a *schistosity* or *schistose foliation* to the rock that commonly produces a wavy type of parting when split. Schistosity is common in low- to high-grade metamorphic environments. Because a schist's mineral grains can be readily identified, each type is known by its most conspicuous mineral or minerals, for example, mica schist, chlorite schist, and talc schist.

Gneiss is a metamorphic rock that is streaked or has segregated bands of light and dark minerals. Gneisses are composed mostly of granular minerals such as quartz and/or feldspar with lesser percentages of platy or elongated minerals such as micas or amphiboles (⯈ Figure 7-13). Quartz and feldspar are the principal light-colored minerals, while biotite and hornblende are the typical dark-colored minerals. Most gneiss breaks in an irregular manner, much like coarsely crystalline nonfoliated rocks.

Most gneiss probably results from recrystallization of clay-rich sedimentary rocks during regional metamorphism. Gneiss also can form from igneous rocks such as granite or older metamorphic rocks.

Another fairly common foliated metamorphic rock is *amphibolite*. It is a dark colored rock, and composed mainly of hornblende and plagioclase. The alignment of the hornblende crystals produces a slightly foliated texture. Many amphibolites result from medium- to high-grade metamorphism of such ferromagnesian mineral–rich igneous rocks as basalt.

In some areas of regional metamorphism, exposures of "mixed rocks" having both igneous and high-grade metamorphic characteristics are present. In these rocks, called *mig-*

(a)

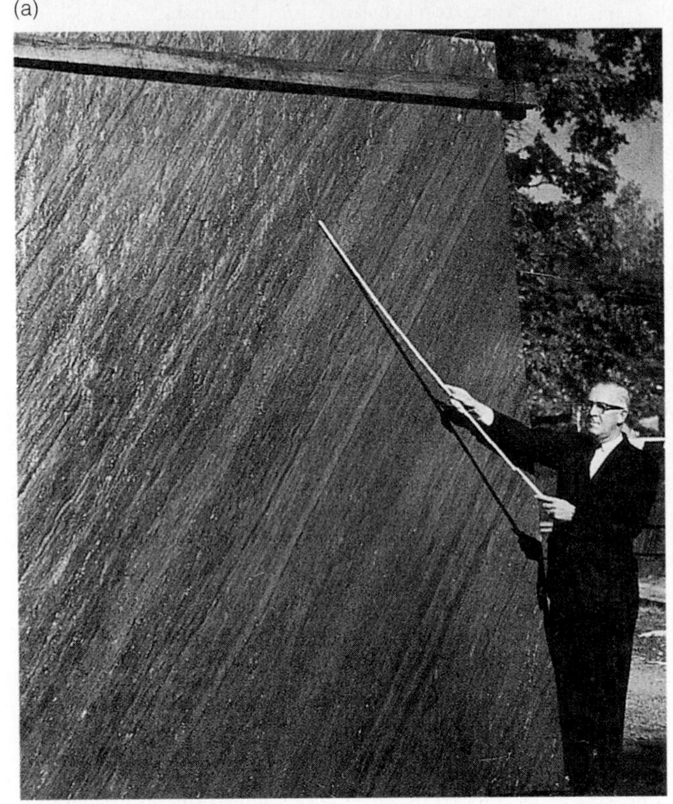

(b)

➢ FIGURE 7-11 (*a*) Hand specimen of slate. (*b*) This panel of Arvonia Slate from Albemarne Slate Quarry, Virginia, shows bedding (upper right to lower left) at an angle to the slaty cleavage. (Photo (*a*) courtesy of Sue Monroe; photo (*b*) courtesy of R. V. Dietrich.)

matites, streaks or lenses of granite are usually intermixed with high-grade ferromagnesian-rich metamorphic rocks, thereby imparting a wavy appearance to the rocks (➢ Figure 7-14).

Most migmatites are thought to be the product of extremely high-grade metamorphism, and several models for their origin have been proposed. Part of the problem in determining the origin of migmatites is explaining how the

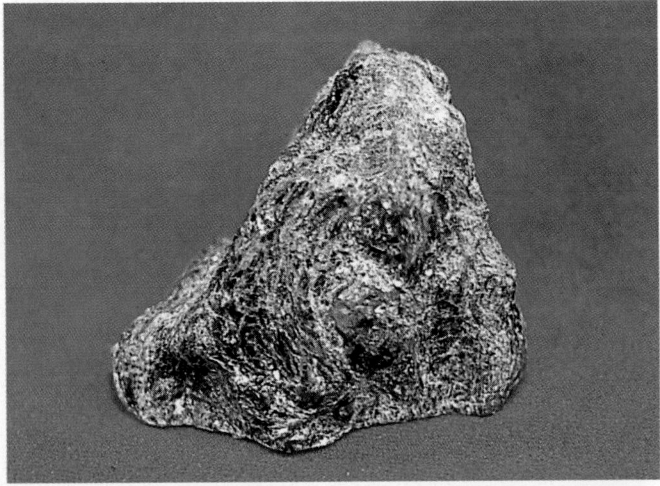

(a)

(b)

➢ FIGURE 7-12 Schist. (*a*) Garnet-mica schist. (*b*) Hornblende-mica-garnet schist. (Photos courtesy of Sue Monroe.)

➢ FIGURE 7-13 Gneiss is characterized by segregated bands of light and dark minerals. This folded gneiss is exposed near Wawa, Ontario, Canada.

⋗ **FIGURE 7-14** Migmatites consist of high-grade metamorphic rock intermixed with streaks or lenses of granite. This migmatite is exposed at Thirty Thousand Islands of Georgian Bay, Lake Huron, Ontario, Canada. (Photo by Ed Bartram, courtesy of R. V. Dietrich.)

granitic component formed. According to one model, the granitic magma formed in place by the partial melting of rock during intense metamorphism. Such an origin is possible providing that the host rocks contained quartz and feldspars and that water was present. Another possibility is that the granitic components formed by the redistribution of minerals by recrystallization in the solid state, that is, by pure metamorphism.

Nonfoliated Metamorphic Rocks

In some metamorphic rocks, the mineral grains do not show a discernible preferred orientation. Instead, these rocks consist of a mosaic of roughly equidimensional minerals and are characterized as **nonfoliated** (⋗ Figure 7-15). Most nonfoliated metamorphic rocks result from contact or regional metamorphism of rocks in which no platy or elognate minerals are present. Frequently, the only indication that a granular rock has been metamorphosed is the large grain size resulting from recrystallization. Nonfoliated metamorphic rocks are generally of two types: those composed mainly of only one mineral, for example, marble or quartzite; and those in which the different mineral grains are too small to be seen without magnification, such as greenstone and hornfels.

Marble is a well-known metamorphic rock composed predominantly of calcite or dolomite; its grain size ranges from fine to coarsely granular (Figs. 7-2 and ⋗ 7-16). Marble results from either contact or regional metamorphism of limestones or dolostones. Pure marble is snowy white or

⋗ **FIGURE 7-15** Nonfoliated textures are characterized by a mosaic of roughly equidimensional minerals as in this photomicrograph of marble.

⊳ **FIGURE 7-16** Marble results from the metamorphism of the sedimentary rock limestone or dolostone. (Photos courtesy of Sue Monroe.)

bluish, but varieties of all colors exist because of the presence of mineral impurities in the parent sedimentary rock. The softness of marble, its uniform texture, and its various colors have made it the favorite rock of builders and sculptors throughout history (see the Prologue).

Quartzite is a hard, compact rock formed from quartz sandstone under medium-to-high-grade metamorphic conditions during contact or regional metamorphism (⊳ Figure 7-17). Because recrystallization is so complete, metamorphic quartzite is of uniform strength and therefore usually breaks across the component quartz grains rather than around them when it is struck. Pure quartzite is white, but iron and other impurities commonly impart a reddish or other color to it. Quartzite is commonly used as foundation material for road and railway beds, because it is very hard and strong.

The name *greenstone* is applied to any compact, dark-green, altered, mafic igneous rock that formed under low-to-high-grade metamorphic conditions. The green color results from the presence of chlorite, epidote, and hornblende.

Hornfels is a fine-grained, nonfoliated metamorphic rock resulting from contact metamorphism; it is composed of various equidimensional mineral grains. The composition of hornfels is directly dependent upon the composition of the parent rock, and many compositional varieties are known. The majority of hornfels, however, are apparently derived from contact metamorphism of clay-rich sedimentary rocks or impure dolostones.

Anthracite is a black, lustrous, hard coal that contains a high percentage of fixed carbon and a low percentage of other elements. It usually forms from the metamorphism of various types of coals by heat and pressure and is thus considered by many geologists to be a metamorphic rock.

 METAMORPHIC ZONES AND FACIES

The first systematic study of metamorphic zones was conducted during the late 1800s by George Barrow and other British geologists working in the Dalradian schists of the southwestern Scottish Highlands. Here clay-rich sedimentary rocks have been subjected to regional metamorphism,

⊳ **FIGURE 7-17** Quartzite results from the metamorphism of quartz sandstone. (Photos courtesy of Sue Monroe.)

Conservation and Museum Science: Geology to Preserve and Understand Our Cultural Heritage

When I first arrived at Haverford College, I went through the catalog and crossed out all the subjects that didn't interest me. I was left with history, political science, and geology. I took courses in all three, but it was geology that caught my imagination. In graduate school I found I had a talent for instrumental analysis and learned to apply geochemical analytical techniques to environmental and art historical research. I also developed a strong interest in how rocks weather.

I was fortunate to find a position that required such skills at the Getty Conservation Institute (GCI), part of the J. Paul Getty Trust. The institute has an ambitious mission: to further the conservation of the world's cultural heritage. The GCI focuses its resources on scientific research, training conservators, and international projects, such as the conservation of Queen Nefertari's tomb in Egypt, Roman mosaics in Cyprus, and the elaborate carvings in Buddhist caves along the silk route in China.

The conservation of our cultural heritage has received much attention in recent years due to public concern over accelerated deterioration and loss caused by increased pollution, development, and tourism. The field of conservation science is relatively new and brings together collaborators from governments, universities, museums, and consultants from many fields. Conservation scientists who study monuments and architecture often have a geological background. Others have training as chemists, engineers, archaeologists, or architects. Museum science has a longer history than conservation science and is concerned with the techniques and materials used to create the objects in our museums as well as with their state of preservation. The related specialty of archaeometry deals with questions of age, authenticity, and provenance. Many of the professionals in these fields have a background in geology. An understanding of organic chemistry is also extremely useful since organic materials and polymeric treatments are common.

Somewhat surprisingly, little is known about many important questions in conservation science. For example, a serious problem in the conservation of monuments is the deterioration of stone and brick caused by the crystallization of salts. Part of my research has been dedicated to better understanding the *microdynamics* of phenomena such as salt crystallization by using a special microscope that allows one to observe at high magnification what actually happens to different materials as they deteriorate.

As part of my research, I have been able to study some of the oldest human cultural artifacts using the latest analytical methods. For example, I recently helped to identify human blood cells in a black pigment sample from a Chumash Indian rock art site near Santa Barbara, California. I also contributed to a research project at the J. Paul Getty Museum on the Kouros statue, a free-standing sculpture of a naked youth (Figure 7-2). The Getty Kouros is thought to be a rare example of Archaic Greek sculpture carved from dolomite marble in the sixth century B.C., although its authenticity has been questioned on stylistic grounds.

It is extremely difficult to prove that a sculpture is authentic; instead one looks for inconsistencies that indicate forgery. The question at hand was whether the marble and its altered surface were consistent with what we know about the sources of marble used for sculpture in Archaic Greek time and with the natural weathering of this type of marble. While we were able to obtain a radiocarbon age of several thousand years for the surface of the Getty Kouros (suggesting authenticity), we were also able to "fake" such a result using laboratory chemicals containing different amounts of carbon 14 (which prevented us from ruling out forgery). However, using a wide range of techniques, including oxygen, carbon, and strontium isotope analysis, we were able to definitively assign the source of the Kouros marble to the Greek island of Thasos. Thasos is located in the northern Aegean, and field work there has documented the presence of ancient quarries of the correct age. Extensive analysis of the altered surface of the Kouros showed that is was composed mostly of whewellite—a mineral produced by the action of lichen and soil fungi. We tried to reproduce the Kouros surface using a wide range of treatments (some from forgers' recipes), but all failed. Thus, all of the data were found to be consistent with authenticity. However, we were unable to rule out the admittedly remote possibility of forgery.

Conservation and museum science operate in an arena where politics, cultural values, and history all play important roles. The future of both fields will depend on increasing public support for museums and conservation projects as well as involing more scientists from universities and industry in collaborative programs.

ERIC DOEHNE is a conservation scientist at the Getty Conservation Institute in Marina del Rey, California, where he works as part of an interdisciplinary group to further the conservation of art, architecture, and monuments. He received his Ph.D. in geology from the University of California, Davis. His research interests include studies of stone weathering mechanisms, the origin of complex patinas on ancient Greek and Roman stone surfaces, and the microdynamics of deterioration phenomena using the environmental scanning electron microscope (E-SEM).

and the resulting metamorphic rocks can be divided into different zones based on the presence of distinctive silicate mineral assemblages. These mineral assemblages, each recognized by the presence of one or more index minerals, indicate different degrees of metamorphism. The index minerals Barrow and his associates chose to represent increasing metamorphic intensity were chlorite, biotite, garnet, staurolite, kyanite, and sillimanite (Table 7-1). Note that these are the metamorphic minerals produced from clay-rich sedimentary rocks. Other mineral assemblages and index minerals are produced from rocks with different original compositions (Table 7-1).

The successive appearance of metamorphic index minerals indicates gradually increasing or decreasing intensity of metamorphism. Going from lower toward higher grade zones, the first appearance of a particular index mineral indicates the location of the minimum temperature and pressure conditions needed for the formation of that mineral. When the locations of the first appearances of that index mineral are connected on a map, the result is a line of equal metamorphic intensity or an **isograd**. The region between isograds is known as a **metamorphic zone**. By noting the occurrence of metamorphic index minerals, geologists can construct a map showing the metamorphic zones of an entire area (▶ Figure 7-18).

Numerous studies of different metamorphic rocks have demonstrated that while the texture and composition of any rock may be altered by metamorphism, the overall chemical composition may be little changed. Thus, the different mineral assemblages found in increasingly higher grade metamorphic rocks derived from the same parent rock result from changes in temperature and pressure (Table 7-1).

A **metamorphic facies** is a group of metamorphic rocks that are characterized by particular mineral assemblages formed under the same broad temperature-pressure conditions (▶ Figure 7-19). Each facies is named after its most characteristic rock or mineral. For example, the green metamorphic mineral chlorite, which forms under relatively low temperatures and pressures, yields rocks said to belong to the *greenschist facies*. Under increasingly higher temperatures and pressures, other metamorphic facies, such as the *amphibolite* and *granulite facies,* develop.

Although usually applied to areas where the original rocks were clay rich, the concept of metamorphic facies can be used with modification in other situations. It cannot, however, be used in areas where the original rocks were pure quartz sandstones or pure limestones or dolostones. Such rocks would yield only quartzites and marbles, respectively.

Metamorphism and Plate Tectonics

Although metamorphism is associated with all three types of plate boundaries (see Figure 1-15), it is most common along convergent plate margins. Metamorphic rocks form at convergent plate boundaries because temperature and pressure increase as a result of plate collisions.

▶ Figure 7-20 illustrates the various temperature-pressure regimes that are produced along an oceanic-

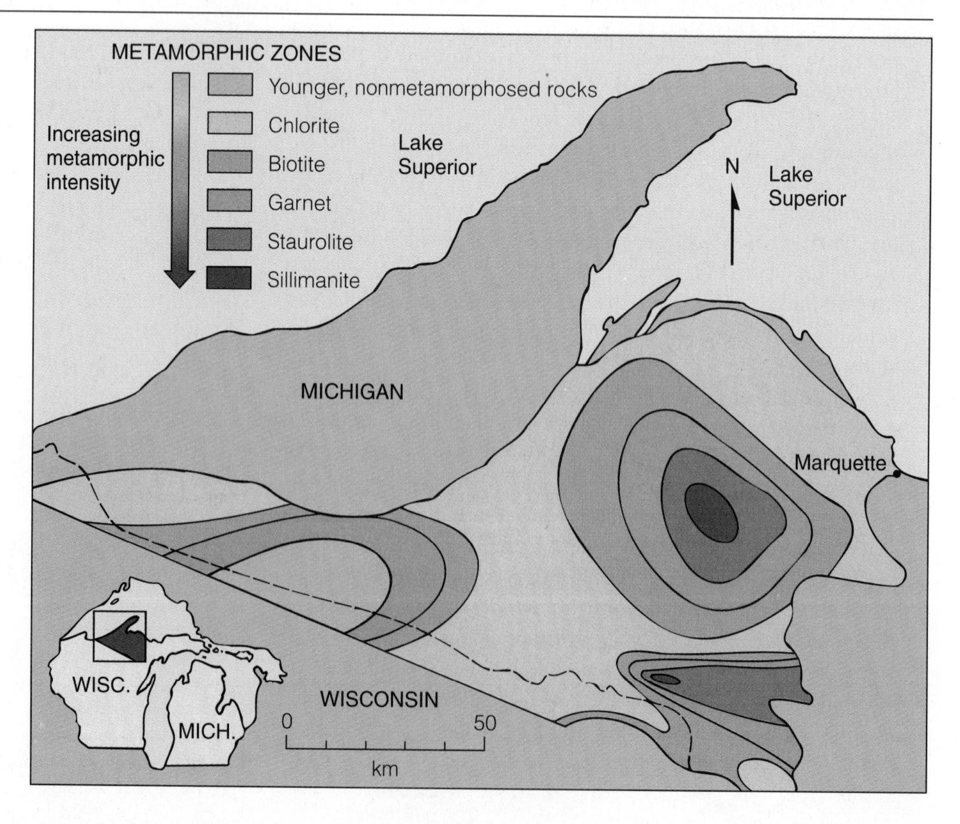

▶ **FIGURE 7-18** Metamorphic zones in the upper peninsula of Michigan. The zones in this region are based on the presence of distinctive silicate mineral assemblages resulting from the metamorphism of sedimentary rocks during an interval of mountain building and minor granitic intrusion during the Proterozoic Eon, about 1.5 billion years ago.

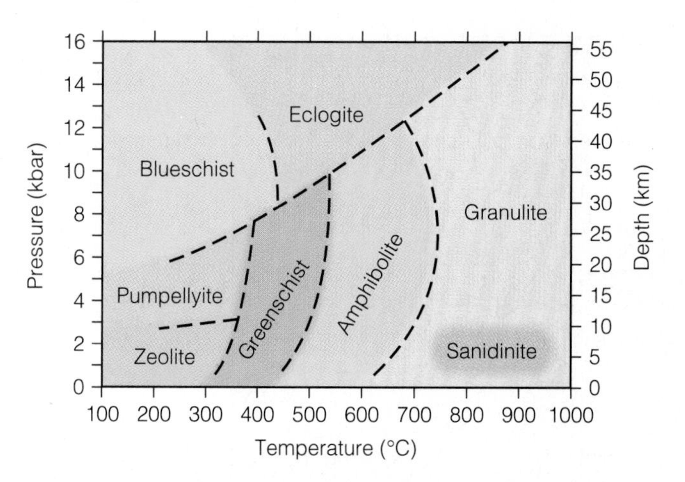

▶ **FIGURE 7-19** A pressure-temperature diagram showing where various metamorphic facies occur. A facies is characterized by a particular mineral assemblage that formed under the same broad temperature-pressure conditions. Each facies is named after its most characteristic rock or mineral.

continental convergent plate boundary and the type of metamorphic facies and rocks that can result. When an oceanic plate collides with a continental plate, tremendous pressure is generated as the oceanic plate is subducted. Because rock is a poor heat conductor, the cold descending oceanic plate heats very slowly, and metamorphism is caused mostly by increasing pressure with depth. Metamorphism in such an environment produces rocks typical of the *blueschist facies* (low temperature, high pressure), which is characterized by the blue-colored amphibole mineral glaucophane (Figure 7-19). Geologists use the occurrence of blueschist facies rocks as evidence of ancient subduction zones. An excellent example of blueschist metamorphism can be found in the California Coast Ranges. Here rocks of the Franciscan Group were metamorphosed under low-temperature, high-pressure conditions that clearly indicate the presence of a former subduction zone (▶ Figure 7-21).

As subduction along the oceanic-continental plate boundary continues, both temperature and pressure increase with depth and can result in high-grade metamorphic rocks. Eventually, the descending plate begins to melt and generates a magma that moves upward. This rising magma may alter the surrounding rock by contact metamorphism, producing migmatites in the deeper portions of the crust and hornfels at shallower depths. Such an environment is characterized by high temperatures and low to medium pressures.

While metamorphism is most common along convergent plate margins, many divergent plate boundaries are characterized by contact metamorphism. Rising magma from mid-oceanic ridges heats the adjacent rocks, producing contact metamorphic minerals and textures. In addition to contact metamorphism, fluids emanating from the rising magma—and from the reaction of the magma and seawater—very commonly produce hydrothermal solutions that may precipitate minerals of great economic value.

Metamorphism and Natural Resources

Many metamorphic rocks and minerals are valuable natural resources. While these resources include various types of ore deposits, the two most familiar and widely used metamorphic rocks, as such, are marble and slate, which, as previously discussed, have been used for centuries in a variety of ways.

Many ore deposits result from contact metamorphism during which hot, ion-rich fluids migrate from igneous

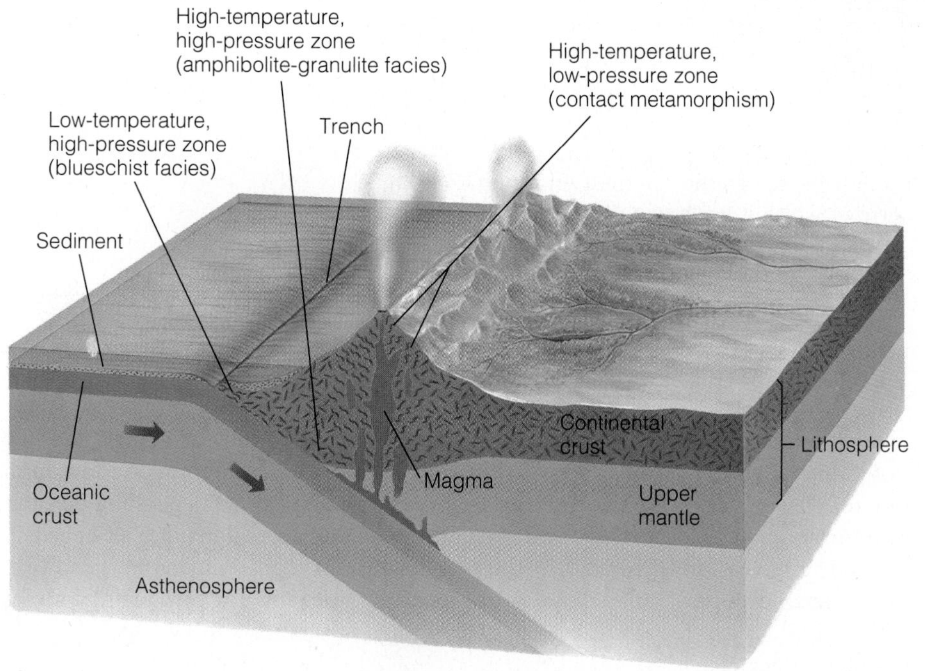

▶ **FIGURE 7-20** Metamorphic facies resulting from various temperature-pressure conditions produced along an oceanic-continental convergent plate boundary.

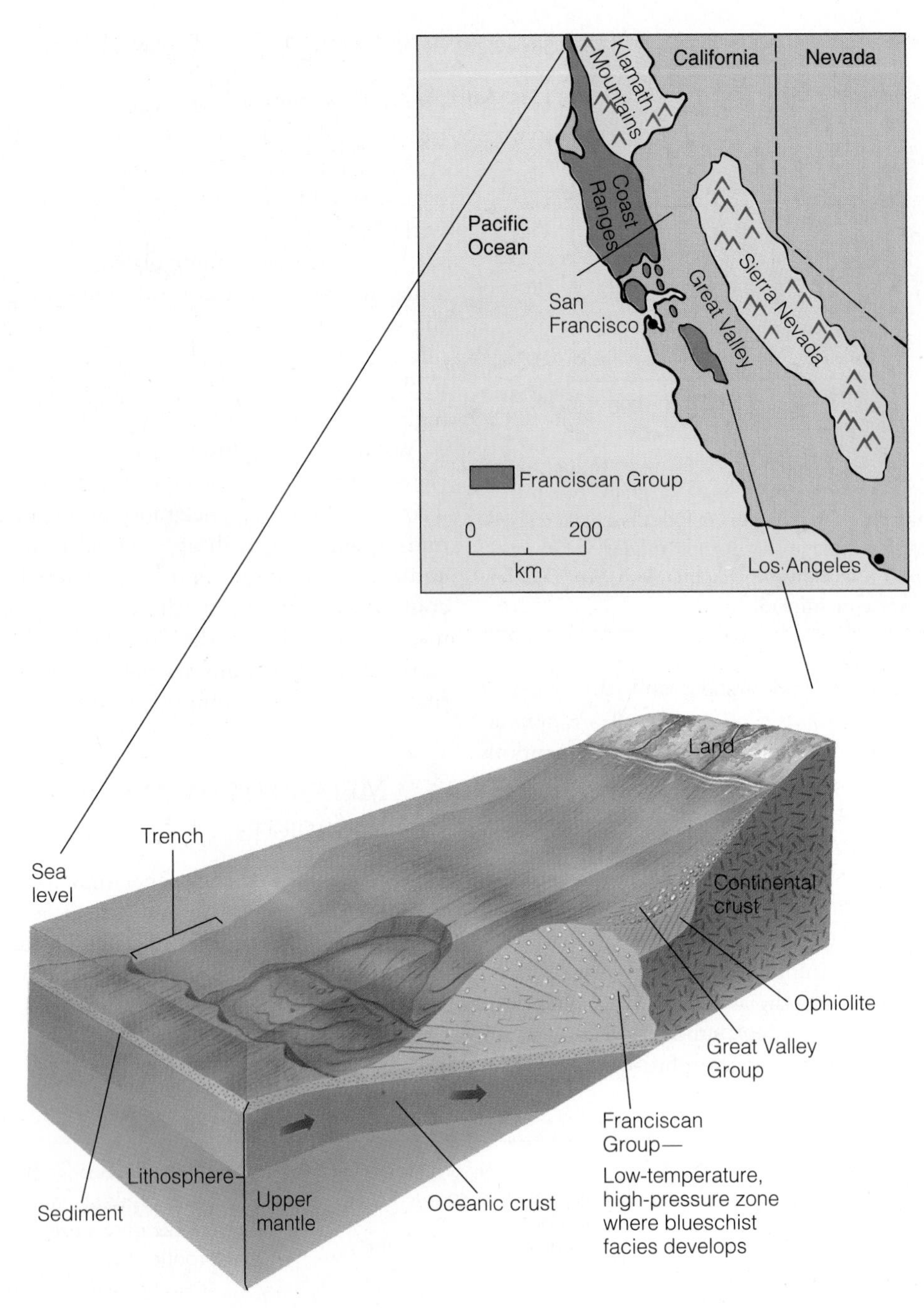

➤ **FIGURE 7-21** Index map of California showing the location of the Franciscan Group and a diagrammatic reconstruction of the environment in which it was regionally metamorphosed under low-temperature, high-pressure subduction conditions approximately 150 million years ago. The red line on the index map shows the orientation of the reconstruction to the current geography.

intrusions into the surrounding rock, thereby producing rich ore deposits. The most common sulfide ore minerals associated with contact metamorphism are bornite, chalcopyrite, galena, pyrite, and sphalerite, while two common oxide ore minerals are hematite and magnetite. Tin and tungsten are also important ores associated with contact metamorphism (◉ Table 7-3).

Other economically important metamorphic minerals include talc for talcum powder; graphite for pencils and dry lubricants (see Perspective 7-2); garnets and corundum, which are used as abrasives or gemstones, depending on their quality; and andalusite, kyanite, and sillimanite, all of which are used in the manufacture of high-temperature porcelains and temperature-resistant minerals for products such as sparkplugs and the linings of furnaces.

Perspective 7-2

GRAPHITE

Graphite (from the Greek *grapho* meaning write) is a soft, gray to black mineral that has a greasy feel and is composed of the element carbon. Graphite occurs in two varieties: crystalline, which consists of thin, flat, nearly pure black flakes; and massive, an impure variety found in compact masses.

Graphite has the same composition as diamond (see Perspective 2-2), but its carbon atoms are strongly bonded together in sheets, with the sheets weakly held together by van der Waals bonds (see Figure 2-5). Because the sheets are loosely held together, they easily slide over one another, giving graphite its ability to mark paper and serve as a dry lubricant.

Graphite occurs mainly in metamorphic rocks produced by contact and regional metamorphism. It is found in marble, quartzite, schist, gneiss, and even in anthracite. Contact metamorphism of impure limestones by igneous intrusions produces some of the graphite found in marbles. The graphite resulting from regional metamorphism of sedimentary rocks probably came from organic matter present in the sediments. Some evidence, however, indicates that the graphite in Precambrian-aged rocks (>570 million years) may be the result of the reduction of calcium carbonate ($CaCO_3$) by an inorganic process. Graphite is also found in igneous rocks, pegmatite dikes, and veins; it is thought to have formed in these environments from the primary constituents of the magma or from the hot fluids and vapors released by the cooling magma.

Major producers of graphite are China, Russia, South Korea, India, and Mexico. In the United States, graphite has been mined in 27 states, but production is now generally limited to Alabama and New York.

Graphite is used for many purposes. The oldest use is in pencil leads, where it is finely ground, mixed with clay, and baked. The amount of clay and the baking time give pencil leads their desired hardness. Other important uses include batteries, brake linings, carbon brushes, crucibles, foundry facings, lubricants, refractories, and steel making.

Synthetic graphite can be produced from anthracite coal or petroleum coke and now accounts for most graphite production. Its extreme purity (99% to 99.5% pure) makes it especially valuable where high purity is required as in the rods that slow down the reaction rates in nuclear reactors.

TABLE 7-3 The Main Ore Deposits Resulting from Contact Metamorphism

Ore Deposit	Major Mineral	Formula	Use
Copper	Bornite Chalcopyrite	Cu_5FeS_4 $CuFeS_2$	Important sources of copper, which is used in various aspects of manufacturing, transportation, communications, and construction.
Iron	Hematite Magnetite	Fe_2O_3 Fe_3O_4	Major sources of iron for manufacture of steel, which is used in nearly every form of construction, manufacturing, transportation, and communications.
Lead	Galena	PbS	Chief source of lead, which is used in batteries, pipes, solder, and elsewhere where resistance to corrosion is required.
Tin	Cassiterite	SnO_2	Principal source of tin, which is used for tin plating, solder, alloys, and chemicals.
Tungsten	Scheelite Wolframite	$CaWO_4$ $(Fe, Mn)WO_4$	Chief sources of tungsten, which is used in hardening metals and manufacturing carbides.
Zinc	Sphalerite	$(Zn, Fe)S$	Major source of zinc, which is used in batteries and in galvanizing iron and making brass.

Chapter Summary

1. Metamorphic rocks result from the transformation of other rocks, usually beneath the Earth's surface, as a consequence of one or a combination of three agents: heat, pressure, and fluid activity.
2. The geothermal gradient and intrusive magmas provide most of the heat for metamorphism. Pressure is either lithostatic or differential. Fluids trapped in sedimentary rocks or emanating from intruding magmas can enhance chemical changes and the formation of new minerals.
3. The three major types of metamorphism are contact, dynamic, and regional.
4. Metamorphic rocks are classified primarily according to their texture. In a foliated texture, platy minerals have a preferred orientation. A nonfoliated texture does not exhibit any discernible preferred orientation of the mineral grains.
5. Foliated metamorphic rocks can be arranged in order of grain size and/or perfection of their foliation. Slate is very fine grained, followed by phyllite and schist; gneiss displays segregated bands of minerals. Amphibolite is another fairly common foliated metamorphic rock.
6. Marble, quartzite, greenstone, and hornfels are common nonfoliated metamorphic rocks.
7. Metamorphic rocks can be arranged into metamorphic zones based on the conditions of metamorphism. Individual metamorphic facies are characterized by particular minerals that formed under specific metamorphic conditions. Such facies are named for a characteristic rock or mineral.
8. Metamorphism can occur along all three kinds of plate boundaries, but most commonly occurs at convergent plate margins.
9. Metamorphic rocks formed near the Earth's surface along an oceanic-continental plate boundary result from low-temperature, high-pressure conditions. As a subducted oceanic plate descends, it is subjected to increasingly higher temperatures and pressures that result in higher-grade metamorphism.
10. Many metamorphic rocks and minerals, such as marble, slate, graphite, talc, and asbestos, are valuable natural resources.

Important Terms

aureole
contact metamorphism
differential pressure
dynamic metamorphism
fluid activity
foliated texture
heat
index minerals
isograd
lithostatic pressure
metamorphic facies
metamorphic rock
metamorphic zone
nonfoliated texture
regional metamorphism

Review Questions

1. The metamorphic rock formed from limestone is:
   a. ____ quartzite;
   b. ____ hornfels;
   c. ____ marble;
   d. ____ slate;
   e. ____ greenstone.
2. From which of the following rock groups can metamorphic rocks form?
   a. ____ plutonic;
   b. ____ sedimentary;
   c. ____ metamorphic;
   d. ____ volcanic;
   e. ____ all of these.
3. Which of the following is not an agent of metamorphism?
   a. ____ foliation;
   b. ____ heat;
   c. ____ pressure;
   d. ____ fluid activity;
   e. ____ none of these.
4. Pressure exerted equally in all directions on an object is:
   a. ____ differential;
   b. ____ directional;
   c. ____ lithostatic;
   d. ____ shear;
   e. ____ none of these.
5. In which type of metamorphism are magmatic heat and fluids the primary agents of change?
   a. ____ contact;
   b. ____ dynamic;
   c. ____ regional;
   d. ____ local;
   e. ____ thermodynamic.
6. Concentric zones surrounding an igneous intrusion are:
   a. ____ metamorphic layers;
   b. ____ thermodynamic rings;
   c. ____ aureoles;
   d. ____ hydrothermal regions;
   e. ____ none of these.
7. Metamorphic rocks resulting from pure dynamic metamorphism are:
   a. ____ fault breccias;
   b. ____ quartzites;
   c. ____ greenstones;
   d. ____ mylonites;
   e. ____ hornfels.

8. Which type of metamorphism produces the majority of metamorphic rocks?
   a. ____ contact;
   b. ____ dynamic;
   c. ____ regional;
   d. ____ lithostatic;
   e. ____ lithospheric.
9. Which of the following metamorphic rocks displays a foliated texture?
   a. ____ marble;
   b. ____ quartzite;
   c. ____ greenstone;
   d. ____ schist;
   e. ____ hornfels.
10. What is the correct metamorphic sequence of increasingly coarser grain size?
    a. ____ phyllite → slate → gneiss → schist;
    b. ____ slate → phyllite → schist → gneiss;
    c. ____ gneiss → phyllite → slate → schist;
    d. ____ schist → gneiss → phyllite → slate;
    e. ____ slate → schist → gneiss → phyllite.
11. Mixed rocks with the characteristics of both igneous and high-grade metamorphic rocks are:
    a. ____ mylonites;
    b. ____ migmatites;
    c. ____ amphibolites;
    d. ____ hornfels;
    e. ____ greenstones.
12. Metamorphic zones:
    a. ____ are characterized by distinctive mineral assemblages;
    b. ____ are separated from each other by isograds;
    c. ____ reflect a metamorphic grade;
    d. ____ all of these;
    e. ____ none of these.
13. To which metamorphic facies do metamorphic rocks formed under low-temperature, low-pressure conditions belong?
    a. ____ granulite;
    b. ____ greenschist;
    c. ____ amphibolite;
    d. ____ blueschist;
    e. ____ eclogite.
14. Along what type of plate boundary is metamorphism most common?
    a. ____ convergent;
    b. ____ divergent;
    c. ____ transform;
    d. ____ mantle plume;
    e. ____ static.
15. Metamorphic rocks form a significant proportion of:
    a. ____ shields;
    b. ____ the cores of mountain ranges;
    c. ____ oceanic crust;
    d. ____ answers (a) and (b);
    e. ____ answers (b) and (c).
16. Name the three agents of metamorphism, and explain how each contributes to metamorphism.
17. What are the two types of pressure? What type of metamorphic textures does each produce?
18. Where does contact metamorphism occur, and what type of changes does it produce?
19. What are aureoles? How can they be used to determine the effects of metamorphism?
20. What is regional metamorphism, and under what conditions does it occur?
21. Describe the two types of metamorphic texture, and explain how they may be produced.
22. Starting with a shale, what metamorphic rocks would be produced by increasing heat and pressure?
23. Name the three common nonfoliated rocks, and describe their characteristics.
24. What is a metamorphic zone?
25. Name some common metamorphic rocks or minerals that are economically valuable, and describe their uses.

POINTS TO PONDER

1. What specific features about foliated metamorphic rocks make them unsuitable as the foundation rock for a dam? Are there any metamorphic rocks that would make a good foundation? Why?
2. If you were in charge of the EPA, how would you formulate a policy that balances the risks and the benefits of removing asbestos from public buildings? What role would geologists play in this policy?
3. If there was no internal mechanism to drive the plates, would metamorphism exist? If so, what types of metamorphism would occur and why? When astronauts land on Mars, do you think they will find metamorphic rocks? Why?
4. Using Figure 7-9, what mineral would form at a pressure of 3 kilobars and 500° C? At a pressure of 6 kilobars and 750° C?
5. At a tempterature of 500° C, what metamorphic facies occurs at 14 kilobars pressure? At 8 kilobars? At a depth of 5 km?

ADDITIONAL READINGS

Abelson, F. H. 1990. The asbestos fiasco. *Science* 247 (4946): 1017.

Best, M. G. 1982. *Igneous and metamorphic petrology*. San Francisco, Calif.: W. H. Freeman and Co.

Bowes, D. R., ed. 1989. *The encyclopedia of igneous and metamorphic petrology*. New York: Van Nostrand Reinhold.

Gillen, C. 1982. *Metamorphic geology*. London: George Allen & Unwin.

Gunter, M. E. 1994. Asbestos as a metaphor for teaching risk perception. *Journal of Geological Education* 42 (1): 17–24.

Hyndman, D. W. 1985. *Petrology of igneous and metamorphic rocks.* 2d ed. New York: McGraw-Hill Book Co.

Kokkou, A. 1993. *The Getty kouros colloquium.* Athens, Greece. Kapon Editions.

Margolis, S. V. 1989. Authenticating ancient marble sculpture. *Scientific American* 260, no. 6: 104–11.

Turner, F. J. 1981. *Metamorphic petrology.* 2d ed. New York: McGraw-Hill Book Co.

Chapter 8

GEOLOGIC TIME

OUTLINE

Three Sisters, in New South Wales, Australia. The Three Sisters are distinctive erosional remnants that formed over a long period of geologic time. They are a popular tourist attraction in Australia's Blue Mountains.

PROLOGUE

What is time? We seem obsessed with it, and organize our lives around it with the help of clocks, calendars, and appointment books. Yet most of us feel we don't have enough of it—we are always running "behind" or "out of time." According to biologists and psychologists, children less than two years old and animals exist in a "timeless present," where there is no past or future. They have no conscious concept of time. Some scientists think that our early ancestors may also have lived in a state of timelessness with little or no perception of a past or future. According to Buddhist, Taoist, and Mayan beliefs, time is circular, and like a circle, all things are destined to return to where they once were. Thus, in these belief systems, there is no beginning or end, but rather a cyclicity to everything.

For most people though, time is linear and moves like a flowing stream. It can be measured and subdivided. We place events in a chronology in which past events have a history and there are expectations for the future.

Albert Einstein, however, changed that view in 1905 with his special theory of relativity. Einstein showed that time is a dimension and is not absolute. In fact, like space, it is bent by gravity. The greater the mass of an object, the greater its gravitational attraction, and thus the slower that time moves relative to an object of lesser mass. For example, if you had two identical clocks and placed one on Jupiter and one on Earth, the clock on Jupiter would run detectably slower than the clock on Earth because Jupiter has 318 times the mass of the Earth and exerts a greater gravitational attraction. Therefore time is unique to any particular location in the universe.

In some respects, time is defined by the methods used to measure it. Many prehistoric monuments were oriented to detect the summer solstice. Sundials were used to divide the day into measurable units. As civilization advanced, mechanical devices were invented to measure time, the earliest being the water clock, first used by the ancient Egyptians and further developed by the Greeks and Romans. The pendulum clock was invented in the seventeenth century and provided the most accurate timekeeping for the next two and a half centuries.

Today the quartz watch is the most popular timepiece. Powered by a battery, a quartz crystal vibrates approximately 100,000 times per second. An integrated circuit counts these vibrations and converts them into a digital or dial reading on your watch face. An inexpensive quartz watch today is more accurate than the best mechanical watch, and precision-manufactured quartz clocks are accurate to within one second per 10 years.

Precise timekeeping is important in our technological world. Ships and aircraft plot their locations by satellite, relying on an extremely accurate time signal. Deep-space probes such as the *Voyagers* (see Chapter 20) require radio commands timed to billionths of a second, while physicists exploring the motion inside the nucleus of an atom deal in trillionths of a second as easily as we talk about minutes.

To achieve such accuracy, scientists use atomic clocks. First developed in the 1940s, these clocks rely on an atom's oscillating electrons, a rhythm so regular that they are accurate to within a few thousandths of a second per day. Recently, an atomic clock thought to be accurate to within one second per three million years was installed at the National Institute of Standards and Technology (NIST). Named the NIST-7, this clock is a refinement over previous atomic clocks because it uses lasers instead of a magnetic field to stimulate cesium 133 atoms. Cesium 133 is used in atomic clocks because its electrons oscillate at a very predictable rate (9,192,631,770 vibrations per second). In fact, a cesium clock was used to prove Einstein's prediction that a clock will slow down as its speed increases.

While physicists deal with incredibly short intervals of time, astronomers and geologists are concerned with geologic time measured in millions or billions of years. When astronomers look at a distant galaxy, they are seeing what it looked like billions of years ago. When geologists investigate rocks in the walls of the Grand Canyon, they are deciphering events that occurred over an interval of 2 billion years. Geologists can measure decay rates of such radioactive elements as uranium, thorium, and rubidium to determine how long ago an igneous rock formed. Furthermore, geologists know that the Earth's rotational velocity has been slowing down a few thousandths of a second per century as a result of the frictional effects of tides, ocean currents, and varying thicknesses of polar ice. Five hundred million years ago a day was only 20 hours long, and at the current rate of slowing, 200 million years from now a day will be 25 hours long.

Time is a fascinating topic that has been the subject of numerous essays and books. And while we can comprehend concepts like milliseconds and understand how a quartz watch works, deep time, or geologic time, is still very difficult for most people to comprehend.

INTRODUCTION

Time is what sets geology apart from most of the other sciences, and an appreciation of the immensity of geologic time is fundamental to understanding both the physical and biological history of our planet (▶ Figure 8-1). Most people have difficulty comprehending geologic time because they tend to view time from the perspective of their own existence. Ancient history is what occurred hundreds or even thousands of years ago, but when geologists talk in terms of ancient geologic history, they are referring to events that happened millions or even billions of years ago!

▶ **FIGURE 8-1** Geologic time is depicted in this spiral history of the Earth from the time of its formation 4.6 billion years ago to the present. (B.Y. = billion years; M.Y. = million years.)

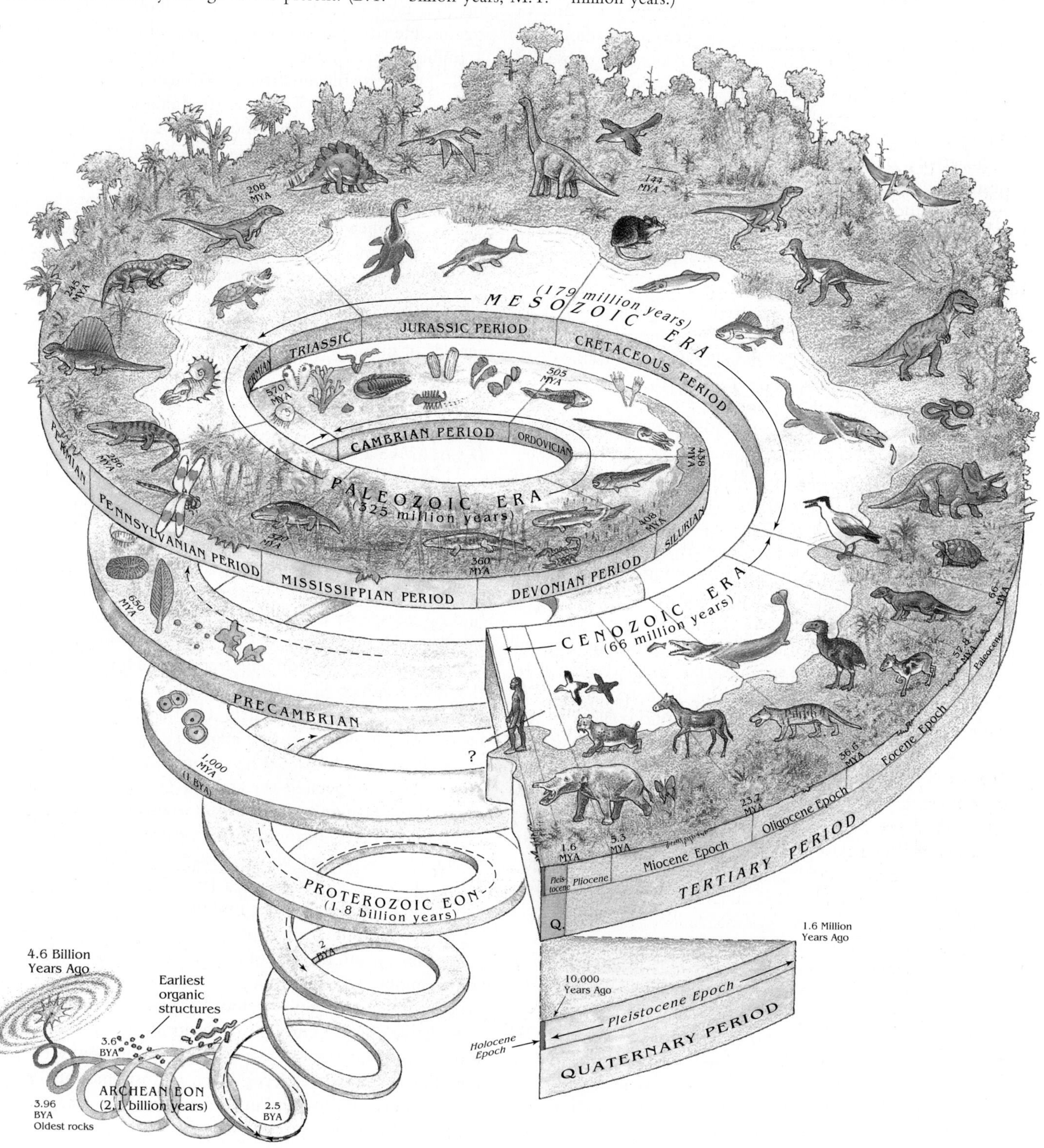

Geologists use two different frames of reference when speaking of geologic time. **Relative dating** involves placing geologic events in a sequential order as determined from their position in the geologic record. Relative dating will not tell us how long ago a particular event occurred, only that one event preceded another. The various principles used to determine relative dating were discovered hundreds of years ago, and since then they have been used to construct the *relative geologic time scale* (▶ Figure 8-2). These principles are still widely used today.

Absolute dating results in specific dates for rock units or events expressed in years before the present. Radiometric dating is the most common method of obtaining absolute age dates. Such dates are calculated from the natural rates of decay of various radioactive elements occurring in trace amounts in some rocks. It was not until the discovery of radioactivity near the end of the nineteenth century that absolute ages could be accurately applied to the relative geologic time scale. Today the geologic time scale is really a dual scale: a relative scale based on rock sequences with radiometric dates expressed as years before the present added to it (Figure 8-2).

Early Concepts of Geologic Time and the Age of the Earth

The concept of geologic time and its measurement have changed through human history. Many early Christian scholars and clerics tried to establish the date of creation by analyzing historical records and the genealogies found in Scripture. Based on their analyses, they generally believed that the Earth and all of its features were no more than about 6,000 years old. The idea of a very young Earth provided the basis for most Western chronologies of Earth history prior to the eighteenth century.

During the eighteenth and nineteenth centuries, several attempts were made to determine the age of the Earth on the basis of scientific evidence rather than revelation. The French zoologist Georges Louis de Buffon (1707–1788) assumed that the Earth gradually cooled to its present condition from a molten beginning. To simulate this history, he melted iron balls of various diameters and allowed them to cool to the surrounding temperature. By extrapolating their cooling rate to a ball the size of the Earth, he determined that the Earth was at least 75,000 years old. While this age was much older than that derived from Scripture, it was still vastly younger than we now know the Earth to be.

Other scholars were equally ingenious in attempting to calculate the Earth's age. For example, geologists reasoned that by determining deposition rates for various sediments, they could calculate how long it would take to deposit any rock layer. Furthermore, they could then extrapolate how old the Earth was from the total thickness of sedimentary rock in the Earth's crust. Rates of deposition vary, however, even for the same type of rock. Furthermore, it is impossible to estimate how much rock has been removed by erosion, or how much a rock sequence has been reduced by compaction. As a result of these variables, estimates ranged from less than a million years to more than a billion years.

Another attempt to determine the Earth's age involved ocean salinity. Scholars assumed that the Earth's ocean waters were originally fresh and that their present salinity was the result of dissolved salt being carried into the ocean basins by streams. John Joly, a nineteenth-century Irish geologist, knew the volume of ocean water and its salinity, and measured the amount of salt currently in the world's streams. He then calculated that it would have taken at least 90 million years for the oceans to reach their present salinity level. This was still much younger than the now accepted age of 4.6 billion years for the Earth, mainly because Joly had no way of calculating how much salt had been recycled or the amount of salt stored in continental salt deposits and sea-floor clay deposits.

In addition to these efforts, the naturalists of the eighteenth and nineteenth centuries were also formulating some of the fundamental geologic principles that are used in deciphering Earth history. From the evidence preserved in the geologic record, it was clear to them that the Earth is very old and that geologic processes have operated over long periods of time.

James Hutton and the Recognition of Geologic Time

The Scottish geologist James Hutton (1726–1797) is considered by many to be the father of modern geology. He relied on known processes to account for Earth history and concluded that the Earth must be very old, stating that "we find no vestige of a beginning, and no prospect of an end."

Observing the processes of wave action, erosion by running water, and sediment transport, Hutton concluded that given enough time these processes could account for the geologic features of the Earth. He thought that "the past history of our globe must be explained by what can be seen to be happening now." This assumption that present-day processes have operated throughout geologic time was the basis for the **principle of uniformitarianism** (see Chapter 1).

Unfortunately, Hutton was not a particularly good writer, so his ideas were not widely disseminated or accepted. In 1830 Charles Lyell published a landmark book, *Principles of Geology*, in which he championed Hutton's concept of uniformitarianism. Instead of relying on catastrophic events to explain various features of the Earth, Lyell recognized that imperceptible changes brought about by present-day processes could, over long periods of time, have tremendous cumulative effects. Through his writings, Lyell firmly established uniformitarianism as the guiding philosophy of geology. Furthermore, the recognition of vast amounts of time was also necessary for, and instrumental in, the acceptance of Darwin's 1859 theory of evolution.

After finally establishing that present-day processes have operated over vast periods of time, geologists were nevertheless nearly forced to accept a very young age for the Earth when a highly respected English physicist, Lord Kelvin (1824–1907), claimed in 1866 to have destroyed the uniformitarian foundation of geology. Starting with the generally accepted belief that the Earth was originally molten, Kelvin assumed that the Earth has gradually been losing heat and that, by measuring this heat loss, he could determine the age of the Earth.

Kelvin knew from deep mines that the Earth's temperature increases with depth, and he reasoned that the Earth is therefore losing heat from its interior. By knowing the melting temperatures of the Earth's rocks, the size of the Earth, and the rate of heat loss, Kelvin was able to calculate the age at which the Earth was entirely molten. From these calculations, he concluded that the Earth could not be older than 100 million years or younger than 20 million years. This wide range in age reflected uncertainties over average temperature increases with depth and the various melting points of the Earth's constituent materials.

After finally establishing that the Earth was very old, and showing how present-day processes operating over long periods of time can explain geological features, geologists were in a quandary. They either had to accept Kelvin's dates and squeeze events into a shorter time frame or reject his calculations.

While Kelvin's reasoning and calculations were sound, his

▶ **FIGURE 8-2** The geologic time scale. Some of the major biological and geological events are indicated along the right-hand margin.

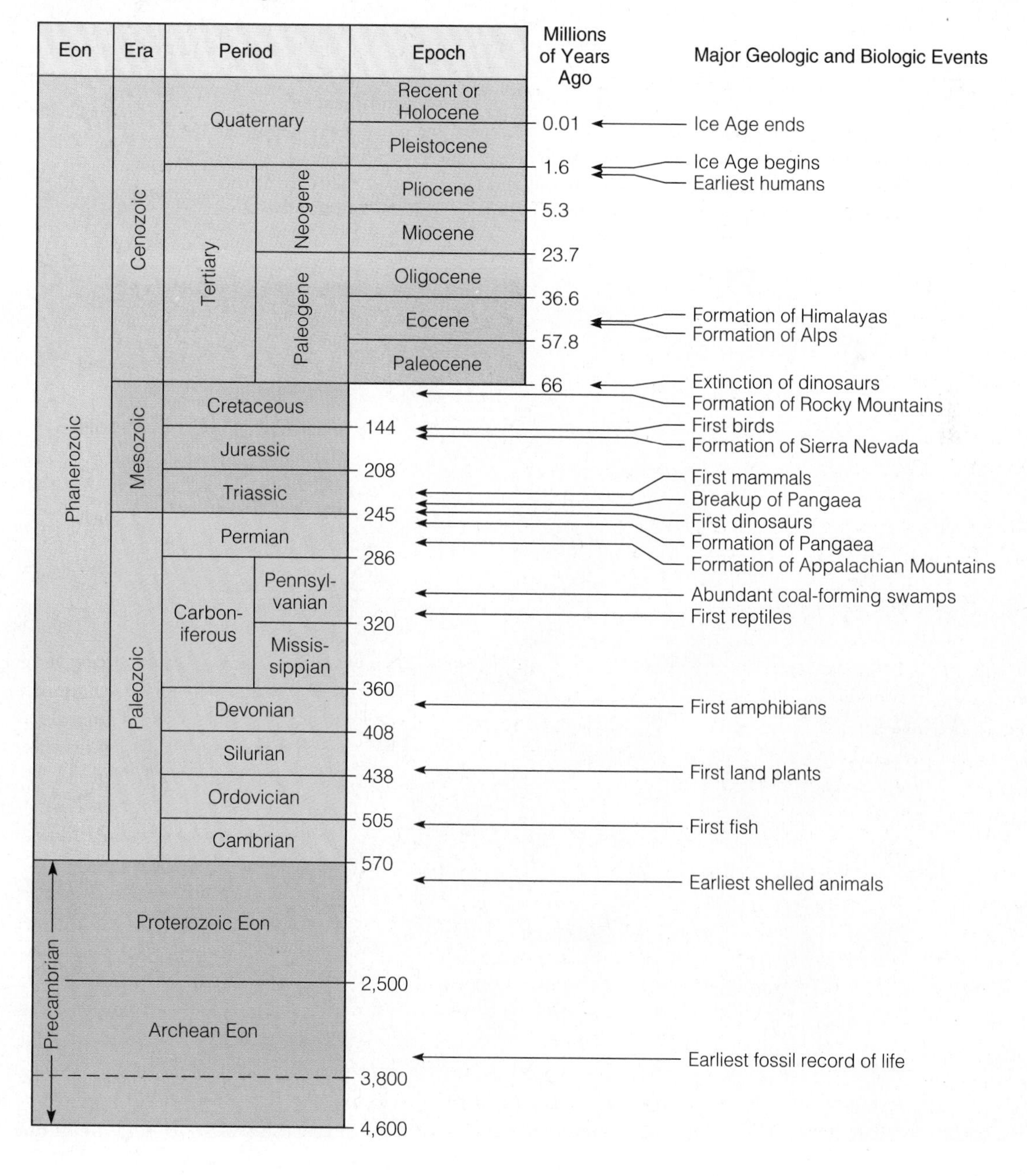

basic premises were false, thereby invalidating his conclusions. Kelvin was unaware that the Earth has an internal heat source, radioactivity, that has allowed it to maintain a fairly constant temperature through time.* His 40-year campaign for a young Earth ended with the discovery of radioactivity near the end of the nineteenth century. His calculations were therefore no longer valid, and his proof for a geologically young Earth collapsed. Moreover, while the discovery of radioactivity destroyed Kelvin's arguments, it provided geologists with a clock that could measure the Earth's age and validate what geologists had been saying all along, namely, that the Earth was indeed very old!

* Actually, the Earth's temperature has decreased through time because the original amount of radioactive materials has been decreasing and thus is not supplying as much heat. However, the temperature is decreasing at a rate considerably slower than would be required to lend any credence to Kelvin's calculations.

Relative Dating Methods

Before the development of radiometric dating techniques, geologists had no reliable means of absolute age dating and depended solely on relative dating methods. These methods only allow events to be placed in sequential order and do not tell us how long ago an event took place. Though the principles of relative dating may now seem self-evident, their discovery was an important scientific achievement because they provided geologists with a means to interpret geologic history and develop a relative geologic time scale.

Fundamental Principles of Relative Dating

Six fundamental geologic principles are used in relative dating: *superposition, original horizontality, lateral continuity, cross-cutting relationships, inclusions*, and *fossil succession*.

FIGURE 8-3 The Grand Canyon of Arizona illustrates three of the six fundamental principles of relative dating. The sedimentary rocks of the Grand Canyon were originally deposited horizontally in a variety of marine and continental environments (principle of original horizontality). The oldest rocks are at the bottom of the canyon, and the youngest rocks are at the top, forming the rim (principle of superposition). The exposed rock layers extend laterally for some distance (principle of lateral continuity).

The seventeenth century was an important time in the development of geology as a science because of the widely circulated writings of the Danish anatomist, Nicolas Steno (1638–1686). Steno observed that during flooding, streams spread out across their floodplains and deposit layers of sediment that bury organisms dwelling on the floodplain. Subsequent flooding events produce new layers of sediments that are deposited or superposed over previous deposits. When lithified, these layers of sediment become sedimentary rock. Thus, in an undisturbed succession of sedimentary rock layers, the oldest layer is at the bottom and the youngest layer is at the top. This **principle of superposition** is the basis for relative age determinations of strata and their contained fossils (▶ Figure 8-3).

Steno also observed that because sedimentary particles settle from water under the influence of gravity, sediment is deposited in essentially horizontal layers, thus illustrating the **principle of original horizontality** (Figure 8-3). Therefore, a sequence of sedimentary rock layers that is steeply inclined from the horizontal must have been tilted after deposition and lithification.

Steno's third principle, the **principle of lateral continuity,** states that sediment extends laterally in all directions until it thins and pinches out or terminates against the edge of the depositional basin (Figure 8-3).

James Hutton is credited with discovering the **principle of cross-cutting relationships.** Based on his detailed studies and observations of rock exposures in Scotland, Hutton recognized that an igneous intrusion or fault must be younger than the rocks it intrudes or cuts (▶ Figure 8-4).

While this principle illustrates that an intrusive igneous structure is younger than the rocks it intrudes, the association of sedimentary and igneous rocks may cause problems in relative dating. Buried lava flows and intrusive igneous bodies such as sills look very similar in a sequence of strata (▶ Figure 8-5). A buried lava flow, however, is older than the rocks above it (principle of superposition), while a sill is younger than all the beds below it and younger than the bed immediately above it.

(a)

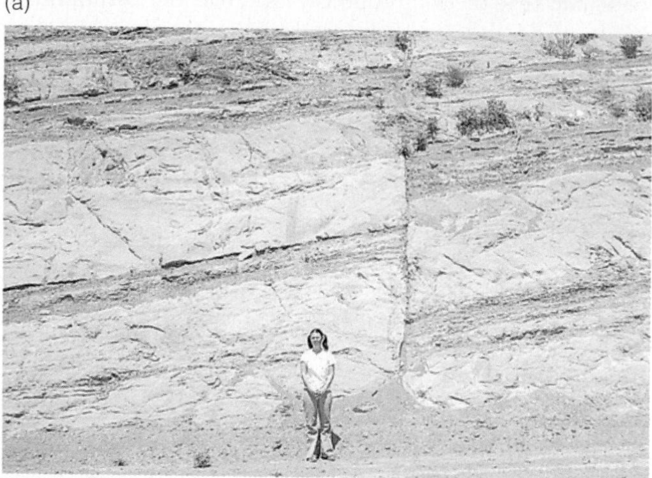

(b)

▶ **FIGURE 8-4** The principle of cross-cutting relationships. (*a*) A dark-colored dike has been intruded into older light-colored granite along the north shore of Lake Superior, Ontario, Canada. (*b*) A fault displacing tilted beds along Templin Highway, Castaic, California.

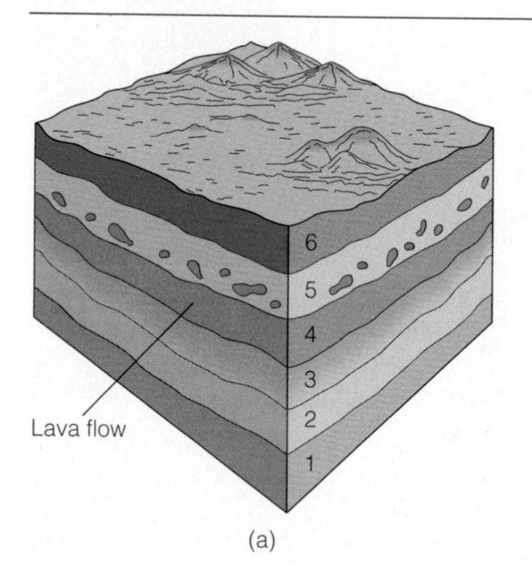

(a)

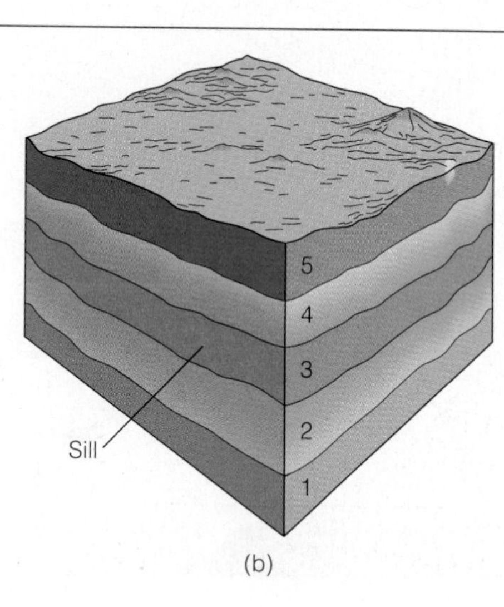

(b)

▶ **FIGURE 8-5** Relative ages of lava flows, sills, and associated sedimentary rocks may be difficult to determine. (*a*) A buried lava flow (4) baked the underlying bed, and bed 5 contains inclusions of the lava flow. The lava flow is younger than bed 3 and older than beds 5 and 6. (*b*) The rock units above and below the sill (3) have been baked, indicating that the sill is younger than beds 2 and 4, but its age relative to bed 5 cannot be determined.

To resolve such relative age problems as these, geologists look to see if the sedimentary rocks in contact with the igneous rocks show signs of baking or alteration by heat (see Chapter 7, Contact Metamorphism). A sedimentary rock showing such effects must be older than the igneous rock with which it is in contact. In Figure 8-5, for example, a sill produces a zone of baking immediately above and below it because it intruded into previously existing sedimentary rocks. A lava flow, on the other hand, bakes only those rocks below it.

Another way to determine relative ages is by using the **principle of inclusions.** This principle holds that inclusions, or fragments of one rock contained within a layer of another, are older than the rock layer itself. The batholith shown in ➤ Figure 8-6a contains sandstone inclusions, and the sandstone unit shows the effects of baking. Accordingly, we conclude that the sandstone is older than the batholith. In Figure 8-6b, however, the sandstone contains granite rock fragments, indicating that the batholith was the source rock for the inclusions and is therefore older than the sandstone.

Fossils have been known for centuries (see Chapter 6), yet their utility in relative dating and geologic mapping was not fully appreciated until the early nineteenth century. William Smith (1769–1839), an English civil engineer involved in surveying and building canals in southern England, independently recognized the principle of superposition by reason-

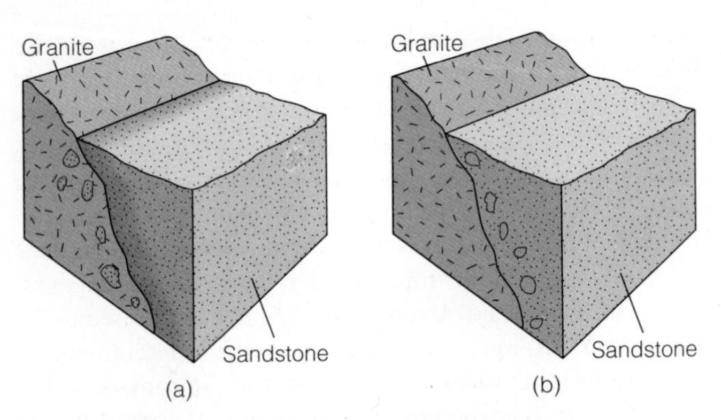

➤ **FIGURE 8-6** (*a*) The batholith is younger than the sandstone because the sandstone has been baked at its contact with the granite and the granite contains sandstone inclusions. (*b*) Granite inclusions in the sandstone indicate that the batholith was the source of the sandstone and therefore is older.

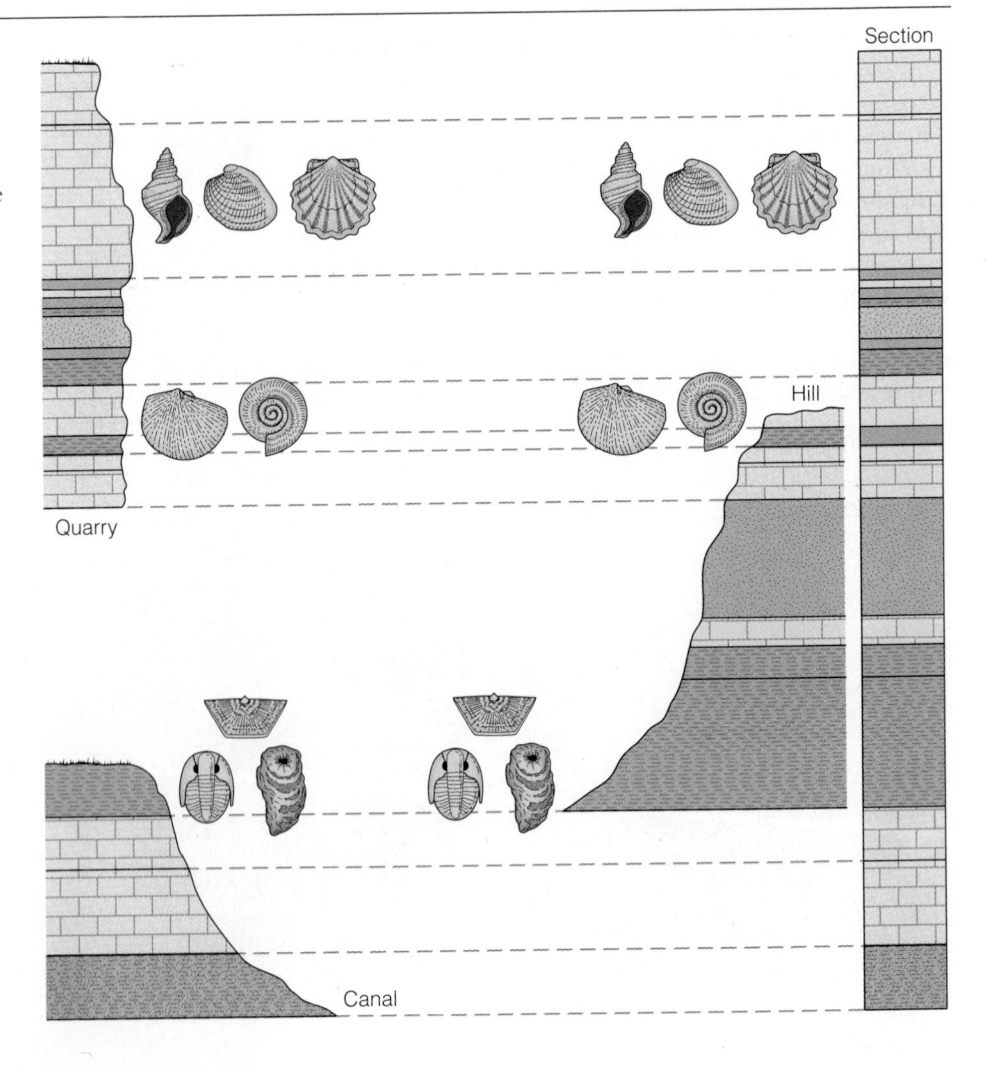

➤ **FIGURE 8-7** This generalized diagram shows how William Smith used fossils to identify strata of the same age in different areas (principle of fossil succession). The composite section on the right shows the relative ages of all strata in this area.

ing that the fossils at the bottom of a sequence of strata are older than those at the top of the sequence. This recognition served as the basis for the **principle of fossil succession** or the *principle of faunal and floral succession* as it is sometimes called (▶ Figure 8-7). According to this principle, fossil assemblages succeed one another through time in a regular and predictable order.

Unconformities

Our discussion so far has been concerned with conformable sequences of strata, sequences in which no depositional breaks of any consequence occur. A sharp bedding plane (see Figure 6-20) separating strata may represent a depositional break of minutes, hours, years, or even tens of years, but it is inconsequential when considered in the context of geologic time.

Surfaces that encompass significant amounts of geologic time are **unconformities,** and any interval of geologic time not represented by strata in a particular area is a *hiatus* (▶ Figure 8-8). Thus, an unconformity is a surface of nondeposition or erosion that separates younger strata from older rocks. As such, it represents a break in our record of geologic time. The famous 12-minute gap in the Watergate tapes of Richard Nixon's presidency is somewhat analogous. Just as we have no record of the conversations that were occurring during this period of time, we have no record of the events that occurred during a hiatus.

Three types of unconformities are recognized. A **disconformity** is a surface of erosion or nondeposition between younger and older beds that are parallel with one another (▶ Figure 8-9). Unless a well-defined erosional surface separates the older from the younger parallel beds, the disconformity frequently resembles an ordinary bedding plane. Accordingly, many disconformities are difficult to recognize and must be identified on the basis of fossil assemblages.

An **angular unconformity** is an erosional surface on tilted or folded strata over which younger strata have been deposited (▶ Figure 8-10). Both younger and older strata may dip, but if their dip angles are different (generally the older strata dip more steeply), an angular unconformity is present.

The angular unconformity illustrated in Figure 8-10b is probably the most famous in the world. It was here at Siccar Point, Scotland, that James Hutton realized that severe upheavals had tilted the lower rocks and formed mountains that were then worn away and covered by younger, flat-lying rocks. The erosional surface between the older tilted rocks and the younger flat-lying strata meant that there was a significant gap in the rock record. Although Hutton did not use the term unconformity, he was the first to understand and explain the significance of such discontinuities in the rock record.

The third type of unconformity is a **nonconformity.** Here an erosion surface cut into metamorphic or igneous rocks is covered by sedimentary rocks (▶ Figure 8-11).

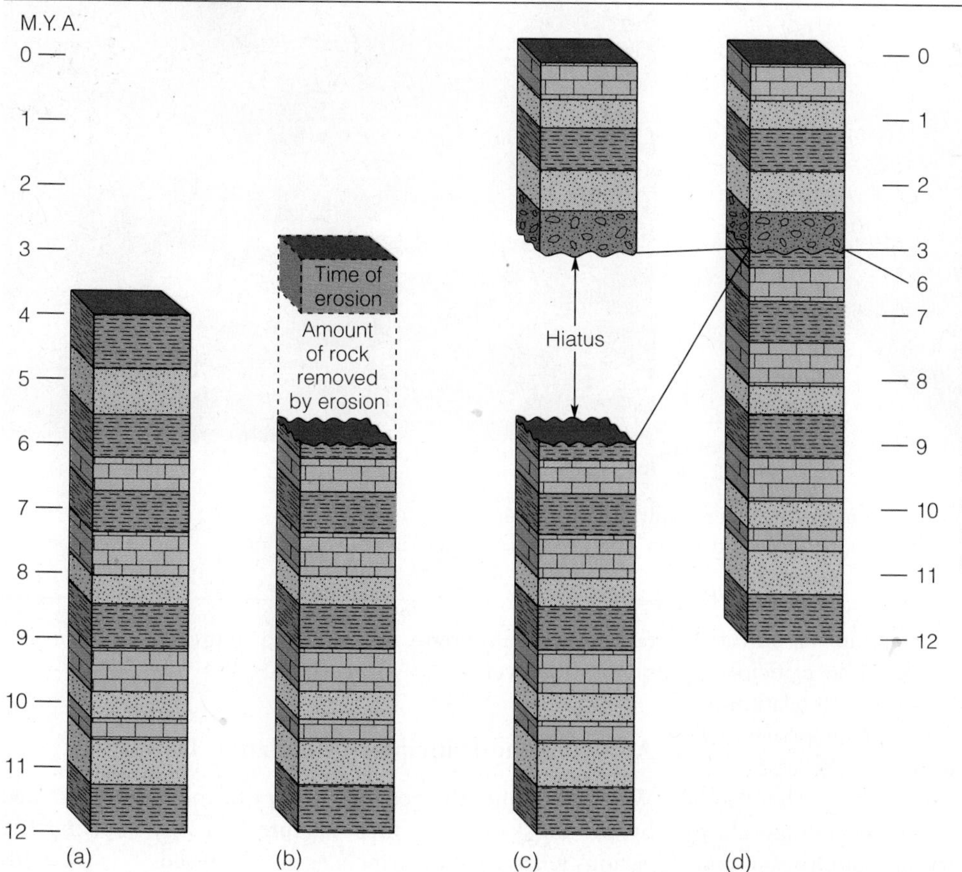

▶ FIGURE 8-8 A simplified diagram showing the development of an unconformity and a hiatus. (*a*) Deposition began 12 million years ago (M.Y.A.) and continued more or less uninterrupted until 4 M.Y.A. (*b*) A 1-million-year episode of erosion occurred, and during that time strata representing 2 million years of geologic time were eroded. (*c*) A hiatus of 3 million years exists between the older strata and the strata that formed during a renewed episode of deposition that began 3 M.Y.A. (*d*) The actual stratigraphic record. The unconformity is the surface separating the strata and represents a major break in our record of geologic time.

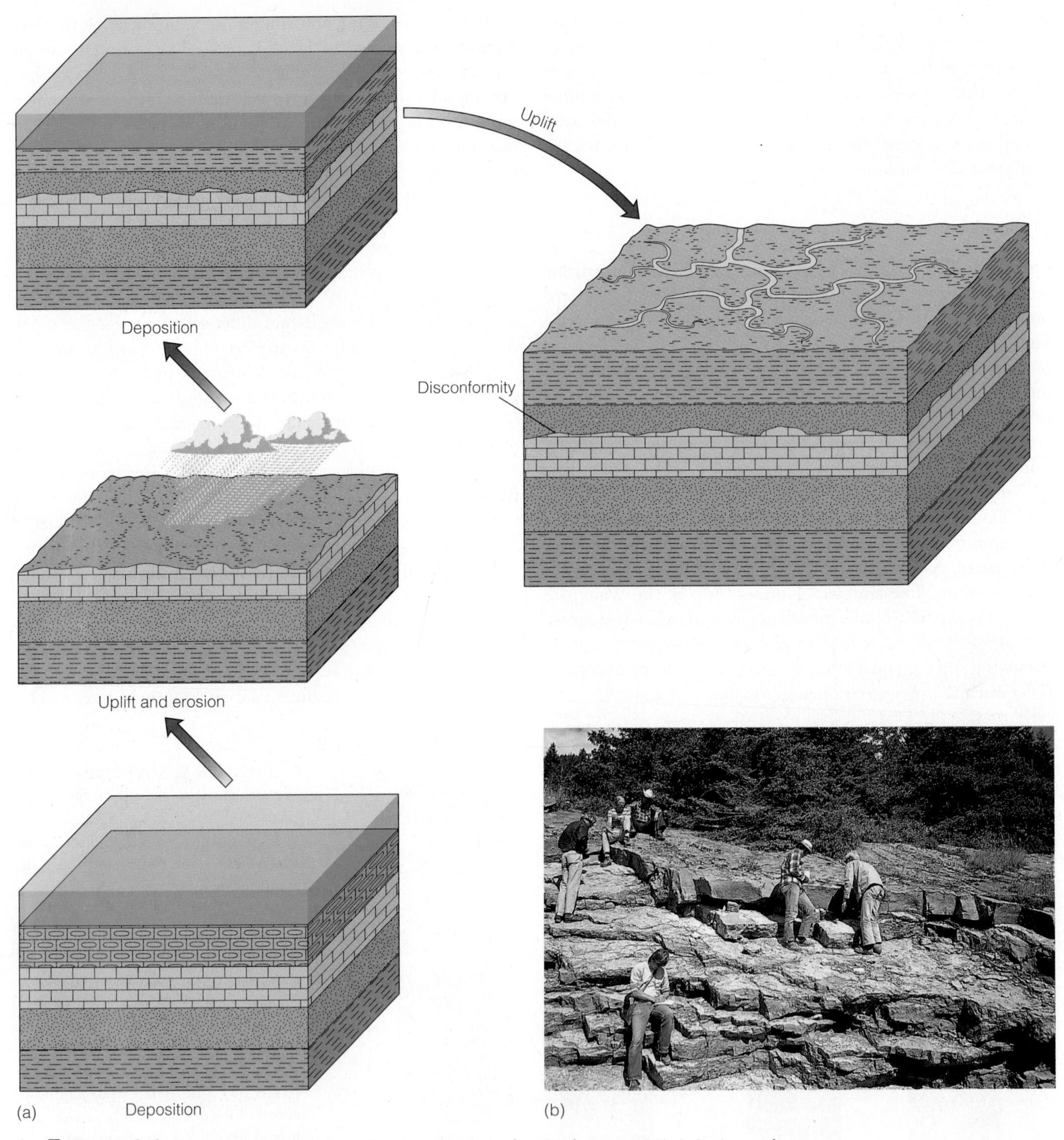

➢ **FIGURE 8-9** (*a*) Formation of a disconformity. (*b*) Disconformity between Mississippian and Jurassic strata in Montana. The geologist at the upper left is sitting on Jurassic strata, and his right foot is resting upon Mississippian rocks.

This type of unconformity closely resembles an intrusive igneous contact with sedimentary rocks. The principle of inclusions is helpful in determining whether the relationship between the igneous rocks and overlying sedimentary rocks is the result of an intrusion or erosion. In the case of an intrusion, the igneous rocks are younger, but in the case of erosion, the sedimentary rocks are younger. Being able to distinguish between a nonconformity and an intrusive contact is very important because they represent different sequences of events.

Applying the Principles of Relative Dating

We can decipher the geologic history of the area represented by the block diagram in ➢ Figure 8-12 by applying the various relative dating principles just discussed. The methods

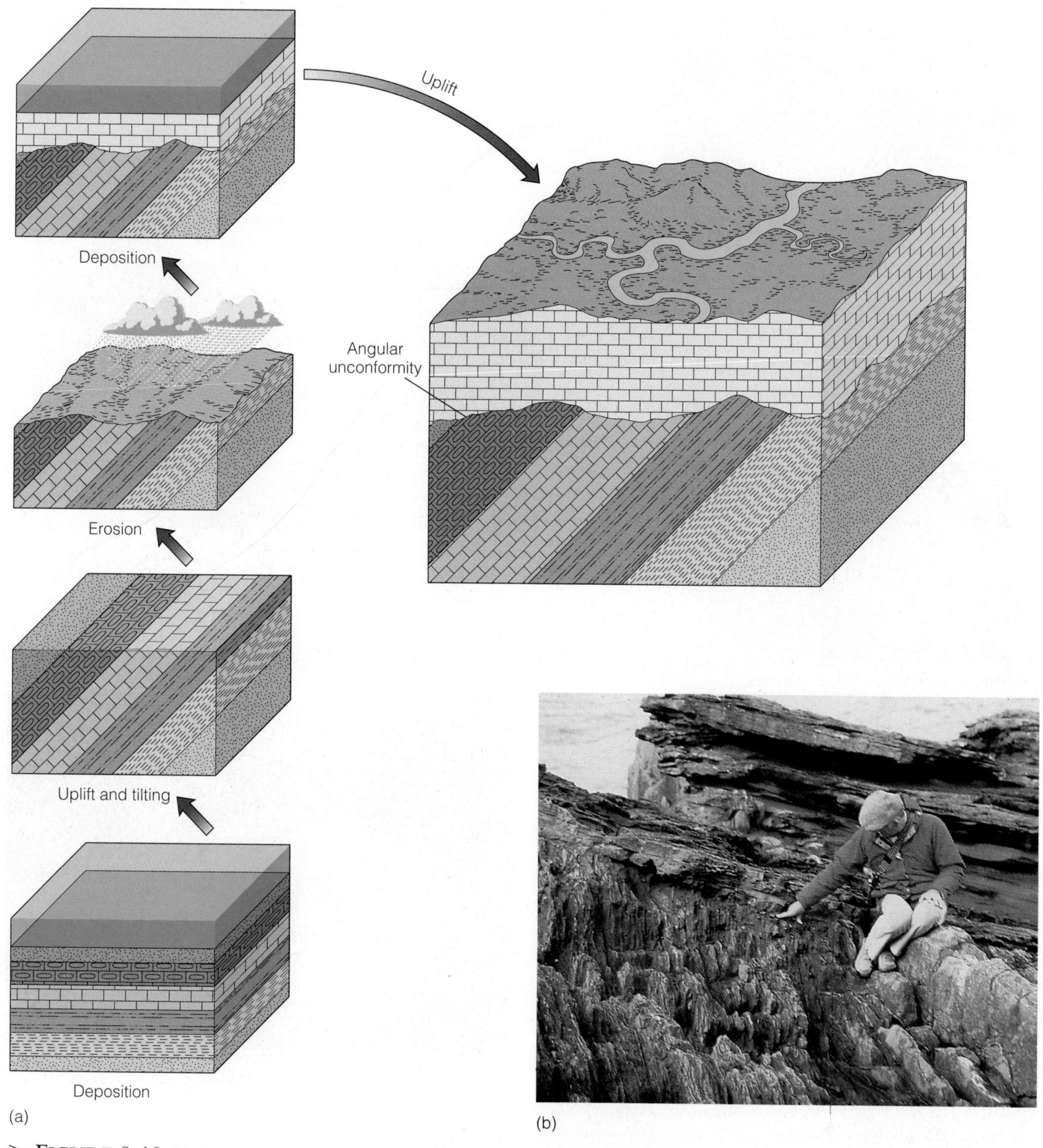

▶ FIGURE 8-10 (*a*) Formation of an angular unconformity. (*b*) Angular unconformity at Siccar Point, Scotland. (Photo courtesy of Dorothy L. Stout.)

and logic used in this example are the same as those applied by nineteenth-century geologists in constructing the geologic time scale.

According to the principles of superposition and original horizontality, beds A, B, C, D, E, F, and G were deposited horizontally; then they were either tilted, faulted (H), and eroded, or after deposition, they were faulted (H), tilted, and then eroded (▶ Figure 8-13a, b, and c). Because the fault cuts beds A–G, it must be younger than the beds according to the principle of cross-cutting relationships.

Beds J, K, and L were then deposited horizontally over this erosional surface producing an angular unconformity (I) (Figure 8-13d). Following deposition of these three beds, the entire sequence was intruded by a dike (M), which,

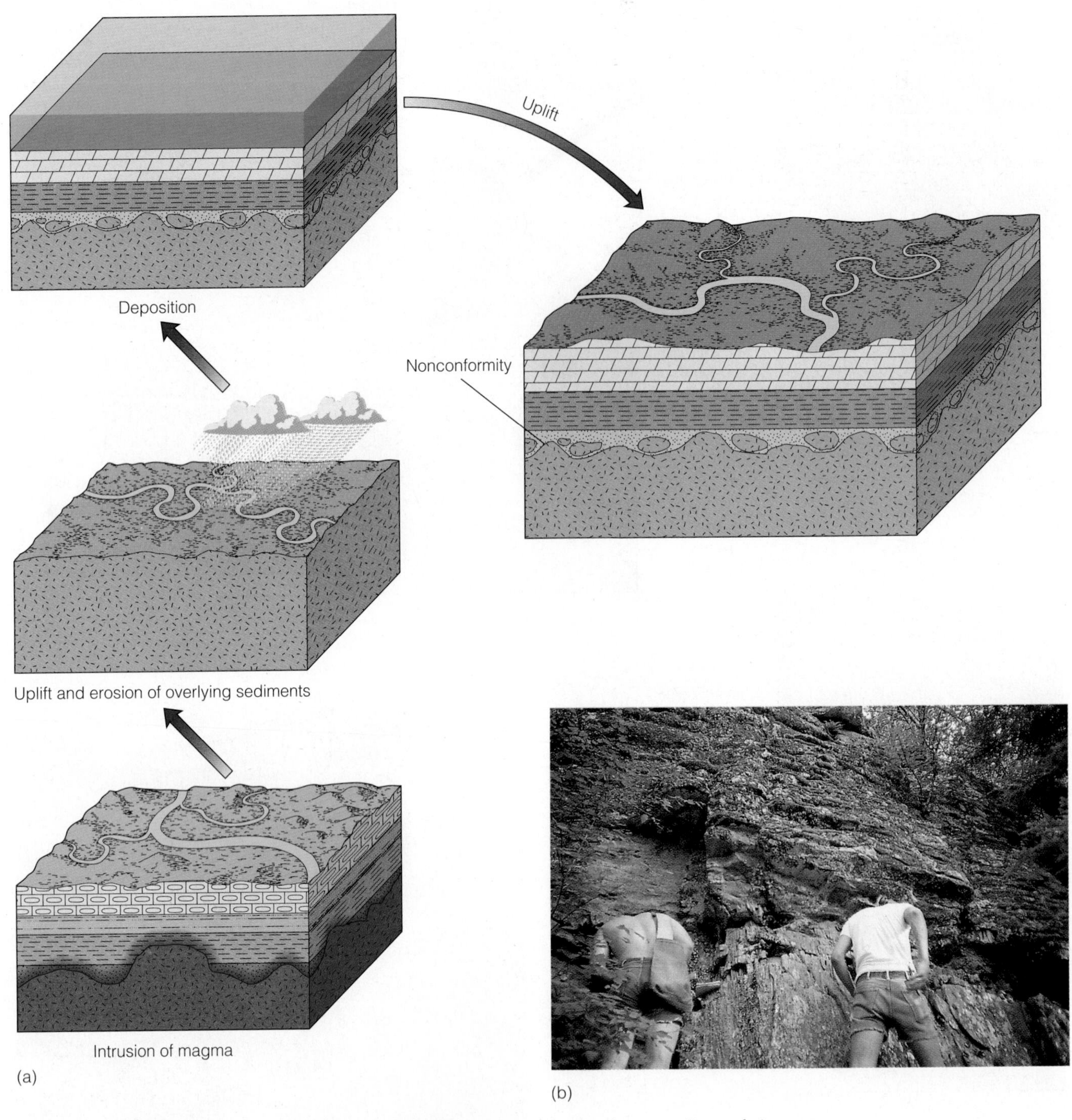

▶ **FIGURE 8-11** (*a*) Formation of a nonconformity. (*b*) Nonconformity between Precambrian granite and the overlying Cambrian-age Deadwood Formation, Wyoming.

according to the principle of cross-cutting relationships, must be younger than all the rocks it intrudes (Figure 8-13e).

The entire area was then uplifted and eroded; next beds P and Q were deposited, producing a disconformity (N) between beds L and P and a nonconformity (O) between the igneous intrusion M and the sedimentary bed P (Figure 8-13f and g). We know that the relationship between igneous intrusion M and the overlying sedimentary bed P is a nonconformity because of the presence of inclusions of M in P (principle of inclusions).

At this point, there are several possibilities for reconstructing the geologic history of this area. According to the principle of cross-cutting relationships, dike R must be younger than bed Q because it intrudes into it. It can have intruded anytime *after* bed Q was deposited; however, we cannot determine whether R was formed right after Q, right after S, or after T was formed. For purposes of this history,

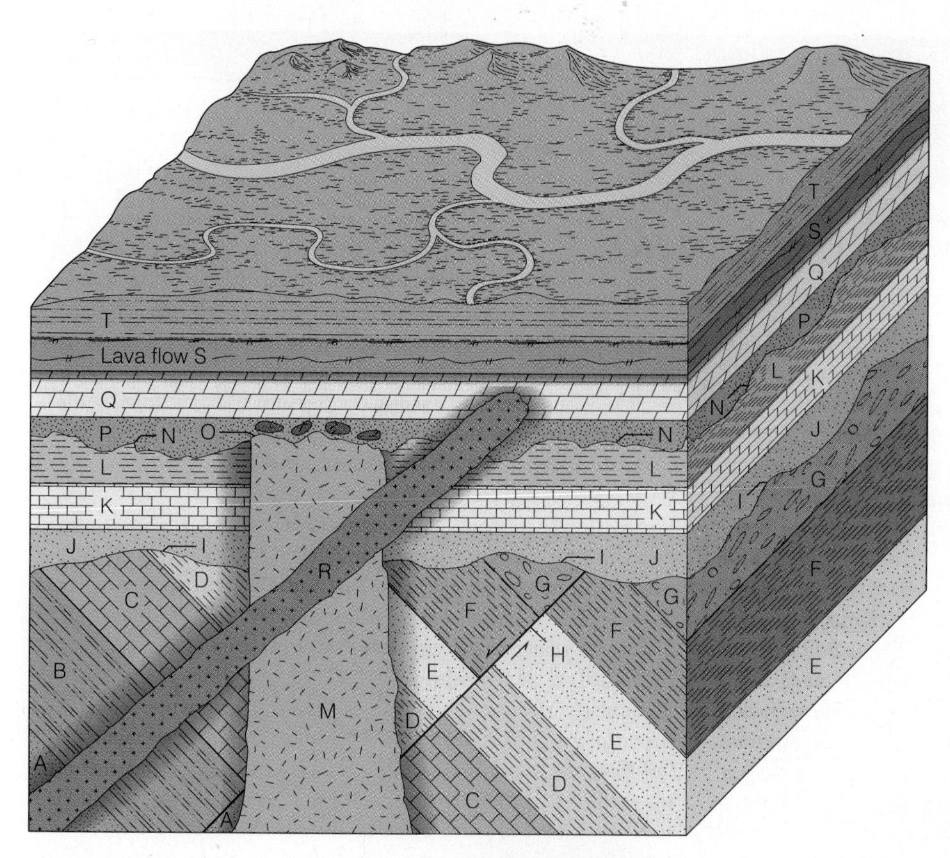

▶ **FIGURE 8-12 (left)** A block diagram of a hypothetical area whose geologic history can be reconstructed by applying the various relative dating principles.

▶ **FIGURE 8-13 (below)** (*a*) Beds A, B, C, D, E, F, and G are deposited. (*b*) The preceding beds are tilted and faulted. (*c*) Erosion. (*d*) Beds J, K, and L are deposited, producing an angular unconformity. (*e*) The entire sequence is intruded by a dike. (*f*) The entire sequence is uplifted and eroded. (*g*) Beds P and Q are deposited, producing a disconformity (N) and a nonconformity (O). (*h*) Dike R intrudes. (*i*) Lava (S) flows over bed Q, baking it. (*j*) Bed T is deposited.

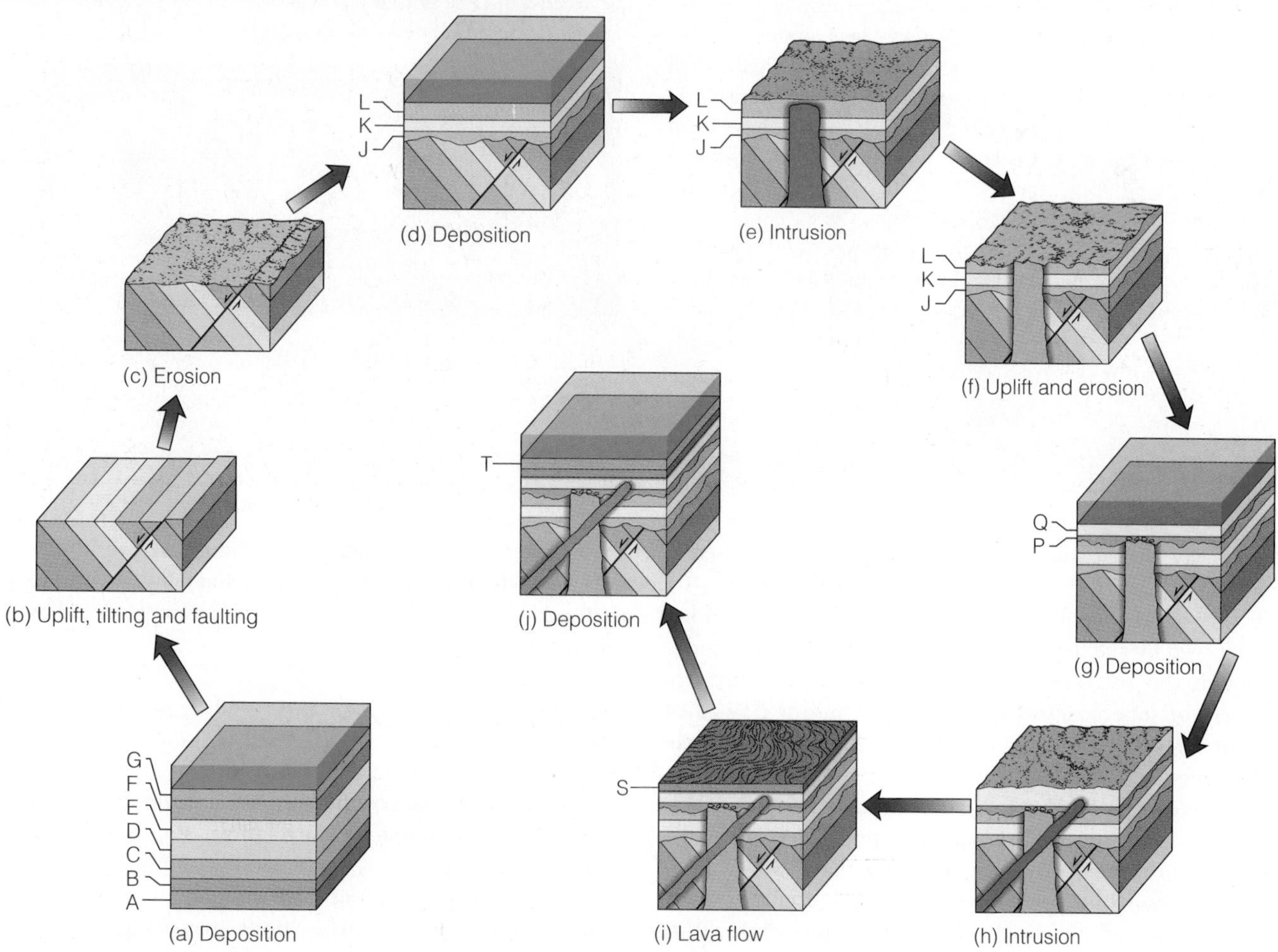

Perspective 8-1

Subsurface Correlation and the Search for Oil and Natural Gas

During the early years of the petroleum industry, geologists searching for oil and gas relied almost exclusively on surface studies and constructed maps showing rocks and geologic structures such as folds and faults. Interpretation of such maps sometimes revealed subsurface structures, such as those in Figure 6-31, in which oil and natural gas might be trapped. Surface methods are still important in petroleum geology, particularly in unexplored regions, but most exploration is now done using subsurface methods.

Subsurface geology is the acquisition and interpretation of data about geologic features beneath the Earth's surface. Drilling operations have provided a wealth of subsurface data. When drilling for oil or natural gas, cores or rock chips called *well cuttings* are commonly recovered from the drill hole. These samples are studied under the microscope and reveal such important information as rock type, porosity (the amount of pore space) and permeability (the ability to transmit fluids), and the presence of oil stains. The samples can also be processed for microfossils that can aid in determining the geologic age of the rocks (▶ Figure 1).

Geophysical instruments may be lowered down the drill hole to record such rock properties as electrical resistivity, density, and radioactivity, thus providing a lithologic record or *well log* of the rocks penetrated. Cores, well cuttings, and well logs are all extremely useful in making subsurface lithostratigraphic correlations (▶ Figure 2).

Subsurface rock units may also be detected and traced by the study of seismic profiles. Energy pulses, such as those from explosions, travel through rocks at a velocity determined by rock density, and some of this energy is reflected from various horizons (contacts between contrasting layers) back to the surface, where it is recorded (▶ Figure 3). Seismic stratigraphy is particularly useful in tracing units in areas such as the continental shelves where it is very expensive to drill holes and other techniques have limited use.

In petroleum exploration, the purpose of most well correlations is to map the structure to see if it has the potential for trapping oil and gas. The choice of subsurface correlation methods depends on the information geologists are seeking, the general geology of the area, and the cost and time available to run different logs.

▶ **FIGURE 1** Microscopic one-celled animals called foraminifera can be used to determine the age of the rock they are found in and can be used to correlate rock units between walls. (Scanning electron micrograph by Dee Breger, Lamont–Doherty Geological Observatory.)

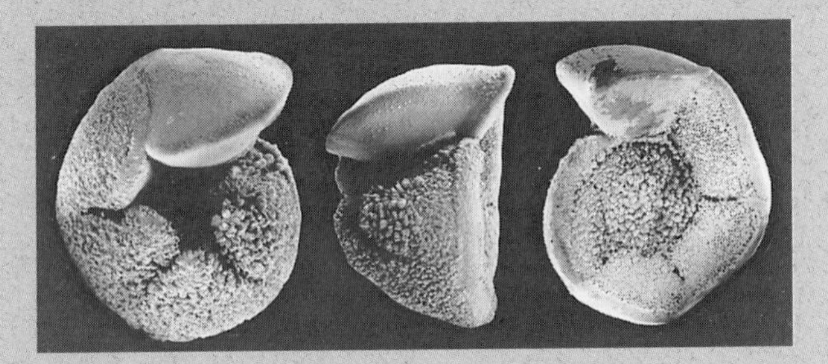

we will say that it intruded after the deposition of bed Q (Figure 8-13g and h).

Following the intrusion of dike R, lava S flowed over bed Q, followed by the deposition of bed T (Figure 8-13i and j). Although the lava flow (S) is not a sedimentary unit, the principle of superposition still applies because it flowed on the Earth's surface, just as sediments are deposited on the Earth's surface.

We have established a relative chronology for the rocks and events of this area by using the principles of relative dating. Remember, however, that we have no way of knowing how many years ago these events occurred unless we can obtain radiometric dates for the igneous rocks. With these dates we can establish the range of absolute ages between which the different sedimentary units were deposited and also determine how much time is represented by the unconformities.

 Correlation

If geologists are to decipher Earth history, they must demonstrate the time equivalency of rock units in different areas. This process is known as **correlation.** Rock units can be correlated in various ways, including similar rock type, position in a sequence, and fossil content.

If exposures are adequate, units may simply be traced

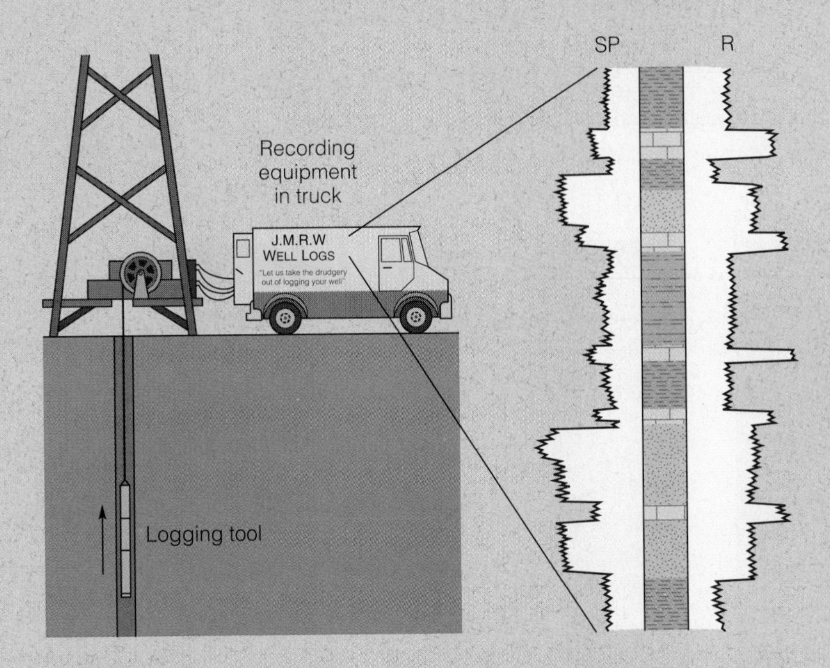

▶ **FIGURE 2** A schematic diagram showing how well logs are made. As the logging tool is withdrawn from the drill hole, data are transmitted to the surface where they are recorded and printed as a well log. The curve labeled SP in this diagrammatic electric log is a plot of self-potential (electrical potential caused by different conductors in a solution that conducts electricty) with depth. The curve labeled R is a plot of electrical resistivity with depth. Electric logs yield information about the rock type and fluid content of subsurface formations. Electric logs are also used to correlate wells.

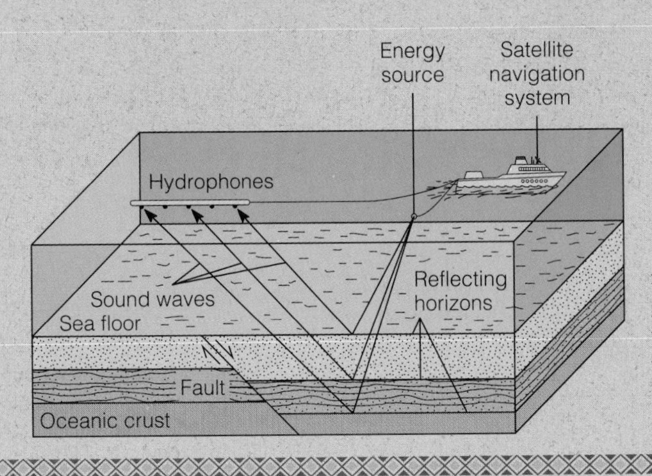

▶ **FIGURE 3** A diagram showing the use of seismic reflections to detect buried rock units at sea. Sound waves are generated at the energy source. Some of the energy of these waves is reflected from various horizons back to the surface where it is detected by hydrophones. Buried rock units can also be detected on land, but here explosive charges are detonated as an energy source.

laterally (principle of lateral continuity), even if occasional gaps exist (▶ Figure 8-14). Other criteria used to correlate units are similarity of rock type, position in a sequence, and *key beds.* Key beds are units, such as coal beds or volcanic ash layers, that are sufficiently distinctive to allow the same unit to be identified in different areas (Figure 8-14). In addition to surface correlation, geologists frequently use well logs, cores, or cuttings to correlate subsurface rock units when exploring for minerals, coal, and petroleum (see Perspective 8-1).

Generally, no single location in a region has a geologic record of all the events that occurred during its history; therefore, to decipher the complete geologic history of the region, geologists must correlate from one area to another. An excellent example is provided by the history of the Colorado Plateau (▶ Figure 8-15). A record of events occuring over approximately 2 billion years of Earth history is present in this region. Because of the forces of erosion, the entire record is not preserved at any single location. Within the walls of the Grand Canyon are rocks of the Precambrian and Paleozoic Era, while Paleozoic and Mesozoic Era rocks are found in Zion National Park, and Mesozoic and Cenozoic Era rocks are exposed in Bryce Canyon (Figure 8-15). By correlating the uppermost rocks at one location with the lowermost equivalent rocks of another area, the history of the entire region can be deciphered.

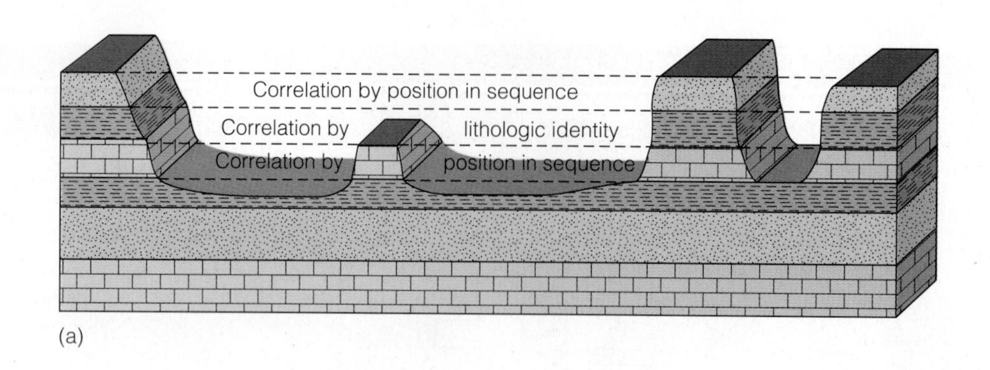

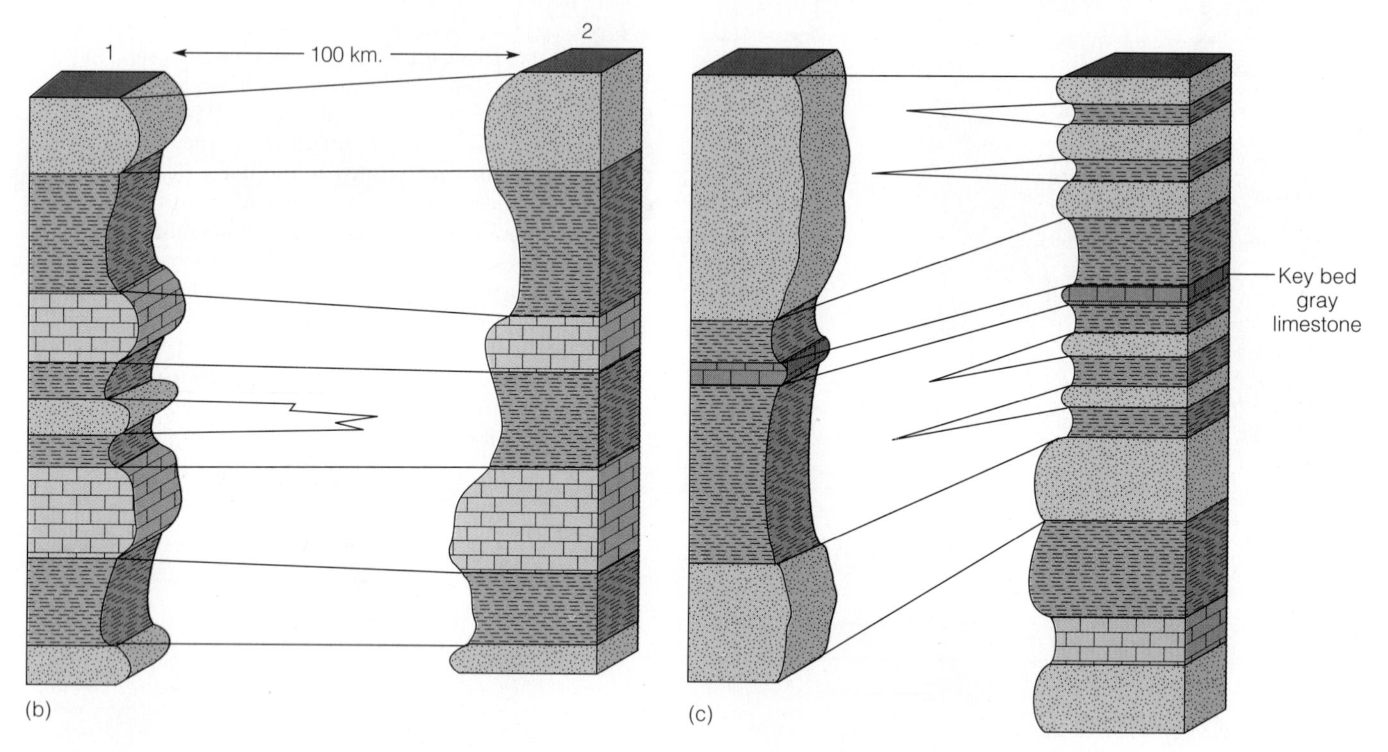

➤ **FIGURE 8-14** Correlation of rock units. (*a*) In areas of adequate exposures, rock units can be traced laterally even if occasional gaps exist. (*b*) Correlation by similarities in rock type and position in a sequence. The sandstone in section 1 is assumed to intertongue or grade laterally into the shale at section 2. (*c*) Correlation using a key bed, a distinctive gray limestone.

Although geologists can match up rocks on the basis of similar rock type and stratigraphic position, correlation of this type can only be done in a limited area where beds can be traced from one site to another. In order to correlate rock units over a large area or to correlate age-equivalent units of different composition, fossils and the principle of fossil succession must be used.

Fossils are useful as time indicators because they are the remains of organisms that lived for a certain length of time during the geologic past. Easily identified fossils that are geographically widespread and existed for a rather short geologic time are particularly useful. Such fossils are called **guide fossils** or *index fossils* (➤ Figure 8-16). The trilobite *Isotelus* and the clam *Inoceramus* meet all of these criteria and are therefore good guide fossils. In contrast, the brachiopod *Lingula* is easily identified and widespread, but its geologic range of Ordovician to Recent makes it of little use in correlation.

Because most fossils have fairly long geologic ranges, geologists construct *assemblage range zones* to determine the age of the sedimentary rocks containing the fossils. Assemblage range zones are established by plotting the overlapping geologic ranges of different species of fossils. The first and last occurrences of two species are used to establish an assemblage zone's boundaries (➤ Figure 8-17). Correlation of assemblage zones generally yields correlation lines that are considered time equivalent. In other words, the strata encompassed by the correlation lines are thought to be the same age.

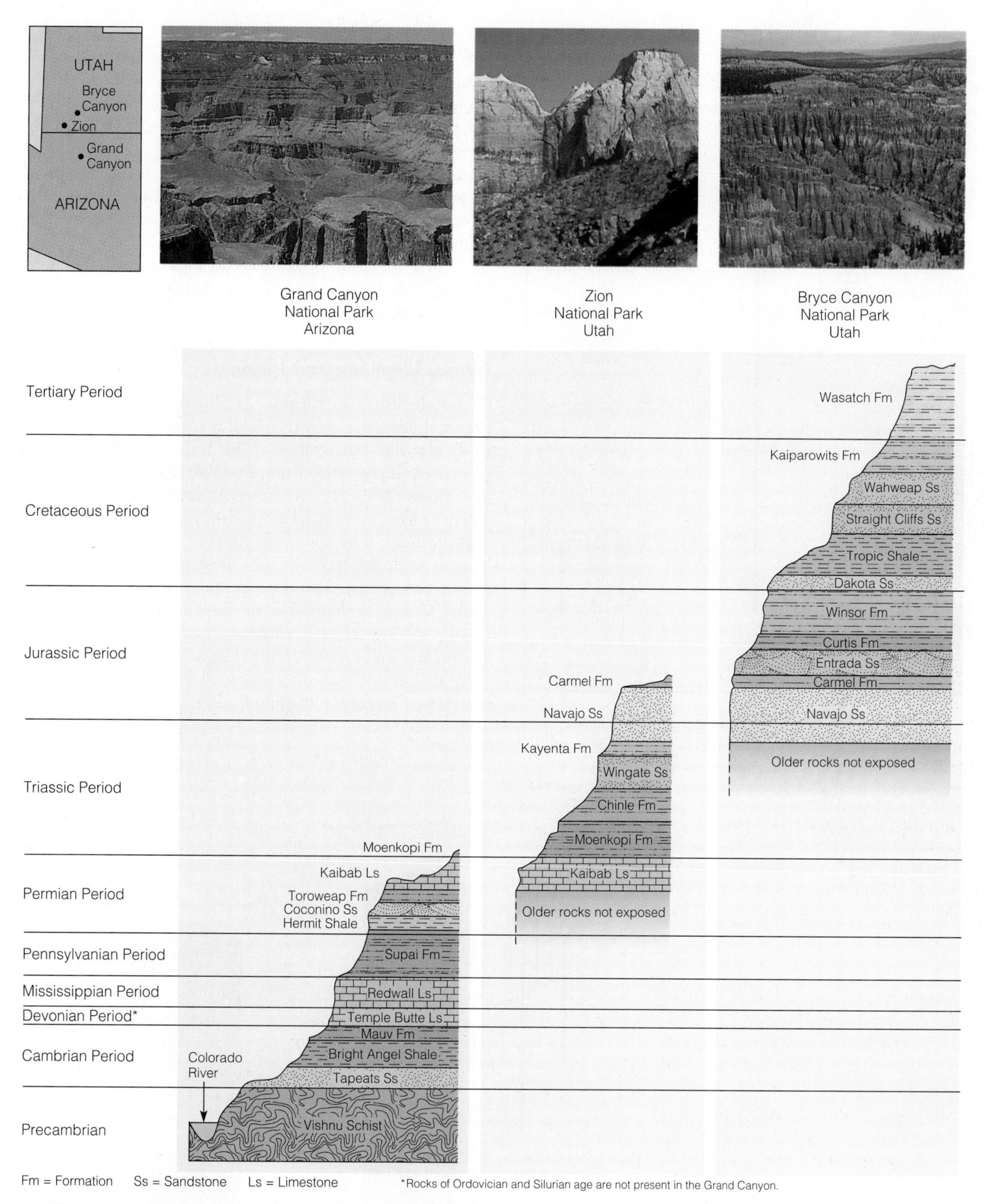

➢ **FIGURE 8-15** Correlation of rocks within the Colorado Plateau. By correlating the rocks from various locations, the history of the entire region can be deciphered.

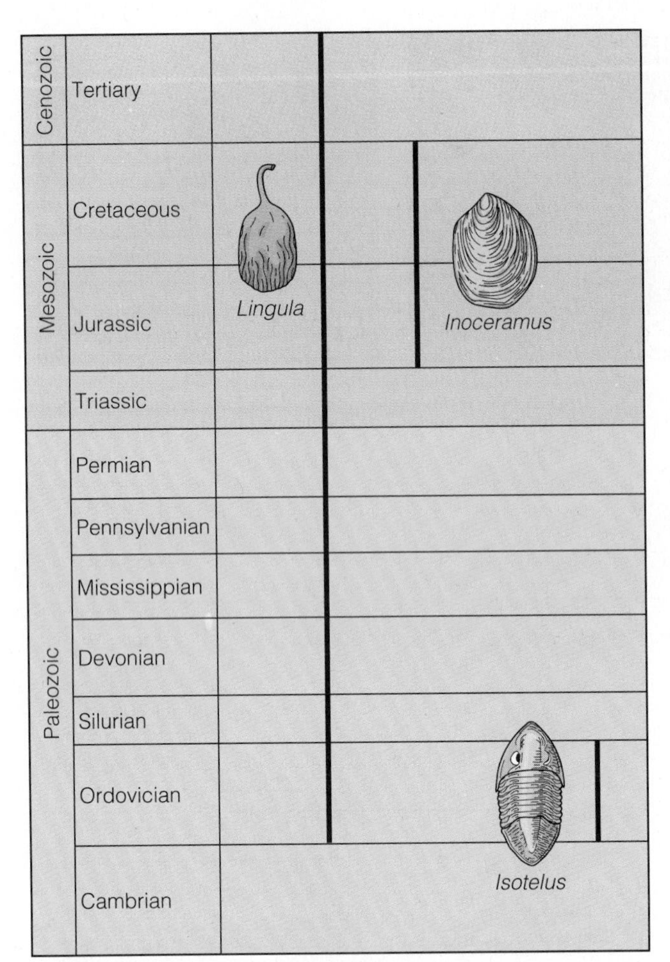

▶ **FIGURE 8-16** The geologic ranges of three marine invertebrates. The brachiopod *Lingula* is of little use in correlation because of its long geologic range. The trilobite *Isotelus* and the bivalve *Inoceramus* are good guide fossils because they are geographically widespread, are easily identified, and have short geologic ranges.

ABSOLUTE DATING METHODS

Although most of the isotopes of the 92 naturally occurring elements are stable, some are radioactive and spontaneously decay to other, more stable isotopes of elements, releasing energy in the process. The discovery, in 1903 by Pierre and Marie Curie, that radioactive decay produces heat as a by-product meant that geologists finally had a mechanism for explaining the internal heat of the Earth that did not rely on residual cooling from a molten origin. Furthermore, geologists had a powerful tool to date geologic events accurately, and verify the long time periods postulated by Hutton, Lyell, and Darwin.

Atoms, Elements, and Isotopes

As we discussed in Chapter 2, all matter is made up of chemical elements, each of which is composed of extremely small particles called *atoms*. The nucleus of an atom is composed of *protons* and *neutrons* with *electrons* encircling it (see Figure 2-2). The number of protons defines an element's *atomic number* and helps determine its properties and characteristics. The combined number of protons and neutrons in an atom is its *atomic mass number*. Not all atoms of the same element have the same number of neutrons in their nuclei. These variable forms of the same element are called *isotopes* (see Figure 2-3). Most isotopes are stable, but some are unstable and spontaneously decay to a more stable form. It is the decay rate of unstable isotopes that geologists measure to determine the absolute age of rocks.

▶ **FIGURE 8-17** Correlation of two sections by using assemblage range zones. These zones are established by the overlapping ranges of fossils A through E.

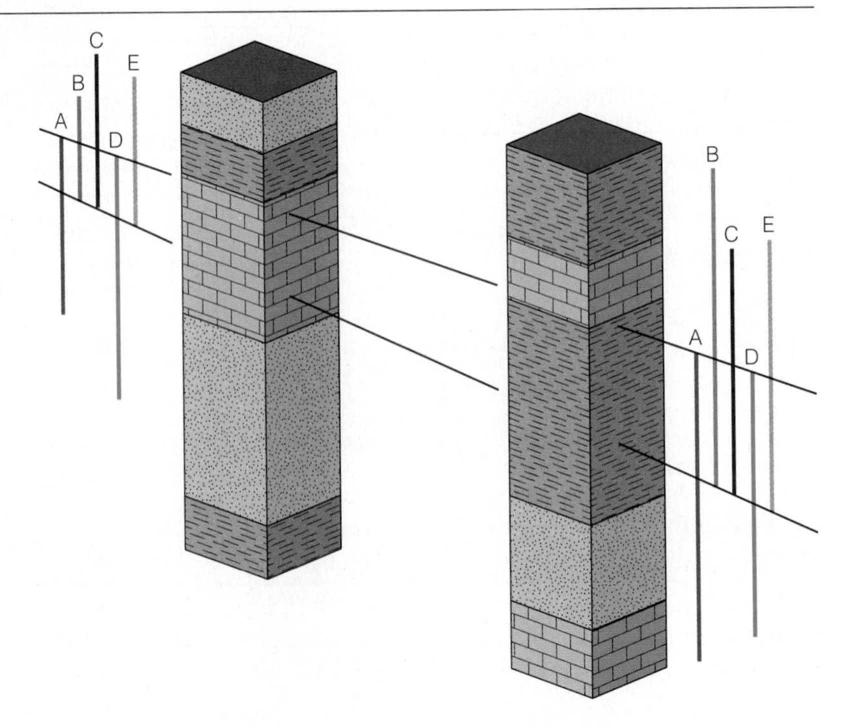

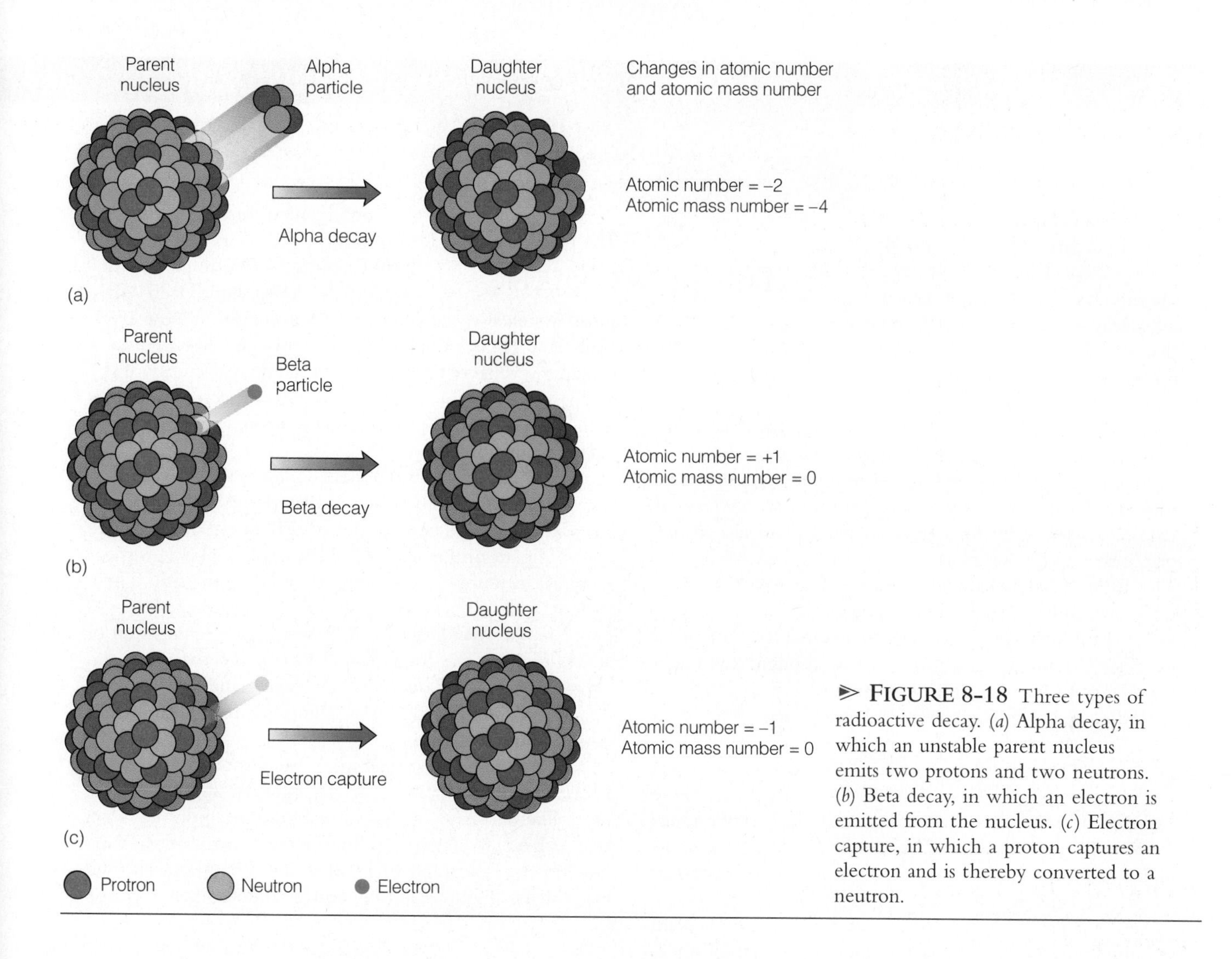

▶ **FIGURE 8-18** Three types of radioactive decay. (*a*) Alpha decay, in which an unstable parent nucleus emits two protons and two neutrons. (*b*) Beta decay, in which an electron is emitted from the nucleus. (*c*) Electron capture, in which a proton captures an electron and is thereby converted to a neutron.

Radioactive Decay and Half-Lives

Radioactive decay is the process whereby an unstable atomic nucleus is spontaneously transformed into an atomic nucleus of a different element. Three types of radioactive decay are recognized, all of which result in a change of atomic structure (▶ Figure 8-18). In **alpha decay,** two protons and two neutrons are emitted from the nucleus, resulting in the loss of two atomic numbers and four atomic mass numbers. In **beta decay,** a fast-moving electron is emitted from a neutron in the nucleus, changing that neutron to a proton and consequently increasing the atomic number by one, but not affecting the atomic mass number. **Electron capture** occurs when a proton captures an electron from an electron shell and thereby converts to a neutron, resulting in the loss of one atomic number but not changing the atomic mass number.

Some elements undergo only one decay step in the conversion from an unstable form to a stable form. For example, rubidium 87 decays to strontium 87 by a single beta emission, and potassium 40 decays to argon 40 by a single electron capture. Other radioactive elements undergo several decay steps (see Perspective 8-2). Uranium 235 decays to lead 207 by seven alpha and six beta steps, while uranium 238 decays to lead 206 by eight alpha and six beta steps (▶ Figure 8-19).

When discussing decay rates, it is convenient to refer to them in terms of half-lives. The **half-life** of a radioactive element is the time it takes for one-half of the atoms of the original unstable **parent element** to decay to atoms of a new, more stable **daughter element.** The half-life of a given radioactive element is always constant and can be precisely measured in the laboratory. Half-lives of various radioactive elements range from less than a billionth of a second to 49 billion years.

Radioactive decay occurs at a geometric rate rather than a linear rate. Therefore, a graph of the decay rate produces a curve rather than a straight line (▶ Figure 8-20). For example, an element with *1,000,000* parent atoms will have *500,000* parent atoms and *500,000* daughter atoms after one half-life. After two half-lives, it will have *250,000* parent atoms (one-half of the previous parent atoms, which is equivalent to one-fourth of the original parent atoms) and *750,000* daughter atoms. After three half-lives, it will have *125,000* parent atoms (one-half of the previous parent atoms or one-eighth of the original parent atoms) and *875,000*

Perspective 8-2

RADON: THE SILENT KILLER

Radon is a colorless, odorless, naturally occurring radioactive gas that has a 3.8-day half-life. It is part of the uranium 238–lead 206 radioactive decay series (Figure 8-19) and occurs in any rock or soil that contains uranium 238. Radon concentrations are reported in picocuries per liter (pCi/L) of air (a curie is the standard measure of radiation, and a picocurie is one-trillionth of a curie, or the equivalent of the decay of about two radioactive atoms per minute). Outdoors, radon escapes into the atmosphere where it is diluted and dissipates to harmless levels (0.2 pCi/L is the ambient outdoor level of radon). Radon levels for indoor air range from less than 1 pCi/L to about 3,000 pCi/L, but average about 1.5 pCi/L. The EPA considers radon levels exceeding 4 pCi/L to be unhealthy and recommends that remedial action be taken to lower them. Continued exposure to elevated levels of radon over an extended period of time is thought by many researchers to increase the risk of lung cancer.

Radon is one of the natural decay products of uranium 238. It rapidly decays by the emission of an alpha particle, producing two short-lived radioactive isotopes—polonium 218 and polonium 214 (Figure 8-19). Both of these isotopes are solid and can become trapped in your lungs every time you breath. When polonium decays, it emits alpha and beta particles that can damage lung cells and cause lung cancer.

Your chances of being adversely affected by radon depend on numerous interrelated factors such as geographic location, the geology of the area, the climate, how buildings are constructed, and the amount of time you spend in your house. Because radon is a naturally occurring gas, contact with it is unavoidable. However, atmospheric concentrations are probably harmless. Only when concentrations of radon build up in poorly ventilated structures does it become a potential health risk.

Concern about the health risks posed by radon first arose during the 1960s when the news media revealed that some homes in the West had been built with uranium mine tailings. Since then, geologists have found that high indoor radon levels can be caused by natural uranium in minerals of the rock and soil on which buildings are constructed. In response to the high cost of energy during the 1970s and 1980s, old buildings were insulated, and new buildings were constructed to be as energy efficient and airtight as possible. Ironically, these energy-saving measures also sealed in radon.

Radon enters buildings through dirt floors, cracks in the floor or walls, joints between floors and walls, floor drains, sumps, and utility pipes as well as through any cracks or pores in hollow block walls (▶ Figure 1). Radon can also be released into a building whenever the water is turned on, particularly if the water comes from a private well. Municipal water is generally safe because it has usually been aerated before it gets to your home.

To find out if your home has a radon problem, you must test for it with commercially available, relatively inexpensive, simple home testing devices. If radon readings are above the recommended EPA levels of 4 pCi/L, several remedial measures can be taken to reduce your risk. These include sealing up all cracks in the foundation, pouring a concrete slab over a dirt floor, increasing the circulation of air throughout the house, especially in the basement and crawl space, providing filters for drains and other utility openings, and limiting time spent in areas with higher concentrations of radon.

It is important to remember that although the radon hazard encompasses most of the country, some areas are more likely to have higher natural concentrations of radon than others (▶ Figure 2). Such rocks as uranium-bearing granites, metamorphic rocks of granitic composition, and black shales (high carbon content) are quite likely to cause

▶ **FIGURE 1** Some of the common entry points where radon can enter a house.

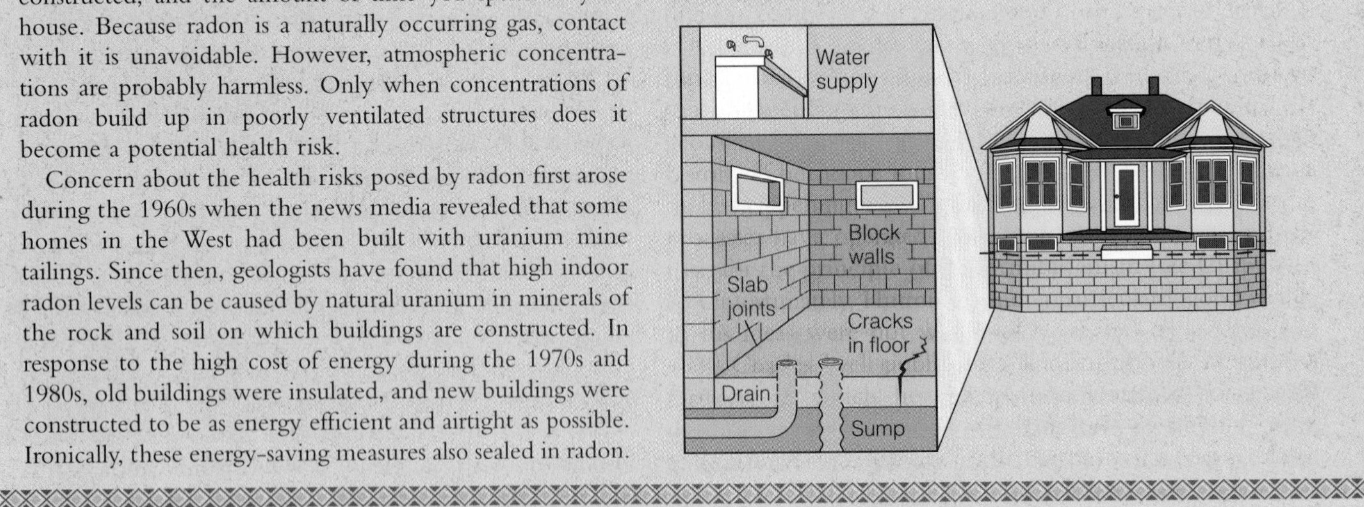

daughter atoms, and so on until the number of parent atoms remaining is so few that they cannot be accurately measured by present-day instruments.

By measuring the parent-daughter ratio and knowing the half-life of the parent (determined in the laboratory), geologists can calculate the age of a sample containing the indoor radon problems. Other rocks such as marine quartz sandstone, noncarbonaceous shales and siltstones, most volcanic rocks, and igneous and metamorphic rocks rich in iron and magnesium typically do not cause radon problems. The permeability of the soil overlying the rock can also affect indoor levels of radon gas. Some soils are more permeable than others and allow more radon to escape into the overlying structures.

The climate and type of construction affect not only how much radon enters a structure, but how much escapes. Concentrations of radon are highest during the winter in northern climates because buildings are sealed as tightly as possible. Homes with basements are more likely to have higher radon levels than those built on concrete slabs. While research continues into the sources of indoor radon and ways of controlling it, the most important thing people can do is to test their home, school, or business for radon.

There is currently a heated debate among scientists concerning the large-scale health hazards from radon exposure and how much money should be spent for its remediation. On the one hand, the EPA and Surgeon General estimate that exposure to high levels of indoor radon causes between 5,000 and 20,000 lung cancer deaths each year. Other scientists dispute these figures because of the difficulty of attributing mortality rates for lung cancer directly to radon, particularly when so many other factors, such as smoking, are involved. Central to this debate are two questions: What concentration levels of indoor radon are acceptable, and exactly how serious is the risk from radon at those levels? Unfortunately, the data for making these determinations are simply not available at this time.

FIGURE 2 Areas in the United States where granite, phosphate-bearing rocks, carbonaceous shales, and uranium occur. These rocks are all potential sources of radon gas.

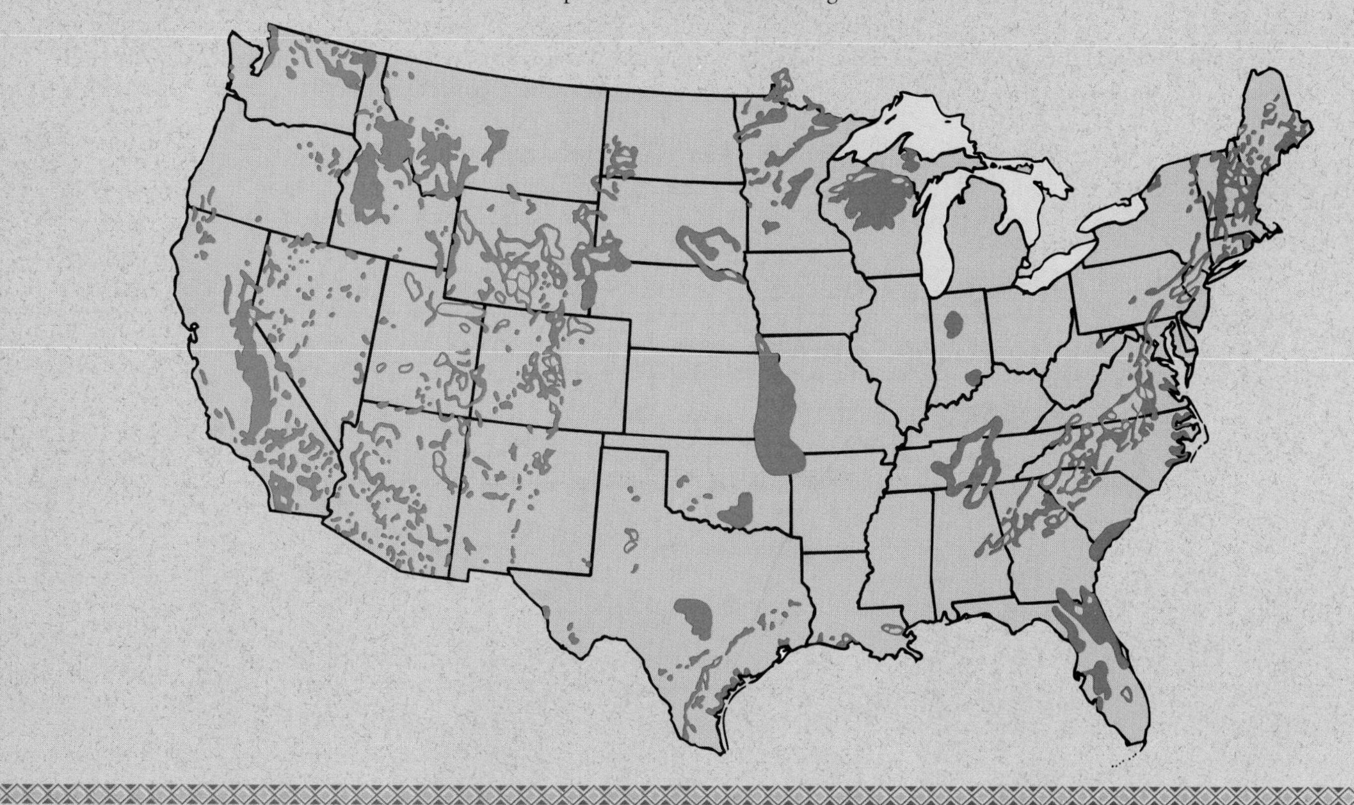

radioactive element. The parent-daughter ratio is usually determined by a *mass spectrometer*, an instrument that measures the proportions of elements of different masses.

Sources of Uncertainty. The most accurate radiometric dates are obtained from igneous rocks. As a magma cools and begins to crystallize, radioactive parent atoms are separated

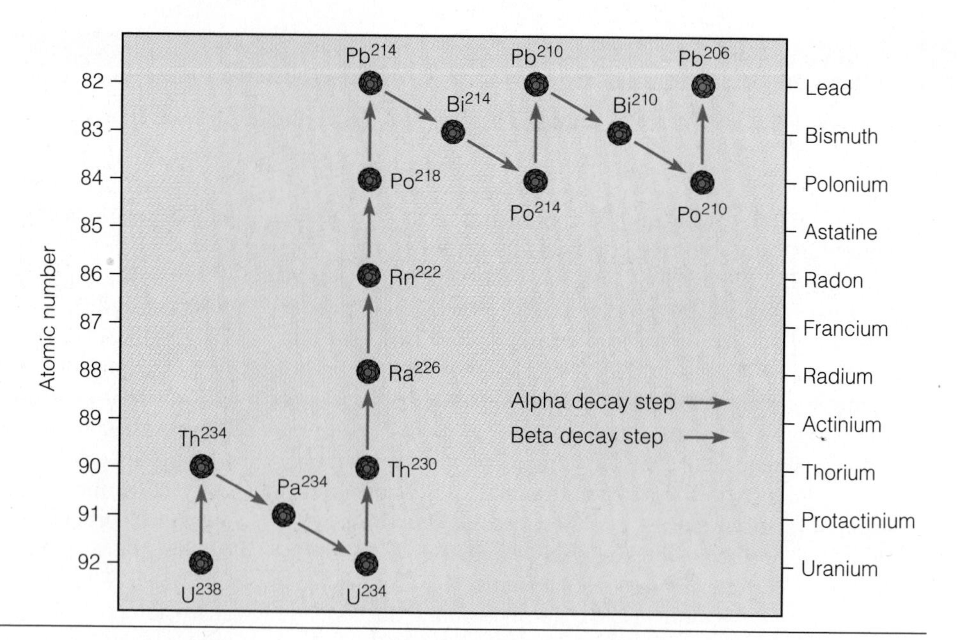

▶ **FIGURE 8-19** Radioactive decay series for uranium 238 to lead 206. Radioactive uranium 238 decays to its stable end product, lead 206, by eight alpha and six beta decay steps. A number of different isotopes are produced as intermediate steps in the decay series.

from previously formed daughter atoms. Because they are the right size, some radioactive parent atoms are incorporated into the crystal structure of certain minerals. The stable daughter atoms, however, are a different size than the radioactive parent atoms and consequently cannot fit into the crystal structure of the same mineral as the parent atoms. Therefore, a mineral crystallizing in a cooling magma will contain radioactive parent atoms but no stable daughter atoms (▶ Figure 8-21). Thus, the time that is being measured is the time of crystallization of the mineral containing the radioactive atoms, and not the time of formation of the radioactive atoms.

Except in unusual circumstances, sedimentary rocks cannot be radiometrically dated, because one would be measuring the age of a particular mineral rather than the time that it was deposited as a sedimentary particle. One of the few instances in which radiometric dates can be obtained on sedimentary rocks is when the mineral glauconite is present. Glauconite is a greenish mineral containing radioactive potassium 40, which decays to argon 40 (◉ Table 8-1). It forms in certain marine environments as a result of chemical reactions with clay minerals during the conversion of sediments to sedimentary rock. Thus, glauconite forms when the sedimentary

▶ **FIGURE 8-20** (*a*) Uniform, linear relationship is characteristic of many familiar processes. (*b*) Geometric radioactive decay curve, in which each time unit represents one half-life, and each half-life is the time it takes for one-half of the parent element to decay to the daughter element.

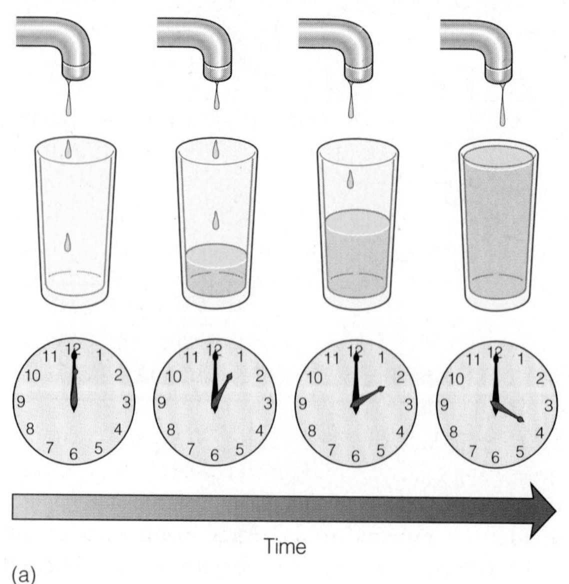

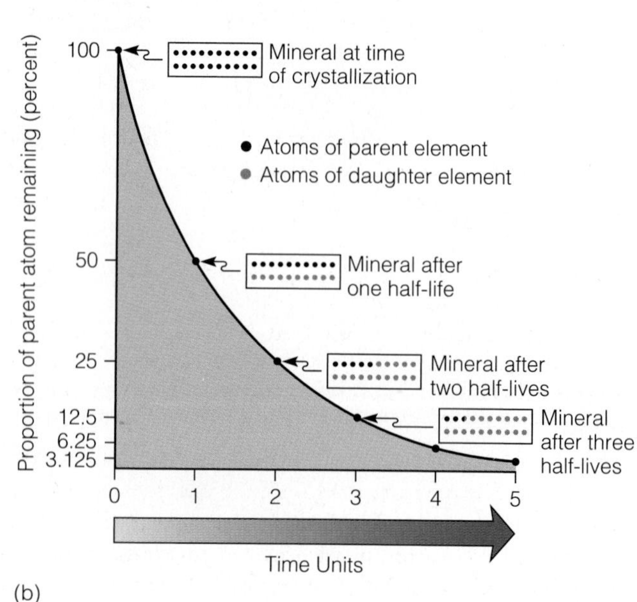

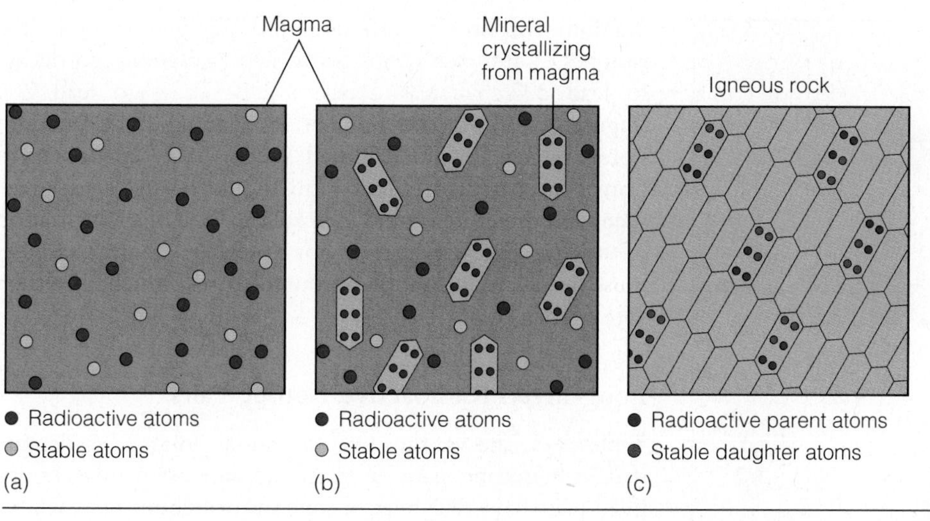

▶ FIGURE 8-21 (*a*) A magma contains both radioactive and stable atoms. (*b*) As the magma cools and begins to crystallize, some radioactive atoms are incorporated into certain minerals because they are the right size and can fit into the crystal structure. Therefore, at the time of crystallization, the mineral will contain 100% radioactive parent atoms and 0% stable daughter atoms. (*c*) After one half-life, 50% of the radioactive parent atoms will have decayed to stable daughter atoms.

TABLE 8-1 Five of the Principal Long-Lived Radioactive Isotope Pairs Used in Radiometric Dating

Isotopes		Half-Life of Parent (Years)	Effective Dating Range (Years)	Minerals and Rocks That Can Be Dated
Parent	*Daughter*			
Uranium 238	Lead 206	4.5 billion	10 million to 4.6 billion	Zircon Uraninite
Uranium 235	Lead 207	704 million		
Thorium 232	Lead 208	14 billion		
Rubidium 87	Strontium 87	48.8 billion	10 million to 4.6 billion	Muscovite Biotite Potassium feldspar Whole metamorphic or igneous rock
Potassium 40	Argon 40	1.3 billion	100,000 to 4.6 billion	Glauconite Muscovite Biotite Hornblende Whole volcanic rock

rock forms, and a radiometric date indicates the time of the sedimentary rock's origin. Being a gas, however, the daughter product argon can easily escape from a mineral. Therefore, any date obtained from glauconite, or any other mineral containing the potassium 40–argon 40 pair, must be considered a minimum age.

To obtain accurate radiometric dates, geologists must be sure that they are dealing with a closed system, meaning that neither parent nor daughter atoms have been added or removed from the system since crystallization and that the ratio between them results only from radioactive decay. Otherwise, an inaccurate date will result. If daughter atoms have leaked out of the mineral being analyzed, the calculated age will be too young; if parent atoms have been removed, the calculated age will be too great.

Leakage may occur if the rock is heated as occurs during metamorphism. If this happens, some of the parent or daughter atoms may be driven from the mineral being analyzed, resulting in an inaccurate age determination. If the daughter product was completely removed, then one would be measuring the time since metamorphism (a useful measurement itself), and not the time since crystallization of the mineral (▶ Figure 8-22). Because heat affects the parent-daughter ratio, metamorphic rocks are difficult to age date accurately. Remember that while the parent-daughter ratio may be affected by heat, the decay rate of the parent element remains constant, regardless of any physical or chemical changes.

In addition to the problem of leakage, some error is also inherent in measuring the minute amounts of the different

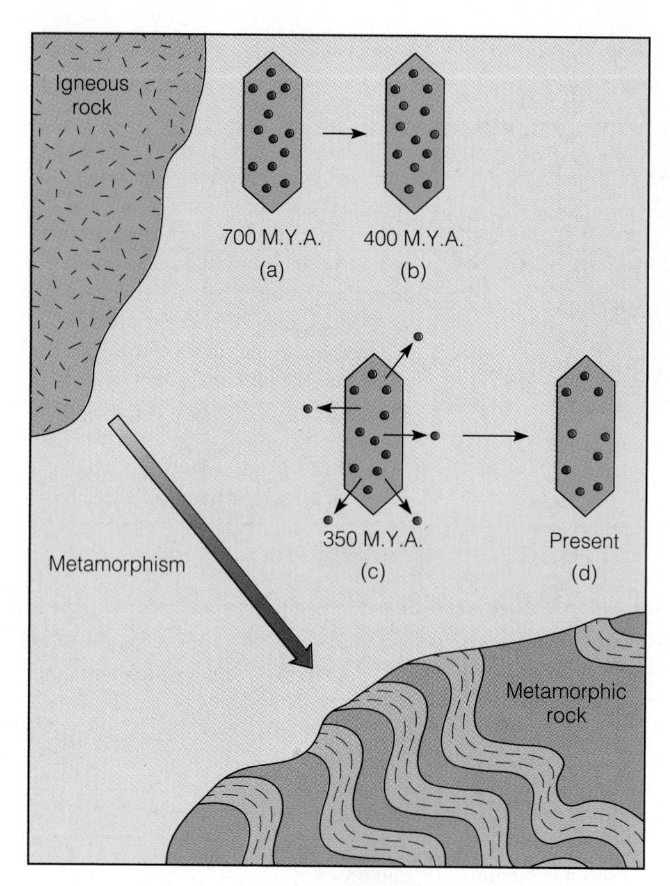

▶ **FIGURE 8-22** The effect of metamorphism in driving daughter atoms out of a mineral that crystallized 700 million years ago (M.Y.A.). The mineral is shown immediately after crystallization (*a*), then at 400 million years (*b*), when some of the parent atoms have decayed to daughter atoms. Metamorphism at 350 M.Y.A. (*c*) drives the daughter atoms out of the mineral into the surrounding rock. (*d*) Assuming the rock has remained a closed chemical system throughout its history, dating the mineral today yields the time of metamorphism, while dating the whole rock provides the time of its crystallization, 700 M.Y.A.

elements and isotopes used for age dating. Measurements by mass spectrometers are ±0.2 to 2.0% accurate. For a rock 10 million years old, a possible error of 20,000 to 200,000 years exists. For a one-billion-year-old rock, this is a 2-million to 20-million-year difference, which is more than geologists would like, but is essentially trivial in rocks of such great age. Therefore, when a radiometric age is given, a plus or minus factor in number of years is usually appended to the age to indicate the limits of error in the dating technique.

To obtain an accurate radiometric date, geologists must make sure that the sample is fresh and unweathered and that it has not been subjected to high temperatures or intense pressures after crystallization. Furthermore, it is sometimes possible to cross-check the date obtained by measuring the parent-daughter ratio of two different radioactive elements in the same mineral. For example, naturally occurring uranium consists of both uranium 235 and uranium 238 isotopes. Through various decay steps, uranium 235 decays to lead 207, whereas uranium 238 decays to lead 206 (Figure 8-19). If the minerals containing both uranium isotopes have remained closed systems, the ages obtained from each parent-daughter ratio should be in close agreement and therefore should indicate the time of crystallization of the magma. If the ages do not closely agree, other samples must be taken and ratios measured to see which, if either, date is correct.

Long-Lived Radioactive Isotope Pairs

Table 8-1 shows the five common, long-lived parent-daughter isotope pairs used in radiometric dating. Long-lived pairs have half-lives of millions or billions of years. All of these were present when the Earth formed and are still present in measurable quantities. Other shorter-lived radioactive isotope pairs have decayed to the point that only small quantities near the limit of detection remain.

The most commonly used isotope pairs are the uranium-lead and thorium-lead series, which are used principally to date ancient igneous intrusives, lunar samples, and some meteorites. The rubidium-strontium pair is also used for very old samples and has been effective in dating the oldest rocks on Earth as well as meteorites. The potassium-argon method is typically used for dating fine-grained volcanic rocks from which individual crystals cannot be separated; hence the whole rock is analyzed. However, argon is a gas, so great care must be taken to assure that the sample has not been subjected to heat, which would allow argon to escape; such a sample would yield an age that is too young. Other long-lived radioactive isotope pairs exist, but they are rather rare and are used only in special situations.

Fission Track Dating

The emission of atomic particles resulting from the spontaneous decay of uranium within a mineral damages its crystal structure. The damage appears as microscopic linear tracks that are visible only after the mineral is etched with hydrofluoric acid. The age of the sample is determined by the number of fission tracks present and the amount of uranium the sample contains. The older the sample, the greater the number of tracks (▶ Figure 8-23).

Fission track dating is of particular interest to geologists because the technique can be used to date samples ranging from only a few hundred to hundreds of millions of years in age. It is most useful for dating samples between about 40,000 and one million years ago, a period for which other dating techniques are not particularly suitable. One of the problems in fission track dating occurs when the rocks have later been subjected to high temperatures. If this happens, the damaged crystal structures are repaired by annealing, and consequently the tracks disappear. In such instances, the calculated age will be younger than the actual age.

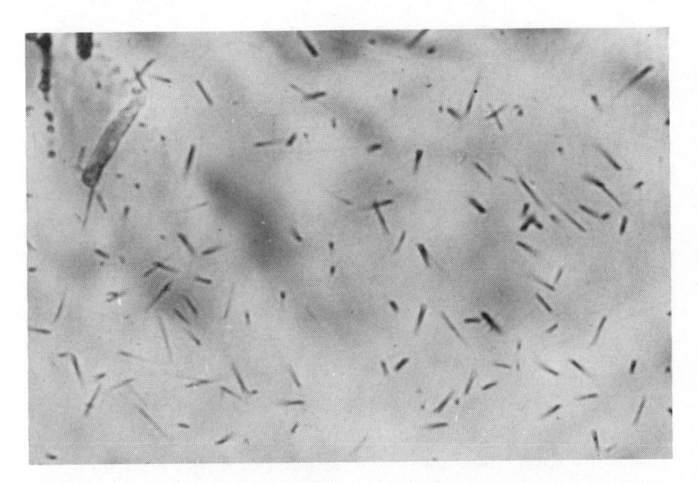

▶ FIGURE 8-23 Each fission track (about 16 microns in length) in this apatite crystal is the result of the radioactive decay of a uranium atom. To make the fission tracks visible, the apatite crystal has been etched with hydrofluoric acid. This apatite crystal comes from one of the dikes of Shiprock, New Mexico, and indicates a calculated age of 27 million years. (Photo courtesy of Charles W. Naeser, U.S. Geological Survey.)

Radiocarbon and Tree-Ring Dating Methods

Carbon is an important element in nature and is one of the basic elements found in all forms of life. It has three isotopes; two of these, carbon 12 and 13, are stable, whereas carbon 14 is radioactive. Carbon 14 has a half-life of 5,730 years plus or minus 30 years. The **carbon 14 dating technique** is based on the ratio of carbon 14 to carbon 12 and is generally used to date once-living material.

The short half-life of carbon 14 makes this dating technique practical only for specimens younger than about 70,000 years. Consequently, the carbon 14 dating method is especially useful in archaeology and has greatly aided in unraveling the events of the latter portion of the Pleistocene Epoch.

Carbon 14 is constantly formed when cosmic rays, which are high-energy particles (mostly protons), strike the atoms of upper-atmospheric gases, splitting their nuclei into protons and neutrons. When a neutron strikes the nucleus of a nitrogen atom (atomic number 7, atomic mass number 14), it may be absorbed into the nucleus and a proton emitted. Thus, the atomic number of the atom decreases by one, while the atomic mass number stays the same. Because the atomic number has changed, a new element, carbon 14 (atomic number 6, atomic mass number 14), is formed. The newly formed carbon 14 is rapidly assimilated into the carbon cycle and, along with carbon 12 and 13, is absorbed in a nearly constant ratio by all living organisms (▶ Figure 8-24). When an organism dies, however, carbon 14 is not replenished, and the ratio of carbon 14 to carbon 12 decreases as carbon 14 decays back to nitrogen by a single beta decay step (Figure 8-24).

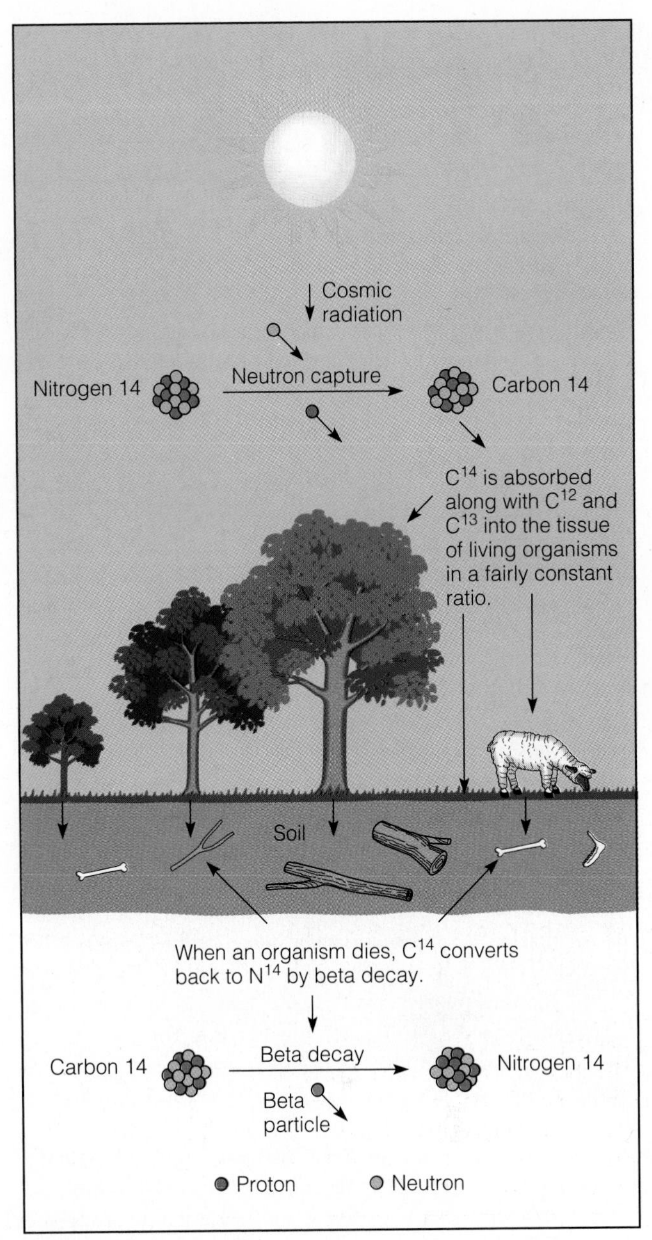

▶ FIGURE 8-24 The carbon cycle showing the formation, dispersal, and decay of carbon 14.

Currently, the ratio of carbon 14 to carbon 12 is remarkably constant in both the atmosphere and living organisms. There is good evidence, however, that the production of carbon 14, and thus the ratio of carbon 14 to carbon 12, has varied somewhat over the past several thousand years, in part, because the amount of CO_2 has varied. This variation was determined by comparing ages established by carbon 14 dating of wood samples against those established by counting annual tree rings in the same samples (▶ Figure 8-25). As a result, carbon 14 ages have been corrected to reflect such variations in the past.

Tree-ring dating is another useful method for dating geologically recent events. The age of a tree can be deter-

PALEONTOLOGY: TRACING LIFE THROUGH TIME

Even as a child, I recall being interested in rocks and fossils. I know now that I was attracted to them for the same reasons that I still enjoy teaching and doing research in historical geology. For one thing, rocks and fossils are a constant reminder that time did not begin with my existence. This knowledge leads to a more relaxed view of what I—and the human species for that matter—am doing here. One's self-importance is continually diminished when you work with fossils that are millions of years old. In fact, some social commentators think that many of today's environmental problems arise because so many people lack an awareness of "deep time." They live only in the present, without regard to the future inhabitants of the planet who must live here for (hopefully) millions of years to come.

Another part of my motivation, however, comes from the "detective" work involved in historical geology. Like a police detective, the historical geologist trys to reconstruct past events from fragmentary evidence. Whether as a sedimentologist trying to determine when an oil basin formed, or a paleontologist trying to find the ancestors of modern mammals, the challenge is to use whatever limited information is available. This can be frustrating, but as with many puzzles, the moment when ideas "come together" is very satisfying. Furthermore, new evidence is always being found so new puzzles always arise and old answers often prove inadequate. Most satisfying of all is the knowledge that the work is more than idle amusement: you are contributing to our understanding of how the Earth and its life came to be what they are today.

Besides being fun, the study of fossils and sedimentary rocks has many practical applications. Our society is built on ores and energy (such as fossil fuels) that come from the Earth. By studying the history of the Earth, we learn how and, more importantly, where these materials formed. For example, many paleontologists work for oil companies, examining microfossils in rock cores brought up by drilling rigs. Historical geologists specializing in sedimentology and stratigraphy are also employed in the search for oil and minerals; they examine the physical characteristics of the rock cores and correlate rock layers. Environmental firms are currently the major employers of geologists, and environmental careers are among the fastest growing fields in the United States.

I have never worked in industry although I was offered jobs by two oil companies when I completed my master's degree in geology. Instead, I chose to go on for a Ph.D. After receiving my degree, I joined the faculty at the University of Tennessee where I teach undergraduate and graduate courses. I'm glad I made this choice because it allows me to carry out research projects of my own choosing. I am helped by a number of graduate students who work in my laboratory, doing research in their own particular areas. For instance, one is making highly sophisticated measurements of fossil shapes by using a television camera connected to a computer. Much of this work is supported by grants from agencies such as the National Science Foundation. Funding from these agencies is very competitive, and the grants usually last only a couple of years. Therefore, scientists must often spend a significant amount of time writing and submitting grant applications if their research is costly.

Some of my own favorite research is currently aimed at finding information on the many extinctions seen in the fossil record. This research has much relevance today, when species are becoming extinct at an alarming rate. Since over 99% of all species that have ever existed have died out, the fossil record contains a vast amount of useful data about extinctions. For instance, we have already learned that some species of animals (such as mammals) are generally more likely to become extinct than others (such as clams). We have also discovered that habitat destruction has been the main cause of extinction throughout geologic time, just as it is today. The only difference is that today humans destroy the habitats, whereas in the past changes in climate, meteorite impacts, and other natural phenomena caused the destruction.

MICHAEL L. MCKINNEY is an associate professor in the Geology and Ecology Programs at the University of Tennessee, Knoxville. He has published four books and many technical articles on evolution, paleontology, and environmental topics.

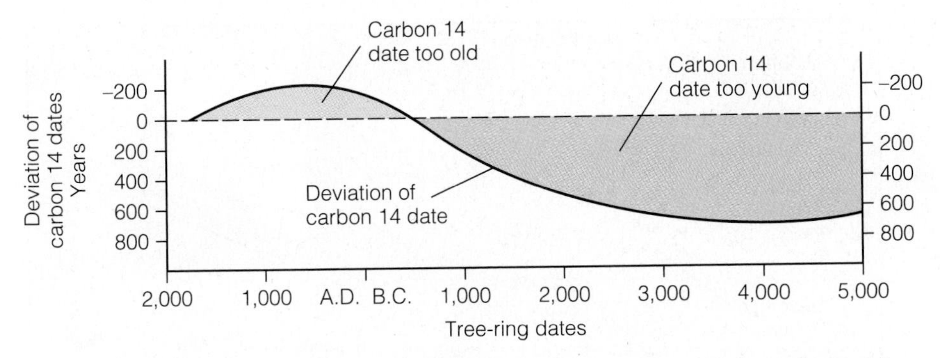

▶ **FIGURE 8-25** Discrepancies exist between carbon 14 dates and those obtained by counting annual tree rings. Back to about 600 B.C., carbon 14 dates are too old, and those from about 600 B.C. to about 5,000 B.C. are too young. Consequently, corrections must be made to the carbon 14 dates for this time period.

mined by counting the growth rings in the lower part of the stem. Each ring represents one year's growth, and the pattern of wide and narrow rings can be compared among trees to establish the exact year in which the rings were formed. The procedure of matching ring patterns from numerous trees and wood fragments in a given area is referred to as cross-dating. By correlating distinctive tree-ring sequences from living to nearby dead trees, a time scale can be constructed that extends back to about 14,000 years ago (▶ Figure 8-26). By matching ring patterns to the composite ring scale, wood samples whose ages are not known can be accurately dated.

The applicability of tree-ring dating is somewhat limited because it can only be used where continuous tree records are found. It is therefore most useful in arid regions, particularly the southwestern United States.

▶ **FIGURE 8-26** In the cross-dating method, tree-ring patterns from different woods are matched against each other to establish a ring-width chronology backward in time.

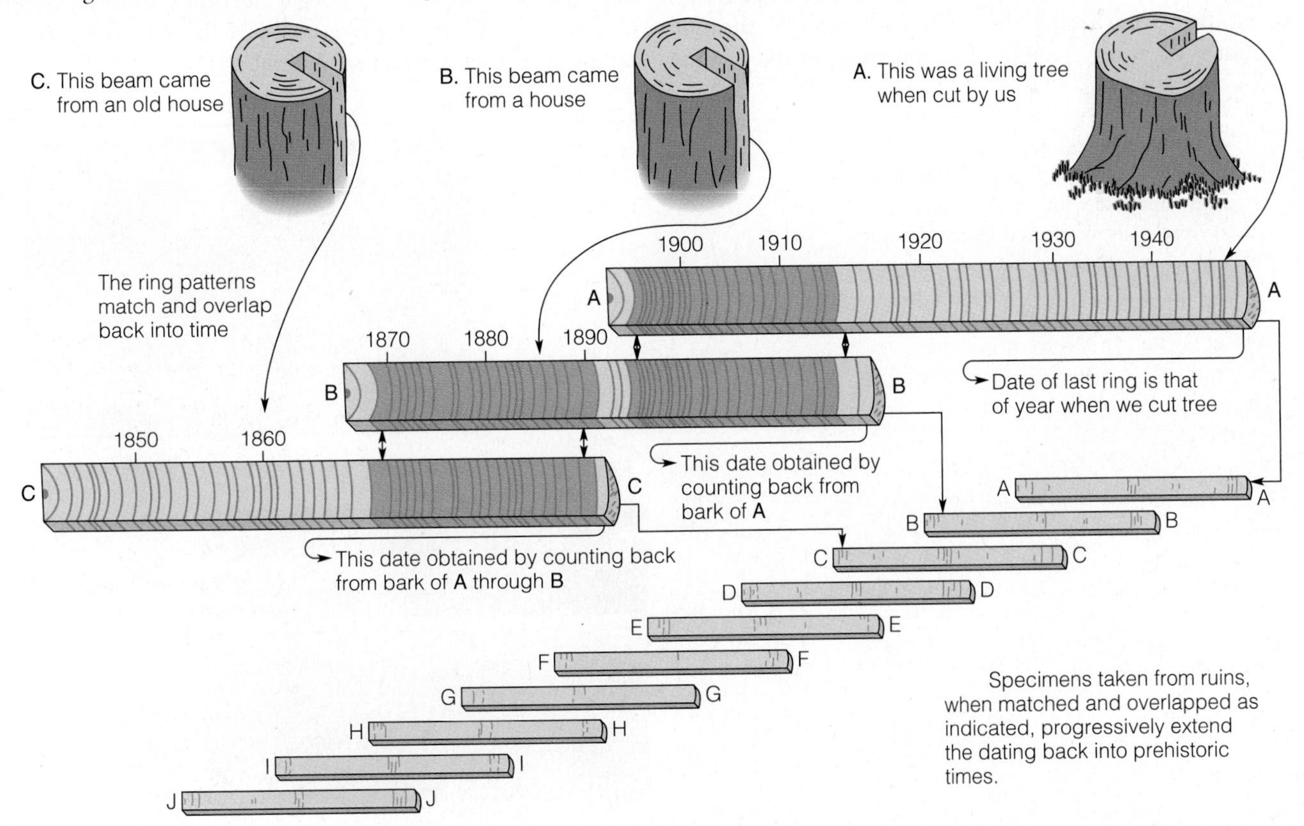

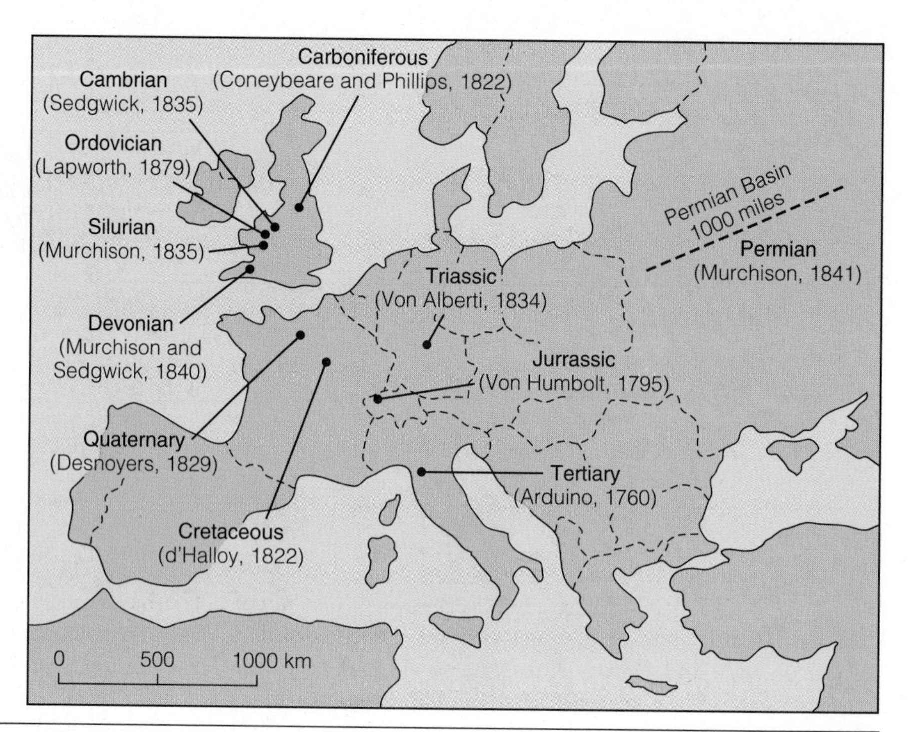

▶ **FIGURE 8-27** The names of the time periods of the geologic time scale were based on areas in England and Europe where the rock units were originally described. Note that the Carboniferous, which is recognized in Europe, is represented in North America by the Mississippian and Pennsylvanian.

THE DEVELOPMENT OF THE GEOLOGIC TIME SCALE

The geologic time scale is a hierarchical scale in which the 4.6-billion-year history of the Earth is divided into time units of varying duration (Figure 8-2). The geologic time scale was not developed by any one individual, but rather evolved, primarily during the nineteenth century, through the efforts of many people. By applying relative dating methods to rock outcrops, geologists in England and western Europe defined the major geologic time units without the benefit of radiometric dating techniques (▶ Figure 8-27). Using the principles of superposition and fossil succession, they were able to correlate various rock exposures and piece together a composite geologic section. This composite section is, in effect, a relative time scale because the rocks are arranged in their correct sequential order.

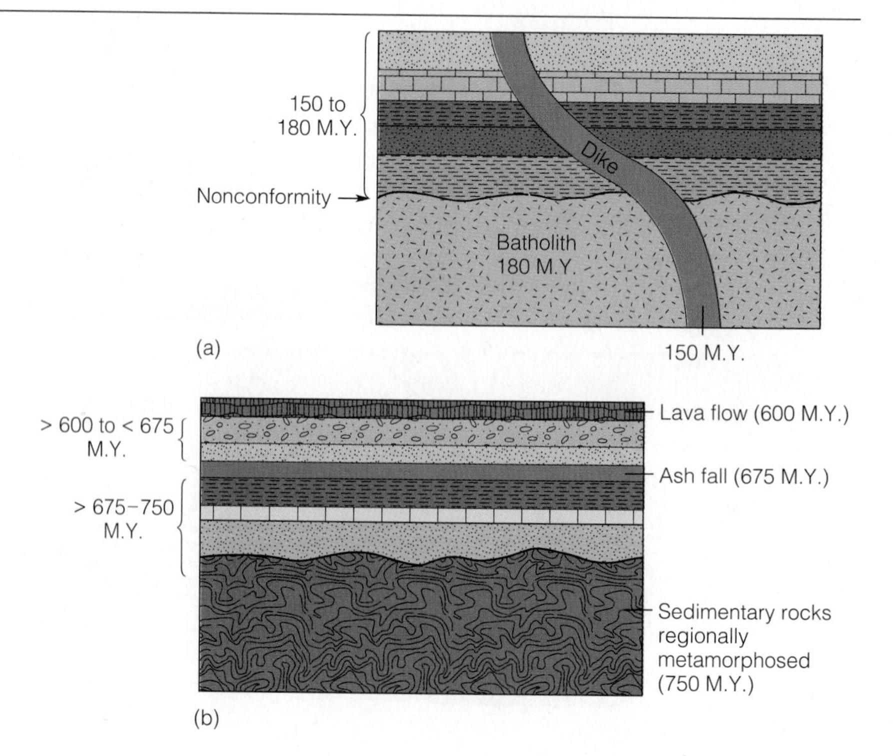

▶ **FIGURE 8-28** Absolute ages of sedimentary rocks can be determined by dating associated igneous rocks. In (*a*) and (*b*), sedimentary rocks are bracketed by rock bodies for which absolute ages have been determined.

By the beginning of the twentieth century, geologists had developed a relative geologic time scale, but did not yet have any absolute dates for the various time unit boundaries. Following the discovery of radioactivity near the end of the last century, radiometric dates were added to the relative geologic time scale (Figure 8-2).

Because sedimentary rocks, with rare exceptions, cannot be radiometrically dated, geologists have had to rely on interbedded volcanic rocks and igneous intrusions to apply absolute dates to the boundaries of the various subdivisions of the geologic time scale (▶ Figure 8-28). An ash fall or lava flow provides an excellent marker bed that is a time-equivalent surface, supplying a minimum age for the sedimentary rocks below and a maximum age for the rocks above. Ash falls are particularly useful because they may fall over both marine and nonmarine sedimentary environments and can provide a connection between these different environments.

Thousands of absolute ages are now known for sedimentary rocks of known relative ages, and these absolute dates have been added to the relative time scale. In this way, geologists have been able to determine the absolute ages of the various geologic periods and to determine their durations (Figure 8-2).

CHAPTER SUMMARY

1. Relative dating involves placing geologic events in a sequential order as determined from their position in the geologic record. Absolute dating results in specific dates for events, expressed in years before the present.
2. During the eighteenth and nineteenth centuries, attempts were made to determine the age of the Earth based on scientific evidence rather than revelation. While some attempts were quite ingenious, they yielded a variety of ages that now are known to be much too young.
3. James Hutton thought that present-day processes operating over long periods of time could explain all the geologic features of the Earth. His observations were instrumental in establishing the basis for the principle of uniformitarianism.
4. Uniformitarianism, as articulated by Charles Lyell, soon became the guiding principle of geology. It holds that the laws of nature have been constant through time and that the same processes operating today have operated in the past, although not necessarily at the same rates.
5. In addition to uniformitarianism, the principles of superposition, original horizontality, lateral continuity, cross-cutting relationships, inclusions, and fossil succession are basic for determining relative geologic ages and for interpreting the geologic history of the Earth.
6. Surfaces of discontinuity that encompass significant amounts of geologic time are common in the geologic record. Such surfaces are unconformities and result from times of nondeposition, erosion, or both.
7. Correlation is the stratigraphic practice of demonstrating equivalency of units in different areas. Time equivalence is most commonly demonstrated by correlating strata containing similar fossils.
8. Radioactivity was discovered during the late nineteenth century, and soon thereafter radiometric dating techniques allowed geologists to determine absolute ages for geologic events.
9. Absolute age dates for rock samples are usually obtained by determining how many half-lives of a radioactive parent element have elapsed since the sample originally crystallized. A half-life is the time it takes for one-half of the radioactive parent element to decay to a stable daughter element.
10. The most accurate radiometric dates are obtained from long-lived radioactive isotope pairs in igneous rocks. The most reliable dates are those obtained by using at least two different radioactive decay series in the same rock.
11. Carbon 14 dating can be used only for organic matter such as wood, bones, and shells and is effective back to about 70,000 years ago. Unlike the long-lived isotope pairs, the carbon 14 dating technique determines age by the ratio of radioactive carbon 14 to stable carbon 12.
12. Through the efforts of many geologists applying the principles of relative dating, a relative geologic time scale was established.
13. Most absolute ages of sedimentary rocks and their contained fossils are obtained indirectly by dating associated metamorphic or igneous rocks.

IMPORTANT TERMS

absolute dating
alpha decay
angular unconformity
beta decay
carbon 14 dating technique
correlation
daughter element
disconformity

electron capture
fission track dating
guide fossil
half-life
nonconformity
parent element
principle of cross-cutting relationships
principle of fossil succession
principle of inclusions
principle of lateral continuity
principle of original horizontality
principle of superposition
principle of uniformitarianism
radioactive decay
relative dating
tree-ring dating
unconformity

Review Questions

1. Placing geologic events in sequential order as determined by their position in the geologic record is called:
   a. ____ absolute dating;
   b. ____ uniformitarianism;
   c. ____ relative dating;
   d. ____ correlation;
   e. ____ historical dating.
2. If a rock is heated during metamorphism and the daughter atoms migrate out of a mineral that is subsequently radiometrically dated, an inaccurate date will be obtained. This date will be ________ the actual date.
   a. ____ younger than;
   b. ____ older than;
   c. ____ the same as;
   d. ____ it cannot be determined;
   e. ____ none of these.
3. Which of the following methods can be used to demonstrate age equivalency of rock units?
   a. ____ lateral tracing;
   b. ____ radiometric dating;
   c. ____ guide fossils;
   d. ____ position in a sequence;
   e. ____ all of these.
4. Which fundamental geologic principle states that the oldest layer is on the bottom of a succession of sedimentary rocks and the youngest is on top?
   a. ____ lateral continuity;
   b. ____ fossil succession;
   c. ____ original horizontality;
   d. ____ superposition;
   e. ____ cross-cutting relationships.
5. In which type of radioactive decay are two protons and two neutrons emitted from the nucleus?
   a. ____ alpha;
   b. ____ beta;
   c. ____ electron capture;
   d. ____ fission track;
   e. ____ radiocarbon.
6. Which of the following is not a long-lived radioactive isotope pair?
   a. ____ uranium-lead;
   b. ____ thorium-lead;
   c. ____ potassium-argon;
   d. ____ carbon-nitrogen;
   e. ____ none of these.
7. What is being measured in radiometric dating?
   a. ____ the time when the radioactive isotope formed;
   b. ____ the time of crystallization of a mineral containing an isotope;
   c. ____ the amount of the parent isotope only;
   d. ____ when the dated mineral became part of a sedimentary rock;
   e. ____ when the stable daughter isotope was formed.
8. If a radioactive element has a half-life of 4 million years, the amount of parent material remaining after 12 million years of decay will be what fraction of the original amount?
   a. ____ $1/32$;
   b. ____ $1/16$;
   c. ____ $1/8$;
   d. ____ $1/4$;
   e. ____ $1/2$.
9. In carbon 14 dating, which ratio is being measured?
   a. ____ the parent to daughter isotope;
   b. ____ C^{14}/N^{14};
   c. ____ C^{12}/C^{13};
   d. ____ C^{12}/N^{14};
   e. ____ C^{12}/C^{14}.
10. How many half-lives are required to yield a mineral with 625 atoms of U^{238} and 19,375 atoms of Pb^{206}?
    a. ____ 4;
    b. ____ 5;
    c. ____ 6;
    d. ____ 8;
    e. ____ 10.
11. What is the difference between relative and absolute dating of geologic events?
12. What are the six fundamental principles used in relative age dating? Why are they so important in deciphering Earth history?
13. Define the three types of unconformities. Why are unconformities important in relative age dating?
14. Explain how a geologist would determine the relative ages of a granite batholith and an overlying sandstone formation.
15. Why is the principle of uniformitarianism important to geologists?
16. Are volcanic eruptions, earthquakes, and storm deposits geologic events encompassed by uniformitarianism?
17. What is radon, and why is it so dangerous to humans?
18. What are assemblage range zones? How can such zones be used to demonstrate time equivalency of strata in widely separated areas?
19. If you wanted to calculate the absolute age of an intrusive body, what information would you need?
20. Assume a hypothetical radioactive isotope with an atomic number of 150 and an atomic mass number of 300 emits five alpha decay particles and three beta decay particles and undergoes two electron capture steps. What is the atomic number and atomic mass number of the resulting stable daughter product?
21. What are some of the potential sources of error in radiometric dating?
22. How can geologists be sure that the absolute age dates they obtain from igneous rocks are accurate?
23. Why is it difficult to date sedimentary and metamorphic rocks radiometrically?
24. How does the carbon 14 dating technique differ from the uranium-lead dating method?

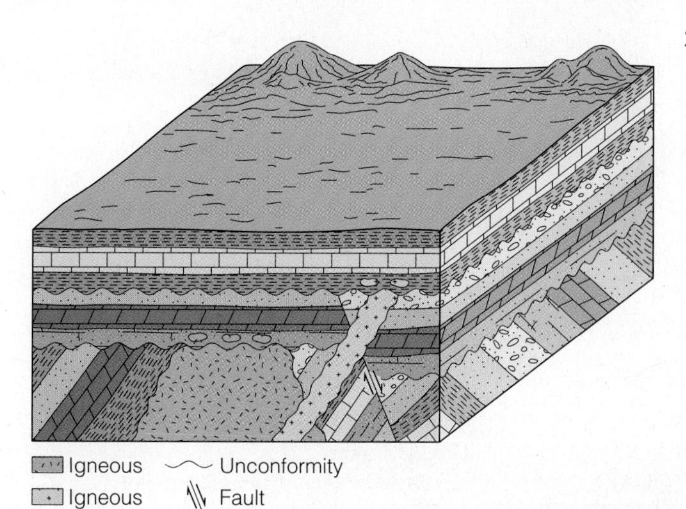

25. Using the principles of relative dating, give the geologic history for this diagram.

Points to Ponder

1. Discuss some of the reasons why the amount of CO_2 has varied during the past several thousand years and thus affects the ages determined by carbon 14 dating.
2. The highly publicized controversies surrounding radon and asbestos (see Perspective 7-1) are similar in that massive amounts of money are being spent to remedy environmental hazards that may not be as serious as policymakers have been led to believe. Discuss how you might objectively assess the potential danger to the public posed by these substances and how you would determine what constitutes an acceptable risk for them.
3. Can the same principles geologists use to interpret the geologic history of the Earth be used by astronauts to determine the geologic history of other planets in our solar system?
4. A zircon mineral in an igneous rock contains both uranium 235 and uranium 238. In measuring the parent-daughter ratio for both uranium isotopes in the zircon, it was discovered that uranium 238 had undergone 2 half-lives in decaying to lead 206, while uranium 235 had undergone 10 half-lives during its decay to lead 207. Using Table 8-1, determine the age of the igneous rock based on each long-lived radioactive isotope pair. Are they the same age? If not, why? What explanations can you give for the difference in geologic age between the two samples.
5. Calculate the age of a rock containing the following atoms of radioactive parent element A and stable daughter element B. Radioactive parent element A: 1,125,000 atoms, stable daughter element B: 34,875,000 atoms. The half life of radioactive parent element A is 6.25 million years. What is the age of the rock containing these parent and daughter elements?

Additional Readings

Albritton, C. C., Jr. 1980. *The abyss of time*. San Francisco, Calif.: Freeman, Cooper and Co.

______ . 1984. Geologic time. *Journal of Geological Education* 32, no. 1: 29–37.

Berry, W. B. N. 1987. *Growth of a prehistoric time scale*. 2d ed. Palo Alto, Calif.: Blackwell Scientific Publications.

Boslough, J. 1990. The enigma of time. *National Geographic* 177, no. 3: 109–32.

Faure, G. 1986. *Principles of isotope geology*. 2d ed. New York: John Wiley & Sons.

Geyh, M. A., and H. Schleicher. 1990. *Absolute age determination*. New York: Springer-Verlag.

Gould, S. J. 1987. *Time's arrow, time's cycle*. Cambridge, Mass.: Harvard University Press.

Gunderson, L. C. S. 1992. Hidden hazards of radon. *Earth* 1, no. 6: 55–61.

Harland, W. B., R. L. Armstrong, A. V. Cox, L. E. Craig, A. G. Smith, and D. G. Smith. 1990. *A geologic time scale 1989*. New York: Cambridge University Press.

Itano, W. M., and N. F. Ramsey. 1993. Accurate measurement of time. *Scientific American* 269, no. 1: 56–57.

Ramsey, N. F. 1988. Precise measurement of time. *American Scientist* 76, no. 1: 42–49.

Wetherill, G. W. 1982. Dating very old objects. *Natural History* 91, no. 9: 14–20.

Chapter 9

EARTHQUAKES

OUTLINE

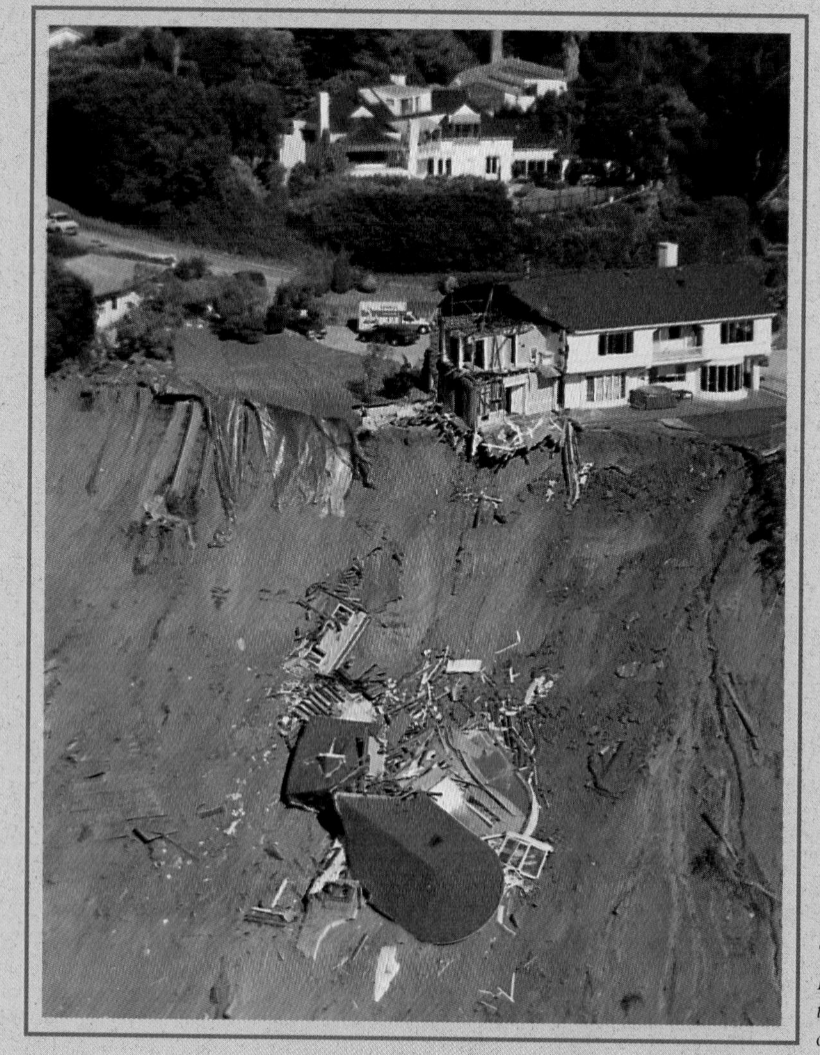

This house in Pacific Palisades was destroyed by the Northridge earthquake of January 17, 1994.

PROLOGUE

In the early morning hours of January 17, 1994, southern California was rocked by a devasting earthquake. The epicenter of this earthquake, which measured 6.7 on the Richter Magnitude Scale, was located in the city of Northridge, in the San Fernando Valley area of southern California (▶ Figure 9-1a). For slightly more than 40 seconds, the earth shook, and when the initial shock was over, 61 people had been killed, thousands were injured, and an oil main and at least 250 gas lines had ruptured, igniting numerous fires (Figure 9-1b), nine freeways were destroyed (Figure 9-1c), and thousands of homes and buildings were destroyed or damaged (Figures 9-1d and e). In addition, so many power lines were knocked down and circuits blown that 3.1 million people were without electricity, and at least 40,000 were without water due to broken water mains. More than 1,000 aftershocks followed the main earthquake, many of them with a magnitude greater than 4, further contributing to the already considerable damage, estimated by some to be as high as $15 to $30 billion. While this was not the "Big One" that Californians have long been expecting, the destruction left in its wake was immense.

Geologists know that the southern California area is riddled with active faults capable of producing strong earthquakes. Yet it was a previously unknown fault 15 km beneath Northridge that caused the tragic January 17 earthquake. While most people are aware that movement along the San Andreas fault and its subsidiary faults has been responsible for many of the major earthquakes California has experienced, geologists are now coming to realize that a network of possibly interconnected hidden faults may be causing most of the earthquakes in the Los Angeles area. These hidden faults form when pressure builds up in the crust as a result of movement between the North American and Pacific plates. What makes these faults so dangerous is that they are hidden and generally show no evidence of their existence at the surface. Geologists are usually unaware of these faults until they break, causing an earthquake. Hidden faults in southern California have been responsible for seven earthquakes of magnitude 4.5 or greater during the past six years. As the Northridge earthquake demonstrated, earthquake prediction will remain an elusive reality until more is learned about these hidden faults and how, and if, they are interconnected.

The 1989 Loma Prieta earthquake in the San Francisco area had taught Californians many important lessons about earthquake preparedness and the factors that affect the damage an area is likely to suffer. For example, as was so dramatically demonstrated, the underlying material and type of building construction are probably the two most important factors determining the amount of damage. None of the structures in San Francisco that were constructed in compliance with current building codes collapsed. Furthermore, areas built on poorly consolidated material or bay mud generally suffered greater damage than those built on bedrock.

The Loma Prieta earthquake also strongly reinforced the importance of careful planning and preparation in earthquake-prone areas. Within hours after the earthquake, shelters were open and emergency relief services were in place and operating smoothly. This was due, in part, to the numerous rehearsals that various agencies had conducted in preparation for just such an emergency.

Considering what had been learned from the 1989 Loma Prieta and other recent large earthquakes, how did the Los Angeles basin area fare after this most recent earthquake? Older, reinforced masonry buildings and more modern wood-frame apartments built over ground-floor garages generally sustained the most damage. Structures built to the tougher building standards during the past five years typically escaped unscathed or with only minor damage.

Caltrans, the state transportation department, began a program of reinforcing bridges and freeway overpasses soon after the 1971 Sylmar earthquake and began a second round of reinforcing structures following the 1989 Loma Prieta earthquake. Most of the reinforced structures held up quite well in the Northridge earthquake, but several of those awaiting reinforcing, including a portion of the vital east-west Santa Monica Freeway, collapsed.

Laws designed to protect utility lines have not yet been implemented, in part, because of the cost. As a consequence, numerous power and gas lines were ruptured, and three water aqueducts were severed, cutting off water and power to thousands of residents.

In terms of rescue operations and emergency preparations, things went fairly well. Shelters were established, people fed and clothed, and disaster relief offices opened in the area shortly after the earthquake. Applying the lessons learned from the 1971 Sylmar earthquake, rescue agencies had invested in better rescue equipment including high-pressure air bags that can lift up to 72 tons, fiber-optics search cameras, and specially trained dogs that can sniff out buried victims. All of this equipment helped in locating and rescuing trapped victims.

The Northridge earthquake was tragic but it was not the "Big One" that Californians have been waiting for. And even though rescue and relief agencies operated efficiently after the earthquake, it reminds us that much still remains to be done in terms of earthquake protection and preparedness.

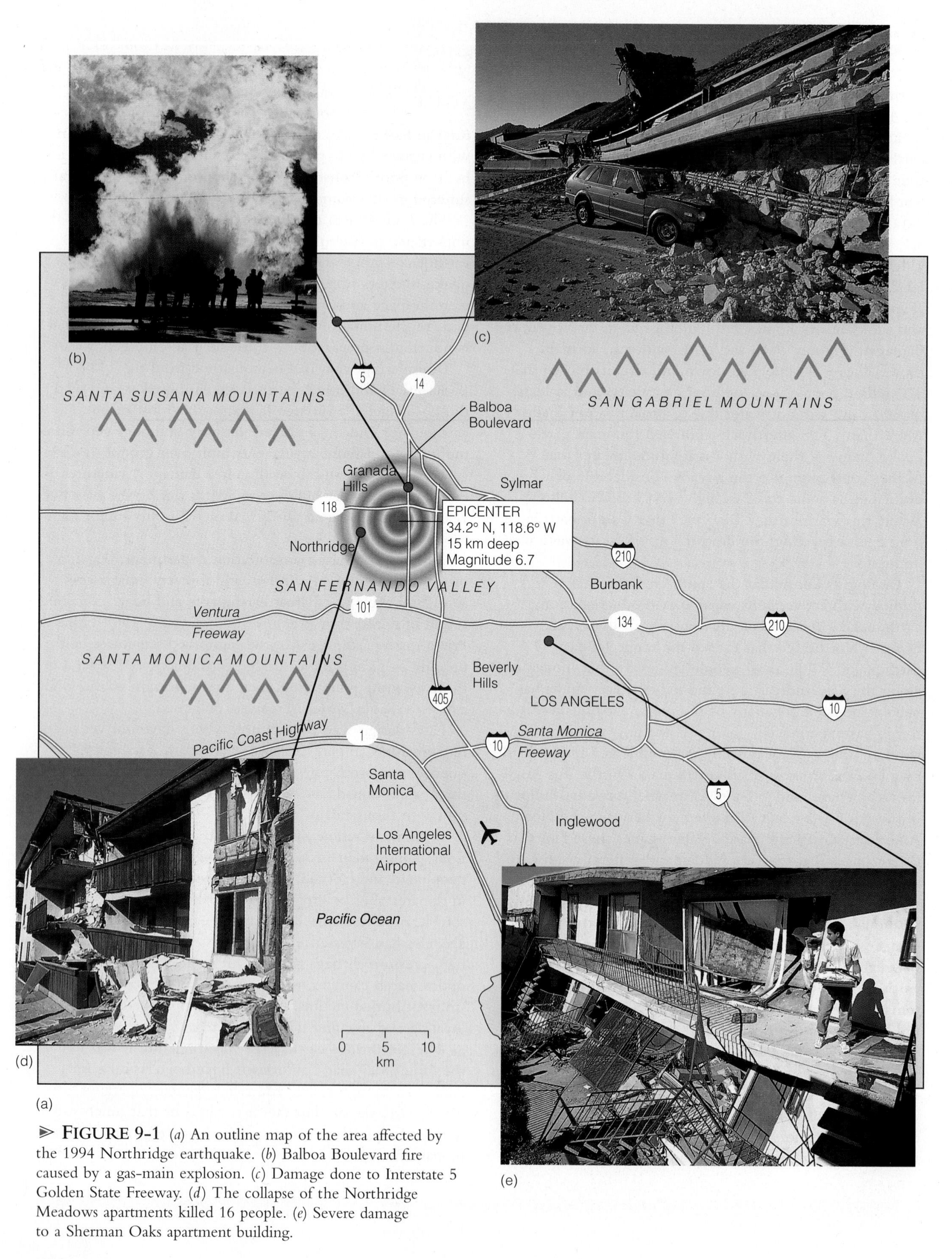

➢ **FIGURE 9-1** (*a*) An outline map of the area affected by the 1994 Northridge earthquake. (*b*) Balboa Boulevard fire caused by a gas-main explosion. (*c*) Damage done to Interstate 5 Golden State Freeway. (*d*) The collapse of the Northridge Meadows apartments killed 16 people. (*e*) Severe damage to a Sherman Oaks apartment building.

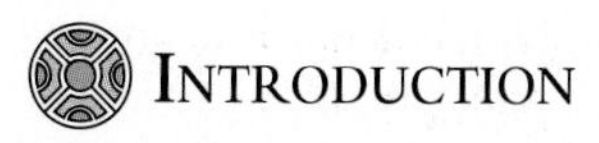

Introduction

Earthquakes are violent and usually unpredictable; typically, they produce a feeling of helplessness. As one of nature's most frightening and destructive phenomena, they have always aroused a sense of fear. Even when an earthquake begins, there is no way to tell how strong the shaking will be or how long it will last.

It is estimated that more than 13 million people have died as a result of earthquakes during the past 4,000 years, and approximately 1 million of these deaths occurred during the last century (◉ Table 9-1).

If you have never experienced an earthquake, try to imagine that as you are reading this book, the ground suddenly and without any warning starts shaking and everything around you begins to sway. If the shock waves are severe enough, you might be knocked down and have trouble standing up. The first thought that would probably go through your mind is, "How long is the shaking going to last and is it going to get any stronger?" You want to do something, but you don't know exactly what to do. If the shaking is severe, windows may break, and if you are in a building, objects will fall from the ceiling and walls, and

TABLE 9-1 Significant Earthquakes of the World

Year	Location	Magnitude (Estimated before 1935)	Deaths (Estimated)
893	India		180,000
1138	Syria		100,000
1556	China (Shanxi Province)	8.0	1,000,000
1668	China (Shandong Province)	8.5	50,000
1730	Japan (Hokkaidō Prefecture)		137,000
1755	Portugal (Lisbon)	8.6	70,000
1811–12	USA (New Madrid, Missouri)	7.5	20
1835	Chile (Concepción)	8.5	?
1857	USA (Fort Tejon, California)	8.3	1
1868	Chile and Peru	8.5	25,000
1872	USA (Owens Valley, California)	8.5	27
1886	USA (Charleston, South Carolina)	7.0	60
1905	India (Punjab-Kashmir region)	8.6	19,000
1906	USA (San Francisco, California)	8.3	700
1908	Italy (Messina)	7.5	83,000
1920	China (Gansu)	8.6	100,000
1923	Japan (Tokyo)	8.3	143,000
1950	India (Assam) and Tibet	8.6	1,530
1960	Chile	8.5	4,000+
1964	USA (Alaska)	8.6	131
1970	Peru (Chimbote)	7.8	25,000
1971	USA (San Fernando, California)	6.6	65
1975	China (Haicheng)	7.3	Some
1976	Guatemala	7.5	23,000
1976	China (Tangshan)	8.0	242,000
1985	Mexico (Mexico City)	8.1	9,500
1988	Armenia	7.0	25,000
1989	USA (Loma Prieta, California)	7.1	63
1990	Iran	7.3	40,000
1992	Turkey	6.8	570
1992	USA (Landers, California)	Five earthquakes varying from 4.6 to 7.5 and numerous aftershocks occurred from April through June	At least 150 injured, but only 1 death.
1992	Egypt	5.9	500+
1993	India	6.4	30,000+
1994	USA (Northridge, California)	6.7	61
1994	Bolivia	8.2	?

there will be loud creaking and groaning noises as the building sways.

In most cases the shaking will stop almost as suddenly as it began, and you will realize that you have survived one of nature's most terrifying natural disasters. What seemed like eternity was probably only tens of seconds or less. Depending on the circumstances, you also may experience a gentle rolling motion as the slowest of the four types of earthquake waves pass below you. You also may feel numerous aftershocks, which typically are not as strong as the main shock.

Having described what it is like to experience an earthquake, we should ask, how do geologists define an earthquake? An **earthquake** is the vibration of the Earth caused by the sudden release of energy, usually as a result of displacement of rocks along fractures, or faulting, beneath the Earth's surface.

Following an earthquake, adjustments along a fault commonly generate a series of earthquakes referred to as **aftershocks.** Most of these are smaller than the main shock, but they can cause considerable damage to already weakened structures. Indeed, much of the destruction from the 1755 earthquake in Lisbon, Portugal, was caused by aftershocks. After a small earthquake, aftershock activity usually ceases within a few days, but it may persist for months following a large earthquake.

Early humans and cultures had much more imaginative and colorful explanations of earthquakes than this scientific explanation. For example, many cultures believed that the Earth rested on some type of organism whose movements caused the Earth to shake. In Japan, it was a giant catfish (▶ Figure 9-2); in Mongolia, a giant frog; in China, an ox; in India, a giant mole; in parts of South America, a whale; and to the Algonquin Indians of North America, an immense tortoise.

Many people believed earthquakes were a punishment or warning to the unrepentant. This view was strongly reinforced by the great Lisbon, Portugal earthquake that occurred on November 1, 1755 (All Saints' Day), when many people were attending church services. So strong was this earthquake that buildings shook all over Europe and chandeliers rattled in parts of the United States. Approximately 70,000 people were killed by a combination of collapsing buildings, a giant seismic sea wave that devastated the waterfront, and a fire that burned throughout the city.

The Greek philosopher Aristotle (384-322 B.C.) offered what he considered to be a natural explanation for earthquakes. He believed that atmospheric winds were drawn into the Earth's interior where they caused fires and swept around the various subterranean cavities trying to escape. It was this movement of underground air that caused earthquakes and occasional volcanic eruptions. Today, geologists know that the majority of earthquakes result from faulting associated with plate movements.

 ELASTIC REBOUND THEORY

Based on studies conducted after the 1906 San Francisco earthquake, H. F. Reid of Johns Hopkins University formulated the **elastic rebound theory** to explain how earthquakes occur. Reid studied three sets of measurements taken across the portion of the San Andreas fault that had broken during the 1906 earthquake. The measurements revealed that points on opposite sides of the fault had moved 3.2 m during the 50-year period prior to breakage in 1906, with the west side moving northward (▶ Figure 9-3).

▶ **FIGURE 9-2** This painting from the Edo period shows people trying to subdue a giant catfish. According to Japanese legend, earthquakes are caused by the movement of a giant catfish.

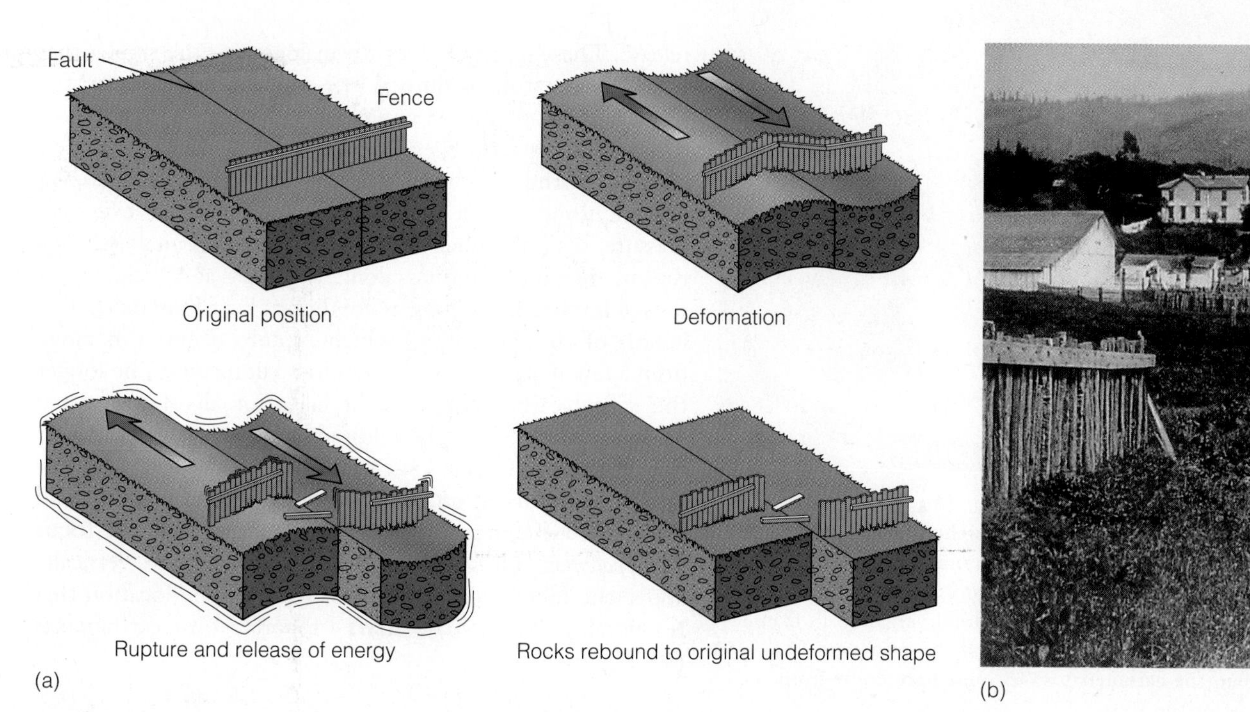

▶ **FIGURE 9-3** (*a*) According to the elastic rebound theory, when rocks are deformed, they store energy and bend. When the inherent strength of the rocks is exceeded, they rupture, releasing the energy in the form of earthquake waves that radiate outward in all directions. Upon rupture, the rocks rebound to their former undeformed shape. (*b*) During the 1906 San Francisco earthquake, this fence in Marin County was displaced 2.5 m.

According to Reid, rocks on opposite sides of the San Andreas fault had been storing energy and bending slightly for at least 50 years before the 1906 earthquake. Any straight line such as a fence or road that crossed the San Andreas fault would gradually be bent, as rocks on one side of the fault moved relative to rocks on the other side (Figure 9-3). Eventually, the strength of the rocks was exceeded, and rupture occurred. When this happened, the rocks on opposite sides of the fault rebounded or "snapped back" to their former undeformed shape, and the energy stored was released as earthquake waves radiating outward from the break (Figure 9-3).

Additional field and laboratory studies conducted by Reid and others have confirmed that elastic rebound is the mechanism by which earthquakes are generated. In laboratory studies, rocks subjected to forces equivalent to those occurring in the Earth's crust initially change their shape. As more force is applied, however, they resist further deformation until their internal strength is exceeded. At that point, they break and snap back to their original undeformed shape, releasing the energy that had been internally stored.

The energy stored in rocks undergoing elastic deformation is analogous to the energy stored in a watch spring that is tightly wound. The tighter the spring is wound, the more energy is stored, and thus, the more energy is available for release. If the spring is wound so tightly that it breaks, then the stored energy is released as the spring rapidly unwinds and partially regains its original shape.

SEISMOLOGY

Seismology, the study of earthquakes, began emerging as a true science around 1880 with the development of instruments that effectively recorded earthquake waves, but people had been studying earthquakes for much longer. The earliest earthquake detector was invented by the Chinese scholar Chang Heng sometime around A.D. 132 (▶ Figure 9-4). A story is told that one day Chang Heng's instrument indicated that there had been an earthquake, but no one in the area had felt a tremor. Most people regarded his experiment as a failure. A few days later, however, a rider arrived with the news of an earthquake that had occurred in a distant province in the direction indicated by Chang's instrument.

Over the succeeding centuries, other instruments were invented to study earthquakes, but it was not until the late nineteenth century that the first effective seismograph was developed. A **seismograph** is an instrument that detects, records, and measures the various vibrations produced by an earthquake (▶ Figure 9-5). The record made by a seismograph is a *seismogram*. Although modern seismographs are very sophisticated instruments that electronically record the motion onto a seismogram or enter it directly into a computer, they still follow the basic principles of operation that were used in the earliest seismographs.

When an earthquake occurs, energy in the form of *seismic waves* radiates outward in all directions from the point of

▶ FIGURE 9-4 The world's first earthquake detector was invented by Chang Heng sometime around A.D. 132. Movement of the vase dislodged a ball from a dragon's mouth into the waiting mouth of a frog below. For example, if the balls from the dragons on the east and west sides of the vase were dislodged, then the eathquake waves must have come from either the east or the west.

release. These seismic waves are analogous to the ripples that result when a stone is thrown into a quiet body of water; the ripples move outward in concentric circles from the point of the stone's impact.

Most earthquakes result when rocks in the Earth's crust rupture along a fault because of the buildup of excessive pressure, which is usually caused by plate movement. Once a rupture begins, it moves along the fault at a velocity of several km/sec for as long as conditions for failure exist. The length of the fault along which rupture occurs can range from a few meters to several hundred kilometers. The longer the rupture, the more time it takes for all of the stored energy in the rocks to be released, and therefore the longer the ground will shake.

The location within the crust where rupture initiates, and thus where the energy is released, is referred to as the **focus** or *hypocenter*. The point on the Earth's surface vertically above the focus is the **epicenter**, which is the location that is usually given in news reports on earthquakes (▶ Figure 9-6).

(a)

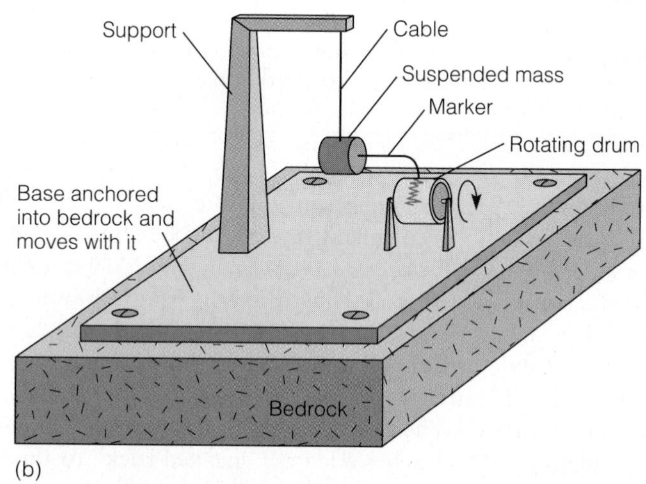

(b)

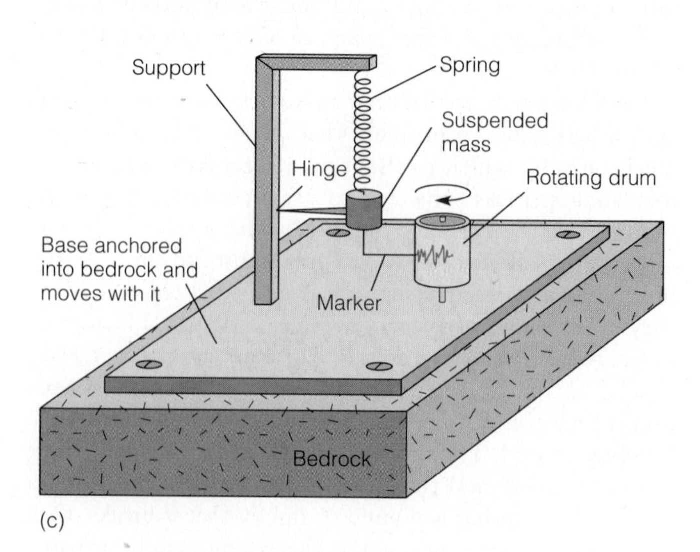

(c)

▶ FIGURE 9-5 (*a*) Modern seismographs record earthquake waves electronically. A geophysicist points to the trace of an earthquake recorded by a seismograph at the National Earthquake Information Service, Golden, Colorado. (*b*) A horizontal-motion seismograph. Because of its inertia, the heavy mass that contains the marker will remain stationary while the rest of the structure moves along with the ground during an earthquake. As long as the length of the arm is not parallel to the direction of ground movement, the marker will record the earthquake waves on the rotating drum. (*c*) A vertical-motion seismograph. This seismograph operates on the same principle as a horizontal-motion instrument and records vertical ground movement.

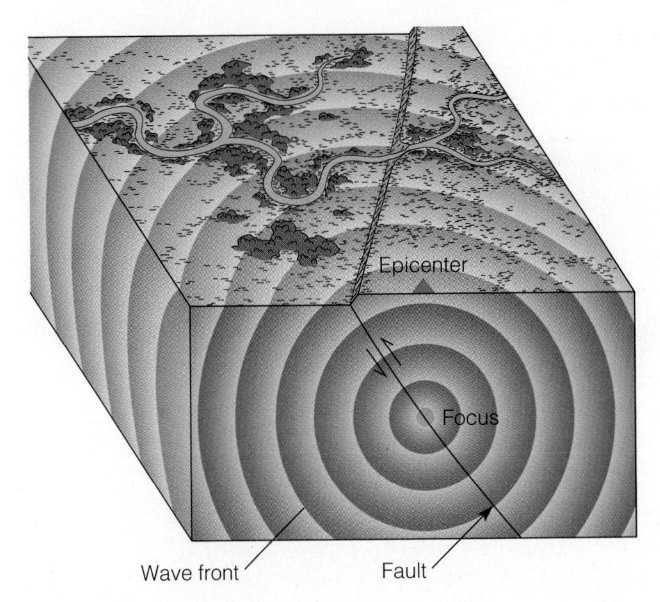

▶ FIGURE 9-6 The focus of an earthquake is the location where rupture begins and energy is released. The place on the Earth's surface vertically above the focus is the epicenter.

Seismologists recognize three categories of earthquakes based on the depth of their foci. *Shallow-focus* earthquakes have a focal depth of less than 70 km. Earthquakes with foci between 70 and 300 km are referred to as *intermediate focus,* and those with foci at a depth greater than 300 km are called *deep focus.* Earthquakes are not evenly distributed among these three categories. Approximately 90% of all earthquake foci occur at a depth of less than 100 km.

Shallow-focus earthquakes are, with few exceptions, the most destructive. All of the known large earthquakes in California have been shallow focus, and most have originated within the upper 10 km of the Earth's crust. The 1964 Alaska earthquake, the strongest yet recorded in the United States, had a focal depth near 30 km.

There is an interesting relationship between earthquake foci and plate margins. Earthquakes generated along divergent or transform plate boundaries are always shallow focus, while almost all intermediate- and deep-focus earthquakes occur within the circum-Pacific belt along convergent margins (▶ Figure 9-7). Furthermore, a pattern emerges when the focal depths of earthquakes near island arcs and their adjacent ocean trenches are plotted. Notice in ▶ Figure 9-8 that the focal depth increases beneath the Tonga Trench in a narrow, well-defined zone that dips approximately 45°.

▶ FIGURE 9-7 The relationship between the distribution of earthquake epicenters and plate boundaries. Approximately 80% of earthquakes occur within the circum-Pacific belt, 15% within the Mediterranean-Asiatic belt, and the remaining 5% within the interiors of plates or along oceanic spreading ridge systems. Each dot represents a single earthquake epicenter.

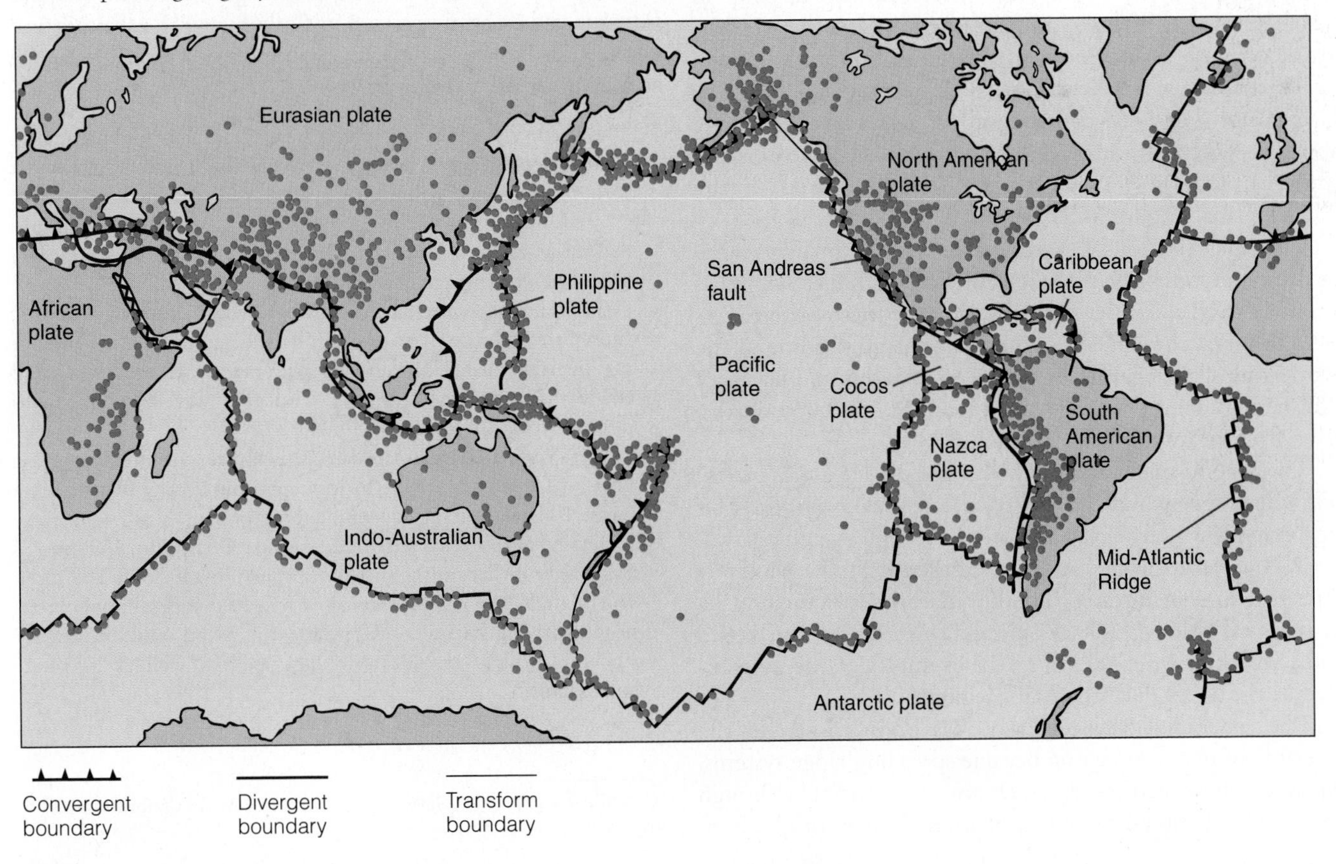

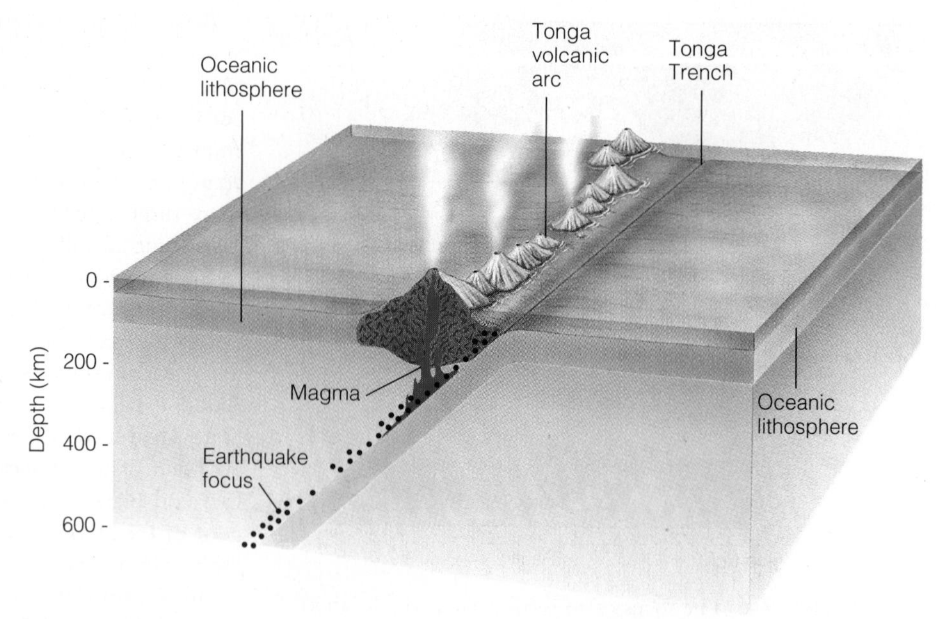

➢ FIGURE 9-8 Focal depth increases in a well-defined zone that dips approximately 45° beneath the Tonga volcanic arc in the South Pacific. Dipping seismic zones are common features of island arcs and deep-ocean trenches.

Dipping seismic zones, called *Benioff zones*, are a feature common to island arcs and deep-ocean trenches. Such zones indicate the angle of plate descent along a convergent plate boundary.

 THE FREQUENCY AND DISTRIBUTION OF EARTHQUAKES

Most earthquakes (almost 95%) occur in seismic belts that correspond to plate boundaries where stresses develop as plates converge, diverge, and slide past each other. Earthquake activity distant from plate margins is minimal, but can be devastating when it occurs. The relationship between plate margins and the distribution of earthquakes is readily apparent when the locations of earthquake epicenters are superimposed on a map showing the boundaries of the Earth's plates (Figure 9-7).

The majority of all earthquakes (approximately 80%) occur in the *circum-Pacific belt,* a zone of seismic activity that encircles the Pacific Ocean basin. Most of these earthquakes are a result of convergence along plate margins. Some of the world's most devastating earthquakes, resulting in billions of dollars of property damage and more than 500,000 deaths, have occurred within this belt (Table 9-1).

The second major seismic belt is the *Mediterranean-Asiatic belt* where approximately 15% of all earthquakes occur. This belt extends westerly from Indonesia through the Himalayas, across Iran and Turkey, and westerly through the Mediterranean region of Europe. The devastating 1990 earthquake in Iran that killed 40,000 people and the 1993 earthquake in India that killed 30,000 are recent examples of the destructive earthquakes that strike this region.

The remaining 5% of earthquakes occur mostly in the interiors of plates and along oceanic spreading ridge systems. Most of these earthquakes are not very strong although several major intraplate earthquakes are worthy of mention. For example, the 1811 and 1812 earthquakes near New Madrid, Missouri, killed some 20 people and nearly destroyed the town of New Madrid. So strong were these earthquakes that they were felt from the Rocky Mountains to the Atlantic Ocean and from the Canadian border to the Gulf of Mexico. In addition, the earthquake caused church bells to ring as far away as Boston, Massachusetts (1,600 km). Within the immediate area, numerous buildings were destroyed and forests were flattened; the land sank several meters in some areas, causing flooding; and the Mississippi River is said to have reversed its flow during the shaking and changed its course slightly. Eyewitnesses described the scene at New Madrid as follows:

> The earth was observed to roll in waves a few feet high with visible depressions between. By and by these swells burst throwing up large volumes of water, sand, and coal.
>
> . . . Undulations of the earth resembling waves, increasing in elevations as they advanced, and when they attained a certain fearful height the earth would burst.
>
> The shocks were clearly distinguishable into two classes, those in which the motion was horizontal and those in which it was perpendicular. The latter was attended with explosions and the terrible mixture of noises, . . . but they were by no means as destructive as the former.
>
> Cpt. Sarpy tied up at this [small] island on the evening of the 15th of December, 1811. In looking around they found that a party of river pirates occupied part of the island and were expecting Sarpy with the intention of robbing him. As soon as Sarpy found that out he quietly dropped lower down the river. In the night the earthquake came and next morning when the accompanying haziness disappeared the island could no longer be seen. It had been utterly destroyed as well as its pirate inhabitants.*

*C. Officer and J. Page, *Tales of the Earth* (New York: Oxford University Press 1993), pp. 49-50.

Guest Essay   *Kate Hutton*

Seismology: A Link between the Public and Science

Earth, as a planet, is unique in several ways. Not the least of its special features is that we live on it. Because we live on it, we know more about it than about any other planet. We are also deeply embroiled in its affairs, so much so that we sometimes "lose sight of the forest for the trees" and forget that nature has natural processes of its own, independent of human concerns.

When these natural processes are too much at cross-purposes with our own agenda, they are termed "natural disasters." Many disasters, however, are caused more by human disregard for natural processes than by those processes themselves. For example, for the most part, earthquakes do not kill people—buildings falling down in earthquakes do.

Seismic risk is only one example of how we sometimes ignore natural processes to our detriment. Our failure to consider the impact of our activities on the environment is another important example, though on a somewhat slower time scale. Science education, especially earth science education, is an important link between people and their planet. In other words, taking this course can make you a more informed and responsible citizen with regard to issues as wide ranging as land use, building codes, environmental protection, and even nuclear disarmament. The need to provide accurate and usably brief information (even "sound bites") for the mass media is another important link that many scientists tend to ignore.

My own career is probably not the best to use as an example of forethought and planning. In my junior and senior high school days, I had a series of passionate interests in many different branches of natural science, ranging from physics and astronomy through geology and biology. Then one year, I felt my first earthquake. It occurred around midnight. There was damage, but not locally, and only our nerves were seriously affected. When I finally did go back to bed, I took the encyclopedia with me. Later that year, I built a model of a seismograph as a science project. It actually worked, until I took it to school on the bus. I remember thinking that it would be fun to run a seismograph station—little did I know.

By the time I went to college, however, chance had placed me far from earthquake country, and astronomy was my big interest. I stayed with that for eight years, until I earned my Ph.D., with only occasional thoughts of switching majors (to, for example, computer science). My dissertation research was with the radio astronomy group at the University of Maryland and the Goddard Space Flight Center in very long baseline interferometry, or VLBI. VLBI is an attempt to use various radio telescopes located around the globe as one big instrument and thereby gain higher resolution of very compact celestial radio sources. In my case, the compact radio sources were the cores of quasars and Seyfert galaxies. In the course of my research, I found myself working beside geophysicists who were using the same data to measure the motion of tectonic plates on the Earth. I was intrigued, and the prospect of earthquake prediction, which at that time was thought to be a possibility, was fascinating. Since the job market in astronomy looked dim, I joined the geophysics side of the VLBI program at Goddard after I finished my degree. A year later, I took a staff position at the California Institute of Technology processing earthquake data, and I've been here ever since.

The earth scientists that I know perform a wide variety of different kinds of work. Some spend a lot of time in the field, some none. Some do fairly abstract mathematical modeling, some dig holes. Being in charge of data processing for the Southern California Seismographic Network, I spend most of my time on the computer. My routine work involves everything from programming to supervision of the data analysts. In addition, because I may be counted on to be here soon after any moderately large earthquake and, unlike some others, I am willing to talk to the press, I've become somewhat of a spokesperson. I find explaining science to the general public through the news media to be an enjoyable challenge. It is very exciting to be right in the middle of the response to one of nature's little "disasters."

It has been said that the fascination of flying comes from its unique combination of boredom and panic. The same can be said of seismology, I suppose. Each day, the chances are fairly good that I will be doing the work that I had planned for the day, but I might not be. I might be scrambling for disk space to record a major aftershock sequence. I might be facing a small forest of microphones, trying to say something intelligent about data we've barely had time to look at.

My job may not be as philosophically satisfying as learning the origin of the universe, but is is certainly relevant to the human condition. At least, it is in California.

DR. HUTTON earned her B.S. in astronomy at Penn State University in 1971 and her M.S. and Ph.D. in 1973 and 1976 at the University of Maryland, also in astronomy. After spending one year in the Geophysics Branch at Gooddard Space Flight Center, she came to Caltech in 1977. Widely known as "the Earthquake Lady," she currently oversees data processing for the Southern California Seismographic Network and often informs the press and public following an earthquake. She lives in Pasadena in a house bolted to the foundation, and she does keep emergency supplies.

▶ **FIGURE 9-9** Damage done to Charleston, South Carolina, by the earthquake of August 31, 1886. This earthquake is the largest reported in the eastern United States.

Another major intraplate earthquake struck Charleston, South Carolina, on August 31, 1886, killing 60 people and causing $23 million in property damage (▶ Figure 9-9). More recently, a large intraplate earthquake struck near Tennant Creek in Australia's Northern Territory in December 1988.

The cause of intraplate earthquakes is not well understood, but geologists think they arise from localized stresses caused by the compression that most plates experience along their margins. The release of these stresses and hence the resulting intraplate earthquakes are due to local factors. Interestingly, many intraplate earthquakes are associated with very ancient and presumed inactive faults that are reactivated at various intervals.

More than 150,000 earthquakes strong enough to be felt by someone are recorded every year by the worldwide network of seismograph stations. In addition, it has been estimated that about 900,000 earthquakes occur annually that are recorded by seismographs, but are too small to be individually cataloged. These small earthquakes result from the energy released as continual adjustments between the Earth's various plates occur.

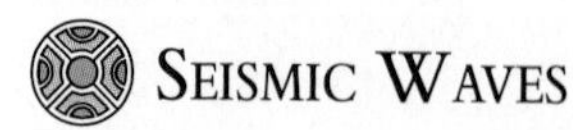

SEISMIC WAVES

The shaking and destruction resulting from earthquakes are caused by two different types of seismic waves: *body waves,* which travel through the Earth and are somewhat like sound waves; and *surface waves,* which travel only along the ground surface and are analogous to ocean waves.

Body Waves

An earthquake generates two types of body waves: P-waves and S-waves (▶ Figure 9-10). **P-waves** or *primary waves* are the fastest seismic waves and can travel through solids, liquids, and gases. P-waves are compressional, or push-pull, waves and are similar to sound waves in that they move material forward and backward along a line in the same direction that the waves themselves are moving (Figure 9-10b). Thus, the material through which P-waves travel is expanded and compressed as the wave moves through it and returns to its original size and shape after the wave passes by. In fact, some P-waves emerging from within the Earth are transmitted into the atmosphere as sound waves that can be heard by humans and animals at certain frequencies.

S-waves or *secondary waves* are considerably slower than P-waves and can only travel through solids. S-waves are *shear waves* because they move the material perpendicular to the direction of travel, thereby producing shear stresses in the material they move through (Figure 9-10c). Because liquids (as well as gases) are not rigid, they have no shear strength and S-waves cannot be transmitted through them.

The velocities of P- and S-waves are determined by the density and elasticity of the materials through which they travel. *Elasticity* is a property of solids, such as rocks, and means that once they have been deformed by an applied force, they return to their original shape when the force is no longer present. Both the density and elasticity of rocks increase with depth, but elasticity increases faster than density, resulting in a general increase in the velocity of

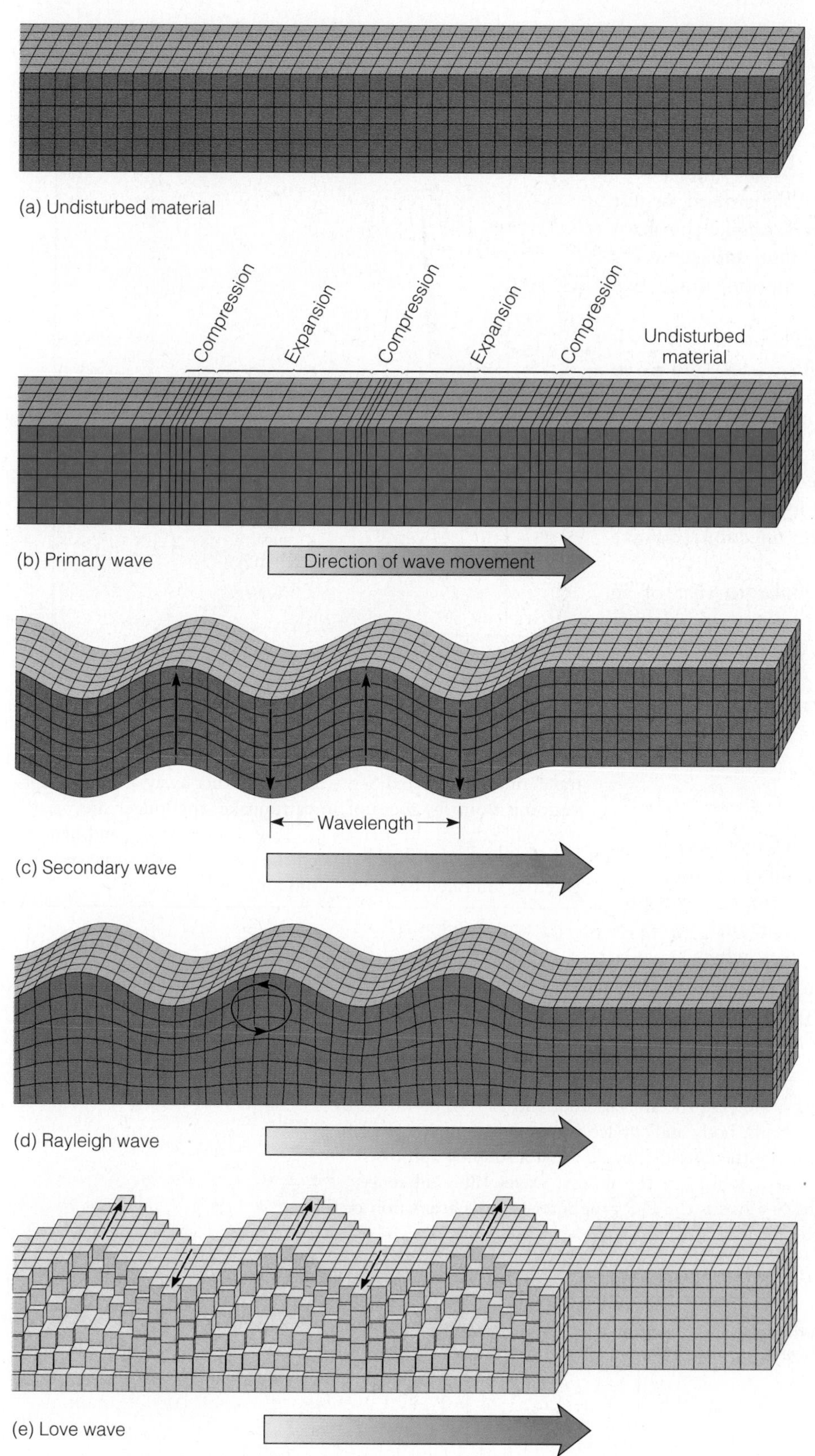

FIGURE 9-10 Seismic waves. (*a*) Undisturbed material. (*b*) Primary waves (P-waves) compress and expand material in the same direction as the wave movement. (*c*) Secondary waves (S-waves) move material perpendicular to the direction of wave movement. (*d*) Rayleigh waves (R-waves) move material in an elliptical path within a vertical plane oriented parallel to the direction of wave movement. (*e*) Love waves (L-waves) move material back and forth in a horizontal plane perpendicular to the direction of wave movement.

seismic waves. Because P-wave velocity is greater than S-wave velocity in all materials, P-waves always arrive at seismic stations first.

Surface Waves

Surface waves travel along the surface of the ground, or just below it, and are slower than body waves. Unlike the sharp jolting and shaking that body waves cause, surface waves generally produce a rolling or swaying motion, much like the experience of being on a boat.

Several types of surface waves are recognized. The two most important are **Rayleigh waves (R-waves)** and **Love waves (L-waves),** named after the British scientists who discovered them, Lord Rayleigh and A. E. H. Love. Rayleigh waves are generally the slower of the two and behave like water waves in that they move forward while the individual particles of material move in an elliptical path within a vertical plane oriented in the direction of wave movement (Figure 9-10d).

The motion of a Love wave is similar to that of an S-wave, but the individual particles of the material only move back and forth in a horizontal plane perpendicular to the direction of wave travel (Figure 9-10e). This type of lateral motion can be particularly damaging to building foundations.

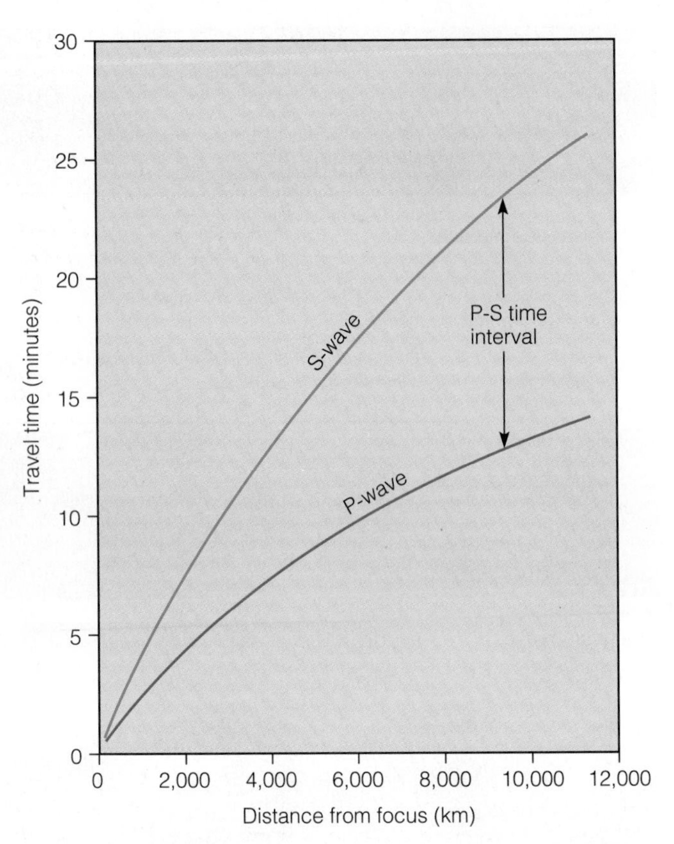

▶ FIGURE 9-12 A time-distance graph showing the average travel times for P- and S-waves. The farther away a seismograph station is from the focus of an earthquake, the longer the interval between the arrivals of the P- and S-waves, and hence the greater the distance between the curves on the time-distance graph as indicated by the P-S time interval.

LOCATING AN EARTHQUAKE

The various seismic waves travel at different speeds and therefore arrive at a seismograph at different times. As ▶ Figure 9-11 illustrates, the different seismic waves produce distinctive seismogram patterns. The first waves to arrive, and thus the fastest, are the P-waves, which travel at nearly twice the velocity of the S-waves that follow. Both the P- and S-waves travel directly from the focus to the seismograph through the interior of the Earth. The last

▶ FIGURE 9-11 A schematic seismogram showing the arrival order and pattern produced by P-, S-, and L-waves. When an earthquake occurs, body and surface waves radiate outward from the focus at the same time. Because P-waves are the fastest, they arrive at a seismograph first, followed by S-waves, and then by surface waves, which are the slowest waves. The difference between the arrival times of the P- and the S-waves is the P-S time interval; it is a function of the distance of the seismograph station from the focus.

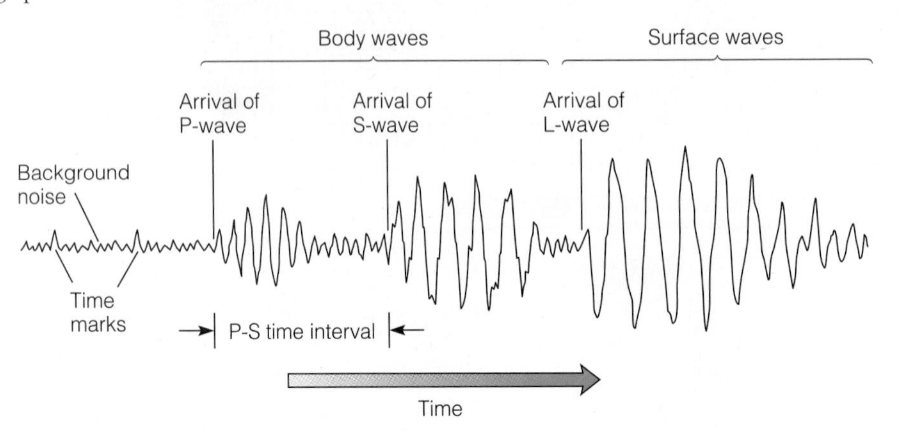

waves to arrive are the L- and R-waves, which are the slowest and also travel the longest route along the Earth's surface.

By accumulating a tremendous amount of data over the years, seismologists have determined the average travel times of P- and S-waves for any specific distance. These P- and S-wave travel times are published as **time-distance graphs** and illustrate that the difference between the arrival times of the P- and S-waves is a function of the distance of the seismograph from the focus; that is, the farther the waves travel, the greater the time between the arrivals of the P- and S-waves (⯈ Figure 9-12).

As ⯈ Figure 9-13 demonstrates, the epicenter of any earthquake can be determined by using a time-distance graph and knowing the arrival times of the P- and S-waves at any three seismograph locations. Subtracting the arrival time of the first P-wave from the arrival time of the first S-wave gives the time interval between the arrivals of the two waves for each seismograph location. Each time interval is then plotted on the time-distance graph, and a line is drawn straight down to the distance axis of the graph, indicating how far away each station is from the focus of the earthquake. Then a circle whose radius equals the distance shown on the time-distance graph from each of the three seismograph locations is drawn on a map (Figure 9-13). The intersection of the three circles is the location of the earthquake's epicenter. A minimum of three locations is needed because two locations will provide two possible epicenters and one location will provide an infinite number of possible epicenters. It should be noted that computers are now used to determine the epicenter of an earthquake, and many seismic stations are used for redundancy and to determine the most accurate location.

MEASURING EARTHQUAKE INTENSITY AND MAGNITUDE

Geologists measure the strength of an earthquake in two different ways. The first, *intensity,* is a qualitative assessment of the kinds of damage done by an earthquake. The second,

⯈ **FIGURE 9-13** Three seismograph stations are needed to locate the epicenter of an earthquake. The P-S time interval is plotted on a time-distance graph for each seismograph station to determine the distance that station is from the epicenter. A circle with that radius is drawn from each station, and the intersection of the three circles is the epicenter of the earthquake.

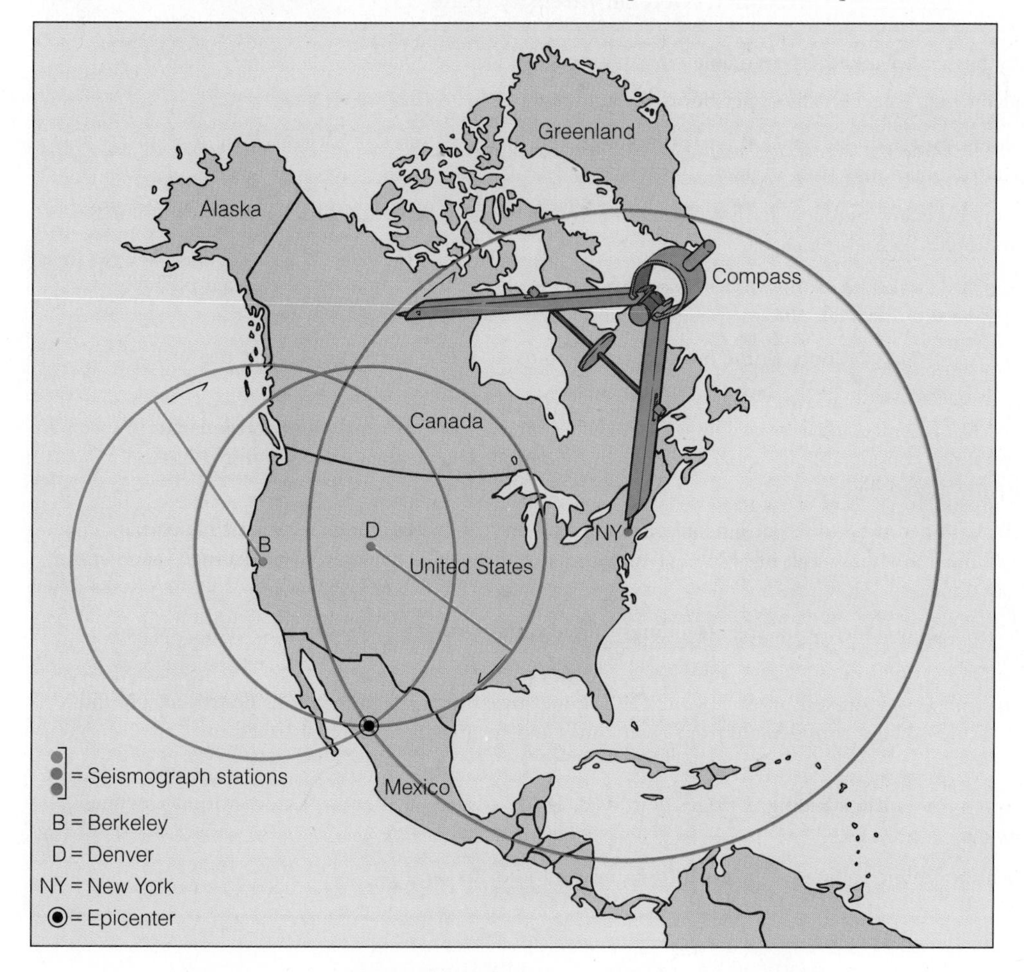

magnitude, is a quantitative measurement of the amount of energy released by an earthquake. Each method provides geologists with important data about earthquakes and their effects. This information can then be used to prepare for future earthquakes.

Intensity

Intensity is a subjective measure of the kind of damage done by an earthquake as well as people's reaction to it. Since the mid-nineteenth century, geologists have used intensity as a rough approximation of the size and strength of an earthquake. The most common intensity scale used in the United States is the **Modified Mercalli Intensity Scale**, which has values ranging from I to XII (◉ Table 9-2). This scale was originally developed by the Italian seismologist Giuseppe Mercalli in 1902 and was later modified for use in the United States by H. O. Wood and F. Neumann of the California Institute of Technology Seismological Laboratory in 1931.

After an assessment of the earthquake damage is made, *isoseismal lines* (lines of equal intensity) are drawn on a map, dividing the affected region into various intensity zones. The intensity value given for each zone is the maximum intensity that the earthquake produced for that zone. Even though intensity maps are not precise because of the subjective nature of the measurements, they do provide geologists with a rough approximation of the location of the earthquake, the kind and extent of the damage done, and the effects of local geology and types of building construction (▶ Figure 9-14). Because intensity is a measure of the kind of damage done by an earthquake, insurance companies still classify earthquakes on the basis of intensity.

While it is generally true that a large earthquake will produce greater intensity values than a small earthquake, many other factors besides the amount of energy released by an earthquake affect its intensity. These include the distance from the epicenter, the focal depth of the earthquake, the population density and local geology of the area, the type of building construction employed, and the duration of shaking.

A comparison of the intensity map for the 1906 San Francisco earthquake and a geologic map of the area shows a strong correlation between the amount of damage done and the underlying rock and soil conditions (▶ Figure 9-15). Damage was greatest in those areas underlain by poorly consolidated material or artificial fill because the effects of

TABLE 9-2 Modified Mercalli Intensity Scale

I	Not felt except by a very few under especially favorable circumstances.
II	Felt only by a few people at rest, especially on upper floors of buildings.
III	Felt quite noticeably indoors, especially on upper floors of buildings, but many people do not recognize it as an earthquake. Standing automobiles may rock slightly.
IV	During the day felt indoors by many, outdoors by few. At night some awakened. Sensation like heavy truck striking building, standing automobiles rocked noticeably.
V	Felt by nearly everyone, many awakened. Some dishes, windows, etc. broken, a few instances of cracked plaster. Disturbance of trees, poles, and other tall objects sometimes noticed.
VI	Felt by all, many frightened and run outdoors. Some heavy furniture moved, a few instances of fallen plaster or damaged chimneys. Damage slight.
VII	Everybody runs outdoors. Damage negligible in buildings of good design and construction; slight to moderate in well-built ordinary structures; considerable in poorly built or badly designed structures; some chimneys broken. Noticed by people driving automobiles.
VIII	Damage slight in specially designed structures; considerable in normally constructed buildings with possible partial collapse; great in poorly built structures. Fall of chimneys, monuments, walls. Heavy furniture overturned. Sand and mud ejected in small amounts.
IX	Damage considerable in specially designed structures. Buildings shifted off foundations. Ground noticeably cracked. Underground pipes broken.
X	Some well-built wooden structures destroyed; most masonry and frame structures with foundations destroyed; ground badly cracked. Rails bent. Landslides considerable from river banks and steep slopes. Water splashed over river banks.
XI	Few, if any (masonry) structures remain standing. Bridges destroyed. Broad fissures in ground. Underground pipelines completely out of service.
XII	Damage total. Waves seen on ground surfaces. Objects thrown upward into the air.

SOURCE: United States Geological Survey.

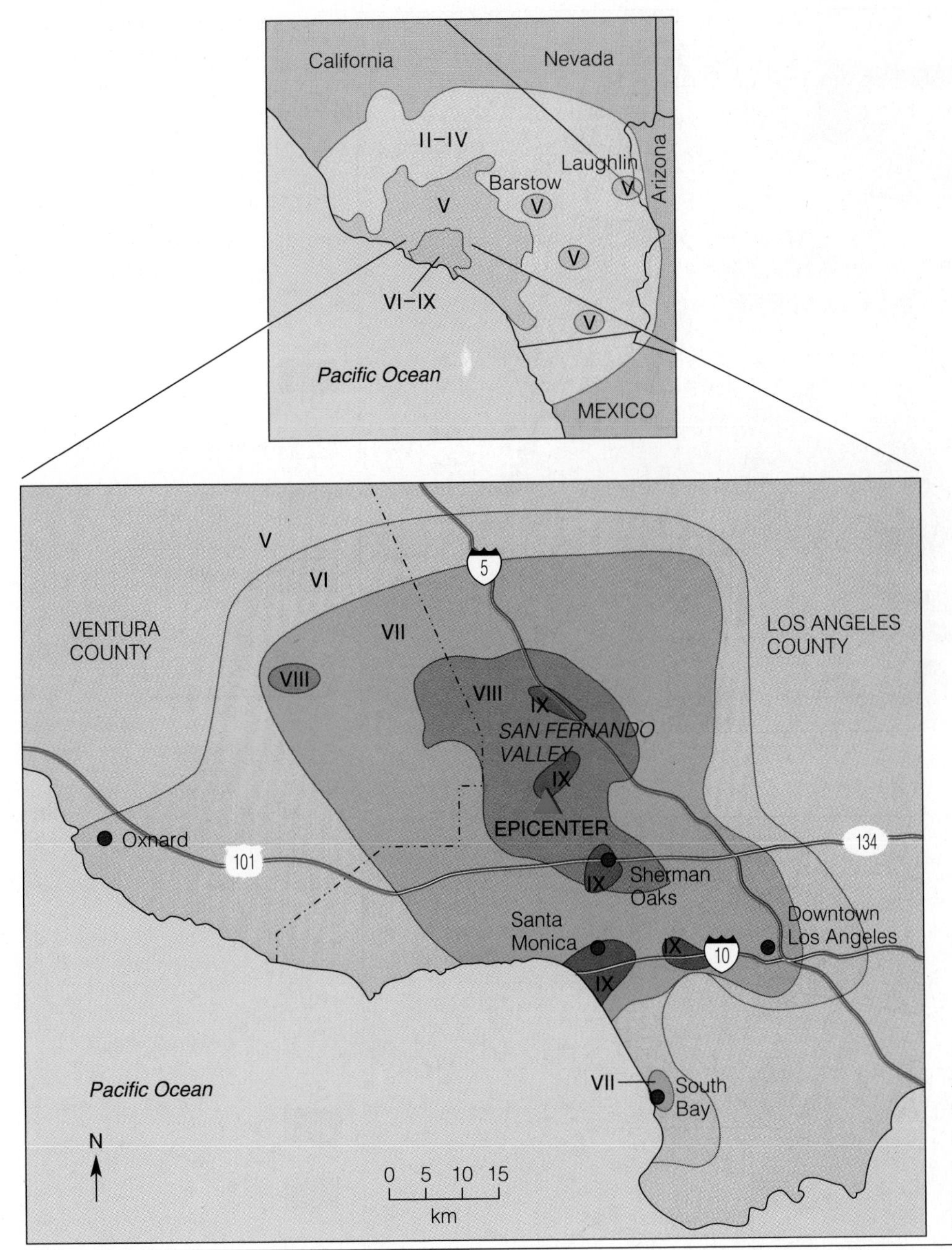

▶ **FIGURE 9-14** Preliminary Modified Mercalli Intensity map for the 1994 Northridge, California earthquake showing the region divided into intensity zones based on the kind of damage done. This earthquake had a magnitude of 6.7.

shaking are amplified in these materials, whereas damage was rather low in areas of solid bedrock. The correlation between the geology and the amount of damage done by an earthquake was further reinforced by the 1989 Loma Prieta earthquake when many of the same areas that were extensively damaged in the 1906 earthquake were once again heavily damaged.

Magnitude

If earthquakes are to be compared quantitatively, we must use a scale that measures the amount of energy released and is independent of intensity. Such a scale was developed in 1935 by Charles F. Richter, a seismologist at the California Institute of Technology. The **Richter Magnitude Scale** measures earthquake **magnitude**, which is the total amount of energy released by an earthquake at its source. It is an open-ended scale with values beginning at 1. The largest magnitude recorded has been 8.6, and though values greater than 9 are theoretically possible, they are highly improbable because rocks are not able to withstand the buildup of pressure required to release that much energy.

The magnitude of an earthquake is determined by measuring the amplitude of the largest seismic wave as recorded on a seismogram (▶ Figure 9-16). To avoid large numbers, Richter used a conventional base-10 logarithmic scale to convert the amplitude of the largest recorded seismic wave to a numerical magnitude value (Figure 9-16). Therefore,

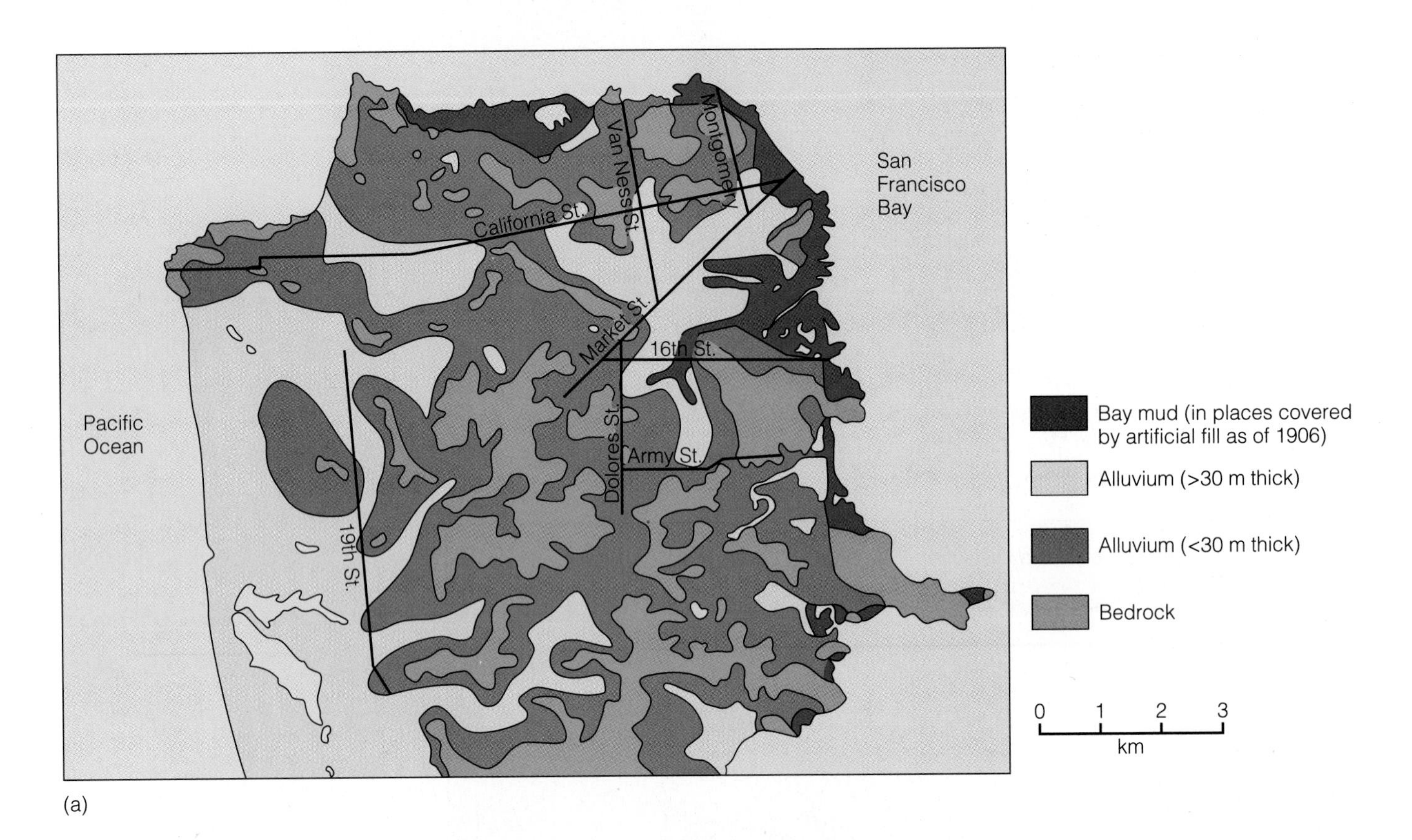

(a)

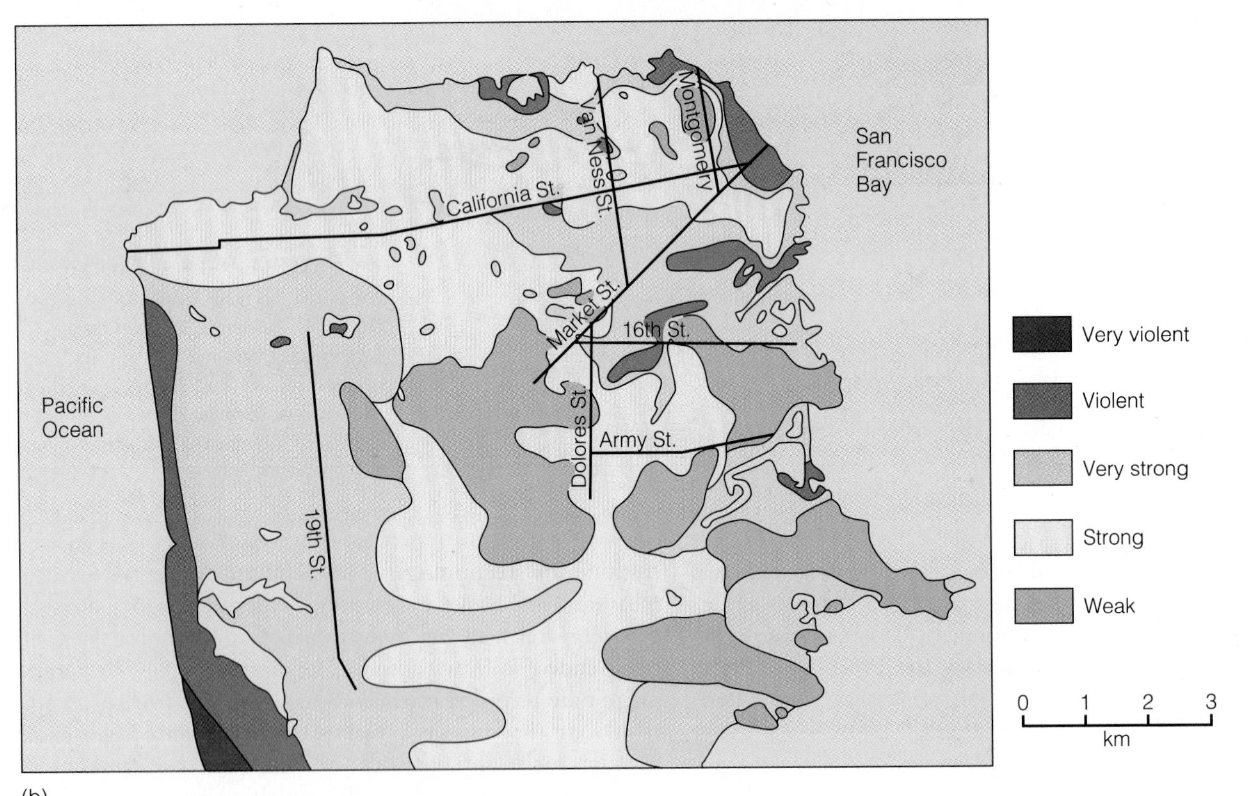

(b)

▶ **FIGURE 9-15** A comparison between (*a*) the general geology of the San Francisco peninsula and (*b*) a Modified Mercalli Intensity map of the same area for the 1906 San Francisco earthquake. Notice the close correlation between the geology and the intensity. Areas of bedrock correspond to the lowest intensity values, while areas of poorly consolidated material (alluvium) or bay mud have the highest intensity values.

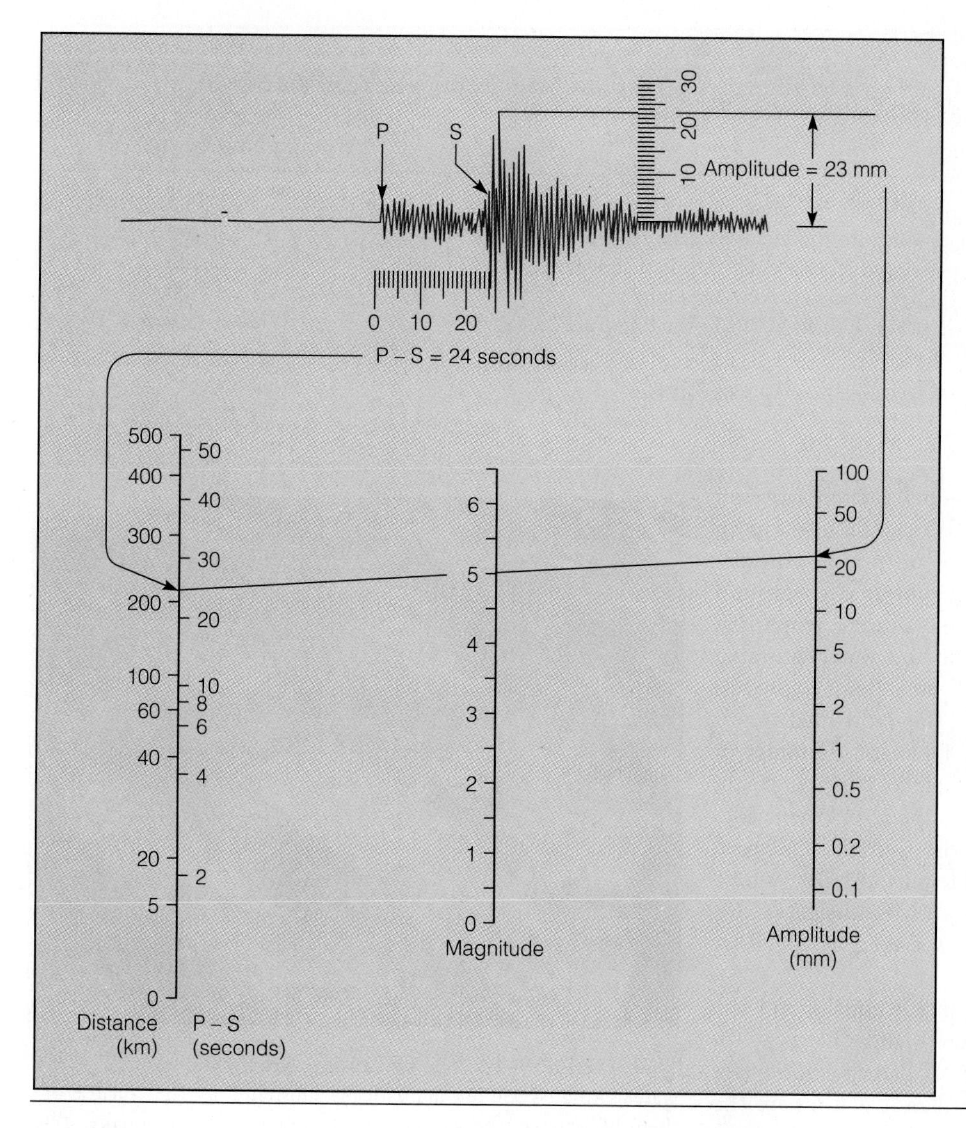

▶ **FIGURE 9-16** The Richter Magnitude Scale measures magnitude, which is the total amount of energy released by an earthquake at its source. The magnitude is determined by measuring the maximum amplitude of the largest seismic wave and marking it on the right-hand scale. The difference between the arrival times of the P- and S-waves (recorded in seconds) is marked on the left-hand scale. When a line is drawn between the two points, the magnitude of the earthquake is the point at which the line crosses the center scale.

each integer increase in magnitude represents a 10-fold increase in wave amplitude. For example, the amplitude of the largest seismic wave for an earthquake of magnitude 6 is 10 times that produced by an earthquake of magnitude 5, 100 times as large as a magnitude 4 earthquake, and 1,000 times that of an earthquake of magnitude 3 (10 × 10 × 10 = 1,000).

While each increase in magnitude represents a 10-fold increase in wave amplitude, each magnitude increase corresponds to a roughly 30-fold increase in the amount of energy released. The 1964 Alaska earthquake with a magnitude of 8.6 released about 900 times the energy of the 1994 Northridge earthquake of magnitude 6.7.

We have already mentioned that more than 900,000 earthquakes are recorded around the world each year. These figures can be placed in better perspective by reference to ◉ Table 9-3, which shows that the vast majority of earthquakes have a Richter magnitude of less than 2.5, and that great earthquakes (those with a magnitude greater than 8.0) occur, on average, only once every five years.

THE DESTRUCTIVE EFFECTS OF EARTHQUAKES

The destructive effects of earthquakes are many and varied and include such phenomena as ground shaking, fire, seismic sea waves, and landslides, as well as disruption of vital services, panic, and psychological shock. The amount of property damage, loss of life, and injury depends on the time of day an earthquake occurs, its magnitude, the distance from the epicenter, the geology of the area, the type of construction of the various structures, the population density, and the duration of shaking. Generally speaking, earthquakes occurring during working and school hours in densely populated urban areas are the most destructive.

Ground Shaking

Ground shaking usually causes more damage and results in more loss of life and injuries than any other earthquake hazard. Structures built on solid bedrock generally suffer less

TABLE 9-3 Average Number of Earthquakes of Various Magnitudes per Year Worldwide

Magnitude	Effects	Average Number per Year
<2.5	Typically not felt, but recorded.	900,000
2.5–6.0	Usually felt. Minor to moderate damage to structures.	31,000
6.1–6.9	Potentially destructive, especially in populated areas.	100
7.0–7.9	Major earthquakes. Serious damage results.	20
>8.0	Great earthquakes. Usually result in total destruction.	1 every 5 years

SOURCE: Data from *Earthquake Information Bulletin* and Gutenberg and Richter (1949).

damage than those built on poorly consolidated material such as water-saturated sediments or artificial fill (▶ Figure 9-17; see Perspective 9-1). Structures on poorly consolidated or water-saturated material are subjected to ground shaking of longer duration and greater S-wave amplitude than those on bedrock. In addition, fill and water-saturated sediments tend to liquefy, or behave as a fluid, a process known as *liquefaction*. When shaken, the individual grains lose cohesion and the ground flows. Dramatic examples of damage resulting from liquefaction include Niigata, Japan, where large apartment buildings were tipped on their sides after the water-saturated soil of the hillside collapsed (▶ Figure 9-18), and Turnagain Heights, Alaska, where many homes were destroyed when the Bootlegger Cove Clay lost all of its strength when it was shaken by the 1964 earthquake (see Figure 14-18).

In addition to the magnitude of an earthquake and the underlying geology, the material used and the type of construction also affect the amount of damage done (see Perspective 9-1). Adobe and mud-walled structures are the weakest of all and almost always collapse during an earthquake. Unreinforced brick structures and poorly built concrete structures are also particularly susceptible to collapse.

▶ FIGURE 9-18 The effects of ground shaking on water-saturated soil are dramatically illustrated by the collapse of these buildings in Niigata, Japan, during a 1964 earthquake. The buildings were designed to be earthquake-resistant and fell over on their sides intact.

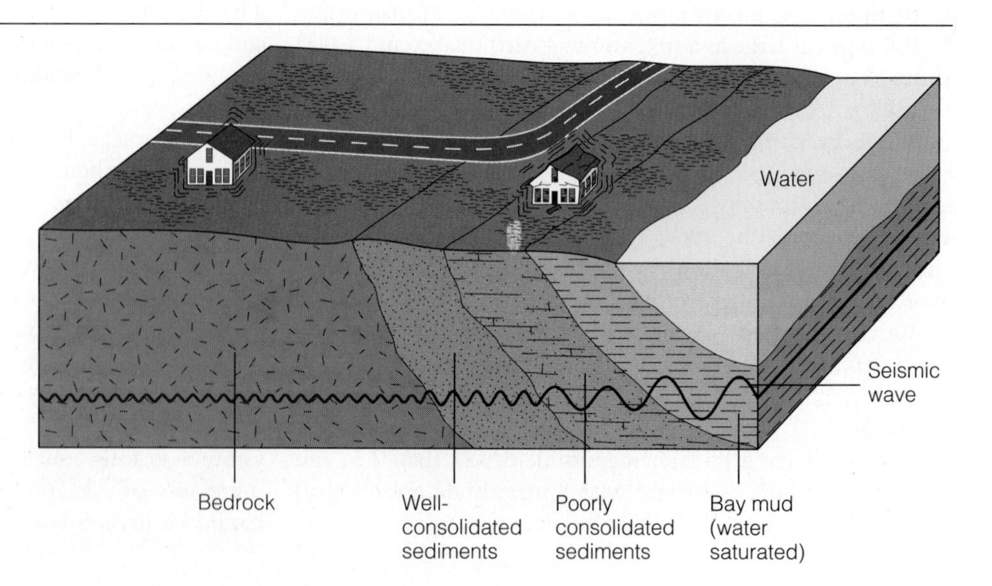

▶ FIGURE 9-17 The amplitude, duration, and period of seismic waves vary in different types of materials. The amplitude and duration of the waves generally increase as they pass from bedrock to poorly consolidated or water-saturated material. Structures built on weaker material typically suffer greater damage than similar structures built on bedrock.

▶ **FIGURE 9-19** Many of the approximately 242,000 people who died in the 1976 earthquake in Tangshan, China, were killed by collapsing structures. Many of the buildings were constructed from unreinforced brick, which has no flexibility, and quickly fell down during the earthquake.

For example, the 1976 earthquake in Tangshan, China, completely leveled the city because hardly any of the structures were built to resist seismic forces. In fact, most of them had unreinforced brick walls, which have no flexibility, and consequently they collapsed during the shaking (▶ Figure 9-19). During the 1906 San Francisco earthquake, 10% of the buildings were destroyed as a direct result of ground shaking. Many of the buildings at the time were constructed of brick and were not designed to withstand the violent shaking unleashed by an earthquake (▶ Figure 9-20). The 6.4 magnitude earthquake that struck India in 1993 killed about 30,000 whereas the 6.7

▶ **FIGURE 9-20** Approximately 10% of the total destruction during the 1906 San Francisco earthquake was the direct result of ground shaking. Many buildings were constructed of brick or masonry and quickly collapsed. The City Hall dome remained standing because it was supported by a steel framework, but the walls and rest of the building collapsed.

Perspective 9-1

Designing Earthquake-Resistant Structures

One way to reduce property damage, injuries, and loss of life is to design and build structures as earthquake-resistant as possible. Many things can be done to improve the safety of current structures and of new buildings as well.

California has a Uniform Building Code that sets minimum standards for building earthquake-resistant structures and is used as a model around the world. The California code is far more stringent than federal earthquake building codes and requires that structures be able to withstand a 25-second main shock. Unfortunately, many earthquakes are of far longer duration. The main shock of the 1964 Alaskan earthquake lasted approximately three minutes and was followed by numerous aftershocks. Although many of the buildings that were extensively damaged in this earthquake had been built according to the California code, they were not designed to withstand shaking of such long duration. Nevertheless, in California and elsewhere in the world, structures built since the California code went into effect have fared much better during moderate to major earthquakes than those built before its implementation.

The major objective in designing earthquake-resistant structures is minimizing the loss of life, injuries, and damage. To achieve this goal, engineers must understand the dynamics and mechanics of earthquakes including the type and duration of the ground motion that occurs and how rapidly the ground accelerates during an earthquake. An understanding of the area's geology is also very important because certain ground materials such as water-saturated sediments or landfill can lose their strength and cohesiveness during an earthquake. Such materials should be avoided if at all possible. Finally, engineers must be aware of how different structures behave under different earthquake conditions.

With the level of technology currently available, a well-designed, properly constructed building should be able to withstand small, short-duration earthquakes of less than 5.5 magnitude with little or no damage. In moderate earthquakes (5.5 to 7.0 magnitude), the damage suffered should not be serious and should be repairable. In a major earth-

➤ **FIGURE 1** This diagram shows some of the things a homeowner can do to reduce the potential damage to a building because of ground shaking during an earthquake.

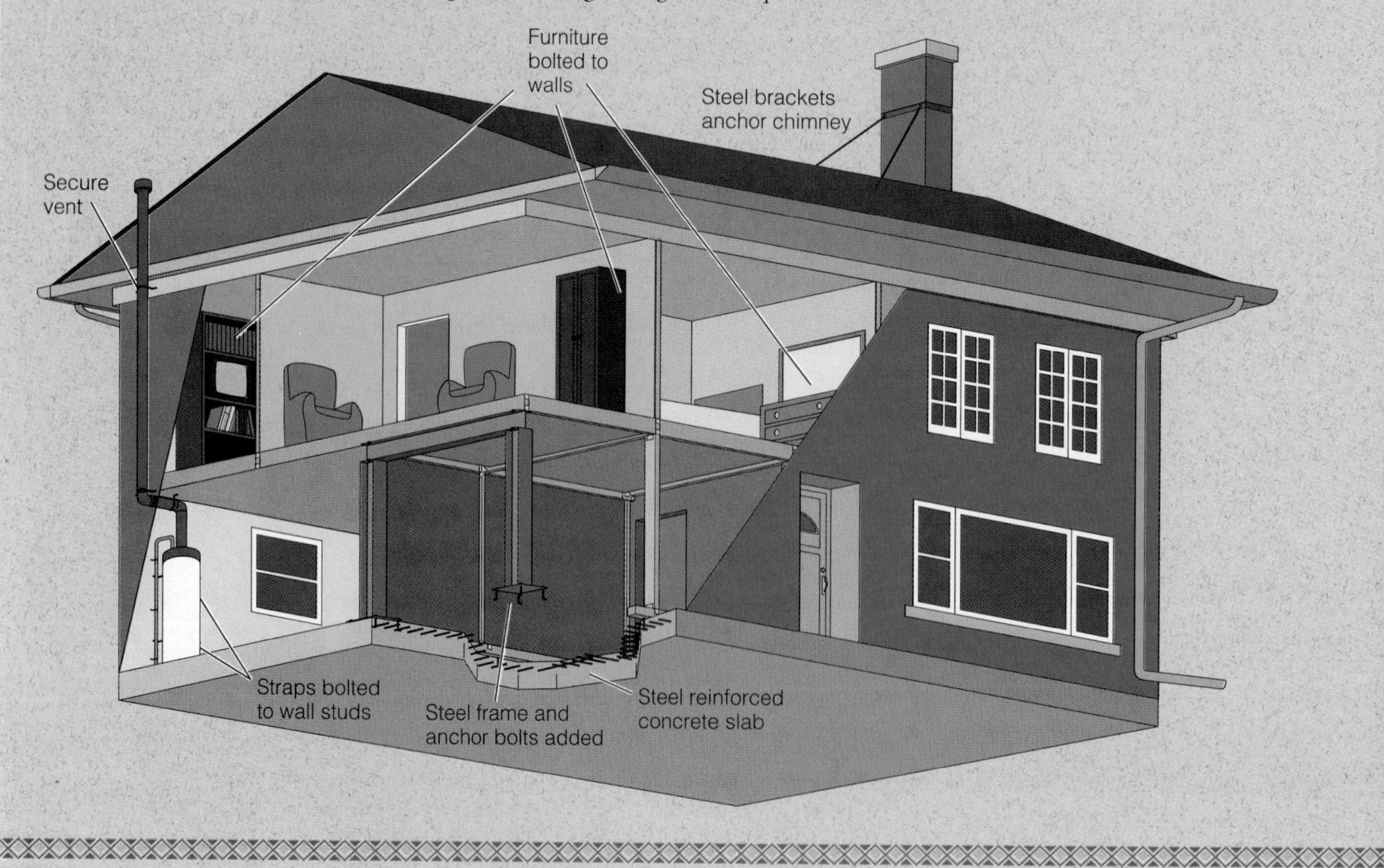

quake of greater than 7.0 magnitude, the building should not collapse, although it may later have to be demolished.

Many factors enter into the design of an earthquake-resistant structure, but the most important is that the building be tied together; that is, the foundation, walls, floors, and roof should all be joined together to create a structure that can withstand both horizontal and vertical shaking (▶ Figure 1). Almost all of the structural failures resulting from earthquake ground movement have occurred at weak connections, where the various parts of a structure were not securely tied together (▶ Figure 2).

The size and shape of a building can also affect its resistance to earthquakes (▶ Figure 3). Rectangular box-shaped buildings are inherently stronger than those of irregular size or shape because different parts of an irregular building may sway at different rates, increasing the stress and likelihood of structural failure (Figure 3b). Buildings with

continued

▶ **FIGURE 2** During the 1971 San Fernando, California earthquake, the Olive View Hospital's stair tower broke away from the main building. The hospital was built to federal earthquake standards, but still suffered major damage.

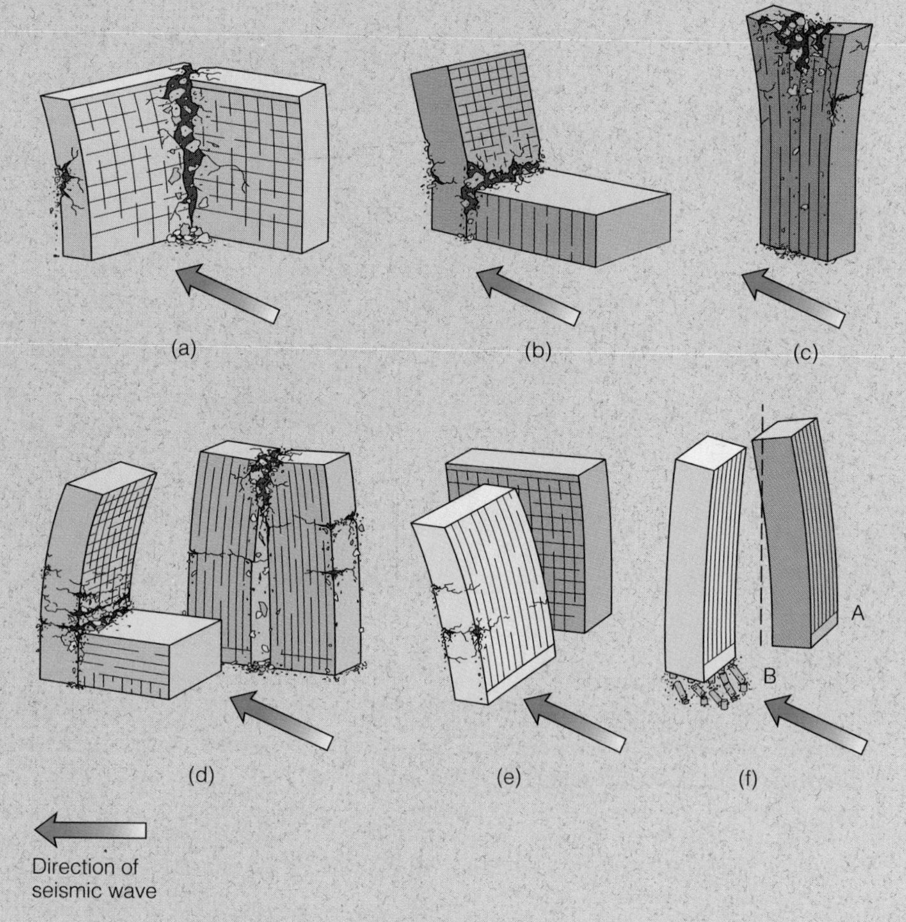

▶ **FIGURE 3** The effects of ground shaking on various tall buildings of differing shapes. (*a*) Damage will occur if two wings of a building are joined at right angles and experience different motions. (*b*) Buildings of different heights will sway differently leading to damage at the point of connection. (*c*) Shaking increases with height and is greatest at the top of a building. (*d*) Closely spaced buildings may crash into each other due to swaying. (*e*) A building whose long axis is parallel to the direction of the seismic waves will sway less than a building whose axis is perpendicular. (*f*) Two buildings of different design will behave differently even when subjected to the same shaking conditions. Building A sways as a unit and remains standing while building B whose first story is composed of tall columns collapses because most of the swaying takes place in the "soft" first story.

continued

open or unsupported first stories are particularly susceptible to damage. Some reinforcement must be done or collapse is a distinct possibility.

Tall buildings, such as skyscrapers, must be designed so that a certain amount of swaying or flexing can occur, but not so much that they touch neighboring buildings during swaying (Figure 3d). If a building is brittle and does not give, it will crack and fail. In addition to designed flexibility, engineers must make sure that a building does not vibrate at the same frequency as the ground does during an earthquake. When that happens, the force applied by the seismic waves at ground level is multiplied several times by the time they reach the top of the building (Figure 3c). This condition is particularly troublesome in areas of poorly consolidated sediment (▶ Figure 4). Fortunately, buildings can be designed so that they will sway at a different frequency than the ground.

What about structures built many years ago? Almost every city and town has older single and multistory structures, constructed of unreinforced brick masonry, poor-quality concrete, and rotting or decaying wood. Just as in new buildings, the most important thing that can be done to increase the stability and safety of older structures is to tie the different components of each building together. This can be done by adding a steel frame to unreinforced parts of a building such as a garage, bolting the walls to the foundation, adding reinforced beams to the exterior, and using beam and joist connectors whenever possible. Although such modifications are expensive, they are usually cheaper than having to replace a building that was destroyed by an earthquake.

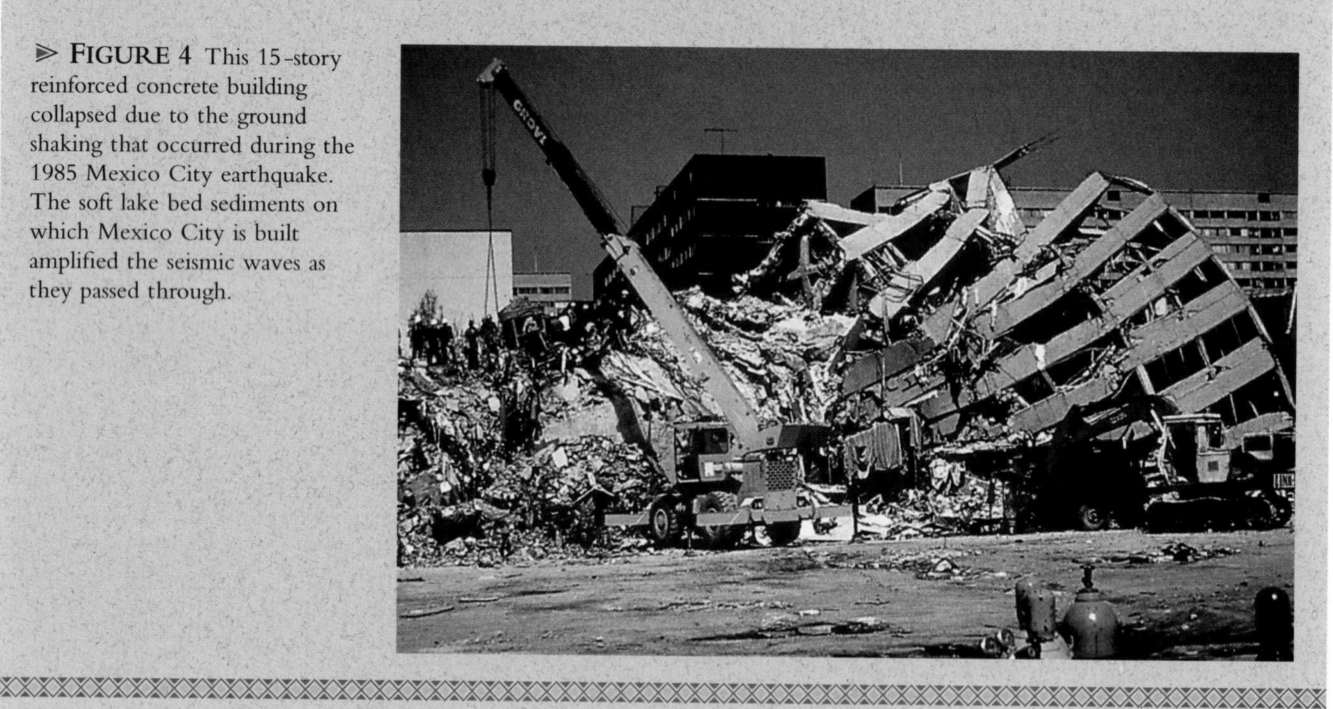

▶ **FIGURE 4** This 15-story reinforced concrete building collapsed due to the ground shaking that occurred during the 1985 Mexico City earthquake. The soft lake bed sediments on which Mexico City is built amplified the seismic waves as they passed through.

magnitude Northridge earthquake resulted in only 61 deaths. Both earthquakes occurred in densely populated regions, but in India the brick and stone buildings could not withstand ground shaking; most collapsed entombing their occupants.

Fire

In many earthquakes, particularly in urban areas, fire is a major hazard. Almost 90% of the damage done in the 1906 San Francisco earthquake was caused by fire. The shaking severed many of the electrical and gas lines, which touched off flames and started numerous fires all over the city. Because water mains were ruptured by the earthquake, there was no effective way to fight the fires. Hence, they raged out of control for three days, destroying much of the city. During the 1989 Loma Prieta earthquake, a fire broke out in the Marina district of San Francisco (▶ Figure 9-21) but was contained within a small area. In contrast to 1906, in

▶ **FIGURE 9-21** San Francisco Marina district fire caused by broken gas lines during the 1989 Loma Prieta earthquake.

1989 San Francisco had a system of valves throughout its water and gas pipeline system so that lines could be isolated from breaks.

During the September 1, 1923, earthquake in Japan, fires destroyed 71% of the houses in Tokyo and practically all the houses in Yokohama. In all, a total of 576,262 houses were completely destroyed by fire, and 143,000 people died, many as a result of the fire. A horrible example occurred in Tokyo where thousands of people gathered along the banks of the Sumida River to escape the raging fires. Suddenly, a firestorm swept over the area, killing more than 38,000 people. The fires from this earthquake were so devastating because most of the buildings were constructed of wood; many fires were started by chemicals and fanned by 20 km/hr winds.

Tsunami

Seismic sea waves or **tsunami** are destructive sea waves that are usually produced by earthquakes but can also be caused by submarine landslides or volcanic eruptions (see the Prologue to Chapter 1). Tsunami are popularly called tidal waves, although they have nothing to do with tides. Instead, most tsunami result from the sudden movement of the sea floor, which sets up waves within the water that travel outward, much like the ripples that form when a stone is thrown into a pond.

Tsunami travel at speeds of several hundred km/hr and are commonly not felt in the open ocean because their wave height is usually less than 1 m and the distance between wave crests is typically several hundred kilometers. When tsunami approach shorelines, however, the waves slow down and water piles up to heights of up to 65 m (▶ Figure 9-22).

Following a 1946 tsunami that killed 154 people and caused $25 million in property damage in Hawaii, the U.S. Coast and Geodetic Survey established a Tsunami Early Warning System in Honolulu, Hawaii, in an attempt to minimize tsunami devastation. This system combines seismographs and other instruments that can detect earthquake-generated waves. Whenever a strong earthquake occurs anywhere within the Pacific basin, its location is determined, and instruments are checked to see if a tsunami has been generated. If it has, a warning is sent out to evacuate people from low-lying areas that may be affected (▶ Figure 9-23).

Ground Failure

Earthquake-triggered landslides are particularly dangerous in mountainous regions and have been responsible for tremendous damage and many deaths. The 1959 earthquake in Madison Canyon, Montana, for example, generated a major rock slide (▶ Figure 9-24), while the 1970 Peru earth-

▶ **FIGURE 9-22** As a tsunami crashes into the street behind them, residents of Hilo, Hawaii, run for their lives. This tsunami was generated by an earthquake in the Aleutian Islands and resulted in massive property damage to Hilo and the deaths of 154 people.

▶ **FIGURE 9-23** Tsumani travel times within the Pacific Ocean basin to Honolulu, Hawaii.

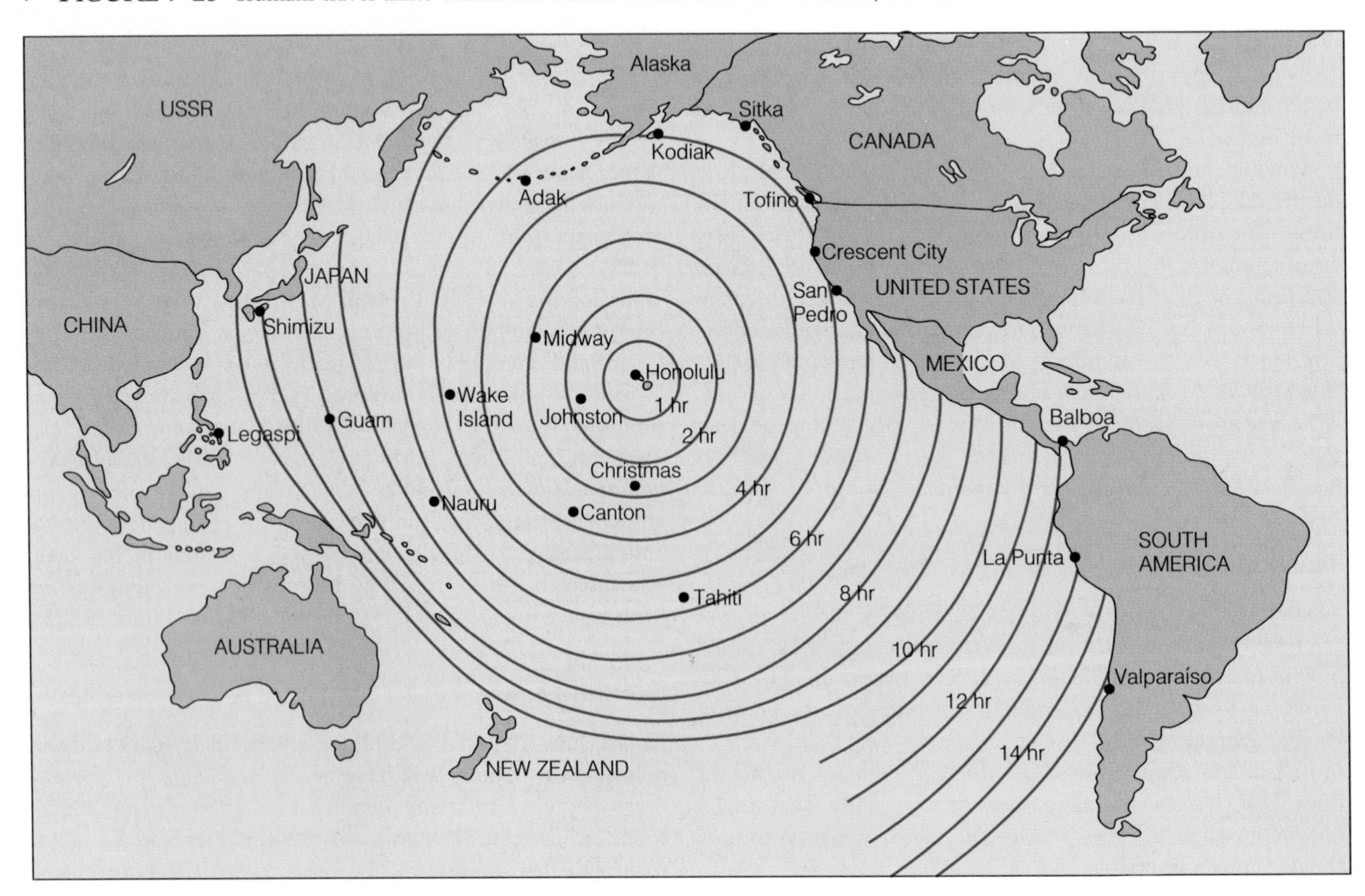

Perspective 9-2

PALEOSEISMOLOGY

Paleoseismology is the study of prehistoric earthquakes and has been an emerging field in geology during the past 15 years, particularly in North America. Data from a variety of sources has convinced an interdisciplinary group of scientists that a shallow-focus earthquake of at least magnitude 7 occurred beneath Seattle, Washington, less than 1,100 years ago. In a point not lost on officials, they noted the catastrophic effects a similar-sized earthquake would have if it occurred in the same area today.

The first link in the chain of evidence for a paleoearthquake came from the discovery of a marine terrace that had been uplifted some 7 m at Restoration Point, 5 km west of Seattle. Carbon 14 analysis of peat within sediments of the terrace indicates that uplift occurred between 500 and 1,700 years ago. Carbon 14 dating of other sites to the north, south, and east also indicates a sudden uplift in the area within the same time period. The amount of uplift suggests to geologists a magnitude 7 or greater earthquake.

Because earthquakes cause tsunami, geologists looked for evidence that a tsunami had occurred, and found it in the form of unusual sand layers in nearby tidal marsh deposits. Carbon 14 dating of organic matter associated with the sands yielded an age between 850 and 1,250 years ago, well within the time period in which the terraces were uplifted.

Geologists also found evidence of rock avalanches in the Olympic Mountains that dammed streams, thereby forming lakes. Drowned trees in the lakes were dated as having died between 1,000 and 1,300 years ago, again fitting in nicely with the date of the paleoearthquake.

One of the final pieces of evidence is the deposits found on the bottom of Lake Washington. The earthquake apparently caused the bottom sediment on the lake to be resuspended and move downslope as a turbidity current. Dating of these sediments indicates they were deposited between 940 and 1,280 years ago, consistent with their being caused by an earthquake.

All of the evidence points to a large (magnitude 7 or greater) shallow-focus earthquake occurring in the Seattle area about 1,000 years ago. As John Adams of the Geological Survey of Canada pointed out, there has not been time to assess the implications of an earthquake of that size occurring today in the Seattle area, but it should be noted that very little has been done to date in planning for a major earthquake in the area. Metropolitan Seattle has a population of 2.5 million people, and its entire port area is built on fill that would probably be hard hit by an earthquake. Furthermore, most of Seattle's schools, hospitals, utilities, and fire and police stations are not built to withstand a strong earthquake.

FIGURE 9–24 On August 17, 1959, an earthquake started a landslide that blocked the Madison River in Montana and created Earthquake Lake. The slide began on one side of the valley, demolished a campsite at the valley bottom, killing approximately 26 people, completely filled the river forming an earthen dam, and continued up the opposite valley slope. This view shows the slide in the background and Earthquake Lake in the foreground.

quake caused an avalanche that destroyed the town of Yungay (see the Prologue to Chapter 14). Most of the 100,000 deaths from the 1920 earthquake in Gansu, China, resulted when cliffs composed of loess (wind-deposited silt) collapsed. More than 20,000 people were killed when two-thirds of the town of Port Royal, Jamaica, slid into the sea following an earthquake on June 7, 1692.

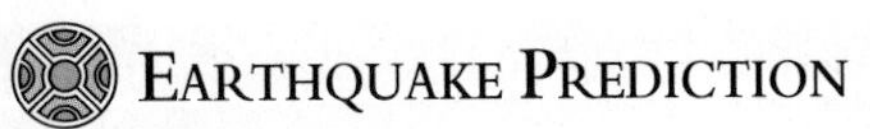

EARTHQUAKE PREDICTION

Can earthquakes be predicted? A successful prediction must include a time frame for the occurrence of an earthquake, its location, and its strength. In spite of the tremendous amount of information geologists have gathered about the cause of earthquakes, successful predictions are still quite

➢ **FIGURE 9-25** (*a*) A 1969 seismic risk map for the United States based on intensity data collected by the U.S. Coast and Geodetic Survey. (*b*) A 1976 seismic risk map based on a 90% probability that the indicated horizontal acceleration of the ground during an earthquake is not likely to be exceeded in 50 years. The horizontal ground acceleration numbers shown in the key are in m/sec^2. Therefore, the higher the number, the greater the hazard.

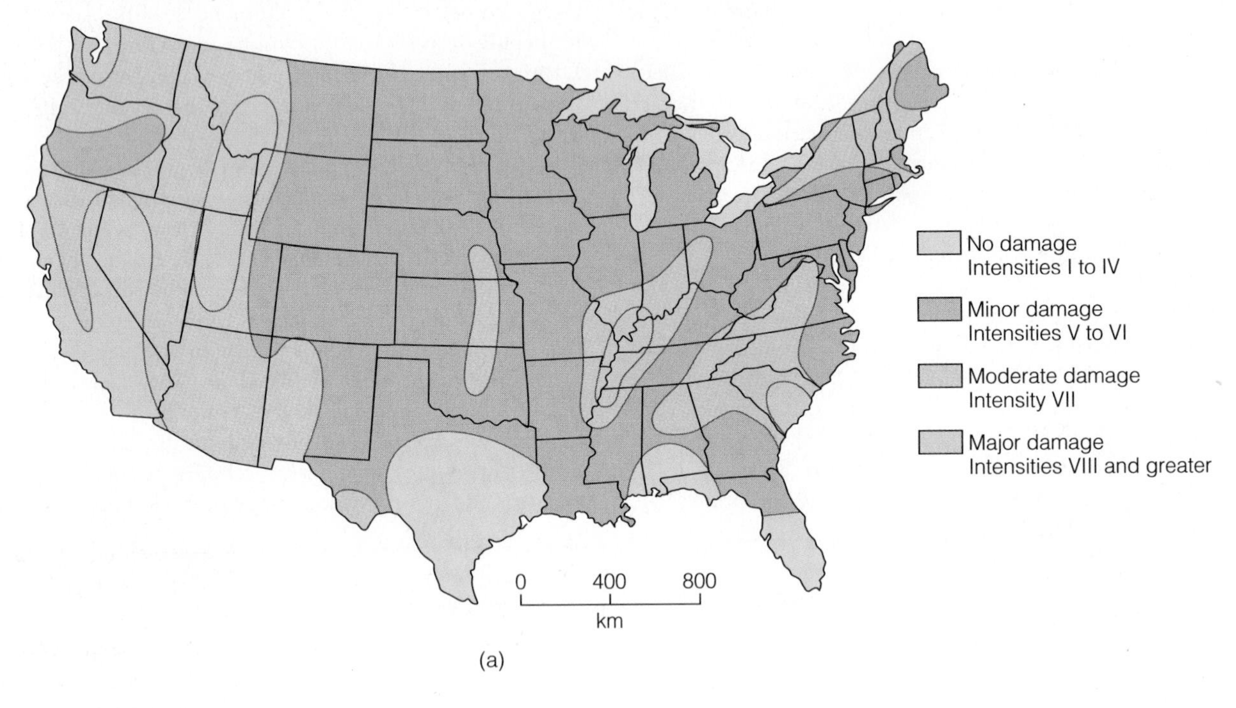

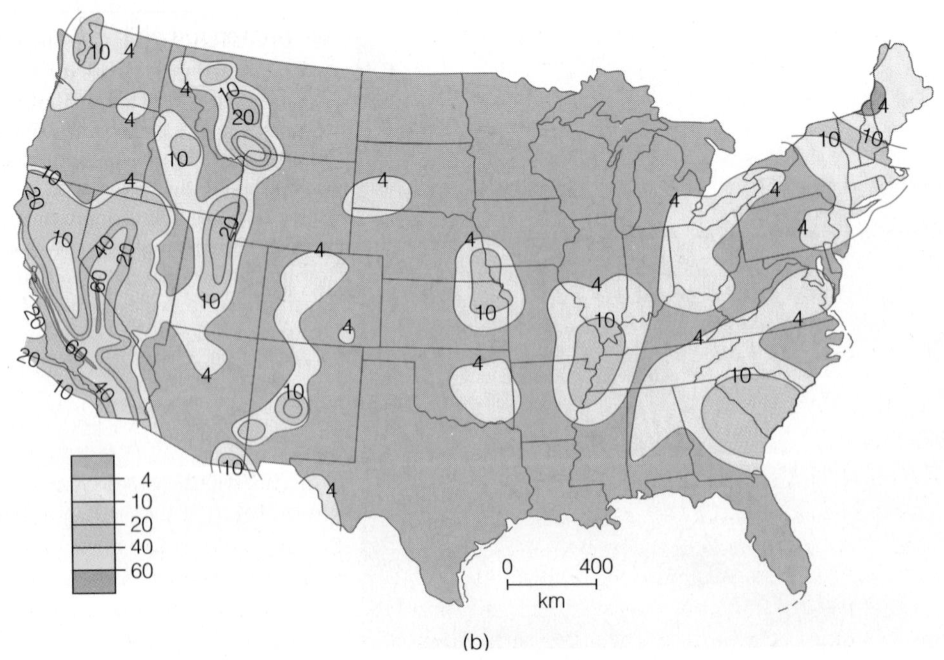

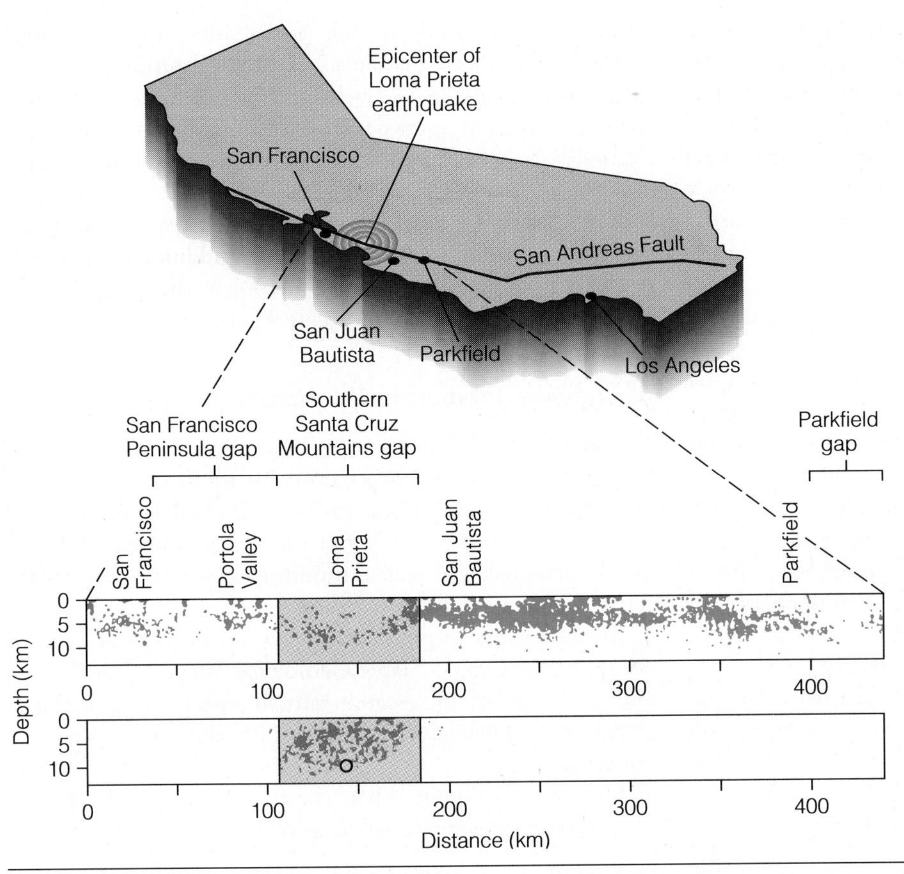

▶ **FIGURE 9-26** Three seismic gaps are evident in this cross section along the San Andreas fault from north of San Francisco to south of Parkfield. The first is between San Francisco and Portola Valley, the second near Loma Prieta Mountain, and the third southeast of Parkfield. The top section shows the epicenters of earthquakes between January 1969 and July 1989. The bottom section shows the southern Santa Cruz Mountains gap after it was filled by the October 17, 1989, Loma Prieta earthquake (open circle) and its aftershocks.

rare. Nevertheless, if reliable predictions can be made, they can greatly reduce the number of deaths and injuries.

From an analysis of historic records and the distribution of known faults, **seismic risk maps** can be constructed that indicate the likelihood and potential severity of future earthquakes based on the intensity of past earthquakes (▶ Figure 9-25). Although such maps cannot predict when the next major earthquake will occur, they are useful in helping people plan for future earthquakes (see Perspective 9-2).

Earthquake Precursors

Studies conducted over the past several decades indicate that most earthquakes are preceded by both short-term and long-term changes within the Earth. Such changes are called **precursors**.

One long-range prediction technique used in seismically active areas involves plotting the location of major earthquakes and their aftershocks to detect areas that have had major earthquakes in the past but are currently inactive. Such regions are locked and not releasing energy. Nevertheless, pressure is continuing to accumulate in these regions due to plate motions, making these *seismic gaps* prime locations for future earthquakes. Several seismic gaps along the San Andreas fault have the potential for future major earthquakes (▶ Figure 9-26). A major earthquake that damaged Mexico City in 1985 occurred along a seismic gap in the convergence zone along the west coast of Mexico (see the Prologue to Chapter 12).

Changes in elevation and tilting of the land surface have frequently preceded earthquakes and may be warnings of impending quakes. Extremely slight changes in the angle of the ground surface can be measured by tiltmeters. Tiltmeters have been placed on both sides of the San Andreas fault to measure tilting of the ground surface that is thought to result from increasing pressure in the rocks. Data from measurements in central California indicate significant tilting occurred immediately preceding small earthquakes. Furthermore, extensive tiltmeter work performed in Japan prior to the 1964 Niigata earthquake clearly showed a relationship between increased tilting and the main shock. Though more research is needed, such changes appear to be useful in making short-term earthquake predictions.

Other earthquake precursors include fluctuations in the water level of wells and local changes in the Earth's magnetic field and the electrical resistance of the ground. These fluctuations are thought to result from changes in the amount of pore space in rocks due to increasing pressure. A change in animal behavior prior to an earthquake also is frequently mentioned. It may be that animals are sensing small and subtle changes in the Earth prior to a quake that humans simply do not sense.

The Chinese used all of the precursors just mentioned, except seismic gaps, to successfully predict a large earthquake in Haicheng on February 4, 1975. The earthquake had a magnitude of 7.3 and destroyed hundreds of buildings but claimed very few lives because most people had been evacuated from the buildings and were outdoors when it occurred. Although this was not the first successful earthquake prediction, it was the first to predict a major earthquake and thus saved thousands of lives.

Another possible earthquake precursor was discovered following the 1989 Loma Prieta earthquake. Electrical engineers at Stanford University, California, noticed that the amplitude of ultralow frequency radio waves increased about three hours before the earthquake. Furthermore, they noticed that the background radio noise for all frequencies abruptly increased 12 days before the earthquake and then suddenly decreased one day before the tremor hit. At this time it is not known why such a change should occur, but it is hoped that it may prove useful in short-term prediction of future earthquakes.

Dilatancy Model. Many of the precursors just discussed can be related to the **dilatancy model,** which is based on changes occurring in rocks subjected to very high pressures. Laboratory experiments have shown that rocks undergo an increase in volume, known as dilatancy, just before rupturing. As pressure builds in rocks along faults, numerous small cracks are produced that alter the physical properties of the rocks. Water enters the cracks and increases the fluid pressure; this further increases the volume of the rocks and decreases their inherent strength until failure eventually occurs, producing an earthquake.

The dilatancy model is consistent with many earthquake precursors (▶ Figure 9-27). Although additional research is needed, it appears that this model has the potential for predicting earthquakes under certain circumstances.

Earthquake Prediction Programs

Currently, only four nations—the United States, Japan, Russia, and China—have government-sponsored earthquake prediction programs. These programs include laboratory and field studies of the behavior of rocks before, during, and after major earthquakes as well as monitoring activity along major active faults.

Most earthquake prediction work in the United States is done by the United States Geological Survey (USGS) and involves a variety of research into all aspects of earthquake-related phenomena. One of the more ambitious programs undertaken by the USGS was the Parkfield earthquake prediction experiment. Over the past 130 years, moderate-sized earthquakes have occurred on an average of every 21

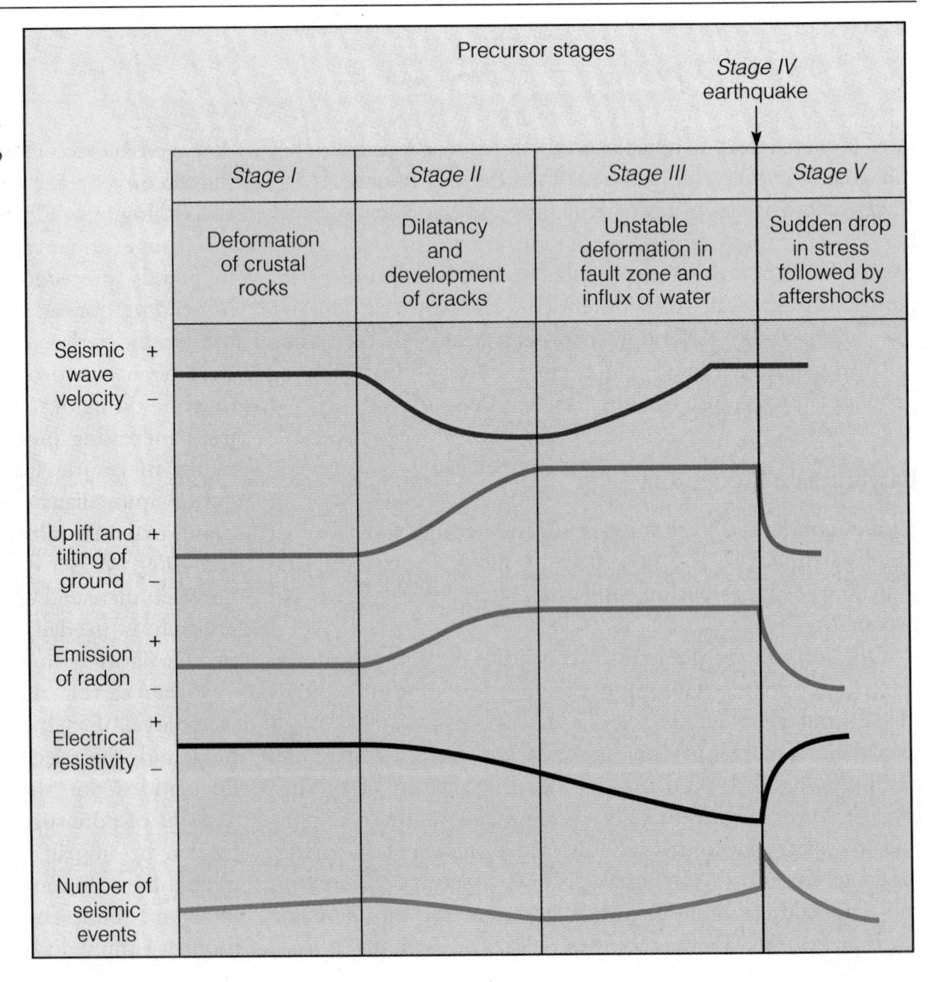

▶ **FIGURE 9-27** The relationship between dilatancy and various other earthquake precursors. The onset of dilatancy matches a change in each of the precursors illustrated. For example, a drop in seismic wave velocity corresponds to the onset of dilatancy and the development of cracks.

to 22 years along a 24 km segment of the San Andreas fault at Parkfield, California. Based on the regularity of these earthquakes and the fact they have all been very similar, the USGS forecast that another moderate-sized earthquake would occur in this region by January 1, 1993. To study earthquake precursors and assess the possibility of short-term predictions of moderate-sized earthquakes, the USGS installed a variety of instruments to monitor conditions along the San Andreas fault in the Parkfield area. When the predicted deadline passed without the promised earthquake occurring, some geologists questioned whether earthquakes could ever be predicted with any certainty. What the Parkfield failure showed, however, was that the Earth's interior is far more complex than had previously been assumed and that accurate predictions of earthquakes are going to be more difficult than geologists originally thought.

The Chinese earthquake prediction program may be one of the most ambitious in the world, which is understandable considering their long history of destructive earthquakes. The Chinese program on earthquake prediction was initiated soon after two large earthquakes occurred at Xingtai (300 km southwest of Beijing) in 1966. The Chinese program includes extensive study and monitoring of all possible earthquake precursors. In addition, the Chinese also emphasize changes in phenomena that can be observed by seeing and hearing without the use of sophisticated instruments. The Chinese have had remarkable success in predicting earthquakes, particularly in the short term, such as the 1975 Haicheng earthquake. They failed, however, to predict the devastating 1976 Tangshan earthquake that killed at least 242,000 people.

Great strides are being made toward dependable, accurate earthquake predictions, and studies are underway to assess public reactions to long-, medium-, and short-term earthquake warnings. Unless short-term warnings are actually followed by an earthquake, most people will probably ignore the warnings as they frequently do now for hurricanes, tornadoes, and tsunami (see Perspective 9-3). Perhaps the best we can hope for is that people will take measures to minimize their risk from the next major earthquake (◉ Table 9-4).

TABLE 9-4 What You Can Do to Prepare for an Earthquake

Anyone who lives in an area that is subject to earthquakes or who will be visiting or moving to such an area can take certain precautions to reduce the risks and losses resulting from an earthquake.

Before an earthquake:

1. Become familiar with the geologic hazards of the area where you live and work.
2. Make sure your house is securely attached to the foundation by anchor bolts and that the walls, floors, and roof are all firmly connected together.
3. Heavy furniture such as bookcases should be bolted to the walls; semiflexible natural gas lines should be used so that they can give without breaking; water heaters and furnaces should be strapped and the straps bolted to wall studs to prevent gas-line rupture and fire. Brick chimneys should have a bracket or brace that can be anchored to the roof.
4. Maintain a several-day supply of fresh water and canned foods, and keep a fresh supply of flashlight and radio batteries as well as a fire extinguisher.
5. Maintain a basic first-aid kit, and have a working knowledge of first-aid procedures.
6. Learn how to turn off the various utilities at your house.
7. Above all, have a course of action planned for when an earthquake strikes.

During an earthquake:

1. Remain calm and avoid panic.
2. If you are indoors, get under a desk or table if possible, or stand in an interior doorway or room corner as these are the structurally strongest parts of a room; avoid windows and falling debris.
3. In a tall building, do not rush for the stairwells or elevators.
4. In an unreinforced or other hazardous building, it may be better to get out of the building rather than stay in it. Be on the alert for fallen power lines and the possibility of falling debris.
5. If you are outside, get to an open area away from buildings if possible.
6. If you are in an automobile, stay in the car, and avoid tall buildings, overpasses, and bridges if possible.

After an earthquake:

1. If you are uninjured, remain calm and assess the situation.
2. Help anyone who is injured.
3. Make sure there are no fires or fire hazards.
4. Check for damage to utilities and turn off gas valves if you smell gas.
5. Use your telephone only for emergencies.
6. Do not go sightseeing or move around the streets unnecessarily.
7. Avoid landslide and beach areas.
8. Be prepared for aftershocks.

A Predicted Earthquake That Didn't Occur

December 3, 1990, passed without incident when a major earthquake that had been predicted publicly for a portion of the Midwest failed to materialize. For months, a five-state region overlying the New Madrid fault zone braced for a potentially devastating earthquake (▶ Figure 1).

During the months leading up to December 3, insurance salespeople did a brisk business selling earthquake insurance to homeowners and businesses; entrepreneurs cashed in on the sale of such earthquake-related items as T-shirts, survival kits, and gas-line shutoff safety devices; seminars on earthquake preparedness and survival drew large crowds; and public officials reviewed disaster plans and coordinated emergency relief efforts, while schools practiced earthquake drills. So great was the concern that an earthquake would occur as predicted, many school districts canceled classes and numerous businesses closed for several days.

The reason for such massive preparation and media attention was a prediction made by Iben Browning. He has a Ph.D. in physiology, genetics, and bacteriology and was previously best known for his work on climates, claims to have correctly predicted the dates of several major earthquakes and volcanic eruptions.

Based on the apparent accuracy of his previous predictions, Browning's New Madrid prediction was taken very seriously and received wide media coverage. He predicted that there was a 50% chance of a magnitude 6.5 to 7.5 earthquake occurring somewhere within the New Madrid fault zone on December 3, 1990 (plus or minus a day). He also predicted for the same time that there was a lesser chance of a similar earthquake occurring along California's San Andreas or Hayward faults and an even greater chance of an 8.2 magnitude earthquake striking Tokyo. None of the predicted earthquakes occurred.

All of Browning's predictions are based on tidal forces. When the Earth, Moon, and Sun are aligned in a straight line, they exert greater than normal gravitational forces (although the forces are still relatively weak) that some think could trigger fault movement. Such a hypothesis is not new. Whenever the various planets are aligned as a result of their orbits around the Sun, doomsayers are always predicting some type of natural disaster will occur, and it never does.

As for Browning's claimed accuracy for predicting previous earthquakes, it seems that none of his predictions can be verified, or his predictions were so vague as to be of no value. In the case of the 1985 Loma Prieta earthquake, for example, 10 days before the earthquake occurred, all he said was that "There will probably be several earthquakes around the world. Richter 6-plus magnitude. . . ."* As it turns out, Browning's predictions of earthquakes corresponding to tidal forces rate no better than random guessing. Earthquakes are the result of complex interactions within the Earth and occur under varied geologic conditions. Consequently, no one factor can be used to predict when and where an earthquake will occur.

Seismologists do admit that based on past earthquake activity in the New Madrid fault zone area, there is a high probability of another major earthquake in the area within the foreseeable future. Yet exactly when that will be, no geoscientist is willing to predict because far too many complex variables are involved.

*Quoted in Spence, W. *et al.* 1993. Responses to Iben Browning's prediction of a 1990 New Madrid, Missouri, earthquake. *U.S. Geological Survey Circular 1083.*

▶ **FIGURE 1** A devastating earthquake was predicted to occur on December 3, 1990, somewhere within the five-state region that overlies the New Madrid fault zone. Luckily, the earthquake did not happen.

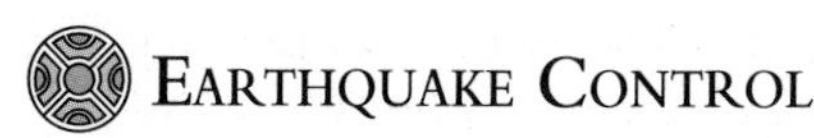

Earthquake Control

If earthquake prediction is still in the future, can anything be done to control earthquakes? Because of the tremendous forces involved, humans are certainly not going to be able to prevent earthquakes. But there may be ways to dissipate the destructive energy of major earthquakes by releasing it in small amounts that will not cause extensive damage.

During the early to mid-1960s, Denver experienced numerous small earthquakes. This was surprising because Denver had not been prone to earthquakes in the past. In 1962, David M. Evans, a geologist, suggested that the earthquakes in Denver were directly related to the injection of contaminated waste water into a disposal well 3,674 m deep at the Rocky Mountain Arsenal, northeast of Denver. The U.S. Army initially denied that there was any connection, but a USGS study concluded that the pumping of waste fluids into the disposal well was the cause of the earthquakes.

▶ Figure 9-28 shows the relationship between the average number of earthquakes in Denver per month and the average amount of waste fluids injected into the disposal well per month. Obviously, a high degree of correlation between the two exists, and the correlation is particularly convincing considering that during the time when no waste fluids were injected, earthquake activity decreased dramatically. The geology of the area consists of highly fractured gneiss overlain by sedimentary rocks. When water was pumped into these fractures, it decreased the friction on opposite sides of the fractures and, in essence, lubricated them so that movement occurred, causing the earthquakes that Denver experienced.

Experiments conducted in 1969 at an abandoned oil field near Rangely, Colorado, confirmed the arsensal hypothesis.

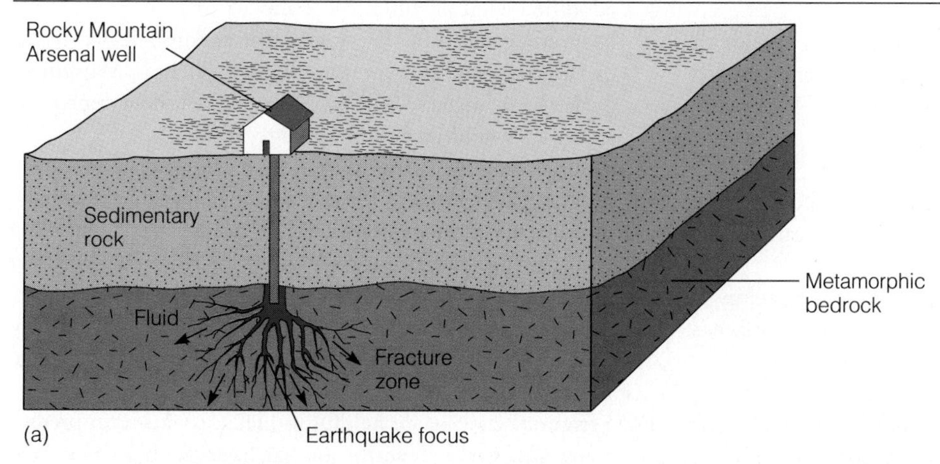

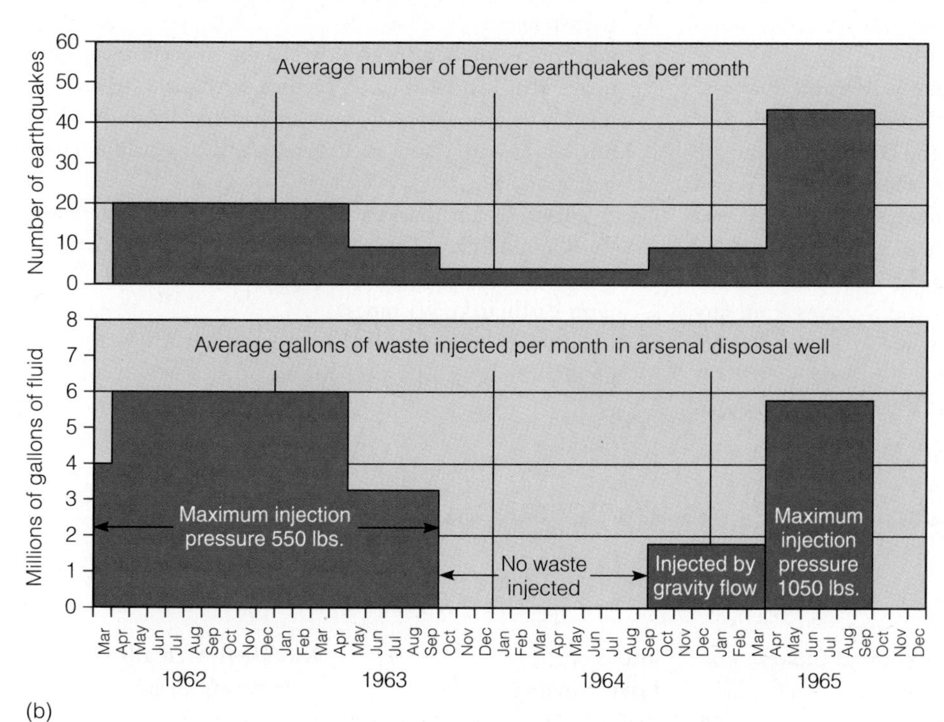

▶ **FIGURE 9-28** (*a*) A block diagram of the Rocky Mountain Arsenal well and the underlying geology. (*b*) A graph showing the relationship between the amount of waste injected into the well per month and the average number of Denver earthquakes per month. There have not been any significant earthquakes in Denver since injection of waste water into the disposal well ceased in 1965.

Water was pumped in and out of abandoned oil wells, the pore-water pressure in these wells was measured, and seismographs were installed in the area to measure any seismic activity. Monitoring showed that small earthquakes were occurring in the area when fluid was injected and that earthquake activity declined when the fluids were pumped out. What the geologists were doing was starting and stopping earthquakes at will, and the relationship between pore-water pressures and earthquakes was established.

Based upon these results, some geologists have proposed that fluids be pumped into the locked segments of active faults to cause small- to moderate-sized earthquakes. They think that this would relieve the pressure on the fault and prevent a major earthquake from occurring. While this plan is intriguing, it also has many potential problems. For instance, there is no guarantee that only a small earthquake might result. Instead a major earthquake might occur, causing tremendous property damage and loss of life. Who would be responsible? Certainly, a great deal more research is needed before such an experiment is performed, even in an area of low population density.

It appears that until such time as earthquakes can be accurately predicted or controlled, the best means of defense is careful planning and preparation (Table 9-4).

Chapter Summary

1. Earthquakes are vibrations of the Earth caused by the sudden release of energy, usually along a fault.
2. The elastic rebound theory states that pressure builds in rocks on opposite sides of a fault until the inherent strength of the rocks is exceeded and rupture occurs. When the rocks rupture, stored energy is released as they snap back to their original position.
3. Seismology is the study of earthquakes. Earthquakes are recorded on seismographs, and the record of an earthquake is a seismogram.
4. The focus of an earthquake is the point where energy is released. Vertically above the focus on the Earth's surface is the epicenter.
5. Most earthquakes occur within seismic belts. Approximately 80% of all earthquakes occur in the circum-Pacific belt, 15% within the Mediterranean-Asiatic belt, and the remaining 5% mostly in the interior of plates or along oceanic spreading ridge systems.
6. The two types of body waves are P-waves and S-waves. Both travel through the Earth, although S-waves do not travel through liquids. P-waves are the fastest waves and are compressional, while S-waves are shear. Surface waves travel along or just below the Earth's surface. The two types of surface waves are Rayleigh and Love waves.
7. The epicenter of an earthquake can be located by the use of a time-distance graph of the P- and S-waves from any given distance. Three seismographs are needed to locate the epicenter of an earthquake.
8. Intensity is a measure of the kind of damage done by an earthquake and is expressed by values from I to XII in the Modified Mercalli Intensity Scale.
9. Magnitude measures the amount of energy released by an earthquake and is expressed in the Richter Magnitude Scale. Each increase in the magnitude number represents about a 30-fold increase in energy released.
10. Ground shaking is the most destructive of all earthquake hazards. The amount of damage done by an earthquake depends upon the geology of the area, the type of building construction, the magnitude of the earthquake, and the duration of shaking.
11. Tsunami are seismic sea waves that are usually produced by earthquakes. They can do a tremendous amount of damage to coastlines.
12. Seismic risk maps are helpful in making long-term predictions about the severity of earthquakes based on past occurrences.
13. Earthquake precursors are any changes preceding an earthquake that can be used to predict earthquakes. Precursors include seismic gaps, changes in surface elevation, tilting, fluctuations in water well levels, and anomalous animal behavior.
14. A variety of earthquake research programs are underway in the United States, Japan, Russia, and China. Studies indicate that most people would probably not heed a short-term earthquake warning.
15. Fluid injection into locked segments of an active fault holds great promise as a possible means of earthquake control.

Important Terms

aftershock
dilatancy model
earthquake
elastic rebound theory
epicenter
focus
intensity
Love wave (L-wave)
magnitude
Modified Mercalli Intensity Scale
precursor

P-wave
Rayleigh wave (R-wave)
Richter Magnitude Scale
seismic risk map
seismograph
seismology
S-wave
time-distance graph
tsunami

REVIEW QUESTIONS

1. According to the elastic rebound theory:
   a. ____ earthquakes originate deep within the Earth;
   b. ____ earthquakes originate in the asthenosphere where rocks are plastic;
   c. ____ earthquakes occur where the strength of the rock is exceeded;
   d. ____ rocks are elastic and do not rebound to their former position;
   e. ____ none of these.
2. With few exceptions, the most destructive earthquakes are:
   a. ____ shallow focus;
   b. ____ intermediate focus;
   c. ____ deep focus;
   d. ____ answers (a) and (b);
   e. ____ answers (b) and (c).
3. The majority of all earthquakes occur in the:
   a. ____ Mediterranean-Asiatic belt;
   b. ____ interior of plates;
   c. ____ circum-Atlantic belt;
   d. ____ circum-Pacific belt;
   e. ____ along spreading ridges.
4. Body waves are:
   a. ____ P-waves;
   b. ____ S-waves;
   c. ____ Rayleigh waves;
   d. ____ answers (b) and (c);
   e. ____ answers (a) and (b).
5. The fastest of the four seismic waves is:
   a. ____ P;
   b. ____ S;
   c. ____ Rayleigh;
   d. ____ Love;
   e. ____ tsunami.
6. An epicenter is:
   a. ____ the location where rupture begins;
   b. ____ the point on the Earth's surface vertically above the focus;
   c. ____ the same as the hypocenter;
   d. ____ the location where energy is released;
   e. ____ none of these.
7. What is the minimum number of seismographs needed to determine an earthquake's epicenter?
   a. ____ 1;
   b. ____ 2;
   c. ____ 3;
   d. ____ 4;
   e. ____ 5.
8. A qualitative assessment of the kinds of damage done by an earthquake is expressed by:
   a. ____ seismicity;
   b. ____ dilatancy;
   c. ____ magnitude;
   d. ____ intensity;
   e. ____ none of these.
9. How much more energy is released by a magnitude 5 earthquake than by one of magnitude 2?
   a. ____ 2.5 times:
   b. ____ 3 times;
   c. ____ 30 times;
   d. ____ 900 times;
   e. ____ 27,000 times.
10. Which of the following usually causes the greatest amount of damage and loss of life?
    a. ____ fire;
    b. ____ tsunami;
    c. ____ ground shaking;
    d. ____ liquefaction;
    e. ____ landslides.
11. A tsunami is a:
    a. ____ measure of the energy released by an earthquake;
    b. ____ seismic sea wave;
    c. ____ precursor to an earthquake;
    d. ____ locked portion of a fault;
    e. ____ seismic gap.
12. How does the elastic rebound theory explain how energy is released during an earthquake?
13. Describe how a seismograph works.
14. What is the difference between body waves and surface waves?
15. What is the difference between the focus and the epicenter of an earthquake?
16. How is the epicenter of an earthquake determined?
17. What is the relationship between plate boundaries and earthquakes?
18. What is the relationship between plate boundaries and focal depth?
19. Explain the difference between intensity and magnitude and between the Modified Mercalli Intensity Scale and the Richter Magnitude Scale.
20. Why is ground shaking so destructive during an earthquake?
21. Explain how tsunami are produced and why they are so destructive.
22. Why are seismic risk maps useful to planners?
23. How can earthquake precursors be used to predict earthquakes?
24. What is the dilatancy model? How does it help explain how earthquake precursors are related?

POINTS TO PONDER

1. Some geologists think that by pumping fluids into locked segments of active faults, small- to moderate-sized earthquakes can be generated. These earthquakes would relieve the built-up pressure along the fault and thus prevent a major earthquake from occurring. Discuss whether this idea is really feasible. Are there any guarantees that the earthquake will be limited to only the locked segments along the fault? Can geologists be sure that the resulting earthquake will not be a major one, causing tremendous property damage and loss of life? Considering the social, political, and economic consequences, do you think such an experiment will ever be tested on an active fault?
2. From the following arrival times of P- and S-waves and the graph in Figure 9-12, calculate how far away from each seismograph station the earthquake occurred. How would you determine the epicenter of this earthquake?

	Arrival Time of P-Wave	Arrival Time of S-Wave
Station A:	2:59:03 P.M.	3:04:03 P.M.
Station B:	2:51:16 P.M.	3:01:16 P.M.
Station C:	2:48:25 P.M.	2:55:55 P.M.

Distance from the earthquake:

Station A: ____________

Station B: ____________

Station C: ____________

3. One of the goals of earthquake prediction is to prevent loss of life and injury by evacuating people from an area before an earthquake occurs. Discuss the social, economic, and political problems associated with such evacuations. For example, who would pay for lost earnings to businesses and individuals because they would not be able to operate or work during the time of evacuation? Who is responsible if looting occurs in the evacuated area? What will happen to public confidence if the predicted earthquake fails to occur?
4. Using the graph in Figure 9-16, answer the following question. A seismograph in Berkeley, California, records the arrival time of an earthquake's P-waves at 6:59:54 P.M. and the S-waves at 7:00:02 P.M. The maximum amplitude of the S-waves as recorded on the seismogram was 75 mm. What was the magnitude of the earthquake and how far away from Berkeley did it occur?
5. Discuss why insurance companies use the qualitative Modified Mercalli Intensity Scale instead of the quantitative Richter Magnitude Scale in classifying earthquakes.
6. On June 8, 1994, a deep-focus (600 km) 8.2 magnitude earthquake occurred beneath Bolivia and was felt as far away in North America as Seattle, Washington, Minneapolis, Minnesota, and Toronto, Canada. While some damage was done in the area of the epicenter, no deaths were reported. Discuss why this particular earthquake should be felt so far away, while other equally strong earthquakes are not felt much beyond the area of the epicenter. Why do you think so little damage was done and no deaths occurred from such a strong earthquake? Lastly, what do you think was the cause of this deep-focus earthquake, and what can it tell geologists about the interior of the Earth?

ADDITIONAL READINGS

Bolt, B. A. 1988. *Earthquakes*. New York: W. H. Freeman and Co.

Canby, T. Y. 1990. California earthquake—prelude to the big one? *National Geographic* 177, no. 5: 76–105.

Davidson, K. 1994. Predicting earthquakes. *Earth* 3, no. 3: 56–63.

______ . 1994. Learning from Los Angeles. *Earth* 3, no. 5: 40–47

Dvorak, J., and T. Peek. 1993. Swept away. *Earth* 2, no. 4: 52–59.

Evans, D. M. 1966. Man-made earthquakes in Denver. *Geotimes* 10, no. 9: 11–18.

Fischman, J. 1992. Falling into the gap: A new theory shakes up earthquake predictions. *Discover* October 1992: 56–63.

Green, H. W., II. 1994. Solving the paradox of deep earthquakes. *Scientific American* 271, no. 3: 64–71.

Hanks, T. C. 1985. *National earthquake hazard reduction program: Scientific status*. U.S. Geological Survey Bulletin 1659.

Johnston, A. 1992. New Madrid: The rift, the river, and the earthquake. *Earth* 1, no. 1: 34–43.

Johnston, A. C., and L. R. Kanter. 1990. Earthquakes in stable continental crust. *Scientific American* 262, no. 3: 68–75.

McCredie, S. 1994. Tsunamis: The wrath of Poseidon. *Smithsonian* 24, no. 12: 28–39.

Monastersky, R. 1993. Lessons from Landers. *Earth* 2, no. 2: 40–47.

Stein, R. S., and R. S. Yeats. 1989. Hidden earthquakes. *Scientific American* 260, no. 6: 48–59.

Chapter 10

THE INTERIOR OF THE EARTH

OUTLINE

Probing the Earth's interior. The world's deepest hole, more than 12 km deep, is on the Kola Pennisula in northwestern Russia. The 30-story building in this image houses the drill rig.

PROLOGUE

During most of historic time, the Earth's interior was perceived as an underground world of vast caverns, heat, and sulfur gases, populated by demons (▶ Figure 10-1). By the 1800s, scientists had some sketchy ideas about the Earth's structure, but outside scientific circles, all kinds of bizarre ideas were proposed. In 1869, for example, Cyrus Reed Teed claimed that the Earth was hollow and that humans lived on the inside. As recently as 1913, Marshall B. Gardner held that the Earth is a large hollow sphere with a 1,300-km-thick outer shell surrounding a central sun.

Although making no claim to present a reliable picture of the Earth's interior, Jules Verne's 1864 novel *A Journey to the Center of the Earth* described the adventures of Professor Hardwigg, his nephew, and an Icelandic guide as they descended into the Earth through the crater of Mount Sneffels in Iceland. During their travels, they followed a labyrinth of passageways until they finally arrived 140 km below the surface. Here, they encountered a vast cavern containing "the central sea" illuminated by some electrical phenomenon related to the northern lights. Along the margins of the sea, they saw forests of prehistoric ferns and palms and a herd of mastodons complete with a gigantic human shepherd. Dwelling in the central sea were Mesozoic-aged marine reptiles and gigantic turtles. Their adventure ended when they were carried upward to the surface on a raft by a rising plume of water.

Scientists in 1864 knew what the average density of the Earth was and that pressure and temperature increase with depth. They also knew that the fabled passageways followed by Professor Hardwigg could not exist, but little else was known, even though humans had probed the Earth through mines and wells for centuries. Even the deepest mines (the gold mines in South Africa) descend only about 3 km below the surface, barely penetrating the uppermost part of the Earth's crust. The deepest drill hole reaches a depth of a little more than 12 km. A drill hole 12 km deep is impressive, but it is less than 0.2% of the distance to the Earth's center. Indeed, if the Earth were the size of an apple, this drill hole would be roughly equivalent to a pinprick penetrating less than halfway through the apple's skin!

In 1958, an ambitious project to drill through the oceanic crust to the mantle was launched. Known as *Mohole,* the project attracted considerable attention and support from scientists and the public, and in 1962 Congress appropriated more than $40 million to finance it. Mohole

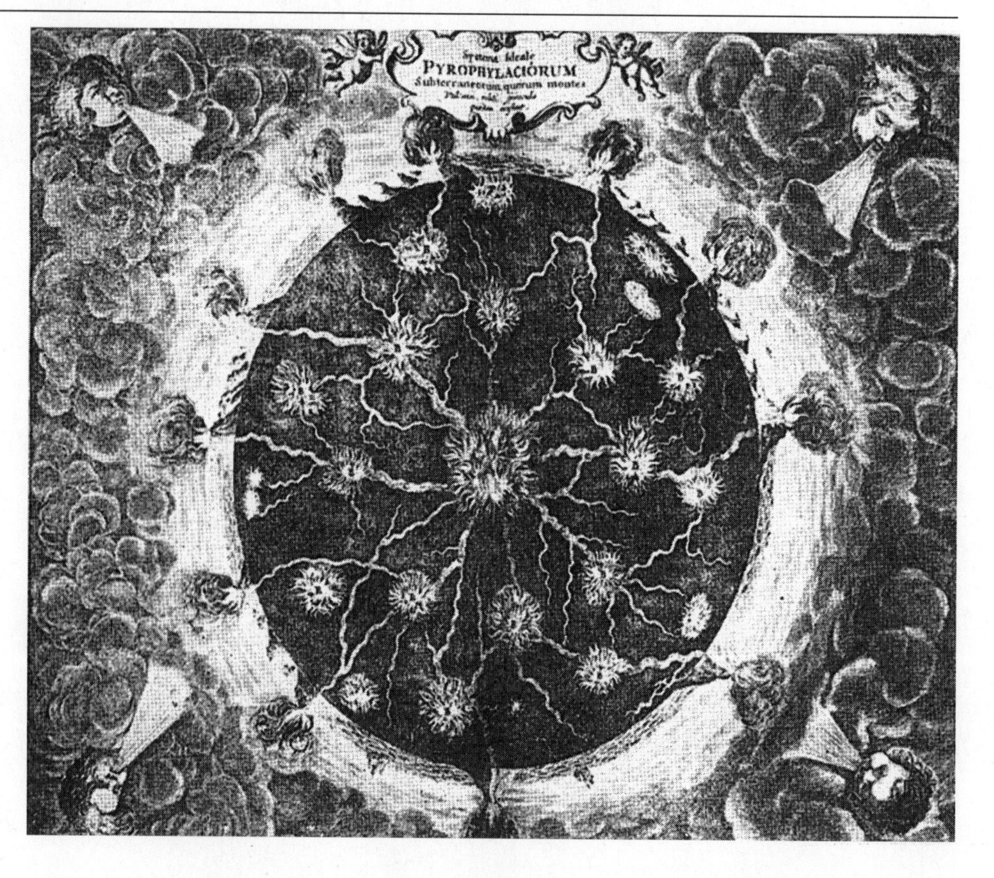

▶ **FIGURE 10-1** In 1678, Athanasius Kircher (1602–1680) published *Mundus Subterraneus,* which contained this drawing showing what he described as the "ideal system of subterranean fire cells from which volcanic mountains arise, as it were, like vents."

encountered technological difficulties, however, and by 1966 funding had dried up; soon thereafter, project Mohole was quickly forgotten.

Because the Earth's interior is hidden from direct observation, it is more inaccessible than the surfaces of the Moon and Mars. Nevertheless, scientists have a reasonably good idea of the Earth's internal structure and composition. No vast openings or passageways exist as in Jules Verne's story; even the deepest known caverns extend to depths of less than 1,500 m. Even at the modest depths to which Professor Hardwigg and his companions are supposed to have descended, the pressure and temperature are so great that rock actually flows even though it remains solid. In deep mines the rock is under such tremendous pressure that rock bursts and popping are constant problems (see Perspective 5-1). In short, the behavior of solids at depth where the temperature and pressure are great is very different from their brittle behavior at the surface.

 INTRODUCTION

The Earth's interior is so inaccessible that most people think little about it. One can appreciate the stunning beauty of the northern lights and yet be completely unaware that they exist because of the interaction between the magnetic field that is generated within the Earth and the solar wind, a continuous stream of electrically charged particles emanating from the Sun. Much of the Earth's geologic activity including volcanism, earthquakes, the movements of plates, and the origin of mountains is caused by internal heat. A continuous slow exchange of material occurs as magma rises from within the Earth, and solid Earth materials are subducted and returned to the interior.

Scientists have known for more than 200 years that the Earth's interior is not homogeneous. Sir Isaac Newton (1642–1727) noted in a study of the planets that the Earth's average density is 5.0 to 6.0 g/cm^3 (water has a density of 1 g/cm^3). In 1797, Henry Cavendish calculated a density value very close to the 5.5 g/cm^3 now accepted. The Earth's average density is considerably greater than that of surface rocks, most of which range from 2.5 to 3.0 g/cm^3. Thus, in order for the average density to be 5.5 g/cm^3, much of the interior must consist of materials with a density greater than the Earth's average density.

The Earth is generally depicted as consisting of concentric layers differing in composition and density that are separated from adjacent layers by rather distinct boundaries (▶ Figure 10-2). Recall that the outermost layer, or the **crust**, is the very thin skin of the Earth. Below the crust and extending about halfway to the Earth's center is the **mantle**, which comprises more than 80% of the Earth's volume (◉ Table 10-1). The central part of the Earth is the **core**, which is divided into a solid inner core and a liquid outer part (Figure 10-2).

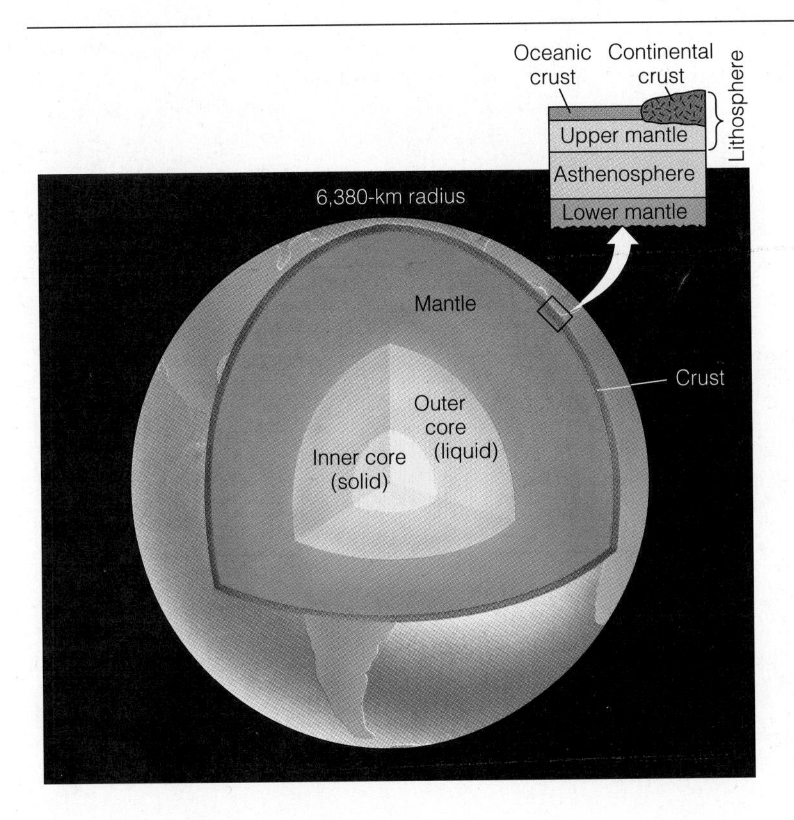

▶ **FIGURE 10-2** The internal structure of the Earth.

TABLE 10-1 Data on the Earth

	Volume (Thousands of km^3)	Percentage of Total Volume	Mass (Trillions of Metric Tons)	Percentage of Total Mass
Inner core	7,512,800	0.70%	19,000,000,000	31.79%
Outer core	169,490,000	15.68		
Mantle	896,990,000	83.02	40,500,000,000	67.77
Continental crust	4,760,800	0.44	250,000,000	0.42
Oceanic crust	1,747,200	0.16		
Atmosphere, water, ice			14,351,000	0.02

Because no direct observations of the Earth's interior can be made, this model of the Earth's internal structure is based on indirect evidence, mostly from the study of seismic waves. Nevertheless, the model is widely accepted by scientists and is becoming increasingly refined as more sophisticated methods of probing what some call "inner space" are developed.

SEISMIC WAVES

The behavior and travel times of P- and S-waves within the Earth provide geologists with much information about its internal structure. Seismic waves travel outward as wave fronts from the source areas, although it is most convenient to depict them as *wave rays,* which are lines showing the direction of movement of small parts of wave fronts (➤ Figure 10-3). Any disturbance such as a passing train or construction equipment can cause seismic waves, but only those generated by large earthquakes, explosive volcanism, asteroid impacts, and nuclear explosions can travel completely through the Earth.

As we noted in Chapter 9, the velocities of P- and S-waves are determined by the density and elasticity of the materials through which they travel. Both the density and elasticity of rocks increase with depth, but elasticity increases faster than density, resulting in a general increase in the velocity of seismic waves. P-waves travel faster than S-waves through all materials, but unlike P-waves, S-waves cannot

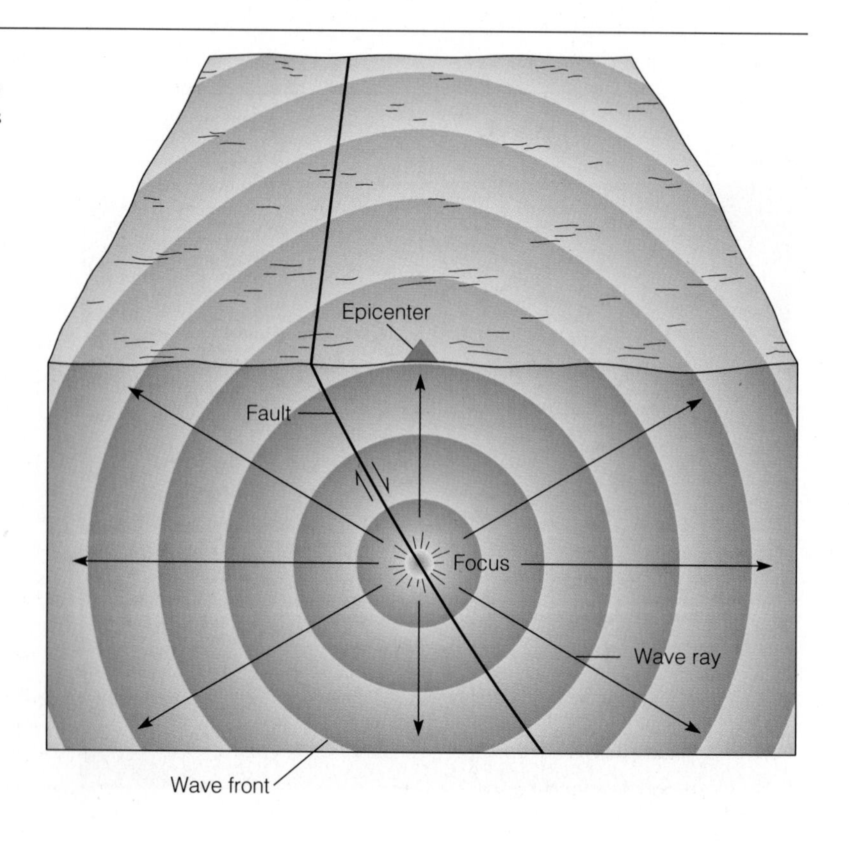

➤ **FIGURE 10-3** Seismic wave fronts move outward in all directions from their source, the focus of an earthquake in this example. Wave rays are lines drawn perpendicular to wave fronts.

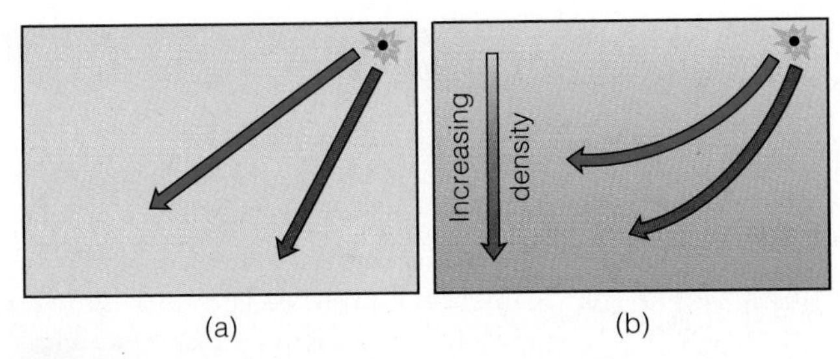

▶ **FIGURE 10-4** (*a*) If the Earth had the same composition and density throughout, seismic wave rays would follow straight paths. (*b*) Because density increases with depth, wave rays are continually refracted so that their paths are curved.

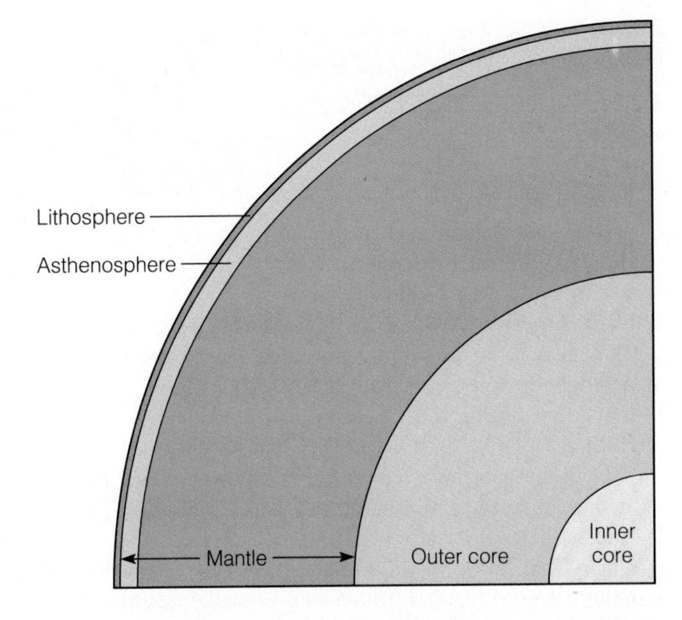

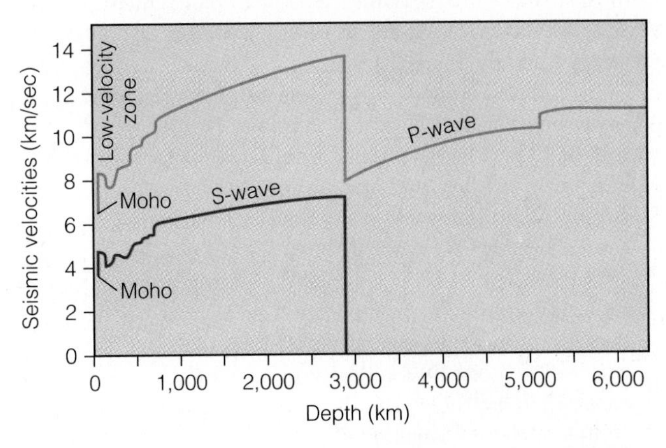

▶ **FIGURE 10-6** Profiles showing seismic wave velocities versus depth. Several discontinuities are shown across which seismic wave velocities change rapidly.

be transmitted through a liquid because liquids have no shear strength (rigidity)—they simply flow in response to a shear stress.

If the Earth were a homogeneous body, P- and S-waves would travel in straight paths as shown in ▶ Figure 10-4a. But as a seismic wave travels from one material into another of different density and elasticity, its velocity and direction of travel change. That is, the wave is bent, a phenomenon known as **refraction** (Figure 10-4b). Because seismic waves pass through materials of differing density and elasticity, they are continually refracted so that their paths are curved; wave rays only travel in a straight line and are not refracted when the direction of their travel is perpendicular to a boundary (▶ Figure 10-5).

In addition to refraction, seismic rays are also **reflected,** much as light is reflected from a mirror. When seismic waves encounter a boundary separating materials of different density or elasticity with the earth, some of the rays' energy is *reflected* back to the Earth's surface (Figure 10-5). If we know the wave velocity and the time required for it to travel from its source to the boundary and back to the surface, we can calculate the depth of the reflecting boundary. Such information is useful in determining not only the depths of the various layers within the Earth, but also the depths of sedimentary rocks that may contain petroleum. Seismic reflection is a common tool used in petroleum exploration (see Perspective 8-1).

▶ **FIGURE 10-5** Refraction and reflection of P-waves. When seismic waves pass through a boundary separating Earth materials of different density or elasticity, they are refracted, and some of their energy is reflected back to the surface.

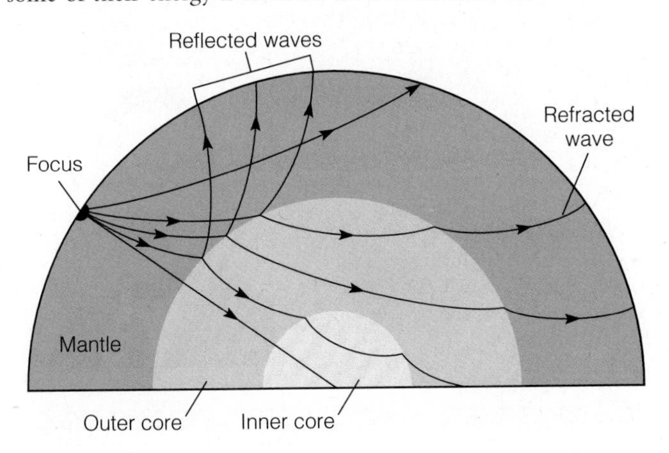

Although changes in seismic wave velocity occur continuously with depth, P-wave velocity increases suddenly at the base of the crust and decreases abruptly at a depth of about 2,900 km (▶ Figure 10-6). These marked changes in the velocity of seismic waves indicate a boundary called a **discontinuity** across which a significant change in Earth materials or their properties occurs. Such discontinuities are the basis for subdividing the Earth's interior into concentric layers.

The contribution of seismology to the study of the Earth's interior cannot be overstated. Beginning in the early 1900s, scientists recognized the utility of seismic wave studies and, between 1906 and 1936, largely worked out the internal structure of the Earth on the basis of these studies.

GEOLOGY: AN UNEXPECTED BUT REWARDING CAREER

From a study conducted by both private and federal agencies during the early 1970s, I discovered that I was one of a very small cadre of five black Americans who had doctorate degrees in geology. This year, I am beginning my 47th year as a professional geologist-geophysicist, and I can say, without hesitation, that it has been a most exciting and rewarding professional career.

Geology was not even in my vocabulary when I graduated from a segregated high school in Cumberland, Maryland, which is located at the eastern margin of the Appalachian Coal Basin. It was certainly not one of my interests when I served as one of the Tuskegee airmen during World War II. But, after leaving the service and while attending the University of Michigan, I did take a historical geology course, merely as an elective.

Ironically, I joined the geological profession due to my experiences in seeking other employment during the summer of 1948. I had graduated from Howard University with a major in mathematics and a minor in physics. Thus, I first attempted to find work in my field as a mathematician with the U.S. Naval Research Laboratory.

On the last of several repeat visits to the laboratory to refile my employment application, which had been "accidentally misplaced" several times, I picked up a page from the Washington, D.C. editon of an Afro-American newspaper that had been left on a trolley car seat. That page contained a very small ad stating that the U.S. Geological Survey was seeking applicants for positions as mathematicians, physicists, and chemists. Married and desperately in need of employment, I quickly transferred to a connecting trolley car and went to the U.S. Geological Survey. I filed an application and within three or four days was hired as an exploration geophysicist with the newly formed airborne magnetometer group in the Geophysics Section of the U.S. Geological Survey.

After several years with the U.S. Geological Survey conducting some of the first airborne geophysical surveys flown on the Colorado Plateau and in other parts of the United States, I returned to school, part-time and at night, completed my undergraduate and graduate requirements in geology, and received my master's degree from the American University in Washington, D.C. I then earned my Ph.D. degree in geology from Johns Hopkins University where I studied with Ernst Cloos, Francis J. Pettijohn, and Clifford A. Hopson, who introduced me to the Appalachian geology underlying my birthplace.

After nearly 20 years in exploration geophysics with the U.S. Geological Survey, I joined the faculty at the University of Massachusetts. After holding a number of administrative positions, I plan to complete my active professional career by returning to teaching and research. My main research interests are in airborne geophysical surveys, and applications of shallow-refraction seismic, gravity, electrical resistivity and conductivity surveys in hydrological and engineering programs.

The irony associated with that earlier act of discrimination and the chance finding of that newspaper page on a trolley car in Washington, D.C., over 40 years ago, is that it deflected me into a new and unfamiliar career that has been both personally and professionally a most rewarding experience. My career in geology reached its most significant point when my professional peers elected me president of the Geological Society of America in 1989.

As I fast approach retirement, I reflect with much pride on the many undergraduate and graduate students I have taught and learned from over the last quarter of a century; the nearly two decades of geological and geophysical consulting work I conducted in several west and central African countries; and my early career with the U.S. Geological Survey.

The economic and environmental future of our country and this planet will be shaped to a large extent by those of us in the various geological professions. The United States now holds the leadership position in research and applications of groundwater and surface-water exploration and development techniques, Earth resources exploration and development, and waste disposal and environmental planning and protection technology. Geology will continue to play a key role in these critically important areas, and minorities and women are now finding a place in this professional workforce.

RANDOLPH W. BROMERY is currently Commonwealth Professor of Geophysics at the University of Massachusetts. In addition to his teaching duties, he has been department chair, vice chancellor and chancellor of the Amherst campus, executive vice president and senior vice president of the university system, interim president of Westfield State College, and interim chancellor of the Massachusetts Board of Regents for Higher Education. In 1989, he was elected president of the Geological Society of America.

Discovery of the Earth's Core

In 1906, R. D. Oldham of the Geological Survey of India discovered that seismic waves arrived later than expected at seismic stations more than 130° from an earthquake focus. He postulated the existence of a core that transmits seismic waves at a slower rate than shallower Earth materials. We now know that P-wave velocity decreases markedly at a depth of 2,900 km, indicating a major discontinuity now recognized as the core-mantle boundary (Figure 10-6).

The sudden decrease in P-wave velocity at the core-mantle boundary causes P-waves entering the core to be refracted in such a way that very little P-wave energy reaches the Earth's surface in the area between 103° and 143° from an earthquake focus (▶ Figure 10-7). This area in which little P-wave energy is recorded by seismometers is a **P-wave shadow zone;** it was discovered by the German seismologist Beno Gutenberg in 1914.

The P-wave shadow zone is not a perfect shadow zone. That is, some weak P-wave energy reaches the surface within the zone. Several hypotheses were proposed to explain this phenomenon, but all were rejected by the Danish seismologist Inge Lehmann (▶ Figure 10-8a), who in 1936 postulated that the core is not entirely liquid. She demonstrated that reflection from a solid inner core could account for the arrival of weak P-waves in the P-wave shadow zone (Figure 10-8b). Lehmann's proposal of a solid inner core was quickly accepted by seismologists.

In 1926, the British physicist Harold Jeffreys realized that S-waves were not simply slowed by the core, but were completely blocked by it. So in addition to a P-wave shadow zone, a much larger and more complete **S-wave shadow zone** exists (▶ Figure 10-9). At locations greater than 103° from an earthquake focus, no S-waves are recorded, indicating that S-waves cannot be transmitted through the core. S-waves will not pass through a liquid, so it seems that the outer core must be liquid or behave as a liquid. The inner core, however, is thought to be solid because P-wave velocity increases at the base of the outer core.

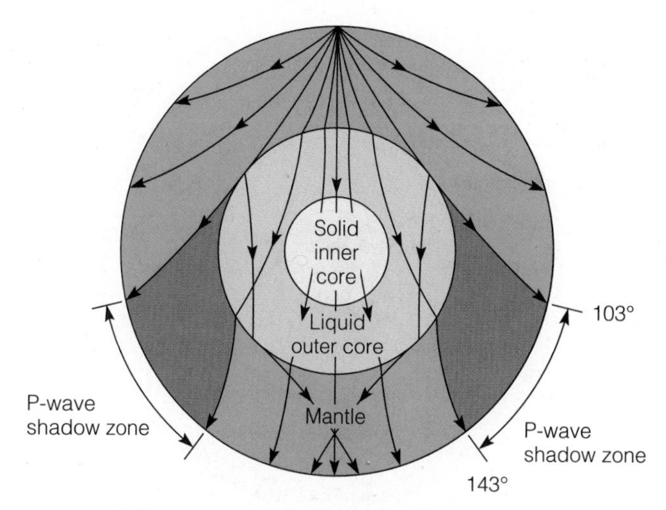

▶ **FIGURE 10-7** P-waves are refracted so that no direct P-wave energy reaches the Earth's surface in the P-wave shadow zone.

Density and Composition of the Core

The core lies below the mantle at a depth of 2,900 km and extends to the center of the Earth. It has a diameter of 6,960 km, about the same as the total diameter of Mars (Figure 10-2). The core constitutes 16.4% of the Earth's volume and nearly one-third of its mass (Table 10-1).

We can estimate the core's density and composition by using seismic evidence and laboratory experiments. Furthermore, meteorites, which are thought to represent remnants of the material from which the solar system formed, can be used to make estimates of density and composition. For

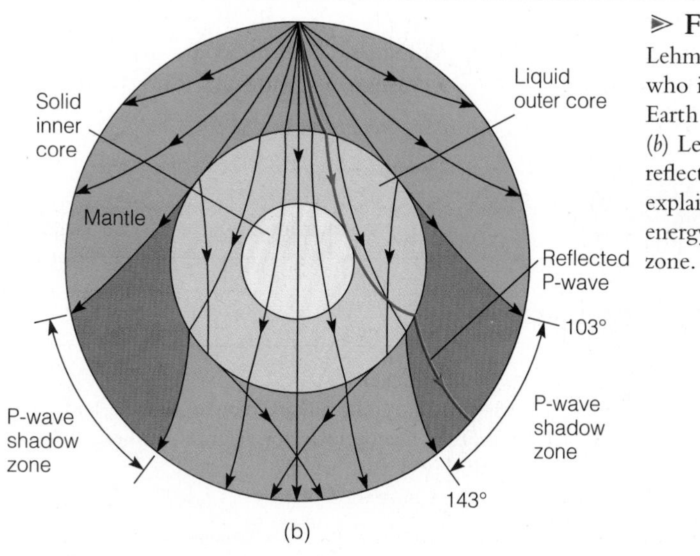

▶ **FIGURE 10-8** (*a*) Inge Lehmann, the Danish seismologist who in 1936 postulated that the Earth has a solid inner core. (*b*) Lehmann proposed that reflection from an inner core could explain the arrival of weak P-wave energy in the P-wave shadow zone.

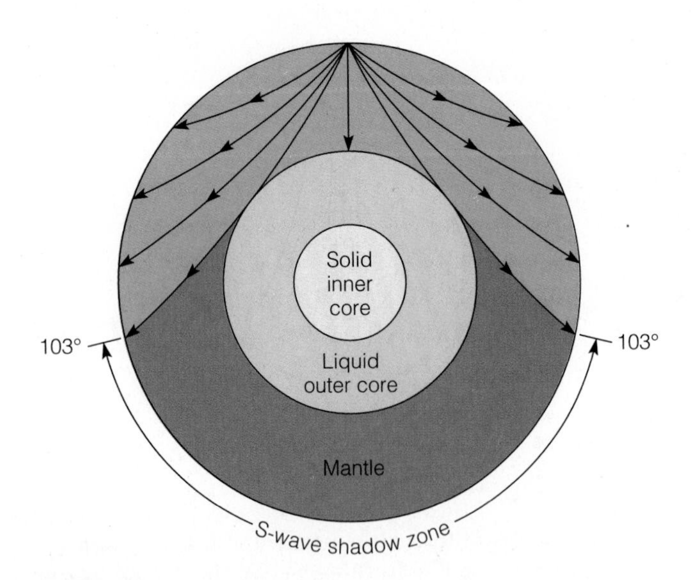

⫸ **FIGURE 10-9** The presence of an S-wave shadow zone indicates that S-waves are being blocked within the Earth.

example, the irons, meteorites composed of iron and nickel alloys, may represent the differentiated interiors of large asteroids and approximate the density and composition of the Earth's core. The density of the outer core varies from 9.9 to 12.2 g/cm^3, and that of the inner core ranges from 12.6 to 13.0 g/cm^3 (◉ Table 10-2). At the Earth's center, the pressure is equivalent to about 3.5 million times normal atmospheric pressure.

The core cannot be composed of the minerals most common at the Earth's surface, because even under the tremendous pressures at great depth they would still not be dense enough to yield an average density of 5.5 g/cm^3 for the Earth. Both the outer and inner core are thought to be composed largely of iron, but pure iron is too dense to be the sole constituent of the outer core. It must be "diluted" with elements of lesser density. Laboratory experiments and comparisons with iron meteorites indicate that perhaps 12% of the outer core consists of sulfur and possibly some silicon, oxygen, nickel, and potassium (Table 10-2).

In contrast, pure iron is not dense enough to account for the estimated density of the inner core. Most geologists think that perhaps 10 to 20% of the inner core also consists of nickel. These metals form an iron-nickel alloy that is thought to be sufficiently dense under the pressure at that depth to account for the density of the inner core.

Any model of the core's composition and physical state must explain not only the variations in density, but also why the outer core is liquid while the inner core is solid, and how the magnetic field is generated within the core (discussed later in this chapter). When the core formed during early Earth history, it was probably completely molten and has since cooled to the point that its interior has crystallized. The temperature at the core-mantle boundary is estimated at 3,500° to 5,000°C, yet the high pressure within the inner core prevents melting. In contrast, the outer core is subject to less pressure. Even more important than the differences in pressure, though, are the compositional differences between the inner and outer core. The sulfur content of the outer core helps depress its melting temperature. An iron-sulfur mixture melts at a lower temperature than does pure iron, or an iron-nickel alloy, so despite the high pressure, the outer core is molten.

 THE MANTLE

Another significant discovery about the Earth's interior was made in 1909 when the Yugoslavian seismologist Andrija Mohorovičić detected a discontinuity at a depth of about 30 km. While studying arrival times of P-waves from Balkan earthquakes, Mohorovičić noticed that seismic stations a few hundred kilometers from an earthquake's epicenter were recording two distinct sets of P- and S-waves.

From his observations Mohorovičić concluded that a sharp boundary separating rocks with different properties exists at a depth of about 30 km. He postulated that P-waves below this boundary travel at 8 km/sec, whereas those above the boundary travel at 6.75 km/sec. When an earthquake occurs, some waves travel directly from the focus to the seismic station, while others travel through the deeper layer and some of their energy is refracted back to the surface (⫸ Figure 10-10). Waves traveling through the deeper layer travel farther to a seismic station, but they do so more rapidly than those in the shallower layer. The boundary identified by Mohorovičić separates the crust from the mantle and is now called the **Mohorovičić discontinuity**, or

TABLE 10-2 Composition and Density of the Earth

	Composition	Density (g/cm^3)
Inner core	Iron with 10 to 20% nickel	12.6–13.0
Outer core	Iron with perhaps 12% sulfur, silicon, oxygen, nickel, and potassium	9.9–12.2
Mantle	Peridotite (composed mostly of ferrogmagnesian silicates)	3.3–5.7
Oceanic crust	Upper part basalt, lower part gabbro	~3.0
Continental crust	Average composition of granodiorite	~2.7

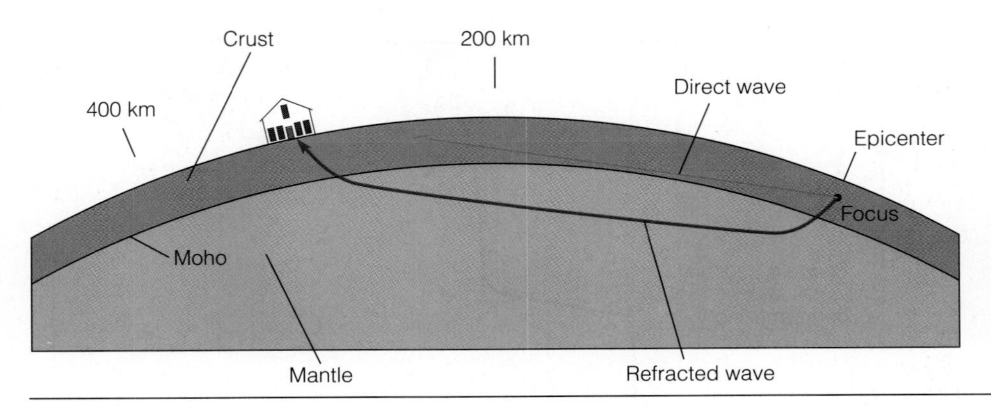

▶ **FIGURE 10-10** Andrija Mohorovičić studied seismic waves and detected a seismic discontinuity at a depth of about 30 km. The deeper, faster seismic waves arrive at seismic stations first, even though they travel farther.

simply the **Moho.** It is present everywhere except beneath spreading ridges, but its depth varies: beneath the continents it ranges from 20 to 90 km, with an average of 35 km; beneath the sea floor it is 5 to 10 km deep (▶ Figure 10-11).

Structure and Composition of the Mantle

Although seismic wave velocity in the mantle generally increases with depth, several discontinuities also exist. Between depths of 100 and 250 km, both P- and S-wave velocities decrease markedly (▶ Figure 10-12). This layer between 100 and 250 km deep is the **low-velocity zone;** it corresponds closely to the **asthenosphere,** a layer in which the rocks are close to their melting point and are less elastic; this decrease in elasticity accounts for the observed decrease in seismic wave velocity. The asthenosphere is an important zone because it may be where some magmas are generated. Furthermore, it lacks strength and flows plastically and is thought to be the layer over which the plates of the outer, rigid **lithosphere** move.

Even though the low-velocity zone and the asthenosphere closely correspond, they are still distinct. The asthenosphere appears to be present worldwide, but the low-velocity zone is not. In fact, the low-velocity zone appears to be poorly defined or even absent beneath the ancient shields of continents.

Other discontinuities have been detected at deeper levels within the mantle. But unlike those between the crust and mantle or between the mantle and core, these probably represent structural changes in minerals rather than compositional changes. In other words, geologists think the mantle is composed of the same material throughout, but the structural states of minerals such as olivine change with depth. At a depth of 400 km, seismic wave velocity increases slightly as a consequence of such changes in mineral structure (Figure 10-12). Another velocity increase occurs at 640 to 720 km where the minerals break down into metal oxides, such as FeO (iron oxide) and MgO (magnesium oxide), and silicon dioxide (SiO_2). A third discontinuity exists at about 1,050 km where P-waves once again increase in velocity. These three discontinuities are within what is called a *transition zone* separating the upper mantle from the lower mantle (Figure 10-12).

Although the mantle's density, which varies from 3.3 to 5.7g/cm^3, can be inferred rather accurately from seismic

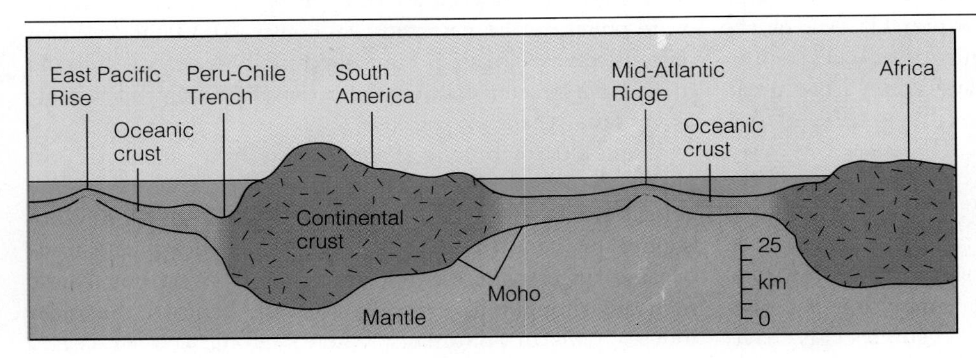

▶ **FIGURE 10-11** The Moho is present everywhere except beneath spreading ridges such as the East Pacific Rise and the Mid-Atlantic Ridge. However, the depth of the Moho varies considerably.

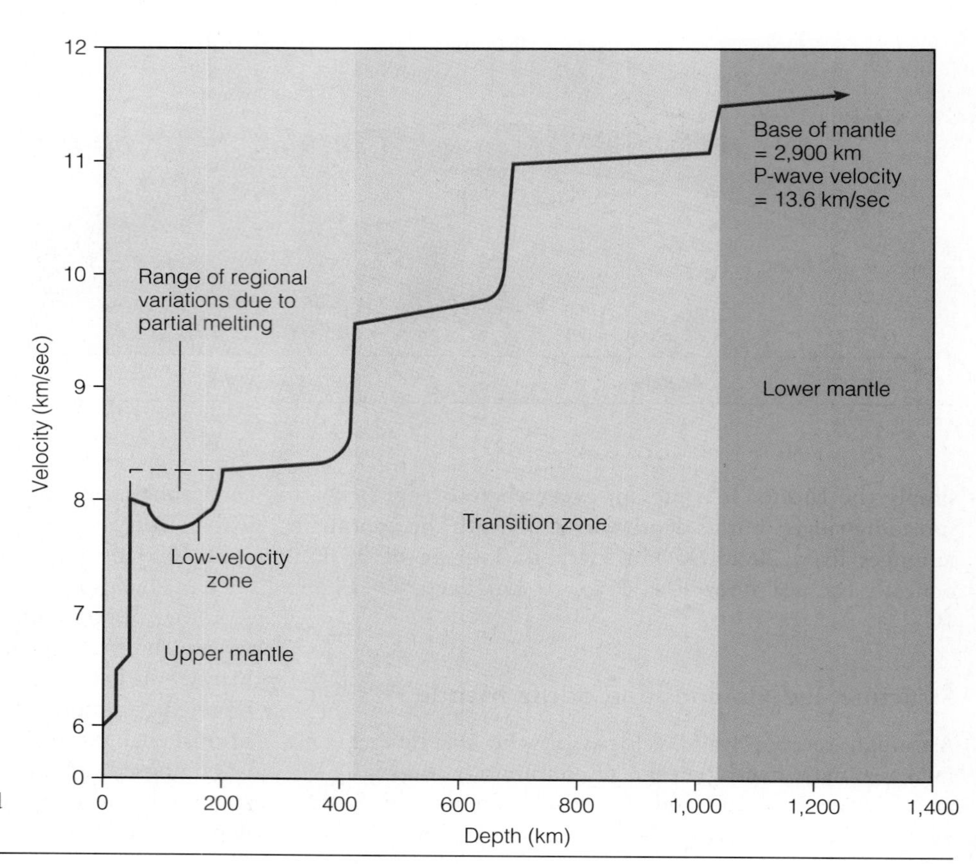

▶ **FIGURE 10-12** Variations in P-wave velocity in the upper mantle and transition zone.

waves, its composition is less certain. The igneous rock *peridotite* is considered the most likely component. Peridotite contains mostly ferromagnesian minerals (60% olivine and 30% pyroxene) with about 10% feldspars (see Figure 3-13a). Peridotite is considered the most likely candidate for three reasons. First, laboratory experiments indicate that it possesses physical properties that would account for the mantle's density and observed rates of seismic wave transmissions. Second, peridotite forms the lower parts of igneous rock sequences thought to be fragments of the oceanic crust and upper mantle emplaced on land (see Chapter 11). And third, peridotite occurs as inclusions in volcanic rock bodies such as *kimberlite pipes* that are known to have come from great depths. These inclusions are thought to be pieces of the mantle (see Perspective 10-1).

Seismic Tomography

The model of the Earth's interior consisting of an iron-rich core and a rocky mantle is probably accurate, but it is also rather imprecise. Recently, however, geophysicists have developed a technique called *seismic tomography* that allows them to develop three-dimensional models of the Earth's interior. In seismic tomography numerous crossing seismic waves are analyzed in much the same way radiologists analyze CAT (computerized axial tomography) scans. In CAT scans, X-rays penetrate the body, and a two-dimensional image of the inside of a patient is formed. Repeated CAT scans, each from a slightly different angle, are computer analyzed and stacked to produce a three-dimensional picture.

In a similar fashion geophysicists use seismic waves to probe the interior of the Earth. From the time of arrival and distance traveled, the velocity of a seismic ray is computed at a seismic station. Only average velocity is determined, rather than variations in velocity. In seismic tomography numerous wave rays are analyzed so that "slow" and "fast" areas of wave travel can be detected (▶ Figure 10-13). Recall that seismic wave velocity is controlled partly by elasticity; cold rocks have greater elasticity and therefore transmit seismic waves faster than hot rocks.

Using this technique, geophysicists have detected areas within the mantle at a depth of about 150 km where seismic velocities are slower than expected. These anomalously hot regions lie beneath volcanic areas and beneath the mid-oceanic ridges, where convection cells of rising hot mantle rock are thought to exist. In contrast, beneath the older interior parts of continents, where tectonic activity ceased hundreds of millions or billions of years ago, anomalously cold spots are recognized. In effect, tomographic maps and three-dimensional diagrams show heat variations within the Earth.

Seismic tomography has also yielded additional and sometimes surprising information about the core. For example,

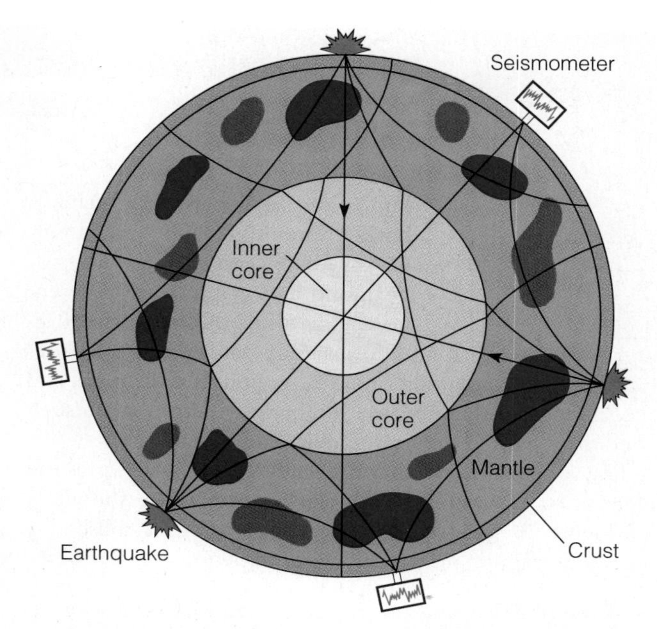

▶ **FIGURE 10-13** Numerous earthquake waves are analyzed to detect areas within the Earth that transmit seismic waves faster or slower than adjacent areas. Areas of fast wave travel correspond to "cold" regions (blue), whereas "hot" regions (red) transmit seismic waves more slowly.

the core-mantle boundary is not a smooth surface, but has broad depressions and rises extending several kilometers into the mantle. Of course, the base of the mantle possesses the same features in reverse; geophysicists have termed these features "anticontinents" and "antimountains." It appears that the surface of the core is continually deformed by sinking and rising masses of mantle material.

As a result of seismic tomography, a much clearer picture of the Earth's interior is emerging. It has already given us a better understanding of complex convection within the mantle, including upwelling convection currents thought to be responsible for the movement of the Earth's lithospheric plates (see Chapter 12).

 THE EARTH'S CRUST

The Earth's crust is the most accessible and best studied of its concentric layers, but it is also the most complex both chemically and physically. Whereas the core and mantle seem to vary mostly in a vertical dimension, the crust shows considerable vertical and lateral variation. (More lateral variation exists in the mantle than was once thought.) The crust along with that part of the upper mantle above the low-velocity zone constitutes the *lithosphere* of plate tectonic theory.

Two types of crust are recognized—continental crust and oceanic crust—both of which are less dense than the underlying mantle. **Continental crust** is the more complex, consisting of a wide variety of igneous, sedimentary, and metamorphic rocks. It is generally described as "granitic," meaning that its overall composition is similar to that of granitic rocks. Specifically, its overall composition corresponds closely to that of granodiorite, an igneous rock having a chemical composition between granite and diorite (see Figure 3-12).

Continental crust varies in density depending on rock type, but with the exception of metal-rich rocks, such as iron ore deposits, most rocks have densities of 2.0 to 3.0 g/cm^3, and the overall density is about 2.70 g/cm^3 (Table 10-2). P-wave velocity in the continental crust is about 6.75 km/sec; at the base of the crust, P-wave velocity abruptly increases to about 8 km/sec.

The continental crust averages about 35 km thick, but is much thinner in such areas as the Rift Valleys of East Africa and a large area called the Basin and Range Province in the western United States. The crust in these areas is being stretched and thinned in what appear to be the early stages of rifting and is as little as 20 km thick. In contrast, continental crust beneath mountain ranges is much thicker and projects deep into the mantle. Beneath the Himalayas of Asia, the crust is as much as 90 km thick. Crustal thickening beneath mountain ranges is an important point that will be discussed in "The Principle of Isostasy" later in the chapter.

Although variations also occur in **oceanic crust,** they are not as distinct as those for the continental crust. Oceanic crust varies from 5 to 10 km thick, being thinnest at spreading ridges. It is denser than continental crust, averaging about 3.0 g/cm^3, and transmits P-waves at about 7 km/sec. Just as beneath the continental crust, P-wave velocity increases at the Moho. The P-wave velocity of oceanic crust is what one would expect if it were composed of basalt. Direct observations of oceanic crust from submersibles and deep-sea drilling confirm that its upper part is indeed composed of basalt. The lower part of the oceanic crust is composed of gabbro, the intrusive equivalent of basalt (see Chapter 11 for a more detailed description of the oceanic crust).

 THE EARTH'S INTERNAL HEAT

During the nineteenth century, scientists realized that the Earth's temperature in deep mines increases with depth. Indeed, very deep mines must be air conditioned so that the miners can survive. More recently, the same trend has been observed in deep drill holes, but even in these we can measure temperatures directly down to a depth of only a few kilometers. The temperature increase with depth, or **geothermal gradient,** near the surface is about 25°C/km, although it varies from area to area. In areas of active or recently active volcanism, the geothermal gradient is greater than in adjacent nonvolcanic areas, and temperature rises faster beneath spreading ridges than elsewhere beneath the sea floor.

Unfortunately, the geothermal gradient is not useful for estimating temperatures deep in the Earth. If we were simply to extrapolate from the surface downward, the temperature

Perspective 10-1

Kimberlite Pipes—Windows to the Mantle

Diamonds have been economically important throughout history, yet prior to 1870, they had been found only in river gravels, where they occur as the result of weathering, transport, and deposition. In 1870, the source of diamonds in South Africa was traced to cone-shaped igneous bodies called *kimberlite pipes* found near the town of Kimberley (▶ Figure 1). Kimberlite pipes are the source rocks for most diamonds.

The greatest concentrations of kimberlite pipes are in southern Africa and Siberia, but they occur in many other areas as well. In North America they have been found in the Canadian Arctic, Colorado, Wyoming, Missouri, Montana, Michigan, and Virginia, and one at Murfreesboro, Arkansas, was briefly worked for diamonds. Diamonds discovered in glacial deposits in some midwestern states indicate that kimberlite pipes are present farther north. The precise source of these diamonds has not been determined, although some kimberlite pipes have recently been identified in northern Michigan.

Kimberlite pipes are composed of dark gray or blue igneous rock called *kimberlite,* which contains olivine, a potassium- and magnesium-rich mica, serpentines, and calcite and silica. Some of these rocks contain inclusions of *peridotite* that are thought to represent pieces of the mantle brought to the surface during the explosive volcanic eruptions that form kimberlite pipes.

If peridotite inclusions are, in fact, pieces of the mantle, they indicate that the magma in kimberlite pipes originated at a depth of at least 30 km. Indeed, the presence of diamonds and the structural form of the silica in the kimberlite can be used to establish both minimum and maximum depths for the origin of the magma. Diamond and graphite are different crystalline forms of carbon (see Figure 2-5), but diamond forms only under high-pressure, high-temperature conditions. The presence of diamond and the absence of graphite in kimberlite indicate that such conditions existed where the magma originated.

The calculated geothermal gradient and the pressure increase with depth beneath the continents are shown in ▶ Figure 2. Laboratory experiments have established a diamond-graphite inversion curve showing the pressure-

▶ **FIGURE 1** Generalized cross section of a kimberlite pipe. Most kimberlite pipes measure less than 500 m in diameter at the surface.

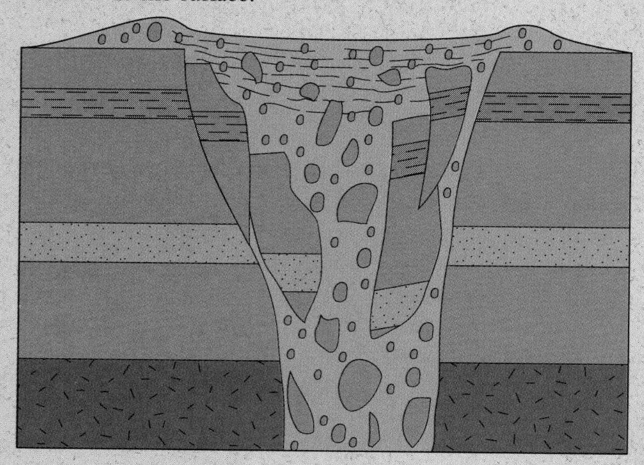

at 100 km would be so high that in spite of the great pressure, all known rocks would melt. Yet except for pockets of magma, it appears that the mantle is solid rather than liquid because it transmits S-waves. Accordingly, the geothermal gradient must decrease markedly.

Current estimates of the temperature at the base of the crust are 800° to 1,200°C. The latter figure seems to be an upper limit: if it were any higher, melting would be expected. Furthermore, fragments of mantle rock in kimberlite pipes (see Perspective 10-1), thought to have come from depths of about 100 to 300 km, appear to have reached equilibrium at these depths and at a temperature of about 1,200°C. At the core-mantle boundary, the temperature is probably between 3,500° and 5,000°C; the wide spread of values indicates the uncertainties of such estimates. If these figures are reasonably accurate, the geothermal gradient in the mantle is only about 1°C/km.

Because the core is so remote and its composition so uncertain, only very general estimates of its temperature can be made. Based on various experiments, the maximum temperature at the center of the core is estimated to be 6,500°C, very close to the estimated temperature for the surface of the Sun!

Heat Flow

Even though rocks are poor conductors of heat, detectable amounts of heat from the Earth's interior escape at the surface by **heat flow.** The amount of heat lost from within the Earth is small and can be detected only by sensitive temperature conditions at which graphite is favored over diamond (Figure 2). According to the data in Figure 2, the intersection of the diamond-graphite inversion curve with the geothermal gradient indicates that kimberlite magma came from a minimum depth of about 100 km.

Diamond can establish only a minimum depth for kimberlite because it is stable at any pressure greater than that occurring at a depth of 100 km. The silica found in kimberlite, on the other hand, is a form that indicates a maximum depth of about 300 km. Quartz is the form of silica found under low-pressure, low-temperature conditions. Under great pressure, however, the crystal structure of quartz changes to its high-pressure equivalent called coesite, and at even greater pressure it changes to stishovite.* Kimberlite pipes contain coesite but no stishovite, indicating that the kimberlite magma must have come from a depth of less than 300 km as indicated by the intersection of the coesite-stishovite inversion curve with the geothermal gradient (Figure 2).

*Coesite and stishovite are also known from other high-pressure environments such as meteorite impact sites.

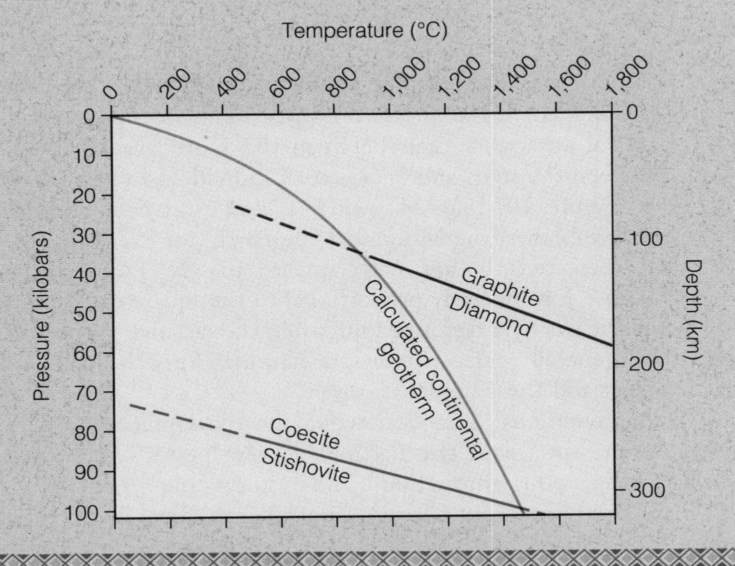

▶ **FIGURE 2** The forms of carbon and silica in kimberlite pipes provide information on the depth at which the magma formed. The presence of diamond and coesite in kimberlite indicates that the magma probably formed between 100 and 300 km as shown by the intersection of the calculated continental geotherm with the graphite-diamond and coesite-stishovite inversion curves.

instruments. Heavy, cylindrical probes are dropped into soft sea-floor sediments, and temperatures are measured at various depths along the cylinder. On the continents, temperature measurements are made at various depths in drill holes and mines.

As one would expect, heat flow is greater in areas of active or recently active volcanism. For example, greater heat flow occurs at spreading ridges, and lower than average values are recorded at subduction zones (▶ Figure 10-14). Higher values are also recorded in areas of continental volcanism, such as in Yellowstone National Park in Wyoming, Lassen National Park in California, and near Mount St. Helens in Washington. Any area possessing higher than average heat flow values is a potential area for the development of geothermal energy (see Chapter 16).

More than 70% of the total heat lost by the Earth is lost through the sea floor, but heat flow values for both ocean basins and continents decrease with increasing age. In the ocean basins heat flow is higher through younger oceanic crust. This result is expected, because high heat flow values occur at spreading ridges where oceanic crust is continually formed by igneous activity. Spreading ridges are also the sites of hydrothermal vents where considerable heat is transported upward by hydrothermal convection. Heat flow through the continental crust is not as well understood, but it too shows lower values for older crust.

It should be apparent that if heat is escaping from within the Earth, then its interior must be cooling unless a mechanism exists to replenish it. Radioactive decay generates heat continuously, but the quantity of radioactive isotopes (except

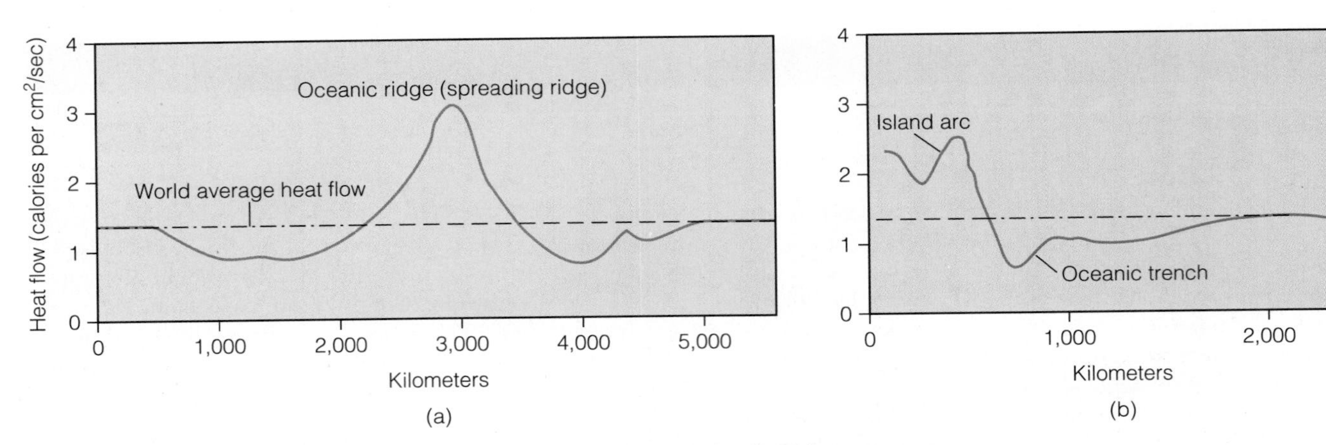

▶ **FIGURE 10-14** Variations in heat flow. (*a*) Higher than average heat flow occurs at spreading ridges and island arcs, both of which are characterized by volcanism. (*b*) Oceanic trenches show lower than average heat flow.

carbon 14) decreases through time as they decay to a stable state. In Perspective 5-1 we discussed the fact that ultramafic lava flows occurred during early Earth history, but largely ceased forming by the end of the Archean Eon about 2.5 billion years ago. Our conclusion was that the Earth possessed more internal heat during its early history and has been cooling continuously since then (see Perspective 3-1).

Measuring Gravity

Sir Isaac Newton (1642–1727) formulated the law of universal gravitation in which the force of gravitational attraction (F) between two masses (m_1 and m_2) is directly proportional to the products of their mass and inversely proportional to the square of the distance (D) between their centers of mass:

$$F = G\frac{m_1 \times m_2}{D^2}$$

G in this equation is the universal gravitational constant. This law applies to any two bodies. Consequently, a gravitational attraction exists between the Earth and its moon, between the stars, and between an individual and the Earth (▶ Figure 10-15a). However, F is greater between two massive bodies, the Moon and the Earth, for example, than between two bodies with smaller masses. Furthermore, because F is inversely proportional to the square of distance, it decreases by a factor of four when the distance is doubled. We generally refer to the gravitational force between an object and the Earth as its *weight*.

Gravitational attraction would be the same everywhere on the surface of the Earth if it were perfectly spherical, homogeneous throughout, and not rotating. As a consequence of rotation, however, a centrifugal force is generated

▶ **FIGURE 10-15** (*a*) The gravitational attraction of the Earth pulls all objects toward its center of mass. Objects 1 and 2 are the same distance from the Earth's center of mass, but the gravitational attraction on 1 is greater because it is more massive. Objects 2 and 3 have the same mass, but the gravitational attraction on 3 is four times less than on 2 because it is twice as far from the Earth's center of mass. (*b*) The Earth's rotation generates a centrifugal force that partly counteracts the force of gravity. Centrifugal force is zero at the poles and maximum at the equator.

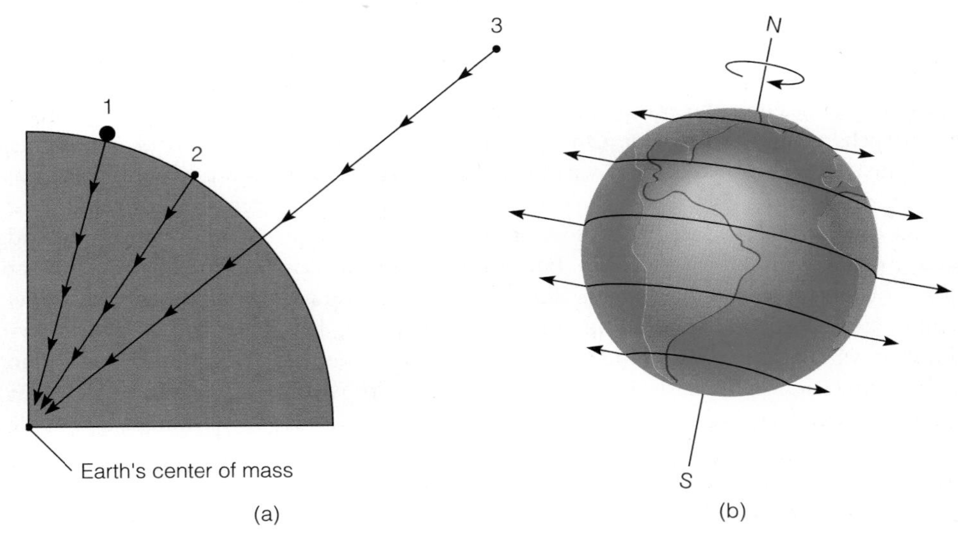

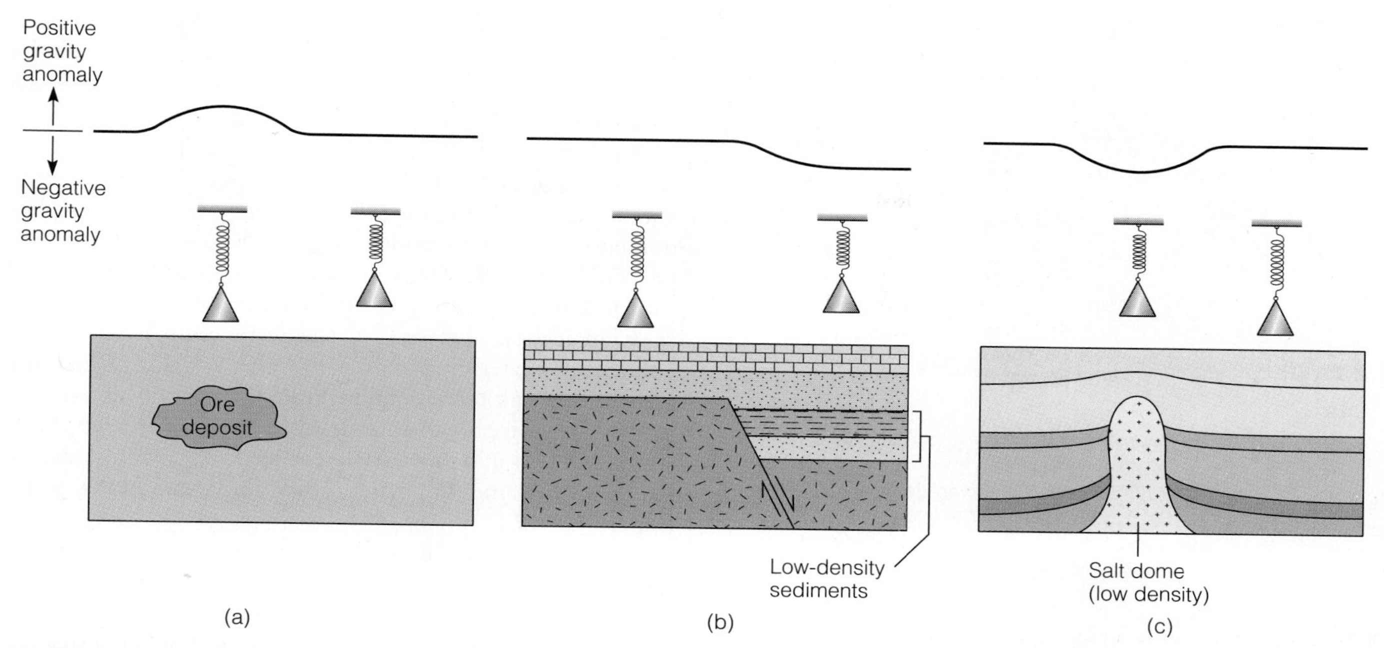

➢ **FIGURE 10-16** (*a*) The mass suspended from a spring in the gravimeter, shown diagrammatically, is pulled downward more over the dense body of ore than it is in adjacent areas, indicating a positive gravity anomaly. (*b*) A negative gravity anomaly over a buried structure. (*c*) Rock salt is less dense than most other types of rocks. A gravity survey over a salt dome shows a negative gravity anomaly.

that partly counteracts the force of gravity (Figure 10-15b). An object at the equator weighs slightly less than the same object would at the poles. The force of gravity also varies with distance between the centers of masses, so an object would weigh slightly less above the surface than if it were at sea level (Figure 10-15a).

Geologists use a sensitive instrument called a *gravimeter* to measure variations in the force of gravity. A gravimeter is simple in principle; it contains a weight suspended on a spring that responds to variations in gravity (➢ Figure 10-16). Gravimeters have been used extensively in exploration for hydrocarbons and mineral resources. Long ago geologists realized that gravity anomalies should exist over buried bodies of ore minerals and salt domes and that geologic structures such as faulted strata could be located by surface gravity surveys (Figure 10-16).

Gravity measurements would be higher over an iron ore deposit than over unconsolidated sediment because of the ore's greater density (Figure 10-16a). Such departures from the expected force of gravity are **gravity anomalies.** In other words, the measurement over the body of iron ore indicates an excess of dense material, or simply a *mass excess,* between the surface and the center of the Earth and is considered to be a **positive gravity anomaly.** A **negative gravity anomaly** indicating a *mass deficiency* exists over low-density sediments because the force of gravity is less than the expected average (Figure 10-16b). Large negative gravity anomalies also exist over salt domes (Figure 10-16c) and at subduction zones, indicating that the crust is not in equilibrium.

 THE PRINCIPLE OF ISOSTASY

More than 150 years ago, British surveyors in India detected a discrepancy of 177 m when they compared the results of two measurements between points 600 km apart. Even though this discrepancy was small, it was an unacceptably large error. The surveyors realized that the gravitational attraction of the nearby Himalaya Mountains probably deflected the plumb line (a cord with a suspended weight) of their surveying instruments from the vertical, thus accounting for the error. Calculations revealed, however, that if the Himalayas were simply thicker crust piled on denser material, the error should have been greater than that observed (➢ Figure 10-17).

In 1865, Sir George Airy proposed that in addition to projecting high above sea level, the Himalayas—and other mountains as well—also project far below the surface and thus have a low-density root (Figure 10-17b). In effect, he was saying that mountains float on denser rock at depth. Their excess mass above sea level is compensated for by a mass deficiency at depth, which would account for the observed deflection of the plumb line during the British survey (Figure 10-17).

Gravity studies have revealed that mountains do indeed have a low-density "root" projecting deep into the mantle. If it were not for this low-density root, a gravity survey across a mountainous area would reveal a huge positive gravity anomaly. The fact that no such anomaly exists indicates that a mass excess is not present, so some of the

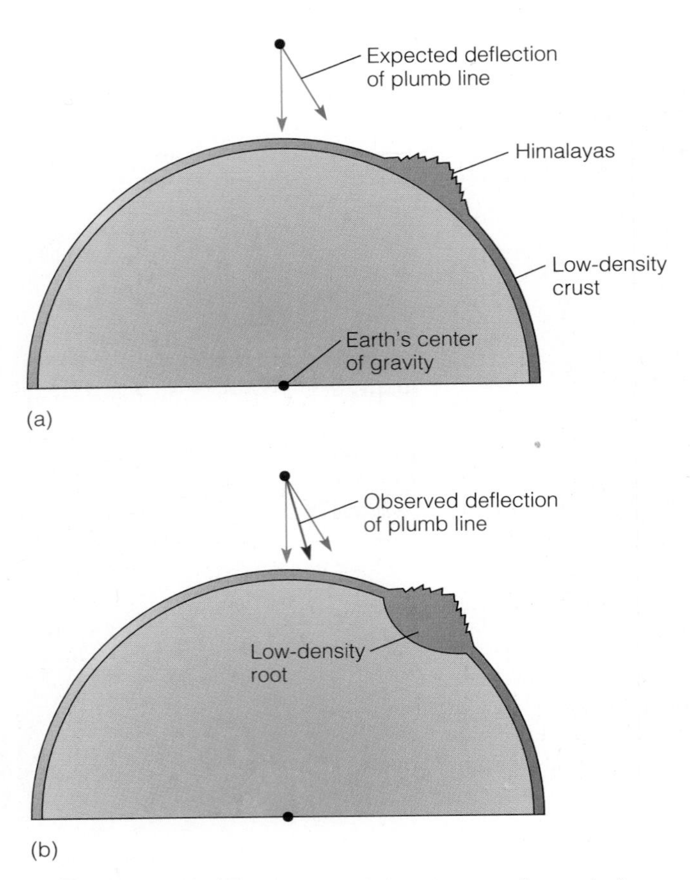

▶ FIGURE 10-17 (*a*) A plumb line is normally vertical, pointing to the Earth's center of gravity. Near a mountain range, one would expect the plumb line to be deflected as shown if the mountains were simply thicker, low-density material resting on denser material. (*b*) The actual deflection of the plumb line during the survey in India was less than expected. It was explained by postulating that the Himalayas have a low-density root.

dense mantle at depth must be displaced by lighter crustal rocks as shown in ▶ Figure 10-18. (Seismic wave studies also confirm the existence of low-density roots beneath mountains.)

Airy's proposal is now called the **principle of isostasy.** According to this principle, the Earth's crust is in floating equilibrium with the more dense mantle below. This phenomenon is easy to understand by analogy to an iceberg (▶ Figure 10-19). Ice is slightly less dense than water, and thus it floats. According to Archimedes' principle of buoyancy, an iceberg will sink in the water until it displaces a volume of water that equals its total weight. When the iceberg has sunk to an equilibrium position, only about 10% of its volume is above water level. If some of the ice above water level should melt, the iceberg will rise in order to maintain the same proportion of ice above and below water (Figure 10-19).

The Earth's crust is similar to the iceberg, or a ship, in that it sinks into the mantle to its equilibrium level. Where the crust is thickest, as beneath mountain ranges, it sinks further down into the mantle but also rises higher above the equilibrium surface (Figure 10-18). Continental crust being thicker and less dense than oceanic crust stands higher than the ocean basins. Should the crust be loaded, as where widespread glaciers accumulate, it responds by sinking further into the mantle to maintain equilibrium (▶ Figure 10-20). In Greenland and Antarctica the surface of the crust has been depressed below sea level by the weight of glacial ice. The crust also responds isostatically to widespread erosion and sediment deposition (▶ Figure 10-21).

Unloading of the Earth's crust causes it to respond by rising upward until equilibrium is again attained. This

▶ FIGURE 10-18 (*a*) Gravity measurements along the line shown would indicate a positive gravity anomaly over the excess mass of the mountains if the mountains were simply thicker crust resting on denser material below. (*b*) An actual gravity survey across a mountain region shows no departure from the expected and thus no gravity anomaly. Such data indicate that the mass of the mountains above the surface must be compensated for at depth by low-density material displacing denser material.

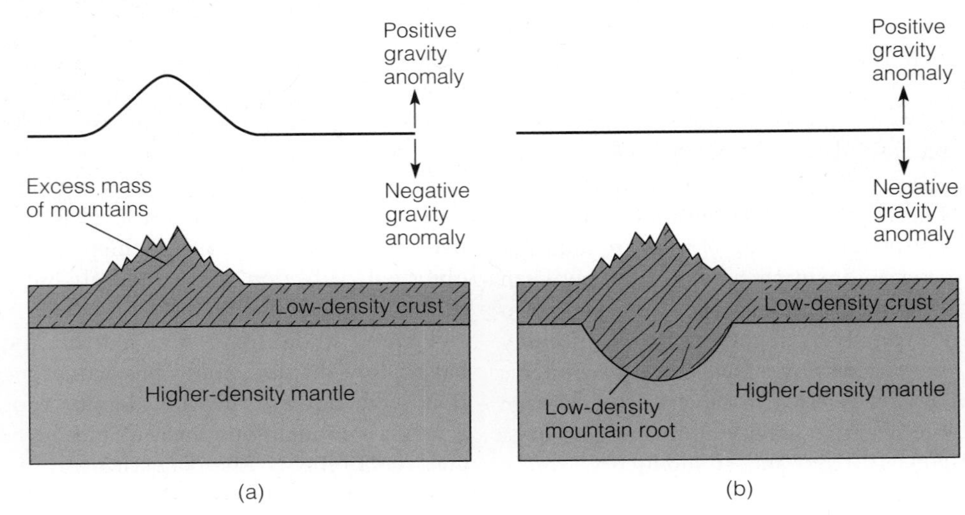

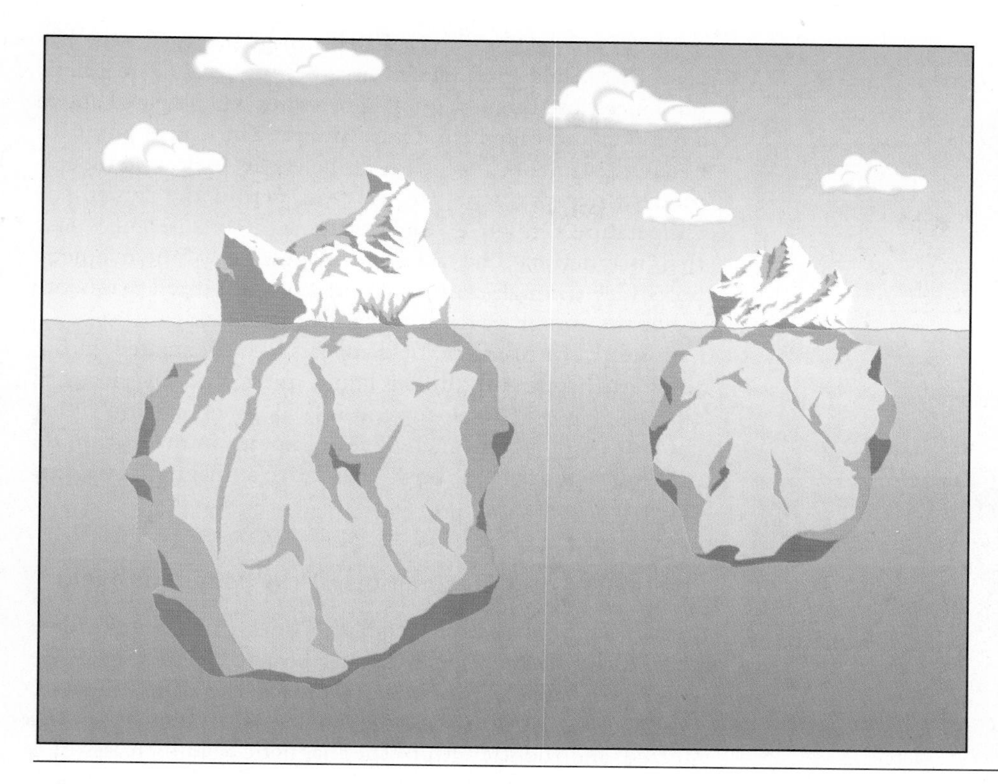

▶ **FIGURE 10-19** An iceberg sinks to an equilibrium position with about 10% of its mass above water level. The larger iceberg sinks farther below and rises higher above the water surface than does the smaller one. If some of the ice above water level should melt, the icebergs will rise to maintain the same proportion of ice above and below water level. The Earth's crust floating in more dense material below is analogous to this example.

phenomenon, known as **isostatic rebound,** occurs in areas that are deeply eroded and in areas that were formerly glaciated. Scandinavia, which was covered by a vast ice sheet until about 10,000 years ago, is still rebounding isostatically at a rate of up to 1 m per century (▶ Figure 10-22a). Coastal cities in Scandinavia have been uplifted rapidly enough that docks constructed several centuries ago are now far from shore. Isostatic rebound has also occurred in eastern Canada where the land has risen as much as 100 m during the last 6,000 years (Figure 10-22b).

If the principle of isostasy is correct, it implies that the mantle behaves as a liquid. In preceding discussions, however, we said that the mantle must be solid because it transmits S-waves, which will not move through a liquid. How can this apparent paradox be resolved? When considered in terms of the short time necessary for S-waves to pass through it, the mantle is indeed solid. But when subjected to stress over long periods of time, it will yield by flowage and at these time scales can be considered a viscous liquid. A familiar substance that has the properties of a solid or a liquid depending on how rapidly deforming forces are applied is

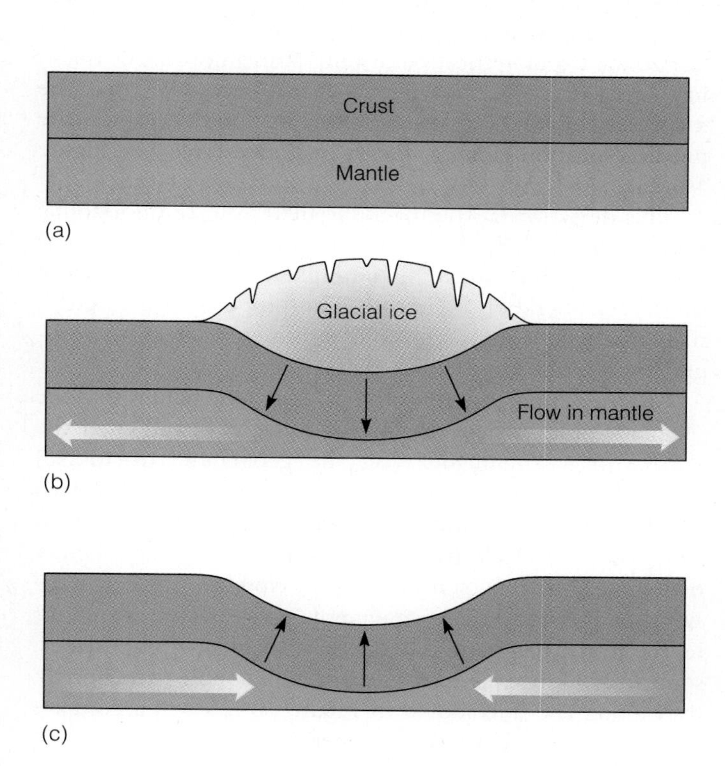

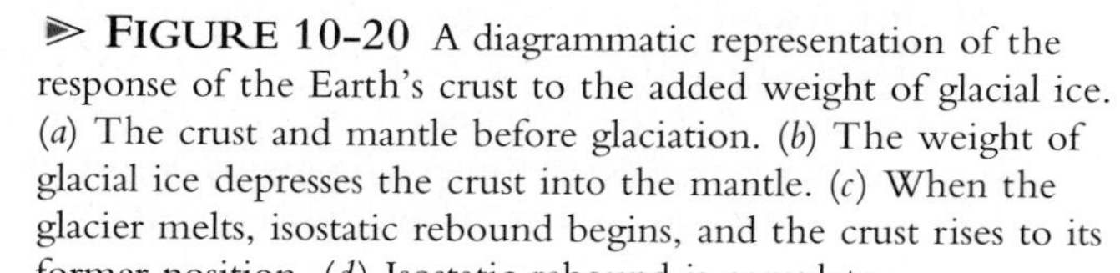

▶ **FIGURE 10-20** A diagrammatic representation of the response of the Earth's crust to the added weight of glacial ice. (*a*) The crust and mantle before glaciation. (*b*) The weight of glacial ice depresses the crust into the mantle. (*c*) When the glacier melts, isostatic rebound begins, and the crust rises to its former position. (*d*) Isostatic rebound is complete.

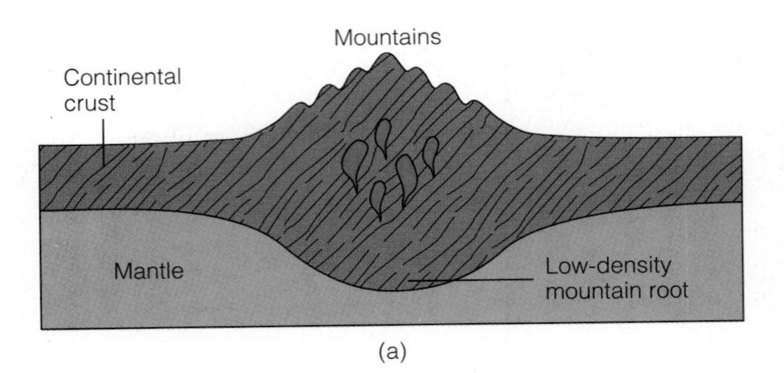

(a)

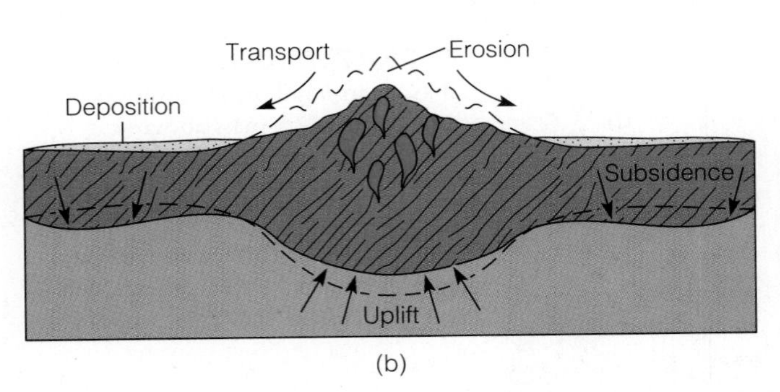

(b)

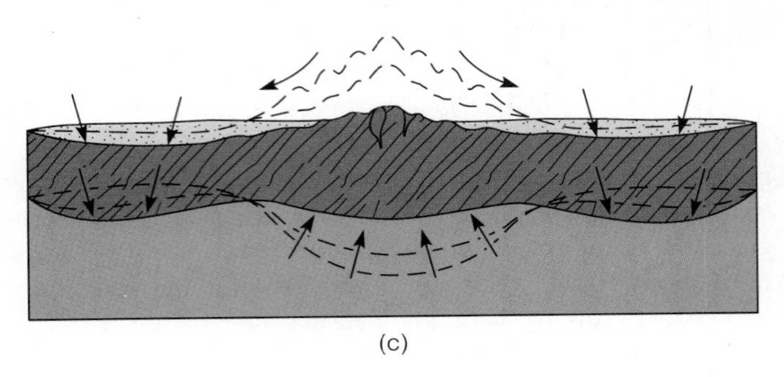

(c)

FIGURE 10-21 A diagrammatic representation showing the isostatic response of the crust to erosion (unloading) and widespread deposition (loading).

silly putty. It will flow under its own weight if given sufficient time, but shatters as a brittle solid if struck a sharp blow.

 THE EARTH'S MAGNETIC FIELD

A simple bar magnet has a **magnetic field**, an area in which magnetic substances are affected by lines of magnetic force radiating from the magnet (▶ Figure 10-23). The magnetic field shown in Figure 10-23 is *dipolar*, meaning that it possesses two unlike magnetic poles referred to as the north and south poles. The Earth possesses a dipolar magnetic field that resembles, on a large scale, that of a bar magnet (▶ Figure 10-24).

What is the source of this magnetic field? A number of naturally occurring minerals are magnetic, with *magnetite* being the most common and most magnetic. It is very unlikely, however, that the Earth's magnetic field is generated by a body of buried magnetite because magnetic substances lose their magnetic properties when heated above a temperature called the **Curie point.** The Curie point for magnetite is 580°C, which is far below its melting temperature. At a depth of 80 to 100 km within the Earth, the temperature is high enough that magnetic substances lose their magnetism. The fact that the locations of the magnetic poles vary through time also indicates that buried magnetite is not the source of the Earth's magnetic field.

Instead, the magnetic field appears to be generated within the Earth, apparently in the liquid outer core, by electrical currents (an electrical current is a flow of electrons that always generates a magnetic field). Experts on magnetism do not fully understand how the Earth's magnetic field is generated, however.

Inclination and Declination of the Magnetic Field

Notice in Figure 10-24 that the lines of magnetic force around the Earth parallel the Earth's surface only near the equator. As the lines of force approach the poles, they are oriented at increasingly large angles with respect to the surface, and the strength of the magnetic field increases; it is weakest at the equator and strongest at the poles. Accordingly, a compass needle mounted so that it can rotate both horizontally and vertically not only points north, but is also inclined with respect to the Earth's surface, except at the magnetic equator. The degree of inclination depends on the needle's location along a line of magnetic force (▶ Figure 10-25).

This deviation of the magnetic field from the horizontal is called **magnetic inclination.** To compensate for this, compasses used in the Northern Hemisphere have a small weight on the south end of the needle. This property of the Earth's magnetic field is important in determining the ancient geographic positions of tectonic plates (see Chapter 12).

Another important aspect of the magnetic field is that the magnetic poles, where the lines of force leave and enter the Earth, do not coincide with the geographic (rotational) poles. At present, an 11½° angle exists between the two (▶ Figure 10-26). Studies of the Earth's magnetic field show that the locations of the magnetic poles vary slightly over time, but they still correspond closely on the average with the locations of the geographic poles. A compass points to the north magnetic pole in the Canadian Arctic islands, some 1,290 km away from the geographic pole (true north); only along the line shown in Figure 10-26 will a compass needle point to both the magnetic and geographic north poles. From any other location, an angle called **magnetic declination** exists between lines drawn from the compass position to the magnetic pole and the geographic pole (Figure 10-26). Magnetic declination must be taken into account during surveying and navigation because, for most places on Earth, compass needles point east or west of true north.

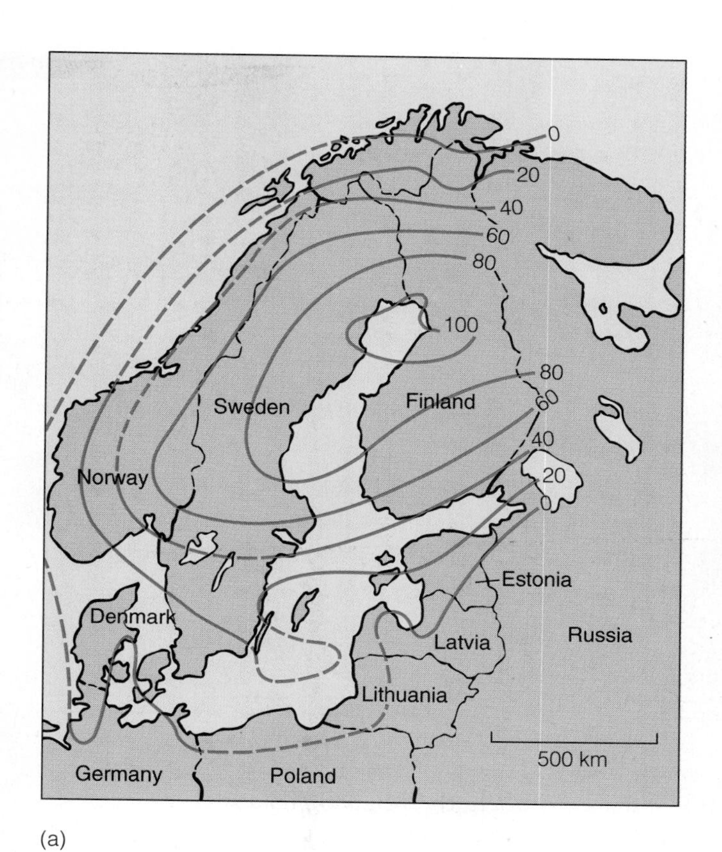

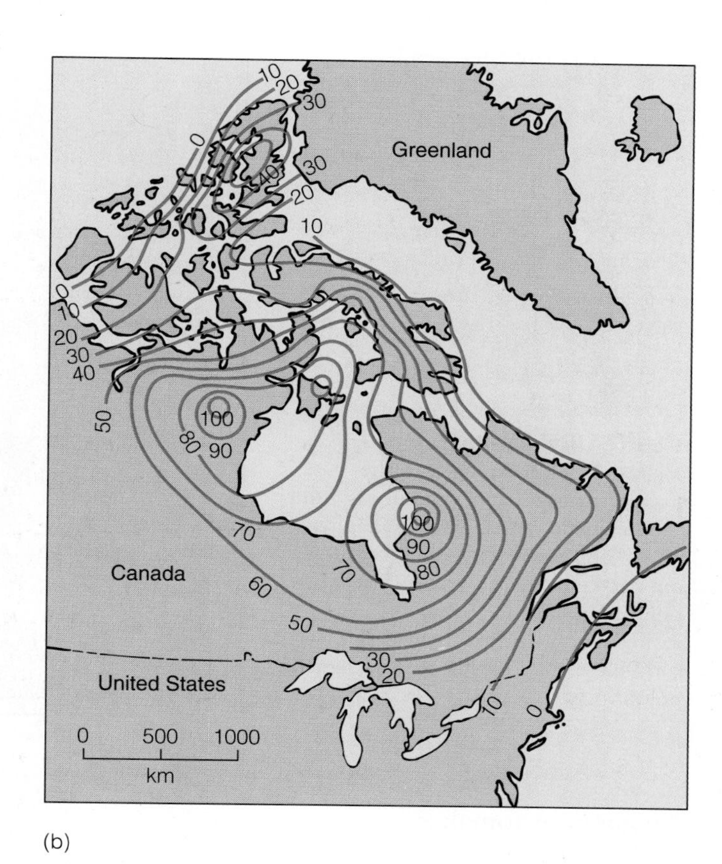

▶ **FIGURE 10-22** (*a*) Isostatic rebound in Scandinavia. The lines show rates of uplift in centimeters per century. (*b*) Isostatic rebound in eastern Canada in meters during the last 6,000 years.

▶ **FIGURE 10-23** Iron filings align themselves along the lines of magnetic force radiating from a magnet.

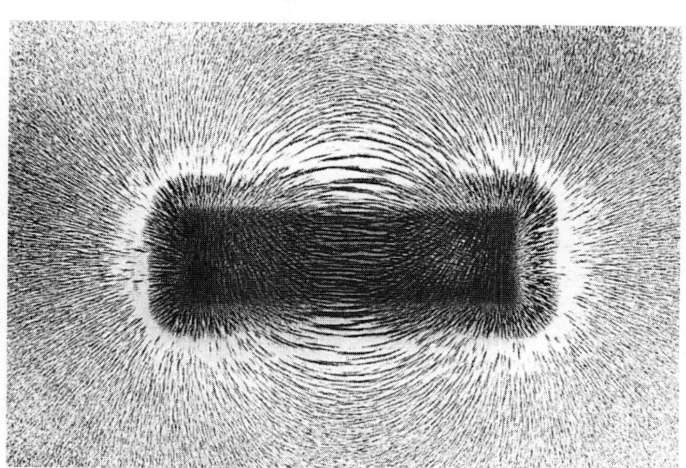

▶ **FIGURE 10-24** The magnetic field of the Earth has lines of force just like those of a bar magnet.

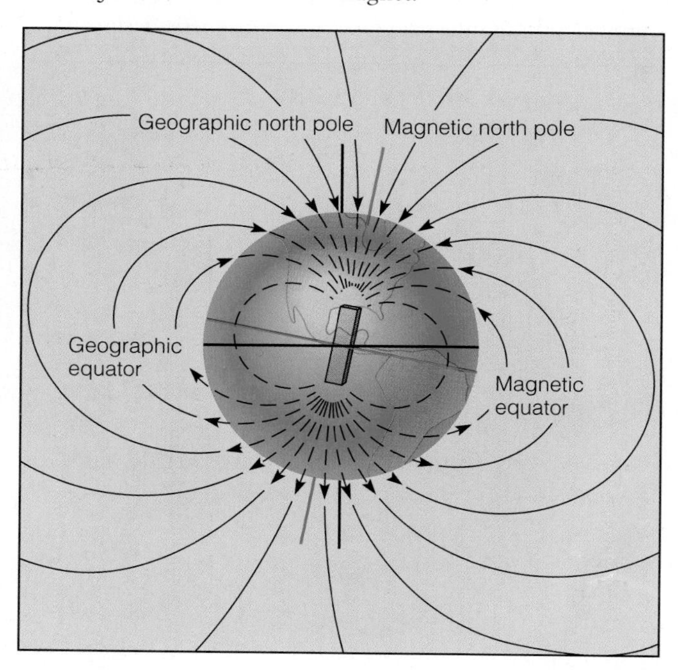

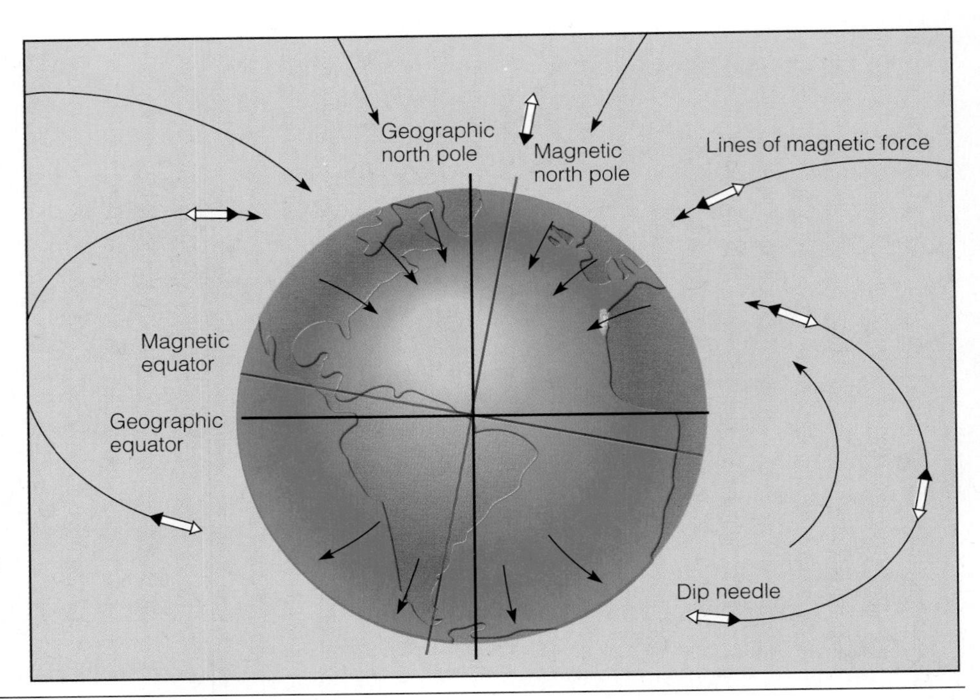

▶ **FIGURE 10-25** Magnetic inclination. The strength of the magnetic field changes uniformly from the magnetic equator to the magnetic poles. This change in strength causes a dip needle to parallel the Earth's surface only at the magnetic equator, whereas its inclination with respect to the surface increases to 90° at the magnetic poles.

Magnetic Anomalies

Variations in the strength of the Earth's magnetic field occur on both regional and local scales. Such variations from the normal are **magnetic anomalies.** Regional variations are most likely related to the complexities of convection within the outer core where the magnetic field is probably generated. Local variations can be accounted for by lateral or vertical variations in rock types within the crust.

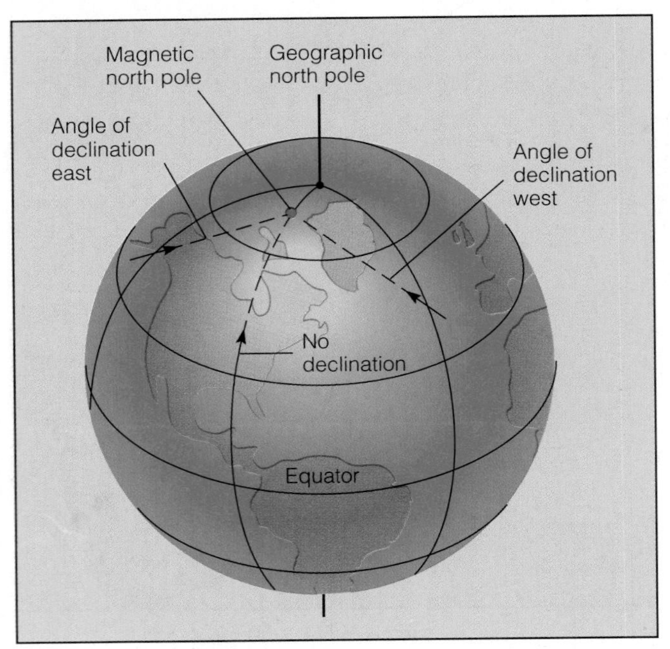

▶ **FIGURE 10-26** Magnetic declination. A compass needle points to the magnetic north pole rather than the geographic pole (true north). The angle formed by the lines from the compass position to the two poles is the magnetic declination.

An instrument called a *magnetometer* can detect slight variations in the strength of the magnetic field, and deviations from the normal are characterized as *positive* or *negative.* A **positive magnetic anomaly** exists in areas where the rocks contain more iron-bearing minerals than elsewhere. In the Great Lakes region of the United States and Canada, huge iron ore deposits containing hematite and magnetite add their magnetism to that of the Earth's magnetic field; the result is a positive magnetic anomaly (▶ Figure 10-27a). Positive magnetic anomalies also exist where extensive basaltic volcanism has occurred because basalt contains appreciable quantities of iron-bearing minerals (Figure 10-27b). Areas underlain by basalt lava flows, such as the Columbia River basalts of the northwestern United States (see Figure 4-18), possess positive magnetic anomalies, whereas an adjacent area underlain by sedimentary rocks shows a **negative magnetic anomaly** (Figure 10-27b).

Geologists have used magnetometers for magnetic surveys for decades because iron-bearing rocks can be easily detected by a positive magnetic anomaly even if they are deeply buried. In addition, magnetometers can detect a variety of buried geologic structures, such as salt domes, which show negative magnetic anomalies (▶ Figure 10-28); these can be detected by gravity surveys as well.

Magnetic Reversals

When a magma cools through the Curie point, its iron-bearing minerals gain their magnetization and align them-

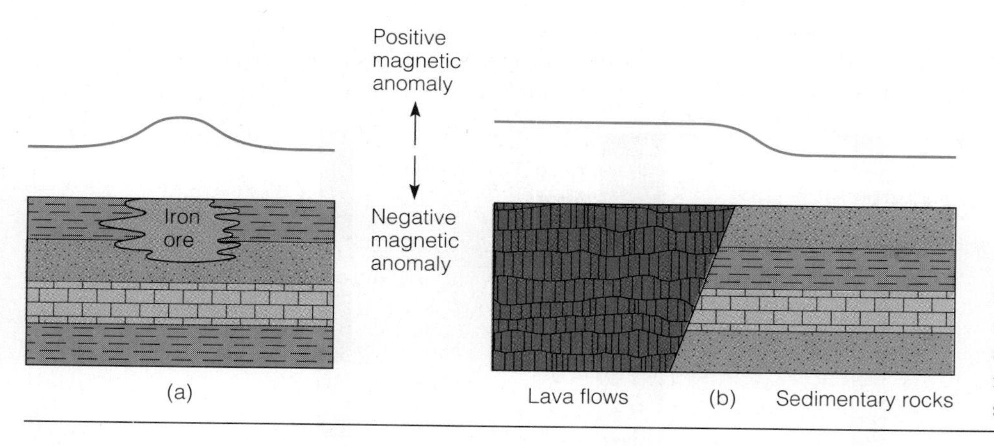

▶ **FIGURE 10-27** (*a*) A positive magnetic anomaly over an iron ore deposit. (*b*) A positive magnetic anomaly over lava flows and a negative anomaly over adjacent sedimentary rocks.

selves with the Earth's magnetic field, recording both its direction and strength. As long as the rock is not subsequently heated above the Curie point, it will preserve that magnetism. If the rock is heated above the Curie point, however, the original magnetism is lost, and when the rock subsequently cools, the iron-bearing minerals will align with the current magnetic field.

The iron-bearing minerals of some sedimentary rocks (especially those that formed on the deep sea floor) are also oriented parallel to the Earth's magnetic field as the sediments are deposited. These rocks also preserve a record of the Earth's magnetic field at the time of their formation. Such information preserved in lava flows and some sedimentary rocks can be used to determine the directions to the Earth's magnetic poles and the latitude of the rock when it was formed.

Paleomagnetism is simply the remanent magnetism in ancient rocks that records the direction and strength of the Earth's magnetic field at the time of their formation. Geologists refer to the Earth's present magnetic field as normal, that is, with the north and south magnetic poles located roughly at the north and south geographic poles. As early as 1906, though, rocks were discovered that showed reversed magnetism. Paleomagnetic studies initially conducted on continental lava flows have clearly shown that the Earth's magnetic field has completely reversed itself numerous times during the geologic past (▶ Figure 10-29). When these **magnetic reversals** occur, the Earth's magnetic polarity is reversed, so that the north arrow on a compass would point south rather than north.

Rocks that have a record of magnetism the same as the present magnetic field are described as having **normal polarity**, whereas rocks with the opposite magnetism have **reversed polarity.** The ages of the normal and reversed polarity events for the past several million years have been determined by applying absolute dating techniques to continental lava flows and have been used to construct a magnetic reversal time scale (▶ Fig 10-30). These same patterns of normal and reversed polarity were soon discovered in the oceanic crust (see Chapter 12).

▶ **FIGURE 10-28** A negative magnetic anomaly over a salt dome.

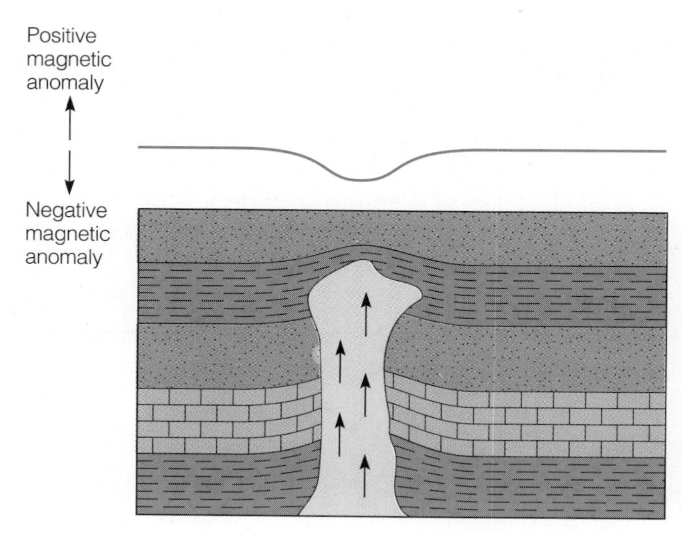

▶ **FIGURE 10-29** Magnetic reversals recorded in a succession of lava flows are shown diagrammatically by red arrows, and the record of normal polarity events is shown by black arrows. The lava flows containing a record of these magnetic-polarity events can be radiometrically dated so that a magnetic time scale as in Figure 10-30 can be constructed.

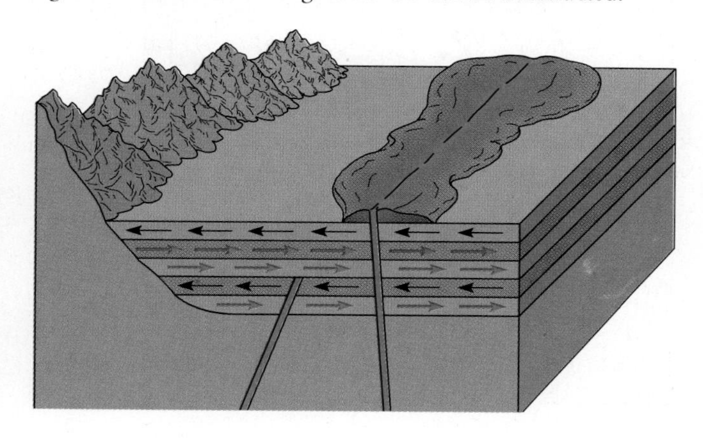

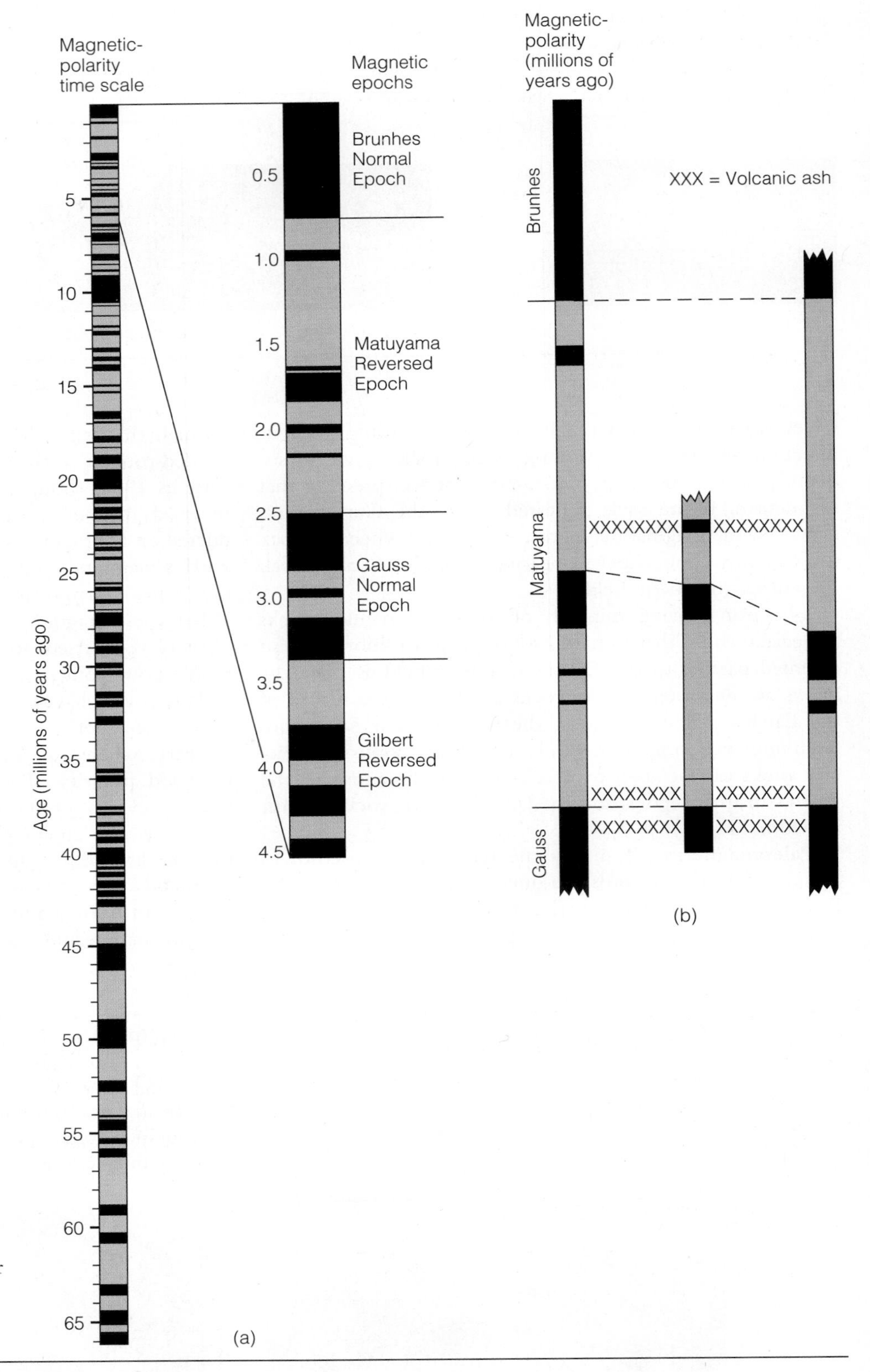

▶ **FIGURE 10-30** (*a*) Normal (black) and reversed polarity events for the last 66 million years. (*b*) Rocks in northern Pakistan correlated with the magnetic-polarity time scale.

The cause of magnetic reversals is not completely known, although they appear to be related to changes in the intensity of the Earth's magnetic field. Calculations indicate that the magnetic field has weakened about 5% during the last century. If this trend continues, there will be a period during the next few thousand years when the magnetic field will be nonexistent and then will reverse. After the reversal occurs, the magnetic field will rebuild itself with opposite polarity.

Chapter Summary

1. The Earth is concentrically layered into oceanic crust and continental crust, a rocky mantle, and an iron-rich core with a solid inner part and a liquid outer part.
2. Much of the information about the Earth's interior has been derived from studies of P- and S-waves that travel through the Earth. Laboratory experiments, comparisons with meteorites, and studies of inclusions in volcanic rocks provide additional information.
3. The Earth's interior is subdivided into concentric layers on the basis of changes in seismic wave velocities at discontinuities.
4. Density and elasticity of Earth materials determine the velocity of seismic waves. Seismic waves are refracted when their direction of travel changes. Wave reflection occurs at boundaries across which the properties of rocks change.
5. The behavior of P- and S-waves within the Earth and the presence of P- and S-wave shadow zones allow geologists to estimate the density and composition of the Earth's interior and to estimate the size and depth of the core and mantle.
6. The Earth's inner core is thought to be composed of iron and nickel, whereas the outer core is probably composed mostly of iron with 10 to 20% other substances. Peridotite is the most likely component of the mantle.
7. The oceanic and continental crusts are of basaltic and granitic composition, respectively. The boundary between the crust and the mantle is the Mohorovičić discontinuity.
8. The geothermal gradient of 25°C/km cannot continue to great depths, otherwise most of the Earth would be molten. The geothermal gradient for the mantle and core is probably about 1°C/km. The temperature at the Earth's center is estimated to be 6,500°C.
9. Detectable amounts of heat escape at the Earth's surface by heat flow. Most of the Earth's internal heat is generated by radioactive decay.
10. According to the principle of isostasy, the Earth's crust is floating in equilibrium with the denser mantle below. Continental crust stands higher than oceanic crust because it is thicker and less dense.
11. Positive and negative gravity anomalies can be detected where excesses and deficiencies of mass occur, respectively. Gravity surveys are useful in exploration for minerals and hydrocarbons.
12. The Earth's magnetic field is thought to be generated by electrical currents in the outer core.
13. The Earth is surrounded by lines of magnetic force similar to those of a bar magnet. The lines of magnetic force are inclined with respect to the Earth's surface, except at the equator, accounting for the phenomenon of magnetic inclination.
14. Although the magnetic poles are close to the geographic poles, they do not coincide exactly. For most places on Earth, an angle called magnetic declination exists between lines drawn from a compass location to the magnetic and geographic north poles.
15. A magnetometer can detect departures from the normal magnetic field, which can be either positive or negative.
16. Although the cause of magnetic reversal is not fully understood, it is clear that the polarity of the magnetic field has completely reversed itself many times during the past.

Important Terms

asthenosphere
continental crust
core
crust
Curie point
discontinuity
geothermal gradient
gravity anomaly (positive and negative)
heat flow
isostatic rebound
lithosphere
low-velocity zone
magnetic anomaly (positive and negative)
magnetic declination
magnetic field
magnetic inclination
magnetic reversal
mantle
Mohorovičić discontinuity (Moho)
normal polarity
oceanic crust
paleomagnetism
principle of isostasy
P-wave shadow zone
reflection
refraction
reversed polarity
S-wave shadow zone

Review Questions

1. The average density of the Earth is ______ g/cm^3.
   a. ____ 12.0;
   b. ____ 5.5;
   c. ____ 6.75;
   d. ____ 1.0;
   e. ____ 2.5.
2. A line showing the direction of movement of a small part of a wave front is a:
   a. ____ seismic discontinuity;
   b. ____ P-wave reflection;
   c. ____ wave ray;
   d. ____ particle beam;
   e. ____ seismic gradient.
3. When seismic waves travel through materials having different properties, their direction of travel changes. This phenomenon is wave:
   a. ____ elasticity;
   b. ____ energy dissipation;
   c. ____ refraction;
   d. ____ deflection;
   e. ____ reflection.
4. A major seismic discontinuity at a depth of 2,900 km is the:
   a. ____ core-mantle boundary;
   b. ____ oceanic crust–continental crust boundary;
   c. ____ Moho;
   d. ____ inner core–outer core boundary;
   e. ____ lithosphere-asthenosphere boundary.
5. The Earth's core is probably composed mostly of:
   a. ____ sulfur;
   b. ____ silica;
   c. ____ nickel;
   d. ____ potassium;
   e. ____ iron.
6. Which of the following provides evidence for the composition of the core?
   a. ____ inclusions in volcanic rocks;
   b. ____ diamonds;
   c. ____ meteorites;
   d. ____ peridotite;
   e. ____ S-wave shadow zone.
7. The seismic discontinuity at the base of the crust is the:
   a. ____ magnetic anomaly;
   b. ____ Moho;
   c. ____ geothermal gradient;
   d. ____ high-velocity zone;
   e. ____ transition zone.
8. Continental crust has an overall composition corresponding closely to that of:
   a. ____ basalt;
   b. ____ sandstone;
   c. ____ granodiorite;
   d. ____ iron-nickel alloy;
   e. ____ gabbro.
9. Oceanic crust is:
   a. ____ 20 to 90 km thick;
   b. ____ thinnest at spreading ridges;
   c. ____ granitic in composition;
   d. ____ less dense than continental crust;
   e. ____ the primary source of magma.
10. Most of the Earth's internal heat is generated by:
    a. ____ moving plates;
    b. ____ volcanism;
    c. ____ earthquakes;
    d. ____ radioactive decay;
    e. ____ meteorite impacts.
11. According to the principle of isostasy:
    a. ____ more heat escapes from oceanic crust than from continental crust;
    b. ____ the Earth's crust is floating in equilibrium with the more dense mantle below;
    c. ____ the Earth's crust behaves both as a liquid and a solid;
    d. ____ much of the asthenosphere is molten;
    e. ____ magnetic anomalies result when the crust is loaded by glacial ice.
12. The magnetic field is thought to be generated by:
    a. ____ the tilt of the Earth's rotational axis;
    b. ____ the solar wind;
    c. ____ electrical currents in the outer core;
    d. ____ deformation of the asthenosphere;
    e. ____ a large deposit of magnetite at the North Pole.
13. Except at the magnetic equator, a compass needle in the Northern Hemisphere points to the magnetic north pole and downward from the horizontal. This phenomenon is:
    a. ____ magnetic declination;
    b. ____ magnetic reflection;
    c. ____ magnetic reversal;
    d. ____ magnetic polarity;
    e. ____ magnetic inclination.
14. Iron-bearing minerals in a magma gain their magnetism and align themselves with the magnetic field when they cool through the:
    a. ____ negative magnetic anomaly;
    b. ____ Curie point;
    c. ____ isostasy curve;
    d. ____ magnetic-polarity field;
    e. ____ magnetic declination.
15. What determines the velocity of P- and S-waves?
16. Explain how seismic waves are refracted and reflected.
17. What is the significance of the S-wave shadow zone?
18. Why is the inner core thought to be solid whereas the outer core is probably liquid?
19. Several seismic discontinuities exist within the mantle. What accounts for these discontinuities?
20. Explain the reasoning used by Mohorovičić to determine that a discontinuity, now called the Moho, exists between the crust and the mantle.
21. How do oceanic and continental crust differ in composition and thickness?
22. What is the geothermal gradient? Why must it decrease within the Earth?
23. If the continental crust is deeply eroded in one area and loaded by widespread, thick sedimentary deposits in another, how will it respond isostatically at each location?
24. What is meant by positive and negative gravity anomalies? Give examples of where each type of anomaly might occur.
25. What is the magnetic field, and how is it thought to be generated?
26. Explain the phenomenon of magnetic inclination.
27. Illustrate how a vertical succession of ancient lava flows preserves a record of magnetic reversals.

Points to Ponder

1. Use the law of universal gravitation to explain how negative and positive gravity anomalies arise.
2. If the Earth were completely solid and had the same composition and density throughout, how would P- and S-waves behave as they traveled through the Earth?
3. What factors account for higher-than-average heat flow values at spreading ridges? Also, how is heat flow related to the age of crustal rocks?

Additional Readings

Anderson, D. L., and A. M. Dziewonski. 1984. Seismic tomography. *Scientific American* 251, no. 4: 60–68.

Bolt, B. A. 1982. *Inside the Earth: Evidence from earthquakes.* San Francisco: W. H. Freeman and Co.

Bonatti, E. 1994. The Earth's mantle below the oceans. *Scientific American* 270, no. 3: 44–51.

Brown, G. C. 1981. *The inaccessible Earth.* London: George Allen & Unwin.

Fowler, C. M. R. 1990. *The solid Earth.* New York: Cambridge University Press.

Heppenheimer, T. A. 1987. Journey to the center of the Earth. *Discover* 8, no. 10: 86–93.

Jeanloz, R. 1990. The nature of the Earth's crust. *Annual Review of Earth Sciences* 18: 357–86.

Jeanloz, R., and T. Lay. 1993. The core-mantle boundary. *Scientific American* 268, no. 5: 48–55.

McKenzie, D. P. 1983. The Earth's mantle. *Scientific American* 249, no. 3: p. 66–78.

Monastersky, R. 1988. Inner space. *Science News* 136: 266–268.

Monastersky, R. 1994. Scrambled Earth: Researchers look deep to learn how the planet cools its heat. *Science News* 145, no. 15: 235–237.

Chapter 11

THE SEA FLOOR

OUTLINE

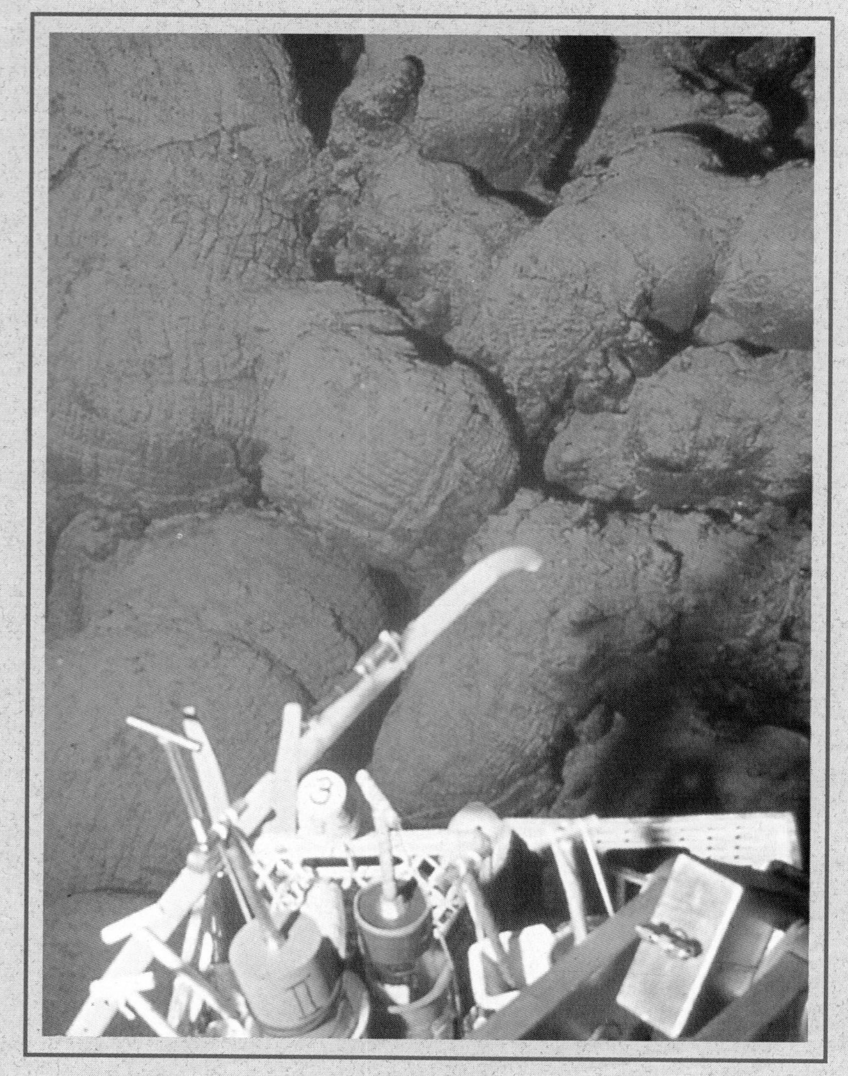

Pillow lava on the floor of the Pacific Ocean near the Galápagos Islands.

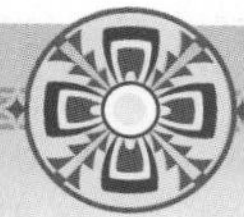

PROLOGUE

In 1979, researchers aboard the submersible *Alvin* descended about 2,500 m to the Galapagos Rift in the eastern Pacific Ocean basin and observed hydrothermal vents on the sea floor (▶ Figure 11-1). Such vents occur near spreading ridges where seawater seeps down into the oceanic crust through cracks and fissures, is heated by the hot rocks, and then rises and is discharged onto the sea floor as hot springs. During the 1960s, hot metal-rich brines apparently derived from hydrothermal vents were detected and sampled in the Red Sea (▶ Figure 11-2). These dense brines were concentrated in pools along the axis of the sea; beneath them thick deposits of metal-rich sediments were found. During the early 1970s, researchers observed hydrothermal vents on the Mid-Atlantic Ridge about 2,900 km east of Miami, Florida, and in 1978 moundlike mineral deposits were sampled from the East Pacific Rise just south of the Gulf of California.

When the submersible *Alvin* descended to the Galapagos Rift in 1979, mounds of metal-rich sediments were observed. Near these mounds the researchers saw what they called black smokers (chimneylike vents) discharging plumes of hot, black water (Figure 11-1). Since 1979 similar vents have been observed at or near spreading ridges in several other areas.

Submarine hydrothermal vents are interesting for several reasons. Near the vents live communities of organisms, including bacteria, crabs, mussels, starfish, and tubeworms, many of which had never been seen before (Figure 11-1). In most biological communities, photosyn-

▶ FIGURE 11-1 The submersible *Alvin* sheds light on hydrothermal vents at the Galapagos Rift, a branch of the East Pacific Rise. Seawater seeps down through the oceanic crust, becomes heated, and then rises and builds chimneys composed of anhydrite ($CaSO_4$) and sulfides of iron, copper, and zinc. The plume of "black smoke" is simply heated water saturated with dissolved minerals. Communities of organisms, including tubeworms, giant clams, crabs, and several types of fish, live near the vents.

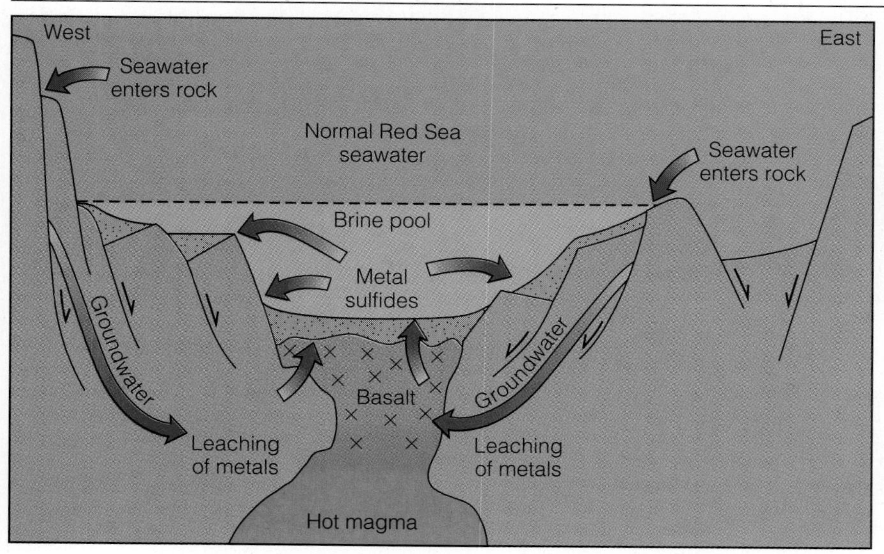

▶ FIGURE 11-2 Cross section of the Red Sea. Water seeps down along fractures, becomes heated, and dissolves metals from rocks of the oceanic crust. The hot water then rises, but because it is denser than seawater, it is trapped in the Atlantis II Deep where metals are precipitated from solution.

thesizing organisms form the base of the food chain and provide nutrients for the herbivores and carnivores. In vent communities, though no sunlight is available for photosynthesis, and the base of the food chain consists of bacteria that practice chemosynthesis; they oxidize sulfur compounds from the hot vent waters, thus providing their own nutrients and the nutrients for other members of the food chain.

Another interesting aspect of these submarine hydrothermal vents is their economic potential. When seawater circulates downward through the oceanic crust, it is heated to as much as 400°C. The hot water then reacts with the crust and is transformed into a metal-bearing solution. As the hot solution rises and discharges onto the sea floor, it cools, precipitating iron, copper, and zinc sulfides and other minerals that accumulate to form a chimneylike vent. These vents are ephemeral, however; one observed in 1979 was inactive six months later. When their activity ceases, the vents eventually collapse and are incorporated into a moundlike mineral deposit.

The economic potential of hydrothermal vent deposits is tremendous. The deposits in the Atlantis II Deep of the Red Sea contain an estimated 100 million tons of metals, including iron, copper, zinc, silver, and gold. These deposits are fully as large as the major sulfide deposits mined on land, such as the Troodos Massif on Cyprus which is thought to have formed on the sea floor by hydrothermal vent activity.

Hydrothermal vent sulfide deposits have formed throughout geologic time. None are currently being mined, but the technology to exploit them exists. In fact, the Saudi Arabian and Sudanese governments have determined that it is feasible to recover these deposits from the Red Sea and are making plans to do so.

INTRODUCTION

About 71% of the Earth's surface is covered by an interconnected body of saltwater we refer to as oceans and seas. Four areas of this saltwater body are distinct enough to be recognized as the Pacific, Atlantic, Indian, and Arctic oceans (► Figure 11-3). The Pacific is by far the largest, containing almost 53% of all water on Earth (◉ Table 11-1). *Sea* is a term designating a smaller body of water, usually a marginal part of an ocean (Figure 11-3). Sea is also used for some bodies of water completely enclosed by land, such as the Dead Sea and the Caspian Sea, but these are actually lakes.

During its very earliest history, the Earth was probably hot and airless and had no surface water. Volcanic activity was more common than it is at present because the Earth possessed more heat, and this activity is thought to have been responsible for the formation of the atmosphere and surface water. As we noted in Chapter 4, volcanoes emit a variety of gases, the most abundant of which is water vapor. The atmosphere and surface waters are thought to have been derived from within the Earth and been emitted at the surface by volcanoes in a process called *outgassing* (► Figure 11-4). As the Earth cooled, water vapor began condensing and fell as rain, which accumulated to form the surface waters. Geologic evidence clearly indicates that an extensive ocean was present more than 3.5 billion years ago.

During most of historic time, people knew little of the oceans and, until fairly recently, thought that the sea floor was flat and featureless. Although the ancient Greeks had determined the size of the Earth rather accurately, Western Europeans were not aware of the vastness of the oceans until the fifteenth and sixteenth centuries when various explorers sought new trade routes to the Indies. When Christopher Columbus set sail on August 3, 1492, in an attempt to find a route to the Indies, he greatly underestimated the width of the Atlantic Ocean. Contrary to popular belief, Columbus was not attempting to demonstrate that the Earth is a sphere—the Earth's spherical shape was well accepted by then. The controversy was over the Earth's circumference

TABLE 11-1 Numerical Data for the Oceans

Ocean*	Surface Area (Million km^2)	Water Volume (Million km^3)	Average Depth (km)	Maximum Depth (km)
Pacific	180	700	4.0	11.0
Atlantic	93	335	3.6	9.2
Indian	77	285	3.7	7.5
Arctic	15	17	1.1	5.2

*Excludes adjacent seas, such as the Caribbean Sea and Sea of Japan, which are marginal parts of oceans.
SOURCE: Pinet, P. R. 1992. *Oceanography.*

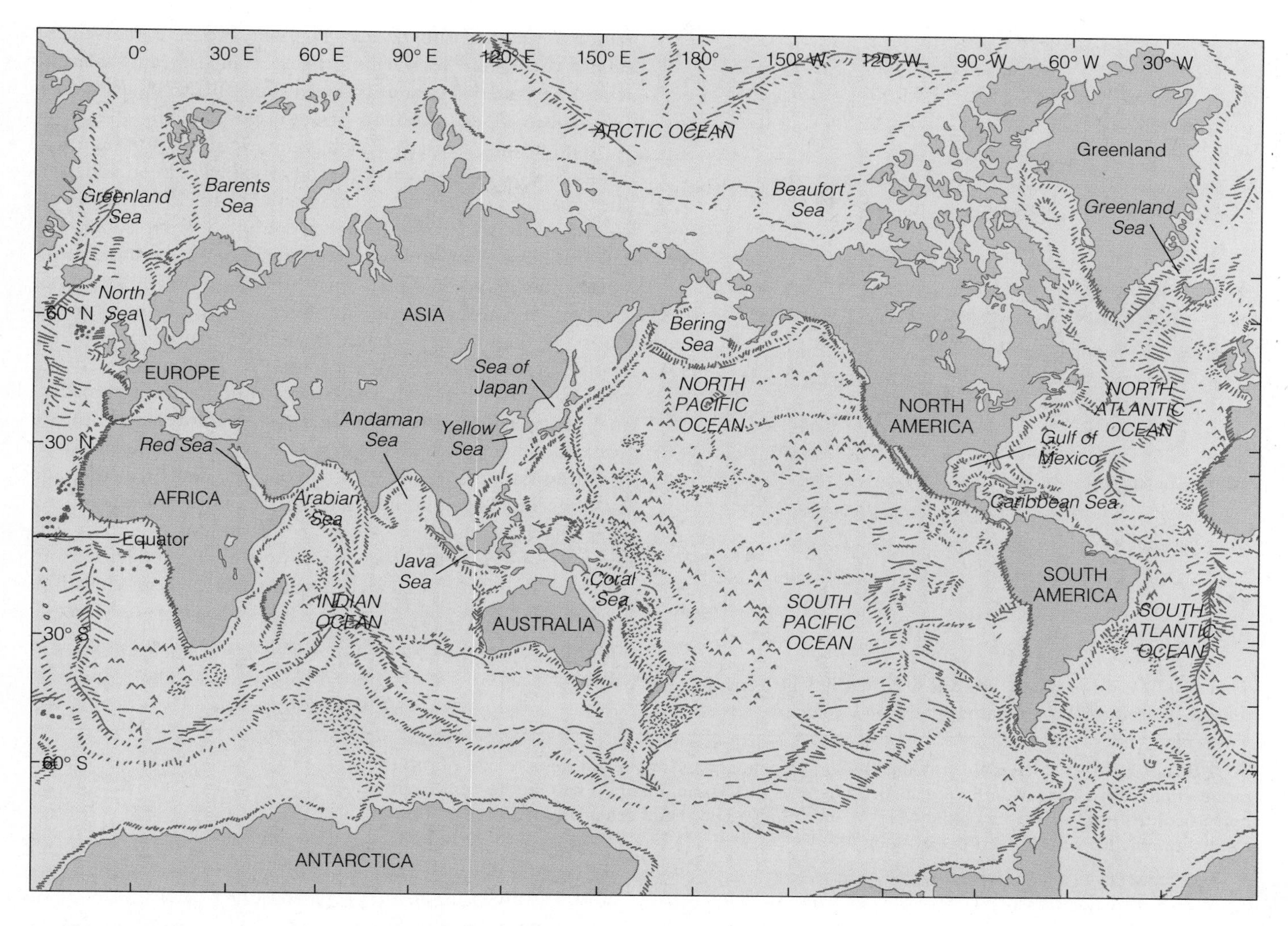

➤ **FIGURE 11-3** This map shows the geographic limits of the four major oceans and many of the various seas of the world.

and what was the shortest route to China. During these and subsequent voyages, Europeans sailed to the Americas, the Pacific Ocean, Australia, New Zealand, the Hawaiian Islands, and many other islands previously unknown to them.

Such voyages added considerably to our knowledge of the oceans, but truly scientific investigations did not begin until the late 1700s. Great Britain was the dominant maritime power, and in order to maintain that dominance, the British sought to increase their knowledge of the oceans. The earliest British scientific voyages were led by Captain James Cook in 1768, 1772, and 1777. In 1872, the converted British warship H.M.S. *Challenger* began a four-year voyage, during which seawater was sampled and analyzed, oceanic depths were determined at nearly 500 locations, rock and sediment samples were recovered from the sea floor, and more than 4,000 new marine species were classified.

Continuing exploration of the oceans revealed that the sea floor is not flat and featureless as formerly believed. Indeed, scientists discovered that the sea floor possesses varied topography including oceanic trenches, submarine ridges, broad plateaus, hills, and vast plains (➤ Figure 11-5). Some people have suggested that some of these features are remnants of the mythical lost continent of Atlantis (see Perspective 11-1).

 OCEANOGRAPHIC RESEARCH

The Deep Sea Drilling Project, an international program sponsored by several oceanographic institutions and funded by the National Science Foundation, began in 1968. Its first research vessel, the *Glomar Challenger*, was capable of drilling in water more than 6,000 m deep. It was equipped to drill into and recover long cores of sea-floor sediment and the oceanic crust. During the next 15 years, the *Glomar Challenger* drilled more than 1,000 holes in the sea floor.

The Deep Sea Drilling Project came to an end in 1983 when the *Glomar Challenger* was retired. However, an inter-

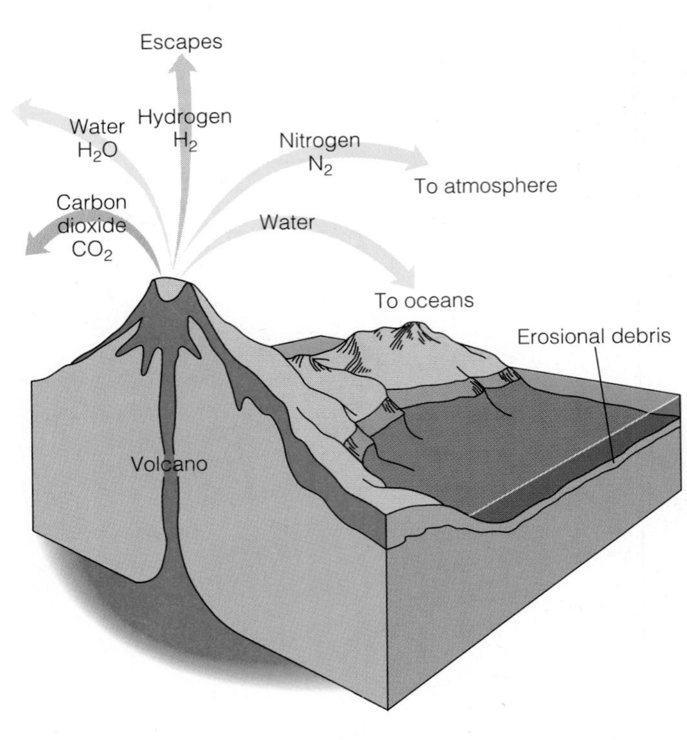

▶ **FIGURE 11-4** Gases derived from within the Earth by outgassing formed the early atmosphere and surface waters.

national project, the Ocean Drilling Program, continued where the Deep Sea Drilling Project left off, and a larger, more advanced research vessel, the JOIDES* *Resolution* (▶ Figure 11-6), made its first voyage in 1985.

In addition to surface vessels, submersibles, some remotely controlled and others carrying scientists, have been added to the research arsenal of oceanographers. In 1985 the *Argo*, towed by a surface vessel and equipped with sonar and television systems, provided the first views of the British ocean liner R.M.S. *Titanic* since it sank in 1912. The U.S. Geological Survey is using a towed device to map the sea floor (▶ Figure 11-7). The system uses sonar to produce images resembling aerial photographs. Researchers aboard the submersible *Alvin* have observed submarine hydrothermal vents (see the Prologue) and have explored parts of the oceanic ridge system.

The first measurements of the oceanic depths were made by lowering a weighted line to the sea floor and measuring the length of the line. Now an instrument called an *echo sounder* is used. Sound waves from a ship are reflected from the sea floor and detected by instruments on the ship,

*JOIDES is an acronym for Joint Oceanographic Institutions for Deep Earth Sampling.

▶ **FIGURE 11-5** This map showing the varied topography of the sea floor resulted from the pioneering work of Maurice Ewing, Bruce Heezen, and Marie Tharp. Their studies and maps helped confirm the existence of the Mid-Atlantic Ridge and made other important contributions to our knowledge of the deep-ocean floor.

➢ **FIGURE 11-6** The *JOIDES Resolution* is an oceanographic research vessel 300 m long.

yielding a continuous profile of the sea floor. Depth is determined by knowing the velocity of sound waves in water and the time it takes for the waves to reach the sea floor and return to the ship.

Seismic profiling is similar to echo sounding but even more useful. Strong waves are generated at an energy source, the waves penetrate the layers beneath the sea floor, and some of the energy is reflected back to the surface from various geologic horizons (see Figure 3, Perspective 8-1). Recall from Chapter 10 that seismic waves are reflected from boundaries where the properties of Earth materials change. Seismic profiling has been particularly useful in mapping the structure of the oceanic crust beneath sea-floor sediments.

➢ **FIGURE 11-7** The sonar system used by the U.S. Geological Survey for sea-floor mapping.

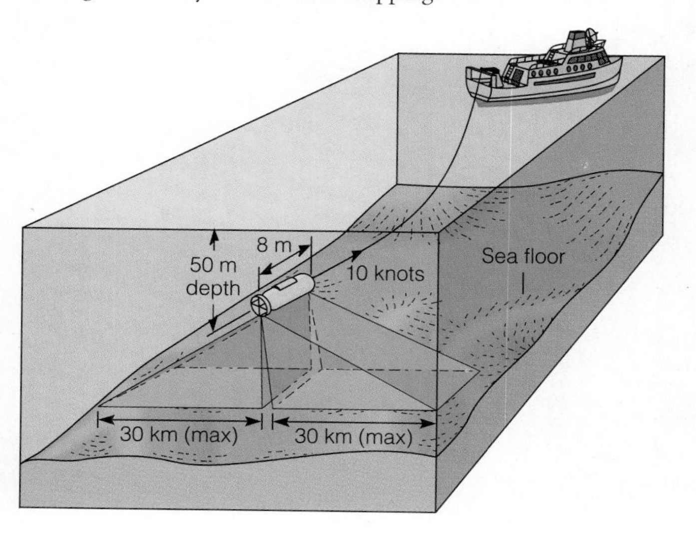

Oceanographers also use gravity surveys to detect gravity anomalies. Salt domes beneath the continental margins are recognized by negative gravity anomalies, and oceanic trenches also exhibit negative gravity anomalies. Magnetic surveys have also provided important information about the sea floor (see Chapter 12).

Although scientific investigations of the oceans have been yielding important information for more than two hundred years, much of our current knowledge has been acquired since World War II. This statement is particularly true with respect to the sea floor, because only in recent decades has instrumentation been available to study this largely hidden domain. The data collected are not only important in their own right but also have provided much of the evidence supporting plate tectonic theory (see Chapter 12).

 CONTINENTAL MARGINS

Continental margins are zones separating the part of a continent above sea level from the deep-sea floor. The conti-

Perspective 11-1

LOST CONTINENTS

Most people have heard of the mythical lost continent of Atlantis, but few are aware of the source of the Atlantis legend or the evidence that is cited for the former existence of this continent. Only two known sources of the Atlantis legend exist, both written in about 350 B.C. by the Greek philosopher Plato. In two of his philosophical dialogues, the *Timaeus* and the *Critias*, Plato tells of Atlantis, a large island continent that, according to him, was located in the Atlantic Ocean west of the Pillars of Hercules, which we now call the Strait of Gibraltar (▶ Figure 1). Plato also wrote that following the conquest of Atlantis by Athens, the continent disappeared:

> . . . there were violent earthquakes and floods and one terrible day and night came when . . . Atlantis . . . disappeared beneath the sea. And for this reason even now the sea there has become unnavigable and unsearchable, blocked as it is by the mud shallows which the island produced as it sank.*

If one assumes that the destruction of Atlantis was a real event, rather than one conjured up by Plato to make a philosophical point, he nevertheless lived long after it was supposed to have occurred. According to Plato, Solon, an Athenian who lived about 200 years before Plato, heard the story from Egyptian priests who claimed the event had occurred 9,000 years before their time. Solon told the story to his grandson, Critias, who in turn told it to Plato.

Present-day proponents of the Atlantis legend generally cite two types of evidence to support their claim that Atlantis did indeed exist. First, they point to supposed cultural similarities on opposite sides of the Atlantic Ocean basin, such as the similar shapes of the pyramids of Egypt and those of Central and South America. They contend that these similarities are due to cultural diffusion from the highly developed civilization of Atlantis. According to archaeologists, however, few similarities actually exist, and those that do can be explained as the independent development of analogous features by different cultures.

Secondly, supporters of the legend assert that remnants of the sunken continent can be found. No "mud shallows" exist in the Atlantic as Plato claimed, but the Azores, Bermuda, the Bahamas, and the Mid-Atlantic Ridge are alleged to be remnants of Atlantis. If a continent had actually sunk in the Atlantic it could be easily detected by a gravity survey. Recall that continental crust has a granitic composition and a lower density than oceanic crust. So if a continent were actually present beneath the Atlantic Ocean, there would be a huge negative gravity anomaly, but no such anomaly has been detected. Furthermore, the crust beneath the Atlantic has been drilled in many places, and all the samples recovered indicate that its composition is the same as that of oceanic crust elsewhere.

In short, there is no geological evidence for Atlantis. Nevertheless, some archaeologists think that the legend may be based on a real event. About 1390 B.C., a huge volcanic eruption destroyed the island of Thera in the Mediterranean Sea, which was an important center of early Greek civilization. The eruption was one of the most violent during historic time, and much of the island disappeared when it subsided to form a caldera (▶ Figure 2). Most of the island's inhabitants escaped (▶ Figure 3), but the eruption probably contributed to the demise of the Minoan culture on Crete. At least 10 cm of ash fell on parts of Crete, and the coastal areas of the island were probably devastated by tsunami. It is possible that Plato used an account of the destruction of Thera, but fictionalized it for his own purposes, thereby giving rise to the Atlantis legend.

▶ **FIGURE 1** According to Plato, Atlantis was a large continent west of the Pillars of Hercules, which we now call the Strait of Gibraltar.

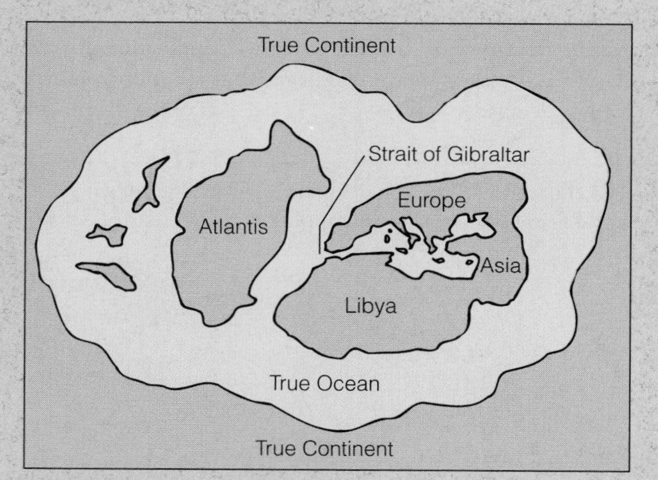

*From the *Timaeus*. Quoted in E. W. Ramage, ed., *Atlantis: Fact or Fiction?* (Bloomington, Ind.: Indiana University Press, 1978), p. 13.

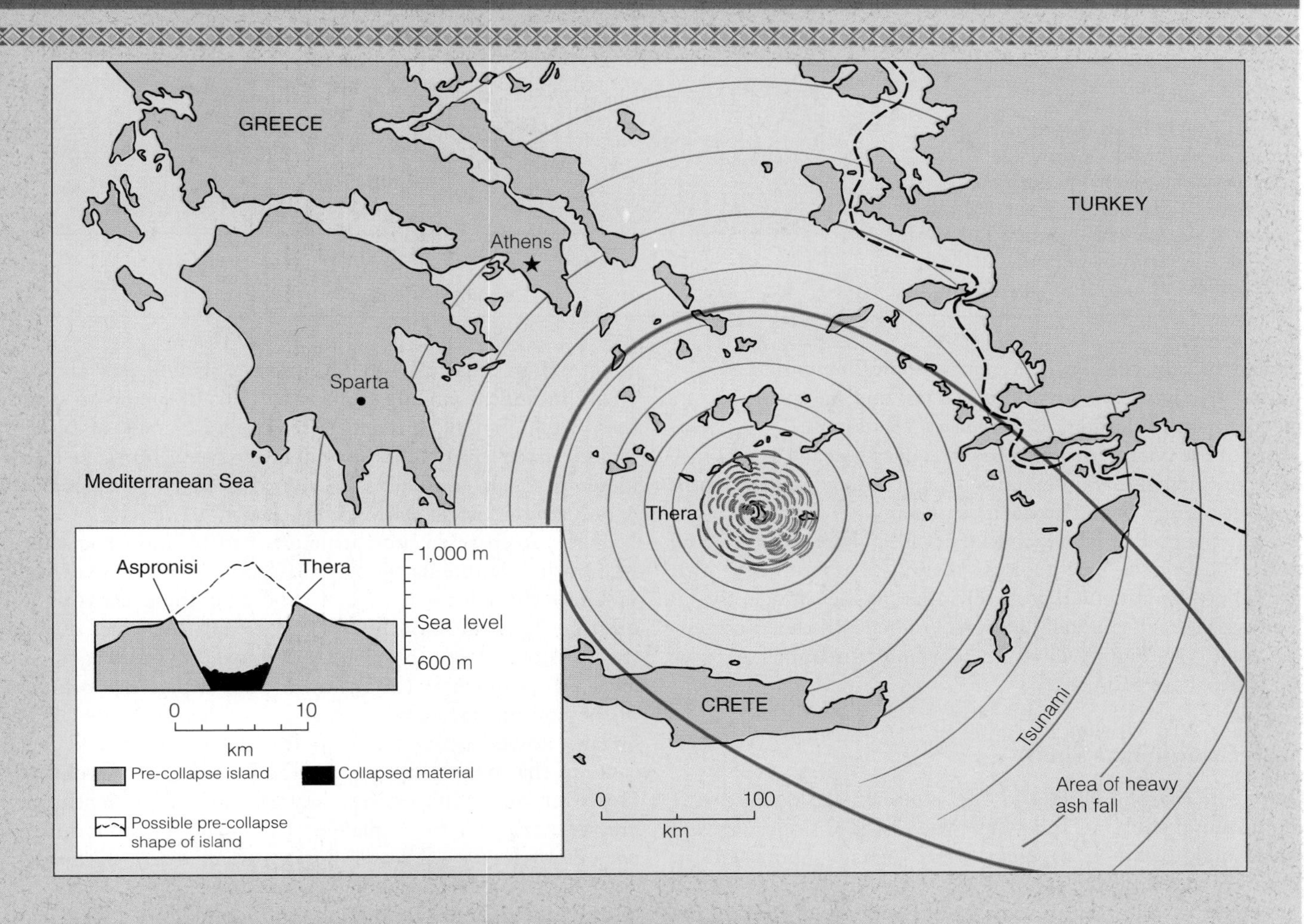

FIGURE 2 (above) The island of Thera was destroyed by a huge eruption in about 1390 B.C. Ash was carried more than 950 km to the southeast, and tsunami probably devastated nearby coastal areas. The inset shows the possible profile of the island before the eruption and its shape immediately after the caldera formed.

FIGURE 3 (left) An artist's rendition of the volcanic eruption on Thera that destroyed most of the island in about 1390 B.C. Most of the island's inhabitants escaped the devastation.

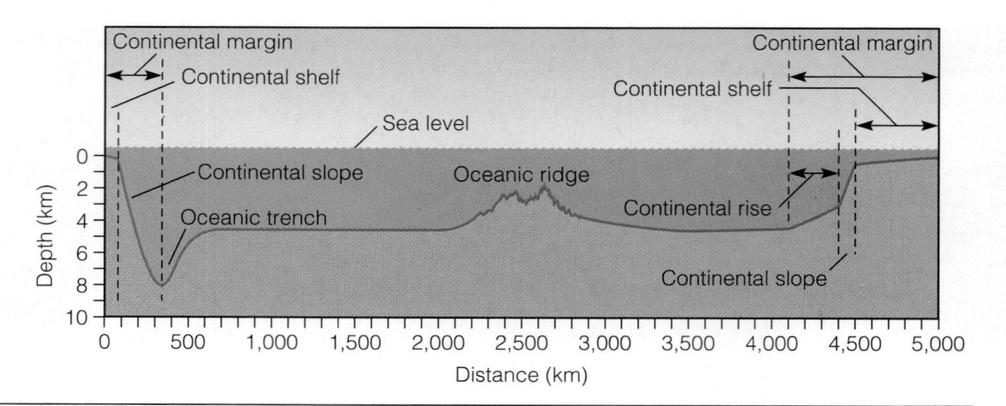

▶ **FIGURE 11-8** A generalized profile of the sea floor showing features of the continental margins. The vertical dimensions of the features in this profile are greatly exaggerated because the vertical and horizontal scales differ.

nental margin consists of a gently sloping *continental shelf*, a more steeply inclined *continental slope*, and, in some cases, a deeper, gently sloping *continental rise* (▶ Figure 11-8). Seaward of the continental margin is the deep-ocean basin. Thus, the continental margin extends to increasingly greater depths until it merges with the deep-sea floor.

Most people perceive continents as land areas outlined by sea level. But the true geologic margin of a continent—that is, where continental crust changes to oceanic crust—is below sea level, generally somewhere beneath the continental slope (▶ Figure 11-9). Accordingly, marginal parts of continents are submerged.

The Continental Shelf

Between the shoreline and the continental slope lies the **continental shelf,** an area where the sea floor slopes seaward at an angle of much less than 1° (on average, about 0.1° or about 2 m/km). The outer edge of the continental shelf is generally taken to correspond to the point at which the inclination of the sea floor increases rather abruptly to several degrees; this *shelf-slope break* occurs at an average depth of about 135 m (Figure 11-8). Continental shelves vary considerably in width, ranging from a few tens of meters to more than 1,000 km. The shelf along the east coast of North America, for example, is as much as several hundred kilometers across in some places, whereas along the west coast it is only a few kilometers wide.

Deep, steep-sided submarine canyons are most characteristic of the continental slope, but some of them extend well up onto the continental shelf. Some of these canyons lie offshore from the mouths of large streams. At times during the Pleistocene Epoch (1,600,000 to 10,000 years ago), sea level was as much as 130 m lower than at present, so much of the continental shelves was above sea level (Figure 11-9). Streams flowed across these exposed shelves and eroded deep canyons that were subsequently flooded when sea level rose. However, most submarine canyons extend to depths far greater than can be explained by stream erosion during periods of lower sea level. Furthermore, many submarine canyons are not associated with streams on land. They are discussed more fully in a following section.

As a consequence of lower sea level during the Pleistocene Epoch, much of the sediment on continental shelves

▶ **FIGURE 11-9** The transition from continental to oceanic crust, and hence the geologic margin of a continent, occurs beneath the continental slope. At times of lower sea level during the Pleistocene Epoch, large parts of the continental shelves were exposed. Accordingly, much of the sediment deposited during these times accumulated in various continental environments.

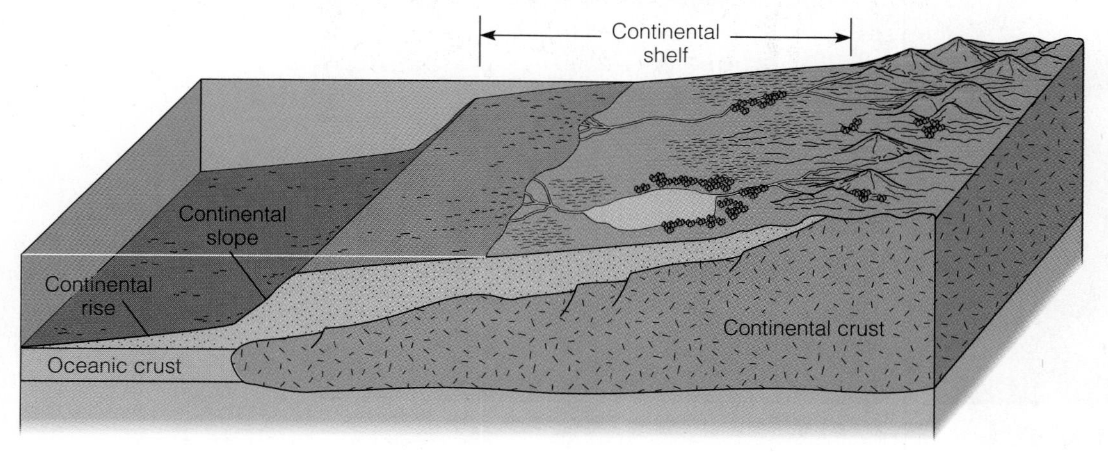

accumulated in stream channels and on floodplains (Figure 11-9). In fact, in areas such as northern Europe and parts of North America, glaciers extended onto the exposed shelves and deposited gravel, sand, and mud. Since the Pleistocene Epoch, sea level has risen submerging the shelf sediments, which are now being reworked by marine processes. That these sediments were, in fact, deposited on land is indicated by evidence of human settlements and fossils of mammoths and mastodons (extinct elephants) and other land-dwelling animals.

The Continental Slope and Rise

The seaward margin of the continental shelf is marked by the *shelf-slope break* (at an average depth of 135 m) where the relatively steep **continental slope** begins (Figure 11-8). Continental slopes average about 4°, but range from 1° to 25°. In many places, especially around the margins of the Atlantic, the continental slope merges with the more gently sloping **continental rise**. In other places, such as around the Pacific Ocean, slopes commonly descend directly into an oceanic trench, and a continental rise is absent (Figure 11-8).

The shelf-slope break is a very important feature in terms of sedimentation. Landward from the break, the shelf is affected by waves and tidal currents. Seaward of the break, the bottom sediments are completely unaffected by surface processes, and their transport onto the slope and rise is controlled by gravity. The continental slope and rise system is the area where most of the sediment derived from continents is eventually deposited. Much of this sediment is transported by turbidity currents through submarine canyons.

Turbidity Currents, Submarine Canyons, and Submarine Fans

Turbidity currents are sediment-water mixtures denser than normal seawater that flow downslope to the deep-sea floor (see Figure 6-21). An individual turbidity current flows onto the relatively flat sea floor where it slows and begins depositing sediment; the coarsest particles are deposited first, followed by progressively smaller particles, thus forming graded bedding (see Figure 6-21). These deposits accumulate as a series of overlapping **submarine fans**, which constitute a large part of the continental rise (➢ Figure 11-10). At their seaward margins, these fans grade into the deposits of the deep-ocean basins.

No one has ever observed a turbidity current in progress, so for many years there was considerable debate about their existence. In 1971 abnormally turbid water was sampled just above the sea floor in the North Atlantic, indicating that a turbidity current had occurred recently. Furthermore, sea-floor samples from many areas show a succession of graded beds and the remains of shallow-water organisms that were apparently displaced into deeper water.

Perhaps the most compelling evidence for the existence of turbidity currents is the pattern of trans-Atlantic cable breaks that occurred south of Newfoundland on November 18, 1929 (➢ Figure 11-11). Initially, it was assumed that an earthquake that occurred on that date had ruptured several trans-Atlantic telephone and telegraph cables. However, while the breaks on the continental shelf near the epicenter occurred when the earthquake struck, cables farther seaward were broken later and in succession. The last cable to break was 720 km from the source of the earthquake, and it did not snap until 13 hours after the first break occurred (Figure

➢ **FIGURE 11-10** Submarine fans formed by the deposition of sediments carried down submarine canyons by turbidity currents. Much of the continental rise is composed of overlapping submarine fans.

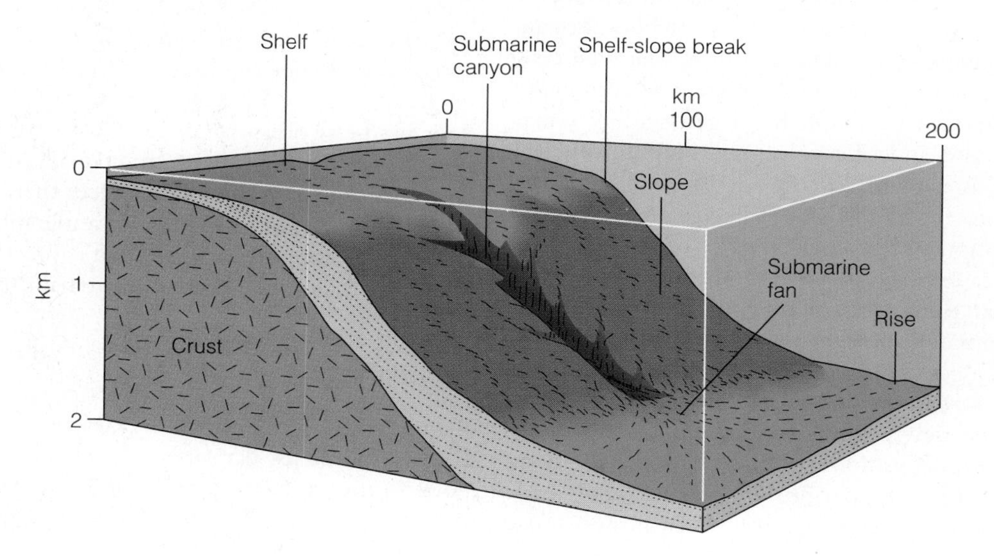

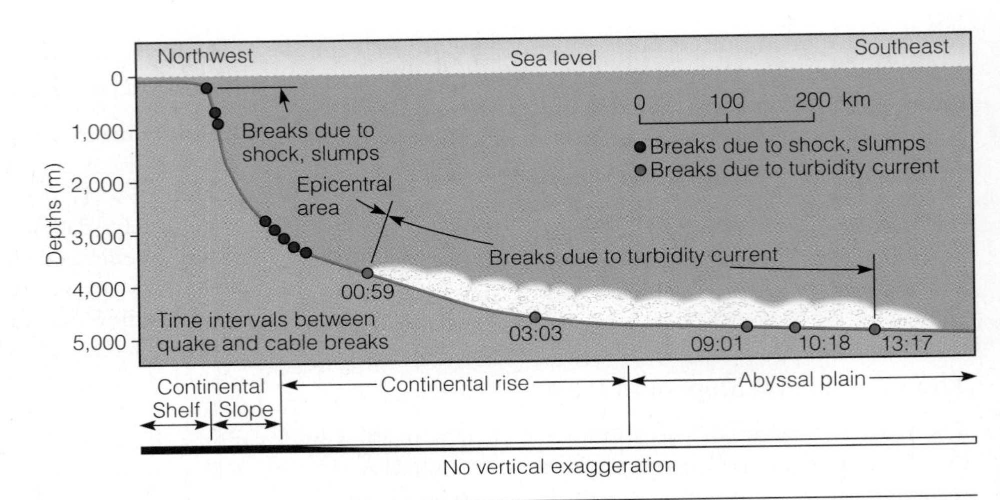

▶ **FIGURE 11-11** Submarine cable breaks caused by an earthquake-generated turbidity current south of Newfoundland. This profile of the sea floor shows the locations of the cables and the times at which they were severed. The vertical dimension in this profile is highly exaggerated. The profile labeled "no vertical exaggeration" shows what the sea floor actually looks like in this area.

11-11). In 1949, geologists realized that the earthquake had generated a turbidity current that moved downslope, breaking the cables in succession. The precise time each cable broke was known, so it was a simple matter to calculate the velocity of the turbidity current. It apparently moved at about 80 km/hr on the continental slope, but slowed to about 27 km/hr when it reached the continental rise.

Deep, steep-sided **submarine canyons** occur on the continental shelves, but they are best developed on continental slopes (Figure 11-10). Some submarine canyons can be traced across the shelf to associated streams on land; apparently, they formed as stream valleys when sea level was lower during the Pleistocene. Many have no association with streams, and their origin is not fully understood. It is known that strong currents move through submarine canyons and perhaps play some role in their origin. Furthermore, turbidity currents periodically move through these canyons and are now thought to be the primary agent responsible for their erosion.

 Types of Continental Margins

Two types of continental margins are generally recognized, *active* and *passive*. An **active continental margin** develops at the leading edge of a continental plate where oceanic lithosphere is subducted (▶ Figure 11-12a). The west coast of South America is a good example. Here, the continental margin is characterized by seismicity, a geologically young mountain range, and andesitic volcanism. Additionally, the continental shelf is narrow, and the continental slope descends directly into the Peru-Chile Trench.

The configuration and geologic activity of the continental margins of eastern North and South America differ considerably from their western margins. In the east, the continental margins developed as a result of the rifting of the supercontinent Pangaea. The continental crust was stretched, thinned, and fractured as rifting proceeded. As plate separation occurred, the newly formed continental margins became the sites of deposition of land-derived sediments. These **passive continental margins** are on the trailing edge of a continental plate (Figure 11-12b). They possess broad continental shelves and a continental slope and rise; vast, flat *abyssal plains* are commonly present adjacent to the rises (Figure 11-12b). Furthermore, passive continental margins lack the intense seismic and volcanic activity characteristic of active continental margins, and an oceanic trench is not present.

Active continental margins obviously lack a continental rise because the slope descends directly into an oceanic trench. Just as on passive continental margins, sediment is transported down the slope by turbidity currents, but it simply fills the trench rather than forming a rise. The proximity of the trench to the continent also explains why the continental shelf is so narrow. In contrast, the continental shelf of a passive continental margin is much wider because land-derived sedimentary deposits build outward into the ocean.

 The Deep-Ocean Basin

Considering that the oceans average more than 3.8 km deep, most of the sea floor lies far below the depth of sunlight penetration, which is rarely more than 100 m. Accordingly, most of the sea floor is completely dark, no plant life exists, the temperature is generally just above 0°C, and the pressure varies from 200 to more than 1,000 atmospheres depending on depth. Submersibles have carried scientists to the greatest oceanic depths, so some of the sea floor has been observed directly. Nevertheless, much of the deep-ocean basin has been studied only by echo sounding, seismic profiling, and remote devices that have descended in excess of 11,000 m. Although oceanographers know considerably more about the deep-ocean basins than they did even a few years ago, many questions remain unanswered.

Abyssal Plains

Beyond the continental rises of passive continental margins are **abyssal plains**, flat surfaces covering vast areas of the sea floor. In some areas they are interrupted by peaks rising more than 1 km, but in general they are the flattest, most

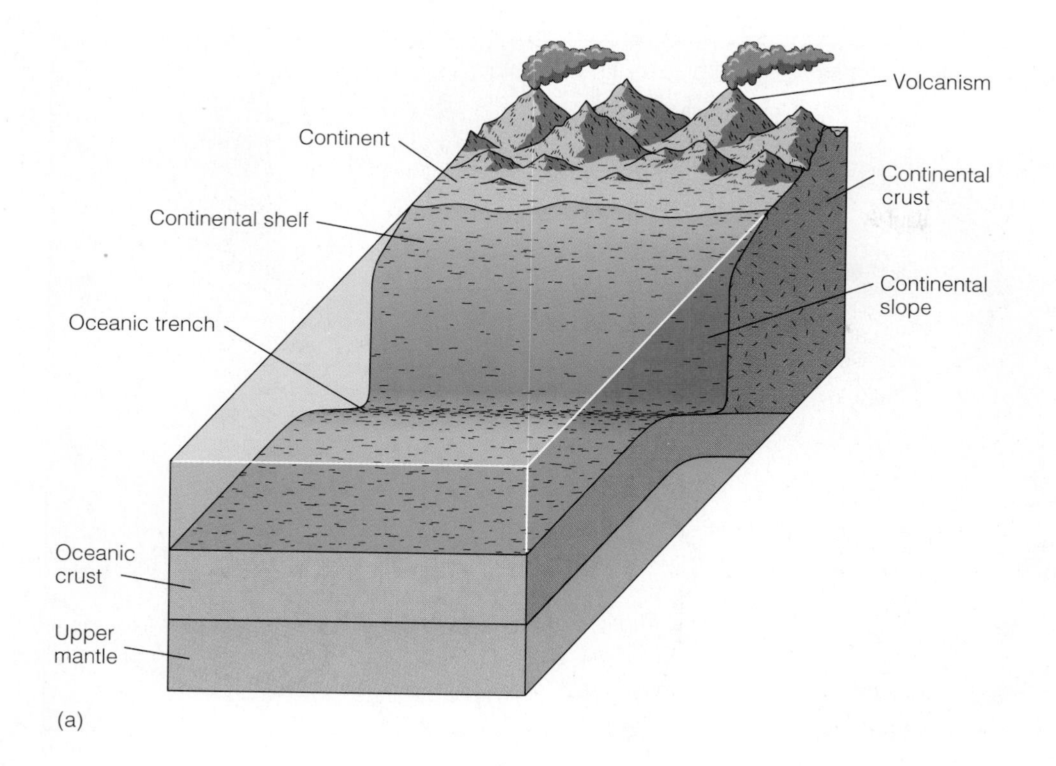

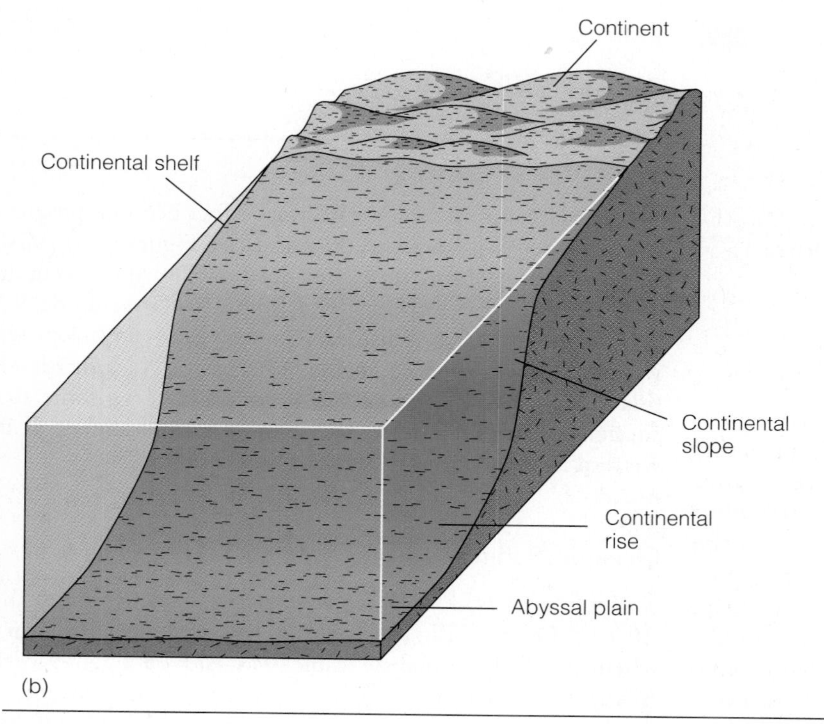

▶ **FIGURE 11-12** Diagrammatic views of (*a*) an active continental margin and (*b*) a passive continental margin.

featureless areas on Earth (▶ Figure 11-13). The flat topography is a result of sediment deposition; where sediment accumulates in sufficient quantities, the rugged sea floor is buried beneath thick layers of sediment (▶ Figure 11-14).

Seismic profiles and sea-floor samples reveal that the abyssal plains are covered with fine-grained sediment derived mostly from the continents and deposited by turbidity currents. Abyssal plains are invariably found adjacent to the continental rises, which are composed mostly of overlapping submarine fans that owe their origin to deposition by turbidity currents (Figure 11-10). Along active continental margins, sediments derived from the shelf and slope are trapped in an oceanic trench, and abyssal plains fail to

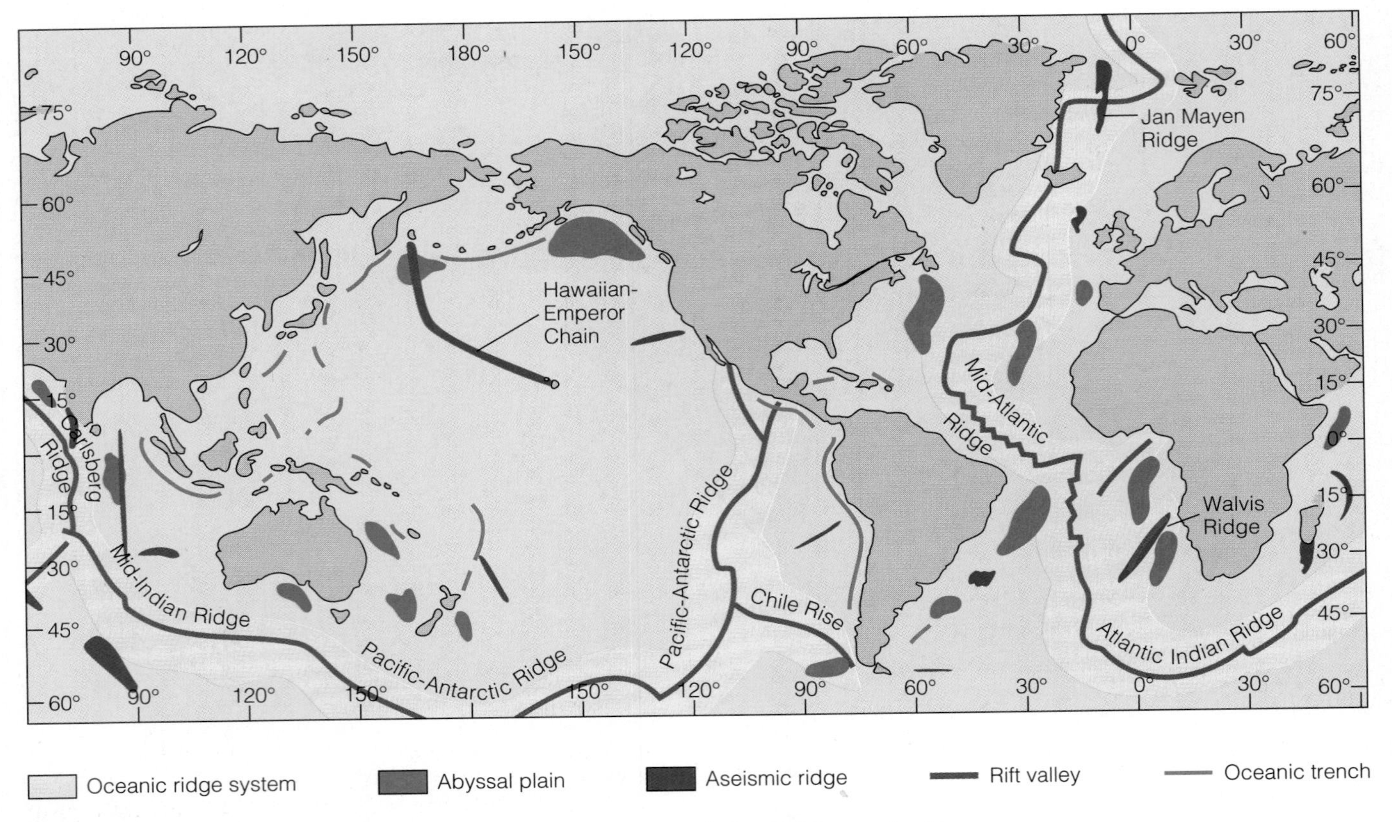

➢ **FIGURE 11-13** The distribution of oceanic trenches, abyssal plains, the oceanic ridge system, and some of the aseismic ridges.

develop. Accordingly, abyssal plains are common in the Atlantic Ocean basin, but rare in the Pacific Ocean basin (Figure 11-13).

Oceanic Trenches

Although **oceanic trenches** constitute a small percentage of the sea floor, they are very important, for it is here that lithospheric plates are consumed by subduction (see Chapter 12). Oceanic trenches are long, narrow features restricted to active continental margins so they are common around the margins of the Pacific Ocean basin (Figure 11-13). (The Peru-Chile Trench west of South America is 5,900 km long, but only 100 km wide. It is more than 8,000 m deep.) On the landward side of oceanic trenches, the continental slope descends at angles of up to 25° (Figure 11-12a). Oceanic trenches are also the sites of the greatest oceanic depths; a depth of more than 11,000 m has been recorded in the Challenger Deep of the Marianas Trench.

Oceanic trenches show anomalously low heat flow compared to the rest of the oceanic crust; it appears that the crust here is cooler and slightly denser than elsewhere. Furthermore, gravity surveys reveal that trenches show a huge negative gravity anomaly, indicating that the crust is held down and is not in isostatic equilibrium. Seismic activity also occurs at or near trenches. In fact, trenches are characterized by Benioff zones in which earthquake foci become progressively deeper in a landward direction (see Figure 9-8). Most of the Earth's intermediate and deep earthquakes occur in such zones as, for example, the June 1994 magnitude 8.2 earthquake in Boliva. Finally, oceanic trenches are associated with volcanoes, either as an arcuate chain of volcanic islands (island arc) or as a chain of volcanoes on land (volcanic arc) adjacent to a trench along the margin of a continent as in western South America (Figure 11-12a).

Oceanic Ridges

A feature called the Telegraph Plateau was discovered in the Atlantic Ocean basin during the late nineteenth century when the first submarine cable was laid between North America and Europe. Following the 1925–1927 voyage of the German research vessel *Meteor*, scientists proposed that this plateau was actually a continuous feature extending the length of the Atlantic Ocean basin. Subsequent investigations revealed that this proposal was correct, and we now call this feature the Mid-Atlantic Ridge (Figure 11-13).

The Mid-Atlantic Ridge is more than 2,000 km wide and rises about 2.5 km above the sea floor adjacent to it. It is part of a much larger **oceanic ridge** system of submarine mountainous topography that runs from the Arctic Ocean through the middle of the Atlantic, curves around South Africa, and

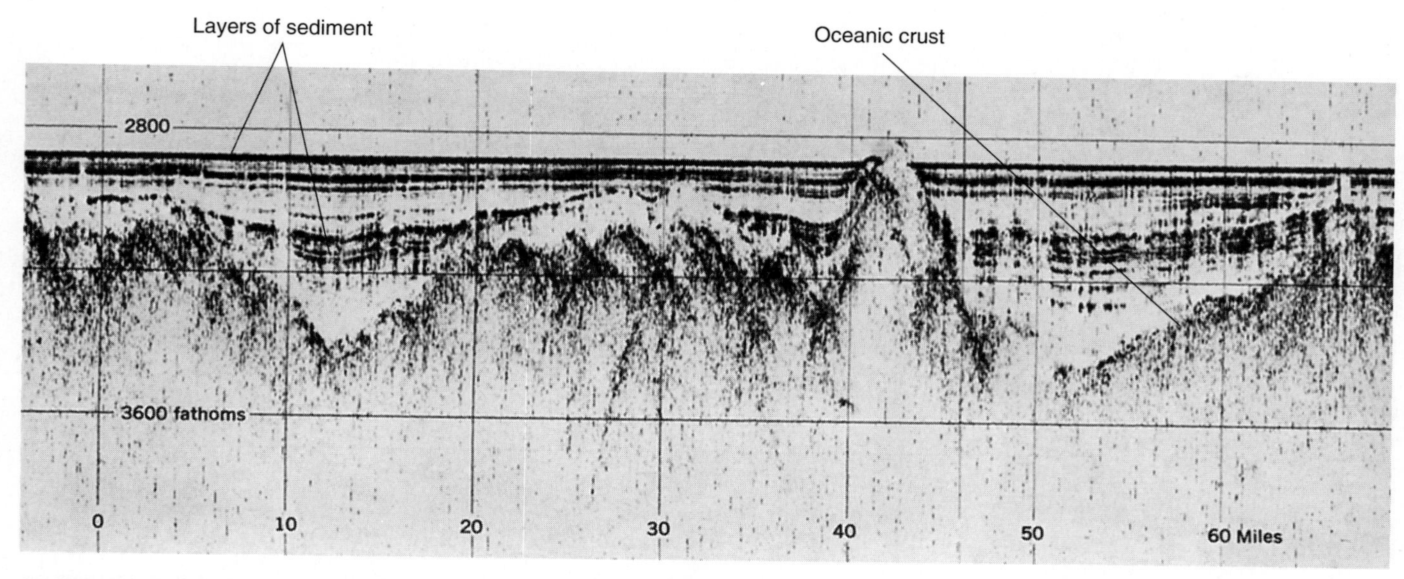

➢ **FIGURE 11-14** Seismic profile showing the burial of rugged sea-floor topography by sediments of the Northern Madeira Abyssal Plain in the Atlantic Ocean.

passes into the Indian Ocean, continuing from there into the Pacific Ocean basin (Figure 11-13). The system extends for at least 65,000 km, surpassing the length of the largest mountain range on land. Mountain ranges on land are typically composed of granitic and metamorphic rocks and sedimentary rocks that have been folded and fractured by compressional forces. Oceanic ridges, on the other hand, are composed of volcanic rocks (mostly basalt) and have features produced by tensional forces.

Running along the crests of some ridges is a rift that appears to have opened up in response to tensional forces (➢ Figure 11-15), although portions of the East Pacific Rise lack this feature. These rifts are commonly 1 to 2 km deep and several kilometers wide. Rifts open as sea-floor spreading occurs (discussed in Chapter 12); ridges are characterized by shallow-focus earthquakes, basaltic volcanism, and high heat flow.

Direct observation of the ridges and their rift valleys began in 1974. As a part of Project FAMOUS (French-American Mid-Ocean Undersea Study), submersible craft descended into the rift of the Mid-Atlantic Ridge, and more recent dives have investigated other rifts. Although no active volcanism was observed, the researchers did see pillow lavas (see Figure 4-10), lava tubes, and sheet lava flows, some of which appear to have formed very recently. In addition, hydrothermal vents such as black smokers have been observed (see the Prologue).

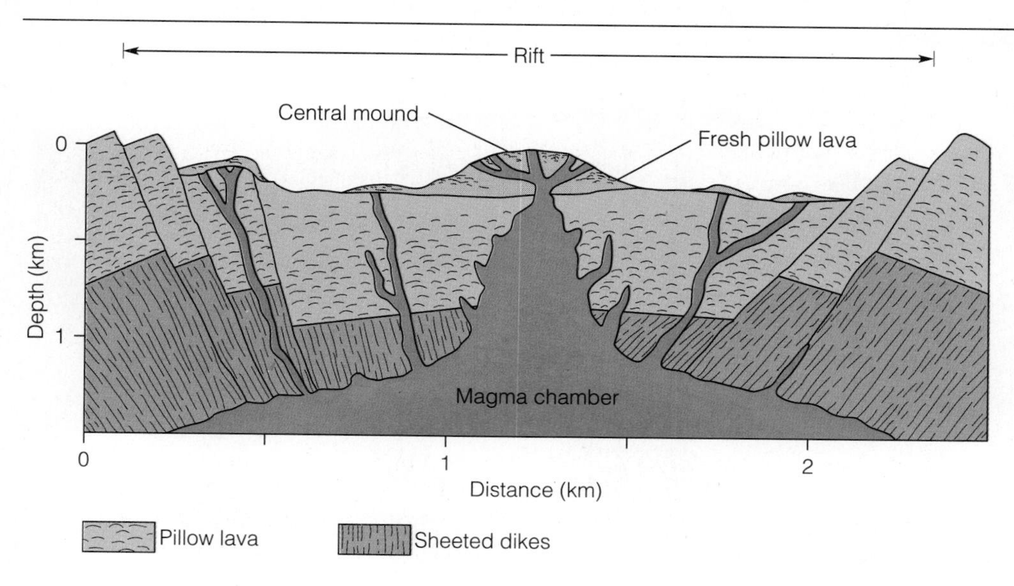

➢ **FIGURE 11-15** Studies during the early 1970s revealed that recent moundlike accumulations of volcanic rocks, mostly basaltic pillow lavas, occurred in a rift along the axis of the Mid-Atlantic Ridge.

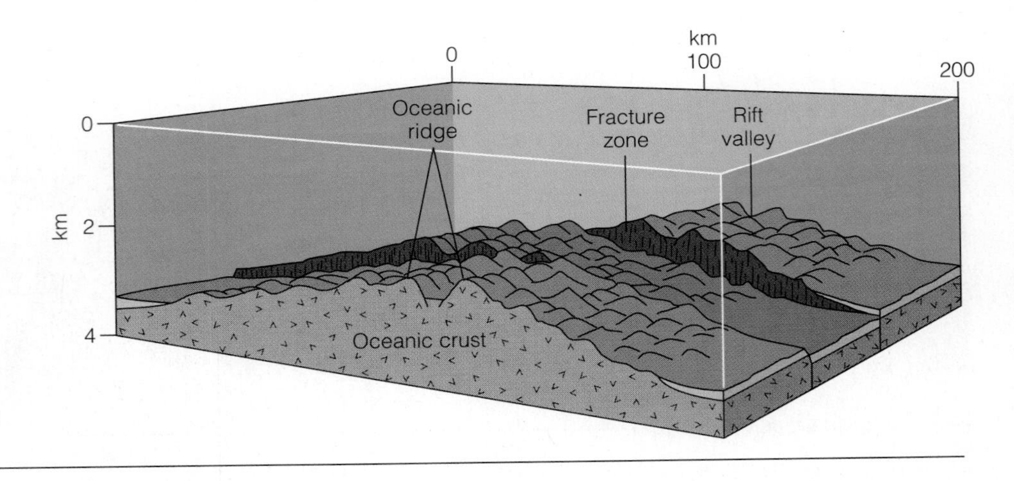

▶ **FIGURE 11-16** Diagrammatic view of a fracture offsetting a ridge. Earthquakes occur only in the segments between offset ridge crests.

Fractures in the Sea Floor

Oceanic ridges are not continuous features winding without interruption around the globe. They abruptly terminate where they are offset along major fractures oriented more or less at right angles to ridge axes (▶ Figure 11-16). Such large-scale fractures run for hundreds of kilometers, although they are difficult to trace where they are buried beneath sea-floor sediments. Many geologists are convinced that some geologic features on the continents can best be accounted for by the extension of such fractures into continents.

Where these fractures offset oceanic ridges, they are characterized by shallow seismic activity only in the area between the displaced ridge segments (Figure 11-16). Furthermore, because ridges are higher than the sea floor adjacent to them, the offset segments yield vertical relief on the sea floor. Nearly vertical escarpments 3 or 4 km high develop, as illustrated in Figure 11-16. We will have more to say about such fractures, called transform faults, in Chapter 12.

Seamounts, Guyots, and Aseismic Ridges

As noted previously, the sea floor is not a flat, featureless plain, except for the abyssal plains, and even these are underlain by rugged topography (Figure 11-14). In fact, a large number of volcanic hills, seamounts, and guyots rise above the sea floor. These features are present in all ocean basins, but are particularly abundant in the Pacific. All are of volcanic origin and differ from one another mostly in size. **Seamounts** rise more than 1 km above the sea floor; if they are flat topped, they are called **guyots** rather than seamounts (▶ Figure 11-17). Guyots are volcanoes that originally extended above sea level. But, as the plate upon which they were situated continued to move, they were carried away from a spreading ridge, and the oceanic crust cooled and descended to greater oceanic depths. So what was once an island was eroded by waves as it slowly sank beneath the sea, giving it the typical flat-topped appearance.

Many other volcanic features, most of them are smaller than seamounts, are also present on the sea floor, but probably originated in the same way. These so-called *abyssal*

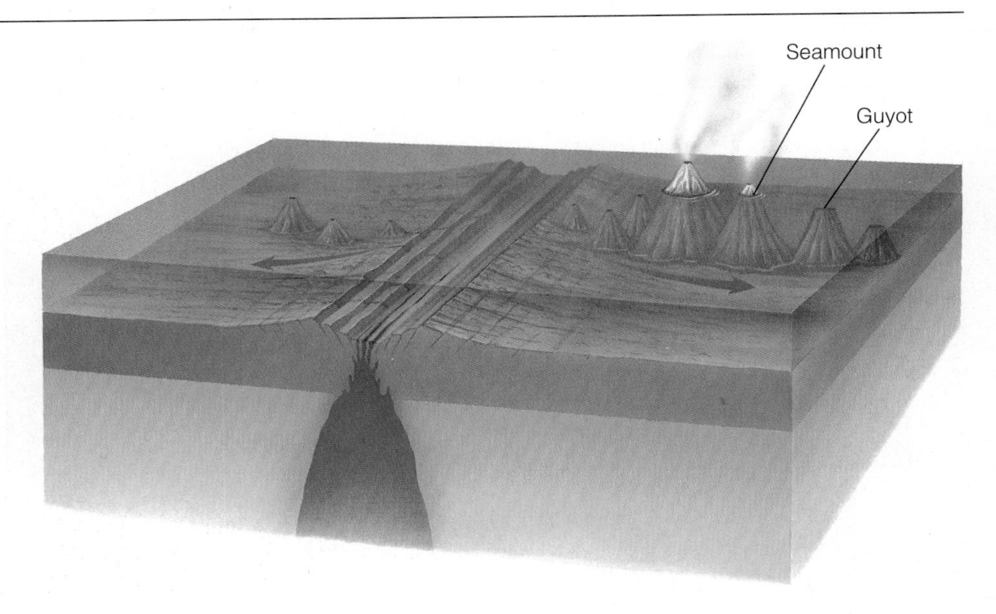

▶ **FIGURE 11-17** Submarine volcanoes may build up above sea level to form seamounts. As the plate upon which these volcanoes rest moves away from a spreading ridge, the volcanoes sink beneath sea level and become guyots.

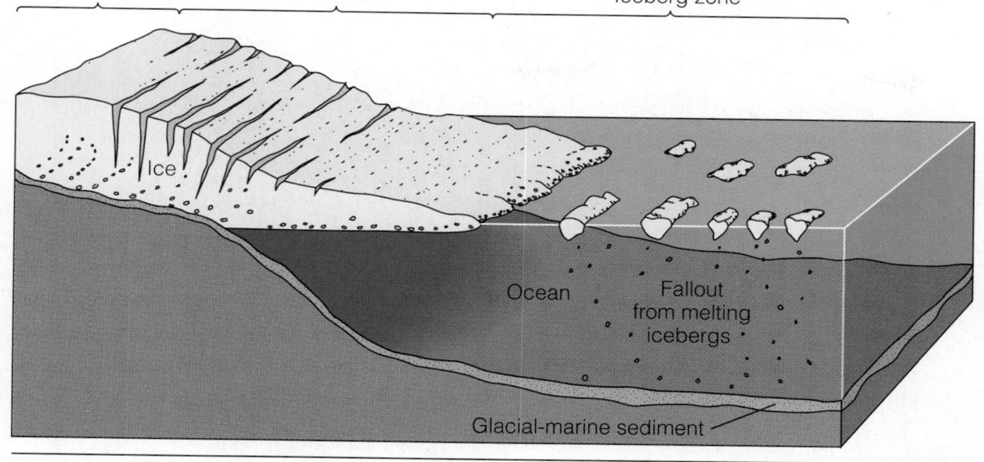

▶ **FIGURE 11-18** The formation of glacial-marine sediments by ice rafting, a process in which icebergs transport sediment into the ocean basins, releasing it as they melt.

hills average only about 250 m high. They are common on the sea floor and underlie thick sediments on the abyssal plains.

Other common features in the ocean basins are long, linear ridges and broad plateaulike features rising as much as 2 to 3 km above the surrounding sea floor. They are known as **aseismic ridges** because they lack seismic activity. A few of these ridges are thought to be small fragments separated from continents during rifting. These fragments, referred to as *microcontinents*, are represented by such features as the Jan Mayen Ridge in the North Atlantic (Figure 11-13).

Most aseismic ridges form as a linear succession of hot spot volcanoes. These may develop at or near an oceanic ridge, but each volcano so formed is carried laterally with the plate upon which it originated. The net result of this activity is a sequence of seamounts/guyots extending from an oceanic ridge (Figure 11-17); the Walvis Ridge in the South Atlantic is a good example (Figure 11-13). Aseismic ridges also form over hot spots unrelated to ridges. The Hawaiian-Emperor chain in the Pacific formed in such a manner (Figure 11-13).

 Deep-Sea Sedimentation

Deep-sea sediments consist mostly of fine-grained deposits because few mechanisms exist that can transport coarse-grained sediment (sand and gravel) far from land. Icebergs can transport coarse sediment into the ocean basins, though, and a broad band of glacial-marine sediment has been deposited adjacent to Antarctica and Greenland (▶ Figure 11-18).

Most of the fine-grained sediment in the deep sea is windblown dust and volcanic ash from the continents and oceanic islands and the shells of microscopic organisms that live in the near-surface waters of the oceans. Other sources of sediment include cosmic dust and deposits resulting from chemical reactions in seawater. The manganese nodules that are fairly common in all the ocean basins are a good example of the latter (▶ Figure 11-19). They are composed mostly of manganese and iron oxides, but also contain copper, nickel, and cobalt. These nodules may be an important source of some metals in the future; the United States, which imports most of its manganese and cobalt, is particularly interested in this potential resource.

The contribution of cosmic dust to deep-sea sediment is negligible. Even though some researchers estimate that as much as 40,000 metric tons of cosmic dust may fall to Earth each year, this is a trivial quantity compared to the volume of sediments derived from other sources.

The bulk of the sediments on the deep-sea floor are *pelagic*, meaning that they settled from suspension far from land. Two categories of pelagic sediment are recognized: pelagic clay and ooze (▶ Figure 11-20). **Pelagic clay** covers most of the deeper parts of the ocean basins. It is generally brown or reddish and is composed of clay-sized particles derived from the continents and oceanic islands. **Ooze**, on the other hand, is composed mostly of shells of

▶ **FIGURE 11-19** Manganese nodules on the sea floor south of Australia.

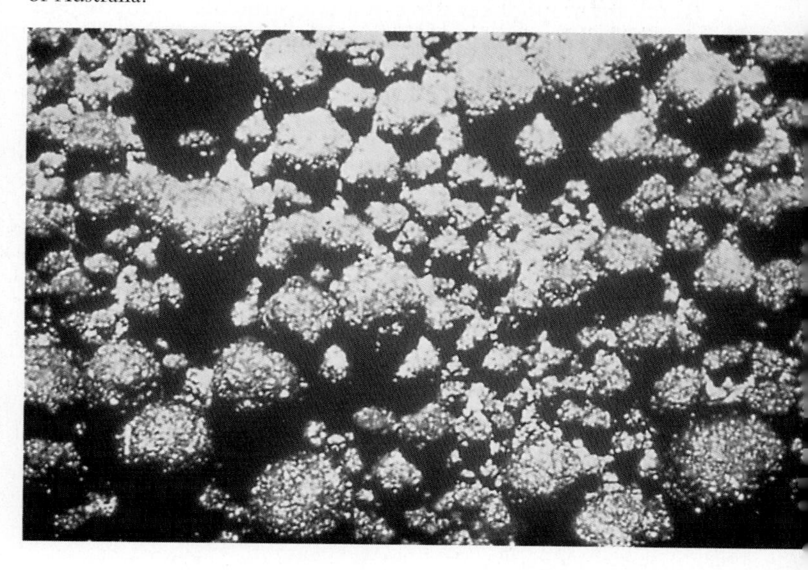

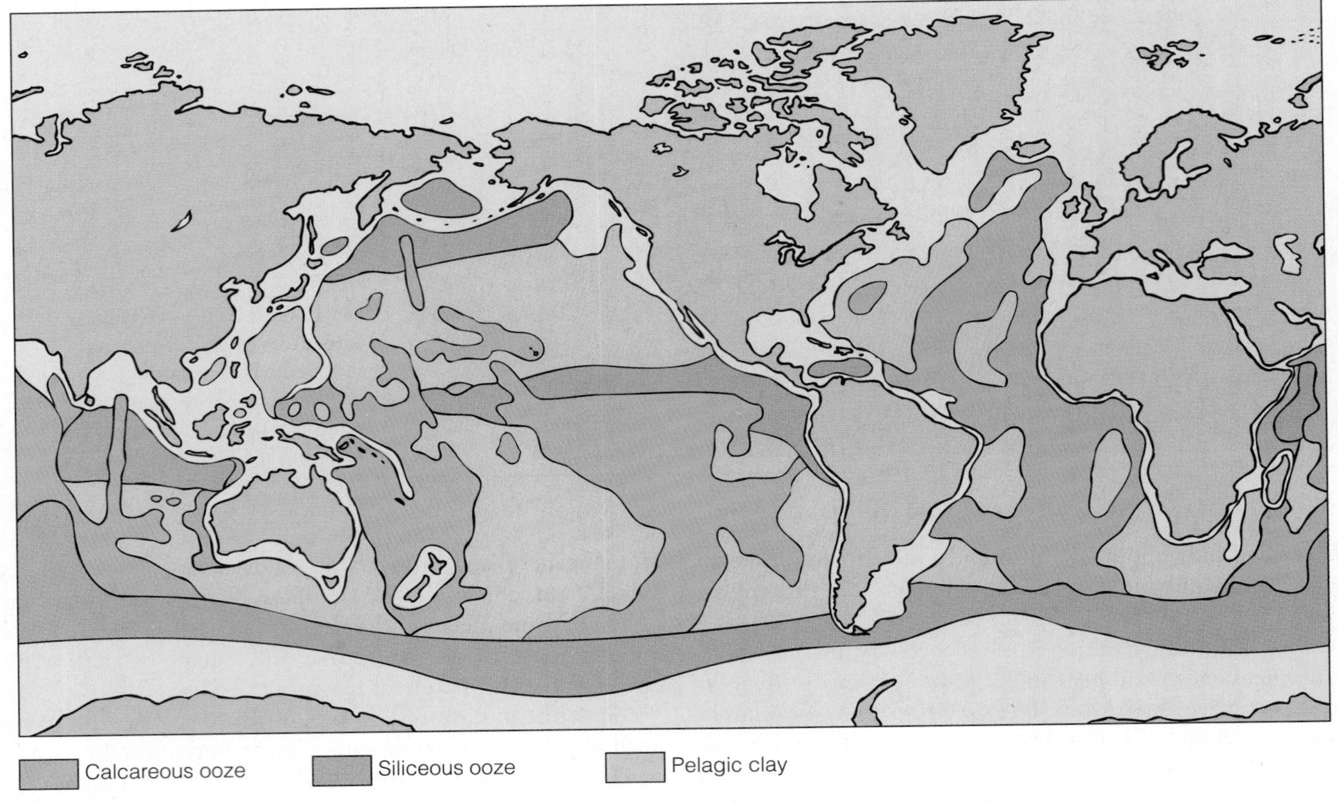

▶ **FIGURE 11-20** The distribution of sediments on the deep-sea floor.

microscopic marine animals and plants. It is characterized as *calcareous ooze* if it contains mostly calcium carbonate ($CaCO_3$) skeletons of tiny marine organisms such as foraminifera. *Siliceous ooze* is composed of the silica (SiO_2) skeletons of such single-celled organisms as radiolarians (animals) and diatoms (plants) (see Figure 6-14).

REEFS

Reefs are moundlike, wave-resistant structures composed of the skeletons of organisms (▶ Figure 11-21). They are commonly called coral reefs, but many other organisms in addition to corals make up reefs. A reef consists of a solid framework of skeletons of corals, clams, and such encrusting organisms as algae and sponges. Reefs grow to a depth of about 45 or 50 m and are restricted to shallow tropical seas where the water is clear, and the temperature does not fall below about 20°C.

Three types of reefs are recognized: fringing, barrier, and atoll (▶ Figure 11-22). *Fringing reefs* are solidly attached to the margins of an island or continent. They have a rough, tablelike surface, are as much as 1 km wide, and, on their seaward side, slope steeply down to the sea floor. *Barrier reefs* are similar to fringing reefs, except that they are separated from the mainland by a lagoon. Probably the best-known barrier reef in the world is the Great Barrier Reef of Australia, which is 2,000 km long.

An *atoll* is a circular to oval reef surrounding a lagoon (▶ Figure 11-23). Atolls form around volcanic islands that subside below sea level as the plate upon which they rest is carried progressively farther from an oceanic ridge (Figure 11-17). As subsidence occurs, the reef organisms construct the reef upward so that the living part of the reef remains in shallow water. The island eventually subsides below sea level, leaving a circular lagoon surrounded by a more-or-less continuous reef (Figure 11-22). Atolls are particularly common in the western Pacific Ocean basin. Many of these began as fringing reefs, but as subsidence occurred, they evolved first to barrier reefs and finally to atolls.

This particular scenario for the evolution of reefs from fringing to barrier to atoll was proposed more than 150 years ago by Charles Darwin while he was serving as a naturalist on the ship H.M.S. *Beagle*. Drilling into atolls has revealed that they do indeed rest upon a basement of volcanic rocks, confirming Darwin's hypothesis.

STRUCTURE AND COMPOSITION OF THE OCEANIC CRUST

Most of the oceanic crust is consumed at subduction zones. However, a tiny amount of it is not subducted, and it along with pieces of the underlying upper mantle, is emplaced in

Guest Essay *Robert D. Ballard*

THE FUTURE BENEATH THE SEA

My childhood dreams always dealt with the sea. When I was growing up, the landmasses of the world were largely explored—the final frontiers were in space and underwater. Space fascinated me, and still does, but the ocean fascinated me even more, possibly because one of my greatest heroes was Captain Nemo in *20,000 Leagues under the Sea.* I had the good fortune to grow up in San Diego where the sea was always part of my life. I can vividly remember walking along the sandy beaches of southern California searching for treasure washed ashore by the tide. Eventually, I graduated from walking along the beach to body surfing and scuba diving. For some reason, I never was interested in the top of the sea or, for that matter, in the sea itself. It was the land beneath the sea that fascinated me. Every time I put on an air tank, I headed straight for the bottom.

While in high school, I wrote a letter to the Scripps Institution of Oceanography in La Jolla, which I visited many times. A kind scientist at Scripps answered my letter and told me how I could apply for a summer scholarship. I was 17 years old, and that summer of 1959 gave me my first great adventures with the sea. On the first cruise, we were hit by a great storm and had to limp back to shore. On the second, our ship was almost sunk by a great wave that knocked out the windows in the bridge and exploded the portholes in the galley. I was hooked.

On that curise I met another kind scientist who encouraged me to attend the University of California at Santa Barbara where he was teaching geology. Not knowing exactly what aspect of the sea I wanted to concentrate on, I majored in all the physical sciences: physics, math, chemistry, and geology. That proved to be my most important decision. This broad-based education in science and technology has made it possible for me to follow the action in my field. I think it is a mistake to focus your interests too narrowly. The broader your experience, the more you can go with the flow as time goes on.

Today, as an undersea explorer at the Woods Hole Oceanographic Institution, I am still fascinated by the sea. Given our exploding population, our diminished interest in the promises of the space program, and the continued development of advanced technology, I truly believe the twenty-first century will usher in an explosion in human activity in the sea. I am convinced the next generation will explore more of the Earth, that is, the 71% that lies underwater, than all previous generations combined.

Just as Lewis and Clark's explorations led to the settling of the West, the exploration of the sea will lead to its colonization. The gathering and hunting of the living resources of the sea, an activity characteristic of primitive societies on land, will be replaced at sea by farming and herding. High-tech barbwire in the form of acoustic, thermal, or other barrier techniques will emerge to control and manage the living resources of the sea.

Oil and gas exploration and exploitation will continue moving into deeper and deeper depths. We have already discovered and mapped oil and gas reserves down to 12,000 feet, which represents the average depth of the ocean, and each year the oil industry brings production wells on line in waters deeper than the previous year.

In recent years, we have discovered major mineral deposits in the deep sea similar to those mined for centuries on Cyprus. They contain high concentrations of copper, lead, and sulfur as well as silver and gold. Their formation continues today in the vast hydrothermal vent systems of the Mid-Ocean Ridge. These mineral deposits will be processed using the very geothermal energy that drives the crustal processes that lead to their formation. Some of these magnificent vent areas will also become the Yellowstone Parks of the deep sea, leading to future arguments over their commerical versus tourist value.

The unique chemosynthetic life-forms that process the toxic material associated with the vent communities will hopefully be bioengineered to convert a portion of our waste products into less harmful or even commercially valuable by-products. These exotic creatures will also help us understand the early origin of life on our planet as well as the potential for life on other planets we once ruled out for their lack of a friendly nearby Sun.

Whether all this occurs during the next generation's time on Earth, only time will tell. But the seeds can be found in programs already under way. With this exploration will come better understanding of the ocean and the land surface beneath it, which is critical to our understanding of the planet as a whole.

ROBERT D. BALLARD is the director of the Center for Marine Exploration and senior scientist in the Department of Applied Physics and Engineering at Woods Hole Oceanographic Institution. He is also the founder and chairman of the Jason Foundation for Education. Ballard earned a B.S. in physical science at the University of California, Santa Barbara and a Ph.D. in marine geology and geophysics at the University of Rhode Island, Graduate School in Oceanography.

mountain ranges on land, usually by moving along large fractures called thrust faults. These preserved slivers of upper mantle and oceanic crust are known as **ophiolites.**

Detailed studies reveal that an ideal ophiolite consists of a layer of deep-sea sedimentary rocks underlain successively by a layer of pillow basalts and sheet lava flows, and a sheeted dike complex, a zone consisting mostly of vertical basaltic dikes (➢ Figure 11-24). Further downward in an ophiolite is massive gabbro, and below that is layered gabbro that probably cooled at the top of a magma chamber. Beneath the gabbro is peridotite representing the upper mantle; this layer is sometimes altered by metamorphism to an assemblage containing serpentine. Thus, a complete ophiolite sequence consists of deep-sea sedimentary rock, oceanic crust, and upper mantle rocks (Figure 11-24).

Sampling and direct observations of the oceanic ridges and deep-sea drilling reveal that the upper part of the oceanic crust is indeed composed of a layer of pillow lavas and sheet flows underlain by a sheeted dike complex, exactly as predicted from ophiolites. It was not until 1989, that direct confirmation was made of what lay below the sheeted dike complex. In that year, scientists in a submersible descended to the walls of a sea-floor fracture (transform fault) in the North Atlantic. There, rocks of the upper mantle and lower oceanic crust were close to the sea floor and, for the first time, scientists observed the peridotite and gabbro parts of the sequence. In short, the structure and composition of oceanic crust were inferred from fragments of presumed oceanic crust on land, and these inferences were later verified by observations.

➢ **FIGURE 11-21** Reefs such as this one in the Red Sea are wave-resistant structures composed of the skeletons of organisms.

RESOURCES FROM THE SEA

Seawater contains many elements in solution, some of which are extracted for various industrial and domestic uses. In many places sodium chloride (table salt) is produced by the evaporation of seawater, and a large proportion of the world's magnesium is produced from seawater. Numerous

➢ **FIGURE 11-22** Three-stage development of an atoll. In the first stage, a fringing reef forms, but as the island sinks, the barrier reef becomes separated from the island by a lagoon. As the island disappears beneath the sea, the barrier reef continues to grow upward, forming an atoll. An oceanic island carried into deeper water by plate movement can account for this sequence.

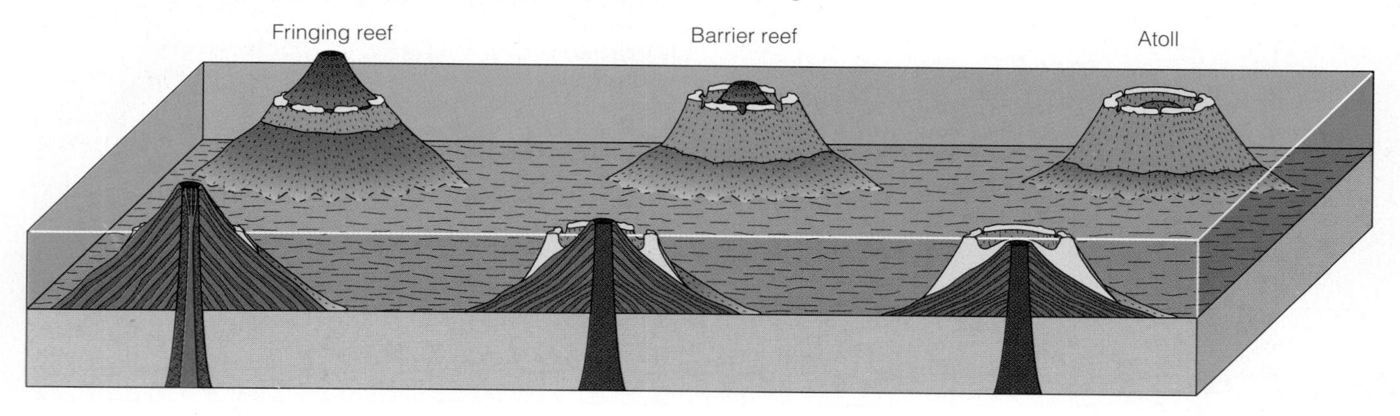

▶ **FIGURE 11-23** View of an atoll in the Pacific Ocean.

other elements and compounds can be extracted from seawater, but for many, such as gold, the cost is prohibitive.

In addition to substances in seawater, deposits on the sea floor or within sea-floor sediments are becoming increasingly important. Many of these potential resources lie well beyond the continental margins, so their ownership is a political and legal problem that has not yet been resolved. Most nations bordering an ocean claim those resources occurring within their adjacent continental margin. The United States, by a presidential proclamation issued on March 10, 1983, claims sovereign rights over an area designated as the **Exclusive Economic Zone (EEZ).** The EEZ extends seaward 200 nautical miles (371 km) from the coast, giving the United States jurisdiction over an area

▶ **FIGURE 11-24** New oceanic crust consisting of the layers shown here forms as magma rises beneath oceanic ridges. The composition of the oceanic crust is known from ophiolites, sequences of rock on land consisting of deep-sea sediments, oceanic crust, and upper mantle.

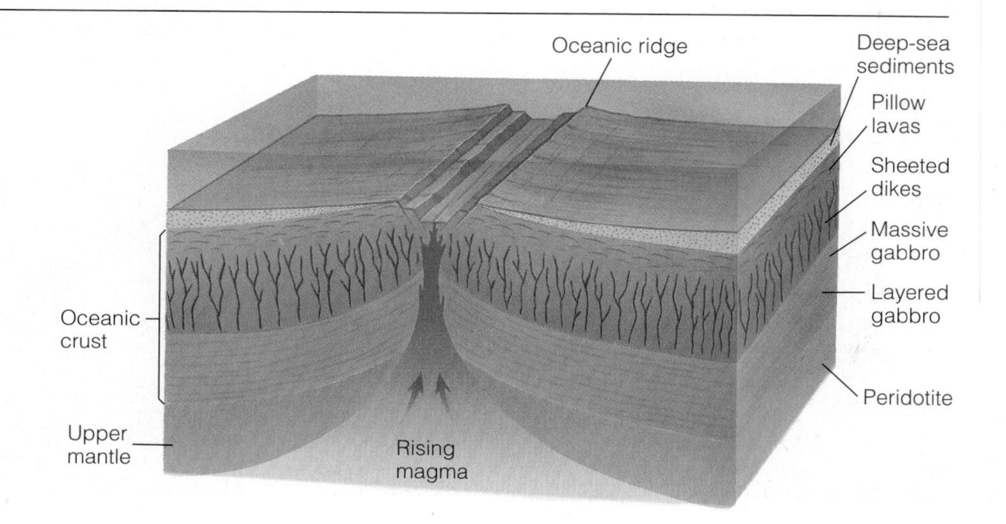

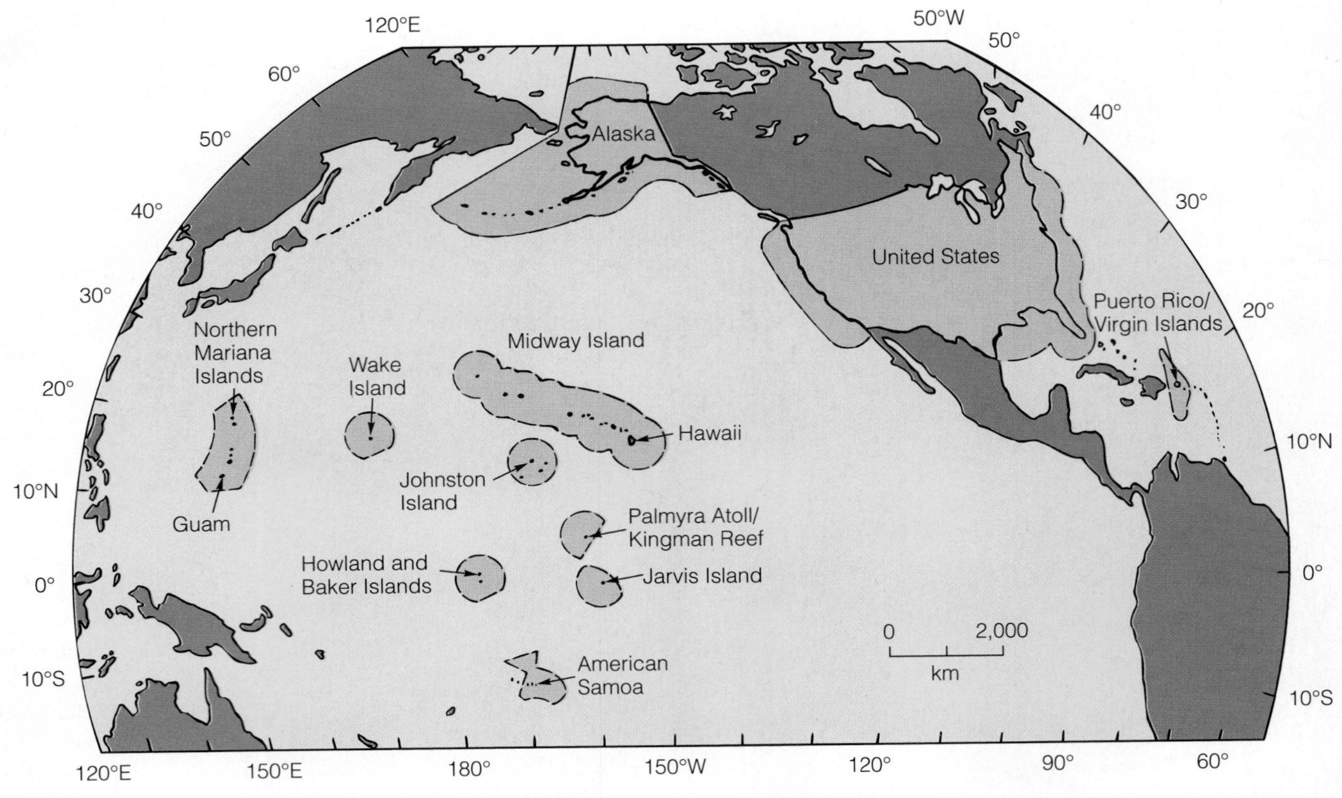

▶ **FIGURE 11-25** The Exclusive Economic Zone (EEZ) includes a vast area adjacent to the United States and its possessions.

about 1.7 times larger than its land area (▶ Figure 11-25).* Also included within the EEZ are the areas adjacent to U.S. territories, such as Guam, American Samoa, Wake Island, and Puerto Rico (Figure 11-25). In short, the United States claims a huge area of the sea floor and any resources on or beneath it.

Numerous resources occur within the EEZ, some of which have been exploited for many years. Sand and gravel for construction are mined from the continental shelf in several areas. About 17% of U.S. oil and natural gas production comes from wells on the continental shelf. Some 30 sedimentary basins occur within the EEZ, several of which are known to contain hydrocarbons whereas others are areas of potential hydrocarbon production.

Other resources of interest include the massive sulfide deposits that form by submarine hydrothermal activity at spreading ridges (see the Prologue). Such deposits containing iron, copper, zinc, and other metals have been identified within the EEZ at the Gorda Ridge off the coasts of California and Oregon; similar deposits occur at the Juan de Fuca Ridge within the Canadian EEZ (▶ Figure 11-26).

Other potential resources include the manganese nodules discussed previously (Figure 11-19), and metalliferous oxide

*A number of other nations also claim sovereign rights to resources within 200 nautical miles of their coasts.

▶ **FIGURE 11-26** Massive sulfide deposits formed by submarine hydrothermal activity have been identified on the Gorda Ridge within the U.S. Exclusive Economic Zone and on the Juan de Fuca Ridge in the Canadian EEZ.

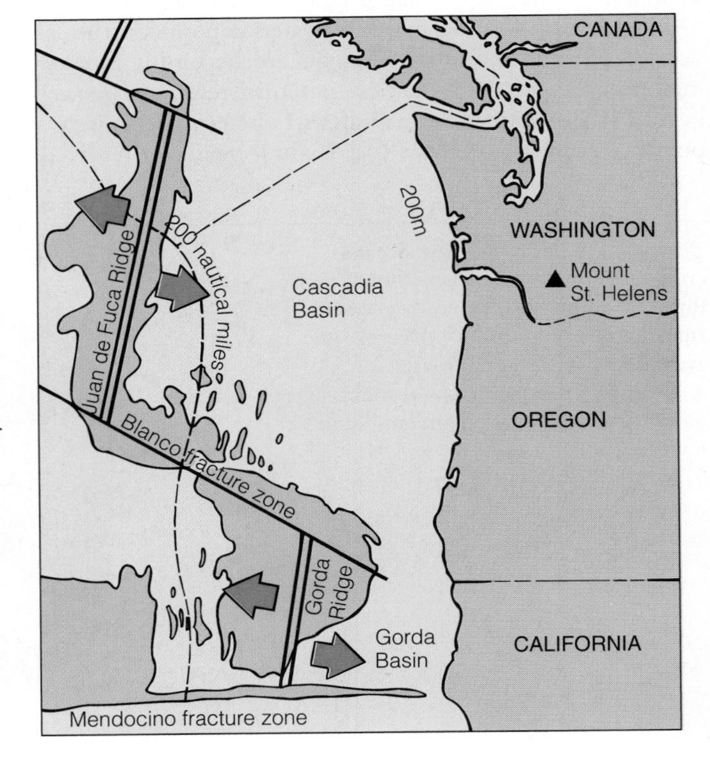

crusts found on seamounts. Manganese nodules contain manganese, cobalt, nickel, and copper; the United States is heavily dependent on imports of the first three of these elements (see Table 2-7). Within the EEZ, manganese nodules occur near Johnston Island in the Pacific Ocean and on the Blake Plateau off the east coast of South Carolina and Georgia. In addition, seamounts and seamount chains within the EEZ in the Pacific are known to have metalliferous oxide crusts several centimeters thick from which cobalt and manganese could be mined.

Chapter Summary

1. Scientific investigations of the oceans began during the late 1700s. Present-day research vessels are equipped to investigate the sea floor by drilling, echo sounding, and seismic profiling.
2. Continental margins separate the continents above sea level from the deep-ocean basin. They consist of a continental shelf, continental slope, and in some cases a continental rise.
3. Continental shelves slope gently in a seaward direction and vary from a few tens of meters to more than 1,000 km wide.
4. The continental slope begins at an average depth of 135 m where the inclination of the sea floor increases rather abruptly from less than 1° to several degrees.
5. Submarine canyons are characteristic of the continental slope, but some of them extend well up onto the shelf and lie offshore from large streams. Stream erosion of the shelf during the Pleistocene Epoch may account for some submarine canyons, but many have no association with streams on land and were probably eroded by turbidity currents.
6. Turbidity currents commonly move through submarine canyons and deposit an overlapping series of submarine fans that constitute a large part of the continental rise.
7. Active continental margins are characterized by a narrow shelf and a slope that descends directly into an oceanic trench with no rise present. Such margins are also characterized by seismic activity and volcanism.
8. Passive continental margins lack volcanism and exhibit little seismic activity. The continental shelf along these margins is broad, and the slope merges with a continental rise. Abyssal plains are commonly present seaward beyond the rise.
9. Oceanic trenches are long, narrow features where oceanic crust is subducted. They are characterized by low heat flow, negative gravity anomalies, and the greatest oceanic depths.
10. Oceanic ridges consisting of mountainous topography are composed of volcanic rocks, and many ridges possess a large rift caused by tensional forces. Basaltic volcanism and shallow-focus earthquakes occur at ridges. Oceanic ridges nearly encircle the globe, but they are interrupted and offset by large fractures in the sea floor.
11. Other important features on the sea floor include seamounts that rise more than a kilometer high and guyots, which are flat-topped seamounts. Many aseismic ridges are oriented more-or-less perpendicular to oceanic ridges and consist of a chain of seamounts and/or guyots.
12. Deep-sea sediments consist mostly of fine-grained particles derived from continents and oceanic islands and the microscopic shells of organisms. The primary types of deep-sea sediments are pelagic clay and ooze.
13. Reefs are wave-resistant structures composed of animal skeletons, particularly corals. Three types of reefs are recognized: fringing, barrier, and atoll.
14. Deep-sea drilling and the study of fragments of sea floor in mountain ranges on land reveal that the oceanic crust is composed in descending order of pillow lava, sheeted dikes, and gabbro.
15. The United States has claimed rights to all resources occurring within 200 nautical miles (371 km) of its shorelines. Numerous resources including various metals occur within this Exclusive Economic Zone.

Important Terms

abyssal plain
active continential margin
aseismic ridge
continental margin
continental rise
continental shelf
continental slope
Exclusive Economic Zone
guyot
oceanic ridge
oceanic trench
ooze
ophiolite
passive continental margin
pelagic clay
reef
seamount
seismic profiling
submarine canyon
submarine fan
turbidity current

REVIEW QUESTIONS

1. Much of the continental rise is composed of:
   a. ____ calcareous ooze;
   b. ____ submarine fans;
   c. ____ fringing reefs;
   d. ____ sheeted dikes;
   e. ____ ophiolite.
2. The greatest oceanic depths occur at:
   a. ____ aseismic ridges;
   b. ____ guyots;
   c. ____ the shelf-slope break;
   d. ____ oceanic trenches;
   e. ____ passive continental margins.
3. Abyssal plains are most common:
   a. ____ around the margins of the Atlantic;
   b. ____ adjacent to the East Pacific Rise;
   c. ____ along the west coast of South America;
   d. ____ in the rift valley of the Mid-Atlantic Ridge;
   e. ____ on continental shelves.
4. A circular reef enclosing a lagoon is a(n):
   a. ____ barrier reef;
   b. ____ seamount;
   c. ____ aseismic ridge;
   d. ____ guyot;
   e. ____ atoll.
5. Submarine canyons are most characteristic of the:
   a. ____ continental shelves;
   b. ____ abyssal plains;
   c. ____ continental slopes;
   d. ____ rift valleys;
   e. ____ fractures in the sea floor.
6. The Earth's surface waters probably originated through the process of:
   a. ____ dewatering;
   b. ____ subduction;
   c. ____ outgassing;
   d. ____ crustal fracturing;
   e. ____ erosion.
7. Continental shelves:
   a. ____ are composed of pelagic sediments;
   b. ____ lie between continental slopes and rises;
   c. ____ descend to an average depth of 1,500 m;
   d. ____ slope gently from the shoreline to the shelf-slope break;
   e. ____ are widest along active continental margins.
8. The flattest, most featureless areas on Earth are the:
   a. ____ oceanic ridges;
   b. ____ abyssal plains;
   c. ____ continental slopes;
   d. ____ aseismic ridges;
   e. ____ continental margins.
9. Sediment that settles from suspension far from land is:
   a. ____ abyssal;
   b. ____ pelagic;
   c. ____ volcanic;
   d. ____ generally coarse grained;
   e. ____ characterized by graded bedding.
10. Which of the following statements is correct?
    a. ____ most of the continental margins around the Atlantic are passive;
    b. ____ oceanic ridges are composed largely of deformed sedimentary rocks;
    c. ____ the deposits of turbidity currents consist of calcareous ooze;
    d. ____ most of the Earth's intermediate and deep earthquakes occur at or near oceanic ridges;
    e. ____ oceanic crust is thicker than continental crust.
11. Massive sulfide deposits form:
    a. ____ on passive continental margins;
    b. ____ as accumulations of microscopic shells on the sea floor;
    c. ____ by precipitation of minerals near hydrothermal vents;
    d. ____ from sediments derived from continents;
    e. ____ in oceanic trenches.
12. The most useful method of determining the structure of the oceanic crust beneath continental shelf sediments is:
    a. ____ echo sounding;
    b. ____ observations from submersible research vessels;
    c. ____ dredging;
    d. ____ seismic profiling;
    e. ____ underwater photography.
13. Which of the following is *not* characteristic of an active continental margin?
    a. ____ volcanism;
    b. ____ earthquakes;
    c. ____ oceanic trench;
    d. ____ volcanic arc;
    e. ____ continental rise.
14. What is the average slope of the continental slope?
    a. ____ 25°;
    b. ____ 1°;
    c. ____ 4°;
    d. ____ 0.1°;
    e. ____ 40°.
15. Graded bedding is a characteristic of:
    a. ____ continental shelves;
    b. ____ turbidity current deposits;
    c. ____ pelagic clay;
    d. ____ siliceous ooze;
    e. ____ manganese nodules.
16. How do sulfide mineral deposits form on the sea floor?
17. What are the characteristics of a passive continental margin? How does such a continental margin originate?
18. Describe the continental rise, and explain why a rise occurs at some continental margins and not at others.
19. Summarize the evidence indicating that turbidity currents transport sediment from the continental shelf onto the slope and rise.
20. Where do abyssal plains most commonly develop? Describe their composition.
21. What is the significance of oceanic trenches, and where are they found?
22. How do mid-oceanic ridges differ from mountain ranges on land?
23. Describe how an aseismic ridge forms.
24. What are the sources of deep-sea sediments?
25. Describe the sequence of events leading to the origin of an atoll.
26. Illustrate and label an ideal sequence of rocks in an ophiolite.
27. What is the Exclusive Economic Zone? What types of metal deposits occur within it?

Points to Ponder

1. Why does the Pacific Ocean basin have so many oceanic trenches whereas the Atlantic has very few?
2. The most distant part of an aseismic ridge is 1,000 km from an oceanic ridge system, and its age is 30 million years. How fast, on the average, did the plate with this ridge move?
3. During the Pleistocene Epoch, or what is called the Ice Age, sea level was as much as 130 m lower than at present. What effect did this lower sea level have on streams? Is there any evidence from the continental shelves that might bear upon this question?

Additional Readings

Anderson, R. N. 1986. *Marine geology*. New York: John Wiley & Sons.

Bishop, J. M. 1984. *Applied oceanography*. New York: John Wiley & Sons.

Davis, R. A. 1987. *Oceanography: An introduction to the marine environment*. Dubuque, Iowa: W. C. Brown.

Edmond, J. M., and K. Von Damm. 1983. Hot springs on the ocean floor. *Scientific American* 248, no. 4: 78–93.

Gass, I. G. 1982. Ophiolites. *Scientific American* 247, no. 2: 122–31.

Hutchinson, D. 1992–93 (Winter). Continental margins: windows into the Earth's history, *Oceanus* 35, no. 4: 34–44.

Kennett, J. P. 1982. *Marine geology*. Englewood Cliffs, N.J.: Prentice-Hall.

Mark, K. 1976. Coral reefs, seamounts, and guyots. *Sea Frontiers* 22, no. 3: 143–49.

Pinet, P. 1992. *Oceanography: An introduction to the planet oceanus*. St. Paul, Minn.: West Publishing Co.

Rona, P. A. 1986. Mineral deposits from sea-floor hot springs. *Scientific American* 254, no. 1: 84–93.

Ross, D. A. 1988. *Introduction to oceanography*. Englewood Cliffs, N.J.: Prentice-Hall.

Thurman, H. V. 1988. *Introductory oceanography*. 5th ed. Columbus, Ohio: Merrill Publishing Co.

Tolmazin, D. 1985. *Elements of dynamic oceanography*. Boston, Mass.: Allen & Unwin.

Chapter 12

PLATE TECTONICS: A UNIFYING THEORY

OUTLINE

View looking down the Great Rift Valley of Africa. Little Magadi, seen in the background, is one of numerous soda lakes forming in the valley. Because of high evaporation rates and lack of any drainage outlets, these lakes are very saline. The Great Rift Valley is part of the system of rift valleys resulting from stretching of the crust as plates move away from each other in eastern Africa.

PROLOGUE

Two tragic events that occurred during 1985 serve to remind us of the dangers of living near a convergent plate margin. On September 19, a magnitude 8.1 earthquake killed more than 9,000 people in Mexico City. Two months later and 3,200 km to the south, a minor eruption of Colombia's Nevado del Ruiz volcano partially melted its summit glacial ice, causing a mudflow that engulfed Armero and several other villages and killed more than 23,000 people. These two tragedies resulted in more than 32,000 deaths, tens of thousands of injuries, and billions of dollars in property damage.

Both of these events occurred along the eastern portion of the Ring of Fire, a chain of intense seismic and volcanic activity that encircles the Pacific Ocean basin (▶ Figure 12-1). Some of the world's greatest disasters occur along this ring because of volcanism and earthquakes generated by plate convergence. The Mexico City earthquake resulted from subduction of the Cocos plate at the Middle America Trench (Figure 12-1). Sudden movement of the Cocos plate beneath Central America generated seismic waves that traveled outward in all directions. The violent shaking experienced in Mexico City, 350 km away, and elsewhere was caused by these seismic waves.

The strata underlying Mexico City consist of unconsoli dated sediment deposited in a large ancient lake. Such sediment amplifies the shaking during earthquakes with the unfortunate consequence that buildings constructed there are commonly more heavily damaged than those built on solid bedrock (see Perspective 9-1, Figure 4).

Less than two months after the Mexico City earthquake, Colombia experienced its greatest recorded natural disaster. Nevado del Ruiz is one of several active volcanoes resulting from the rise of magma generated where the Nazca plate is subducted beneath South America (Figure 12-1). A minor eruption of Nevado del Ruiz partially melted the glacial ice on the mountain; the meltwater rushed down the valleys, mixed with the sediment, and turned it into a deadly viscous mudflow.

The city of Armero, Colombia, lies in the valley of the Lagunilla River, one of several river valleys inundated by mudflows. Twenty thousand of the city's 23,000 inhabitants died, and most of the city was destroyed (▶ Figure 12-2). Another 3,000 people were killed in nearby valleys. A geologic hazard assessment study completed one month before the eruption showed that Armero was in a high-hazard mudflow area!

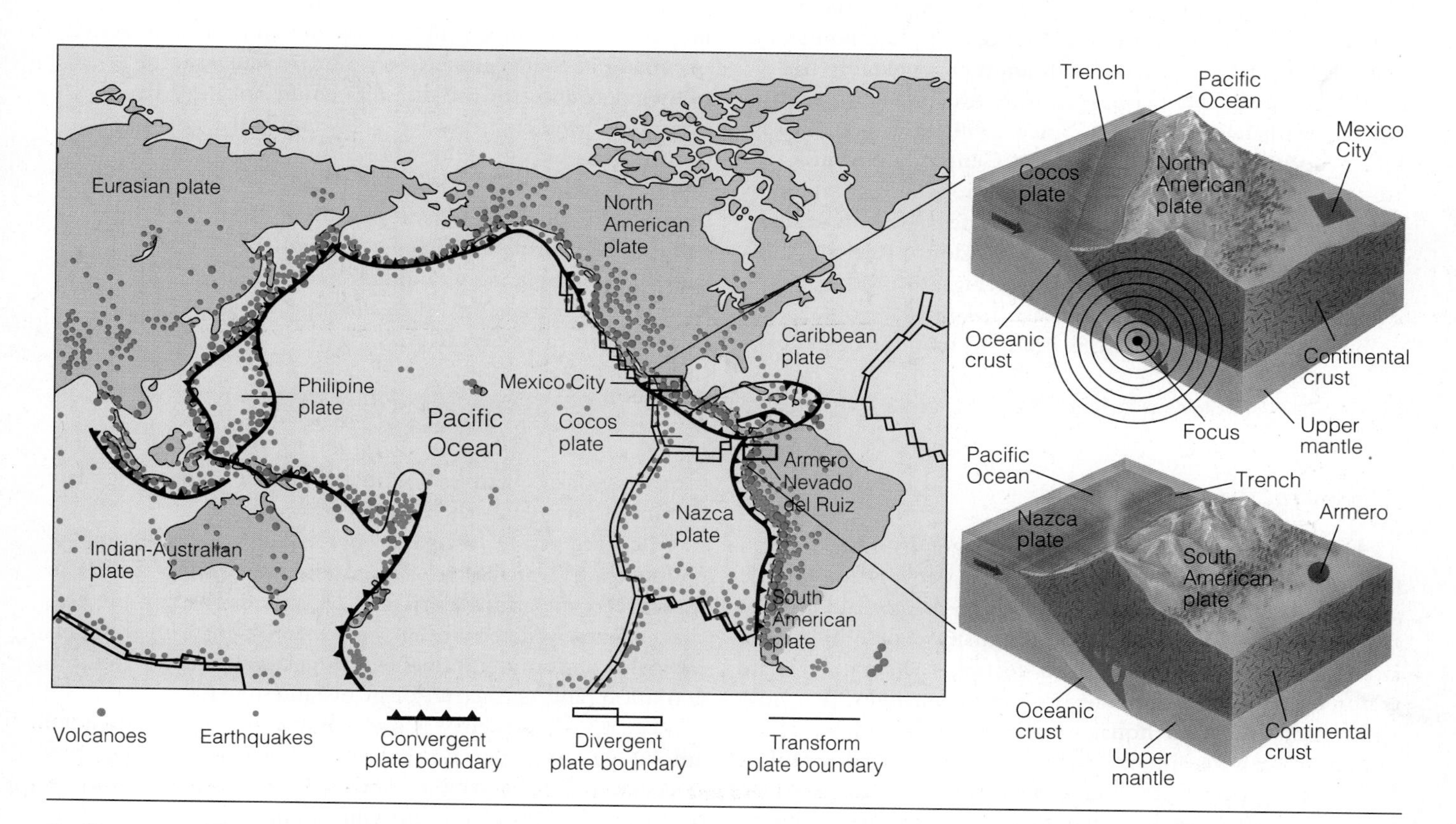

▶ **FIGURE 12-1** The Ring of Fire is a zone of intense earthquake and volcanic activity that encircles the Pacific Ocean basin. Most of this activity results from plate convergence as illustrated by the two insets.

➢ **FIGURE 12-2** The 1985 eruption of Nevado del Ruiz in Colombia melted some of its glacial ice. The meltwater mixed with sediments and formed a huge mudflow that destroyed the city of Armero and killed 20,000 of its inhabitants.

These two examples vividly illustrate some of the dangers of living in proximity to a convergent plate boundary. Subduction of one plate beneath another repeatedly triggers large earthquakes, the effects of which are frequently felt far from their epicenters. Since 1900, earthquakes have killed more than 112,000 people in Central and South America alone. Though volcanic eruptions in this region have not caused nearly as many casualties as earthquakes, they have, nevertheless, caused tremendous property damage and have the potential for triggering devastating events such as the 1985 Colombian mudflow.

Because the Ring of Fire is home to millions of people, can anything be done to decrease the devastation that inevitably results from the earthquake and volcanic activity occurring in that region? Given our present state of knowledge, most of the disasters could not have been accurately predicted, but better planning and advance preparations by the nations bordering the Ring of Fire could have prevented much tragic loss of life. As long as people live near convergent plate margins, there will continue to be disasters. However, by studying and understanding geologic activity along convergent as well as divergent and transform plate margins, geologists can help to minimize the destruction.

 INTRODUCTION

The recognition that the Earth's geography has changed continuously through time has led to a revolution in the geological sciences, forcing geologists to greatly modify the way they view the Earth. Although many people have only a vague notion of what plate tectonic theory is, plate tectonics has a profound effect on all of our lives. It is now realized that most earthquakes and volcanic eruptions occur near plate margins and are not merely random occurrences. Furthermore, the formation and distribution of many important natural resources, such as metallic ores, are related to plate boundaries, and geologists are now incorporating plate tectonic theory into their prospecting efforts.

The interaction of plates determines the location of continents, ocean basins, and mountain systems, which in turn affects the atmospheric and oceanic circulation patterns that ultimately determine global climates. Plate movements have also profoundly influenced the geographic distribution, evolution, and extinction of plants and animals.

Plate tectonic theory is now almost universally accepted among geologists, and its application has led to a greater understanding of how the Earth has evolved and continues to do so. This powerful, unifying theory accounts for many apparently unrelated geologic events, allowing geologists to view such phenomena as part of a continuing story rather than as a series of isolated incidents.

We will first review the various hypotheses that preceded plate tectonic theory, examining the evidence that led some people to accept the idea of continental movement and others to reject it. Because plate tectonic theory has evolved from numerous scientific inquiries and observations, only the more important ones will be covered.

Early Ideas About Continental Drift

The idea that the Earth's geography was different during the past is not new. During the fifteenth century, Leonardo da Vinci observed that "above the plains of Italy where flocks of birds are flying today fishes were once moving in large schools." In 1620, Sir Francis Bacon commented on the similarity of the shorelines of western Africa and eastern South America but did not make the connection that the Old and New Worlds might once have been sutured together. Alexander von Humboldt made the same observation in 1801, although he attributed these similarities to erosion rather than the splitting apart of a larger continent.

One of the earliest specific references to continental drift is in the 1858 book *Creation and Its Mysteries Revealed* by Antonio Snider-Pellegrini. He suggested that all of the continents were linked together during the Pennsylvanian Period and later split apart. He based his conclusions on the similarities between plant fossils in the Pennsylvanian-aged coal beds of Europe and North America, and attributed the separation of the continents to the biblical deluge.

During the late nineteenth century, the Austrian geologist Edward Suess noted the similarities between the Late Paleozoic plant fossils of India, Australia, Africa, Antarctica, and South America as well as evidence of glaciation in the rock sequences of these southern continents. In 1885 he proposed the name Gondwanaland (or **Gondwana** as we will use here) for a supercontinent composed of these southern landmasses. Gondwana is a province in east-central India where evidence exists for extensive glaciation as well as abundant fossils of the ***Glossopteris* flora** (▶ Figure 12-3), an association of Late Paleozoic plants found only in India and the Southern Hemisphere continents. Suess thought the distribution of plant fossils and glacial deposits was a consequence of extensive land bridges that once connected the continents and later sank beneath the ocean.

In 1910, the American geologist Frank B. Taylor published a pamphlet presenting his own theory of continental drift. He explained the formation of mountain ranges as a result of the lateral movement of continents. He also envisioned the present-day continents as parts of larger polar continents that eventually broke apart and migrated toward the equator after the Earth's rotation supposedly slowed due to gigantic tidal forces. According to Taylor, these tidal forces were generated when the Earth captured the Moon about 100 million years ago.

(a)

(b)

▶ **FIGURE 12-3** Representative members of the *Glossopteris* flora. Fossils of these plants are found on all five of the Gondwana continents. *Glossopteris* leaves from (*a*) the Upper Permian Dunedoo Formation and (*b*) the Upper Permian Illawarra Coal Measures, Australia. (Photos courtesy of Patricia G. Gensel, University of North Carolina.)

Although we now know that Taylor's mechanism is incorrect, one of his most significant contributions was his suggestion that the Mid-Atlantic Ridge, discoverd by the British H.M.S. *Challenger* expeditions of 1872–1876, might mark the site along which an ancient continent broke apart to form the present-day Atlantic Ocean.

Alfred Wegener and the Continental Drift Hypothesis

Alfred Wegener, a German meteorologist (▶ Figure 12-4), is generally credited with developing the hypothesis of **continental drift.** In a 1912 lecture before the German Geological Association in Frankfurt, Germany, Wegener first presented his ideas for moving continents. In his monumental book, *The Origin of Continents and Oceans* (first published in 1915), Wegener proposed that all of the landmasses were

▶ **FIGURE 12-4** Alfred Wegener, a German meteorologist, proposed the continental drift hypothesis in 1912 based on a tremendous amount of geological, paleontological, and climatological evidence. He is shown here waiting out the Arctic winter in an expedition hut.

originally united into a single supercontinent that he named **Pangaea,** from the Greek meaning "all land." Wegener portrayed his grand concept of continental movement in a series of maps showing the breakup of Pangaea and the movement of the various continents to their present-day locations. Wegener had amassed a tremendous amount of geological, paleontological, and climatological evidence in support of continental drift, but the initial reaction of scientists to his then-heretical ideas can best be described as mixed.

Opposition to Wegener's ideas became particularly widespread in North America after 1928 when the American Association of Petroleum Geologists held an international symposium to review the hypothesis of continental drift. After each side had presented its arguments, the opponents of continental drift were clearly in the majority, even though the evidence in support of continental drift, most of which came from the Southern Hemisphere, was impressive and difficult to refute. One problem with the hypothesis, however, was its lack of a mechanism to explain how continents, composed of granitic rocks, could seemingly move through the denser basaltic oceanic crust.

Nevertheless, the eminent South African geologist Alexander du Toit further developed Wegener's arguments and gathered more geological and paleontological evidence in support of continental drift. In 1937, du Toit published *Our Wandering Continents,* in which he contrasted the glacial deposits of Gondwana with coal deposits of the same age found in the continents of the Northern Hemisphere. To resolve this apparent climatological paradox, du Toit moved the Gondwana continents to the South Pole and brought the northern continents together such that the coal deposits were located at the equator. He named this northern landmass **Laurasia.** It consisted of present-day North America, Greenland, Europe, and Asia (except for India).

In spite of what seemed to be overwhelming evidence, most geologists still refused to accept the idea that continents moved. It was not until the 1960s when oceanographic research provided convincing evidence that the continents had once been joined together and subsequently separated that the hypothesis of continental drift finally became widely accepted.

 The Evidence for Continental Drift

The evidence used by Wegener, du Toit, and others to support the hypothesis of continental drift includes the fit of the shorelines of continents; the appearance of the same rock sequences and mountain ranges of the same age on continents now widely separated; the matching of glacial deposits and paleoclimatic zones; and the similarities of many extinct plant and animal groups whose fossil remains are found today on widely separated continents.

Continental Fit

Wegener, like some before him, was impressed by the close resemblance between the coastlines of continents on opposite sides of the Atlantic Ocean, particularly between South America and Africa. He cited these similarities as partial evidence that the continents were at one time joined together as a supercontinent that subsequently split apart. As his critics pointed out, though, the configuration of coast-

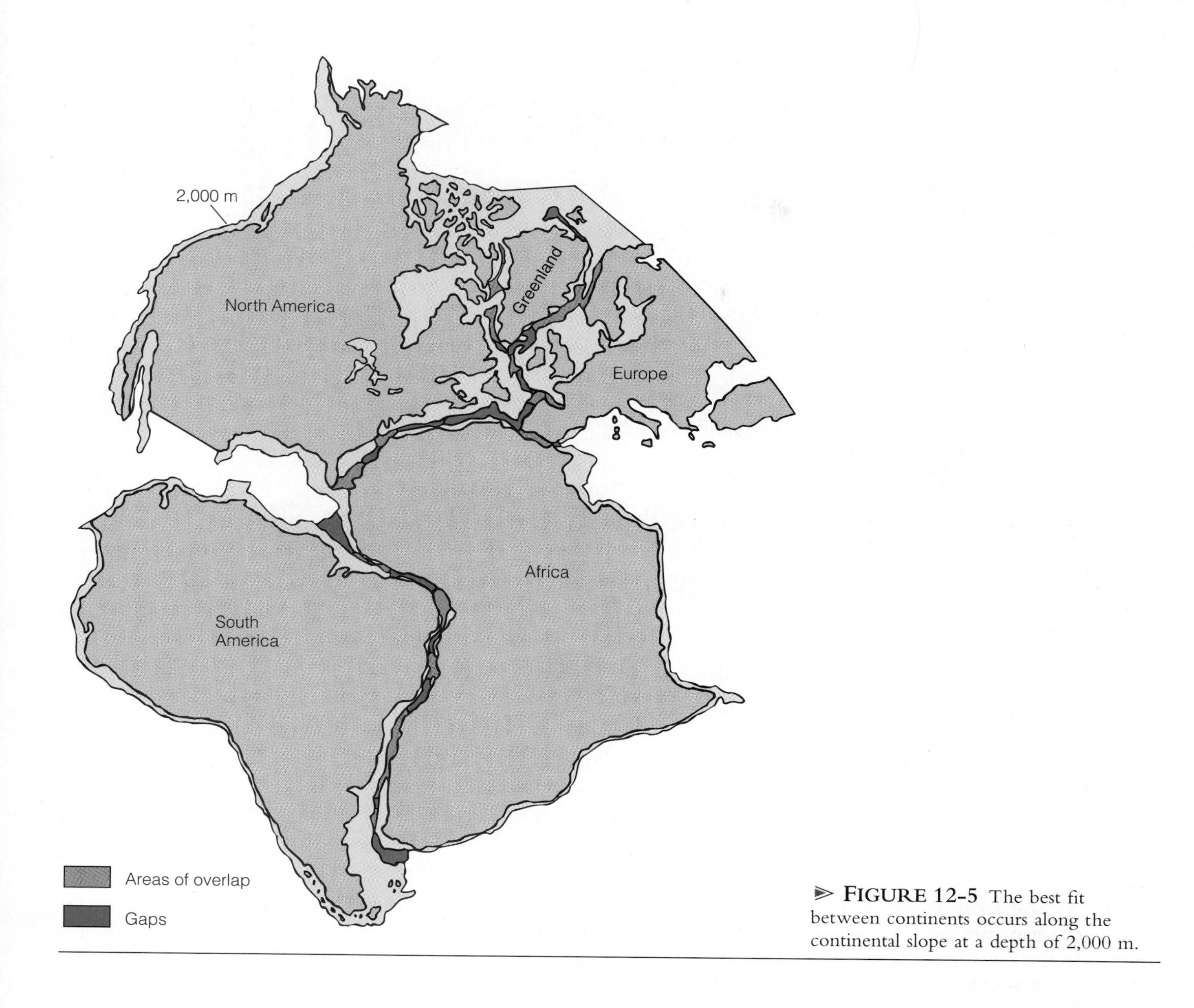

▶ **FIGURE 12-5** The best fit between continents occurs along the continental slope at a depth of 2,000 m.

lines results from erosional and depositional processes and therefore is continually being modified. So even if the continents had separated during the Mesozoic Era, as Wegener proposed, it is not likely that the coastlines would fit exactly.

A more realistic approach is to fit the continents together along the continental slope where erosion would be minimal. Recall from Chapter 11 that the true margin of a continent—that is, where continental crust changes to oceanic crust—is beneath the continental slope. In 1965 Sir Edward Bullard, an English geophysicist, and two associates showed that the best fit between the continents occurs along the continental slope at a depth of about 2,000 m (▶ Figure 12-5). Since then, other reconstructions using the latest ocean basin data have confirmed the close fit between continents when they are reassembled to form Pangaea.

Similarity of Rock Sequences and Mountain Ranges

If the continents were at one time joined together, then the rocks and mountain ranges of the same age in adjoining locations on the opposite continents should closely match. Such is the case for the Gondwana continents (▶ Figure 12-6). Marine, nonmarine, and glacial rock sequences of Pennsylvanian to Jurassic age are almost identical for all five Gondwana continents, strongly indicating that they were joined together at one time.

The trends of several major mountain ranges also support the hypothesis of continental drift. These mountain ranges seemingly end at the coastline of one continent only to apparently continue on another continent across the ocean. The folded Appalachian Mountains of North America, for

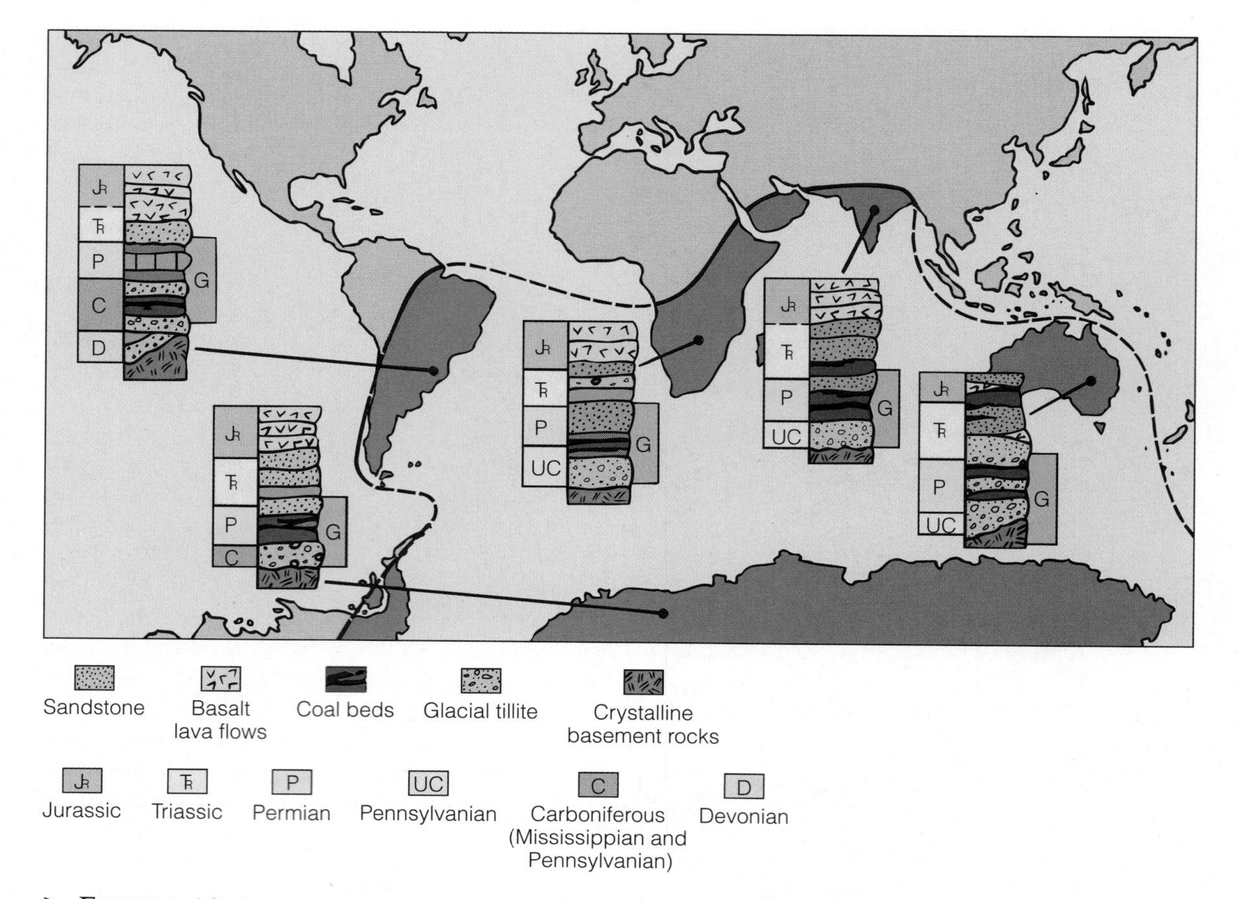

▶ **FIGURE 12-6** Marine, nonmarine, and glacial rock sequences of Pennsylvanian to Jurassic age are nearly the same for all Gondwana continents. Such close similarity strongly suggests that they were joined together at one time. The range indicated by G is that of the *Glossopteris* flora.

example, trend northeastward through the eastern United States and Canada and terminate abruptly at the Newfoundland coastline (▶ Figure 12-7a). Mountain ranges of the same age and deformational style occur in eastern Greenland, Ireland, Great Britain, and Norway. Even though these mountain ranges are currently separated by the Atlantic Ocean, they form an essentially continuous mountain range when the continents are positioned next to each other (Figure 12-7b).

Glacial Evidence

During the Late Paleozoic Era, massive glaciers covered large continental areas of the Southern Hemisphere. Evidence for this glaciation includes layers of till (sediments deposited by glaciers) and striations (scratch marks) in the bedrock beneath the till. Fossils and sedimentary rocks of the same age from the Northern Hemisphere, however, give no indication of glaciation. Fossil plants found in coals indicate that the Northern Hemisphere had a tropical climate during the time that the Southern Hemisphere was glaciated.

All of the Gondwana continents except Antarctica are currently located near the equator in subtropical to tropical climates. Mapping of glacial striations in bedrock in Australia, India, and South America indicates that the glaciers moved from the areas of the present-day oceans onto land (▶ Figure 12-8a). This would be highly unlikely because large continental glaciers flow outward from their central area of accumulation toward the sea.

If the continents did not move during the past, one would have to explain how glaciers moved from the oceans onto land and how large-scale continental glaciers formed near the equator. But if the continents are reassembled as a single landmass with South Africa located at the south pole, the direction of movement of Late Paleozoic continental glaciers makes sense. Furthermore, this geographic arrangement places the northern continents nearer the tropics, which is consistent with the fossil and climatological evidence from Laurasia (Figure 12-8b).

Fossil Evidence

Some of the most compelling evidence for continental drift comes from the fossil record. Fossils of the *Glossopteris* flora are found in equivalent Pennsylvanian- and Permian-aged coal deposits on all five Gondwana continents. The *Glossop-*

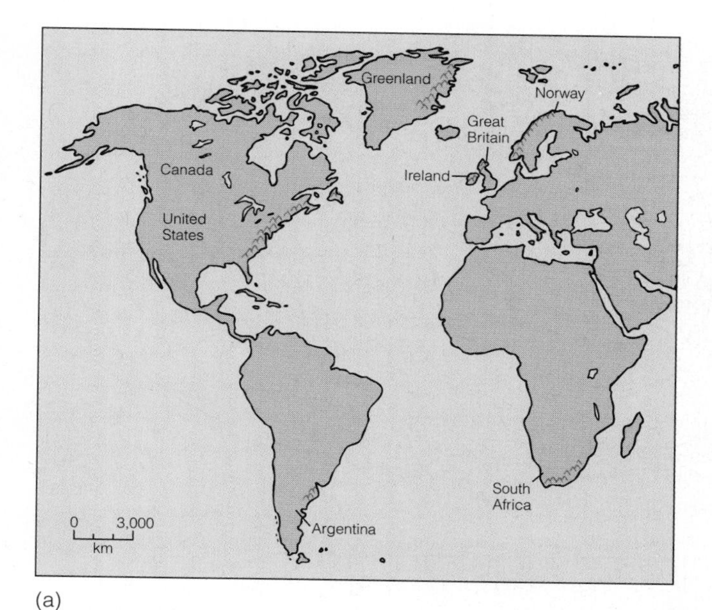

(a)

(b)

➢ FIGURE 12-7 (*a*) Various mountain ranges of the same age and style of deformation are currently widely separated by oceans. (*b*) When the continents are brought together, however, a single continuous mountain range is formed. Such evidence indicates the continents were at one time joined together and were subsequently separated.

teris flora is characterized by the seed fern *Glossopteris* (Figure 12-3) as well as by many other distinctive and easily identifiable plants. Plant pollen and spores can be dispersed over great distances by wind, but *Glossopteris*-type plants produced seeds that were too large to have been carried by winds. Even if the seeds had floated across the ocean, they probably would not have remained viable for any length of time in salt water.

The present-day climates of South America, Africa, India, Australia, and Antarctica range from tropical to polar and are much too diverse to support the type of plants that compose the *Glossopteris* flora. Wegener reasoned therefore that these continents must once have been joined such that these widely separated localities were all in the same latitudinal climatic belt (➢ Figure 12-9).

The fossil remains of animals also provide strong evidence for continental drift. One of the best examples is *Mesosaurus,* a freshwater reptile whose fossils are found in Permian-aged rocks in certain regions of Brazil and South Africa and nowhere else in the world (Figure 12-9). Because the physiology of freshwater and marine animals is completely different, it is hard to imagine how a freshwater reptile could have swum across the Atlantic Ocean and found a freshwater environment nearly identical to its former habitat. Moreover, if *Mesosaurus* could have swum across the ocean, its fossil remains should be widely dispersed. It is more logical to assume that *Mesosaurus* lived in lakes in what are now adjacent areas of South America and Africa, but were then united into a single continent.

Lystrosaurus and *Cynognathus* are both land-dwelling reptiles that lived during the Triassic Period; their fossils are found only on the present-day continental fragments of Gondwana (Figure 12-9). Because they are both land animals, they certainly could not have swum across the oceans currently separating the Gondwana continents. Therefore, the continents must once have been connected.

The evidence favoring continental drift seemed overwhelming to Wegener and his supporters yet the lack of a suitable mechanism to explain continental movement prevented its widespread acceptance. Not until new evidence from studies of the Earth's magnetic field and oceanographic research showed that the ocean basins were geologically young features did renewed interest in continental drift occur.

Paleomagnetism and Polar Wandering

Some of the most convincing evidence for continental drift came from the study of paleomagnetism, a relatively new discipline during the 1950s. During that time, some geologists were researching past changes of the Earth's magnetic field in order to better understand the present-day magnetic field. As so often happens in science, these studies led to other discoveries. In this case, they led to the discovery that the ocean basins are geologically young features, and that the continents have indeed moved during the past, just as Wegener and others had proposed.

Recall from Chapter 10 that the Earth's magnetic poles correspond closely to the location of the geographic poles (see Figure 10-24). When a magma cools, the iron-bearing minerals align themselves with the Earth's magnetic field when they reach the Curie point, thus recording both the direction and the intensity of the magnetic field. This

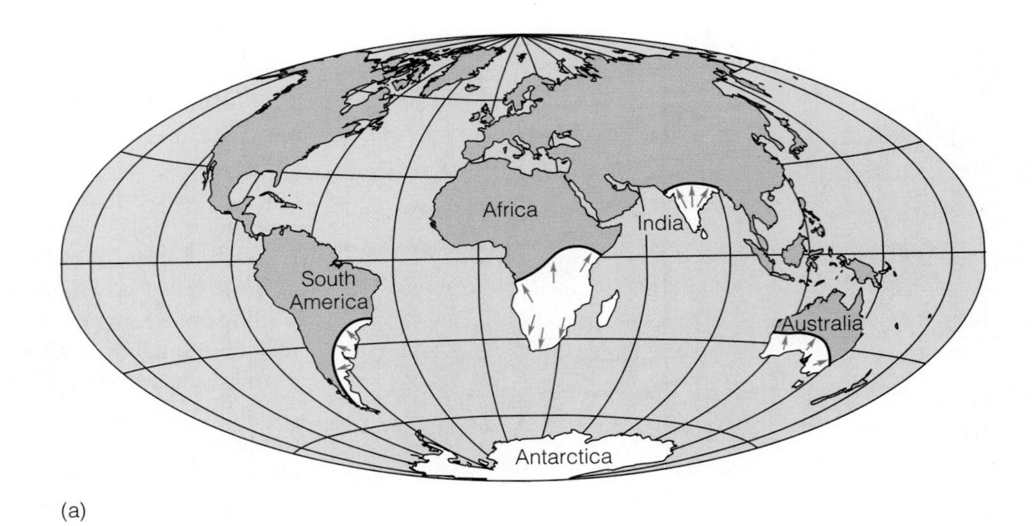

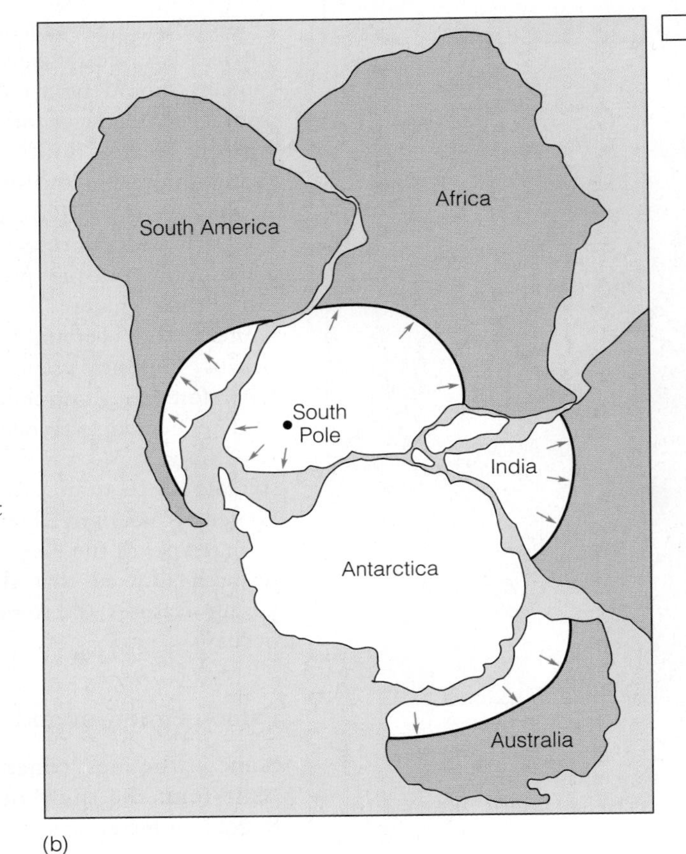

▶ **FIGURE 12-8** (*a*) If the continents did not move in the past, then Late Paleozoic glacial striations preserved in bedrock in Australia, India, and South America indicate that glacial movement for each continent was from the oceans onto land within a subtropical to tropical climate. Such an occurrence is highly unlikely. (*b*) If the continents are brought together so that South Africa is located at the South Pole, then the glacial movement indicated by the striations makes sense. In this situation, the glacier, located in a polar climate, moved radially outward from a thick central area toward its periphery.

information can be used to determine the location of the Earth's magnetic poles and the latitude of the rock when it formed.

Research conducted during the 1950s by the English geophysicist S. K. Runcorn and his associates showed that the location of the paleomagnetic pole, as measured by the paleomagnetism in European lava flows of different ages, varied widely. They found that during the past 500 million years, the north magnetic pole has apparently wandered from the Pacific Ocean northward through eastern and then northern Asia to its present-day location near the geographic north pole (▶ Figure 12-10). This paleomagnetic evidence from Europe could be interpreted in three ways: the continent remained fixed and the north magnetic pole moved; the north magnetic pole stood still and the continent moved; or both the continent and the north magnetic pole moved.

When paleomagnetic readings from numerous lava flows

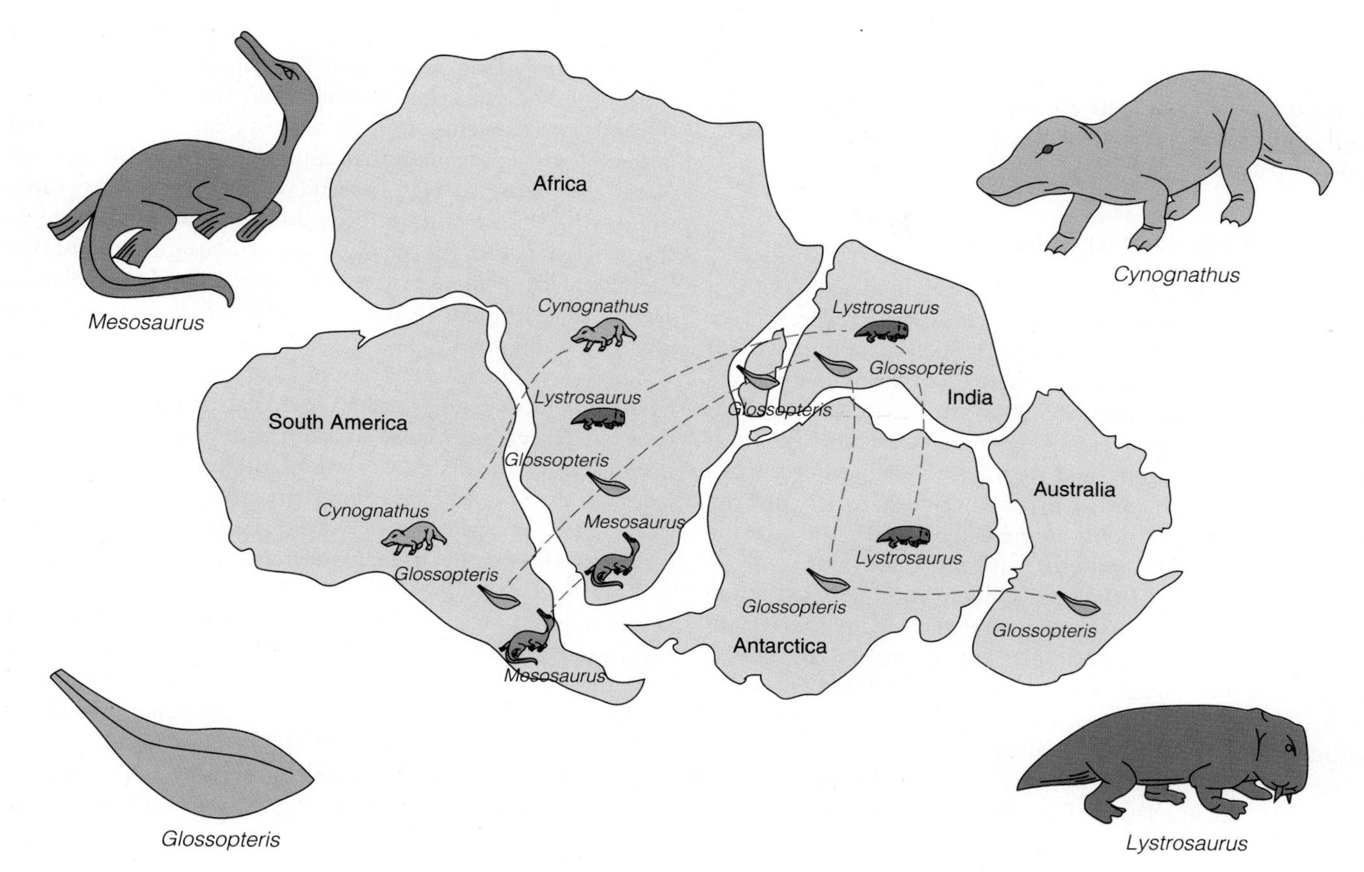

► **FIGURE 12-9** Some of the animals and plants whose fossils are found today on the widely separated continents of South America, Africa, India, Australia, and Antarctica. These continents were joined together during the Late Paleozoic to form Gondwana, the southern landmass of Pangaea. *Glossopteris* and similar plants are found in Pennsylvanian- and Permian-aged deposits on all five continents. *Mesosaurus* is a freshwater reptile whose fossils are found in Permian-aged rocks in Brazil and South Africa. *Cynognathus* and *Lystrosaurus* are land reptiles who lived during the Early Triassic Period. Fossils of *Cynognathus* are found in South America and Africa, while fossils of *Lystrosaurus* have been recovered from Africa, India, and Antarctica.

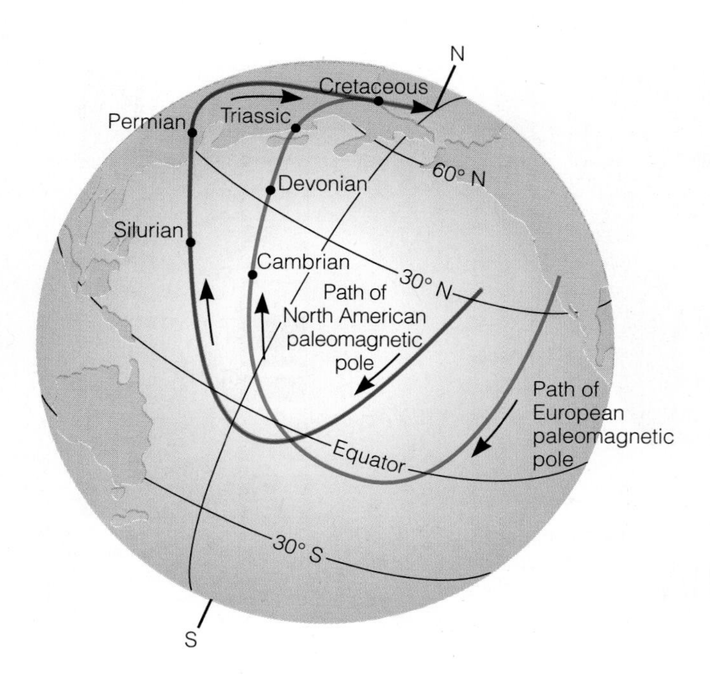

► **FIGURE 12-10** The apparent paths of polar wandering for North America and Europe. The apparent location of the north magnetic pole is shown for different periods on each continent's polar wandering path.

of different ages in North America were plotted on a map, they pointed to different magnetic pole locations than did flows of the same ages in Europe (Figure 12-10). Furthermore, analysis of lava flows from all continents indicated that each continent had its own series of magnetic poles! Does this mean that each continent had a different north magnetic pole? That would be highly unlikely and difficult to reconcile with the laws of physics and what we know about how the Earth's magnetic field is generated (see Chapter 10).

Therefore, the best explanation for the apparent wandering of the magnetic poles is that they have remained at their present locations near the geographic poles and the continents have moved. When the continents are fitted together so that the paleomagnetic data point to only one magnetic pole, we find, just as Wegener did, that the rock sequences, mountain ranges, and glacial deposits match, and that the fossil and climatic evidence is consistent with the reconstructed paleogeography.

 SEA-FLOOR SPREADING

In addition to the paleomagnetic research in the 1950s, a renewed interest in oceanographic research led to extensive mapping of the world's ocean basins. Such mapping revealed that the Mid-Atlantic Ridge is part of a worldwide oceanic ridge system more than 65,000 km long. It was also discovered that oceanic ridges are characterized by high heat flow, basaltic volcanism, and seismicity. Furthermore, magnetic reversals, as recorded in oceanic-crust rocks, and the age of deep-sea sediments immediately above the oceanic crust occur in distinct patterns with respect to ridges.

Harry H. Hess of Princeton University conducted much of his oceanographic research while serving in the central Pacific during World War II. His discovery of guyots (submerged, flat-topped volcanic islands) provided geologists with evidence that the sea floor is moving away from the oceanic ridges (see Figure 11-17).

As a result of his discovery of guyots and other research conducted during the 1950s, Hess published a landmark paper in 1962 in which he proposed the hypothesis of **sea-floor spreading** to account for continental movement. Hess proposed that the continents do not move across or through oceanic crust, but rather that the continents and oceanic crust move together and are both parts of large plates. According to Hess, oceanic crust separates at oceanic ridges where new crust is formed by upwelling magma. As the magma cools, the newly formed oceanic crust moves laterally away from the ridge, thus explaining how volcanic islands that formed at or near ridge crests later become guyots (see Figure 11-17).

Hess revived the idea (proposed in the 1930s and 1940s by Arthur Holmes and others) of a heat transfer system—or **thermal convection cells**—within the mantle as a mechanism to move the plates. According to Hess, hot magma rises from the mantle, intrudes along rift zone fractures defining oceanic ridges, and forms new crust. Cold crust is subducted back into the mantle at deep-sea trenches where it is heated and recycled.

How could Hess's hypothesis be confirmed? If new crust is forming at oceanic ridges and the Earth's magnetic field is periodically reversing itself, then these magnetic reversals should be preserved as magnetic anomalies in the rocks of the oceanic crust (➤ Figure 12-11).

Around 1960, magnetic data gathered by scientists from the Scripps Institution of Oceanography in California indi-

➤ **FIGURE 12-11** The sequence of magnetic anomalies preserved within the oceanic crust on both sides of an oceanic ridge is identical to the sequence of magnetic reversals already known from continental lava flows. Magnetic anomalies are formed when basaltic magma intrudes into oceanic ridges; when the magma cools below the Curie point, it records the Earth's magnetic polarity at the time. Sea-floor spreading splits the previously formed crust in half, so that it moves laterally away from the oceanic ridge. Repeated intrusions record a symmetrical series of magnetic anomalies that reflect periods of normal and reversed polarity. The magnetic anomalies are recorded by a magnetometer, which measures the strength of the magnetic field.

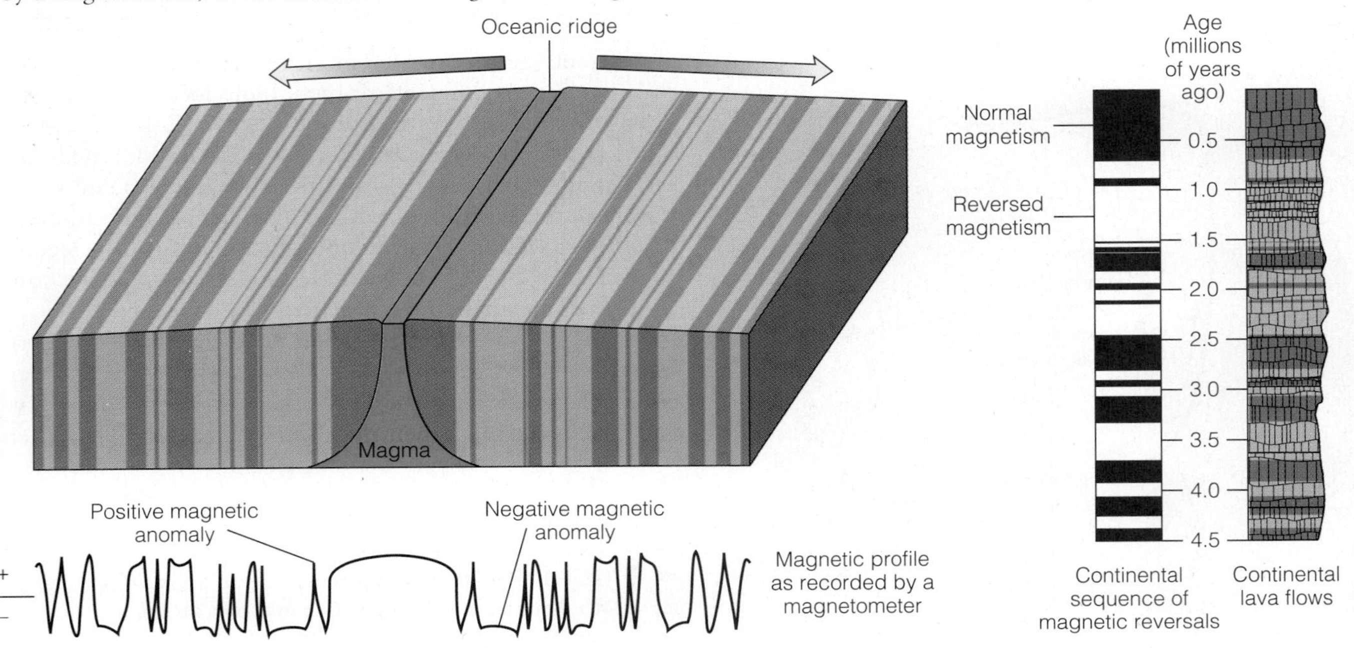

cated an unusual pattern of alternating positive and negative magnetic anomalies for the Pacific ocean floor off the west coast of North America. The pattern consisted of a series of roughly north-south parallel stripes, but they were broken and offset by essentially east-west fractures. It was not until 1963 that F. Vine and D. Matthews of Cambridge University and L. W. Morley, a Canadian geologist, independently arrived at a model that explained this pattern of magnetic anomalies.

These three geologists proposed that when basaltic magma intruded along the crests of oceanic ridges, it would record the magnetic polarity at the time it cooled. As the ocean floor moved away from these oceanic ridges, repeated intrusions would form a symmetrical series of magnetic stripes, recording periods of normal and reverse polarity (Figure 12-11). Shortly thereafter, the Vine, Matthews, and Morley proposal was supported by evidence from magnetic readings across the Reykjanes Ridge, part of the Mid-Atlantic Ridge south of Iceland. A group from the Lamont-Doherty Geological Observatory at Columbia University found that magnetic anomalies in this area did form stripes that were distributed parallel to and symmetrical about the oceanic ridge. By the end of the 1960s, comparable magnetic anomaly patterns were found surrounding most oceanic ridges.

Magnetic surveys for most of the ocean floor have now been completed (▶ Figure 12-12). They demonstrate that the youngest oceanic crust is adjacent to the spreading ridges and that the age of the crust increases with distance from the ridge axis, as would be expected according to the sea-floor spreading hypothesis. Furthermore, the oldest oceanic crust is less than 180 million years old, whereas the oldest continental crust is 3.96 billion years old; this difference in age provides confirmation that the ocean basins are geologically young features whose openings and closings are partially responsible for continental movement.

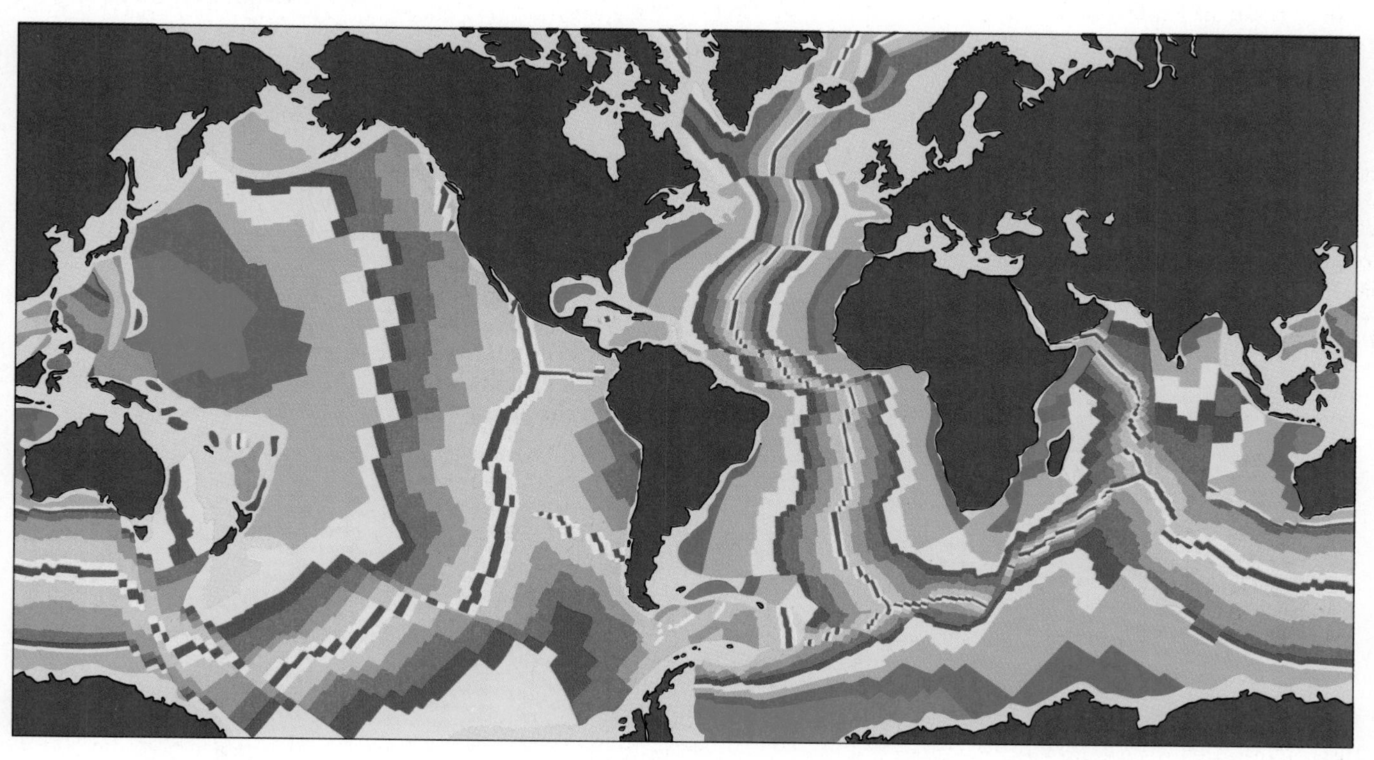

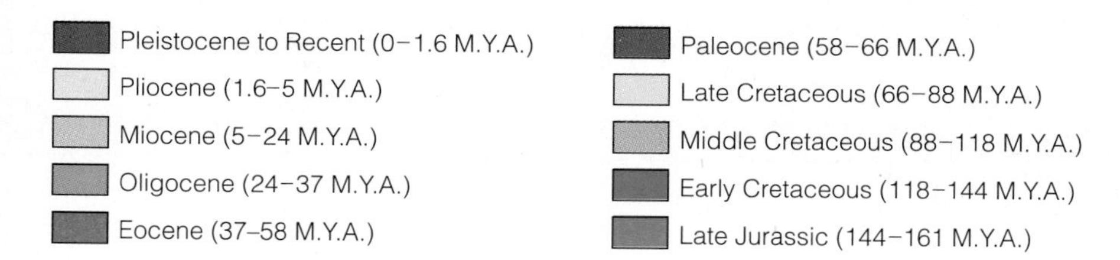

▶ **FIGURE 12-12** The age of the world's ocean basins established from magnetic anomalies demonstrates that the youngest oceanic crust is adjacent to the spreading ridges and that its age increases away from the ridge axis.

Perspective 12-1

THE SUPERCONTINENT CYCLE

At the end of the Paleozoic Era, all continents were amalgamated into the supercontinent Pangaea. Pangaea began fragmenting during the Triassic Period and continues to do so, thus accounting for the present distribution of continents and oceans. It now appears that another supercontinent existed at the end of the Proterozoic Eon, and some evidence indicates that even earlier supercontinents existed. Recently, it has been proposed that supercontinents consisting of all or most of the Earth's landmasses form, break up, and re-form in a cycle spanning about 500 million years.

The supercontinent cycle hypothesis is an expansion on the ideas of the Canadian geologist J. Tuzo Wilson. During the early 1970s, Wilson proposed a cycle (now known as the Wilson cycle) that includes continental fragmentation, the opening and closing of an ocean basin, and finally reassembly of the continent. According to the supercontinent cycle hypothesis, heat accumulates beneath a supercontinent because rocks of continents are poor conductors of heat. As a result of the heat accumulation, the supercontinent domes upward and fractures. Basaltic magma rising from below fills the fractures. As a basalt-filled fracture widens, it begins subsiding and forms a long narrow ocean such as the present-day Red Sea. Continued rifting eventually forms an expansive ocean basin such as the Atlantic.

According to proponents of the supercontinent cycle, one of the most convincing arguments for their hypothesis is the "surprising regularity" of mountain building caused by compression during continental collisions. These mountain-building episodes occur about every 400 to 500 millions years, and are followed by an episode of rifting about 100 million years later. In other words, a supercontinent fragments and its individual plates disperse following a rifting episode, an interior ocean forms, and then the dispersed fragments reassemble to form another supercontinent (▶ Figure 1).

In addition to explaining the dispersal and reassembly of supercontinents, the supercontinent cycle hypothesis also explains two distinct types of orogens (linear parts of the Earth's crust that were deformed during mountain building): *interior orogens* resulting from compression induced by continental collisions and *peripheral orogens* that form at continental margins in response to subduction and igneous activity. Following the fragmentation of a supercontinent, an interior ocean forms and widens as plates separate. After about 200 million years, plate separation ceases as the oceanic crust becomes cooler and denser and begins to subduct at the margins of the interior ocean basin, thus transforming inward-facing passive continental margins into active continental margins. Rising magma resulting from subduction forms plutons and volcanoes along these continental margins, and an interior orogeny develops when plates converge, causing compression-induced deformation, crustal thickening, and metamorphism. The present-day Appalachian Mountains of the eastern United States and Canada originally formed as an interior orogen (see Figure 13-36).

Peripheral orogenies are caused by convergence of oceanic plates with the margins of a continent, igneous activity resulting from subduction, and collisions of island arcs with the continent. Subduction-related volcanism in peripheral orogens is rather continuous, but collisions of island arcs with the supercontinent are episodic. Peripheral orogenies can develop at any time during the supercontinent cycle. A good example of a currently active peripheral orogeny is the Andes of western South America where the Nazca plate is being subducted beneath the South American plate (see Figure 13-29).

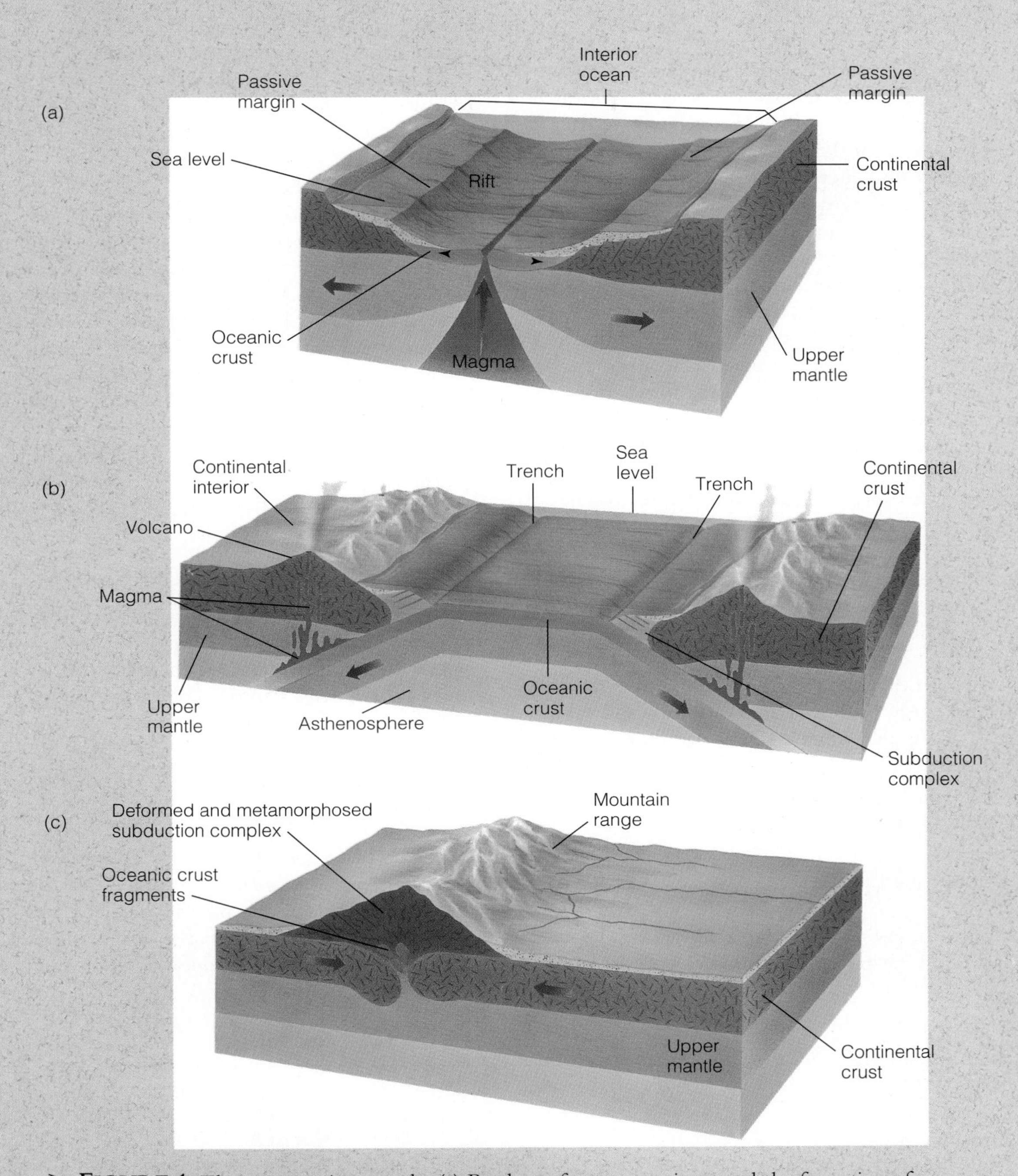

▶ **FIGURE 1** The supercontinent cycle. (*a*) Breakup of a supercontinent and the formation of an interior ocean basin. (*b*) Subduction along the margins of the interior ocean basin begins approximately 200 million years later, resulting in volcanic activity and deformation along an active oceanic-continental plate boundary. (*c*) Continental collisions and the formation of a new supercontinent result when all of the oceanic crust of the interior ocean basin is subducted.

Deep-Sea Drilling and the Confirmation of Sea-Floor Spreading

To many geologists, the paleomagnetic data amassed in support of continental drift and sea-floor spreading were convincing. Results from the Deep-Sea Drilling Project (see Chapter 11) have confirmed the interpretations made by earlier paleomagnetic studies. Cores of deep-sea sediments and seismic profiles obtained by the *Glomar Challenger* and other research vessels have provided much of the data that support the sea-floor spreading hypothesis.

According to this hypothesis, oceanic crust is continuously forming at mid-oceanic ridges, moving away from these ridges by sea-floor spreading, and being consumed at subduction zones. If this is the case, oceanic crust should be youngest at the ridges and become progressively older with increasing distance away from them. Moreover, the age of the oceanic crust should be symmetrically distributed about the ridges. As we have just noted, paleomagnetic data confirm these statements. Furthermore, fossils from sediments overlying the oceanic crust and radiometric dating of rocks found on oceanic islands both substantiate this predicted age distribution.

Sediments in the open ocean accumulate, on average, at a rate of less than 0.3 cm per 1,000 years. If the ocean basins were as old as the continents, we would expect deep-sea sediments to be several kilometers thick. However, data from numerous drill holes indicate that deep-sea sediments are at most only a few hundred meters thick and are thin or absent at oceanic ridges. Their near-absence at the ridges should come as no surprise because these are the areas where new crust is continuously produced by volcanism and sea-floor spreading. Accordingly, sediments have had little time to accumulate at or very close to spreading ridges where the oceanic crust is young, but their thickness increases with distance away from the ridges (➢ Figure 12-13).

 PLATE TECTONIC THEORY

Plate tectonic theory is based on a simple model of the Earth. The rigid lithosphere, consisting of both oceanic and continental crust, as well as the underlying upper mantle, consists of numerous variable-sized pieces called **plates** (➢ Figure 12-14). The plates vary in thickness; those composed of upper mantle and continental crust are as much as 250 km thick, whereas those composed of upper mantle and oceanic crust are up to 100 km thick.

The lithosphere overlies the hotter and weaker semiplastic asthenosphere. It is thought that movement resulting from some type of heat transfer system within the asthenosphere causes the overlying plates to move. As plates move over the asthenosphere, they separate, mostly at oceanic ridges, while in other areas such as at oceanic trenches, they collide and are subducted back into the mantle.

Most geologists accept plate tectonic theory, in part because the evidence for it is overwhelming, and also because it is a unifying theory that accounts for a variety of apparently unrelated geologic features and events. Consequently, geologists now view many geologic processes, such as mountain building, seismicity, and volcanism, from the perspective of plate tectonics (see Perspective 12-1). Furthermore, because all of the inner planets have had a similar origin and early history, geologists are interested in determining whether plate tectonics is unique to Earth or if it operates in the same way on other planets (see Perspective 12-2).

 PLATE BOUNDARIES

Plates move relative to one another such that their boundaries can be characterized as *divergent, convergent,* and *transform.* Interaction of plates at their boundaries accounts for most of the Earth's seismic and volcanic activity and, as will be apparent in the next chapter, the origin of mountain systems.

Divergent Boundaries

Divergent plate boundaries or **spreading ridges** occur where plates are separating and new oceanic lithosphere is forming. Divergent boundaries are places where the crust is being extended, thinned, and fractured as magma (derived from the partial melting of the mantle) rises to the surface. The

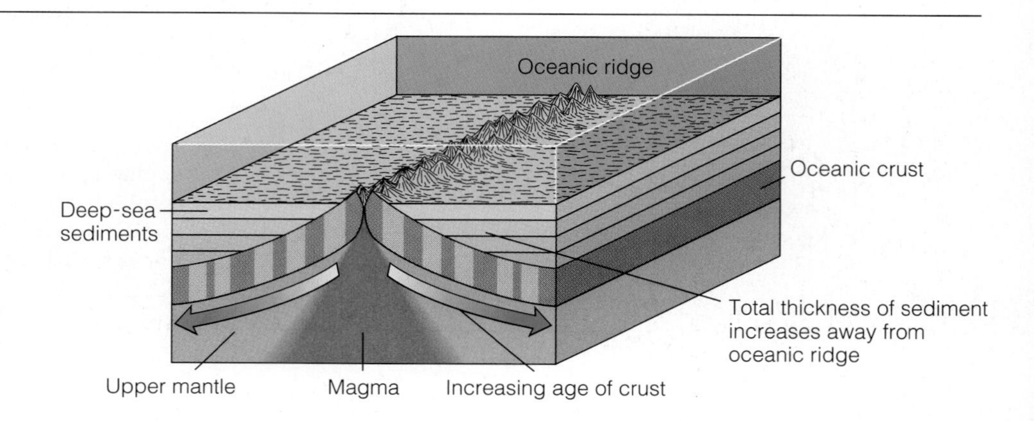

➢ **FIGURE 12-13** The total thickness of deep-sea sediments increases away from oceanic ridges. This is because oceanic crust becomes older away from oceanic ridges, and there has been more time for sediment to accumulate.

Perspective 12-2

Tectonics of the Terrestrial Planets

The four inner or terrestrial planets—Mercury, Venus, Earth, and Mars—all had a similar early history involving accretion, differentiation into a metallic core and silicate mantle and crust, and formation of an early atmosphere by outgassing. Their early history was marked by widespread volcanism and meteorite impacts, both of which helped modify their surfaces. The volcanic and tectonic activity and resultant surface features (other than meteorite craters) of these planets are clearly related to the way they transport heat from their interiors to their surfaces.

The Earth appears to be unique in that its surface is broken up into a series of plates. The creation and destruction of these plates at spreading ridges and subduction zones transfer the majority of the Earth's internally produced heat. In addition, movement of the plates, together with lifeforms, the formation of sedimentary rocks, and water, is responsible for the cycling of carbon dioxide between the atmosphere and lithosphere and thus the maintenance of a habitable climate on Earth (see Perspective 20-2).

Heat is transferred between the interior and surface of both Mercury and Mars mainly by lithospheric conduction. This method is sufficient for these planets because both are significantly smaller than Earth or Venus. Because both Mercury and Mars have a single, globally continuous plate, they have exhibited fewer types of volcanic and tectonic activity than has the Earth. The initial interior warming of Mercury and Mars produced tensional features such as normal faults (see Chapter 13) and widespread volcanism, while their subsequent cooling produced folds and faults resulting from compressional forces, as well as volcanic activity.

Mercury's surface is heavily cratered and shows little in the way of primary volcanic structures. It does, however, have a global system of lobate scarps (see Figure 20-10). These have been interpreted as evidence that Mercury shrank a little soon after its crust hardened, resulting in crustal cracking.

Mars has numerous features that indicate an extensive early period of volcanism. These include Olympus Mons, the solar system's largest volcano (see Figure 20-12), lava flows, and uplifted regions thought to have resulted from mantle convection. In addition to volcanic features, Mars also displays abundant evidence of tensional tectonics, including numerous faults and large fault-produced valley structures. While Mars was tectonically active during the past, there is no evidence that plate tectonics comparable to that on Earth has ever occurred there.

Venus underwent essentially the same early history as the other terrestrial planets, including a period of volcanism, but it is more Earth-like in its tectonics than either Mercury or Mars. Initial radar mapping in 1990 by the *Magellan* spacecraft revealed a surface of extensive lava flows, volcanic domes, folded mountain ranges, and an extensive and intricate network of faults, all of which attest to an internally active planet. (➢ Figure 1).

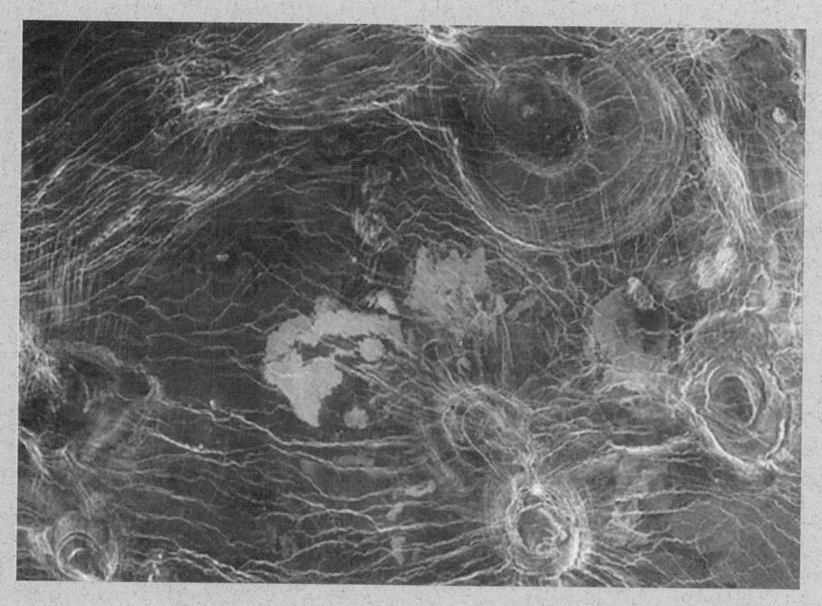

➢ **FIGURE 1** This radar image of Venus made by the *Magellan* spacecraft reveals circular and oval-shaped volcanic features. A complex network of cracks and fractures extends outward from the volcanic features. Geologists think these features were created by blobs of magma rising from the interior of Venus with dikes filling some of the cracks.

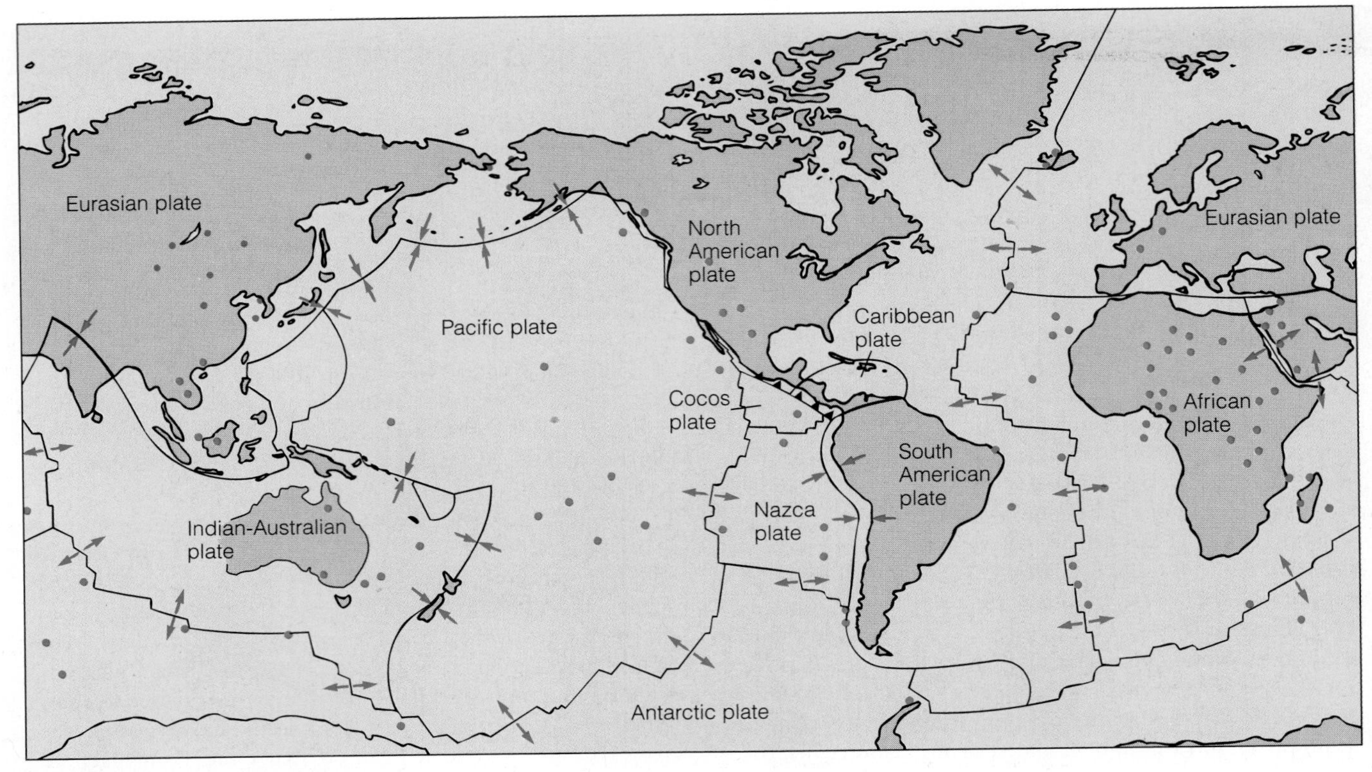

➤ **FIGURE 12-14** A map of the world showing the plates, their boundaries, direction of movement, and hot spots.

magma is almost entirely basaltic and intrudes into vertical fractures to form dikes and lava flows (see Figure 4-10). As successive injections of magma cool and solidify, they form new oceanic crust and record the intensity and orientation of the Earth's magnetic field (Figure 12-11). Divergent boundaries most commonly occur along the crests of oceanic ridges, for example, the Mid-Atlantic Ridge. Oceanic ridges are thus characterized by rugged topography with high relief resulting from displacement of rocks along large fractures, shallow-focus earthquakes, high heat flow, and basaltic flows or pillow lavas.

Divergent boundaries also occur under continents during the early stages of continental breakup (➤ Figure 12-15). When magma wells up beneath a continent, the crust is initially elevated, extended, and thinned (Figure 12-15a), producing fractures and rift valleys. During this stage, magma typically intrudes into the faults and fractures forming sills, dikes, and lava flows; the latter often cover the rift valley floor (Figure 12-15b). The East African rift valleys are an excellent example of this stage of continental breakup (➤ Figure 12-16).

If spreading proceeds, some rift valleys will continue to lengthen and deepen until they form a narrow linear sea separating two continental blocks (Figure 12-15c). The Red Sea separating the Arabian Peninsula from Africa (Figure 12-16) and the Gulf of California, which separates Baja California from mainland Mexico, are good examples of this more advanced stage of rifting.

As a newly created narrow sea continues enlarging, it may eventually become an expansive ocean basin such as the Atlantic, which separates North and South America from Europe and Africa by thousands of kilometers (Figure 12-15d). The Mid-Atlantic Ridge is the boundary between these diverging plates; the American plates are moving westward, and the Eurasian and African plates are moving eastward.

Convergent Boundaries

Because new lithosphere is formed at divergent plate boundaries, older lithosphere must be destroyed and recycled in order for the entire surface area of the Earth to remain constant. Otherwise, we would have an expanding Earth. Such plate destruction occurs at **convergent plate boundaries** where two plates collide.

At a convergent boundary, the leading edge of one plate descends beneath the margin of the other by **subduction.** A dipping plane of earthquake foci, referred to as a Benioff

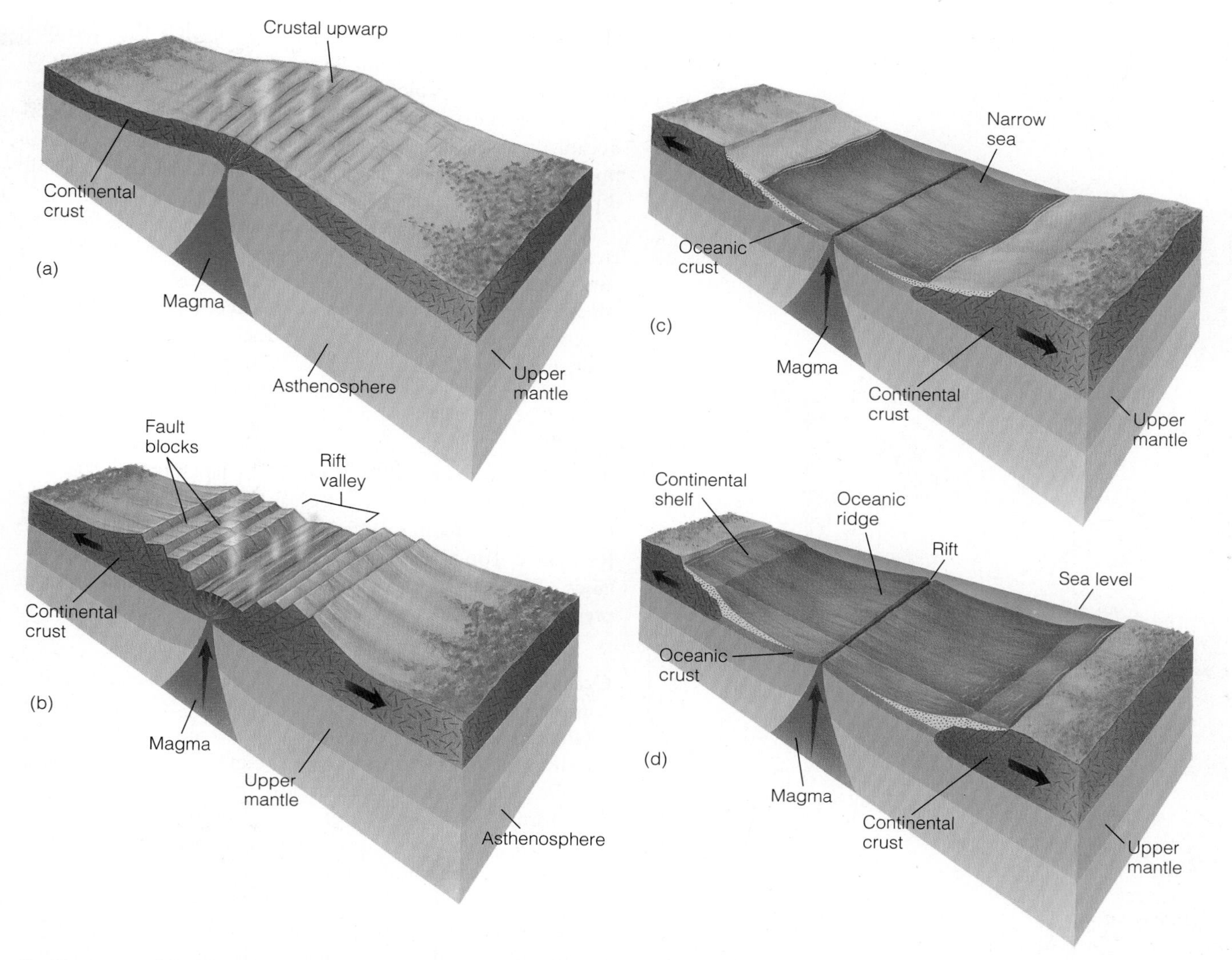

➢ **FIGURE 12-15** History of a divergent plate boundary. (*a*) Rising magma beneath a continent pushes the crust up, producing numerous cracks and fractures. (*b*) As the crust is stretched and thinned, rift valleys develop, and lava flows onto the valley floors. (*c*) Continued spreading further separates the continent until a narrow seaway develops. (*d*) As spreading continues, an oceanic ridge system forms, and an ocean basin develops and grows.

zone, defines subduction zones (see Figure 9-8). Most of these planes dip from oceanic trenches beneath adjacent island arcs or continents, marking the surface of slippage between the converging plates. As the subducting plate moves down into the asthenosphere, it is heated and eventually incorporated into the mantle. When both of the converging plates are continental, subduction does not occur because continental crust is not dense enough to be subducted into the mantle.

Convergent boundaries are characterized by deformation, volcanism, mountain building, metamorphism, seismicity, and important mineral deposits. Three types of convergent plate boundaries are recognized: *oceanic-oceanic, oceanic-continental,* and *continental-continental.*

Oceanic-Oceanic Boundaries. When two oceanic plates converge, one of them is subducted beneath the other along

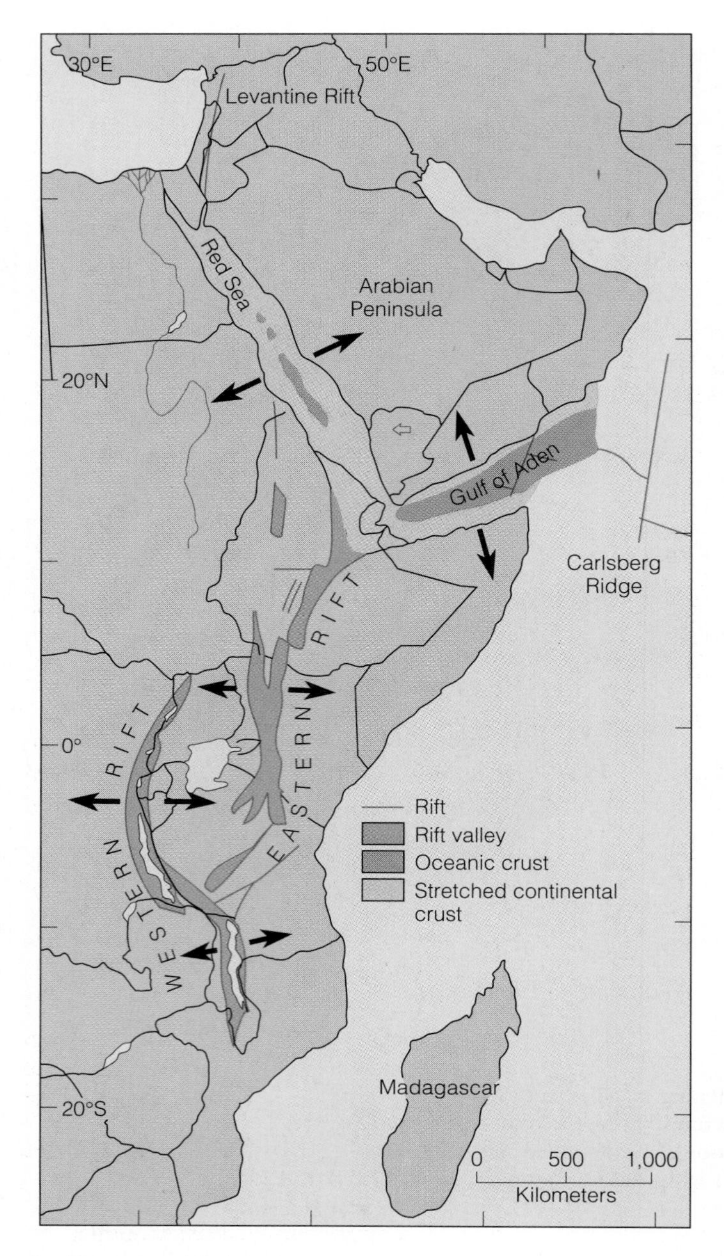

▶ **FIGURE 12-16** The East African rift valley is being formed by the separation of eastern Africa from the rest of the continent along a divergent plate boundary. The Red Sea represents a more advanced stage of rifting, in which two continental blocks are separated by a narrow sea.

an **oceanic-oceanic plate boundary** (▶ Figure 12-17). The subducting plate bends downward to form the outer wall of an oceanic trench. A *subduction complex*, composed of wedge-shaped slices of highly folded and faulted marine sediments and oceanic lithosphere scraped off the descending plate, forms along the inner wall. As the subducting plate descends into the asthenosphere, it is heated and partially melted, generating magma, commonly of andesitic composition. This magma is less dense than the surrounding mantle rocks and rises to the surface through the nonsubducting or overriding plate where it forms a curved chain of volcanoes called a **volcanic island arc** (any plane intersecting a sphere makes an arc). This arc is nearly parallel to the oceanic trench and is separated from it by up to several hundred kilometers—the distance depends on the angle of dip of the subducting plate (Figure 12-17).

In those areas where the rate of subduction is faster than the forward movement of the overriding plate, the lithosphere on the landward side of the volcanic island arc may be subjected to tensional stress and stretched and thinned, resulting in the formation of a *back-arc basin*. This back-arc basin may grow by spreading if magma breaks through the thin crust and forms new oceanic crust (Figure 12-17). A good example of a back-arc basin associated with an oceanic-oceanic plate boundary is the Sea of Japan between the Asian continent and the islands of Japan.

Most present-day active volcanic island arcs are in the Pacific Ocean basin and include the Aleutian Islands, the Kermadec-Tonga arc, and the Japanese and Philippine Islands. The Scotia and Antillean (Caribbean) island arcs are present in the Atlantic Ocean basin.

Oceanic-Continental Boundaries. When an oceanic and a continental plate converge, the denser oceanic plate is subducted under the continental plate along an **oceanic-continental plate boundary** (▶ Figure 12-18). Just as at oceanic-oceanic plate boundaries, the descending oceanic plate forms the outer wall of an oceanic trench.

As the cold, wet, and slightly denser oceanic plate descends into the hot asthenosphere, melting occurs and magma is generated. This magma rises beneath the overriding continental plate and can extrude at the surface, producing a chain of andesitic volcanoes (also called a volcanic arc), or intrude into the continental margin as plutons, especially batholiths. An excellent example of an oceanic-continental plate boundary is the Pacific coast of South America where the oceanic Nazca plate is currently being subducted under South America.

Continental-Continental Boundaries. When two continental plates converge along a **continental-continental plate boundary,** one plate may partially slide under the other, but neither plate will be subducted because of their low and equal densities and great thickness (▶ Figure 12-19). These continents are initially separated from one another by oceanic crust that is being subducted under one of the continents. The edge of that continent will display the characteristics of an oceanic-continental boundary with the development of a deep-sea trench and volcanic arc (Figure 12-18). Eventually, the oceanic crust is totally consumed and the two continents collide. The Himalayas, the world's youngest and highest mountain system, resulted from the collision between India and Asia that began about 40 to 50 million years ago and is still continuing (see Figure 13-31).

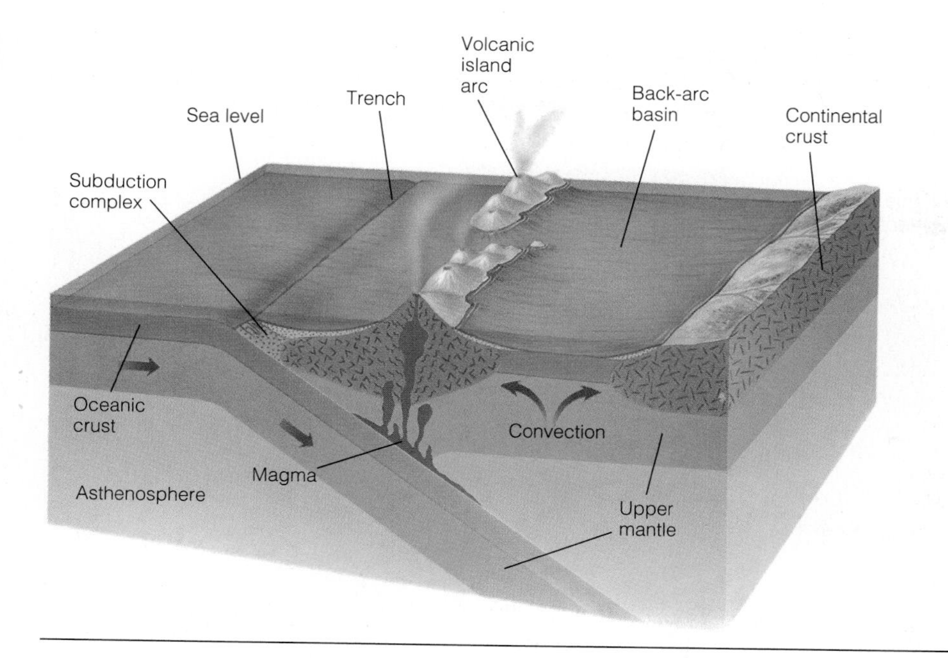

▶ **FIGURE 12-17** Oceanic-oceanic plate boundary. An oceanic trench forms where one oceanic plate is subducted beneath another. On the nonsubducted plate, a volcanic island arc forms from the rising magma generated from the subducting plate.

Transform Boundaries

The third type of plate boundary is a **transform boundary.** These occur along transform faults where plates slide laterally past one another roughly parallel to the direction of plate movement. Although lithosphere is neither created nor destroyed along a transform boundary, the movement between plates results in a zone of intensely shattered rock and numerous shallow-focus earthquakes.

Transform faults are particular types of faults that "transform" or change one type of motion between plates into another type of motion. The majority of transform faults connect two oceanic ridge segments, but they can also connect ridges to trenches and trenches to trenches (▶ Figure 12-20). Though the majority of transform faults occur in oceanic crust and are marked by distinct fracture zones, they may also extend into continents.

One of the best-known transform faults is the San Andreas fault in California. It separates the Pacific plate from the North American plate and connects spreading ridges in the Gulf of California and off the coast of northern California (▶ Figure 12-21). Many of the earthquakes affecting California are the result of movement along this fault.

 PLATE MOVEMENT AND MOTION

How fast and in what direction are the Earth's various plates moving, and do they all move at the same rate? Rates of plate movement can be calculated in several ways. The least

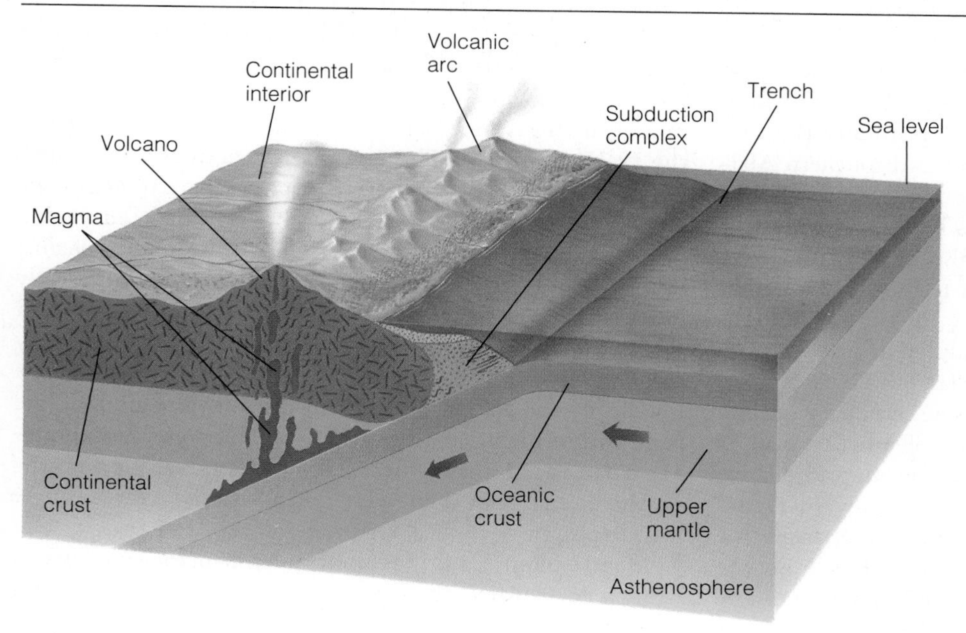

▶ **FIGURE 12-18** Oceanic-continental plate boundary. When an oceanic plate is subducted beneath a continental plate, an andesitic volcanic mountain range is formed on the continental plate as a result of rising magma.

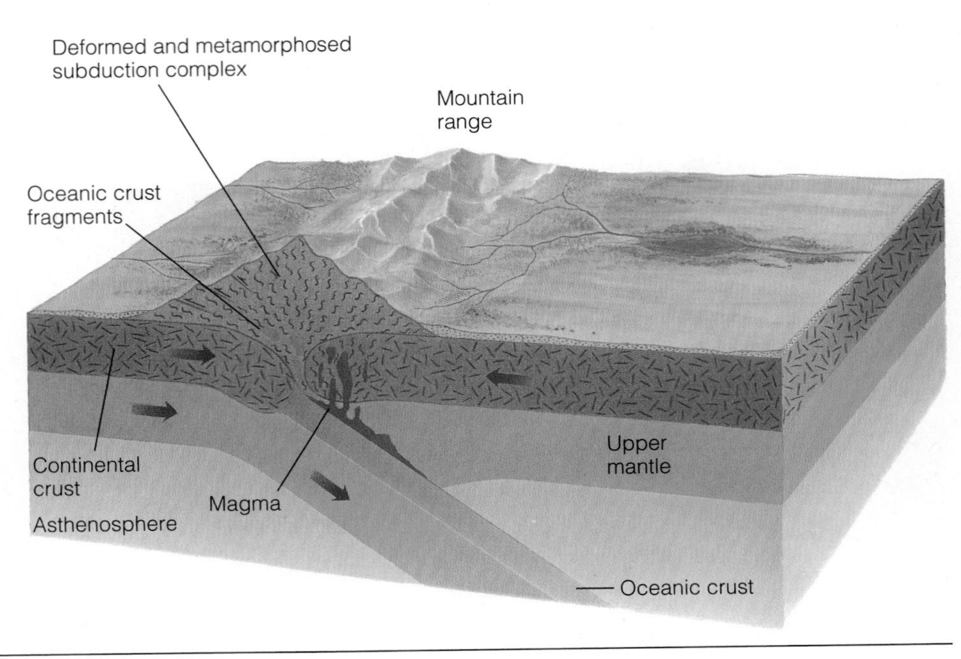

⊳ FIGURE 12-19 Continental-continental plate boundary. When two continental plates converge, neither is subducted because of their great thickness and low and equal densities. As the two continental plates collide, a mountain range is formed in the interior of a new and larger continent.

accurate method is to determine the age of the sediments immediately above any portion of the oceanic crust and divide that age by the distance from the spreading ridge. Such calculations give an average rate of movement.

A more accurate method of determining both the average rate of movement and relative motion is by dating the magnetic reversals in the crust of the sea floor. The distance from an oceanic ridge axis to any magnetic reversal indicates the width of new sea floor that formed during that time interval. Thus, for a given interval of time, the wider the strip of sea floor, the faster the plate has moved. In this way not only can the present average rate of movement and relative motion be determined (⊳ Figure 12-22), but the average rate of movement during the past can also be calculated by dividing the distance between reversals by the amount of time elapsed between reversals.

From the information in Figure 12-22, it is obvious that the rate of movement varies among plates. The southeastern part of the Pacific plate and the Cocos plates are the two fastest moving plates, while the Arabian and southern African plates are the slowest.

The average rate of movement as well as the relative motion between any two plates can also be determined by satellite laser ranging techniques. Laser beams from a station on one plate are bounced off a satellite (in geosynchronous orbit) and returned to a station on a different plate. As the plates move away from each other, the laser beam takes more time to go from the sending station to the stationary satellite and back to the receiving station. This difference in elapsed time is used to calculate the rate of movement and relative motion between plates. In addition, rates of movement and relative motion have also been calculated by measuring the difference between arrival times of radio signals from the same quasar to receiving stations on different plates. The rates of plate movement determined by these two techniques correlate closely with those determined from magnetic reversals.

Hot Spots and Absolute Motion

Plate motions derived from magnetic reversals, satellites, and lasers give only the relative motion of one plate with respect to another. To determine absolute motion, we must have a fixed reference from which the rate and direction of plate movement can be determined. **Hot spots,** which may provide reference points, are locations where stationary columns of magma, originating deep within the mantle (mantle plumes), slowly rise to the Earth's surface and form volcanoes or flood basalts (Figure 12-14).

One of the best examples of hot spot activity is that over which the Emperor Seamount–Hawaiian Island chain formed (⊳ Figure 12-23). Currently, the only active volcanoes in this island chain are on the island of Hawaii and the Loihi Seamount. The rest of the islands and seamounts of the chain are also of volcanic origin and are progressively older west-northwestward along the Hawaiian chain and north-northwestward along the Emperor Seamount chain.

The reason these islands and seamounts are progressively older as one moves toward the north and northwest is that the Pacific plate has moved over an apparently stationary mantle plume. Thus, a line of volcanoes was formed near the middle of the Pacific plate, marking the direction of the plate's movement. In the case of the Emperor Seamount–Hawaiian Island chain, the Pacific plate moved first north-northwesterly and then west-northwesterly over a single mantle plume.

Mantle plumes and hot spots are useful to geologists in helping to explain some of the geologic activity occurring

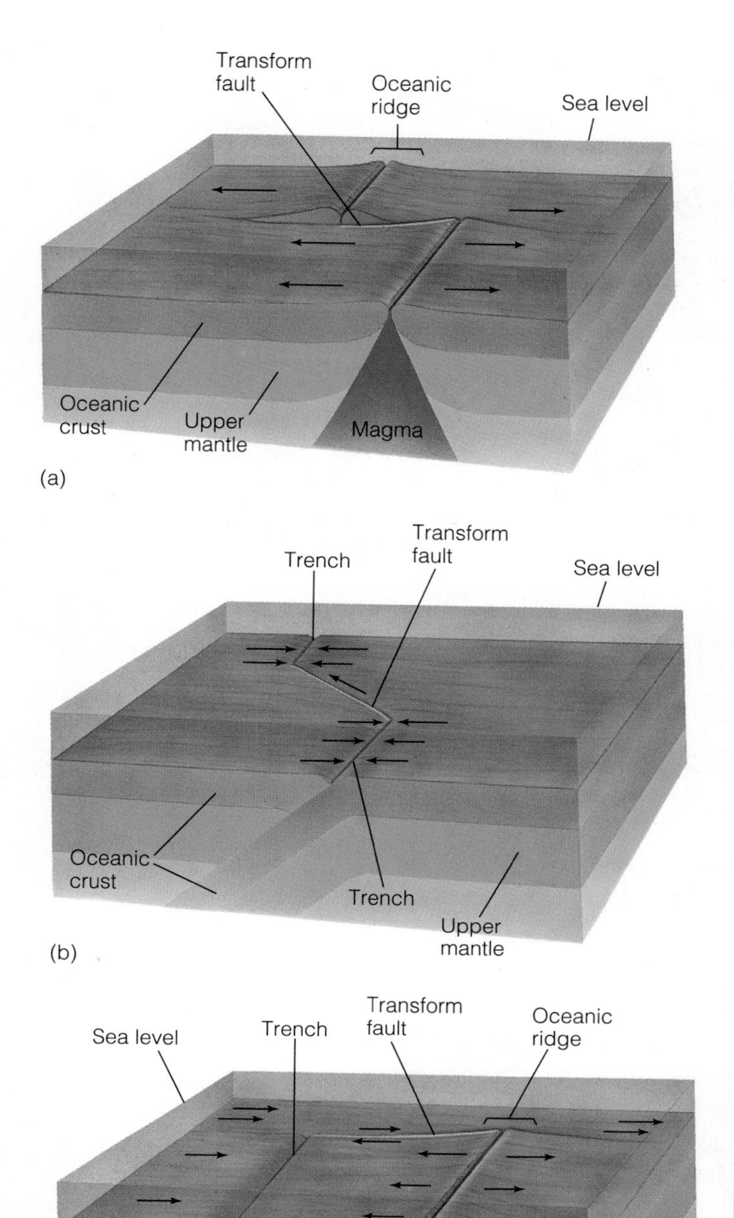

➢ **FIGURE 12-20** Horizontal movement between plates occurs along a transform fault. (*a*) The majority of transform faults connect two oceanic ridge segments. Note that relative motion between the plates only occurs between the two ridges. (*b*) A transform fault connecting two trenches. (*c*) A transform fault connecting a ridge and a trench.

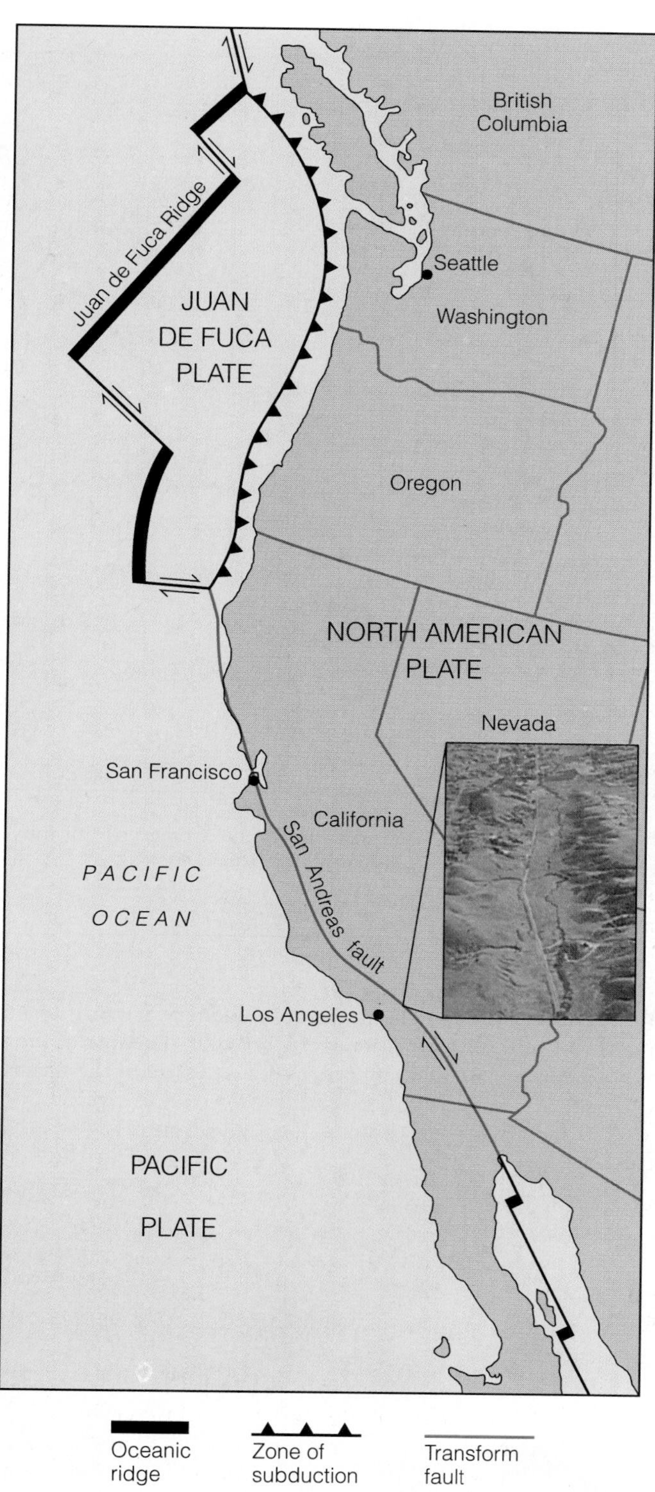

➢ **FIGURE 12-21** Transform plate boundary. The San Andreas fault is a transform fault separating the Pacific plate from the North American plate. Movement along this fault has caused numerous earthquakes. The photograph shows a segment of the San Andreas fault as it cuts through the Leona Valley, California. (Photo courtesy of Eleanora I. Robbins, U.S. Geological Survey.)

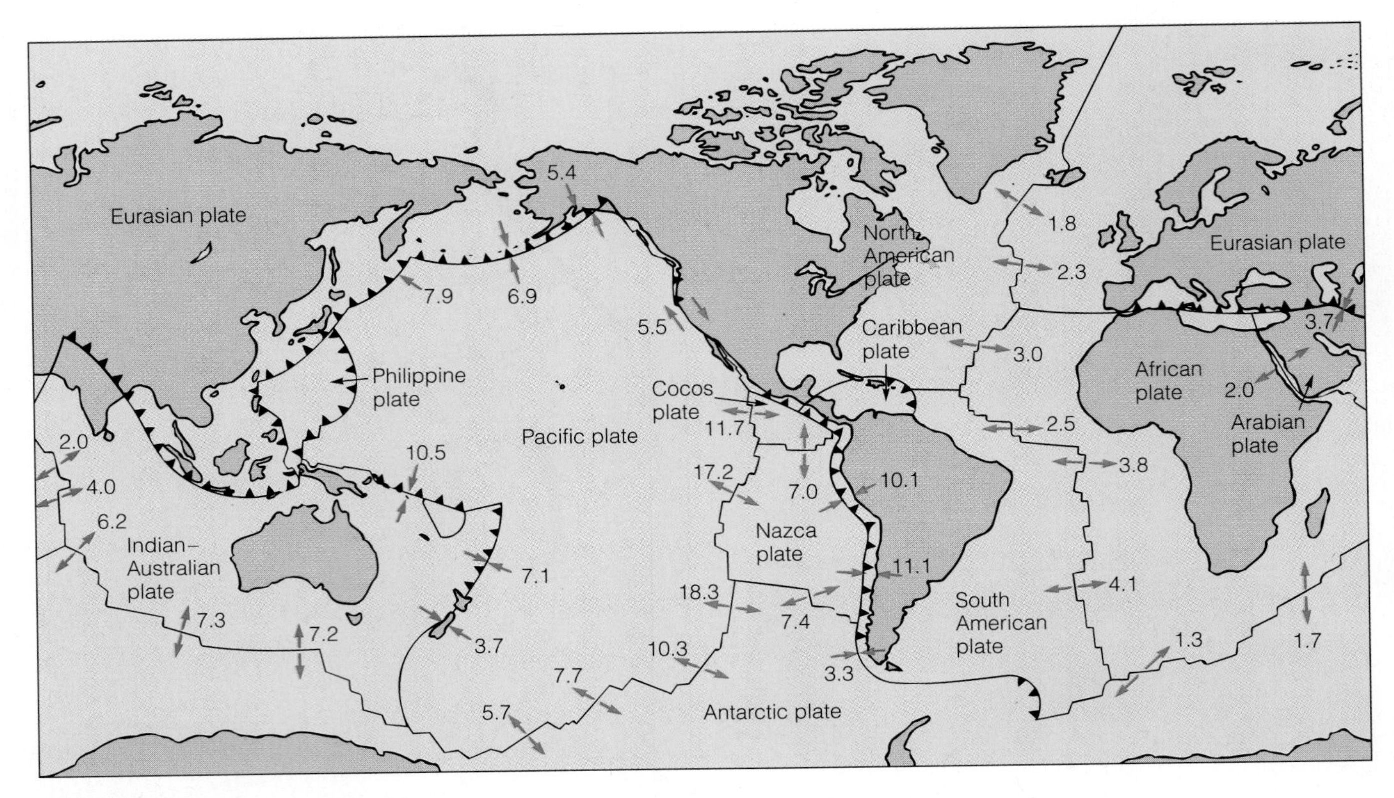

▶ **FIGURE 12-22** This map shows the average rate of movement in centimeters per year and relative motion of the Earth's plates.

▶ **FIGURE 12-23** The Emperor Seamount–Hawaiian Island chain formed as a result of movement of the Pacific plate over a hot spot. The line of the volcanic islands traces the direction of plate movement. The numbers indicate the age of the islands in millions of years.

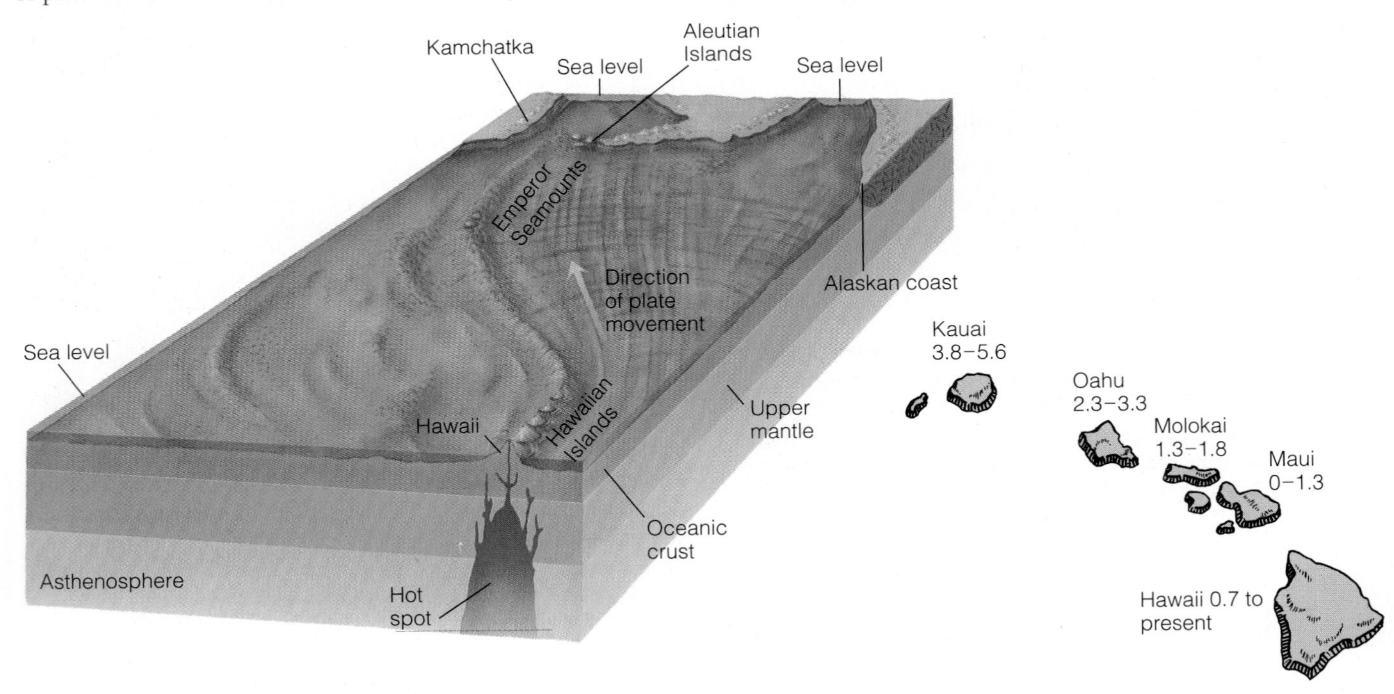

Guest Essay *Daniel Sarewitz*

Geology Meets Public Policy

I had always assumed that I would be a novelist, so naturally I started college intending to major in English. During my freshman year, it occurred to me that I could read and write fiction regardless of my profession, but that I would never understand the origins of mountains and oceans unless I spent some time learning about geology. At some point during my education, I realized that my vague aspirations to write fiction were overshadowed by the fact that I had become—without ever planning to do so—a geologist.

Over a period of 10 years, I did field research on the tectonic evolution of mountain belts in western Idaho, the central Philippines, the Argentine Andes, and Central Asia. As time passed, I began to consider my future options. One possibility was to become a professor, but I doubted that I was suited to academic life. The obvious alternative was to work for an oil or mineral company. This set of options—academia or industry—seemed unacceptably circumscribed, but when I looked for examples of geologists who had gone on to nontraditional careers, I couldn't find any.

As an alternative, I came to Washington, D.C., as part of a fellowship program that places about 25 scientists a year in congressional staff positions. The fellowship gave me entry to a wrold that would otherwise have been inaccessible. Although most fellows return to academia after their year in Washington, some choose to stay on, as I have done.

I began working as a fellow the week before the October 1989 earthquake in Loma Prieta, California, and spent a good part of the next year attempting to transform the publicity generated by the earthquake into a renewed federal commitment to research on earthquake hazard reduction. Much of my work was educational: congressional staff and members of Congress alike needed to understand that bigger, more damaging earthquakes were inevitable in the future; that earthquakes occurred throughout the United States, not just in California; and that federal funding for earthquake research could save lives and money.

Congressional action often comes only on the heels of a crisis. With no major U.S. earthquakes in almost 20 years, funding for the federal earthquake program had declined significantly. In the wake of the Loma Prieta event, however, Congress voted to more than double funding over a period of four years.

This victory was short-lived. One year after the earthquake, the president asked Congress to cut earthquake research funding at the U.S. Geological Survey (USGS) back to pre–Loma Prieta levels. Some said this request was simply part of the attempt to reduce federal spending and balance the budget. Others suggested that political rifts within the Department of Interior, which administers the USGS, led to the requested cuts. But, regardless of why the decision was made, it illustrates that federal science policy is commonly based not on science, but on politics and fiscal concerns.

At times, the quality of congressional debate over scientific issues is astonishingly ill-informed. But even if elected officials were well versed in science, legitimate tradeoffs would still have to be made between political and scientific considerations. For example, a member of Congress from a state whose economy depends on high-sulfur coal production may feel obliged to vote against regulations that prohibit using such coal to generate electricity, even though he or she understands that burning this coal contributes to acid rain. All the same, to make wise decisions, members of Congress must be able to weigh the relative importance of political pressures and scientific data. They cannot do so without the advice of staff who are scientifically literate.

In the coming years, Congress will be increasingly faced with complex decisions that are intimately related to the geosciences. Global warming, energy policy, water supply, nuclear and solid waste disposal, exploitation of the resources of the sea floor and the Antarctic, and federal funding of academic research facilities are a few of the issues that will be on the national agenda. Only two members of the House of Representatives have degrees in science or engineering. Few congressional staff members have scientific backgrounds.

As scientific and technical expertise become increasingly critical to the legislative process, the need for geoscientists on "The Hill" will increase. Thus, the geoscience community should recognize that it can make an important contribution to the formulation of public policy and that careers in public policy represent a legitimate—and growing—area of professional opportunity.

DANIEL SAREWITZ is a science policy analyst for the Committee on Science, Space, and Technology of the U.S. House of Representatives. He earned his Ph.D. in geological sciences from Cornell University in 1985 and served as a Geological Society of America Congressional Science Fellow from September 1989 to August 1990.

within plates as opposed to that occurring at or near plate boundaries. In addition, if mantle plumes are essentially fixed with respect to the Earth's rotational axis—although some evidence suggests they might not be—they may prove useful as reference points for determining paleolatitude.

The Driving Mechanism of Plate Tectonics

A major obstacle to the acceptance of continental drift was the lack of a driving mechanism to explain continental movement. When it was shown that continents and ocean floors moved together and not separately and that new crust formed at spreading ridges by rising magma, most geologists accepted some type of convective heat system as the basic process responsible for plate motion. However, the question of exactly what drives the plates still remains.

Two models involving thermal convection cells have been proposed to explain plate movement (➢ Figure 12-24). In one model, thermal convection cells are restricted to the asthenosphere, whereas in the second model the entire mantle is involved. In both models spreading ridges mark the ascending limbs of adjacent convection cells, while trenches occur where the convection cells descend back into the Earth's interior. The locations of spreading ridges and trenches are therefore determined by the convection cells themselves, and the lithosphere is considered to be the top of the thermal convection cell. Each plate thus corresponds to a single convection cell.

Although most geologists agree that the Earth's internal heat plays an important role in plate movement, problems are inherent in both models. The major problem associated with the first model is the difficulty in explaining the source of heat for the convection cells and why they are restricted to the asthenosphere. In the second model, the source of heat comes from the Earth's outer core, but it is still not known how heat is transferred from the outer core to the mantle. Nor is it clear how convection can involve both the lower mantle and the asthenosphere.

Some geologists think that in addition to thermal convection within the Earth, plate movement also occurs, in part, because of a mechanism involving "slab-pull" or "ridge-push" (➢ Figure 12-25). Both of these mechanisms are gravity driven, but still depend on thermal differences within the Earth. In "slab-pull," the subducting cold slab of lithosphere, being denser than the surrounding warmer asthenosphere, pulls the rest of the plate along with it as it descends into the asthenosphere. As the lithosphere moves downward, there is a corresponding upward flow back into the spreading ridge.

Operating in conjunction with "slab-pull" is the "ridge-push" mechanism. As a result of rising magma, the oceanic ridges are higher than the surrounding oceanic crust. It is thought that gravity pushes the oceanic lithosphere away from the higher spreading ridges and toward the trenches.

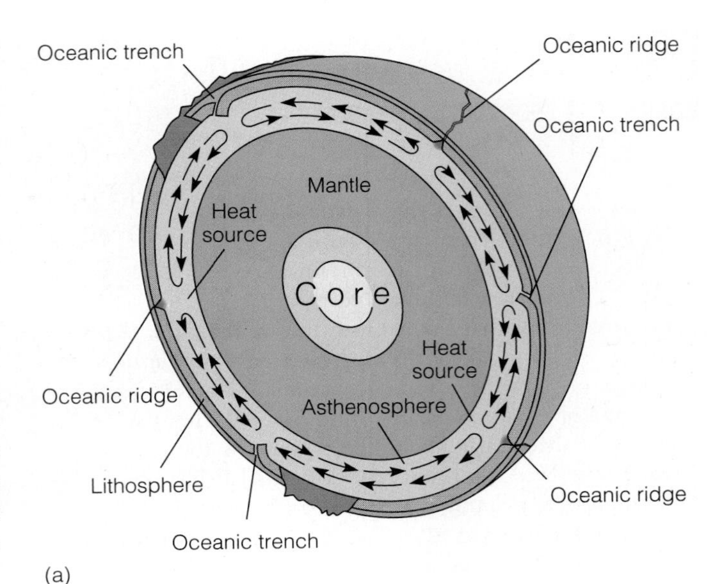

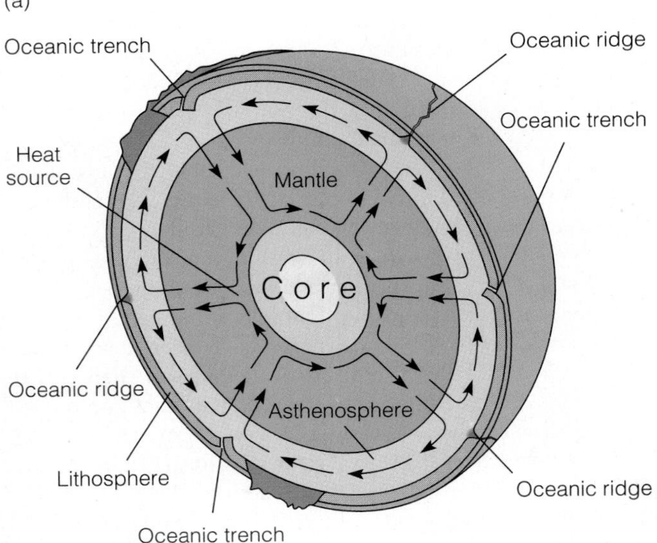

➢ **FIGURE 12-24** Two models involving thermal convection cells have been proposed to explain plate movement. (*a*) In one model, thermal convection cells are restricted to the asthenosphere. (*b*) In the other model, thermal convection cells involve the entire mantle.

Currently, geologists are fairly certain that some type of convective system is involved in plate movement, and the extent to which other mechanisms such as "slab-pull" and "ridge-push" are involved is still unresolved. Consequently, a comprehensive theory of plate movement has not yet been developed, and much still remains to be learned about the Earth's interior.

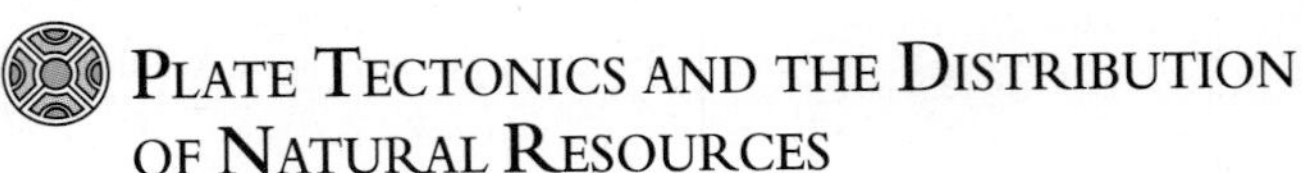

Plate Tectonics and the Distribution of Natural Resources

In addition to being responsible for the major features of the Earth's crust, plate movements also affect the formation and

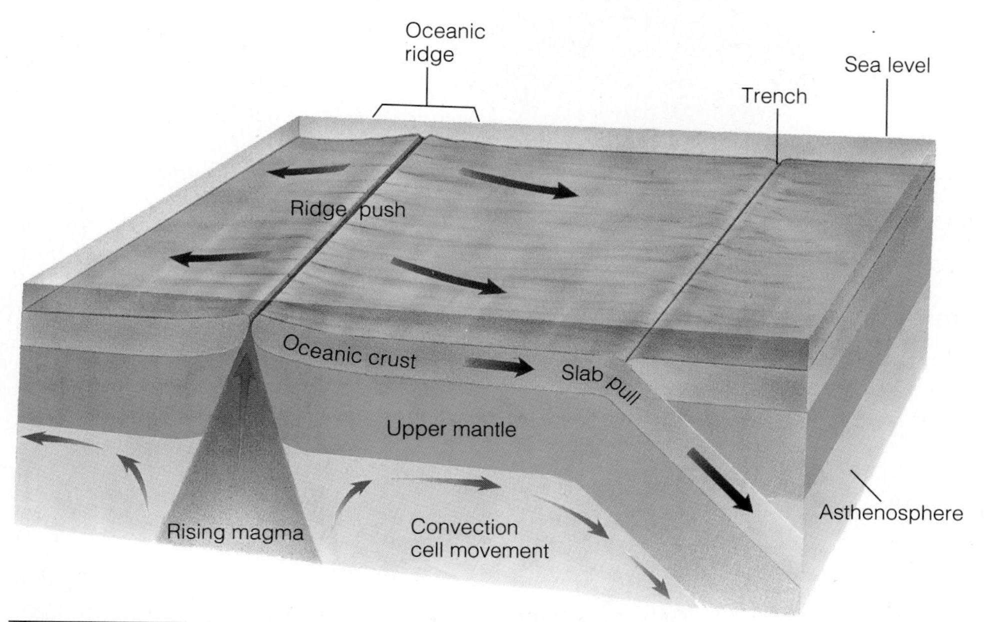

➢ FIGURE 12-25 Plate movement is also thought to occur because of gravity-driven "slab-pull," or "ridge-push" mechanisms. In "slab-pull," the edge of the subducting plate descends into the Earth's interior, and the rest of the plate is pulled downward. In "ridge-push," rising magma pushes the oceanic ridges higher than the rest of the oceanic crust. Gravity thus pushes the oceanic lithosphere away from the ridges and toward the trenches.

distribution of some natural resources. Consequently, geologists are using plate tectonic theory in their search for new mineral deposits and in explaining the occurrence of known deposits.

Many metallic mineral deposits such as copper, gold, lead, silver, tin, and zinc are related to igneous and associated hydrothermal activity, so it is not surprising that a close relationship exists between plate boundaries and the occurrence of these valuable deposits.

The magma generated by partial melting of a subducting plate rises toward the Earth's surface, and as it cools, it precipitates and concentrates various metallic sulfide ores. Many of the world's major metallic ore deposits (such as copper and molybdenum) are associated with convergent plate boundaries including those in the Andes, Rockies, the Coast Ranges of North and South America, Japan, the Philippines, Russia, and a zone extending from the eastern Mediterranean region to Pakistan. In addition, the majority of the world's gold is associated with sulfide deposits located at ancient convergent plate boundaries in such areas as South Africa, Canada, California, Alaska, Venezuela, Brazil, southern India, Russia, and western Australia.

The copper deposits of western North and South America are an excellent example of the relationship between convergent plate boundaries and the distribution, concentration, and exploitation of valuable metallic ores (➢ Figure 12-26). The world's largest copper deposits are found along this belt. The majority of the copper deposits in the Andes and the southwestern United States were formed less than 60 million years ago when oceanic plates were subducted under the North and South American plates. The rising magma and associated hydrothermal fluids carried minute amounts of copper, which was originally widely disseminated but eventually became concentrated in the cracks and fractures of the surrounding andesites. These

➢ FIGURE 12-26 Important copper deposits are located along the west coasts of North and South America.

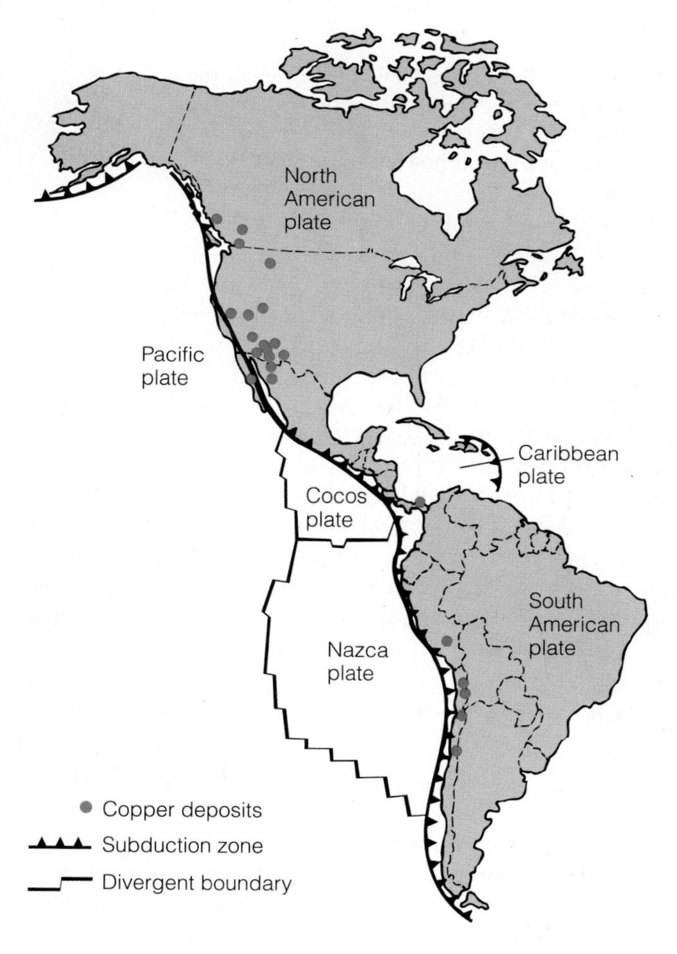

➢ FIGURE 12-27 Bingham Mine in Utah is a huge open-pit copper mine with reserves estimated at 1.7 billion tons. More than 400,000 tons of rock are removed each day. (Photo courtesy of R. V. Dietrich.)

low-grade copper deposits contain from 0.2 to 2% copper and are extracted from large open-pit mines (➢ Figure 12-27).

Divergent plate boundaries also yield valuable resources. Hydrothermal vents are the sites of much metallic mineral precipitation (see the Prologue in Chapter 11). The island of Cyprus in the Mediterranean is rich in copper and has been supplying all or part of the world's needs for the last 3,000 years. The concentration of copper on Cyprus formed as a result of precipitation adjacent to hydrothermal vents. This deposit was brought to the surface when the copper-rich sea floor collided with the European plate, warping the sea floor and forming Cyprus.

Studies indicate that minerals of such metals as copper, gold, iron, lead, silver, and zinc are currently forming as sulfides in the Red Sea. The Red Sea is opening as a result of plate divergence and represents the earliest stage in the growth of an ocean basin.

It is becoming increasingly clear that if we are to keep up with the continuing demands of a global industrialized society, the application of plate tectonic theory to the origin and distribution of mineral resources is essential.

CHAPTER SUMMARY

1. The concept of continental movement is not new. The earliest maps showing the similarity between the east coast of South America and the west coast of Africa provided people with the first evidence that the continents may once have been united and subsequently separated from each other.
2. Alfred Wegener is generally credited with developing the hypothesis of continental drift. He provided abundant geological and paleontological evidence to show that the continents were once united into one supercontinent he named Pangaea. Unfortunately, Wegener could not explain

how the continents moved, and most geologists ignored his ideas.

3. The hypothesis of continental drift was revived during the 1950s when paleomagnetic studies indicated the presence of multiple magnetic north poles instead of just one as there is today. This paradox was resolved by moving the continents into different positions, making the paleomagnetic data consistent with a single magnetic north pole.
4. Magnetic surveys of the oceanic crust reveal magnetic anomalies in the rocks indicating that the Earth's magnetic field has reversed itself in the past. Since the anomalies are parallel and form symmetric belts adjacent to the oceanic ridges, new oceanic crust must have formed as the sea floor was spreading.
5. Sea-floor spreading has been confirmed by dating the sediments overlying the oceanic crust and by radiometric dating of rocks on oceanic islands. Such dating reveals that the oceanic crust becomes older with distance from spreading ridges.
6. Plate tectonic theory became widely accepted by the 1970s because of the overwhelming evidence supporting it and because it provides geologists with a powerful theory for explaining such phenomena as volcanism, seismicity, mountain building, global climatic changes, past and present animal and plant distribution, and the distribution of mineral resources.
7. Three types of plate boundaries are recognized: divergent boundaries, where plates move away from each other; convergent boundaries, where two plates collide; and transform boundaries, where two plates slide past each other.
8. The average rate of movement and relative motion of plates can be calculated in several ways. The results of these different methods all agree and indicate that the plates move at different average velocities.
9. Absolute motion of plates can be determined by the movement of plates over mantle plumes. A mantle plume is an apparently stationary column of magma that rises to the Earth's surface where it becomes a hot spot and forms a volcano.
10. Although a comprehensive theory of plate movement has yet to be developed, geologists think that some type of convective heat system is involved.
11. A close relationship exists between the formation of some mineral deposits and plate boundaries. Furthermore, the formation and distribution of some natural resources are related to plate movements.

IMPORTANT TERMS

continental-continental plate boundary
continental drift
convergent plate boundary
divergent plate boundary
***Glossopteris* flora**
Gondwana
hot spot
Laurasia
oceanic-continental plate boundary
oceanic-oceanic plate boundary
Pangaea
plate
plate tectonic theory
sea-floor spreading
spreading ridge
subduction
thermal convection cell
transform boundary
transform fault
volcanic island arc

REVIEW QUESTIONS

1. The man who is credited with developing the continental drift hypothesis is:
   a. ____ Wilson;
   b. ____ Hess;
   c. ____ Vine;
   d. ____ Wegener;
   e. ____ du Toit.
2. The southern part of Pangaea, consisting of South America, Africa, India, Australia, and Antarctica, is called:
   a. ____ Gondwana;
   b. ____ Laurasia;
   c. ____ Atlantis;
   d. ____ Laurentia;
   e. ____ Pacifica.
3. Which of the following has been used as evidence for continental drift?
   a. ____ continental fit;
   b. ____ fossil plants and animals;
   c. ____ similarity of rock sequences;
   d. ____ paleomagnetism;
   e. ____ all of these.
4. Magnetic surveys of the ocean basins indicate that:
   a. ____ the oceanic crust is oldest adjacent to spreading ridges;
   b. ____ the oceanic crust is youngest adjacent to the continents;
   c. ____ the oceanic crust is youngest adjacent to spreading ridges;
   d. ____ the oceanic crust is the same age in all ocean basins;
   e. ____ answers (a) and (b).
5. Plates:
   a. ____ are the same thickness everywhere;
   b. ____ vary in thickness;
   c. ____ include the crust and upper mantle;

d. ____ answers (a) and (c);
e. ____ answers (b) and (c).

6. Divergent boundaries are the areas where:
   a. ____ new continental lithosphere is forming;
   b. ____ new oceanic lithosphere is forming;
   c. ____ two plates come together;
   d. ____ two plates slide past each other;
   e. ____ answers (b) and (d).
7. Along what type of plate boundary does subduction occur?
   a. ____ divergent;
   b. ____ transform;
   c. ____ convergent;
   d. ____ answers (a) and (b);
   e. ____ answers (b) and (c).
8. The west coast of South America is an example of a(n) ________ plate boundary.
   a. ____ divergent;
   b. ____ continental-continental;
   c. ____ oceanic-oceanic;
   d. ____ oceanic-continental;
   e. ____ transform.
9. Back-arc basins are associated with ________ plate boundaries.
   a. ____ divergent;
   b. ____ convergent;
   c. ____ transform;
   d. ____ answers (a) and (b);
   e. ____ answers (b) and (c).
10. The San Andreas fault is an example of a(n) ________ boundary.
   a. ____ divergent;
   b. ____ convergent;
   c. ____ transform;
   d. ____ oceanic-continental;
   e. ____ continental-continental.
11. Which of the following will allow you to determine the absolute motion of plates?
   a. ____ hot spots;
   b. ____ the age of the sediment directly above any portion of the ocean crust;
   c. ____ magnetic reversals in the sea-floor crust;
   d. ____ satellite laser ranging techniques;
   e. ____ all of these.
12. The island of Hawaii and the Loihi Seamount formed as the result of:
   a. ____ oceanic-oceanic plate boundaries;
   b. ____ hot spots;
   c. ____ divergent plate boundaries;
   d. ____ transform boundaries;
   e. ____ oceanic-continental plate boundaries.
13. The driving mechanism of plate movement is thought to be:
   a. ____ isostasy;
   b. ____ magnetism;
   c. ____ thermal convection cells;
   d. ____ rotation of the Earth;
   e. ____ none of these.
14. The formation and distribution of copper deposits are associated with ________ boundaries.
   a. ____ divergent;
   b. ____ convergent;
   c. ____ transform;
   d. ____ answers (a) and (b);
   e. ____ answers (b) and (c).

Name the type of plate boundary indicated in the illustration below.

15. ________________.
16. ________________.
17. ________________.
18. ________________.
19. What evidence convinced Wegener that the continents were once joined together and subsequently broke apart?
20. Why can't the similarity between the coastlines of continents alone be used to prove they were once joined together?
21. What is the significance of polar wandering in relation to continental drift?
22. How can magnetic anomalies be used to show that the sea floor has been spreading?
23. What evidence besides magnetic anomalies convinced geologists of sea-floor spreading?
24. Why is plate tectonics such a powerful unifying theory?
25. Summarize the geologic features characterizing the three different types of plate boundaries.
26. What are mantle plumes and hot spots? How can they be used to determine the direction and rate of movement of plates?

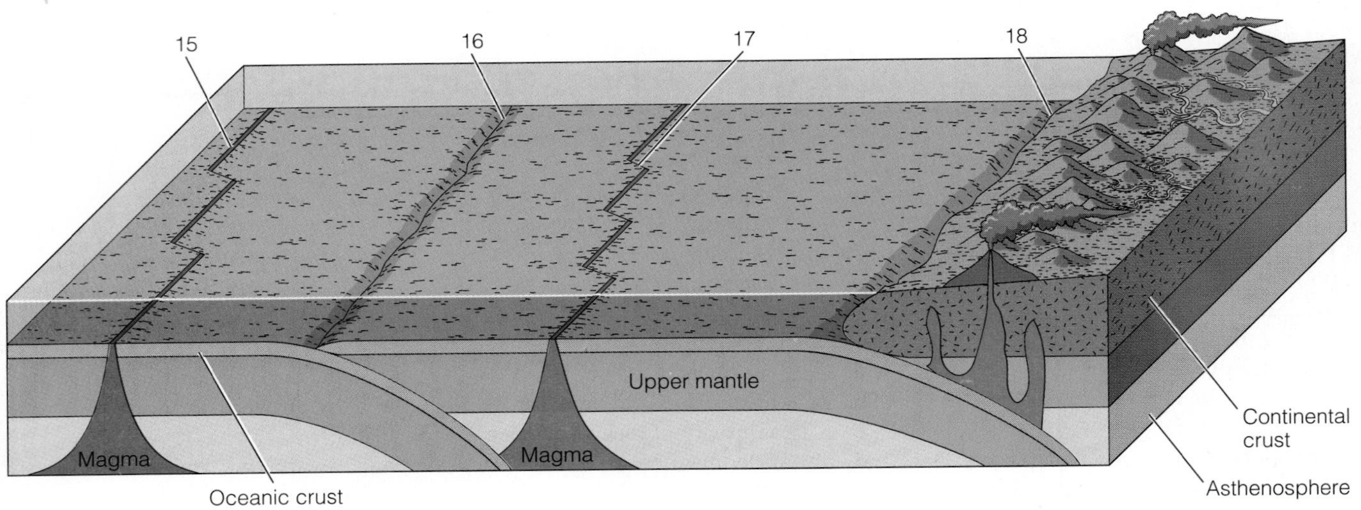

Points to Ponder

1. What features would an astronaut look for on the Moon or another planet to find out if plate tectonics is currently active or if it was active during the past?
2. If movement along the San Andreas fault, which separates the Pacific plate from the North American plate, averages 5.5 cm per year, how long will it take before Los Angeles is opposite San Francisco?
3. Using the age for each of the Hawaiian Islands in Figure 12-23, calculate the average rate of movement per year for the Pacific plate since each island formed. Is the average rate of movement the same for each island? Would you expect it to be? Explain why it may not be.
4. The average rate of movement away from the Mid-Atlantic Ridge between North America and Africa is 3.0 cm per year (Figure 12-22), and the average sedimentation rate in the open ocean is 0.275 cm per 1,000 years. Calculate the age of the oceanic crust adjacent to Norfolk, Virginia, and the thickness of the sediment overlying the oceanic crust at that location. Does your calculation agree with the age of the oceanic crust at that location as shown in Figure 12-12?

Additional Readings

Allégre, C. 1988. *The behavior of the Earth.* Cambridge, Mass.: Harvard University Press.

Bonatti, E. 1987. The rifting of continents. *Scientific American* 256, no. 3: 96–103.

Brimhall, G. 1991. The genesis of ores. *Scientific American* 264, no. 5: 84–91.

Condie, K. 1989. *Plate tectonics and crustal evolution.* 3d ed. New York: Pergamon Press.

Cox, A., and R. B. Hart. 1986. *Plate tectonics: How it works.* Palo Alto, Calif.: Blackwell Scientific Publishers.

Cromie, W. J. 1989. The roots of midplate volcanism. *Mosaic* 20, no. 4: 19–25.

Jordan, T. H., and J. B. Minster. 1988. Measuring crustal deformation in the American west. *Scientific American* 259, no. 2: 48–59.

Kearey, P., and F. J. Vine. 1990. *Global tectonics.* Palo Alto, Calif.: Blackwell Scientific Publishers.

Klein, G. D. ed. 1994. Pangea: Paleoclimate, tectonics, and sedimentation during accretion, zenith, and breakup of a supercontinent. *Geological Society of America Special Paper 288.* Boulder, Colo.: Geological Society of America.

Murphy, J. B., and R. D. Nance. 1992. Mountain belts and the supercontinent cycle. *Scientific American* 266, no. 4: 84–91.

Nance, R. D., T. R. Worsley, and J. B. Moody. 1988. The supercontinent cycle. *Scientific American* 259, no. 1: 72–79.

Parks, N. 1994. Exploring Loihi: The next Hawaiian Island. *Earth* 3, no. 5: 56–63.

Vink, G. E., W. J. Morgan, and P. R. Vogt. 1985. The Earth's hot spots. *Scientific American* 252, no. 4: 50–57.

Chapter 13

DEFORMATION, MOUNTAIN BUILDING, AND THE EVOLUTION OF CONTINENTS

OUTLINE

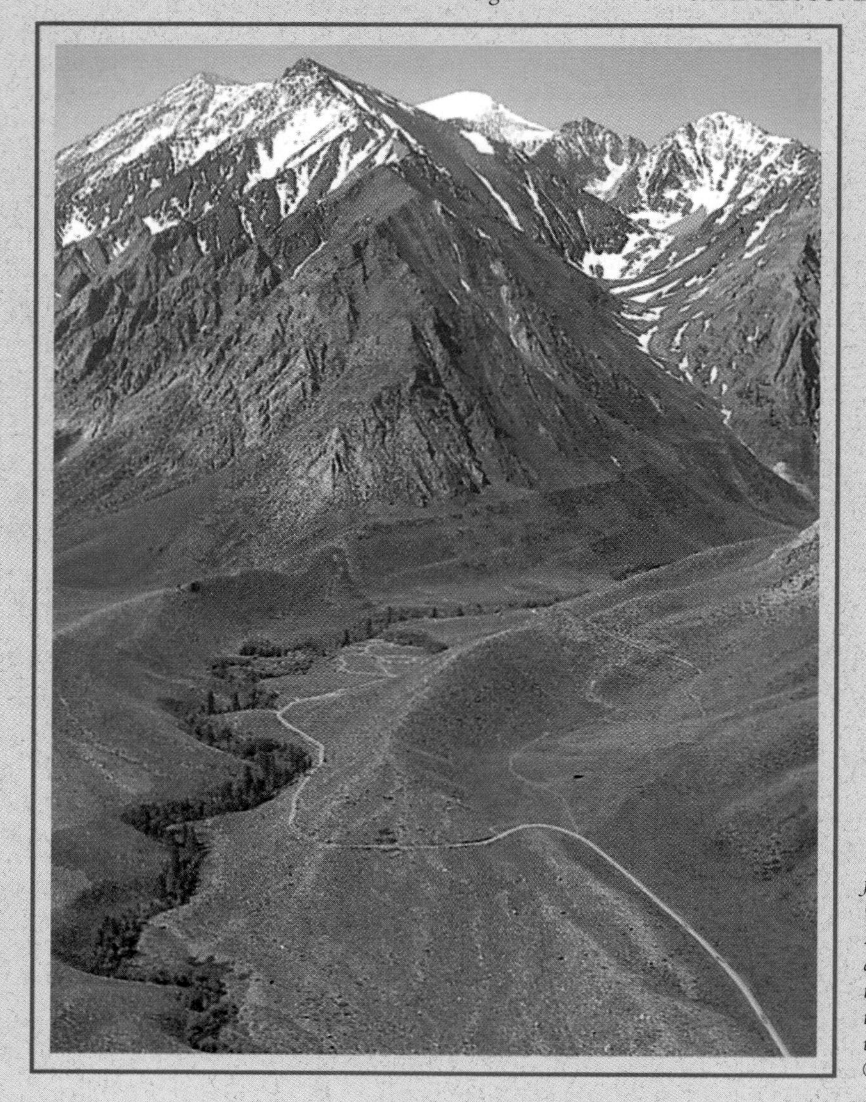

Owens Valley and the east face of the Sierra Nevada, California. The Sierra Nevada has been uplifted along a huge fracture so that it now stands more than 3,000 m above the valleys to the east. (Photo © Peter Kresan.)

PROLOGUE

North America has many scenic mountain ranges, but few are as grand as the Teton Range of northwestern Wyoming (▶ Figure 13-1). The Native Americans of the region called these mountains Teewinot, meaning many pinnacles, which is an appropriate name because the Teton Range consists of numerous jagged peaks. The loftiest of these, the Grand Teton, rises 4,190 m above sea level (Figure 13-1). Higher and larger mountain ranges exist in North America, but none rise so abruptly as the Tetons, which ascend nearly vertically more than 2,100 m above the floor of Jackson Hole, the valley to their east. In 1929, when Grand Teton National Park was established, it was largely limited to the peaks within the Teton Range, but in 1950 it was enlarged to include Jackson Hole and now covers 1,260 km^2.

Mountains began forming in the Teton region about 90 million years ago, but these early mountains were quite different from the present ones. They formed when the Earth's crust was contorted and folded, and they were oriented northwest-southeast. The present-day, north-south–trending Teton Range began forming only about 10 million years ago when a large part of the crust was uplifted along a large fracture called the Teton fault (▶ Figure 13-2).

The Teton Range is made up of a variety of rocks, but most of those exposed in the range are Precambrian-aged metamorphic and plutonic rocks that formed at great depth. These rocks were overlain by a thick sequence of younger sedimentary rocks that are now present only on the west flank of the range (Figure 13-2). Movement on the Teton fault resulted in uplift of the Teton block relative to Jackson Hole, the block to the east; the total displacement on this fault is about 6,100 m. As the Teton block was uplifted, the overlying sedimentary rocks were eroded, exposing the underlying metamorphic and plutonic rocks. The fault is along the east side of the Teton block, so as uplift proceeded, the block was tilted ever more steeply toward the west (Figure 13-2). Displacement of recent sedimentary deposits along the east flank of the Teton Range shows that uplift is continuing today.

Deformation of the Earth's crust was responsible for the origin of the Teton Range, but its spectacular, rugged scenery developed rather recently and in response to a completely different process—glaciation. Currently, the

▶ **FIGURE 13-1** View of the Teton Range in Wyoming. The highest peak visible is the Grand Teton.

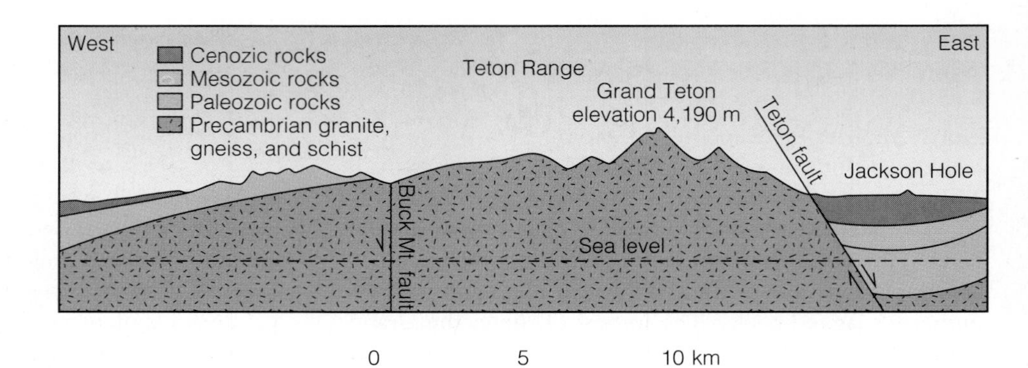

▶ **FIGURE 13-2** A cross section of the Teton Range, Wyoming.

range supports about a dozen small glaciers, but periodically during the last 200,000 years it had many more and much larger glaciers. Glaciers are particularly effective agents of erosion; they deeply scoured the valleys and intricately sculpted the uplifted Teton block, producing excellent examples of glacial landforms. The Grand Teton, which is a glacial feature known as a horn peak, is one of the most prominent of these (see Chapter 17).

 INTRODUCTION

Deformed rocks are a manifestation of the dynamic nature of the Earth. Many ancient rocks are fractured or highly contorted, clearly indicating that forces within the Earth caused deformation during the past. Such deformation is not restricted to the past, however; seismic activity and continuing deformation at convergent, divergent, and transform plate boundaries indicate that deforming forces remain active.

Mountains can form in a variety of ways, some of which involve little or no deformation, but in most mountains the rocks have been complexly deformed by compressive forces at convergent plate boundaries. The Alps of Europe, the Appalachians of North America, the Himalayas of Asia, and many others owe their existence, and in some cases continuing evolution, to plate convergence. In short, deformation and mountain building are closely related phenomena.

A large part of this chapter is devoted to a review of the various types of geologic structures, their descriptive terminology, and the forces responsible for them. The study of deformed rocks has several applications. For instance, geologic structures, such as folds and fractures resulting from deformation, provide a record of the kinds and intensities of forces that operated during the past. By interpreting these structures, geologists can make inferences about Earth history that enable us to satisfy our curiosity about the past, or search more efficiently for various natural resources. Understanding the nature of geologic structures helps geologists find and recover resources such as petroleum and natural gas (see Figure 6-31). Local geologic structures must also be considered when selecting sites for dams, large bridges, and nuclear power plants, especially if the sites are in areas of active deformation.

 DEFORMATION

Fractured and contorted rocks such as those in ▶ Figure 13-3 are said to be **deformed;** that is, their original shape or volume or both have been altered by **stress,** which is the result of force applied to a given area of rock. If the intensity of the stress is greater than the internal strength of the rock, it will undergo **strain,** which is deformation caused by stress.

Three types of stress are recognized: compression, tension, and shear. **Compression** results when rocks are squeezed or compressed by external forces directed toward one another. Rock layers subjected to compression are

▶ **FIGURE 13-3** Rock layers deformed by folding and fracturing.

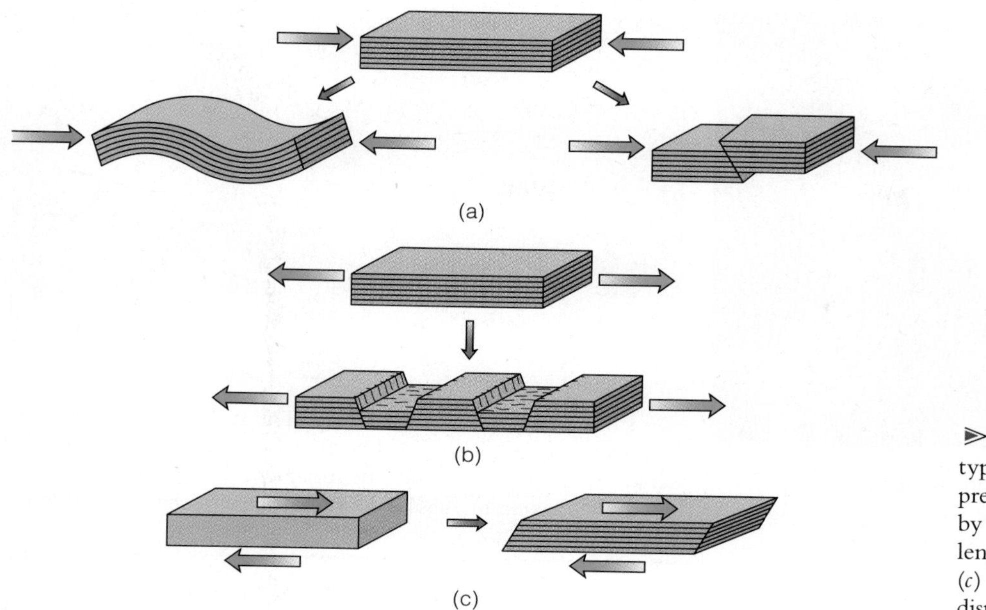

▶ **FIGURE 13-4** Stress and possible types of resulting deformation. (*a*) Compression causes shortening of rock layers by folding or faulting. (*b*) Tension lengthens rock layers and causes faulting. (*c*) Shear stress causes deformation by displacement along closely spaced planes.

commonly shortened in the direction of stress by folding or faulting (▶ Figure 13-4a). **Tension** results from forces acting in opposite directions along the same line (Figure 13-4b). Such stress tends to lengthen rocks or pull them apart. In **shear stress,** forces act parallel to one another but in opposite directions, resulting in deformation by displacement of adjacent layers along closely spaced planes (Figure 13-4c).

Strain is characterized as **elastic** if a deformed object returns to its original shape when the stresses are relaxed. Squeezing a tennis ball causes strain, but once the stress is released, the tennis ball returns to its original shape. Most rocks are not very elastic, but the Earth's crust responds by elastic strain when it is loaded by glacial ice and is depressed into the mantle. Recall from Chapter 10 that the crust has responded elastically by isostatic rebound in large areas of Scandinavia and Canada following the last Ice Age (see Figure 10-22).

When stress is applied to rocks, they respond first by elastic strain, but when strained beyond their elastic limit, they cannot recover their original shape. Such rocks deform by **plastic strain** as when they are folded, or they behave as brittle solids and are **fractured** (▶ Figure 13-5).

Many rocks show the effects of plastic deformation that must have occurred deep within the Earth's crust where the temperature and pressure are high. Recall from Chapters 7 and 10 that the behavior of rock materials under these conditions is very different from their behavior near the surface. At or near the surface, they behave as brittle solids, whereas under conditions of high temperature and high pressure, they more commonly deform plastically rather than fracture. The foci of most earthquakes are at depths of less than 30 km, indicating that deformation by fracturing becomes increasingly difficult with depth, and no fracturing is known to occur at depths greater than 700 km.

The type of strain that occurs depends on the kind of stress applied, the amount of pressure, the temperature, the rock type, and the length of time the rock is subjected to the stress. A small stress applied over a long period of time, as on the slab shown in ▶ Figure 13-6, will cause plastic defor-

▶ **FIGURE 13-5** Strain or deformation in response to stress. Rocks initially respond to stress by elastic deformation and return to their original shape when the stress is released. If the elastic limit is exceeded as in curve A, rocks deform plastically, which is permanent deformation. The amount of plastic deformation rocks exhibit before fracturing depends on their ductility: if they are ductile, they show considerable plastic deformation (curve A), but if they are brittle, they show little or no plastic deformation before failing by fracture (curve B).

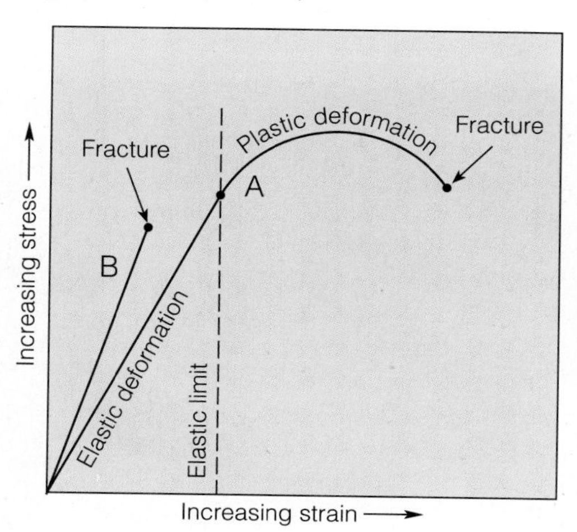

⋗ **FIGURE 13-6** This marble slab in the Rock Creek Cemetery, Washington, D.C., bent under its own weight in about 80 years.

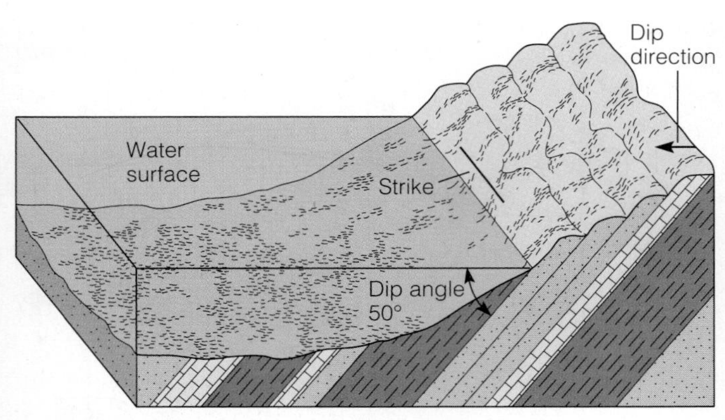

⋗ **FIGURE 13-8** Strike and dip. The strike is formed by the intersection of a horizontal plane (the water surface) with the surface of an inclined plane (the surface of the rock layer). The dip is the maximum angular deviation of the inclined plane from horizontal.

mation. By contrast, a large stress applied rapidly to the same object, as when it is struck by a hammer, will probably result in fracture. Rock type is important because not all rocks respond to stress in the same way. Rocks are considered to be either *ductile* or *brittle* depending on the amount of plastic strain they exhibit. Brittle rocks show little or no plastic strain before fracture, whereas ductile rocks exhibit a great deal (Figure 13-5).

Strike and Dip

According to the principle of original horizontality, when sediments are deposited, they accumulate in nearly horizontal layers (⋗ Figure 13-7a). Thus, sedimentary rock layers that are steeply inclined must have been tilted following deposition and lithification (Figure 13-7b). Some igneous rocks, especially ash falls and many lava flows, also form nearly horizontal layers. To describe the orientation of deformed rock layers, geologists use the concept of *strike and dip*.

Strike is the direction of a line formed by the intersection of a horizontal plane with an inclined plane, such as a rock layer. In ⋗ Figure 13-8, the surface of any of the tilted rock layers constitutes an inclined plane, and the intersection of a horizontal plane with any of these inclined planes forms a line, the direction of which is the strike. The strike line's

⋗ **FIGURE 13-7** (*a*) These rock layers at Dead Horse State Park, Utah, have been uplifted, but they are still horizontal as when they were deposited and lithified. (*b*) These sedimentary rocks in Utah are inclined from horizontal, so we can infer that they were tilted after deposition and lithification. (Photo courtesy of David J. Matty.)

(a)

(b)

orientation is determined by using a compass to measure its angle with respect to north. **Dip** is a measure of the maximum angular deviation of an inclined plane from horizontal, so it must be measured perpendicular to the strike direction (Figure 13-8).

Geologic maps indicate strike and dip by using a long line oriented in the strike direction and a short line perpendicular to the strike line and pointing in the dip direction (➤ Figure 13-9a). The number adjacent to the strike and dip symbol indicates the dip angle.

Folds

If you place your hands on a tablecloth and move them toward one another, the tablecloth is deformed by compression into a series of up- and down-arched folds. Similarly, rock layers within the Earth's crust commonly respond to compression by folding. Unlike the tablecloth, though, folding in rock layers is permanent; that is, the rocks have been strained plastically so once folded, they stay folded. Most folding probably occurs deep within the crust where rocks are ductile.

Monoclines, Anticlines, and Synclines. A **monocline** is a simple bend or flexure in otherwise horizontal or uniformly dipping rock layers (Figure 13-9a). The large monocline in Figure 13-9b formed when the Bighorn Mountains of Wyoming were uplifted along a large fault. This fault did not penetrate to the surface, so as uplift occurred, the near-surface layers of rock were bent so that they appear to be draped over the margin of the uplifted block.

An **anticline** is an up-arched fold, while a **syncline** is a down-arched fold (➤ Figure 13-10). Both anticlines and synclines have an *axial plane* that divides them into halves; each half is a *limb* (➤ Figure 13-11). Because folds most commonly occur as a series of anticlines alternating with synclines, a limb is generally shared by an anticline and an adjacent syncline.

It is important to remember that anticlines and synclines are defined by the orientation of rock layers and not by the configuration of the Earth's surface. Thus, folds may or may not correspond to mountains and valleys (➤ Figure 13-12) and may, in fact, underlie areas where the Earth's surface is rather flat. Indeed, folds are commonly exposed to view in areas that have been deeply eroded. But even where folds have been eroded, anticlines and synclines can easily be distinguished from each other by strike and dip and by the relative ages of the folded rocks. As ➤ Figure 13-13 shows, in an eroded anticline, each limb dips outward or away from the center of the fold, where the oldest rocks are. In eroded

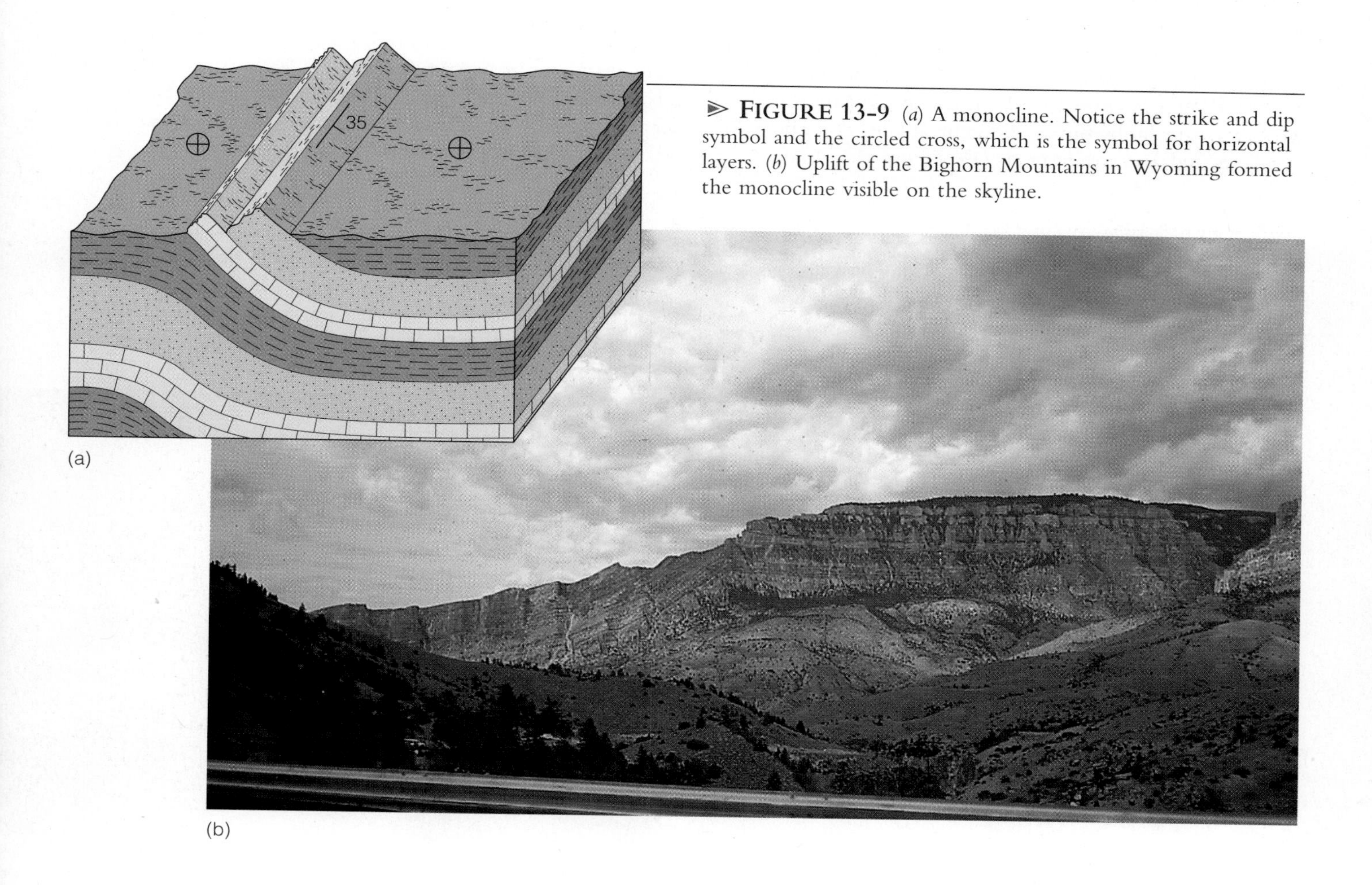

➤ **FIGURE 13-9** (*a*) A monocline. Notice the strike and dip symbol and the circled cross, which is the symbol for horizontal layers. (*b*) Uplift of the Bighorn Mountains in Wyoming formed the monocline visible on the skyline.

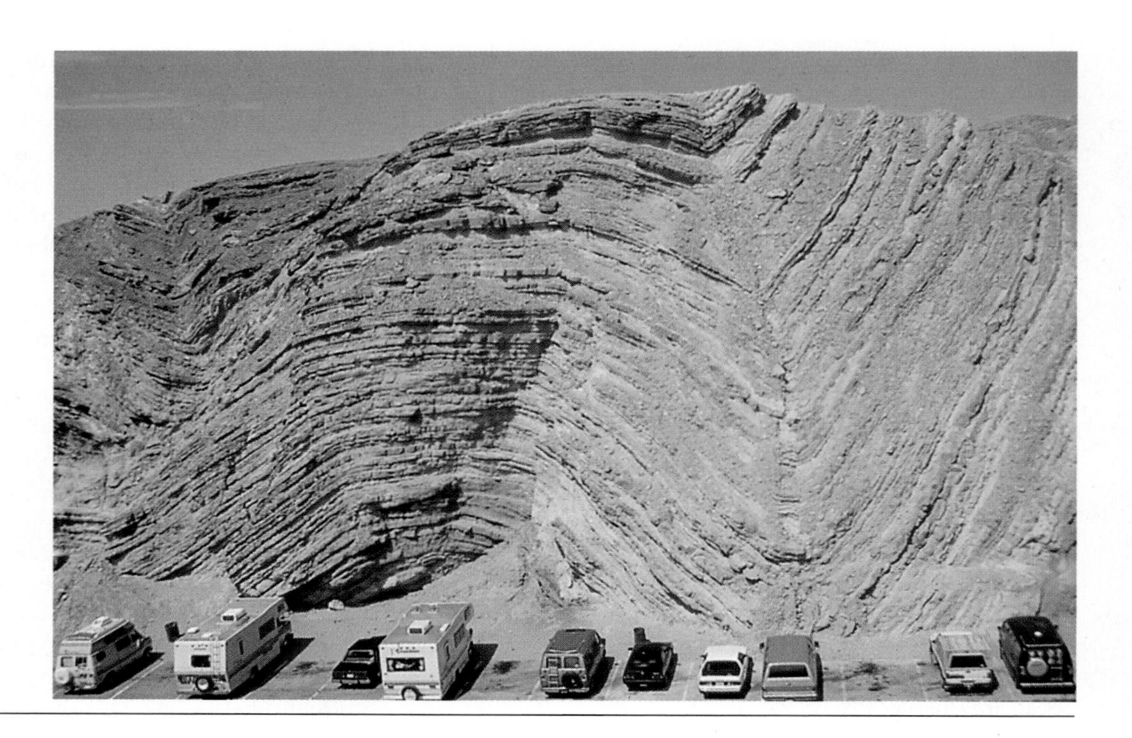

➢ **FIGURE 13-10** Anticline and syncline in the Calico Mountains of southeastern California.

synclines, on the other hand, each limb dips inward toward the fold's center, and the youngest rocks coincide with the center of the fold.

Thus far, we have described *symmetrical,* or upright, folds in which the axial plane is vertical, and each fold limb dips at the same angle (Figure 13-11). If the axial plane is inclined, however, the limbs dip at different angles, and the fold is characterized as *asymmetrical* (➢ Figure 13-14a). In an *overturned fold,* both limbs dip in the same direction. In other words, one fold limb has been rotated more than 90° from its original position so that it is now upside down (Figure 13-14b). Folds in which the axial plane is horizontal are referred to as *recumbent* (Figure 13-14c). Overturned and recumbent folds are particularly common in mountain ranges that formed by compression at convergent plate boundaries (discussed later in this chapter).

➢ **FIGURE 13-11** Syncline and anticline showing the axial plane, axis, and fold limbs.

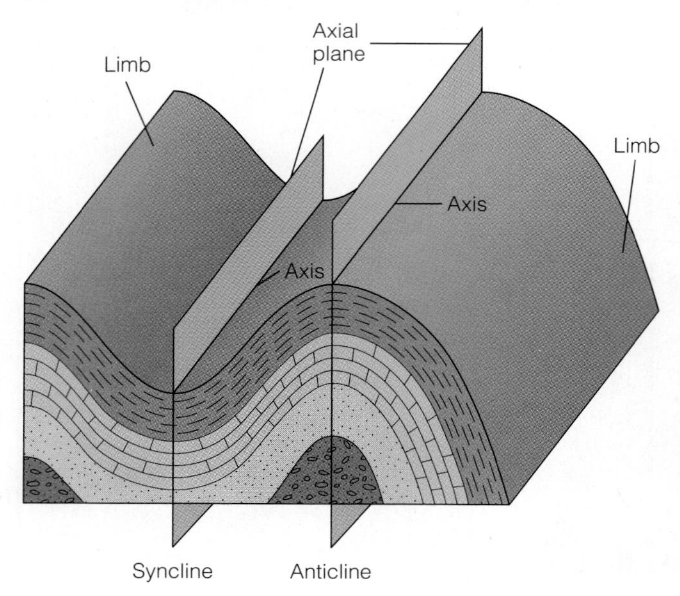

➢ **FIGURE 13-12** These folded rocks in Kootenay National Park, British Columbia, Canada, illustrate that anticlines and synclines do not necessarily correspond to mountains and valleys, respectively. Note that the fold at the mountain peak is a syncline.

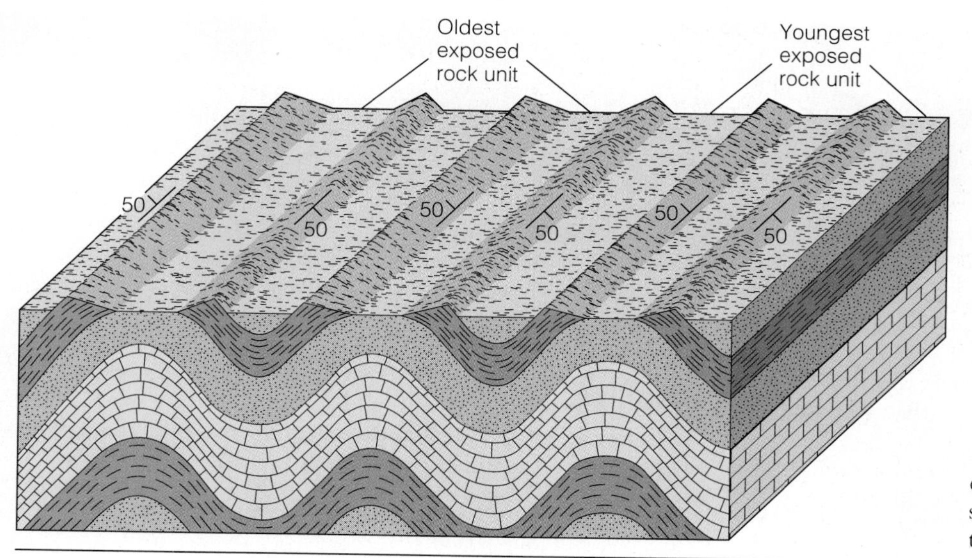

▶ **FIGURE 13-13** Identifying eroded anticlines and synclines by strike and dip and the relative ages of the folded rock layers.

▶ **FIGURE 13-14** (*a*) An asymmetrical fold. The axial plane is not vertical, and the fold limbs dip at different angles. (*b*) Overturned folds. Both fold limbs dip in the same direction, but one limb is inverted. Notice the special strike and dip symbol to indicate overturned beds. (*c*) Recumbent folds.

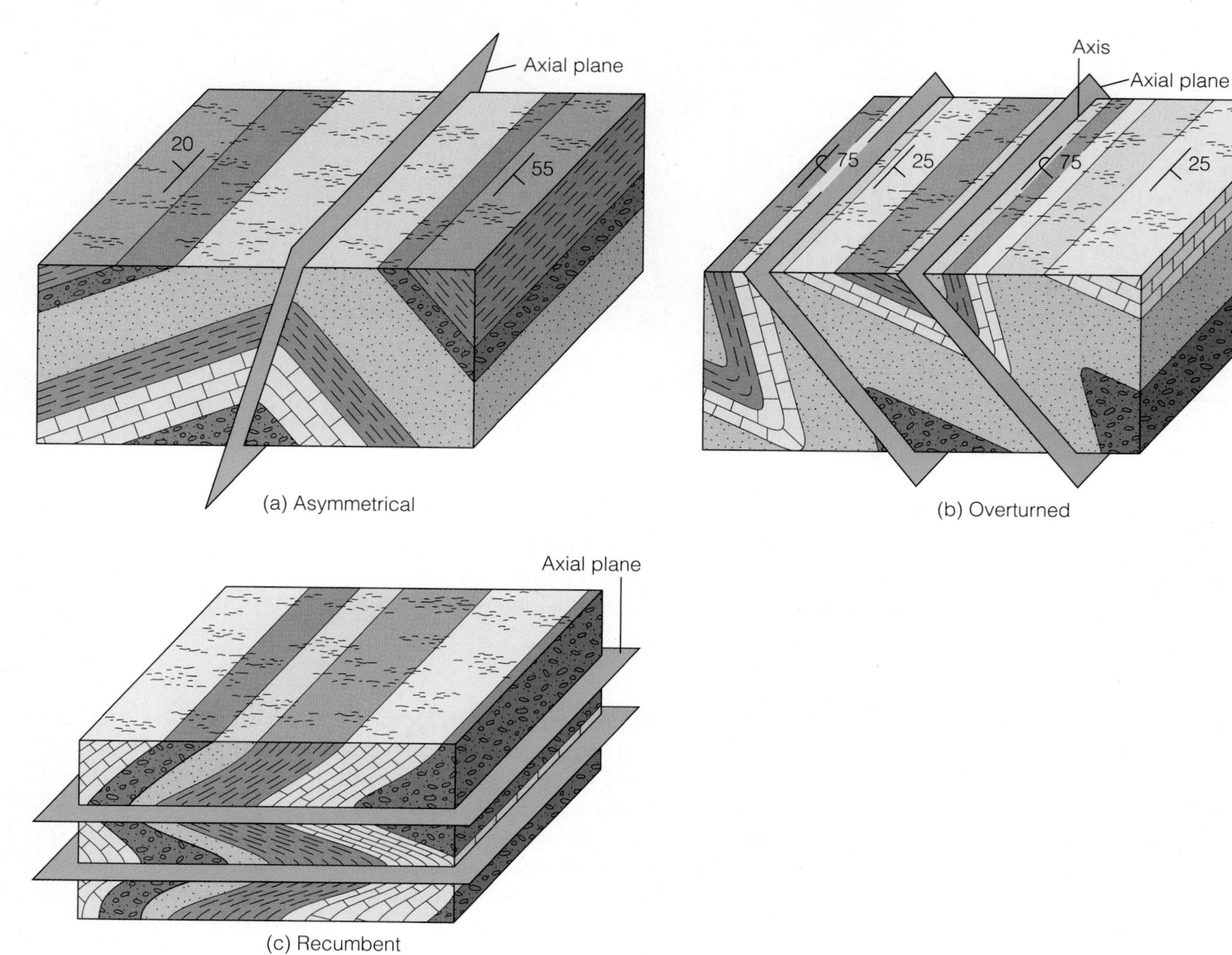

Perspective 13-1

Folding, Joints, and Arches

Arches National Park in eastern Utah is noted for its panoramic vistas, which include such landforms as Delicate Arch, Double Arch, Landscape Arch, and many others (▶ Figure 1). Unfortunately, the term *arch* is used for a variety of geologic features of different origin, but here we will restrict the term to mean an opening through a wall of rock that is formed by weathering and erosion.

The arches of Arches National Park continue to form as a result of weathering and erosion of the folded and jointed Entrada Sandstone, the rock underlying much of the park. Accordingly, geologic structures play a significant role in the origin of arches. Where the Entrada Sandstone was folded into anticlines, it was stretched so that parallel, vertical joints formed. Weathering and erosion occur most vigorously along joints because these processes can attack the exposed rock from both the top and the sides, whereas only the top is attacked in unjointed rocks.

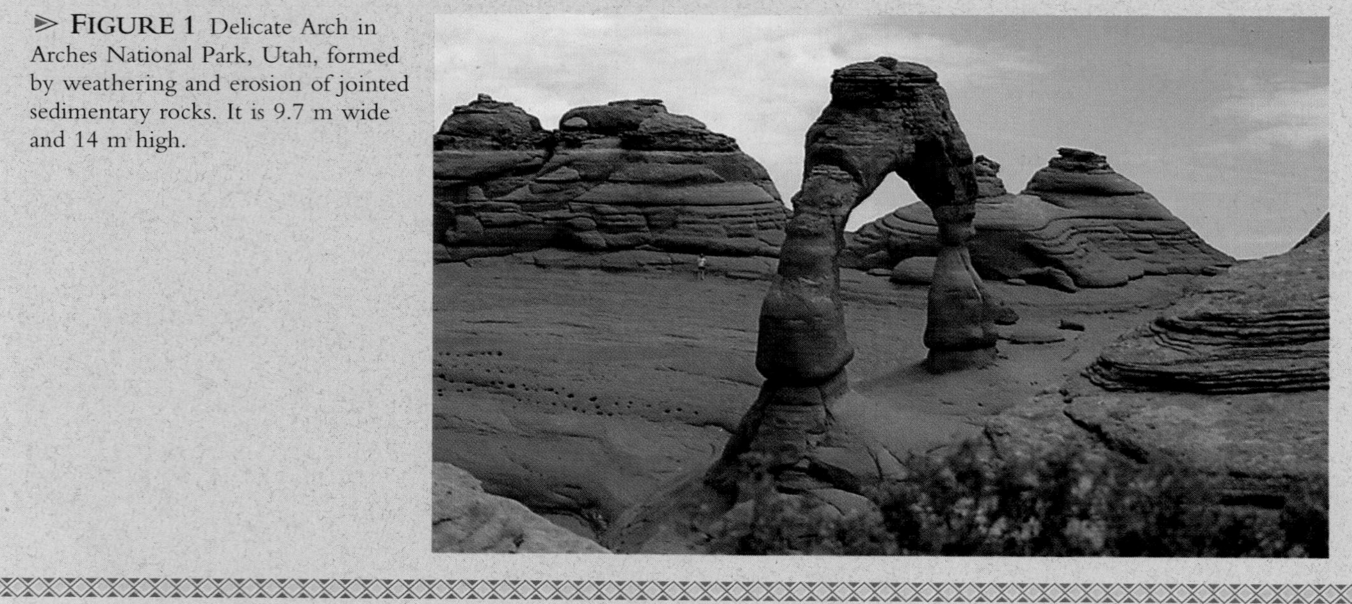

▶ **FIGURE 1** Delicate Arch in Arches National Park, Utah, formed by weathering and erosion of jointed sedimentary rocks. It is 9.7 m wide and 14 m high.

Plunging Folds. Folds may be further characterized as *nonplunging* or *plunging*. In the former, the fold *axis*, a line formed by the intersection of the axial plane with the folded beds, is horizontal (Figure 13-11). However, it is much more common for the axis to be inclined so that it appears to plunge beneath the surrounding rocks; folds possessing an inclined axis are **plunging folds** (▶ Figure 13-15). To differentiate plunging anticlines from plunging synclines, geologists use exactly the same criteria used for nonplunging folds: that is, all rocks dip away from the fold axis in plunging anticlines, whereas in plunging synclines all rocks dip inward toward the axis. The oldest exposed rocks are in the center of an eroded plunging anticline, whereas the youngest exposed rocks are in the center of an eroded plunging syncline (Figure 13-15b).

In Chapter 6 we noted that anticlines form one type of structural trap for petroleum and natural gas (see Figure 6-31). As a matter of fact, most of the world's petroleum production comes from anticlinal traps, although several other types are also important. Accordingly, geologists are particularly interested in correctly identifying the geologic structures in areas of potential petroleum and natural gas production.

Domes and Basins. **Domes** and **basins** are the circular to oval equivalents of anticlines and synclines, respectively (▶ Figure 13-16). They tend to be rather equidimensional whereas anticlines and synclines are elongate structures. Essentially the same criteria used in recognizing anticlines and synclines are used to differentiate domes and basins. In an eroded dome, the oldest exposed rocks are in the center, whereas in a basin the opposite is true. All of the rocks in a dome dip away from a central point (as opposed to dipping away from a fold

Erosion along joints causes them to enlarge, thereby forming long slender fins of rock between adajcent joints (▶ Figure 2). Some parts of these fins are more susceptible to weathering and erosion than others, and as the sides are attacked, a recess may form. If it does, eventually pieces of the unsupported rock above the recess will fall away, forming an arch as the original recess is enlarged (Figure 2). Thus, arches are remnants of fins formed by weathering and erosion along joints.

Historical observations show that arches continue to form today. In 1940, for example, Skyline Arch was enlarged when a large block fell from its underside. The park also contains many examples of arches that collapsed during prehistoric time. When arches collapse, they leave isolated pinnacles and spires. Arches National Park is well worth visiting, the pinnacles, spires, and arches are impressive features indeed.

▶ **FIGURE 2** Hole-in-the-Wall or "Baby Arch" shows the early development of an arch in a fin of rock. It measures 7.6 m wide and 4.5 m high. (Photo courtesy of Sue Monroe.)

axis, which is a line). By contrast, all the rocks in a basin dip inward toward a central point (Figure 13-16).

Some domes and basins are small structures that are easily recognized by their surface exposure patterns, but many are so large that they can be visualized only on geologic maps or aerial photographs. Many of these large-scale structures formed in the continental interior, not by compression, but as a result of vertical uplift of parts of the Earth's crust with little additional folding and faulting.

The Black Hills of South Dakota are a large eroded dome with a core of ancient rocks surrounded by progressively younger rocks. The rocks dip outward rather uniformly from the Black Hills, which have been uplifted so that they now stand more than 2,000 m above the adjacent plains. One of the best-known large basins in the United States is the Michigan basin. Most of the Michigan basin is buried beneath younger rocks so it is not directly observable at the surface. Nevertheless, strike and dip of exposed rocks near the basin margin and thousands of drill holes for oil and gas clearly show that the rock layers beneath the surface are deformed into a large structural basin.

Joints

Joints are fractures along which no movement has occurred or movement is perpendicular to the fracture surface (▶ Figure 13-17). In other words, a joint may open up, but the rocks on opposite sides of the fracture show no movement parallel to the fracture. This lack of movement parallel to joint surfaces is what distinguishes joints from *faults,* which do show movement parallel to the fracture surface. Coal miners originally used the term "joint" long

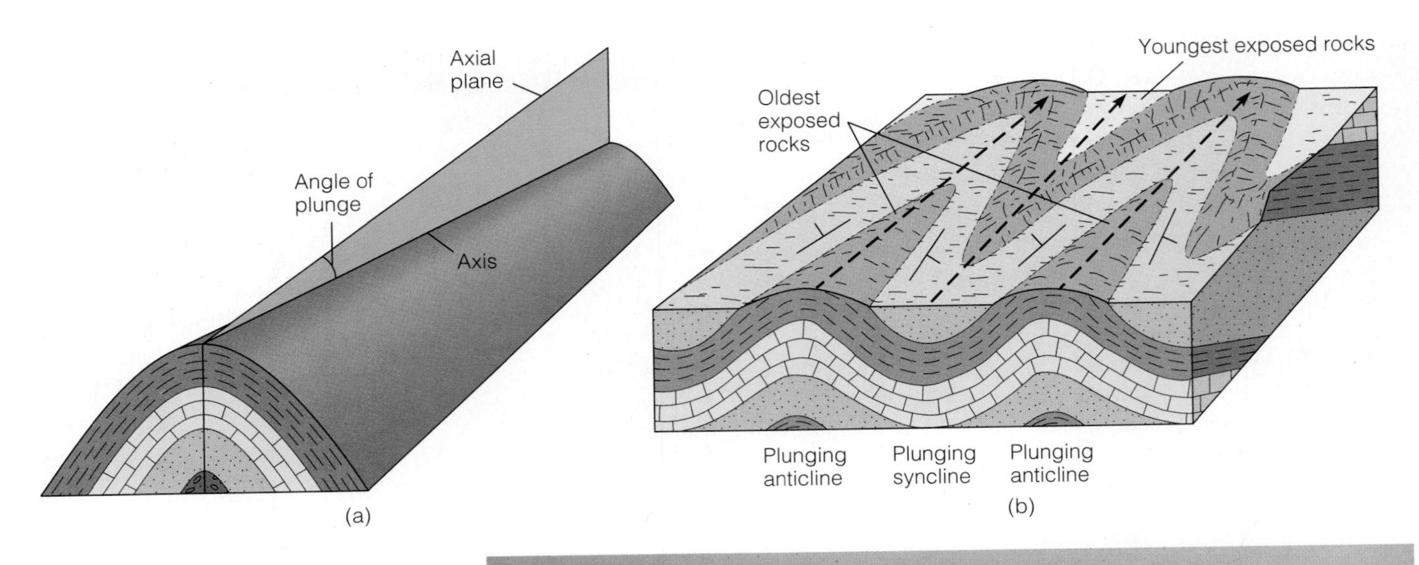

➢ **FIGURE 13-15** Plunging folds. (*a*) A schematic illustration of a plunging fold. (*b*) A block diagram showing surface and cross-sectional views of plunging folds. The long arrow at the center of each fold shows the direction of plunge. (*c*) Surface view of the eroded, plunging Sheep Mountain anticline in Wyoming. This anticline plunges toward the observer.

➢ **FIGURE 13-16** Block diagrams of (*a*) a dome and (*b*) a basin. Note that in a dome the oldest exposed rocks are in the center and all rocks dip away from a central point, whereas in a basin the youngest exposed rocks are in the center and all rocks dip inward toward a central point.

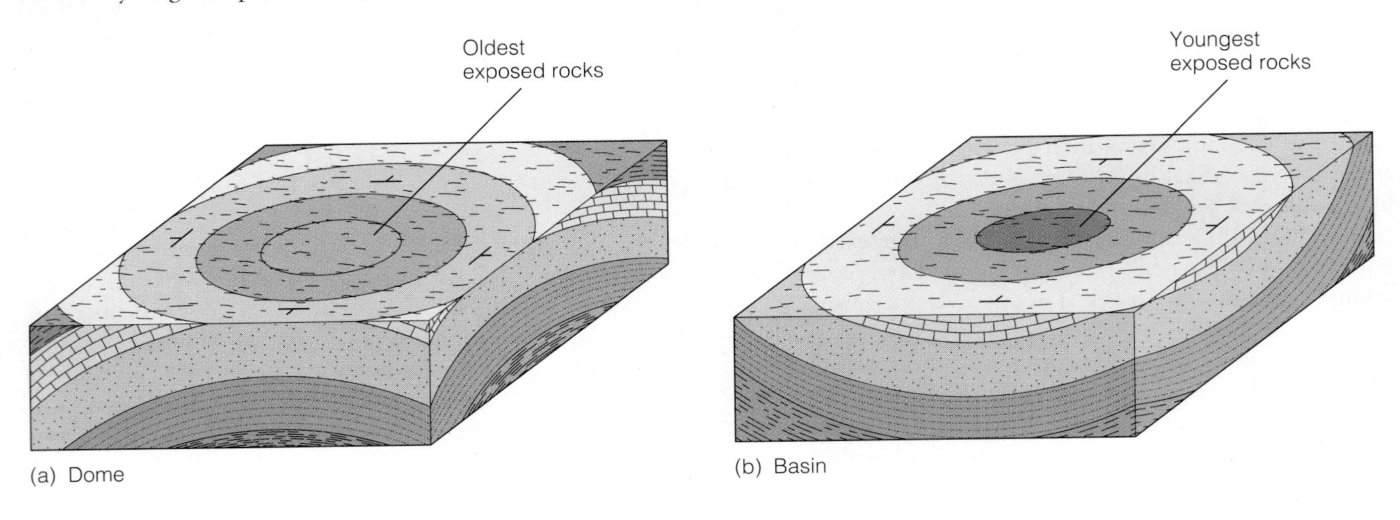

ago for cracks in rock that appeared to be surfaces where adjacent blocks were "joined" together.

Joints are the commonest structures in rocks. Recall that rocks near the Earth's surface are brittle and therefore commonly fail by fracturing when subjected to stresses (Figure 13-5). Hence nearly all near-surface rocks are jointed to some degree. Joints can form in response to compression, tension, and shearing. They vary from minute fractures to those of regional extent and are often arranged in parallel or nearly parallel sets. It is common for a region to have two or perhaps three prominent sets. Regional mapping reveals that joints and joint sets are usually related to other geologic structures such as faults and large folds. Weathering and erosion of jointed rocks in Utah have produced the spectacular scenery of Arches National Park (see Perspective 13-1).

It may seem odd that joints, which indicate brittle behavior, are so common in folded rocks that have been deformed plastically. Of course, some joints form before folding when the rocks are near the surface where they are brittle. But even rocks that exhibit considerable plastic deformation, such as occurs during folding, can be fractured (Figure 13-5). The crest of an anticline provides a good example of how this might occur. Although anticlines are produced by compression, the rock layers are arched such that tension occurs perpendicular to fold crests, and joints form parallel to the long axis of the fold in the upper part of a folded layer (➢ Figure 13-18).

We have already discussed two other types of joints in earlier chapters: columnar joints and sheet jointing. Columnar joints form in some lava flows and in some plutons. Recall from Chapters 3 and 4 that as cooling magma

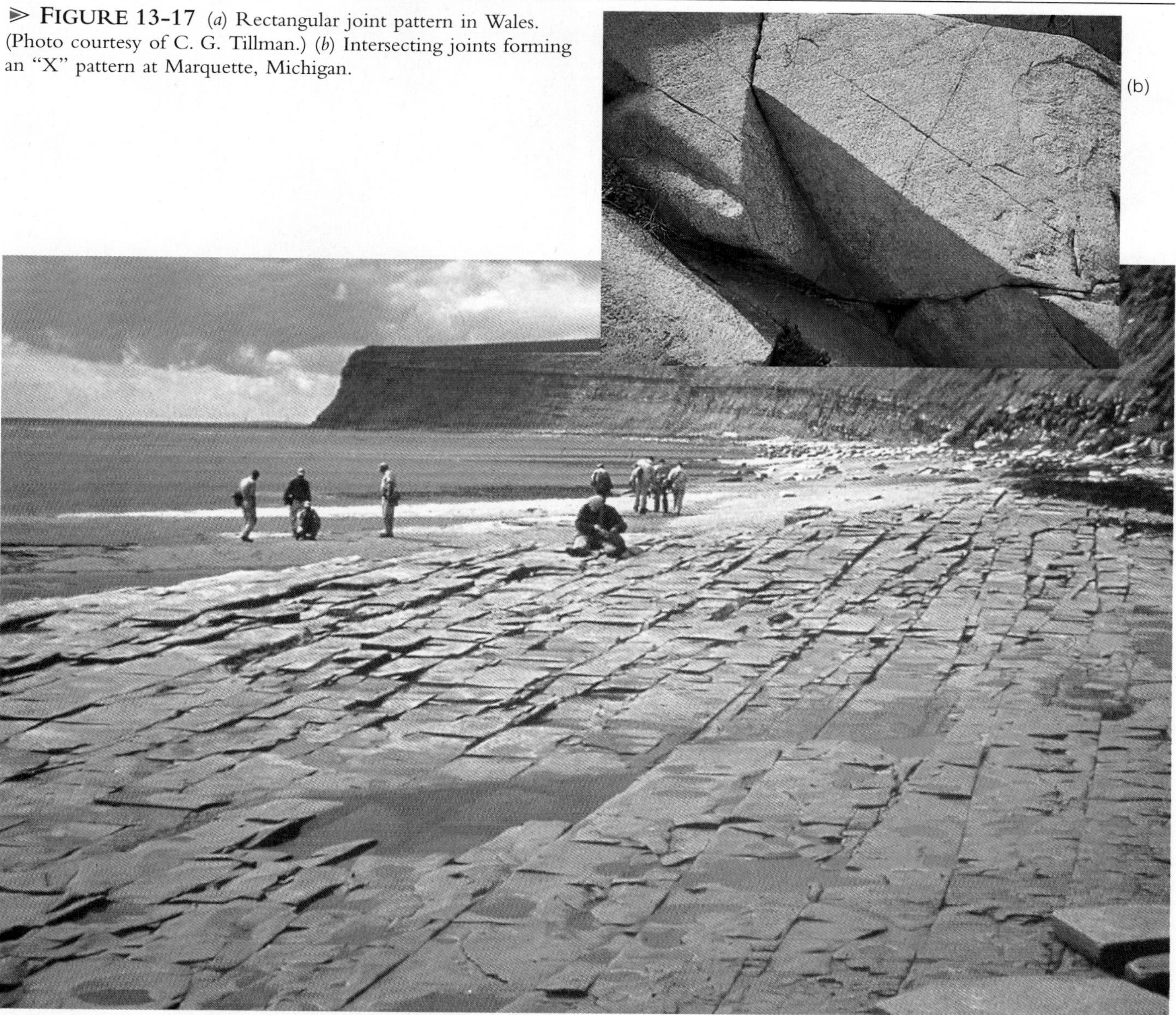

➢ **FIGURE 13-17** (*a*) Rectangular joint pattern in Wales. (Photo courtesy of C. G. Tillman.) (*b*) Intersecting joints forming an "X" pattern at Marquette, Michigan.

contracts, it develops tensional stresses that form polygonal fracture patterns (see Figures 3-1 and 4-9). Sheet jointing forms in response to pressure release (see Figure 5-7).

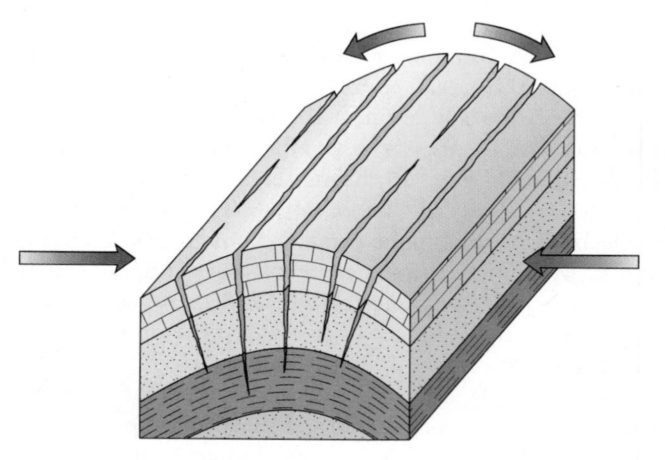

➢ FIGURE 13-18 Folding and the formation of joints parallel to the crest of an anticline.

Faults

A **fault** is a fracture along which blocks on opposite sides of the fracture move parallel with the fracture surface, which is a **fault plane** (➢ Figure 13-19a and b). One of the manifestations of faulting is a *fault scarp,* or a cliff formed as a result of vertical movement (Figure 13-19b). Fault scarps are usually quickly modified by erosion and obscured. When rocks on opposite sides of a fault plane move past one another, they may be scratched and polished by friction along the fault plane (Figure 13-19b) or crushed and shattered into angular blocks forming a *fault breccia* (Figure 13-19c).

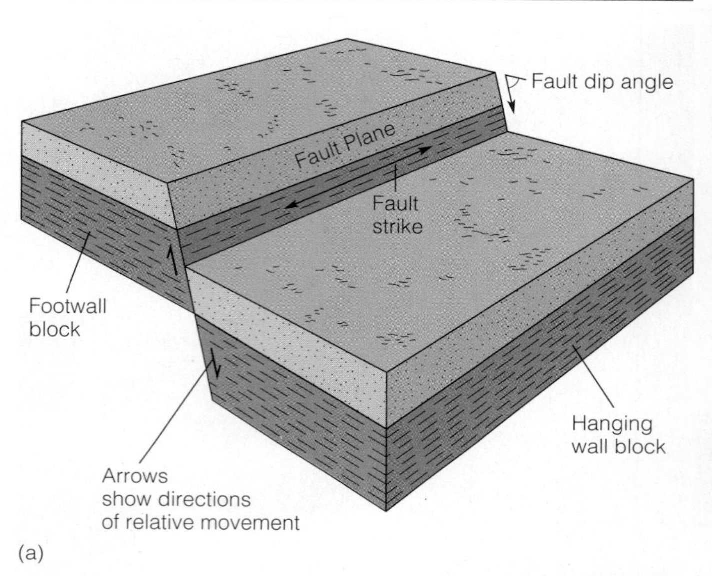

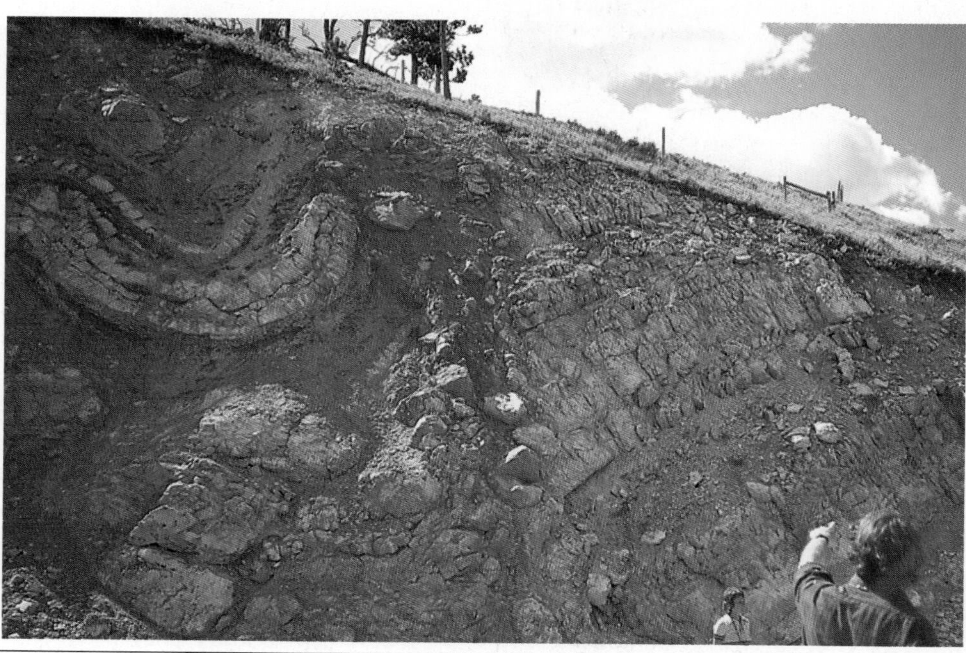

➢ FIGURE 13-19 (*a*) Fault terminology. (*b*) Polished, scratched fault plane and fault scarp near Klamath Falls, Oregon. (Photo courtesy of David J. Matty.) (*c*) Fault breccia, the zone of rubble directly above the geologist, along a fault in the Bighorn Mountains, Wyoming.

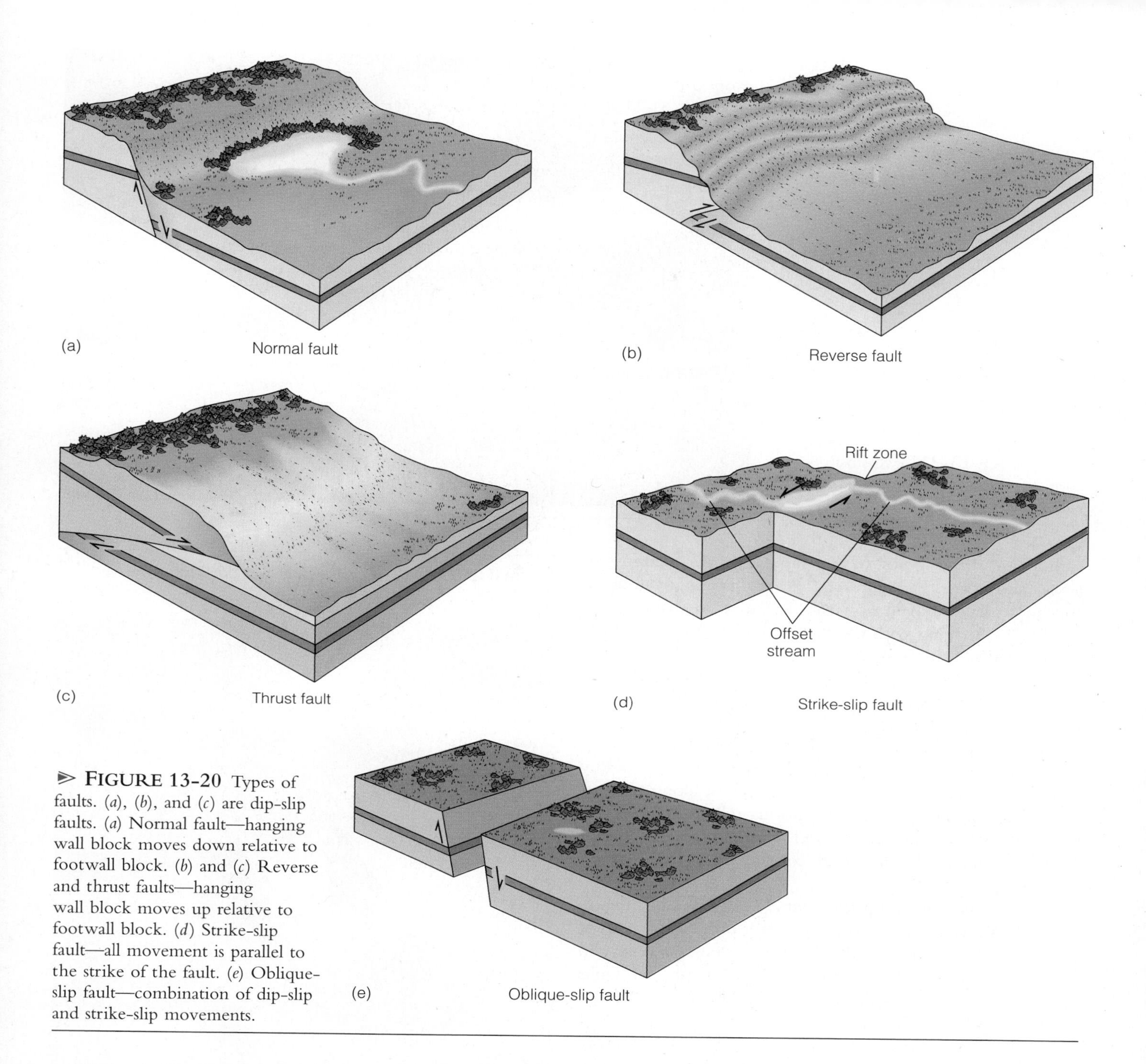

▶ FIGURE 13-20 Types of faults. (*a*), (*b*), and (*c*) are dip-slip faults. (*a*) Normal fault—hanging wall block moves down relative to footwall block. (*b*) and (*c*) Reverse and thrust faults—hanging wall block moves up relative to footwall block. (*d*) Strike-slip fault—all movement is parallel to the strike of the fault. (*e*) Oblique-slip fault—combination of dip-slip and strike-slip movements.

Notice in Figure 13-19a that the blocks adjacent to the fault plane are designated *hanging wall block* and *footwall block*. The **hanging wall block** is the block of rock overlying the fault, whereas the **footwall block** lies beneath the fault plane. These two blocks can be recognized on any fault except one that is vertical.

To differentiate among the various types of faults, one must be able to identify hanging wall and footwall blocks and understand the concept of relative movement. Geologists refer to *relative movement* because one usually cannot tell which block actually moved or if both blocks moved. In Figure 13-19a, for example, it cannot be determined whether the hanging wall block moved up, the footwall block moved down, or both blocks moved. Nevertheless, the hanging wall block appears to have moved downward relative to the footwall block.

Like rock layers, fault planes can also be characterized by their strike and dip (Figure 13-19). Two basic types of faults are recognized according to whether the blocks on opposite sides of the fault plane have moved parallel to the direction of dip or along the direction of strike.

Dip-Slip Faults. In **dip-slip faults,** all movement is parallel with the dip of the fault plane (▶ Figure 13-20a, b, and c). In other words, all movement is such that one block moves up or down relative to the block on the opposite side of the

▶ **FIGURE 13-21** Two small normal faults cutting through layers of volcanic ash in Oregon.

fault plane. Two types of dip-slip faults are recognized: *normal* and *reverse*. In Figure 13-20a, the hanging wall block appears to have moved downward relative to the footwall block; dip-slip faults with this type of relative movement are **normal faults** (▶ Figure 13-21).

Normal faults are caused by tensional stresses, such as those that occur when the Earth's crust is stretched and thinned by rifting. The mountain ranges in the Basin and Range Province, a large area in the western United States and northern Mexico, are bounded on one or both sides by large normal faults. A normal fault is present along the east side of the Sierra Nevada in California where uplift of the block west of the fault has elevated the mountains more than 3,000 m above the lowlands to the east (see Chapter Opening photo).

The second type of dip-slip fault is a **reverse fault** (Figure 13-20b). A reverse fault involving a fault plane with a dip of less than 45° is a **thrust fault** (Figure 13-20c). Reverse and thrust faults are easily distinguished from normal faults because the hanging wall block moves up relative to the footwall block (▶ Figure 13-22).

Reverse and thrust faults are caused by compression. Many large faults of these varieties are present in mountain ranges that formed by compression at convergent plate boundaries (discussed later in this chapter). A well-known thrust fault is the Lewis overthrust of Montana, where a large slab of Precambrian-aged rocks moved at least 75 km eastward on the fault and now rests upon much younger rock of Cretaceous age (▶ Figure 13-23).

Strike-Slip Faults. Shearing forces are responsible for **strike-slip faulting,** a type of faulting involving horizontal movement in which blocks on opposite sides of a fault plane slide sideways past one another (Figure 13-20d). In other words, all movement is in the direction of the fault plane's strike.

▶ **FIGURE 13-22** Reverse fault in welded tuff, Mojave Desert, California.

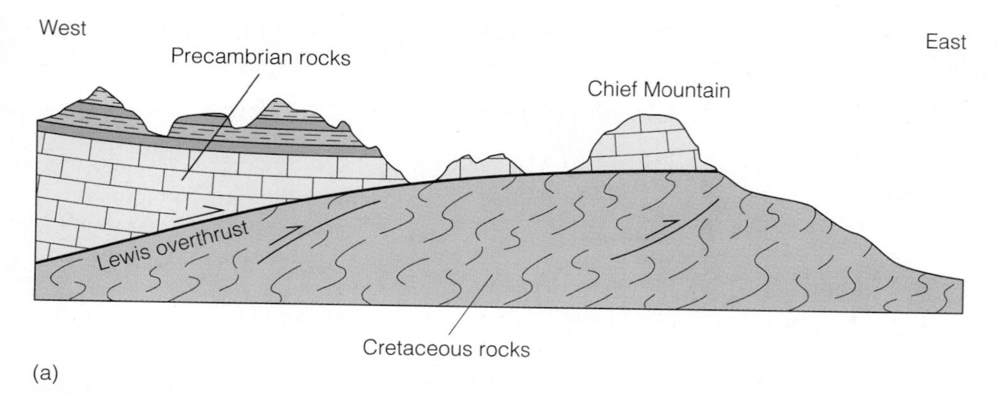

(b)

(c)

▶ **FIGURE 13-23** The Lewis overthrust fault in Glacier National Park, Montana. (*a*) Cross section showing the fault. As the slab of Precambrian rocks moved east along the fault, it deformed the rocks below. Chief Mountain is an erosional remnant of a more extensive slab of rock. (*b*) The trace of the fault is the light line on the side of the mountain. (*c*) Chief Mountain.

Geologic Maps

A geologic map uses lines, symbols, and colors to depict the distribution of various types of rocks, show the age relationships among rock units, and delineate geologic structures (▶ Figure 1). Thus, these maps convey information about the identity of bedrock formations such as granite, sandstone, and limestone; indicate the surface distribution of sediment deposited by streams, glaciers, wind, and landslides; and show geologic structures such as faults, regional joint patterns, and folds. In short, they provide in graphic form considerable information about the structure, composition, and distribution of geologic materials in a given region.

Geologic maps are constructed in a rather straightforward manner, using information gathered during field studies (▶ Figure 2). Surface exposures of rocks are studied and pertinent information recorded, such as strike and dip, composition, and relative age relationships. Notice in Figure 2 that surface exposures, or outcrops, of the various rocks are discontinuous, the most common situation encountered by geologists. Nevertheless, the exposures can be used to infer what lies beneath the surface between the outcrops. Thus, the geologic map shows the rocks of this area as if there were no soil cover (Figure 2c).

Geologists, of course, are the primary users of geologic maps. Once geologic structures and the types and relative ages of rocks in an area are known, geologists can interpret geologic processes and the geologic history. Furthermore, because geologic maps use strike and dip symbols and other symbols depicting geologic structures, cross sections of mapped areas can be constructed to illustrate three-dimensional relationships among rock units and geologic structures (Figure 1). This step is particularly important for many economic ventures because cross sections may indi-

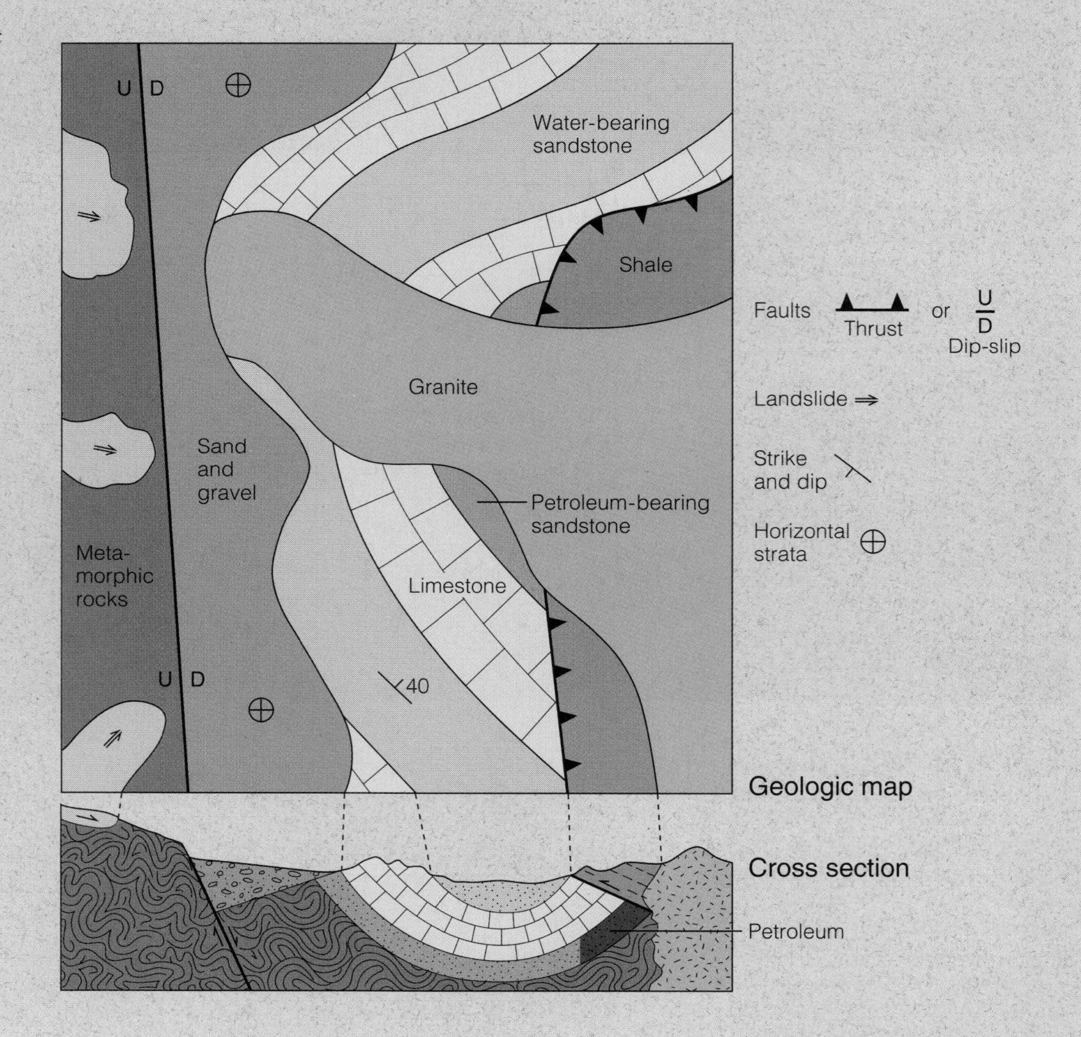

▶ **FIGURE 1** Geologic map and cross section showing the rocks and geologic structures of an area.

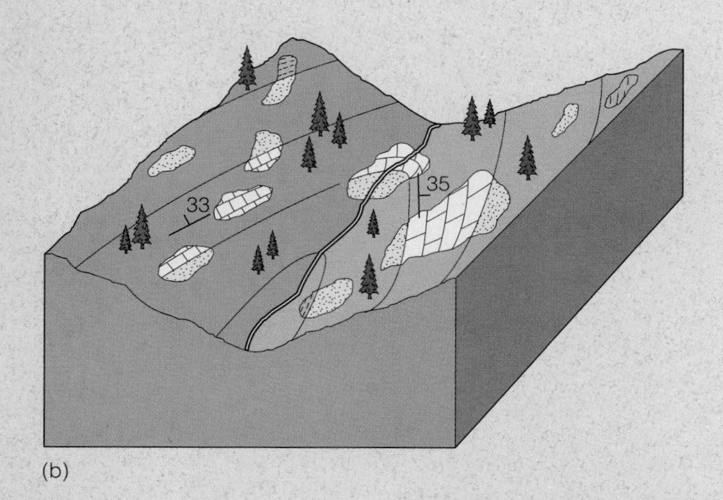

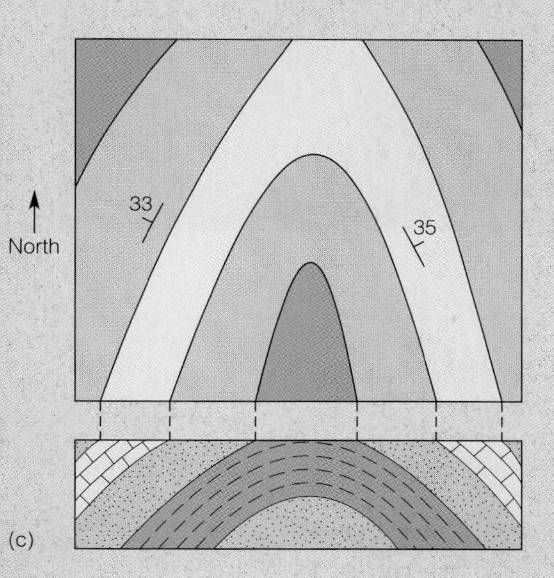

▶ **FIGURE 2** Construction of a geologic map and cross section from surface rock exposures (outcrops). (*a*) Valley with outcrops. In much of the area, the rocks are covered by soil. (*b*) Data from the outcrops are used to infer what is present in the covered areas between outcrops. The lines shown represent boundaries between different types of rock. (*c*) A geologic map (top) showing the area as if the soil had been removed. Strike and dip would be recorded at many places, but only two are shown here. The orientation of the rocks as recorded by strike and dip is used to construct the cross section (bottom).

cate where an oil well should be drilled or where other resources might be present beneath the surface.

In addition to geologists, people in many other disciplines use geologic maps. You, for example, may one day be a member of a safety-planning board of a city with known active faults. Geologic maps might therefore be useful in developing zoning regulations and construction codes. Or perhaps you will become a land-use planner or a member of a county commission charged with selecting a site for a sanitary landfill. Geologic maps would almost certainly be important in making these decisions.

Geologic maps are also used extensively in planning and constructing dams, choosing the best routes for highways through mountainous regions, and selecting sites for nuclear reactors (see Figure 13-7b). All such structures must be situated on stable foundations. For example, building a nuclear reactor in an area of active faulting or constructing a dam across a valley with an active fault would be unwise.

In areas of continuing volcanic activity, geologic maps provide essential information about the kinds of volcanic processes that might occur, such as lava flows, ash falls, or mudflows. Recall from Chapter 4 that one way in which volcanic hazards are assessed is by mapping and determining the geologic history of a region. Studies and maps prepared by geologists of the U.S. Geological Survey two years before the eruption of Mount St. Helens in Washington in 1980 allowed local officials to deal more effectively with the consequences of the eruption when it occurred.

Figure 1, a geologic map of a hypothetical area, depicts not only bedrock and various geologic structures, but also recent landslide and stream deposits. Given the position of the fault, which is still active, and the landslide-prone area, zoning this area for housing developments would not seem prudent, although it might be perfectly satisfactory for agriculture or some other kind of land use. The map also shows the distribution of unconsolidated layers of sand, and these might be important sources of groundwater. The identification of a petroleum-bearing sandstone might be important to the economic development of this region. In short, this map provides a wealth of information that can be used for a variety of purposes.

▶ **FIGURE 13-24** Right-lateral offset of a gully by the San Andreas fault in southern California. The gully is offset about 21 m.

One of the best-known strike-slip faults is the San Andreas fault of California.* Movement on this fault caused the October 29, 1989 earthquake that damaged much of Oakland, San Francisco, and several communities to the south and resulted in a 10-day delay of the World Series.

Strike-slip faults can be characterized as right-lateral or left-lateral, depending on the apparent direction of offset. In Figure 13-20d, for example, an observer looking at the block on the opposite side of the fault determines whether it appears to have moved to the right or to the left. In this example, movement appears to have been to the left, so the fault is characterized as a *left-lateral strike-slip fault.* Had this been a *right-lateral strike-slip fault,* the block across the fault from the observer would appear to have moved to the right. The San Andreas fault is a right-lateral strike-slip fault (see Figures 9-3b and ▶ 13-24).

Oblique-Slip Faults. It is possible for movement on a fault to show components of both dip-slip and strike-slip. Strike-slip movement may be accompanied by a dip-slip component giving rise to a combined movement that includes left-lateral and reverse, or right-lateral and normal. Faults having components of both dip-slip and strike-slip movement are **oblique-slip faults.** The fault shown in Figure 13-20e has a combined movement that includes left-lateral and normal.

*Recall from Chapter 12 that the San Andreas fault is also called a transform fault in plate tectonics terminology.

Various faults as well as folds and joints are commonly depicted on *geologic maps.* Geologists, engineers, city and regional planners, and people in various other professions use these maps for numerous purposes (see Perspective 13-2).

MOUNTAINS

The term *mountain* refers to any area of land that stands significantly higher than the surrounding country. Some mountains are single, isolated peaks, but much more commonly they are parts of a linear association of peaks and/or ridges called *mountain ranges* that are related in age and origin. A *mountain system* is a mountainous region consisting of several or many mountain ranges such as the Rocky Mountains and Appalachians. Mountain systems are complex linear zones of intense deformation and crustal thickening characterized by many of the geologic structures previously discussed.

Mountain systems are indeed impressive features and represent the effects of dynamic processes operating within the Earth. The forces necessary to elevate the Himalayas of Asia to nearly 9 km above sea level are difficult to comprehend, yet when compared with the size of the Earth, even the loftiest mountains are very small features. In fact, the greatest difference in elevation on Earth is about 20 km; if we depicted this to scale on a globe 1 m in diameter, its relief would be less than 2 mm. From the human perspective, however, mountain systems are large-scale manifestations of tremendous forces that have produced folded, faulted, and thickened parts of the crust. Furthermore, in some mountain systems, such as the Andes of South America and the Himalayas of Asia, the mountain-building processes remain active today.

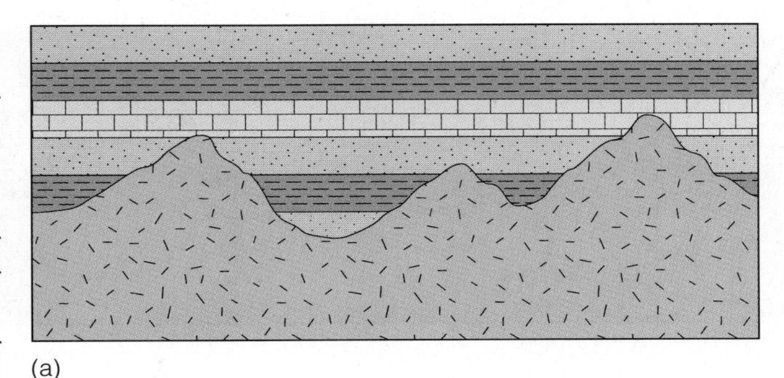

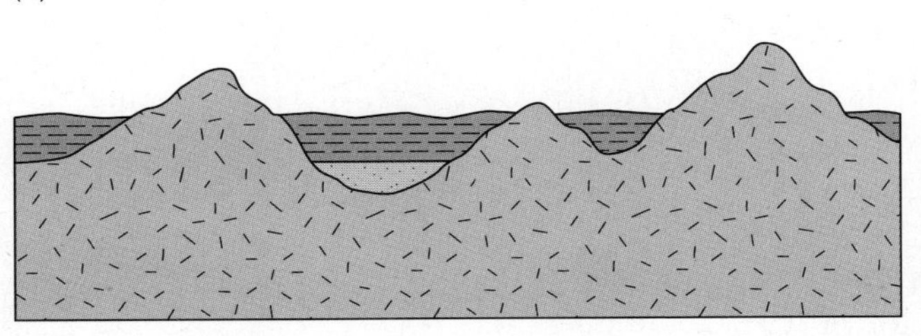

➤ **FIGURE 13-25** (*a*) Pluton intruded into sedimentary rocks. (*b*) Erosion of the softer overlying rocks reveals the pluton and forms small mountains.

Types of Mountains

Mountainous topography can develop in a variety of ways, some of which involve little or no deformation of the Earth's crust. A single volcanic mountain can develop over a hot spot, but more commonly a series of volcanoes develops as a plate moves over the hot spot, as in the case of the Hawaiian Islands (see Figure 12-23).

Mountainous topography also forms where the crust has been intruded by batholiths that are subsequently uplifted and eroded (➤ Figure 13-25). The Sweetgrass Hills of northern Montana consist of resistant plutonic rocks exposed following uplift and erosion of the softer overlying sedimentary rocks.

Block-faulting is yet another way mountains are formed. Block-faulting involves movement on normal faults so that one or more blocks are elevated relative to adjacent areas (➤ Figure 13-26). A classic example is the large-scale block-faulting currently occurring in the Basin and Range Province of the western United States, a large area centered on Nevada but extending into several adjacent states and northern Mexico. In the Basin and Range Province, the Earth's crust is being stretched in an east-west direction and tensional stresses produce north-south oriented, range-bounding faults. Differential movement on these faults has yielded uplifted blocks called *horsts* and down-dropped blocks called *grabens* (Figure 13-26). Horsts and grabens are bounded on both sides by parallel normal faults. Erosion of the horsts has yielded the mountainous topography now present, and the grabens have filled with sediments eroded from the horsts.

The processes discussed above can certainly yield mountains. However, the truly large mountain systems of the continents, such as the Alps of Europe and the Appalachians in North America, were produced by compression along convergent plate margins.

Mountain Building: Orogenesis

During an episode of mountain building, termed an **orogeny,** intense deformation occurs, generally accompanied by metamorphism and the emplacement of plutons, especially batholiths. *Orogenesis,* the processes whereby mountains form, is still not fully understood, but it is known to be related to plate movements. In fact, the advent of plate tectonic theory has completely changed the way geologists view the origin of mountain systems.

Any theory accounting for orogenesis must adequately explain the characteristics of mountain systems such as their geometry and location; they tend to be long and narrow and to be located at or near plate margins. Mountain systems also show intense deformation, especially compression-induced overturned and recumbent folds and reverse and thrust faults. Furthermore, the deeper, interior parts or cores of mountain systems are characterized by granitic plutons and regional metamorphism. The presence of deformed shallow and deep marine sedimentary rocks that have been elevated far above sea level is another feature.

Plate Boundaries and Orogenesis

Most of the Earth's geologically recent and present-day orogenic activity is concentrated in two major zones or belts: the *Alpine-Himalayan orogenic belt* and the *circum-Pacific orogenic belt* (➤ Figure 13-27). Both belts consist of a number of smaller segments known as *orogens,* many of which are areas of active orogenesis today.

Most orogenesis occurs in response to compressive stresses at convergent plate boundaries. Recall from Chapter 12 that three varieties of convergent plate boundaries are recognized: oceanic-oceanic, oceanic-continental, and continental-continental.

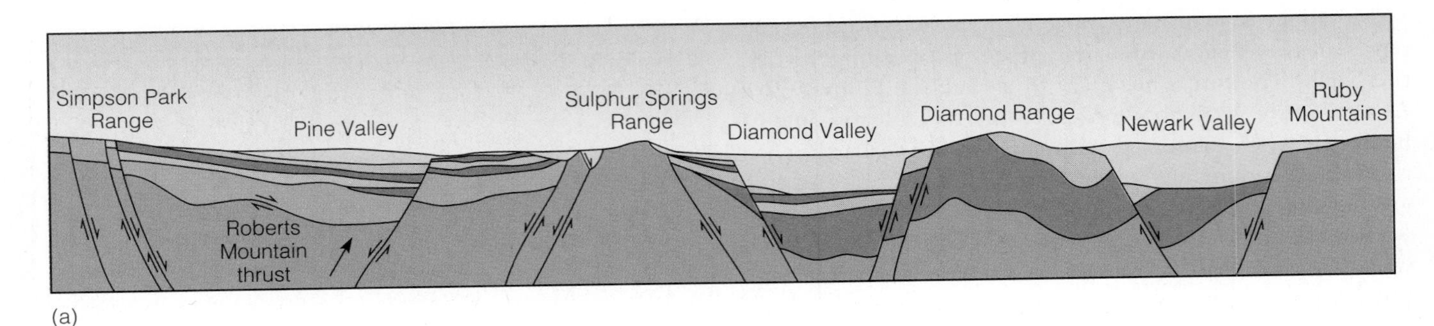

(a)

(b)

➢ FIGURE 13-26 Block-faulting and the origin of horsts and grabens. (*a*) Cross section of part of the Basin and Range Province in Nevada. The ranges (horsts) and valleys (grabens) are bounded by normal faults. (*b*) View of the Humboldt Range in Nevada.

Orogenesis at Oceanic-Oceanic Plate Boundaries. Orogenies occurring where oceanic lithosphere is subducted beneath oceanic lithosphere are characterized by the formation of a volcanic island arc and by deformation, igneous activity, and metamorphism. The subducted plate forms the outer wall of an oceanic trench, and the inner wall of the trench consists of a subduction complex or *accretionary wedge* composed of wedge-shaped slices of highly folded and faulted marine sedimentary rocks and oceanic lithosphere scraped from the descending plate (➢ Figure 13-28). This subduction complex is elevated as a result of uplift along faults as subduction continues. In addition, plate convergence results in low-temperature, high-pressure metamorphism characteristic of the blueschist facies (see Figure 7-20).

Deformation also occurs in the island arc system where it is caused largely by the emplacement of plutons of intermediate and felsic composition, and many rocks show evidence of high-temperature, low-pressure metamorphism. As a result, the overriding oceanic plate is thickened as it is intruded by plutons and becomes more continental. The overall effect of island arc orogenesis is the origin of two more-or-less parallel orogenic belts consisting of a deformed volcanic island arc underlain by batholiths and a seaward belt of deformed trench rocks (Figure 13-28). The Japanese Islands are a good example of this type of deformation.

In the back-arc basin, volcanic rocks derived from the island arc and sediments eroded from the island arc and the adjacent continent are also deformed as the plates continue to converge. They are intensely folded and displaced toward the adjacent continent along low-angle thrust faults. Eventually, the entire island arc complex with its core of metamorphic and plutonic rocks is fused to the edge of the continent, and the back-arc basin sediments are thrust onto the continent and form a thick stack of thrust sheets (Figure 13-28).

Orogenesis at Oceanic-Continental Plate Boundaries. Several mountain systems such as the Alps of Europe and the Andes of South America formed at oceanic-continental plate boundaries where oceanic lithosphere is subducted. The Andes of western South America are perhaps the best

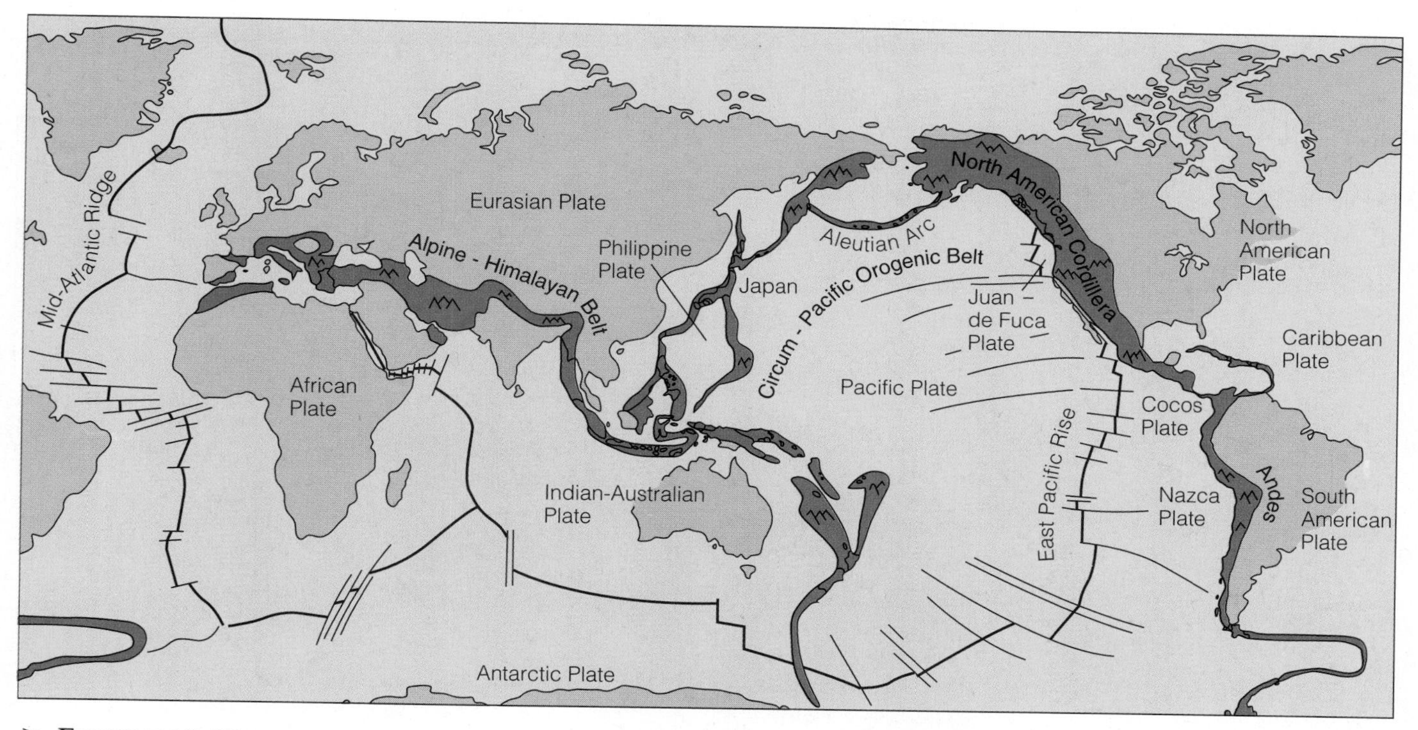

▶ **FIGURE 13-27** Most of the Earth's geologically recent and present-day orogenic activity is concentrated in the circum-Pacific and Alpine-Himalayan orogenic belts.

example of such continuing orogeny (▶ Figure 13-29). Among the ranges of the Andes are the highest mountain peaks in the Americas; 49 of these peaks are more than 6,000 m high. The Andes also include many active volcanoes, and the western part of South America is an extremely active segment of the circum-Pacific earthquake belt. Furthermore, one of the Earth's great oceanic trench systems, the Peru-Chile Trench, lies just off the west coast (see Figure 11-13).

Prior to 200 million years ago, the western margin of South America was a passive continental margin, where sediments accumulated on the continental shelf, slope, and rise much as they currently do along the east coast of North America. When Pangaea began fragmenting in response to rifting along what is now the Mid-Atlantic Ridge, the South American plate moved westward, and eastward-moving oceanic lithosphere began subducting beneath the continent (Figure 13-29). What had been a passive continental margin was now an active one.

As subduction proceeded, rocks of the continental margin and trench were folded and faulted and are now part of an accretionary wedge along the west coast of South America (Figure 13-29). Accretionary wedges here and elsewhere commonly contain fragments of oceanic crust and upper mantle called ophiolites (see Figure 11-24). A well-known accretionary wedge in North America is the Franciscan Group, a 7,000 m thick, complex assemblage of various rock types exposed in the Coast Ranges of California (▶ Figure 13-30).

Subduction also resulted in partial melting of the descending plate, producing an andesitic volcanic arc of composite volcanoes at the edge of the continent. More viscous felsic magmas, mostly of granitic composition, were emplaced as large plutons beneath the volcanic arc (Figure 13-29). The coastal batholith of Peru, for example, consists of perhaps 800 individual plutons that were emplaced over several tens of millions of years.

As a result of the events just described, the Andes Mountains consist of a central core of granitic rocks capped by andesitic volcanoes. To the west of this central core along the coast are the deformed rocks of the accretionary wedge. And to the east of the central core are sedimentary rocks that have been intensely folded and thrust eastward onto the continent (Figure 13-29). Present-day subduction, volcanism, and seismicity along South America's west coast indicate that the Andes Mountains are still actively forming.

Orogenesis at Continental-Continental Plate Boundaries. The Himalayas of Asia began forming when India collided with Asia about 40 to 50 million years ago. Prior to that time, India was far south of Asia and separated from it by an ocean basin (▶ Figure 13-31a). As the Indian plate moved northward, a subduction zone formed along the southern margin of Asia where oceanic lithosphere was consumed

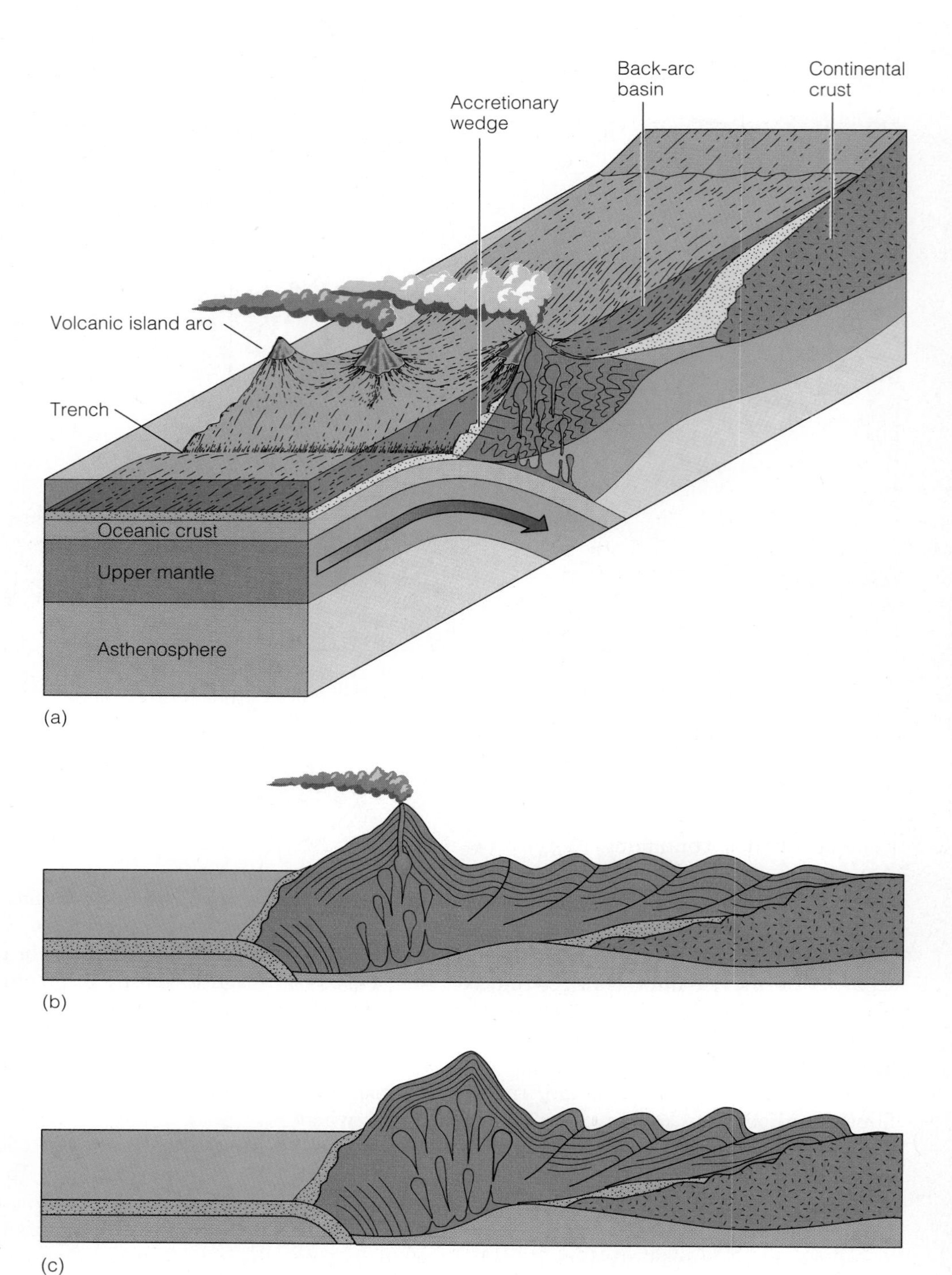

➢ **FIGURE 13-28** Orogenesis and the origin of a volcanic island arc at an oceanic-oceanic plate boundary. (*a*) Subduction of an oceanic plate beneath an island arc. (*b*) Continued subduction, emplacement of plutons, and beginning of deformation of back-arc basin sediments. (*c*) Thrusting of back-arc basin sediments onto the adjacent continent and suturing of the island arc to the continent.

(Figure 13-31a). Partial melting generated magma, which rose to form a volcanic arc, and large granite plutons were emplaced into what is now Tibet. At this stage, the activity along Asia's southern margin was similar to what is now occurring along the west coast of South America.

The ocean separating India from Asia continued to close, and India eventually collided with Asia (Figure 13-31b). As a result, two continental plates became welded, or sutured, together. Thus, the Himalayas are now located within a continent rather than along a continental margin (Figures 13-27 and 13-31b). The exact time of India's collision with Asia is uncertain, but between 40 and 50 million years ago, India's rate of northward drift decreased abruptly—from 15 to 20 cm per year to about 5 cm per year. Because continental lithosphere is not dense enough to be subducted, this decrease in rate seems to mark the time of collision and

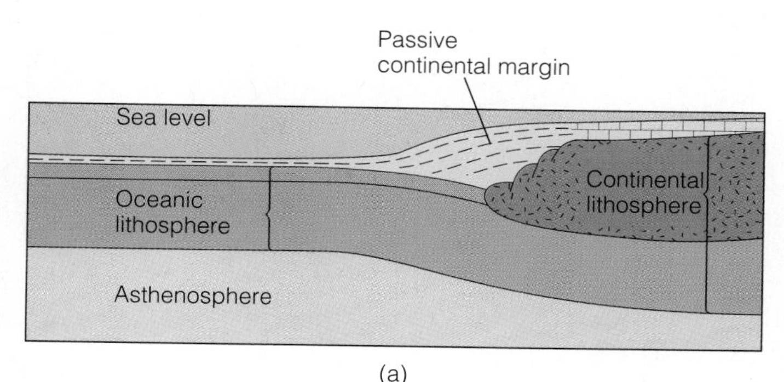

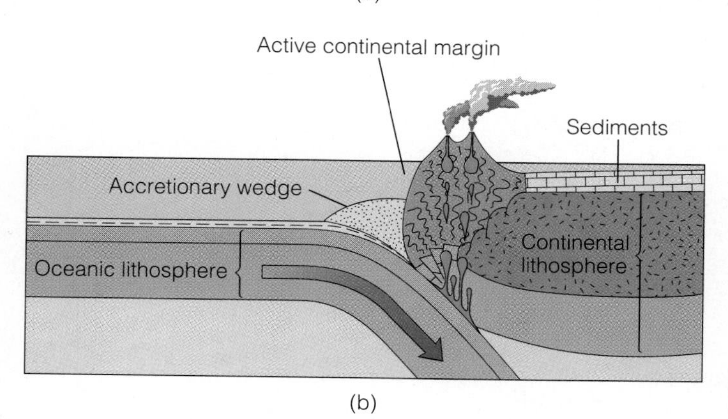

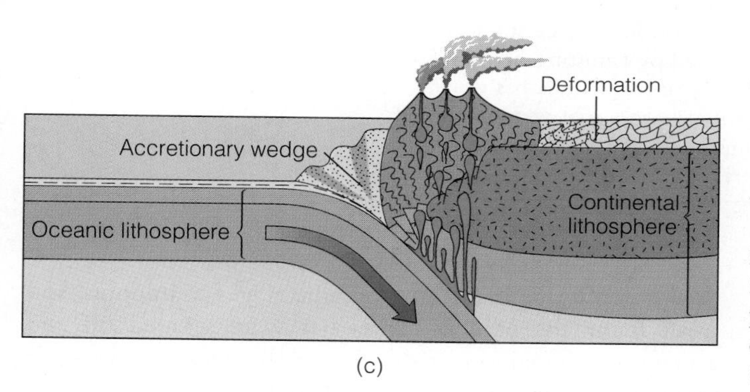

India's resistance to subduction. Consequently, India's leading margin was thrust beneath Asia, causing crustal thickening, thrusting, and uplift. Sedimentary rocks that had been deposited in the sea south of Asia were thrust northward, and two major thrust faults carried rocks of Asian origin onto the Indian plate (Figure 13-31c and d). Rocks deposited in the shallow seas along India's northern margin now form the higher parts of the Himalayas.

As the Himalayas were uplifted, they were also eroded, but at a rate insufficient to match the uplift. Much of the debris shed from the rising mountains was transported to the south and deposited as a vast blanket of sediment on the Ganges Plain and as huge submarine fans in the Arabian Sea and the Bay of Bengal. Since its collision with Asia, India has been thrust about 2,000 km beneath Asia. Currently, India is moving north at a rate of about 5 cm per year.

A number of other mountain systems also formed as a result of collisions between two continental plates. The Urals in Russia and the Appalachians of North America formed by such collisions. Also, west of the Himalayas, the Arabian plate is colliding with Asia along the Zagros Mountains of Iran.

MICROPLATE TECTONICS AND MOUNTAIN BUILDING

In the preceding sections, we discussed orogenies along convergent plate boundaries resulting in continental accre-

➢ **FIGURE 13-29** Generalized diagrams showing three stages in the development of the Andes of South America. (*a*) Prior to 200 million years ago, the west coast of South America was a passive continental margin. (*b*) Orogenesis began when the west coast of South America became an active continental margin. (*c*) Continued deformation, volcanism, and plutonism.

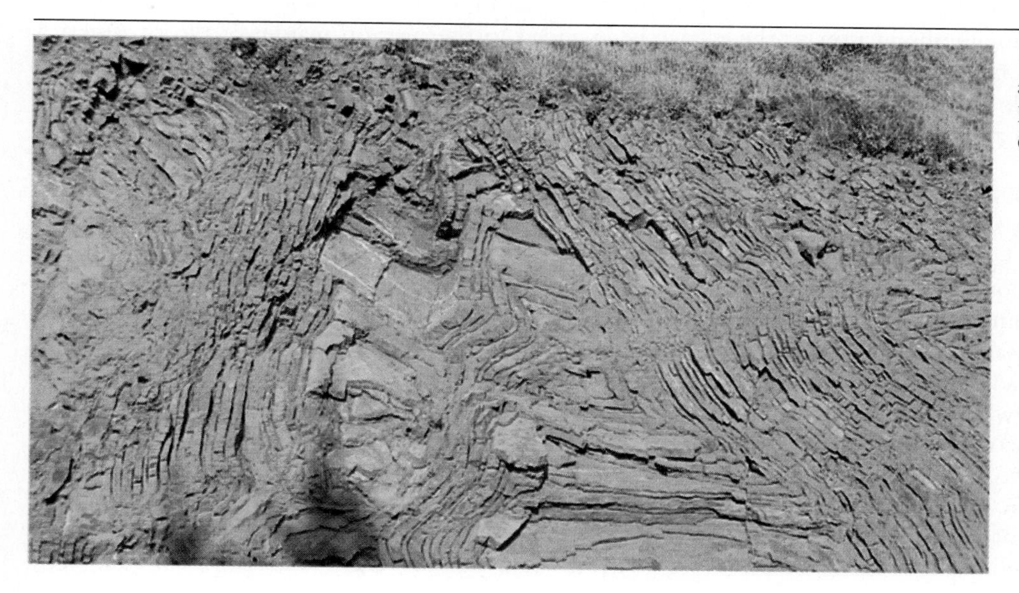

➢ **FIGURE 13-30** Deformed accretionary wedge rocks of the Franciscan Group in Marin County, California.

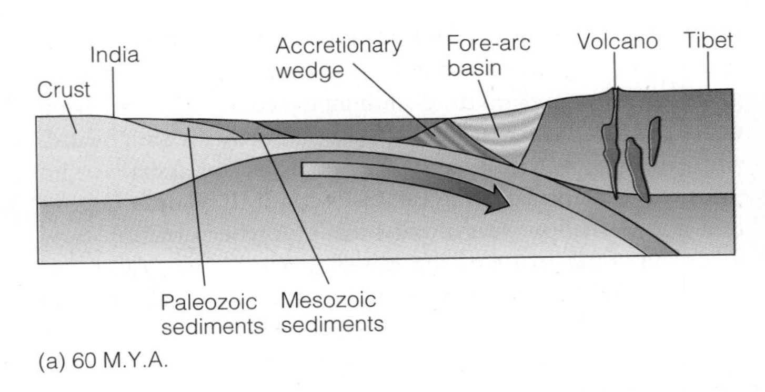

(a) 60 M.Y.A.

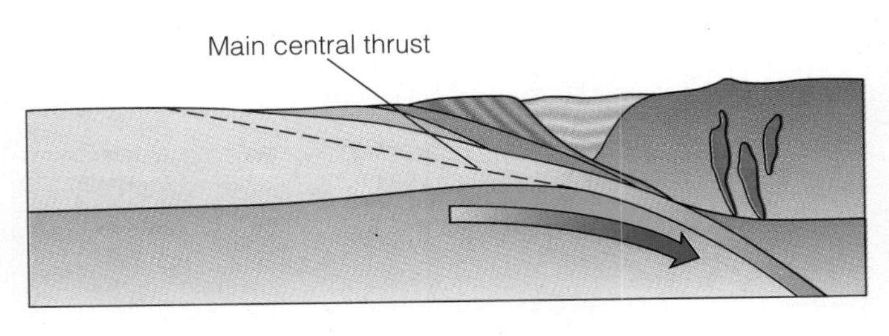

(b) 40–50 M.Y.A.

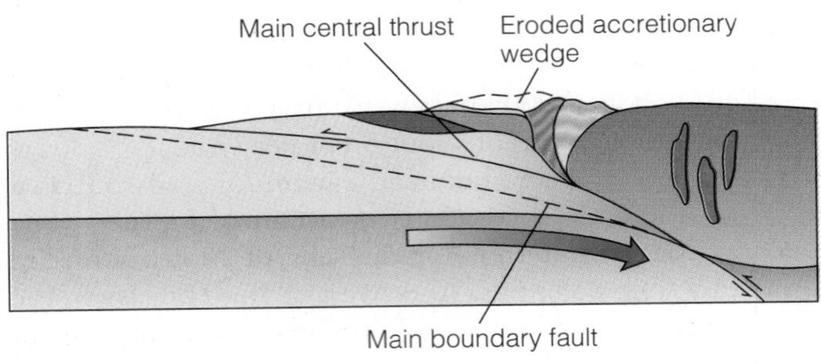

(c) 20–40 M.Y.A.

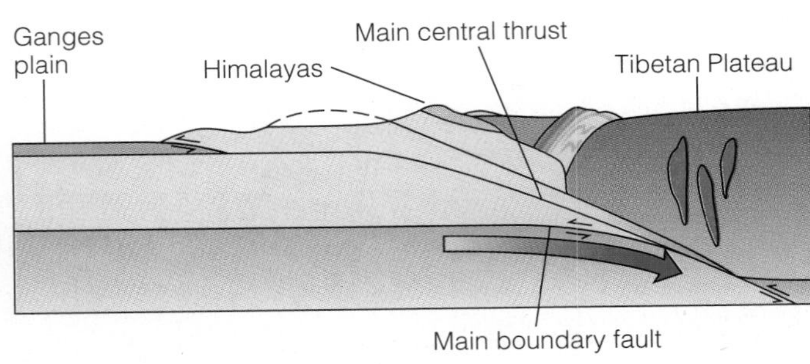

(d) 20–0 M.Y.A.

➤ **FIGURE 13-31** Simplified cross sections showing the collision of India with Asia and the origin of the Himalayas. (*a*) The northern margin of India before its collision with Asia. Subduction of oceanic lithosphere beneath southern Tibet as India approached Asia. (*b*) About 40 to 50 million years ago, India collided with Asia, but because India was too light to be subducted, it was underthrust beneath Asia. (*c*) Continued convergence accompanied by thrusting of rocks of Asian origin onto the Indian subcontinent. (*d*) Since about 10 million years ago, India has moved beneath Asia along the main boundary fault. Shallow marine sedimentary rocks that were deposited along India's northern margin now form the higher parts of the Himalayas. Sediment eroded from the Himalayas has been deposited on the Ganges Plain.

tion. Much of the material accreted to continents during these events is simply eroded older continental crust, but a significant amount of new material is added to continents as well—igneous rocks that formed as a consequence of subduction and partial melting and the suturing of an island arc to a continent, for example. Although subduction is the predominant influence on tectonic history in many regions of orogenesis, other processes are also involved in mountain building and continental accretion, especially the accretion of microplates.

During the late 1970s and 1980s, geologists discovered that portions of many mountain systems are composed of small accreted lithospheric blocks that are clearly of foreign origin. These **microplates** differ completely in their fossil content, structural trends, and paleomagnetic properties from the rocks of the surrounding mountain system. In fact, these microplates are so different from adjacent rocks that most geologists think they formed elsewhere and were carried great distances as parts of other plates until they collided with other microplates or continents.

Geologic evidence indicates that more than 25% of the entire Pacific coast from Alaska to Baja California consists of accreted microplates. The accreting microplates are composed of volcanic island arcs, oceanic ridges, seamounts, and small fragments of continents that were scraped off and accreted to the continent's margin as the oceanic plate with which they were carried was subducted under the continent. It is estimated that more than 100 different-sized microplates have been added to the western margin of North America during the last 200 million years (➤ Figure 13-32).

The basic plate tectonic reconstruction of orogenies and continental accretion remains unchanged, but the details of these reconstructions are decidedly different in view of microplate tectonics. For example, growth along active continental margins is faster than along passive continental margins because of the accretion of microplates. Furthermore, these accreted microplates are often new additions to a continent rather than reworked older continental material.

Most of the microplates identified so far are in mountains of the North American Pacific coast region, but a number of such plates are suspected to be present in other mountain systems as well. They are more difficult to recognize in older mountain systems, such as the Appalachians, because of greater deformation and erosion. Nevertheless, about a dozen microplates have been identified in the Appalachians, although their boundaries are hard to discern. Micro-

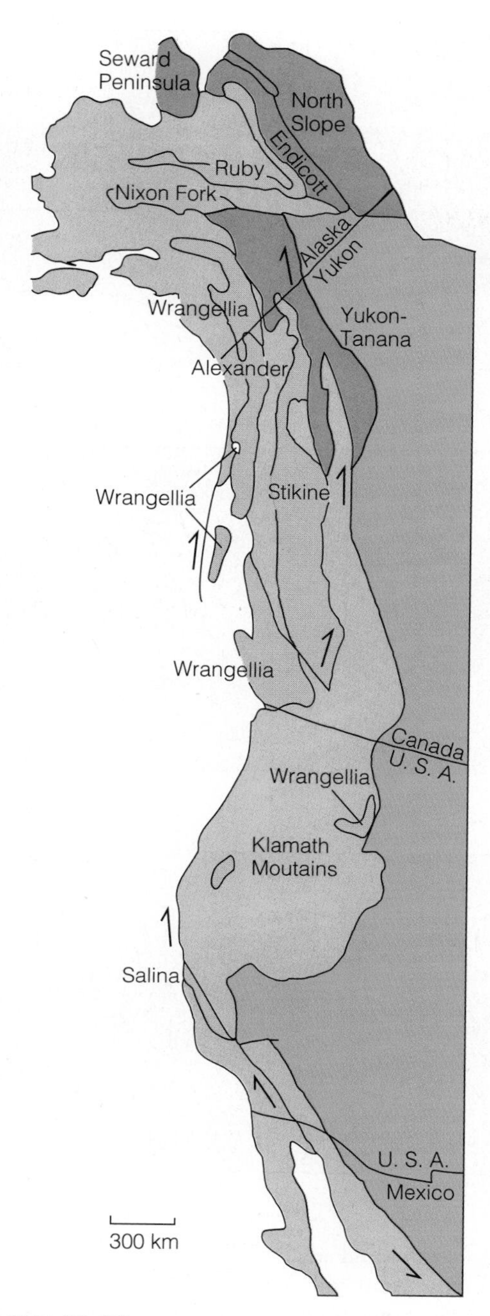

➤ **FIGURE 13-32** Some of the accreted lithospheric blocks called microplates that form the western margin of North America. The light brown blocks probably originated as parts of continents other than North America. The reddish brown blocks are possibly displaced parts of North America.

plate tectonics provides a new way of viewing the Earth and of gaining a better understanding of the geologic history of the continents.

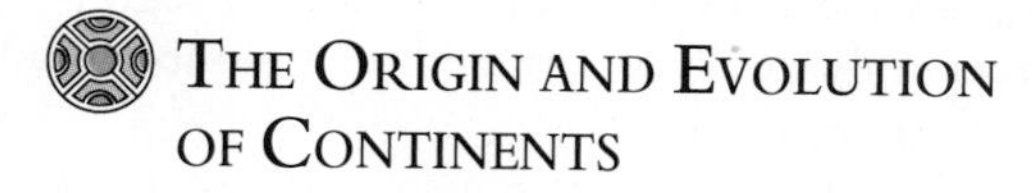

THE ORIGIN AND EVOLUTION OF CONTINENTS

Rocks 3.8 billion years old that are thought to represent continental crust are known from several areas, including Minnesota, Greenland, and South Africa. Most geologists agree that even older continental crust probably existed, and, in fact, rocks dated at 3.96 billion years were recently discovered in Canada.

According to one model for the origin of continents, the earliest crust was thin and unstable and was composed of ultramafic igneous rock. This early ultramafic crust was disrupted by upwelling basaltic magmas at ridges and was consumed at subduction zones. Ultramafic crust would therefore have been destroyed because its density was great enough to make recycling by subduction very likely. Apparently, only crust of a more granitic composition, which has a lower density, is resistant to destruction by subduction.

A second stage in crustal evolution began when partial melting of earlier formed basaltic crust resulted in the formation of andesitic island arcs, and partial melting of lower crustal andesites yielded granitic magmas that were emplaced in the crust that had formed earlier. As plutons were emplaced in these island arcs they became more like continental crust. By 3.96 to 3.8 billion years ago, plate motions accompanied by subduction and collisions of island arcs had formed several granitic continental nuclei (➤ Figure 13-33).

Shields, Cratons, and the Evolution of Continents

Each continent is characterized by one or more areas of exposed ancient rocks called a **shield** (see Figure 7-3). Extending outward from these shields are broad platforms of ancient rocks buried beneath younger sediments and sedimentary rocks. The shields and buried platforms are collectively called **cratons,** so shields are simply the exposed parts of cratons. Cratons are considered to be the stable interior parts of continents.

In North America, the *Canadian Shield* includes much of Canada; a large part of Greenland; parts of the Lake Superior region in Minnesota, Wisconsin, and Michigan; and parts of the Adirondack Mountains of New York (➤ Figure 13-34). In general, the Canadian Shield is a vast area of subdued topography, numerous lakes, and exposed ancient metamorphic, volcanic, plutonic, and sedimentary rocks.

Each continent evolved by accretion along the margins of ancient cratons. To this extent, all continents developed similarly, but the details of each continent's history differ. Here we shall concentrate on the evolution of North America. Several cratons that would eventually become part of the North American craton had formed by 2.5 billion years ago, but these were independent minicontinents that were later assembled into a larger craton. The Slave, Hearn, and Superior cratons in ➤ Figure 13-35, for instance, were separate landmasses until a larger craton formed.

A major episode in the Precambrian evolution of North America took place between 2.0 and 1.8 billion years ago when several major orogens developed. These orogens are zones of complexly deformed rocks, many of which have been metamorphosed and intruded by plutons and thus represent areas of ancient mountain building. Smaller cratons

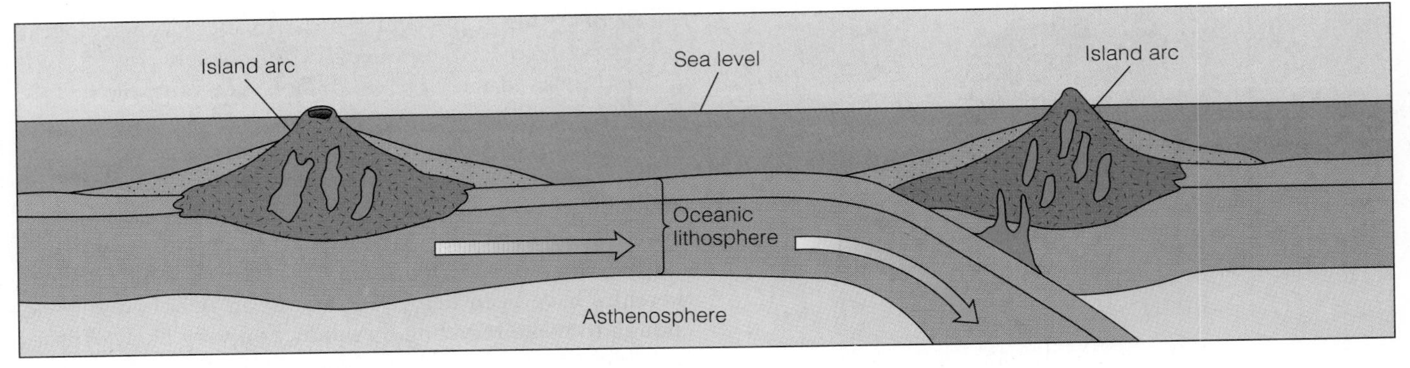

(a)

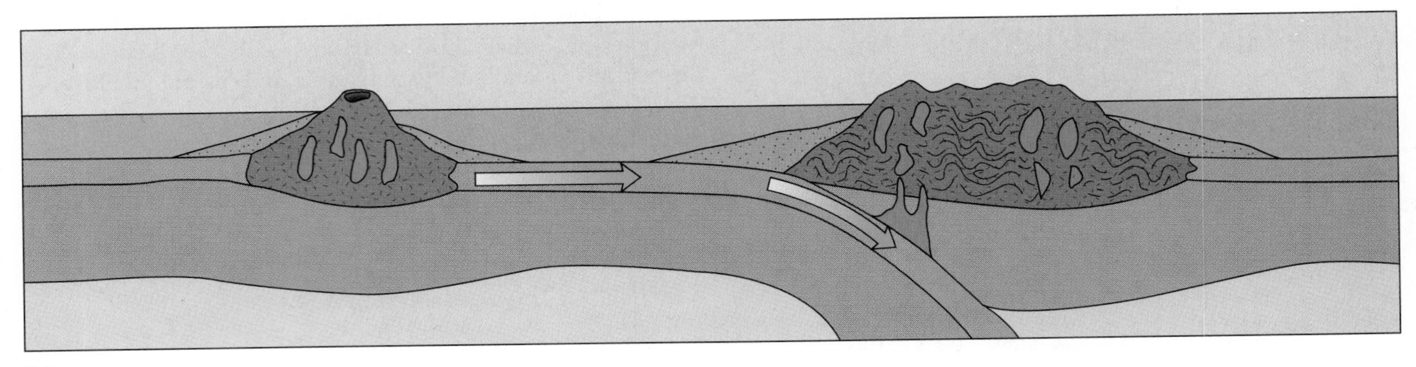

(b)

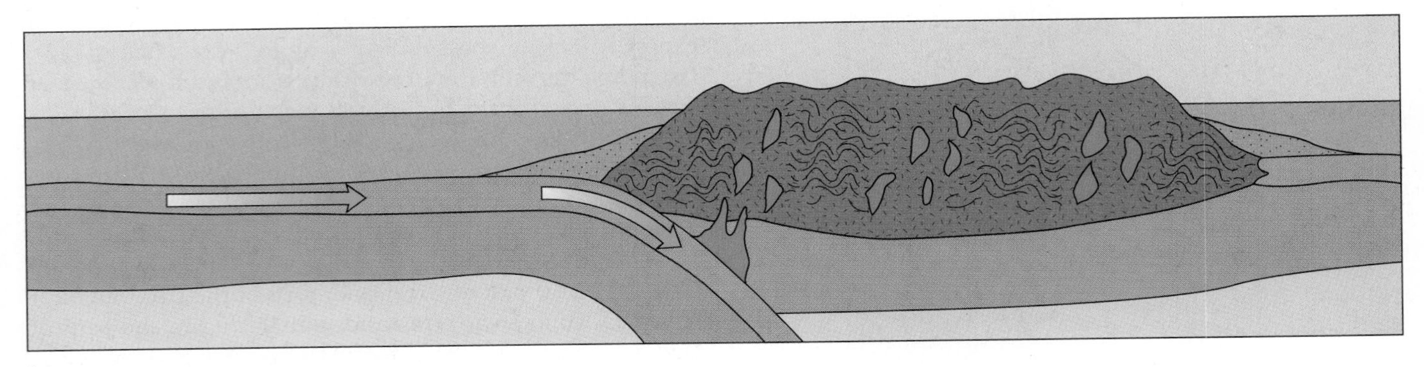

(c)

▶ **FIGURE 13-33** Three stages in the origin of granitic continental crust. Andesitic island arcs formed by the partial melting of basaltic oceanic crust are intruded by granitic magmas. As a result of plate movements, island arcs collide and form larger units or cratons. (*a*) Two island arcs on separate plates move toward one another. (*b*) The island arcs shown in (*a*) collide, forming a small craton, and another island arc approaches this craton. (*c*) The island arc shown in (*b*) collides with the craton.

were sutured along these orogenic belts so that by 1.8 billion years ago much of what is now Greenland, central Canada, and the north-central United States formed a large craton (Figure 13-35a).

Following this initial stage of North America's evolution, continental accretion occurred along the southern and eastern margins of the craton (Figure 13-35b and c). By the end of this time, about 1 billion years ago, the size and shape of North America were approximately as shown in Figure 13-35c. No further episode of continental accretion occurred until the Paleozoic Era.

Orogeny and accretion during the last 570 million years occurred mostly along the eastern, southern, and western margins of the craton, giving rise to the present configuration of North America. During this time, the craton itself has been remarkably stable. It has periodically been invaded by the seas during marine transgressions followed by regressions, but it has been only mildly deformed into a number of large basins and domes.

In the east and south, the Appalachian and Ouachita mountains formed during the Paleozoic Era in response to compression generated by the closure of an ocean basin

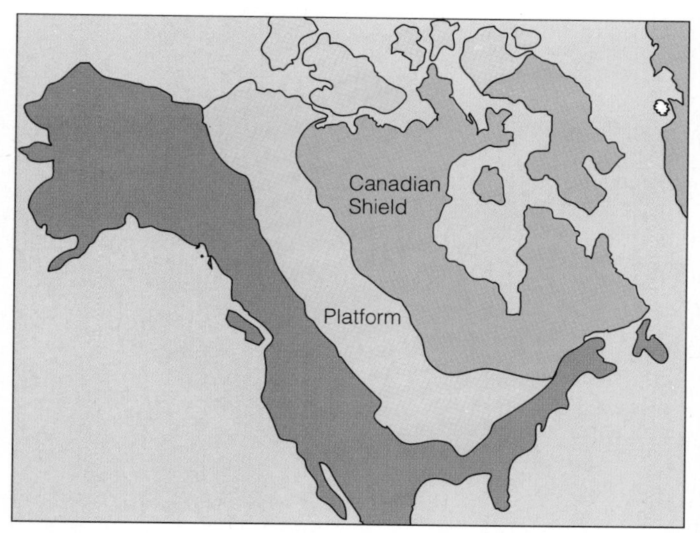

➢ **FIGURE 13-34 (above)** The North American craton. The Canadian Shield is a large area of exposed Precambrian-aged rocks. Extending from the shield are platforms of buried Precambrian rocks. The shield and platforms collectively make up the craton.

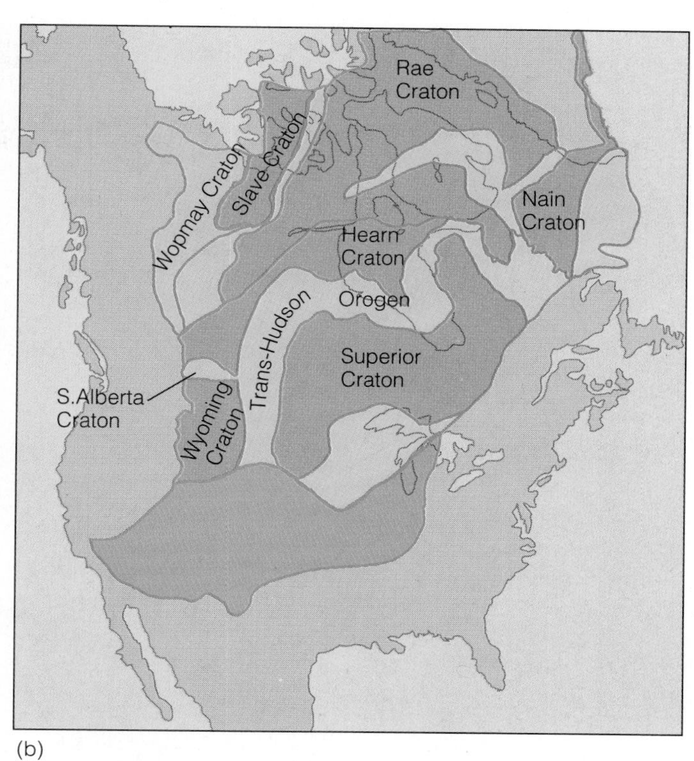

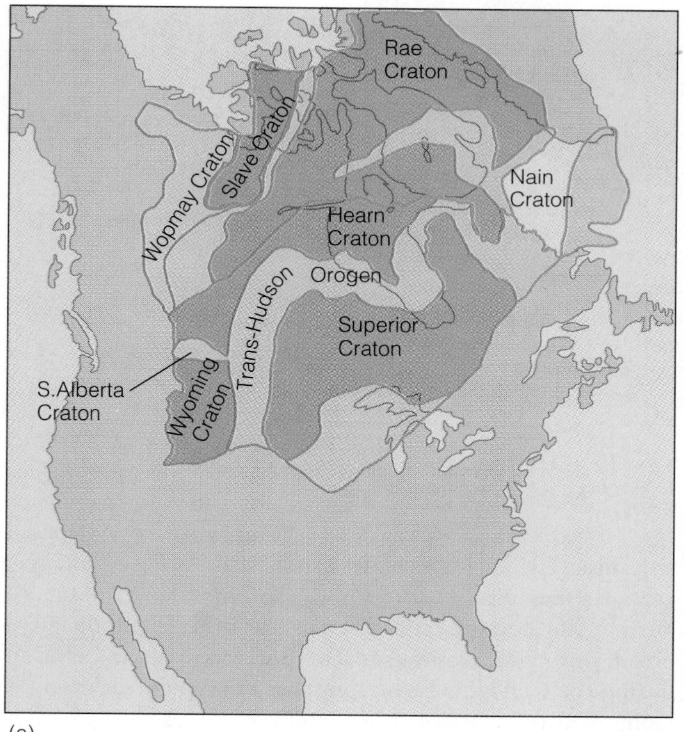

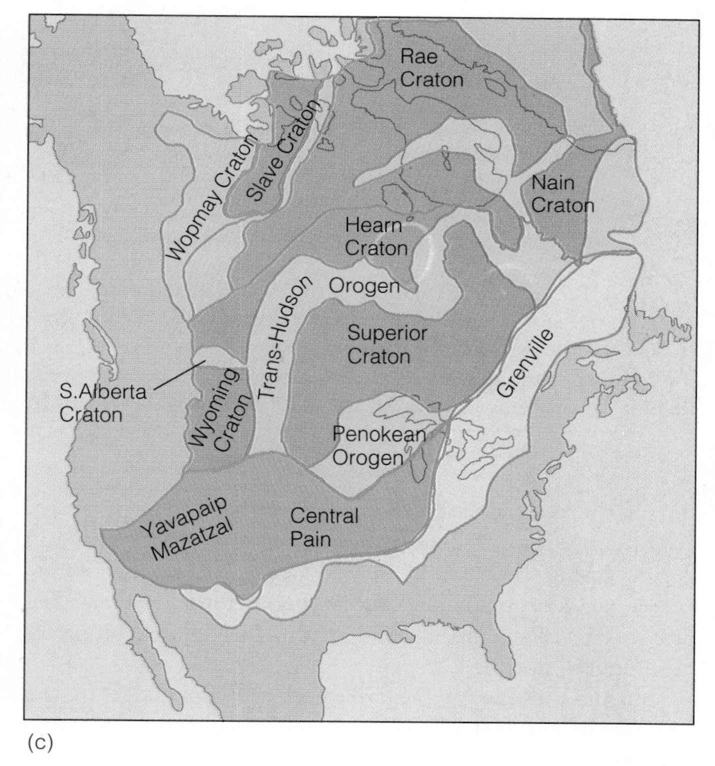

➢ **FIGURE 13-35** Three stages in the early evolution of the North American craton. (*a*) By about 1.8 billion years ago, North America consisted of the elements shown here. The various orogens formed when older cratons collided to form a larger craton. (*b*) and (*c*) Continental accretion along the southern and eastern margins of North America. By about 1 billion years ago, North America had the size and shape shown diagrammatically in (*c*).

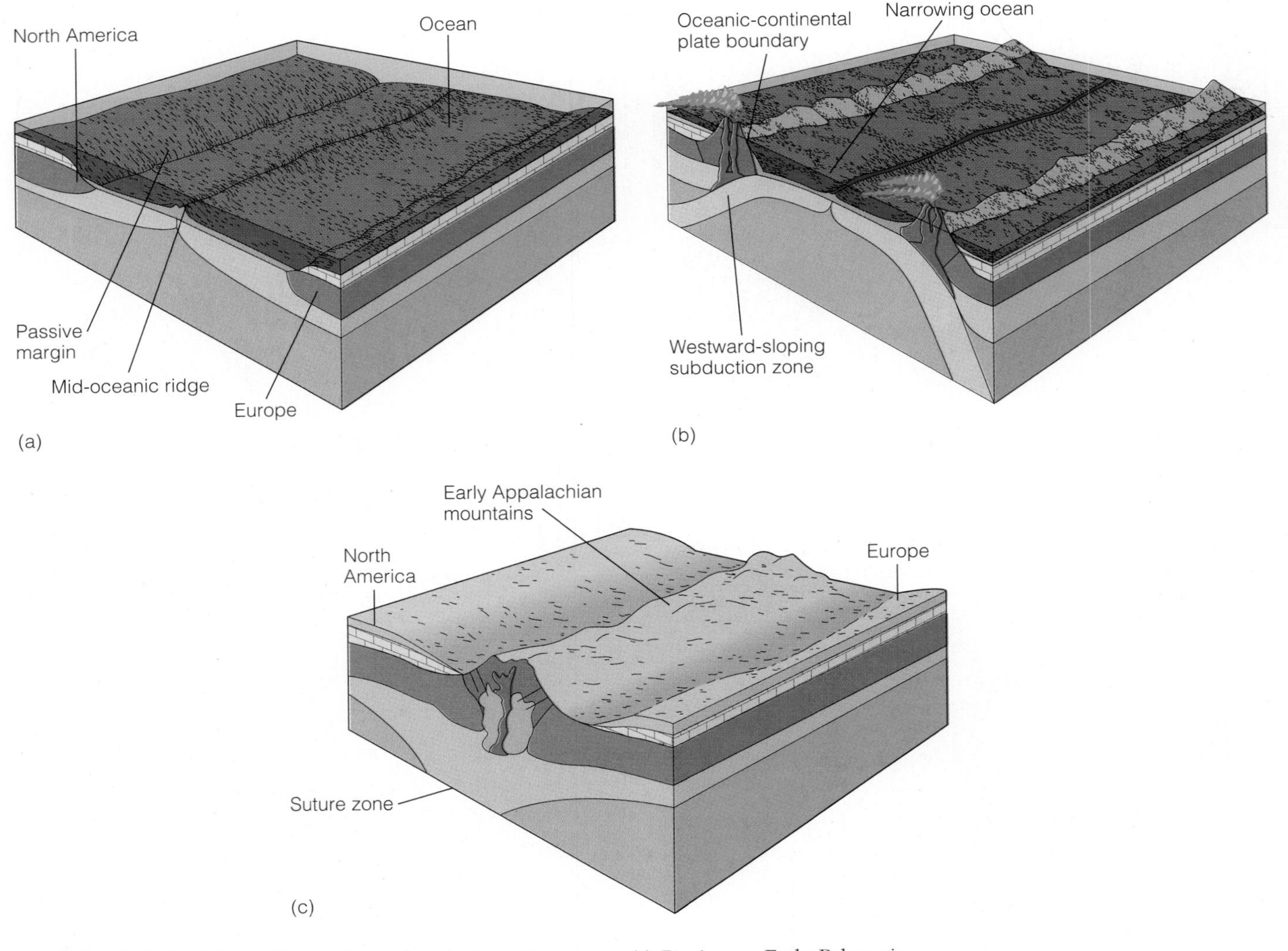

➤ **FIGURE 13-36** Evolution of the Appalachian Mountains. (*a*) During an Early Paleozoic stage, an ocean was opening up along a divergent boundary. Both the east coast of North America and the west coast of Europe were passive plate margins. (*b*) Beginning in the Ordovician Period, the North American and European passive margins became oceanic-continental plate boundaries, resulting in orogenic activity. (*c*) By the Late Paleozoic, North America and Europe collided.

during the amalgamation of Pangaea. What had been passive continental margins became active ones as the ocean basin began closing. By the end of the Paleozoic Era, a continent-continent collision occurred, causing deformation in what are now the Appalachian and Ouachita Mountains (➤ Figure 13-36).

During the Late Triassic Period, the first stage in the breakup of Pangaea began, with North America separating from Eurasia and North Africa. Along the east coast from Nova Scotia to North Carolina, block-faulting occurred and formed numerous ranges with intervening valleys much like those of the present-day Basin and Range Province of the western United States (Figure 13-26).

The valleys resulting from block-faulting filled with poorly sorted red-colored nonmarine detrital sediments, some of which are well known for dinosaur footprints (➤ Figure 13-37). Rifting was accompanied by widespread volcanism, which resulted in extensive lava flows and numerous dikes and sills. Erosion of the block-fault mountains during the Jurassic and Cretaceous periods produced a broad, low-lying erosion surface; renewed uplift and erosion during the Cenozoic Era account for the present-day topography of the Appalachian Mountains.

The North American Cordillera is a complex mountainous region in western North America, extending from Alaska into central Mexico (➤ Figure 13-38). It has a long, complex geologic history involving accretion of island arcs along the continental margin, orogeny at an oceanic-continental boundary, vast outpourings of basaltic lavas, and block-faulting. Although the Cordillera has a long history of

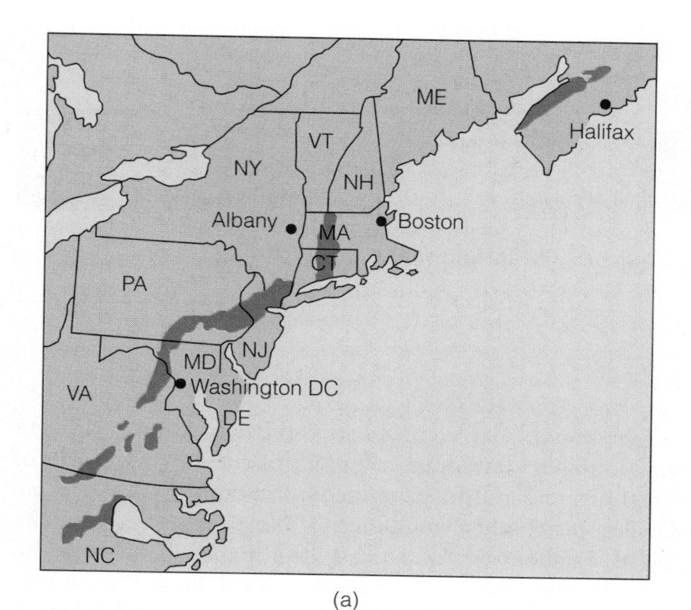

(a)

(b)

▶ **FIGURE 13-37** (*a*) Areas where Triassic block-fault basin deposits are exposed in eastern North America. (*b*) Reptile tracks in Triassic basin rocks were uncovered during the excavation for a new state building in Hartford, Connecticut. Because the tracks were so spectacular, the building site was moved, and the excavation was designated as a state park. (Photo courtesy of Dinosaur State Park.)

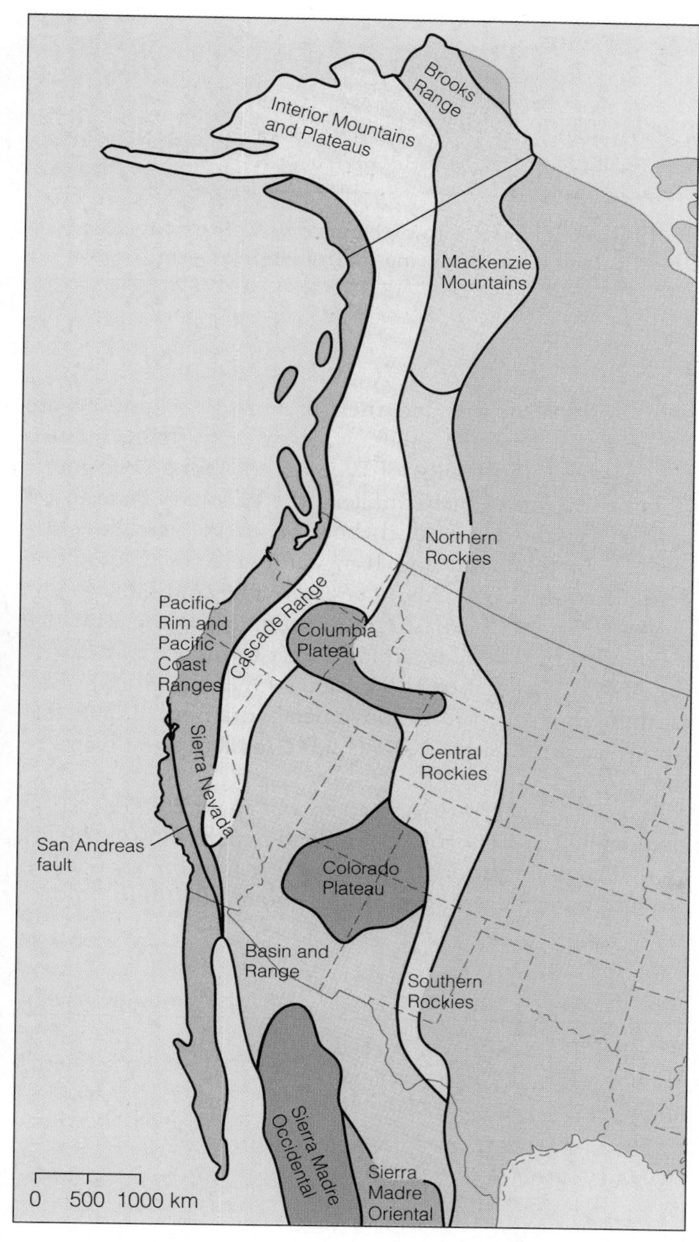

▶ **FIGURE 13-38** The North American Cordillera is a complex mountainous region extending from Alaska into central Mexico. It consists of a number of elements as shown here.

deformation, the most recent episode of large-scale deformation was the Laramide orogeny, which began 85 to 90 million years ago. Like many other orogenies, it occurred along an oceanic-continental boundary. The main Laramide orogeny was centered in the Rocky Mountains of present-day Colorado and Wyoming, but deformation occurred far to the north in Canada and Alaska and as far south as Mexico City. The Lewis overthrust of Montana resulted from Laramide compression (Figure 13-23), and in the Canadian Rockies, thrust sheets piled one upon another (▶ Figure 13-39).

The Laramide orogeny ceased about 40 million years ago, but since that time the Rocky Mountains have continued to evolve. The mountain ranges formed during the orogeny were eroded, and the valleys between ranges filled with sediments. Many of the ranges were nearly buried in their

▶ **FIGURE 13-39** Cross section of part of the Canadian Rocky Mountains. The folding and thrust faulting occurred during the Laramide orogeny.

own erosional debris, and their present-day elevations are the result of renewed uplift, which is continuing in such areas as the Teton Range of Wyoming (see the Prologue).

In other parts of the Cordillera, the Colorado Plateau was uplifted far above sea level, but the rocks were little deformed. In the Basin and Range Province, block-faulting began during the Middle Cenozoic and continues to the present (Figure 13-26). At its western edge, the province is bounded by a large escarpment that forms the east face of the Sierra Nevada (see the Chapter Opening photo). This escarpment resulted from movement on a normal fault that has elevated the Sierra Nevada 3,000 m above the basins to the east.

In the Pacific Northwest, an area of about 200,000 km^2 is covered by the Cenozoic Columbia River basalts (see Figure 4-18). Issuing from long fissures, these flows overlapped to produce an aggregate thickness of about 1,000 m.

The present-day elements of the Pacific coast section of the Cordillera developed as a result of the westward drift of North America, the partial consumption of the oceanic Farallon plate, and the collision of North America with the Pacific-Farallon ridge. During the Early Cenozoic, the entire Pacific coast was bounded by a subduction zone that stretched from Mexico to Alaska (▶ Figure 13-40). Most of the Farallon plate was consumed at this subduction zone, and now only two small remnants exist—the Juan de Fuca and Cocos plates (Figure 13-40). As discussed earlier, the continuing subduction of these small plates accounts for seismicity and volcanism in the Cascade Range of the Pacific Northwest and Central America, respectively. Westward drift of the North American plate also resulted in its collision with the Pacific-Farallon ridge and the origin of the Queen Charlotte and San Andreas transform faults (Figure 13-40).

▶ **FIGURE 13-40** (*a*), (*b*), and (*c*) Three stages in the westward drift of North America and its collision with the Pacific-Farallon ridge. As the North American plate overrode the ridge, its margin became bounded by transform faults rather than a subduction zone.

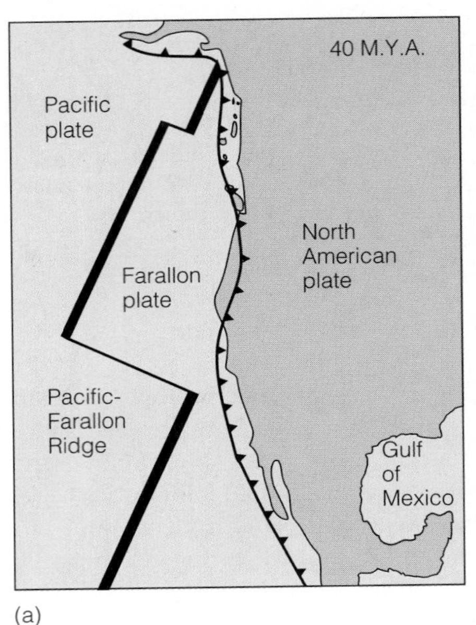

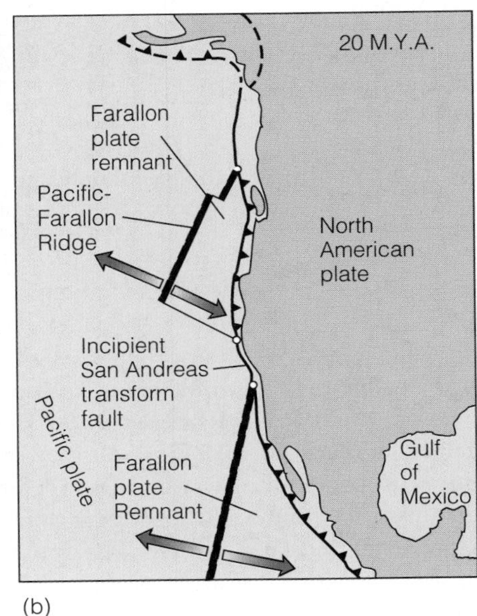

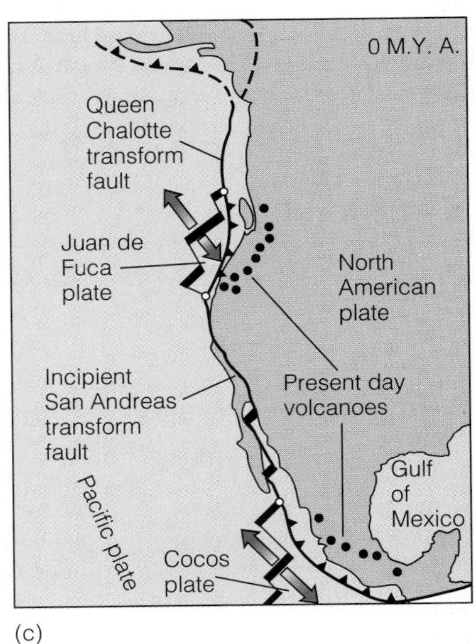

Chapter Summary

1. Contorted and fractured rocks have been deformed or strained by applied stresses.
2. Stresses are characterized as compression, tension, or shear. Elastic strain is not permanent, meaning that when the stress is removed, the rocks return to their original shape or volume. Plastic strain and fracture are both permanent types of deformation.
3. The orientation of deformed layers of rock is described by strike and dip.
4. Rock layers that have been buckled into up- and down-arched folds are anticlines and synclines, respectively. They can be identified by the strike and dip of the folded rocks and by the relative age of the rocks in the center of eroded folds.
5. Domes and basins are the circular to oval equivalents of anticlines and synclines, but are commonly much larger structures.
6. Two types of structures resulting from fracturing are recognized: joints are fractures along which the only movement, if any, is perpendicular to the fracture surface, and faults are fractures along which the blocks on opposite sides of the fracture move parallel to the fracture surface.
7. Joints, which are the commonest geologic structures, form in response to compression, tension, and shear.
8. On dip-slip faults, all movement is in the dip direction of the fault plane. Two varieties of dip-slip faults are recognized: normal faults form in response to tension, while reverse faults are caused by compression.
9. Strike-slip faults are those on which all movement is in the direction of strike of the fault plane. They are characterized as right-lateral or left-lateral depending on the apparent direction of offset of one block relative to the other.
10. Some faults show components of both dip-slip and strike-slip; they are called oblique-slip faults.
11. Mountains can form in a variety of ways, some of which involve little or no folding or faulting. Mountain systems consisting of several mountain ranges result from deformation related to plate movements.
12. Most orogenies occur where plates converge and one plate is subducted beneath another or where two continental plates collide.
13. A volcanic island arc, deformation, igneous activity, and metamorphism characterize orogenies occurring at oceanic-oceanic plate boundaries. Subduction of oceanic lithosphere at an oceanic-continental plate boundary also results in orogeny.
14. Some mountain systems, such as the Himalayas, are within continents far from a present-day plate boundary. Such mountains formed when two continental plates collided and became sutured.
15. Geologists now realize that mountain building also occurs when microplates collide with continents.
16. A craton is the stable core of a continent. Broad areas in which the cratons of continents are exposed are called shields; each continent has at least one shield area.
17. Cratons formed as a result of accretion, a process involving the addition of eroded continental material, igneous rocks, and island arcs to the margin of a craton during orogenesis.
18. In North America, the craton evolved during Precambrian time by collisions of smaller cratons along belts of deformation known as orogens and by accretion along its southern and eastern margins. Since then, orogenies have resulted in continental accretion along the craton's margins.

Important Terms

anticline
basin
compression
craton
deformation
dip
dip-slip fault
dome
elastic strain
fault
fault plane
footwall block
fracture
hanging wall block
joint
microplate
monocline
normal fault
oblique-slip fault
plastic strain
plunging fold
reverse fault
shear stress
shield
strain
stress
strike
strike-slip fault
syncline
tension
thrust fault

REVIEW QUESTIONS

1. Rocks that show a large amount of plastic strain are said to be:
   a. ____ brittle;
   b. ____ fractured;
   c. ____ ductile;
   d. ____ sheared;
   e. ____ all of these.
2. Most folding results from:
   a. ____ fracturing;
   b. ____ compaction;
   c. ____ rifting;
   d. ____ convection;
   e. ____ compression.
3. An elongate fold in which all the rock layers dip in toward the center is a(n):
   a. ____ dome;
   b. ____ monocline;
   c. ____ basin;
   d. ____ syncline;
   e. ____ anticline.
4. An overturned fold is one in which:
   a. ____ both limbs dip in the same direction;
   b. ____ the axial plain is vertical;
   c. ____ the axis is inclined;
   d. ____ the rock layers in one limb are horizontal;
   e. ____ the rocks are faulted as well as folded.
5. An oval to circular fold with all rocks dipping outward from a central point is a(n):
   a. ____ plunging anticline;
   b. ____ dome;
   c. ____ overturned syncline;
   d. ____ recumbent syncline;
   e. ____ basin.
6. A fault on which the hanging wall block appears to have moved down relative to the footwall block is a ________ fault.
   a. ____ thrust;
   b. ____ strike-slip;
   c. ____ normal;
   d. ____ reverse;
   e. ____ joint.
7. Strike-slip faults:
   a. ____ are low-angle reverse faults;
   b. ____ have mainly vertical displacement;
   c. ____ have mainly horizontal movement;
   d. ____ are faults on which no movement has yet occurred;
   e. ____ are characterized by uplift of the footwall block.
8. Solids that have been deformed by movement along closely spaced slippage planes are said to have been:
   a. ____ sheared;
   b. ____ folded;
   c. ____ subjected to tension;
   d. ____ elastically strained;
   e. ____ overturned.
9. Fractures along which no movement has occurred are:
   a. ____ joints;
   b. ____ monoclines;
   c. ____ transform faults;
   d. ____ axial planes;
   e. ____ fold limbs.
10. The intersection of an inclined plane with a horizontal plane is the definition of:
    a. ____ horizontal rock layers;
    b. ____ dip-slip movement;
    c. ____ folded rock layers;
    d. ____ strike;
    e. ____ joint.
11. In mountain systems that form at continental margins:
    a. ____ the Earth's crust is thicker than average;
    b. ____ most deformation is caused by tensional stresses;
    c. ____ little or no volcanic activity occurs;
    d. ____ stretching and thinning of the continental crust occur;
    e. ____ most deformation results from rifting.
12. Sediments deposited in an oceanic trench and then deformed and scraped off against the landward side of the trench during an orogeny form a(n):
    a. ____ divergent margin complex;
    b. ____ accretionary wedge;
    c. ____ back-arc basin deposit;
    d. ____ island arc system;
    e. ____ orogenic continental margin complex.
13. An excellent example of a mountain system now forming as a result of a continent-continent collision is the:
    a. ____ Andes;
    b. ____ Rocky Mountains;
    c. ____ Himalayas;
    d. ____ Alps;
    e. ____ Appalachians.
14. What types of evidence indicate that stress remains active within the Earth?
15. How do compression, tension, and shear differ from one another?
16. What is meant by the elastic limit of rocks?
17. Explain how rock type, time, temperature, and pressure influence the type of strain in rocks.
18. Draw a simple geologic map showing a plunging anticline and an adjacent plunging syncline. Assume that these folds plunge to the east.
19. Draw a simple cross section showing the displacement on a normal fault.
20. Draw a simple sketch map showing the displacement on a left-lateral strike-slip fault.
21. Discuss two ways in which mountains can form with little or no folding and faulting.
22. Cite two examples of mountain systems in which mountain-building processes remain active.
23. How do geologists account for mountain systems within continents, such as the Urals in Russia?
24. Explain how continents "grow" by accretion.
25. What is the difference between a reverse fault and a thrust fault?

Points to Ponder

1. Two recumbent folds are shown in Figure 13-14c. What kind of evidence would you need to determine which of these folds is an anticline and which is a syncline?
2. During the Paleozoic Era, eastern North America experienced considerable deformation whereas during the Mesozoic and Cenozoic eras deformation has occurred mostly in the western part of the continent. What accounts for this shifting pattern of deformation?
3. Over 5 million years, rocks are displaced 6,000 m along a normal fault. What was the average yearly movement on this fault? Is this average likely to represent the actual rate of displacement on this fault? Explain.
4. How is it possible for the same kind of rock to behave both elastically and plastically?
5. What kinds of evidence would indicate that mountain building occurred in the Canadian Shield where no mountains are now present?

Additional Readings

Davis, G. H. 1984. *Structural geology of rocks and regions*. New York: John Wiley & Sons.

Dennis, J. G. 1987. *Structural geology: An introduction*. Dubuque, Iowa: Wm. C. Brown.

Hatcher, R. D., Jr. 1990. *Structural geology: Principles, concepts, and problems*. Columbus, Ohio: Merrill Publishing Co.

Howell, D. G. 1985. Terranes. *Scientific American* 253, no. 5: 116–25.

____ . 1989. *Tectonics of suspect terranes: Mountain building and continental growth*. London: Chapman and Hall.

Jones, D. L., A. Cox, P. Coney, and M. Beck. 1982. The growth of western North America. *Scientific American* 247, no. 5: 70–84.

Lisle, R. J. 1988. *Geological structures and maps: A practical guide*. New York: Pergamon Press.

Miyashiro, A., K. Aki, and A. M. C. Segnor. 1982. *Orogeny*. New York: John Wiley & Sons.

Molnar, P. 1986. The geologic history and structure of the Himalaya. *American Scientist* 74, no. 2: 144–54.

____ . 1986. The structure of mountain ranges. *Scientific American* 255, no. 1: 70–79.

Spencer, E. W. 1988. *Introduction to the structure of the Earth*. New York: McGraw-Hill Book Company.

Chapter 14

MASS WASTING

OUTLINE

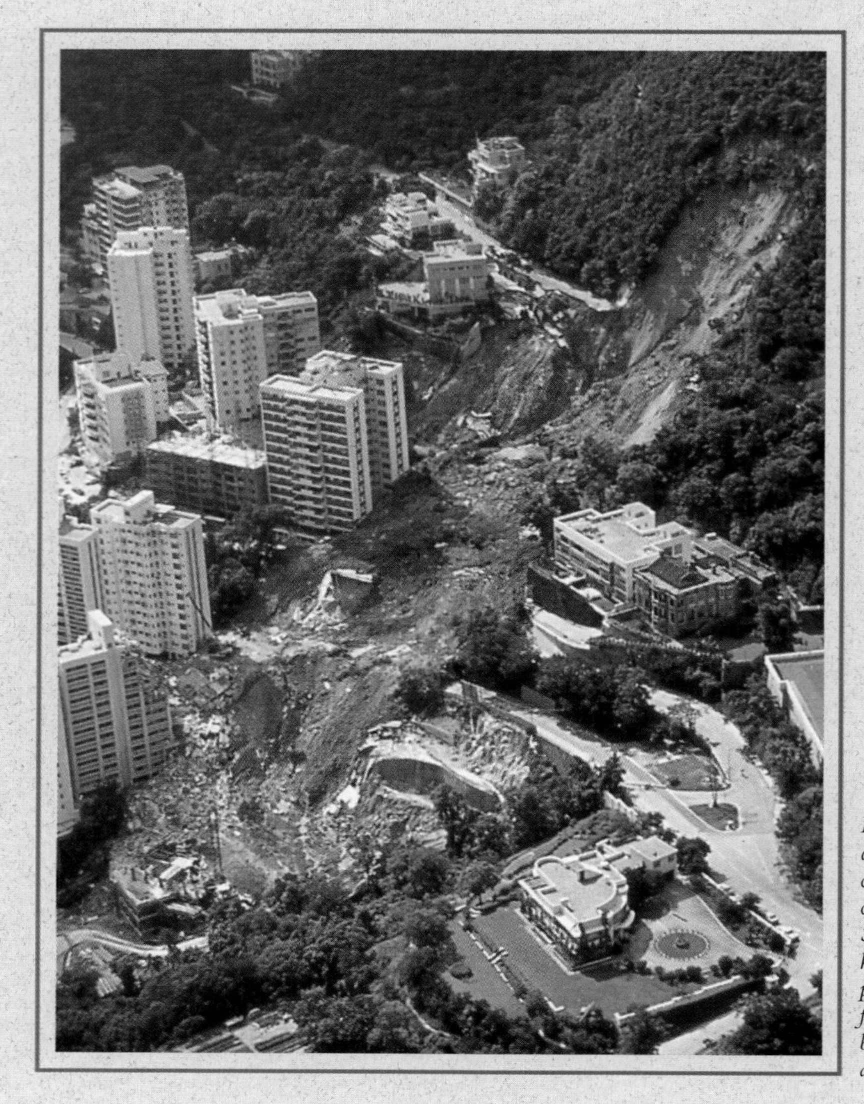

Hong Kong's most destructive landslide occurred on Po Shan road on June 18, 1972. Sixty-seven people were killed when a 68 m wide portion of this steep hillside failed, destroying a 4-story building and a 13-story apartment block.

Prologue

On May 31, 1970, a devastating earthquake occurred about 25 km west of Chimbote, Peru. High in the Peruvian Andes, about 65 km to the east, the violent shaking from the earthquake tore loose a huge block of snow, ice, and rock from the north peak of Nevado Huascarán (6,654 m), setting in motion one of this century's worst landslides. Free-falling for about 1,000 m, this block of material smashed to the ground, displacing thousands of tons of rock and generating a gigantic debris avalanche (▶ Figure 14-1). Hurtling down the mountain's steep glacial valley at speeds up to 320 km per hour, the avalanche, consisting of more than 50 millon m^3 of mud, rock, and water, flowed over ridges 140 m high, obliterating everything in its path.

About 3 km east of the town of Yungay, part of the debris flow overrode the valley walls and within seconds buried Yungay, instantly killing more than 20,000 of its residents (Figure 14-1). The main mass of the flow continued down the valley, overwhelming the town of Ranrahirca and several other villages, burying about 5,000 more people. When the flow reached the bottom of the valley, its momentum carried it across the Rio Santa and some 60 m up the opposite bank. In a span of roughly four minutes from the time of the initial ground shaking, approximately 25,000 people died, and most of the area's transportation, power, and communication networks were destroyed.

Ironically, the only part of Yungay that was not buried was Cemetery Hill, where 92 people survived by running to its top (▶ Figure 14-2). A Peruvian geophysicist who was giving a French couple a tour of Yungay provided a vivid eyewitness account of the disaster:

> As we drove past the cemetery the car began to shake. It was not until I had stopped the car that I realized that we were experiencing an earthquake. We immediately got out of the car and observed the effects of the earthquake around us. I saw several homes as well as a small bridge crossing a creek near Cemetery Hill collapse. It was, I suppose, after about one-half to three-quarters of a minute when the earthquake shaking began to subside. At that time I heard a great roar coming from Huascarán. Looking up, I saw what appeared to be a cloud of dust and it looked as though a large mass of rock and ice was breaking loose from the north peak. My immediate reaction was to run for the high ground of Cemetery Hill, situated about 150 to 200 m away. I began running and noticed that there were many others in Yungay who were also running toward Cemetery Hill. About half to three-quarters of the way up the hill, the wife of my friend stumbled and fell and I turned to help her back to her feet.
>
> The crest of the wave had a curl, like a huge breaker coming in from the ocean. I estimated the wave to be at least 80 m high. I observed hundreds of people in Yungay running in all directions and many of them toward Cemetery Hill. All the while, there was a continuous loud roar and rumble. I reached the upper level of the cemetery near the top just as the debris flow struck the base of the hill and I was probably only 10 seconds ahead of it.
>
> At about the same time, I saw a man just a few meters down hill who was carrying two small children toward the hilltop. The debris flow caught him and he threw the two children toward the hilltop, out of the path of the flow, to safety, although the debris flow swept him down the valley, never to

▶ FIGURE 14-1 An earthquake 65 km away triggered a landslide on Nevado Huascarán, Peru, that destroyed the towns of Yungay and Ranrahirca and killed more than 25,000 people.

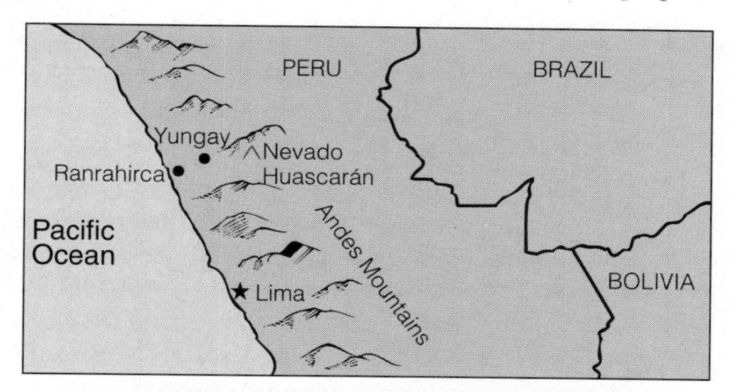

▶ **FIGURE 14-2** Cemetery Hill was the only part of Yungay to escape the 1970 landslide that destroyed the rest of the town. Only 92 people survived the destruction by running to the top of the hill.

be seen again. I also remember two women who were no more than a few meters behind me and I never did see them again. Looking around, I counted 92 persons who had also saved themselves by running to the top of the hill. It was the most horrible thing I have ever experienced and I will never forget it.*

As tragic and devastating as this debris avalanche was, it was not the first time a destructive landslide had swept down this valley. In January 1962, another large chunk of snow, ice, and rock broke off from the main glacier and generated a large debris avalanche that buried several villages and killed about 4,000 people.

 INTRODUCTION

Geologists use the term *landslide* in a general sense to cover a wide variety of mass movements that may cause loss of life, property damage, or a general disruption of human activities. In 218 B.C., avalanches in the European Alps buried 18,000 people; an earthquake-generated landslide in Hsian, China, killed an estimated 1,000,000 people in 1556; and 7,000 people died when mudflows and avalanches destroyed Huaraz, Peru, in 1941. What makes these mass movements so terrifying, and yet so fascinating, is that they almost always occur with little or no warning and are over in a very short time, leaving behind a legacy of death and destruction (◉ Table 14-1).

Every year about 25 people are killed by landslides in the United States alone, while the total annual cost of damages from them exceeds $1 billion. Almost all of the major landslides have natural causes, yet many of the smaller ones are the result of human activity and could have been prevented or their damage minimized.

Mass wasting (also called *mass movement*) is defined as the downslope movement of material under the direct influence of gravity. Most types of mass wasting are aided by weathering and usually involve surficial material. The material moves at rates ranging from almost imperceptible, as in the case of creep, to extremely fast as in a rockfall or slide. Though water can play an important role, the relentless pull of gravity is the major force behind mass wasting.

Mass wasting is an important geologic process that can occur at any time and almost any place. Although most people associate mass wasting with steep and unstable slopes, it can also occur on near-level land, given the right geologic conditions. Furthermore, while the rapid types of mass wasting, such as avalanches and mudflows, typically get the most publicity, the slow, imperceptible types, such as creep, usually do the greatest amount of property damage.

 FACTORS INFLUENCING MASS WASTING

When the gravitational force acting on a slope exceeds its resisting force, slope failure (mass wasting) occurs. The resisting forces helping to maintain slope stability include the slope material's strength and cohesion, the amount of internal friction between grains, and any external support of the

*B. A. Bolt et al., *Geological Hazards* (New York: Springer-Verlag 1977), pp. 37–39.

TABLE 14-1 Selected Landslides, Their Cause, and the Number of People Killed

Date	Location	Type	Deaths
218 B.C.	Alps (European)	Avalanche—destroyed Hannibal's army	18,000
1512	Alps (Biasco)	Landslide—temporary lake burst	>600
1556	China (Hsian)	Landslides—earthquake triggered	1,000,000
1689	Austria (Montaton Valley)	Avalanche	>300
1806	Switzerland (Goldau)	Rock glide	457
1881	Switzerland (Elm)	Rockfall	115
1892	France (Haute-Savoie)	Icefall, mudflow	150
1903	Canada (Frank, Alberta)	Rock glide	70
1920	China (Kansu)	Landslides—earthquake triggered	~200,000
1936	Norway (Loen)	Rockfall into fiord	73
1941	Peru (Huaraz)	Avalanche and mudflow	7,000
1959	USA (Madison Canyon, Montana)	Landslide—earthquake triggered	26
1962	Peru (Mt. Huascarán)	Ice avalanche and mudflow	~4,000
1963	Italy (Vaiont Dam)	Landslide—subsequent flood	~2,000
1966	Brazil (Rio de Janeiro)	Landslides	279
1966	United Kingdom (Aberfan, South Wales)	Debris flow—collapse of mining waste tip	144
1970	Peru (Mt. Huascarán)	Rockfall and debris avalanche—earthquake triggered	25,000
1971	Canada (St. Jean-Vianney, Quebec)	Quick clays	31
1972	USA (West Virginia)	Landslide and mudflow—collapse of mining waste tip	400
1974	Peru (Mayunmarca)	Rock glide and debris flow	430
1978	Japan (Myoko Kogen Machi)	Mudflow	12
1979	Indonesia (Sumatra)	Landslide	80
1980	USA (Washington)	Avalanche and mudflow	63
1981	Indonesia (West Irian)	Landslide—earthquake triggered	261
1981	Indonesia (Java)	Mudflow	252
1983	Iran (Northern area)	Landslide and avalanche	90
1987	El Salvador (San Salvador)	Landslide	1,000
1988	Chile (Tupungatito area)	Mudflow	41
1989	Tadzhikistan	Mudflow—earthquake triggered	274
1989	Indonesia (West Irian)	Landslide—earthquake triggered	120
1991	Guatemala (Santa Maria)	Landslide	33
1994	Colombia (Paez River Valley)	Avalanche—earthquake triggered	>300

SOURCE: Data from J. Whittow, *Disasters: The Anatomy of Environmental Hazards* (Athens, Ga.: University of Georgia Press, 1979) and *Geotimes*.

slope (▶ Figure 14-3). These factors collectively define a slope's **shear strength**.

Opposing a slope's shear strength is the force of gravity. Gravity operates vertically but has a component acting parallel to the slope, thereby causing instability (Figure 14-3). The greater a slope's angle, the greater the component of force acting parallel to the slope, and the greater the chance for mass wasting. The steepest angle that a slope can maintain without collapsing is its *angle of repose*. At this angle, the shear strength of the slope's material exactly counterbalances the force of gravity. For unconsolidated material, the angle of repose normally ranges from 25° to 40°. Slopes steeper than 40° usually consist of unweathered solid rock.

All slopes are in a state of dynamic equilibrium, which means that they are constantly adjusting to new conditions. While we tend to view mass wasting as a disruptive and usually destructive event, it is one of the ways that a slope adjusts to new conditions. Whenever a building or road is constructed on a hillside, the equilibrium of that slope is affected. The slope must then adjust, perhaps by mass wasting, to this new set of conditions.

Many factors can cause mass wasting: a change in slope gradient, weakening of material by weathering, increased water content, changes in the vegetation cover, and overloading. Although most of these are interrelated, we will examine them separately for ease of discussion, but will also show how they individually and collectively affect a slope's equilibrium.

Slope Gradient

Slope gradient is probably the major cause of mass wasting. Generally speaking, the steeper the slope, the less stable it is. Therefore, steep slopes are more likely to experience mass wasting than gentle ones.

A number of processes can oversteepen a slope. One of the most common is undercutting by stream or wave action

▶ FIGURE 14-3 A slope's shear strength depends on the slope material's strength and cohesiveness, the amount of internal friction between grains, and any external support of the slope. These factors promote slope stability. The force of gravity operates vertically but has a component acting parallel to the slope. When this force, which promotes instability, exceeds a slope's shear strength, slope failure occurs.

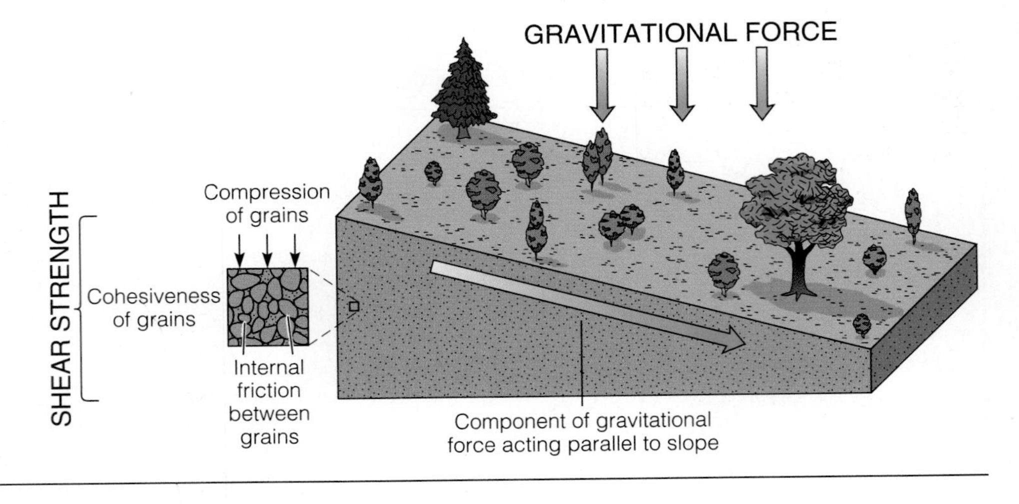

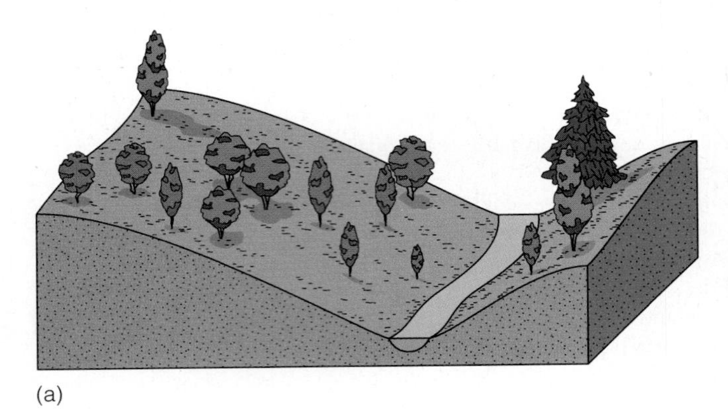

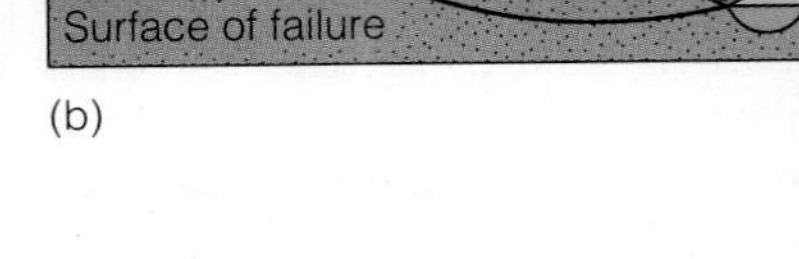

▶ FIGURE 14-4 Undercutting by stream erosion (*a*) removes a slope's base, which increases the slope angle and (*b*) can lead to slope failure and slumping. (*c*) Undercutting by stream erosion caused slumping along this stream near Weidman, Michigan.

(▶ Figure 14-4). This removes the slope's base, increases the slope angle, and thereby increases the gravitational force acting parallel to the slope. Wave action, especially during storms, often results in mass movements along the shores of oceans or large lakes.

Excavations for road cuts and hillside building sites are another major cause of slope failure (▶ Figure 14-5). Grading the slope too steeply, or cutting into its side, increases the stress in the rock or soil until it is no longer strong enough to remain at the steeper angle and mass movement ensues. Such action is analogous to undercutting by streams or waves and has the same result, thus explaining why so many mountain roads are plagued by frequent mass movements. Fortunately, many of the slope failures associ-

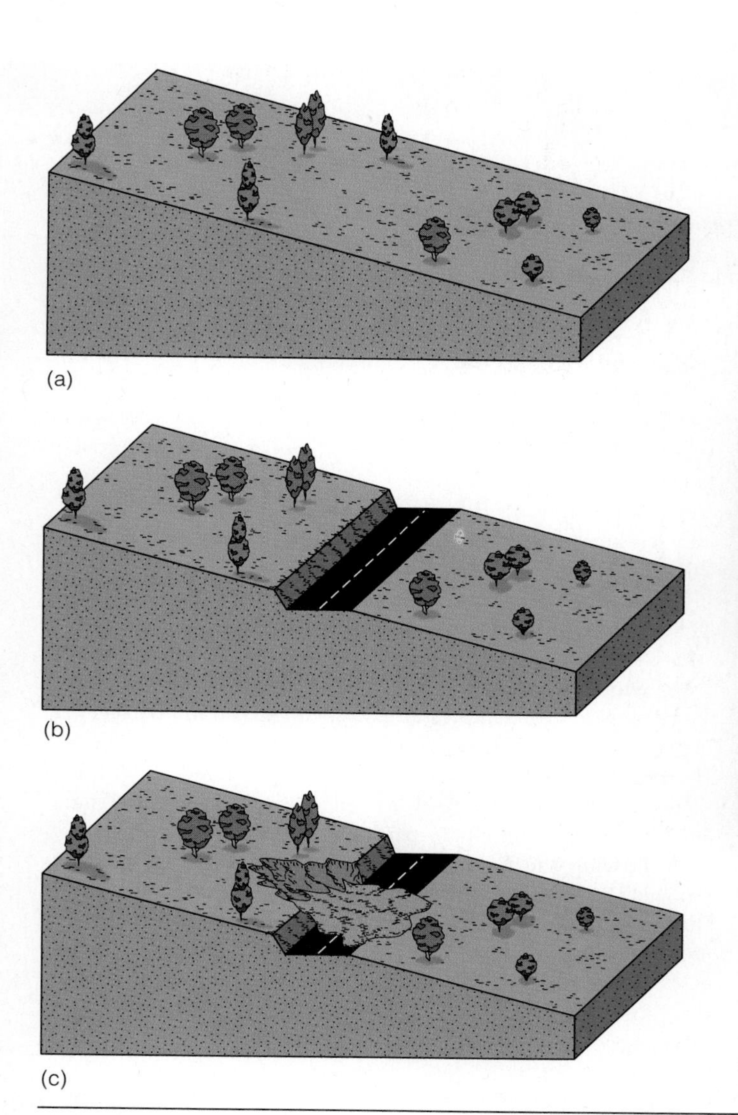

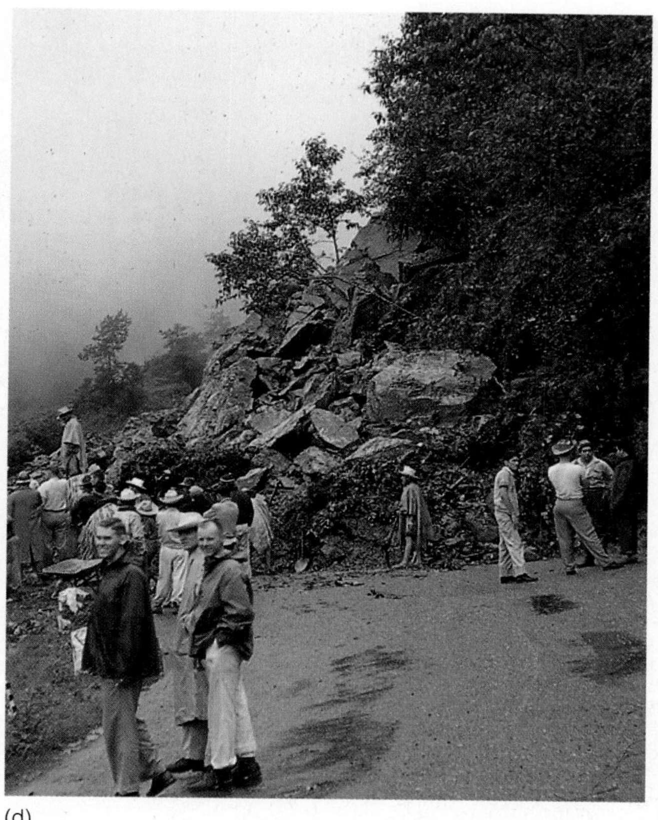

▶ FIGURE 14-5 (*a*) Highway excavations disturb the equilibrium of a slope by (*b*) removing a portion of its support as well as oversteepening it at the point of excavation. (*c*) Such action can result in frequent landslides. (*d*) Cutting into the hillside to construct this portion of the Pan American Highway in Mexico resulted in a rockfall that completely blocked the road. (Photo courtesy of R. V. Dietrich.)

ated with hillside road cuts and building construction can be avoided or greatly minimized by better understanding of the factors involved.

Weathering and Climate

Mass wasting is more likely to occur in loose or poorly consolidated slope material than in solid bedrock. As soon as solid rock is exposed at the Earth's surface, weathering begins to disintegrate and decompose it, thereby reducing its shear strength and increasing its susceptibility to mass wasting. The deeper the weathering zone extends, the greater the likelihood of some type of mass movement.

Recall from Chapter 5 that some rocks are more susceptible to weathering than others and that climate plays an important role in the rate and type of weathering. In the tropics, where temperatures are high and considerable rainfall occurs, the effects of weathering extend to depths of several tens of meters, and rapid mass movements most commonly occur in the deep weathering zone. In arid and semiarid regions, the weathering zone is usually considerably shallower. Nevertheless, localized and intense cloudbursts can drop large quantities of water on an area in a short time. With little vegetation to absorb this water, runoff is rapid and frequently results in mudflows. In high mountains, rockfalls are common because of frost action (see Chapter 5).

Water Content

The amount of water in rock or soil influences slope stability. Large quantities of water from melting snow or heavy storms greatly increase the likelihood of slope failure. The additional weight that water adds to a slope can be enough to cause mass movement. Furthermore, water percolating through a slope's material helps to decrease friction between grains, contributing to a loss of cohesion. For example, slopes composed of dry clay are usually quite stable, but when wetted, they quickly lose cohesiveness and internal friction and become an unstable slurry. This occurs because

Perspective 14-1

THE TRAGEDY AT ABERFAN, WALES

The debris brought out of underground coal mines in southern Wales typically consists of a wet mixture of various sedimentary rock fragments. This material is usually dumped along the nearest valley slope where it builds up into large waste piles called tips. A tip is fairly stable as long as the material composing it is relatively dry and its sides are not too steep.

Between 1918 and 1966, seven large tips composed of mine debris had been built at various elevations on the valley slopes above the small coal-mining village of Aberfan. Shortly after 9:00 A.M. on October 21, 1966, the 250 m high, rain-soaked Tip No. 7 collapsed, and a black sludge flowed down the valley with the roar of a loud train (▶ Figure 1). Before it came to a halt 800 m from its starting place, the flow had destroyed two farm cottages, crossed a canal, and buried Pantglas Junior School, suffocating virtually all the children of Aberfan. A total of 144 people died in the flow, among them 116 children who had gathered for morning assembly in the school.

After the disaster, everyone asked, "Why did this tragedy occur and could it have been prevented?" The subsequent investigation revealed that no stability studies had ever been made on the tips and that repeated warnings about potential failure of the tips, as well as previous slides, had all been ignored.

In 1939, 8 km to the south, a tip constructed under conditions almost identical to those of Tip No. 7 collapsed. Luckily, no one was injured, but unfortunately, the failure was soon forgotten and the Aberfan tips continued to grow. In 1944 Tip No. 4 failed, and again no one was injured.

In 1958 Tip No. 7 was sited solely on the basis of available space, with no regard to the area's geology. Although springs and seeps were known to be a potential source of instability, the tip was still located over a spring. In spite of previous tip failures and warnings of slope failure by tip workers and others, mine debris was being piled onto Tip No. 7 until the day of the disaster.

What exactly caused Tip No. 7 and the others to fail? The official investigation revealed that the foundation of the tips had become saturated with water from the springs over which they were built. In the case of the collapsed tips, pore pressure from the water exceeded the friction between grains, and the entire mass liquefied like a "quicksand." Behaving as a liquid, the mass quickly moved downhill spreading out laterally. As it flowed, water escaped from the mass, and the sedimentary particles regained their cohesion.

Following the inquiry, it was recommended that a National Tip Safety Committee be established to assess the dangers of existing tips and advise on the construction of new tip sites.

clay, which can hold large quantities of water, consists of platy particles that easily slide over each other when wet. For this reason, clay beds are frequently the slippery layer along which overlying rock units slide downslope (see Perspective 14-1).

Vegetation

Vegetation affects slope stability in several ways. By absorbing the water from a rainstorm, vegetation decreases water saturation of a slope's material and the resultant loss of shear strength that frequently leads to mass wasting. In addition, the root system of vegetation helps to stabilize a slope by binding soil particles together and holding the soil to bedrock.

The removal of vegetation by either natural or human activity is a major cause of many mass movements. Summer brush and forest fires in southern California, for example, frequently leave the hillsides bare of vegetation. Fall rainstorms saturate the ground causing mudslides that do tremendous damage and cost millions of dollars to clean up (▶ Figure 14-6). The soils of many hillsides in New Zealand are sliding because deep-rooted native bushes have been replaced by shallow-rooted grasses used for sheep grazing. When heavy rains saturate the soil, the shallow-rooted grasses cannot hold the slope in place, and parts of it slide downhill.

Overloading

Overloading is almost always the result of human activity and typically results from dumping, filling, or piling up of material. Under natural conditions, a material's load is carried by its grain-to-grain contacts, and a slope is thus

▶ **FIGURE 1** Location map and aerial view of the Aberfan tip disaster in which 144 people died.

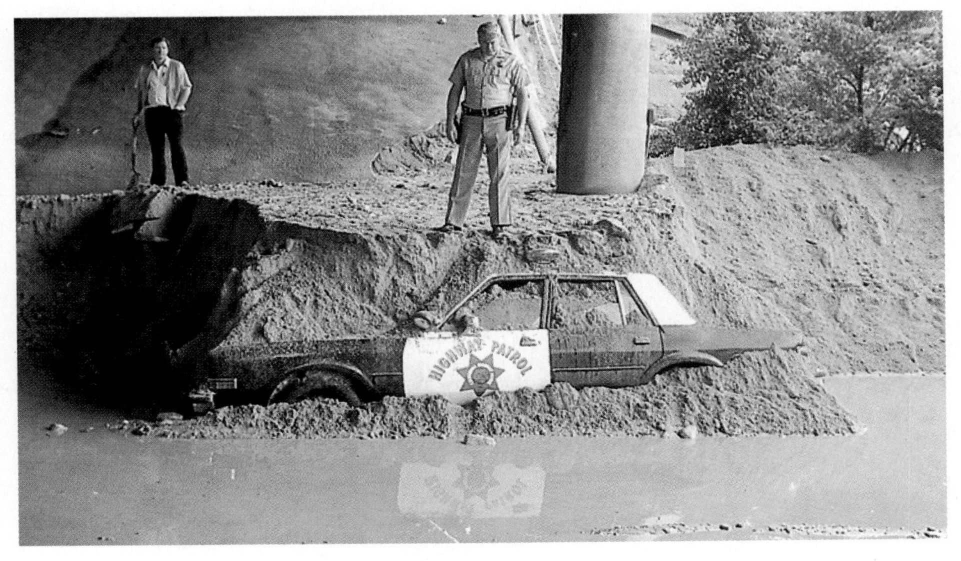

▶ **FIGURE 14-6** A California Highway Patrol officer stands on top of a 2 m high wall of mud that rolled over a patrol car near the Golden State Freeway on October 23, 1987. Flooding and mudslides also trapped other vehicles and closed the freeway.

maintained by the friction between the grains. The additional weight created by overloading increases the water pressure within the material, which in turn decreases its shear strength, thereby weakening the slope material. If enough material is added, the slope will eventually fail, sometimes with tragic consequences.

Geology and Slope Stability

The relationship between topography and the geology of an area is important in determining slope stability (▶ Figure 14-7). If the rocks underlying a slope dip in the same direction as the slope, mass wasting is more likely to occur than if the rocks are horizontal or dip in the opposite direction. When the rocks dip in the same direction as the slope, water can percolate along the various bedding planes and decrease the cohesiveness and friction between adjacent rock layers (Figure 14-7a). This is particularly true when there are interbedded clay layers because clay becomes very slippery when wet.

Even if the rocks are horizontal or dip in a direction opposite to that of the slope, joints may dip in the same direction as the slope. Water migrating through them weathers the rock and expands these openings until the weight of the overlying rock causes it to fall (Figure 14-7b).

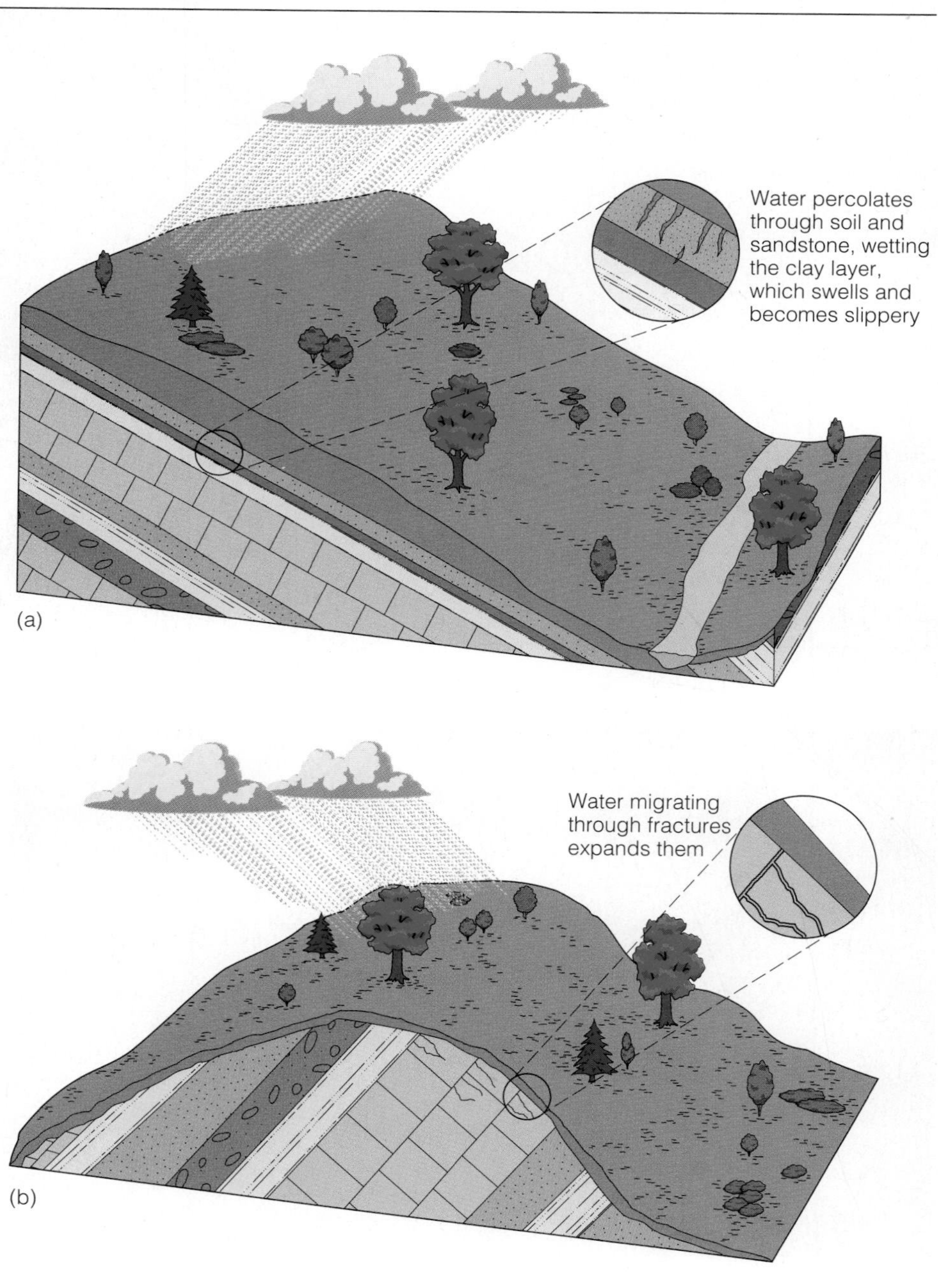

▶ **FIGURE 14-7** (*a*) Rocks dipping in the same direction as a hill's slope are particularly susceptible to mass wasting. Undercutting of the base of the slope by a stream removes support and steepens the slope at the base. Water percolating through the soil and into the underlying rock increases its weight and, if clay layers are present, wets the clay making them slippery. (*b*) Fractures dipping in the same direction as a slope are enlarged by chemical weathering, which can remove enough material to cause mass wasting.

Triggering Mechanisms

While the factors previously discussed all contribute to slope instability, most—though not all—rapid mass movements are triggered by a force that temporarily disturbs slope equilibrium. The most common triggering mechanisms are strong vibrations from earthquakes and excessive amounts of water from a winter snow melt or a heavy rainstorm.

Volcanic eruptions, explosions, and even loud claps of thunder may also be enough to trigger a landslide if the slope is sufficiently unstable. Many *avalanches*, which are rapid movements of snow and ice down steep mountain slopes, are triggered by the sound of a loud gunshot or, in rare cases, even a person's shout.

Rapid mass movements involve a visible movement of material. Such movements usually occur quite suddenly, and the material moves very quickly downslope. Rapid mass movements are potentially dangerous and frequently result in loss of life and property damage. Most rapid mass movements occur on relatively steep slopes and can involve rock, soil, or debris.

Slow mass movements advance at an imperceptible rate and are usually only detectable by the effects of their movement such as tilted trees and power poles or cracked foundations. Although rapid mass movements are more dramatic, slow mass movements are responsible for the downslope transport of a much greater volume of weathered material.

 TYPES OF MASS WASTING

Geologists recognize a variety of mass movements (◉ Table 14-2). Some are of one distinct type, while others are a combination of different types. It is not uncommon for one type of mass movement to change into another along its course. A landslide, for example, may start out as a slump at its head and, with the addition of water, become an earthflow at its base. Even though many slope failures are combinations of different materials and movements, it is still convenient to classify them according to their dominant behavior.

Mass movements are generally classified on the basis of three major criteria (Table 14-2): (1) rate of movement (rapid or slow); (2) type of movement (primarily falling, sliding, or flowing); and (3) type of material involved (rock, soil, or debris).

Falls

Rockfalls are a common type of extremely rapid mass movement in which rocks of any size fall through the air (► Figure 14-8). Rockfalls occur along steep canyons, cliffs, and road cuts and build up accumulations of loose rocks and rock fragments at their base called *talus* (see Figure 5-6).

Rockfalls result from failure along joints or bedding planes in the bedrock and are commonly triggered by natural or human undercutting of slopes, or by earthquakes. Many rockfalls in cold climates are the result of frost wedging. Chemical weathering caused by water percolating through the fissures in carbonate rocks (limestone, dolostone, and marble) is also responsible for many rockfalls.

Rockfalls range in size from small rocks falling from a cliff to massive falls involving millions of cubic meters of debris

TABLE 14-2 Classification of Mass Movements and Their Characteristics

Type of Movement	Subdivision	Characteristics	Rate of Movement
Falls	Rockfall	Rocks of any size fall through the air from steep cliffs, canyons, and road cuts	Extremely rapid
Slides	Slump	Movement occurs along a curved surface of rupture; most commonly involves unconsolidated or weakly consolidated material	Extremely slow to moderate
	Rock glide	Movement occurs along a generally planar surface	Rapid to very rapid
Flows	Mudflow	Consists of at least 50% silt- and clay-sized particles and up to 30% water	Very rapid
	Debris flow	Contains larger-sized particles and less water than mudflows	Rapid to very rapid
	Earthflow	Thick, viscous, tongue-shaped mass of wet regolith	Slow to moderate
	Quick clays	Composed of fine silt and clay particles saturated with water; when disturbed by a sudden shock, lose their cohesiveness and flow like a liquid	Rapid to very rapid
	Solifluction	Water-saturated surface sediment	Slow
	Creep	Downslope movement of soil and rock	Extremely slow
Complex movements		Combination of different movement types	Slow to extremely rapid

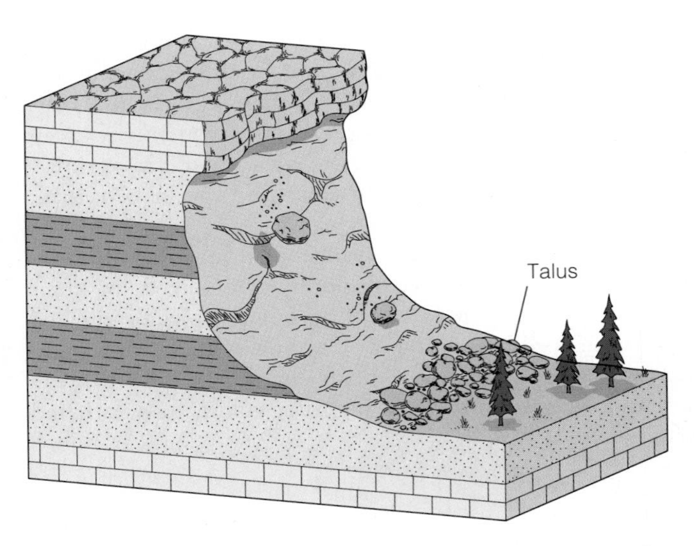

▶ FIGURE 14-8 Rockfalls result from failure along cracks, fractures, or bedding planes in the bedrock and are common in areas of steep cliffs.

that destroy buildings, bury towns, and block highways (▶ Figure 14-9). Rockfalls are a particularly common hazard in mountainous areas where roads have been built by blasting and grading through steep hillsides of bedrock. Anyone who has ever driven through the Appalachians, the Rocky Mountains, or the Sierra Nevada is familiar with the "Watch for Falling Rocks" signs posted to warn drivers of the danger. Slopes particularly prone to rockfalls are sometimes covered with wire mesh in an effort to prevent dislodged rocks from falling to the road below. Another tactic is to put up wire mesh fences along the base of the slope to catch or slow down bouncing or rolling rocks.

Slides

A **slide** involves movement of material along one or more surfaces of failure. The type of material may be soil, rock, or a combination of the two, and it may break apart during movement or remain intact. A slide's rate of movement can vary from extremely slow to very rapid (Table 14-2).

Two types of slides are generally recognized: (1) slumps or rotational slides, in which movement occurs along a curved surface; and (2) rock or block glides, which move along a more-or-less planar surface.

A **slump** involves the downward movement of material along a curved surface of rupture and is characterized by the backward rotation of the slump block (▶ Figure 14-10).

▶ FIGURE 14-9 Rockfall in Jefferson County, Colorado. All eastbound traffic and part of the westbound lane of Interstate 70 were blocked by the rockfall. Heavy rainfall and failure along joints and foliation planes in Precambrian gneiss caused this rockfall.

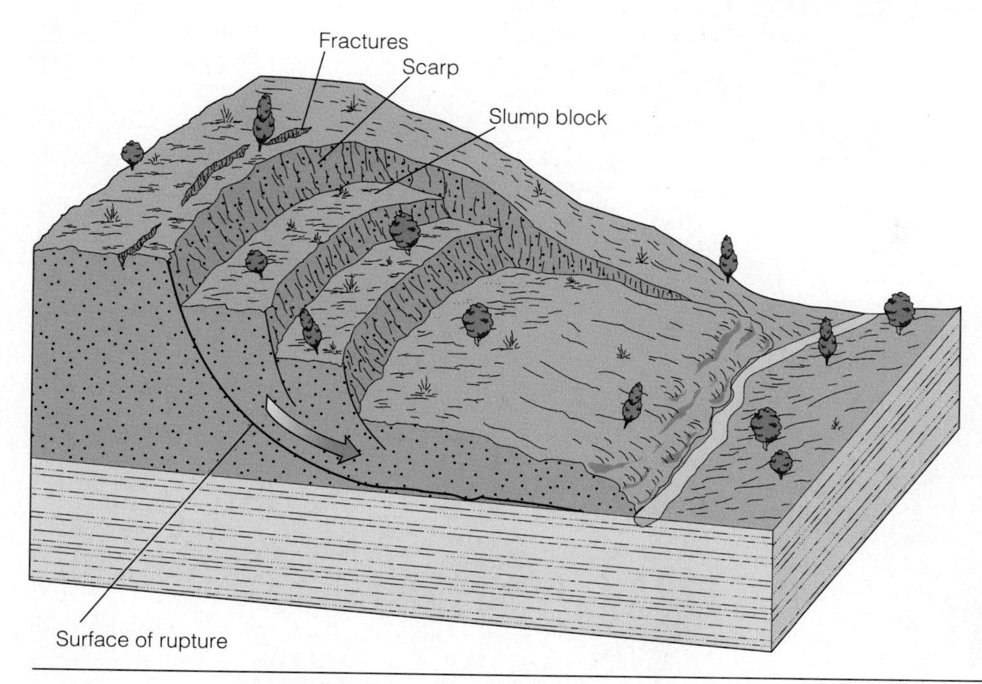

➢ **FIGURE 14-10** In a slump, material moves downward along the curved surface of a rupture, causing the slump block to rotate backward. Most slumps involve unconsolidated or weakly consolidated material and are typically caused by erosion along the slope's base.

Slumps occur most commonly in unconsolidated or weakly consolidated material and range in size from small individual sets, such as occur along stream banks, to massive, multiple sets that affect large areas and cause considerable damage.

Slumps can be caused by a variety of factors, but the most common is erosion along the base of a slope, which removes support for the overlying material. This local steepening may be caused naturally by stream erosion along its banks (Figure 14-10) or by wave action at the base of a coastal cliff. Slope oversteepening can also be caused by human activity, such as the construction of highways and housing developments. Slumps are particularly prevalent along highway cuts where they are generally the most frequent type of slope failure observed.

While many slumps are merely a nuisance, large-scale slumps involving populated areas and highways can cause extensive damage. Such is the case in coastal southern California where slumping and sliding have been a constant problem. Many areas along the coast are underlain by poorly to weakly consolidated silts, sands, and gravels interbedded with clay layers, some of which are weathered ash falls. In addition, southern California is tectonically active so that many of these deposits are cut by faults and joints, which allow the infrequent rains to percolate downward rapidly, wetting and lubricating the clay layers.

Southern California lies in a semiarid climate and is dry most of the year. When it does rain, typically between November and March, large amounts of rain can fall in a short time. Thus, the ground quickly becomes saturated, leading to landslides along steep canyon walls as well as along coastal cliffs (➢ Figure 14-11). Most of the slope failures along the southern California coast are the result of slumping. These slumps have destroyed many expensive homes and forced numerous roads to be closed and relocated.

A **rock** or *block* **glide** occurs when rocks move downslope along a more-or-less planar surface. Most rock glides occur because the local slopes and rock layers dip in the same direction (➢ Figure 14-12), although they can also occur along fractures parallel to a slope.

The largest known rock glide in the world is the prehistoric Saidmarreh landslide in southwestern Iran (➢ Figure 14-13). A slab of limestone 305 m thick, 14 km long, and 5 km wide became detached from the Kabir Kuh ridge and slid down and across the adjacent 8 km wide Saidmarreh Valley with enough momentum to climb over a ridge 460 m high before stopping nearly 18 km from its source! The volume of the slipped material was about 21 km^3, and it weighed approximately 50 billion tons. When the debris from the rock glide finally settled, it covered an area of 166 km^2.

The causes of this rock glide probably involved three factors: (1) the massive limestone dipped in the same direction as the local slope; (2) the limestone was underlain by a weak claystone; and (3) its base was undercut by the Karkheh River. In addition, the area is seismically active, and it is thought an earthquake probably triggered the slide.

Rock glides are also common occurrences along the southern California coast. At Point Fermin, seaward-dipping rocks with interbedded slippery clay layers are undercut by waves causing numerous glides (➢ Figure 14-14a).

Farther south in the town of Laguna Beach, startled residents watched as a rock glide destroyed or damaged 50 homes on October 2, 1978 (Figure 14-14b). Just as at Point Fermin, the rocks at Laguna Beach dip about 25° in the same

➢ **FIGURE 14-11** Undercutting of steep sea cliffs by wave action resulted in massive slumping in the Pacific Palisades area of southern California on March 31 and April 3, 1958. Highway 1 was completely blocked. Note the heavy earth-moving equipment for scale.

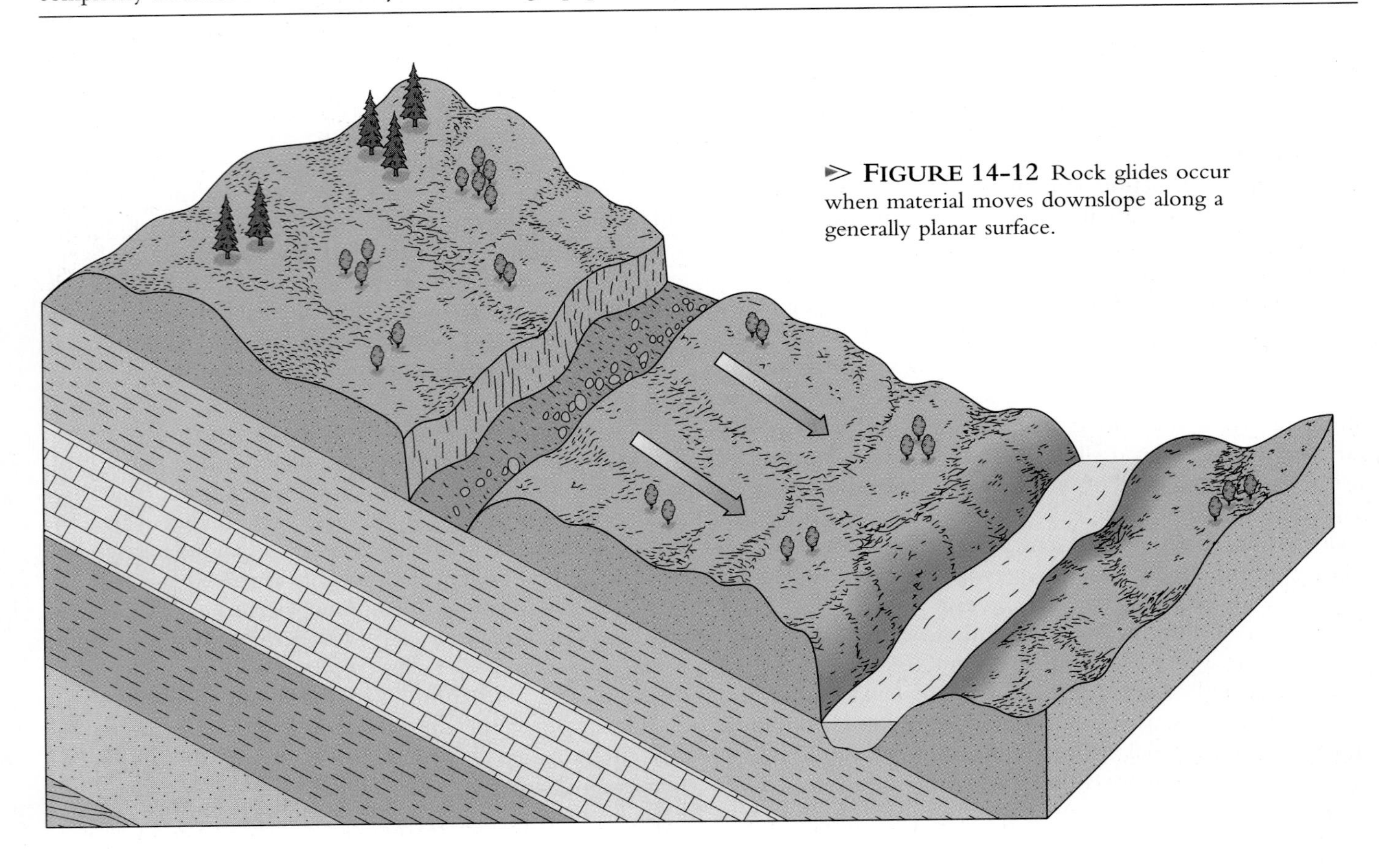

➢ **FIGURE 14-12** Rock glides occur when material moves downslope along a generally planar surface.

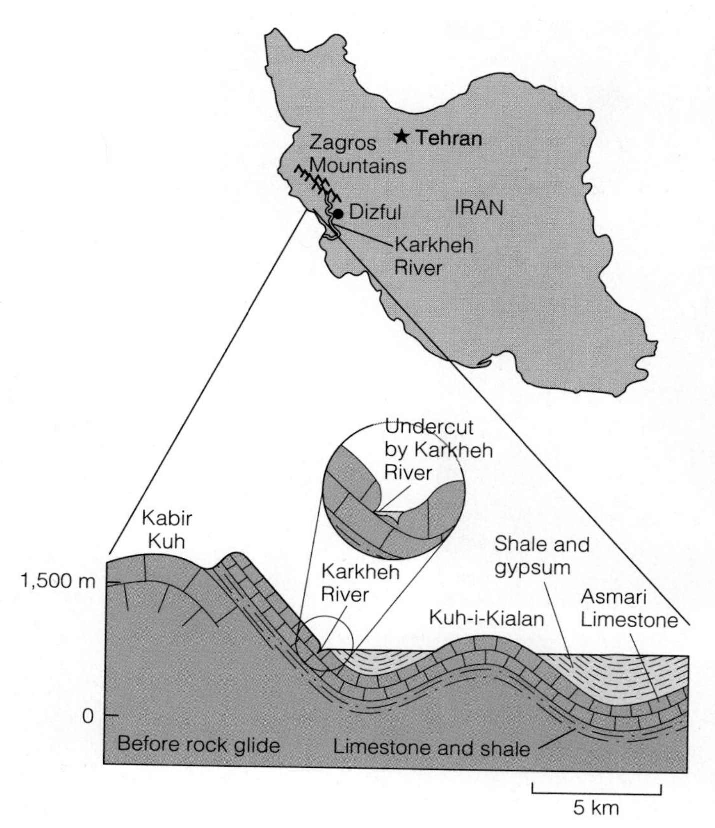

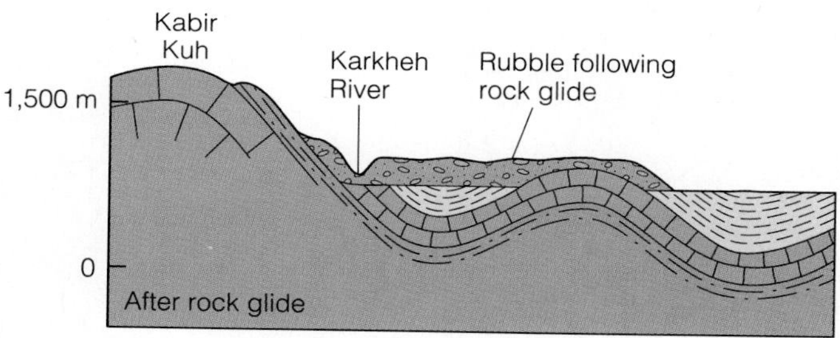

▶ **FIGURE 14-13** The world's largest known rock glide occurred in the Saidmarreh Valley, some 96 km northwest of Dizful, Iran. An earthquake is thought to have triggered this massive prehistoric slide that covered an area of 166 km^2.

direction as the slope of the canyon walls and contain clay beds that "lubricate" the overlying rock layers, causing the rocks and the houses built on them to glide. In addition, percolating water from the previous winter's heavy rains wet a subsurface clayey siltstone, thus reducing its shear strength and helping to activate the glide. Although the 1978 glide covered only about 5 acres, it was part of a larger ancient slide complex.

Not all rock glides are the result of rocks dipping in the same direction as a hill's slope. The rock glide at Frank, Alberta, Canada, on April 29, 1903, illustrates how nature and human activity can combine to create a situation with tragic results (▶ Figure 14-15).

It would appear at first glance that the coal-mining town of Frank, lying at the base of Turtle Mountain, was in no danger from a landslide (Figure 14-15). After all, many of the rocks dipped away from the mining valley, unlike the situations at Saidmarreh, Point Fermin, and Laguna Beach. The joints in the massive limestone composing Turtle Mountain, however, dip steeply toward the valley and are essentially parallel with the slope of the mountain itself. Furthermore, Turtle Mountain is supported by weak limestones, shales, and coal layers that underwent slow plastic deformation from the weight of the overlying massive limestone. Coal mining along the base of the valley also contributed to the stress on the rocks by removing some of the underlying support. All of these factors, as well as frost action and chemical weathering that widened the joints, finally resulted in a massive rock glide. Almost 40 million m^3 of rock slid down Turtle Mountain along joint planes, killing 70 people and partially burying the town of Frank.

Flows

Mass movements in which material flows as a viscous fluid or displays plastic movement are termed *flows*. Their rate of movement ranges from extremely slow to extremely rapid (Table 14-2). In many cases, mass movements may begin as falls, slumps, or slides and change into flows further downslope.

Mudflows are the most fluid of the major mass movement types. They consist of at least 50% silt- and clay-sized material combined with a significant amount of water (up to 30%). Mudflows are common in arid and semiarid environments where they are triggered by heavy rainstorms that quickly saturate the regolith, turning it into a raging flow of mud that engulfs everything in its path. Mudflows can also occur in mountain regions (▶ Figure 14-16) and in areas covered by volcanic ash where they can be particularly destructive (see Chapter 4). Because mudflows are so fluid, they generally follow preexisting channels until the slope decreases or the channel widens, at which point they fan out.

Mudflows are very dangerous because they typically form quickly, usually move very rapidly (at speeds up to 80 km per hour), and are capable of transporting all different sizes of objects. As urban areas in arid and semiarid climates continue to expand, mudflows and the damage they create are becoming problems. For example, mudflows are very common in the steep hillsides around Los Angeles where they have damaged or destroyed many homes.

In addition to the damage they cause on hillsides, mudflows are also a hazard to structures built along the bases of steep mountain fronts. Any building, highway, or railroad tracks in the path of the mudflow will be quickly moved or buried. A mudflow in Cajon Pass near Los Angeles, for example, carried a locomotive a distance of more than 600 m before burying it.

Debris flows are composed of larger-sized particles than those in mudflows and do not contain as much water. Consequently, they are usually more viscous than mudflows,

(a)

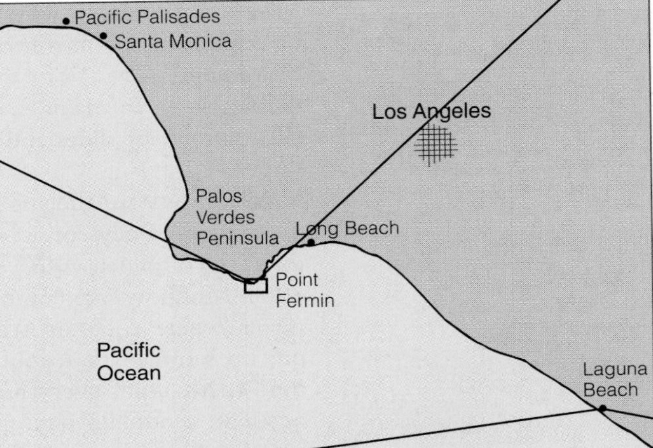

➢ **FIGURE 14-14** (*a*) A combination of interbedded clay beds that become slippery when wet, rocks dipping in the same direction as the slope of the sea cliffs, and undercutting of the sea cliffs by wave action has caused numerous rock glides and slumps at Point Fermin, California. (*b*) The same combination of factors apparently activated a rock glide farther south at Laguna Beach that destroyed numerous homes and cars on October 2, 1978. (Photo (*a*) courtesy of Eleanora I. Robbins, U. S. Geological Survey.)

(b)

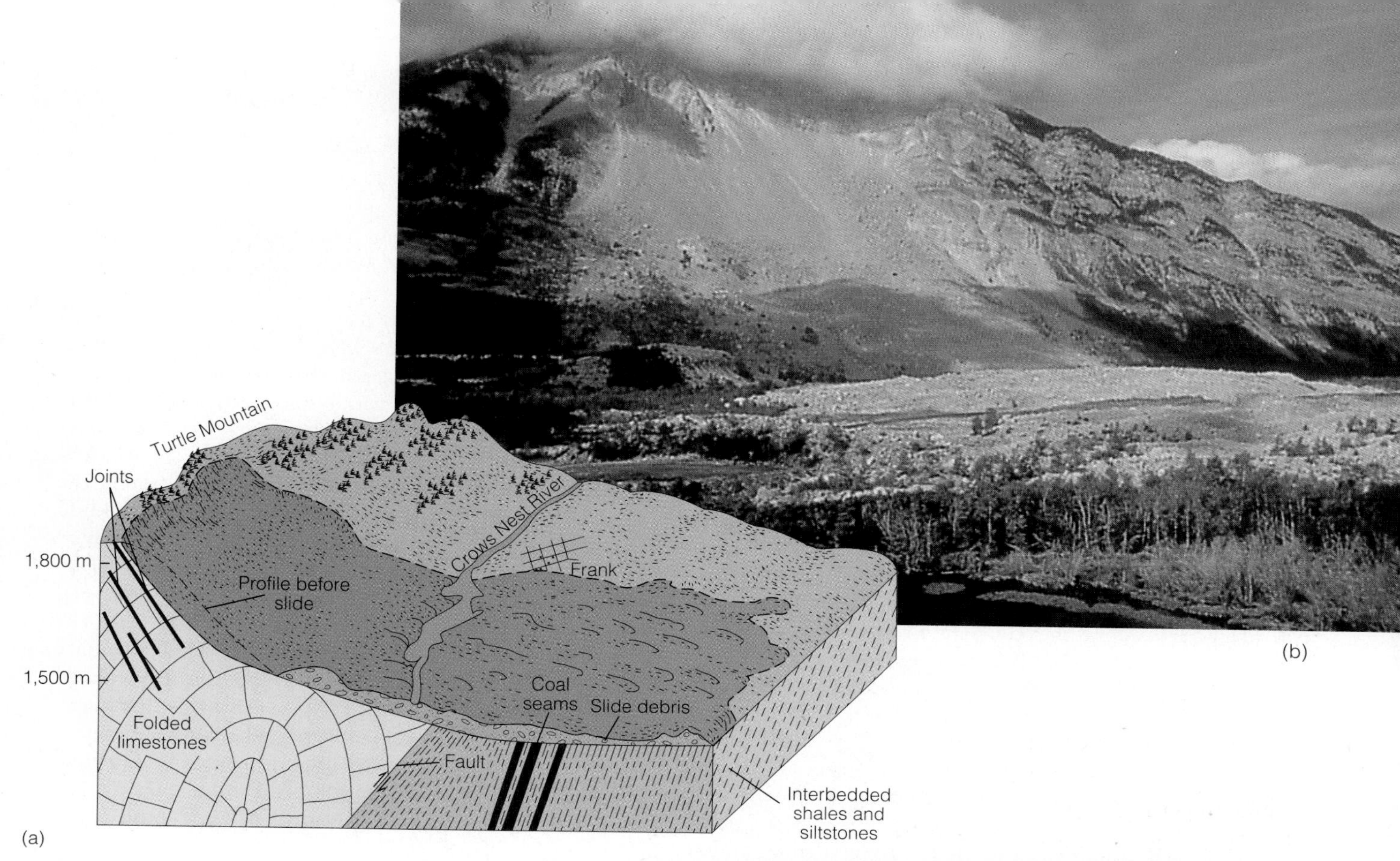

➢ **FIGURE 14-15** (*a*) The tragic Turtle Mountain rock glide that killed 70 people and partially buried the town of Frank, Alberta, Canada, on April 29, 1903, was caused by a combination of factors. These included joints that dipped in the same direction as the slope of Turtle Mountain, a fault partway down the mountain, weak shale and siltstone beds underlying the base of the mountain, and mined-out coal seams. (*b*) Results of the 1903 rock glide at Frank.

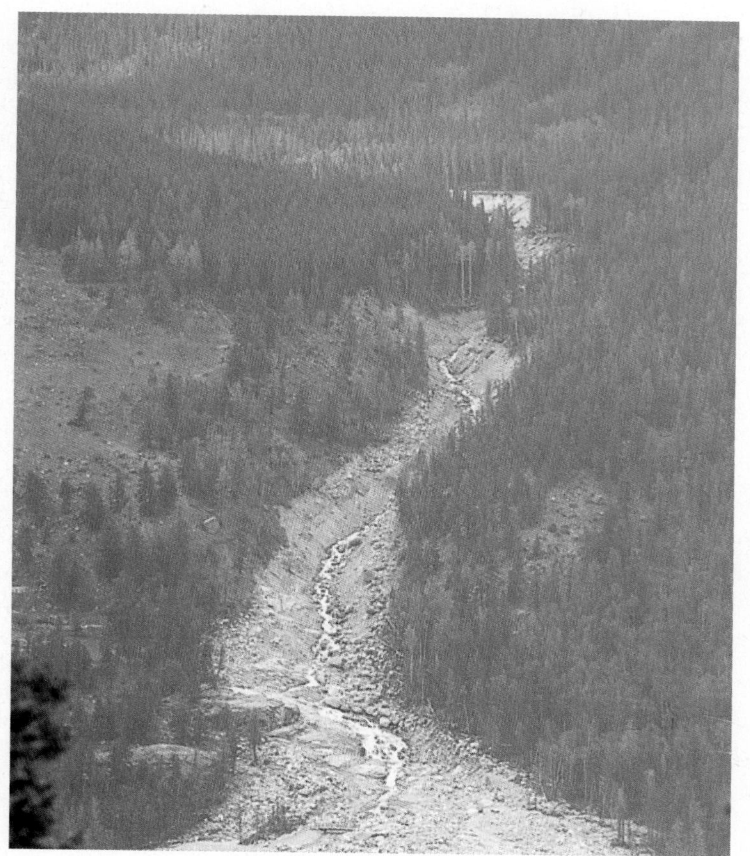

➢ **FIGURE 14-16** A mudflow near Estes Park, Colorado.

Engineering Geology: Using Geological Principles for Practical Solutions

I grew up in southern California in the 1960s and 1970s. My parents both worked, and busy though they were, our family went on trips to the mountains, desert, and up and down the coast. I remember packing up the station wagon with the tent and camping gear up top and our dog, Whitey, sharing the back with the metal cooler. When we arrived at our destination, we camped, hiked, explored, and, of course, collected momentos such as rocks. Rock collecting never became an obsession for me, but the exposure to the natural world that started when I was very young has continued.

After a bit of trial and error at California State University, I finally arrived in the Geology Department, and once there, I was hooked. My B.S. degree in geology has turned out to be a door that, once opened, leads in many directions.

Currently, I am a consulting engineering geologist, applying basic geologic principles to solve problems related to construction. The varied nature of the projects I work on is very simulating, but the time frame in which private work is completed is more accelerated than academic studies would allow. High-quality technical work is top priority and is integrated with good consulting service and sound business practice.

A residential project site that was underlain by layered sedimentary rock provides a good example of how basic geologic principles are used to solve a practical problem. During the preliminary investigation, a typical stratigraphic section was determined based on data collected during background research, aerial photograph review, field mapping, and subsurface exploration (drilling and trenching). During analysis, the stratigraphic section was utilized to predict where certain "problem" layers would occur, and remedial recommendations were presented prior to mass grading. During grading, the duty of the field geologist is to confirm the preliminary interpretations and refine them, if necessary. On this particular project, a potentially active fault transected the site. At one location, this fault intersected a landslide. Exposures during grading revealed that the landslide actually cut off the fault, indicating the fault was older than the landslide. Within the landslide graben, an old organic soil horizon (paleosol) was carbon dated to be 12,500 years old. Since this soil layer was buried in the graben of the landslide after the failure occurred, the fault must be older than 12,500 years and not active according to the definition of a Recent fault.

On one large planned community project, I encountered a variety of geological problems that impacted different aspects of the development. The problems included landslides, deeply incised alluvial-filled channels, shallow sandstone layers, expansive claystones, and diatomaceous siltstones. The topographic relief at the site was over 400 feet. Storm drain outlet structures were built that had to function beneath 20-year flood levels. They have been tested twice by nature since completion and performed as designed. Residential structures were built over compacted fills as thick as 200 vertical feet. A 30-inch-diameter major water trunk line was relocated around the development. The electrical transmission lines crossing the site were removed from the towers and put in an underground vault. The cemented rock encountered was disposed of by burying it in deep fills or crushing it to gravel size and using it to bridge removal bottoms in saturated alluvium. Sand, the other by-product of crushing, was used as backfill for utility trenches. Hard rock in cut areas was undercut and replaced with compacted fill to achieve design elevations to mitigate problems during foundation, utility, and pool excavation. Select grading techniques were utilized to minimize the waiting period for settlement of deep fills prior to building and exclusion of highly expansive clays at finish grade in fill areas.

Many short-term projects have exposed me to various types of challenges including the following: investigation of earthquake-damaged homes in the Northridge area (seismic shaking, liquefaction); investigation of a landslide in San Clemente that destroyed five homes and closed the Atchison, Topeka, and Santa Fe Railroad for two weeks and the Pacific Coast Highway for more than one year; investigation of a distressed property in the Newport Harbor area where a seawall was in jeopardy of failure; investigation for a major transportation corridor planned across hillside terrain in southern Orange County; environmental impact study and report for proposed improvements for a 23-mile segment of a major highway in Orange County; numerous groundwater seepage investigations in occupied residential areas; numerous forensic geotechnical investigations involving distressed residential structures.

After 12 years working as a professional geologist, I believe I am fulfilling what is best suited for me personally and will continue to see what opportunities will arise next.

WILLIAM GOODMAN is the principal geologist and co-owner of a geotechnical consulting firm working primarily in California. He graduated with a B.S. degree in geology from California State University, Fullerton in 1983. He is currently a Registered Geologist in California and Arizona and a Certified Engineering Geologist.

typically do not move as rapidly, and rarely are confined to preexisting channels. Debris flows can, however, be just as damaging because they can transport large objects.

Earthflows move more slowly than either mudflows or debris flows. An earthflow slumps from the upper part of a hillside, leaving a scarp, and flows slowly downslope as a thick, viscous, tongue-shaped mass of wet regolith (▶ Figure 14-17). Like mudflows and debris flows, earthflows can be of any size, and are frequently destructive. They occur most commonly in humid climates on grassy soil-covered slopes following heavy rains.

Some clays spontaneously liquefy and flow like water when they are disturbed. Such **quick clays** have caused serious damage and loss of lives in Sweden, Norway, eastern Canada, and Alaska (Table 14-1). Quick clays are composed of silt and clay particles made by the grinding action of glaciers. Geologists think these fine sediments were originally deposited in a marine environment where their pore space was filled with salt water. The ions in the salt water helped establish strong bonds between the clay particles, thus stabilizing and strengthening the clay. When the clays were subsequently uplifted above sea level, however, the salt water was flushed out by fresh groundwater, reducing the effectiveness of the ionic bonds between the clay particles and thereby reducing the overall strength and cohesiveness of the clay. Consequently, when the clay is disturbed by a sudden shock or shaking, it essentially turns to a liquid and flows.

An example of the damage that can be done by quick clays occurred in the Turnagain Heights area of Anchorage, Alaska, in 1964 (▶ Figure 14-18). Underlying most of the Anchorage area is the Bootlegger Cove Clay, a massive clay unit of poor permeability. Because the Bootlegger Cove Clay forms a barrier preventing groundwater from flowing through the adjacent glacial deposits to the sea, considerable hydraulic pressure builds up behind the clay. Some of this water has flushed out the salt water in the clay and also has saturated the lenses of sand and silt associated with the clay beds. When the 8.5-magnitude Good Friday earthquake struck on March 27, 1964, the shaking turned parts of the Bootlegger Cove Clay into a quick clay and precipitated a series of massive slides in the coastal bluffs that destroyed most of the homes in the Turnagain Heights subdivision.

Solifluction is the slow downslope movement of water-saturated surface sediment. Solifluction can occur in any climate where the ground becomes saturated with water, but

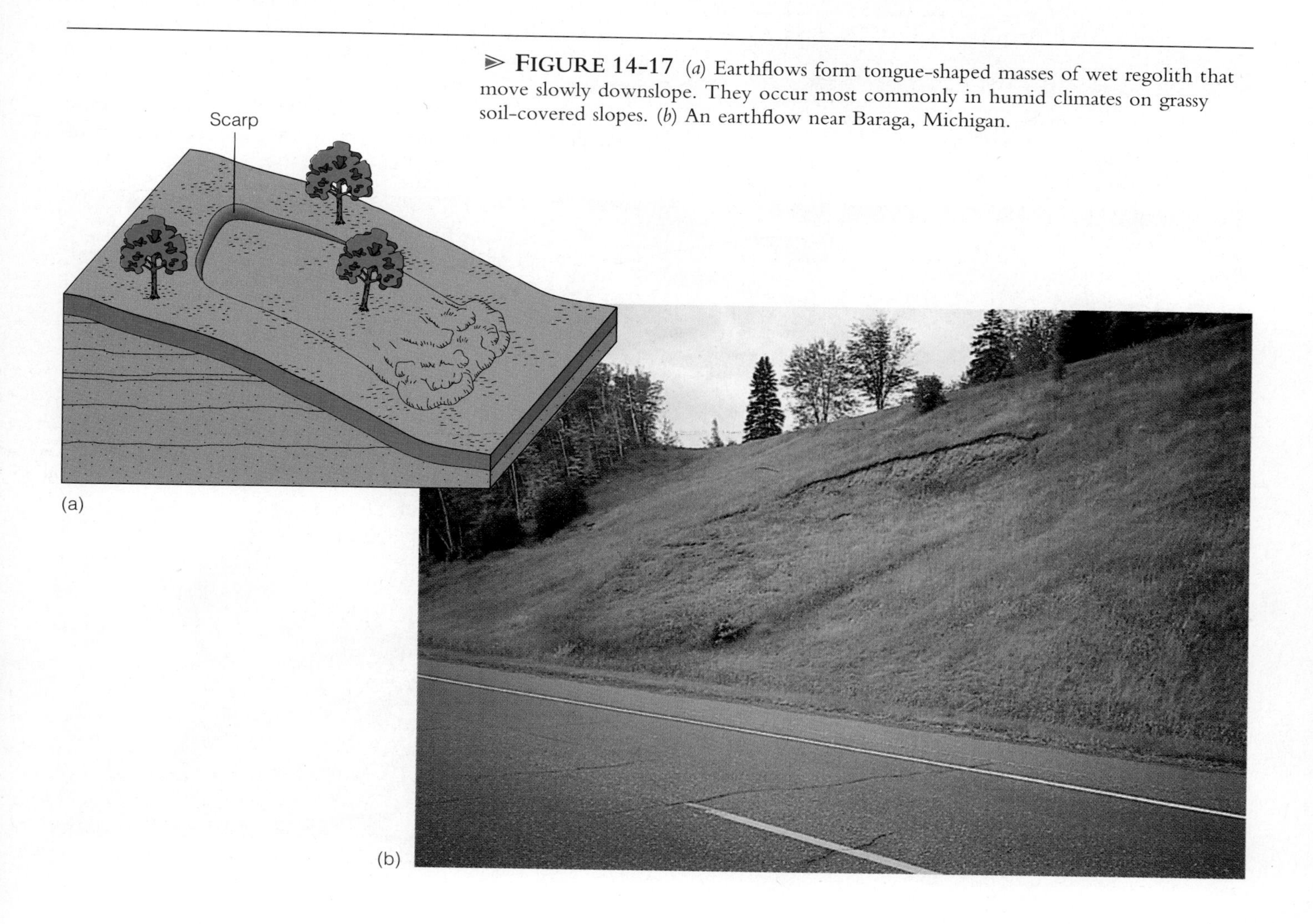

▶ **FIGURE 14-17** (*a*) Earthflows form tongue-shaped masses of wet regolith that move slowly downslope. They occur most commonly in humid climates on grassy soil-covered slopes. (*b*) An earthflow near Baraga, Michigan.

▶ **FIGURE 14-18** (*a*) Groundshaking by the 1964 Alaska earthquake turned parts of the Bootlegger Cove Clay into a quick clay, causing numerous slides (*b*) that destroyed many homes in the Turnagain Heights subdivision of Anchorage.

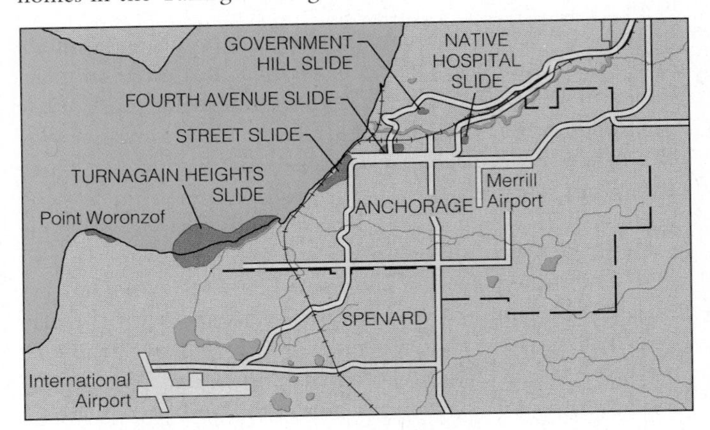

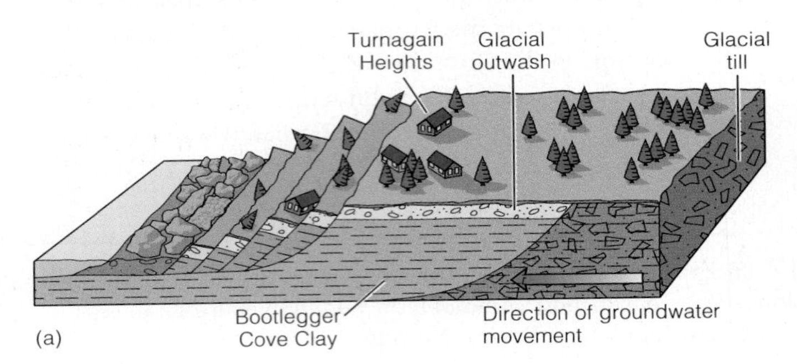

is most common in cold climates where the upper surface periodically thaws and freezes. **Permafrost** is ground that remains permanently frozen. It covers nearly 20% of the world's land surface (▶ Figure 14-19a). During the warmer season when the upper portion of the permafrost thaws, water and surface sediment form a soggy mass that flows by solifluction and produces a characteristic lobate topography (Figure 14-19b).

As might be expected, many problems are associated with construction in a permafrost environment. A good example is what happens when an uninsulated building is constructed directly on permafrost. In this instance, heat escapes through the floor, thaws the ground below, and turns it into a soggy, unstable mush. Because the ground is no longer solid, the building settles unevenly into the ground, and numerous structural problems result (▶ Figure 14-20).

Construction of the Alaska pipeline from the oil fields in Prudhoe Bay to the ice-free port of Valdez raised numerous concerns over the effect it might have on the permafrost and the potential for solifluction. Some thought that oil flowing through the pipeline would be warm enough to melt the permafrost, causing the pipeline to sink further into the ground and possibly rupture. After numerous studies were conducted, scientists concluded that the pipeline, completed in 1977, could safely be buried for more than half of its 1,280 km length; where melting of the permafrost might cause structural problems to the pipe, it was insulated and installed above ground.

Creep is the slowest type of flow. It is also the most widespread and significant mass wasting process in terms of

(b)

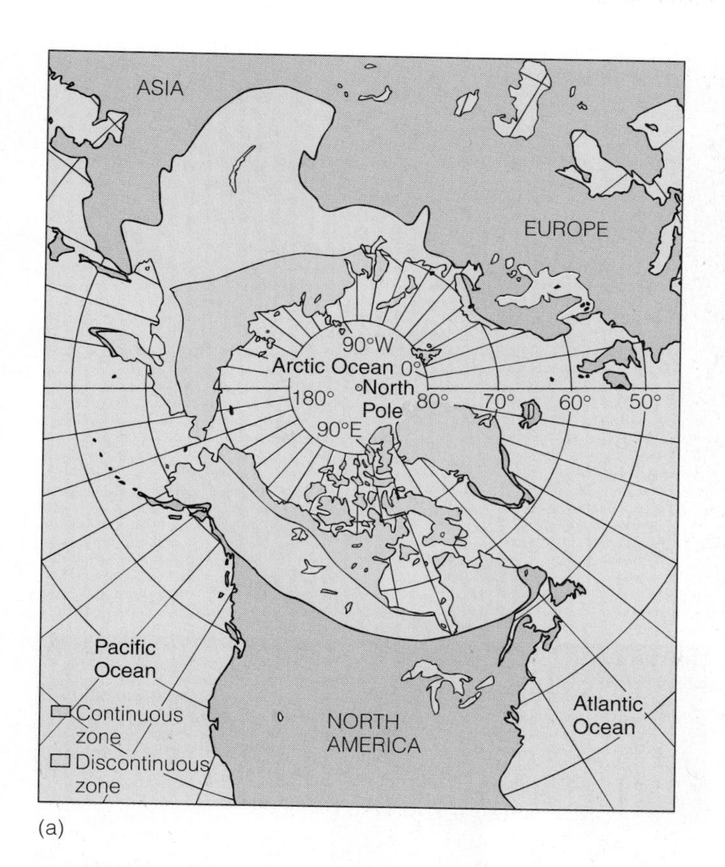

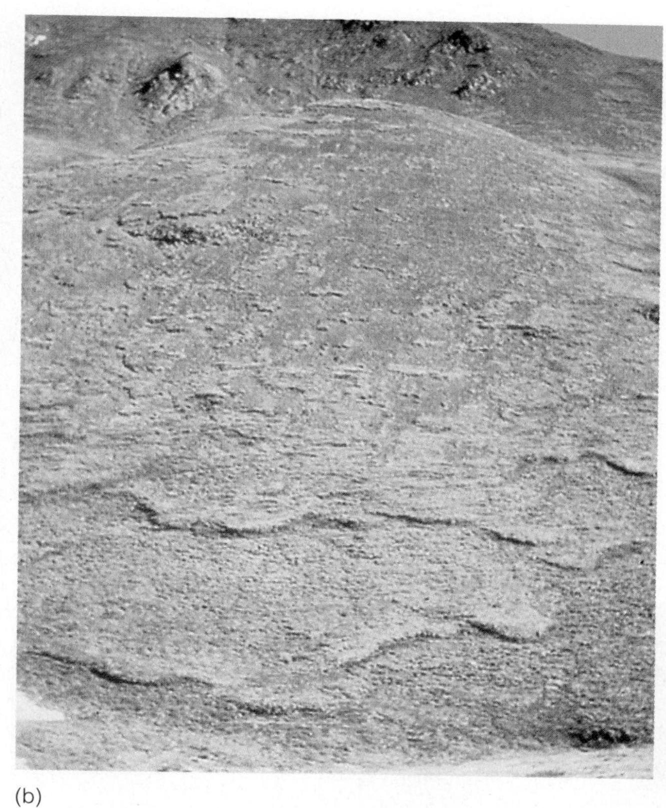

➢ **FIGURE 14-19** (*a*) Distribution of permafrost areas in the Northern Hemisphere. (*b*) Solifluction flows near Suslositna Creek, Alaska, show the typical lobate topography that is characteristic of solifluction conditions.

➢ **FIGURE 14-20** This house, located south of Fairbanks, Alaska, has settled unevenly because the underlying permafrost has thawed.

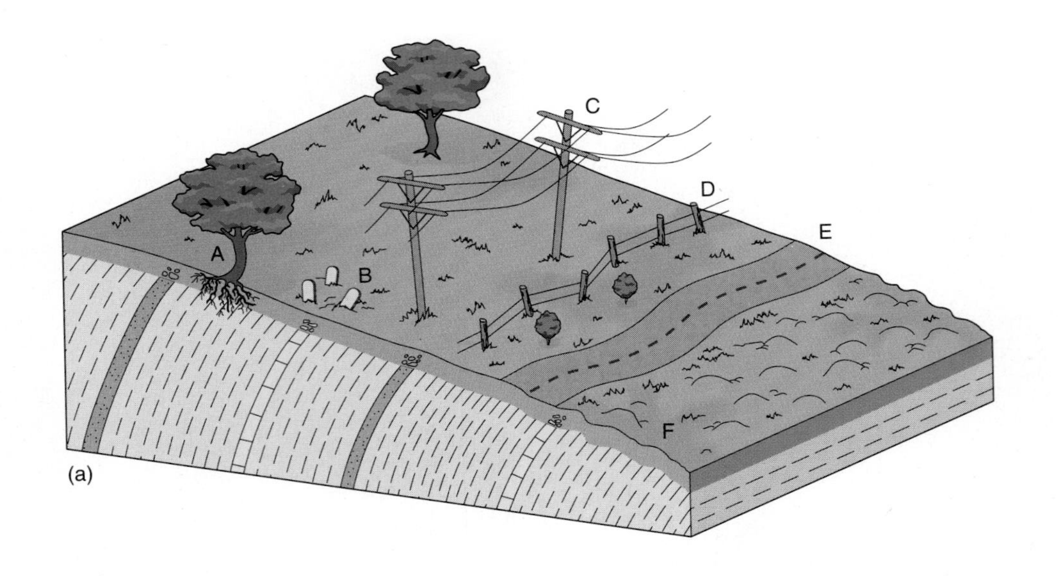

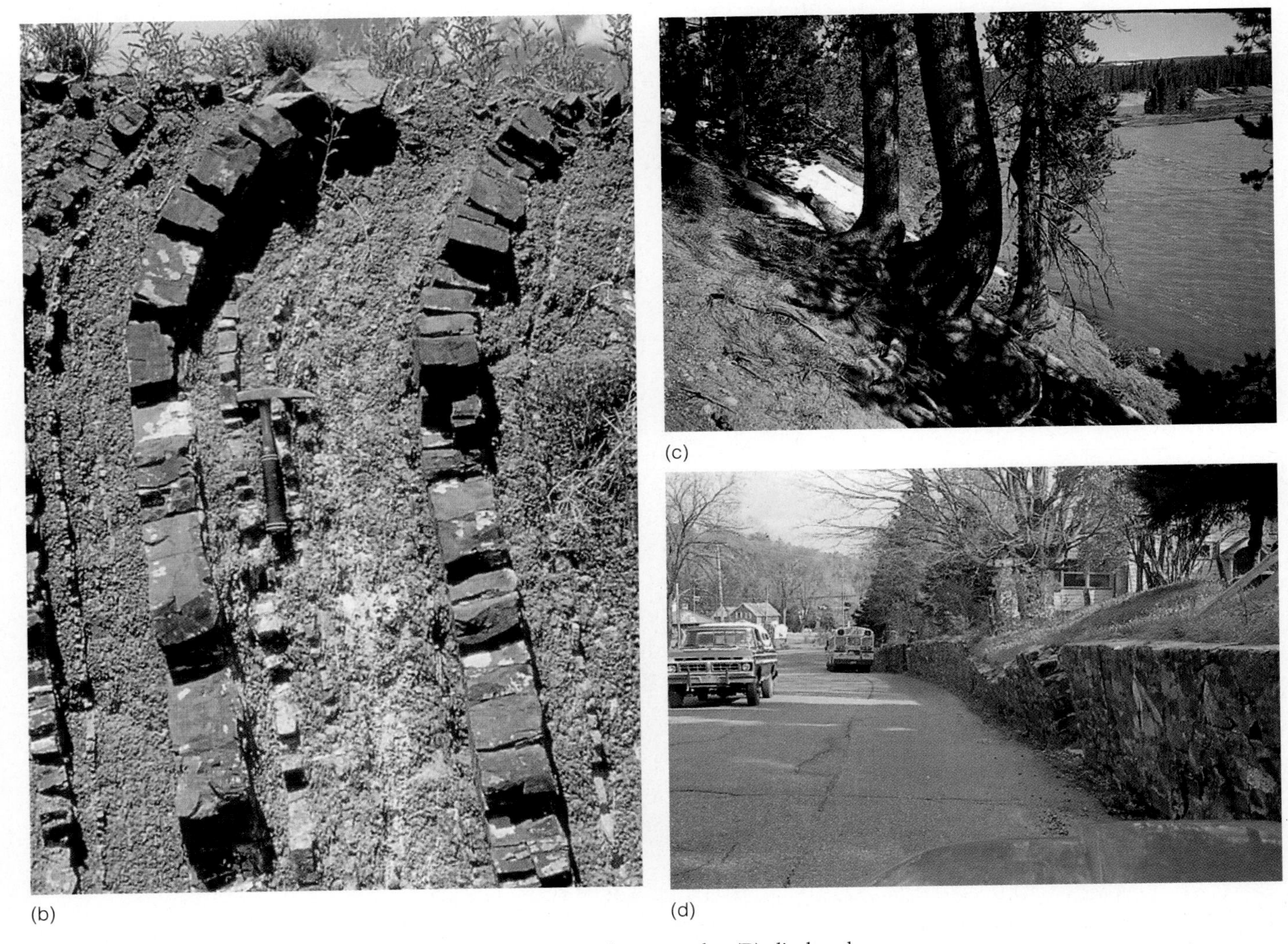

➢ **FIGURE 14-21** (*a*) Some evidence of creep: (A) curved tree trunks; (B) displaced monuments; (C) tilted power poles; (D) displaced and tilted fences; (E) roadways moved out of alignment; (F) hummocky surface. (*b*) Creep has bent these sandstone and shale beds of the Haymond Formation near Marathon, Texas. (*c*) Trees bent by creep, Wyoming. (*d*) Stone wall tilted due to creep, Champion, Michigan. (Photo courtesy of David J. Matty.)

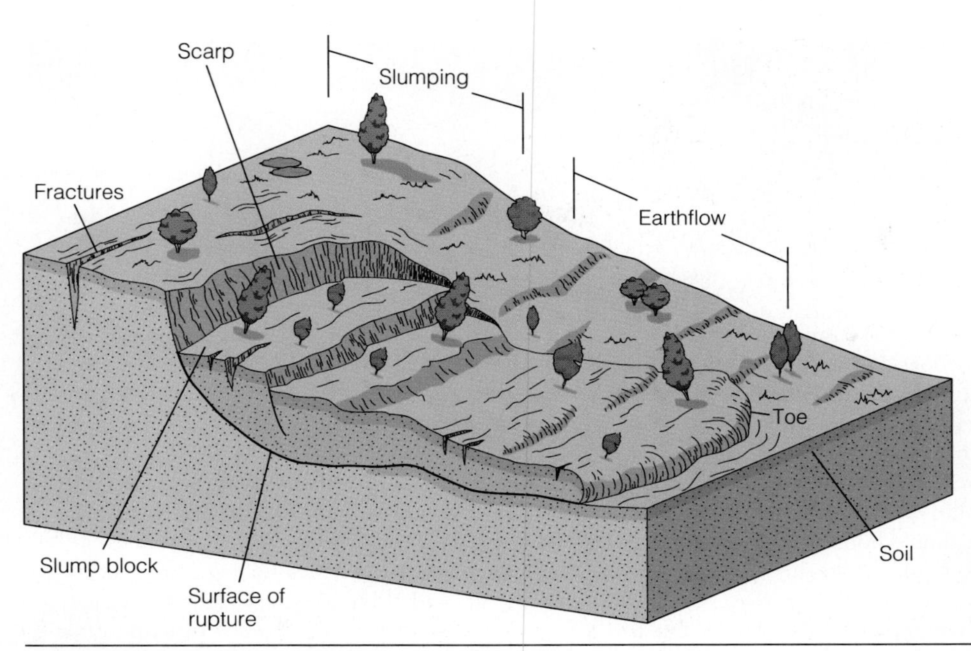

⊳ **FIGURE 14-22** A complex movement in which slumping occurs at the head followed by an earthflow.

the total amount of material moved downslope and the monetary damage that it does annually. Creep involves extremely slow downhill movement of soil or rock under the inexorable pull of gravity. Although it can occur anywhere and in any climate, it is most effective and significant as a geologic agent in humid regions. In fact, it is the most common form of mass wasting in the southeastern United States and the southern Appalachian Mountains.

Because the rate of movement is essentially imperceptible, we are frequently unaware of creep's existence until we notice its effects: tilted trees and power poles, broken streets and sidewalks, and cracked retaining walls or foundations (⊳ Figure 14-21). Creep usually involves the whole hillside and probably occurs, to some extent, on any weathered or soil-covered, sloping surface.

Not only is creep difficult to recognize, it is difficult to control. Although engineers can sometimes slow or stabilize creep, many times the only course of action is to simply avoid the area if at all possible or, if the zone of creep is relatively thin, design structures that can be anchored into the solid bedrock.

Complex Movements

Recall that many mass movements are combinations of different movement types. When one type is dominant, the movement can be classified as one of the movements described thus far. If several types are more-or-less equally involved, it is called a **complex movement**.

The most common type of complex movement is the slide-flow in which there is sliding at the head and then some type of flowage farther along its course. Most slide-flow landslides involve well-defined slumping at the head, followed by a debris flow or earthflow (⊳ Figure 14-22). Any combination of different mass movement types can be classified as a complex movement.

A **debris avalanche** is a complex movement that often occurs in very steep mountain ranges. Debris avalanches typically start out as rockfalls when large quantities of rock, ice, and snow are dislodged from a mountainside, frequently as a result of an earthquake. The material then slides or flows down the mountainside, picking up additional surface material and increasing in speed. The 1970 Peru earthquake set in motion the debris avalanche that destroyed the town of Yungay (see the Prologue).

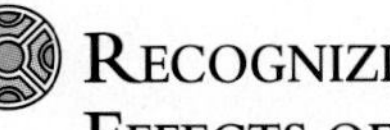

Recognizing and Minimizing the Effects of Mass Movements

The most important factor in eliminating or minimizing the damaging effects of mass wasting is a thorough geologic investigation of the region in question. In this way, former landslides and areas susceptible to mass movements can be identified and perhaps avoided (see Perspective 14-2). By assessing the risks of possible mass wasting before construction begins, steps can be taken to eliminate or minimize the effects of such events.

Identifying areas with a high potential for slope failure is important in any hazard assessment study; these studies include identifying former landslides as well as sites of potential mass movement. Scarps, open fissures, displaced or tilted objects, a hummocky surface, and sudden changes in vegetation are some of the features indicating former landslides or an area susceptible to slope failure. The effects of weathering, erosion, and vegetation, however, may obscure the evidence of previous mass wasting.

Soil and bedrock samples are also studied, both in the field and laboratory, to assess such characteristics as compo-

Perspective 14-2

The Vaiont Dam Disaster

On October 9, 1963, more than 240 million m^3 of rock and soil slid into the Vaiont Reservoir, in Italy, triggering a destructive flood that killed nearly 3,000 people (▶ Figure 1). To fully appreciate the immensity of this catastrophe, consider the following: Within a period of 15 to 30 seconds, the slide filled the reservoir with a mass of debris 2 km long and as high as 175 m above the reservoir level. The impact of the debris created a wave of water that overflowed the dam by 100 m and was still more than 70 m high 1.6 km downstream. The slide also set off a blast of wind that shook houses, broke windows, and even lifted the roof off one house in the town of Casso, which is 260 m above the reservoir on the opposite side of the slide; it also set off shock waves recorded by seismographs throughout Europe. Considering the forces generated by the slide, it is a tribute to the designer and construction engineer that the dam itself survived the disaster (▶ Figure 2)!

The dam was built in a glacial valley underlain by thick layers of folded and faulted limestones and clay layers that were further weakened by jointing (▶ Figure 3). Signs of previous slides in the area were obvious, and the few boreholes in the valley slopes revealed clay layers and small-scale slide planes. In spite of the geological evidence of previous mass wasting and objections to the site by some of the early investigators, construction of the 265.5 m high Vaiont Dam began.

A combination of both adverse geological features and conditions resulting from the dam construction contributed to the massive landslide. Among the geological causes were the rocks themselves, which were weak to begin with and dipped in the same direction as the valley walls of the reservoir. Fractured limestones make up the bulk of the rocks and are interbedded with numerous clay beds that are particularly prone to slippage. Active solution of the limestones by slightly acid groundwater further weakened them by developing and expanding an extensive network of cracks, joints, fissures, and other openings.

During the two weeks before the slide occurred, heavy rains saturated the ground, adding extra weight and reducing the shear strength of the rocks. In addition to water from the rains, water from the reservoir infiltrated the rocks of the lower valley walls, further reducing their strength.

Soon after the dam was completed, a relatively small slide of one million m^3 of material occurred on the south side of the reservoir. Following this slide, it was decided to limit the amount of water in the reservoir and to install monitoring devices throughout the potential slide area. Between 1960 and 1963, the eventual slide area moved an average of about 1 cm per week. On September 18, 1963, numerous monitoring stations reported movement had increased to about 1 cm per day. It was assumed that these were individual blocks moving, but it was actually the entire slide area!

Heavy rains fell between September 28 and October 9,

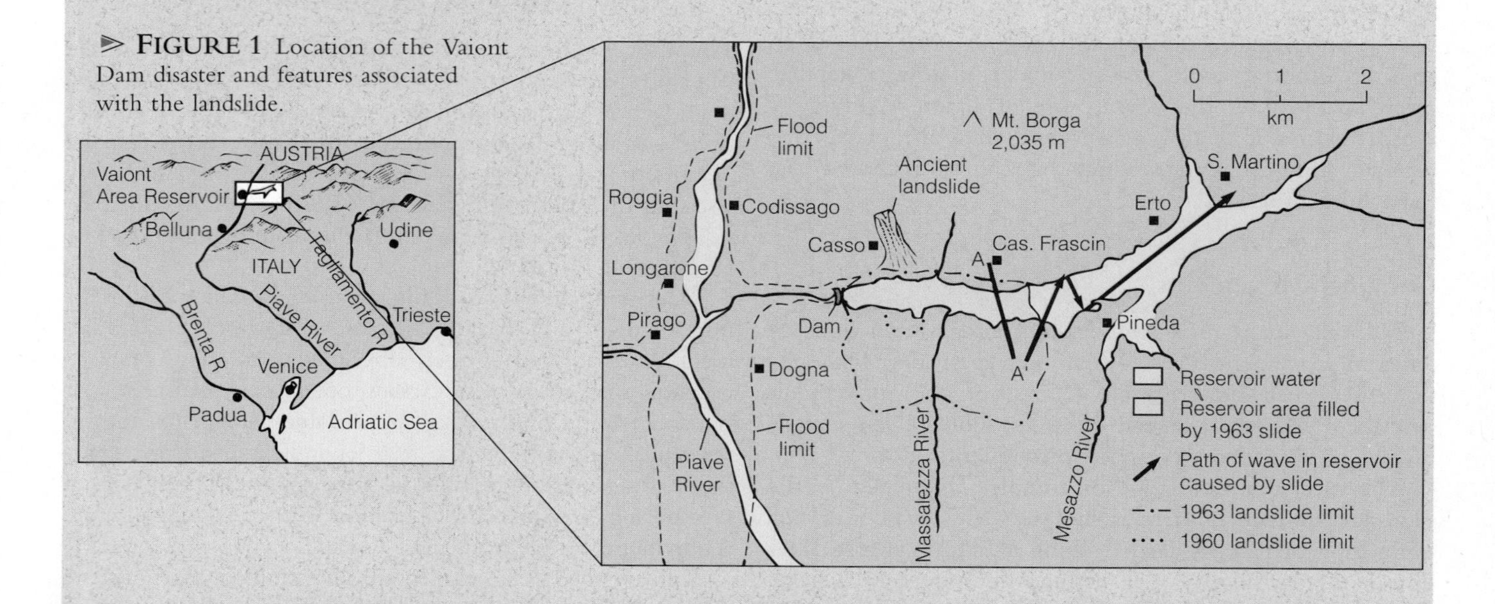

▶ **FIGURE 1** Location of the Vaiont Dam disaster and features associated with the landslide.

increasing the amount of subsurface water. By October 8, the creep rate had increased to almost 39 cm per day, and engineers finally realized that the entire slide area was moving, and quickly began lowering the reservoir level. On October 9, the rate of movement in the slide area had increased still further, in some locations up to 80 cm per day, and there were reports that the reservoir level was actually rising. This was to be expected if the south bank was moving into the reservoir and displacing water. Finally, at 10:41 P.M. that night, during yet another rainstorm, the south bank of the Vaiont valley slid into the reservoir.

The lesson to be learned from this disaster is that before construction on any dam begins, a complete and systematic appraisal of the area must be conducted. Such a study should examine the geology of the area, identify past mass movements, assess their potential for recurrence, and evaluate the effects that the project will have on the rocks, including how it will alter their shear strength over time. Without these precautions, similar disasters will occur and lives will needlessly be lost.

▶ **FIGURE 2** Aerial view of the Vaiont Dam.

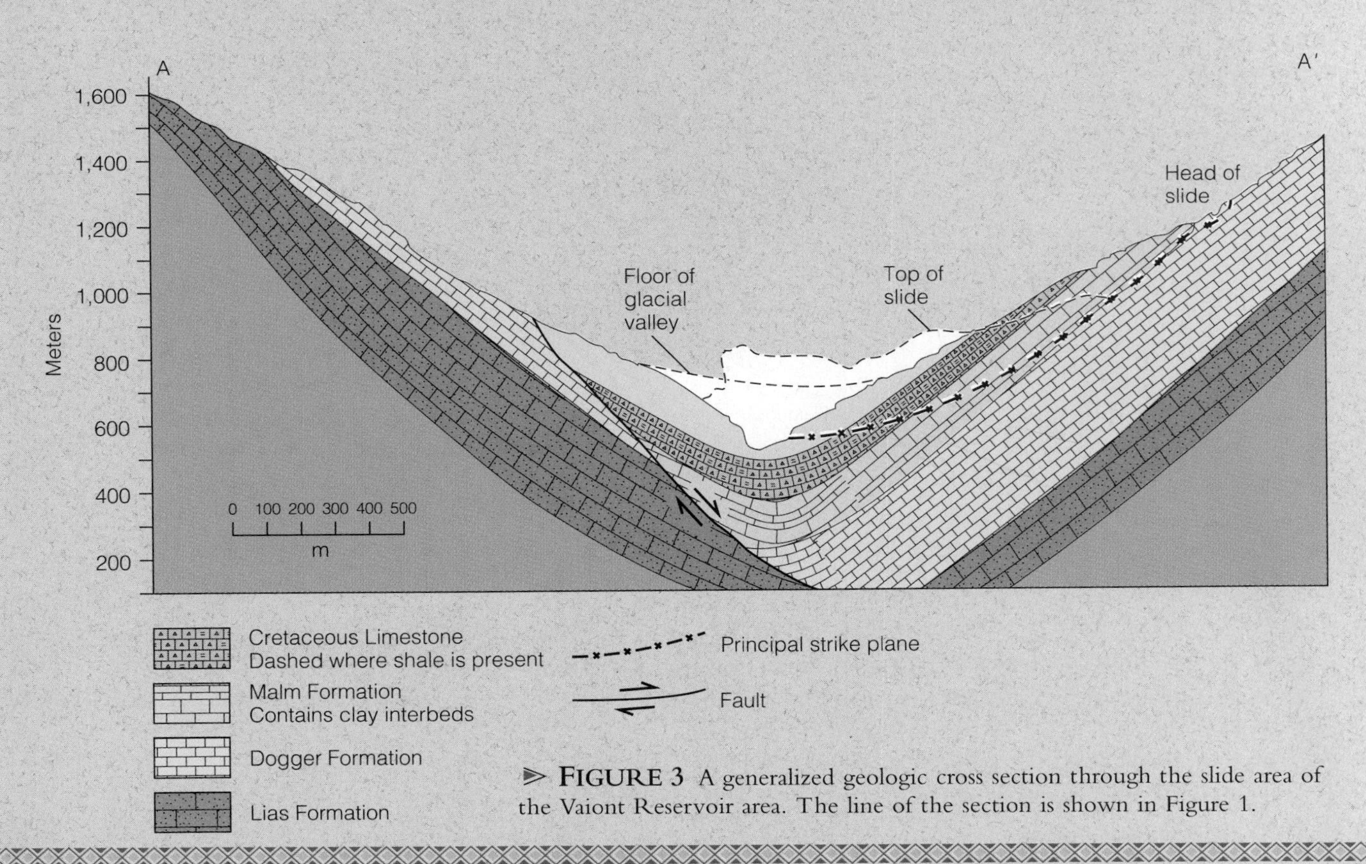

▶ **FIGURE 3** A generalized geologic cross section through the slide area of the Vaiont Reservoir area. The line of the section is shown in Figure 1.

sition, susceptibility to weathering, cohesiveness, and ability to transmit fluids. These studies help geologists and engineers predict slope stability under a variety of conditions.

The information derived from a hazard assessment study can be used to produce *slope stability maps* of the area (▶ Figure 14-23). These maps allow planners and developers to make decisions about where to site roads, utility lines, and housing or industrial developments based on the relative stability or instability of a particular location. In addition, the maps also indicate how extensive an area's landslide problem is and the type of mass movement that may occur. This information is important for grading slopes or building structures to prevent or minimize slope failure damage.

Although most large mass movements usually cannot be prevented, geologists and engineers can employ various methods to minimize the danger and damage resulting from them. Because water plays such an important role in many landslides, one of the most effective and inexpensive ways to reduce the potential for slope failure or to increase existing slope stability is through surface and subsurface drainage of a hillside. Drainage serves two purposes. It reduces the weight of the material likely to slide and increases the shear strength of the slope material by lowering pore pressure.

Surface waters can be drained and diverted by ditches, gutters, or culverts designed to direct water away from slopes. Drainpipes perforated along one surface and driven into a hillside can help remove subsurface water (▶ Figure 14-24). Finally, planting vegetation on hillsides helps stabilize slopes by holding the soil together and reducing the amount of water in the soil.

Another way to help stabilize a hillside is to reduce its slope. Recall that overloading or oversteepening by grading is a common cause of slope failure. By reducing the gradient of a hillside, the potential for slope failure is decreased. Two common methods are generally employed to reduce a slope's gradient. In the *cut-and-fill* method, material is removed from the upper part of the slope and used as fill at the base, thus providing a flat surface for construction and reducing the slope (▶ Figure 14-25). The second method, which is called *benching*, involves cutting a series of benches or steps into a hillside. This process reduces the average slope, and the benches serve as collecting sites for small landslides or rockfalls that might occur. Benching is most commonly used on steep hillsides in conjunction with a system of surface drains to divert runoff.

In some situations, retaining walls can be constructed to

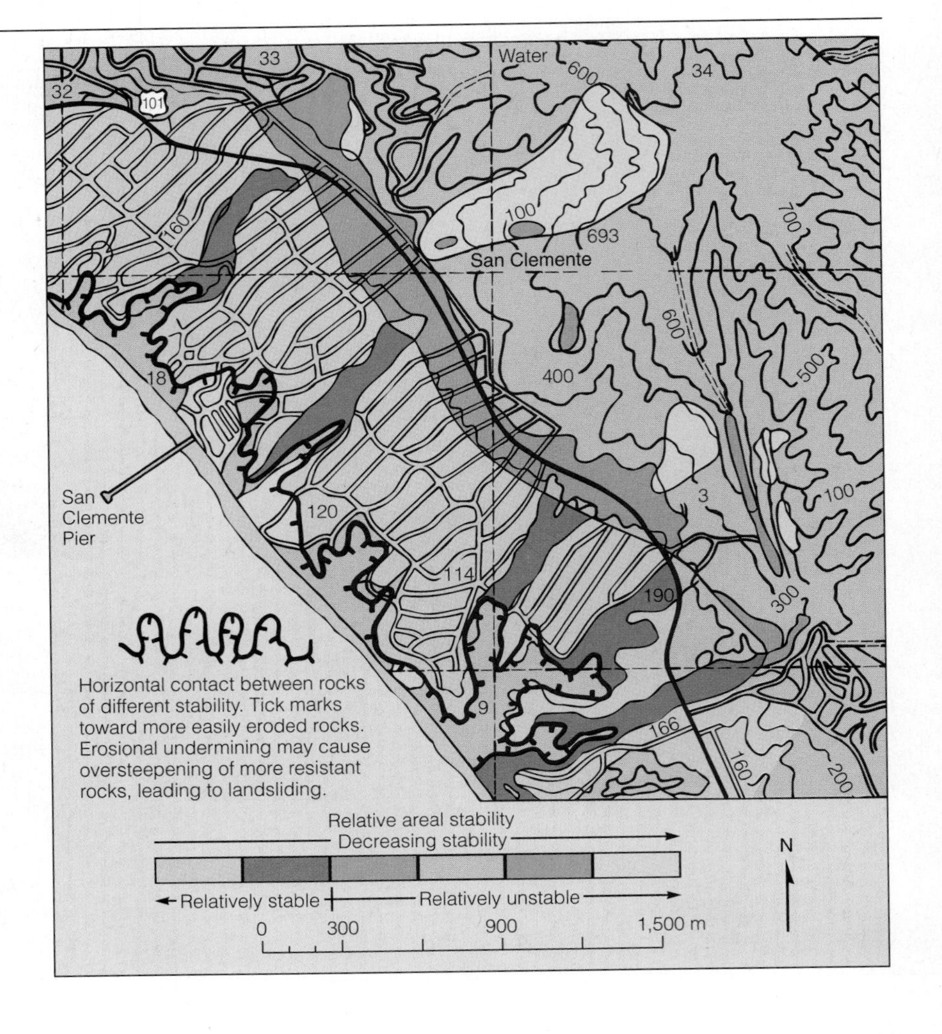

▶ **FIGURE 14-23** Relative slope stability map of part of San Clemente, California, showing areas delineated according to relative stability.

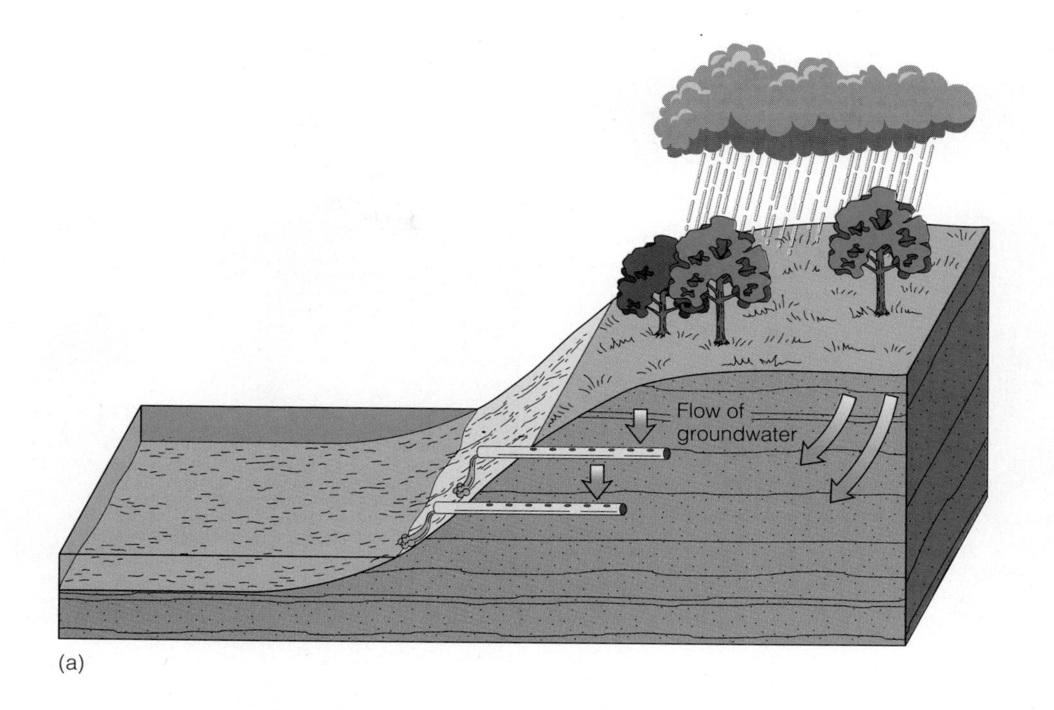

(a)

(b)

➢ **FIGURE 14-24** (*a*) Driving drainpipes perforated on one side into a hillside with the perforated side up can remove some subsurface water and help stabilize the hillside. (*b*) A drainpipe driven into the hillside at Point Fermin, California, helps remove subsurface water and stabilize the slope.

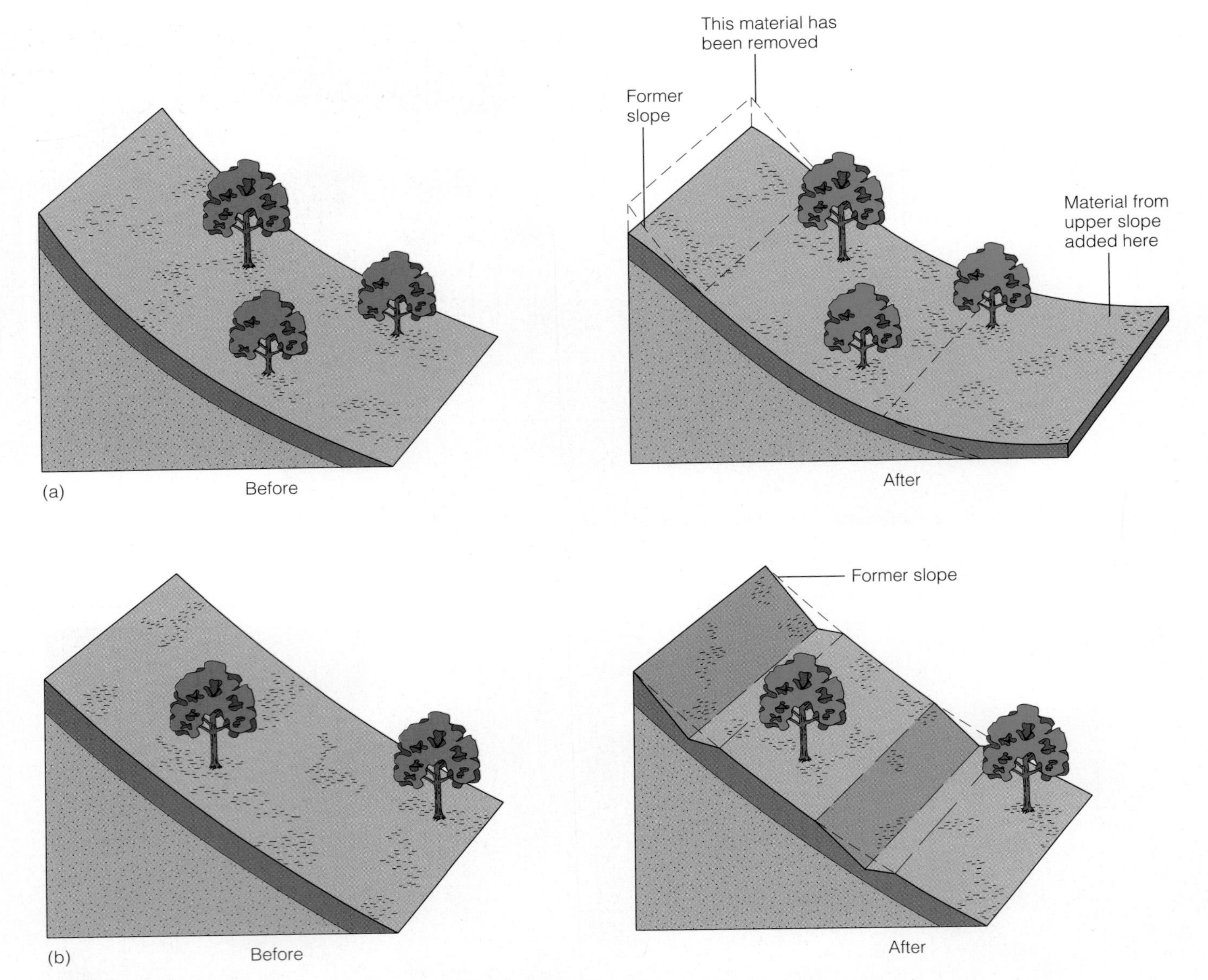

▶ **FIGURE 14-25** Two common methods used to help stabilize a hillside and reduce its slope. (*a*) In the cut-and-fill method, material from the steeper upper part of the hillside is removed, thereby reducing the slope angle, and is used to fill in the base. This provides some additional support at the base of the slope. (*b*) Benching involves making several cuts along a hillside to reduce the overall slope.

provide support for the base of the slope (▶ Figure 14-26). These are usually anchored well into bedrock, backfilled with crushed rock, and provided with drain holes to prevent the buildup of water pressure in the hillside.

Rock bolts, similar to those employed in tunneling and mining, can sometimes be used to fasten potentially unstable rock masses into the underlying stable bedrock (▶ Figure 14-27). This technique has been used successfully on the hillsides of Rio de Janeiro, Brazil, and to help secure the slopes at the Glen Canyon Dam on the Colorado River.

Recognition, prevention, and control of landslide-prone areas is expensive, but not nearly as expensive as the damage can be when such warning signs are ignored or not recognized. The collapse of Tip No. 7 at Aberfan, Wales (see Perspective 14-1) and the Vaiont Dam disaster (see Perspective 14-2) are two tragic examples where the warning signs of impending disaster were ignored.

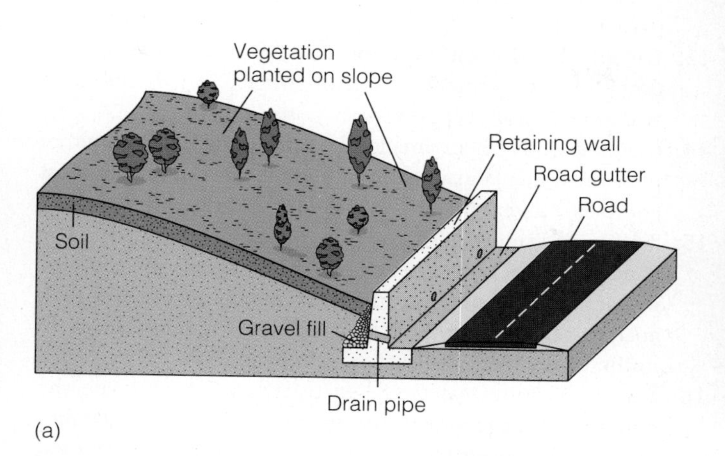

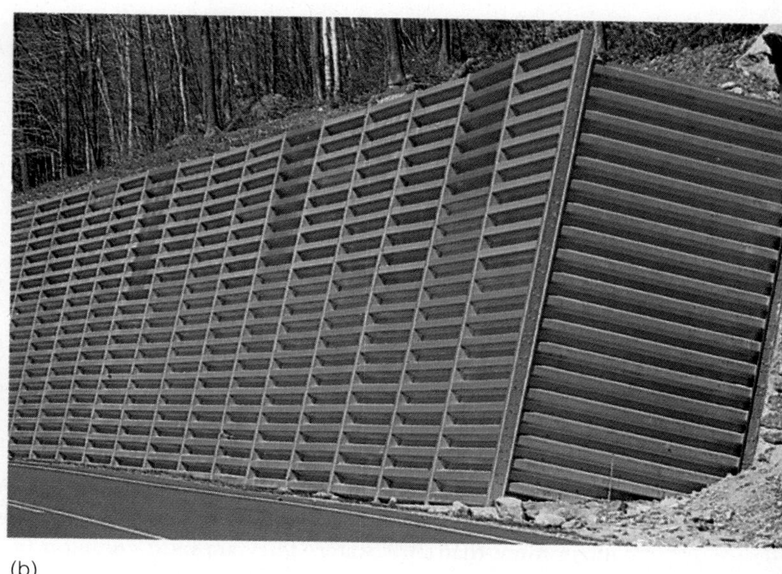

FIGURE 14-26 (*a*) Retaining walls anchored into bedrock, backfilled with gravel, and provided with drainpipes can support a slope's base and reduce landslides. (*b*) Steel retaining wall built to stabilize the slope and keep falling and sliding rocks off the highway.

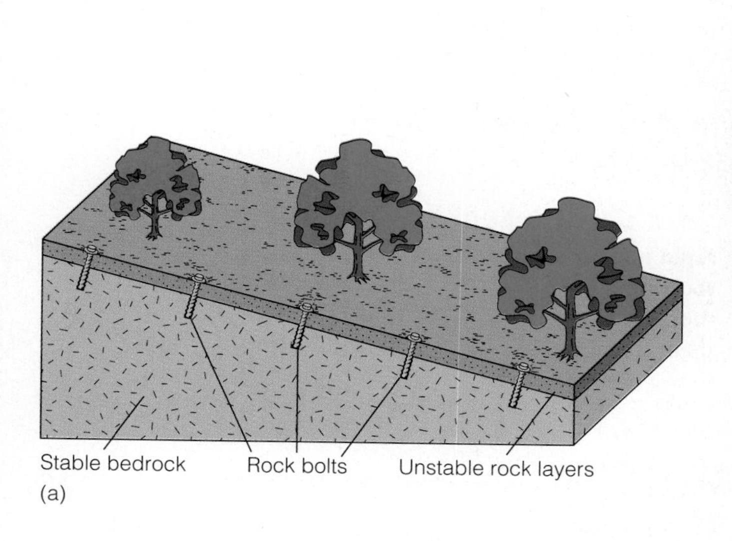

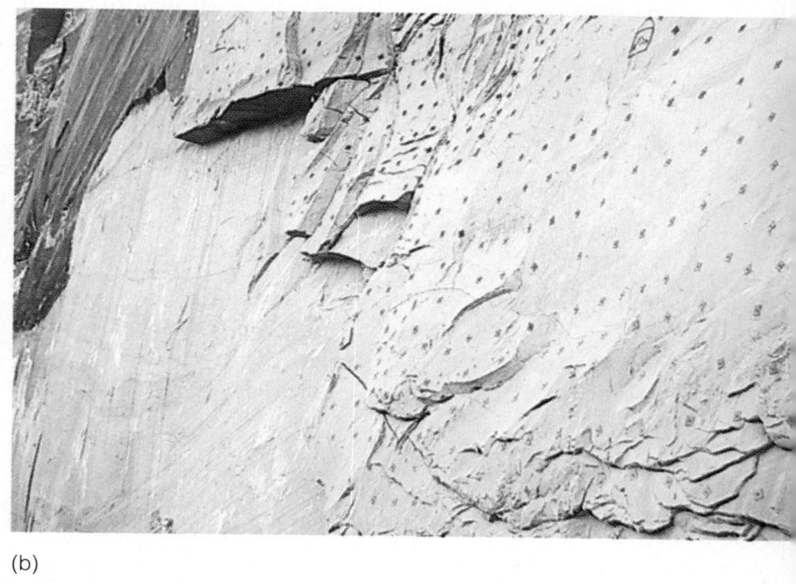

FIGURE 14-27 (*a*) Rock bolts secured in bedrock can help stabilize a slope and reduce landslides. (*b*) Rock bolts are used to help secure rock above the outlet of the west diversion tunnel of the Glen Canyon Dam. As can be seen, however, some portions of rock still broke away.

CHAPTER SUMMARY

1. Mass wasting is the downslope movement of material under the influence of gravity. It occurs when the gravitational force acting parallel to a slope exceeds the slope's strength.
2. Mass wasting frequently results in the loss of life, as well as causing millions of dollars in damage annually.
3. Mass wasting can be caused by many factors including slope gradient, weathering of slope material, water content,

overloading, and removal of vegetation. Usually, several of these factors in combination contribute to slope failure.

4. Mass movements are generally classified on the basis of their rate of movement (rapid vs. slow), type of movement (falling, sliding, or flowing), and type of material (rock, soil, or debris).
5. Rockfalls are a common mass movement in which rocks free-fall.
6. Two types of slides are recognized. Slumps are rotational slides involving movement along a curved surface; they are most common in poorly consolidated or unconsolidated material. Rock glides occur when movement takes place along a more-or-less planar surface; they usually involve solid pieces of rock.
7. Several types of flows are recognized on the basis of their rate of movement, type of material, and amount of water.
8. Mudflows consist of mostly clay- and silt-sized particles and contain more than 30% water. They are most common in semiarid and arid environments and generally follow pre-existing channels.
9. Debris flows are composed of larger-sized particles and contain less water than mudflows. They are more viscous and do not flow as rapidly as mudflows.
10. Earthflows move more slowly than either debris flows or mudflows; they move downslope as thick, viscous, tongue-shaped masses of wet regolith.
11. Quick clays are clays that spontaneously liquefy and flow like water when they are disturbed.
12. Solifluction is the slow downslope movement of water-saturated surface material and is most common in areas of permafrost.
13. Creep, the slowest type of flow, is the imperceptible downslope movement of soil or rock. Creep is the most widespread of all types of mass wasting.
14. Complex movements are combinations of different types of mass movements in which one type is not dominant. Most complex movements involve sliding and flowing.
15. The most important factor in reducing or eliminating the damaging effects of mass wasting is a thorough geologic investigation of the area including mapping, soil and rock analysis, and the construction of slope stability maps to outline areas susceptible to mass movements.
16. Several measures can be taken to reduce or eliminate the potential damage from mass wasting. First, slope stability maps can be constructed to outline areas susceptible to mass wasting, and secondly, slopes can be stabilized by retaining walls, draining excess water, regrading slopes, planting vegetation, and using rock bolts.

IMPORTANT TERMS

complex movement
creep
debris avalanche
debris flow
earthflow
mass wasting
mudflow
permafrost
quick clay
rapid mass movement
rockfall
rock glide
shear strength
slide
slow mass movement
slump
solifluction

REVIEW QUESTIONS

1. Shear strength includes:
   a. ____ the strength and cohesion of material;
   b. ____ the amount of internal friction between grains;
   c. ____ gravity;
   d. ____ all of these;
   e. ____ answers (a) and (b).
2. Which of the following is not a factor influencing mass wasting?
   a. ____ gravity;
   b. ____ weathering;
   c. ____ slope gradient;
   d. ____ water content;
   e. ____ none of these.
3. Which of the following factors can actually enhance slope stability?
   a. ____ water content;
   b. ____ vegetation;
   c. ____ overloading;
   d. ____ rocks dipping in the same direction as the slope;
   e. ____ none of these.
4. Mass wasting can occur:
   a. ____ on gentle slopes;
   b. ____ on steep slopes;
   c. ____ in flat-lying areas;
   d. ____ all of these;
   e. ____ none of these.
5. A type of mass wasting common in mountainous regions in which talus accumulates is:
   a. ____ creep;
   b. ____ solifluction;
   c. ____ rockfalls;
   d. ____ slides;
   e. ____ mudflows.
6. Movement of material along a surface or surfaces of failure is a:

a. ____ slide;
b. ____ fall;
c. ____ flow;
d. ____ solifluction;
e. ____ none of these.

7. Downslope movement along an essentially planar surface is a(n):
a. ____ slump;
b. ____ rockfall;
c. ____ earthflow;
d. ____ landslide;
e. ____ rock glide.

8. Which of the following helps to stabilize a slope?
a. ____ surface and subsurface drainage;
b. ____ planting vegetation;
c. ____ reducing its slope;
d. ____ rock bolts;
e. ____ all of these.

9. Which of the following are the most fluid of mass movements?
a. ____ earthflows;
b. ____ debris flows;
c. ____ mudflows;
d. ____ solifluction;
e. ____ slumps.

10. The most widespread and costly type of mass wasting in terms of total material moved and monetary damage is:
a. ____ creep;
b. ____ solifluction;
c. ____ mudflow;
d. ____ debris flow;
e. ____ slumping.

11. Which of the following features indicate former landslides or areas susceptible to slope failure?
a. ____ displaced objects;
b. ____ scarps;
c. ____ hummocky surfaces;
d. ____ open fissures;
e. ____ all of these.

12. What are the forces that help to maintain slope stability?
13. What roles do climate and weathering play in mass wasting?
14. How does vegetation affect slope stability? Give several examples.
15. What is overloading and why is it dangerous?
16. Give several examples of how the relationship between topography and the underlying geology affects slope stability.
17. What are rockfalls? Where are they most common and why?
18. What is the difference between a slump and a rock glide? Why are slumps particularly common along road cuts?
19. Differentiate between a mudflow, debris flow, and earthflow.
20. Why are quick clays so dangerous?
21. What precautions must be taken when building in permafrost areas?
22. How can creep be controlled once it has started?
23. What are complex movements, and how do they differ from other types of mass movements?
24. What are some of the indications of previous mass wasting? How can you recognize areas that are susceptible to mass movement?

Points to Ponder

1. What features would you look for to determine if the site where you wanted to build a house was safe?
2. If the slab of limestone that became detached from the Kabir Kuh ridge in Iran causing the Saidmarreh landslide was 305 m thick and covered an area of 70 km^2, and the resulting debris from the rock glide covered an area of 166 km^2, how thick was this layer of debris?
3. Discuss how slope stability maps used in conjunction with seismic risk maps could be extremely useful to planners, developers, and the government agencies responsible for overseeing development and growth in an area.
4. Do you think it will ever be possible to predict mass wasting events? Explain.

Additional Readings

Brabb, E. E., and B. L. Harrod, eds. 1989. *Landslides: Extent and economic significance.* Brookfield, Va.: A. A. Balkema.

Crozier, M. J. 1989. *Landslides: Causes, consequences, and environment.* Dover, N.H.: Croom Helm.

Fleming, R. W., and F. A. Taylor. 1980. *Estimating the cost of landslide damage in the United States.* U.S. Geological Survey Circular 832.

Kiersch, G. A. 1964. Vaiont reservoir disaster. *Civil Engineering* 34: 32–39.

McPhee, J. 1989. *The control of nature.* New York: Farrar, Straus & Giroux.

Small, R. J., and M. J. Clark. 1982. *Slopes and weathering.* New York: Cambridge University Press.

Zaruba, Q., and V. Mencl. 1982. *Landslides and their control.* 2d ed. Amsterdam, The Netherlands: Elsevier Publishing Co.

Chapter 15

Running Water

Outline

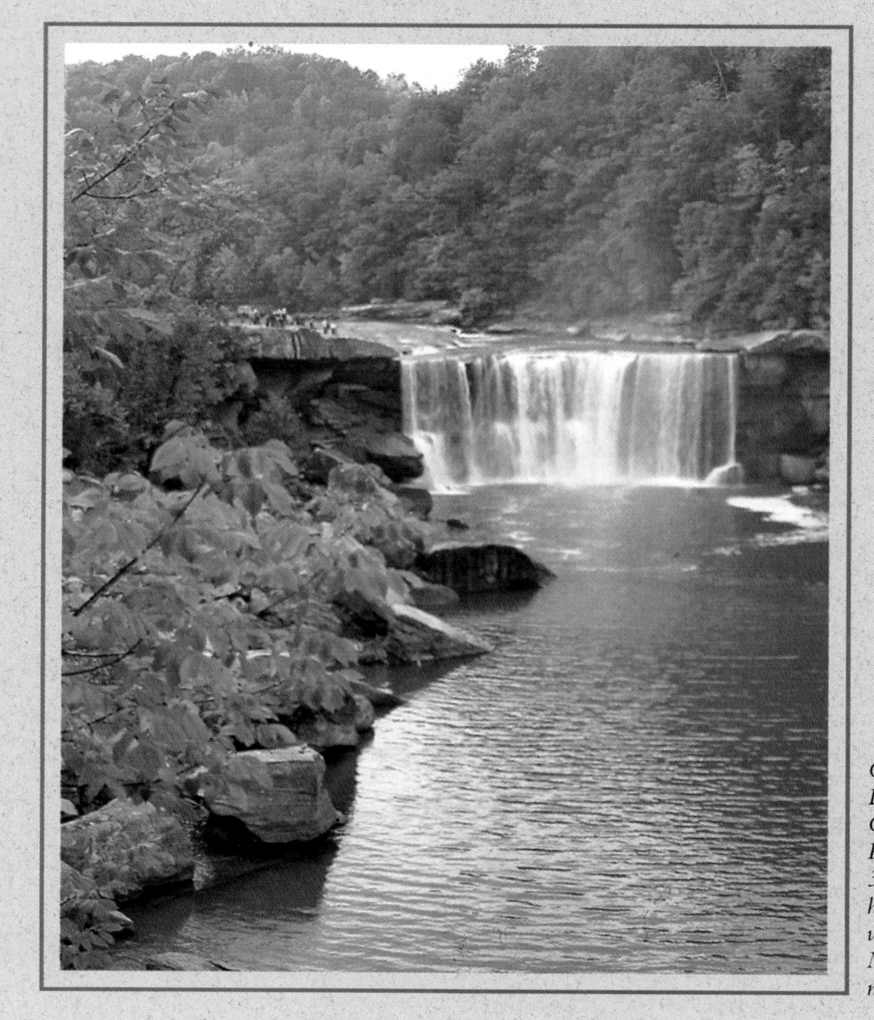

Cumberland Falls on the Big South Fork River in Cumberland Falls State Resort Park, Kentucky. At 38 m wide and 18 m high, it is the second largest waterfall east of the Rocky Mountains and one of the most impressive.

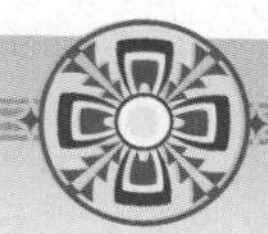

PROLOGUE

According to one news report, the flooding between late June and August 1993 in the midwestern United States caused an estimated $15 to $20 billion in property damage. Fifty people died as a result of the flooding, 70,000 were left homeless, and more than 200 counties in several states including every county in Iowa were declared disaster areas. In addition to Iowa, flooding also occurred in South Dakota, North Dakota, Minnesota, Wisconsin, Nebraska, Missouri, Illinois, and Kansas.

One of the areas flooded earliest was Davenport, Iowa, where the Mississippi River crested on July 7 at 1.8 m above flood stage. By July 13 it crested almost 10 m above flood stage at Quincy, Illinois. The Des Moines River crested at 8 m above flood stage on July 11 and flooded Des Moines, Iowa, and surrounding areas. By the end of the first week in July, at least 10 million acres of farmland had been flooded, with more rain expected; by August a total of 23 million acres had been flooded.

Despite the heroic efforts of thousands of volunteers and the National Guard to stabilize levees by sandbagging them, levees failed everywhere or were simply overtopped by the rising floodwaters. Grafton, Illinois, at the juncture of the Mississippi and Illinois rivers was 80% underwater. Grafton, which lies on the floodplain and has no protective levee system, has been flooded six times in the last 20 years. In the aftermath of the flood, the mayor and many of Grafton's residents decided to move much of the town to higher ground at an estimated cost of $25 million.

Much of St. Charles County, Missouri, at the confluence of the Mississippi and Missouri rivers was so extensively flooded that 8,000 people were evacuated from this county alone. In the small town of Portage des Sioux in St. Charles County, only eight homes were not flooded or were only slightly damaged by flooding (⊳ Figure 15-1). Even the 5-m-high pedestal of a statue near the Mississippi was swamped by the floodwaters.

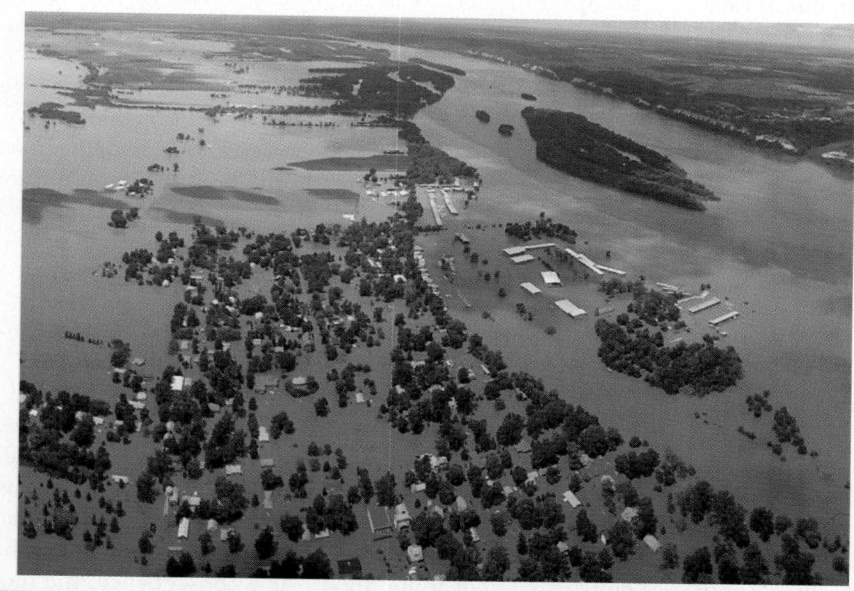

(a)

⊳ **FIGURE 15-1** (*a*) Portage des Sioux, St. Charles County, Missouri, on July 16, 1993. The channel of the Mississippi River is at the far right. (*b*) Portage des Sioux on November 7 after the flood waters had subsided.

(b)

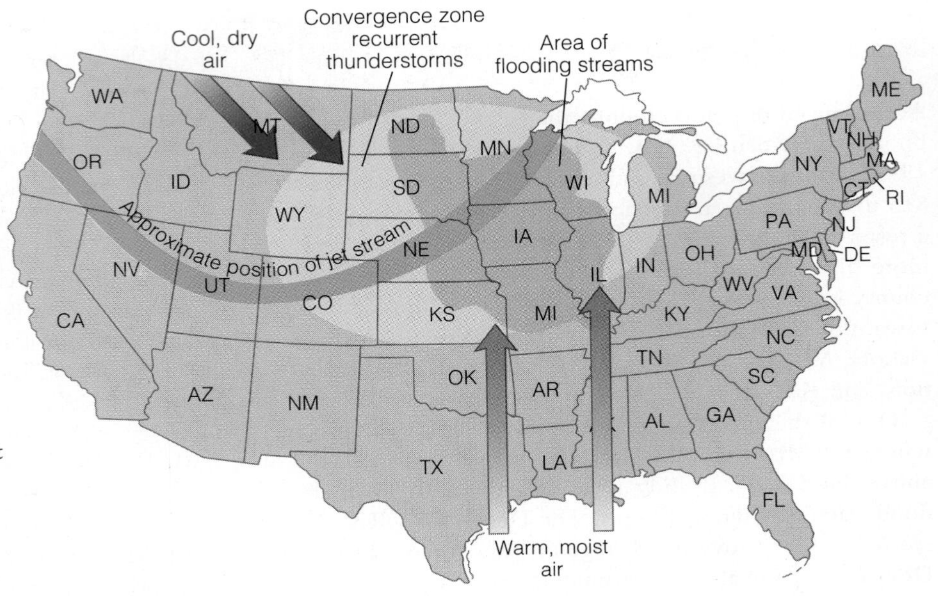

▶ **FIGURE 15-2** The dominant weather pattern for June and July 1993. The jet stream remained over the Midwest during the summer rather than shifting north over Canada as it usually does. Thunderstorms developed in the convergence zone where warm, moist air and cool, dry air met.

By mid-July flooding was extensive and still more rain was expected. On July 16, an additional 17.8 cm of rain fell in North Dakota and Minnesota, and 15.2 cm more fell in South Dakota further swelling already flooding rivers. Obviously, the direct cause of flooding was too much water for the Mississippi and Missouri rivers and their tributaries to handle. But the reason so much water was present was the unusual behavior of the jet stream.

The *jet stream* is a narrow band of strong winds in the atmosphere. It is usually over the Midwest during the spring and then shifts north into Canada during the summer. In 1993 it remained over the Midwest, and as hot, moisture-laden air moved up from the south, thunderstorms developed over the Midwest (▶ Figure 15-2). In short, these storms are usually distributed over a much larger region and spread to the northeast, but in this case they simply dumped their precipitation over a much smaller region, resulting in as much as 1½ to 2 times the normal amount.

One important lesson learned from the Flood of '93 is that despite our best efforts at flood control, some floods will occur anyway. The 29 dams on the Mississippi and 36 reservoirs on upstream tributaries, as well as about 5,800 km of levees, had little effect on holding the floodwaters in check. Reservoirs have a limited capacity and when full are of no further use in controlling floods. In fact, some reservoirs are kept at high levels for recreation, which is inconsistent with their role in flood control.

No doubt debate will now focus on the utility of levees in flood control. Levees are effective in protecting many areas during floods, yet in some cases they actually exacerbate the problem by restricting the flow that would have formerly spread over a floodplain. They are certainly expensive to build and maintain, and their overall effectiveness has been and will continue to be questioned.

The object of most criticism is the U.S. Army Corps of Engineers, which has spent about $25 billion in this century to build 500 dams and more than 16,000 km of levees. No one doubts that some of these projects have been successful, at least within the limits of their design. But critics charge that such flood-control projects actually make the problem of flooding worse, particularly because development in flood-prone areas commonly follows the completion of flood-control projects, and nothing can be done to prevent some floods. As a consequence, the most destructive and most costly type of natural disaster continues to be flooding.

 INTRODUCTION

Among the terrestrial planets, the Earth is unique in having abundant liquid water. Both Mercury and the Earth's moon are too small to retain any water, and Venus, because of its runaway greenhouse effect, is too hot to retain surface water. Mars has only some frozen water and trace amounts of water vapor in its atmosphere. In marked contrast, 71% of the Earth's surface is covered by water, and a small but important quantity of water vapor is present in its atmosphere.

The volume of water on Earth is estimated at 1.36 billion km^3, most of which (97.2%) is in the oceans. About 2% is frozen in glaciers, and the remaining 0.8% constitutes all the water in streams, lakes, swamps, groundwater, and the atmosphere (◉ Table 15-1). Thus, only a tiny portion of the total water on Earth is in streams, but running water is nevertheless the most important erosional agent modifying the Earth's surface.

Despite the importance of running water as an agent of erosion, sediment transport, and deposition, its role is limited in some areas. In areas covered by glacial ice, such as Greenland and Antarctica, running water is currently not important. Some parts of deserts are also little affected by running water. Even in most desert regions, though, the effects of running water are manifest, although the channels are dry most of the time (▶ Figure 15-3).

In addition to its significance as a geologic agent, running water is important for many other reasons as well. It is a source of fresh (nonsaline) water for industry, domestic use, and agriculture, and about 8% of the electricity used in North America is generated by falling water at hydroelectric stations (▶ Figure 15-4). Streams have been, and continue to be, important avenues of commerce and much of the interior of North America was first explored by following such large streams as the St. Lawrence, Mississippi, and Missouri rivers.

Much of this discussion of running water is necessarily descriptive, but one should always be aware that streams are dynamic systems that must continually respond to change. For example, paving in urban areas increases surface runoff to streams, while other human actions such as building dams and impounding reservoirs also alter the dynamics of a stream system. Natural changes, too, affect stream dynamics. When more rain falls in a stream's drainage area due to a long-term climatic change, more water flows in the stream's channel, and greater energy is available for erosion and transport of sediments. In short, streams adjust to any change occurring within their overall systems.

THE HYDROLOGIC CYCLE

Although the quantity of water in streams is small at any one time, during the course of a year very large volumes of water move through stream channels. In fact, water is continually recycled from the oceans, through the atmosphere, to the continents, and back to the oceans. This continuous recycling of water is called the **hydrologic cycle** (▶ Figure 15-5). (The hydrologic cycle will also be relevant to our discussions of groundwater in Chapter 16 and glaciers in Chapter 17.)

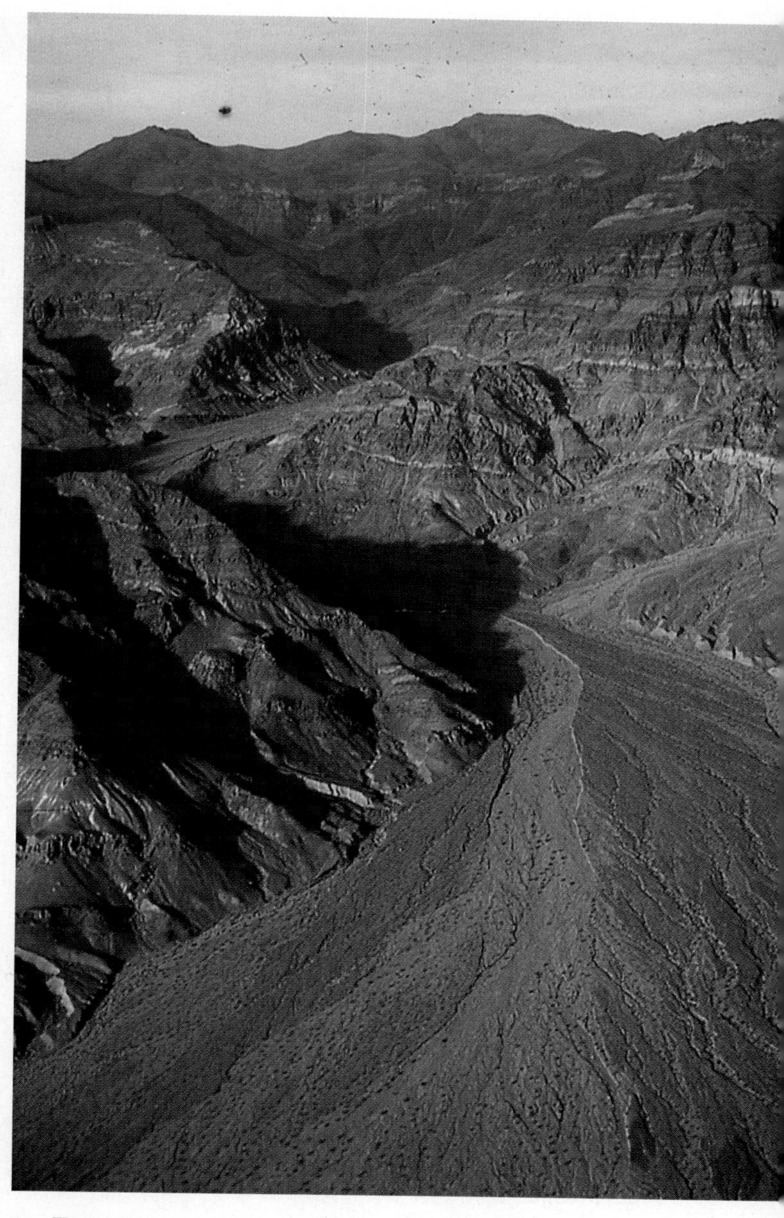

▶ **FIGURE 15-3** Although stream channels in deserts are dry most of the time, evidence for running water is common even in such arid areas as Death Valley, California.

TABLE 15-1 Water on Earth

Location	Volume (km^3)	Percentage of Total
Oceans	1,327,500,000	97.20
Icecaps and glaciers	29,315,000	2.15
Groundwater	8,442,580	0.625
Freshwater and saline lakes and inland seas	230,325	0.017
Atmosphere at sea level	12,982	0.001
Average in stream channels	1,255	0.0001

➢ **FIGURE 15-4** About 8% of the electricity used in North America is generated by falling water at hydroelectric stations such as this one on the New River in Virginia. (Photo courtesy of R. V. Dietrich.)

➢ **FIGURE 15-5** During the hydrologic cycle, water evaporates from the oceans and rises as water vapor to form clouds that release their precipitation either over the oceans or over land. Much of the precipitation falling on land returns to the oceans by surface runoff, thus completing the cycle.

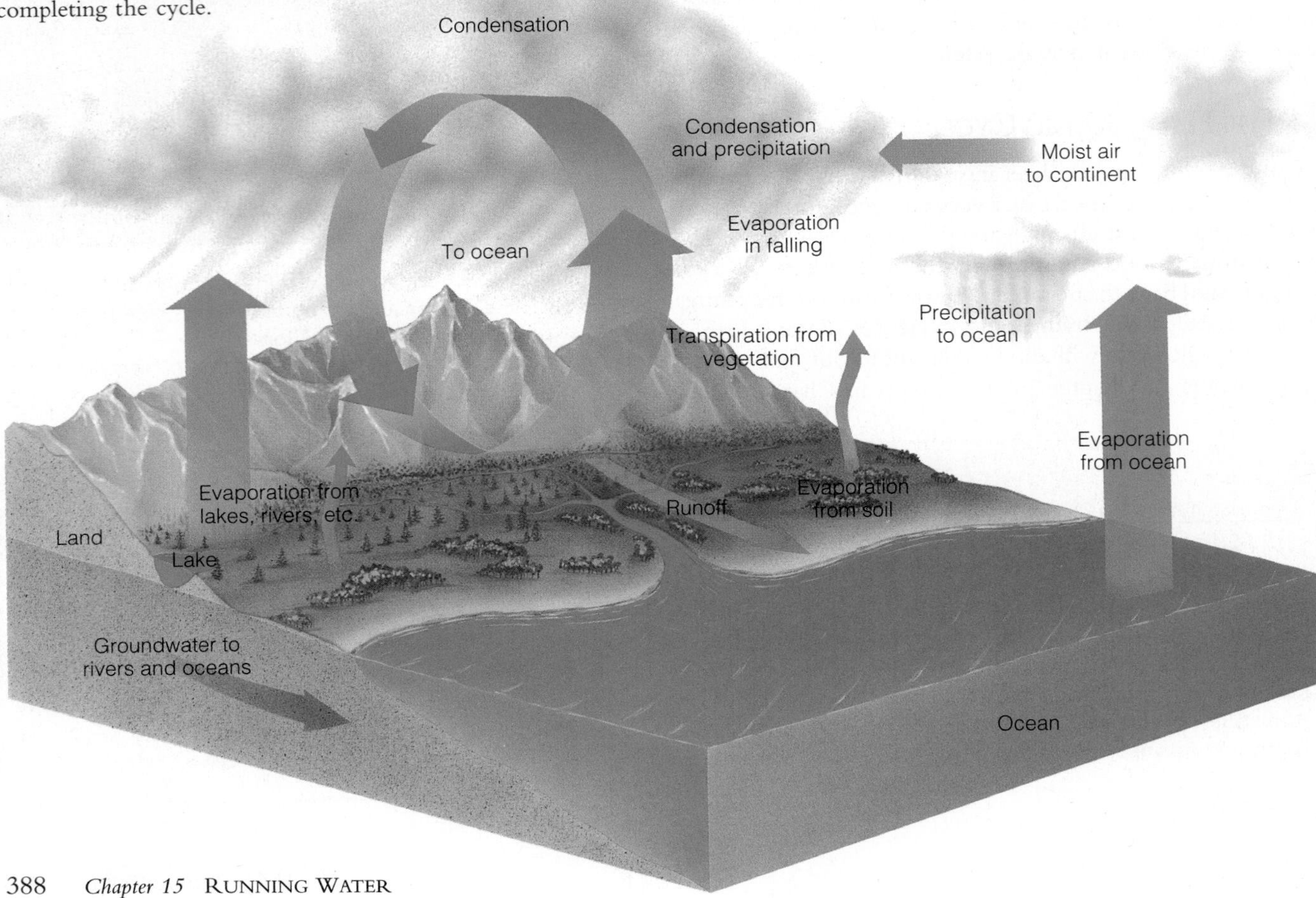

The hydrologic cycle, which is powered by solar radiation, is possible because water changes phases easily under Earth surface conditions. Huge quantities of water evaporate from the oceans each year as the surface waters are heated by solar energy. The amount of ocean water evaporated yearly corresponds to a layer about 1 m thick from all the oceans. Approximately 85% of all water that enters the atmosphere is derived from the oceans; the remaining 15% comes from evaporation of water on land.

When water evaporates, the vapor rises into the atmosphere where the complex processes of condensation and cloud formation occur. About 80% of the precipitation falls directly into the oceans, in which case the hydrologic cycle is limited to a three-step process of evaporation, condensation, and precipitation.

About 20% of all precipitation falls on land as rain and snow. In this case, the hydrologic cycle involves more steps: evaporation, condensation, movement of water vapor from the oceans to the continents, precipitation, and runoff and infiltration. Some of the precipitation evaporates as it falls and reenters the hydrologic cycle as vapor; water evaporated from lakes and streams also reenters the cycle as vapor as does moisture evaporated from plants by *transpiration* (Figure 15-5).

Each year about 36,000 km^3 of the precipitation falling on land returns to the oceans by **runoff,** the surface flow of streams. Of course, the amount of precipitation and runoff varies widely across the continents; some areas are extremely arid, whereas others receive hundreds of centimeters of precipitation yearly. In any case, the water returning to the oceans by runoff enters the Earth's ultimate reservoir where it begins the hydrologic cycle again.

Some of the precipitation falling on land is temporarily stored in lakes, snow fields, and glaciers or seeps below the surface where it is temporarily stored as groundwater. This water is effectively removed from the system for up to thousands of years, but eventually, glaciers melt, lakes feed streams, and groundwater flows into streams or directly into the oceans (Figure 15-5). Our concern here is with the comparatively small quantity returning to the oceans as runoff, for the energy of running water is responsible for a great many surface features.

RUNNING WATER

Water possesses no strength so it will flow on any slope no matter how slight. The flow of water, or any other fluid, can be characterized as *laminar* or *turbulent*. In laminar flow, lines of flow called streamlines are all parallel with one another. In other words, all flow occurs in parallel layers with no mixing between layers (➢ Figure 15-6a). By contrast, in turbulent flow, the streamlines intertwine, causing a complex mixing of the fluid (Figure 15-6b).

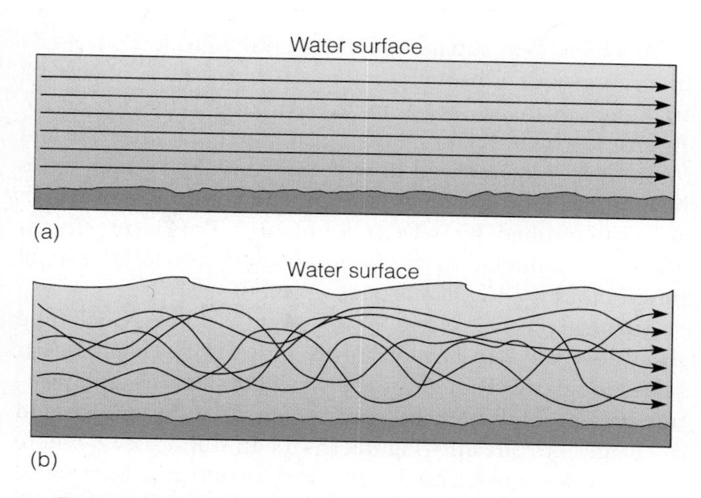

➢ FIGURE 15-6 (*a*) In laminar flow, streamlines are all parallel to one another, and little or no mixing occurs between adjacent layers in the fluid. (*b*) In turbulent flow, streamlines are complexly intertwined, indicating mixing between adjacent layers in the fluid. Most flow in streams is turbulent.

Laminar flow is most easily observed in viscous fluids such as cold motor oil or syrup. One can also see laminar flow in parking lots where a thin film of water containing oil moves slowly over the surface. Turbulent flow, on the other hand, occurs in all streams. The primary control on the type of flow is velocity; roughness of the surface over which flow occurs also plays a role. Laminar flow occurs when water flows very slowly, as when groundwater moves through the tiny pores in sediments and soil. In streams, however, the flow is usually fast enough and the channel walls and bed rough enough so that flow is fully turbulent. Laminar flow is so slow, and generally so shallow, that it causes little or no erosion. Turbulent flow is much more energetic and is capable of considerable erosion and sediment transport.

Sheet Flow versus Channel Flow

The amount of runoff in any area during a rainstorm depends on the **infiltration capacity,** the maximum rate at which soil or other surface materials can absorb water. Infiltration capacity depends on several factors, including the intensity and duration of rainfall. Loosely packed, dry soils absorb water faster than tightly packed, wet soils. Hard, dry surfaces, such as those that develop during droughts, also have low infiltration capacities. Therefore, when they do receive rain, there is still considerable runoff.

If rain is absorbed as fast as it falls, no surface runoff occurs. Should the infiltration capacity be exceeded, or should surface materials become saturated, excess water collects on the surface and, if a slope exists, moves downhill. Even on steep slopes such flow is initially slow, and hence little or no erosion occurs. As the water moves downslope, however, it accelerates and may move by *sheet flow,* a more-or-less continuous film of water flowing over the surface. Sheet flow is not confined to depressions, and it accounts for *sheet erosion,* a particular problem on some agricultural lands (see Chapter 5).

In *channel flow,* surface runoff is confined to long, trough-like depressions. Channels vary in size from rills containing a trickling stream of water to the Amazon River of South America, which is 6,450 km long and up to 2.4 km wide and 90 m deep. Channelized flow is described by various terms including rill, brook, creek, stream, and river, most of which are distinguished by size and volume. The term **stream** carries no connotation of size and is used here to refer to all runoff confined to channels regardless of size.

Streams receive water from several sources, including sheet flow and rain falling directly into stream channels. Far more important, though, is the water supplied by soil moisture and groundwater, both of which flow downslope and discharge into streams (Figure 15-5). In humid areas where groundwater is plentiful, streams may maintain a fairly stable flow year round, even during dry seasons, because they are continually supplied by groundwater. In contrast, the amount of water in streams of arid and semiarid regions fluctuates widely because these streams depend more on infrequent rainstorms and surface runoff for their water supply.

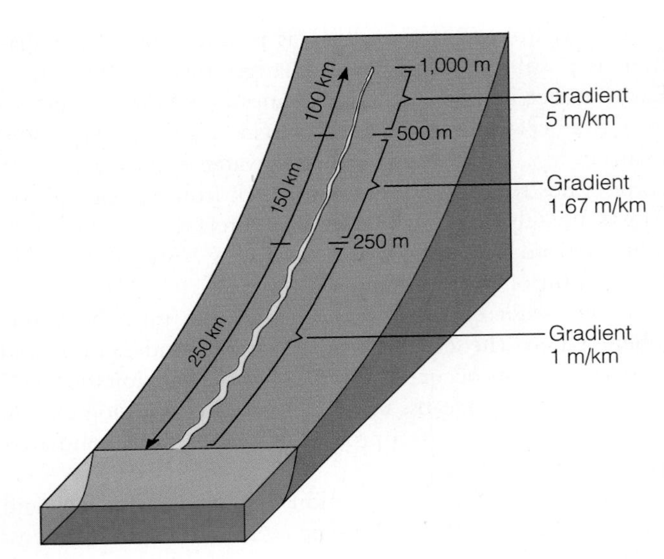

➢ **FIGURE 15-7** The average gradient of this stream is 2 m/km. Gradient can be calculated for any segment of a stream as shown in this example. Notice that the gradient is steepest in the headwaters area and decreases in a downstream direction.

Stream Gradient

Streams flow downhill from a source area to a lower elevation where they empty into another stream, a lake, or the sea.* The slope over which a stream flows is its **gradient.** For example, if the source (headwaters) of a stream is 1,000 m above sea level and the stream flows 500 km to the sea, it drops 1,000 m vertically over a horizontal distance of 500 km (➢ Figure 15-7). Its gradient is calculated by dividing the vertical drop by the horizontal distance; in this example, it is 1,000 m/500 km = 2 m/km.

Gradients vary considerably, even along the course of a single stream. Generally, streams are steeper in their upper reaches where their gradients may be tens of meters per kilometer, but in their lower reaches the gradient may be as little as a few centimeters per kilometer. Some streams in mountainous regions have particularly steep gradients of several hundred meters per kilometer.

Velocity and Discharge

Stream velocity and discharge are closely related variables. **Velocity** is simply a measure of the downstream distance traveled per unit of time. Velocity is usually expressed in feet per second (ft/sec) or meters per second (m/sec) and varies considerably among streams and even within the same stream.

Variations in flow velocity occur not only with distance along a stream channel but also across a channel's width. For example, because of friction, flow velocity is slower and more turbulent near a stream's bed and banks than it is farther from these boundaries. The bed and banks cause frictional resistance to flow, whereas the water some distance away is unaffected by friction and has a higher velocity (➢ Figure 15-8).

Other controls on velocity include channel shape and roughness. Broad, shallow channels and narrow, deep channels have proportionally more water in contact with their perimeters than do channels with semicircular cross sections (➢ Figure 15-9). Consequently, the water in semicircular

➢ **FIGURE 15-8** In a stream, flow velocity varies as a result of friction with the banks and bed. The maximum flow velocity is near the center and top of a stream in a straight channel. The lengths of the arrows in this illustration are proportional to the velocity.

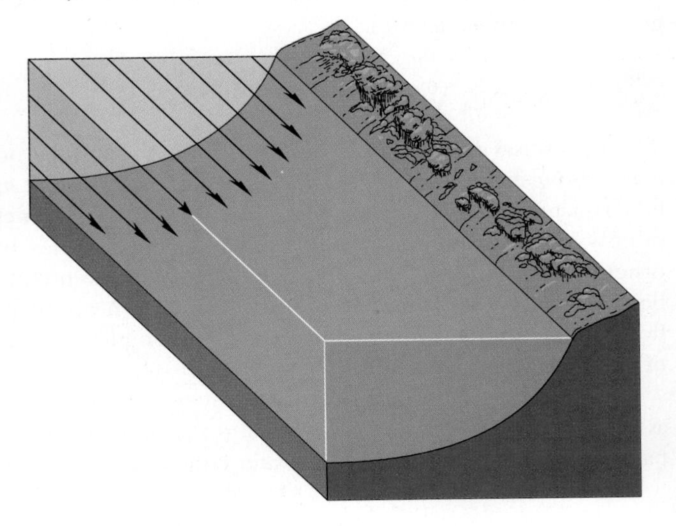

*The flow in certain desert streams diminishes in a downstream direction by evaporation and infiltration until the streams disappear. Some streams in regions with numerous caverns may disappear below the ground.

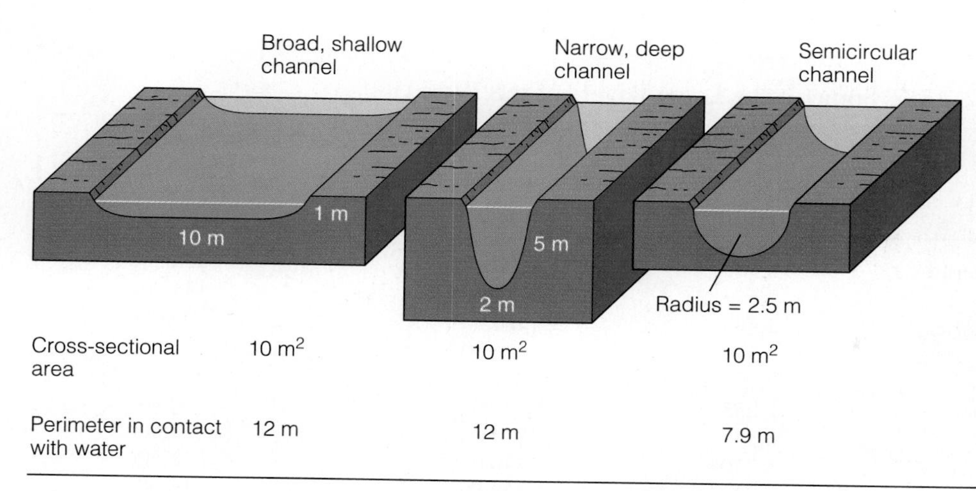

▶ **FIGURE 15-9** All three of these channels have the same cross-sectional area, but each has a different shape. The semicircular channel has the least perimeter in contact with the water and causes the least frictional resistance to flow. If other variables, such as channel roughness, are the same in all of these channels, flow velocity will be greatest in the semicircular channel.

channels flows more rapidly because it encounters less frictional resistance. In many streams the maximum flow velocity occurs near the surface at the center of the channel; it occurs slightly below the surface because of frictional resistance from the air above. In sinuous (meandering) channels, however, the line of maximum flow velocity switches from one side of the channel to the other and corresponds to the channel center only along straight reaches (▶ Figure 15-10).

Channel roughness is a measure of the frictional resistance within a channel. Frictional resistance to flow is greater in a channel containing large boulders than in one with banks and a bed composed of sand or clay. In channels with abundant vegetation, flow is slower than in barren channels of comparable size.

The most obvious control on velocity is gradient, and one might think that the steeper the gradient, the greater the flow velocity. In fact, the average velocity generally increases in a downstream direction, even though the gradient decreases in the same direction. Three factors contribute to this: First, velocity increases continuously, even as gradient decreases, in response to the acceleration of gravity unless other factors retard flow. Secondly, in their upstream reaches, streams commonly have boulder-strewn, broad, shallow channels, so flow resistance is high and velocity is correspondingly slower. Downstream, however, channels generally become more semicircular, and the bed and banks are usually composed of finer-grained materials, thus reducing the effects of friction. Thirdly, the number of tributary streams joining a larger stream increases in a downstream direction. Thus, the total volume of water (discharge) increases, and increasing discharge results in increased velocity.

▶ **FIGURE 15-10** In a sinuous (meandering) channel, flow velocity varies from one side of the channel to the other. As the water flows around curves, it flows fastest near the outer bank. The dashed line in this illustration follows the path of maximum flow velocity.

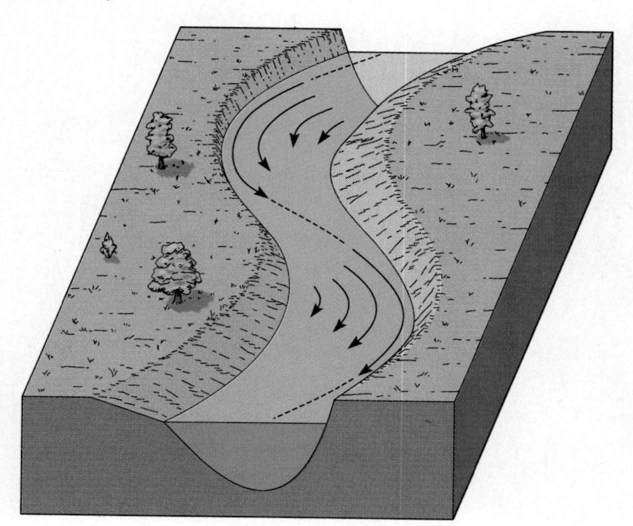

Discharge is the total volume of water in a stream moving past a particular point in a given period of time. To determine discharge, one must know the dimensions of a channel—that is, its cross-sectional area (A)—and its flow velocity (V). Discharge (Q) can then be calculated by the formula $Q = VA$, and it is generally expressed in cubic feet per second (ft^3/sec) or cubic meters per second (m^3/sec). The Mississippi River has an average discharge of 18,000 m^3/sec (◉ Table 15-2), but, as for all streams, the discharge varies, being greatest during floods and lowest during long, dry spells.

STREAM EROSION

Streams possess two kinds of energy, potential and kinetic. *Potential energy* is the energy of position, such as that possessed by water behind a dam or at a high elevation. In stream flow, potential energy is converted to *kinetic energy*, the energy of motion. Most of this kinetic energy is dissipated as heat within streams by fluid turbulence, but a small amount, perhaps 5%, is available to erode and transport sediment. Erosion involves the physical removal of dissolved substances and loose particles of soil and rock from a source

TABLE 15-2 Some of the Large Rivers of the World

River	Continent	Length (km)	Drainage Area (km^2)	Average Discharge at Mouth (m^3/sec)
Amazon	South America	6,450	6,150,000	200,000
Zaire (Congo)	Africa	4,675	3,720,000	44,000
Orinoco	South America	2,580	970,000	36,000
Ganges and Brahmaputra	Asia	2,500 2,900	1,480,000	31,000
Rio Negro	South America	2,255	1,000,000	30,000
Yangtze	Asia	5,530	1,810,000	29,000
Mississippi	North America	3,770	3,240,000	18,000
Zambezi	Africa	2,740	1,420,000	16,000
Mekong	Asia	4,190	786,000	15,000
St. Lawrence	North America	1,290	1,030,000	9,900
Mackenzie	North America	4,245	1,810,000	9,700
Ohio	North America	2,110	526,000	8,000
Columbia	North America	2,000	850,000	7,500
Yukon	North America	3,190	850,000	6,400
Nile	Africa	6,705	3,030,000	1,000
Murray	Australia	2,580	1,060,000	700

area. Thus, the sediment transported in a stream consists of both dissolved materials and solid particles.

Because the *dissolved load* of a stream is invisible, it is commonly overlooked, but it is an important part of the total sediment load. Some of it is acquired from the stream bed and banks where soluble rocks such as limestone and dolostone are exposed, but much of it is carried into streams by sheet flow and by groundwater.

The solid sediment carried in streams ranges from clay-sized particles to large boulders (▶ Figure 15-11). Much of this sediment is supplied to streams by mass wasting (▶ Figure 15-12), but some is derived directly from the stream bed and banks. The power of running water, called **hydraulic action,** is sufficient to set particles in motion. Everyone has seen the results of hydraulic action, although perhaps not in streams. For example, if the flow from a

▶ FIGURE 15-11 The solid load of stream varies depending on its energy and may consist of particles up to the size of boulders. (Photo courtesy of B. M. C. Pape.)

➢ **FIGURE 15-12** Streams such as the Snake River in Idaho receive some of their sediment load by mass wasting processes, frost wedging in this case. (Photo courtesy of R. V. Dietrich.)

garden hose is directed onto loose soil, a hole is soon gouged out by hydraulic action.

Another process of erosion in streams is **abrasion,** in which exposed rock is worn and scraped by the impact of solid particles. If running water contains no sediment, little or no erosion of solid rock surfaces will result, but if it is transporting sand and gravel, the impact of these particles abrades exposed rock surfaces. *Potholes* in the beds of streams are one obvious manifestation of abrasion (➢ Figure 15-13). These circular to oval holes occur where eddying currents containing sand and gravel swirl around and erode depressions into solid rock.

➢ **FIGURE 15-13** Potholes in the bed of the Chippewa River in Ontario, Canada.

Guest Essay E. Burton Kemp III

Managing Our Water Resources

My interest in geology began when I was a senior in high school. At that time, the petroleum industry was at a high point. In particular, overseas exploration and development were beginning to expand dramatically. The possibility of extensive travel to foreign countries, coupled with the potential for a comfortable livelihood, was attractive to me. At the same time, I had always had an interest in the natural and physical sciences. Therefore, geology seemed to be a perfect choice for me: it was a natural physical science, and it offered me the opportunity to travel, seek my fortune, and satisfy my interest in naturally occurring phenomena.

My original intent was to pursue a career in the petroleum industry; however, at the time of my graduation from undergraduate school, the oil and gas industry was in one of the cyclic downturns to which it is prone. After much thought, I decided that a change in direction would be prudent. I shifted my emphasis toward more engineering-related areas of study and began a career as an engineering geologist with the U.S. Army Corps of Engineers. I also elected to pursue graduate studies with emphasis on engineering geology and its many related fields such as geomorphology, hydrogeology, environmental geology, and the application of geology to engineering structures.

During my career, I have been privileged to participate in a number of investigations, studies, projects, and activities that have been extremely rewarding, from both a professional and a personal viewpoint. Specifically, I participated in the initial site investigations and final design for a number of large projects such as dams, control structures, and navigation locks. Cooper Dam in northeast Texas; Caddo Dam in northwest Louisiana; Old River Control Auxiliary Structure in central Louisiana; Lock and Dam Numbers 1 through 5 on the Red River Waterway in north-central Louisiana; and the Lake Pontchartrain & Vicinity Hurricane Protection Project in southeast Louisiana are some examples of these projects. These structures are used to control floods and hurricane tidal surges, thus reducing property damage and the potential for loss of life. They also help maintain our nation's waterways so that waterborne commerce can move efficiently.

I have also been engaged in several ongoing groundwater studies involving foundation effects, groundwater levels and supply potential, and hazardous toxic waste contamination. These studies have led to such projects as the evaluation and design of an emergency dewatering system to prevent excessive uplift during dewatering of a critical control structure in eastern New Orleans; groundwater evaluation and development programs in east and southwest Texas; design of a groundwater monitoring program for a hazardous holding pond in north-central Louisiana; detailed analysis of the subsurface for hazardous toxic waste analysis at several industrial facilities in Louisiana, Arkansas, and South Carolina; and evaluation of the potential for groundwater contamination at a proposed High Level Nuclear Waste Repository in central Mississippi. These studies included studying, examining, and monitoring the effects that changes in groundwater levels and chemistry have on our environment.

Finally, through participation in detailed studies of the Louisiana Coastal Zone marshes and wetlands, I have been able to help define land loss rates, make predictions on future areas where land loss problems will increase or diminish, and isolate areas that have minimal losses and offer the best potential to reduce future losses or create new marshes. These coastal zone studies will make it possible to reduce the losses within our valuable wetlands and help preserve these natural resources.

These studies and investigations, as well as participation on numerous committees, boards, review panels, study groups, and seminars, have afforded me the opportunity to work closely with many other disciplines in conserving, protecting, and improving our natural environment. I feel strongly that geology is a very dynamic and rewarding field and will continue to be equally exciting and challenging in the future.

E. BURTON KEMP III earned an M.S. in geology from Tulane University Graduate School. He has worked in the area of engineering geology for more than 30 years and is currently a district geologist with the U.S. Army Corps of Engineers for the New Orleans District.

Transport of Sediment Load

Streams transport a solid load of sedimentary particles and a **dissolved load** consisting of ions taken into solution by chemical weathering. Sedimentary particles are transported either as suspended load or as bed load. **Suspended load** consists of the smallest particles, such as silt and clay, which are kept suspended by fluid turbulence (▶ Figure 15-14). The Mississippi River transports nearly 200 million metric tons of suspended load past Vicksburg, Mississippi, each year, and the Yellow River of China carries almost four times as much suspended load per year. Particles transported in suspension are deposited only where turbulence is minimal as in lakes or the quiet offshore waters of the sea.

Bed load consists of the coarser particles such as sand and gravel (Figure 15-14). Fluid turbulence is insufficient to keep such large particles suspended, so they move along the stream bed. Part of the bed load can be suspended temporarily as when an eddying current swirls across a stream bed and lifts sand grains into the water. These particles move forward at approximately the flow velocity, but at the same time they settle toward the stream bed where they come to rest, to be moved again later by the same process. This process of intermittent bouncing and skipping along the stream bed is called *saltation* (Figure 15-14).

Particles too large to be suspended even temporarily are transported by rolling or sliding (Figure 15-14). Obviously, greater flow velocity is required to move particles of these sizes. The maximum-sized particles that a stream can carry define its *competence,* a factor related to flow velocity. ▶ Figure 15-15 shows the velocities required to erode, transport, and deposit particles of various sizes. As expected, high velocity is necessary to erode and transport gravel-sized particles, whereas sand is eroded and transported at lower velocities. Notice, though, that high velocity is needed to erode clay because clay deposits are very cohesive: the tiny clay particles adhere to one another and can be disrupted only by energetic flow conditions. Once eroded, however, very little energy is needed to keep the clay particles in motion.

Capacity is a measure of the total load a stream can carry. It varies as a function of discharge; with greater discharge, more sediment can be carried. Capacity and competence may seem quite similar, but they are actually related to different aspects of stream transport. For instance, a small, swiftly flowing stream may have the competence to move gravel-sized particles but not to transport a large volume of sediment, so it has a low capacity. A large, slow-flowing stream, on the other hand, has a low competence, but may have a very large suspended load, and hence a large capacity.

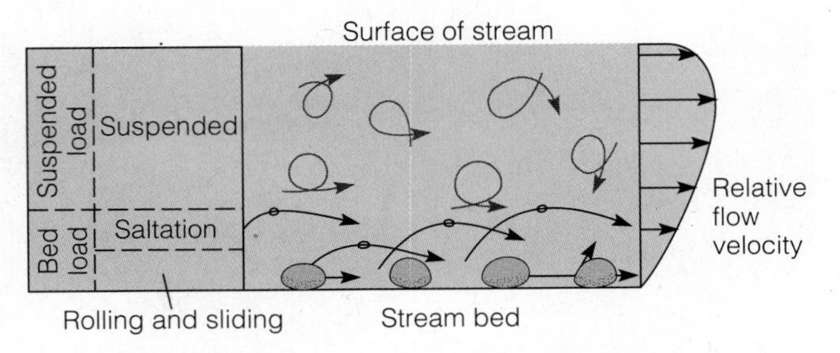

▶ **FIGURE 15-14** Methods of sediment transport by running water. The arrows in the velocity profile at the right are proportional to flow velocity, indicating that the water flows fastest near the surface and slowest along the stream bed.

Stream Deposition

Streams can transport sediment a considerable distance from the source area. Some of the sediments deposited in the Gulf of Mexico by the Mississippi River came from such distant sources as Pennsylvania, Minnesota, and southern Alberta, Canada. Along the way, deposition may occur in a variety of environments, such as stream channels, the floodplains adjacent to channels, and the points where streams flow into lakes or the seas or flow from mountain valleys onto adjacent lowlands.

Streams do most of their erosion, sediment transport, and deposition when they flood. Consequently, stream deposits, collectively called **alluvium,** do not represent the continuous day-to-day activity of streams, but rather those periodic, large-scale events of sedimentation associated with flooding.

Braided Streams and Their Deposits

Braided streams possess an intricate network of dividing and rejoining channels (▶ Figure 15-16). Braiding develops when a stream is supplied with excessive sediment, which over time is deposited as sand and gravel bars within its channel. During high-water stages, these bars are submerged, but during low-water stages, they are exposed and divide a single channel into multiple channels (Figure

▶ **FIGURE 15-15** Sediment erosion, transport, and deposition by running water are related to particle size and flow velocity.

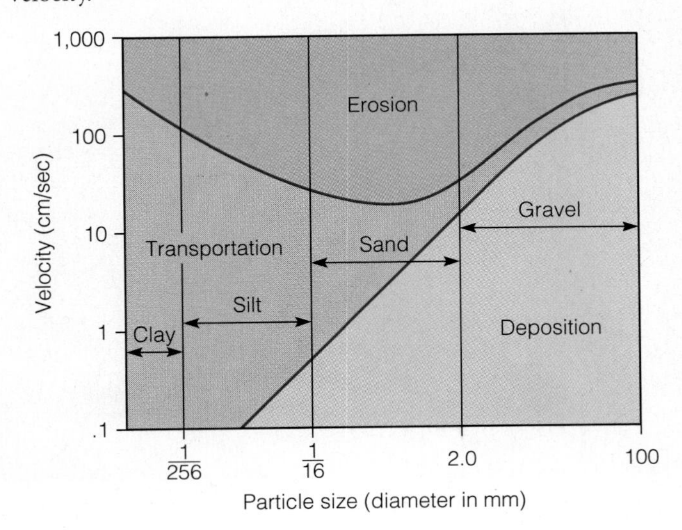

▶ **FIGURE 15-16** A braided stream near Santa Fe, New Mexico. The deposits in this stream are composed entirely of sand.

▶ **FIGURE 15-17** Aerial view of a meandering stream. The broad, flat area adjacent to the stream channel is the floodplain. Notice the crescent-shaped lakes—these are cutoff meanders, which are known as oxbow lakes.

➤ **FIGURE 15-18** The cut bank of a meandering stream.

15-16). Braided streams have broad, shallow channels. They are generally characterized as bed load transport streams, and their deposits are composed mostly of sheets of sand and gravel (Figure 15-16).

Braided streams are common in arid and semiarid regions where there is little vegetation and erosion rates are high. Streams with easily eroded banks are also likely to become braided. In fact, a stream that is braided where its banks are easily eroded may have a single, sinuous or meandering channel when it flows into an area of more resistant materials. Streams fed by melting glaciers are also commonly braided because the melting glacial ice yields so much sediment (see Chapter 17).

Meandering Streams and Their Deposits

Meandering streams have a single, sinuous channel with broadly looping curves called *meanders* (➤ Figure 15-17). Such stream channels are semicircular in cross section along straight reaches, but at meanders they are markedly asymmetric, being deepest near the outer bank, which commonly descends vertically into the channel. The outer bank is called the *cut bank* because flow velocity and turbulence are greatest on that side of the channel where it is eroded (➤ Figure 15-18). In contrast, flow velocity is at a minimum near the inner bank, which slopes gently into the channel.

As a consequence of the unequal distribution of flow velocity across meanders, the cut bank is eroded, and deposition occurs along the opposite side of the channel. The net effect is that a meander migrates laterally, and the channel maintains a more or less constant width because erosion on the cut bank is offset by an equal amount of deposition on the opposite side of the channel. The deposit formed in this manner is a **point bar;** it consists of cross-bedded sand or, in some cases, gravel (➤ Figure 15-19). Point bars are the characteristic deposits that accumulate within meandering stream channels.

It is not uncommon for meanders to become so sinuous that the thin neck of land separating adjacent meanders is eventually cut off during a flood. The valley floors of meandering streams are commonly marked by crescent-shaped **oxbow lakes**, which are actually cutoff meanders (➤ Figures 15-17 and 15-20). These oxbow lakes may persist as lakes for some time, but are eventually filled with organic matter and fine-grained sediment carried by floods. Once filled, oxbow lakes are called *meander scars*.

One immediate effect of meander cutoff is an increase in flow velocity; following the cutoff, the stream abandons part of its old course and flows a shorter distance, thereby increasing its gradient. Numerous cutoffs would, of course, significantly shorten a meandering stream, but streams usually establish new meanders elsewhere when old ones are cut off.

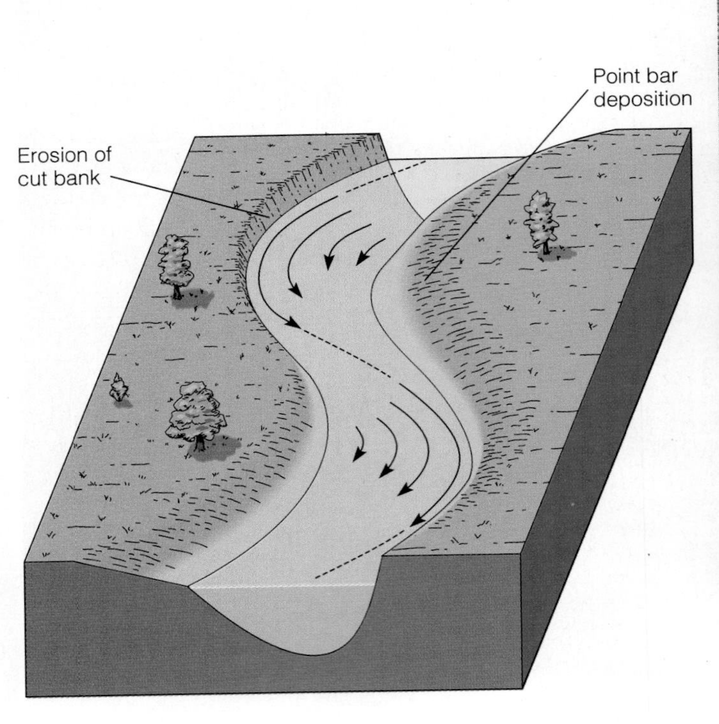

(a)

(b)

➢ FIGURE 15-19 (*a*) In a meandering channel, flow velocity is greatest near the outer bank. The dashed line follows the path of maximum flow velocity. Because of varying flow velocity across the channel, the outer or cut bank is eroded, and a point bar is deposited on the inner side of the meander. (*b*) Two small point bars in a meandering stream.

➢ FIGURE 15-20 Four stages in the origin of an oxbow lake. In (*a*) and (*b*), the meander neck becomes narrower. (*c*) The meander neck is cut off, and part of the channel is abandoned. (*d*) When it is completely isolated from the main channel, the abandoned meander is an oxbow lake.

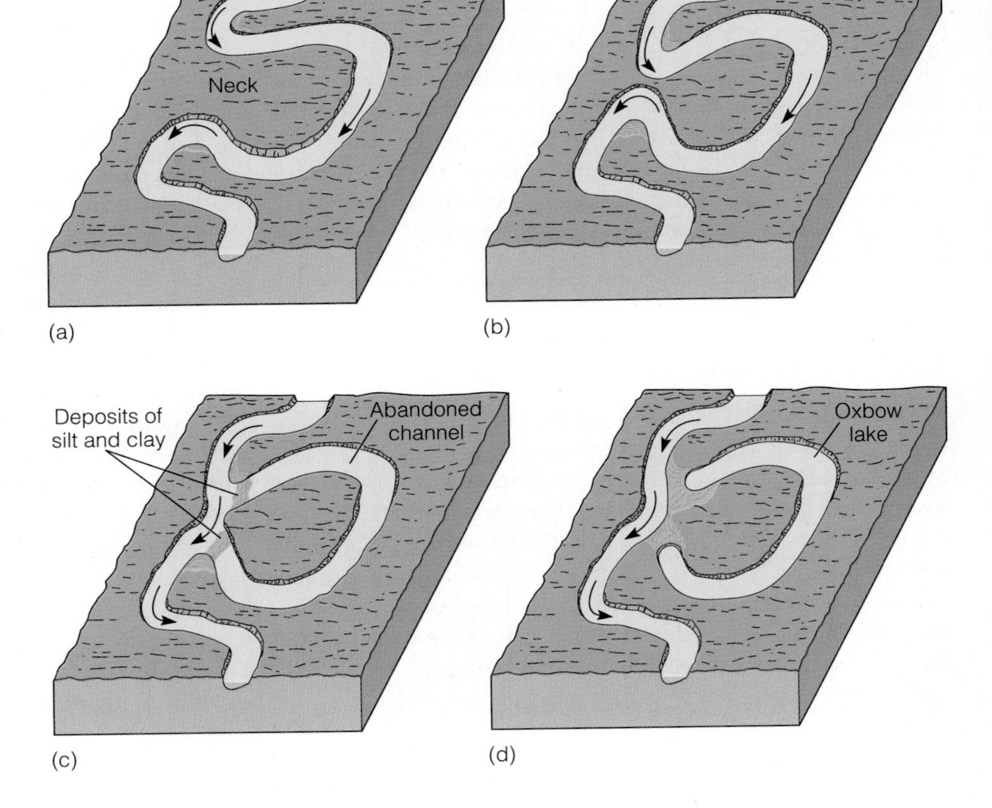

Floods and Floodplain Deposits

Most streams periodically receive more water than their channel can carry, so they spread across low-lying, relatively flat areas called **floodplains** adjacent to their channels (Figure 15-17) (see Perspective 15-1). Even small streams commonly have a floodplain, but this feature is usually proportional to the size of the stream; thus, small streams have narrow floodplains, whereas the lower Mississippi and other large streams have floodplains many kilometers wide. Streams restricted to deep, narrow valleys usually have little or no floodplain.

Some floodplains are composed mostly of sand and gravel that were deposited as point bars. When a meandering stream erodes its cut bank and deposits on the opposite bank, it migrates laterally across its floodplain. As lateral migration occurs, a succession of point bars develops by *lateral accretion* (⊳ Figure 15-21). That is, the deposits build laterally as a result of repeated episodes of sedimentation on the inner banks of meanders.

Many floodplains are dominated by *vertical accretion* of fine-grained sediments. When a stream overflows its banks and floods, the velocity of the water spilling onto the floodplain diminishes rapidly because of greater frictional resistance to flow as the water spreads out as a broad, shallow sheet. In response to the diminished velocity, ridges of sandy alluvium called **natural levees** are deposited along the margins of the stream channel (⊳ Figure 15-22). Natural levees are built up by repeated deposition of sediment during numerous floods. These natural levees separate most of the floodplain from the stream channel, so floodplains are commonly poorly drained and swampy. In fact, tributary streams may parallel the main stream for many kilometers until they find a way through the natural levee system (Figure 15-22).

The floodwaters spilling from a main channel carry large quantities of silt- and clay-sized sediment beyond the natural

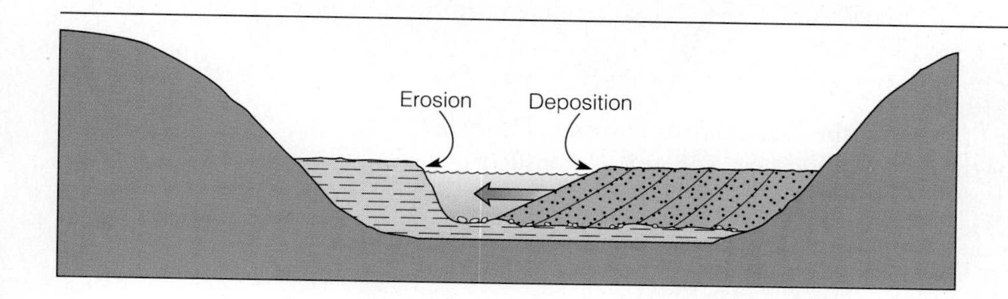

⊳ **FIGURE 15-21** Floodplain deposits forming by lateral accretion of point bars.

⊳ **FIGURE 15-22** Three stages in the formation of vertical accretion deposits on a floodplain. (*a*) Stream at low-water stage. (*b*) Flooding stream and deposition of natural levees. The levees form after many such episodes of flooding. (*c*) After flooding. Notice the tributary stream, which parallels the main stream until it finds a way through the natural levees.

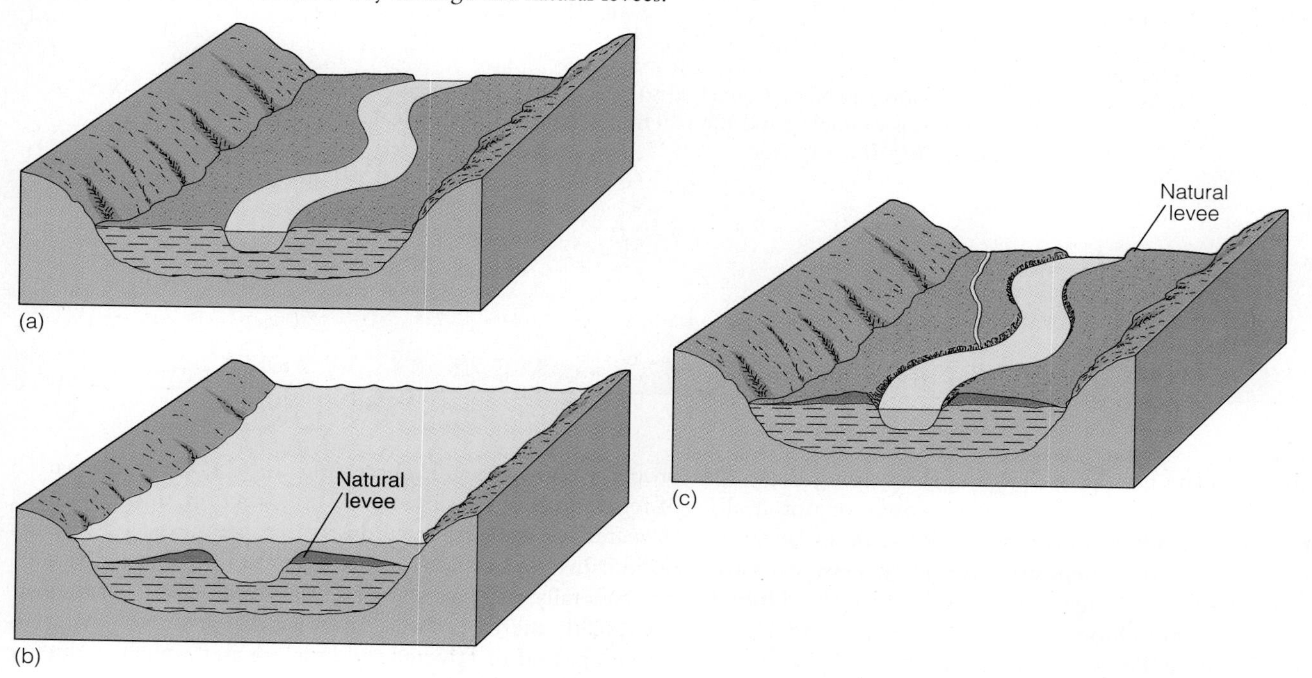

Perspective 15-1

PREDICTING AND CONTROLLING FLOODS

Occasionally, a stream receives more water than its channel can handle, and it floods, occupying part or all of its floodplain. To monitor stream behavior, the U.S. Geological Survey maintains more than 11,000 stream gauging stations, and various state agencies also monitor streams. Data collected at gauging stations can be used to construct a *hydrograph* showing how a stream's discharge varies over time (➤ Figure 1). Hydrographs are useful in planning irrigation and water supply projects, and they give planners a better idea of what to expect during floods.

Stream gauge data are also used to construct *flood-frequency curves* (➤ Figure 2). To construct such a curve, the peak discharges for a stream are first arranged in order of volume; the flood with the greatest discharge has a magnitude rank of 1, the second largest is 2, and so on (◉ Table 1). The *recurrence interval*—that is, the time period during which a flood of a given magnitude or larger can be expected over an average of many years—is determined by the equation shown in Table 1. For this stream, floods with magnitude ranks of 1 and 23 have recurrence intervals of 77.00 and 3.35 years, respectively. Once the recurrence interval has been calculated, it is plotted against discharge, and a line is drawn through the data points (Figure 2).

According to Figure 2, the 10-year flood for the Rio Grande near Lobatos, Colorado, has a discharge of 245 m^3/sec. This means that, on average, we can expect one flood of this size or greater to occur within a 10-year interval. One cannot, however, predict that such a flood

TABLE 1 Some of the Data and Recurrence Intervals for the Rio Grande near Labatos, Colorado

Year	Discharge (m^3/sec)	Rank	Recurrence Interval
1900	133	23	3.35
1901	103	35	2.20
1902	16	69	1.12
1903	362	2	38.50
1904	22	66	1.17
1905	371	1	77.00
1906	234	10	7.70
1907	249	7	11.00
1908	61	45	1.71
1909	211	13	5.92
.			
.			
.			
1974	22	64	1.20
1975	68	43	1.79

The greatest yearly discharge is given a magnitude rank (m) ranging from 1 to N ($N = 76$ in this example), and the recurrence interval (R) is calculated by the equation $R = (N + 1)/m$.

SOURCE: U.S. Geological Survey Open-File Report 79–681.

levees and onto the floodplain. During the waning stages of a flood, the floodwaters may flow very slowly or not at all, and the suspended silt and clay eventually settle as layers of mud that build upward by deposition during successive floods.

Annual property damage from flooding in the United States exceeds \$100 million. And in spite of the completion of more and more flood-control projects, the amount of property damage is not decreasing. In fact, the combination of fertile soils, level surfaces for construction, and proximity to water for agriculture, industry, and domestic uses makes floodplains popular sites for settlement. These human activities generally increase the potential for flooding. Urbanization greatly increases surface runoff because surface materials are compacted or covered by concrete and asphalt, reducing

▶ **FIGURE 1** Hydrograph for Sycamore Creek near Ashland City, Tennessee, for the February 1989 flood. (From U.S. Geological Survey Water-Resources Investigations Report 89–4207.)

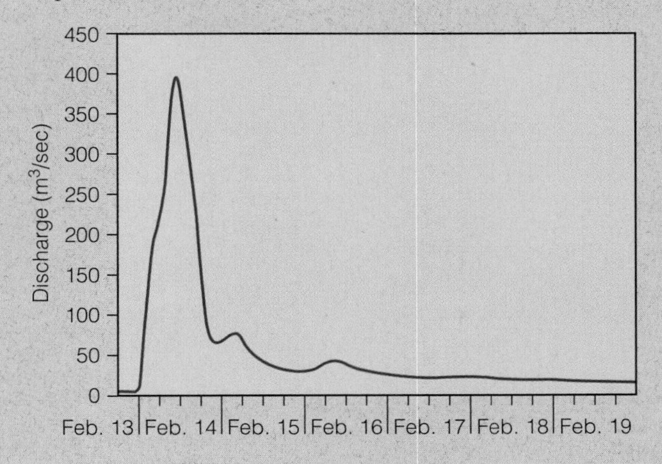

▶ **FIGURE 2** Flood-frequency curve for the Rio Grande near Lobatos, Colorado. The curve was constructed from the data in Table 1.

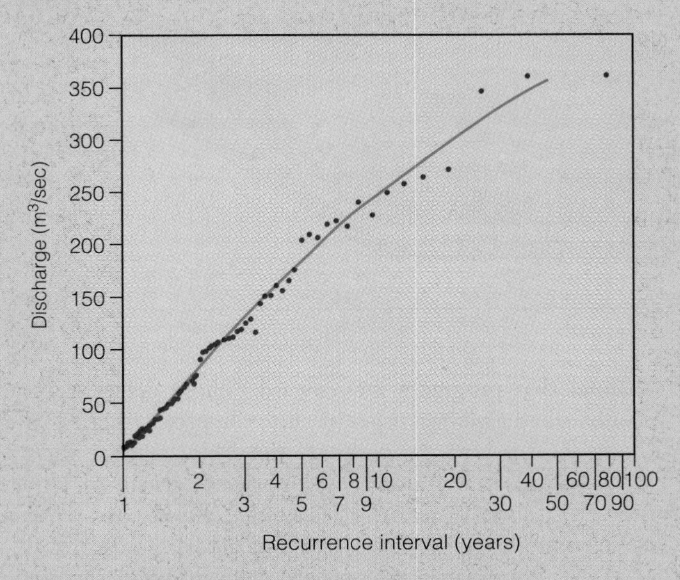

will occur in any particular year, only that it has a probability of 1 in 10 (1/10) of occurring in any year. Furthermore, 10-year floods are not necessarily separated by 10 years. That is, two such events could occur in the same year or in successive years, but over a period of centuries their average occurrence would be once every 10 years.

Unfortunately, stream gauge data in the United States have generally been available for only a few decades, and rarely for more than a century. Accordingly, we have a good idea of stream behavior over short periods, the 2-year and 5-year floods, for example, but our knowledge of long-term behavior is limited by the short period of record keeping. Accordingly, predictions of 50-year or 100-year floods from Figure 2 are unreliable. In fact, the largest magnitude flood shown in Figure 2 may have been a unique event for this stream that will never be repeated. On the other hand, it may actually turn out to be a magnitude 2 or 3 flood when data for a longer time are available.

Although flood-frequency curves have limited applicability, they are nevertheless helpful in making decisions regarding flood control. Careful mapping of floodplains can identify areas at risk for floods of a given magnitude. For a particular stream, planners must decide what magnitude of flood to protect against because the cost goes up faster than the increasing sizes of floods would indicate.

Federal, state, and local agencies and land-use planners use flood-frequency analyses to develop recommendations and regulations concerning construction on and use of floodplains. Geologists and engineers are interested in such analyses for planning appropriate flood-control projects. They must decide, for example, where dams and basins should be constructed to contain the excess water of floods.

When flood-control projects are well planned and constructed, they are functional. What many people fail to realize is that these projects are designed to contain floods of a given size; should larger floods occur, streams spill onto floodplains anyway. Furthermore, dams occasionally collapse and reservoirs eventually fill with sediment unless dredged. In short, flood-control projects are not only initially expensive, but they require constant, costly maintenance. Such costs must be weighed against the cost of damage if no control projects were undertaken.

their infiltration capacity. Storm drains in urban areas quickly carry water to nearby streams, many of which flood much more commonly than they did in the past.

Deltas

The fundamental process of delta formation is rather simple: when a stream flows into another body of water, its flow velocity decreases rapidly and deposition occurs. As a result of such deposition, a **delta** forms, causing the local shoreline to build out, or *prograde* (▶ Figure 15-23). Deltas in lakes are common, but marine deltas are much larger and far more complex.

The simplest prograding deltas exhibit a characteristic vertical sequence in which *bottomset beds* are successsively overlain by *foreset beds* and *topset beds* (Figure 15-23a). These

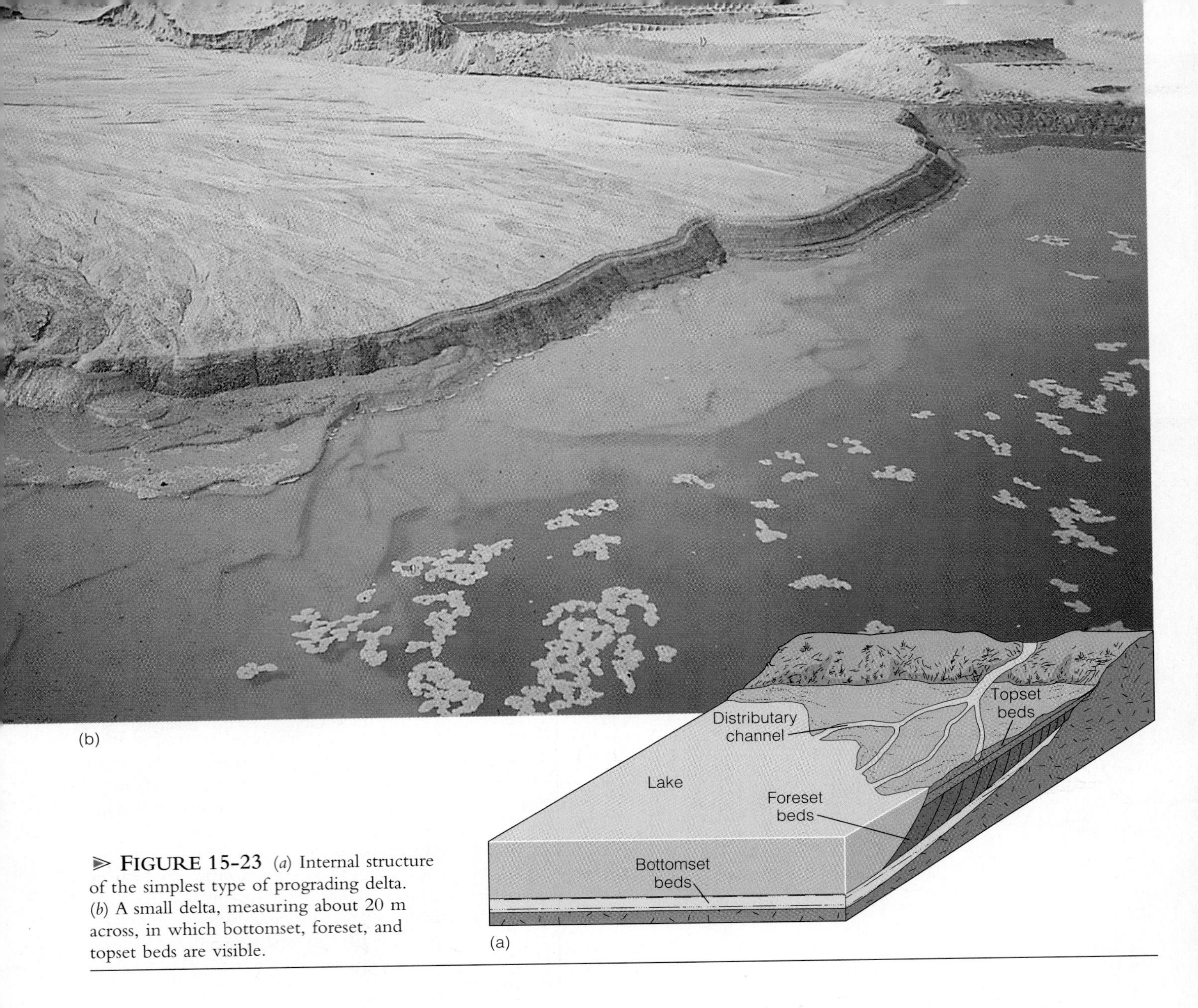

▶ **FIGURE 15-23** (*a*) Internal structure of the simplest type of prograding delta. (*b*) A small delta, measuring about 20 m across, in which bottomset, foreset, and topset beds are visible.

sequences develop when a stream enters another body of water, and the finest sediments are carried some distance beyond the stream's mouth, where they settle from suspension and form bottomset beds. Nearer the stream's mouth, foreset beds are formed as sand and silt are deposited in gently inclined layers. The topset beds consist of coarse-grained sediments deposited in a network of *distributary channels* traversing the top of the delta. In effect, streams lengthen their channels as they extend across prograding deltas (Figure 15-23).

Many small deltas in lakes have the three-part division described above, but large marine deltas are usually much more complex. Depending on the relative importance of stream, wave, and tidal processes, three major types of marine deltas are recognized (▶ Figure 15-24). *Stream-dominated deltas,* such as the Mississippi River delta, consist of long fingerlike sand bodies, each deposited in a distributary channel that progrades far seaward. These deltas are commonly called *bird's-foot deltas* because the projections resemble the toes of a bird. In contrast, the Nile delta of Egypt is *wave-dominated,* although it also possesses distributary channels; the seaward margin of the delta consists of a series of barrier islands formed by reworking of sediments by waves, and the entire margin of the delta progrades seaward. *Tide-dominated deltas,* such as the Ganges-Brahmaputra of Bangladesh, are continually modified into tidal sand bodies that parallel the direction of tidal flow (Figure 15-24).

Coal can form in several depositional environments, such as the fresh water marshes between distributary channels of deltas (Figure 15-24a). These marshes are dominated by nonwoody plants whose remains accumulate to form peat, the first stage in the origin of coal (see Chapter 6). If peat is buried, the volatile components of the plants are driven off leaving mostly carbon that eventually forms coal.

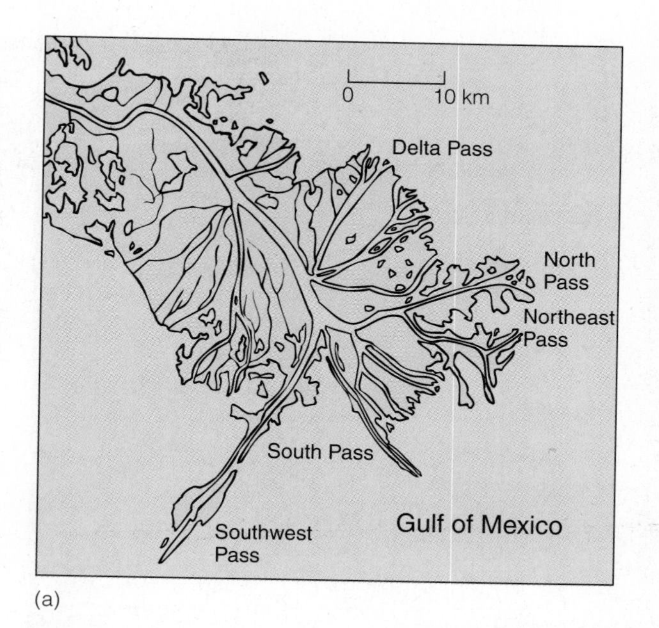

(a)

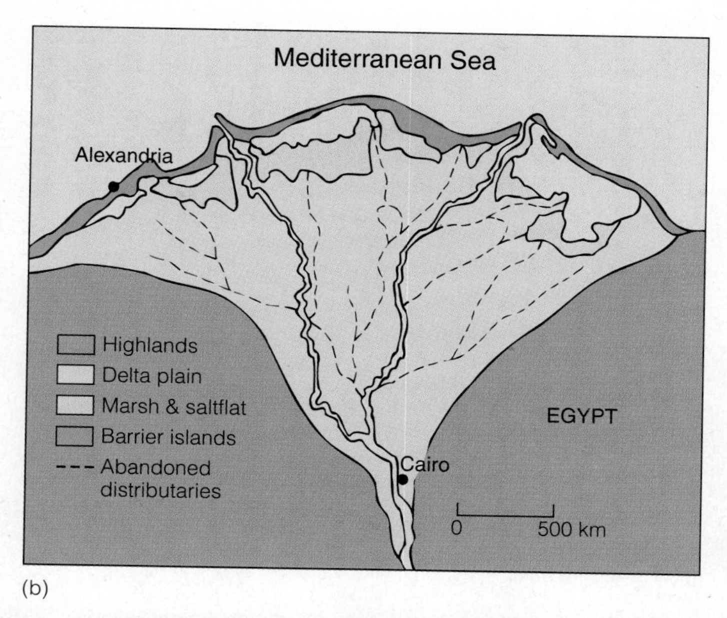

(b)

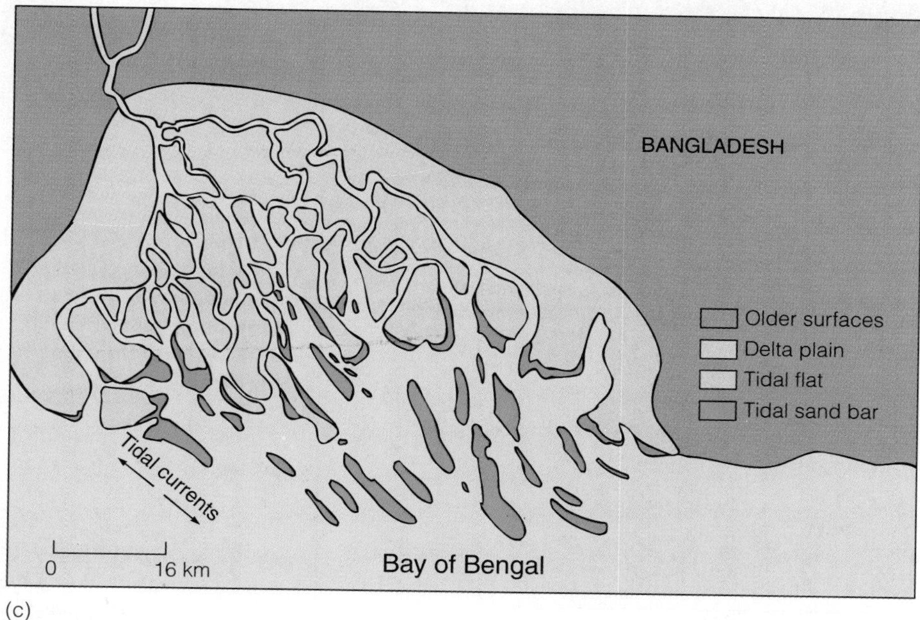

(c)

▶ **FIGURE 15-24** (*a*) The Mississippi River delta of the U.S. Gulf Coast is stream-dominated, and (*b*) the Nile delta of Egypt is wave-dominated. (*c*) The Ganges-Brahmaputra delta of Bangladesh is tide-dominated.

Delta progradation is one way that potential reservoirs for oil and gas form. Because of their porosity and permeability and association with organic-rich marine sediments, distributary sand bodies commonly contain oil and gas. Much of the oil and gas production of the Gulf Coast of Texas comes from buried delta deposits. Some of the older deposits of the Niger River delta of Africa and the Mississippi River delta are also known to contain vast reserves of oil and gas.

Alluvial Fans

Alluvial fans are lobate deposits on land (▶ Figure 15-25). They form best on lowlands adjacent to highlands in arid and semiarid regions where little or no vegetation exists to stabilize surface materials. When periodic rainstorms occur, surface materials are quickly saturated and runoff begins. During a particularly heavy rain, all of the surface flow in a drainage area is funneled into a mountain canyon leading to an adjacent lowland. The stream is confined in the mountain canyon so that it cannot spread laterally. But as it discharges from the canyon onto the lowland area, it quickly spreads out, its velocity diminishes, and deposition ensues.

The alluvial fans that develop by the process just described are mostly accumulations of sand and gravel, a large proportion of which is deposited by streams. In some cases the water flowing through a mountain canyon picks up so much sediment that it becomes a viscous mudflow (see

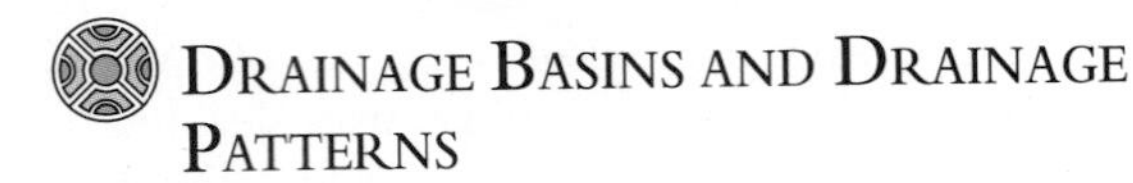

▶ **FIGURE 15-25** (*a*) Alluvial fans form where a stream discharges from a mountain canyon onto an adjacent lowland. (*b*) Alluvial fan adjacent to the Paniment Range on the margin of Death Valley, California.

Chapter 14). Consequently, mudflow deposits make up a large part of some alluvial fans.

DRAINAGE BASINS AND DRAINAGE PATTERNS

Thousands of streams, most of which are parts of larger drainage systems, flow either directly or indirectly into the oceans. A stream such as the Mississippi River consists of a main stream and all of the smaller *tributary streams* that supply water to it. The Mississippi and all of its tributaries, or any other drainage system for that matter, carry surface runoff from an area known as the **drainage basin** (Table 15-2). Individual drainage basins are separated from adjacent ones by topographically higher areas called **divides** (▶ Figures 15-26 and 15-27). Some divides are rather modest rises, such as that separating the Great Lakes' drainage basin from that of the Mississippi River, whereas others, such as the Continental Divide along the crest of the Rocky Mountains, are more impressive.

Various **drainage patterns** are recognized based on the regional arrangement of channels in a drainage system. The most common is *dendritic drainage,* which consists of a network of channels resembling tree branching (▶ Figure 15-28a). Dendritic drainage develops on gently sloping surfaces where the materials respond more or less homogeneously to erosion. Areas of flat-lying sedimentary rocks and some terrains of igneous or metamorphic rocks usually display a dendritic drainage pattern.

In marked contrast to dendritic drainage in which tributaries join larger streams at various angles, *rectangular drainage* is characterized by channels with right angle bends and tributaries that join larger streams at right angles (Figure 15-28b). The positions of the channels are strongly con-

➢ **FIGURE 15-26** Small drainage basins separated from one another by divides.

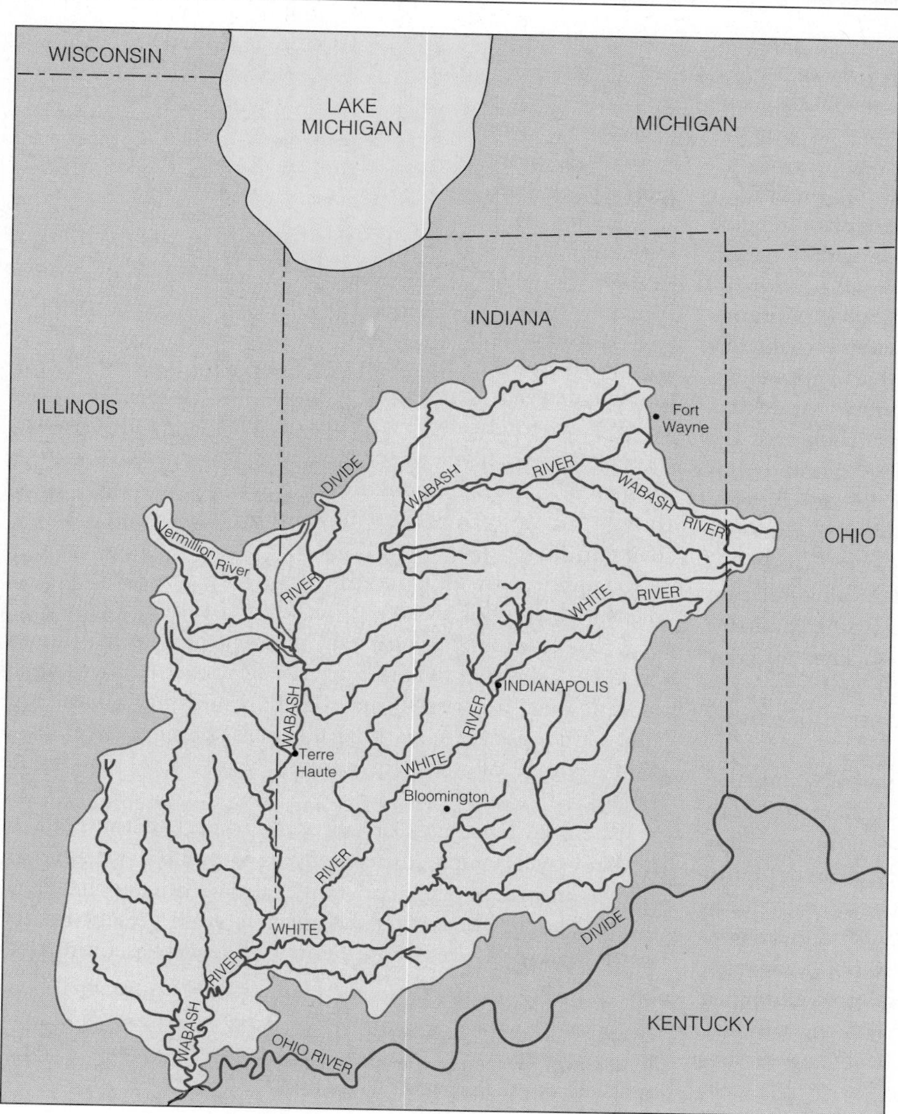

➢ **FIGURE 15-27** The drainage basin of the Wabash River, which is one of the tributaries of the Ohio River. This drainage basin covers about 85,500 km^2, mostly in Indiana. All of the streams within the drainage basin, such as the Vermillion River, have their own smaller drainage basins. Divides are shown by brown lines.

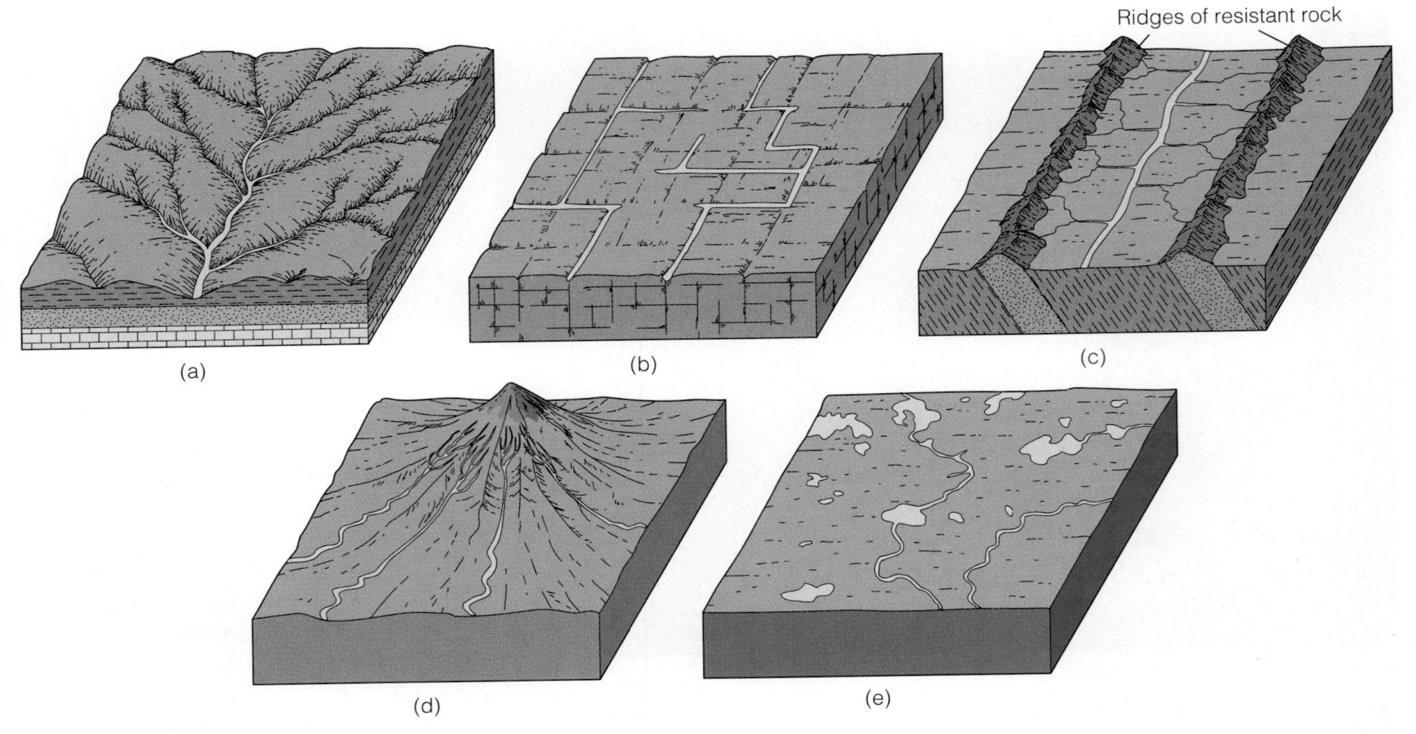

▶ **FIGURE 15-28** Examples of drainage patterns. (*a*) Dendritic drainage. (*b*) Rectangular drainage. (*c*) Trellis drainage. (*d*) Radial drainage. (*e*) Deranged drainage.

trolled by geologic structures, particularly regional joint systems that intersect at right angles. Rectangular drainage develops because streams more easily erode and establish channels along the traces of joints.

In some parts of the eastern United States, such as Virginia and Pennsylvania, erosion of folded sedimentary rocks develops a landscape of alternating parallel ridges and valleys. The ridges consist of more resistant rocks such as sandstone, whereas the valleys overlie less resistant rocks such as shale. Main streams follow the trends of the valleys. Short tributaries flowing from the adjacent ridges join the main stream at nearly right angles, hence the name *trellis drainage* (Figure 15-28c).

In *radial drainage,* streams flow outward in all directions from a central high area (Figure 15-28d). Radial drainage develops on large, isolated volcanic mountains, such as Mount Shasta in California (see Figure 4-6b), and where the Earth's crust has been arched up by the intrusion of plutons such as laccoliths.

In some areas streams flow in and out of swamps and lakes with irregular flow directions. Drainage patterns characterized by such irregularity are called *deranged* (Figure 15-28e). The presence of deranged drainage indicates that the drainage system developed recently and has not yet formed an organized system. In areas in Minnesota, Wisconsin, and Michigan that were glaciated until about 10,000 years ago, the previously established drainage systems were obliterated by glacial ice. Following the final retreat of the glaciers, drainage systems became established, but have not yet become fully organized.

BASE LEVEL

Streams require a slope in order to flow, so they can erode downward only to the level of the body of water into which they discharge. A stream flowing into the sea, for example, cannot erode its valley lower than sea level—if it could, it would have to flow uphill to reach the sea.* The lower limit to which streams can erode is called **base level** (▶ Figure 15-29). Theoretically, a stream could erode its entire valley to very near sea level, so sea level is commonly referred to as *ultimate base level.* Streams never reach ultimate base level, though, because they must have some gradient in order to maintain flow.

In addition to ultimate base level, streams have *local* or *temporary base levels.* A lake or another stream can serve as a local base level for the upstream segment of a stream (Figure 15-29b). Likewise, where a stream flows across particularly resistant rock, a waterfall may develop, forming a local base level. The escarpment Niagara Falls plunges over is a good example of a local or temporary base level (▶ Figure 15-30).

When sea level rises or falls with respect to the land, or the land over which a stream flows is uplifted or subsides, changes in base level occur. For example, during the Pleistocene Epoch when extensive glaciers were present on the Northern Hemisphere continents, sea level was more than

*Streams flowing into depressions below sea level, such as Death Valley in California, have a base level corresponding to the lowest point of the depression and are not limited by sea level.

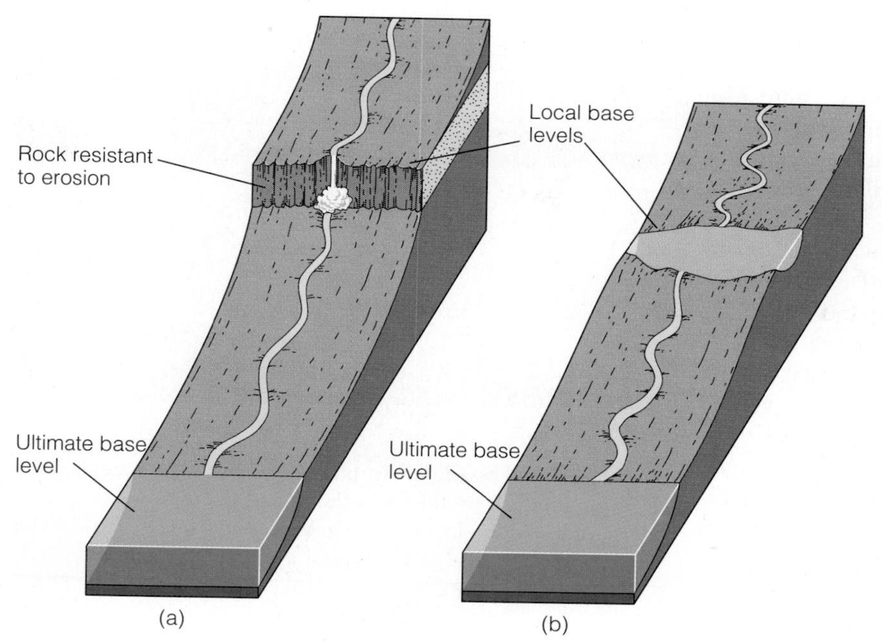

▶ **FIGURE 15-29** In both (*a*) and (*b*), sea level is ultimate base level. In (*a*) a resistant rock layer forms a local base level, while in (*b*) a lake is a local base level.

▶ **FIGURE 15-30** Niagara Falls on the border between New York and Ontario, Canada. The resistant rock forming the escarpment over which the falls plunge is a local base level. The more impressive falls in the foreground are in Canada, and those in the background are in the United States.

100 m lower than at present. Accordingly, streams deepened their valleys by adjusting to a new, lower base level. In addition, many streams extended their valleys onto the exposed continental shelves. Rising sea level at the end of the Pleistocene caused base level to rise, and the streams responded by depositing sediments and backfilling previously formed valleys.

Streams adjust to human intervention, but not always in anticipated or desirable ways. Geologists and engineers are well aware that the process of building a dam and impounding a reservoir creates a local base level (➢ Figure 15-31a). Where a stream enters a reservoir, its flow velocity diminishes rapidly and deposition occurs, so reservoirs are eventually filled with sediment unless they are dredged. Another consequence of building a dam is that the water discharged at the dam is largely sediment-free, but it still possesses energy to transport sediment. Commonly, such streams simply acquire a new sediment load by vigorously eroding downstream from the dam.

Draining a lake along a stream's course may seem like a small change that is well worth the time and expense to expose dry land for agriculture or commercial development. But unless one anticipates the stream's probable response, dire consequences can result. Remember that a lake is a temporary base level, so draining it lowers the base level for that part of the stream above the lake, and the stream will very likely respond by rapid downcutting (Figure 15-31b).

 The Graded Stream

A stream's *longitudinal profile* shows the elevations of a channel along its length as viewed in cross section (➢ Figure 15-32). The longitudinal profiles of many streams show a number of irregularities such as lakes and waterfalls, which are local base levels (Figure 15-32a). Over time such irregularities tend to be eliminated by stream processes; where the gradient is steep, erosion decreases it, and where the gradient is too low to maintain sufficient flow velocity for sediment transport, deposition occurs, steepening the gradient. In short, streams tend to develop a smooth, concave longitudinal profile of equilibrium, meaning that all parts of the system are dynamically adjusting to one another (Figure 15-32b).

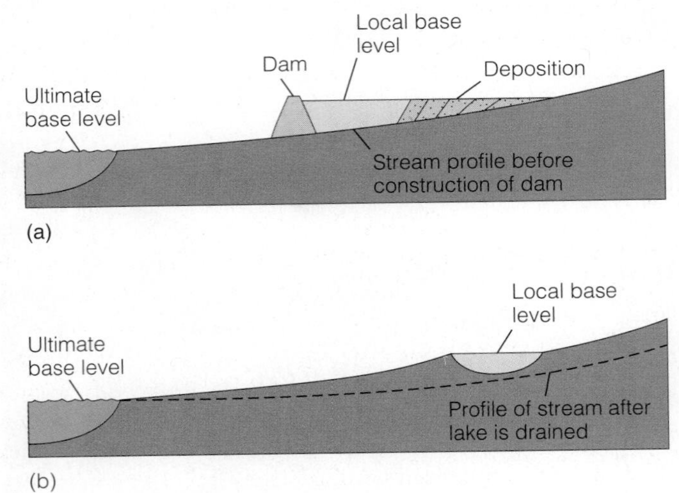

➢ **FIGURE 15-31** (*a*) The process of constructing a dam and impounding a reservoir creates a local base level. A stream deposits much of its sediment load where it flows into a reservoir. (*b*) A stream adjusts to a lower base level when a lake is drained.

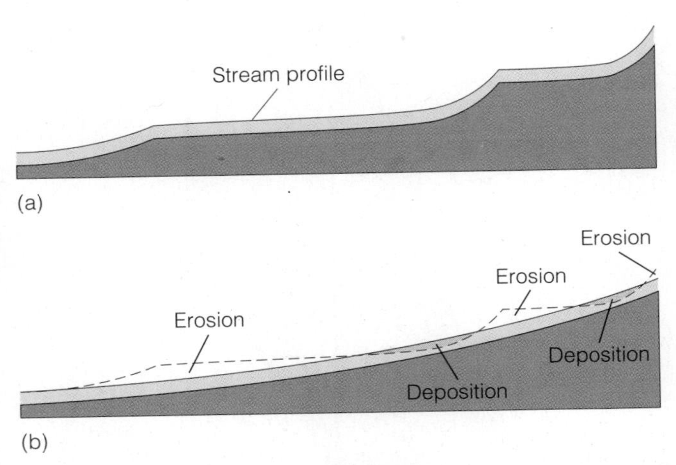

➢ **FIGURE 15-32** (*a*) An ungraded stream has irregularities in its longitudinal profile. (*b*) Erosion and deposition along the course of a stream eliminate irregularities and cause it to develop the smooth, concave profile typical of a graded stream.

Streams possessing an equilibrium profile are said to be **graded streams;** that is, a delicate balance exists between gradient, discharge, flow velocity, channel characteristics, and sediment load such that neither significant erosion nor deposition occurs within the channel. Such a delicate balance is rarely attained, so the concept of a graded stream is an ideal. Nevertheless, many streams do indeed approximate the graded condition, although not along their entire courses and usually only temporarily.

Even though the concept of a graded stream is an ideal, we can generally anticipate the responses of a graded stream to changes altering its equilibrium. A change in base level would cause a stream to adjust as previously discussed. Increased rainfall in a stream's drainage basin would result in greater discharge and flow velocity. In short, the stream would now possess greater energy—energy that must be dissipated within the stream system by, for example, a change in channel shape. A change from a semicircular to a broad, shallow channel would dissipate more energy by friction. On the other hand, the stream may respond by active downcutting in which it erodes a deeper valley and effectively reduces its gradient until it is once again graded.

Vegetation inhibits erosion by having a stabilizing effect on soil and other loose surface materials. So a decrease in vegetation in a drainage basin might lead to higher erosion

▶ **FIGURE 15-33** The Grand Canyon in Arizona is a vast system of canyons eroded by the Colorado River and its tributaries. (Photo courtesy of L. E. Andrews.)

rates, causing more sediment to be washed into a stream than it can effectively carry. Accordingly, the stream may respond by deposition within its channel, which increases the stream's gradient until it is sufficiently steep to transport the greater sediment load.

 Development of Stream Valleys

Valleys are common landforms, and with few exceptions they form and evolve as a consequence of stream erosion, although other processes, especially mass wasting, also contribute. The shapes and sizes of valleys vary considerably; some are small, steep-sided *gullies,* whereas others are broad and have gently sloping valley walls. Some steep-walled, deep valleys of vast proportions are called *canyons.* The Grand Canyon of Arizona, for example, is an interconnected system of canyons eroded by the Colorado River and its tributaries (▶ Figure 15-33). Particularly narrow and deep valleys are *gorges* (▶ Figure 15-34).

A valley may begin where runoff has sufficient energy to dislodge surface materials and excavate a small rill. Once formed, a rill collects more surface runoff and becomes deeper and wider until a full-fledged valley develops (▶ Figure 15-35). Several processes are involved in the origin and evolution of valleys, including downcutting, lateral erosion, mass wasting, sheet wash, and headward erosion.

Downcutting occurs when a stream possesses more energy than it requires to transport its sediment load, so some of its excess energy cuts its valley deeper (Figure 15-35a). If downcutting were the only process operating, valleys would be narrow and steep sided as in Figure 15-34. In most cases, however, the valley walls are undercut by the stream. Such undermining, termed *lateral erosion,* creates unstable conditions so that part of a bank or valley wall may move downslope by any one or a combination of mass wasting processes (Figure 15-12). Furthermore, sheet wash and erosion of rill and gully tributaries carry materials from the valley walls into the main stream.

In addition to becoming deeper and wider, stream valleys are commonly lengthened as well. Valleys are lengthened in an upstream direction by *headward erosion* as drainage divides are eroded by entering runoff water (▶ Figure 15-36a). In some cases headward erosion eventually breaches the drainage divide and diverts part of the drainage of another stream by a process called *stream piracy* (Figure 15-36b). Once stream piracy has occurred, both drainage systems must adjust; one now has more water, greater discharge, and greater potential to erode and transport sediment, whereas the other is diminished in all of these aspects.

➢ **FIGURE 15-34** The Black Canyon of the Gunnison River in Colorado is a gorge because it is steep walled and deep.

Superposed Streams

Streams flow downhill in response to gravity, so their courses are determined by preexisting topography. Yet a number of streams seem, at first glance, to have defied this fundamental control. For example, the Delaware, Potomac, and Susquehanna rivers in the eastern United States have valleys that cut directly through ridges lying in their paths. All of these streams are **superposed.** In order to understand how superposition occurs, it is necessary to know the geologic histories of these streams.

During the Mesozoic Era, the Appalachian Mountain region was eroded to a sediment-covered plain crossed by numerous streams generally flowing eastward. During the

➢ **FIGURE 15-35** Valley development. (*a*) If valleys formed mostly by downcutting, they would be narrow and steep sided. (*b*) Valleys are deepened by downcutting, but most of them are also widened by lateral erosion, mass wasting, and sheet wash.

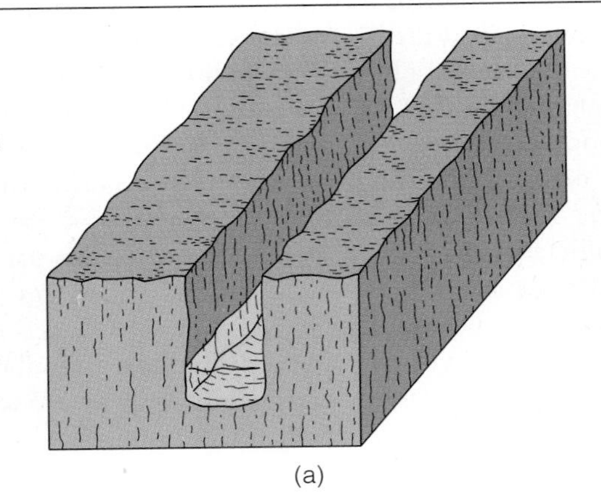

(a)

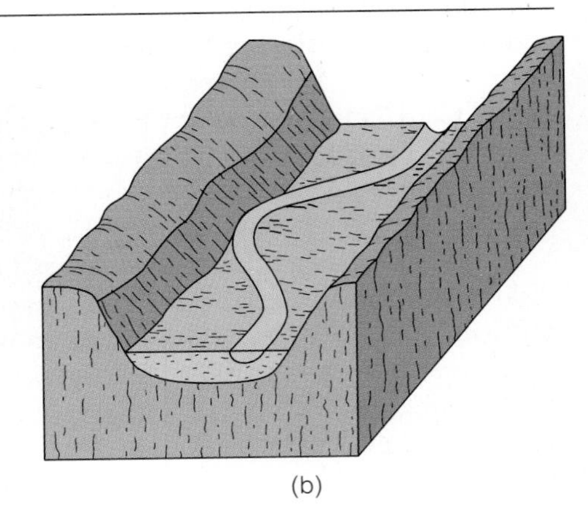

(b)

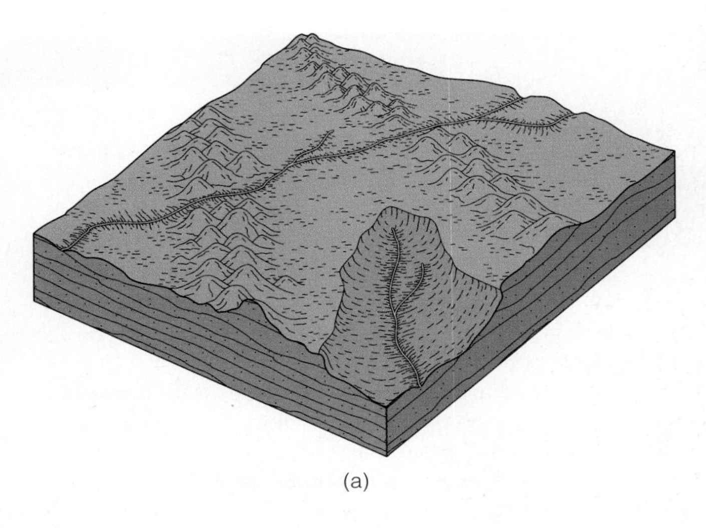

(a)

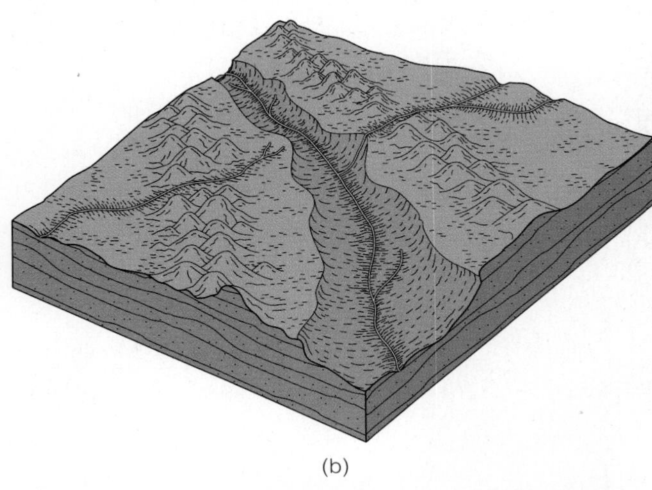

(b)

▶ **FIGURE 15-36** Two stages in stream piracy. (*a*) In the first stage, the stream at the lower elevation extends its channel by headward erosion. In (*b*) it has captured some of the drainage of the stream flowing at the higher elevation.

Cenozoic Era, regional uplift commenced, and the streams eroded downward and were superposed directly upon resistant rock. Instead of changing course, they cut narrow, steep-walled canyons called *water gaps* (▶ Figure 15-37).

STREAM TERRACES

Adjacent to many streams are erosional remnants of floodplains formed when the streams were flowing at a higher level. These erosional remnants are **stream terraces.** They consist of a fairly flat upper surface and a steep slope descending to the level of the lower, present-day floodplain (▶ Figure 15-38). In some cases, a stream has several steplike surfaces above its present-day floodplain, indicating that stream terraces formed several times.

Although all stream terraces result from erosion, they are preceded by an episode of floodplain formation and deposition of sediment. Subsequent erosion causes the stream to cut downward until it is once again graded (▶ Figure 15-39). Once the stream again becomes graded, it begins eroding laterally and establishes a new floodplain at a lower level. Several such episodes account for the multiple terrace levels seen adjacent to some streams (Figures 15-38 and 15-39).

Stream terraces are commonly cut into previously deposited sediment, but some are cut into solid bedrock. Where they are cut into bedrock, the terrace surface is generally covered by a thin veneer of sediment. In many stream valleys, terraces are paired, meaning that they occur at the same elevation on opposite sides of the channel (Figure 15-39b).

Renewed erosion and the formation of stream terraces are usually attributed to a change in base level. Either uplift of the land over which a stream flows or lowering of sea level yields a steeper gradient and increased flow velocity, thus initiating an episode of downcutting. When the stream reaches a level at which it is once again graded, downcutting ceases.

▶ **FIGURE 15-37** The origin of a superposed stream. (*a*) A stream begins cutting down into horizontal strata. (*b*) The horizontal layer is removed by erosion, exposing the underlying structure. The stream cuts a narrow valley in resistant rocks that form the ridges.

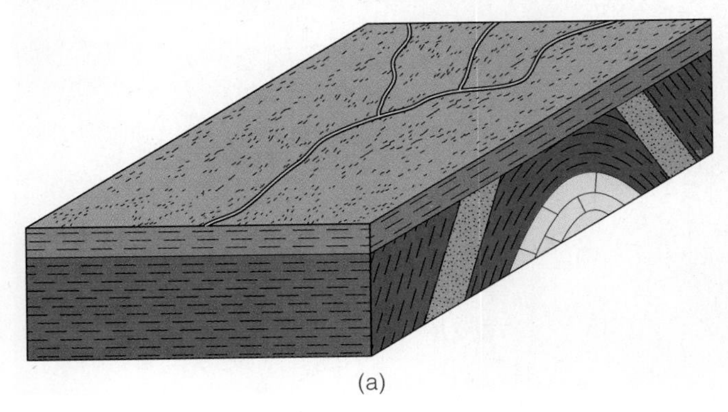

(a)

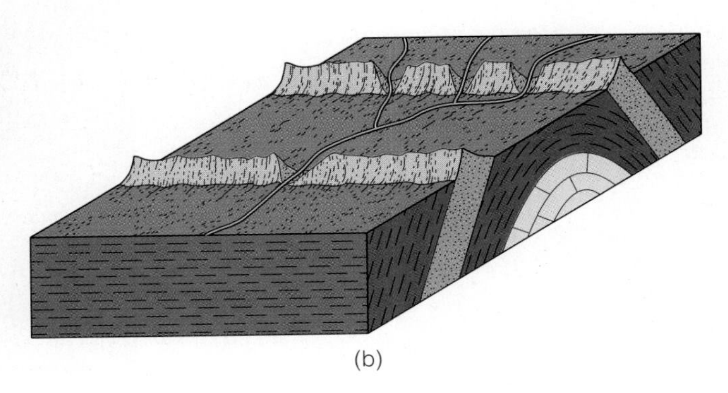

(b)

Perspective 15-2

Natural Bridges

The term *natural bridge* has been used to describe a variety of features including spans of rock resulting from wave erosion, the partial collapse of cavern roofs, and weathering and erosion along closely spaced, parallel joints as in Arches National Park in Utah (see Perspective 13-1). Here we are concerned only with natural bridges that span a valley eroded by running water.

The best place to observe this type of natural bridge is in Natural Bridges National Monument in southwestern Utah. Three natural bridges are present within the monument, and all originated in the same way. Of these three, Sipapu Bridge is the largest (▶ Figure 1); it stands 67 m above White Canyon and has a span of 81.5 m.

The process by which these natural bridges were formed is well understood, and, as a matter of fact, it is still going on. In the first stage, a meandering stream was incised into solid bedrock (▶ Figure 2). In Natural Bridges National

▶ **FIGURE 1** Sipapu Bridge in Natural Bridges National Monument, Utah. (Photo courtesy of Sue Monroe.)

▶ **FIGURE 15-38** The elevated flat areas adjacent to the Madison River in Montana are stream terraces.

Monument, this rock unit is the Cutler Formation, which consists of sandstone formed from windblown sand deposited during the Permian Period. When local meandering streams became incised, lateral erosion created a thin wall of rock between adjacent meanders that was eventually breached (Figure 2). As the breach was subsequently enlarged, the stream abandoned its old meander, and the stream flow was diverted. As we discussed previously, oxbow lakes are formed by a similar process (Figure 15-20). The only significant difference is that the streams that form natural bridges are incised.

Natural bridges are temporary features. Once formed, they are destroyed by other processes. Rocks fall from the undersides of bridges, their surfaces are weathered and eroded, and eventually they collapse. The monument contains several examples of collapsed bridges, but new ones are in the process of forming.

▶ FIGURE 2 Origin of a natural bridge. (*a*) A meandering stream flows across a gently sloping surface. (*b*) Incised meanders develop as the stream erodes down into solid rock. (*c*) A thin wall of rock between meanders is eventually breached, forming a natural bridge.

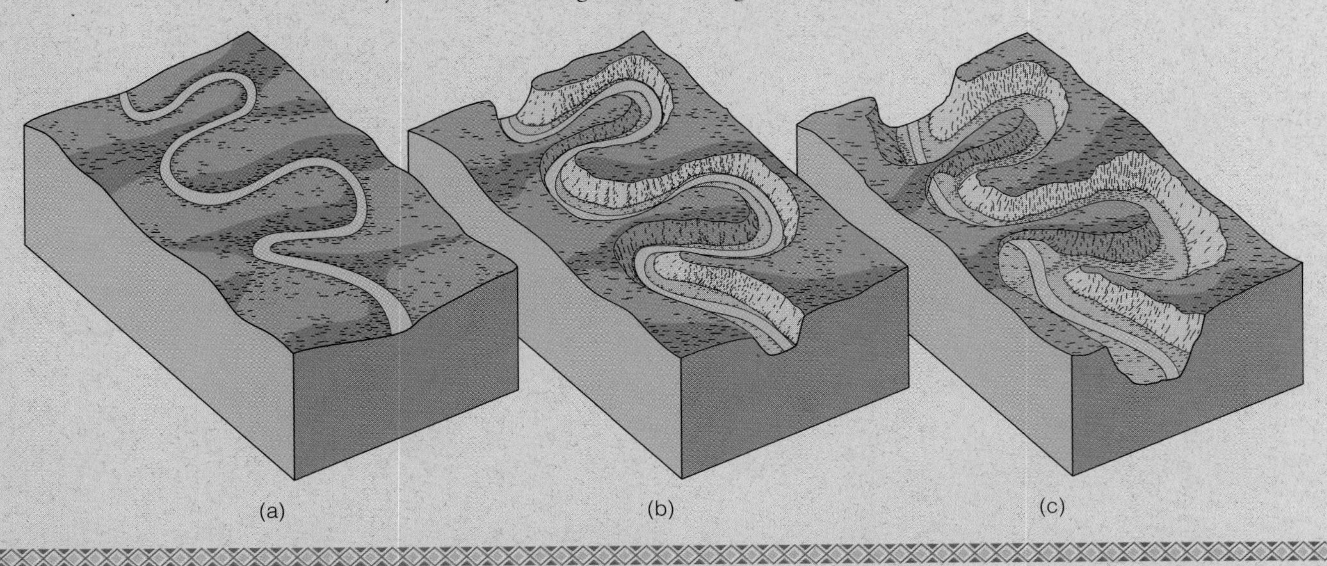

Although changes in base level no doubt account for many stream terraces, greater runoff in a stream's drainage basin can also result in the formation of terraces. Recall that one of the variables controlling velocity is discharge, so a stream can erode downward and form terraces with no change in base level.

 INCISED MEANDERS

Some streams are restricted to deep, meandering canyons cut into solid bedrock, where they form features called **incised meanders.** The San Juan River in Utah, for example, occupies a meandering canyon more than 390 m deep (▶ Figure 15-40). Such streams, being restricted by solid rock walls, are generally ineffective in eroding laterally; thus, they lack a floodplain and occupy the entire width of the canyon floor. Some incised meandering streams do erode laterally, thereby cutting off meanders and producing natural bridges (see Perspective 15-2).

It is not difficult to understand how a stream can cut downward into solid rock, but forming a meandering pattern in bedrock is another matter. Because lateral erosion is inhibited once downcutting begins, one must infer that the meandering course was established when the stream flowed across an area covered by alluvium. For instance, suppose that a stream near base level has established a meandering pattern. If the land over which the stream flows is uplifted, erosion is initiated, and the meanders become incised into the underlying bedrock.

Uplift does not account for all incised meanders. A stream far above base level can establish a meandering pattern provided that it flows over a gently sloping surface. The meandering pattern was already established before erosion into bedrock occurred.

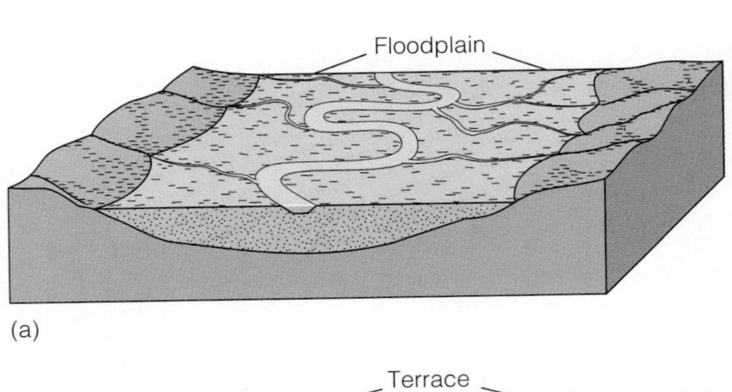

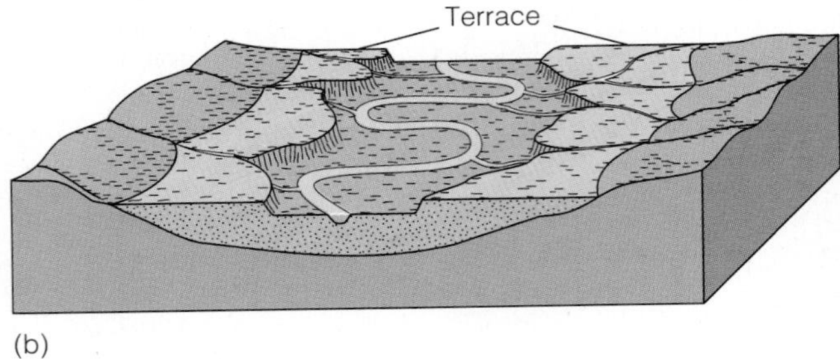

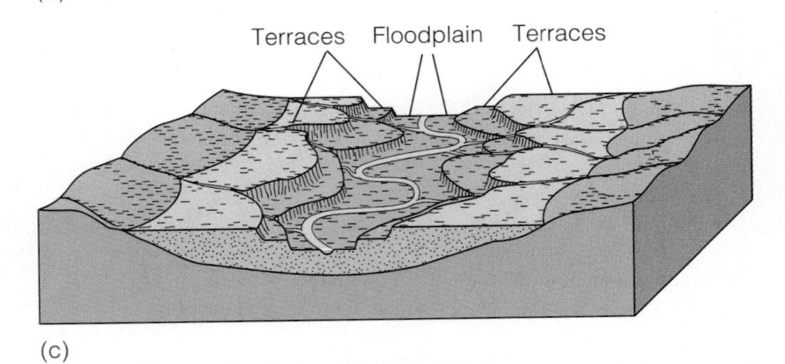

➢ **FIGURE 15-39 (right)** Origin of stream terraces. (*a*) A stream has a broad floodplain adjacent to its channel. (*b*) The stream erodes downward and establishes a new floodplain at a lower level. Remnants of its old floodplain are stream terraces. (*c*) Another level of terraces originates as the stream erodes downward again.

➢ **FIGURE 15-40 (below)** The Goose Necks of the San Juan River are incised meanders.

Chapter Summary

1. Water is continuously evaporated from the oceans, rises as water vapor, condenses, and falls as precipitation. About 20% of all precipitation falls on land and eventually returns to the oceans, mostly by surface runoff.
2. Running water moves by either laminar or turbulent flow. In the former, streamlines parallel one another, whereas in the latter they are complexly intertwined. Most flow in streams is turbulent.
3. Runoff can be characterized as either sheet flow or channel flow. Channels of all sizes are called streams.
4. Gradient generally varies from steep to gentle along the course of a stream, being steep in upper reaches and gentle in lower reaches.
5. Flow velocity and discharge are related. A change in one of these parameters causes the other to change as well.
6. A stream and its tributaries carry runoff from its drainage basin. Drainage basins are separated from one another by divides.
7. Streams erode by hydraulic action, abrasion, and dissolution of soluble rocks.
8. The coarser part of a stream's sediment load is transported as bed load, and the finer part as suspended load. Streams also transport a dissolved load of ions in solution.
9. Competence is a measure of the maximum-sized particles that a stream can carry and is related to velocity. Capacity is a function of discharge and is a measure of the total load transported by a stream.
10. Braided streams are characterized by a complex of dividing and rejoining channels. Braiding occurs when sediment transported by the stream is deposited within channels as sand and gravel bars.
11. Meandering streams have a single, sinuous channel with broad looping curves. Meanders migrate laterally as the cut bank is eroded and point bars form on the inner bank. Oxbow lakes are cutoff meanders in which fine-grained sediments and organic matter accumulate.
12. Floodplains are rather flat areas paralleling stream channels. They may be composed mostly of point bar deposits formed by lateral accretion or mud accumulated by vertical accretion during numerous floods.
13. Deltas are alluvial deposits at a stream's mouth. Many small deltas in lakes conform to the three-part division of bottomset, foreset, and topset beds, but large marine deltas are more complex. Marine deltas are characterized as stream-, wave-, or tide-dominated.
14. Alluvial fans are lobate alluvial deposits on land consisting mostly of sand and gravel. They form best in arid and semiarid regions where erosion rates are high.
15. Sea level is ultimate base level, the lowest level to which streams can erode. However, streams commonly have local base levels such as lakes, other streams, or the points where they flow across particularly resistant rocks.
16. Streams tend to eliminate irregularities in their channels so that they develop a smooth, concave profile of equilibrium. Such streams are graded. In a graded stream, a balance exists between gradient, discharge, flow velocity, channel characteristics, and sediment load so that little or no erosion or deposition occurs within the channel.
17. Stream valleys develop by a combination of processes including downcutting, lateral erosion, mass wasting, sheet wash, and headward erosion.
18. Many streams flowing through valleys cut into ridges directly in their paths are superposed, meaning that they once flowed on a higher surface and eroded downward into resistant rocks.
19. Renewed downcutting by a stream possessing a floodplain commonly results in the formation of stream terraces, which are remnants of an older floodplain at a higher level.
20. Incised meanders are generally attributed to renewed downcutting by a meandering stream so that it now occupies a deep, meandering valley.

Important Terms

abrasion
alluvial fan
alluvium
base level
bed load
braided stream
delta
discharge
dissolved load
divide
drainage basin
drainage pattern
floodplain
graded stream
gradient
hydraulic action
hydrologic cycle
incised meander
infiltration capacity
meandering stream
natural levee
oxbow lake
point bar
runoff
stream
stream terrace
superposed stream
suspended load
velocity

Review Questions

1. Trellis drainage develops on:
   a. ____ natural levees;
   b. ____ granite;
   c. ____ fractured basalt;
   d. ____ tilted sedimentary rock layers;
   e. ____ horizontal layers of volcanic rocks.
2. Mounds of sediment deposited on the margin of a stream are:
   a. ____ natural levees;
   b. ____ oxbow lakes;
   c. ____ bottomset beds;
   d. ____ incised meanders;
   e. ____ alluvial fans.
3. The direct impact of running water is:
   a. ____ bed load;
   b. ____ saltation;
   c. ____ hydraulic action;
   d. ____ meander cutoff;
   e. ____ base level.
4. Most of the fresh water on Earth is in:
   a. ____ the groundwater system;
   b. ____ the atmosphere;
   c. ____ lakes;
   d. ____ streams;
   e. ____ glaciers.
5. The vertical drop of a stream in a given horizontal distance is its:
   a. ____ discharge;
   b. ____ gradient;
   c. ____ velocity;
   d. ____ base level;
   e. ____ drainage pattern.
6. Sediment transport by intermittent bouncing and skipping along a stream bed is:
   a. ____ saltation;
   b. ____ dissolved load;
   c. ____ capacity;
   d. ____ suspended load;
   e. ____ alluvium.
7. The capacity of a stream is a measure of its:
   a. ____ volume of water;
   b. ____ velocity;
   c. ____ total load of sediment;
   d. ____ discharge;
   e. ____ ability to erode.
8. A meandering stream is one having:
   a. ____ numerous sand and gravel bars in its channel;
   b. ____ a single, sinuous channel;
   c. ____ a broad, shallow channel;
   d. ____ a deep, narrow valley;
   e. ____ long, straight reaches and waterfalls.
9. Which of the following is a local base level?
   a. ____ lake;
   b. ____ ocean;
   c. ____ floodplain
   d. ____ point bar;
   e. ____ alluvial fan.
10. Erosional remnants of floodplains that are higher than the current level of a stream are:
    a. ____ oxbow lakes;
    b. ____ cut banks;
    c. ____ stream terraces;
    d. ____ incised meanders;
    e. ____ natural bridges.
11. All of the sediment carried by saltation and rolling and sliding along a stream bed is the:
    a. ____ suspended load;
    b. ____ drainage capacity;
    c. ____ stream profile;
    d. ____ bed load;
    e. ____ channel pattern.
12. Infiltration capacity is the:
    a. ____ rate at which a stream erodes;
    b. ____ distance a stream flows from its source to the ocean;
    c. ____ maximum rate that surface materials can absorb water;
    d. ____ vertical distance a stream can erode below sea level;
    e. ____ variation in flow velocity across a stream channel.
13. A drainage pattern in which streams flow in and out of lakes with irregular flow directions is:
    a. ____ radial;
    b. ____ longitudinal;
    c. ____ deranged;
    d. ____ rectangular;
    e. ____ graded.
14. How do solar radiation, the changing phases of water, and runoff cause the recycling of water from the oceans to the atmosphere and back to the oceans?
15. What is the difference between laminar and turbulent flow, and why is flow in streams usually turbulent?
16. Explain what infiltration capacity is and why it is important in considering runoff.
17. How do channel shape and roughness control flow velocity?
18. Is the statement "the steeper the gradient, the greater the flow velocity" correct? Explain.
19. What do braided streams look like, and what do they transport and deposit?
20. How is it possible for a meandering stream to erode laterally yet maintain a more or less constant channel width?
21. How do oxbow lakes and meander scars form?
22. What are alluvial fans and where are they best developed?
23. Sea level is ultimate base level for most streams. If sea level drops with respect to the land, how would a stream respond?
24. What is a graded stream, and why are streams rarely graded except temporarily?
25. How do headward erosion and stream piracy lengthen a stream channel?

Points to Ponder

1. A stream 2,000 m above sea level at its source flows 1,500 km to the sea. What is the stream's gradient? Do you think the gradient you calculated will be correct for all segments of this stream? Explain.
2. What long-term changes may occur in the hydrologic cycle? How might human activities bring about such changes?
3. According to one estimate, 10.76 km^3 of sediment is eroded from the continents each year, much of it by running water. Given that the volume of the continents above sea level is 92,832,194 km^3, they should be eroded to sea level in only 8,627,527 years. Although the calculation is correct, there is something seriously wrong with this line of reasoning. What is it?
4. Calculate the daily discharge for a stream 148 m wide, 2.6 m deep, with a flow velocity of 0.3 m/sec.
5. Why is the Earth the only planet with abundant liquid water?

Additional Readings

Beven, K., and P. Carling, eds. 1989. *Floods*. New York: John Wiley & Sons.

Chorley, R. J., ed. 1971. *Introduction to fluvial processes*. London: Methuen.

Crickmay, C. H. 1974. *The work of the river.* London: Macmillan.

Frater, A., ed. 1984. *Great rivers of the world*. Boston: Little, Brown.

Graves, W., ed. 1993. Water: The power, promise, and turmoil of North America's fresh water. *National Geographic*. Special edition, October.

Knighton, D. 1984. *Fluvial forms and processes*. London: Edward Arnold.

Leopold, L. B., M. G. Wolman, and J. P. Miller. 1964. *Fluvial processes in geomorphology*. San Francisco: W. H. Freeman & Co.

McPhee, J. 1989. *The control of nature*. New York. Farrar, Straus, & Giroux.

Morisawa, M. 1968. *Streams: Their dynamics and morphology*. New York: McGraw-Hill.

Petts, G., and I. Foster. 1985. *Rivers and landscape*. London: Edward Arnold.

Rachocki, A. 1981. *Alluvial fans*. New York: John Wiley & Sons.

Schumm, S. A. 1977. *The fluvial system*. New York: John Wiley & Sons.

Chapter 16

GROUNDWATER

OUTLINE

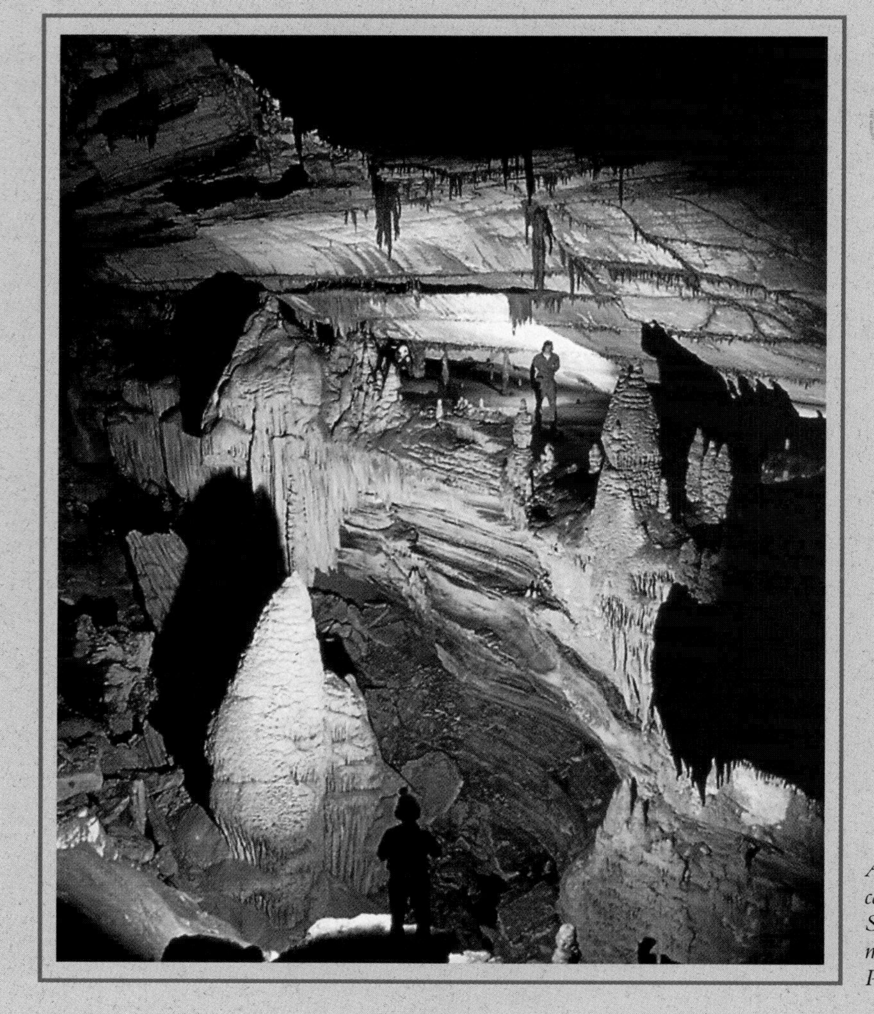

A variety of cave deposits can be seen in Blanchard Springs Caverns, one of the many caves in the Ozark Plateau.

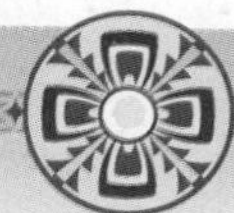

PROLOGUE

For more than two weeks in February 1925, Floyd Collins, an unknown farmer and cave explorer, became a household word (⯈ Figure 16-1). News about the attempts to rescue him from a narrow subsurface fissure near Mammoth Cave, Kentucky, captured the attention of the nation.

The saga of Floyd Collins is rooted in what is known as the Great Cave War of Kentucky. The western region of Kentucky is riddled with caves formed by groundwater weathering and erosion. Many of them were developed as tourist attractions to help supplement meager farm earnings. The largest and best known is Mammoth Cave (see Perspective 16-1). So spectacular is Mammoth, with its numerous caverns, underground rivers, and dramatic cave deposits, that it soon became the standard by which all other caves were measured.

As Mammoth Cave drew more and more tourists, rival cave owners became increasingly bold in attempting to lure visitors to their caves and curio shops. Signs pointing the way to Mammoth Cave frequently disappeared, while "official" cave information booths redirected unsuspecting tourists away from Mammoth Cave. It was in this environment that Floyd Collins grew up.

Seven years before his tragic death, Collins had discovered Crystal Cave on the family farm and opened it up for visitors. But like most of the caves in the area, Crystal Cave attracted few tourists—they visited Mammoth Cave instead. Perhaps it was the thought of discovering a cave rivaling Mammoth or even connecting to it that drove Collins to his fateful exploration of Sand Cave on January 30, 1925.

As Collins inched his way back up through the narrow fissure he had crawled down, he dislodged a small oblong piece of limestone from the ceiling that immediately pinned his left ankle. Try as he might, he was trapped in total darkness 17 m below ground. As he lay half on his left side, Collins's left arm was partially wedged under him, while his right arm was held fast by an overhanging ledge. During his struggles to free himself, he dislodged enough silt and small rocks to bury his legs, further immobilizing him and adding to his anguish.

The next day several neighbors reached Collins and were able to talk to him, feed him, encourage him, and

(b)

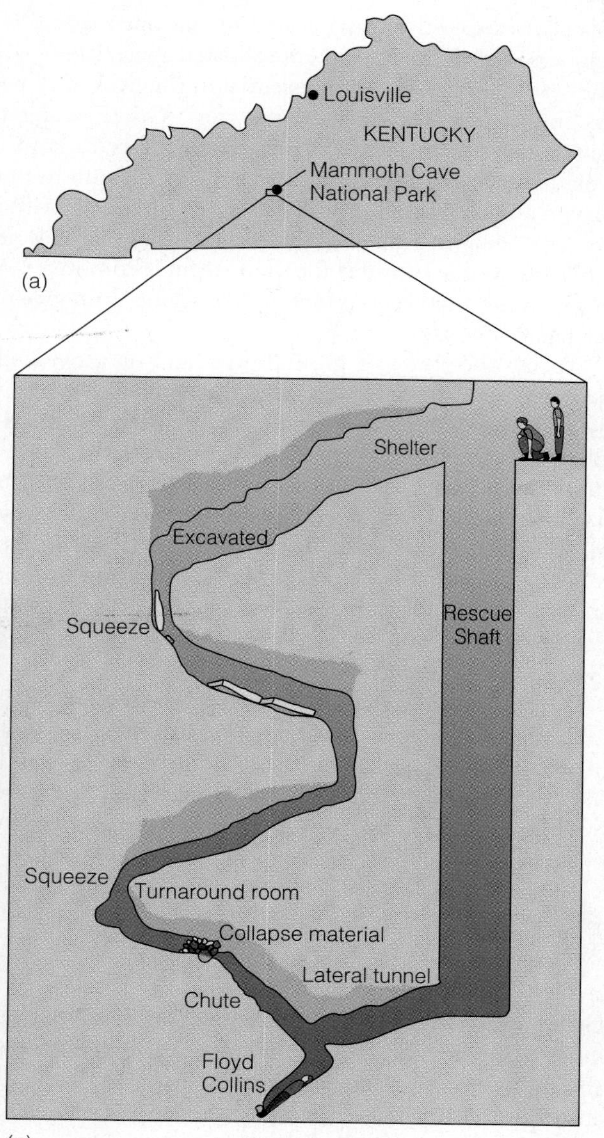

⯈ FIGURE 16-1 (*a*) Location of the cave in which Floyd Collins was trapped. (*b*) Collins looking out of a fissure near the cave where he ultimately died. (*c*) Cross section showing the fissure where Collins was trapped, the rescue shaft that was sunk, and the lateral tunnel that finally reached him.

try to make him more comfortable, but they could not get him out. Word of his plight quickly spread and the area soon swarmed with reporters. Eventually, volunteers were able to excavate an area around Collins's upper body, but could not free his pinned legs. While an anxious country waited, rescue attempts led by Floyd's brother Homer continued.

Three days after he had become trapped, a harness was put around Collins's chest and rescuers tried to pull him free. After numerous attempts to yank him out, workers had to abandon that plan because Collins was unable to bear the pain. Meanwhile at the surface, a carnival-like atmosphere had developed as some 20,000 people converged on the scene, and the National Guard had to be called out to maintain order.

Two days after the attempt to pull Collins out of the fissure failed, part of the passageway used by the rescuers collapsed, sealing Collins's fate. The only hope now was to dig a vertical relief shaft from which a lateral tunnel could be dug to reach Collins. On February 16, rescuers finally reached the chamber where Collins's lifeless body lay entombed. After his body was brought out, he was buried near Crystal Cave, where his grave is appropriately marked by a beautiful stalagmite and pink granite headstone.

INTRODUCTION

Groundwater—the water stored in the open spaces within underground rocks and unconsolidated material—is a valuable natural resource that is essential to the lives of all people. Its importance to humans is not new. Groundwater rights have always been important in North America, and many legal battles have been fought over them. Groundwater also played a crucial role in the development of the U.S. railway system during the nineteenth century when railroads needed a reliable source of water for their steam locomotives. Much of the water used by the locomotives came from groundwater tapped by wells.

Today, the study of groundwater and its movement has become increasingly important as the demand for fresh water by agricultural, industrial, and domestic users has reached an all-time high. More than 65% of the groundwater used in the United States each year goes for irrigation, with industrial use second, followed by domestic needs. These demands have severely depleted the groundwater supply in many areas and led to such problems as ground subsidence and saltwater contamination. In other areas, pollution from landfills, toxic waste, and agriculture has rendered the groundwater supply unsafe.

As the world's population and industrial development expand, the demand for water, particularly groundwater, will increase. Not only is it important to locate new groundwater sources, but, once found, these sources must be protected from pollution and managed properly to ensure that users do not withdraw more water than can be replenished.

GROUNDWATER AND THE HYDROLOGIC CYCLE

Groundwater represents approximately 22% (8.4 million km^3) of the world's supply of fresh water. This amount is about 36 times greater than the total for all of the streams and lakes of the world (see Chapter 15) and equals about one-third the amount locked up in the world's ice caps (see Chapter 17). If the world's groundwater were spread evenly over the Earth's surface, it would be about 10 m deep.

Groundwater is one reservoir of the hydrologic cycle. The major source of groundwater is precipitation that infiltrates the ground and moves through the soil and pore spaces of rocks (see Figure 15-5). Other sources include water infiltrating from lakes and streams, recharge ponds, and wastewater treatment systems. As the groundwater moves through soil, sediment, and rocks, many of its impurities, such as disease-causing microorganisms, are filtered out. Not all soils and rocks are good filters, however, and some serious pollutants are not removed. Groundwater eventually returns to the surface reservoir when it enters lakes, streams, or the ocean.

POROSITY AND PERMEABILITY

Porosity and *permeability* are important physical properties of Earth materials and are largely responsible for the amount, availability, and movement of groundwater. Water soaks into the ground because the soil, sediment, or rock has open spaces or pores. **Porosity** is the percentage of a material's total volume that is pore space. While porosity most often consists of the spaces between particles in soil, sediments, and sedimentary rocks, other types of porosity can include cracks, fractures, faults, and vesicles in volcanic rocks (▶ Figure 16-2).

Porosity varies among different rock types and is dependent on the size, shape, and arrangement of the material composing the rock (◉ Table 16-1). Most igneous and metamorphic rocks as well as many limestones and dolostones have very low porosity because they are composed of tightly interlocking crystals. Their porosity can be increased, however, if they have been fractured or weathered by groundwater. This is particularly true for massive limestone and dolostone whose fractures can be enlarged by acidic groundwater.

By contrast, detrital sedimentary rocks composed of

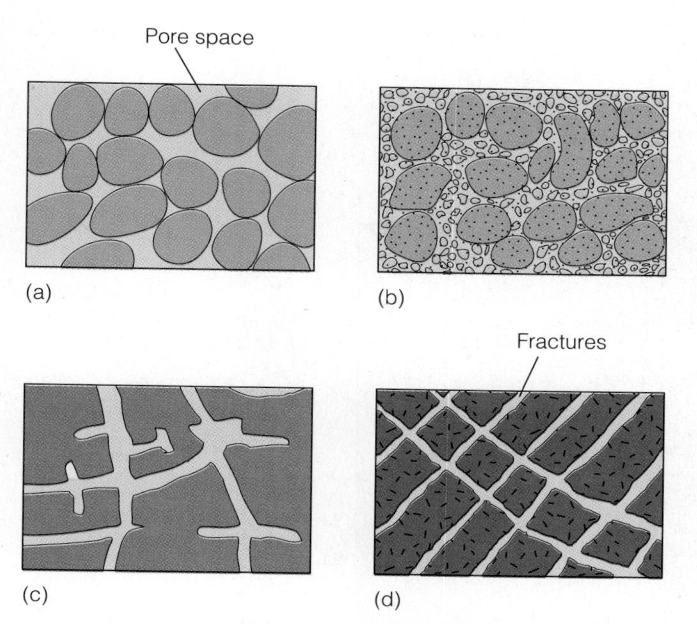

▶ **FIGURE 16-2** A rock's porosity is dependent on the size, shape, and arrangement of the material composing the rock. (*a*) A well-sorted sedimentary rock has high porosity while (*b*) a poorly sorted one has low porosity. (*c*) In soluble rocks such as carbonates, porosity can be increased by solution, while (*d*) crystalline rocks can be rendered porous by fracturing.

well-sorted and well-rounded grains can have very high porosity because any two grains touch only at a single point, leaving relatively large open spaces between the grains (Figure 16-2a). Poorly sorted sedimentary rocks, on the other hand, typically have low porosity because finer grains fill in the space between the larger grains, further reducing porosity (Figure 16-2b). In addition, the amount of cement between grains can also decrease porosity.

TABLE 16-1 Porosity Values for Different Materials

Material	Percentage Porosity
Unconsolidated sediment	
Soil	55
Gravel	20–40
Sand	25–50
Silt	35–50
Clay	50–70
Rocks	
Sandstone	5–30
Shale	0–10
Solution activity in limestone, dolostone	10–30
Fractured basalt	5–40
Fractured granite	10

SOURCE: U.S. Geological Survey, Water Supply Paper 2220 (1983) and others.

Although porosity determines the amount of groundwater a rock can hold, it does not guarantee that the water can be extracted. The capacity of a material for transmitting fluids is its **permeability**. Permeability is dependent not only on porosity, but also on the size of the pores or fractures and their interconnections. For example, deposits of silt or clay are typically more porous than sand or gravel. Nevertheless, shale has low permeability because the pores between its clay particles are very small, and the molecular attraction between the clay and the water is great, thereby preventing movement of the water. In contrast, the pore spaces between grains in sandstone and conglomerate are much larger, and the molecular attraction on the water is therefore low. Chemical and biochemical sedimentary rocks, such as limestone and dolostone, and many igneous and metamorphic rocks that are highly fractured can also be very permeable provided that the fractures are interconnected.

A permeable layer transporting groundwater is called an **aquifer**, from the Latin *aqua* meaning water. The most effective aquifers are deposits of well-sorted and well-rounded sand and gravel. Limestones in which fractures and bedding planes have been enlarged by solution are also good aquifers. Shales and many igneous and metamorphic rocks make poor aquifers because they are typically impermeable. Rocks such as these and any other materials that prevent the movement of groundwater are called **aquicludes**.

THE WATER TABLE

When precipitation occurs over land, some of it evaporates, some is carried away by runoff in streams, and the remainder seeps into the ground. As this water moves down from the surface, some of it adheres to the material that it is moving through and halts its downward progress. This region is the **zone of aeration,** and its water is called *suspended water* (▶ Figure 16-3). The pore spaces in this zone contain both water and air. Extending irregularly upward a few centimeters to several meters from the zone of saturation below is the **capillary fringe**. Water moves upward in this region because of surface tension, much as water moves upward through a paper towel.

Beneath the zone of aeration lies the **zone of saturation** where all of the pore spaces are filled with groundwater (Figure 16-3). The base of the zone of saturation varies from place to place, but usually extends to a depth where an impermeable layer is encountered or to a depth where confining pressure closes all open space.

The surface separating the zone of aeration from the underlying zone of saturation is the **water table** (Figure 16-3). In general, the configuration of the water table is a subdued replica of the overlying land surface; that is, it has its highest elevations beneath hills and its lowest elevations in valleys. In most arid and semiarid regions, however, the water table is quite flat and is below the level of river valleys.

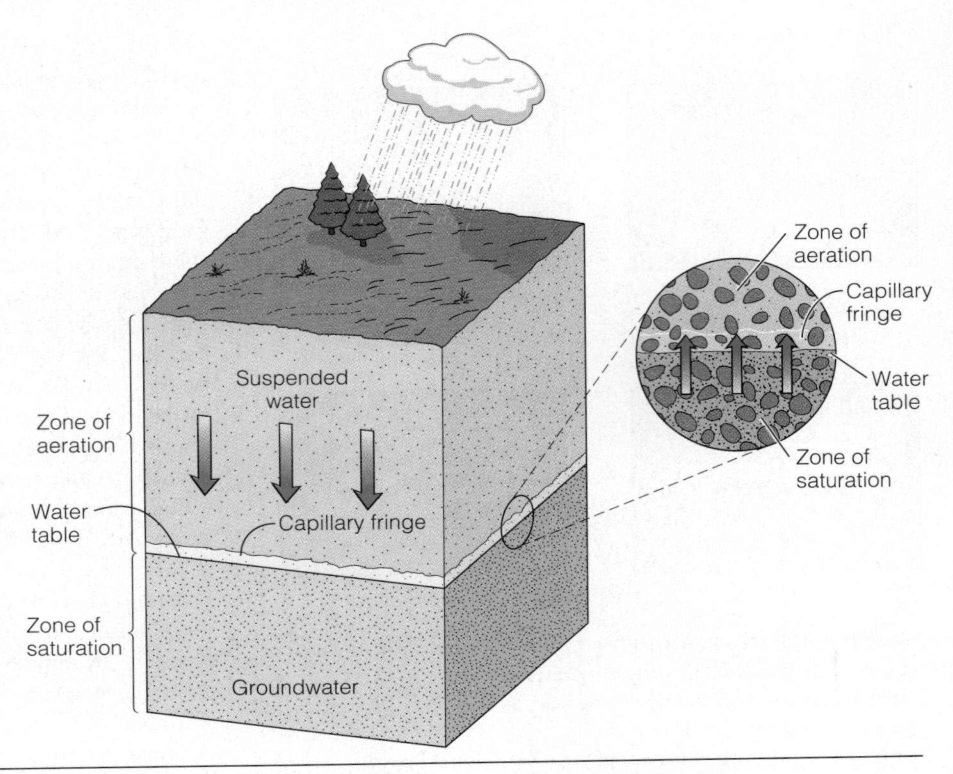

▶ **FIGURE 16-3** The zone of aeration contains both air and water within its open space, while all of the open space in the zone of saturation is filled with groundwater. The water table is the surface separating the zones of aeration and saturation. Within the capillary fringe, water rises upward by surface tension from the zone of saturation into the zone of aeration.

Several factors contribute to the surface configuration of a region's water table. These include regional differences in the amount of rainfall, permeability, and the rate of groundwater movement. During periods of high rainfall, groundwater tends to rise beneath hills because it cannot flow fast enough into the adjacent valleys to maintain a level surface. During droughts, the water table falls and tends to flatten out because it is not being replenished.

 GROUNDWATER MOVEMENT

Groundwater velocity varies greatly and depends on many factors. Velocities ranging from 250 m per day in some extremely permeable material to less than a few centimeters per year in nearly impermeable material have been measured. In most ordinary aquifers, the average velocity of groundwater is a few centimeters per day.

Gravity provides the energy for the downward movement of groundwater. Water entering the ground moves through the zone of aeration to the zone of saturation (▶ Figure 16-4). When water reaches the water table, it continues to move through the zone of saturation from areas where the water table is high toward areas where it is lower, such as at streams, lakes, or swamps (Figure 16-4). Only some of the water follows the direct route along the slope of the water table. Most of it takes longer curving paths downward and then enters a stream, lake, or swamp from below. This occurs because groundwater moves from areas of high pressure toward areas of lower pressure within the saturated zone.

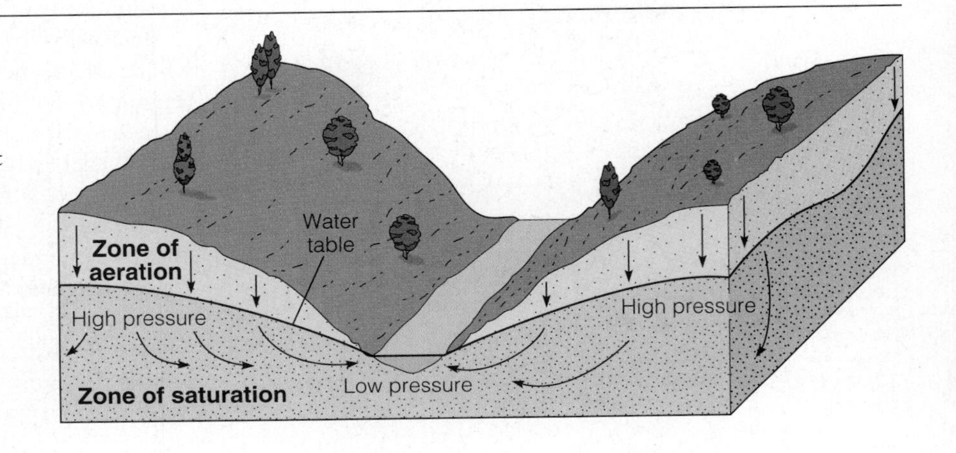

▶ **FIGURE 16-4** Groundwater moves downward due to the force of gravity. It moves through the zone of aeration to the zone of saturation where some of it moves along the slope of the water table and the rest of it moves through the zone of saturation from areas of high pressure toward areas of low pressure.

➢ **FIGURE 16-5** Storm Water Basin No. 129, Garden City Park, Long Island, New York, is one of many recharge basins operated by the Nassau County Department of Public Works.

Springs, Water Wells, and Artesian Systems

Adding water to the zone of saturation is called **recharge**, and it causes the water table to rise. Water may be added by natural means, such as rainfall or melting snow, or artificially at recharge basins or wastewater treatment plants (➢ Figure 16-5). If groundwater is discharged without sufficient replenishment, the water table drops. Groundwater discharges naturally whenever the water table intersects the ground surface as at a spring or along a stream, lake, or swamp. Groundwater can also be discharged artificially by pumping water from wells.

Springs

A **spring** is a place where groundwater flows or seeps out of the ground. Springs have always fascinated people because the water flows out of the ground for no apparent reason and from no readily identifiable source. It is not surprising that springs have long been regarded with superstition and revered for their supposed medicinal value and healing powers. Nevertheless, there is nothing mystical or mysterious about springs.

Although springs can occur under a wide variety of geologic conditions, they all form in basically the same way (➢ Figure 16-6). When percolating water reaches the water table or an impermeable layer, it flows laterally, and if this flow intersects the Earth's surface, the water discharges onto the surface as a spring (➢ Figure 16-7). The Mammoth Cave area in Kentucky, for example, is underlain by fractured limestones that have been enlarged into caves by solution activity (see Perspective 16-1). In this geologic environment, springs occur where the fractures and caves intersect the ground surface allowing groundwater to exit onto the surface. Springs most commonly occur along valley walls where streams have cut valleys below the regional water table.

Springs can also develop wherever a perched water table intersects the Earth's surface (➢ Figure 16-8). A **perched water table** may occur wherever a local aquiclude occurs within a larger aquifer, such as a lens of shale within a sandstone. As water migrates through the zone of aeration, it is stopped by the local aquiclude, and a localized zone of saturation "perched" above the main water table is created. Water moving laterally along the perched water table may intersect the Earth's surface to produce a spring.

Water Wells

A **water well** is made by digging or drilling into the zone of saturation. Once the zone of saturation is reached, water percolates into the well and fills it to the level of the water table. Most wells must be pumped to bring the groundwater to the surface.

When a well is pumped, the water table in the area around the well is lowered, because water is removed from the aquifer faster than it can be replenished. A **cone of depression** thus forms around the well, varying in size

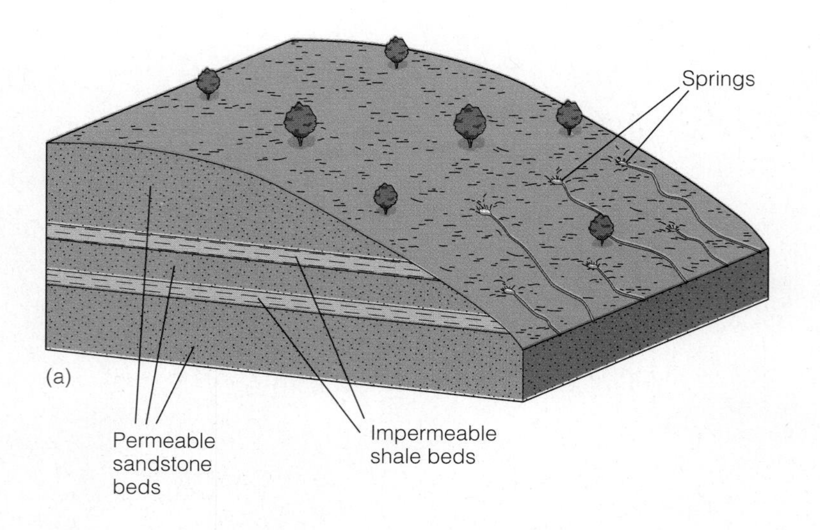

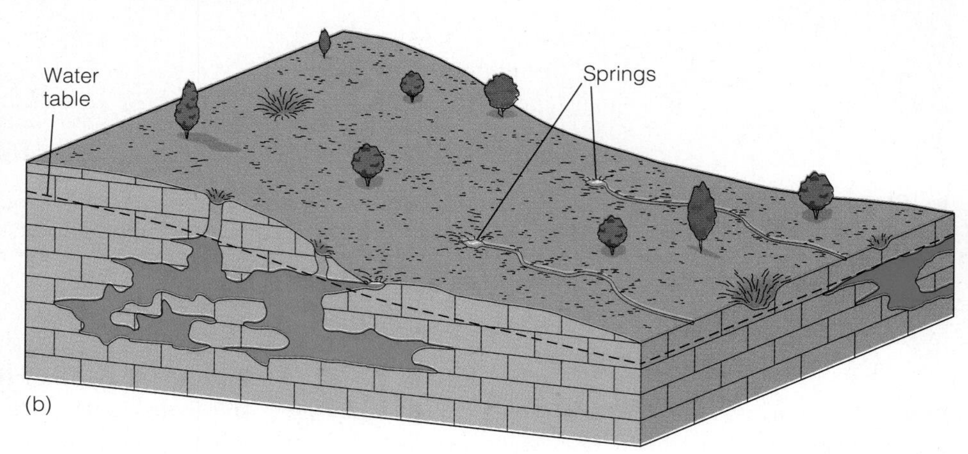

➢ **FIGURE 16-6** Springs form wherever laterally moving groundwater intersects the Earth's surface. (*a*) Most commonly, they form when percolating water reaches an impermeable layer and migrates laterally until it seeps out at the surface. (*b*) Springs also can occur in areas underlain by fractured soluble rocks such as limestones where groundwater moves freely through underground cavities until it reaches the surface and flows out.

➢ **FIGURE 16-7**
Periodic Spring, near Afton, Wyoming.

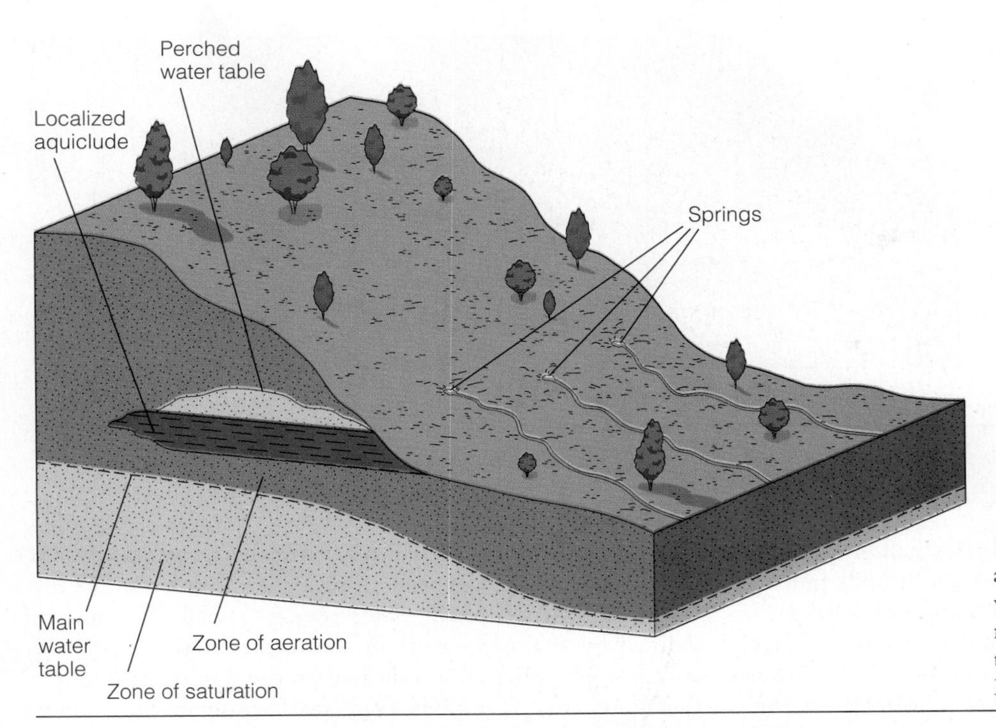

▶ **FIGURE 16-8** If a localized aquiclude, such as a shale layer, occurs within an aquifer, a perched water table may result with springs occurring where the perched water table intersects the Earth's surface.

according to the rate and amount of water being withdrawn (▶ Figure 16-9). If water is pumped out of a well faster than it can be replaced, the cone of depression grows until the well goes dry. This lowering of the water table normally does not pose a problem for the average domestic well, provided that the well is drilled sufficiently deep into the zone of saturation. The tremendous amounts of water used by industry and irrigation, however, may create a large cone of depression that lowers the water table sufficiently to cause shallow wells in the immediate area to go dry (Figure 16-9). This situation is not uncommon and frequently results in lawsuits by the owners of the shallow dry wells. Furthermore, lowering of the regional water table is becoming a serious problem in many areas, particularly in the southwestern United States where rapid growth has placed tremendous demands on the groundwater system. Unrestricted withdrawal of groundwater cannot continue indefinitely, and the rising costs and decreasing supply of groundwater should soon limit the growth of this region of the United States.

People in rural areas and those without access to a municipal water system are well aware of the problems of locating an adequate groundwater supply. The distribution and type of rocks present, their porosity and permeability, fracture patterns, and so on are all factors that determine whether a water well will be successful (▶ Figure 16-10).

Artesian Systems

The word **artesian** comes from the French town and province of Artois (called Artesium during Roman times) near Calais, where the first European artesian well was drilled in A.D. 1126 and is still flowing today. The term

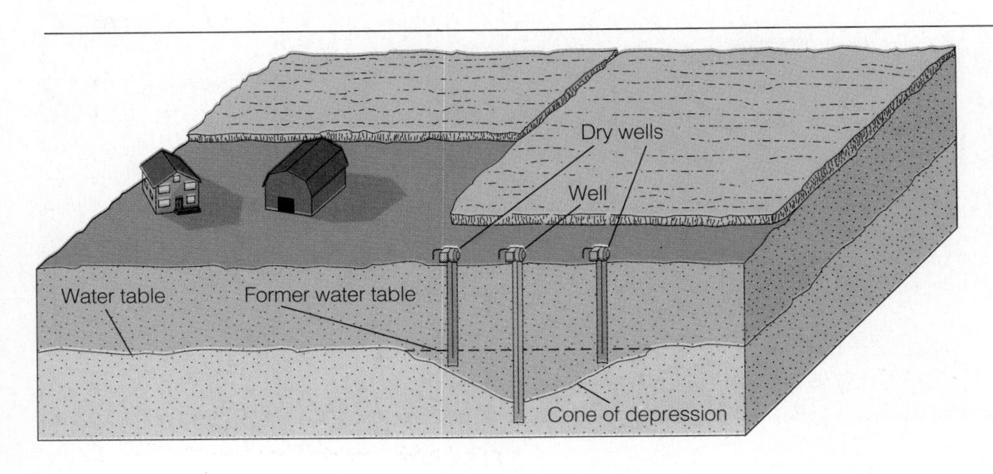

▶ **FIGURE 16-9** A cone of depression forms whenever water is withdrawn from a well. If water is withdrawn faster than it can be replenished, the cone of depression will grow in depth and circumference, lowering the water table in the area and causing nearby shallow wells to go dry.

▶ **FIGURE 16-10** Many factors determine whether a water well will be successful. Wells A and E were drilled to the same depth. Well A was successful because it tapped a perched water table, whereas well E did not. To be successful, it will have to be drilled below the water table like well C. Well B tapped a fracture below the water table and was successful, whereas well D missed the fractures and was dry.

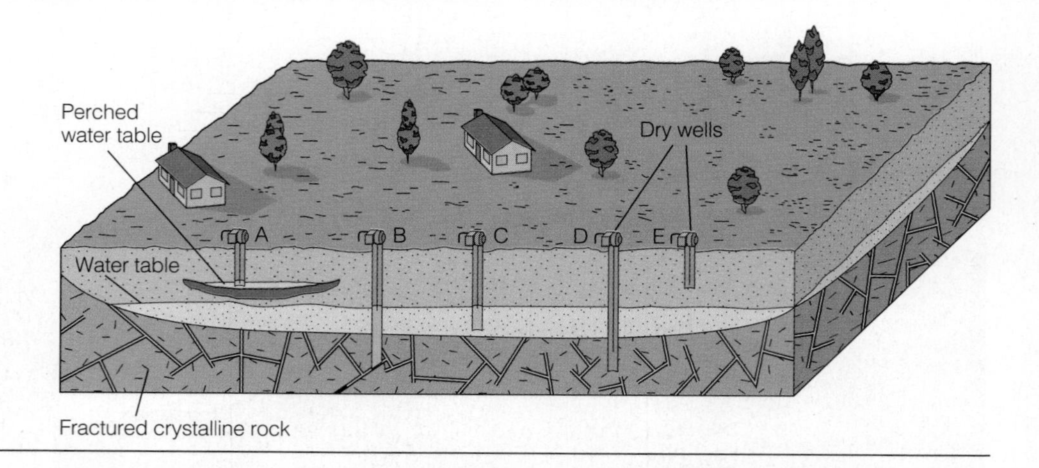

artesian, can be applied to any system in which groundwater is confined and builds up high hydrostatic (fluid) pressure. Water in such a system is able to rise above the level of the aquifer if a well is drilled through the confining layer, thereby reducing the pressure and forcing the water upward (▶ Figure 16-11). For an artesian system to develop, three geologic conditions must be present (▶ Figure 16-12): (1) the aquifer must be confined above and below by aquicludes to prevent water from escaping; (2) the rock sequence is usually tilted and exposed at the surface, enabling the aquifer to be recharged; and (3) there is sufficient precipitation in the recharge area to keep the aquifer filled.

The elevation of the water table in the recharge area and the distance of the well from the recharge area determine the height to which artesian water rises in a well. The surface defined by the water table in the recharge area, called the *artesian-pressure surface*, is indicated by the sloping dashed line in Figure 16-12. If there were no friction in the aquifer, well water from an artesian aquifer would rise exactly to the elevation of the artesian-pressure surface. Friction, however, slightly reduces the pressure of the aquifer water and consequently the level to which artesian water rises. This is why the pressure surface slopes.

An artesian well will flow freely at the ground surface only if the wellhead is at an elevation below the artesian-

▶ **FIGURE 16-11** Artesian well at Deep Well Ranch, South Fork of the Madison River, Gallatin County, Montana.

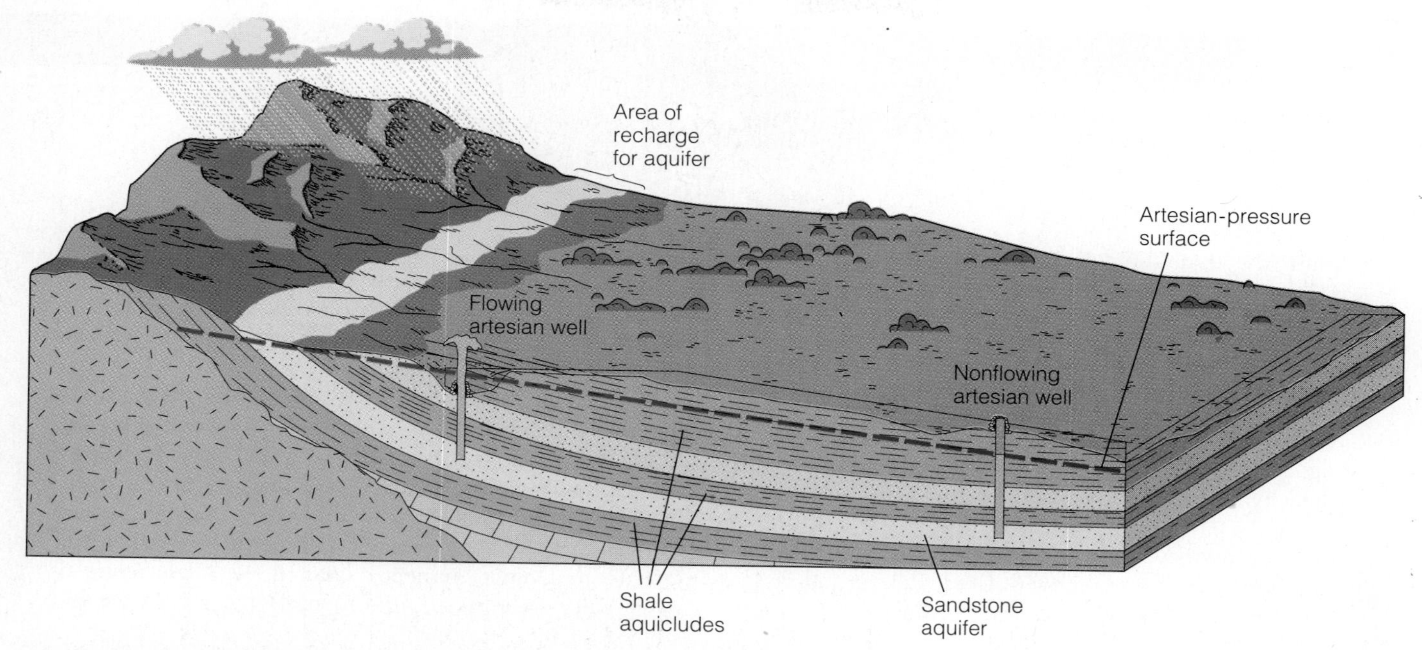

➢ **FIGURE 16-12** An artesian system must have an aquifer confined above and below by aquicludes, the aquifer must be exposed at the surface, and there must be sufficient precipitation in the recharge area to keep the aquifer filled. The elevation of the water table in the recharge area, which is indicated by a sloping dashed line (the artesian-pressure surface), defines the highest level to which well water can rise. If the elevation of a wellhead is below the elevation of the artesian-pressure surface, the well will be free-flowing because the water will rise toward the artesian-pressure surface, which is at a higher elevation than the wellhead. If the elevation of a wellhead is at or above that of the artesian-pressure surface, the well will be nonflowing.

pressure surface. In this situation, the water flows out of the well because it rises toward the artesian-pressure surface, which is at a higher elevation than the wellhead (Figure 16-11). In a nonflowing artesian well, the wellhead is above the artesian-pressure surface, and the water will rise in the well only as high as the artesian-pressure surface.

In addition to artesian wells, many artesian springs also exist. Such springs can occur if a fault or fracture intersects the confined aquifer allowing water to rise above the aquifer. Oases in deserts are commonly artesian springs.

Because the geologic conditions necessary for artesian water can occur in a variety of ways, artesian systems are quite common in many areas of the world underlain by sedimentary rocks. One of the best-known artesian systems in the United States underlies South Dakota and extends southward to central Texas. The majority of the artesian water from this system is used for irrigation. The aquifer of this artesian system, the Dakota Sandstone, is recharged where it is exposed along the margins of the Black Hills of South Dakota. The hydrostatic pressure in this system was originally great enough to produce free-flowing wells and to operate waterwheels. The extensive use of water for irrigation over the years, has reduced the pressure in many of the wells so that they are no longer free-flowing and the water must be pumped.

Another example of an important artesian system is the Floridan aquifer system. Here Tertiary-aged carbonate rocks are riddled with fractures, caves, and other openings that have been enlarged and interconnected by solution activity. These carbonates are exposed at the surface in the northwestern and central parts of the state where they are recharged, and they dip toward both the Atlantic and Gulf coasts where they are covered by younger sediments. The carbonates are interbedded with shales forming a series of confined aquifers and aquicludes. This artesian system is tapped in the southern part of the state where it is an important source of fresh water and one that is being rapidly depleted.

 GROUNDWATER EROSION AND DEPOSITION

When rainwater begins seeping into the ground, it immediately starts to react with the minerals it contacts, weathering them chemically. In an area underlain by soluble rock, groundwater is the principal agent of erosion and is responsible for the formation of many major features of the landscape.

Limestone, a common sedimentary rock composed primarily of the mineral calcite ($CaCO_3$), underlies large areas

Perspective 16-1

Mammoth Cave National Park, Kentucky

Within the limestone region of western Kentucky lies the largest cave system in the world. In 1941, approximately 51,000 acres were set aside and designated as Mammoth Cave National Park. In 1981 it became a World Heritage Site. Recently, the National Park Service has been considering closing Mammoth Cave because of the health hazard created by raw sewage and contaminated groundwater in the area.

From ground level, the topography of the area is unimposing with numerous sinkholes, lakes, valleys, and disappearing streams. Beneath the surface, however, are more than 230 km of interconnecting passageways whose spectacular geologic features have been enjoyed by numerous cave explorers and tourists alike.

Based on carbon 14 dates from some of the many artifacts found in the cave (such as woven cord and wooden bowls), Mammoth Cave had been explored and used by Native Americans for more than 3,000 years prior to its rediscovery in 1799 by a bear hunter named Robert Houchins. During the War of 1812, approximately 180 metric tons of saltpeter (a potassium nitrate mineral), used in the manufacture of gunpowder, were mined from Mammoth Cave. At the end

➢ **FIGURE 1** Frozen Niagara is a spectacular example of massive travertine flowstone deposits.

of the Earth's surface (➢ Figure 16-13). Although limestone is practically insoluble in pure water, it readily dissolves if a small amount of acid is present. Carbonic acid (H_2CO_3) is a weak acid that forms when carbon dioxide combines with water ($H_2O + CO_2 \rightarrow H_2CO_3$) (see Chapter 5). Because the atmosphere contains a small amount of carbon dioxide (0.03%), and carbon dioxide is also produced in soil by the decay of organic matter, most groundwater is slightly acidic. When groundwater percolates through the various openings in limestone, the slightly acidic water readily reacts with the calcite to dissolve the rock by forming soluble calcium bicarbonate, which is carried away in solution (see Chapter 5).

of the war, the saltpeter market collapsed, and Mammoth Cave was developed as a tourist attraction, easily overshadowing the other caves in the area. Over the next 150 years, the discovery of new passageways and caverns helped establish Mammoth Cave as the world's premier cave and the standard against which all others were measured (see the Prologue).

Mammoth Cave formed in much the same way as all other caves (Figure 16-17). Groundwater flowing through the St. Genevieve Limestone eroded a complex network of openings, passageways, and caverns. Flowing through the various caverns is the Echo River, a system of subsurface streams that eventually joins the Green River at the surface.

The colorful cave deposits are the primary reason millions of tourists have visited Mammoth Cave. Here can be seen numerous cave deposits, as well as spectacular travertine flowstone deposits (▶ Figure 1). Other attractions include the Giant's Coffin, a 15 m collapse block of limestone, and giant rooms such as Mammoth Dome, which is about 58 m high (▶ Figure 2). The cave is also home to more than 200 species of insects and other animals, including about 45 blind species; some of these can be seen on the Echo River Tour, which conveys visitors 5 km along the underground stream.

▶ FIGURE 2 Looking up Mammoth Dome, the largest room in Mammoth Cave, Kentucky.

Sinkholes and Karst Topography

In regions underlain by soluble rock, the ground surface may be pitted with numerous depressions that vary in size and shape. These depressions, called **sinkholes** or merely *sinks*, mark areas where the underlying rock has been dissolved (▶ Figure 16-14). Sinkholes form in one of two ways. The first is when the soluble rock below the soil is dissolved by seeping water. Natural openings in the rock are enlarged and filled in by the overlying soil. As the groundwater continues to dissolve the rock, the soil is eventually removed, leaving depressions that are typically shallow with gently sloping sides.

Sinkholes also form when a cave's roof collapses, usually producing a steep-sided crater. Sinkholes formed in this way

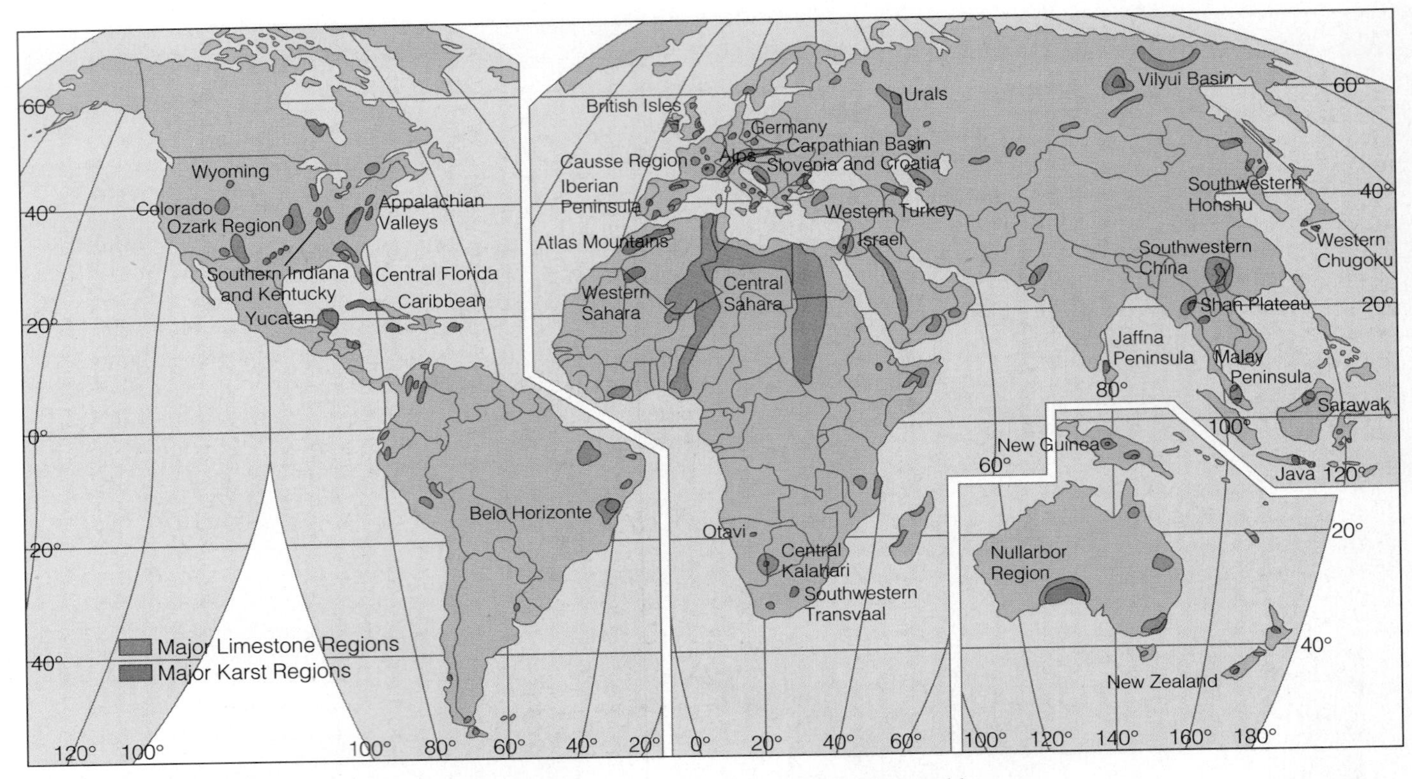

▶ **FIGURE 16-13** The distribution of the major limestone and karst areas of the world.

are a serious hazard, particularly in populated areas. In regions prone to sinkhole formation, the depth and extent of underlying cave systems must be mapped before any development to ensure that the underlying rocks are thick enough to support planned structures.

A **karst topography** is one that has developed largely by groundwater erosion (▶ Figure 16-15). The name *karst* is derived from the plateau region of western Slovenia, western Croatia, and northeastern Italy where this type of topography is well developed. In the United States, regions of karst topography include large areas of southwestern Illinois, southern Indiana, Kentucky, Tennessee, northern Missouri, Alabama, and central and northern Florida (Figure 16-13).

Karst topography is characterized by numerous caves,

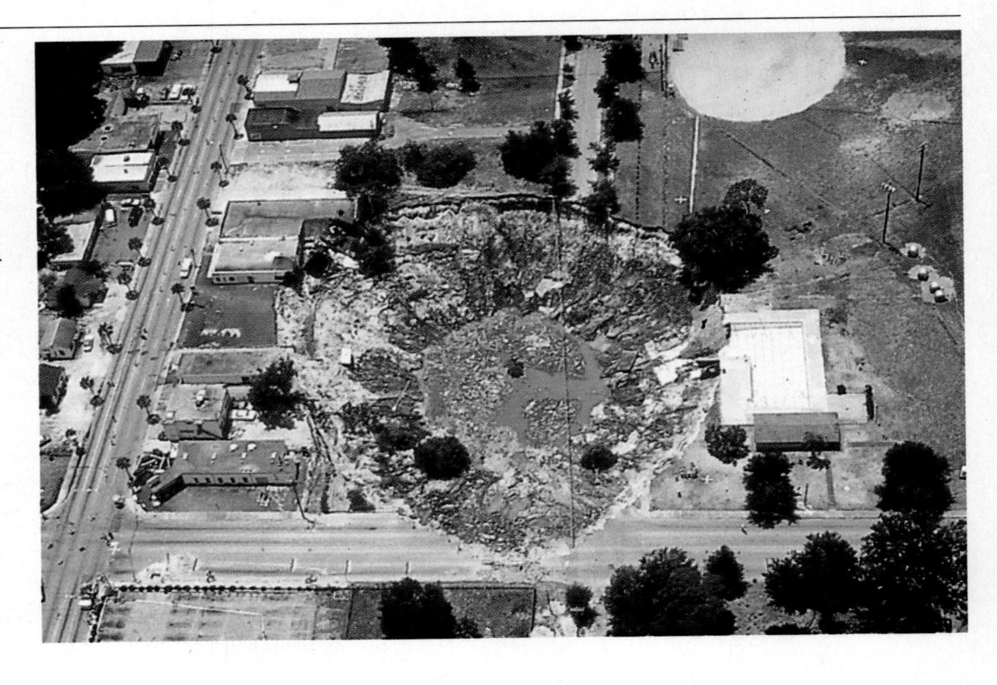

▶ **FIGURE 16-14** This sinkhole formed on May 8 and 9, 1981, in Winter Park, Florida, due to a drop in the water table after prior dissolution of the underlying limestone. The sinkhole destroyed a house, numerous cars, and the municipal swimming pool. It has a diameter of 100 m and a depth of 35 m.

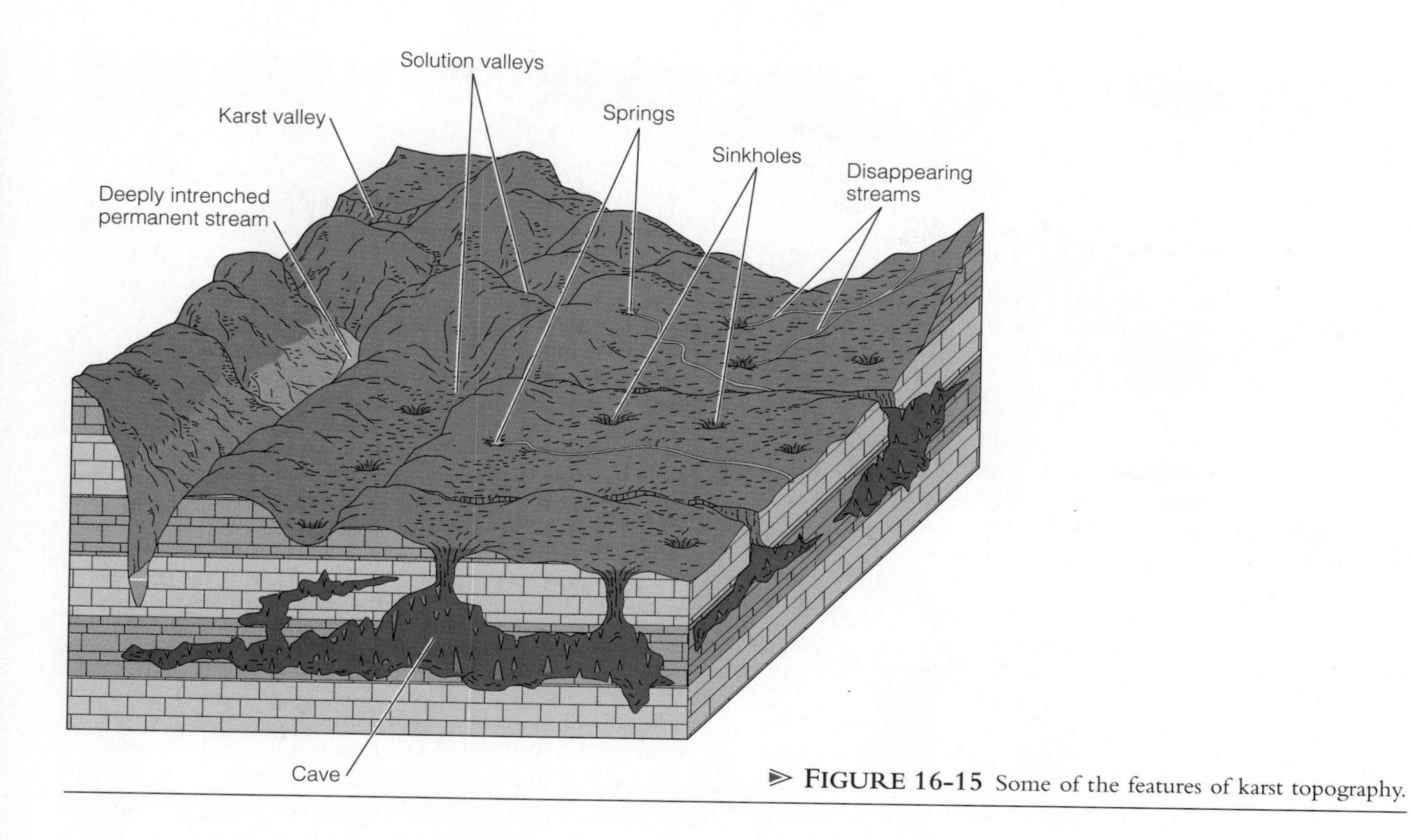

▶ **FIGURE 16-15** Some of the features of karst topography.

springs, sinkholes, solution valleys, and disappearing streams (Figure 16-15). When adjacent sinkholes merge, they form a network of larger, irregular, closed depressions called *solution valleys. Disappearing streams* are another feature of areas of karst topography. They are so named because they typically flow only a short distance at the surface and then disappear into a sinkhole. The water continues flowing underground through various fractures or caves until it surfaces again at a spring or other stream.

Karst topography can range from the spectacular high-relief landscapes of China to the subdued and pockmarked landforms of Kentucky (▶ Figure 16-16). What is com-

▶ **FIGURE 16-16** (*a*) The Stone Forest, 126 km southeast of Kunming, People's Republic of China, is a high-relief karst landscape formed by the dissolution of carbonate rocks. (*b*) Solution valleys, sinkholes, and sinkhole lakes dominate the subdued karst topography east of Bowling Green, Kentucky.

(a)

(b)

Guest Essay *Christopher C. Wieland*

RESTORING THE EARTH AS AN ENVIRONMENTAL CONSULTANT

I came to geology by accident. I was introduced to earth science through a physical geography course. After completing a masters in physical geography, I decided to continue in geology. I studied sedimentary geology, soil science, geomorphology, and hydrogeology, and like many people in the late 1970s, I wanted to work in the environmental field.

As a start, I did my master's thesis on peat deposits in the northern Everglades. These deposits are composed of the preserved remains of aquatic plants. Everglades ecosystems developed as a direct response to water depth and flow patterns. Water depth was initially controlled by lake-bottom topography, but was later controlled by the peat deposits themselves. Consequently, these peat deposits record about 6,000 years of ecosystem formation and development in a former lake.

Jobs in environmental research were hard to find in the early 1980s, but the petroleum industry offered some exciting work. I started with Texaco in 1982 as a development geologist, later moving to an exploration group. Exploration geologists look for new oil and gas fields by examining the stratigraphy and structure of an area for possible traps and then drill to test their theories. Development geologists take over new discoveries and determine the most efficient way to extract the most oil or gas. It was always exciting to propose a well location, drill for oil or gas, and test your predictions.

Later, I moved into Texaco's applied research laboratories where we examined sedimentary rocks using microscopes, scanning electron microscopes, and X-ray diffraction. The information obtained was used to plan well stimulation and, with micropaleontologists, to identify the depositional environments of key rock units. Because many petroleum reservoirs are not controlled by rock structure, knowing the depositional environment of a rock formation helps to determine if stratigraphic traps might exist. Development geologists also use this information in planning well locations for new fields. One of my most interesting projects was the early exploration of the continental slope in the Gulf of Mexico. This oceanographic region is geologically very complex, but contains important reserves of oil and gas. Odd structures related to the movement of thick Louann Formation salt deposits and sands require new methods and concepts of exploration. Our job was to figure out what techniques might prove useful.

I eventually left the petroleum industry to take a position with an environmental consulting firm in Oak Ridge, Tennessee. This gave me the opportunity to pursue my interests in environmental restoration. Environmental restoration requires teamwork with other disciplines. Geologists and hydrogeologists investigate the various attributes of the soils and rocks affected by contaminants. Chemists analyze samples for contaminant concentrations and help geologists and hydrogeologists interpret the data. Biologists examine vegetation and animal life for adverse effects (particularly to threatened or endangered species) and, with hydrogeologists, assess the effects of contaminants on streams and lakes. After investigating the nature, extent, and effects of contaminants, geologists and hydrogeologists team with toxicologists, biologists, and other environmental scientists to determine whether immediate steps need to be taken to protect people and to what extent contaminants pose long-term hazards to humans, animals, and plants. Computer modeling of contaminant fate and transport in the subsurface often plays a major role in risk assessments, because it offers a means of quantitatively estimating the behavior of the contaminants over many years.

Once the distribution of contaminants and the risks associated with them are understood, the site is ready to be remediated. Remediation usually means helping the Earth clean up a human-made mess. Restoration may involve extracting groundwater with wells or trenches and treating it to remove contaminants, excavating and treating soils, or treating soils in place using microbes or chemicals. Computer models of groundwater flow are developed to aid in designing well fields and the intersection of drawdown cones around the wells. Geochemical models may be used to determine whether the contaminants or chemicals used to extract contaminants might interact with soil and rock minerals. Earth scientists also help to determine the volume of contaminated soil or groundwater that must be treated.

Environmental consulting obviously requires good technical skills and the ability to work in teams. It also requires negotiating skills, because part of the consultant's job is to find a way to clean up pollutants that meets with the approval of government regulators, citizens and environmental groups, and clients. On top of that, some business and management skills are also needed, because consulting is essentially a business.

CHRISTOPHER C. WIELAND received an M.A. degree in physical geography/geomorphology from the University of Pittsburgh in 1978 and an M.S. in geology from the University of Florida in 1981. He is currently employed by Radian Corporation in Oak Ridge, Tennessee.

mon to all karst topography, though, is that thick-bedded, readily soluble rock is present at the surface or just below the soil, and enough water is present for solution activity to occur. Karst topography is, therefore, typically restricted to humid and temperate climates. Currently, however, some of the best karst topography can be found in arid and semiarid regions such as Bexar County, Texas, and the Carlsbad Caverns region in New Mexico. The examples of karst topography in these regions are relicts that originally formed when the climate was more humid.

Caves and Cave Deposits

Caves are some of the most spectacular examples of the combined effects of weathering and erosion by groundwater. As groundwater percolates through carbonate rocks, it dissolves and enlarges fractures and openings to form a complex interconnecting system of crevices, caves, caverns, and underground streams. A **cave** is usually defined as a naturally formed subsurface opening that is generally connected to the surface and is large enough for a person to enter. A *cavern* is a very large cave or a system of interconnected caves.

More than 17,000 caves are known in the United States. Most of them are small, but some are quite large and spectacular. Some of the more famous caves in the United States are Mammoth Cave, Kentucky (see Perspective 16-1); Carlsbad Caverns, New Mexico; Lewis and Clark Caverns, Montana; Wind Cave and Jewel Cave, South Dakota; Lehman Cave, Nevada; and Meramec Caverns, Missouri, which Jesse James and his outlaw band often used as a hideout. While the United States has many famous caves, the deepest known cave in North America is the 536 m deep Arctomys Cave in Mount Robson Provincial Park, British Columbia, Canada.

Caves and caverns form as a result of the dissolution of carbonate rocks by weakly acidic groundwater (▶ Figure 16-17). Groundwater percolating through the zone of aeration slowly dissolves the carbonate rock and enlarges its fractures and bedding planes. Upon reaching the water table, the groundwater migrates toward the region's surface streams (Figure 16-4). As the groundwater moves through the zone of saturation, it continues to dissolve the rock and gradually forms a system of horizontal passageways through which the dissolved rock is carried to the streams. As the surface streams erode deeper valleys, the water table drops in response to the lower elevation of the streams. The water that flowed through the system of horizontal passageways now percolates down to the lower water table where a new system of passageways begins to form. The abandoned channelways now form an interconnecting system of caves and caverns. Caves eventually become unstable and collapse, littering the floor with fallen debris.

When most people think of caves, they think of the seemingly endless variety of colorful and bizarre-shaped deposits found in them. Although a great many different types of cave deposits exist, most form in essentially the same manner and are collectively known as **dripstone**. As water seeps through a cave, some of the dissolved carbon dioxide in the water escapes, and a small amount of calcite is precipitated. In this manner, the various dripstone deposits are formed.

Stalactites are icicle-shaped structures hanging from cave ceilings that form as a result of precipitation from dripping water (▶ Figure 16-18). With each drop of water, a thin layer of calcite is deposited over the previous layer, forming a cone-shaped projection that grows downward from the ceiling.

The water that drips from a cave's ceiling also precipitates a small amount of calcite when it hits the floor. As additional calcite is deposited, an upward-growing projection called a **stalagmite** forms (Figure 16-18). If a stalactite and stalagmite meet, they form a **column**. Groundwater seeping from a crack in a cave's ceiling may form a vertical sheet of rock called a *drip curtain*, while water flowing across a cave's floor may produce *travertine terraces* (Figure 16-17).

Modifications of the Groundwater System and Their Effects

Groundwater is a valuable natural resource that is rapidly being exploited with little regard to the effects of overuse and misuse. Currently, about 20% of all water used in the United States is groundwater. This percentage is increasing, and unless this resource is used more wisely, sufficient amounts of clean groundwater will not be available in the future. Modifications of the groundwater system may have many consequences including (1) lowering of the water table, which causes wells to dry up; (2) loss of hydrostatic pressure, which causes once free-flowing wells to require pumping; (3) saltwater encroachment; (4) subsidence; and (5) contamination of the groundwater supply.

Lowering of the Water Table

Withdrawing groundwater at a significantly greater rate than it is replaced by either natural or artificial recharge can have serious effects. For example, the High Plains aquifer is one of the most important aquifers in the United States. Underlying most of Nebraska, large parts of Colorado and Kansas, portions of South Dakota, Wyoming, and New Mexico, as well as the panhandle regions of Oklahoma and Texas, it accounts for approximately 30% of the groundwater used for irrigation in the United States (▶ Figure 16-19). Irrigation from the High Plains aquifer is largely responsible for the high agricultural productivity of this region, where a significant percentage of the nation's corn, cotton, and wheat is grown, and half of U.S. beef cattle are raised. Large areas of land (more than 14 million acres) are currently irrigated with water pumped from the High Plains aquifer. Irrigation is

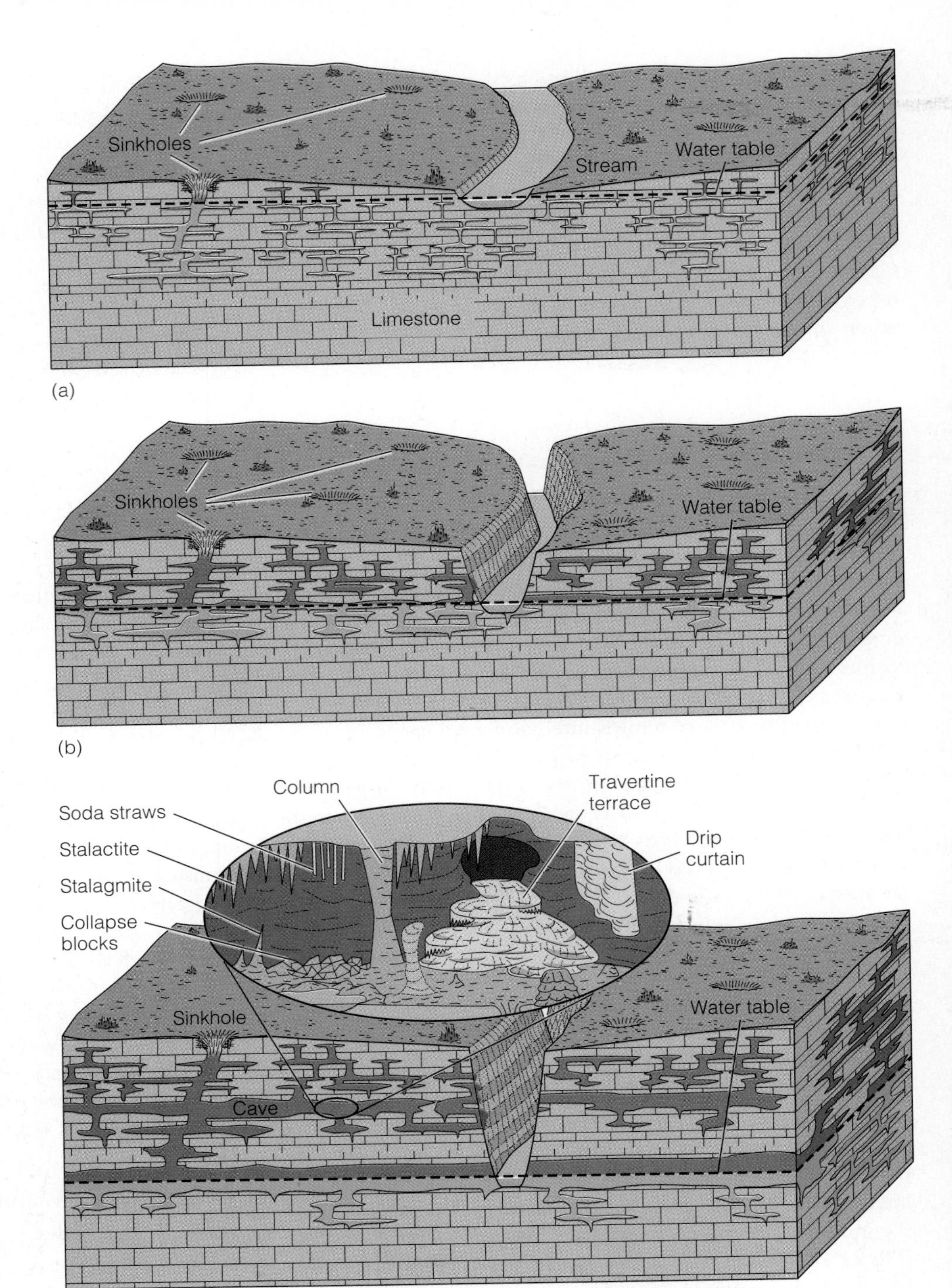

▶ **FIGURE 16-17** The formation of caves. (*a*) As groundwater percolates through the zone of aeration and flows through the zone of saturation, it dissolves the carbonate rocks and gradually forms a system of passageways. (*b*) Groundwater moves along the surface of the water table, forming a system of horizontal passageways through which dissolved rock is carried to the surface streams, thus enlarging the passageways. (*c*) As the surface streams erode deeper valleys, the water table drops, and the abandoned channelways form an interconnecting system of caves and caverns.

popular because yields from irrigated lands can be triple what they would be without irrigation.

While the High Plains aquifer has contributed to the high productivity of the region, it cannot continue providing the quantities of water that it has in the past. In some parts of the High Plains, from 2 to 100 times more water is being pumped annually than is being recharged. Consequently, water is being removed from the aquifer faster than it is being replenished, causing the water table to drop significantly in many areas (Figure 16-19).

What will happen to this region's economy if long-term withdrawal of water from the High Plains aquifer greatly exceeds its recharge rate such that it can no longer supply the quantities of water necessary for irrigation? Solutions range from going back to farming without irrigation to diverting water from other regions such as the Great Lakes. Farming without irrigation would result in greatly decreased yields and higher costs and prices for agricultural products, while the diversion of water from elsewhere would cost billions of dollars and the price of agricultural products would still rise.

➢ **FIGURE 16-18** Stalactites are the icicle-shaped structures seen hanging from the ceiling, while the upward-pointing structures on the cave floor are stalagmites. Several columns are present where the stalactites and stalagmites have met in this chamber of Luray Caves, Virginia.

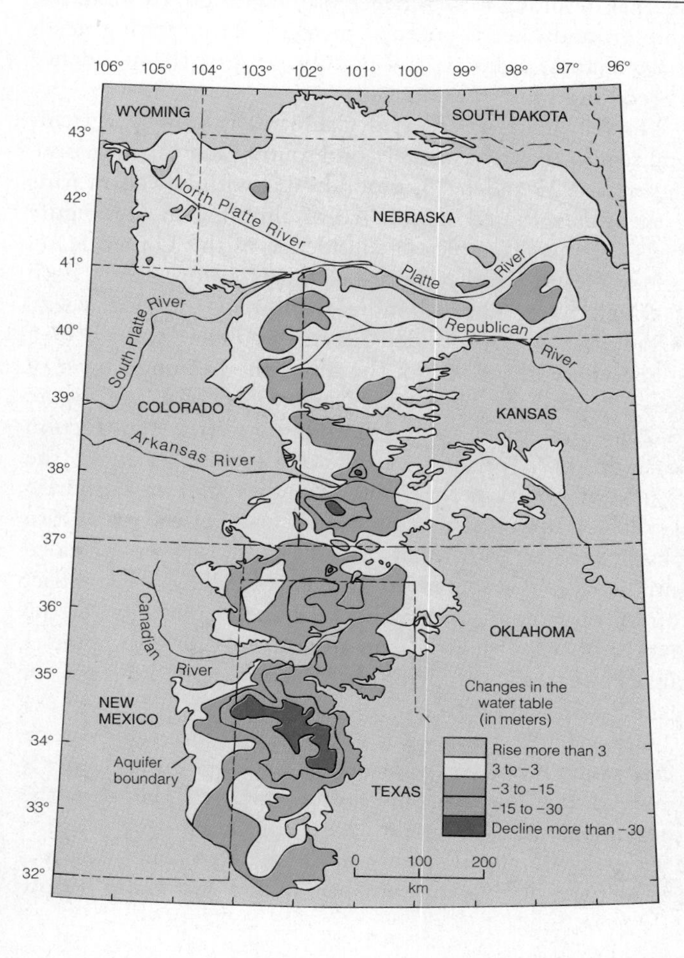

➢ **FIGURE 16-19** Areal extent of the High Plains aquifer and changes in the water table, predevelopment to 1980.

Saltwater Incursion

The excessive pumping of groundwater in coastal areas can result in **saltwater incursion** such as occurred on Long Island, New York, during the 1960s. Along coastlines where permeable rocks or sediments are in contact with the ocean, the fresh groundwater, being less dense than seawater, forms a lens-shaped body above the underlying salt water (➢ Figure 16-20a). The weight of the fresh water exerts pressure on the underlying salt water. As long as rates of recharge equal rates of withdrawal, the contact between the fresh groundwater and the seawater will remain the same. If excessive pumping occurs, however, a deep cone of depression forms in the fresh groundwater (Figure 16-20b). Because some of the pressure from the overlying fresh water has been removed, salt water migrates upward to fill the pore space that formerly contained fresh water. When this occurs, wells become contaminated with salt water and remain contaminated until recharge by fresh water restores the former level of the fresh groundwater water table.

Saltwater incursion is a major problem in many rapidly growing coastal communities. As the population in these areas grows, greater demand for groundwater creates an even greater imbalance between recharge and withdrawal.

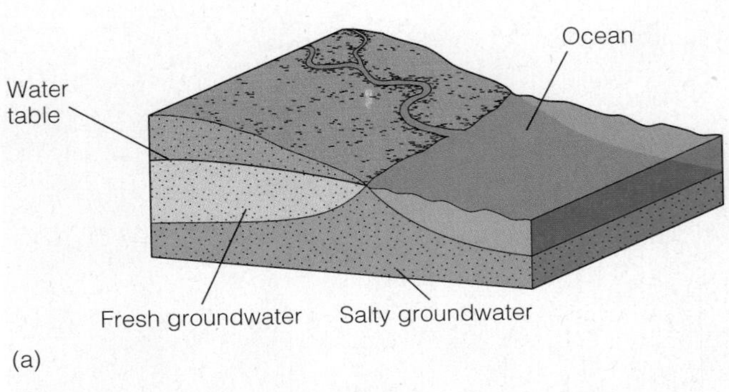

(a)

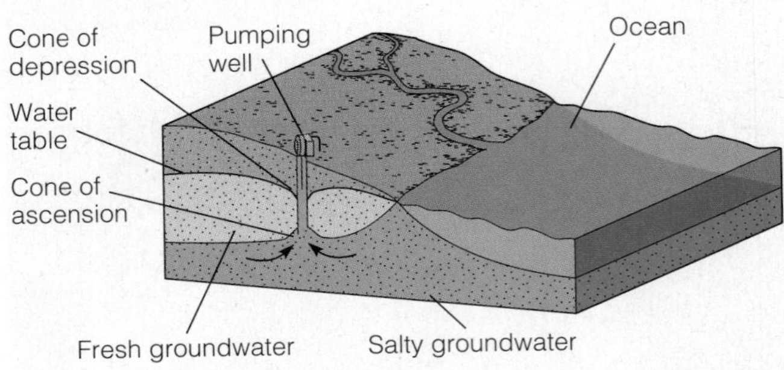

(b)

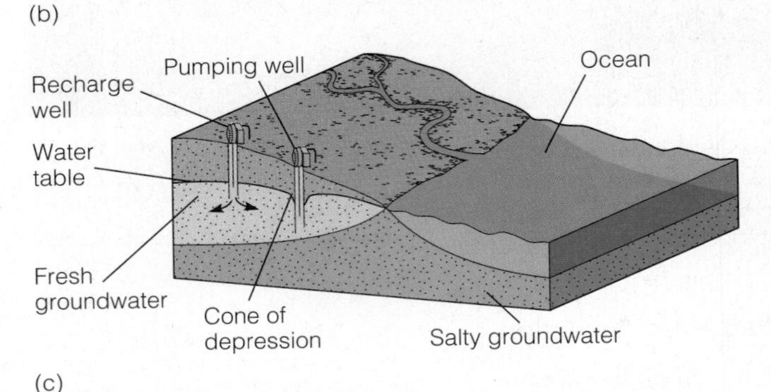

(c)

➤ **FIGURE 16-20** Saltwater incursion. (*a*) Because fresh water is not as dense as salt water, it forms a lens-shaped body above the underlying salt water. (*b*) If excessive pumping occurs, a cone of depression develops in the fresh groundwater, and a cone of ascension forms in the underlying salty groundwater that may result in saltwater contamination of the well. (*c*) Pumping water back into the groundwater system through recharge wells can help lower the interface between the fresh groundwater and the salty groundwater and reduce saltwater incursion.

Not only is saltwater incursion a major concern for some coastal communities, it is also becoming a problem in the Salinas Valley, California, which produces fruits and vegetables valued at about $1.7 billion annually. Here, in an area encompassing about 160,000 acres of rich farmland 161 km south of San Francisco, saltwater incursion caused by overpumping of several shallow-water aquifers is threatening the groundwater supply that is used for irrigation. Because of the drought in recent years and increased domestic needs caused by a burgeoning population, overpumping has resulted in increased seepage of salt water into the groundwater system such that large portions of some of the aquifers are now too salty even for irrigation. At some locations in the Salinas Valley, seawater has migrated more than 11 km inland during the past 13 years. Left unchecked, the farmlands of the valley could become too salty to support most agriculture.

Farmers currently use 85% of all the groundwater removed, and unless some curbs are placed on the amount removed for irrigation, saltwater contamination of the shallow aquifers will continue to accelerate until the water is too salty for irrigation. Uncontaminated deep aquifers could be tapped to replace the contaminated shallow groundwater, but doing so is too expensive for most farmers and small towns.

To counteract the effects of saltwater incursion, recharge wells are often drilled to pump water back into the groundwater system (Figure 16-20c). Recharge ponds that allow large quantities of fresh surface water to infiltrate the groundwater supply may also be constructed (Figure 16-5). Both of these methods are successfully used on Long Island, which has had a saltwater incursion problem for several decades.

Subsidence

As excessive amounts of groundwater are withdrawn from poorly consolidated sediments and sedimentary rocks, the water pressure between grains is reduced, and the weight of the overlying materials causes the grains to pack closer together, resulting in subsidence of the ground. As more and more groundwater is pumped to meet the increasing needs of agriculture, industry, and population growth, subsidence is becoming more prevalent.

The San Joaquin Valley of California is a major agricultural region that relies largely on groundwater for irrigation. Between 1925 and 1975, groundwater withdrawals in parts of the valley caused subsidence of almost 9 m (➤ Figure 16-21). Other examples of subsidence in the United States include New Orleans, Louisiana, and Houston, Texas, both of which have subsided more than 2 m, and Las Vegas, Nevada, which has subsided 8.5 m (◉ Table 16-2).

Elsewhere in the world, the tilt of the Leaning Tower of Pisa is partly due to groundwater withdrawal (➤ Figure 16-22). The tower started tilting soon after construction began in 1173 because of differential compaction of the foundation. During the 1960s, the city of Pisa withdrew ever larger amounts of groundwater, causing the ground to subside further; as a result, the tilt of the tower increased until it was considered in danger of falling over. Strict control of groundwater withdrawal and stabilization of the foundation have reduced the amount of tilting to about 1 mm per year, thus ensuring that the tower should stand for several more centuries.

A spectacular example of subsidence occurred in Mexico City, which is built on a former lake bed. As groundwater is removed for the ever increasing needs of the city, the fine-grained lake sediments are compacting, and Mexico City is slowly and unevenly subsiding. Its opera house has settled more than 3 m, and half of the first floor is now

▶ **FIGURE 16-21** The dates on this power pole dramatically illustrate the amount of subsidence the San Joaquin Valley has undergone since 1925. Due to withdrawal of groundwater for agricultural needs and the ensuing compaction of sediment, the ground subsided almost 9 m between 1925 and 1975.

TABLE 16-2 Subsidence of Cities and Regions Due Primarily to Groundwater Removal

Location	Maximum Subsidence (m)	Area Affected (km^2)
Mexico City, Mexico	8.0	25
Long Beach and Los Angeles, California	9.0	50
Taipei Basin, Taiwan	1.0	100
Shanghai, China	2.6	121
Venice, Italy	0.2	150
New Orleans, Louisiana	2.0	175
London, England	0.3	295
Las Vegas, Nevada	8.5	500
Santa Clara Valley, California	4.0	600
Bangkok, Thailand	1.0	800
Osaka and Tokyo, Japan	4.0	3,000
San Joaquin Valley, California	9.0	9,000
Houston, Texas	2.7	12,100

SOURCE: Data from R. Dolan and H. G. Goodell, "Sinking Cities," *American Scientist* 74 (1986): 38–47; and J. Whittow, *Disasters: The Anatomy of Environmental Hazards* (Athens, Ga.: University of Georgia Press, 1979).

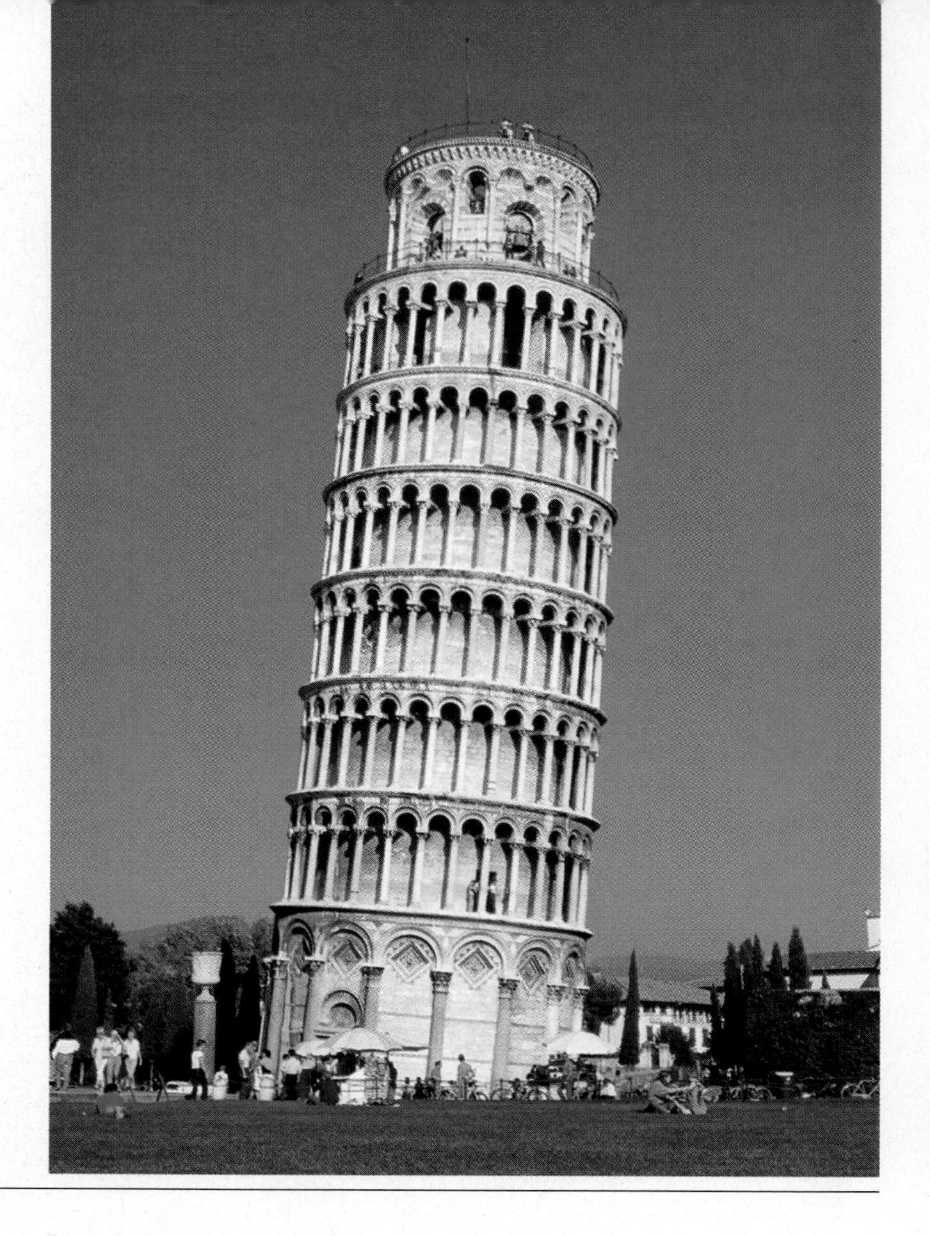

➢ **FIGURE 16-22** The Leaning Tower of Pisa, Italy. The tilting is partly the result of subsidence due to the removal of groundwater.

below ground level. Other parts of the city have subsided more than 6 m, creating similar problems for other structures (➢ Figure 16-23).

The extraction of oil can also cause subsidence. Long Beach, California, has subsided 9 m as a result of 34 years of oil production. More than $100 million of damage was done to the pumping, transportation, and harbor facilities in this area because of subsidence and encroachment of the sea (➢ Figure 16-24). Once water was pumped back into the oil reservoir, subsidence virtually stopped.

Groundwater Contamination

A major problem facing our society is the safe disposal of the numerous pollutant by-products of an industrialized economy. We are becoming increasingly aware that our streams, lakes, and oceans are not unlimited reservoirs for waste, and that we must find new safe ways to dispose of pollutants.

The most common sources of groundwater contamination are sewage, landfills, toxic waste disposal sites (see Perspective 16-2), and agriculture. Once pollutants get into the groundwater system, they will spread wherever groundwater travels, which can make containment of the contamination difficult. Furthermore, because groundwater moves very slowly, it takes a very long time to cleanse a groundwater reservoir once it has become contaminated.

In many areas, septic tanks are the most common way of disposing of sewage. A septic tank slowly releases sewage into the ground where it is decomposed by oxidation and microorganisms and filtered by the sediment as it percolates through the zone of aeration. In most situations, by the time the water from the sewage reaches the zone of saturation, it has been cleansed of any impurities and is safe to use (➢ Figure 16-25a). If the water table is very close to the surface or if the rocks are very permeable, water entering the zone of saturation may still be contaminated and unfit to use.

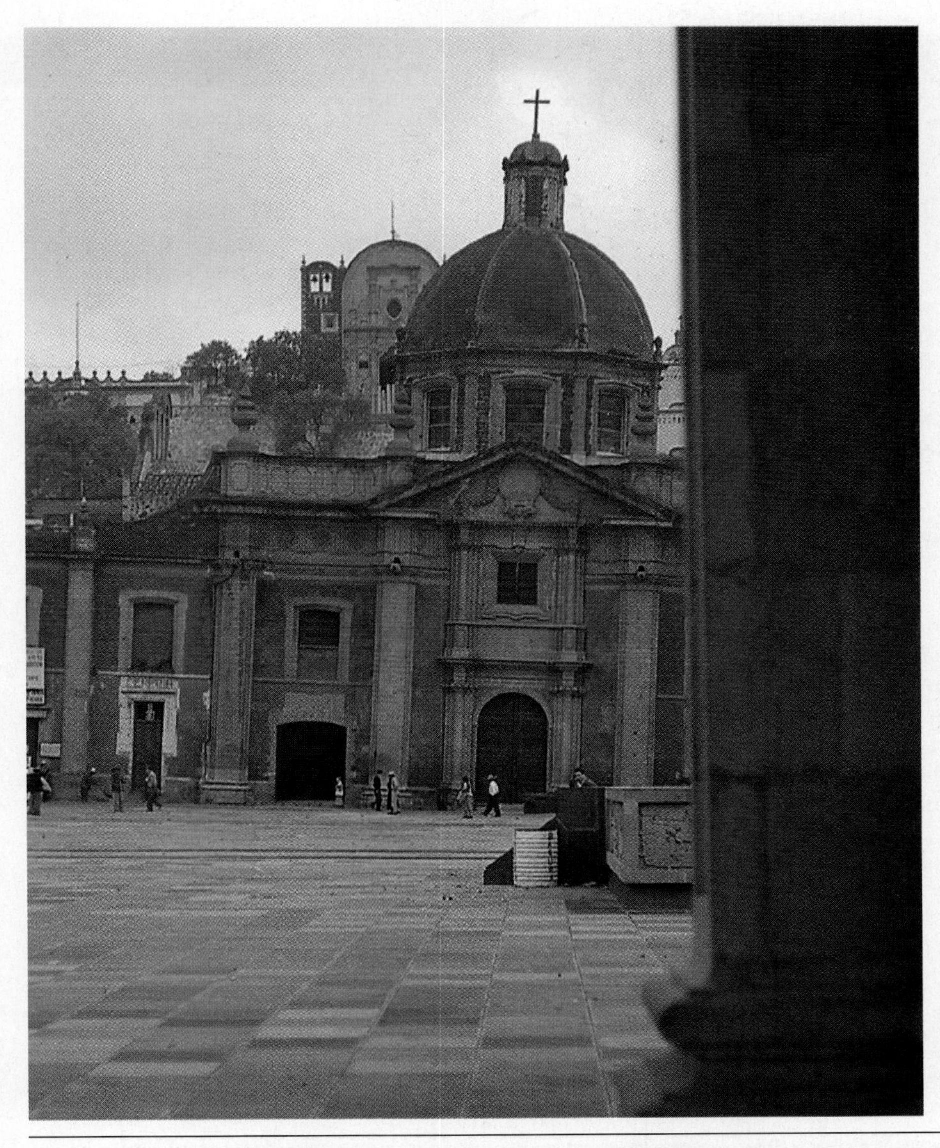

➢ **FIGURE 16-23** The right side of this church (Our Lady of Guadalupe) in Mexico City has settled slightly more than a meter. (Photo courtesy of R. V. Dietrich.)

➢ **FIGURE 16-24** The withdrawal of petroleum from the oil field in Long Beach, California, resulted in up to 9 m of ground subsidence because of sediment compaction. It was not until water was pumped back into the reservoir to replace the petroleum that ground subsidence essentially ceased. (2 to 29 feet = 0.6 to 8.8 meters)

Perspective 16-2

Radioactive Waste Disposal

One of the problems of the nuclear age is finding safe storage sites for the radioactive waste from nuclear power plants, the manufacture of nuclear weapons, and the radioactive by-products of nuclear medicine. Radioactive waste can be grouped into two categories: low-level and high-level waste. Low-level wastes are low enough in radioactivity that, when properly handled, they do not pose a significant environmental threat. Most low-level wastes can be safely buried in controlled dump sites where the geology and groundwater system are well known and careful monitoring is provided.

High-level radioactive waste, such as the spent uranium fuel assemblies used in nuclear reactors and the material used in nuclear weapons, is extremely dangerous because of high amounts of radioactivity; it therefore presents a major environmental problem. Currently, some 22,500 metric tons of spent uranium fuel are awaiting disposal, and the Department of Energy (DOE) estimates that by the year 2000 the nation will have produced almost 50,000 metric tons of highly radioactive waste that must be disposed of safely.

In 1986, Congress chose Yucca Mountain in southern Nevada as the only candidate to house the nation's ever increasing amounts of civilian high-level radioactive waste (▶ Figure 1). Congress also authorized the DOE to study the suitability of the site. Such a facility must be able to isolate high-level waste from the environment for at least 10,000 years, which is the minimum time the waste will remain dangerous. The Yucca Mountain site will have a capacity of 70,000 metric tons of waste and will not be completely filled until around the year 2030, at which time its entrance shafts will be sealed and backfilled (▶ Figure 2).

The canisters holding the waste are designed to remain leakproof for at least 300 years, so there is some possibility that leakage could occur over the next 10,000 years. The DOE thinks, however, that the geology of the area will prevent radioactive isotopes from entering the groundwater system. Under an Environmental Protection Agency (EPA) regulation, a radioactive dump site must be located so that the travel time for groundwater from the site to the outside environment is at least 1,000 years.

The radioactive waste at the Yucca Mountain repository will be buried in a volcanic tuff at a depth of about 300 m. The water table in the area will be an additional 200 to 420 m below the dump site. Thus, the canisters will be stored in the zone of aeration, which was one of the reasons Yucca Mountain was selected. Only about 15 cm of rain fall in this area per year, and only a small amount of this percolates into the ground. Most of the water that does seep into the

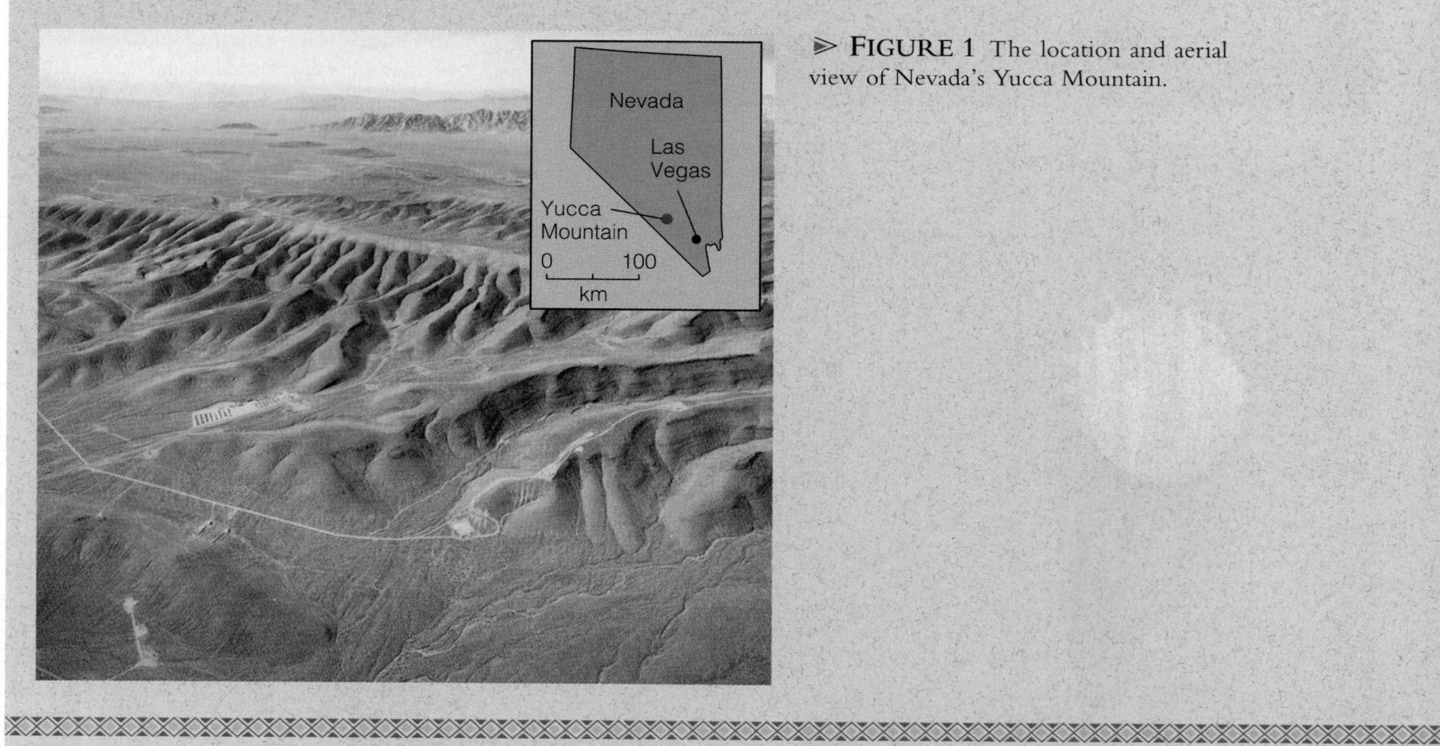

▶ **FIGURE 1** The location and aerial view of Nevada's Yucca Mountain.

ground evaporates before it migrates very far. Thus, the rock at the depth the canisters are buried will be very dry, helping prolong the lives of the canisters.

Geologists think that the radioactive waste at Yucca Mountain is most likely to contaminate the environment if it is in liquid form; if liquid, it could seep into the zone of saturation and enter the groundwater supply. But because of the low moisture in the zone of aeration, there is little water to carry the waste downward, and it will take well over 1,000 years to reach the zone of saturation. In fact, the DOE estimates that the waste will take longer than 10,000 years to move from the repository to the water table.

Some geologists are concerned that the climate will change during the next 10,000 years. If the region should become more humid, more water will percolate through the zone of aeration. This will increase the corrosion rate of the canisters and could cause the water table to rise, thereby decreasing the travel time between the repository and the zone of saturation. This area of the country was much more humid during the Ice Age, 1.6 million to 10,000 years ago (see Chapter 17).

Another concern is that the area is seismically active. It is, in fact, riddled with faults. At least 27 earthquakes with magnitudes greater than 3 occurred in the area between 1852 and 1991. The most recent, which occurred on June 29, 1992, had a magnitude of 5.6 and an epicenter only 32 km from Yucca Mountain. Nevertheless, the DOE is convinced that earthquakes pose little danger to the underground repository itself because the disruptive effects of an earthquake are usually confined to the surface.

Finally, some suggest that the DOE has not thoroughly evaluated the economic potential of the area. Exploration is occurring around the Yucca Mountain site, and some Nevada government officials think that there is geologic evidence for various metals and possibly oil and gas in the area. Should human intrusion occur during the thousands of years that the site is supposed to be isolated, dangerous radiation could be released into the environment. Others think the economic potential is being sufficiently evaluated and that the area in question has a low economic potential.

While it appears that Yucca Mountain meets all of the requirements for a safe high-level radioactive waste dump, the site is still controversial, and further studies must be conducted to ensure that the groundwater supply in this area is not rendered unusable by nuclear waste.

➢ **FIGURE 2** Schematic diagram of the proposed Yucca Mountain high-level radioactive waste dump facility.

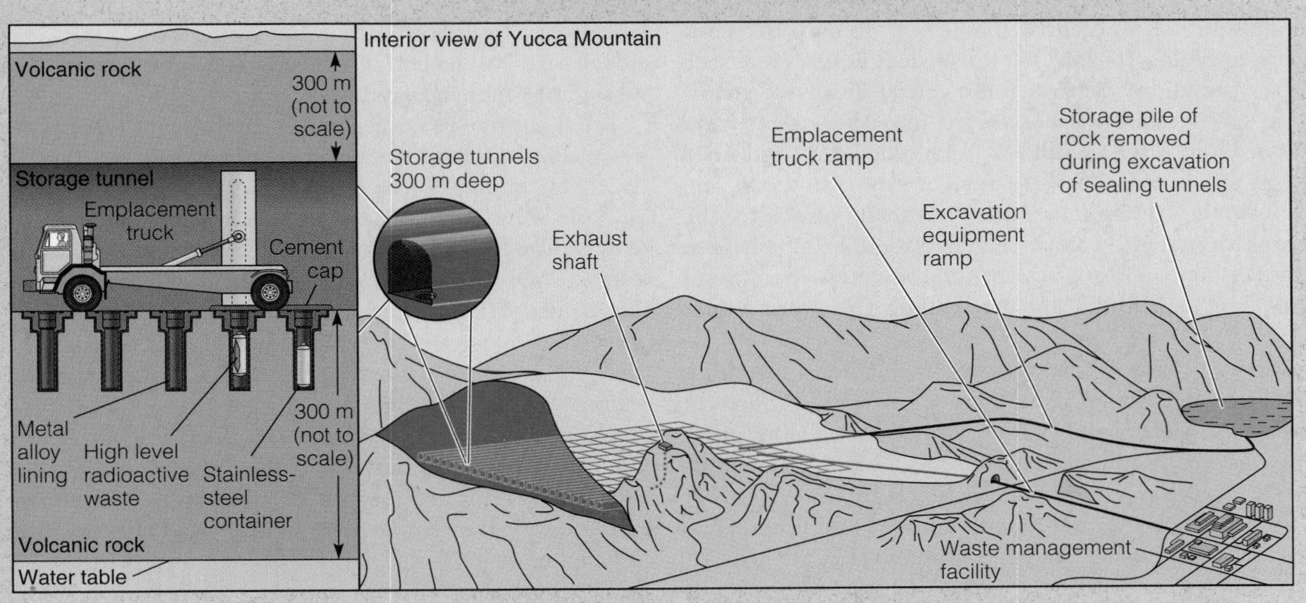

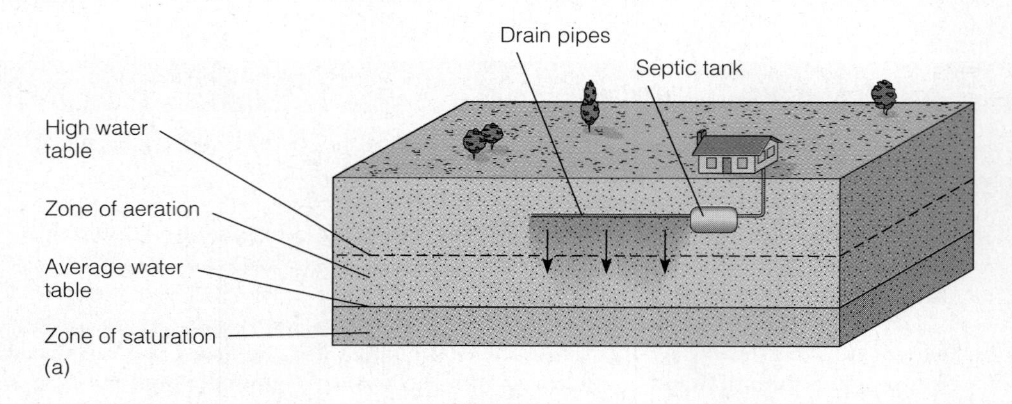

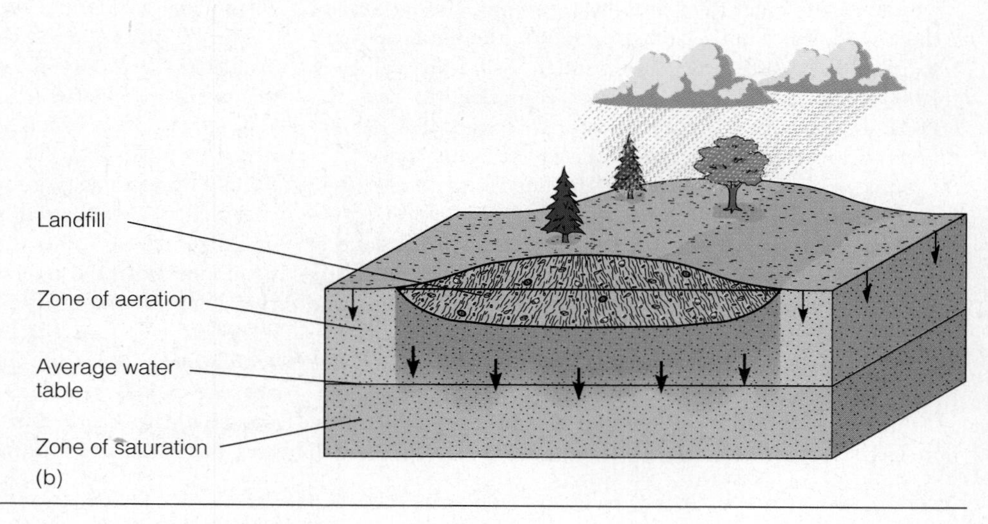

FIGURE 16-25 (*a*) A septic system slowly releases sewage into the zone of aeration. Oxidation, bacterial degradation, and filtering by the sediments usually remove all of the impurities before they reach the water table. If the rocks are very permeable or the water table is too close to the septic system, contamination of the groundwater can result. (*b*) Unless there is an impermeable barrier between a landfill and the water table, pollutants can be carried into the zone of saturation and contaminate the groundwater supply.

Landfills are also potential sources of groundwater contamination (Figure 16-25b). Not only does liquid waste seep into the ground, but rainwater also carries dissolved chemicals and other pollutants downward into the groundwater reservoir. Unless the landfill is carefully designed and lined below by an impermeable layer such as clay, many toxic and cancer-causing compounds will find their way into the groundwater system. Paints, solvents, cleansers, pesticides, and battery acid are just a few of the toxic household items that end up in landfills and can pollute the groundwater supply.

Toxic waste sites where dangerous chemicals are either buried or pumped underground are an increasing source of groundwater contamination. The United States alone must dispose of several thousand metric tons of hazardous chemical waste per year. Unfortunately, much of this waste has been, and still is being, improperly dumped and is contaminating the surface water, soil, and groundwater.

 HOT SPRINGS AND GEYSERS

The subsurface rocks in regions of recent volcanic activity usually stay hot for thousands of years. Groundwater percolating through these rocks is heated and, if returned to the surface, forms *hot springs* or *geysers*. Yellowstone National Park in the United States, Rotorua, New Zealand, and Iceland are all famous for their hot springs and geysers. They are all sites of recent volcanism, and consequently their subsurface rocks and groundwater are very hot.

A **hot spring** (also called a thermal spring or warm spring) is a spring in which the water temperature is warmer than the temperature of the human body (37°C) (Figure 16-26). Some hot springs, though, are much hotter, with temperatures ranging up to the boiling point in many instances. Of the approximately 1,100 known hot springs in the United States, more than 1,000 are in the Far West, while the rest are in the Black Hills of South Dakota, the Ouachita region of Arkansas, Georgia, and the Appalachian region.

Hot springs are also common in other parts of the world. One of the most famous is at Bath, England, where shortly after the Roman conquest of Britain in A.D. 43, numerous bathhouses and a temple were built around the hot springs (Figure 16-27).

The heat for most hot springs comes from magma or cooling igneous rocks. The geologically recent igneous activity in the western United States accounts for the large number of hot springs in that region. The water in some hot springs, however, is circulated deep into the Earth, where it is warmed by the normal increase in temperature, the geothermal gradient. For example, the spring water of Warm Springs, Georgia, is heated in this manner. This hot spring

➢ **FIGURE 16-26 (above)** Hot springs are springs with a water temperature greater than 37°C. This hot spring is in West Thumb Geyser Basin, Yellowstone National Park, Wyoming.

was a health and bathing resort long before the Civil War; later, with the establishment of the Georgia Warm Springs Foundation, it was used to help treat polio victims.

Geysers are hot springs that intermittently eject hot water and steam with tremendous force. The word comes from the Icelandic *geysir,* which means to gush or rush forth. One of the most famous geysers in the world is Old Faithful in Yellowstone National Park in Wyoming (➢ Figure 16-28). With a thunderous roar, it erupts a column of hot water and steam every 30 to 90 minutes. Other well-known geyser areas are found in Iceland and New Zealand.

Geysers are the surface expression of an extensive underground system of interconnected fractures within hot igneous rocks (➢ Figure 16-29). Groundwater percolating down into the network of fractures is heated as it comes into contact with the hot rocks. Because the water near the bottom of the fracture system is under greater pressure than

➢ **FIGURE 16-27 (right)** One of the many bathhouses in Bath, England, that were built around hot springs shortly after the Roman conquest in A.D. 43.

▶ **FIGURE 16-28** Old Faithful Geyser in Yellowstone National Park, Wyoming, is one of the world's most famous geysers, erupting approximately every 30 to 90 minutes.

that near the top, it must be heated to a higher temperature before it will boil. Thus, when the deeper water is heated to very near the boiling point, a slight rise in temperature or a drop in pressure, such as from escaping gas, will cause it to change instantly to steam. The expanding steam quickly pushes the water above it out of the ground and into the air, thereby producing a geyser eruption. After the eruption, relatively cool groundwater starts to seep back into the fracture system where it is heated to near its boiling temperature and the eruption cycle begins again. Such a process explains how geysers can erupt with some regularity.

Hot spring and geyser water typically contains large quantities of dissolved minerals because most minerals dissolve more rapidly in warm water than in cold water. Due to this high mineral content, the waters of many hot springs are believed by some to have medicinal properties. Numerous spas and bathhouses have been built at hot springs throughout the world to take advantage of these supposed healing properties.

When the highly mineralized water of hot springs or geysers cools at the surface, some of the material in solution is precipitated, forming various types of deposits. The amount and type of precipitated mineral depend on the solubility and composition of the material through which the groundwater flows. If the groundwater contains dissolved calcium carbonate ($CaCO_3$), then *travertine* or *calcareous tufa* (both of which are varieties of limestone) are precipitated. Spectacular examples of hot spring travertine deposit are found at Pamukhale in Turkey and at Mammoth Hot Springs in Yellowstone National Park (▶ Figure 16-30). Groundwater containing dissolved silica will, upon reaching the surface, precipitate a soft, white, hydrated mineral called *siliceous sinter* or *geyserite*, which can accumulate around a geyser's opening (▶ Figure 16-31).

Geothermal Energy

Energy harnessed from steam and hot water trapped within the Earth's crust is called **geothermal energy**. It is a desirable and relatively nonpolluting alternate form of energy. Approximately 1 to 2% of the world's current energy needs could be met by geothermal energy. In those areas where it is plentiful, geothermal energy can supply most, if not all, of the energy needs, sometimes at a fraction of the cost of other types of energy. Unfortunately, many of the areas of geothermal power are remote and far from population centers. Furthermore, many are in national parks, such as Yellowstone National Park, and thus cannot be exploited. Some of the countries currently using geothermal energy in one form or another include Iceland, the United States, Mexico, Italy, New Zealand, Japan, the Philippines, and Indonesia.

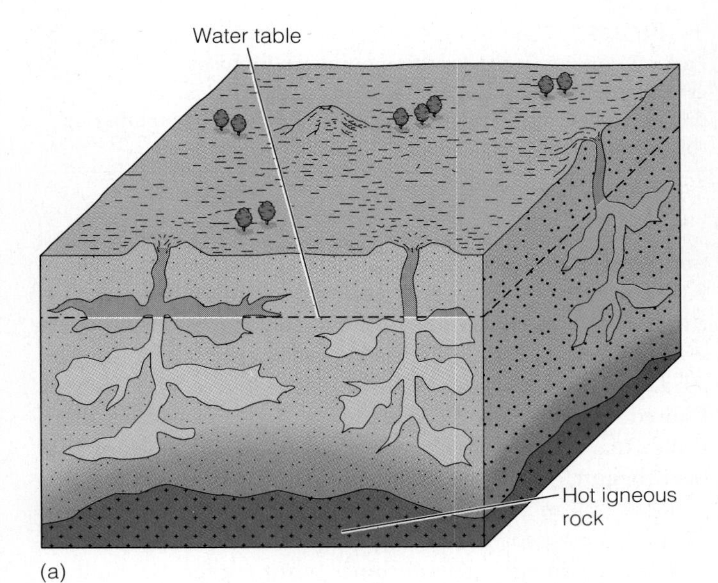

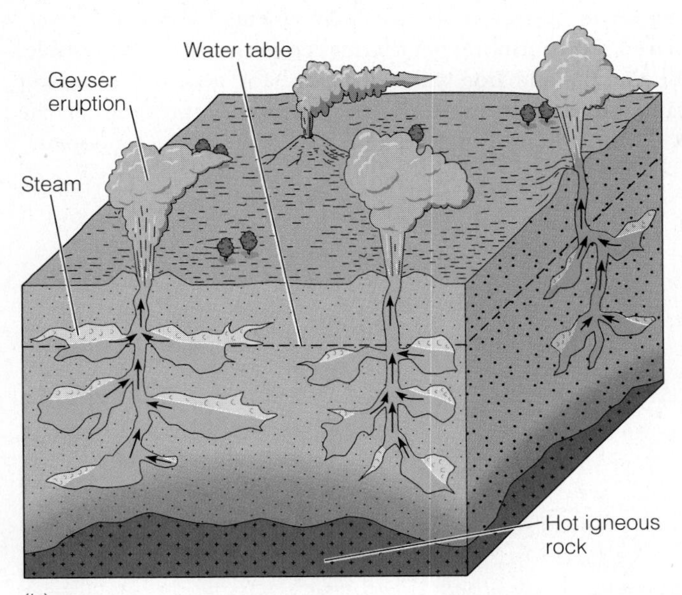

▶ **FIGURE 16-29 (left)** The formation of a geyser. (*a*) Groundwater percolates downward into a network of interconnected openings and is heated by the hot igneous rocks. The water near the bottom of the fracture system is under greater pressure than that near the top and consequently must be heated to a higher temperature before it will boil. (*b*) Any rise in temperature of the water above its boiling point or a drop in pressure will cause the water to change to steam, which quickly pushes the water above it upward and out of the ground, producing a geyser eruption.

Geothermal energy has been successfully used in Iceland since 1928. In Reykjavik, Iceland's capital, steam and hot water from wells drilled in geothermal areas are pumped into buildings for heating and hot water. Fruits and vegetables are grown year-round in hothouses heated from geothermal wells. Direct heating in this manner is significantly cheaper than fuel oil or electrical heating and much cleaner.

The city of Rotorua in New Zealand is world famous for its volcanoes, hot springs, geysers, and geothermal fields. Since the first well was sunk in the 1930s, more than 800 wells have been drilled to tap the hot water and steam below. Geothermal energy in Rotorua is used in a variety of ways, including home, commercial, and greenhouse heating.

In the United States, the first commercial geothermal electrical generating plant was built in 1960 at The Geysers, about 120 km north of San Francisco, California (▶ Figure 16-32). Here, wells were drilled into the numerous near-vertical fractures underlying the region. As pressure on the rising groundwater decreases, the water changes to steam that is piped directly to electricity-generating turbines. The

▶ **FIGURE 16-30 (below)** Minerva Terrace at Mammoth Hot Springs in Yellowstone National Park, Wyoming, formed when calcium carbonate–rich hot spring water cooled, precipitating travertine deposits.

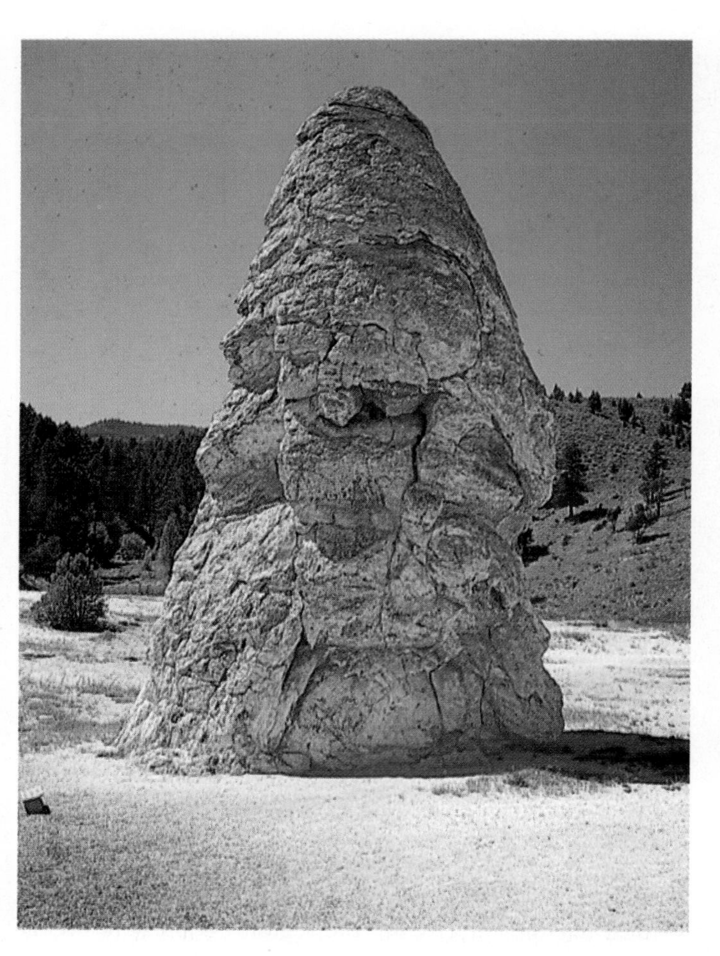

➢ **FIGURE 16-31** Liberty Cap in Yellowstone National Park, Wyoming, is a geyserite mound produced by repeated geyser eruptions. Each eruption of hot silica-rich water precipitated a small amount of geyserite, eventually building up this large mound.

present electrical generating capacity at The Geysers is about 2,000 megawatts, which is enough to supply about two-thirds of the electrical needs of the San Francisco Bay area.

As oil reserves decline, geothermal energy is becoming an attractive alternative, particularly in parts of the western United States, such as the Salton Sea area of southern California, where geothermal exploration and development have begun. While geothermally generated electricity is a generally clean source of power, it can also be expensive because most geothermal waters are acidic and very corrosive. Consequently, the turbines must either be built of expensive corrosion-resistant alloy metals or frequently replaced. Furthermore, geothermal power is not inexhaustible. The steam and hot water removed for geothermal power cannot be easily replaced, and eventually pressure in the wells drops until the geothermal field must be abandoned.

➢ **FIGURE 16-32** The Geysers, Sonoma County, California. Plumes of steam can be seen rising from several steam-generating plants.

Chapter Summary

1. The water stored in the pore spaces of subsurface rocks and unconsolidated material is groundwater.
2. Groundwater is part of the hydrologic cycle and represents approximately 22% of the world's supply of fresh water.
3. Porosity is the percentage of a rock, sediment, or soil consisting of pore space. Permeability is the ability of a rock, sediment, or soil to transmit fluids. A material that transmits groundwater is an aquifer, and one that prevents the movement of groundwater is an aquiclude.
4. The water table is the surface that separates the zone of aeration (in which pore spaces are filled with both air and water) from the zone of saturation (in which all pore spaces are filled with water).
5. Groundwater moves slowly through the pore spaces in the zone of aeration and moves through the zone of saturation to outlets such as streams, lakes, and swamps.
6. A spring occurs wherever the water table intersects the Earth's surface. Some springs are the result of a perched water table, that is, a localized aquiclude within an aquifer and above the regional water table.
7. Water wells are made by digging or drilling into the zone of saturation. When water is pumped out of a well, a cone of depression forms. If water is pumped out faster than it can be recharged, the cone of depression deepens and enlarges and may locally drop to the base of the well, resulting in a dry well.
8. Artesian systems are those in which confined groundwater builds up high hydrostatic pressure. Three conditions must generally be met before an artesian system can form: the aquifer must be confined above and below by aquicludes; the aquifer is usually tilted and exposed at the Earth's surface so it can be recharged; and precipitation must be sufficient to keep the aquifer filled.
9. Karst topography results from groundwater weathering and erosion and is characterized by sinkholes, caves, solution valleys, and disappearing streams.
10. Caves form when groundwater in the zone of saturation weathers and erodes soluble rock such as limestone. Cave deposits, called dripstone, result from the precipitation of calcite.
11. Modifications of the groundwater system can cause serious problems. Excessive withdrawal of groundwater can result in dry wells, loss of hydrostatic pressure, saltwater encroachment, and ground subsidence.
12. Groundwater contamination is becoming a serious problem and can result from sewage, landfills, and toxic waste.
13. Hot springs and geysers may occur where groundwater is heated by hot subsurface volcanic rocks, or by the geothermal gradient. Geysers are hot springs that intermittently eject hot water and steam.
14. Geothermal energy comes from the steam and hot water trapped within the Earth's crust. It is a relatively nonpolluting form of energy that is used as a source of heat and to generate electricity.

Important Terms

aquiclude
aquifer
artesian system
capillary fringe
cave
column
cone of depression
dripstone
geothermal energy
geyser
groundwater
hot spring
karst topography
perched water table
permeability
porosity
recharge
saltwater incursion
sinkhole
spring
stalactite
stalagmite
water table
water well
zone of aeration
zone of saturation

Review Questions

1. What is the correct order, from highest to lowest, of groundwater usage in the United States?
   a. ____ agricultural, industrial, domestic;
   b. ____ industrial, domestic, agricultural;
   c. ____ domestic, agricultural, industrial;
   d. ____ agricultural, domestic, industrial;
   e. ____ industrial, agricultural, domestic.

2. What percentage of the world's supply of fresh water is represented by groundwater?
   a. ____ 5;
   b. ____ 18;
   c. ____ 22;
   d. ____ 43;
   e. ____ 50.
3. The capacity of a material to transmit fluids is:
   a. ____ porosity;
   b. ____ permeability;
   c. ____ solubility;
   d. ____ aeration quotient;
   e. ____ saturation.
4. The water table is a surface separating the:
   a. ____ zone of porosity from the underlying zone of permeability;
   b. ____ capillary fringe from the underlying zone of aeration;
   c. ____ capillary fringe from the underlying zone of saturation;
   d. ____ zone of aeration from the underlying zone of saturation;
   e. ____ zone of saturation from the underlying zone of aeration.
5. Groundwater:
   a. ____ moves slowly through the pore spaces of Earth materials;
   b. ____ moves fastest through the central area of a material's pore space;
   c. ____ can move upward against the force of gravity;
   d. ____ moves from areas of high pressure toward areas of low pressure;
   e. ____ all of these.
6. A perched water table:
   a. ____ occurs wherever there is a localized aquiclude within an aquifer;
   b. ____ is frequently the site of springs;
   c. ____ lacks a zone of aeration;
   d. ____ answers (a) and (b);
   e. ____ answers (b) and (c).
7. An artesian system is one in which:
   a. ____ water is confined;
   b. ____ water can rise above the level of the aquifer when a well is drilled;
   c. ____ water must be pumped;
   d. ____ answers (a) and (c);
   e. ____ answers (a) and (b).
8. Which of the following is not an example of groundwater erosion?
   a. ____ karst topography;
   b. ____ stalactites;
   c. ____ sinkholes;
   d. ____ caves;
   e. ____ caverns.
9. What percentage of the water used in the United States is provided by groundwater?
   a. ____ 50;
   b. ____ 40;
   c. ____ 30;
   d. ____ 20;
   e. ____ 10.
10. Rapid withdrawal of groundwater can result in:
   a. ____ a cone of depression;
   b. ____ ground subsidence;
   c. ____ saltwater incursion;
   d. ____ loss of hydrostatic pressure;
   e. ____ all of these.
11. Karst topography is characterized by:
   a. ____ caves;
   b. ____ springs;
   c. ____ sinkholes;
   d. ____ solution valleys;
   e. ____ all of these.
12. The water in hot springs and geysers:
   a. ____ must be warmer than 37°C;
   b. ____ is noncorrosive;
   c. ____ contains large quantities of dissolved minerals;
   d. ____ answers (a) and (b);
   e. ____ answers (a) and (c).
13. An area in the United States famous for its hot springs and geysers is:
   a. ____ Lansing, Michigan;
   b. ____ Cedar Rapids, Iowa;
   c. ____ Omaha, Nebraska.
   d. ____ Peoria, Illinois;
   e. ____ Yellowstone National Park, Wyoming.
14. Which of the following is not a cave deposit?
   a. ____ stalagmite;
   b. ____ room;
   c. ____ dripstone;
   d. ____ stalactite;
   e. ____ none of these.
15. Discuss the role of groundwater in the hydrologic cycle.
16. How can a rock be porous and yet not be permeable?
17. What types of materials make good aquifers and aquicludes?
18. Why is the water table a subdued replica of the surface topography? What causes the water table level to fluctuate?
19. Why does groundwater move so much slower than surface water?
20. Where are springs likely to occur?
21. What is a cone of depression and why is it important?
22. Why are some artesian wells free-flowing while others must be pumped?
23. How does groundwater weather and erode?
24. How do caves and their various features form?
25. Discuss the various effects that excessive groundwater removal may have on a region. Give some examples.
26. Discuss the various ways that a groundwater system may become contaminated.
27. What is the difference between a thermal spring and a geyser?

Points to Ponder

1. Describe some of the ways of quantitatively measuring the rate of groundwater movement.
2. One of the concerns that geologists have about using Yucca Mountain as a repository for nuclear waste is that the climate may become more humid during the next 10,000 years, thus allowing more water to percolate through the zone of aeration. What would the average rate of groundwater movement have to be during the next 10,000 years to reach the canisters containing radioactive waste buried at a depth of 300 m?
3. Why isn't geothermal power a virtually unlimited source of energy?
4. Why should we be concerned about how fast the groundwater supply is being depleted in some areas?

Additional Readings

Bryan, T.S. 1992. Valley of Geysers. *Earth* 1, no. 4: 20–29.

Courbon, P., C. Chabert, P. Bosted, and K. Lindslay. 1989. *Atlas of the great caves of the world*. St. Louis: Cave Books.

Dietrich, R.V. 1993. How are caves formed? *Rocks and Minerals* 68, no. 4:264–68.

Dolan, R., and H. G. Goodell. 1986. Sinking cities. *American Scientist* 74, no. 1: 38–47.

Fetter, C. W. 1988. *Applied hydrogeology*. 2d ed. Columbus, Ohio: Merrill Publishing Co.

Fincher, J. 1990. Dreams of riches led Floyd Collins to a nightmarish end. *Smithsonian* 21, no. 2: 137–49.

Grossman, D., and S. Shulman. 1994. Verdict at Yucca Mountain. *Earth* 3, no. 2:54–63.

Jennings, J. N. 1983. Karst landforms. *American Scientist* 71, no. 6: 578–86.

———. 1985. *Karst geomorphology*. 2d ed. Oxford, England: Basil Blackwell.

Monastersky, R. 1988. The 10,000-year test. *Science News* 133: 139–41.

Price, M. 1985. *Introducing groundwater*. London: Allen & Unwin.

Rinehart, J. S. 1980. *Geysers and geothermal energy*. New York: Springer-Verlag.

Sloan, B., ed. 1977. *Caverns, caves, and caving*. New Brunswick, N.J.: Rutgers University Press.

Chapter 17

GLACIERS AND GLACIATION

OUTLINE

Glacier on Mount Foresta, Alaska.

PROLOGUE

Following the Great Ice Age, which ended about 10,000 years ago, a general warming trend occurred that was periodically interrupted by short relatively cool periods. One such cool period, from about A.D. 1500 to the mid-to-late 1800s, was characterized by the expansion of small glaciers in mountain valleys and the persistence of sea ice at high latitudes for longer periods than had occurred previously. This interval of nearly four centuries is known as the *Little Ice Age.*

The climatic changes leading to the Little Ice Age actually began by about A.D. 1300. During the preceding centuries, Europe had experienced rather mild temperatures, and the North Atlantic Ocean was warmer and more storm-free than it is at the present. During this time, the Vikings discovered and settled Iceland, and by A.D. 1200, about 80,000 people resided there. They also reached Greenland and North America and established two colonies on the former and one on the latter. As the climate deteriorated the North Atlantic became stormier, and sea ice occurred further south and persisted longer each year. As a consequence of poor sea conditions and political problems in Norway, all shipping across the North Atlantic ceased, and the colonies in Greenland and North America eventually disappeared.

During the Little Ice Age, many of the small glaciers in Europe and Iceland expanded and moved far down their valleys, reaching their greatest historic extent by the early 1800s ( Figure 17-1a). A small ice cap formed in Iceland where none had existed previously, and glaciers in Alaska and the mountains of the western United States and Canada also expanded to their greatest limits during historic time. Although glaciers caused some problems in Europe where they advanced across roadways and pastures, destroying some villages in Scandinavia and threatening villages elsewhere, their overall impact on humans was minimal. Far more important from the human perspective was that during much of the Little Ice Age the summers in northern latitudes were cooler and wetter.

Although worldwide temperatures were a little lower during this time, the change in summer conditions rather than cold winters or glaciers caused most of the problems. Particularly hard hit were Iceland and the Scandinavian

(a)

(b)

▶ **FIGURE 17-1** (*a*) During the Little Ice Age, many European glaciers, such as this one in Switzerland, extended much farther down their valleys than they do at present. This painting, called *The Unterer Grindelwald,* was painted in 1826 by Samuel Birmann (1793–1847). (*b*) The canals of Holland are frozen in this mid-seventeenth-century painting, *The Village of Nieukoop in Winter,* by Jan-Abrahamsz Beerstraten. These canals rarely freeze today.

countries, but at times much of northern Europe was affected (Figure 17-1b). Growing seasons were shorter during many years, resulting in food shortages and a number of famines. Iceland's population declined from its high of 80,000 in 1200 to about 40,000 by 1700. Between 1610 and 1870, sea ice was observed near Iceland for as much as three months a year, and each time the sea ice persisted for long periods, poor growing seasons and food shortages followed.

Exactly when the Little Ice Age ended is debatable. Some authorities put the end at 1880, whereas others think it ended as early as 1850. In any case, during the late 1800s, the sea ice was retreating northward, glaciers were retreating back up their valleys, and summer weather became more stable.

INTRODUCTION

Most people have some idea of what a glacier is, but many confuse glaciers with other masses of snow and ice. A **glacier** is a mass of ice composed of compacted and recrystallized snow that flows under its own weight on land. Accordingly, sea ice as in, for example, the north polar region is not glacial ice, nor are drifting icebergs glaciers even though they may have derived from glaciers that flowed into the sea. Snow fields in high mountains may persist in protected areas for years, but these are not glaciers either because they are not actively moving.

At the present time, glaciers cover nearly 15 million km^2, or about one-tenth of the Earth's land surface (◉ Table 17-1). Numerous glaciers exist in the mountains of the western United States, especially Alaska, western Canada, the Andes in South America, the Alps of Europe, the Himalayas of Asia, and other high mountains. Even Mount Kilimanjaro in Africa, although near the equator, is high enough to support glaciers. In fact, Australia is the only continent lacking glaciers. By far the largest existing glaciers are in Greenland and Antarctica; both areas are almost completely covered by glacial ice (Table 17-1).

At first glance, glaciers may appear to be static, but like other geologic agents such as streams, glaciers are dynamic systems that are continually adjusting to changes. For example, just as streams vary their sediment load depending on the available energy, increases or decreases in the amount of ice in a glacier alter its ability to erode and transport sediment. Glaciers are particularly effective agents of erosion, transport, and deposition. They deeply scour the land over which they move, producing a number of easily recognized erosional landforms, and eventually deposit their sediment load just as other agents of erosion and transport do. Although numerous examples of landscapes that originated from recent glaciation can be found, most glacial landscapes developed during the Pleistocene Epoch, or what is commonly called the Ice Age (1.6 million to 10,000 years ago), when glaciers covered much more extensive areas than they do now, particularly on the Northern Hemisphere continents.

TABLE 17-1 Present-Day Ice-Covered Areas

Antarctica	12,653,000 km^2
Greenland	1,802,600
Northeast Canada	153,200
Central Asian ranges	124,500
Spitsbergen group	58,000
Other Arctic islands	54,000
Alaska	51,500
South American ranges	25,000
West Canadian ranges	24,900
Iceland	11,800
Scandinavia	5,000
Alps	3,600
Caucasus	2,000
New Zealand	1,000
USA (other than Alaska)	650
Others	about 800
	14,971,550

Total volume of present ice: 28 to 35 million km^3

SOURCE: C. Embleton and C. A. King, *Glacial Geomorphology* (New York: Halsted Press, 1975).

GLACIERS AND THE HYDROLOGIC CYCLE

About 2.15% of the world's water is contained in glaciers and has been temporarily removed from the hydrologic cycle (See Table 15-1). But many glaciers at high latitudes, as in Alaska, flow directly into the sea where they melt, or where icebergs break off by a process called *calving* and drift out to sea where they eventually melt. At lower latitudes where they can exist only at high elevations, glaciers flow to lower elevations where they melt and the water returns to the oceans by surface runoff. In areas of low precipitation, as in parts of the western United States and Canada, glaciers are important fresh water reservoirs that release water to streams during the dry season.

In addition to melting, glaciers also lose water by sublimation, a process in which ice changes directly to water vapor without an intermediate liquid phase. Water vapor thus derived rises into the atmosphere where it may con-

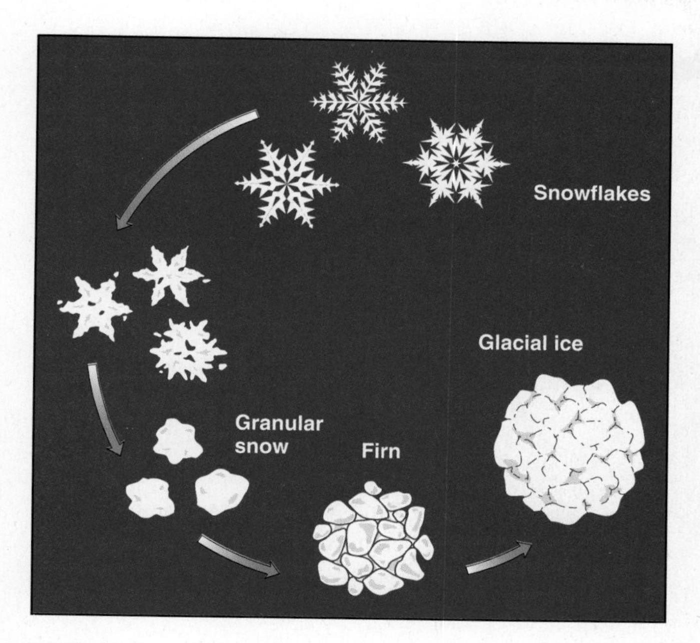

▶ **FIGURE 17-2** The conversion of freshly fallen snow to firn and glacial ice.

dense and fall once again as snow or rain. In any case, it too is eventually returned to the oceans.

 THE ORIGIN OF GLACIAL ICE

Ice is a mineral in every sense of the word; it has a crystalline structure and possesses characteristic physical and chemical properties. Accordingly, geologists consider glacial ice to be rock, although it is a type of rock that is easily deformed. It forms in a fairly straightforward manner (▶ Figure 17-2). When an area receives more winter snow than can melt during the spring and summer seasons, a net accumulation occurs. Freshly fallen snow consists of about 80% air and 20% solids, but it compacts as it accumulates, partly thaws, and refreezes; in the process, the original snow layer is converted to a granular type of ice called **firn.** Deeply buried firn is further compacted and is finally converted to **glacial ice,** consisting of about 90% solids (Figure 17-2). When accumulated snow and ice reach a critical thickness of about 40 m, the pressure on the ice at depth is sufficient to cause deformation and flow, even though it remains solid. Once the critical thickness is reached and flow begins, the moving mass of ice becomes a *glacier.* In polar regions where little snow melts in summer, glaciers can exist at or very near sea level, but at lower latitudes they are found only at higher elevations.

Plastic flow, which causes permanent deformation, occurs in response to pressure and is the primary way that glaciers move. They may also move by **basal slip,** which occurs when a glacier slides over the underlying surface (▶ Figure 17-3). Basal slip is facilitated by the presence of meltwater that reduces frictional resistance between the underlying surface and the glacier.

 TYPES OF GLACIERS

Geologists generally recognize two basic types of glaciers: *valley* and *continental.* A **valley glacier,** as its name implies, is confined to a mountain valley or perhaps to an interconnected system of mountain valleys (▶ Figure 17-4). Large valley glaciers commonly have several smaller tributary glaciers, much as large streams have tributaries. Valley glaciers flow from higher to lower elevations and are invariably small in comparison to continental glaciers, even though some may be more than 100 km long, several kilometers wide, and several hundred meters thick.

Continental glaciers, also called *ice sheets,* cover vast areas (at least 50,000 km^2) and are unconfined by topography (▶ Figure 17-5). In contrast to valley glaciers, which flow downhill within the confines of a valley, continental glaciers flow outward in all directions from a central area of accumulation. Valley glaciers flow in the direction of an existing slope, whereas the direction a continental glacier flows is determined by variations in ice thickness. Currently, only two continental glaciers exist, one in Greenland and the other in Antarctica. Both are more than 3,000 m thick in their central areas, become thinner toward their margins, and cover all but the highest mountains (▶ Figure 17-6). During the Pleistocene Epoch, continental glaciers covered large parts of the Northern Hemisphere continents. Many of the erosional and depositional landforms in much of Canada and the northern tier of the United States formed as a result of Pleistocene glaciation.

▶ **FIGURE 17-3** Movement of a glacier by a combination of plastic flow and basal slip. If a glacier is solidly frozen to the underlying surface, it moves only by plastic flow. Notice that the top of the glacier moves farther in a given time than the bottom does.

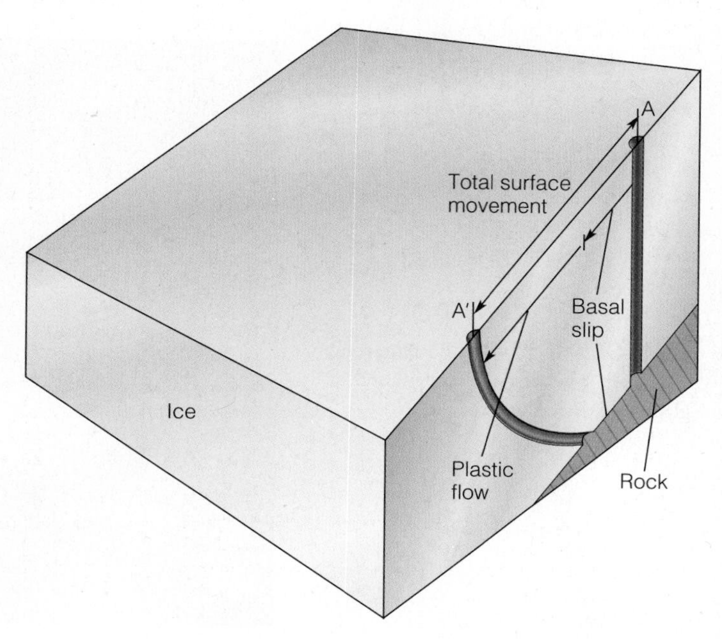

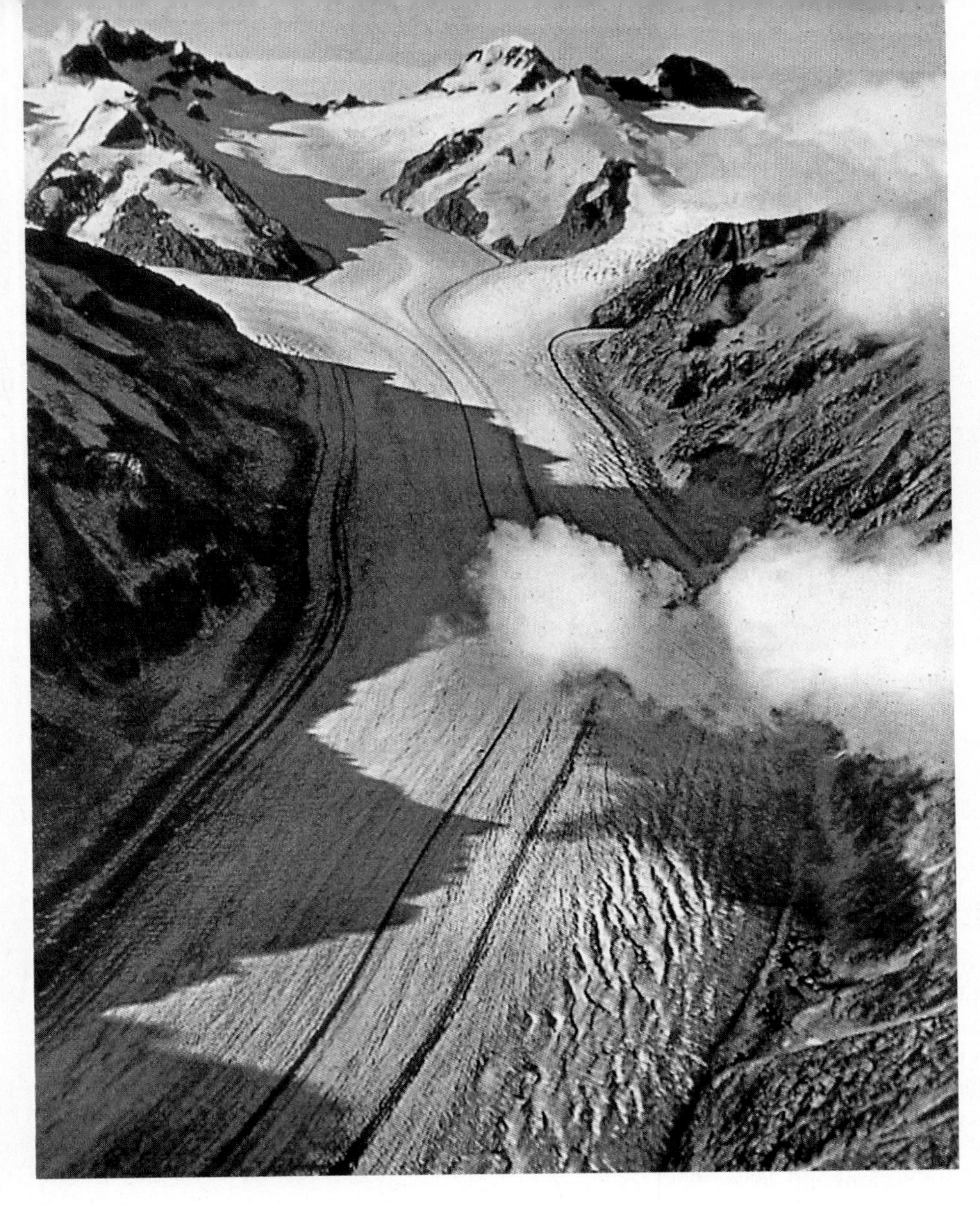

➢ FIGURE 17-4 A large valley glacier in Alaska. Notice the tributaries to the large glacier.

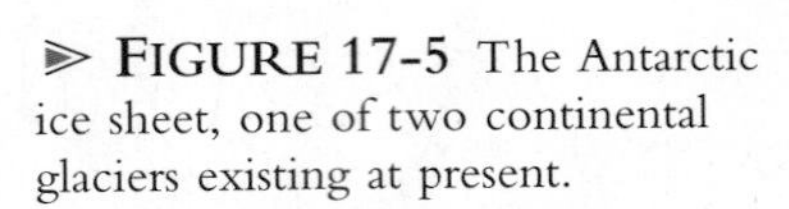

➢ FIGURE 17-5 The Antarctic ice sheet, one of two continental glaciers existing at present.

Although valley and continental glaciers are easily differentiated by their size and location, an intermediate variety called an *ice cap* is also recognized. Ice caps are similar to, but smaller than, continental glaciers and cover less than 50,000 km^2. Some ice caps form when growing valley glaciers overtop the divides and passes between adjacent valleys and coalesce to form a continuous ice cap. They also form on fairly flat terrain including some of the islands of the Canadian Arctic and Iceland.

▶ **FIGURE 17-6** (*a*) Antarctica is almost completely covered by an ice sheet averaging about 2,160 m thick and reaching a maximum thickness of about 4,000 m. (*b*) The Greenland ice sheet has a maxium thickness of about 3,350 m.

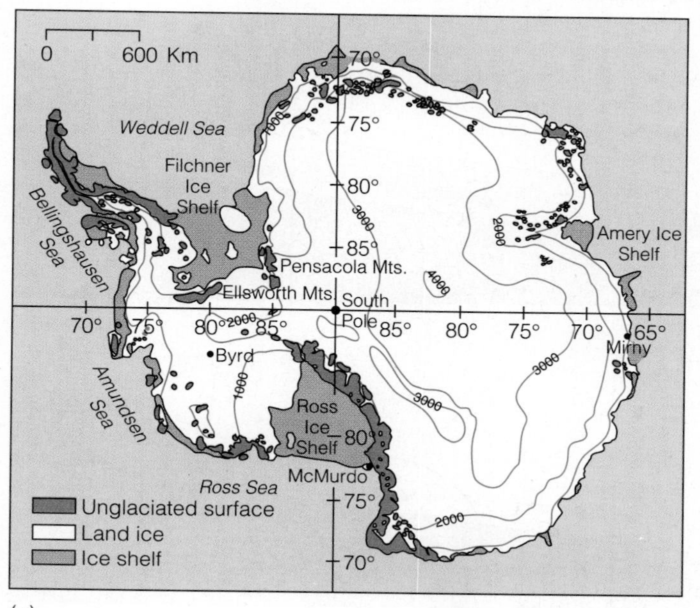

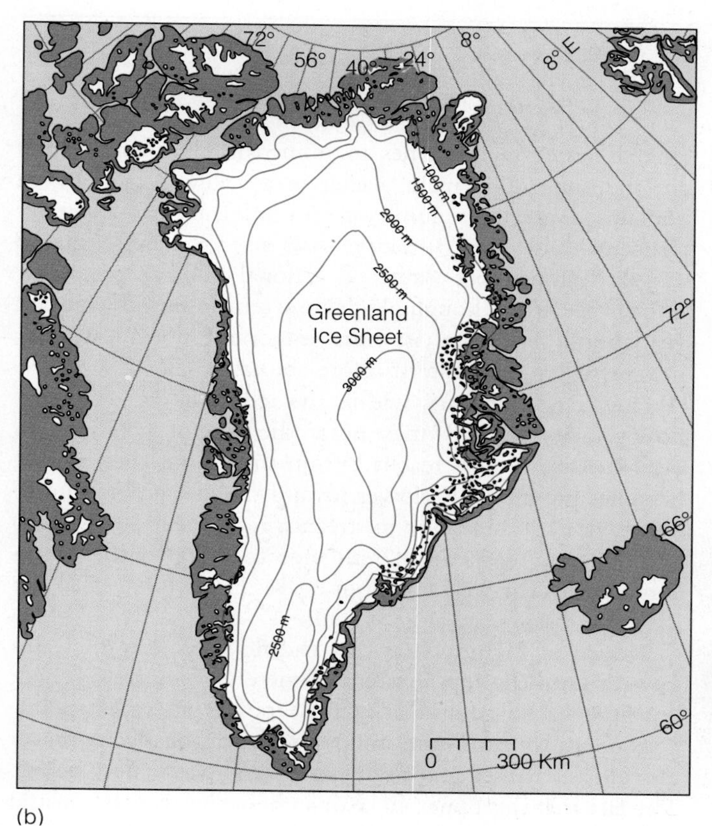

THE GLACIAL BUDGET

Just as a savings account grows and shrinks as funds are deposited and withdrawn, glaciers expand and contract in response to accumulation and wastage. Their behavior can be described in terms of a **glacial budget,** which is essentially a balance sheet of accumulation and wastage. The upper part of a valley glacier is a **zone of accumulation** where additions exceed losses, and the glacier's surface is perennially covered by snow. In contrast, the lower part of the same glacier is a **zone of wastage,** where losses from melting, sublimation, and calving of icebergs exceed the rate of accumulation (▶ Figure 17-7).

At the end of winter, a glacier's surface is usually completely covered with the accumulated seasonal snowfall. During spring and summer, the snow begins to melt, first at lower elevations and then progressively higher up the glacier. The elevation to which snow recedes during a wastage season is called the **firn limit** (Figure 17-7). One can easily identify the zones of accumulation and wastage by noting the position of the firn limit.

Observations of a single glacier reveal that the position of the firn limit usually changes from year to year. If it does not change or shows only minor fluctuations, the glacier is said to have a balanced budget; that is, additions in the zone of accumulation are exactly balanced by losses in the zone of wastage, and the distal end or *terminus* of the glacier remains

▶ **FIGURE 17-7** The glacial budget is the annual balance between additions in the zone of accumulation and losses in the zone of wastage. Ice and rock debris are progressively buried by newly formed ice in the zone of accumulation, but eventually reach the surface in the zone of wastage as the overlying ice melts.

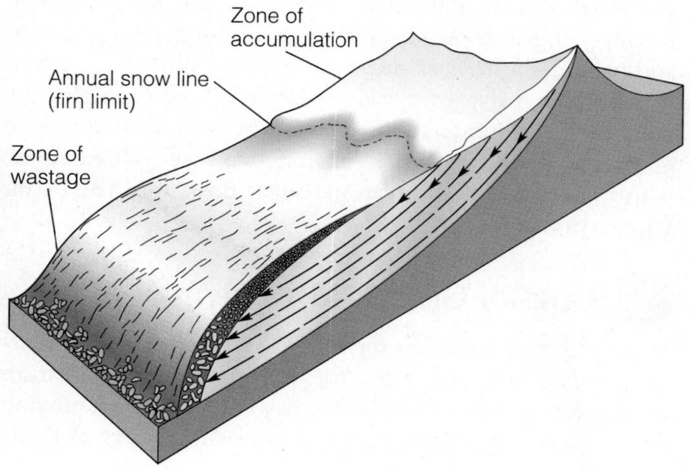

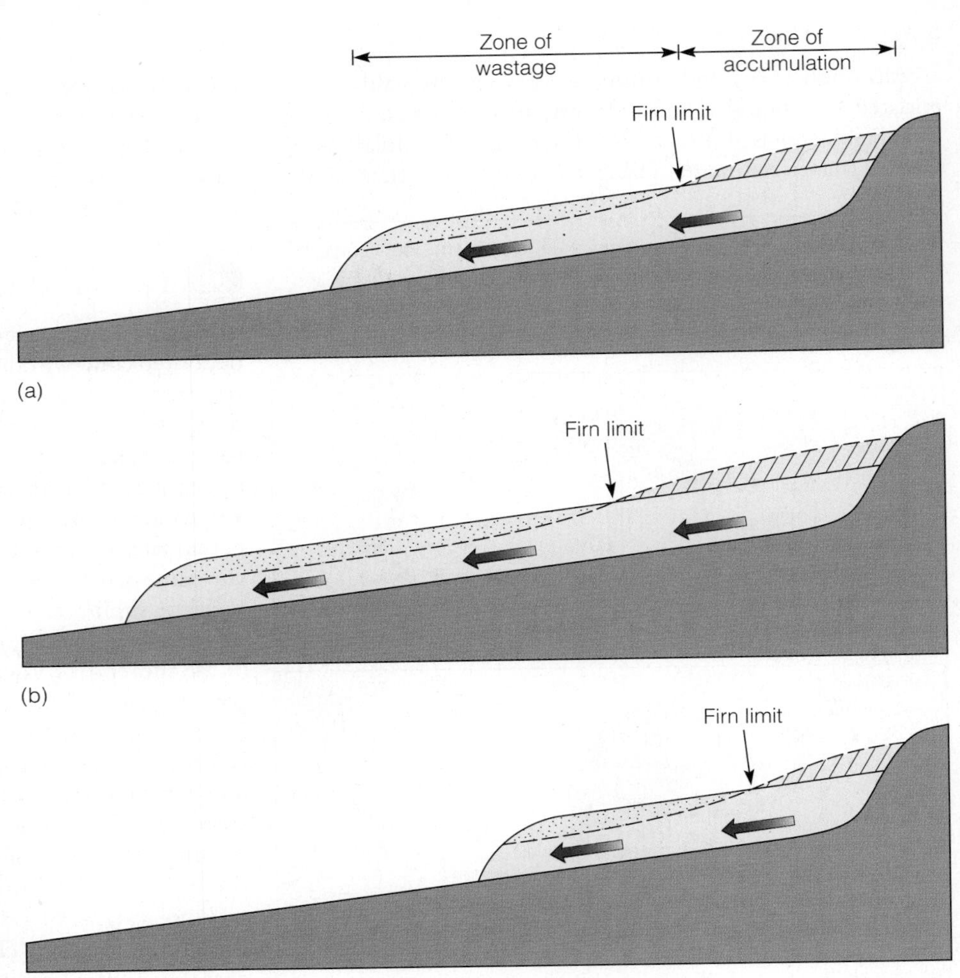

▶ **FIGURE 17-8** Response of a hypothetical glacier to changes in its budget. (*a*) If the losses in the zone of wastage, shown by stippling, equal additions in the zone of accumulation, shown by crosshatching, the terminus of the glacier remains stationary. (*b*) Gains exceed losses, and the glacier's terminus advances. (*c*) Losses exceed gains, and the glacier's terminus retreats, although the glacier continues to flow.

stationary (▶ Figure 17-8a). When the firn limit moves down the glacier, the glacier has a positive budget; its additions exceed its losses, and its terminus advances (Figure 17-8b). If the budget is negative, the glacier recedes—its terminus retreats up the glacial valley (Figure 17-8c). But even though a glacier's terminus may be receding, the glacial ice continues to move toward the terminus by plastic flow and basal slip. If a negative budget persists long enough, however, a glacier recedes and thins until it no longer flows, and becomes a *stagnant glacier.*

Although we used a valley glacier as our example, the same budget considerations control the flow of continental glaciers as well. The entire Antarctic ice sheet, for example, is in the zone of accumulation, but it flows into the ocean where wastage occurs.

 Rates of Glacial Movement

In general, valley glaciers move more rapidly than continental glaciers, but the rates for both vary, ranging from centimeters to tens of meters per day. Valley glaciers moving down steep slopes flow more rapidly than glaciers of comparable size on gentle slopes, assuming that all other variables are the same. The main glacier in a valley glacier system contains a greater volume of ice and thus has a greater discharge and flow velocity than its tributaries (Figure 17-4). Temperature exerts a seasonal control on valley glaciers because although plastic flow remains rather constant year-round, basal slip is more important during warmer months when meltwater is more abundant.

Flow rates also vary within the ice itself. For example, flow velocity generally increases in the zone of accumulation until the firn limit is reached; from that point, the velocity becomes progressively slower toward the glacier's terminus. Valley glaciers are similar to streams, in that the valley walls and floor cause frictional resistance to flow, so the ice in contact with the walls and floor moves more slowly than the ice some distance away (▶ Figure 17-9).

Notice in Figure 17-9 that the flow velocity increases upward until the top few tens of meters of ice are reached, but little or no additional increase occurs after that point. This upper ice constitutes the rigid part of the glacier that is moving as a consequence of basal slip and plastic flow below. The fact that this upper 40 m or so of ice behaves as a brittle

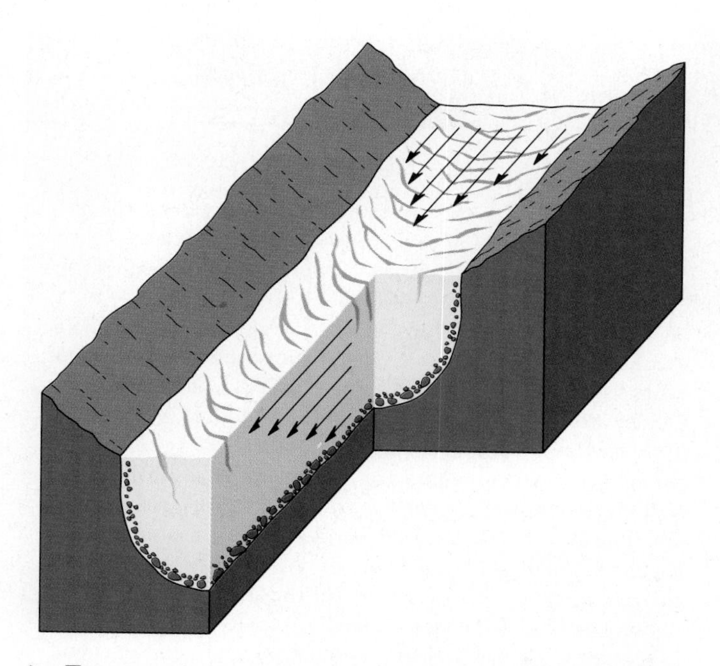

➢ **FIGURE 17-9** Flow velocity in a valley glacier varies both horizontally and vertically. Velocity is greatest at the top-center of the glacier. Friction with the walls and floor of the glacial trough causes the flow to be slower adjacent to these boundaries. The length of the arrows in the figure is proportional to the velocity.

solid is clearly demonstrated by large fractures called *crevasses* that develop when a valley glacier flows over a step in its valley floor where the slope increases or where it flows around a corner (➢ Figure 17-10). In either case, the glacial ice is stretched (subjected to tension), and large crevasses develop, but they extend downward only to the zone of plastic flow. In some cases, a valley glacier descends over such a steep precipice that crevasses break up the ice into a jumble of blocks and spires, and an *ice fall* develops (Figure 17-10).

The flow rates of valley glaciers are also complicated by *glacial surges*, which are bulges of ice that move through a glacier at a velocity several times faster than the normal flow. Although surges are best documented in valley glaciers, they occur in ice caps and continental glaciers as well. During a surge, a glacier's terminus may advance several kilometers during a year. In 1993, the terminus of the Bering Glacier in Alaska, advanced more than 1.5 km in just three weeks.

The causes of surges are not fully understood, but some of them have occurred following a period of unusually heavy precipitation in the zone of accumulation. Others developed when excessive amounts of snow and ice were dislodged from mountain peaks and fell onto the upper parts of glaciers. In either case, rapid changes in the glacial budget occurred.

➢ **FIGURE 17-10** Crevasses and an ice fall in a glacier in Alaska.

Guest Essay Dennis C. Trabant

Understanding Ice: A Career in Glaciology

As a sophomore in engineering at Kansas State University, I took a geology course as an elective. I was surprised to discover how little I knew about the Earth I lived on. I took two courses the next semester and was hooked. During my final semester, I took a seminar directed by my old Geology 101 professor. Among other things, he introduced us to permafrost and high-latitude phenomena. My new interest in ice-related phenomena guided my selection of the University of Alaska for graduate studies. There my major professor was a glaciologist.

I had begun working on a glacier in northern Alaska as an advanced degree project when I joined the U.S. Geological Survey (USGS) in 1973. The USGS studies glaciers to understand glacier-related hydrologic processes for improving predictions of water resources, glacier-related hazards, and the consequences of global change. My first duties involved maintaining the long-term climate and glacier mass-balance observations on two glaciers in Alaska. One of my later projects was to develop a coordinated ice-motion and mass-balance monitoring methodology for research on surging and calving glaciers. For reasons that we do not fully understand, surging glaciers periodically increase their flow speeds by 10 times or more and then return to more normal flow rates for periods of 20 to 50 years. Columbia Glacier is a large calving glacier along the south coast of Alaska. I was a member of the team that predicted the retreat of the glacier and the associated increased iceberg hazard to oil tankers leaving the southern terminal of the Alaska Pipeline. These predictions were partly based on analysis of the team's ice-motion and mass-balance measurements. During the Columbia Glacier study, I introduced the use of beacon radio tracking in glacier research for determining glacier motion and internal temperature. Beacon radios are better known for their use in animal tracking studies.

I have been involved in volcano-glacier hazards analysis since a posteruptive snow and ice hazard assessment of Mount St. Helens shortly after its reawakening in 1980. Following the eruption of Mount St. Helens, heightened interest in volcano-hydrologic hazards lead to glacier ice-volume assessment work in the Cascade Mountains of Washington and Oregon.

The consequences of global change are revealed in the long-term data series from two stations in Alaska. Gulkana Glacier in interior Alaska is approaching a newly reestablished mass-balance equilibrium, while Wolverine Glacier on the coast is growing under the influence of a small climate warming. The association of glacier growth and climate warming is contrary to the generally accepted idea that glacier volume decreases with warming. The measurements at the coastal glacier apply for only a small temperature rise and are caused by the increased storminess associated with warmer waters at the surface of the Gulf of Alaska.

In 1987 I became the staff glaciologist for the USGS's Alaska District where I oversee the snow and glacier-related work. I joined the staff of the USGS's Alaska Volcano Observatory during the eruptions of Redoubt volcano in 1989 and 1990. These eruptions mobilized $113-121 \times 10^6$ m^3 of perennial snow and ice, producing lahars and floods that swept the full 35 km length of the river valley to the sea. The Redoubt work led to understanding that mechanical entrainment of ice and snow in volcanic flows dominates the first phase of the conversion of the snow and ice to the liquid state and flood generation.

Recent work has established a relation between Synthetic Aperture Radar (SAR) image patterns and glacier-bed topography. Although SAR wavelengths do not penetrate to glacier beds, glacier-bed topography is reflected on the ice surface by crevasse size and orientation, which influence SAR image patterns. The indirect relation has proved to be useful in identifying subglacial bedrock structures.

I am currently working on hazard assessments on glacier-clad volcanoes. Part of my work involves documenting the surge of Bering Glacier along the southern coast of Alaska. Bering Glacier is the world's largest temperate-ice surge-type glacier. It began surging during the spring of 1993 and is expected to continue for several years. My interest there is in the relation between ice motion and subglacial hydrology. Four time-lapse cameras are recording ice motion on a daily basis. Other monitors are reporting hourly values from proglacial lakes that are being overridden by the advancing ice.

For those interested in pursuing glaciology as a possible career, glaciology has no fixed curriculum requirements. Most glaciologists have geology or geophysics backgrounds, but others come from subatomic physics, pure mathematics, electrical engineering, and oceanography. My own career has given me an opportunity to travel throughout Alaska, the western United States, and abroad and to spend time in environments that few people are privileged to visit.

DENNIS C. TRABANT received a B.S. in geology from Kansas State University and an M.S. in geology/glaciology from the University of Alaska. He is a glaciologist for the U.S. Geological Survey and is currently on the staff of the Alaska Volcano Observatory.

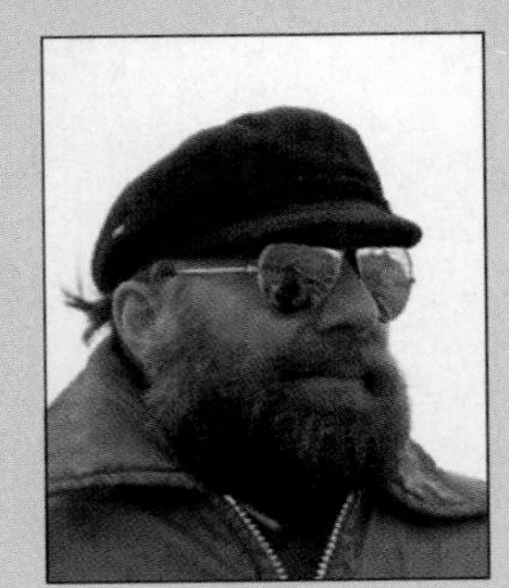

Continental glaciers ordinarily flow at a rate of centimeters to meters per day. Nevertheless, even a rather modest rate of a meter or so per day has a great cumulative effect after several decades. One reason continental glaciers move comparatively slowly is that they exist at higher latitudes and are frozen to the underlying surface most of the time, which limits the amount of basal slip. Some basal slip does occur even beneath the Antarctic ice sheet, but most of its movement is by plastic flow. Nevertheless, some parts of continental glaciers manage to achieve extremely high flow rates. Near the margins of the Greenland ice sheet, the ice is forced between mountains in what are called *outlet glaciers.* In some of these outlets, flow velocities exceeding 100 m per day have been recorded.

▶ **FIGURE 17-11** A glacial erratic near Hammond, New York. (Photo courtesy of R. V. Dietrich.)

 GLACIAL EROSION AND TRANSPORT

Glaciers are currently limited in areal extent, but during the Pleistocene Epoch, they covered much larger areas and were more important than their present distribution would indicate. Glaciers are moving solids that can erode and transport huge quantities of materials, especially unconsolidated sediment and soil. In many areas of Canada and the northern United States, glaciers transported boulders, some of huge proportions, for long distances before depositing them. Such boulders are known as **glacial erratics** (▶ Figure 17-11).

Important erosional processes associated with glaciers include bulldozing, plucking, and abrasion. Bulldozing, although not a formal geologic term, is fairly self-explanatory: a glacier simply shoves or pushes unconsolidated materials in its path. *Plucking,* also called *quarrying,* occurs when glacial ice freezes in the cracks and crevices of a bedrock projection and eventually pulls it loose. One manifestation of plucking is a landform called a *roche moutonnée,* which is French for "rock sheep." As shown in ▶ Figure 17-12, a glacier smooths the "upstream" side of an obstacle, such as a small hill, and plucks pieces of rock from the "downstream" side by repeatedly freezing and pulling away from the obstacle.

Sediment-laden glacial ice can effectively erode by **abrasion.** Bedrock over which sediment-laden glacial ice has moved commonly develops a **glacial polish,** a smooth surface that glistens in reflected light (▶ Figure 17-13a). Abrasion also yields **glacial striations,** consisting of rather straight scratches rarely more than a few millimeters deep on rock surfaces (Figure 17-13b). Abrasion also thoroughly pulverizes rocks so that they yield an aggregate of clay- and silt-sized particles having the consistency of flour—hence the name *rock flour.* Rock flour is so common in streams discharging from glaciers that the water generally has a milky appearance.

▶ **FIGURE 17-12** Origin of a roche moutonnée. As the ice moves over a hill, it smooths the "upstream" side by abrasion and shapes the "downstream" side by plucking.

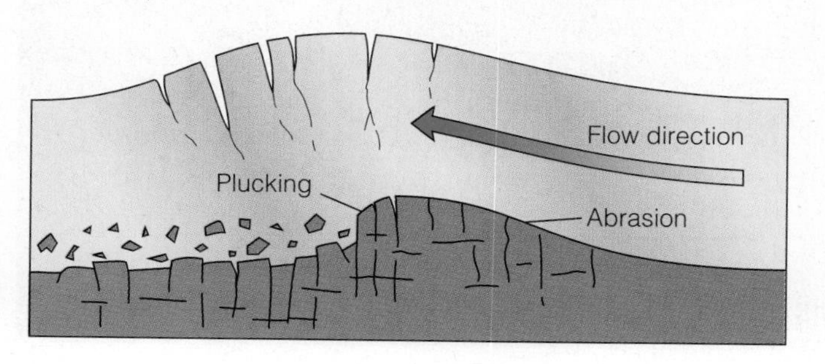

Continental glaciers can derive sediment from mountains projecting through them, and windblown dust settles on their surfaces. Otherwise, most of their sediment is derived from the surface they move over and is transported in the lower part of the ice sheet. In contrast, valley glaciers carry sediment in all parts of the ice, but it is concentrated at the base and along the margins. Some of the marginal sediment is derived by abrasion and plucking, but much of it is supplied by mass wasting processes.

Erosional Landforms of Valley Glaciers

Some of the world's most inspiring scenery is produced by valley glaciers. Many mountain ranges are scenic to begin with, but when modified by valley glaciers, they take on a unique aspect of jagged, angular peaks and ridges in the midst of broad valleys (▶ Figure 17-14). Many landforms resulting from valley glaciation are easily recognized (▶ Figure 17-15). Such features enable us to appreciate the tremendous erosive power of moving ice.

U-Shaped Glacial Troughs. A **U-shaped glacial trough** is one of the most distinctive features of valley glaciation (Figure 17-15c). Mountain valleys eroded by running water are typically V-shaped in cross section; that is, they have

(a)

(b)

▶ FIGURE 17-13 (*a*) Glacial polish on quartzite near Marquette, Michigan. (*b*) Glacial striations in basalt at Devil's Postpile National Monument, California.

valley walls that descend steeply to a narrow valley bottom (Figure 17-15a). In contrast, valleys scoured by glaciers are deepened, widened, and straightened such that they have very steep or vertical walls, but broad, rather flat valley floors; thus, they exhibit a U-shaped profile (▶ Figure 17-16).

Many glacial troughs contain triangular-shaped *truncated spurs*, which are cutoff or truncated ridges that extend into the preglacial valley (Figure 17-15c). Another common feature is a series of steps or rock basins in the valley floor where the glacier eroded rocks of varying resistance; many of the basins now contain small lakes.

During the Pleistocene, when glaciers were extensive, sea level was about 130 m lower than at present, so glaciers flowing into the sea eroded their valleys to much greater depths than they do now. When the glaciers melted at the end of the Pleistocene, sea level rose, and the ocean filled the lower ends of the glacial troughs so that now they are

▶ FIGURE 17-14 The rugged, angular landscape typical of areas eroded by valley glaciers is apparent in Glacier National Park, Montana.

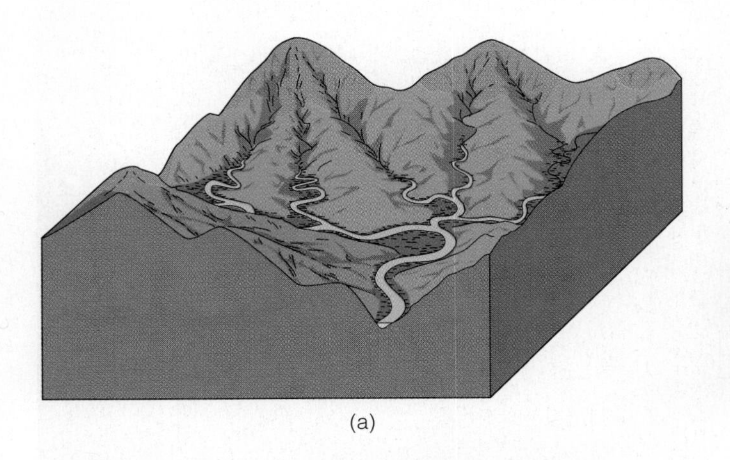

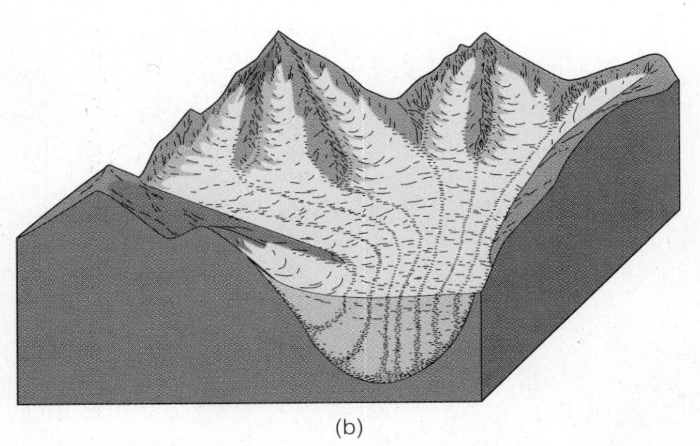

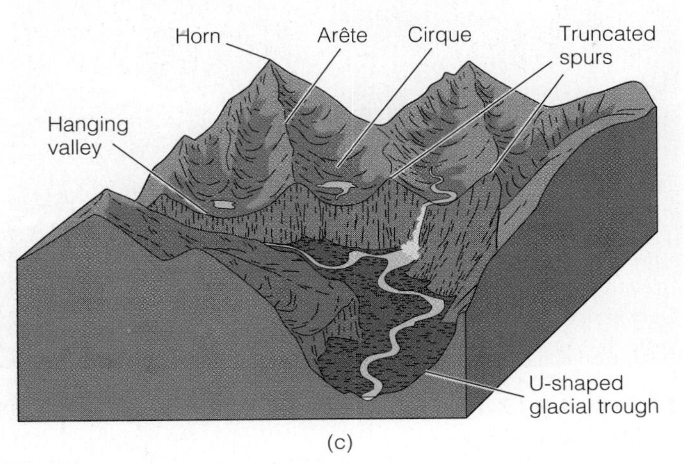

▶ **FIGURE 17-15** Erosional landforms produced by valley glaciers. (*a*) A mountain area before glaciation. (*b*) The same area during the maximum extent of the valley glaciers. (*c*) After glaciation.

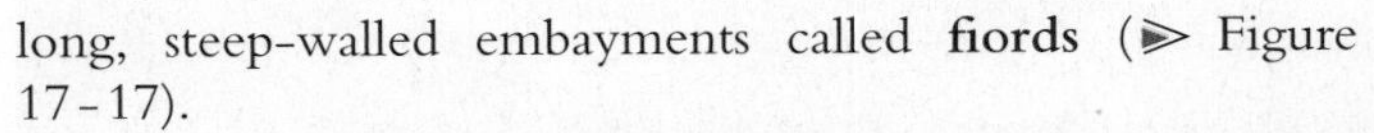

long, steep-walled embayments called **fiords** (▶ Figure 17-17).

Fiords are restricted to high latitudes where glaciers can be maintained even at low elevations, such as Alaska, western Canada, Scandinavia, Greenland, southern New Zealand, and southern Chile. Lower sea level during the Pleistocene was not entirely responsible for the formation of all fiords. Unlike running water, glaciers can erode a considerable distance below sea level. In fact, a glacier 500 m thick can

▶ **FIGURE 17-16** A U-shaped glacial trough in northwestern Montana.

➢ **FIGURE 17-17** Milford Sound, a fiord in New Zealand. (Photo courtesy of George and Linda Lohse.)

stay in contact with the sea floor and effectively erode it to a depth of about 450 m before the buoyant effects of water cause the glacial ice to float! The depth of some fiords is impressive; some in Norway and southern Chile are about 1,300 m deep.

Hanging Valleys. Although waterfalls can form in several ways, some of the world's highest and most spectacular are found in recently glaciated areas. Yosemite Falls in Yosemite National Park, California, plunge 435 m vertically, cascade down a steep slope for another 205 m, and then fall vertically 97 m, for a total descent of 737 m (➢ Figure 17-18). The falls plunge from a **hanging valley,** which is a tributary valley whose floor is at a higher level than that of the main valley. Where the two valleys meet, the mouth of the hanging valley is perched far above the main valley's floor (Figure 17-15c). Accordingly, streams flowing through hanging valleys plunge over vertical or very steep precipices.

Although not all hanging valleys form by glacial erosion, many do. As Figure 17-15 shows, the large glacier in the main valley vigorously erodes, whereas the smaller glaciers in tributary valleys are less capable of large-scale erosion. When the glaciers disappear, the smaller tributary valleys remain as hanging valleys.

Cirques, Arêtes, and Horns. Perhaps the most spectacular erosional landforms in areas of valley glaciation occur at the upper ends of glacial troughs and along the divides separating adjacent glacial troughs. Valley glaciers form and move out from steep-walled, bowl-shaped depressions called **cirques** at the upper end of their troughs (Figure 17-15c). Cirques are typically steep-walled on three sides, but one side is open and leads into the glacial trough. Some cirques slope continuously into the glacial trough, but many have a lip or threshold at their lower end (➢ Figure 17-19).

Although the details of cirque origin are not fully understood, they apparently form by erosion of a preexisting depression on a mountain side. As snow and ice accumulate in the depression, frost wedging and plucking enlarge it until it takes on the typical cirque shape. In cirques with a lip or threshold, the glacial ice apparently not only moves outward but rotates as well, scouring out a depression rimmed by rock. Such depressions commonly contain a small lake known as a *tarn* (Figure 17-19).

Cirques become wider and are cut deeper into mountain sides by headward erosion as a result of abrasion, plucking, and several mass wasting processes. For example, part of a steep cirque wall may collapse, while frost wedging continues to pry loose other rocks that tumble downslope,

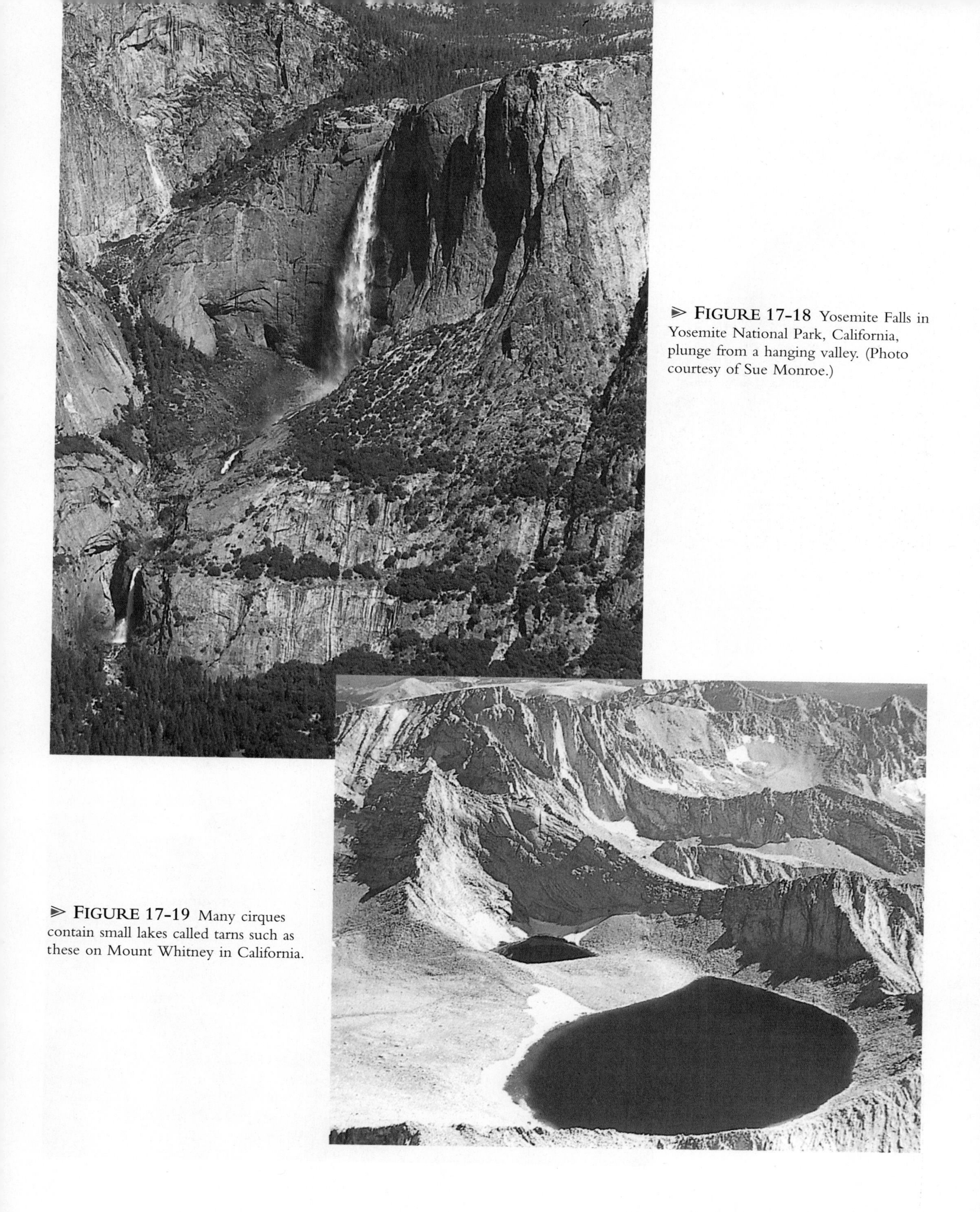

➢ **FIGURE 17-18** Yosemite Falls in Yosemite National Park, California, plunge from a hanging valley. (Photo courtesy of Sue Monroe.)

➢ **FIGURE 17-19** Many cirques contain small lakes called tarns such as these on Mount Whitney in California.

so a combination of processes can erode a small mountainside depression into a large cirque; the largest one known is the Walcott Cirque in Antarctica, which is 16 km wide and 3 km deep.

The fact that cirques expand laterally and by headward erosion accounts for the origin of two other distinctive erosional features, *arêtes* and *horns*. **Arêtes**—narrow, serrated ridges—can form in two ways. In many cases, cirques form on opposite sides of a ridge, and headward erosion reduces the ridge until only a thin partition of rock remains (Figure 17-15c). The same effect occurs when erosion in two parallel glacial troughs reduces the intervening ridge to a thin spine of rock (► Figure 17-20).

The most majestic of all mountain peaks are **horns**; these steep-walled, pyramidal peaks are formed by headward erosion of cirques. In order for a horn to form, a mountain peak must have at least three cirques on its flanks, all of which erode headward (Figure 17-15c). Excellent examples of horns include Mount Assiniboine in the Canadian Rockies, the Grand Teton in Wyoming (see Figure 13-1), and the most famous of all, the Matterhorn in Switzerland (► Figure 17-21).

► **FIGURE 17-20 (left)** This knifelike ridge between glacial troughs in Alaska is an arête. (Photo courtesy of Frank Hanna.)

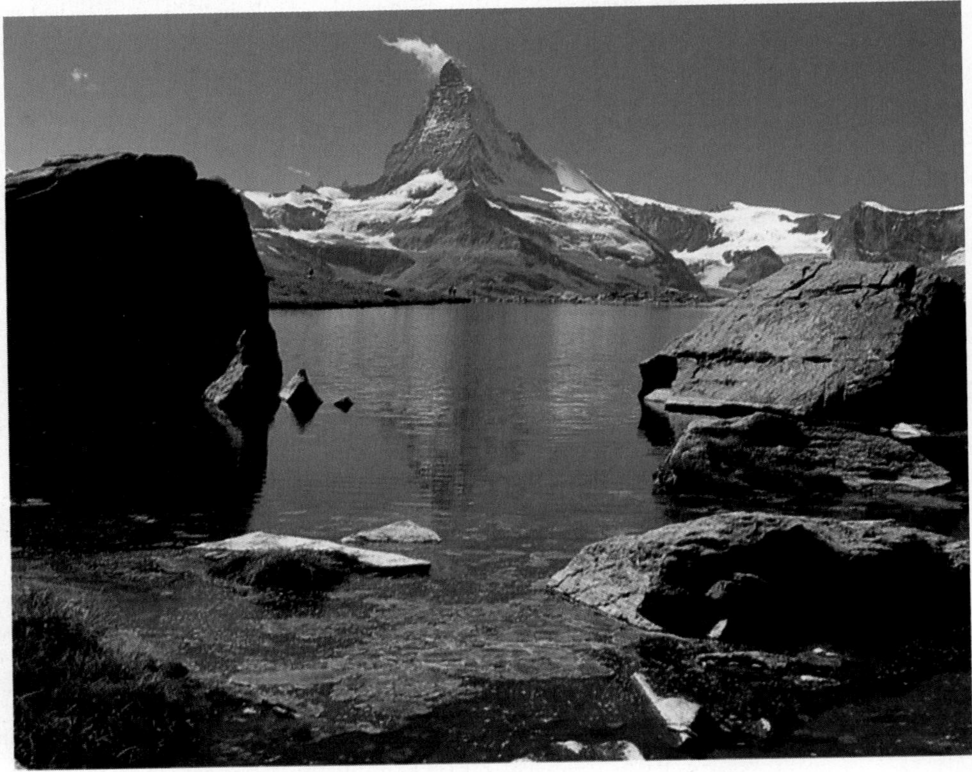

► **FIGURE 17-21 (right)** The Matterhorn in Switzerland is a well-known horn.

Erosional Landforms of Continental Glaciers

Areas eroded by continental glaciers tend to be smooth and rounded because such glaciers bevel and abrade high areas that projected into the ice. Rather than yielding the sharp, angular landforms typical of valley glaciation, they produce a landscape of rather flat, monotonous topography interrupted by rounded hills.

In a large part of Canada, particularly the vast Canadian Shield region, continental glaciation has stripped off the soil and unconsolidated surface sediment, revealing extensive exposures of striated and polished bedrock (▶ Figure 17-22). Similar though smaller bedrock exposures are also widespread in the northern United States from Maine through Minnesota.

Another consequence of erosion in these areas is the complete disruption of drainage that has not yet become reestablished. Much of the area is characterized by deranged drainage (see Figure 15-28e), numerous lakes and swamps, low relief, extensive bedrock exposures, and little or no soil. Such areas are generally referred to as *ice-scoured plains* (Figure 17-22).

▶ **FIGURE 17-22** An ice-scoured plain in the Northwest Territories of Canada.

Glacial Deposits

All sediment deposited as a result of glacial activity is called **glacial drift.** A vast sheet of Pleistocene-aged glacial drift exists in the northern tier of the United States and adjacent parts of Canada (▶ Figure 17-23). Smaller accumulations of similar material are found where valley glaciers existed or remain active. In many areas, glacial deposits are important sources of groundwater and rich soils, and many are exploited for their sand and gravel.

Geologists generally recognize two distinct types of glacial drift, *till* and *stratified drift.* **Till** consists of sediment deposited directly by glacial ice. It is not sorted or stratified; that is, its particles are not separated by size or density, and it does not exhibit any layering. Till deposited by valley glaciers looks much like the till of continental glaciers except that the latter's deposits are much more extensive and have generally been transported much farther.

Stratified drift is sorted by size and density and, as its name implies, is layered. In fact, most of the sediments recognized as stratified drift are braided stream deposits; the streams in which they were deposited received their water and sediment load directly from melting glacial ice.

Landforms Composed of Till

Landforms composed of till include several types of *moraines* and elongated hills known as *drumlins.*

End Moraines. The terminus of either a valley or a continental glacier may become stabilized in one position for some period of time, perhaps a few years or even decades. Stabilization of the ice front does not mean that the glacier has ceased flowing, only that it has a balanced budget (Figure 17-8). When an ice front is stationary, flow within the glacier continues, and the sediment transported within or upon the ice is dumped as a pile of rubble at the glacier's terminus (▶ Figure 17-24). Such deposits are **end moraines,** which continue to grow as long as the ice front is stabilized. End moraines of valley glaciers are commonly crescent-shaped ridges of till spanning the valley occupied by the glacier. Those of continental glaciers similarly parallel the ice front, but are much more extensive.

Following a period of stabilization, a glacier may advance or retreat, depending on changes in its budget. If it advances, the ice front overrides and modifies its former moraine.

▶ **FIGURE 17-23** Exposure of Pleistocene-aged glacial drift near Plymouth, Massachusetts.

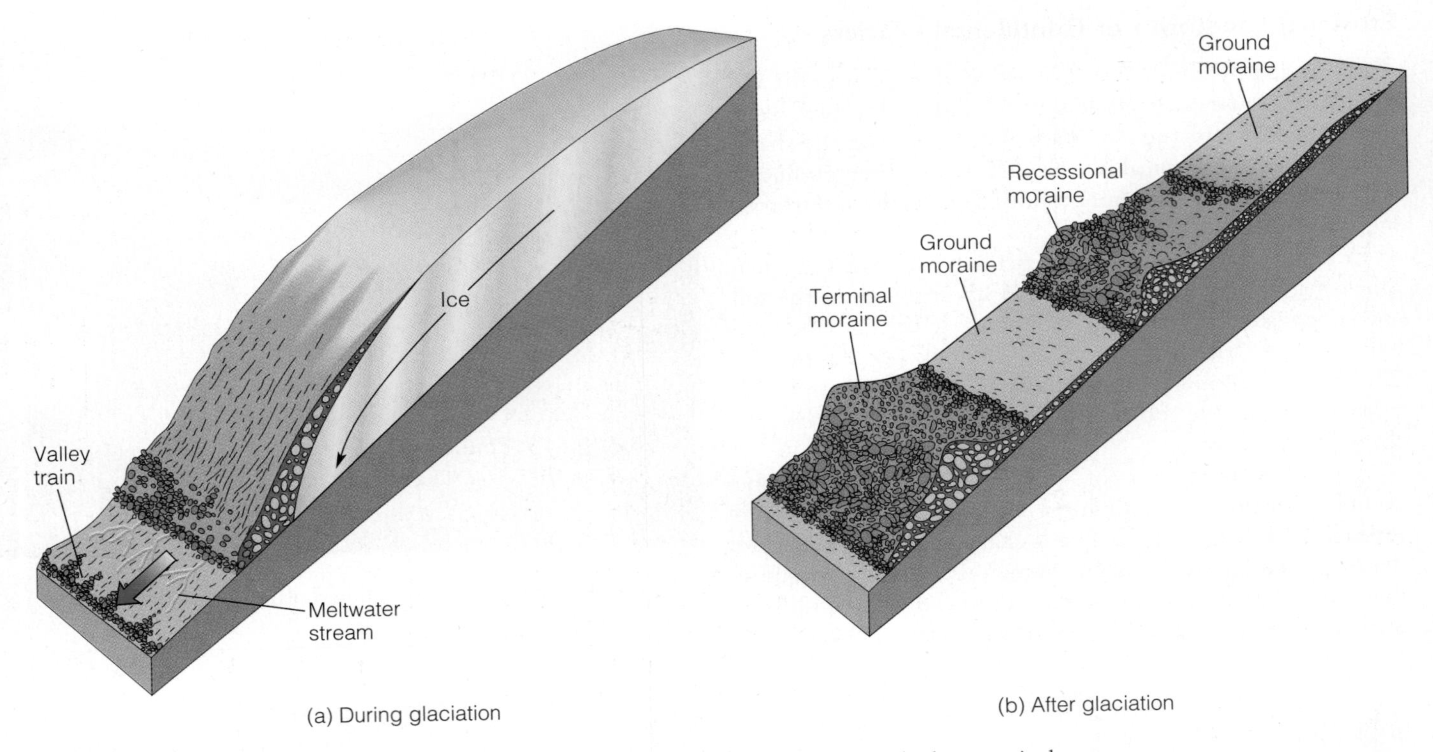

▶ **FIGURE 17-24** (*a*) The origin of an end moraine. (*b*) End moraines are described as terminal moraines or recessional moraines depending on their relative positions with respect to the glacier that produced them.

Should a negative budget occur, however, the ice front retreats toward the zone of accumulation. As the ice front recedes, till is deposited as it is liberated from the melting ice and forms a layer of **ground moraine** (Figure 17-24b). Ground moraine has an irregular, rolling topography, whereas end moraine consists of long ridgelike accumulations of sediment.

After a glacier has retreated for some time, its terminus may once again stabilize, and it will deposit another end moraine. Because the ice front has receded, such moraines are called **recessional moraines** (Figure 17-24b). During the Pleistocene Epoch, continental glaciers in the mid-continent region extended as far south as southern Ohio, Indiana, and Illinois. Their outermost end moraines, marking the greatest extent of the glaciers, go by the special name **terminal moraine** (valley glaciers also deposit terminal moraines). As the glaciers retreated from the positions where their terminal moraines were deposited, they temporarily ceased retreating numerous times and deposited dozens of recessional moraines.

Lateral and Medial Moraines. Valley glaciers transport considerable sediment along their margins. Much of this sediment is abraded and plucked from the valley walls, but a significant amount falls or slides onto the glacier's surface by mass wasting processes. In any case, when a glacier melts, this sediment is deposited as long ridges of till called **lateral moraines** along the margin of the glacier (▶ Figure 17-25).

Where two lateral moraines merge, as when a tributary glacier flows into a larger glacier, a **medial moraine** forms (Figure 17-25). A large glacier will often have several dark stripes of sediment on its surface, each of which is a medial moraine. Although medial moraines are identified by their position on a valley glacier, they are, in fact, formed from the coalescence of two lateral moraines. One can generally determine how many tributaries a valley glacier has by the number of its medial moraines.

Drumlins. In many areas where continental glaciers have deposited till, the till has been reshaped into elongated hills called **drumlins.** Some drumlins measure as much as 50 m high and 1 km long, but most are much smaller. From the side, a drumlin looks like an inverted spoon with the steep end on the side from which the glacial ice advanced, and the gently sloping end pointing in the direction of ice movement (▶ Figure 17-26). Thus, drumlins can be used to determine the direction of ice movement.

Drumlins are most often found in areas of ground moraine that were overridden by an advancing ice sheet. Although no one has fully explained the origin of drumlins, it appears that they form in the zone of plastic flow as glacial ice modifies preexisting till into streamlined hills. Drumlins rarely occur as single, isolated hills; instead they occur in *drumlin fields* that contain hundreds or thousands of drumlins. Drumlin fields are found in several states and Ontario,

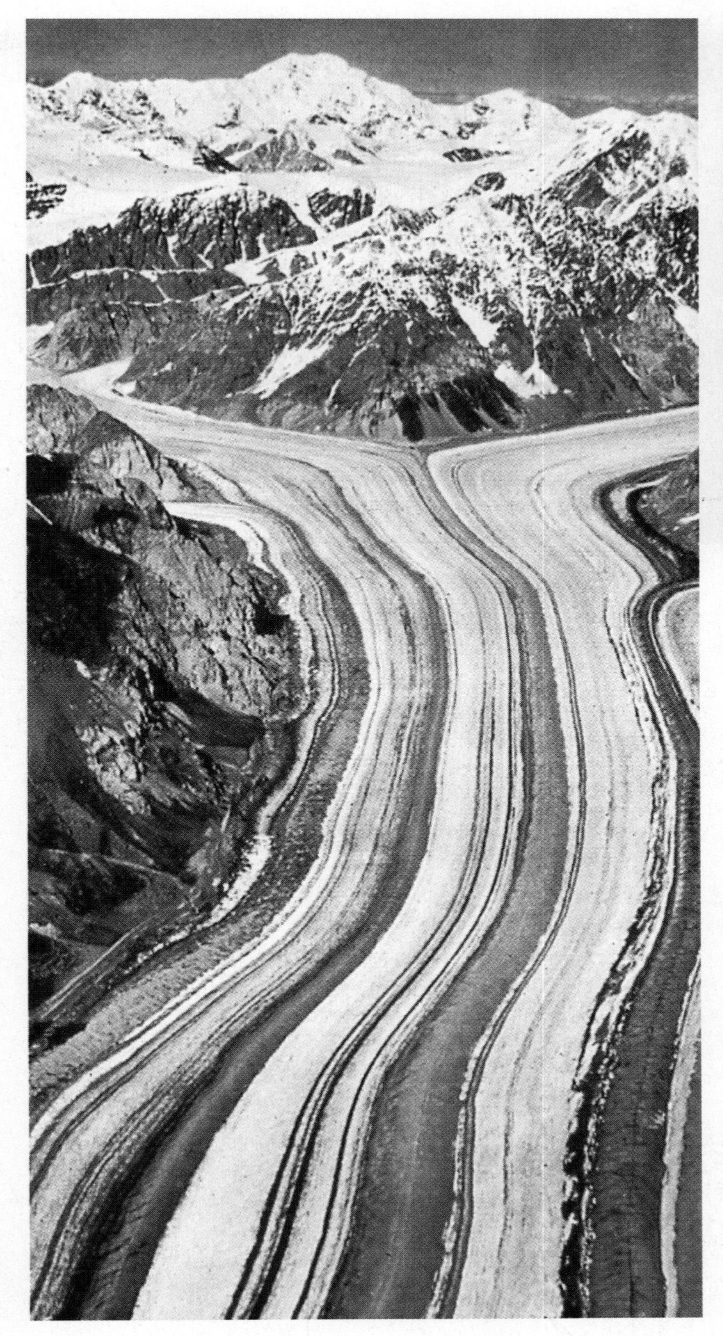

▶ FIGURE 17-25 Lateral and medial moraines on a glacier in Alaska. The material transported and deposited along the margins of a valley glacier is lateral moraine. Where two lateral moraines merge, a medial moraine forms.

Canada, but perhaps the finest example is near Palmyra, New York.

Landforms Composed of Stratified Drift

As already noted, stratified drift is a type of glacial deposit that exhibits sorting and layering, an indication that it was deposited by running water. Stratified drift is associated with both valley and continental glaciers, but as one would expect, it is more extensive in areas of continental glaciation.

▶ FIGURE 17-26 These elongated hills in Antrim County, Michigan are drumlins. (Photo courtesy of B. M. C. Pape.)

Outwash Plains and Valley Trains. Glaciers discharge meltwater laden with sediment most of the time, except perhaps during the coldest months. This meltwater forms a series of braided streams that radiate out from the front of continental glaciers over a wide region. So much sediment is supplied to these streams that much of it is deposited within the channels as sand and gravel bars. The vast blankets of sediments so formed are called **outwash plains** (▶ Figure 17-27a).

Valley glaciers discharge huge amounts of meltwater and, like continental glaciers, have braided streams extending from them. However, these streams are generally confined to the lower parts of glacial troughs, and their long, narrow deposits of stratified drift are known as **valley trains** (Figure 17-27b).

Outwash plains and valley trains commonly contain numerous circular to oval depressions, many of which contain small lakes. These depressions are *kettles*; they form when a retreating ice sheet or valley glacier leaves a block of ice that is subsequently partly or wholly buried (▶ Figure 17-28). When the ice block eventually melts, it leaves a depression; if the depression extends below the water table, it becomes the site of a small lake. So many kettles occur in some outwash plains that they are called *pitted outwash plains*. Although kettles are most common in outwash plains, they can also form in end moraines.

Kames and Eskers. **Kames** are conical hills as much as 50 m high composed of stratified drift (Figures 17-28 and ▶ 17-29a). Many kames form when a stream deposits sediment in a depression on a glacier's surface; as the ice melts, the

(a)

(b)

➢ **FIGURE 17-27** (*a*) Deposits of an outwash plain in Michigan. (Photo courtesy of David J. Matty.) (*b*) Braided streams discharging from a valley glacier, such as these in the Yukon Territory, Canada, deposit the stratified drift of a valley train.

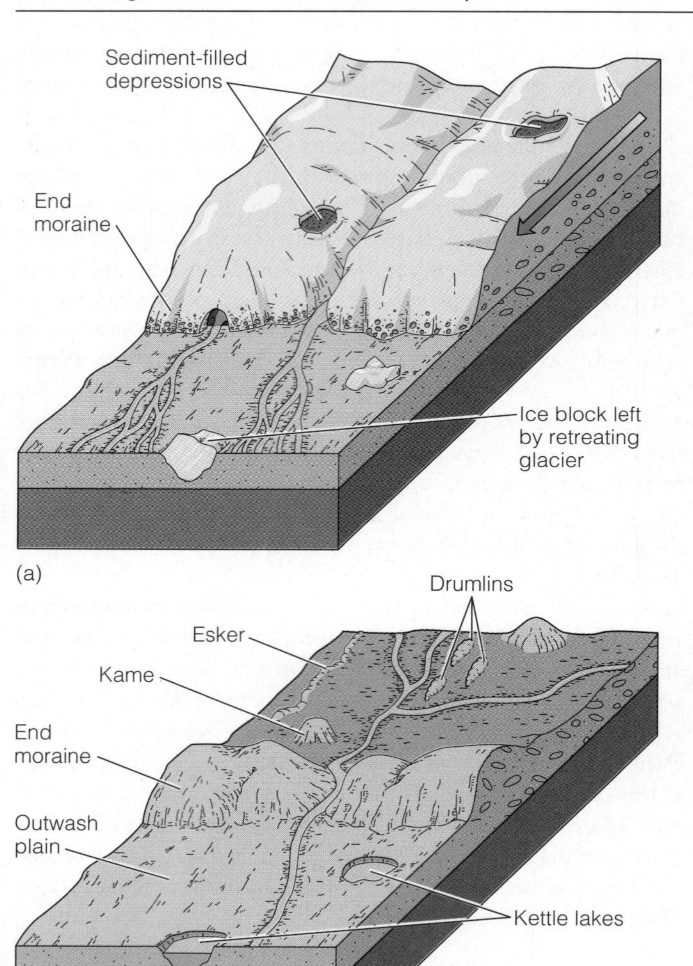

➢ **FIGURE 17-28** Two stages in the origin of kettles, kames, and eskers. (*a*) During glaciation. (*b*) After glaciation.

deposit is lowered to the surface. They also form in cavities within or beneath stagnant ice.

Long sinuous ridges of stratified drift, many of which meander and have tributaries, are called **eskers** (Figures 17-28 and 17-29b). Most eskers have sharp crests and sides that slope at about 30°. Some are quite high, as much as 100 m, and can be traced for more than 100 km. Eskers occur most commonly in areas once covered by continental glaciers, but they are also associated with large valley glaciers. The sorting and stratification of the sediments within eskers clearly indicate deposition by running water. The physical properties of ancient eskers and observations of present-day glaciers indicate that they form in tunnels beneath stagnant ice (Figure 17-28).

Glacial Lake Deposits

Numerous lakes exist in areas of glaciation. Some formed as a result of glaciers scouring out depressions; others occur where a stream's drainage was blocked (see Perspective 17-1); and others are the result of water accumulating behind moraines or in kettles. Regardless of how they formed, glacial lakes, like all lakes, are areas of deposition. Sediment may be carried into them and deposited as small deltas, but of special interest are the fine-grained deposits.

(a)

(b)

▶ **FIGURE 17-29** (*a*) This small, conical hill in Wisconsin is a kame. (Photo courtesy of B. M. C. Pape.) (*b*) An area of ground moraine and an esker.

Mud deposits in glacial lakes are commonly finely laminated, consisting of alternating light and dark layers. Each light-dark couplet is called a *varve* (▶ Figure 17-30). Each varve represents an annual episode of deposition; the light layers form during the spring and summer and consist of silt and clay; the dark layers form during the winter when the smallest particles of clay and organic matter settle from suspension as the lake freezes over. The number of varves indicates how many years a glacial lake has existed.

▶ **FIGURE 17-30** Glacial varves with a dropstone. Each varve, a dark-light couplet, is an annual deposit.

Another distinctive feature of glacial lakes containing varved deposits is the presence of *dropstones* (Figure 17-30). These are pieces of gravel, some of boulder size, in otherwise very fine-grained deposits. The presence of varves indicates that currents and turbulence in such lakes was minimal, otherwise clay and organic matter would not have settled from suspension. How then can we account for dropstones in a low-energy environment? Most of them were probably carried into the lakes by icebergs that eventually melted and released sediment contained in the ice.

PLEISTOCENE GLACIATION

In hindsight, it is hard to believe that so many competent naturalists of the last century were skeptical that widespread glaciers existed on the northern continents during the not-too-distant past. Many naturalists invoked the biblical flood to account for the large boulders throughout Europe that occur far from their sources. Others believed that the boulders were rafted to their present positions by icebergs floating in floodwaters. It was not until 1837 that the Swiss naturalist Louis Agassiz argued convincingly that the displaced boulders, many coarse-grained sedimentary deposits, polished and striated bedrock, and many of the valleys of Europe resulted from huge ice masses moving over the land.

We know today that the Pleistocene Ice Age began about 1.6 million years ago and consisted of several intervals of glacial expansion separated by warmer interglacial periods. At least four major episodes of Pleistocene glaciation have

Perspective 17-1

Glacial Lake Missoula and the Channeled Scablands

The term *scabland* is used in the Pacific Northwest to describe areas where the surface deposits have been scoured, thus exposing the underlying rock. Such an area exists in a large part of eastern Washington where numerous deep and generally dry channels are present. Some of these channels, cut into basalt lava flows, are more than 70 m deep, and their floors are covered by gigantic "ripple marks" as much as 10 m high and 70 to 100 m apart. Additionally, a number of high hills in the area are arranged so that they appear to have been islands in a large braided stream.

In 1923, J Harlan Bretz proposed that the channeled scablands of eastern Washington were formed during a single, gigantic flood. Bretz's unorthodox explanation was rejected by most geologists who preferred a more traditional interpretation based on normal stream erosion over a long period of time. In contrast, Bretz held that the scablands were formed rapidly during a flood of glacial meltwater lasting only a few days.

The problem with Bretz's hypothesis was that he could not identify an adequate source for his floodwater. He knew that glaciers had advanced as far south as Spokane, Washington, but he could not explain how so much ice melted so rapidly. The answer to Bretz's dilemma came from western Montana where an enormous ice-dammed lake (Lake Missoula) had formed. Lake Missoula formed when an advancing glacier plugged the Clark Fork Valley at Ice Cork, Idaho, causing the water to fill the valleys of western Montana (▶ Figure 1). At its highest level, Lake Missoula covered about 7,800 km^2 and contained an estimated 2,090 km^3 of water (about 42% of the volume of present-day Lake Michigan). The shorelines of Lake Missoula are still clearly visible on the mountainsides around Missoula, Montana (▶ Figure 2).

When the ice dam impounding Lake Missoula failed, the water rushed out at tremendous velocity and drained south and southwest across Idaho and into Washington. The maximum rate of flow is estimated to have been nearly 11 million m^3/sec, about 55 times greater than the average discharge of the Amazon River. When these raging floodwaters reached eastern Washington, they stripped away the soil and most of the surface sediment, carving out huge valleys in solid bedrock. The currents were so powerful and turbulent they plucked out and moved pieces of basalt measuring 10 m across. Within the channels, sand and gravel were shaped into huge ridges, the so-called giant ripple marks (▶ Figure 3).

▶ **FIGURE 1** Location of glacial Lake Missoula and the channeled scablands of eastern Washington.

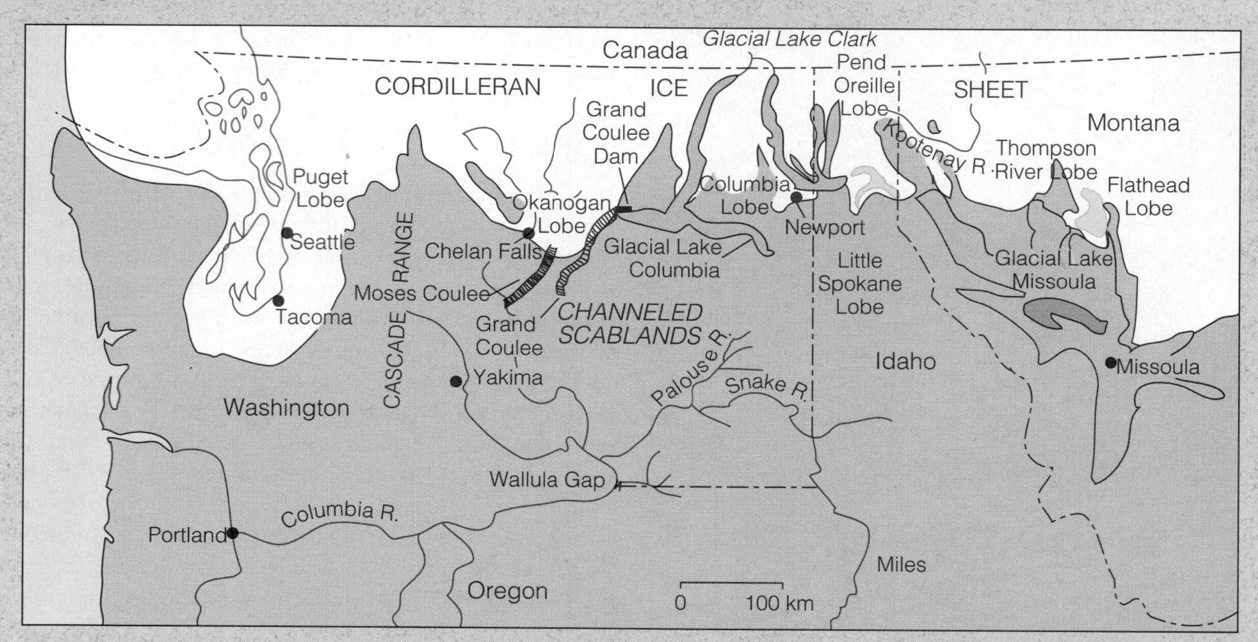

➢ **FIGURE 2** The horizontal lines on Sentinel Mountain at Missoula, Montana, are wave-cut shorelines of glacial Lake Missoula.

➢ **FIGURE 3** These gravel ridges are the so-called giant ripple marks that formed when glacial Lake Missoula drained across this area near Camas Hot Springs, Montana.

Bretz originally thought that one massive flood formed the channeled scablands, but geologists now know that Lake Missoula formed, flooded, and reformed at least four times and perhaps as many as seven times. The largest lake formed 18,000 to 20,000 years ago, and its draining produced the last great flood.

How long did the flood last and did humans witness it? It has been estimated that approximately one month passed from the time the ice dam first broke and water rushed out onto the scablands to the time the scabland streams returned to normal flow. No one knows for sure if anyone witnessed the flood. The oldest known evidence of humans in the region is from the Marmes Man site in southeastern Washington dated at 10,130 years ago, long after the last flood from Lake Missoula. However, it is now generally accepted that Native Americans were present in North America at least 15,000 years ago, and perhaps witnessed one of nature's greatest floods.

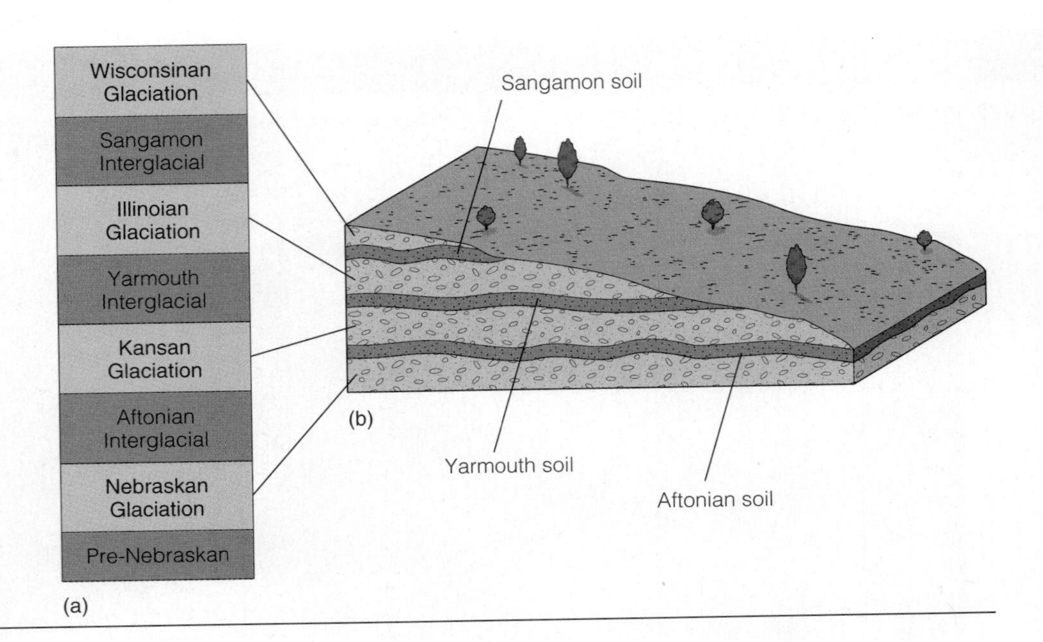

▶ **FIGURE 17-31** (*a*) Standard terminology for Pleistocene glacial and interglacial stages in North America. (*b*) A reconstruction showing an idealized succession of deposits and soils developed during the glacial and interglacial stages.

been recognized in North America (▶ Figure 17-31), and six or seven major glacial advances and retreats are recognized in Europe. It now appears that at least 20 warm-cold cycles can be detected in deep-sea cores. In view of these data, the traditional four-part subdivision of the Pleistocene of North America must be modified. Based on the best available evidence, it appears that the Pleistocene ended about 10,000 years ago. But geologists do not know if the present interglacial period will persist indefinitely, or whether we will enter another glacial interval.

The onset of glacial conditions really began about 40 million years ago when surface ocean waters at high southern latitudes suddenly cooled. By about 38 million years ago, glaciers had formed in Antarctica, but a continuous ice sheet did not develop there until 15 million years ago. Following a brief warming trend during the Late Tertiary Period, ice sheets began forming in the Northern Hemisphere about 2 to 3 million years ago, and the Pleistocene Ice Age was under way. At their greatest extent, Pleistocene glaciers covered about three times as much of the Earth's surface as they do now and were up to 3 km thick (▶ Figure 17-32). Large areas of North America were covered by glacial ice as were Greenland, Scandinavia, Great Britain, Ireland, and a large part of northern Russia. Mountainous areas also experienced an expansion of valley glaciers and the development of ice caps.

Pleistocene Climates

As one would expect, the climatic effects responsible for Pleistocene glaciation were worldwide. Contrary to popular belief, though, the world was not as frigid as it is commonly portrayed in cartoons and movies. During times of glacier growth, those areas in the immediate vicinity of the glaciers experienced short summers and long, wet winters.

Areas outside the glaciated regions experienced varied climates. During times of glacial growth, lower ocean temperatures reduced evaporation so that most of the world was drier than it is today, but some areas that are arid today were much wetter. For example, as the cold belts at high latitudes expanded, the temperate, subtropical, and tropical zones were compressed toward the equator, and the rain that now falls on the Mediterranean shifted so that it fell on the Sahara of North Africa enabling lush forests to grow in what is now desert. California and the arid southwestern United States were also wetter because a high-pressure zone over the northern ice sheet deflected Pacific winter storms southward.

Following the Pleistocene, mild temperatures prevailed between 8,000 and 6,000 years ago. After this warm period, conditions gradually became cooler and moister favoring the growth of valley glaciers on the Northern Hemisphere continents. Careful studies of the deposits at the margins of present-day glaciers reveal that during the last 6,000 years (a time called the *Neoglaciation*), glaciers expanded several times. The last expansion, which occurred between 1500 and the mid-to-late 1800s, was the Little Ice Age (see the Prologue).

Pluvial and Proglacial Lakes

During the Pleistocene, many of the basins in the western United States contained large lakes that formed as a result of greater precipitation and overall cooler temperatures (especially during the summer), which lowered the evaporation rate (▶ Figure 17-33). The largest of these *pluvial lakes*, as they are called, was Lake Bonneville, which attained a maximum size of 50,000 km^2 and a depth of at least 335 m (Figure 17-33). The vast salt deposits of the Bonneville Salt Flats west of Salt Lake City in Utah formed as parts of this

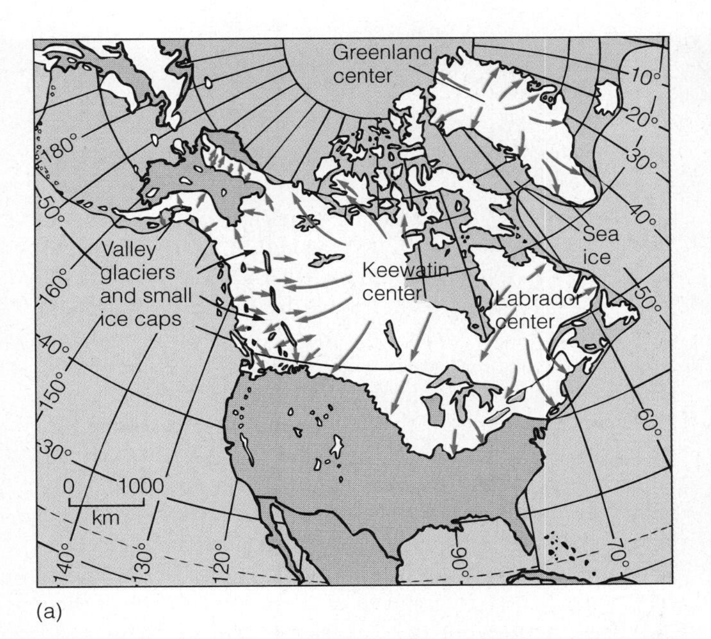

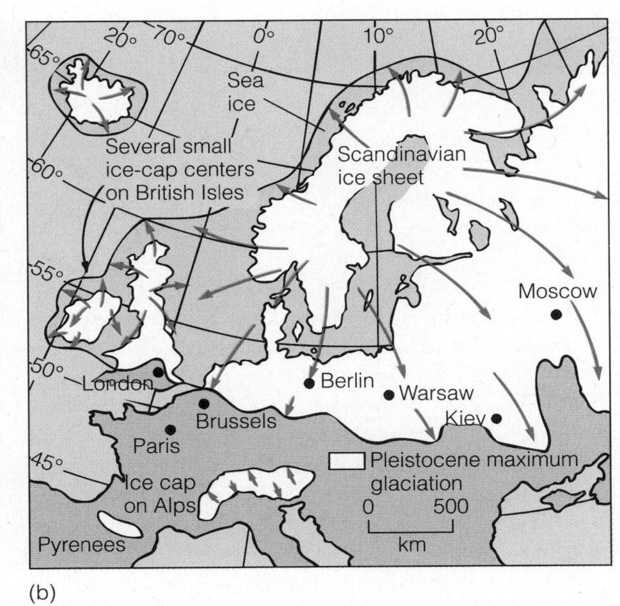

▶ **FIGURE 17-32** Centers of ice accumulation, the maximum extent of Pleistocene glaciation, and directions of ice movement in (*a*) North America and (*b*) Europe.

ancient lake dried up: Great Salt Lake is simply the remnant of this once much larger lake.

Another large pluvial lake existed in Death Valley, California (see Perspective 18-2), which is now the hottest, driest place in North America. During the Pleistocene, that area received enough rainfall to maintain a lake 145 km long and 178 m deep. When the lake evaporated, the dissolved salts were precipitated on the valley floor; some of these evaporite deposits, especially borax, are important mineral resources.

▶ **FIGURE 17-33** Pleistocene pluvial lakes in the western United States.

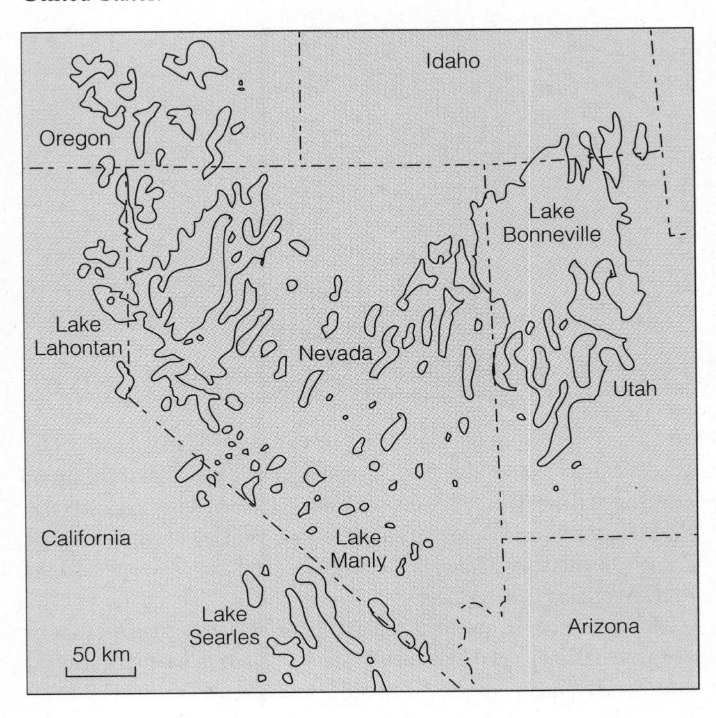

In contrast to pluvial lakes, which form far from glaciers, *proglacial lakes* are formed by the meltwater accumulating along the margins of glaciers. In fact, one shoreline of proglacial lakes is the ice front itself, while the other shorelines consist of moraines. Lake Agassiz, named in honor of the French naturalist Louis Agassiz, was a large proglacial lake covering about 250,000 km^2, mostly in Manitoba, Saskatchewan, and Ontario, Canada, but extending into North Dakota. It persisted until the glacial ice along its northern margin melted, at which time the lake was able to drain northward into Hudson Bay.

Numerous proglacial lakes existed during the Pleistocene, but most of these eventually disappeared. Lake Agassiz and a number of smaller lakes drained when the glaciers disappeared, but some simply filled with sediment. Notable exceptions are the Great Lakes, all of which first formed as proglacial lakes (see Perspective 17-2).

Changes in Sea Level

More than 70 million km^3 of snow and ice covered the continents during the maximum glacial coverage of the Pleistocene. The storage of ocean waters in glaciers lowered sea level 130 m and exposed extensive areas of the present-day continental shelves, which were quickly covered with vegetation. Indeed, a land bridge existed across the Bering Strait from Alaska to Siberia. Native Americans crossed the Bering land bridge, and various animals migrated between the continents; the American bison, for example, migrated from Asia. The British Isles were connected to Europe

Perspective 17-2

A Brief History of the Great Lakes

Before the Pleistocene, no large lakes existed in the Great Lakes region, which was then an area of generally flat lowlands with broad stream valleys draining to the north (➢ Figure 1). As the glaciers advanced southward, they eroded the stream valleys more deeply, forming what were to become the basins of the Great Lakes. During these glacial advances, the ice front moved forward as a series of lobes, some of which flowed into the preexisting lowlands where the ice became thicker and moved more rapidly. As a result, the lowlands were deeply eroded—four of the five Great Lakes basins were eroded below sea level.

At their greatest extent, the glaciers covered the entire Great Lakes region and extended far to the south (Figure 17-32a). As the ice sheet retreated northward during the late Pleistocene, the ice front periodically stabilized, and numerous recessional moraines were deposited. By about 14,000 years ago, parts of the Lake Michigan and Lake Erie basins were ice-free, and glacial meltwater began forming proglacial lakes (➢ Figure 2). As the retreat of the ice sheet continued—although periodically interrupted by minor readvances of the ice front—the Great Lakes basins were uncovered, and the lakes expanded until they eventually reached their present size and configuration (Figure 2). Currently, the Great Lakes contain nearly 23,000 km^3 of water, about 18% of the water in all fresh water lakes.

Although the history of the Great Lakes just presented is generally correct, it is oversimplified. For instance, the areas and depths of the evolving Great Lakes fluctuated widely in response to minor readvances of the ice front. Furthermore, as the lakes filled, they spilled over the lowest parts of their margins, cutting outlets that partly drained them. And finally, as the glaciers retreated northward, isostatic rebound raised the southern parts of the Great Lakes region, greatly altering their drainage systems. We shall have more to say about isostatic rebound in this region in a later section.

The present-day Great Lakes and their St. Lawrence

➢ **FIGURE 1** Theoretical preglacial drainage in the Great Lakes region. The divide separating the preglacial Mississippi and St. Lawrence drainage basins was probably near its present location, indicated by the dashed line. The future sites of the Great Lakes are outlined by dotted lines.

during the glacial intervals because the shallow floor of the North Sea was above sea level. When the glaciers disappeared, these areas were again flooded, drowning the plants and forcing the animals to migrate farther inland.

Lowering of sea level during the Pleistocene also affected the base level of most major streams. When sea level dropped, streams downcut as they sought to adjust to a new lower base level (see Chapter 15). Stream channels in coastal areas were extended and deepened along the emergent continental shelves. When sea level rose at the end of the Pleistocene, the lower ends of river valleys along the east coast of North America were flooded and are now important harbors (see Chapter 19).

A tremendous quantity of water is still stored on land in present-day glaciers (see Table 15-1). If these glaciers should completely melt, sea level would rise about 70 m, flooding

River drainage constitute one of the great commercial waterways of the world. Oceangoing vessels can sail into the interior of North America as far west as Duluth, Minnesota. To do so they must bypass Niagara Falls between Lake Erie and Lake Ontario via a system of locks. Niagara Falls plunges 51 m over the Niagaran escarpment, which consists of resistant dolostone that was exposed during the last glacial retreat. Erosion of softer shale at the base of the falls is causing Niagara Falls to retreat upstream at a rate of about a meter per year (see Figure 15-30). It is estimated that in about 25,000 years the escarpment between the lakes will have been eliminated entirely! As a consequence, Lake Erie will become a small lake adjacent to vast swampy areas, and the upper Great Lakes will be considerably smaller than at present.

▶ **FIGURE 2** Four stages in the evolution of the Great Lakes. As the glacial ice retreated northward, the lake basins began filling with meltwater. The dotted lines indicate the present-day shorelines of the lakes.

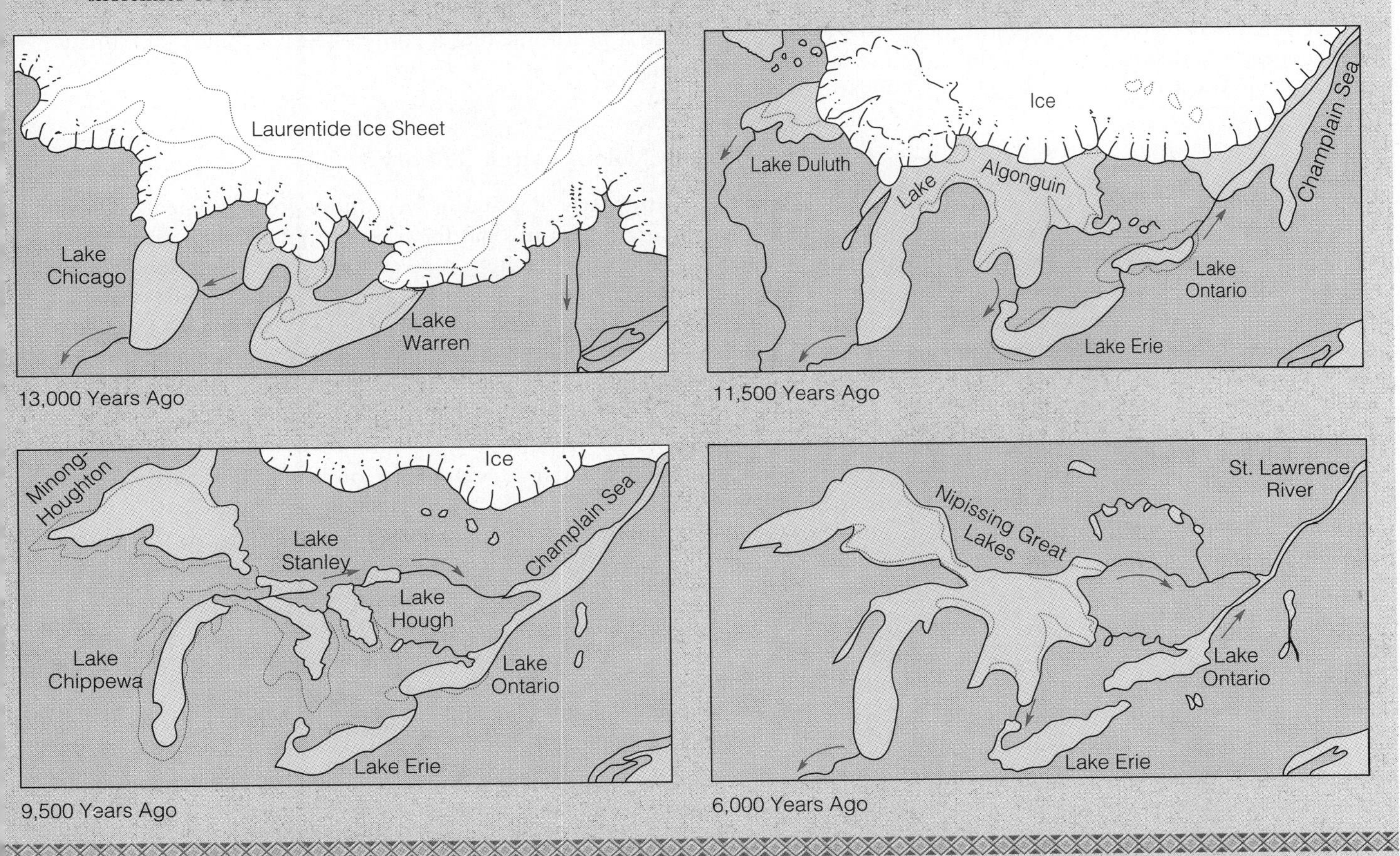

all of the coastal areas of the world where many of the world's large population centers are located (▶ Figure 17-34).

 Glaciers and Isostasy

In Chapter 10 we discussed the concept of isostasy and noted that loading or unloading of the Earth's crust causes it to respond isostatically to an increased or decreased load by subsiding and rising, respectively. There is no question that isostatic rebound has occurred in the areas formerly covered by continental glaciers. In fact, a number of features in such areas can be explained as a consequence of isostatic adjustments of the Earth's crust.

When the Pleistocene ice sheets formed and increased in size, the weight of the ice caused the crust to respond by

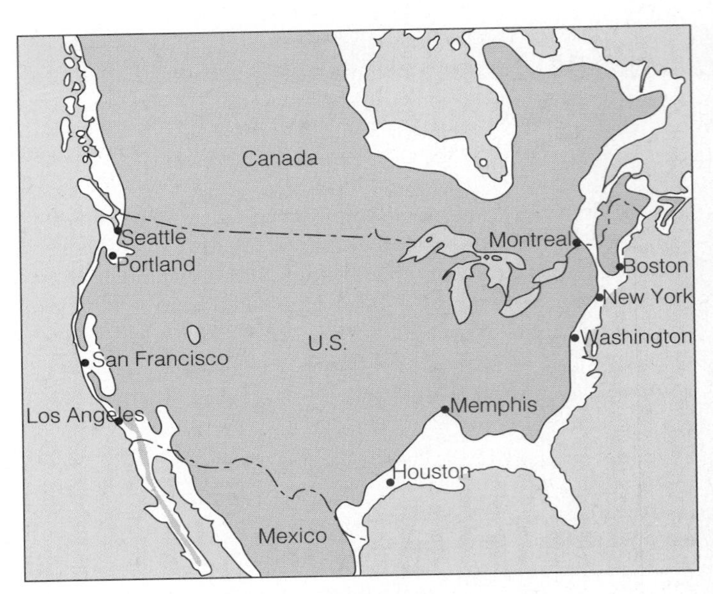

▶ **FIGURE 17-34** Large parts of North America—and all other continents—would be flooded by the 70 m rise in sea level that would occur if all the Earth's glacial ice melted.

slowly subsiding deeper into the mantle. In some places, the Earth's surface was depressed as much as 300 m below preglacial elevations. As the ice sheets disappeared, the downwarped areas gradually rebounded to their former positions. As noted in Chapter 10, parts of Scandinavia are still rebounding at a rate of about 1 m per century (see Figure 10-22a).

In Perspective 17-2 we noted that the Great Lakes evolved as the glaciers retreated to the north. As one would expect, isostatic rebound began as the ice front retreated north. Rebound began first in the southern part of the region because that area was free of ice first. Furthermore, the greatest loading by glaciers, and hence the greatest crustal depression, occurred farther north in Canada in the zones of accumulation. For these reasons, rebound has not been evenly distributed over the entire glaciated area: it increases in magnitude from south to north (see Figure 10-22b). As a result of this uneven isostatic rebound, coastal features in the Great Lakes region, such as old shorelines, are now elevated higher above their former levels in the north and thus slope to the south.

 CAUSES OF GLACIATION

So far we have examined the effects of glaciation, but have not addressed the central questions of what causes large-scale glaciation and why so few episodes of widespread glaciation have occurred. For more than a century, scientists have been attempting to develop a comprehensive theory explaining all aspects of ice ages, but have not yet been completely successful. One reason for their lack of success is that the climatic changes responsible for glaciation, the cyclic occurrence of glacial-interglacial episodes, and short-term events such as the Little Ice Age operate on vastly different time scales.

Only a few periods of glaciation are recognized in the geologic record, each separated from the others by long intervals of mild climate. Such long-term climatic changes probably result from slow geographic changes related to plate tectonic activity. Moving plates can carry continents to high latitudes where glaciers can exist, provided that they receive enough precipitation as snow. Plate collisions, the subsequent uplift of vast areas far above sea level, and the changing atmospheric and oceanic circulation patterns caused by the changing shapes and positions of plates also contribute to long-term climatic change.

Intermediate-term climatic events, such as the glacial-interglacial episodes of the Pleistocene, occur on time scales of tens to hundreds of thousands of years. The cyclic nature of this most recent episode of glaciation has long been a problem in formulating a comprehensive theory of climatic change.

The Milankovitch Theory

A particularly interesting hypothesis for intermediate-term climatic events was put forth by the Yugoslavian astronomer Milutin Milankovitch during the 1920s. He proposed that minor irregularities in the Earth's rotation and orbit are sufficient to alter the amount of solar radiation that the Earth receives at any given latitude and hence can affect climatic changes. Now called the **Milankovitch theory,** it was initially ignored, but has received renewed interest during the last 20 years.

Milankovitch attributed the onset of the Pleistocene Ice Age to variations in three parameters of the Earth's orbit. The first of these is orbital eccentricity, which is the degree to which the orbit departs from a perfect circle (▶ Figure 17-35a). Calculations indicate a roughly 100,000-year cycle between times of maximum eccentricity. This corresponds closely to 20 warm-cold climatic cycles that occurred during the Pleistocene. The second parameter is the angle between the Earth's axis and a line perpendicular to the plane of the ecliptic (Figure 17-35b). This angle shifts about 1.5° from its current value of 23.5° during a 44,000-year cycle. The third parameter is the precession of the equinoxes, which causes the position of the equinoxes and solstices to shift slowly around the Earth's elliptical orbit in a 22,000-year cycle (Figure 17-35c and d).

Continuous changes in these three parameters cause the amount of solar heat received at any latitude to vary slightly over time. The total heat received by the planet, however, is little changed. Milankovitch proposed, and now many scientists agree, that the interaction of these three parameters provides the triggering mechanism for the glacial-interglacial episodes of the Pleistocene.

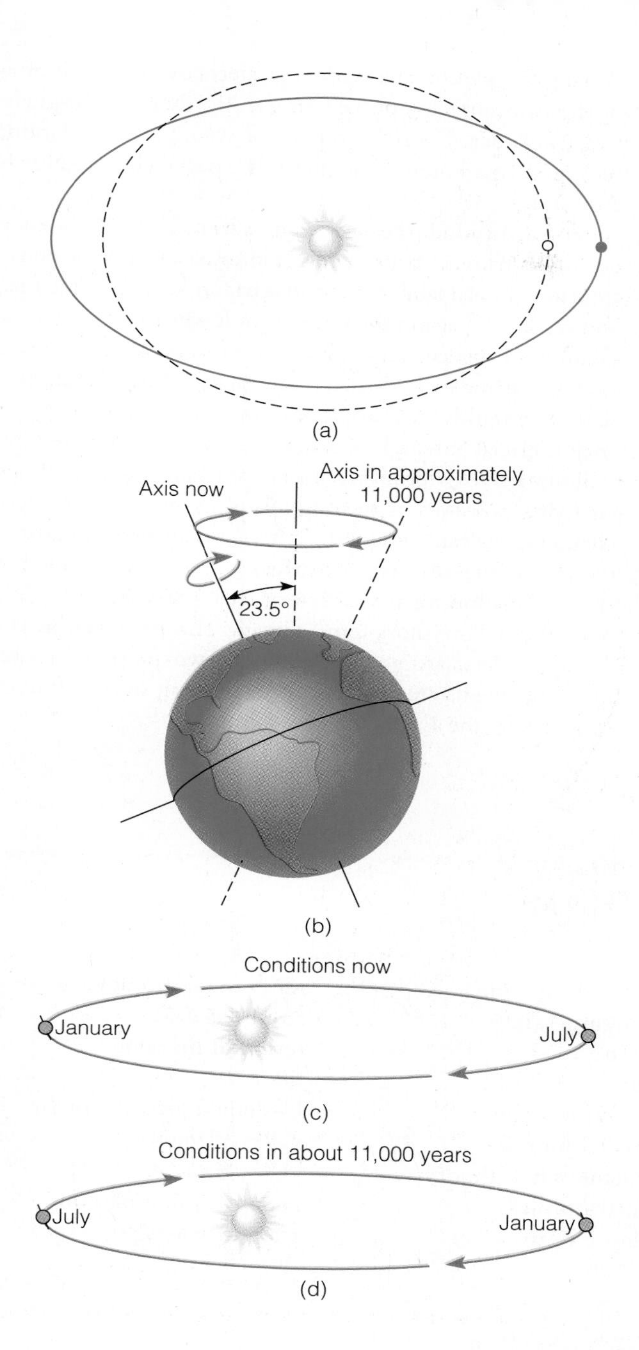

▶ FIGURE 17-35 (*a*) The Earth's orbit varies from nearly a circle (dashed line) to an ellipse (solid line) and back again in about 100,000 years. (*b*) The Earth moves around its orbit while spinning about its axis, which is tilted to the plane of the ecliptic at 23.5° and points toward the North Star. The Earth's axis of rotation slowly moves and traces out the path of a cone in space. (*c*) At present, the Earth is closest to the Sun in January when the Northern Hemisphere experiences winter. (*d*) In about 11,000 years, as a result of precession, the Earth will be closer to the Sun in July, when summer occurs in the Northern Hemisphere.

Short-Term Climatic Events

Climatic events having durations of several centuries, such as the Little Ice Age, are too short to be accounted for by plate tectonics or Milankovitch cycles. Several hypotheses have been proposed, including variations in solar energy and volcanism.

Variations in solar energy could result from changes within the Sun itself or from anything that would reduce the amount of energy the Earth receives from the Sun. The latter could result from the solar system passing through clouds of interstellar dust and gas or from substances in the Earth's atmosphere reflecting solar radiation back into space. Records kept over the past 75 years indicate that during this time the amount of solar radiation has varied only slightly. Although variations in solar energy may influence short-term climatic events, such a correlation has not been demonstrated.

During large volcanic eruptions, tremendous amounts of ash and gases are spewed into the atmosphere where they reflect incoming solar radiation and thus reduce atmospheric temperatures. Recall from Perspective 4-1 that small droplets of sulfur gases remain in the atmosphere for years and can have a significant effect on the climate. Several large-scale volcanic events have been recorded, such as the 1815 eruption of Tambora, and are known to have had climatic effects. However, no relationship between periods of volcanic activity and periods of glaciation has yet been established.

CHAPTER SUMMARY

1. Glaciers are masses of ice on land that move by plastic flow and basal slip. Glaciers currently cover about 10% of the land surface and contain 2.15% of all water on Earth.
2. Valley glaciers are confined to mountain valleys and flow from higher to lower elevations, whereas continental glaciers cover vast areas and flow outward in all directions from a zone of accumulation.
3. A glacier forms when winter snowfall in an area exceeds summer melt and therefore accumulates year after year. Snow is compacted and converted to glacial ice, and when the ice is about 40 m thick, pressure causes it to flow.
4. The behavior of a glacier depends on its budget, which is the relationship between accumulation and wastage. If a glacier possesses a balanced budget, its terminus remains stationary; a positive or negative budget results in advance or retreat of the terminus, respectively.
5. Glaciers move at varying rates depending on the slope, discharge, and season. Valley glaciers tend to flow more rapidly than continental glaciers.

6. Glaciers are powerful agents of erosion and transport because they are solids in motion. They are particularly effective at eroding soil and unconsolidated sediment, and they can transport any size sediment supplied to them.
7. Continental glaciers transport most of their sediment in the lower part of the ice, whereas valley glaciers carry sediment in all parts of the ice.
8. Erosion of mountains by valley glaciers produces several sharp, angular landforms including cirques, arêtes, and horns. U-shaped glacial troughs, fiords, and hanging valleys are also products of valley glaciation.
9. Continental glaciers abrade and bevel high areas, forming a smooth, rounded landscape.
10. Depositional landforms include moraines, which are ridge-like accumulations of till. Several types of moraines are recognized, including terminal, recessional, lateral, and medial moraines.
11. Drumlins are composed of till that was apparently reshaped into streamlined hills by continental glaciers.
12. Stratified drift consists of sediments deposited by meltwater streams issuing from glaciers; it is found in outwash plains and valley trains. Ridges called eskers and conical hills called kames are also composed of stratified drift.
13. During the Pleistocene Epoch, glaciers covered about 30% of the land surface especially in the Northern Hemisphere and Antarctica. Several intervals of widespread glaciation, separated by interglacial periods, occurred in North America.
14. Areas far beyond the ice were affected by Pleistocene glaciation; climate belts were compressed toward the equator, large pluvial lakes existed in what are now arid regions, and sea level was as much as 130 m lower than at present.
15. Loading of the Earth's crust by Pleistocene glaciers caused isostatic subsidence. When the glaciers disappeared, isostatic rebound began and continues in some areas.
16. Major glacial intervals separated by tens or hundreds of millions of years probably occur as a consequence of the changing positions of tectonic plates, which in turn cause changes in oceanic and atmospheric circulation patterns.
17. Currently, the Milankovitch theory is widely accepted as the explanation for glacial-interglacial intervals.
18. The reasons for short-term climatic changes, such as the Little Ice Age, are not understood. Two proposed causes for these events are changes in the amount of solar energy received by the Earth and volcanism.

IMPORTANT TERMS

abrasion
arête
basal slip
cirque
continental glacier
drumlin
end moraine
esker
fiord
firn
firn limit
glacial budget
glacial drift
glacial erratic
glacial ice
glacial polish
glacial striation
glacier
ground moraine
hanging valley
horn
kame
lateral moraine
medial moraine
Milankovitch theory
outwash plain
plastic flow
recessional moraine
stratified drift
terminal moraine
till
U-shaped glacial trough
valley glacier
valley train
zone of accumulation
zone of wastage

REVIEW QUESTIONS

1. Crevasses in glaciers extend down to:
   a. ____ about 300 m;
   b. ____ the base of the glacier;
   c. ____ the zone of plastic flow;
   d. ____ variable depths depending on how thick the ice is;
   e. ____ the outwash layer.
2. If a glacier has a negative budget:
   a. ____ the terminus will retreat;
   b. ____ its accumulation rate is greater than its wastage rate;
   c. ____ all flow ceases;
   d. ____ the glacier's length increases;
   e. ____ crevasses will no longer form.
3. The bowl-shaped depression at the upper end of a glacial trough is a(n):
   a. ____ inselberg;
   b. ____ cirque;
   c. ____ lateral moraine;
   d. ____ drumlin;
   e. ____ till.
4. Which of the following is not an erosional landform?
   a. ____ horn;
   b. ____ arête;
   c. ____ lateral moraine;
   d. ____ U-shaped glacial trough;
   e. ____ roche moutonnée.
5. Headward erosion of a group of cirques on the flanks of a mountain may produce a:
   a. ____ tarn;
   b. ____ varve;
   c. ____ drumlin;
   d. ____ kettle;
   e. ____ horn.

6. Rocks abraded by glaciers develop a smooth surface that shines in reflected light. Such a surface is called glacial ______ :
   a. ___ grooves;
   b. ___ polish;
   c. ___ flour;
   d. ___ striations;
   e. ___ till.
7. A small lake in a cirque is a:
   a. ___ pluvial lake;
   b. ___ proglacial lake;
   c. ___ tarn;
   d. ___ salt lake;
   e. ___ glacial trough lake.
8. Firn is:
   a. ___ freshly fallen snow;
   b. ___ a granular type of ice;
   c. ___ a valley train;
   d. ___ another name for the zone of wastage;
   e. ___ a type of glacial groove.
9. Pressure on ice at depth in a glacier causes it to move by:
   a. ___ rock creep;
   b. ___ fracture;
   c. ___ basal slip;
   d. ___ surging;
   e. ___ plastic flow.
10. A pyramid-shaped peak formed by glacial erosion is a:
    a. ___ fiord;
    b. ___ medial moraine;
    c. ___ horn;
    d. ___ cirque;
    e. ___ hanging valley.
11. Glacial drift is a general term for:
    a. ___ the erosional landforms of continental glaciers;
    b. ___ all the deposits of glaciers;
    c. ___ icebergs floating at sea;
    d. ___ the movement of glaciers by plastic flow and basal slip;
    e. ___ the annual wastage rate of a glacier.
12. The number of medial moraines on a glacier generally indicates the number of its ______ :
    a. ___ tributary glaciers;
    b. ___ terminal moraines;
    c. ___ eskers;
    d. ___ outwash plains;
    e. ___ valley trains.
13. A knifelike ridge separating glaciers in adjacent valleys is a(n):
    a. ___ fiord;
    b. ___ horn;
    c. ___ arête;
    d. ___ cirque;
    e. ___ lateral moraine.
14. How does glacial ice form, and why is it considered to be a rock?
15. Other than size, how do valley glaciers differ from continental glaciers?
16. What is the relative importance of plastic flow and basal slip for glaciers at high and low latitudes?
17. Explain in terms of the glacial budget how a once active glacier becomes a stagnant glacier.
18. What is a glacial surge, and what are the probable causes of surges?
19. Explain how glaciers erode by abrasion and plucking.
20. Describe the processes responsible for the origin of a cirque, U-shaped glacial trough, and hanging valley.
21. How do the erosional landforms of continental glaciers differ from those of valley glaciers?
22. Discuss the processes whereby terminal, recessional, and lateral moraines form.
23. How does a medial moraine form, and how can one determine the number of tributaries a valley glacier has by its medial moraines?
24. Describe drumlins, and explain how they form.
25. How do pluvial lakes differ from proglacial lakes? Give an example of each of these types of lakes.

Points to Ponder

1. In a roadside outcrop, you observe a deposit of alternating light and dark laminated mud containing a few large boulders. Explain the sequence of events responsible for deposition.
2. A glacier has a cross-sectional area of 400,000 m^2 and a flow velocity of 2 m/day. How long would it take for a discharge of 1 km^3 to occur?
3. We can be sure that the ancient shorelines of the Great Lakes were horizontal when they were formed, yet now they are not only elevated above their former level but also tilt toward the south. How do you account for these observations?
4. How might human activities affect the amount of glacial ice on Earth?
5. In North America, valley glaciers are common in Alaska and western Canada, and small ones are present in the mountains of the western United States; none, however, occur east of the Rocky Mountains. How can you explain this distribution of glaciers?

Additional Readings

Broecker, W. S., and G. H. Denton. 1990. What drives glacial cycles? *Scientific American* 262, no. 1: 49–56.

Carozzi, A. V. 1984. Glaciology and the ice age. *Journal of Geological Education* 32: 158–70.

Covey, C. 1984. The Earth's orbit and the ice ages. *Scientific American* 250, no. 2: 58–66.

Drewry, D. J. 1986. *Glacial geologic processes.* London: Edward Arnold.

Grove, J. M. 1988. *The Little Ice Age.* London: Methuen.

Hambrey, M., and J. Alean. 1992. *Glaciers.* New York: Cambridge University Press.

Imbrie, J., and K. P. Imbrie. 1979. *Ice ages: Solving the mystery.* New Jersey: Enslow Press.

John, B. S. 1979. *The winters of the world.* London: David & Charles.

Kurten, B. 1988. *Before the Indians*. New York: Columbia University Press.
Sharp, R. P. 1988. *Living ice: Understanding glaciers and glaciation*. New York: Cambridge University Press.
Williams, R. S., Jr. 1983. *Glaciers: Clues to future climate?* United States Geological Survey.

Chapter 18

THE WORK OF WIND AND DESERTS

OUTLINE

View looking along the ridge of a linear dune in the Rub' al-Khali in Saudi Arabia. The Rub' al-Khali, covering some 570,000m^2, is the largest area of continuous sand in the world.

Prologue

During the last few decades, deserts have been advancing across millions of acres of productive land, destroying rangelands, croplands, and even villages. Such expansion, estimated at 70,000 km^2 per year, has exacted a terrible toll in human suffering. Because of the relentless advance of deserts, hundreds of thousands of people have died of starvation or been forced to migrate as "environmental refugees" from their homelands to camps where the majority are severely malnourished. This expansion of deserts into formerly productive lands is called ***desertification***.

Most regions undergoing desertification lie along the margins of existing deserts. These margins have a delicately balanced ecosystem that serves as a buffer between the desert on one side and a more humid environment on the other. Their potential to adjust to increasing environmental pressures from natural causes or human activity is limited. Currently, such fringe areas include large regions in several parts of the world (▶ Figure 18-1).

Although some gradual expansion and contraction of desert regions occurs as a result of natural processes such as climatic change, much recent desertification has been greatly accelerated by human activities. In many areas, the natural vegetation has been cleared as crop cultivation has expanded into increasingly drier fringes to support the growing population. Because these areas are especially prone to droughts, crop failures are common occurrences, leaving the land bare and susceptible to increased wind and water erosion.

Because grasses constitute the dominant natural vegetation in most fringe areas, raising livestock is a common economic activity. Usually, these areas achieve a natural balance between vegetation and livestock as nomadic herders graze their animals on the available grasses. In many fringe areas, however, livestock numbers have been greatly increasing in recent years, and they now far exceed the land's capacity to support them. As a result, the vegetation cover that protects the soil has diminished, causing the soil to crumble. This leads to further drying of the soil and accelerated soil erosion by wind and water (▶ Figure 18-2).

Drilling water wells also contributes to desertification because human and livestock activity around a well site strips away the vegetation. With its vegetation gone, the topsoil blows away, and the resultant bare areas merge with the surrounding desert. In addition, the water used for irrigation from these wells sometimes contributes to desertification by increasing the salt content of the soil. As the water evaporates, a small amount of salt is deposited in the soil and is not flushed out as it would be in an area that receives more rain. Over time, the salt concentration becomes so high that plants can no longer grow.

▶ **FIGURE 18-1** Desert areas of the world and areas threatened by desertification.

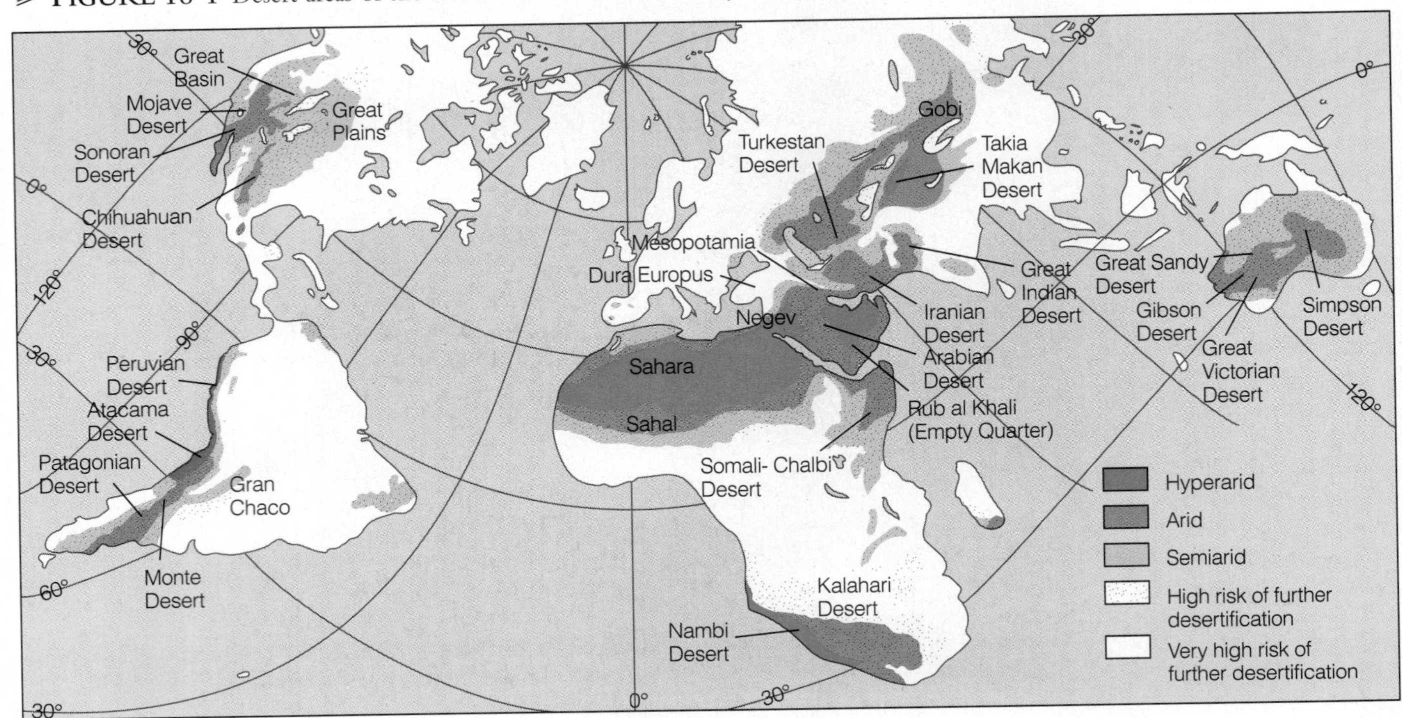

▶ FIGURE 18-2 A sharp line marks the boundary between pasture and an encroaching dune in Niger, Africa. As the goats eat the remaining bushes, the dune will continue to advance, and more land will be lost to desertification.

Desertification resulting from soil salinization is a major problem in North Africa, the Middle East, southwest Asia, and the western United States.

Collecting firewood for heating and cooking is another major cause of desertification, particularly in many less-developed countries where wood is the major fuel source. In the Sahel of Africa (a belt 300 to 1,100 km wide that lies south of the Sahara), the expanding population has completely removed all trees and shrubs in the areas surrounding many towns and cities. Journeys of several days on foot to collect firewood are common there. Furthermore, the use of dried animal dung to supplement firewood has exacerbated desertification because important nutrients in the dung are not returned to the soil.

The Sahel averages between 10 and 60 cm of rainfall per year, 90% of which evaporates when it falls. Because drought is common in the Sahel, the region can support only a limited population of livestock and humans. Traditionally, herders and livestock existed in a natural balance with the vegetation, following the rains north during the rainy season and returning south to greener rangeland during the dry seasons. Some areas were alternately planted and left fallow to help regenerate the soil. During fallow periods, livestock fed off the stubble of the previous year's planting, and their dung helped fertilize the soil.

With the emergence of new nations and increased foreign aid to the Sahel during the 1950s and 1960s, nomads and their herds were restricted, and large areas of grazing land were converted to cash crops such as peanuts and cotton that have a short growing season. Expanding human and animal populations and more intensive agriculture put increasing demands on the land. These developments combined with the worst drought of the century (1968–1973) brought untold misery to the people of the Sahel. Without rains, the crops failed and the livestock denuded the land of what little vegetation remained. As a result, nearly 250,000 people and 3.5 million cattle died of starvation, and the adjacent Sahara expanded southward as much as 150 km.

The tragedy of the Sahel and prolonged droughts in other desert fringe areas serve to remind us of the delicate equilibrium of ecosystems in such regions. Once the fragile soil cover has been removed by erosion, it takes centuries for new soil to form (see Chapter 5).

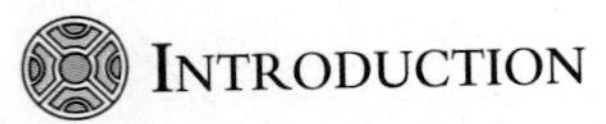

INTRODUCTION

Most people associate the work of wind with deserts. Wind is an effective geologic agent in desert regions, but it also plays an important role wherever loose sediment can be eroded, transported, and deposited, such as along shorelines or on the plains (see Perspective 5-2). Therefore, we will first consider the work of wind in general and then will turn to the distribution, characteristics, and landforms of deserts.

SEDIMENT TRANSPORT BY WIND

Wind is a turbulent fluid and therefore transports sediment in much the same way as running water. Although wind typically flows at a greater velocity than water, it has a lower density and, thus, can carry only clay- and silt-sized particles as *suspended load*. Sand and larger particles are moved along the ground as *bed load*.

Bed Load

Sediments too large or heavy to be carried in suspension by water or wind are moved as bed load either by *saltation* or by rolling and sliding. As we discussed in Chapter 15, saltation is the process by which a portion of the bed load moves by intermittent bouncing along a stream bed. Saltation also occurs on land. The wind starts sand grains rolling and lifts and carries some grains short distances before they fall back to the surface. As the descending sand grains hit the surface, they strike other grains causing them to bounce along by saltation (▶ Figure 18-3). Wind tunnel experiments have shown that once sand grains begin moving, they will continue to move, even if the wind drops below the speed necessary to start them moving! This happens because once saltation begins, it sets off a chain reaction of collisions between sand grains that keeps the grains in constant motion.

Saltating sand usually moves near the surface, and even when winds are strong, grains are rarely lifted higher than about a meter. If the winds are very strong, these wind-whipped grains can cause extensive abrasion. A car's paint can be removed by sandblasting in a short time, and its windshield will become completely frosted and translucent from pitting.

▶ FIGURE 18-3 Most sand is moved near the ground surface by saltation. Sand grains are picked up by the wind and carried a short distance before falling back to the ground where they usually hit other grains, causing them to bounce and move in the direction of the wind.

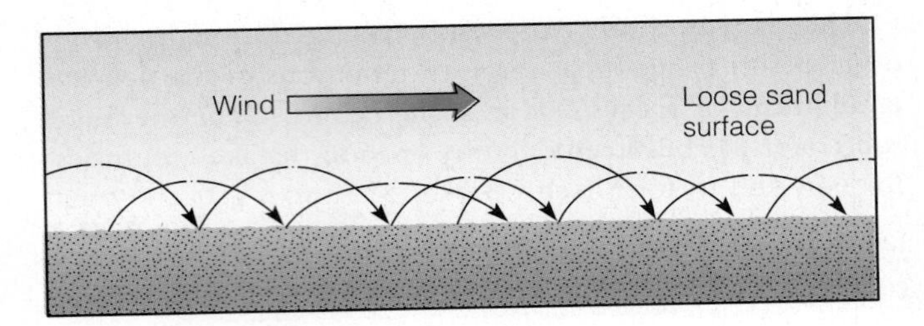

Particles larger than sand can also be moved along the ground by the process of *surface creep*. This type of movement occurs when saltating sand grains strike the larger particles and push them forward along the ground.

Suspended Load

Silt- and clay-sized particles constitute most of a wind's suspended load. Even though these particles are much smaller and lighter than sand-sized particles, wind usually starts the latter moving first. The reason for this phenomenon is that a very thin layer of motionless air lies next to the ground where the small silt and clay particles remain undisturbed. The larger sand grains, however, stick up into the turbulent air zone where they can be moved. Unless the stationary air layer is disrupted, the silt and clay particles remain on the ground providing a smooth surface. This phenomenon can be observed on a dirt road on a windy day. Unless a vehicle travels over the road, little dust is raised even though it is windy. When a vehicle moves over the road, it breaks the calm boundary layer of air and disturbs the smooth layer of dust, which is picked up by the wind and forms a dust cloud in the vehicle's wake.

In a similar manner, when a sediment layer is disturbed, silt- and clay-sized particles are easily picked up and carried in suspension by the wind, creating clouds of dust or even dust storms (see ▶ Figure 18-4). Once these fine particles are lifted into the atmosphere, they may be carried thousands of kilometers from their source. For example, large quantities of fine dust from the southwestern United States were blown eastward and fell on New England during the Dust Bowl of the 1930s (see Perspective 5-2).

WIND EROSION

Recall that streams and glaciers are effective agents of erosion, much more so than wind. Even in deserts, where wind is most effective, running water is still responsible for most erosional landforms, although stream channels are typically dry (see Figure 15-3). Nevertheless, wind action can still produce many distinctive erosional features and is an extremely efficient sorting agent.

Abrasion

Wind erodes material in two ways: abrasion and deflation. **Abrasion** involves the impact of saltating sand grains on an object and is analogous to sandblasting. The effects of abrasion are usually minor because sand, the most common agent of abrasion, is rarely carried more than 1 m above the surface. Rather than creating major erosional features, wind abrasion merely modifies existing features by etching, pitting, smoothing, or polishing. Thus, wind abrasion is most effective on soft sedimentary rocks.

➢ **FIGURE 18-4** A dust storm in Death Valley, California.

Ventifacts are a common product of wind abrasion; these are stones whose surfaces have been polished, pitted, grooved, or faceted by the wind (➢ Figure 18-5). If the wind blows from different directions, or if the stone is moved, the ventifact will have multiple facets. Ventifacts are most common in deserts, yet they can also form wherever stones are exposed to saltating sand grains, as on beaches in humid regions and some outwash plains in New England.

➢ **FIGURE 18-5** (*a*) A ventifact forms when wind-borne particles (1) abrade the surface of a rock (2) forming a flat surface. If the rock is moved, (3) additional flat surfaces are formed. (*b*) Large ventifacts lying on desert pavement in Death Valley National Monument, California.

Perspective 18-1

Evidence of Wind Activity on Mars

Data gathered by the two *Viking* landers (July and September 1976) as well as by the *Viking* and *Mariner 9* orbiters reveal that Mars is a planet with a complex geologic history involving volcanism, running water, and wind (see Chapter 20). Although the majority of Martian topographic features formed early in the planet's history by meteorite impacts, volcanism, and running water, many surface features were created by wind and are still forming.

One of the surprising features of Mars is the general lack of sand and sand dunes. While dunes do exist around the northern Martian ice caps, other areas of the Martian surface resemble a rocky volcanic desert. At the two *Viking* landing sites, the terrain varies from flat to rolling plains littered with rocks (▶ Figure 1). Evidence of wind activity can be seen in the form of linear deposits or streaks of fine-grained material on the lee (downwind) sides of most of the rocks, as well as small dunes and angular-faceted rocks similar to ventifacts and yardangs on Earth (Figure 1).

Large-scale dust storms are seasonal phenomena on Mars (▶ Figure 2) and occur during the southern summer when

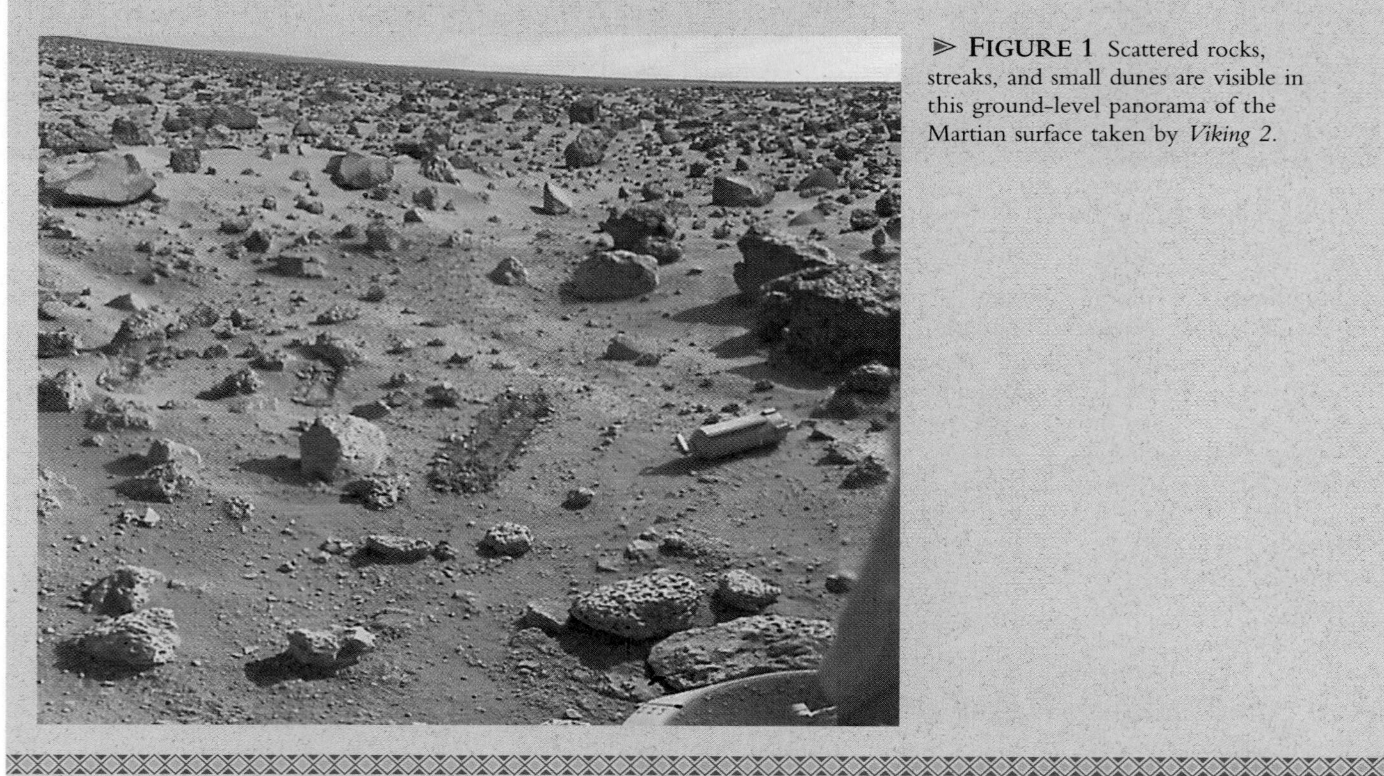

▶ **FIGURE 1** Scattered rocks, streaks, and small dunes are visible in this ground-level panorama of the Martian surface taken by *Viking 2*.

Yardangs are larger features than ventifacts and also result from wind erosion (▶ Figure 18-6). They are elongated and streamlined ridges that look like an overturned ship's hull. They are typically found grouped in clusters aligned parallel to the prevailing winds. They probably form by differential erosion in which depressions, parallel to the direction of wind, are carved out of a rock body, leaving sharp, elongated ridges. These ridges may then be further modified by wind abrasion into their characteristic shape. Although yardangs are fairly common desert features, interest in them was renewed when images radioed back from Mars showed that they are also widespread features on the Martian surface (see Perspective 18-1).

Deflation

Another important mechanism of wind erosion is **deflation,** which is the removal of loose surface sediment by the wind.

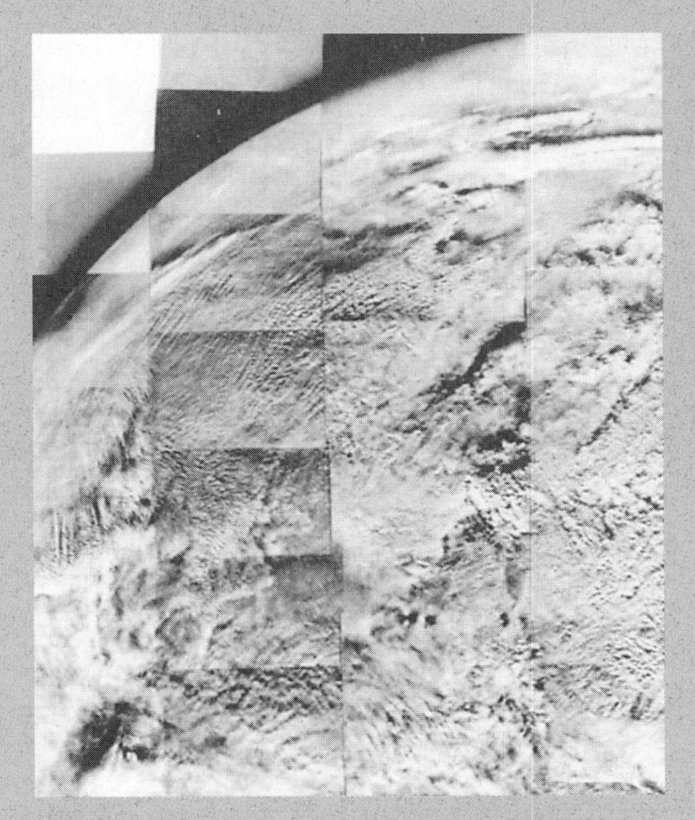

⊳ **FIGURE 2** A planetary dust storm obscured *Mariner 9*'s view of the Martian surface for the first few weeks after it went into orbit around Mars in 1971.

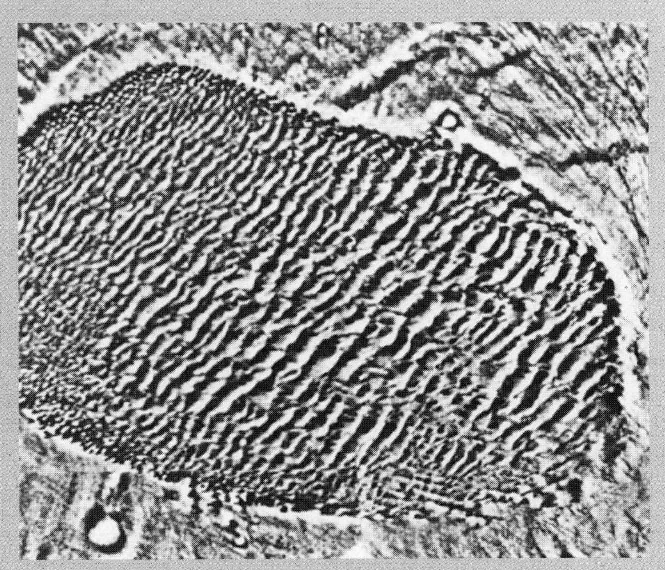

⊳ **FIGURE 3** Large dune fields surrounding the north polar ice cap are testimony to the incessant wind action occurring on Mars.

Mars is closest to the Sun. Once winds lift the fine particles into the atmosphere, they remain suspended for long periods. The dust raised by these storms may obscure the entire planet for weeks at a time. One result of these yearly dust storms is the seasonal change in the dust cover on the ground, which accounts for the seasonal color variations observed from Earth.

Although the atmosphere of Mars is too thin to suspend anything larger than dust-sized particles, sand-sized particles can be moved along the ground by saltation. Large dune fields composed of sand-sized particles have been discovered surrounding the north polar ice cap (⊳ Figure 3). The origin of these dunes is still controversial. Geologists think that most of the debris on the northern plains and the dunes themselves consist of material eroded from the polar deposits. When the deposits of dust-sized particles were removed by the wind, the sand-sized particles were left behind and were transported by saltation to form dunes.

Among the characteristic features of deflation in many arid and semiarid regions are **deflation hollows** (also called *blowouts*). These shallow depressions of variable dimensions result from differential erosion of surface materials (⊳ Figure 18-7). Ranging in size from several kilometers in diameter and tens of meters deep to small depressions only a few meters wide and less than a meter deep, deflation hollows are common in the southern Great Plains region of the United States. In addition, large areas measuring hundreds of square kilometers, such as the Qattara Depression in northwestern Egypt and some of Australia's interior desert basins, may also have formed primarily by deflation.

In many dry regions, the removal of sand-sized and smaller particles by wind leaves a surface of pebbles, cobbles, and boulders. As the wind removes the fine-grained material from the surface, the effects of gravity and occasional floodwaters rearrange the remaining coarse particles into a mosaic of close-fitting rocks called **desert pavement**

➢ **FIGURE 18-6** Profile view of a streamlined yardang in the Roman playa deposits of the Kharga Depression, Egypt. (Photo courtesy of Marion A. Whitney.)

(Figures 18-5b and ➢ 18-8). Once a desert pavement forms, it protects the underlying material from further deflation.

 WIND DEPOSITS

Although wind is of minor importance as an erosional agent, wind deposits can also form impressive structures. These deposits are primarily of two types. The first, called *dunes,* occur in several distinctive types, all of which consist of sand-sized particles that are usually deposited near their source. The second is *loess*, which consists of layers of windblown silt and clay that are deposited over large areas downwind and commonly far from their source.

➢ **FIGURE 18-7** A deflation hollow in Death Valley, California.

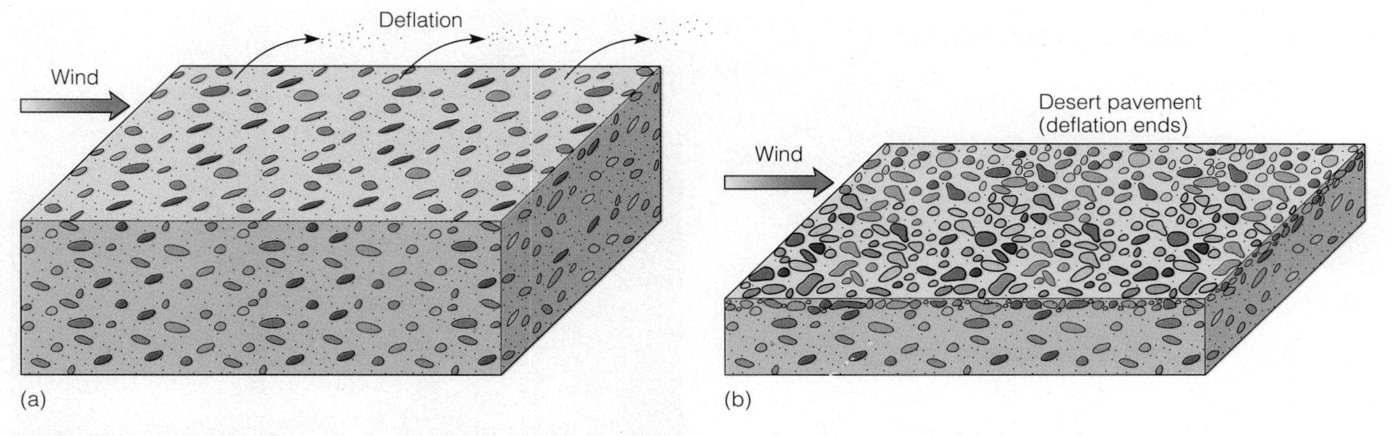

➤ **FIGURE 18-8** (*a*) Desert pavement forms when deflation removes fine-grained material from the ground surface leaving larger-sized particles. (*b*) As deflation continues and more material is removed, the larger particles are concentrated and form a desert pavement, which protects the underlying material from additional deflation.

The Formation and Migration of Dunes

The most characteristic features associated with sand-covered regions are **dunes**, which are mounds or ridges of wind-deposited sand. Dunes form when wind flows over and around an obstruction. This results in two zones of quiet air, called wind shadows, that form immediately in front of and behind the obstruction (➤ Figure 18-9). As saltating sand grains settle in these wind shadows, they begin to accumulate and build up a deposit of sand. As they grow, these sand deposits become self-generating in that they form ever larger wind barriers that further reduce the wind's velocity, forcing it to deposit any sand it carries.

Most dunes have an asymmetrical profile, with a gentle windward slope and a steeper downwind or leeward slope that is inclined in the direction of the prevailing wind (➤ Figure 18-10a). Sand grains move up the gentle windward slope by saltation and accumulate on the leeward side forming an angle between 30° and 34° from the horizontal, which is the angle of repose of dry sand. When this angle is exceeded by accumulating sand, the slope collapses, and the sand slides down the leeward slope, coming to rest at its base. As sand moves from a dune's windward side and periodically slides down its leeward slope, the dune slowly migrates in the direction of the prevailing wind (Figure 18-10b). When preserved in the geologic record, dunes help geologists determine the prevailing direction of ancient winds (➤ Figure 18-11).

➤ **FIGURE 18-9** (*a*) When wind flows around an obstacle, two wind shadows form, one in front of the obstacle and the other behind it. Sand accumulates in both of these wind shadows. (*b*) The accumulating sand forms a mound that may develop into a dune.

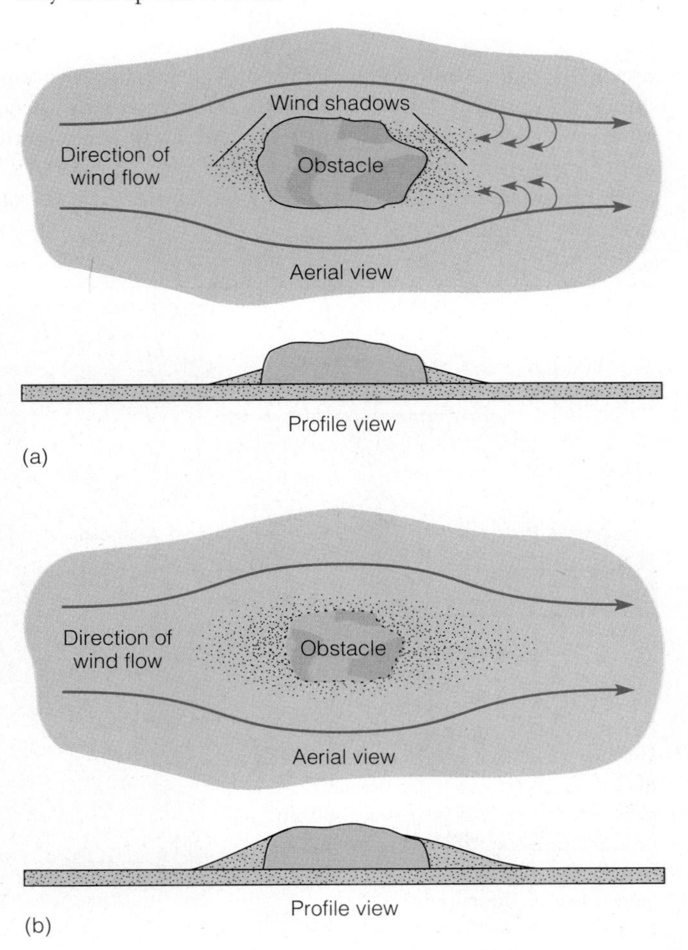

Dune Types

Four major dune types are generally recognized (*barchan*, *longitudinal*, *transverse*, and *parabolic*), although intermediate forms between the major types also exist. The size, shape, and arrangement of dunes result from the interaction of such factors as sand supply, the direction and velocity of the

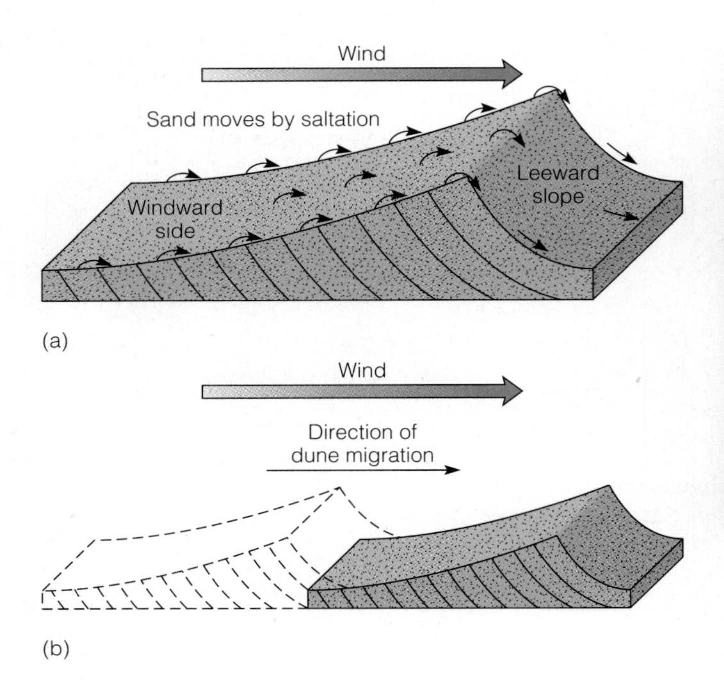

▶ **FIGURE 18-10** (*a*) Profile view of a sand dune. (*b*) Dunes migrate when sand moves up the windward side and slides down the leeward slope. Such movement of the sand grains produces a series of inclined beds that slope in the direction of wind movement.

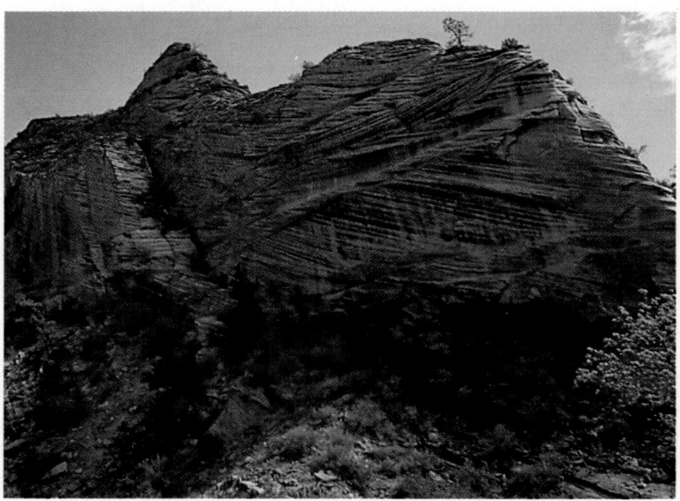

▶ **FIGURE 18-11** Cross-bedding in this sandstone in Zion National Park, Utah, helps geologists determine the prevailing direction of the wind that formed these ancient sand dunes.

prevailing wind, and the amount of vegetation. Although dunes are usually found in deserts, they can also occur wherever there is an abundance of sand such as along the upper parts of many beaches.

Barchan dunes are crescent-shaped dunes whose tips point downwind (▶ Figure 18-12). They form in areas where there is a generally flat, dry surface with little vegetation, a limited supply of sand, and a nearly constant wind direction. Most barchans are small, with the largest reaching about 30 m in height. Barchans are the most mobile of the major dune types, moving at rates that can exceed 10 m per year.

Longitudinal dunes (also called *seif dunes*) are long, parallel ridges of sand aligned generally parallel to the direction of the prevailing winds; they form where the sand

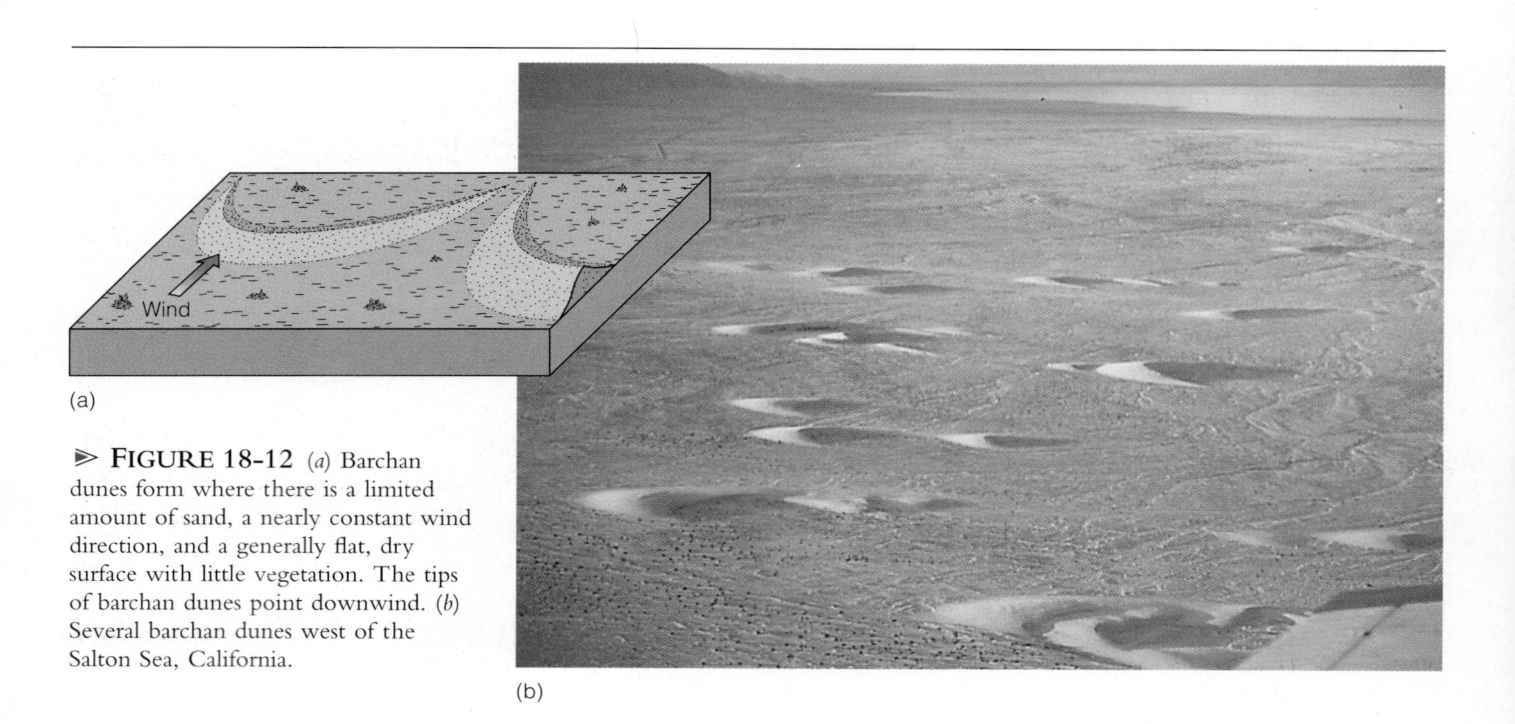

▶ **FIGURE 18-12** (*a*) Barchan dunes form where there is a limited amount of sand, a nearly constant wind direction, and a generally flat, dry surface with little vegetation. The tips of barchan dunes point downwind. (*b*) Several barchan dunes west of the Salton Sea, California.

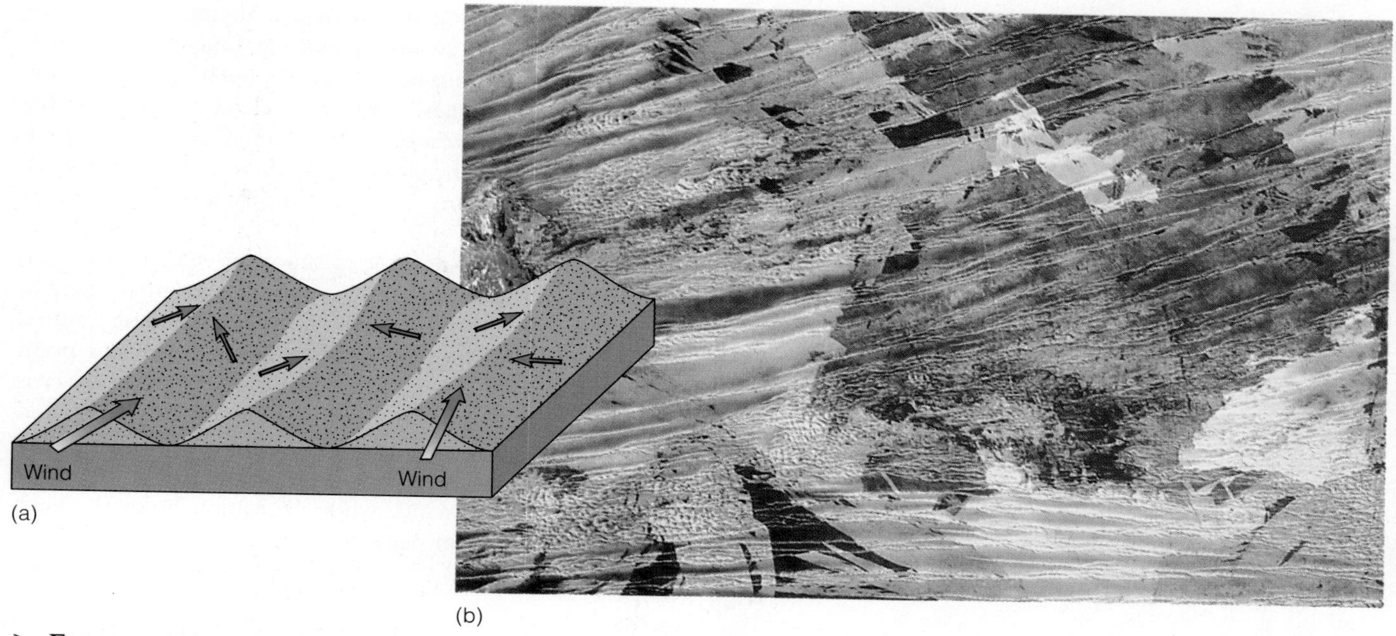

▶ **FIGURE 18-13** (*a*) Longitudinal dunes form long, parallel ridges of sand aligned roughly parallel to the prevailing wind direction. They typically form where sand supplies are limited. (*b*) 15 m-high longitudinal dunes in the Gibson Desert, west central Australia are shown in this image. The bright blue areas between the dunes are shallow pools of rainwater, while the darkest patches are areas where the Aborigines have set fires to encourage the growth of spring grasses.

supply is somewhat limited (▶ Figure 18-13). Longitudinal dunes result when winds converge from slightly different directions to produce the prevailing wind. They range in size from about 3 m to more than 100 m high, and some stretch for more than 100 km. These dunes are especially well developed in central Australia, where they cover nearly one-fourth of the continent. They also cover extensive areas in Saudi Arabia, Egypt, and Iran.

Transverse dunes form long ridges perpendicular to the prevailing wind direction in areas with abundant sand and little or no vegetation (▶ Figure 18-14). When viewed from the air, transverse dunes have a wavelike appearance

▶ **FIGURE 18-14** (*a*) Transverse dunes form long ridges of sand that are perpendicular to the prevailing wind direction in areas of little or no vegetation and abundant sand. (*b*) Aerial view of transverse dunes, Great Sand Dunes National Monument, Colorado.

▶ **FIGURE 18-15** Barchanoid dunes at White Sands National Monument, New Mexico.

and the areas covered by them are therefore sometimes called sand seas. The crests of transverse dunes can be as high as 200 m, and the dunes may be as much as 3 km wide. Some transverse dunes develop a clearly distinguishable barchan form and may separate into individual barchan dunes along the edges of the dune field where there is less sand. Such intermediate-form dunes are known as *barchanoid dunes* (▶ Figure 18-15).

Parabolic dunes are most common in coastal areas with abundant sand, strong onshore winds, and a partial cover of vegetation (▶ Figure 18-16). Although parabolic dunes have a crescent shape like barchan dunes, their tips point upwind. Parabolic dunes form when the vegetation cover is broken and deflation produces a blowout. As the wind transports the sand out of the depression, it builds up on the convex downwind dune crest. The central part of the dune is excavated by the wind, while vegetation holds the ends and sides fairly well in place.

Loess

Windblown silt and clay deposits composed of angular quartz grains, feldspar, micas, and calcite are known as **loess**. The distribution of loess shows that it is derived from three main sources: deserts, Pleistocene glacial outwash deposits, and the floodplains of rivers in semiarid regions. It must be stabilized by moisture and vegetation in order to accumulate. Consequently, loess is not found in deserts, even though they

▶ **FIGURE 18-16** (*a*) Parabolic dunes typically form in coastal areas where there is a partial cover of vegetation, a strong onshore wind, and abundant sand. (*b*) Parabolic dune developed along the Lake Michigan shoreline west of St. Ignace, Michigan.

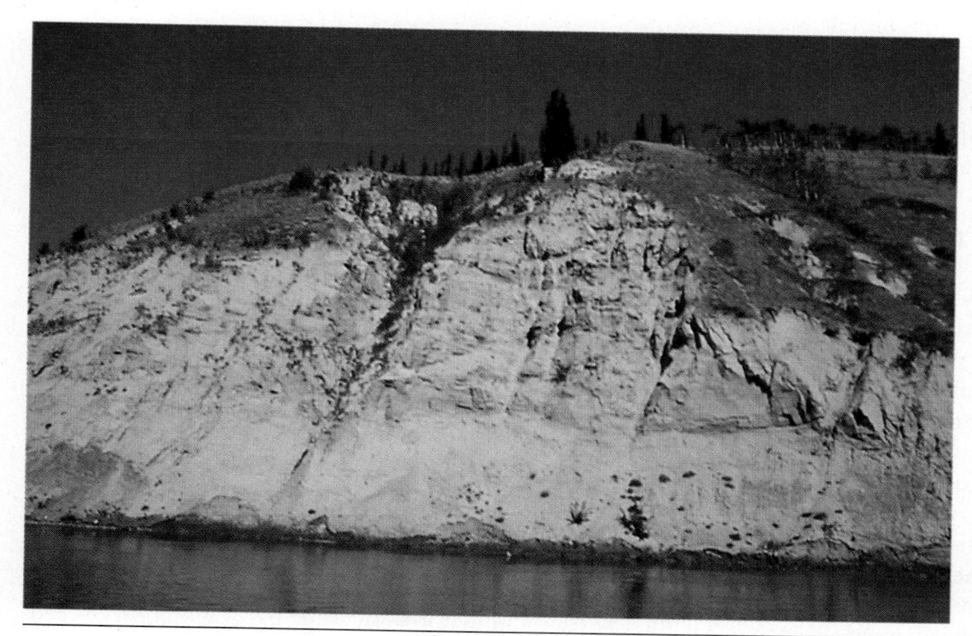

▶ FIGURE 18-17 These steep banks along the Yukon River, Yukon Territory, Canada, are formed of loess.

provide much of its material. Because of its unconsolidated nature, loess is easily eroded, and as a result, eroded loess areas are characterized by steep cliffs and rapid lateral and headward stream erosion (▶ Figure 18-17).

At present, loess deposits cover approximately 10% of the Earth's land surface and 30% of the United States (▶ Figure 18-18). The most extensive and thickest loess deposits occur in northeast China where accumulations greater than 30 m are common. The extensive deserts in central Asia are the source for this loess. Other important loess deposits are on the North European Plain from Belgium eastward to Ukraine, and in Central Asia, and in the pampas of Argentina. In the United States, they occur in the Great Plains, the Midwest, the Mississippi River Valley, and eastern Washington.

▶ FIGURE 18-18 The distribution of the Earth's major loess-covered areas.

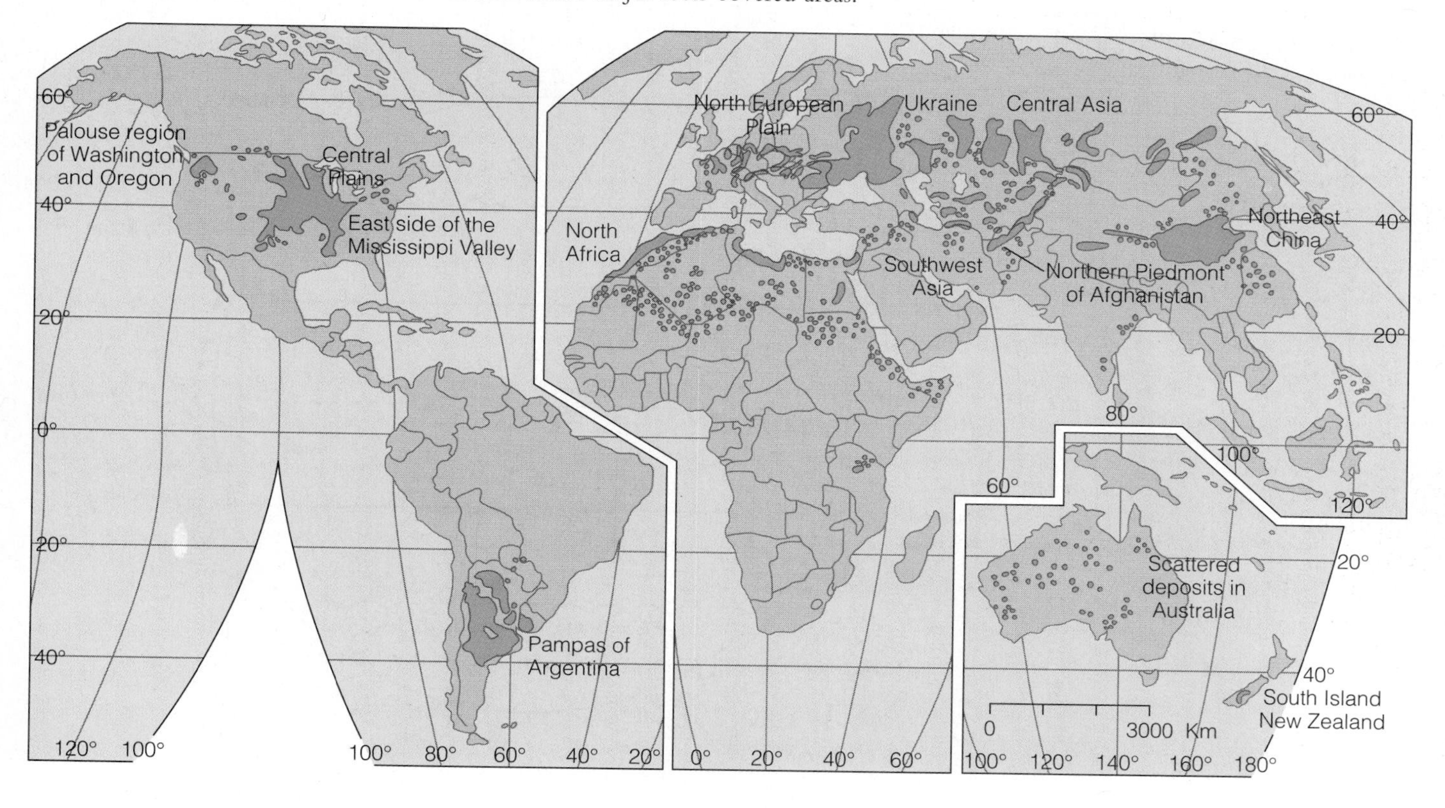

Loess-derived soils are among the most fertile in the world. It is not surprising that the world's major grain-producing regions correspond to the distribution of large loess deposits such as the North European Plain, Ukraine, and the Great Plains of North America.

 AIR PRESSURE BELTS AND GLOBAL WIND PATTERNS

To understand the work of wind and the distribution of deserts, we need to consider the global pattern of air pressure belts and winds, which are responsible for the Earth's atmospheric circulation patterns. Air pressure is the density of air exerted on its surroundings (that is, its weight). When air is heated, it expands and rises, reducing its mass for a given volume and causing a decrease in air pressure. Conversely, when air is cooled, it contracts and air pressure increases. Therefore, those areas of the Earth's surface that receive the most solar radiation, such as the equatorial regions, have low air pressure, while the colder areas, such as the polar regions, have high air pressure.

Air flows from high-pressure zones to low-pressure zones. If the Earth did not rotate, winds would move in a straight line from one zone to another. Because the Earth rotates, winds are deflected to the right of their direction of motion (clockwise) in the Northern Hemisphere and to the left of their direction of motion (counterclockwise) in the Southern Hemisphere. Such deflection of air between latitudinal zones resulting from the Earth's rotation is known as the **Coriolis effect**. Therefore, the combination of latitudinal pressure differences and the Coriolis effect produces a worldwide pattern of east-west–oriented wind belts (▶ Figure 18-19).

The Earth's equatorial zone receives the most solar energy, which heats the surface air, causing it to rise. As the air rises, it cools and releases moisture that falls as rain in the equatorial region (Figure 18-19). The rising air is now much drier as it moves northward and southward toward each pole. By the time it reaches 20° to 30° north and south latitude, the air has become cooler and denser and begins to descend. Compression of the atmosphere warms the descending air mass and produces a warm, dry, high-pressure area, providing the perfect conditions for the formation of the low-latitude deserts of the Northern and Southern hemispheres (▶ Figure 18-20).

THE DISTRIBUTION OF DESERTS

Dry climates occur in the low and middle latitudes. In these climates, the potential loss of water by evaporation exceeds the yearly precipitation (Figure 18-20). Dry climates cover 30% of the Earth's land surface and are subdivided into *semiarid* and *arid* regions. Semiarid regions receive more precipitation than arid regions, yet are moderately dry. Their soils are usually well developed and fertile and support a natural grass cover. Arid regions, generally described as **deserts**, are very dry; they receive, on average, less than 25 cm of rain per year, typically have poorly developed soils, and are mostly or completely devoid of vegetation.

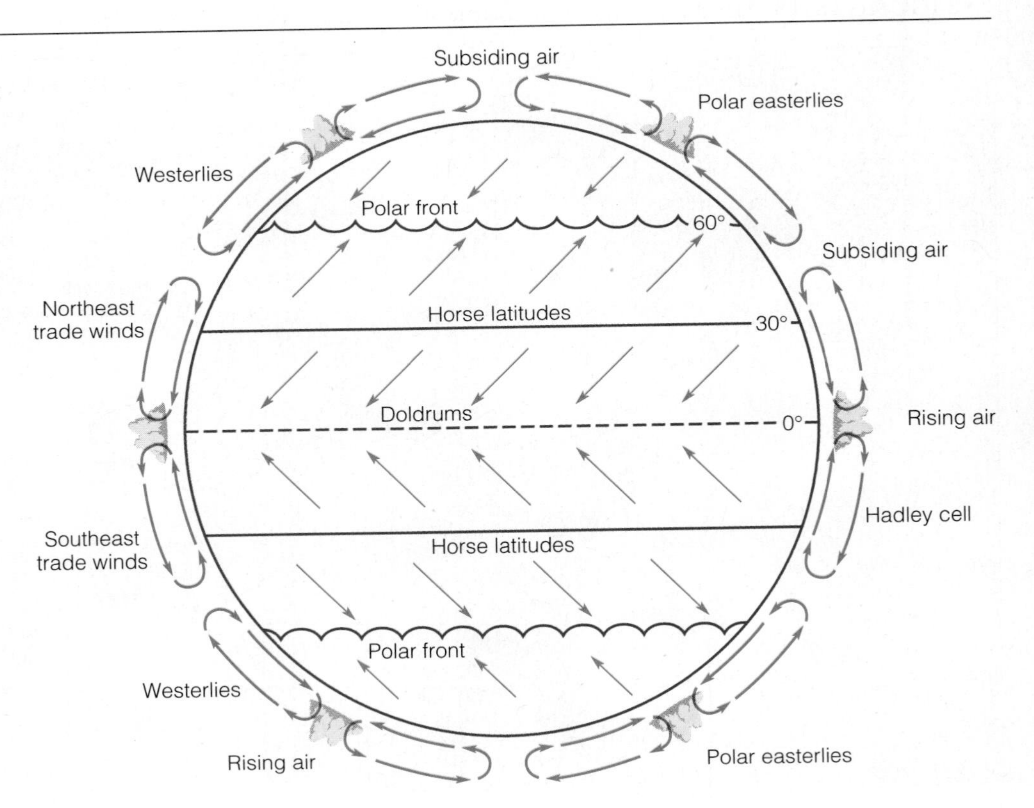

▶ **FIGURE 18-19** The general circulation of the Earth's atmosphere.

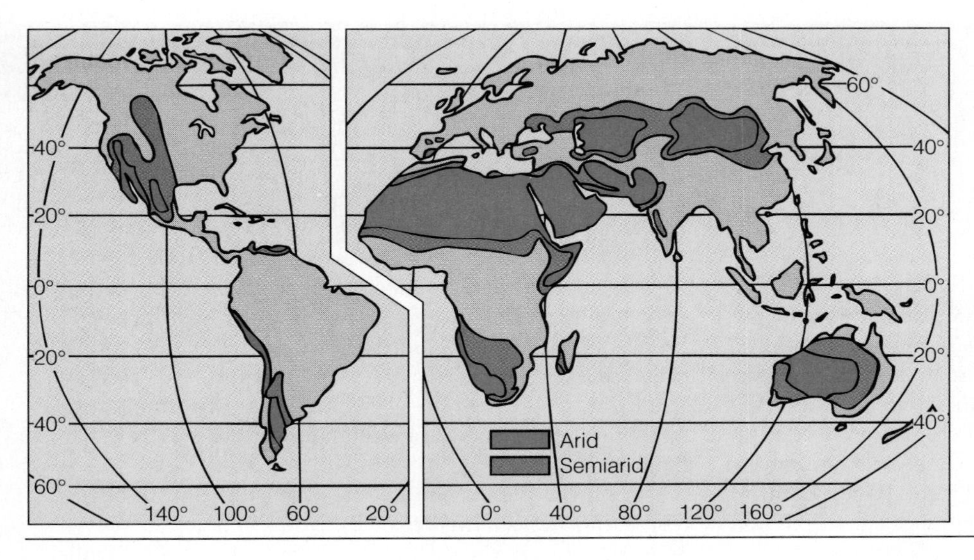

▶ **FIGURE 18-20** The distribution of the Earth's arid and semiarid regions.

The majority of the world's deserts are found in the dry climates of the low and middle latitudes (Figure 18-20). In North America, most of the southwestern United States and northern Mexico are characterized by this hot, dry climate, while in South America this climate is primarily restricted to the Atacama Desert of coastal Chile and Peru. The Sahara in northern Africa and, the Arabian Desert in the Middle East, along with the majority of Pakistan and western India, form the largest essentially unbroken desert environment in the Northern Hemisphere. More than 40% of Australia is desert, and most of the rest of it is semiarid. It is no wonder that it is called the "desert continent."

The remaining dry climates of the world are found in the middle and high latitudes, mostly within continental interiors in the Northern Hemisphere (Figure 18-20). Many of these areas are dry because of their remoteness from moist maritime air and the presence of mountain ranges that produce a **rainshadow desert** (▶ Figure 18-21). When moist marine air moves inland and meets a mountain range, it is forced upward. As the air rises, it cools, forming clouds and producing precipitation that falls on the windward side of the mountains. The air that descends on the leeward side of the mountain range is much warmer and drier, producing a rainshadow desert.

Three widely separated areas are included within the mid-latitude dry climate zone (Figure 18-20). The largest of these is the central part of Eurasia extending from just north of the Black Sea eastward to north-central China. The Gobi Desert in China is the largest desert in this region. The Great Basin area of North America is the second largest mid-latitude dry climate zone and results from the rainshadow produced by the Sierra Nevada (see Perspective 18-2). This region adjoins the southwestern deserts of the United States that formed as a result of the low-latitude subtropical high-pressure zone. The smallest of the mid-latitude dry climate areas is the Patagonian region of southern and western Argentina. Its dryness results from the rainshadow effect of the Andes. The remainder of the world's deserts are found in the cold, but dry high latitudes, such as Antarctica.

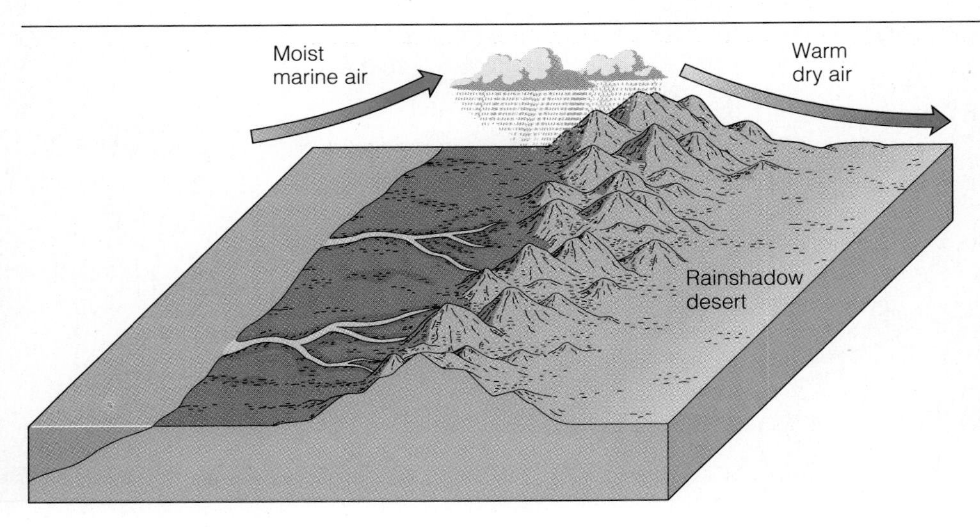

▶ **FIGURE 18-21** Many deserts in the middle and high latitudes are rainshadow deserts, so named because they form on the leeward side of mountain ranges. When moist marine air moving inland meets a mountain range, it is forced upward where it cools and forms clouds that produce rain. This rain falls on the windward side of the mountains. The air descending on the leeward side is much warmer and drier, producing a rainshadow desert.

Perspective 18-2

DEATH VALLEY NATIONAL MONUMENT

Death Valley National Monument was established in 1933 and encompasses 7,700 km^2 of southeastern California and part of western Nevada (▶ Figure 1). The hottest, driest, and lowest of the U.S. National Monuments and Parks, it receives less than 5 cm of rain per year and features normal daytime summer temperatures above 42°C. The highest temperature ever recorded was 57°C in the shade! The topographic relief in Death Valley is impressive. Telescope Peak near the southwestern border is 3,368 m high, while the lowest point in the Western Hemisphere—86 m below sea level—is less than 32 km to the east at Badwater.

Within Death Valley and its bordering mountains are excellent examples of a wide variety of desert landforms and economically valuable evaporite deposits. In addition, numerous folds, faults and landslides, and considerable evidence of volcanic activity can be seen.

The geologic history of Death Valley is complex and is still being worked out, but rocks from every geologic era can be found in the valley or the surrounding mountains. Although the geologic history of the region reaches back to the Precambrian, Death Valley itself formed less than 4 million years ago.

Death Valley formed during the Pliocene Epoch, when the Earth's crust was stretched and rifted, forming various horsts and grabens. Death Valley continues to subside along normal faults and is sinking most rapidly along its western side. This movement has been so great that more than 3,000 m of sediments are buried beneath the present valley floor.

During the Pleistocene Epoch, when the climate of this region was more humid than it is today, numerous pluvial lakes spread over the valley (see Chapter 17). Lake Manly, the largest of these pluvial lakes (145 km long and 178 m deep), dried up about 10,000 years ago, when the climate became arid.

Volcanic activity has also been occurring during the last several thousand years. The most famous volcanic feature in Death Valley is Ubehebe Crater, an explosion crater that formed approximately 2,000 years ago (▶ Figure 2).

In addition to the usual desert features, Death Valley also includes some unusual ones such as the Devil's Golf Course,

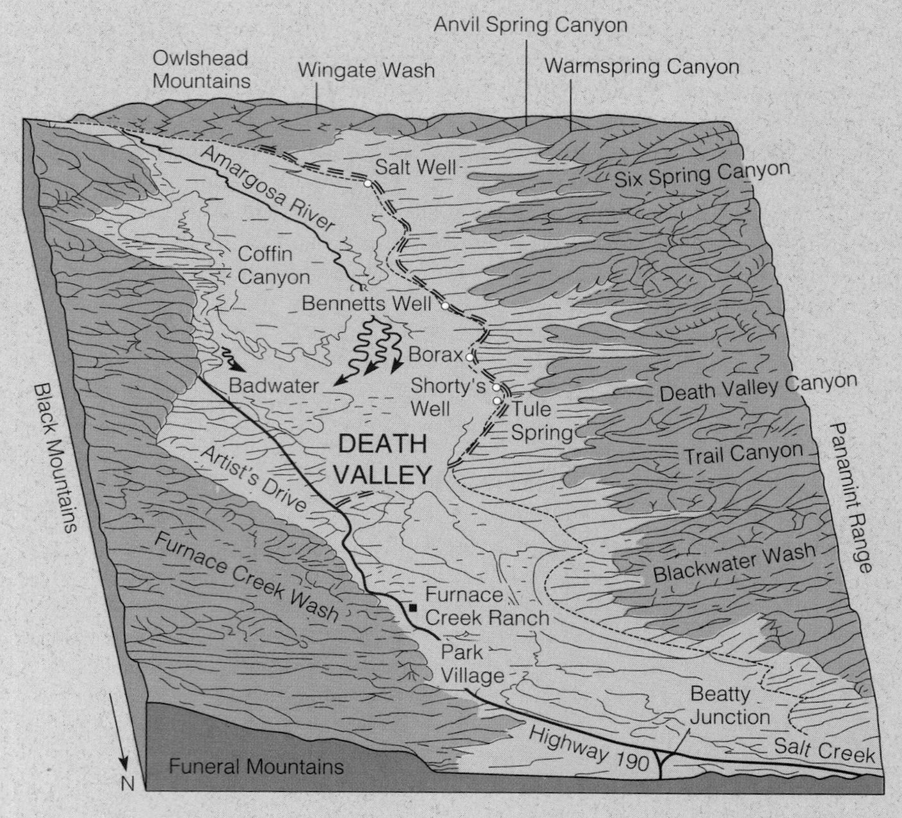

▶ FIGURE 1 Death Valley National Monument, California, encompasses 7,700 km^2 of southeastern California and part of western Nevada.

FIGURE 2 Ubehebe Crater, an explosion crater, last erupted approximately 2,000 years ago.

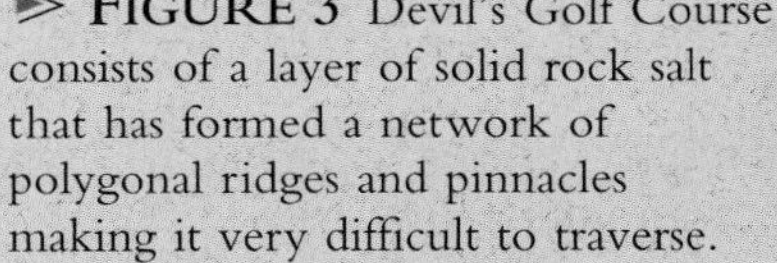

FIGURE 3 Devil's Golf Course consists of a layer of solid rock salt that has formed a network of polygonal ridges and pinnacles making it very difficult to traverse.

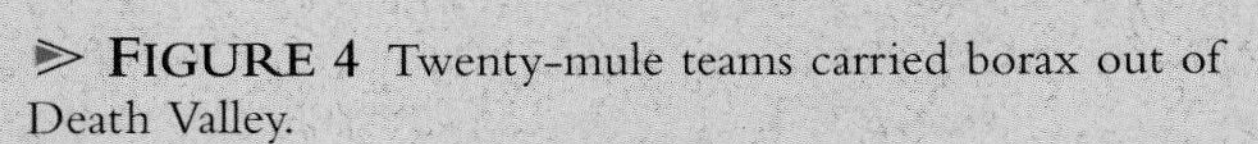

FIGURE 4 Twenty-mule teams carried borax out of Death Valley.

a bed of solid rock salt displaying polygonal ridges and pinnacles that are almost impossible to traverse (Figure 3). The Harmony Borax Works was home to the famous 20-mule teams that hauled out countless wagons of borax (Figure 4). The borax, used for ceramic glazes, fertilizers, glass, solder, and pharmaceuticals, was leached from volcanic ash by hot groundwater and then accumulated in layers of lake sediment.

Besides the numerous geologic features that have made Death Valley famous, it is also home to more than 600 species of plants as well as numerous animal species.

▶ **FIGURE 18-22** Sonoran Desert. Desert vegetation is typically sparse, widely spaced, and characterized by slow growth rates. These cacti are an excellent example of the type of vegetation that has adapted to the harsh desert environment.

 CHARACTERISTICS OF DESERTS

To people who live in humid regions, deserts may seem stark and inhospitable. Instead of a landscape of rolling hills and gentle slopes with an almost continuous cover of vegetation, deserts are dry, have little vegetation, and consist of nearly continuous rock exposures or sand dunes. And yet despite the great contrast between deserts and more humid areas, the same geologic processes are at work, only operating under different climatic conditions.

Temperature, Precipitation, and Vegetation

The heat and dryness of deserts are well known. Many of the deserts of the low latitudes have average summer temperatures that range between 32° and 38°C. It is not uncommon for some low-elevation inland deserts to record daytime highs of 46° to 50°C for weeks at a time. The highest temperature ever recorded was 58°C in El Azizia, Libya, on September 13, 1922.

During the winter months when the angle of the Sun is lower and there are fewer daylight hours, daytime temperatures average between 10° and 18°C. Winter nighttime lows can be quite cold, with frost and freezing temperatures common in the more poleward deserts. Winter daily temperature fluctuations in low-latitude deserts are among the greatest in the world, ranging between 18° and 35°C. Temperatures have been known to fluctuate from below 0°C to more than 38°C in a single day!

The dryness of the low-latitude deserts results primarily from the year-round dominance of the subtropical high-pressure belt, while the dryness of the mid-latitude deserts is due to their isolation from moist marine winds and the rainshadow effect created by mountain ranges. The dryness of both is further accentuated by their high temperatures.

Although deserts are defined as regions receiving, on average, less than 25 cm of rain per year, the amount of rain that falls each year is very unpredictable and unreliable. It is not uncommon for an area to receive more than an entire year's average rainfall in one cloudburst and then to receive very little rain for several years. Thus, yearly rainfall averages can be quite misleading.

Although the driest deserts, or those with large areas of shifting sand, are mostly or completely devoid of vegetation, most deserts can support a sparse plant community in which vegetation is typically small, widely spaced, and characterized by slow growth rates (▶ Figure 18-22). The stems and leaves of desert plants are usually hard and waxy to minimize water loss by evaporation and protect the plant from sand erosion. Most plants have a widespread shallow root system to absorb the dew that forms each morning in all but the driest deserts and to help anchor the plant in what little soil there may be. In extreme cases, many plants lie dormant during particularly dry years and spring to life after the first rain shower with a beautiful profusion of flowers.

Weathering and Soils

Mechanical weathering is dominant in desert regions. Daily temperature fluctuations and frost wedging are the primary forms of mechanical weathering (see Chapter 5). The breakdown of rocks by roots and from salt crystal growth are of minor importance. Some chemical weathering does occur, but its rate is greatly reduced by aridity and the scarcity of organic acids produced by the sparse vegetation. Most chemical weathering takes place during the winter months when more precipitation occurs, particularly in the mid-latitude deserts.

An interesting feature seen in many deserts is a thin, red, brown, or black shiny coating on the surface of many rocks. This coating, called **rock varnish**, is composed of iron and manganese oxides (➢ Figure 18-23). Because many of the varnished rocks contain little or no iron and manganese oxides, the varnish is thought to result from either wind-blown iron and manganese dust that settles on the ground or from the precipitated waste of microorganisms.

Desert soils, if developed, are usually thin and patchy because the limited rainfall and the resultant scarcity of vegetation reduce the efficiency of chemical weathering and hence soil formation. Furthermore, the sparseness of the vegetative cover enhances wind and water erosion of what little soil actually forms.

➢ FIGURE 18-23 The shiny black coating on this rock exposed at Castle Valley, Utah, is rock varnish. It is composed of iron and manganese oxides.

Mass Wasting, Streams, and Groundwater

When traveling through a desert, most people are impressed by such wind-formed features as moving sand, sand dunes, and sand and dust storms. They may also notice the dry washes and dry stream beds. Because of the lack of running water, most people would conclude that wind is the most important erosional agent in deserts. They would be wrong. Running water, even though it occurs infrequently, causes most of the erosion. The dry conditions and sparse vegetation characteristic of deserts enhance water erosion. If you look closely, you will see the evidence of erosion and transportation by running water nearly everywhere.

Recall that most of a desert's average annual rainfall of 25 cm or less comes in brief, heavy, localized cloudbursts. During these times, considerable erosion occurs because the ground cannot absorb all of the rainwater. With so little vegetation to hinder its flow, runoff is rapid, especially on moderately to steeply sloping surfaces, resulting in flash floods and sheetflows. Dry stream channels quickly fill with raging torrents of muddy water and mudflows, which carve out steep-sided gullies and overflow their banks. During these times, a tremendous amount of sediment is rapidly transported and deposited far downstream.

While water is the major erosive agent in deserts today, recall that it was even more important during the Pleistocene Epoch when these regions were more humid (see Chapter 17). During that time, many of the major topographic features of deserts were forming. Today that topography is being modified by wind and infrequently flowing streams.

Most desert streams are poorly integrated and flow only intermittently. Many of them never reach the sea because the water table is usually far deeper than the channels of most streams, so they cannot draw upon groundwater to replace water lost to evaporation and absorption into the ground. This type of drainage in which a stream's load is deposited within the desert is called *internal drainage* and is common in most arid regions.

While the majority of deserts have internal drainage, some deserts have permanent through-flowing streams such as the Nile and Niger rivers in Africa, the Rio Grande and Colorado River in the southwestern United States, and the Indus River in Asia. These streams are able to flow through the desert region because their headwaters are well outside the desert and water is plentiful enough to offset losses resulting from evaporation and infiltration. Demands for greater amounts of water for agriculture and domestic use from the Colorado River, however, are leading to increased salt concentrations in its lower reaches and causing political problems between the United States and Mexico.

The water table in most desert regions is below the stream channels and is only recharged for a short time after a rainfall. In deserts with through-flowing streams, the water table slopes away from the streams. The through-flowing streams help to recharge the groundwater supply and can support vegetation along their banks. Trees, which have high moisture requirements, are rare in deserts, but may occasionally occur along the banks of both ephemeral and permanent streams, where their roots can reach the higher water table.

Wind

Although running water does most of the erosional work in deserts, wind can also be an effective geologic agent capable of producing a variety of distinctive erosional and depositional features. Wind is very effective in transporting and depositing unconsolidated sand, silt, and dust-sized particles. Contrary to popular belief, however, most deserts are not sand-covered wastelands, but rather consist of vast areas of rock exposures. Sand-covered regions, or sandy deserts, constitute less than 25% of the world's deserts. The sand in these areas has accumulated primarily by the action of wind.

 Desert Landforms

Because of differences in temperature, precipitation, and wind, as well as the underlying rocks and recent tectonic events, the landforms in arid regions vary considerably. Although wind is an important geologic agent in deserts, many distinctive landforms are produced and modified by running water.

After an infrequent and particularly intense rainstorm, excess water that is not absorbed by the ground may accumulate in low areas and form **playa lakes** (▶ Figure 18-24a). These lakes are very temporary, lasting from a few hours to several months. Most of them are very shallow and have rapidly shifting boundaries as water flows in or leaves by evaporation and seepage into the ground. Furthermore, the water in playa lakes is often very saline.

When a playa lake evaporates, the dry lake bed is called a **playa** or *salt pan* and is characterized by mudcracks and precipitated salt crystals (Figure 18-24b). Salts in some playas are thick enough to be mined commercially. For example, borates have been mined in Death Valley, California, for more than a hundred years (see Perspective 18-2).

Other common features of deserts, particularly in the Basin and Range region of the United States (see Figure 13-26), are *alluvial fans* and *bajadas*. **Alluvial fans** form after a cloudburst, when sediment-laden streams flowing out from the generally straight, steep mountain fronts deposit their load on the relatively flat desert floor. Because there are no valley walls to contain it, the sediment spreads out laterally, forming a gently sloping and poorly sorted fan-shaped sedimentary deposit (▶ Figure 18-25). Alluvial fans are similar in origin and shape to deltas (see Chapter 15) but are formed entirely on land. Alluvial fans may coalesce to form a **bajada**. This broad alluvial apron typically has an undulating surface resulting from the overlap of adjacent fans (▶ Figure 18-26).

▶ **FIGURE 18-24** (*a*) Playa lake formed after a rainstorm filled Croneis Dry Lake, Mojave Desert, California. (*b*) Racetrack Playa, Death Valley, California. Inyo Mountains can be seen in the background.

➢ **FIGURE 18-25** Aerial view of an alluvial fan, Death Valley, California.

➢ **FIGURE 18-26** Coalescing alluvial fans forming a bajada at the base of the Black Mountains, Death Valley, California.

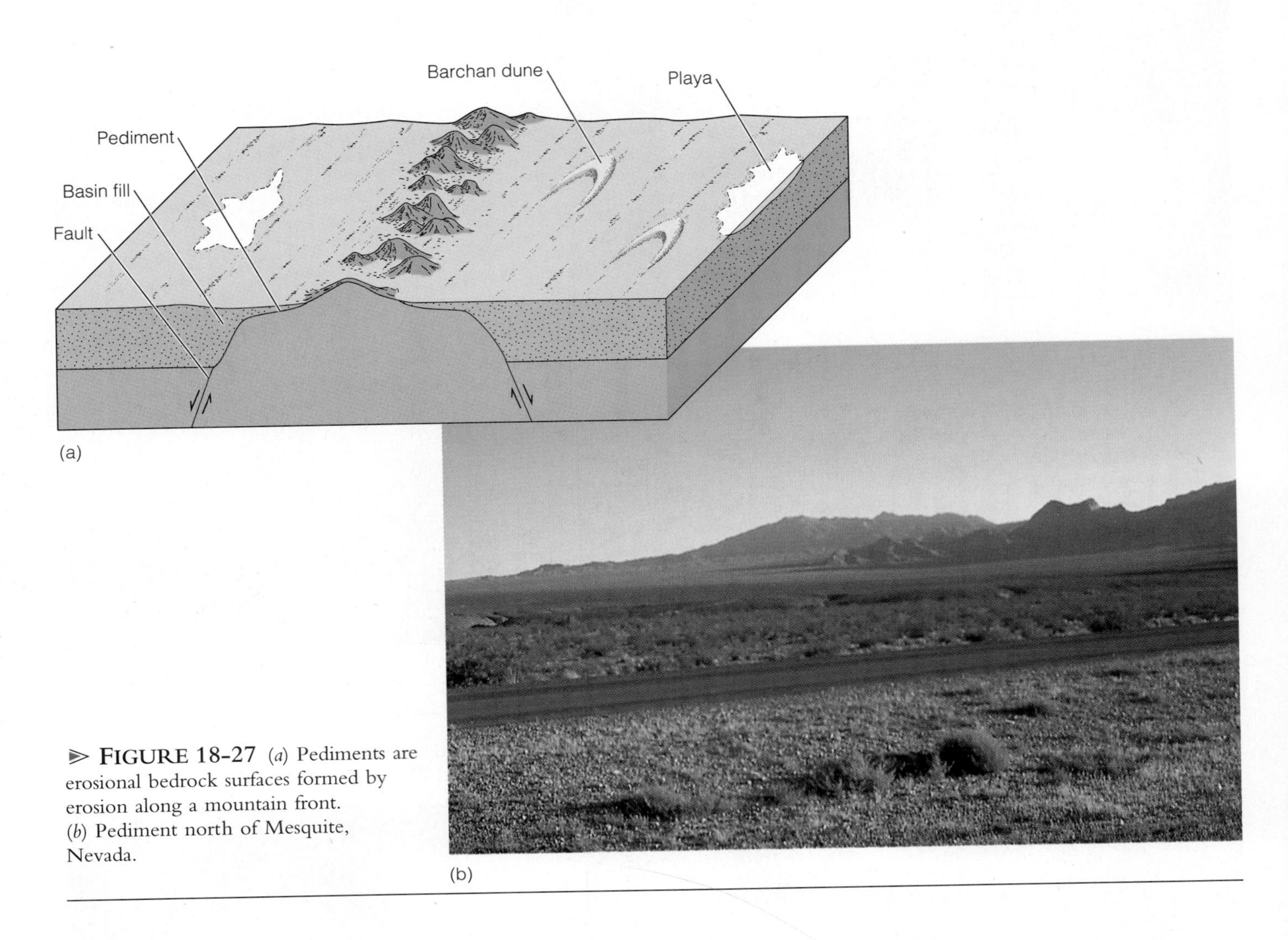

▶ **FIGURE 18-27** (*a*) Pediments are erosional bedrock surfaces formed by erosion along a mountain front. (*b*) Pediment north of Mesquite, Nevada.

Large alluvial fans and bajadas are frequently important sources of groundwater for domestic and agricultural use. Their outer portions are typically composed of fine-grained sediments suitable for cultivation, and their gentle slopes allow good drainage of water. Many alluvial fans and bajadas are also the sites of large towns and cities, such as San Bernardino, California, Salt Lake City, Utah, and Teheran, Iran.

Most mountains in desert regions, including those of the Basin and Range region, rise abruptly from gently sloping surfaces called pediments. **Pediments** are erosional bedrock surfaces of low relief that slope gently away from mountain bases (▶ Figure 18-27). Most pediments are covered by a thin layer of debris or by alluvial fans or bajadas.

The origin of pediments has been the subject of much controversy. Most geologists agree that they are erosional features developed on bedrock in association with the erosion and retreat of a mountain front (Figure 18-27a). The disagreement concerns how the erosion has occurred. Although not all geologists would agree, it appears that pediments are produced by the combined erosional activities of lateral erosion by streams, sheet flooding, and various weathering processes along the retreating mountain front. Thus, pediments grow at the expense of the mountains, and they will continue to expand as the mountains are eroded away or partially buried.

Rising conspicuously above the flat plains of many deserts are isolated steep-sided erosional remnants called **inselbergs**, a German word meaning "island mountain" (▶ Figure 18-28). Inselbergs have survived for a longer period of time than other mountains because of their greater resistance to weathering.

Other easily recognized erosional remnants common to arid and semiarid regions are mesas and buttes (▶ Figure 18-29). A **mesa** is a broad, flat-topped erosional remnant bounded on all sides by steep slopes. Continued weathering and stream erosion will form isolated pillar-like structures known as **buttes**. Buttes and mesas consist of relatively easily weathered sedimentary rocks capped by nearly horizontal, resistant rocks such as sandstone, limestone, or basalt. They form when the resistant rock layer is breached, allowing rapid erosion of the less resistant underlying sediment. One of the best-known areas of mesas and buttes in the United States is Monument Valley on the Arizona-Utah border.

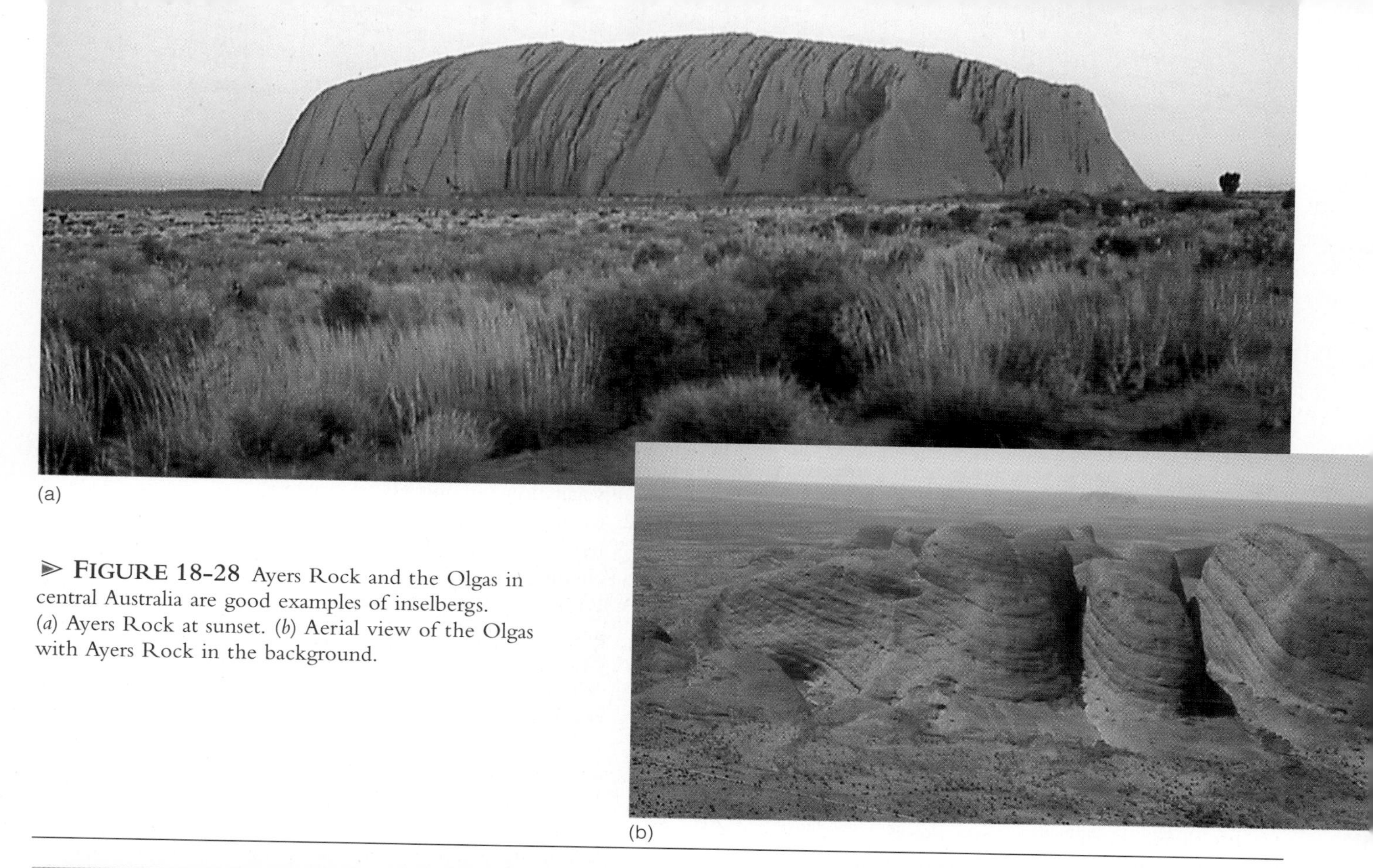

(a)

(b)

➢ **FIGURE 18-28** Ayers Rock and the Olgas in central Australia are good examples of inselbergs. (*a*) Ayers Rock at sunset. (*b*) Aerial view of the Olgas with Ayers Rock in the background.

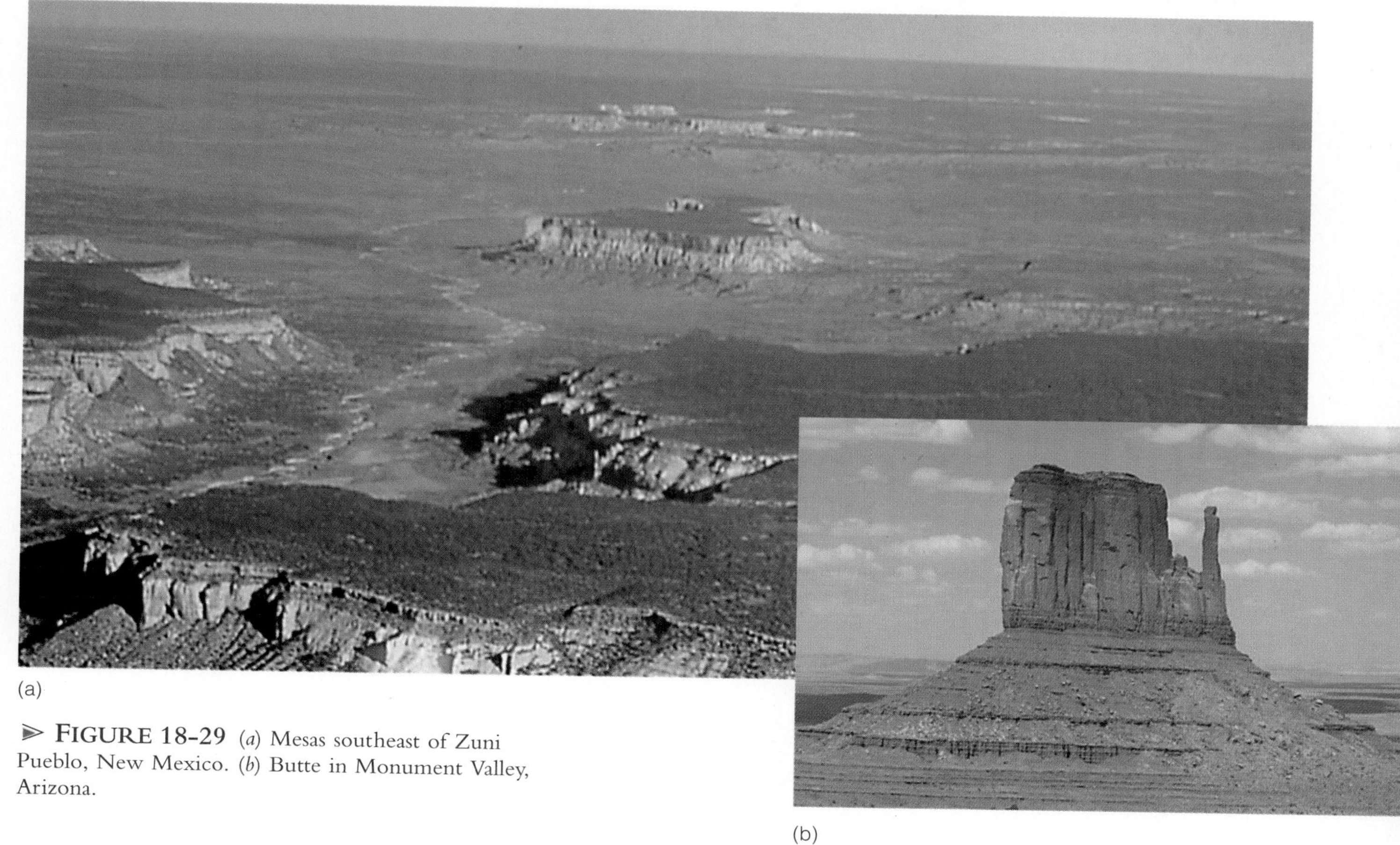

(a)

(b)

➢ **FIGURE 18-29** (*a*) Mesas southeast of Zuni Pueblo, New Mexico. (*b*) Butte in Monument Valley, Arizona.

Chapter Summary

1. Wind can transport sediment in suspension or by saltation and surface creep as part of the bed load.
2. Wind erodes material either by abrasion or deflation. Abrasion is a near-surface effect caused by the impact of saltating sand grains. Ventifacts are common wind-abraded features.
3. Deflation is the removal of loose surface material by the wind. Deflation hollows resulting from differential erosion of surface material are common features of many deserts, as is desert pavement, which effectively protects the underlying surface from additional deflation.
4. The two major wind deposits are dunes and loess. Dunes are mounds or ridges of wind-deposited sand, whereas loess is wind-deposited silt and clay.
5. The four major dune types are barchan, longitudinal, transverse, and parabolic. The amount of sand available, the prevailing wind direction, the wind velocity, and the amount of vegetation present determine which type will form.
6. Loess is derived from deserts, Pleistocene glacial outwash deposits, and river floodplains in semiarid regions. Loess covers approximately 10% of the Earth's land surface and weathers to a rich and productive soil.
7. Deserts are very dry (averaging less than 25 cm rain/year), have poorly developed soils, and are mostly or completely devoid of vegetation.
8. The winds of the major east-west–oriented air pressure belts resulting from rising and cooling air are deflected by the Coriolis effect. These belts help control the world's climate.
9. Dry climates are located in the low and middle latitudes where the potential loss of water by evaporation exceeds the yearly precipitation. Dry climates cover 30% of the Earth's surface and are subdivided into semiarid and arid regions.
10. The majority of the world's deserts are in the low-latitude dry climate zone between 20° and 30° north and south latitudes. Their dry climate results from a high-pressure belt of descending dry air. The remaining deserts are in the middle latitudes, where their distribution is related to the rainshadow effect, and in the dry polar regions.
11. Deserts are characterized by lack of precipitation and high evaporation rates. Furthermore, rainfall is unpredictable and, when it does occur, tends to be very intense and of short duration. As a consequence of such aridity, desert vegetation and animals are scarce.
12. Mechanical weathering is the dominant form of weathering in deserts. The sparse precipitation and slow rates of chemical weathering result in poorly developed soils.
13. Running water is the dominant agent of erosion in deserts and was even more important during the Pleistocene Epoch when present-day deserts were more humid.
14. Wind is an erosional agent in deserts and is very effective in transporting and depositing unconsolidated fine-grained sediments.
15. Important desert landforms include playas, which are dry lake beds; when temporarily filled with water, they form playa lakes. Alluvial fans are poorly sorted, fan-shaped sedimentary deposits that may coalesce to form bajadas.
16. Pediments are low relief erosional bedrock surfaces that gently slope away from mountain bases. The origin of pediments is controversial, although most geologists think that they form by the combined activities of lateral erosion by streams, sheet flooding, and various weathering processes.
17. Inselbergs are isolated steep-sided erosional remnants that rise above the surrounding desert plains. Buttes and mesas are, respectively, pinnacle-like and flat-topped erosional remnants with steep sides.

Important Terms

abrasion
alluvial fan
bajada
barchan dune
butte
Coriolis effect
deflation
deflation hollow
desert
desert pavement
desertification
dry climate
dune
inselberg
loess
longitudinal dune
mesa
parabolic dune
pediment
playa
playa lake
rainshadow desert
rock varnish
transverse dune
ventifact
yardang

Review Questions

1. Between what latitudes in both hemispheres do the driest deserts in the world occur?
   a. ____ 10° and 20°;
   b. ____ 20° and 30°;
   c. ____ 30° and 40°;
   d. ____ 40° and 60°;
   e. ____ 60° and 80°.
2. The Coriolis effect causes wind to be deflected:
   a. ____ to the right in the Northern Hemisphere and the left in the Southern Hemisphere;
   b. ____ to the left in the Northern Hemisphere and the right in the Southern Hemisphere;
   c. ____ only to the left for both hemispheres;
   d. ____ only to the right for both hemispheres;
   e. ____ not at all.
3. The primary process by which bed load is transported is:
   a. ____ suspension;
   b. ____ abrasion;
   c. ____ saltation;
   d. ____ precipitation;
   e. ____ answers (a) and (c).
4. Which particle size constitutes most of a wind's suspended load?
   a. ____ sand;
   b. ____ silt;
   c. ____ clay;
   d. ____ answers (a) and (b);
   e. ____ answers (b) and (c).
5. Which of the following is a feature produced by wind erosion?
   a. ____ playa;
   b. ____ loess;
   c. ____ dune;
   d. ____ yardang;
   e. ____ none of these.
6. What is the approximate angle of repose for dry sand?
   a. ____ 15°;
   b. ____ 25°;
   c. ____ 35°;
   d. ____ 45°;
   e. ____ 55°.
7. Which of the following is a crescent-shaped dune whose tips point downwind?
   a. ____ barchan;
   b. ____ longitudinal;
   c. ____ parabolic;
   d. ____ transverse;
   e. ____ barchanoid.
8. Which of the following dunes form long ridges of sand aligned roughly parallel to the direction of the prevailing wind?
   a. ____ barchan;
   b. ____ longitudinal;
   c. ____ parabolic;
   d. ____ transverse;
   e. ____ barchanoid.
9. Where are the thickest and most extensive loess deposits in the world?
   a. ____ United States;
   b. ____ pampas of Argentina;
   c. ____ Belgium;
   d. ____ Ukraine;
   e. ____ northeast China.
10. What is the primary cause of the dryness of low-latitude deserts?
    a. ____ rainshadow effect;
    b. ____ isolation from moist marine winds;
    c. ____ dominance of the subtropical high-pressure belt;
    d. ____ Coriolis effect;
    e. ____ all of these.
11. The dominant form of weathering in deserts is _____, desert vegetation is _____, and soils are _____:
    a. ____ mechanical, abundant, thick;
    b. ____ mechanical, sparse, thin;
    c. ____ mechanical, abundant, thin;
    d. ____ chemical, diverse, thick;
    e. ____ chemical, diverse, thin.
12. The major agent of erosion in deserts today is _______, and during the Pleistocene Epoch it was _______:
    a. ____ wind, running water;
    b. ____ wind, wind;
    c. ____ running water, wind;
    d. ____ running water, running water;
    e. ____ wind, glaciers.
13. The dry lake beds in many deserts are:
    a. ____ playas;
    b. ____ bajadas;
    c. ____ inselbergs;
    d. ____ pediments;
    e. ____ mesas.
14. Describe how the global distribution of air pressure belts and winds operates.
15. What are the two ways that sediments are transported by wind?
16. Describe the two ways that wind erodes. How effective an erosional agent is wind?
17. What is the difference between a ventifact and yardang? How do both form?
18. Why is desert pavement important in a desert environment?
19. How do sand dunes migrate?
20. Describe the four major dune types and the conditions necessary for their formation.
21. What is loess and why is it important?
22. Why are most of the world's deserts located in the low latitudes?
23. How are temperature, precipitation, and vegetation interrelated in desert environments?
24. What role does groundwater play in the internal drainage system of deserts?
25. Explain the difference between a butte and a mesa, and describe how they form.

Points to Ponder

1. Because much of the recent desertification has been greatly accelerated by human activities, is there anything we can do to slow the process or restore some type of equilibrium or buffer zone between encroaching deserts and adjacent productive lands?
2. Is it possible for the same dune types to form on Mars as form on Earth? What does that tell us about the atmosphere of Mars?
3. Because rainshadow deserts form on the leeward side of mountain ranges, and mountain ranges result from plate interactions, what features in the geologic record would you look for to identify rainshadow deserts?
4. Why are so many desert rock formations red?
5. If deserts are very dry regions in which mechanical weathering predominates, why are so many of their distinctive landforms the result of running water and not wind?

Additional Readings

Agnew, C., and A. Warren. 1990. Sand trap. *The Sciences* March/April: 14–19.

Brookfield, M. E., and T. S. Ahlbrandt. 1983. *Eolian sediments and processes.* New York: Elsevier Publishers.

Cooke, R. A., A. Warren, and A. Goudie. 1993. *Desert geomorphology.* London: UCL Press.

Dorn, R. I. 1991. Rock varnish. *American Scientist* 79, no. 6: 542–53.

Ellis, W. S. 1987. Africa's Sahel: The stricken land. *National Geographic* 172, no. 2: 140–79.

Greeley, R., and J. Iversen. 1985. *Wind as a geologic process.* Cambridge: Cambridge University Press.

Hunt, C. B. 1975. *Death Valley: Geology, ecology, archaeology.* Berkeley, Calif.: University of California Press.

MacKinnon, D. J., and P. S. Chavez, Jr. 1993. Observing dust storms from space. *Earth* 2, no. 3: 60–65.

Sheridan, D. 1981. *Desertification of the United States.* Washington, D.C.: Council on Environmental Quality.

Somerville, D. 1994. Into the Red Center. *Earth* 3, no. 1: 32–41.

Thomas, D. S. G., ed. 1989. *Arid zone geomorphology.* New York: Halsted Press.

Walker, A. S. 1982. Deserts of China. *American Scientist* 70, no. 4: 366–76.

Waters, T. 1993. Dunes. *Earth* 2, no. 1: 44–51.

Wells, S. G., and D. R. Haragan. 1983. *Origin and evolution of deserts.* Albuquerque, N. Mex.: University of New Mexico Press.

Whitney, M. A. 1985. Yardangs. *Journal of Geological Education* 33, no. 2: 93–96.

Chapter 19

SHORELINES AND SHORELINE PROCESSES

OUTLINE

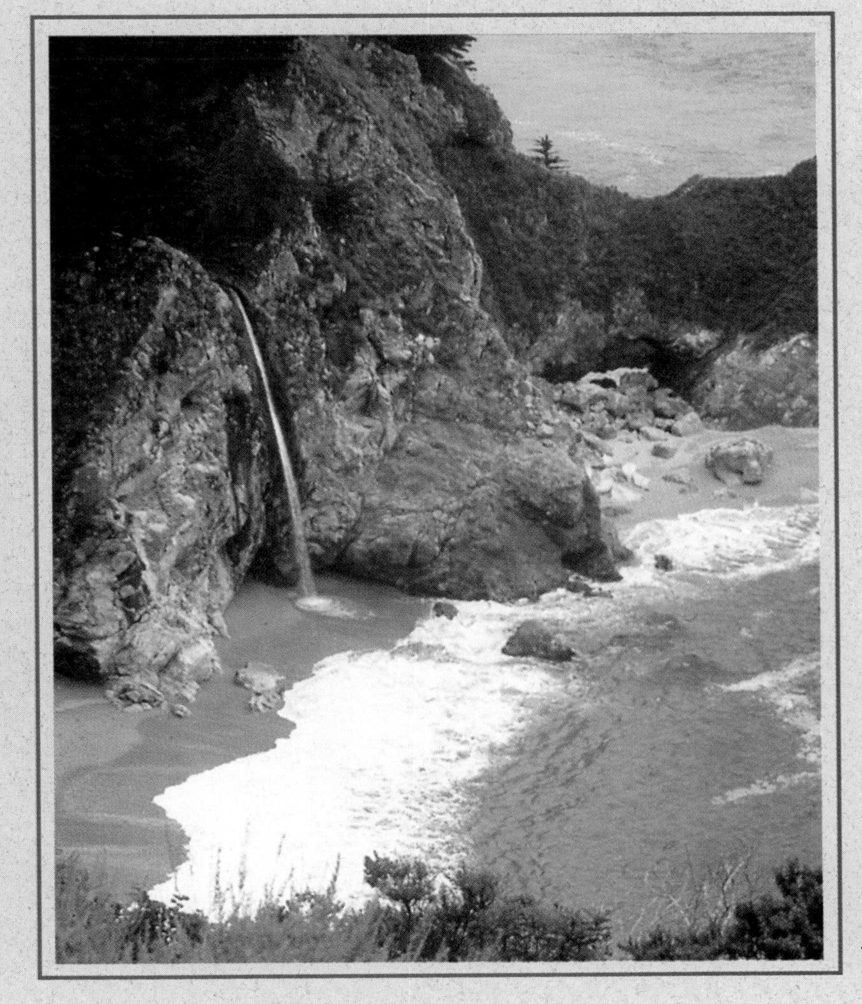

Beach in a small cove along California's Pacific Ocean shoreline. (Photo courtesy of Sue Monroe.)

Prologue

Although wind-generated waves, especially those formed during hurricanes, are responsible for most geologic work on shorelines, tsunami and landslide surges can have disastrous effects. As we explained in Chapter 9, tsunami are generated by fault displacement of the sea floor, submarine slides and slumps, and explosive volcanic eruptions.

In the open sea, tsunami may pass unnoticed, but when they enter shallow coastal waters, their wave height increases to as much as 65 m! The first indication of an approaching tsunami is a rapid withdrawal of the sea from coastal regions, followed a few minutes later by destructive waves. In many cases, tsunami come in as a rapidly rising tide, and their backwash, which undermines structures and carries loose objects out to sea, causes most of the damage and fatalities.

The largest waves occur in restricted bodies of water, such as bays or lakes, when water is suddenly displaced by large landslides or rockfalls. The largest of these so-called *landslide surges* occurred on July 9, 1958, in Lituya Bay, Alaska, when an estimated 30.5 million m^3 of rock plunged into the bay from a height of more than 900 m. The sudden displacement of water caused a surge on the opposite side of the bay that rose 536 m above sea level. The wave then moved out of the harbor into the open ocean where it quickly dissipated.

Of considerably more concern to shoreline dwellers is coastal flooding caused by storm-generated waves. Flooding during hurricanes is caused by large waves being driven onshore and by intense rainfall, more than 60 cm in 24 hours in some cases. In addition, as a hurricane moves over the ocean, low atmospheric pressure beneath the eye of the storm causes the ocean surface to bulge upward as much as 0.5 m. When the eye reaches the shoreline, the bulge coupled with wind-driven waves piles up in a *storm surge* that can rise several meters above normal high tide and inundate areas several kilometers inland.

Several coastal areas in the United States have been devastated by storm surges. In 1900, Galveston, Texas, a town of 38,000 on a long narrow barrier island a short distance from the mainland, was nearly destroyed. Storm waves surged inland and eventually covered the entire island killing between 6,000 and 8,000 people. To protect the city from future flooding, a colossal two-part project was begun in 1902. First, a huge seawall was constructed to protect the city from waves (▶ Figure 19-1). Next, parts of the city were raised to the level of the top of the seawall. Buildings were elevated and supported on jacks while sand fill was pumped beneath them.

More recently, in 1989, Charleston, South Carolina, and nearby areas were flooded by a storm surge generated by Hurricane Hugo (▶ Figure 19-2). This storm surge, which ranged from 2.5 to 6 m high, devastated an extensive area of shoreline, caused 21 deaths, and resulted in more than $7 billion in property damage. In fact, coastal flooding is responsible for most of the damage and about 90% of all fatalities attributed to hurricanes.

One of the greatest natural disasters of the twentieth century occurred in 1970 when a storm surge estimated at 8 to 10 m high flooded the low-lying coastal areas of Bangladesh, drowning 300,000 people. Since 1970, the

▶ **FIGURE 19-1** Construction of this seawall to protect Galveston, Texas, from storm waves began in 1902.

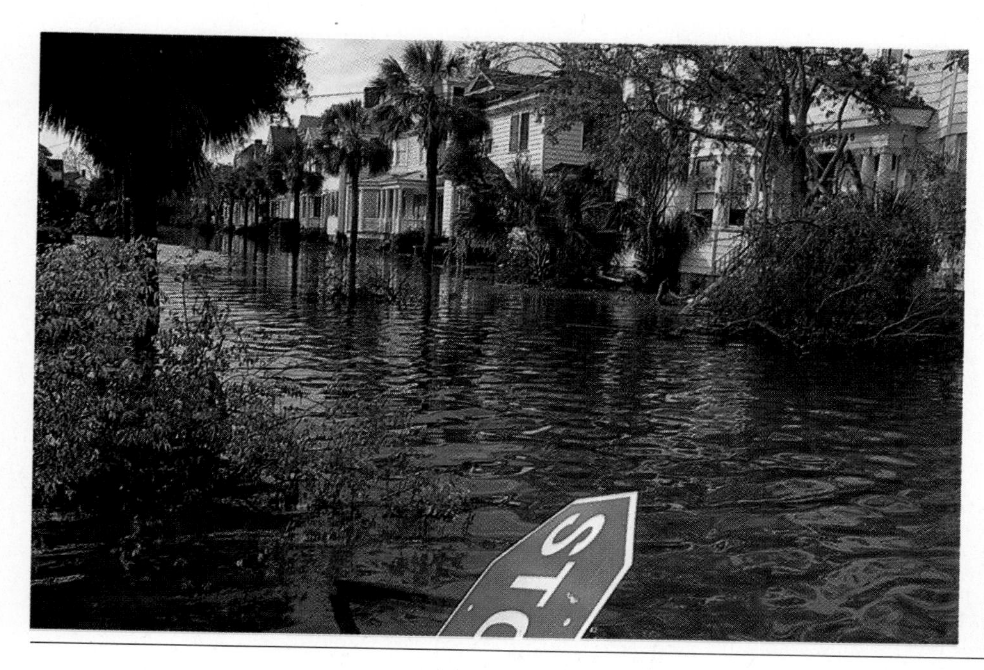

➢ **FIGURE 19-2** Charleston, South Carolina, was flooded by a storm surge produced by Hurricane Hugo on September 22, 1989.

coastal areas of Bangladesh have been flooded several more times, the most recent and most tragic being on April 30, 1991, when an estimated 139,000 people were drowned, and 10 million were left homeless.

The problem of coastal flooding during hurricanes is exacerbated by rising sea level (a topic that will be discussed in more detail in Perspective 19-2). To gain some perspective on the magnitude of the problem, consider this: Of the nearly $2 billion paid out by the federal government's National Flood Insurance Program since 1974, most has gone to owners of beachfront homes.

INTRODUCTION

Shorelines are the areas between low tide and the highest level on land affected by storm waves. In this chapter, we are concerned mostly with ocean shorelines where processes such as waves, nearshore currents, and tides continually modify existing shoreline features. Waves and nearshore currents are also effective geologic agents in large lakes, where the shorelines exhibit many of the same features present along seashores. The most notable differences are that waves and nearshore currents are more energetic on seashores, and even the largest lakes lack appreciable tides.

In contrast to other geologic agents such as running water, wind, and glaciers that operate over vast areas, shoreline processes are restricted to a narrow zone at any particular time. Shorelines migrate landward or seaward, however, depending on changes in sea level and uplift or subsidence of the coastal region. Although shorelines constitute an environment that is constantly changing, their appeal is so strong that about two-thirds of the world's population is concentrated in narrow bands adjacent to them. Many of the world's large cities such as New York, Los Angeles, New Orleans, Tokyo, London, Rio de Janeiro, and Shanghai are coastal cities.

The continents possess more than 400,000 km of shorelines. They vary from rocky, steep shorelines, such as those in Maine and much of the western United States and Canada, to those with broad sandy beaches as in eastern North America from New Jersey southward. Whatever their type, all shorelines involve a continual interplay between the energy levels of shoreline processes and the shoreline materials. In areas where energy levels are particularly high, erosion predominates and the shoreline may retreat landward. Where sediment supply from the land is great, deposition dominates. On shorelines with broad sandy beaches, beach sand is continually shifted from one area to another by waves and nearshore currents.

Living near a shoreline is appealing, but it is not without risks. In many parts of the world, including most of the United States and much of Canada, sea level is rising, and buildings that were built some distance from the ocean are now being undermined and destroyed (➢ Figure 19-3). Slumps and slides are common along rocky, steep shorelines; narrow offshore barrier islands migrate landward by erosion on their seaward sides and deposition on their landward sides; and hurricanes expend much of their fury on shorelines and coastal regions in general.

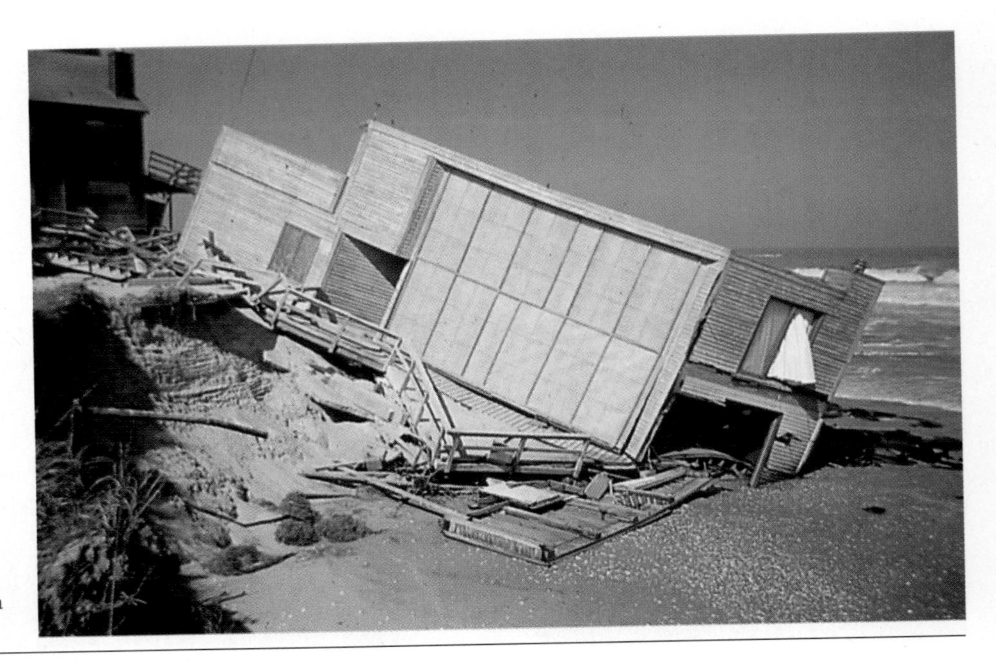

▶ FIGURE 19-3 Building damaged at Nags Head, North Carolina, during a storm in March 1989.

Scientists from several disciplines have contributed to our understanding of shorelines as dynamic systems. Elected officials and city planners of coastal communities must become familiar with shoreline processes so they can develop policies that serve the public as well as protect the fragile shoreline environment. In short, the study of shorelines is not only interesting, but has many practical applications.

WAVE DYNAMICS

Waves are oscillations of a water surface. They occur on all bodies of water, but are most significant in large lakes and the oceans where they serve as agents of erosion, transport, and deposition. Many of the erosional and depositional features of the world's shorelines form and are modified by the energy of incoming waves.

▶ Figure 19-4 shows a typical series of waves in deep water and the terminology applied to them. The highest part of a wave is its **crest,** and the low point between crests is the **trough. Wave length** is the distance between successive wave crests (or troughs), and **wave height** is the vertical distance from trough to crest. The speed at which a wave advances, generally called celerity (C), can be calculated if one knows the wave length (L) and the **wave period** (T), which is the time required for two successive wave crests (or troughs) to pass a given point:

$$C = L/T$$

The speed of wave advance (C) is actually a measure of the velocity of the wave form rather than a measure of the speed of the molecules of water. In fact, the water in waves moves forward and back as a wave passes but has little or no net forward movement. As waves move across a water surface, the water "particles" rotate in circular orbits and transfer energy in the direction of wave advance (Figure 19-4).

The diameters of the orbits followed by water particles in waves diminish rapidly with depth, and at a depth of about one-half wave length ($L/2$), called **wave base,** they are essentially zero (Figure 19-4). At depths exceeding wave base, the water and sea floor, or lake floor, are unaffected by surface waves. The significance of wave base will be explored more fully in later sections.

Wave Generation

Waves can be generated by several processes including displacement of water by landslides, displacement of the sea floor by faulting, and volcanic explosions. Most of the geologic work done on shorelines, however, occurs from wind-generated waves. When wind blows over water, some of its energy is transferred to the water, causing the water surface to oscillate. The mechanism that transfers energy from wind to water is related to the frictional drag resulting from one fluid (air) moving over another (water).

In an area of wave generation, as beneath a storm center at sea, sharp-crested, irregular waves called *seas* develop. Seas are an aggregate of waves of various heights and lengths, and one wave cannot be clearly distinguished from another. As seas move out from the area of wave generation, though, they are sorted into broad *swells* that have rounded, long crests and are all about the same size (▶ Figure 19-5).

As one would expect, the harder and longer the wind blows, the larger are the waves generated. Wind velocity and duration, however, are not the only factors controlling the size of waves. High-velocity wind blowing over a small pond will never generate large waves regardless of how long it blows. In fact, waves occur on ponds and most lakes only while the wind is blowing; once the wind stops, the water

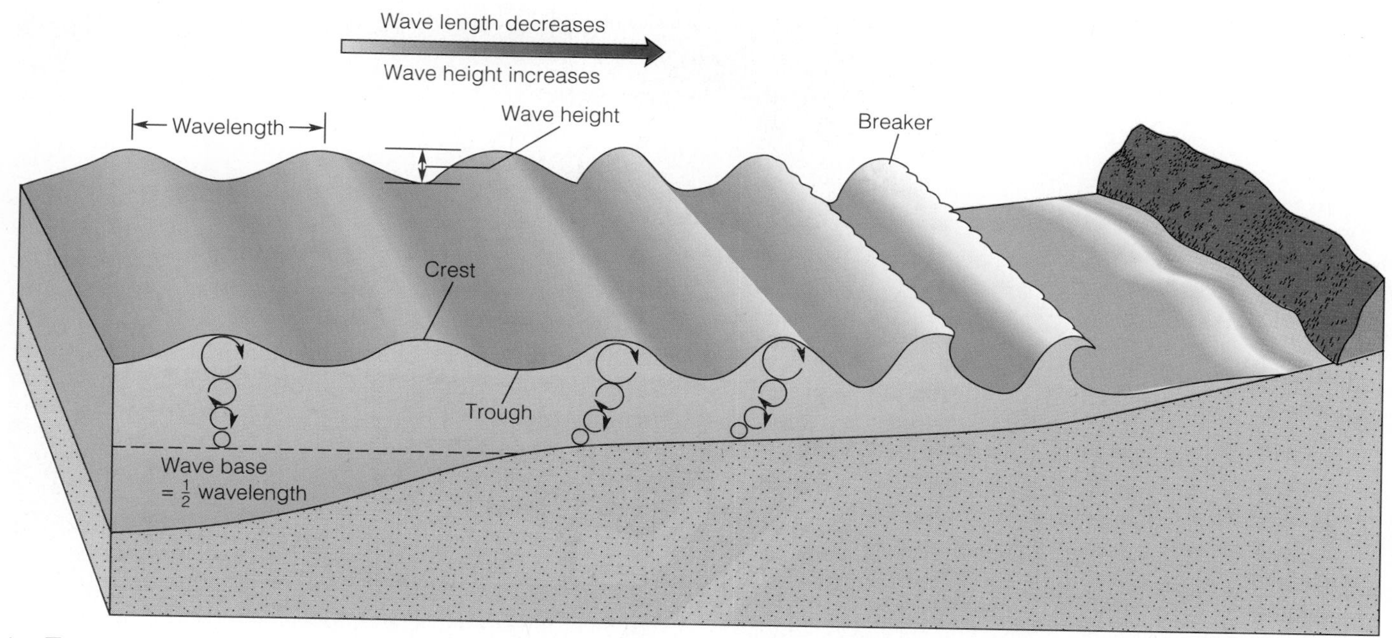

▶ **FIGURE 19-4** Waves and the terminology applied to them. The water in waves moves in circular orbits that decrease in size with depth. At wave base, which is at a depth of one-half wave length, water is not disturbed by surface waves. As deep water waves move toward shore, the orbital motion of water within them is disrupted when they enter water shallower than wave base. Wave length decreases while wave height increases, causing the waves to oversteepen and plunge forward as breakers.

quickly smooths out. In contrast, the surface of the ocean is always in motion, and waves with heights of 34 m have been recorded during storms.

The reason for the disparity between the wave sizes on ponds and lakes and on the oceans is the **fetch,** which is the distance the wind blows over a continuous water surface. The greater the fetch, the greater the size of the waves. Fetch is limited by the available water surface, so on ponds and lakes it corresponds to their length or width, depending on wind direction. A wind blowing the length of Lake Superior, for example, can generate large waves, but even larger ones develop in the oceans. To produce waves of greater length and height, more energy must be transferred from wind to water; hence large waves form beneath large storms at sea.

▶ **FIGURE 19-5** Small swells in the Atlantic Ocean near Massachusetts.

Guest Essay Orrin H. Pilkey

From Abyssal Plains to Beaches: A Geologist's Career

There was no question in my mind, as I wandered around old mine dumps in southern New Mexico, that I would become a geologist. I was nine years old and fascinated by my discovery of bornite, pyrite, turquoise, calcite and quartz crystals, and even native gold. Later in high school, when my family lived in the desert of Washington State, I added agates, petrified wood, and basalt lava flows of unbelievable magnitude to my geologic enthusiasm base.

I followed geology through undergraduate and graduate school, becoming more intrigued than ever with my career choice. I remember in particular some field trips in Montana where we hiked from the Precambrian central core of a mountain range through the whole Paleozoic and Mesozoic section in a single day. One could see the evolution of life for hundreds of millions of years. I used to wonder, why isn't everyone a geologist?

I was there for the revolution. In undergraduate and graduate school, I learned that there are large-scale tensional features in the crust such as the rift valleys and compressional features such as mountain ranges. The explanations for the simultaneous occurrence of such features on the surface of the Earth were tenuous at best, even a bit embarrassing in retrospect. The arrival of sea-floor spreading and plate tectonics in the early 1960s was like a breath of fresh air. Suddenly, the explanations for crustal evolution made sense, and our understanding of many other large-scale geologic phenomena fell into place as well.

I began my career in the early 1960s as a marine geologist at the University of Georgia Marine Institute studying the sediments of the continental shelf. It had just become apparent that the sediments on many continental shelves were relict; that is, the sediment had been left behind by a lowered sea level and was not in equilibrium with present-day sediment sources such as river mouths. At a time when the dominant assumption was that sediment moved from the land to the sea, we discovered that sediment transported from the continental shelf was a very important source of beach and estuarine sediment.

At Duke University I began the study of abyssal plains, which entailed spending an average of a month and a half at sea each year, sampling mostly with piston cores. I mention the amount of time at sea only to note that the romance of the sea for field oceanographers is tempered by long absences from family. Abyssal plains are the flattest areas on the surface of the Earth, made so by turbidity currents that flow catastrophically down continental margins. They pond up in deep closed basins like water behind dams, forming the plains' flat surfaces. One exciting discovery was the size and shape of sand layers resulting from individual flow events on the plains, including one containing more than 100 km^2 of material (compared to 1 km^2 released by Mount St. Helens).

In 1969, my parents' retirement home in Mississippi was destroyed by Hurricane Camille, which led to my eventual drift into coastal geology. I arrived in coastal geology at about the time of a second revolution, albeit a smaller one than plate tectonics. This revolution was the discovery by Robert Dolan and Paul Godfrey that barrier islands are capable of migration. Understanding this made quite a difference in our perspective on barrier island processes including shoreline retreat. And, of course, if an island is capable of migration, there is no real need to stop shoreline retreat except to save houses. In 1975, my father and I wrote a small book, entitled *How to Live with an Island,* about the oceanographic and geologic hazards of a single island in North Carolina. The response was immense, both pro and con, and I began to learn the satisfaction of doing science with real and immediate impact on politics and economics. Since that time I have helped to produce an 18-volume state-specific series of coastal hazards books for public consumption, and my students and I have studied beach replenishment (it's costly and temporary), the impact of seawalls on beaches (seawalls destroy beaches), and the validity of engineering mathematical models of beach behavior (they don't work). We are also examining the role of storms in barrier island evolution and the ramifications in terms of "safe" development on barrier islands. In all of this I find myself in frequent conflict with engineers concerning the viability of engineering structures and with politicians concerning the viability of coastal development. One thing is certain—many more people are interested in the impact of seawalls on beach quality than in the sedimentology of abyssal plains.

ORRIN H. PILKEY is James B. Duke Professor of Geology and director of the Program for the Study of Developed Shorelines in the Department of Geology at Duke University. Along the way he was president of the Society for Sedimentary Geology and the North Carolina Academy of Science and taught briefly at the University of Puerto Rico. In 1987 he received the Shepard medal for marine geology research.

➢ **FIGURE 19-6** Breakers pounding the shoreline near Stinson Beach, California.

Shallow-Water Waves and Breakers

Waves moving out from the area of generation form swells and lose only a small amount of energy as they travel across the ocean. In deep-water swells, the water surface oscillates and water particles move in orbital paths, with little net displacement of water occurring in the direction of wave advance (Figure 19-4). When these waves enter progressively shallower water, the water is displaced in the direction of wave advance (Figure 19-4).

When deep-water waves enter shallow water, they are transformed from broad, undulating swells into sharp-crested waves. This transformation begins at a water depth of wave base; that is, it begins where wave base intersects the sea floor. At this point, the waves "feel" the bottom, and the orbital motions of water particles within waves are disrupted (Figure 19-4). As they move further shoreward, the speed of wave advance and wave length decrease, and wave height increases. In effect, as waves enter shallower water, they become oversteepened; the wave crest advances faster than the wave form, until eventually the crest plunges forward as a **breaker** (➢ Figure 19-6). Breakers are commonly several times higher than deep-water waves, and when they plunge forward, their kinetic energy is expended on the shoreline.

 NEARSHORE CURRENTS

It is convenient to identify the *nearshore zone* as the area extending seaward from the shoreline to just beyond the area where waves break. It includes a *breaker zone* and a *surf zone,* which is where breaking waves rush forward onto the shore followed by seaward movement of the water as backwash (Figure 19-4). The width of the nearshore zone varies depending on the wave length of the approaching waves, because long waves break at a greater depth, and thus farther offshore, than do short waves. Two types of currents are important in the nearshore zone, *longshore currents* and *rip currents*.

Wave Refraction and Longshore Currents

Deep-water waves are characterized by long, continuous crests, but rarely are their crests parallel with the shoreline (➢ Figure 19-7). One part of a wave enters shallow water

➢ **FIGURE 19-7** Wave refraction. These oblique waves are refracted and more nearly parallel the shoreline as they enter progressively shallower water. The refracted waves generate a longshore current that flows in the direction of wave approach, from right to left in this example.

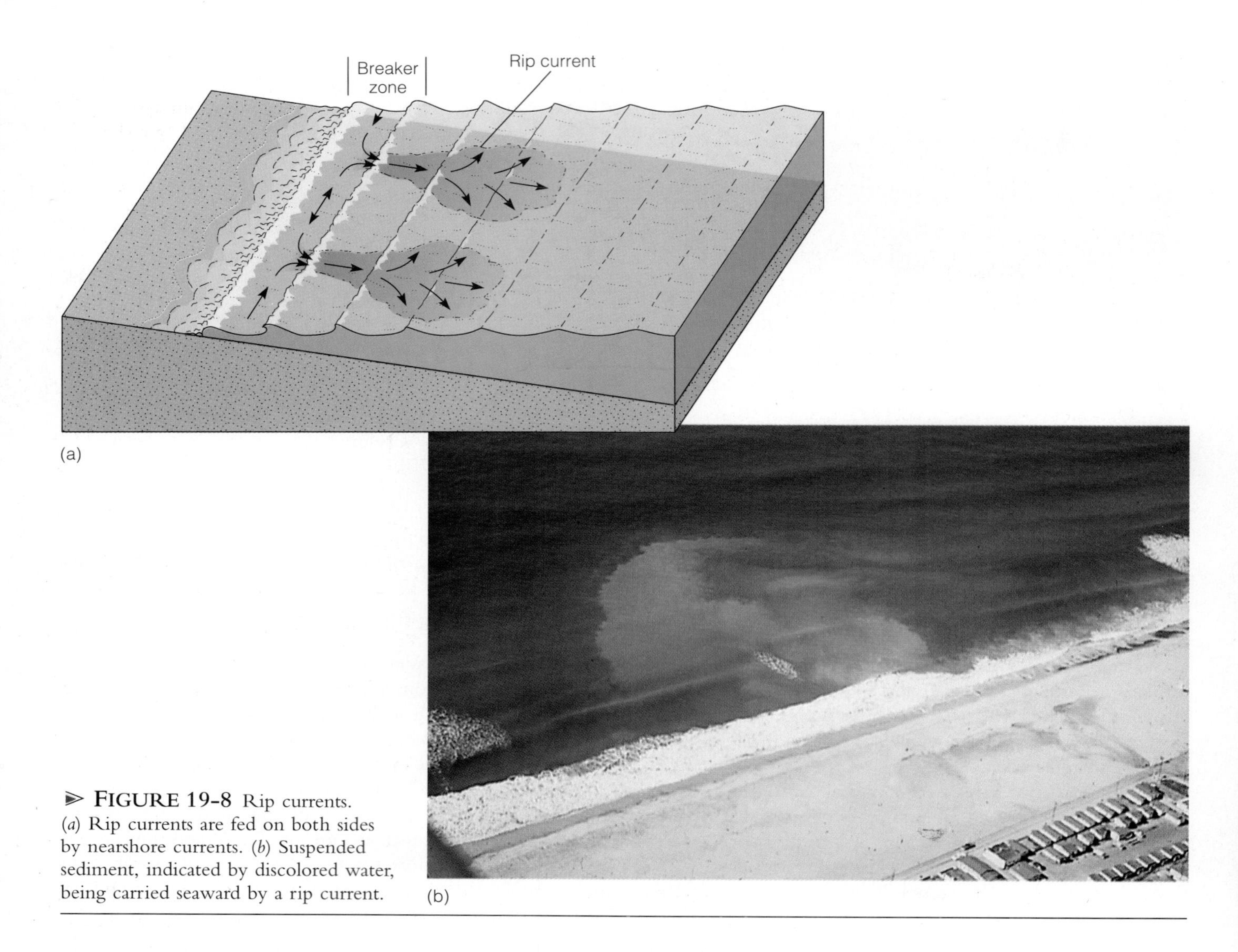

▶ **FIGURE 19-8** Rip currents. (*a*) Rip currents are fed on both sides by nearshore currents. (*b*) Suspended sediment, indicated by discolored water, being carried seaward by a rip current.

where it encounters wave base and begins breaking before other parts of the same wave. As a wave begins breaking, its velocity diminishes, but the part of the wave still in deep water races ahead until it too encounters wave base. The net effect of this oblique approach is that the waves bend so that they more nearly parallel the shoreline (Figure 19-7). Such a phenomenon is called **wave refraction.**

Even though waves are refracted, they still usually strike the shoreline at some angle, causing the water between the breaker zone and the beach to flow parallel to the shoreline. These **longshore currents,** as they are called, are long and narrow and flow in the same general direction as the approaching waves (Figure 19-7). These currents are particularly important agents of transport and deposition in the nearshore zone.

Rip Currents

Waves carry water into the nearshore zone, so there must be a mechanism for mass transfer of water back out to sea. One way in which water moves seaward from the nearshore zone is in **rip currents,** which are narrow surface currents that flow out to sea through the breaker zone (▶ Figure 19-8). Surfers commonly take advantage of rip currents for an easy ride out beyond the breaker zone, but such currents pose a danger to inexperienced swimmers. Some rip currents flow at several kilometers per hour, so if a swimmer is caught in one, it is useless to try to swim directly back to shore. Instead, because rip currents are narrow and usually nearly perpendicular to the shore, one can swim parallel to the shoreline for a short distance and then turn shoreward with no difficulty.

Rip currents can be characterized as circulating cells fed by longshore currents. When waves approach a shoreline, the amount of water builds up until the excess moves out to sea through the breaker zone. The rip currents are fed by nearshore currents that increase in velocity from midway between each rip current (Figure 19-8a).

The configuration of the sea floor plays an important role in determining the location of rip currents. They commonly develop where wave heights are lower than in adjacent areas.

(a)

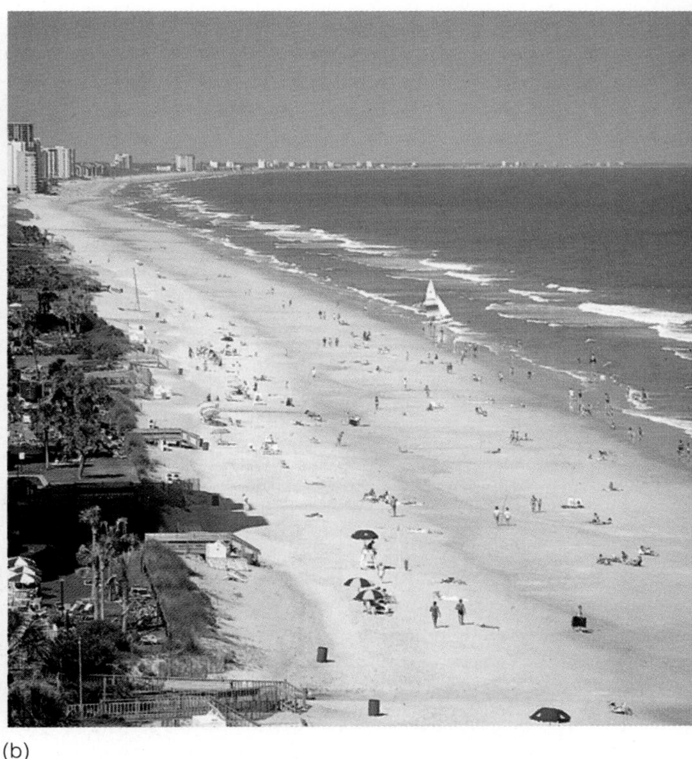

(b)

▶ **FIGURE 19-9** (*a*) A pocket beach in California. (*b*) The Grand Strand of South Carolina, shown here at Myrtle Beach, is 100 km of nearly continuous beach.

Such differences in wave height are commonly controlled by variations in water depth. For example, if waves move over a depression, the height of the wave over the depression tends to be less than in adjacent areas.

SHORELINE DEPOSITION

Depositional features of shorelines include *beaches*, *spits*, *baymouth bars*, and *barrier islands*. The characteristics of beaches are determined by wave energy, and they are continually modified by waves and longshore currents. Spits and baymouth bars both result from deposition by longshore currents, but the origin of barrier islands is controversial. Rip currents play only a minor role in the configuration of shorelines, but they do transport fine-grained sediment seaward through the breaker zone.

Beaches

Beaches are the most familiar of all coastal landforms, attracting millions of visitors each year and providing the economic base for many communities. They consist of a long, narrow strip of unconsolidated sediment, commonly sand, and are constantly changing. Depending on shoreline configuration and wave intensity, beaches may be discontinuous, existing only as *pocket beaches* in protected areas such as embayments, or they may be continuous for long distances (▶ Figure 19-9).

By definition a **beach** is a deposit of unconsolidated sediment extending landward from low tide to a change in topography such as a line of sand dunes, a sea cliff, or the point where permanent vegetation begins (▶ Figure 19-10). Typically, a beach has several component parts including a *backshore* that is usually dry, being covered by water only by storm waves or exceptionally high tides. The backshore consists of one or more **berms,** platforms composed of sediment deposited by waves. Berms are nearly horizontal or

▶ **FIGURE 19-10** Cross section of a typical beach showing its component parts.

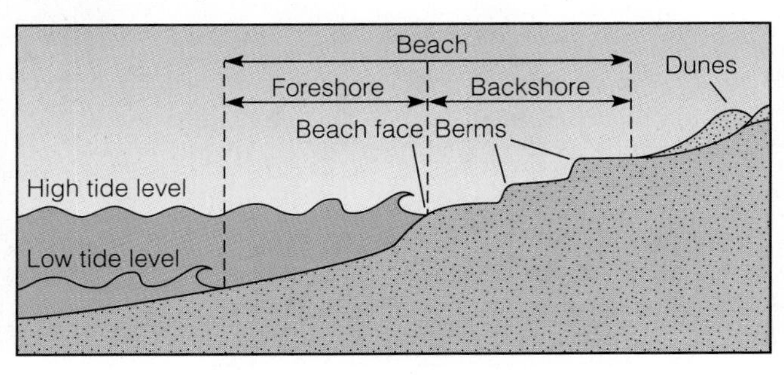

Perspective 19-1

Cape Cod, Massachusetts

Cape Cod, some of which is designated as a National Seashore, resembles a large human arm extending into the Atlantic Ocean from the coast of Massachusetts (▶ Figure 1). It projects 65 km east of the mainland to the "elbow," and then extends another 65 km northward where it resembles a half-curled hand. Cape Cod and the adjacent Elizabeth Islands, Martha's Vineyard, and Nantucket Island all owe their existence to deposition by late Pleistocene glaciers and the subsequent modification of these glacial deposits by waves. The granite bedrock or foundation upon which the cape and nearby islands were built occurs at depths of 90 to 150 m.

Between about 23,000 and 16,000 years ago, during the greatest southward advance of the continental glaciers in this area, the ice front was stabilized in the area of present-day Martha's Vineyard and Nantucket Island (▶ Figure 2a). Recall that a stabilized terminus means the glacier has a balanced budget. Nevertheless, flow continues within the glacier, transporting and depositing sediment as an end moraine. The end moraine that makes up these islands is a terminal moraine because it is the most southerly of all the moraines deposited by this glacier.

As the climate became warmer, the Cape Cod Bay Lobe of this vast glacier began retreating northward. About 14,000 years ago, it once again became stabilized in the area of present-day Cape Cod and the Elizabeth Islands where it deposited a large recessional moraine (Figure 2b). As the ice front continued to retreat northward, meltwater trapped between the ice front and the recessional moraine formed Glacial Lake Cape Cod covering about 1,000 km^2 (Figure 2c). Deposits of mud, silt, and sand accumulated in this lake, and these, too, make up part of present-day Cape Cod.

When the continental glaciers withdrew entirely from this region, Cape Cod looked different than it does now. On its east side were several headlands and embayments, but by 6,000 years ago sea level had risen enough so that waves began smoothing the shoreline by redistributing the sediment. The headlands were eroded whereas the embayments were filled in, and the shoreline has eroded from 1 to 4 km

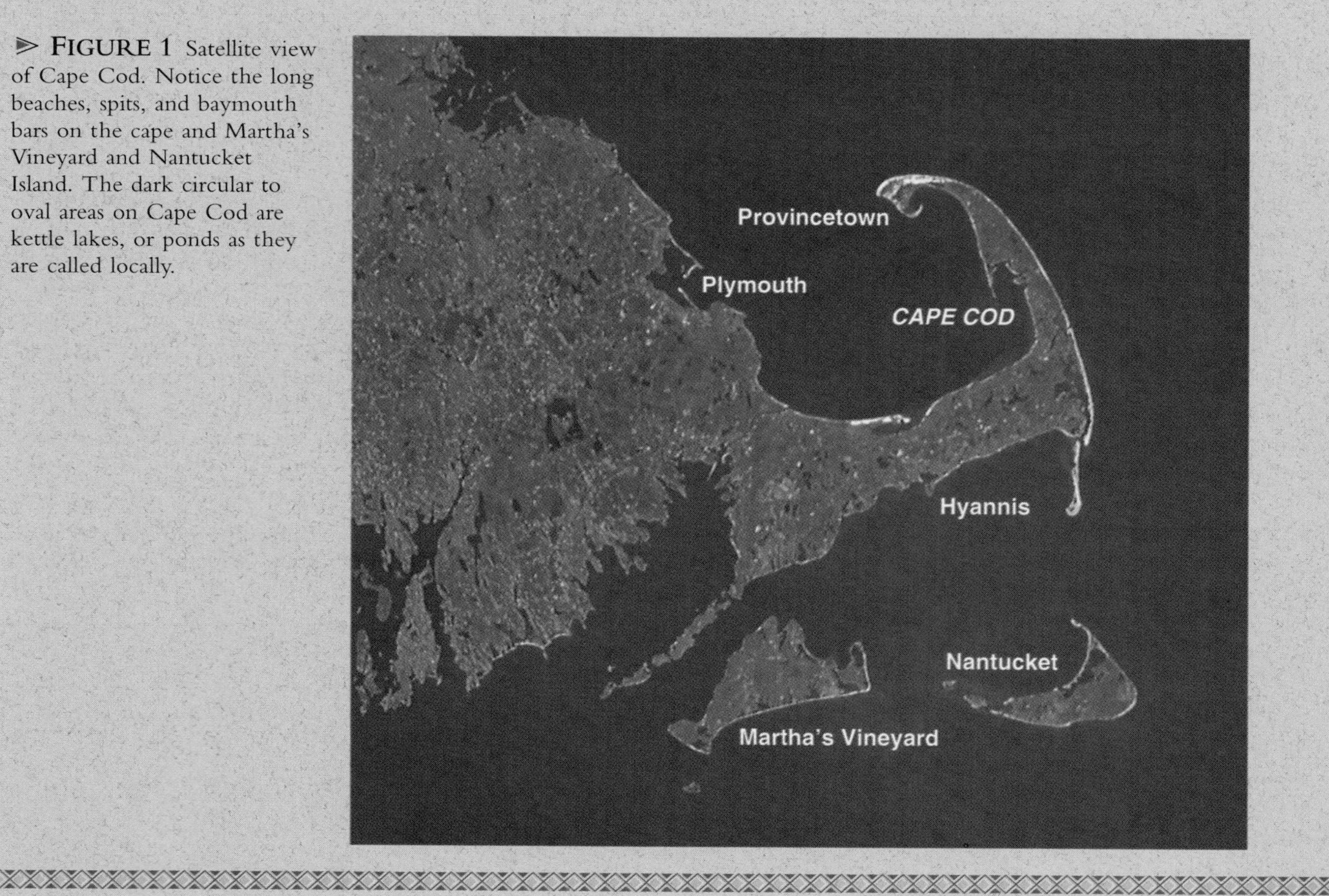

▶ **FIGURE 1** Satellite view of Cape Cod. Notice the long beaches, spits, and baymouth bars on the cape and Martha's Vineyard and Nantucket Island. The dark circular to oval areas on Cape Cod are kettle lakes, or ponds as they are called locally.

landward from its former location. Even today there are many steep wave-cut cliffs on the cape, and those facing east are currently eroding at nearly 1 m per year. Most of this erosion occurs during storms, so cliff retreat is episodic rather than continuous.

Some of the sediment eroded from Cape Cod is transported offshore where it settles beyond the reach of waves, but much of it is transported by longshore currents and redistributed along the cape. Hundreds of thousands of cubic meters of sand are transported along the shores of the cape and deposited as spits and baymouth bars, many of which are still forming. Extending south from the "elbow," is a long spit, and another continues to form at the north where sand is transported around the end of the cape and forms a hook (Figure 1). The wind also transports some of the sand inland where it accumulates as a series of coastal dunes.

Other distinctive features of Cape Cod are its numerous circular to oval ponds. These ponds occupy kettles that formed when large blocks of ice were partly or completely buried by outwash deposits. When these ice masses finally melted, they left depressions measuring up to 0.8 km in diameter that filled with water when the water table rose as a consequence of rising sea level.

Native Americans lived on Cape Cod for thousands of years before the Europeans arrived. Unfortunately, by 1764 these earliest inhabitants had nearly ceased to exist, mostly because of diseases. The first European settlers in this region were the Pilgrims. Despite the persistent myth that they first landed at Plymouth Rock, their first landfall was actually on Cape Cod near what is now Provincetown.

(a)

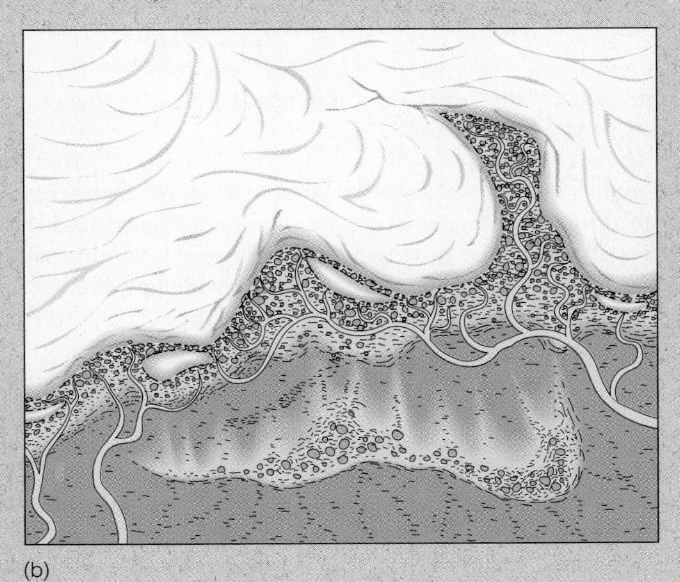

(b)

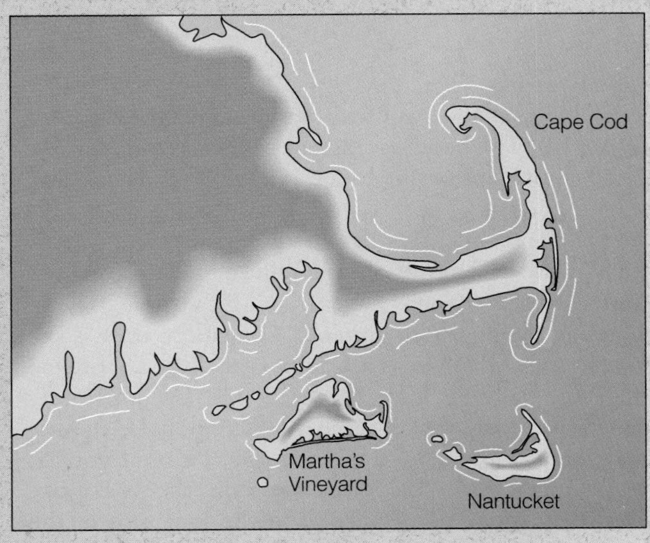

(c)

➤ **FIGURE 2** (*a*) Position of the Cape Cod Lobe of glacial ice 23,000 to 16,000 years ago when it deposited its terminal moraine that would become Martha's Vineyard and Nantucket Island. (*b*) Position of the Cape Cod Lobe when the recessional moraine was deposited that now forms Cape Cod. (*c*) About 5,000 years ago, rising sea level covered the lowlands between the moraines, and beaches and sand spits formed.

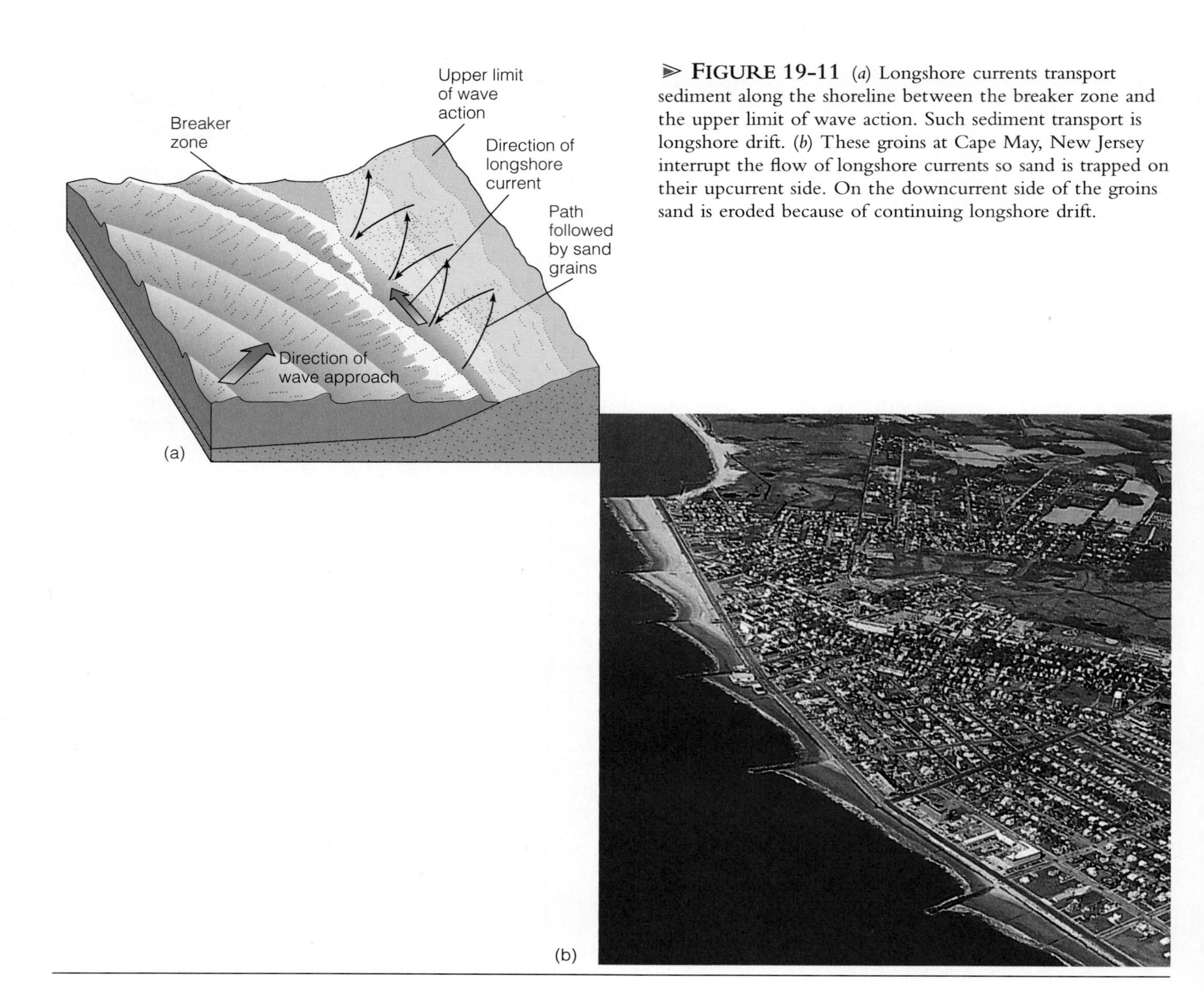

▶ **FIGURE 19-11** (*a*) Longshore currents transport sediment along the shoreline between the breaker zone and the upper limit of wave action. Such sediment transport is longshore drift. (*b*) These groins at Cape May, New Jersey interrupt the flow of longshore currents so sand is trapped on their upcurrent side. On the downcurrent side of the groins sand is eroded because of continuing longshore drift.

slope gently in a landward direction. The sloping area below the berm that is exposed to wave swash is called the *beach face* (Figure 19-10). The beach face is part of the *foreshore*, an area covered by water during high tide but exposed during low tide (Figure 19-10).

Some of the sediment on beaches is derived from weathering and wave erosion of the shoreline, but most of it is transported to the coast by streams and redistributed along the shoreline by longshore currents. **Longshore drift** is the phenomenon by which sand is transported along a shoreline by longshore currents (▶ Figure 19-11). As previously noted, waves usually strike beaches at some angle, causing the sand grains to move up the beach face at a similar angle; as the sand grains are carried seaward in the backwash, however, they move perpendicular to the long axis of the beach, so individual sand grains move in a zigzag pattern in the direction of longshore currents. This movement is not restricted to the beach, but extends seaward to the outer edge of the breaker zone (Figure 19-11a).

In an attempt to widen a beach or prevent erosion, shoreline residents often build *groins*, structures that project seaward at right angles from the shoreline (Figure 19-11b). They interrupt the flow of longshore currents causing sand to be deposited on their upcurrent side, widening the beach at that location. However, erosion inevitably occurs on the downcurrent side of a groin (Figure 19-11b).

Many beaches are sandy, but in areas of particularly vigorous wave activity, they might be gravel covered. Most beach sand is composed of quartz, but a number of other minerals and rock fragments may be present as well. One of the most common accessory minerals in beach sands is magnetite; because of its high specific gravity, magnetite is commonly separated from the other minerals and is visible as thin, black layers.

Although quartz is the most common mineral in most beach sands, there are some notable exceptions. The black sand beaches of Hawaii are composed of sand-sized basalt rock fragments, and some Florida beaches are composed of the fragmented calcium carbonate shells of marine organisms. In short, beaches are composed of whatever material is available; quartz is most abundant simply because it is available in most areas and is the most durable and stable of the common rock-forming minerals (see Chapter 2).

Seasonal Changes in Beaches

A beach is an area where wave energy is dissipated, so the loose grains composing the beach are constantly affected by wave motion. However, the overall configuration of a beach remains unchanged as long as equilibrium conditions persist. The beach profile consisting of a berm or berms and a beach face shown in Figure 19-10 can be thought of as a profile of equilibrium; that is, all parts of the beach are adjusted to the prevailing conditions of wave intensity and nearshore currents.

Tides and longshore currents affect the configuration of beaches to some degree, but by far the most important agent modifying their equilibrium profile is storm waves. In many areas, beach profiles change with the seasons; so we recognize *summer beaches* and *winter beaches*, which have adjusted to the conditions prevailing at these times (➤ Figure 19-12). Summer beaches are generally covered with sand and are characterized by a wide berm, a gently sloping beach face, and a smooth offshore profile. Winter beaches, on the other hand, tend to be coarser grained and steeper; they have a small berm or none at all, and their offshore profiles reveal sand bars paralleling the shoreline (Figure 19-12).

Seasonal changes in beach profiles are related to changing wave intensity. During the winter, energetic storm waves erode the sand from the beach and transport it offshore where it is stored in sand bars (Figure 19-12). The same sand that was eroded from the beach during the winter returns the next summer when it is driven onshore by the more gentle swells that occur during that season. The volume of sand in the system remains more or less constant; it simply moves farther offshore or onshore depending on the energy of waves. The terms *winter* and *summer beach,* although widely used, are somewhat misleading. A "winter beach" profile can develop at any time of the year if a large storm occurs, and likewise a "summer beach" profile can develop during a prolonged calm period in the winter.

Spits and Baymouth Bars

Other than the beach itself, some of the most common depositional landforms on shorelines are *spits* and *baymouth bars,* both of which are variations of the same feature. A **spit** is simply a continuation of a beach forming a point, or "free end," that projects into a body of water, commonly a bay. A

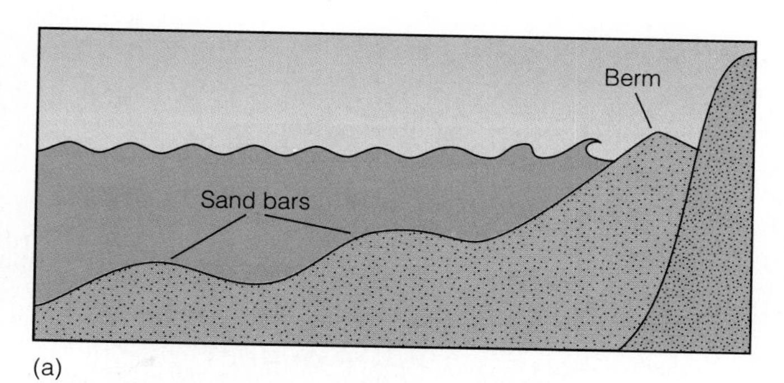

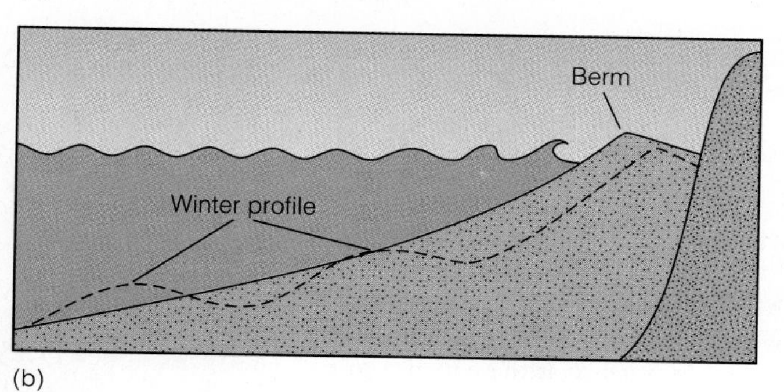

➤ **FIGURE 19-12** Seasonal changes in beach profiles. (*a*) A winter beach showing offshore sand bars. (*b*) A summer beach with its wider berm and more gently sloping beach face.

baymouth bar is a spit that has grown until it completely closes off a bay from the open sea (➤ Figure 19-13).

Both spits and baymouth bars form and grow as a result of longshore drift. Where currents are weak, as in the deeper water at the opening to a bay, longshore current velocity diminishes, and sediment is deposited, forming a sand bar. The free ends of many spits are curved by wave refraction or waves approaching from a different direction. Such spits are called *hooks* or *recurved spits* (Figure 19-13) (see Perspective 9-1).

A rarer type of spit, called a **tombolo,** extends out into the sea and connects an island to the mainland. Tombolos develop on the shoreward sides of islands as shown in ➤ Figure 19-14. Wave refraction around an island causes converging currents that turn seaward and deposit a sand bar connecting the shore with the island. A similar feature may form when an artificial breakwater is constructed offshore (➤ Figure 19-15).

Although spits, baymouth bars, and tombolos are most commonly found on irregular seacoasts, many examples of the same features occur in large lakes (➤ Figure 19-16). Whether along seacoasts or lakeshores, these sand deposits present a continuing problem where bays must be kept open for pleasure boating or commercial shipping. The entrances to these bays must either be regularly dredged or protected. The most common way to protect entrances to bays is to build *jetties*, which are structures extending seaward (or lakeward) that protect the bay from deposition by longshore currents (➤ Figure 19-17).

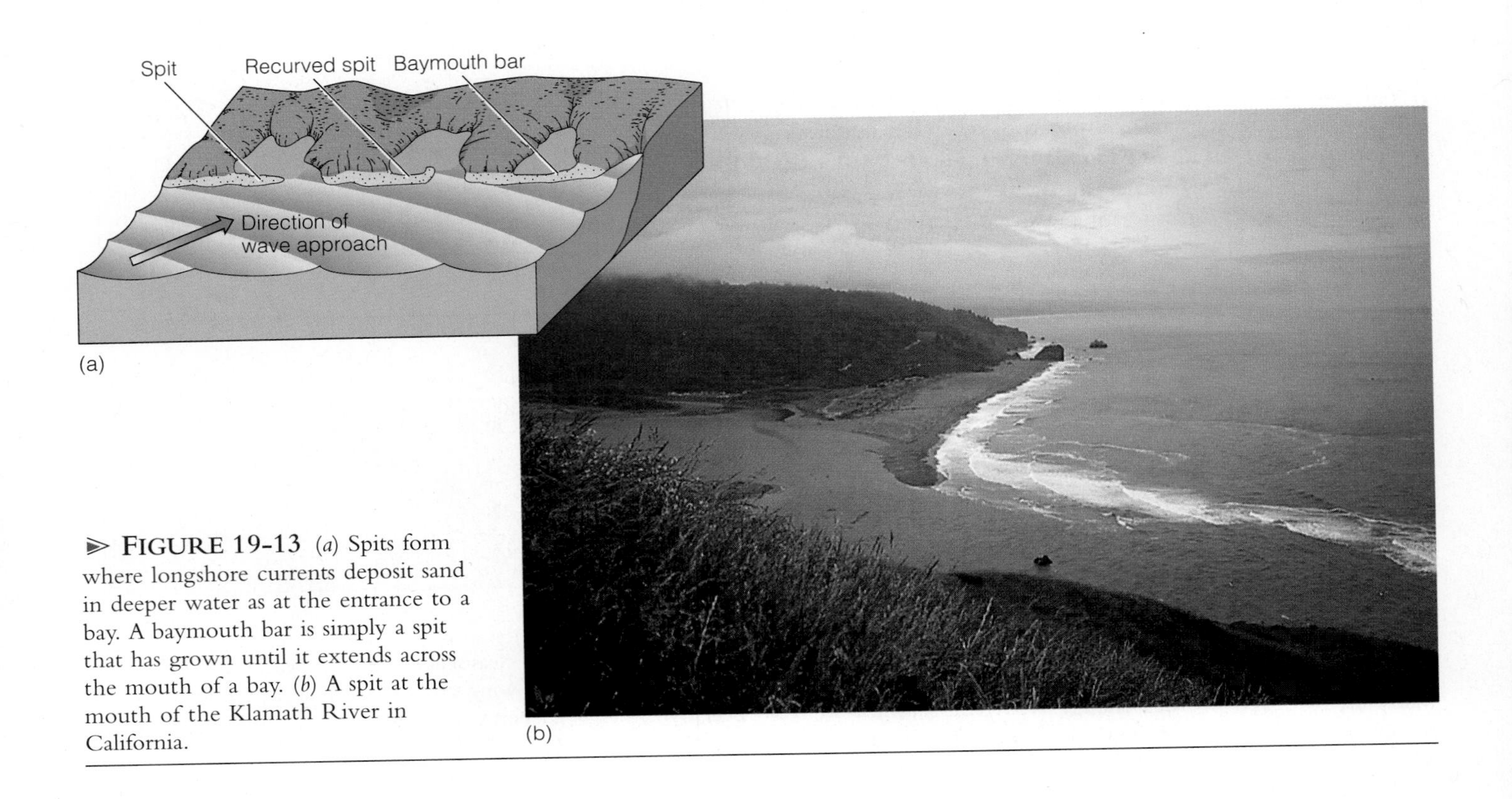

▶ **FIGURE 19-13** (*a*) Spits form where longshore currents deposit sand in deeper water as at the entrance to a bay. A baymouth bar is simply a spit that has grown until it extends across the mouth of a bay. (*b*) A spit at the mouth of the Klamath River in California.

▶ **FIGURE 19-14** (*a*) Origin of a tombolo. Wave refraction around an island causes longshore currents to converge and deposit a sand bar that joins the island with the mainland. (*b*) This small island is connected to the Oregon shoreline by a tombolo.

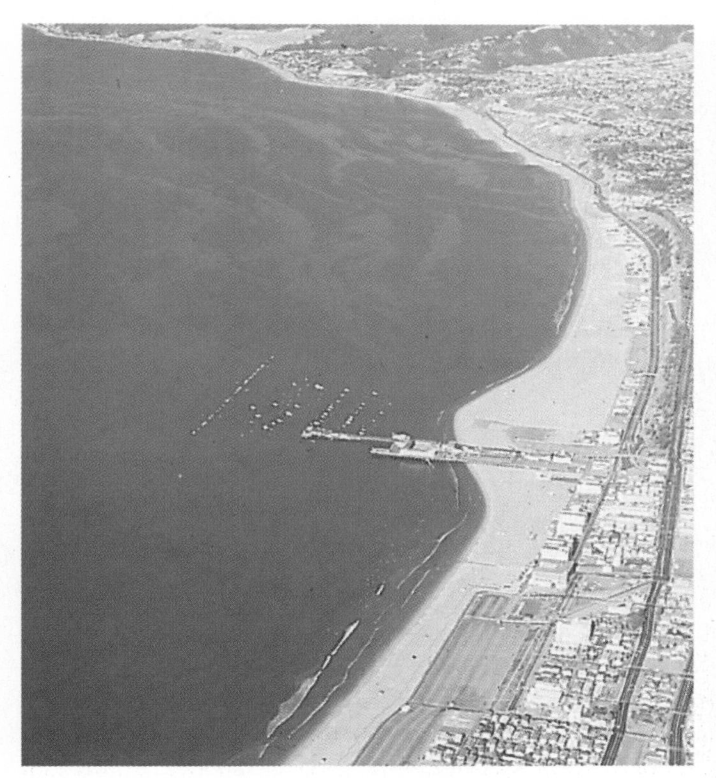

▶ **FIGURE 19-15** Soon after this breakwater was constructed offshore at Santa Monica, California, a bulge appeared in the beach. Wave refraction around the breakwater resulted in the formation of a feature similar to a tombolo.

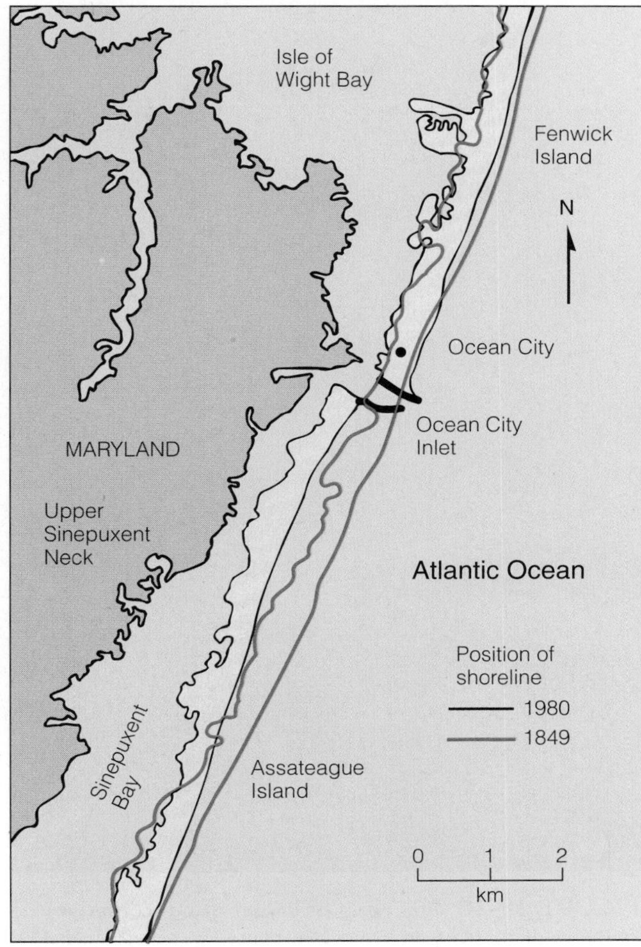

▶ **FIGURE 19-17 (above)** The seaward-projecting heavy, black lines represent jetties that were constructed in the 1930s to protect the Ocean City Inlet at Ocean City, Maryland. The jetties have protected the inlet, but they also disrupted the net southerly longshore drift. As a consequence, Assateague Island has been starved of sediment and migrated about 500 m landward and is now offset from Fenwick Island to the north. Both islands are barrier islands.

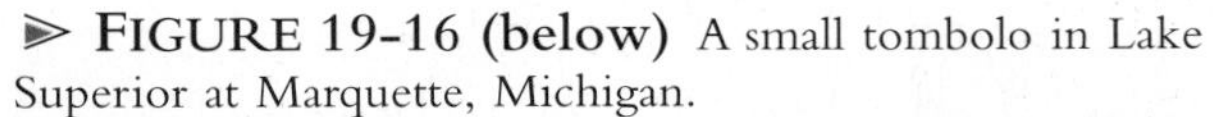

▶ **FIGURE 19-16 (below)** A small tombolo in Lake Superior at Marquette, Michigan.

Barrier Islands

Long, narrow islands composed of sand and separated from the mainland by a lagoon are called **barrier islands** (Figures 9-17 and ➢ 19-18). On their seaward margins, barrier islands are smoothed by waves, but their lagoon sides are irregular. During large storms, waves completely overtop these islands and deposit lobes of sand in the lagoon. Once deposited, these lobes are modified only slightly because they are protected from further wave action. Windblown sand dunes are common on barrier islands and are generally the highest part of these islands.

The origin of barrier islands has been long debated and is still not completely resolved. It is known that they form on gently sloping continental shelves with abundant sand in areas where both tidal fluctuations and wave energy levels are low. Although barrier islands occur in many areas, most of them are along the east coast of the United States from New York to Florida and along the U.S. Gulf Coast. Barrier islands are not nearly as common in Canada, but some are found along the coasts of New Brunswick and Prince Edward Island. According to one model, barrier islands formed as spits that became detached from the land, while another model proposes that they formed as beach ridges on coasts that subsequently subsided (➢ Figure 19-19).

Because sea level is currently rising, most barrier islands are migrating in a landward direction. Such migration is a natural consequence of the evolution of these islands, but it is a problem for the island residents and communities. Barrier islands generally migrate rather slowly, but the rates for many are rapid enough to cause shoreline problems (see Perspective 19-2).

➢ **FIGURE 19-18** This chain of barrier islands comprises the Outer Banks of North Carolina. Cape Hatteras, near the center of the photo, juts the farthest out into the Atlantic.

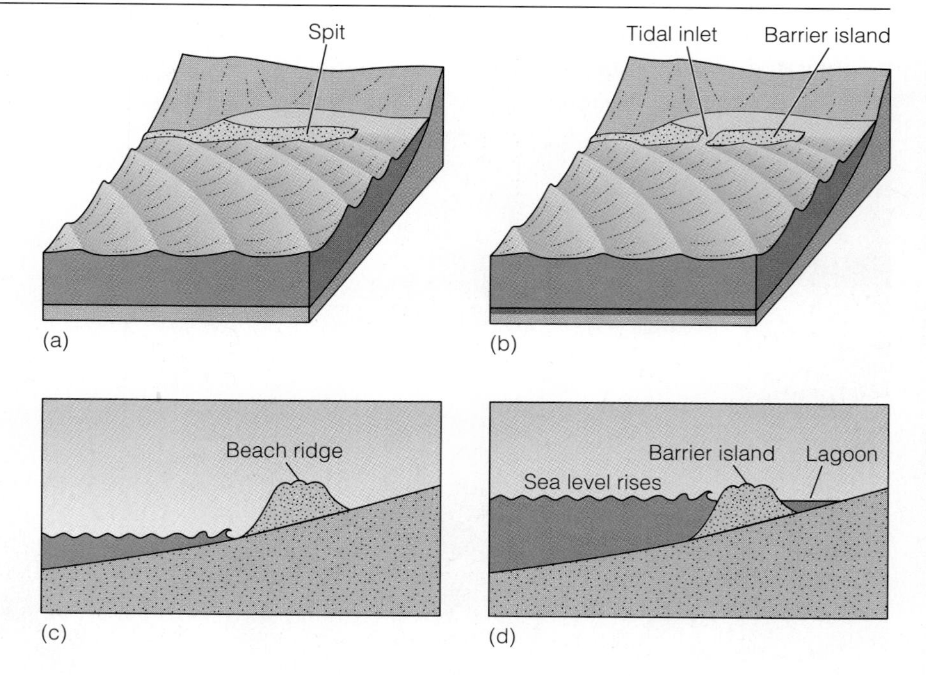

➢ **FIGURE 19-19** Two models for the origin of barrier islands. (*a*) Longshore currents extend a spit along a coast. (*b*) During a storm, the spit is breached, forming a tidal inlet and a barrier island. (*c*) A beach ridge forms on land. (*d*) Sea level rises, partly submerging the beach ridge.

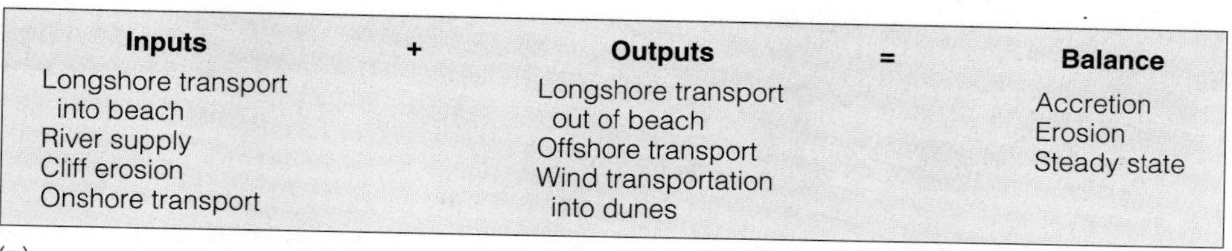

(a)

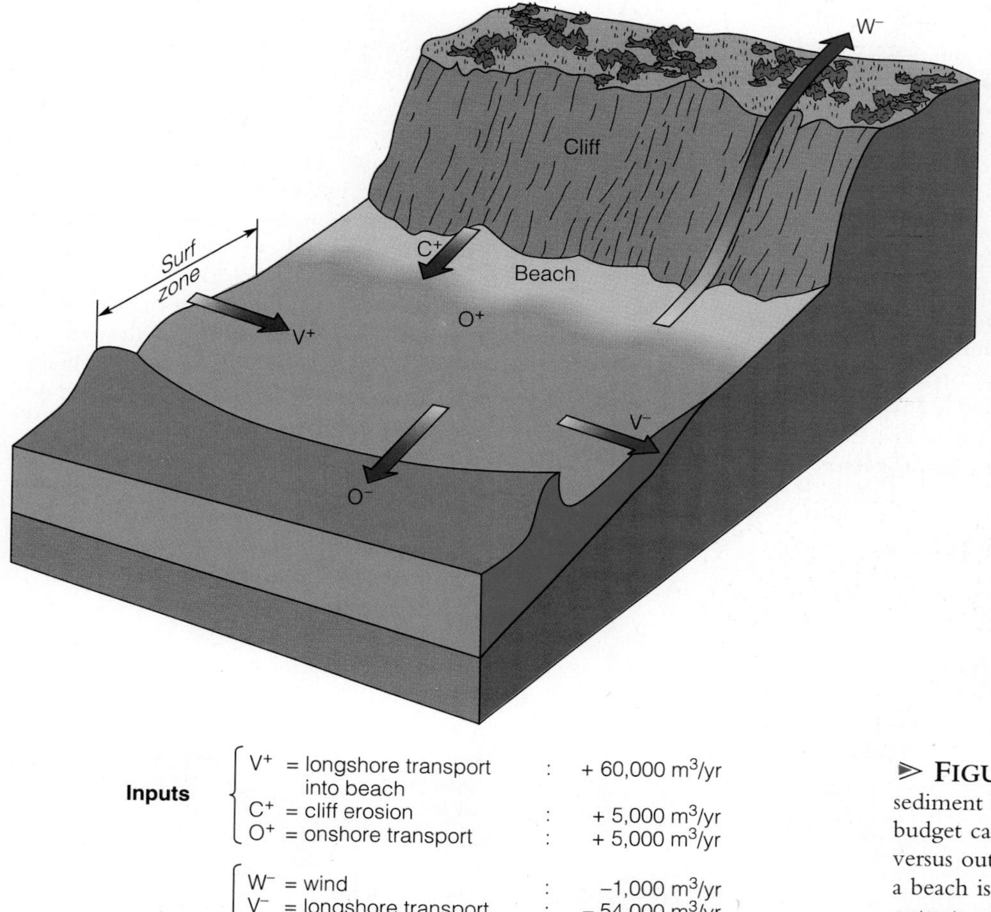

Inputs	V^+ = longshore transport into beach	:	+ 60,000 m^3/yr
	C^+ = cliff erosion	:	+ 5,000 m^3/yr
	O^+ = onshore transport	:	+ 5,000 m^3/yr
Outputs	W^- = wind	:	−1,000 m^3/yr
	V^- = longshore transport out of beach	:	− 54,000 m^3/yr
	O^- = offshore transport (includes transport to submarine canyons)	:	− 20,000 m^3/yr
	Balance	:	− 5,000 m^3/yr (net erosion)

(b)

► FIGURE 19-20 The nearshore sediment budget. (*a*) The long-term sediment budget can be assessed by considering inputs versus outputs. If inputs and outputs are equal, a beach is in a steady state or equilibrium. If outputs exceed inputs, however, the beach has a negative budget and erosion occurs. Accretion occurs when the beach has a positive budget with inputs exceeding outputs. (*b*) A hypothetical example of a negative nearshore sediment budget. In this example the beach is losing 5,000 m^3 a year to erosion.

The Nearshore Sediment Budget

We can think of the gains and losses of sediment in the nearshore zone in terms of a **nearshore sediment budget.** If a nearshore system has a balanced budget, sediment is supplied as fast as it is removed, and the volume of sediment remains more or less constant, although sand may shift offshore and onshore with the changing seasons (Figure 19-12). A positive budget means gains exceed losses, whereas a negative budget results when losses exceed gains. If a negative budget prevails long enough, the nearshore system is depleted and beaches may disappear (► Figure 19-20).

Erosion of sea cliffs provides some sediment to beaches, but in most areas probably no more than 5 to 10% of the total sediment supply is derived from this source. There are exceptions, however; almost all the sediment on the beaches of Maine is derived from the erosion of shoreline rocks. Most of the sediment on typical beaches is transported to the shoreline by streams and then redistributed along the shoreline by longshore drift. Thus, longshore drift also plays a role

Perspective 19-2

Rising Sea Level and Coastal Management

Shorelines in most of the United States and much of Canada are eroding as sea level rises. According to one study, 54% of U.S. shorelines are eroding at rates ranging from millimeters per year to more than 10 m in a few areas (▶ Figure 1). As sea level rises, buildings that were once far from the ocean are now being undermined and destroyed (Figure 19-3). Many other areas of the world are experiencing shoreline problems as well.

During the last century, sea level rose about 12 cm worldwide, and all indications are that it will continue to rise. The absolute rate of sea level rise in a particular shoreline region depends on two factors. The first is the volume of water in the ocean basins, which is increasing as a result of the melting of glacial ice and the thermal expansion of near-surface seawater. Many scientists think that sea level will continue to rise as a consequence of global warming resulting from concentrations of greenhouse gases in the atmosphere.

The second factor controlling sea level is the rate of uplift or subsidence of a coastal area. In some areas, uplift is occurring so fast that sea level is actually falling with respect to the land. In other areas, sea level is rising while the coastal region is simultaneously subsiding, resulting in a net change in sea level of as much as 30 cm per century. Perhaps such a "slow" rate of sea level change seems insignificant; after all it amounts to only a few millimeters per year. However, in gently sloping coastal areas, as in the eastern United States from New Jersey southward, even a slight rise in sea level would eventually have widespread effects.

Many of the nearly 300 barrier islands along the east and Gulf coasts of the United States are migrating landward as a natural consequence of rising sea level. During storms, their beaches are eroded by large waves that carry the beach sand over the islands and into their lagoons. Thus, sand is removed from the seaward sides of barrier islands and deposited on their landward sides, resulting in a gradual landward shift

▶ **FIGURE 1** Shoreline erosion in the United States. No data are available for shoreline areas left uncolored.

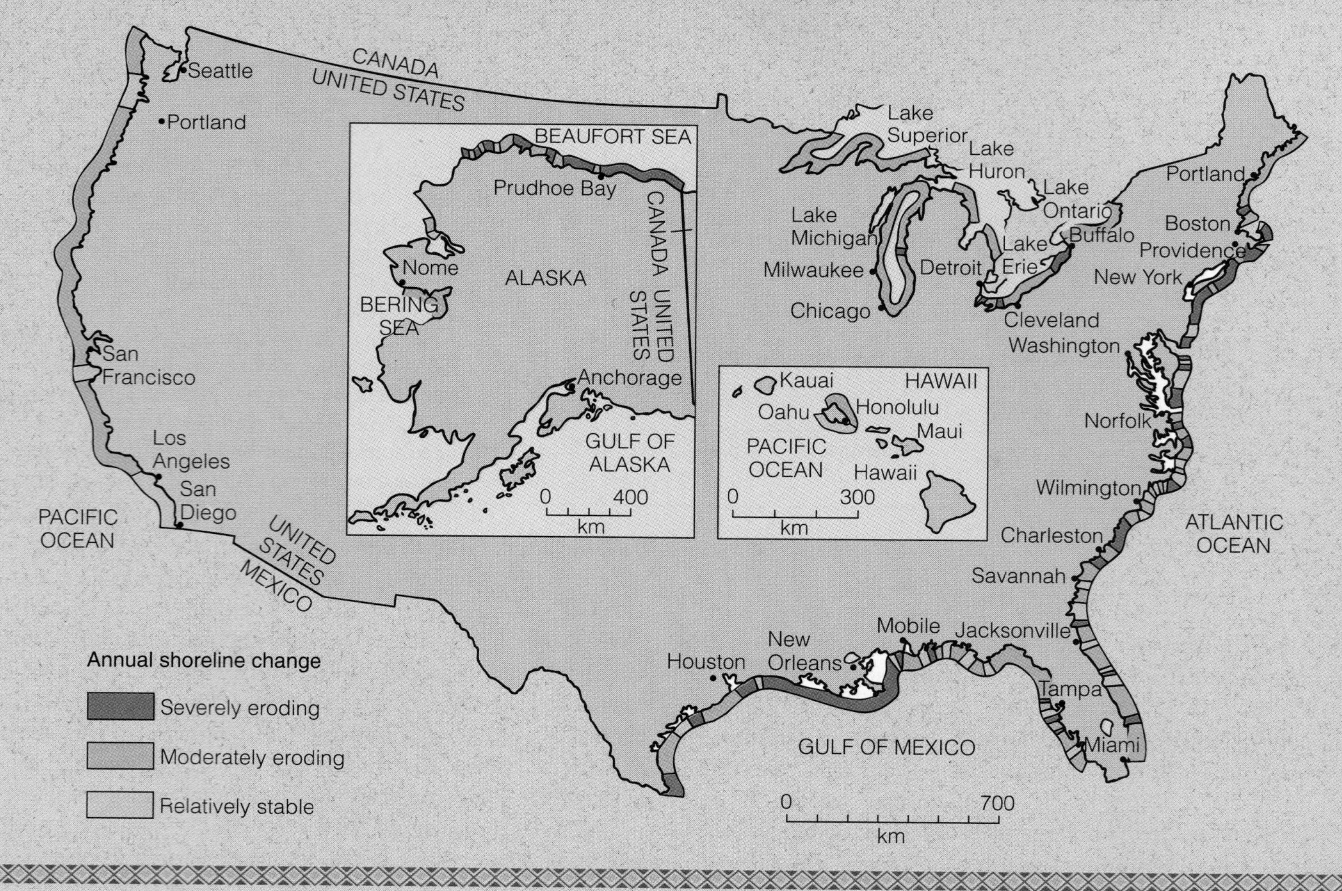

of the entire island complex (➢ Figure 2). During the last 120 years, Hatteras Island, North Carolina, has migrated nearly 500 m landward so that Cape Hatteras lighthouse, which was 460 m from the shoreline when it was built in 1870, now stands on a promontory in the Atlantic Ocean.

Landward migration of barrier islands would pose few problems if it were not for the numerous communities, resorts, and vacation homes located on them. Moreover, barrier islands are not the only threatened areas. For example, Louisiana's coastal wetlands, an important wildlife habitat and seafood-producing area, are currently being lost at a rate of about 90 km^2 per year. Much of this loss results from sediment compaction, but sea level rise exacerbates the problem.

Rising sea level also directly threatens many beaches that communities depend on for revenue. The beach at Miami Beach, Florida, for example, was disappearing at an alarming rate until the Army Corps of Engineers began replacing the eroded beach sand (➢ Figure 3). The problem is even more serious in other countries. A rise in sea level of only 2 m would inundate large areas of the east and Gulf coasts, but would cover 20% of the entire country of Bangladesh. Other problems associated with sea level rise include increased coastal flooding during storms and saltwater incursions that may threaten groundwater supplies (see Chapter 16).

Since nothing can be done to prevent sea level from rising, engineers and scientists must examine what can be done to prevent or minimize the effects of shoreline erosion. At present, only a few viable options exist. One is to put strict controls on coastal development. North Carolina, for example, permits large structures to be sited no closer to the shoreline than 60 times the annual erosion rate. Although a growing awareness of shoreline processes has resulted in similar legislation elsewhere, some states have virtually no restrictions on coastal development.

Regulating coastal development is commendable, but it has no impact on existing structures and coastal communities. A general retreat from the shoreline may be possible, but expensive, for individual dwellings and small communities, but it is impractical for large population centers. Such communities as Atlantic City, New Jersey, Miami Beach, Florida, and Galveston, Texas, have adopted one of two strategies to combat coastal erosion. One is to build protective barriers such as seawalls. Seawalls, such as the one at Galveston, Texas (see the Prologue), can be effective, but they are tremendously expensive to construct and maintain.

(continued on next page)

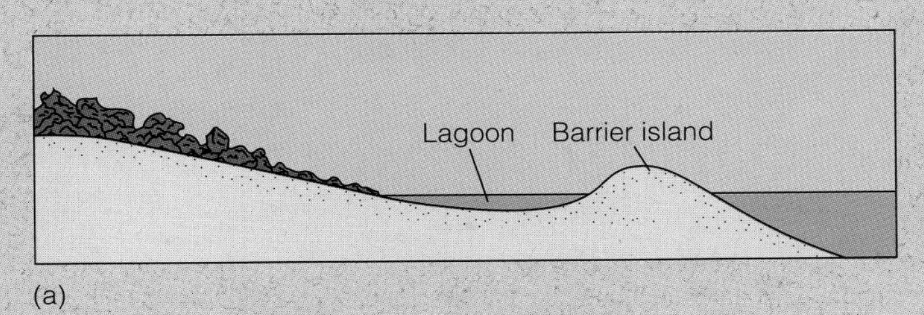

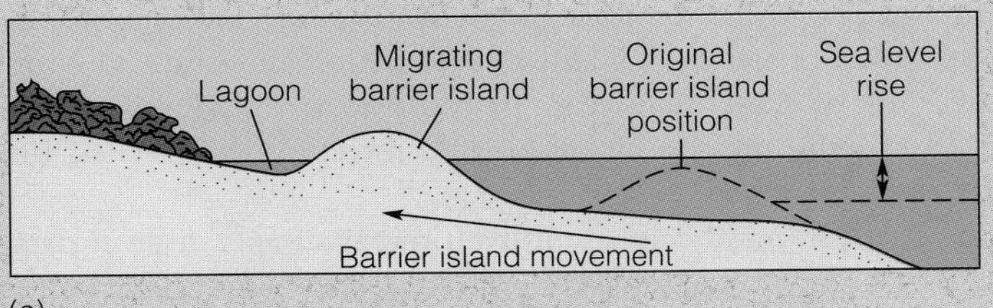

➢ **FIGURE 2** Rising sea level and the landward migration of barrier islands. (*a*) Barrier island before landward migration in response to rising sea level. (*b*) Landward movement occurs when storm waves wash sand from the seaward side of the islands and deposit it in the lagoon. (*c*) Over time, the entire complex migrates landward.

(*continued*)

During the last five years, for example, more than $50 million has been spent to replenish the beach sand and build a protective seawall at Ocean City, Maryland. Unfortunately, seawalls retard erosion only in the area directly behind them; Galveston Island west of the seawall has been eroded back about 45 m.

Another option, adopted by both Atlantic City, New Jersey, and Miami Beach, Florida, is to pump sand onto the beaches to replace that lost to erosion (Figure 3). This, too, is expensive as the sand must be replenished periodically because erosion is a continuing process. In many areas, groins are constructed to preserve beaches, but unless additional sand is artificially supplied to the beaches, longshore currents invariably erode sand from the downcurrent sides of the groins.

Rising sea level has already had a significant economic impact, and all options for dealing with this phenomenon are expensive. Fortifying the shoreline with seawalls, groins, and other structures is initially expensive, requires constant maintenance, and in the long run will be ineffective if sea level continues to rise. A general retreat from the shoreline is simply impractical for most coastal communities. Perhaps the best option is to replace sand lost to erosion by pumping it from elsewhere, usually farther offshore. For some areas, however, little can be done to offset the effects of rising sea level.

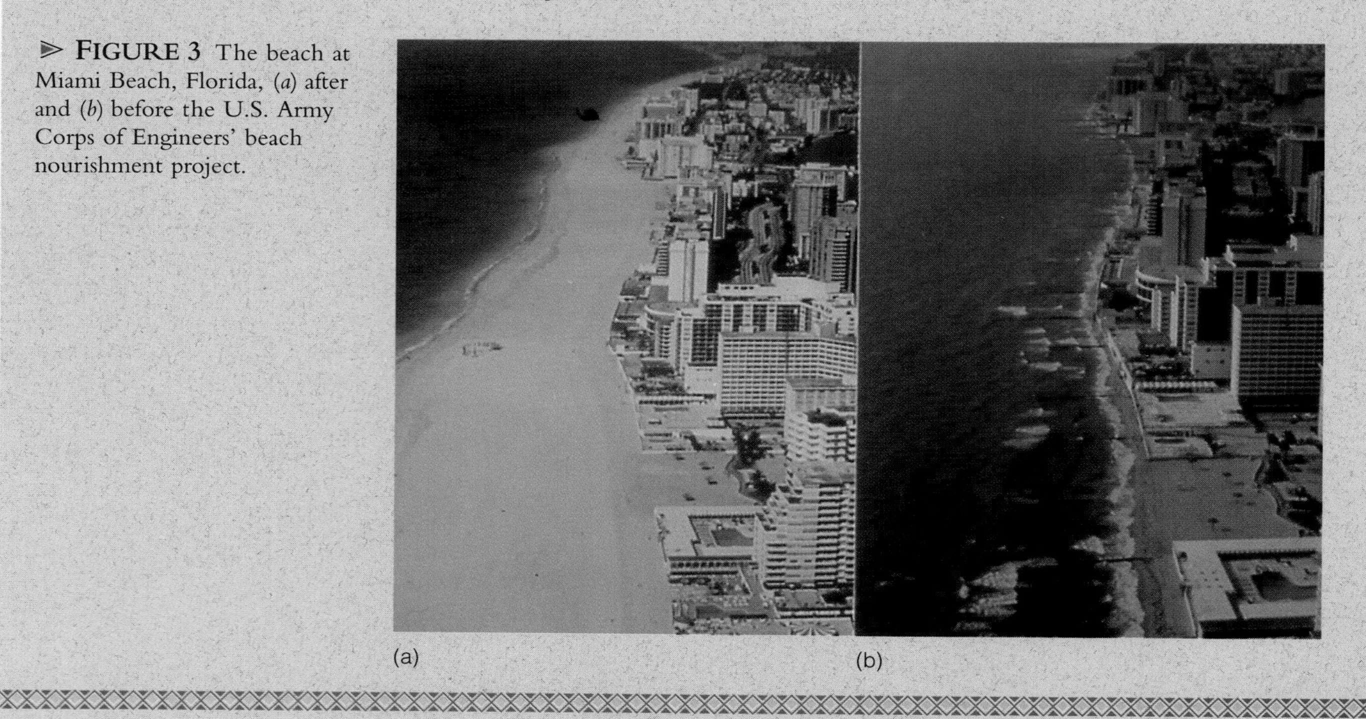

FIGURE 3 The beach at Miami Beach, Florida, (*a*) after and (*b*) before the U.S. Army Corps of Engineers' beach nourishment project.

in the nearshore sediment budget because it continually moves sediment into and away from beach systems (Figure 19-20).

The primary ways in which a nearshore system loses sediment include longshore drift, offshore transport, wind, and deposition in submarine canyons. Offshore transport mostly involves fine-grained sediment that is carried seaward where it eventually settles in deeper water. Wind is an important process because it removes sand from beaches and blows it inland where it commonly piles up as sand dunes.

If the heads of submarine canyons are nearshore, huge quantities of sand are funneled into them and deposited in deeper water. La Jolla and Scripps submarine canyons off the coast of southern California funnel off an estimated 2 million m^3 of sand each year. In most areas, however, submarine canyons are too far offshore to interrupt the flow of sand in the nearshore zone.

It should be apparent from the preceding discussion that if a nearshore system is in equilibrium, its incoming supply of sediment exactly offsets its losses. Such a delicate balance tends to continue unless the system is somehow disrupted. One common change that affects this balance is the construction of dams across the streams supplying sand. The sediment contribution from a stream is proportional to its drainage area. Once dams have been built, all sediment from the upper reaches of the drainage systems is trapped in reservoirs and thus cannot reach the shoreline.

 SHORELINE EROSION

Along seacoasts where erosion rather than deposition predominates, beaches are lacking or poorly developed, and a sea cliff commonly develops ( Figure 19-21). Sea cliffs are erosional features frequently pounded by waves, especially during storms: the cliff face retreats landward as a consequence of *corrosion, hydraulic action,* and *abrasion. Corrosion* is an erosional process involving the wearing away of rock by chemical processes, especially the solvent action of seawater. The force of the water itself, called *hydraulic action,* is a particularly effective erosional process. Waves exert tremendous pressure on shorelines by direct impact, but are most effective on sea cliffs composed of unconsolidated sediment or rocks that are highly fractured. *Abrasion* is an erosional process involving the grinding action of rocks and sand carried by waves.

Wave-Cut Platforms and Associated Landforms

The rate at which sea cliffs are eroded and retreat in a landward direction depends on wave intensity and the resistance of the coastal rocks or sediments. Most sea cliff retreat occurs during storms and, as one would expect, occurs most rapidly in sea cliffs composed of unconsolidated sediment. For example, a sea cliff composed of unconsolidated glacial drift on Cape Cod, Massachusetts, retreats as much as 30 m per century, and some parts of the White Cliffs of Dover in Great Britain are retreating at a rate of more than 100 m per century. By comparison, sea cliffs consisting of dense igneous or metamorphic rocks may retreat at negligible rates.

Sea cliffs retreat mostly as a result of hydraulic action and abrasion at their bases. As a sea cliff is undercut by erosion, the upper part is left unsupported and susceptible to mass wasting processes. Thus, sea cliffs retreat little by little, and as they do, they leave a beveled surface known as a **wave-cut platform** that slopes gently in a seaward direction (Figure 19-21). Broad wave-cut platforms exist in many areas, but invariably the water over them is shallow because the abrasive planing action of waves is only effective to a depth of about 10 m.

Wave-cut platforms are surfaces of sediment transport. The sediment eroded from sea cliffs is transported seaward until it reaches deeper water at the edge of the wave-cut platform. There it is deposited and forms a seaward extension of the wave-cut platform called a *wave-built platform* (Figure 19-21).

Sea cliffs do not retreat uniformly, however, because some of the materials of which they are composed are more

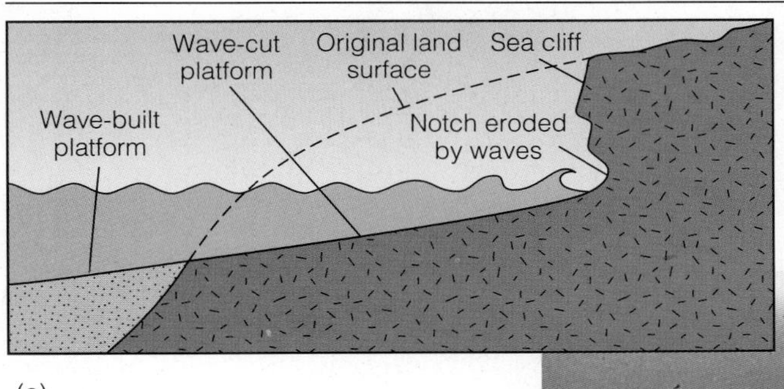

(a)

(b)

▶ **FIGURE 19-21** (*a*) Wave erosion of a sea cliff produces a gently sloping surface called a wave-cut platform. Deposition at the seaward margin of the wave-cut platform forms a wave-built platform. (*b*) Sea cliffs and a wave-cut platform.

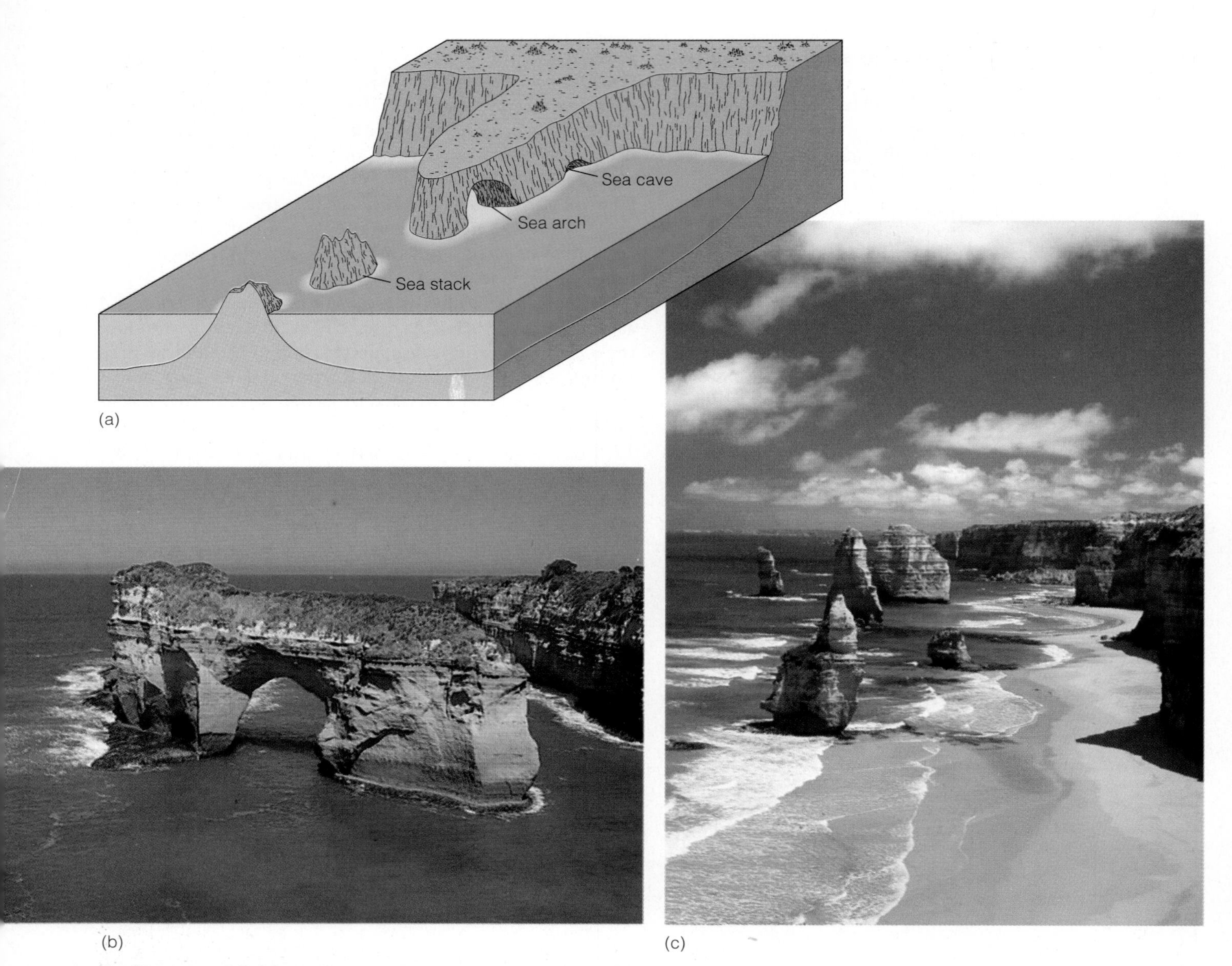

➤ **FIGURE 19-22** (*a*) Erosion of a headland and the origin of sea caves, sea arches, and sea stacks. (*b*) This sea stack in Australia has an arch developed in it. (*c*) Sea stacks on Australia's south coast.

resistant to erosion than others. **Headlands** are seaward-projecting parts of the shoreline that are eroded on both sides due to wave refraction (➤ Figure 19-22a). *Sea caves* may form on opposite sides of a headland, and if these join, they form a *sea arch* (Figure 19-22). Continued erosion generally causes the span of an arch to collapse, yielding isolated *sea stacks* on wave-cut platforms (Figure 19-22). In the long run, shoreline processes tend to straighten an initially irregular shoreline. They do so because wave refraction causes more wave energy to be expended on headlands and less on embayments. Headlands become eroded, and some of the sediment yielded by erosion is deposited in the embayments. The net effect of these processes is to straighten the shoreline (➤ Figure 19-23).

Wave-cut platforms and their associated features are most common along seashores, but they are also present along the shores of large lakes. A number of such features are present in the Great Lakes region, many of which have been raised above lake level by isostatic rebound (➤ Figure 19-24).

Types of Coasts

Coasts can be classified in several ways, but none of them is completely satisfactory because of variations in the factors controlling coastal development and differences in the composition and configuration of coasts. Rather than attempt to categorize all coasts, we shall simply note that two types of coasts have already been discussed, those dominated by

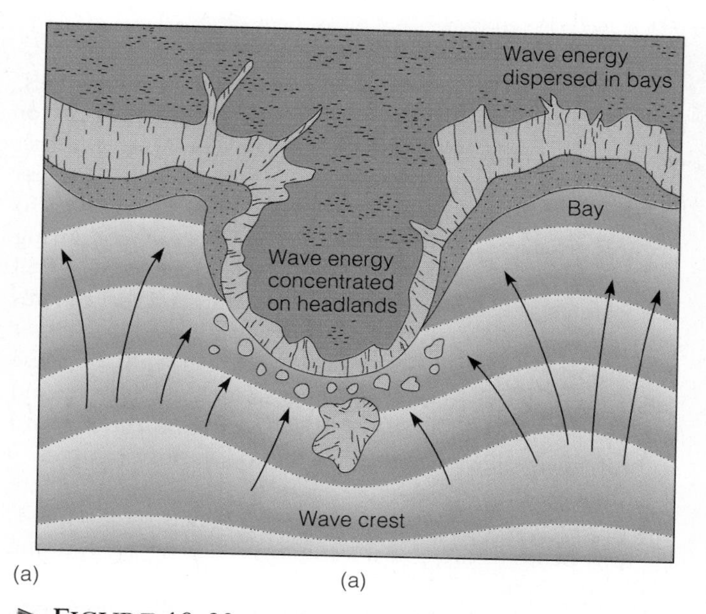

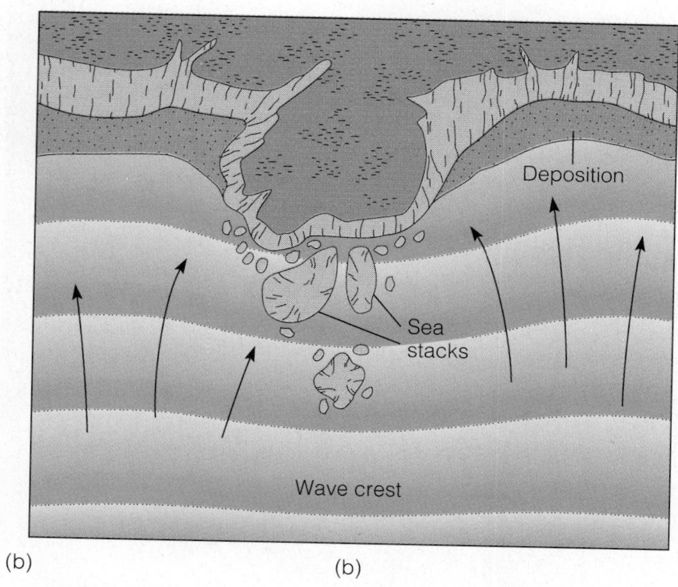

▶ **FIGURE 19-23** (*a*) Wave refraction acts to straighten shorelines by concentrating wave energy on headlands. (*b*) The same shoreline after extensive erosion of the headlands and deposition in the bays.

deposition and those dominated by erosion, and shall look further at the changing relationships between coasts and sea level.

Depositional coasts, such as the U.S. Gulf Coast, are characterized by an abundance of detrital sediment and the presence of such depositional landforms as deltas and barrier islands. Erosional coasts are generally steep and irregular and typically lack well-developed beaches except in protected areas. They are further characterized by erosional features such as sea cliffs, wave-cut platforms, and sea stacks. Many of the beaches along the west coast of North America fall into this category.

The following section examines coasts in terms of their changing relationships to sea level. But note that while some coasts, such as those in southern California, are described as emergent (uplifted), these same coasts may be erosional as well. In other words, coasts commonly possess features allowing them to be classified in several ways.

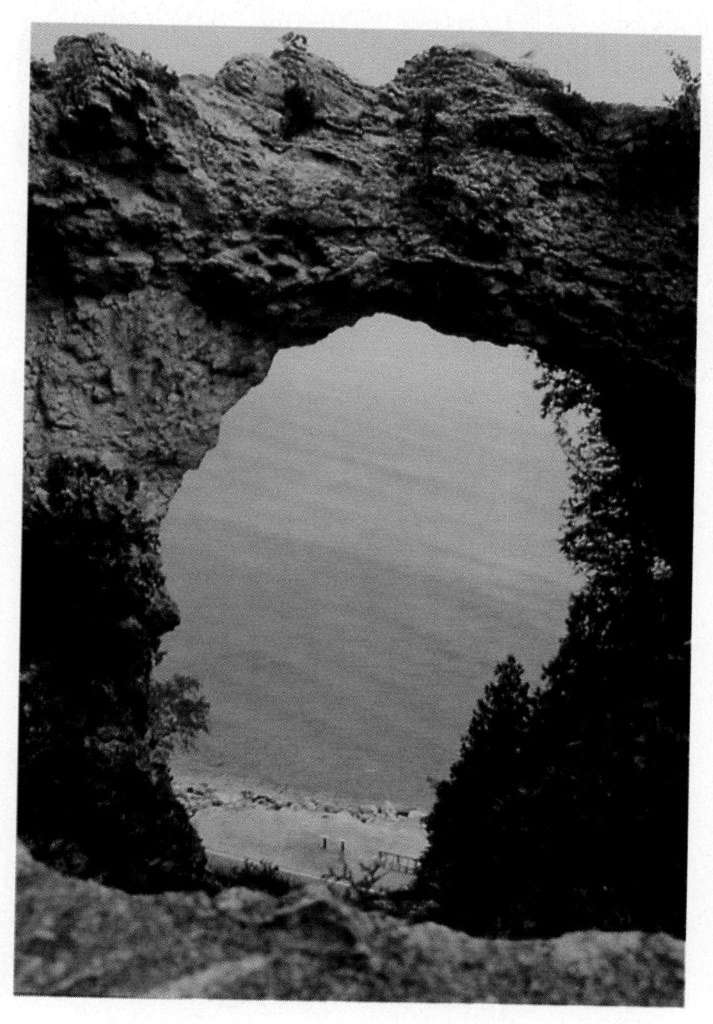

▶ **FIGURE 19-24** Arch Rock on Mackinac Island is a wave erosion feature now above the level of Lake Huron, which is visible through the arch.

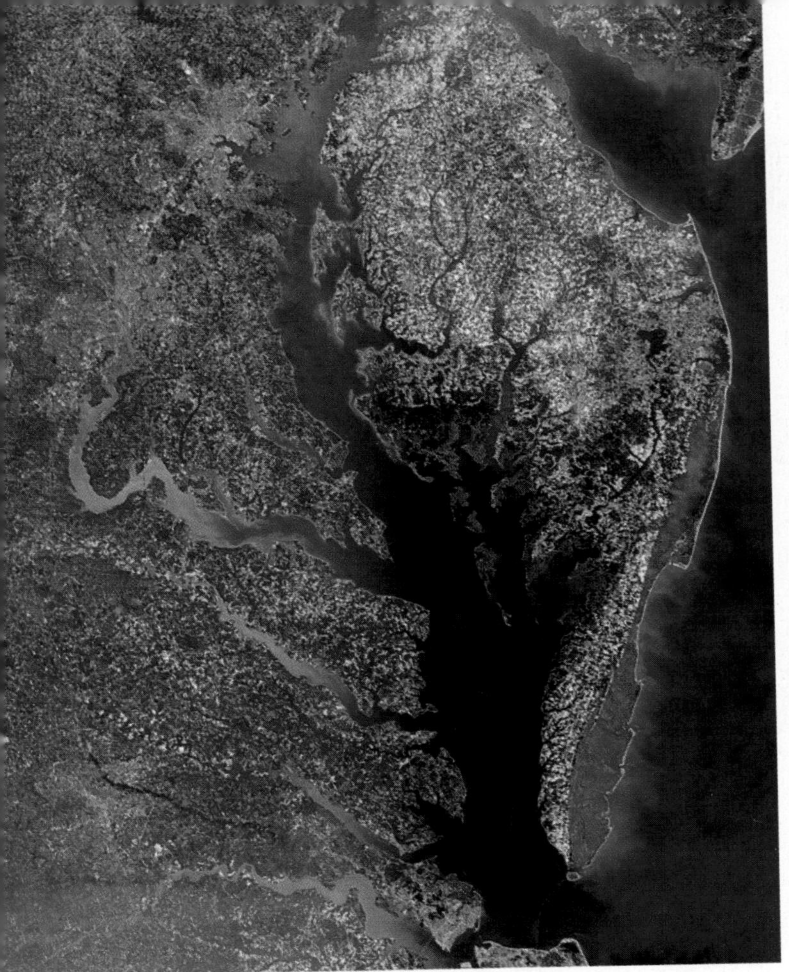

Submergent and Emergent Coasts

If sea level rises with respect to the land or the land subsides, coastal regions are flooded and said to be **submergent** or *drowned* (➢ Figure 19-25). Much of the east coast of North America from Maine southward through South Carolina was flooded during the post-Pleistocene rise in sea level, so that it is now an extremely irregular coast. Recall that during the expansion of glaciers during the Pleistocene, sea level was as much as 130 m lower than at present, and that streams eroded their valleys more deeply as they adjusted to a lower base level. When sea level rose, the lower ends of these valleys were drowned, forming *estuaries* such as Delaware and Chesapeake bays (Figure 19-25). Estuaries are the seaward ends of river valleys where seawater and freshwater mix.

Submerged coasts also occur at higher latitudes where Pleistocene glaciers flowed into the sea. When sea level rose, the lower ends of the glacial troughs were drowned, forming fiords (see Figure 17-17).

Emergent coasts are found where the land has risen with respect to sea level (➢ Figure 19-26). Emergence can occur when water is withdrawn from the oceans as occurred during the Pleistocene expansion of glaciers. At present coasts are emerging as a result of isostasy or tectonism. In northeastern Canada and the Scandinavian countries the coasts are irregular because isostatic rebound is elevating formerly glaciated terrain from beneath the sea.

➢ **FIGURE 19-25** Chesapeake Bay is a large estuary. It formed when the east coast of the United States was flooded as sea level rose following the Pleistocene Epoch.

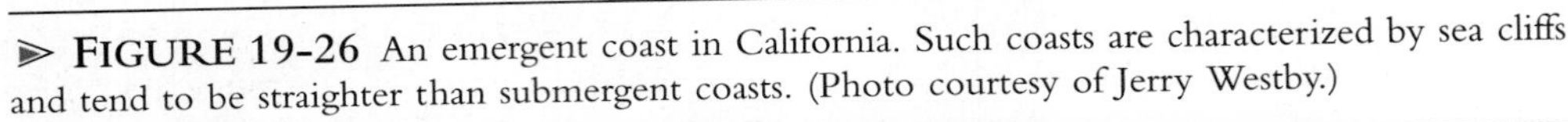

➢ **FIGURE 19-26** An emergent coast in California. Such coasts are characterized by sea cliffs and tend to be straighter than submergent coasts. (Photo courtesy of Jerry Westby.)

▶ **FIGURE 19-27** This gently sloping surface in Ireland is a marine terrace.

Coasts rising in response to tectonism, on the other hand, tend to be straight because the sea-floor topography being exposed as uplift proceeds is smooth. The west coasts of North and South America are rising as a result of plate tectonics. Distinctive features of such coasts include **marine terraces** (▶ Figure 19-27), which are old wave-cut platforms now elevated above sea level. Uplift in such areas appears to be episodic rather than continuous, as indicated by the multiple levels of terraces in some areas. In southern California, for example, several terrace levels are present, each of which probably represents a period of tectonic stability followed by uplift. The highest of these terraces is now about 425 m above sea level.

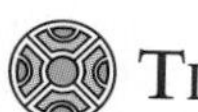

TIDES

Along seacoasts the surface of the ocean rises and falls once or twice daily in response to the gravitational attraction of the Moon and Sun. Such regular fluctuations in the sea's surface are **tides.** Two high tides and two low tides occur daily in most areas as sea level rises and falls anywhere from a few centimeters to more than 15 m (▶ Figure 19-28). During rising or *flood tide,* more and more of the nearshore area is flooded until high tide is reached. *Ebb tide* occurs when currents flow seaward during a decrease in the height of the tide.

Both the Moon and the Sun have sufficient gravitational attraction to exert tide-generating forces strong enough to deform the solid body of the Earth, but they have a much greater influence on the oceans. The Sun is 27 million times more massive than the Moon, but it is 390 times as far from the Earth, and its tide-generating force is only 46% as strong as that of the Moon. Accordingly, the tides are dominated by the Moon, but the Sun does play an important role as well.

If we consider only the Moon acting on a spherical, water-covered Earth, the tide-generating forces produce two bulges on the ocean surface (▶ Figure 19-29a). One bulge extends toward the Moon because it is on the side of the Earth where the Moon's gravitational attraction is greatest. The other bulge occurs on the opposite side of the Earth, where the Moon's gravitational attraction is least. These two bulges always point toward and away from the Moon (Figure 19-29a), so as the Earth rotates and the Moon's position changes, an observer at a particular shoreline location experiences the rhythmic rise and fall of tides twice daily. The heights of two successive high tides may vary depending on the Moon's inclination with respect to the equator.

The Moon revolves around the Earth every 28 days, so its position with respect to any latitude changes slightly each day. That is, as the Moon moves in its orbit and the Earth rotates on its axis, it takes the Moon 50 minutes longer each day to return to the same position it was in the previous day. Thus, an observer would experience a high tide at 1:00 P.M. on one day, for example, and at 1:50 P.M. on the following day.

Tides are also complicated by the combined effects of the Moon and the Sun. Even though the Sun's tide-generating force is weaker than the Moon's, when the Moon and Sun are aligned every two weeks, their forces are added together and generate *spring tides,* which are about 20% higher than average tides (Figure 19-29b). When the Moon and Sun are

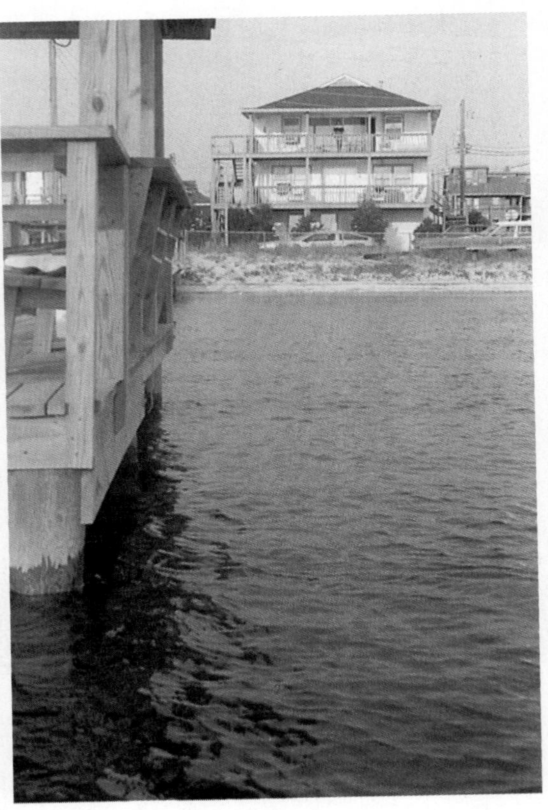

(a) (b)

➢ **FIGURE 19-28** (*a*) Low tide and (*b*) high tide.

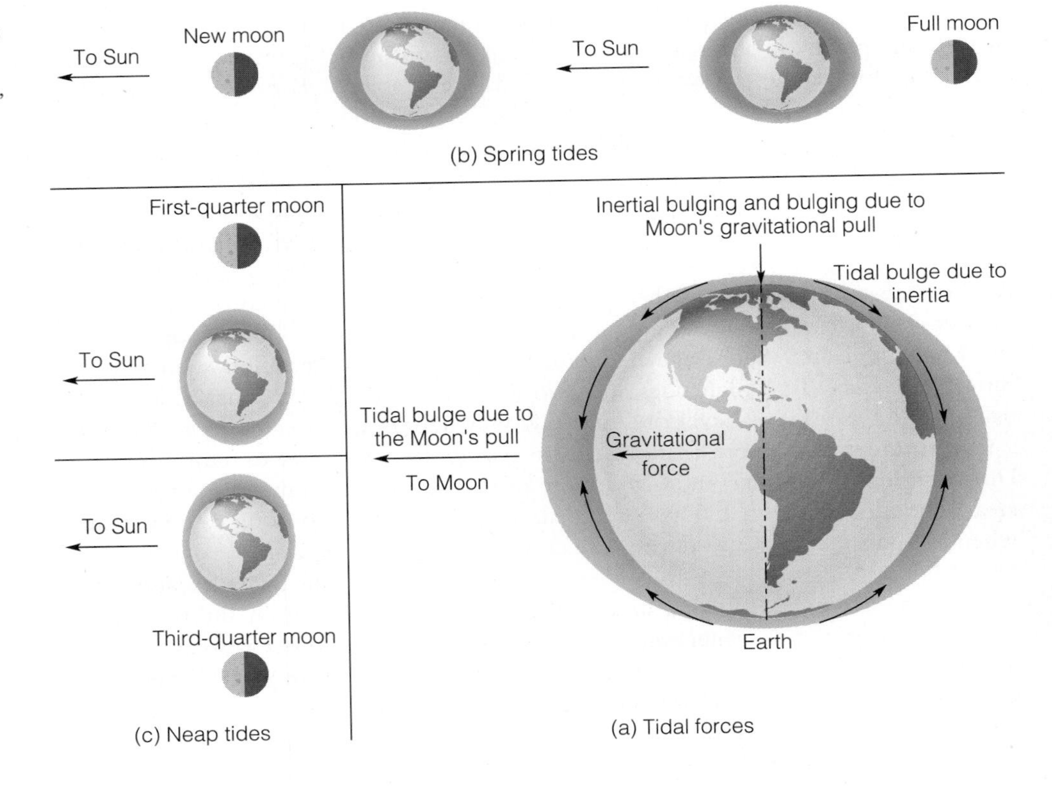

➢ **FIGURE 19-29** (*a*) The tides are caused by the gravitational pull of the Moon and, to a lesser degree, the Sun. The Earth-Moon-Sun alignments at the times of the (*b*) spring and (*c*) neap tides are shown.

at right angles to one another, also at two-week intervals, the Sun's tide-generating force cancels some of that of the Moon, and *neap tides* about 20% lower than average occur (Figure 19-29c).

Tidal ranges are also affected by shoreline configuration. Broad, gently sloping continental shelves as in the Gulf of Mexico have low tidal ranges, whereas steep, irregular shorelines experience a much greater rise and fall of tides. Tidal ranges are greatest in some narrow, funnel-shaped bays and inlets. For example, in the Bay of Fundy in Nova Scotia a tidal range of 16.5 m has been recorded, and ranges greater than 10 m occur in several other areas.

Tides have an important impact on shorelines because the area of wave attack constantly shifts onshore and offshore as the tides rise and fall. Tidal currents themselves have little modifying effect on shorelines, except in narrow passages where tidal current velocity is great enough to erode and transport sediment. Indeed, if it were not for strong tidal currents, some passageways would be blocked by sediments deposited by longshore currents.

Chapter Summary

1. Shorelines are continually modified by the energy of waves and longshore currents and, to a lesser degree, by tidal currents.
2. Waves are oscillations on water surfaces that transmit energy in the direction of wave movement. Surface waves affect the water and sea floor only to wave base, which is equal to one-half the wave length.
3. Little or no net forward motion of water occurs in waves in the open sea. When waves enter shallow water, they are transformed into waves in which water does move in the direction of wave advance.
4. Wind-generated waves, especially storm waves, are responsible for most geologic work on shorelines, but waves can also be generated by faulting, volcanic explosions, and rockfalls.
5. Breakers form where waves enter shallow water and the orbital motion of water particles is disrupted. The waves become oversteepened and plunge forward onto the shoreline, expending their kinetic energy.
6. Waves approaching a shoreline at an angle generate a longshore current. Such currents are capable of considerable erosion, transport, and deposition.
7. Rip currents are narrow surface currents that carry water from the nearshore zone seaward through the breaker zone.
8. Beaches are the most common shoreline depositional features. They are continually modified by nearshore processes, and their profiles generally exhibit seasonal changes.
9. Spits, baymouth bars, and tombolos all form and grow as a result of longshore current transport and deposition.
10. Barrier islands are nearshore sediment deposits of uncertain origin. They parallel the mainland but are separated from it by a lagoon.
11. The volume of sediment in a nearshore system remains rather constant unless the system is somehow disrupted as when dams are built across the streams supplying sand to the system.
12. Many shorelines are characterized by erosion rather than deposition. Such shorelines have sea cliffs and wave-cut platforms. Other features commonly present include sea caves, sea arches, and sea stacks.
13. Submergent and emergent coasts are defined on the basis of their relationships to changes in sea level.
14. The gravitational attraction of the Moon and Sun causes the ocean surface to rise and fall as tides twice daily in most shoreline areas. Most tidal currents have little effect on shorelines.

Important Terms

barrier island
baymouth bar
beach
berm
breaker
crest
emergent coast
fetch
headland
longshore current
longshore drift
marine terrace
nearshore sediment budget
rip current
shoreline
spit
submergent coast
tide
tombolo
trough
wave base
wave-cut platform
wave height
wave length
wave period
wave refraction

Review Questions

1. Which of the following is not a depositional landform?
   a. ____ spit;
   b. ____ tombolo;
   c. ____ baymouth bar;
   d. ____ beach;
   e. ____ sea stack.
2. Wave base is:
   a. ____ the distance offshore that waves break;
   b. ____ the width of a longshore current;
   c. ____ the depth at which the orbital motion in surface waves dies out;
   d. ____ the distance wind blows over a water surface;
   e. ____ the height of storm waves.
3. Waves approaching a shoreline obliquely generate:
   a. ____ flood tides;
   b. ____ longshore currents;
   c. ____ tidal currents;
   d. ____ berms;
   e. ____ marine terraces.
4. Most beach sand is composed of what mineral?
   a. ____ basalt;
   b. ____ calcite;
   c. ____ gravel;
   d. ____ quartz;
   e. ____ feldspar.
5. Erosion of a sea cliff produces a gently sloping surface called a(n):
   a. ____ submergent coast;
   b. ____ wave-cut platform;
   c. ____ beach;
   d. ____ backshore;
   e. ____ emergent coast.
6. Islands composed of sand and separated from the mainland by a lagoon are:
   a. ____ barrier islands;
   b. ____ atolls;
   c. ____ baymouth bars;
   d. ____ sea stacks;
   e. ____ sea arches.
7. The force of waves impacting on shorelines is:
   a. ____ corrosion;
   b. ____ wave oscillation;
   c. ____ hydraulic action;
   d. ____ terracing;
   e. ____ translation.
8. The distance the wind blows over a water surface is the:
   a. ____ fetch;
   b. ____ berm;
   c. ____ spit;
   d. ____ wave period;
   e. ____ wave trough.
9. The bending of waves so that they more nearly parallel the shoreline is:
   a. ____ wave translation;
   b. ____ wave oscillation;
   c. ____ wave deflection;
   d. ____ wave refraction;
   e. ____ wave reflection.
10. The excess water in the nearshore zone returns to the open sea by:
    a. ____ tombolos;
    b. ____ longshore currents;
    c. ____ wave refraction;
    d. ____ emergence;
    e. ____ rip currents.
11. A sand deposit extending into the mouth of a bay is a:
    a. ____ headland;
    b. ____ beach;
    c. ____ spit;
    d. ____ wave-cut platform;
    e. ____ sea stack.
12. Although there are exceptions, most beaches receive most of their sediment from:
    a. ____ wave erosion of sea cliffs;
    b. ____ erosion of offshore reefs;
    c. ____ streams;
    d. ____ breakers;
    e. ____ coastal submergence.
13. Erosional remnants of a shoreline now rising above a wave-cut platform are:
    a. ____ barrier islands;
    b. ____ sea stacks;
    c. ____ beaches;
    d. ____ marine terraces;
    e. ____ spits.
14. Which of the following is a distinctive feature of emergent coasts?
    a. ____ marine terraces;
    b. ____ estuaries;
    c. ____ drowned river valleys;
    d. ____ very high tidal range;
    e. ____ fiords.
15. How do deep- and shallow-water waves differ?
16. What is wave base and how does it affect waves as they enter shallow water?
17. Explain how a longshore current is generated.
18. What is longshore drift?
19. Sketch a north-south shoreline along which several groins have been constructed. Assume that waves approach from the northwest.
20. Explain why quartz is the most common mineral composing beach sands.
21. Sketch the profiles of a summer beach and a winter beach, and explain why they differ.
22. How does a tombolo form?
23. How does a wave-cut platform develop?
24. Explain how an initially irregular shoreline is straightened. A sketch may be helpful.
25. Why does an observer at a shoreline experience two high and two low tides each day?
26. What are the characteristics of a submergent coast?

Points to Ponder

1. An initially straight shoreline is composed of granite flanked on both sides by glacial drift. Diagram and explain this shoreline's probable reponse to erosion.
2. How might burning fossil fuels have some long-term effect on shoreline erosion?
3. Why are long, broad sandy beaches more common in eastern North America than western North America?
4. A hypothetical nearshore area has a balanced sand budget, but a dam is constructed on the stream supplying most of the land-derived sand and a wall is built to protect sea cliffs from erosion. What will likely happen to the beach in this area?

Additional Readings

Abrahamson, D. E., ed. 1989. *The challenge of global warming.* Washington, D.C.: Island Press.

Bird, E. C. F. 1984. *Coasts: An introduction to coastal geomorphology.* New York: Blackwell.

Bird, E. C. F., and M. L. Schwartz. 1985. The world's coastline. New York: Van Nostrand Reinhold Co.

Flanagan, R. 1993. Beaches on the brink. *Earth* 2, no. 6: 24–33.

Fox, W. T. 1983. *At the sea's edge.* Englewood Cliffs, N.J.: Prentice-Hall.

Garrett, C., and L. R. M. Maas. 1993. Tides and their effects. *Oceanus* 36, no. 1: 27–37.

Hecht, J. 1988. America in peril from the sea. *New Scientist* 118:54–59.

Komar, P. D. 1976. *Beach processes and sedimentation.* Englewood Cliffs, N.J.: Prentice-Hall.

______. 1983. *CRC handbook of coastal processes and erosion.* Boca Raton, Fla.: CRC Press.

Pethick, J. 1984. *An introduction to coastal geomorphology.* London: Edward Arnold.

Schneider, S. H. 1990. *Global warming: Are we entering the greenhouse century?* San Francisco, Calif.: Sierra Club Books.

Snead, R. 1982. *Coastal landforms and surface features.* Stroudsburg, Pa.: Hutchinson Ross Publishing Co.

The Open University. 1989. *Waves, tides, and shallow-water processes.* New York: Pergamon Press.

Walden, D. 1990. Raising Galveston. *American Heritage of Invention & Technology* 5:8–18.

Williams, S. J., K. Dodd, and K. K. Gohn. 1990. Coasts in crisis. *U.S. Geological Survey Circular 1075.*

Woods Hole Oceanographic Institute. 1993. Coastal science and policy I. *Oceanus* 36, no. 1.

Chapter 20

A History of the Universe, Solar System, and Planets

Outline

At the stage of development shown here, planetesimals have formed in the inner solar system, and large eddies of gas and dust remain at great distances from the protosun.

Prologue

On August 20 and September 5, 1977, *Voyagers 1* and *2* were launched on an ambitious mission to explore the outer planets. They both flew by Jupiter and Saturn, but *Voyager 1* took a course out of the solar system while *Voyager 2* went on to Uranus and Neptune. Twelve years and 7.13 billion km after it was launched, *Voyager 2* radioed back spectacular images of the blue planet Neptune (▶ Figure 20-1) and its pink and blue mottled moon Triton. Its primary mission completed, and with all but a few of its instruments turned off to conserve power, *Voyager 2*'s last act will be to measure the exotic fields and subatomic particles it passes through on its voyage to infinity.

The discoveries made by these two space probes were truly fantastic and in many cases totally unexpected by scientists. In addition to discovering three new moons of Jupiter, the *Voyagers* found dusty rings encircling the planet, demonstrating that rings are a common feature in the outer solar system. They showed that the Great Red Spot is an enormous, persistent eddy in Jupiter's atmosphere, and they detected lightning discharges that are 10,000 times more powerful than those on Earth. The *Voyagers* sent back images of one of Jupiter's moons, Io, spewing forth hot sulfurous gases 320 km into space. Another Jovian moon, Europa, was revealed to be encrusted with a thick shell of ice covering a liquid ocean several kilometers below its surface (▶ Figure 20-2). This ice is crisscrossed with what appear to be cracks that occasionally open to erupt water and then refreeze.

As the *Voyagers* flew past Saturn, they revealed the spectacular complexity of its 70,000 km wide ring system and sent back images of spiral bands of debris only 35 m thick. The *Voyagers* also discovered that the Saturnian moon Titan has an atmosphere rich in hydrocarbons and nitrogen and may even have oceans consisting primarily of lightweight hydrocarbons.

As *Voyager 2* passed by Uranus on January 24, 1986, it found nine dark, compact rings encircling the planet, discovered 10 new moons, and revealed a corkscrew-shaped magnetic field that extends for millions of kilometers from the planet.

Voyager 2 reached its final target, Neptune, in August 1989 and sent back spectacular images and data that were, for the most part, completely unanticipated. Instead of a quiet, placid planet, Neptune turned out to be a dynamic world cloaked in a thin atmosphere composed predominantly of hydrogen and helium mixed with some methane. Winds up to 2,000 km/hour blow over the planet creating tremendous storms, the largest of which, the Great Dark Spot, is in the southern hemisphere. It is nearly as big as the Earth and is similar to the Great Red Spot on Jupiter. Indeed, one of the mysteries raised by *Voyager 2*'s discovery is where Neptune gets the energy to drive such a storm system.

▶ FIGURE 20-1 A shimmering blue planet set against the black backdrop of space, Neptune reveals itself to *Voyager 2*'s instruments during its August 1989 flyby. Shown here is Neptune's turbulent atmosphere with its Great Dark Spot and various wispy clouds.

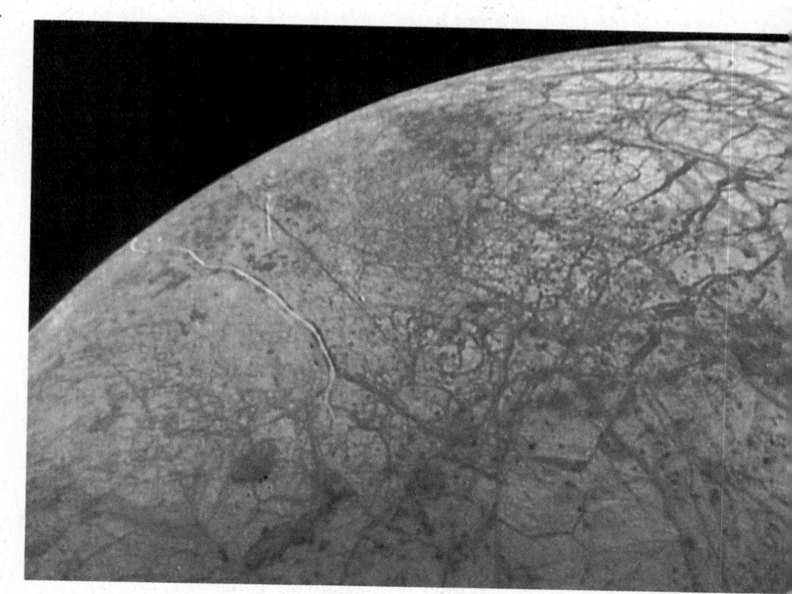

▶ FIGURE 20-2 Europa, the second moon out from Jupiter, is covered by a thick surface layer of ice crisscrossed by numerous fractures. These fractures appear to be rifts where water has risen to the surface and frozen. This system of fractures and the lack of craters are evidence that Europa is a geologically active moon.

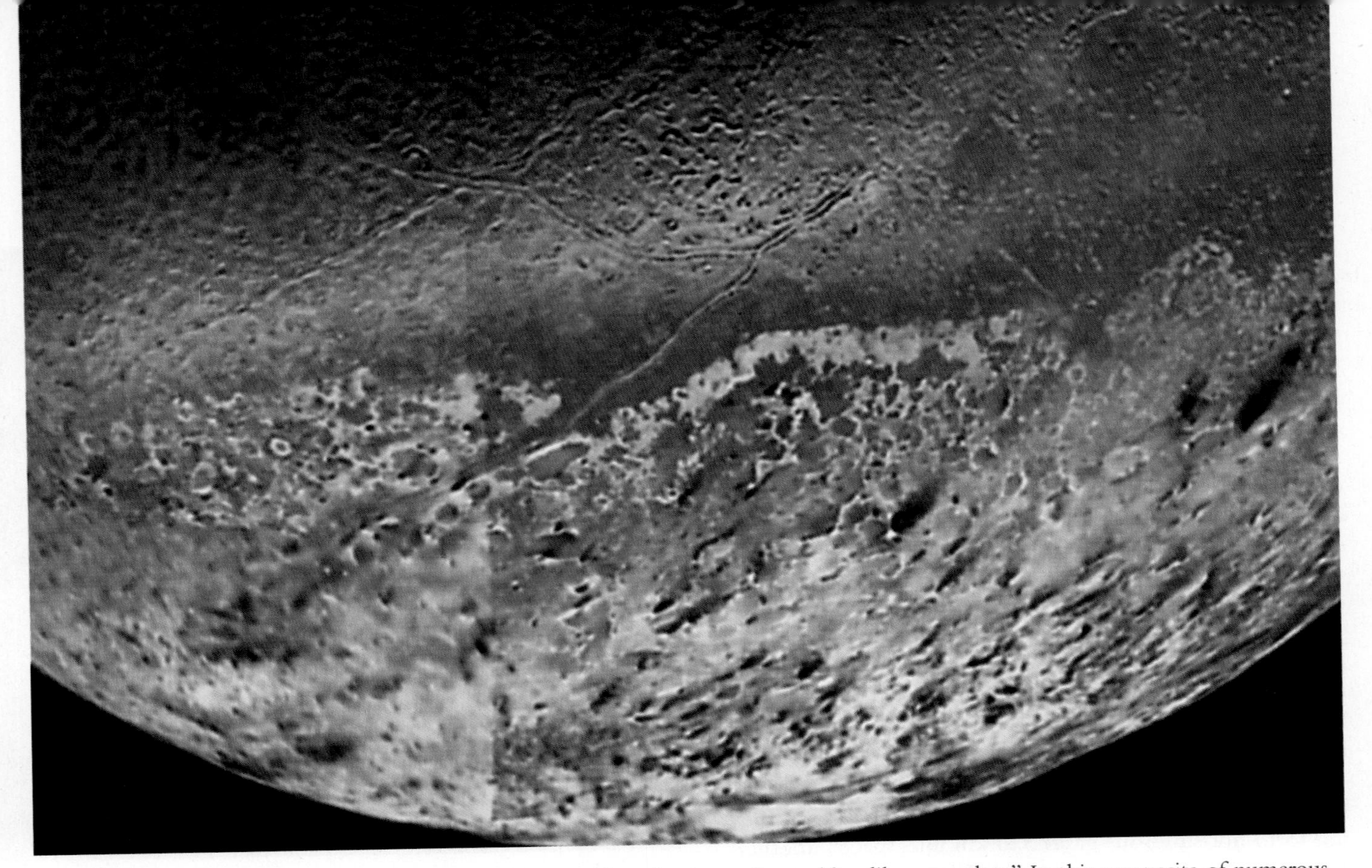

➢ **FIGURE 20-3** Neptune's moon Triton is described by scientists as "a world unlike any other." In this composite of numerous high-resolution images taken by *Voyager 2* during its August 1989 flyby, various features can be seen. The large south polar ice cap at the bottom consists mostly of frozen nitrogen that was deposited during the previous Tritonian winter and is slowly evaporating. The dark plumes in the lower right may be the result of volcanic activity. Smooth plains and fissures in the upper half are evidence of geologic activity in which the surface has been cracked and flooded by slushy ice that refroze.

Equally intriguing were the discoveries of six new Neptunian moons and three rings encircling the planet. The most astonishing discoveries, however, were found on Neptune's largest moon Triton, which has a diameter of 2,720 km, 700 km less than our own moon (➢ Figure 20-3). Triton, with a mottled surface of delicate pinks, reds, and blues, is turning out to be one of the most colorful objects in the solar system. Its surface consists primarily of water ice, with minor amounts of nitrogen and a methane frost. Geysers were also discovered to be erupting carbon-rich material and frozen nitrogen particles some 8 km above its surface. This makes Triton only the second place other than Earth undergoing active volcanism.

Some areas of Triton are smooth while others have a very irregular appearance indicating numerous episodes of deformation. Heavily cratered areas bear witness to bombardment by meteorites or the collapse of its surface. Perhaps the most intriguing aspect of Triton is that it may have once been a planet—much like Pluto, which it resembles in size and possibly composition—that was captured by Neptune's gravitational field soon after the formation of the solar system. Much still needs to be learned about Triton and Neptune's other moons before this hypothesis can be accepted.

INTRODUCTION

Of all the known planets and their moons only one, Earth, has life on it. This unique planet, revolving around the Sun every 365.25 days, is a dynamic and complex body. When viewed from the blackness of space, the Earth is a brilliant, shimmering, bluish planet, wrapped in a veil of swirling white clouds (see Figure 1-4). Beneath these clouds is a surface covered by oceans, seven continents, and numerous islands.

The Earth has not always looked the way it does today. Based on various lines of evidence, many scientists think that it began as a homogeneous mass of rotating dust and gases that contracted, heated, and differentiated during its early history to form a medium-sized planet with a metallic core, a mantle composed of iron- and magnesium-rich rocks, and a thin crust. Overlying this crust is an atmosphere currently composed of 78% nitrogen and 21% oxygen.

As the third planet from the Sun, Earth seems to have formed at just the right distance from the Sun (149,600,000 km) so that it is neither too hot nor too cold to support life as we know it. Furthermore, its size is just right to hold an atmosphere. If the Earth were smaller, its gravity would be so weak that it could retain little, if any, atmosphere.

The Origin of the Universe

Most scientists think that the universe originated between 13 and 20 billion years ago in what is popularly called the "**Big Bang**." In a region infinitely smaller than an atom, both time and space were set at zero. As explained by Einstein's theory of relativity, space and time are unalterably linked to form a space-time continuum. In other words, without space there can be no time. Therefore, there is no "before the Big Bang," only what occurred after it.

Two fundamental phenomena indicate that the Big Bang occurred. The first is the expansion of the universe. When astronomers look beyond the solar system, they observe that everywhere in the universe galaxies are apparently moving away from each other at tremendous speeds. By measuring this expansion rate, they can calculate how long ago the galaxies were all together at a single point. Secondly, a background radiation of 2.7° above absolute zero (absolute zero equals −273°C) permeates the entire universe. This background radiation is thought to be the faint afterglow of the Big Bang.

At the time of the Big Bang, matter as we know it did not exist, and the universe consisted of pure energy. Within the first second after the Big Bang, the four basic forces—gravity, electromagnetic force, strong nuclear force, and weak nuclear force (◉ Table 20-1)—had all separated, and the universe experienced enormous expansion. Matter and antimatter collided and annihilated each other. Fortunately, there was a slight excess of matter left over that would become the universe. When the universe was three minutes old, temperatures were cool enough for protons and neutrons to fuse together to form the nuclei of hydrogen and helium atoms. Approximately 100,000 years later, electrons joined with the previously formed nuclei to make complete atoms of hydrogen and helium. At the same time, photons (the energetic particles of light) separated from matter, and light burst forth for the first time.

As the universe continued expanding and cooling, stars and galaxies formed, and the chemical makeup of the universe changed. Early in its history, the universe was 100% hydrogen and helium, whereas today it is 98% hydrogen and helium by weight.

Over the course of their history, stars undergo many nuclear reactions whereby lighter elements are converted into heavier elements by nuclear fusion in which atomic nuclei combine to form more massive nuclei. Such reactions, which convert hydrogen to helium, occur in the cores of all stars. The subsequent conversion of helium to heavier elements, such as carbon, depends on the mass of the star. When a star dies, often explosively, the heavier elements that were formed in its core are returned to interstellar space and are available for inclusion in new stars. When new stars form, they will have a small amount of these heavier elements, which may be converted to still heavier elements. In this way, the chemical composition of the galaxies, which are made up of billions of stars, is gradually enhanced in heavier elements.

TABLE 20-1 The Four Basic Forces of the Universe

Four forces appear to be responsible for all interactions of matter:

1. **Gravity** is the attraction of one body toward another.
2. The **electromagnetic force** combines electricity and magnetism into the same force and binds atoms into molecules. It also transmits radiation across the various spectra at wavelengths ranging from gamma rays (shortest) to radio waves (longest) through massless particles called *photons*.
3. The **strong nuclear force** binds protons and neutrons together in the nucleus of an atom.
4. The **weak nuclear force** is responsible for the breakdown of an atom's nucleus, producing radioactive decay.

SOURCE: Adapted by permission from Table 7-2, page 175 of *Historical Geology: Evolution of the Earth and Life through Time* by Reed Wicander and James S. Monroe.

The Origin and Early Development of the Solar System

Having briefly examined the origin and history of the universe, we can now examine how our own solar system, which is part of the Milky Way Galaxy, formed. The solar system consists of the Sun, nine planets, 61 known moons, a tremendous number of asteroids—most of which orbit the Sun in a zone between Mars and Jupiter—and millions of comets, meteorites, and interplanetary dust and gases (◉ Table 20-2).

General Characteristics of the Solar System

Any theory that attempts to explain the origin and history of the solar system must take into account several general characteristics (◉ Table 20-3).

All of the planets revolve around the Sun in the same direction, in nearly circular orbits, and in approximately the

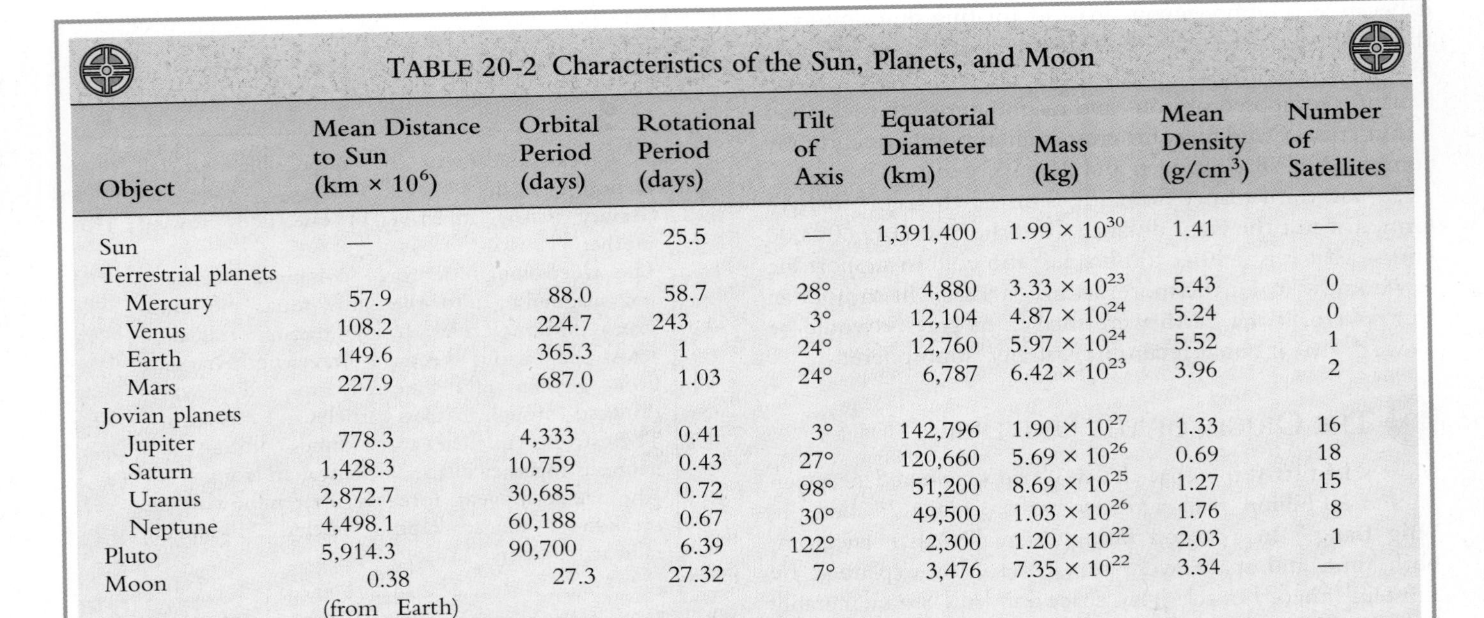

TABLE 20-2 Characteristics of the Sun, Planets, and Moon

Object	Mean Distance to Sun (km × 10^6)	Orbital Period (days)	Rotational Period (days)	Tilt of Axis	Equatorial Diameter (km)	Mass (kg)	Mean Density (g/cm^3)	Number of Satellites
Sun	—	—	25.5	—	1,391,400	1.99×10^{30}	1.41	—
Terrestrial planets								
Mercury	57.9	88.0	58.7	28°	4,880	3.33×10^{23}	5.43	0
Venus	108.2	224.7	243	3°	12,104	4.87×10^{24}	5.24	0
Earth	149.6	365.3	1	24°	12,760	5.97×10^{24}	5.52	1
Mars	227.9	687.0	1.03	24°	6,787	6.42×10^{23}	3.96	2
Jovian planets								
Jupiter	778.3	4,333	0.41	3°	142,796	1.90×10^{27}	1.33	16
Saturn	1,428.3	10,759	0.43	27°	120,660	5.69×10^{26}	0.69	18
Uranus	2,872.7	30,685	0.72	98°	51,200	8.69×10^{25}	1.27	15
Neptune	4,498.1	60,188	0.67	30°	49,500	1.03×10^{26}	1.76	8
Pluto	5,914.3	90,700	6.39	122°	2,300	1.20×10^{22}	2.03	1
Moon	0.38 (from Earth)	27.3	27.32	7°	3,476	7.35×10^{22}	3.34	—

TABLE 20-3 General Characteristics of the Solar System

1. **Planetary orbits and rotation**
   - Planetary and satellite orbits lie in a common plane.
   - Nearly all of the planetary and satellite orbital and spin motions are in the same direction.
   - The rotation axes of nearly all the planets and satellites are roughly perpendicular to the plane of the ecliptic.
2. **Chemical and physical properties of the planets**
   - The terrestrial planets are small, have a high density (4.0 to 5.5 g/cm^3), and are composed of rock and metallic elements.
   - The Jovian planets are large, have a low density (0.7 to 1.7 g/cm^3), and are composed of gases and frozen compounds.
3. **The slow rotation of the Sun**
4. **Interplanetary material**
   - The existence and location of the asteroid belt.
   - The distribution of interplanetary dust.

SOURCE: Reprinted by permission from Table 7-3, page 177 of *Historical Geology: Evolution of the Earth and Life through Time* by Reed Wicander and James S. Monroe.

same plane (called the *plane of the ecliptic*), except for Pluto whose orbit is both highly elliptical and tilted 17° to the orbital plane of the rest of the planets (▶ Figure 20-4b).

All of the planets, except Venus and Uranus, and nearly all the planetary moons rotate counterclockwise when viewed from a point high in space above the Earth's North Pole. Furthermore, the axes of rotation of the planets, except for those of Uranus and Pluto, are nearly perpendicular to the plane of the ecliptic (Figure 20-4b).

The nine planets can be divided into two groups based on their chemical and physical properties. The four inner planets—Mercury, Venus, Earth, and Mars—are all small and have high mean densities (Table 20-2), indicating that they are composed of rock and metallic elements. They are known as the **terrestrial planets** because they are similar to *terra,* which is Latin for Earth.

The next four planets—Jupiter, Saturn, Uranus, and Neptune—are called the **Jovian planets** because they all resemble Jupiter. The Jovian planets are large and have low mean densities, indicating they are composed of lightweight gases such as hydrogen and helium, as well as frozen compounds such as ammonia and methane. The outermost planet, Pluto, is small and has a low mean density of slightly more than 2.0 g/cm^3.

The slow rotation of the Sun is another feature that must be accounted for in any comprehensive theory of the origin of the solar system. If the solar system formed from the collapse of a rotating cloud of gas and dust as is currently accepted, the Sun, which was at the center of that cloud, should have a very rapid rate of rotation, instead of its leisurely 25-day rotation.

Finally, any theory of the origin of the solar system must accommodate the nature and distribution of the various interplanetary objects such as the asteroid belt, comets, and interplanetary gases and dust.

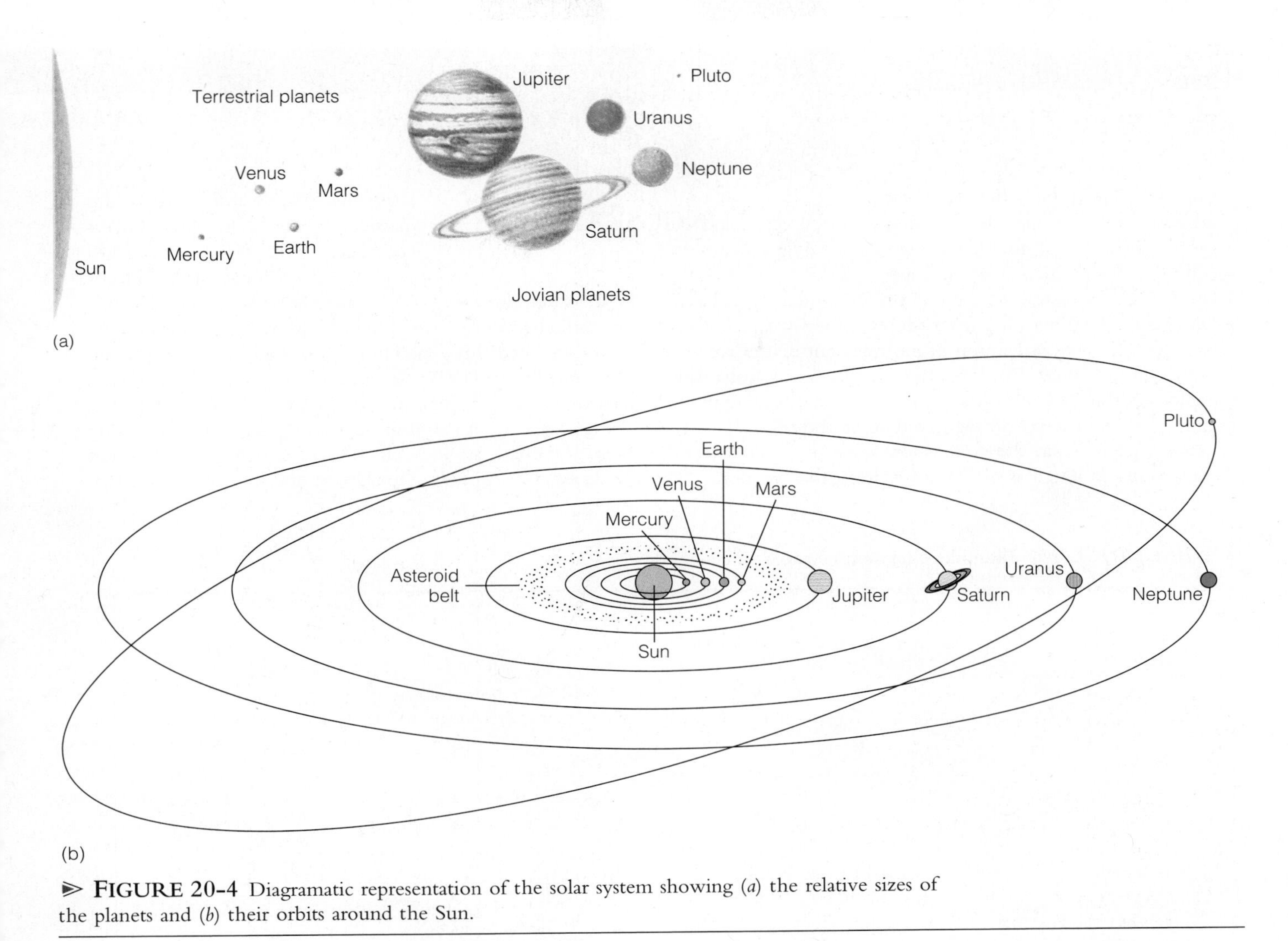

▶ **FIGURE 20-4** Diagramatic representation of the solar system showing (*a*) the relative sizes of the planets and (*b*) their orbits around the Sun.

Current Theory of the Origin and Early History of the Solar System

Various scientific theories for the origin of the solar system have been proposed, modified, and discarded since the French scientist and philosopher René Descartes first proposed in 1644 that the solar system formed from a gigantic whirlpool within a universal fluid. Most theories have involved an origin from a primordial rotating cloud of gas and dust. Through the forces of gravity and rotation, this cloud then shrank and collapsed into a rotating disk. Detached rings within the disk condensed into planets, and the Sun condensed in the center of the disk.

The problem with most of these theories is that they failed to explain the slow rotation of the Sun, which according to the laws of physics should be rotating rapidly. This problem was finally solved with the discovery of solar wind, which is an outflow of ionized gases from the Sun that interact with its magnetic field and slow down its rotation through a magnetic braking process (▶ Figure 20-5).

▶ **FIGURE 20-5** The slow rotation of the Sun is the result of the interaction of its magnetic force lines with ionized gases of the solar nebula. Thus, the rotation is slowed by a magnetic braking process.

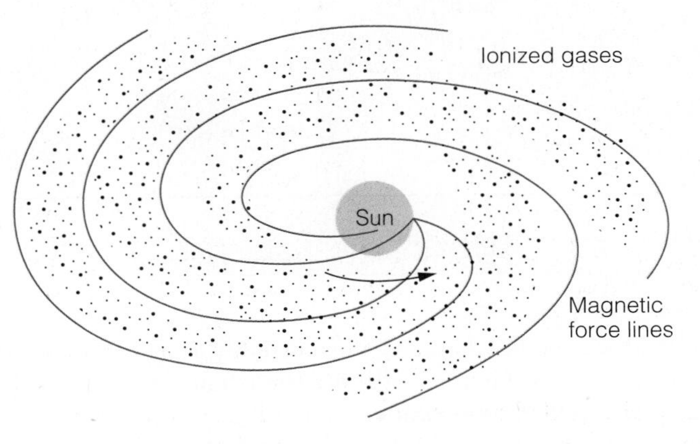

Perspective 20-1

THE TUNGUSKA EVENT

On June 30, 1908, a bright object crossed the morning sky moving from southeast to northwest over central Siberia, and a few seconds later a huge explosion occurred in the Tunguska River basin (▶ Figure 1). The noise from the explosion was heard up to 1,000 km away, a column of incandescent matter rose to a height of about 20 km, the shock wave from the explosion traveled around the world twice, and seismographs around the world registered an earthquake. Eyewitnesses reported that the concussion wave threw people to the ground as much as 60 km away from the blast site.

What the object was that caused this massive explosion remains uncertain. Part of the uncertainty is because the event occurred in an extremely remote area, and an expedition to investigate was not launched until 1921. Unfortunately, illness and exhaustion prevented this expedition from reaching the explosion site. Finally, in 1927, 19 years after the explosion, an expedition led by Leonid Kulik successfully reached the Tunguska basin. A vast peat bog called the Southern Swamp was identified as the site above which the explosion occurred; subsequent investigations and studies indicate that

▶ **FIGURE 1** The Tunguska explosion occurred in central Siberia.

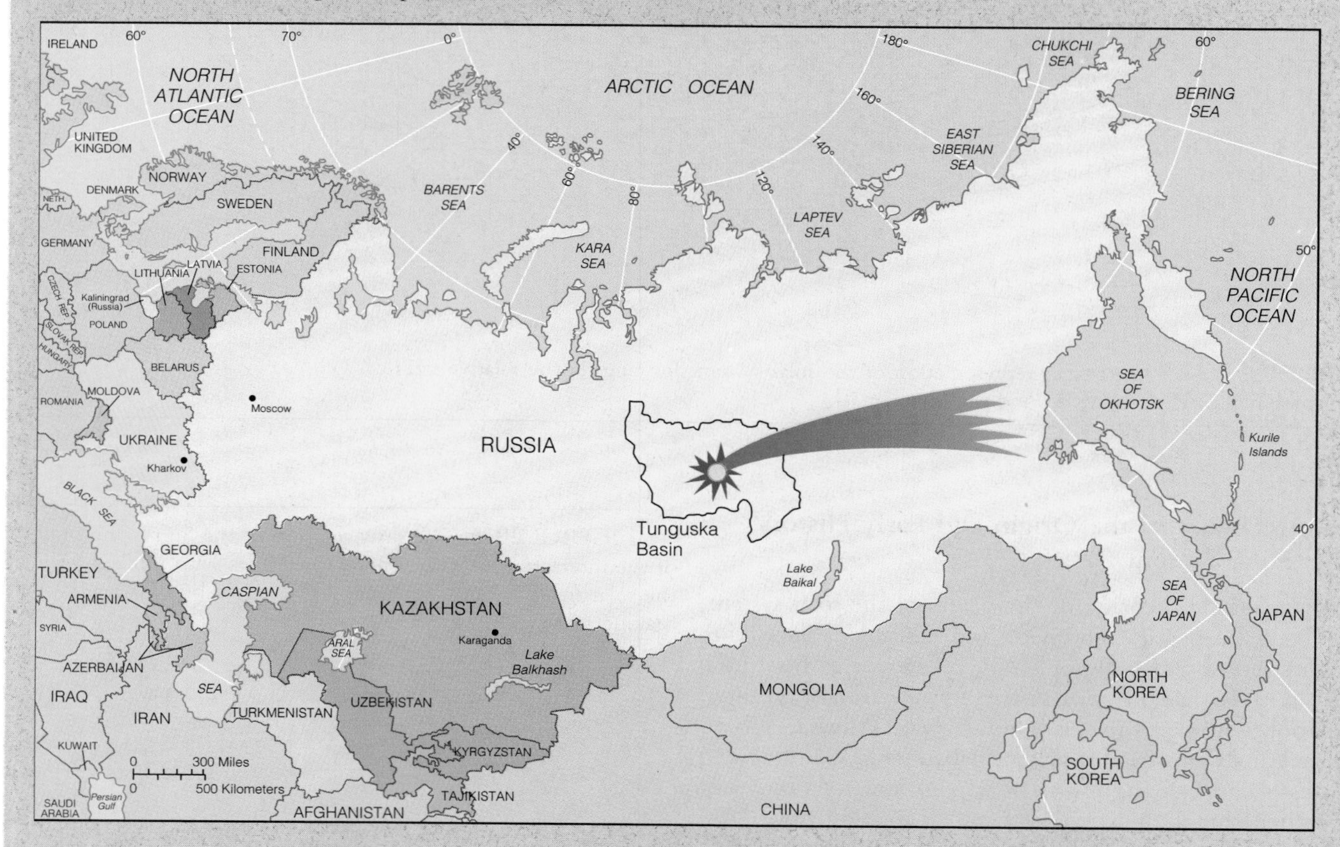

As we discussed in Chapter 1, the currently accepted **solar nebula theory** (see Figure 1-10) for the origin of the solar system involves the condensation and collapse of interstellar material in a spiral arm of the Milky Way Galaxy. The collapse of this cloud of gases and small grains into a counterclockwise rotating disk caused about 90% of the material to be concentrated in the central part of the disk and resulted in the formation of an embryonic Sun, around which swirled a rotating cloud of material called a solar nebula. Within this solar nebula were localized eddies in which gas and solid particles condensed. Collisions between the various gases and particles resulted in accretion into

➢ FIGURE 2 Evidence of the Tunguska event is still apparent in this photograph taken 20 years later. The destruction was caused by some type of explosion in central Siberia in 1908.

the explosion occurred about 8 km above the surface, and it is estimated to have been about 12.5 megatons (equivalent to 12.5 million tons of TNT). More than 1,000 km^2 of forest were leveled by the explosion, and according to earlier accounts, tens of thousands of animals perished, and several nomadic settlements vanished (➢ Figure 2).

Even before the explosion site was reached, scientists had hypothesized that the explosion resulted from a meteorite impact. In fact, part of the incentive for investigation may have been economic; the Soviets believed that a meteor was present and could be mined for its iron content. When investigators finally reached the site, however, no evidence of a meteor crater was ever identified. This led most scientists to conclude that the Tunguska event was caused by a small, icy comet that exploded in the atmosphere. According to this hypothesis, a comet, perhaps 50 m in diameter, entered the atmosphere and began heating up; as this heating occurred, frozen gases were instantaneously converted to the gaseous state and released a tremendous amount of energy, causing a large explosion.

This explanation has recently been challenged. According to computer simulations and studies, the nuclei of comets explode at too high an altitude to account for the pattern and magnitude of destruction found in the Tunguska River basin. Instead, scientists conclude that a stony meteorite about 30 m in diameter, traveling at 15 km per second, would explode at approximately the same altitude as the Tunguska object did, producing the same type of destruction.

These computer simulations and studies rule out not only the possibility that the Tunguska object was a comet, but that it could have been an iron meteorite. Iron meteorites can resist greater pressures than stony meteorites and thus reach the Earth's surface where they would produce a crater. Stony meteorites, however, are intermediate in density between comets and irons and would explode at altitudes similar to the Tunguska object.

planetesimals. As the planetesimals collided and grew in size and mass, they eventually became planets. It is estimated that a planet the size of the Earth formed from planetesimals in as short a time as 1 million years.

The composition and evolutionary history of the planets are indicated, in part, by their distance from the Sun. The terrestrial planets are composed of rock and metallic elements that condensed at the high temperatures of the inner nebula. The Jovian planets, all of which have small central rocky cores compared to their overall size, are composed mostly of hydrogen, helium, ammonia, and methane, which condense at low temperatures. Thus, the farther away from the Sun that condensation occurred, the lower the temperature, and hence the higher the percentage of volatile elements, which condense at low temperatures, relative to refractory elements, which condense at high temperatures.

While the planets were accreting, material that had been pulled into the center of the nebula also condensed, collapsed, and was heated to several million degrees by gravitational compression. The result was the birth of a star, our Sun.

During the early accretionary phase of the solar system's history, collisions between various bodies were common, as indicated by the craters on many planets and moons. An unusually large collision involving Venus could explain why it rotates clockwise rather than counterclockwise, and a collision could also explain why Uranus and Pluto do not rotate nearly perpendicular to the plane of the ecliptic.

It is thought that the *asteroids* probably formed as planetesimals in a localized eddy between what eventually became Mars and Jupiter in much the same way as other planetesimals formed the terrestrial planets. The tremendous gravitational field of Jupiter, however, prevented this material from ever accreting into a planet.

The *comets,* which are interplanetary bodies composed of loosely bound rocky and icy material, are thought to have condensed near the orbits of Uranus and Neptune. Each time the comets pass by Jupiter and Saturn, the gravitational effect of those planets increases their speed, forcing the comets further out into the solar system.

The solar nebula theory of the formation of the solar system thus accounts for the similarities in orbits and rotation of the planets and their moons, the differences in composition between the terrestrial and Jovian planets, the slow rotation of the Sun, and the presence of the asteroid belt. Based on the available data, the solar nebula theory best explains the features of the solar system and provides a logical explanation for its evolutionary history.

METEORITES

Meteorites are thought to be pieces of material that originated during the formation of the solar system 4.6 billion years ago. Early in the history of the solar system, a period of heavy meteorite bombardment occurred as the solar system cleared itself of the many pieces of material that had not yet accreted into planetary bodies or moons. Since then, meteorite activity has greatly diminished. Most of the meteorites that currently reach the Earth are probably fragments resulting from collisions between asteroids.

Meteorites are classified into three broad groups based on their proportions of metals and silicate minerals (▶ Figure 20-6a). About 93% of all meteorites are composed of iron and magnesium silicate minerals and are known as **stones** (Figure 20-6b). There are many varieties of stones, and they provide geologists with much information about the origin and history of the solar system.

Irons, the second group, accounting for about 6% of all meteorites, are composed primarily of a combination of iron and nickel alloys (Figure 20-6c). Their large crystal size and chemical composition indicate that they cooled very slowly in large objects such as asteroids where the hot iron-nickel interior could be insulated from the cold of space. Collisions between such slowly cooling asteroids produced the iron meteorites that we find today.

Stony-irons, the third group, are composed of nearly equal amounts of iron and nickel and silicate minerals; they make up less than 1% of all meteorites (Figure 20-6d). Stony-irons are generally thought to represent fragments from the zone between the silicate and metallic portions of a large differentiated asteroid.

Astronomers have identified at least 40 asteroids more than a kilometer in diameter whose orbits cross the Earth's and estimate that there may be as many as 1,000 such asteroids. A collision between a large asteroid and the Earth formed the famous Meteor Crater in Arizona 25,000 to 50,000 years ago (▶ Figure 20-7). While asteroid-Earth collisions are rare, they do happen and could have devastating results if they occurred in a populated area (▶ Figure 20-8) (see Perspective 20-1). Many scientists think that a collision with a meteorite about 10 km in diameter led to the extinctions of dinosaurs and several other groups of animals 66 million years ago (▶ Figure 20-9). Such a collision would have generated a tremendous amount of dust that would have blocked out the Sun, thereby lowering global temperatures and preventing photosynthesis, which, in turn, would have triggered a collapse of the ecosystem and massive extinctions. We know that the ash and gases released into the atmosphere from volcanic eruptions has affected climates (see Perspective 4-1), and studies indicate that a collision with a large meteorite could produce enough dust to similarly affect global climate.

THE PLANETS

A tremendous amount of information about each planet in the solar system has been derived from Earth-based observations and measurements as well as from the numerous space probes launched during the past 30 years. Such information as a planet's size, mass, density, composition, presence of a magnetic field, and atmospheric composition has allowed scientists to formulate hypotheses concerning the origin and history of the planets and their moons.

The Terrestrial Planets

It appears that all of the terrestrial planets had a similar early history during which volcanism and cratering from meteorite impacts were common. During the accretionary phase of formation and soon thereafter, each planet appears to have undergone differentiation as a result of heating by radioactive decay. The mass, density, and composition of the planets indicate that each formed a metallic core and a silicate mantle-crust during this phase. Images sent back by the various space probes also clearly show that volcanism and cratering by meteorites continued during the differentiation

➢ **FIGURE 20-6** (*a*) Relative proportions of the three groups of meteorites. (*b*) Stony meteorite. (*c*) Iron meteorite. (*d*) Stony-iron meteorite.

▶ **FIGURE 20-7** Meteor Crater, Arizona, is the result of an Earth-asteroid collision that occurred between 25,000 and 50,000 years ago. It produced a crater 1.2 km in diameter and 180 m deep.

▶ **FIGURE 20-8** Artistic rendition of what the moment of impact would look like if the nucleus of a comet, 48 km in diameter, hit northern New Jersey. Everything visible in this picture, including the buildings of lower Manhattan in the foreground, would be vaporized, and a plume of fine material would be ejected into the atmosphere and circulated around the Earth.

(a)

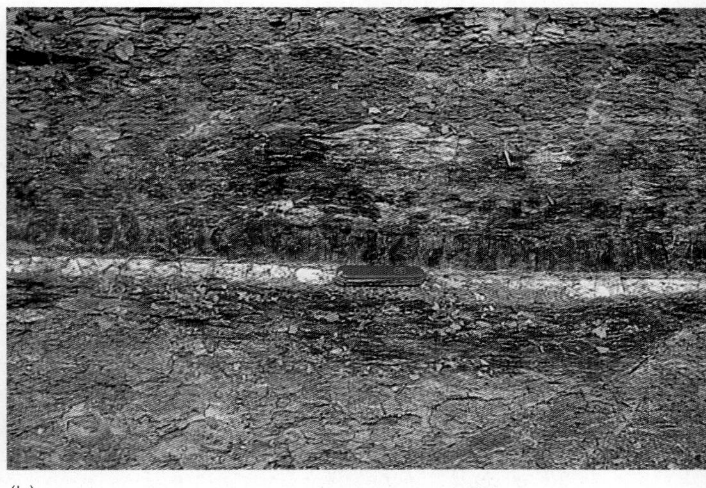

(b)

➢ **FIGURE 20-9** (*a*) View of the Cretaceous-Tertiary boundary in the Raton Basin, Colorado. The boundary, the thin white clay layer, is at the level of the knee of R. Farley Fleming of the U.S. Geological Survey. (*b*) Close-up of the clay layer which represents the preserved layer of dust ejected into the atmosphere when a meteorite struck the Earth 66 million years ago. (Photos courtesy of D. J. Nichols, U.S. Geological Survey.)

phase. Volcanic eruptions produced lava flows, and an atmosphere developed on each planet by **outgassing,** a process whereby light gases from the interior rise to the surface during volcanic eruptions. (see Perspective 20-2).

Mercury. Mercury, the closest planet to the Sun, apparently has changed very little since it was heavily cratered during its early history (➢ Figure 20-10a). Most of what we know about this small (4,880 km diameter) planet comes from

(a)

(c)

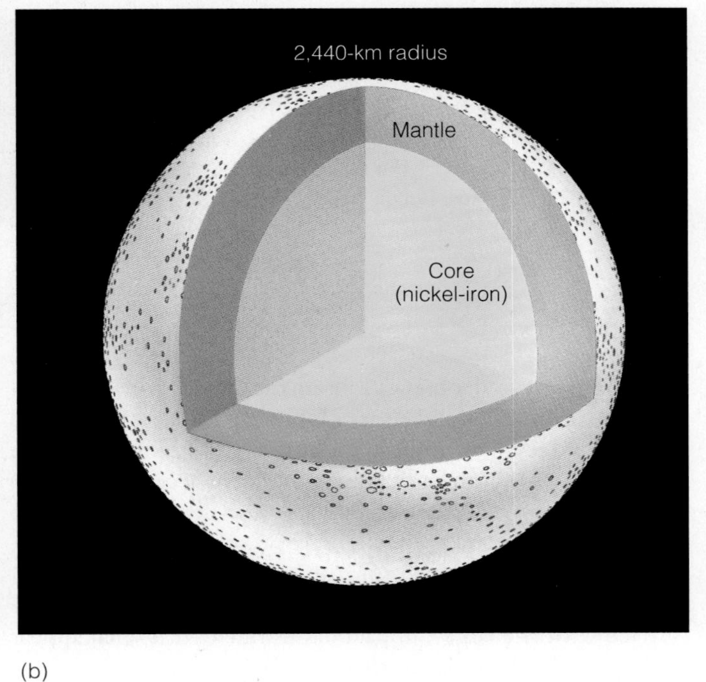

(b)

➢ **FIGURE 20-10** (*a*) Mercury has a heavily cratered surface that has changed very little since its early history. (*b*) Internal structure of Mercury, showing its large solid core relative to its overall size. (*c*) Seven scarps (indicated by arrows) can clearly be seen in this image. It is thought that these scarps formed when Mercury cooled and contracted early in its history.

Perspective 20-2

The Evolution of Climate on the Terrestrial Planets

The origins and early evolution of the terrestrial planets appear to have been similar, yet each of these planets has acquired a dramatically different climate. Why? All four planets were initially alike, with atmospheres high in carbon dioxide and water vapor derived by outgassing. Mercury, because of its small size and proximity to the Sun, lost its atmosphere by evaporation early in its history. Venus, Earth, and Mars, however, all were temperate enough during their early histories to have had fluid water on their surfaces, yet only Earth still has surface water and a climate capable of supporting life.

The reason that these three planets evolved such different climates is related to the recycling of carbon dioxide between the atmosphere and the crust (carbon-silicate geochemical cycle) as well as their distance from the Sun. Carbon dioxide recycling is an important regulator of climates because carbon dioxide, other gases, and water vapor allow sunlight to pass "through" them but trap the heat reflected back from the planet's surface. Heat is thus retained, and the temperature of the atmosphere and surface increases in what is known as the **greenhouse effect**.

Carbon dioxide combines with water in the atmosphere to form carbonic acid. When this slightly acidic rain falls, it decomposes rocks, releasing calcium and bicarbonate ions into streams and rivers and, ultimately, the oceans. In the oceans, marine organisms use some of these ions to construct shells of calcium carbonate. When the organisms die, their shells become part of the total carbonate sediments. During subduction, these carbonate sediments are heated under pressure and release carbon dioxide gas that reenters the atmosphere primarily through volcanic eruptions (▶ Figure 1).

The recycling of carbon dioxide has allowed the Earth to maintain a moderate climate throughout its history, although proponents of the Gaia hypothesis hold a somewhat different view (see Perspective 21-1). For example, when the Earth's surface cools, less water vapor is present in the atmosphere and there is less rain. The amount of carbon dioxide leaving the atmosphere thus decreases, and less decomposition of rocks occurs. Overall, however, in the long term, the amount of carbon dioxide returned to the atmosphere does not change because it is continually replenished by plate subduction and volcanism. This leads to a temporary increase in carbon dioxide in the atmosphere, greater greenhouse warming, and higher surface temperatures.

Just the opposite would happen if the surface temperature should increase. Oceanic evaporation would then increase, leading to greater rainfall and more rapid decomposition of rock; and as a result, carbon dioxide would be removed from the atmosphere. Greenhouse warming would then decrease and surface temperatures would fall.

Venus today is almost completely waterless. Many scientists think that during its early history, when the Sun was dimmer, Venus may have had vast oceans. During this time, which probably lasted no longer than the first billion or so years of Venus's existence, water vapor as well as carbon dioxide was being released into the atmosphere by volcanism. The water vapor condensed and formed oceans, while carbon dioxide cycled (by plate tectonics) just as it does on Earth. As the Sun's energy output increased these oceans eventually evaporated. Once the oceans disappeared, there was no water to return carbon to the crust, and carbon dioxide began accumulating in the atmosphere, creating a greenhouse effect and raising temperatures.

Mars, like Venus and Earth, probably once had a moderate climate and surface water, as indicated by the crisscrossing network of valleys on its oldest terrain. Because Mars is smaller than the Earth, it had less internal heat when it formed and hence cooled rapidly. Eventually, the interior of Mars became so cold that it no longer released carbon dioxide. As a result, the amount of atmospheric carbon

measurements and observations made during the flybys of *Mariner 10* in 1974 and 1975 (Table 20-2). Its high overall density of 5.4 g/cm^3 indicates that it has a large metallic core measuring 3,600 km in diameter; the core accounts for 80% of Mercury's mass (Figure 20-10b). Furthermore, Mercury has a weak magnetic field (about 1% as strong as the Earth's), indicating that the core is probably partially molten.

Images sent back by *Mariner 10* show a heavily cratered surface with the largest impact basins filled with what appear to be lava flows similar to the lava plains on the Moon. The lava plains are not deformed, however, indicating that there has been little or no tectonic activity. Another feature of Mercury's surface is a large number of scarps (Figure 20-10c). It is suggested that these scarps formed when Mercury cooled and contracted.

Because Mercury is so small, its gravitational attraction is insufficient to retain atmospheric gases; any atmosphere that it may have held when it formed probably escaped into space very quickly. Nevertheless, *Mariner 10* detected very small quantities of hydrogen and helium, thought to

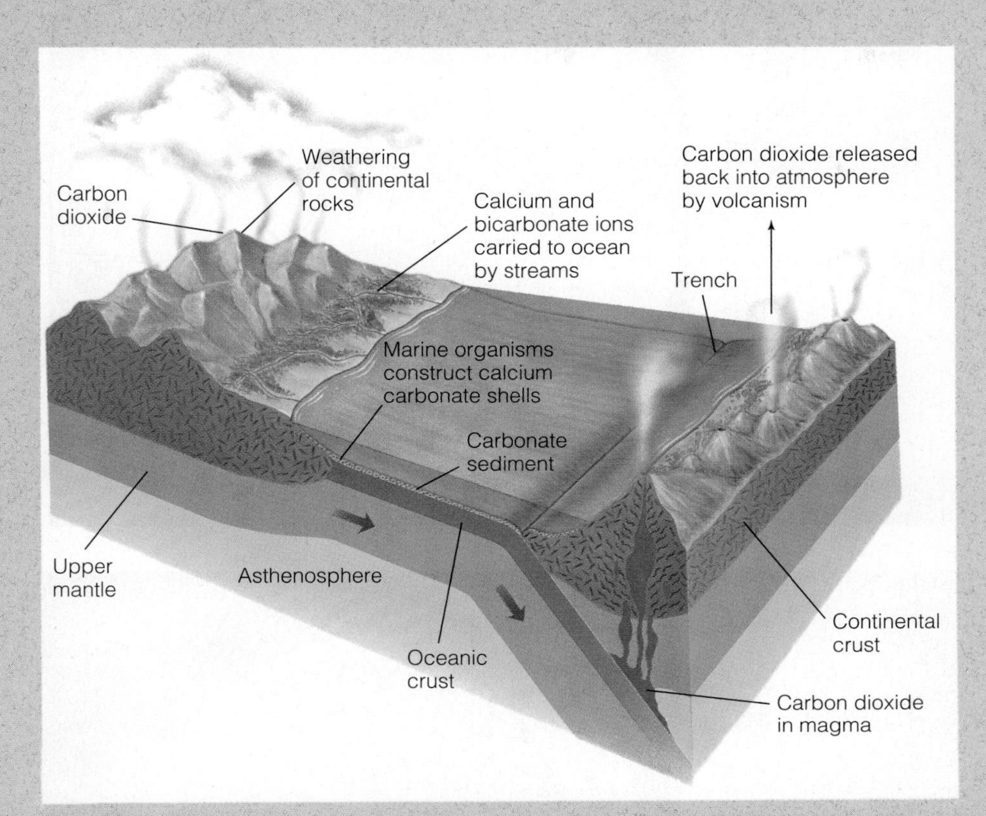

➤ **FIGURE 1** The carbon-silicate geochemical cycle illustrates how carbon dioxide is recycled. Carbon dioxide is removed from the atmosphere by combining with water and forming slightly acidic rain that falls on the Earth's surface and decomposes rocks. This decomposition releases calcium and bicarbonate ions that ultimately reach the oceans. Marine organisms use these ions to construct shells of calcium carbonate. When they die, the shells become part of the carbonate sediments that are eventually subducted. As the sediments are subjected to heat and pressure, they release carbon dioxide gas back into the atmosphere primarily through volcanic eruptions.

dioxide decreased to its current level. The greenhouse effect was thus weakened, and the Martian atmosphere became thin and cooled to its present low temperature.

If Mars had been the size of Earth or Venus, it very likely would have had enough internal heat to continue recycling carbon dioxide, offsetting the effects of low sunlight levels caused by its distance from the Sun. In other words, Mars would still have enough carbon dioxide in its atmosphere so that it could maintain a "temperate climate."

have originated from the solar winds that stream by Mercury.

Venus. Of all the planets, Venus is the most similar in size and mass to the Earth (Table 20-2). It differs, though, in most other respects. Venus is searingly hot with a surface temperature of 475°C and an oppressively thick atmosphere composed of 96% carbon dioxide and 3.5% nitrogen with traces of sulfur dioxide and sulfuric and hydrochloric acid (➤ Figure 20-11a). From information obtained by the various space probes that have passed by, orbited Venus, and descended to its surface, we know that three distinct cloud layers composed of droplets of sulfuric acid envelop the planet. Furthermore, winds up to 360 km/ hour occur at the top of the clouds, while the planet's surface remains calm.

Radar images from both orbiting spacecraft and the Venusian surface reveal a wide variety of terrains (Figures 20-11b), some of which are unlike anything seen elsewhere in the solar system. Even though no active volcanism has been observed on Venus, the presence of volcanic domes,

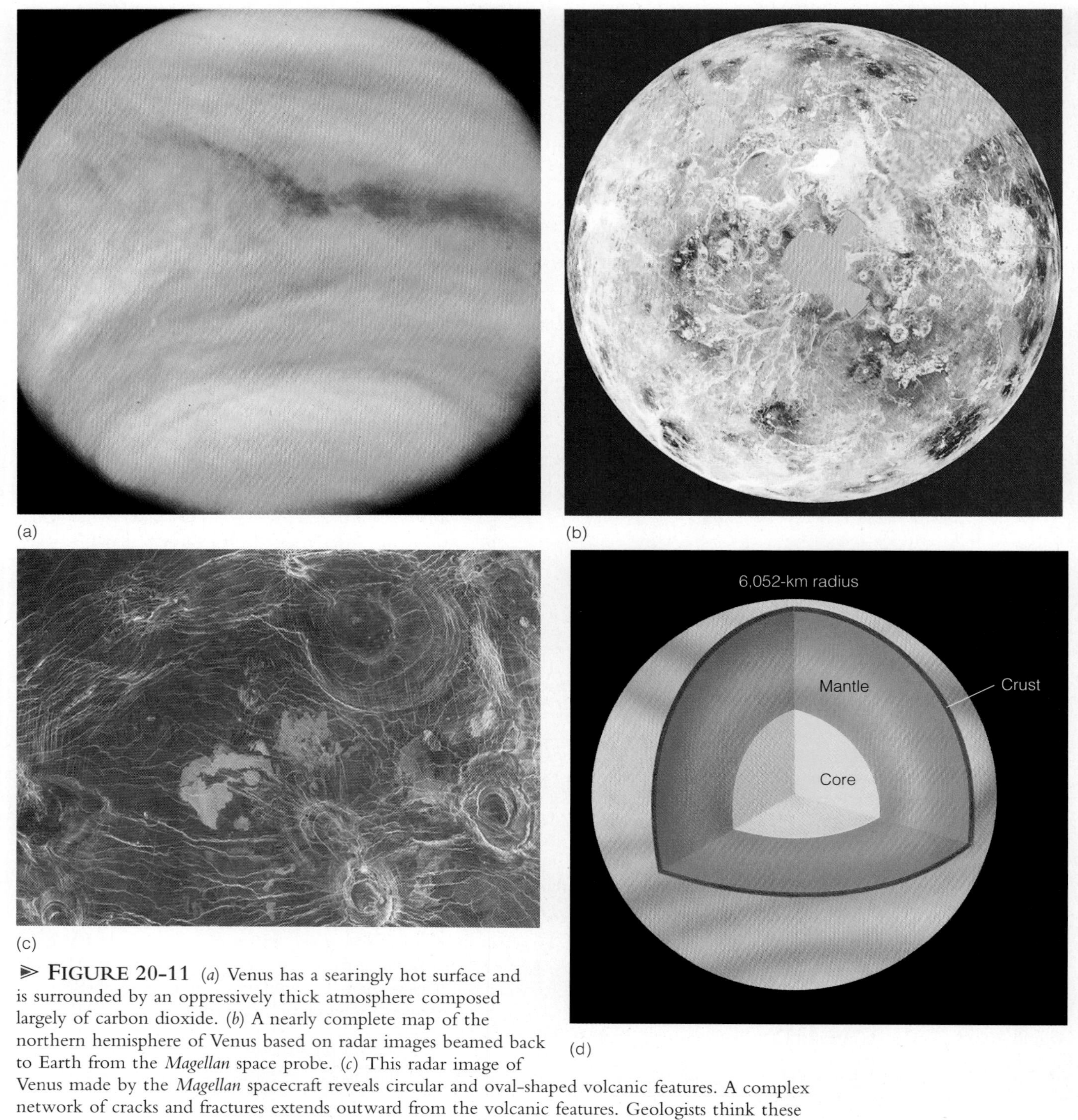

▶ **FIGURE 20-11** (*a*) Venus has a searingly hot surface and is surrounded by an oppressively thick atmosphere composed largely of carbon dioxide. (*b*) A nearly complete map of the northern hemisphere of Venus based on radar images beamed back to Earth from the *Magellan* space probe. (*c*) This radar image of Venus made by the *Magellan* spacecraft reveals circular and oval-shaped volcanic features. A complex network of cracks and fractures extends outward from the volcanic features. Geologists think these features were created by blobs of magma rising from the interior of Venus with dikes filling some of the cracks. (*d*) The internal structure of Venus.

extensive lava flows, folded mountain ranges, a complex network of faults, and what appear to be deep trenches comparable to the oceanic trenches on Earth indicate that internal and surface activity has occurred during the past (Figure 20-11c). There is, however, no evidence for active plate tectonics as on Earth.

Mars. Mars, the red planet, is differentiated, as are all the terrestrial planets, into a metallic core and a silicate mantle and crust (▶ Figure 20-12a). The thin Martian atmosphere consists of 95% carbon dioxide, 2.7% nitrogen, 1.7% argon, and traces of other gases. Mars has distinct seasons during which its polar ice caps of frozen carbon dioxide expand and recede.

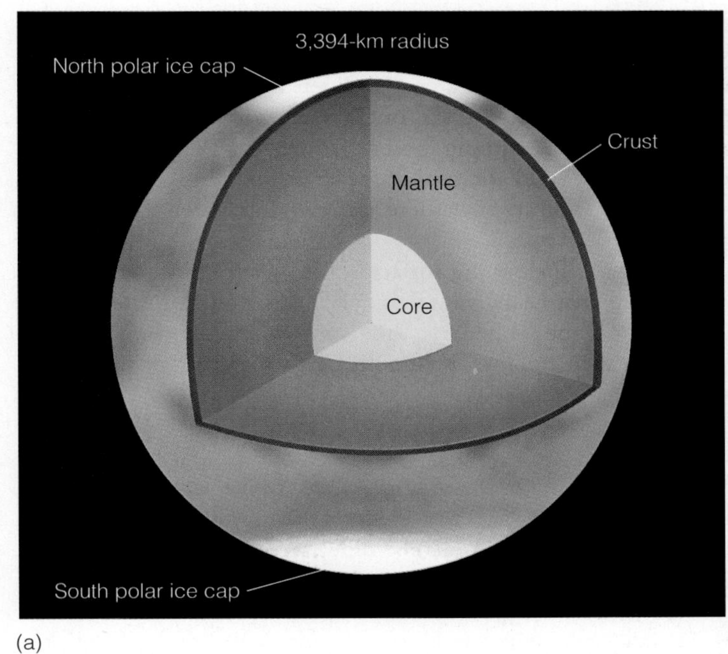

(a)

(c)

(b)

➢ **FIGURE 20-12** (*a*) The internal structure of Mars. (*b*) A striking view of Mars is revealed in this mosaic of 102 *Viking* images. The largest canyon known in the solar system, *Valles Marineris*, can be clearly seen in the center of this image, while three of the planet's volcanoes are visible on the left side of the image. (*c*) Outflow channels extending from areas of chaotic terrain on Mars. Arrows show inferred directions of flow.

Perhaps the most striking aspect of Mars is its surface features, many of which have not yet been satisfactorily explained. Like the surfaces of Mercury and the Moon, the southern hemisphere is heavily cratered, attesting to a period of meteorite bombardment. *Hellas,* a crater with a diameter of 2,000 km, is the largest known impact structure in the solar system and is found in the Martian southern hemisphere.

The northern hemisphere is much different, having large smooth plains, fewer craters, and evidence of extensive volcanism. The largest known volcano in the solar system, *Olympus Mons,* has a basal diameter of 600 km, rises 27 km above the surrounding plains, and is topped by a huge circular crater 80 km in diameter.

The northern hemisphere is also marked by huge canyons that are essentially parallel to the Martian equator. One of these canyons, *Valles Marineris,* is at least 4,000 km long, 250 km wide, and 7 km deep and is the largest yet discovered in the solar system (Figure 20-12b). If it were present on Earth, it would stretch from San Francisco to New York! It is not yet known how these vast canyons formed, although geologists postulate that they may have started as large rift zones that were subsequently modified by running water and wind erosion. Such hypotheses are based on comparison to rift structures found on Earth and topographic features formed by geologic agents of erosion such as water and wind.

Tremendous wind storms have strongly influenced the surface of Mars and led to dramatic dune formations (see Perspective 18-1, Figure 3). Even more stunning than the dunes, however, are the outflow channels that appear to be the result of running water (Figure 20-12c). Mars is currently too cold for surface water to exist, yet the channels strongly indicate that there was running water on the planet during the past.

The fresh-looking surfaces of its many volcanoes strongly suggest that Mars was tectonically active during the past and may still be. There is, however, no evidence that plate movement, such as occurs on Earth, has ever occurred.

The Jovian Planets

The Jovian planets are completely unlike any of the terrestrial planets in size and chemical composition (Table 20-2) and have had completely different evolutionary histories. While they all apparently contain a small core in relation to their overall size, the bulk of a Jovian planet is composed of volatile elements and compounds that condense at low temperatures such as hydrogen, helium, methane, and ammonia.

Jupiter. Jupiter is the largest of the Jovian planets (Table 20-2; ⯈ Figure 20-13a). With its moons, rings, and radiation belts, it is the most complex and varied planet in

(a)

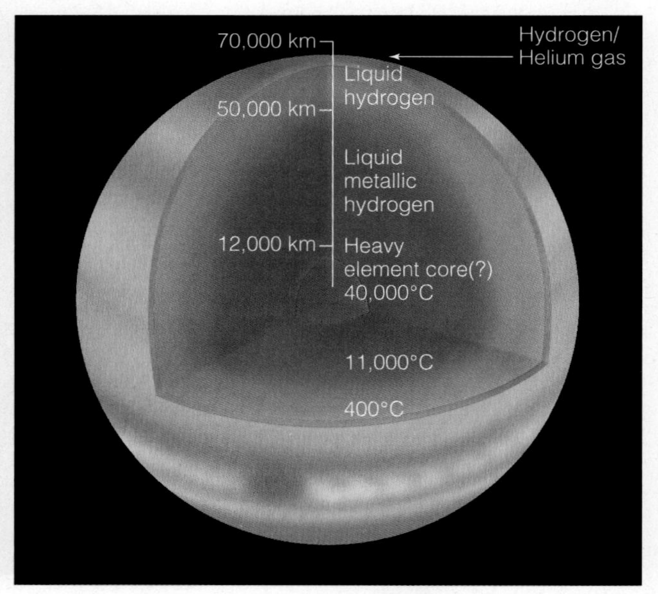

(b)

⯈ **FIGURE 20-13** (*a*) Jupiter, the largest planet in the solar system, is also the most complex with its moons, rings, and radiation belts. Its cloudy atmosphere displays a regular pattern of bands and spots, the largest of which is the Great Red Spot shown in the lower part of this photograph. (*b*) The internal structure of Jupiter.

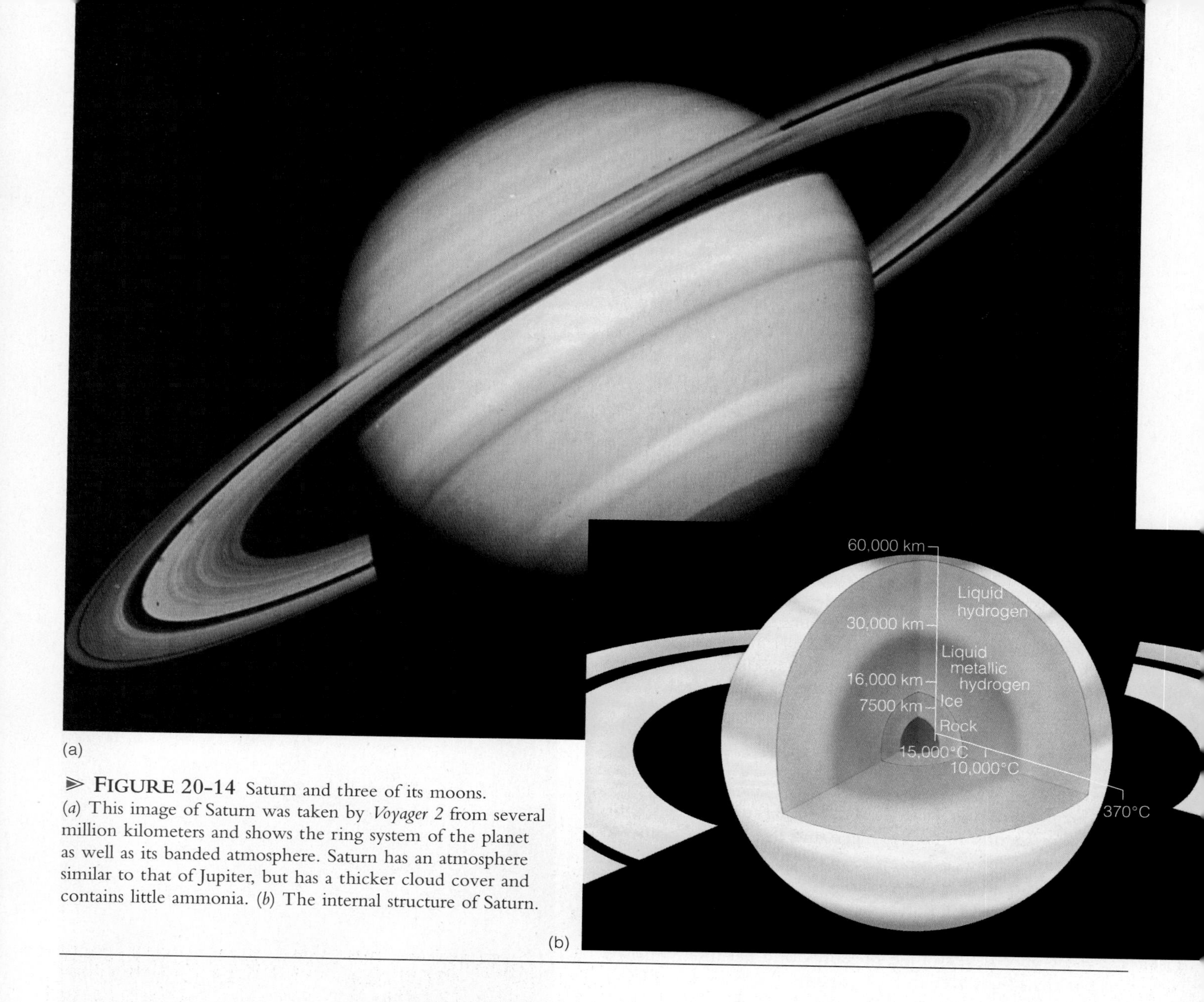

▶ **FIGURE 20-14** Saturn and three of its moons. (*a*) This image of Saturn was taken by *Voyager 2* from several million kilometers and shows the ring system of the planet as well as its banded atmosphere. Saturn has an atmosphere similar to that of Jupiter, but has a thicker cloud cover and contains little ammonia. (*b*) The internal structure of Saturn.

the solar system. Jupiter's density is only one-fourth that of Earth, but because it is so large, it has 318 times the mass (Table 20-2). It is an unusual planet in that it emits almost 2.5 times more energy than it receives from the Sun. One explanation is that most of the excess energy is left over from the time of its formation. When Jupiter formed, it heated up because of gravitational contraction (as did all the planets) and is still cooling. Jupiter's massive size insulates its interior, and hence it has cooled very slowly.

Jupiter has a relatively small central core of solid rocky material formed by differentiation. Above this core is a thick zone of liquid metallic hydrogen followed by a thicker layer of liquid hydrogen; above that is a thin layer of clouds (Figure 20-13b). Surrounding Jupiter are a strong magnetic field and an intense radiation belt.

Jupiter has a dense atmosphere of hydrogen, helium, methane, and ammonia, which some think are the same gases that composed the Earth's first atmosphere. Jupiter's cloudy atmosphere is divided into a series of different colored bands as well as a variety of spots (the Great Red Spot) and other features, all interacting in incredibly complex motions.

Revolving around Jupiter are 16 moons varying greatly in tectonic and geologic activity. Also surrounding Jupiter is a thin, faint ring, a feature shared by all the Jovian planets.

Saturn. Saturn is slightly smaller than Jupiter, about one-third as massive, and about one-half as dense, but has a similar internal structure and atmosphere (Table 20-2). Like Jupiter, Saturn gives off more energy (2.2 times as much) than it gets from the Sun. Saturn's most conspicuous feature is its ring system, consisting of thousands of rippling, spiraling bands of countless particles (▶ Figure 20-14a).

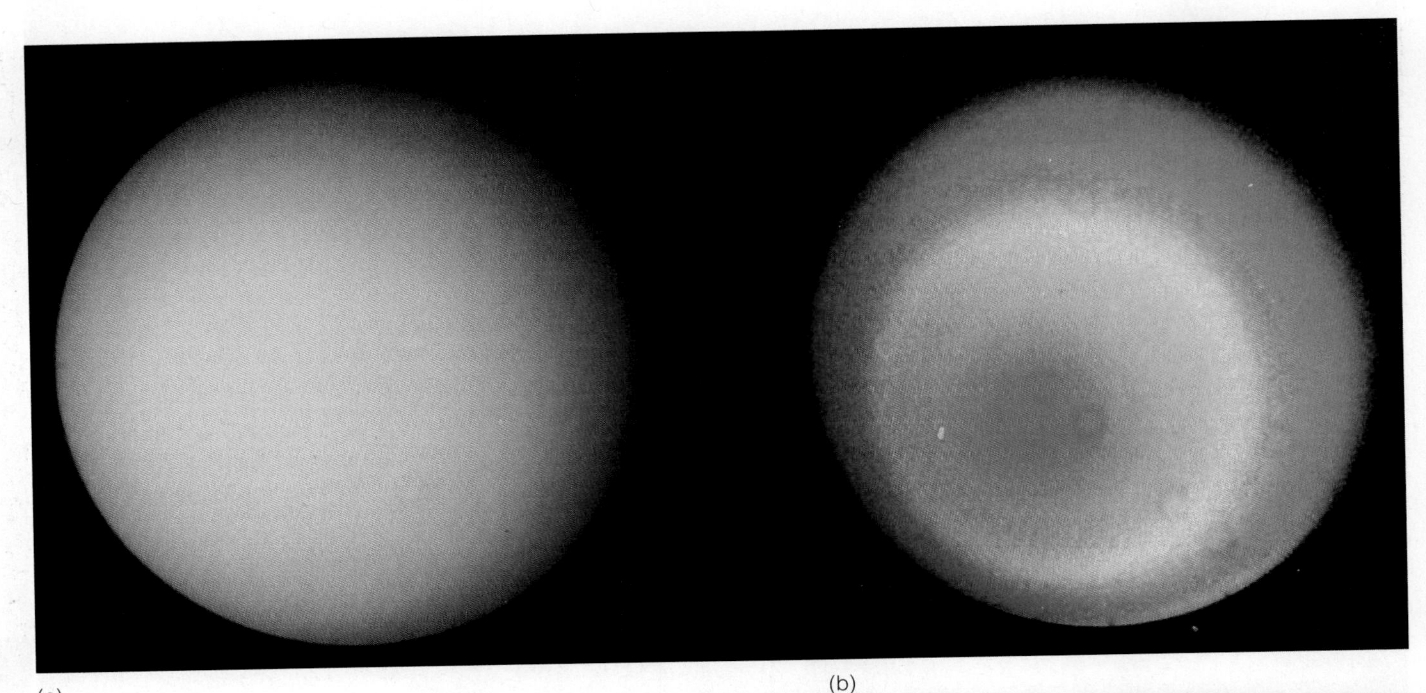

(a) (b)

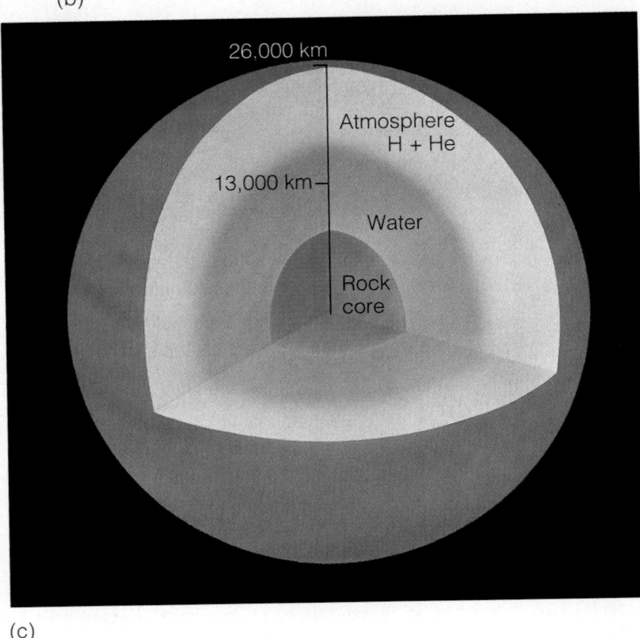

(c)

▶ **FIGURE 20-15** (*a*) Images of Uranus taken by *Voyager 2* under ordinary light show a featureless planet. (*b*) When color is enhanced by computer processing techniques, Uranus is seen to have zonal flow patterns in its atmosphere. (*c*) The internal structure of Uranus.

The composition of Saturn is similar to Jupiter's, but consists of slightly more hydrogen and less helium. Saturn's core is not as dense as Jupiter's, and as in the case of Jupiter, a layer of liquid metallic hydrogen overlies the core, followed by a zone of liquid hydrogen and helium, and, lastly, a layer of clouds (Figure 20-14b). Liquid metallic hydrogen can exist only at very high pressures, and because Saturn is smaller than Jupiter, such high pressures are found at greater depths in Saturn. Therefore, there is less of this conducting material than on Jupiter, and as a consequence, Saturn has a weaker magnetic field.

Even though the atmospheres of Saturn and Jupiter are similar, Saturn's atmosphere contains little ammonia because it is farther from the Sun and therefore is colder. The cloud layer on Saturn is thicker than on Jupiter, but it lacks the contrast between the different bands. Unlike Jupiter, Saturn has seasons because its axis tilts 27°.

Saturn has 18 known moons, most of which are small with low densities. Titan, however, is the second largest moon in the solar system. With its nitrogen and methane atmosphere, it is also the most distinctive and perhaps the most interesting moon in the solar system (see the Prologue).

Uranus. Uranus is much smaller than Jupiter, but their densities are about the same (Table 20-2). It is the only planet that lies on its side (▶ Figure 20-15a and b); that is, its axis of rotation nearly parallels the plane of the ecliptic. Some scientists think that a collision with an Earth-sized body early in its history may have knocked Uranus on its side.

Data gathered by the flyby of *Voyager 2* suggest that Uranus has a water zone beneath its cloud cover. Because the planet's density is greater than if it were composed entirely of hydrogen and helium, it is thought that Uranus must have a dense, rocky core, and this core may be surrounded by a deep global ocean of liquid water (Figure 20-15c).

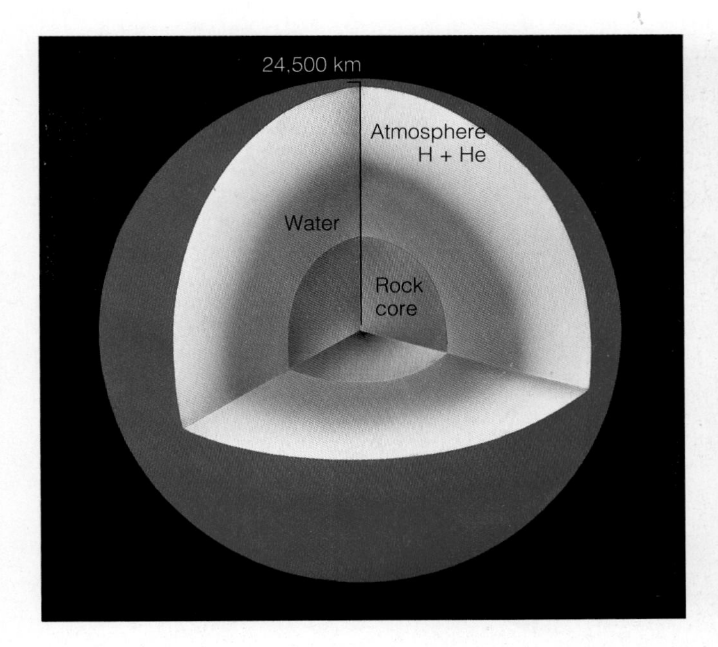

▶ **FIGURE 20-16** The internal structure of Neptune.

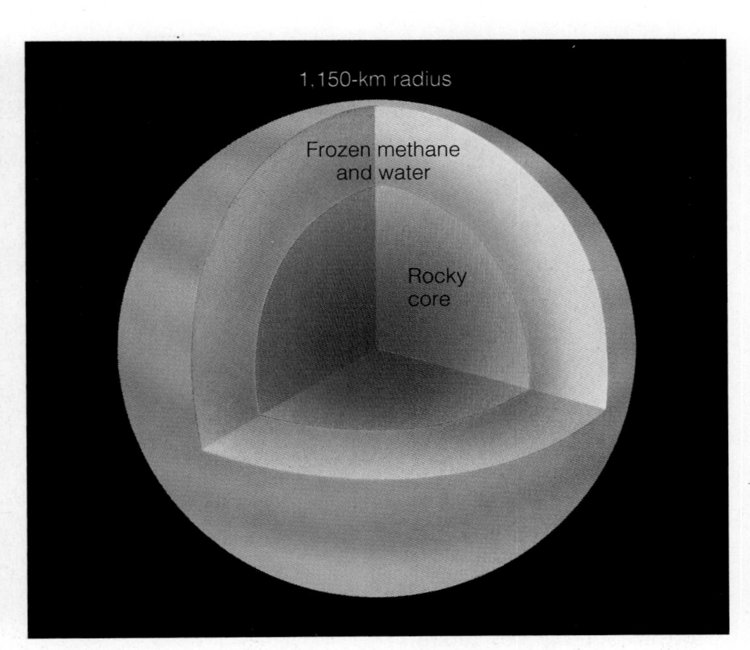

▶ **FIGURE 20-17** The internal structure of Pluto.

The atmospheric composition of Uranus is similar to that of Jupiter and Saturn with hydrogen being the dominant gas, followed by helium and some methane. Uranus also has a banded atmosphere and a circulation pattern much like those of Jupiter and Saturn. Surrounding Uranus is a huge corkscrew-shaped magnetic field that stretches for millions of kilometers into space. Uranus has at least nine thin, faint rings and 15 small moons circling it.

Neptune and Pluto. The flyby of *Voyager 2* in August 1989 provided the first detailed look at Neptune and showed it to be a dynamic, stormy planet (Figure 20-1). Its atmosphere is similar to those of the other Jovian planets, and it exhibits a pattern of zonal winds and giant storm systems comparable to those of Jupiter. The internal structure of Neptune is similar to that of Uranus (Table 20-2); it has a rocky core approximately 17,000 km in diameter surrounded by a semifrozen slush of water and liquid methane (▶ Figure 20-16). Its atmosphere is composed of hydrogen and helium with some methane. Encircling Neptune are three faint rings and eight moons.

With a diameter of only 2,300 km, Pluto is the smallest planet and, strictly speaking, it is not one of the Jovian planets (Table 20-2). Little is known about Pluto, but recent studies indicate it has a rocky core overlain by a mixture of methane gas and ice (▶ Figure 20-17). It has a thin, two-layer atmosphere with a clear upper layer overlying a more opaque lower layer. Pluto's mid-latitude areas are dark, while its north polar region is distinctly brighter. Furthermore, its south polar region is extremely bright, which is interpreted to indicate seasonality.

Pluto differs from all the other planets in that it has a highly elliptical orbit that is tilted with respect to the plane of the ecliptic. It has one known moon, Charon, that is nearly half its size with a surface that appears to differ markedly from Pluto's.

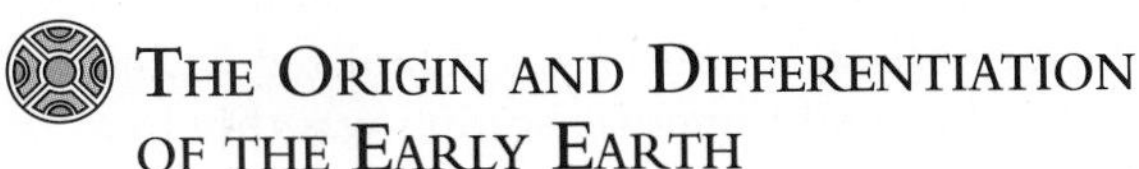

The Origin and Differentiation of the Early Earth

As matter was accreting in the various turbulent eddies that swirled around the early Sun, enough material eventually gathered together in one eddy to form the planet Earth. Recall from Chapter 1 that the Earth consists of concentric layers of different composition and densities (▶ Figure 20-18). This differentiation into concentric layers is a fundamental characteristic of all the terrestrial planets and presumably occurred very early in their history.

Geologists know that the Earth is 4.6 billion years old. The oldest known rocks, however, are 3.96-billion-year-old metamorphic rocks from Canada. Like the younger crustal rocks, these rocks are composed of relatively light silicate minerals. It appears that a crust, a denser mantle, and an iron-nickel core were already present 3.96 billion years ago, or 640 million years after the Earth formed.

The early Earth was probably cool, of generally uniform composition and density throughout, and composed mostly of silicate compounds, iron and magnesium oxides, and smaller amounts of all the other chemical elements (see Figure 1-11a). In order for the dispersed iron and nickel to concentrate in the core, the Earth must have heated up enough for them to melt and sink through the surrounding lighter silicate minerals.

This initial heating could have occurred in three ways. Some heat was no doubt generated by the impact of

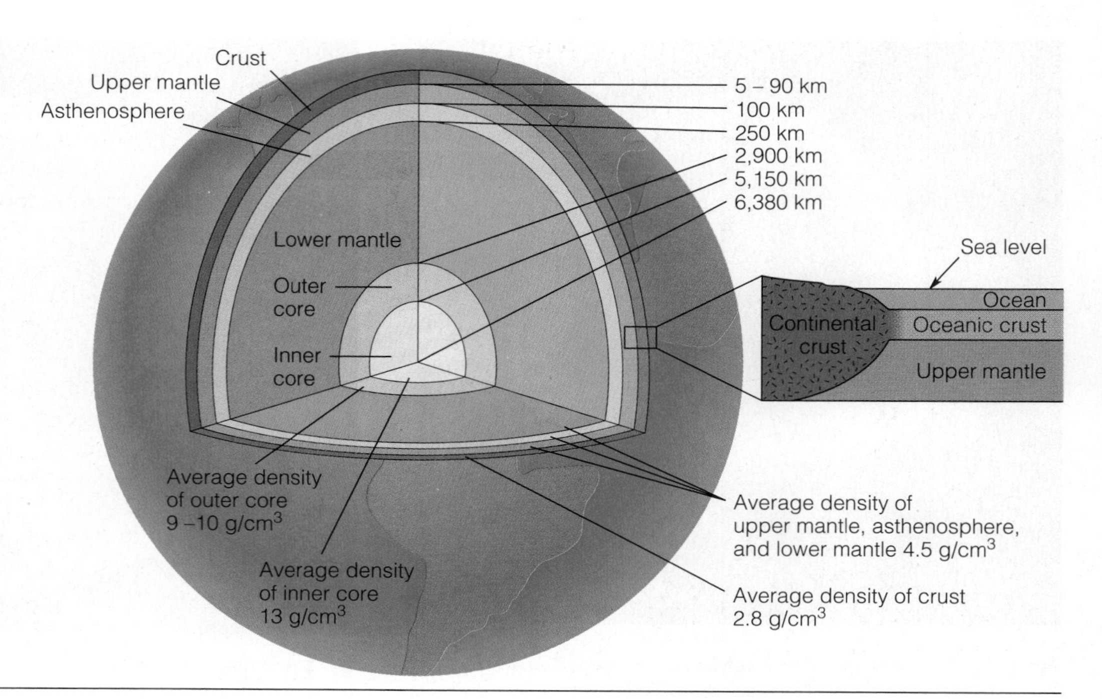

▶ **FIGURE 20-18** Cross section of the Earth showing the various layers and their average density. The crust is divided into a continental and oceanic portion. Continental crust is 20 to 90 km thick; oceanic crust is 5 to 10 km thick.

meteorites; most of this heat was radiated back into space, but some was probably retained by the accreting planet. Heat was also generated within the early Earth as it was reduced to a smaller volume by gravitational compression. Rock is a poor conductor of heat, so this heat accumulated within the Earth. The third cause of internal heating was the decay of radioactive elements such as uranium, thorium, and others. Even though these elements form only a very small portion of the Earth, the heat generated during radioactive decay was absorbed by the surrounding rock.

The combination of meteorite impacts, gravitational compression, and heat from radioactive decay increased the temperature of the early Earth enough to melt iron and nickel, which, being denser than silicate minerals, settled to the center of the Earth and formed the core (see Figure 1-11b). Simultaneously, the lighter silicates slowly flowed upward, beginning the differentiation of the mantle from the core.

Calculations indicate that with a uniform distribution of elements in an early solid Earth, enough heat could be generated to begin melting iron and nickel at depths between 400 and 850 km. Melting would have had to begin at shallow depths because the temperature at which melting begins increases with pressure; therefore the melting point of any material increases toward the Earth's center.

The most significant event in the history of the Earth was its differentiation into a layered planet. This led to the formation of a crust and continents as well as to the outgassing of light volatile elements from the interior that eventually resulted in the formation of the oceans and atmosphere.

 The Origin of the Earth-Moon System

We probably know more about our Moon than any other celestial object except the Earth (▶ Figure 20-19). Nevertheless, even though the Moon has been studied for centuries through telescopes and has been sampled directly, many questions remain unanswered.

The Moon is one-fourth the diameter of the Earth, has a low density (3.3 g/cm^3) relative to the terrestrial planets, and exhibits an unusual chemistry in that it is bone-dry, having been largely depleted of most volatile elements (Table 20-2). The Moon orbits the Earth and rotates on its own axis at the same rate, so we always see the same side. Furthermore, the Earth-Moon system is unique among the terrestrial planets. Neither Mercury nor Venus has a moon, and the two small moons of Mars—Phobos and Deimos—are probably captured asteroids.

The surface of the Moon can be divided into two major parts: the low-lying dark-colored plains, called *maria,* and the light-colored *highlands* (Figure 20-19). The highlands are the oldest parts of the Moon and are heavily cratered, providing striking evidence of the massive meteorite bombardment that occurred in the solar system more than four billion years ago.

Study of the several hundred kilograms of rocks returned by the *Apollo* missions indicates that three kinds of materials dominate the lunar surface: igneous rocks, breccias, and dust. Basalt, a common dark-colored igneous rock on Earth, is one of the several different types of igneous rocks on the Moon and makes up the greater part of the maria. The

▶ **FIGURE 20-19** The side of the Moon as seen from Earth. The light-colored areas are the lunar highlands, which were heavily cratered by meteorite impacts. The dark-colored areas are maria, which formed when lava flowed out onto the surface.

presence of igneous rocks that are essentially the same as those on Earth shows that magmas similar to those on Earth were generated on the Moon long ago.

The lunar surface is covered with a regolith estimated to be 3 to 4 m thick. This gray covering, which is composed of compacted aggregates of rock fragments called breccia, glass spherules, and small particles of dust, is thought to be the result of debris formed by meteorite impacts.

The interior structure of the Moon is quite different from that of the Earth, indicating a different evolutionary history (▶ Figure 20-20). The highland crust is thick (65 to 100 km) and comprises about 12% of the Moon's volume. It was formed about 4.4 billion years ago, immediately following the Moon's accretion. The highlands are composed principally of the igneous rock anorthosite, which is made up of light-colored feldspar minerals that are responsible for their white appearance.

A thin covering (1 to 2 km thick) of basaltic lava fills the maria; lava covers about 17% of the lunar surface, mostly on the side facing the Earth. These maria lavas came from partial melting of a thick underlying mantle of silicate composition. Moonquakes occur at a depth of about 1,000 km, but below that depth seismic shear waves apparently are not transmitted. Because shear waves do not travel through liquid, their lack of transmission implies that the innermost mantle may be partially molten. There is increasing evidence that the Moon has a small (600 km to 1,000 km diameter) metallic core comprising 2 to 5% of its volume.

The origin and earliest history of the Moon are still unclear, but the basic stages in its subsequent development are well understood. It formed some 4.6 billion years ago and shortly thereafter was partially or wholly melted, yielding a silicate melt that cooled and crystallized to form the mineral anorthite. Because of the low density of the anorthite crystals and the lack of water in the silicate melt, the thick anorthosite highland crust formed. The remaining silicate melt cooled and crystallized to produce the zoned mantle, while the heavier metallic elements formed the small metallic core.

The formation of the lunar mantle was completed by about 4.4 to 4.3 billion years ago. The maria basalts, derived from partial melting of the upper mantle, were extruded during great lava floods between 3.8 and 3.2 billion years ago.

Numerous models have been proposed for the origin of the Moon, including capture from an independent orbit, formation with the Earth as part of an integrated two-planet system, breaking off from the Earth during accretion, and formation resulting from a collision between the Earth and a large planetesimal. These various models are not mutually exclusive, and elements of some occur in others. At this time, scientists cannot agree on a single model, as each has some inherent problems. However, the model that seems to account best for the Moon's particular composition and structure involves an impact by a large planetesimal with a young Earth (▶ Figure 20-21).

▶ **FIGURE 20-20** The internal structure of the Moon is different from that of the Earth. The upper mantle is the source for the maria lavas. Moonquakes occur at a depth of 1,000 km. Because seismic shear waves are not transmitted below this depth, it is thought that the innermost mantle is liquid. Below this layer is a small metallic core.

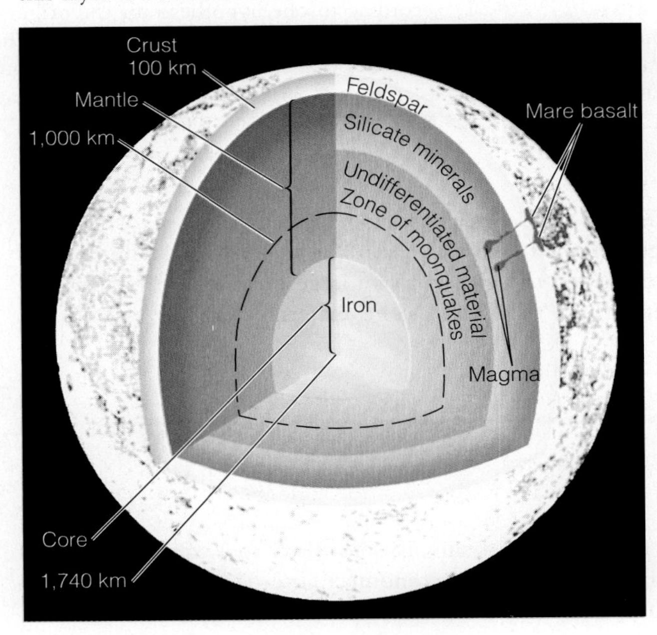

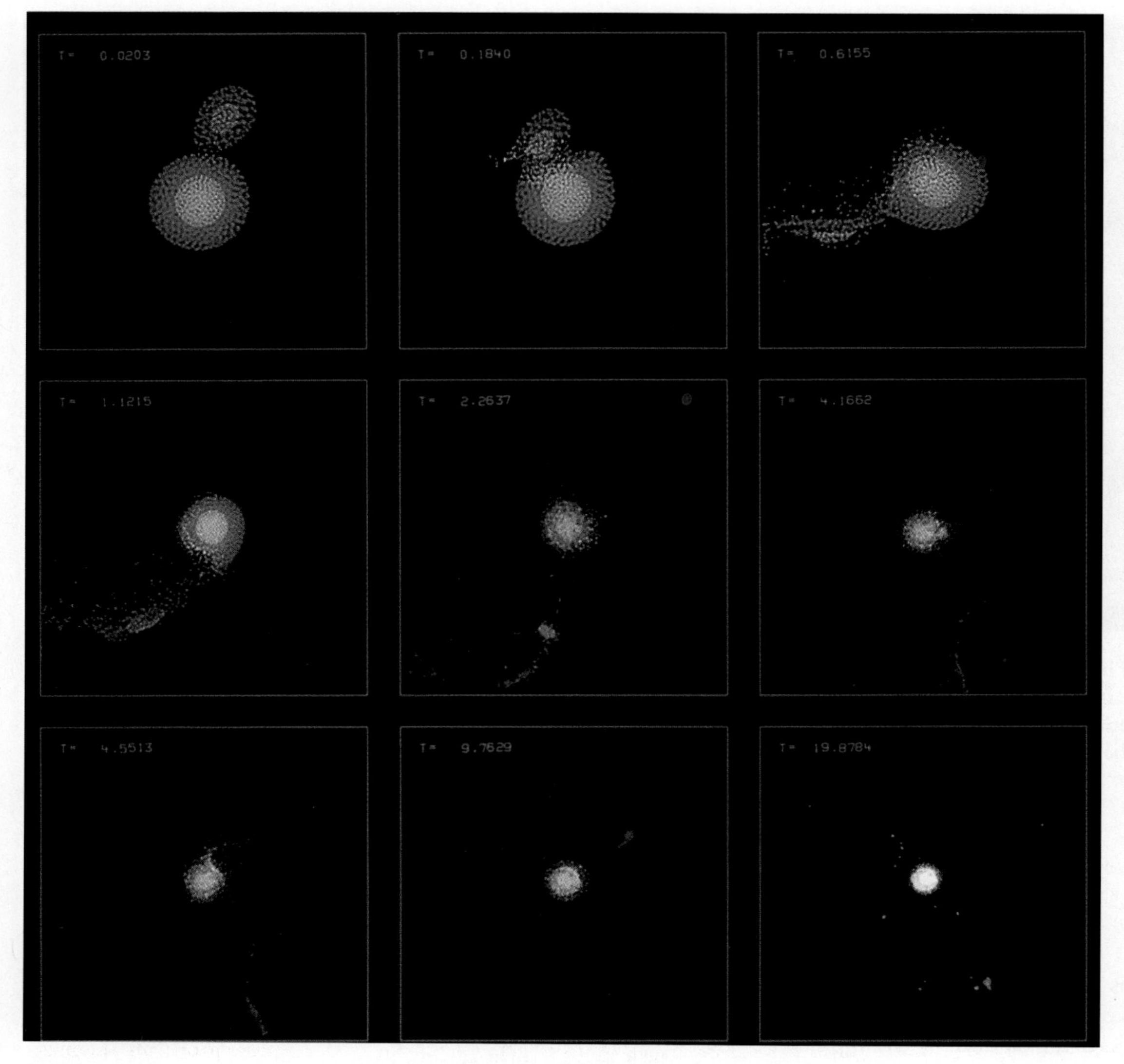

➢ **FIGURE 20-21** According to one hypothesis for the origin of the Moon, a large planetesimal the size of Mars crashed into the Earth 4.6 to 4.4 billion years ago, causing the ejection of a mass of hot material that formed the Moon. This computer simulation shows the formation of the Moon as a result of an Earth-planetesimal collision.

In this model, a giant planetesimal, the size of Mars or larger, crashed into the Earth about 4.6 to 4.4 billion years ago, causing the ejection of a large quantity of hot material that formed the Moon. The material that was ejected was mostly in the liquid and vapor phase and came primarily from the mantle of the colliding planetesimal. As it cooled, the various lunar layers crystallized out in the order we have discussed.

CHAPTER SUMMARY

1. The universe began with a Big Bang approximately 13 to 20 billion years ago. Astronomers have deduced this age from the fact that celestial objects are moving away from each other in what appears to be an ever expanding universe.
2. The universe has a background radiation of 2.7° above absolute zero, representing the cooling remnant of the Big Bang.
3. About 4.6 billion years ago, the solar system formed from a

rotating cloud of interstellar matter. As this cloud condensed, it eventually collapsed under the influence of gravity and flattened into a counterclockwise rotating disk. Within this rotating disk, the Sun, planets, and moons formed from the turbulent eddies of nebular gases and solids.

4. Meteorites provide vital information about the age and composition of the solar system. The three major groups are stones, irons, and stony-irons. Each has a different composition, reflecting a different origin.
5. Temperature as a function of distance from the Sun played a major role in the type of planets that evolved. The inner terrestrial planets are composed of rock and metallic elements that condense at high temperatures. The outer Jovian planets plus Pluto are composed mostly of hydrogen, helium, ammonia, and methane, all of which condense at lower temperatures.
6. All of the terrestrial planets are differentiated into a core, mantle, and crust, and all seem to have had a similar early history during which volcanism and cratering from meteorite impacts were common.
7. The Jovian planets differ from the terrestrial planets in size and chemical composition and followed completely different evolutionary histories. All of the Jovian planets have a small core compared to their overall size; they are mainly composed of volatile elements and compounds.
8. The Earth formed from a swirling eddy of nebular material 4.6 billion years ago, and by at least 3.96 billion years ago, it had differentiated into its present-day structure. It accreted as a solid body and then underwent differentiation during a period of internal heating.
9. The Moon probably formed as a result of a Mars-sized planetesimal crashing into the Earth 4.6 to 4.4 billion years ago and causing it to eject a large quantity of hot material. As the material cooled, the various lunar layers crystallized, forming a zoned body.

IMPORTANT TERMS

Big Bang
greenhouse effect
irons
Jovian planet
meteorite
outgassing
planetesimal
solar nebula theory
stones
stony-irons
terrestrial planet

REVIEW QUESTIONS

1. Which of the following is not one of the four basic forces?
   a. ____ gravity;
   b. ____ electromagnetic;
   c. ____ strong nuclear;
   d. ____ weak nuclear;
   e. ____ photon.
2. The composition of the universe has been changing since the Big Bang. Yet 98% of it by weight still consists of the elements:
   a. ____ hydrogen and carbon;
   b. ____ helium and carbon;
   c. ____ hydrogen and helium;
   d. ____ carbon and nitrogen;
   e. ____ hydrogen and nitrogen.
3. Which of the following is not a terrestrial planet?
   a. ____ Mercury;
   b. ____ Jupiter;
   c. ____ Earth;
   d. ____ Venus;
   e. ____ Mars.
4. The age of the solar system is generally accepted by scientists as:
   a. ____ 4.6 billion years;
   b. ____ 10 billion years;
   c. ____ 15.5 billion years;
   d. ____ 20 billion years;
   e. ____ 50 billion years.
5. The major problem that plagued most early theories of the origin of the solar system involved the:
   a. ____ distribution of elements throughout the solar system;
   b. ____ rotation of the planets around their axes;
   c. ____ slow rotation of the Sun;
   d. ____ revolution of the planets around the Sun;
   e. ____ source of meteorites and asteroids.
6. The most widely accepted theory regarding the origin of the Moon involves:
   a. ____ capture from an independent orbit;
   b. ____ an independent origin from the Earth;
   c. ____ breaking off from the Earth during the Earth's accretion;
   d. ____ formation resulting from a collision between the Earth and a large planetesimal;
   e. ____ none of these.
7. Images radioed back by *Voyagers 1* and *2* revealed that:
   a. ____ all of the Jovian planets have rings;
   b. ____ Neptune is a placid planet;
   c. ____ Uranus has a large spot like those of Jupiter and Neptune;
   d. ____ Pluto has an atmosphere similar to that of Mars;
   e. ____ all of these.

8. The planets can be separated into terrestrial and Jovian primarily on the basis of which property?
   a. ____ size;
   b. ____ atmosphere;
   c. ____ density;
   d. ____ color;
   e. ____ none of these.
9. Which of the following events did all of the terrestrial planets experience early in their history?
   a. ____ accretion;
   b. ____ differentiation;
   c. ____ volcanism;
   d. ____ meteorite impacting;
   e. ____ all of these.
10. Which of the following is not characteristic of Mercury?
   a. ____ a strong magnetic field;
   b. ____ heavy cratering of its surface;
   c. ____ scarps;
   d. ____ numerous lava flows;
   e. ____ small amounts of atmospheric hydrogen and helium.
11. The atmosphere of Venus is:
   a. ____ thick and composed of carbon dioxide;
   b. ____ similar to Earth's;
   c. ____ nonexistent;
   d. ____ thin, like that of Mars;
   e. ____ none of these.
12. The surface of Mars possesses:
   a. ____ huge valleys;
   b. ____ massive volcanoes;
   c. ____ large craters;
   d. ____ smooth plains;
   e. ____ all of these.
13. Which planets give off more energy than they receive?
   a. ____ Jupiter;
   b. ____ Saturn;
   c. ____ Uranus;
   d. ____ answers (a) and (b);
   e. ____ answers (a) and (c).
14. Both Jupiter and Saturn have a relatively small rocky core overlain by a zone of:
   a. ____ helium;
   b. ____ liquid metallic hydrogen;
   c. ____ frozen ammonia;
   d. ____ hydrogen;
   e. ____ carbon dioxide.
15. The only planet whose axis of rotation nearly parallels the plane of the ecliptic is:
   a. ____ Venus;
   b. ____ Saturn;
   c. ____ Uranus;
   d. ____ Neptune;
   e. ____ Pluto.
16. What was the main source of heat for the Earth early in its history?
   a. ____ meteor impact;
   b. ____ radioactivity;
   c. ____ gravitational compression;
   d. ____ an initial molten condition;
   e. ____ spontaneous combustion.
17. What two fundamental phenomena indicate that the Big Bang occurred?
18. How does the solar nebula theory account for the general characteristics of the solar system?
19. What are the three major groups of meteorites?
20. How do the terrestrial planets differ from the Jovian planets?
21. What are the similarities and differences in the origin and history of the four terrestrial planets?
22. Discuss why Venus, Earth, and Mars currently have quite different atmospheres.
23. What are the similarities and differences in the origin and history of the four Jovian planets?
24. Discuss the origin of the Earth-Moon system.
25. Discuss how the *Voyager* space probes have changed our ideas about the planets they have flown by.

POINTS TO PONDER

1. The Sun is thought to have been much dimmer during the early history of the solar system. Because of this, it is thought that Venus had oceans early in its history as did Mars. If the Sun was as bright during its early history as it is now, how might this have affected the evolution of the solar system, particularly the evolution of the terrestrial planets?
2. One of the theories proposed to account for the origin of the Moon is that it broke off from the Earth during accretion. What information about the Earth and Moon would you need to prove this theory? How would you gather such information?
3. Based on our current knowledge about Mars, do you think it is possible for humans to colonize the planet effectively and become self-sufficient? What major problems must be overcome?
4. It is thought that the atmospheres of Jupiter and Saturn are very similar to that of the early Earth. Why have these two planets' atmospheres remained essentially the same during the past 4.6 billion years, while those of the terrestrial planets have evolved?
5. Based on what you know about how volcanic eruptions affect climatic conditions, discuss how the present global ecosystem might be affected by a collision between the Earth and a meteorite about 10 km in diameter.

Additional Readings

Benzel, R. 1990. Pluto. *Scientific American* 262, no. 6: 50–59.

Fernie, J. D. 1993. The Tunguska event. *American Scientist* 81, no. 5: 412–15.

Freedman, W. L. 1992. The expansion rate and size of the universe. *Scientific American* 267, no. 5: 54–61.

Grieve, R. A. F. 1990. Impact cratering on the Earth. *Scientific American* 262, no. 4: 66–73.

Horgan, J. 1990. Universal truths. *Scientific American* 263, no. 4: 108–17.

Ingersoll, A. P. 1987. Uranus. *Scientific American* 256, no. 1: 38–45.

Kasting, J. F., O. B. Toon, and J. B. Pollack. 1988. How climate evolved on the terrestrial planets. *Scientific American* 258, no. 2: 90–97.

Kinoshita, J. 1989. Neptune. *Scientific American* 261, no. 5: 82–91.

Kuhn, K. F. 1994. *In quest of the universe*, 2d ed. St. Paul, Minn.: West Publishing Co.

Luhmann, J. G., J.B. Pollack, and L. Colin. 1994. The Pioneer mission to Venus. *Scientific American* 270, no. 4: 90–97.

Lunine, J. I. 1994. Does Titan have oceans? *American Scientist* 82, no. 2: 134–43.

McSween, H. Y., Jr. 1989. Chondritic meteorites and the formation of planets. *American Scientist* 77, no. 2: 146–53.

Osterback, D. E., J. A. Gwinn, and R. S. Brasher. 1993. Edwin Hubble and the expanding universe. *Scientific American* 269, no. 1: 84–89.

Patrick, R. R., and R. C. Howe. 1994. Volcanism on the terrestrial planets. *Journal of Geologic Education* 42, no. 3: 225–38.

Peterson, I. 1993. A rocky start. *Science News* 143, no. 12: 190–91.

Saunders, R. S. 1990. The surface of Venus. *Scientific American* 263, no. 6: 60–65.

Taylor, G. J. 1994. The scientific legacy of Apollo. *Scientific American* 271, no. 1: 40–47.

Taylor, S. R. 1987. The origin of the Moon. *American Scientist* 75, no. 5: 468–77.

Chapter 21

Geology, the Environment, and Natural Resources

Outline

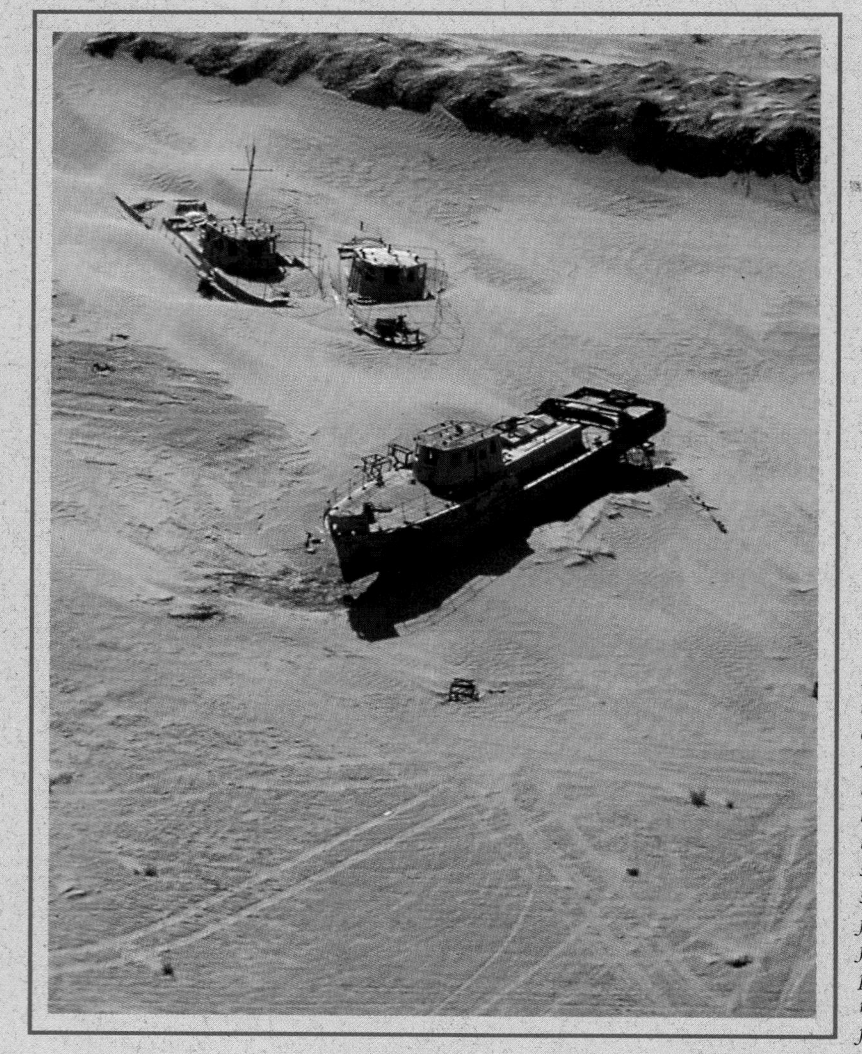

Fishing boats lie abandoned in the dry sea bed that was once part of the Aral Sea. As recently as the 1960s, the Aral Sea supported a large fishing industry, but today, the water is too saline for any fish to survive. Consequently, frozen fish are shipped in from the Pacific for processing in plants that are now more than 20 km from the Aral Sea.

Prologue

The Aral Sea in the Central Asian desert region of Kazakhstan and Uzbekistan (⯈ Figure 21-1) is a continuing environmental and human disaster. In 1960 it was the world's fourth largest lake. Since then it has steadily shrunk so that today it is only the sixth largest. During this time the Aral Sea decreased in size from 67,000 km^2 to about 34,000 km^2, and its volume has fallen from 1,090 km^3 to 300 km^3. Currently, it is divided into two separate basins, a small northern basin and a larger southern basin. At its present rate of reduction, the Aral Sea could disappear within the next 30 years.

What could have caused such a disaster? For thousands of years the Aral Sea was fed by two large rivers, the Amu Dar'ya and Syr Dar'ya. The headwaters of both rivers begin in the mountains more than 2,000 km to the southeast, and flow northward through the Kyzyl Kum and Kara Kum deserts into the Aral Sea. Until the early part of this century, a balance existed between the water supplied by the Amu Dar'ya and Syr Dar'ya rivers and the rate of evaporation of the Aral Sea, which has no outlet. In 1918, it was decreed that waters from the two rivers supplying the Aral Sea would be diverted to irrigate millions of acres of cotton so the Soviet Union could become self-sufficient in cotton production. As a result of this decision, the Aral Sea has become an ecological and environmental nightmare affecting some 35 million people.

To gain an appreciation of the magnitude of this disaster, consider that as late as the 1960s the Aral Sea was home to 24 native species of fish and supported a major fishing industry. The town of Muynak was the major fishing port and produced 3% of the Soviet Union's annual catch. The fishing industry provided about 60,000 jobs, with approximately 10,000 fishermen working out of Muynak. Today, Muynak is more than 20 km from the shoreline of the Aral Sea, and there are no native fish species left in the Aral.

As the Aral Sea shrinks, vast areas of the former sea bottom are exposed. This new sediment, far from being rich and productive, contains large amounts of sodium chloride and sodium sulfate. The concentration of salts is so high that only one plant species grows in it. It is too salty for anything else.

As the winds blow across the near-barren land, salt and dust are picked up and carried throughout the Aral region. These salts cause great damage to the cotton crops and other vegetation of the Aral basin and also exact a heavy toll on humans. The dry, salty dust has resulted in dramatic increases in respiratory and eye diseases during the past 30 years, as well as increased reported cases of throat cancer. In addition, the drinking water supply has become so polluted that many people suffer from intestinal disorders.

Because of the diversion of water from the Amu Dar'ya and Syr Dar'ya rivers and the resulting reduction in the Aral Sea, desertification has become a major problem in the Aral basin. This has caused a change in the weather

⯈ **FIGURE 21-1** Location and 1985 space shuttle view of the Aral Sea.

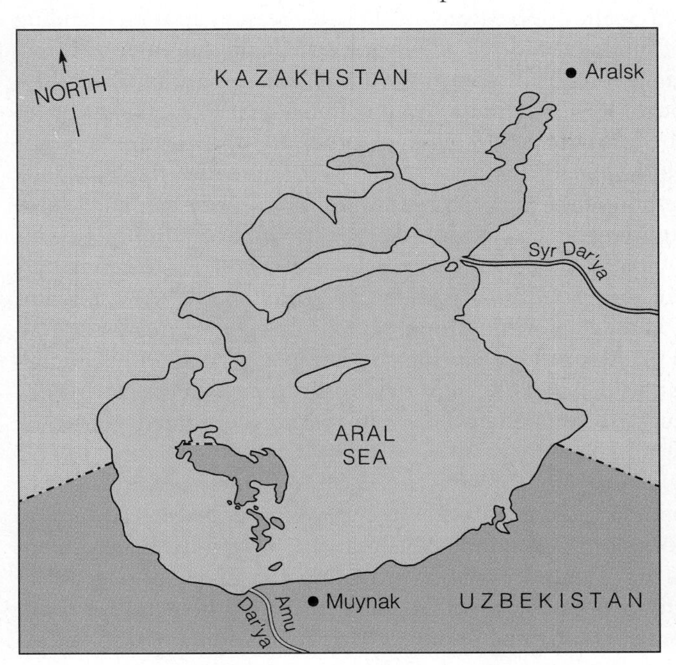

patterns of the region, so that it is now colder in the winter and hotter and drier in the summer.

In spite of the damage done to the region by the dying of the Aral Sea, irrigation for cotton production continues. The soil in many of the fields has become increasingly salty due to lower stream and groundwater levels, so more irrigation is required to maintain the same production levels. This means that even more water is diverted from the Aral Sea, causing it to shrink still further.

Can the Aral Sea be saved? Only by a concerted and cooperative effort by the countries that comprise the Aral basin. Realizing this, Kazakhstan, Uzbekistan, Kyrgyzstan, Tadzhikistan, and Turkmenistan took the first steps in 1992 by signing a formal agreement on sharing the waters of the Amu Dar'ya and Syr Dar'ya. In addition, they entered into discussions to create a council to oversee and coordinate the management of the basin's resources. Though the full restoration of the Aral Sea is probably not possible, maintaining its current surface level and even raising the surface level of the small northern Aral Sea are feasible. If this could be accomplished, salinity levels would be reduced, making it possible to reintroduce native fish species and revive commercial fishing.

INTRODUCTION

A theme of this book is that the Earth is a complex, dynamic planet that has changed continuously since its origin some 4.6 billion years ago. These changes and the present-day features we observe are the result of interactions between the various interrelated internal and external Earth systems and cycles. The rock cycle, the hydrologic cycle, and even the supercontinent cycle are just a few of the various natural cycles and systems we have discussed.

The rock cycle (see Figure 1-16), with its recycling of Earth materials to form the three major rock groups, illustrates the interrelationships between the Earth's internal and external processes. The hydrologic cycle (see Figure 15-5) is the continuous recycling of water from the oceans, to the atmosphere, to the land, and eventually back to the oceans again. Changes within this cycle can have profound effects on the surface of the Earth. For example, when climatic changes result in the formation of continental glaciers, large quantities of water that would normally be returned to the sea are locked up in glaciers. This results in a lower mean sea level, which affects ultimate base level, causing streams to increase their downcutting and shorelines to retreat seaward, opening up new land that can be colonized by plants and animals. During the Pleistocene Epoch, land bridges existed across the Bering Strait from Alaska to Siberia, and from Great Britain to mainland Europe.

We have also seen how intimately intertwined the Earth's various surface systems are. To those unfamiliar with geology, building a dam across a stream to control floodwaters downstream might seem like a good idea. Because of the interconnections between the Earth's various systems, the construction of that dam can have profound effects on the landscape upstream from the dam as well as on beaches along the shoreline where the stream finally drains. The reservoir that accumulates behind the dam results in a new temporary base level, causing the stream to readjust itself and its gradient to the new conditions. Sand that was formerly carried by the stream and deposited in the ocean where it was transported by longshore currents along the shoreline to form a beach is now trapped behind the dam. As a result, the beach is reduced or even eliminated.

On a larger scale, the movement of plates has had a profound effect on the formation of landscapes, the distribution of mineral resources, and atmospheric and oceanic circulation patterns, as well as the evolution and diversification of life.

Whereas an obvious interrelationship exists between the environment and the Earth's biota, some scientists think that an even more fundamental relationship is present, one in which the biosphere itself is so completely self-regulating as to make continued life possible (see Perspective 21-1). Although this view, known as the *Gaia hypothesis,* has not been completely accepted by most scientists, there can be little doubt that the Earth's various components do not act in isolation, but are interconnected such that when one part of a system changes, it has an effect on the other parts of the system.

Accordingly, we must understand that changes we make in the global ecosystem can have wide-ranging effects that we might not be aware of. For this reason, an understanding of geology, and science in general, is of paramount importance so that disruption to the ecosystem is minimal. On the other hand, we must also remember that humans are part of the ecosystem and, like all other life-forms, our presence alone affects the ecosystem. We must therefore act in a responsible manner, based on sound scientific knowledge, so future generations will inherit a habitable environment.

The concept of *sustainable development* has received increasing attention, particularly since the United Nations Conference on Environment and Development met in Rio de Janeiro, Brazil, during the summer of 1992. This important concept links satisfying basic human needs with safeguarding our environment to ensure continued economic development.

If we are to have a world in which poverty is not widespread, then we must develop policies that ensure continuing economic development along with management of our natural resources. Meeting the needs of a growing global population will result in increased demand for food, water, and natural resources, particularly nonrenewable mineral and energy resources. As the demand for these resources increases, geologists will play an increasingly important role in locating them, as well as ensuring protection of the environment for the benefit of future generations.

 SCIENCE AND THE CITIZEN

We live in an age of ever increasing complexity, in which scientific and technological innovations are emerging at an astonishingly rapid rate. New discoveries are announced almost daily. Advances in computer technology have revolutionized the way we live and work. Almost every aspect of modern society involves the application of science and technology.

As jobs become more technologically oriented, it is imperative that everyone know more science and how it affects our lives. Unfortunately, according to a 1985 U.S. National Science Board report, the last time most high school students ever take a math or science course is the tenth grade. Furthermore, students in the United States spend only one-half to one-third as much time learning science as do students in Germany, China, and Japan. If our nation is to compete in the global marketplace, we must have a scientifically literate workforce.

It is becoming a cliché to point out that the American public knows and understands very little science. In 1988, only one American in five knew what DNA was, yet we are debating whether and under what conditions the genetic code of organisms should be purposely altered and spending millions of dollars to map the human genome. Furthermore, about 50% of American adults said they did not understand the concept of radiation. Yet we are asked to vote on measures to build or close down nuclear power plants, debate the merits of building a radioactive waste dump in Nevada's Yucca Mountain (see Perspective 16-2), and decide if we should have our homes checked for high concentrations of radon (see Perspective 8-2).

Based on a study conducted in 1985, John Miller, director of the Public Opinion Laboratory at Northern Illinois University, estimated that only 5% of the U.S. public was scientifically literate. According to Dr. Miller, scientific literacy means understanding the scientific method, knowing the common vocabulary of science, and appreciating the social impact of scientific advances.

The essential point of Dr. Miller's survey was that only 1 in every 20 Americans understands science and the way science works. This does not mean that we must all become scientists. It does point out, however, that as scientifically illiterate consumers and citizens, we run the risk of falling victim to charlatans. How can we make informed decisions about the many critical environmental issues affecting us if we cannot separate fact from fiction and logically follow debates about issues involving science and technology?

It is equally important that the scientific community do a better job of informing and explaining the benefits of their research to society. Science does not exist in a vacuum. It proceeds on the basis of the scientific method (see Chapter 1), and everyone needs at least a rudimentary knowledge of this method and the way science works to make intelligent and informed decisions. Without such a basic understanding of science and its implications for society, we risk losing the many potential benefits science and technology can provide.

One of the benefits of studying geology is that one gains an appreciation of the geological processes that ultimately affect us all. Most readers of this book will not go on to become professional geologists, but they may become involved in geological decisions in a peripheral way, as a member of a planning board, for instance, or as a property owner with mineral rights. In such cases, a knowledge of geology is imperative if one is to make informed decisions. Furthermore, many professionals must deal with geological issues as part of their jobs (see Perspective 13-2). Lawyers, for example, are becoming more involved in issues ranging from ownership of natural resources to environmental impact reports. As government takes a greater role in environmental issues and regulations, there is an even greater need for people to become knowledgeable in the geological sciences.

 GEOLOGY AND THE ENVIRONMENT

Geology is playing an increasingly important role in environmental issues and regulation. While geologists have traditionally been involved in the exploration for mineral and energy resources, they are now also being asked to apply their expertise to many of the environmental problems facing the world today.

The events surrounding the massive underground flooding of the Akzo Salt Mine, located in western New York (and the largest salt mine in the Western Hemisphere), serves to remind us of how interrelated geologic systems and events are. The collapse of a 183 m section of roof within the mine in March, 1994, apparently triggered an initial earthquake of magnitude 3.6, which was followed by other earthquakes. These earthquakes produced a fracture more than 30 m long in the rocks separating an aquifer from the underground mine. This fracture allowed water to pour into the salt mine at a rate of 14,000 gallons per minute, forming a large underground salt lake. The water is now dissolving the salt pillars that support the mine, causing further collapse. Already, huge sinkholes have formed (and are continuing to form) at the surface and are filling in with water. As a result of the subsidence in the area, a bridge along Route 20A has sunk, causing the road to be closed at that location. All production at the mine has had to be shut down, and according to company officials, it isn't likely that the mine will be able to mine enough salt to meet next winter's demand, which will necessitate importing rock salt from Europe.

Probably the greatest challenge to the environment is population growth. It is projected that the world's population will grow by 1.7 billion people during the next two decades, bringing the Earth's human population to 7 billion. Though this may not seem to be a geological problem, we must remember that these people must be fed, housed, and clothed, and all with a minimal impact on the environment.

Perspective 21-1

THE GAIA HYPOTHESIS

In 1785, James Hutton said, "I think the Earth is a superorganism." Nearly 200 years later, in 1972, James Lovelock, a British scientist-inventor, proposed his controversial Gaia hypothesis, which he updated in 1988. According to Lovelock and other proponents of this hypothesis, "the dynamic forces of life so dominate our planet that life has a controlling influence on the oceans and atmosphere." In other words, life has shaped the environment in order to keep it within a comfortable range rather than adapting to an otherwise benign environment.

There is little doubt that interrelationships exist between life-forms and the environment. No one would deny that microorganisms aid in such processes as soil formation and deposition of some rocks, or that green plants obtain carbon dioxide from the atmosphere and release oxygen as a waste product. The destruction of tropical rain forests is much in the news (see the Prologue to Chapter 5), because it results in changes in the relative proportions of gases in the atmosphere—that is, more carbon dioxide and less oxygen. These changes could produce dramatic climatic effects such as global warming.

Although most scientists accept such relationships as those just mentioned, the Gaia hypothesis, named for the Greek Earth goddess, proposes an even more fundamental relationship between organisms and the environment. According to Lovelock, rather than adapting to an evolving environment determined solely by physical and chemical processes, organisms have the capacity to control the environment, especially the atmosphere and oceans, in ways that make continued life possible. It is this idea that the biosphere as a whole is self-regulating that many scientists find hard to accept.

Life exists within a narrow range of physical and chemical conditions, and Lovelock proposes that feedback mechanisms exist that control the environment to suit the needs of organisms. To demonstrate how organisms maintain environmental parameters within narrow limits, Lovelock proposed a mathematical model that he called Daisyworld, in which an imaginary planet is populated only by white and black daisies. If the temperature on Daisyworld rises, the black daisies absorb too much heat and die, thus leaving mostly white daisies that reflect more heat and cool the planet down. When Daisyworld cools sufficiently, black daisies thrive again and absorb more heat. In short, there is a feedback mechanism for temperature control.

Unquestionably, biologic processes are important, but those who accept the Gaia hypothesis also claim that the proportions of various gases in the atmosphere are kept in balance by feedback mechanisms. They point out that the present-day atmosphere is dominated by nitrogen and oxygen, both reactive gases that should have long ago combined with other elements to form nitrates. Furthermore, they claim that without life, carbon dioxide would have become the dominant atmospheric gas, as it is on Venus. Such feedback mechanisms, according to proponents of the Gaia hypothesis, indicate that the Earth is a giant self-regulating body in which there is an intimate connection between the evolution of the living components of the planet and the nonliving.

As some critics point out, however, the composition of the atmosphere has changed through geologic time. If the biosphere was able to regulate the environment to suit itself, as the Gaia hypothesis holds, why have there been periods of biological instability?

Many biologists dismiss the Gaia hypothesis because it is teleological; that is, it appeals to design or purpose in nature and thus cannot be tested. Some geologists point out that plate tectonics alone can control the Earth's temperature through the recycling of carbon dioxide (see Perspective 20-2).

While the Gaia hypothesis is, to say the least, controversial, it remains to be seen whether it will eventually become an acceptable theory. As in any scientific endeavor, new and radical ideas must demonstrate their worth in the competitive field of hypothesis formulating, evidence testing, and prediction. Perhaps Gaia will be supported as scientists investigate its theoretical postulates, or it may be rejected or modified, depending on future discoveries. In any case, Gaia has forced scientists to critically evaluate the relationship between life and the global environment.

Some of this population growth will be in areas that are already at risk from such geological hazards as earthquakes, volcanic eruptions, and mass wasting. Safe and adequate water supplies must be found and kept from being polluted. More oil, gas, coal, and alternative energy resources must be discovered and utilized to provide the energy to fuel the economies of nations with ever increasing populations. New mineral resources must be found. In addition, ways to reduce usage and reuse materials must be found so as to decrease dependency on new sources of these materials.

Thus, there is a great need for well-trained and competent geologists, not only to find new sources of natural resources, but also to ensure that the environment is properly managed. There will also be a great need in the future for people with a strong background in the geological sciences to advise policy makers in drafting laws concerning the environment.

When such environmental issues as acid rain, the greenhouse effect, and the depletion of the ozone layer are discussed and debated, it is important to remember that they are not isolated topics, but are part of a larger system that involves the entire Earth. Furthermore, it is important to remember that the Earth goes through cycles of a much longer duration than the human perspective of time. Although they may have disastrous effects on the human species, global warming and cooling are part of a larger cycle that has resulted in numerous glacial advances and retreats during the past 1.6 million years. In fact, geologists can make important contributions to the debate on global warming because of their geological perspective. Long-term trends can be studied by analyzing deep-sea sediments, changes in sea level during the geologic past, and the distribution of plants and animals through time.

Having examined the various aspects of geology, we now turn our attention to several of the important environmental issues concerning our planet that were only peripherally addressed earlier. While the subjects examined are by no means the only important ones, they help illustrate many of the geological processes and interrelationships we have discussed throughout this book.

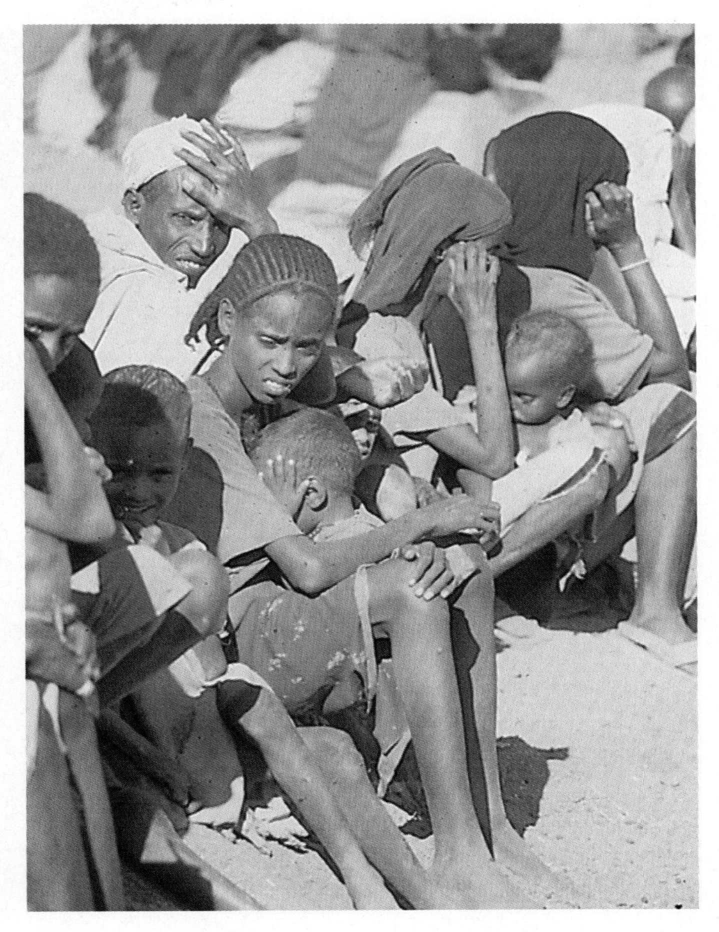

➤ **FIGURE 21-2** Overpopulation is the greatest environmental problem facing the world today. Until the world's growing population is brought under control, scenes like this one will become more common.

Overpopulation

Overpopulation is the greatest environmental problem facing the world today because the human population is exceeding the Earth's ability to support it (➤ Figure 21-2). An ecosystem's carrying capacity, or ability to support a given number of organisms, is limited by at least three factors: its food supply, its resource supply, and its ability to assimilate the pollution produced by the organisms.

As the global human population increases, nations are finding it increasingly difficult to maintain adequate food supplies. Crop yields can only be increased so much, and as the population continues to grow, productive farmlands are being taken over by villages, towns, and cities, which only increases the pressure on the remaining land. To increase food production on the land that remains, farmers must increasingly rely on the use of fertilizers and pesticides, which leads to pollution of water supplies by runoff and depletion of natural minerals in the soil.

In addition to crops, raising livestock as a food source is a common economic activity in many parts of the world even though it is much more energy- and resource-intensive than crop production. Unfortunately, in some of the drier areas the number of animals has increased to the point that they now exceed the land's capacity to support them. As a result, the vegetation cover that protects the soil has been depleted, causing accelerated soil erosion and the loss of valuable grazing land (see Figure 18-2).

Some of the world's greatest population densities occur in countries of the Pacific Rim, or Ring of Fire, so named because of the intense volcanic and seismic activity that encircles the Pacific Ocean (see Figure 12-1). In these countries, productive farmland is frequently destroyed by volcanic eruptions. Many of the populous island nations in the Ring of Fire, however, are located in tropical climates where weathering occurs rapidly, replenishing the nutrients and forming new productive soils.

In addition to food and water, human populations require a great many other resources, such as energy and building materials. Many of these resources, including oil, gas, coal, and minerals, are finite and nonrenewable (see Chapter 2). Exploration for natural resources has been geologists' main activity in the past and will continue to be an important aspect of the science in the future. As natural resources dwindle and become more difficult to find and extract, geologists will need increasingly sophisticated techniques to ensure the world has an adequate supply. Because many of

Guest Essay *Bonnie Robinson*

CLEANSING THE EARTH—WASTE MANAGEMENT*

I remember the moment when I first became interested in geology. My fifth-grade teacher was discussing the theory of continental drift; using a map of the world, she showed us how North and South America could fit against Europe and Africa to form a single giant continent! This intriguing concept made so much sense—it was like putting together the pieces of a giant jigsaw puzzle—and that is how I still view geology.

From that time, I knew that the sciences were my future. It was an unusual pursuit for an African-American girl growing up in an urban environment. I always enjoyed being outdoors and examining maps. I was a rockhound, collecting rocks and fossils wherever I went, especially in Washington's Rock Creek Park. Family trips and the Girl Scouts strengthened my awareness of the natural sciences and the enviornment.

I majored in geology at Oberlin College and broadened my understanding of the field during summer internships at the Smithsonian Institution. Geology was fascinating because it linked all of the natural and physical sciences together with engineering and applied them to the study of the Earth. I learned that geology influenced other fields of endeavor. For example, proper land-use planning requires knowledge of geology, social sciences, and other skills.

After college I worked in environmental geology at the U.S. Geological Survey, followed by graduate studies at the University of California, Santa Cruz. I spent the next 13 years as a petroleum geologist, working on oil and gas exploration and development projects throughout the western United States. My interest in environmental issues affecting the petroleum industry led to my desire to work in the field of waste management.

In my position at the Environmental Protection Agency (EPA), I am involved in the development of programs for improved management of wastes generated during the exploration and production (E&P) of crude oil and natural gas and the mining and processing of ores and minerals. The EPA's Office of Solid Waste is conducting studies of the characteristics of the wastes, waste handling methods, and their impacts on human health and the environment.

Our primary mission is to foster improvements in the management of solid wastes generated in oil and gas and mining operations, largely through voluntary efforts to improve state programs and industry practices. The principal goal is the prevention of adverse impacts to the environment from these wastes through proper waste management practices, particularly source induction.

These programs would have far-reaching implications due to the complexity of the industries involved, the wide range of environmental settings affected, and the variety of state regulatory programs. Oil and gas production is scattered throughout more than 30 states, where over 26,000 companies are involved in the exploration and production of oil and gas. Each year thousands of new wells are drilled and thousands of well sites are abandoned. The major wastes generated at these locations consist of water extracted with the oil and gas, drilling fluids, and a variety of lesser wastes.

Mining and mineral-processing operations occur at sites throughout the United States. Several billion metric tons of wastes are generated annually, including waste rock tailings or material rejected after the valuable minerals have been extracted, mine water, and overburden or unconsolidated soil and rock material underlying the ore body. These wastes often contain varying amounts of potentially hazardous constituents.

One of the key issues facing the EPA is how to determine the most efficient alternatives for improving waste management in these industries without significant adverse impacts on oil and gas and mineral production. Continued domestic production is vital to the nation's interest, but it must be balanced with adequate environmental protection.

The geosciences offer a wide range of opportunities for active involvement in many of the challenging issues facing the world today. Knowledge of science and technology, or science literacy, is essential for intelligent decision making regarding critical national issues. A geological background can also serve as a foundation upon which to build careers in other disciplines, such as law, planning, or politics. Opportunities exist for full participation by minorities and women, who are severely underrepresented in science and technology. It is vital that we encourage, develop, and utilize this pool of talent.

BONNIE ROBINSON earned an A.B. degree in geology from Oberlin College in 1974, followed by graduate studies at the University of California at Santa Cruz. She worked as a petroleum geologist in Denver, Colorado, for 13 years and has been with the U.S. Environmental Protection Agency for more than 4 years.

*Opinions expressed in this paper are solely those of the author and do not necessarily represent those of the U.S. Environmental Protection Agency.

the natural resources essential to society today are finite, we must find economical ways to recycle these resources and use a larger percentage of renewable resources.

The last factor in determining an ecosystem's carrying capacity is its ability to assimilate pollution. We are all aware of air pollution, solid waste disposal, contaminated groundwater, and acid rain, to name only a few of the problems of an increasingly industrialized planet. In natural ecosystems, wastes are usually diluted to harmless levels and reused and recycled through nutrient cycles, so that a natural balance exists between populations and waste production.

Humans produce tremendous amounts of wastes, many of which do not quickly or easily degrade. Furthermore, many of these pollutants are lethal and contaminate the soil, air, and water. According to a study by the Conservation Foundation, every citizen of the United States produces 22,500 kg of domestic, agricultural, and industrial waste and pollution per year. This represents 25% of the world's pollution, yet the United States contains only 6% of the world's population. The problem is certainly not limited to the United States. The industrialized countries of the world make up a minority of the world's total population, yet are responsible for much of the pollution. Some of the worst environmental disasters, such as the nuclear disaster at Chernobyl in the Ukraine, have occurred in the industrialized countries.

The problem of overpopulation and its effects on the global ecosystem are varied. For many of the poor and nonindustrialized countries, the problem is too many people and not enough food. For the more developed and industrialized countries, it is too many people rapidly depleting both the nonrenewable and renewable natural resource base. And in the most industrialized countries, it is people producing more pollutants than the environment can safely recycle on a human time scale. In all cases, it is an environmental imbalance created by a human population exceeding the Earth's carrying capacity.

Global Warming

Carbon dioxide is produced as a by-product of respiration and the burning of organic material. As such, it is a component of the global ecosystem and is constantly being recycled as part of the carbon cycle (see Figure 8-24). The concern in recent years over the increase in atmospheric carbon dioxide has to do with its role in the greenhouse effect. Recall from Perspective 20-2 that the recycling of carbon dioxide between the crust and atmosphere is an important climatic regulator because carbon dioxide, as well as other gases such as methane, nitrous oxide, chlorofluorocarbons, and water vapor, allow sunlight to pass through them but trap the heat reflected back from the Earth's surface. Heat is thus retained, causing the temperature of the Earth's surface and, more importantly, the atmosphere to increase, producing the greenhouse effect.

Until the Industrial Revolution began during the mid-eighteenth century, humans' contribution to the global temperature pattern was negligible. With industrialization and its accompanying burning of tremendous amounts of fossil fuels, carbon dioxide levels in the atmosphere have been steadily increasing since about 1880 (Figure ▶ 21-3). In fact, atmospheric levels of carbon dioxide are currently almost 25% higher than a century ago and at their current yearly rate of increase will double from their present concentrations within the next 60 to 70 years. This increase in atmospheric carbon dioxide is largely attributed to the burning of coal, oil, and natural gas, which releases carbon dioxide as a by-product of combustion.

Recent research also indicates that deforestation of large areas, particularly in the tropics, is another cause of increased levels of carbon dioxide. This is because plants use carbon dioxide in photosynthesis and thus remove it from the atmosphere. With a decrease in the global vegetation cover, less carbon dioxide is removed from the atmosphere.

Carbon dioxide is not the only gas responsible for an

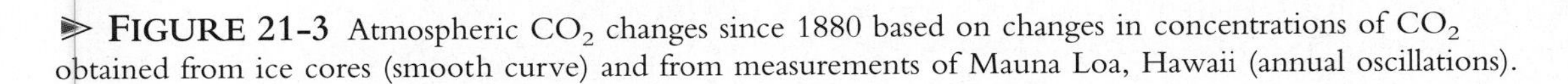

▶ **FIGURE 21-3** Atmospheric CO_2 changes since 1880 based on changes in concentrations of CO_2 obtained from ice cores (smooth curve) and from measurements of Mauna Loa, Hawaii (annual oscillations).

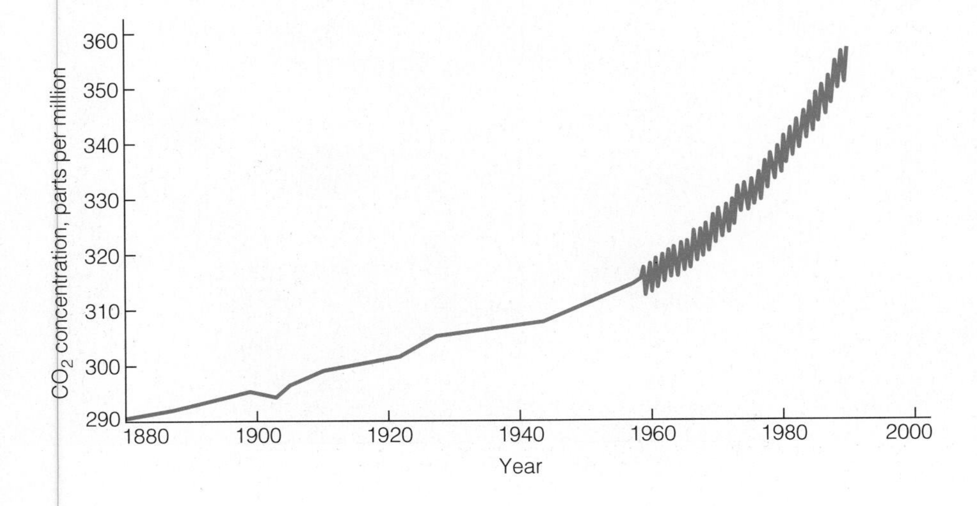

increased greenhouse effect. Methane and chlorofluorocarbons are also major contributors to global warming patterns. Methane is a natural gas that is 20 times as effective as carbon dioxide in trapping heat in the atmosphere. Current atmospheric levels of methane are double those of the preindustrialized period. Methane is a by-product of manure and the decomposition of crops such as rice under oxygen-poor conditions and is also produced by the digestive processes of livestock and termites. Termites are particularly abundant in deforested areas of the world, so deforestation can have the combined effect of increasing methane production and reducing carbon dioxide in the atmosphere when a forest is cut down.

Chlorofluorocarbons are well known for their effect on the degradation of the ozone layer, but also are thought to be responsible for about 20% of the global warming potential. Chlorofluorocarbons are released from air conditioners, refrigerators, and many spray cans and are at least 10,000 times as effective as carbon dioxide in trapping atmospheric heat. As more countries agree to reduce or eliminate chlorofluorocarbons, the contribution of these gases to global warming should decrease.

Because of the increase in human-produced greenhouse gases during the last two hundred years, many scientists are concerned that a global warming trend has already begun and will result in severe global climatic shifts. A graph of global mean temperatures since the late 1800s tends to substantiate this claim. It shows a rise in temperature of 0.6°C during this time, with the six warmest years during the 20th century occurring in the 1980s. Other scientists point out that the warming observed before 1940 may simply be a recovery and adjustment from the effects of the "Little Ice Age" that occurred from about 1500 to the mid- to late-1800s (see the Prologue to Chapter 17).

Most computer models based on the current rate of increase in greenhouse gases show the Earth warming as a whole by 2°C to 5°C during the next 75 years. Such a temperature change will be uneven, however, with the greatest warming occurring in the higher latitudes. As a consequence of this warming, rainfall patterns will shift

FIGURE 21-4 Probable changes in the world's rainfall patterns resulting from global warming.

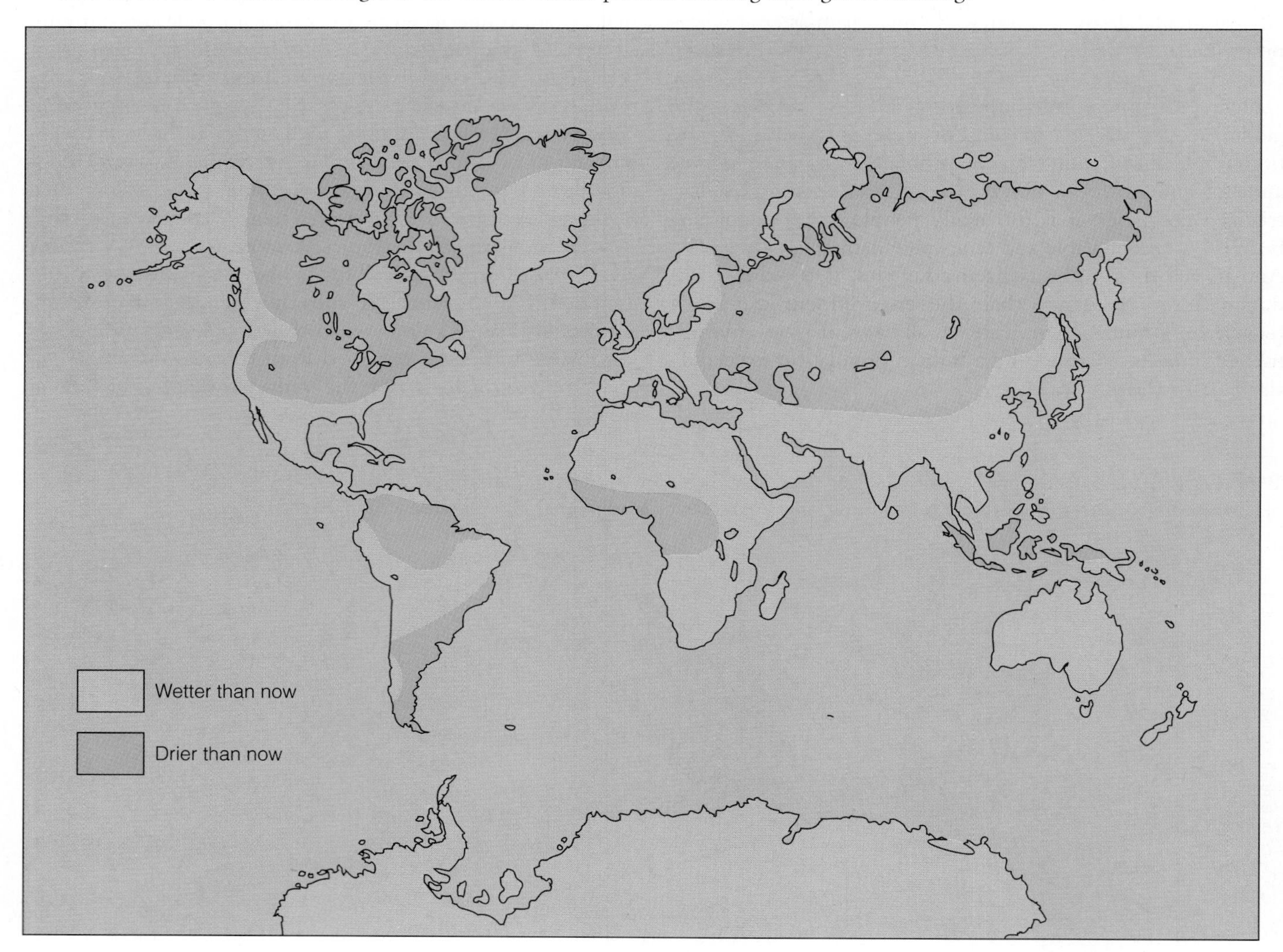

dramatically (▶ Figure 21-4). This will have a major effect on the largest grain-producing areas of the world, such as the American Midwest. Drier and hotter conditions will intensify the severity and frequency of droughts, leading to more crop failures and higher food prices. With such shifts in climate, the Earth may experience an increase in desertification (see the Prologue to Chapter 18).

Glaciers provide important information about climatic change because they respond to such changes by advancing or retreating. Because of their small size and volume of ice, valley glaciers are particularly sensitive to even small climatic perturbations. As ◉ Table 21-1 shows, the number of retreating valley glaciers in various areas has risen dramatically as they respond to higher average global temperatures. The behavior of the Alaskan glaciers did not change significantly because they are much larger (on average more than 100 times larger than valley glaciers studied elsewhere) and a greater lag time occurs before they respond to climatic change.

TABLE 21-1 Response of Valley Glaciers to Climate Changes, 1974–1990

	Number		
	Advancing	*Retreating*	*Stationary*
Switzerland			
1980	70	28	3
1983	46	57	6
1986	42	13	9
1987	35	58	13
1989	19	83	3
North Cascades, Washington			
1974	9	1	0
1984	5	31	11
1988	2	36	9
1990	1	41	5
1992	1	44	2
Norway			
1980	5	7	0
1985	1	10	1
Alaska			
1976	15	18	18
1985	16	24	14
Italy			
1981	25	10	10
1985	25	14	6
1988	26	92	13
1989	11	106	5
1990	9	123	9

SOURCES: Mauri Pelto, Director of the North Cascade Glacier Climate Project; the Swiss Glacier Commission; the Norwegian Water Resources and Electricity Board; the USGS; and *Ice* 98, no. 1: 3.

TABLE 21-2 Measures That Can Reduce Global Warming

Reduce the population growth rate.
Switch from coal-fired and oil-fired power plants to natural gas.
Implement technologies that burn coal more efficiently.
Expand cogeneration—processes that trap waste heat and put it to good use.
Boost automobile efficiency.
Expand mass transit.
Develop alternative liquid fuels for the transportation sector.
Improve the efficiency of industry.
Make new and existing homes more energy-efficient.
Build many new homes that use solar energy for space heating.
Reduce global deforestation.
Begin a massive reforestation effort.
Phase out all chlorofluorocarbons as soon as possible.
Reduce consumption of unnecessary items.
Expand recycling efforts.

With a continued rise in global temperature, not only will valley glaciers retreat, but the ice caps will melt, contributing to rising sea level, which will greatly affect coastal areas where many of the world's large population centers are located (see Figure 17-34). For example, approximately 17% of Bangladesh would be flooded, as would many of the extensive lowland rice-producing regions in other Asian countries. Many low-lying cities such as Houston and New Orleans would have to be ringed with dikes and levees or abandoned.

Can anything be done to reverse this apparent warming trend? The answer is yes, but it will require major shifts in the way people live and changes in the economic structure of the industrialized countries (◉ Table 21-2). Among the ways to reduce the buildup of greenhouse gases are to reduce the wholesale logging of the tropical rainforests and begin reforesting areas that are now denuded of vegetation. Sharp reductions in the consumption of fossil fuels through conservation measures and a switch to alternative energy sources must also occur. While it will take time to reduce the trend of global warming, immediate changes now will have long-term benefits.

We cannot leave the subject of global warming without pointing out that many scientists are not convinced that the global warming trend is the direct result of increased human activity related to industrialization. They point out that while there has been an increase in greenhouse gases, there is still uncertainty about their rate of generation and rate of removal and about whether the 0.6°C rise in global temperature during the past century is the result of normal climatic variations through time or the result of human activity. Furthermore, they point out that even if there is a general global warming during the next 75 years, it is not certain that the dire predictions made by proponents of global warming will come true. The Earth, as we know, is a remarkably complex system, with many feedback mechanisms and interconnections throughout its various subsystems. It is very difficult to predict all of the consequences that global warming would have for atmospheric and oceanic circulation patterns.

Acid Rain

Atmospheric pollution is one of the consequences of industrialization. Several of the most industrialized nations, including the United States, Canada, and the former Soviet Union, have reduced their emissions into the atmosphere, but many developing nations continue to increase theirs. Some of the results of atmospheric pollution include smog, possible disruption of the ozone layer, global warming, and acid rain.

Recall that water and carbon dioxide in the atmosphere react to form carbonic acid that dissociates and yields hydrogen ions and bicarbonate ions (see Chapter 5). The net effect of this reaction is that all rainfall is slightly acidic. Thus, acid rain is the direct consequence of the self-cleansing nature of the atmosphere; that is, many suspended particles of gases in the atmosphere are soluble in water and are removed from the atmosphere during precipitation.

Several natural processes, including volcanism and the metabolism of soil bacteria, introduce gases into the atmosphere that cause acid rain. Human activities, however, produce added atmospheric stress. The burning of fossil fuels, for example, has added excess carbon dioxide to the atmosphere. Nitrogen oxide from internal combustion engines (◉ Table 21-3) and nitrogen dioxide, which is formed in the atmosphere from nitrogen oxide, react to form nitric acid. Although carbon dioxide and nitrogen gases contribute to acid rain, the greatest culprit is sulfur dioxide, which is primarily released by burning coal that contains sulfur (Table 21-3). Once in the atmosphere, sulfur dioxide reacts with oxygen to form sulfuric acid, the main component of acid rain.

TABLE 21-3 U.S. Sources of Sulfur and Nitrogen Oxide Emissions, 1980

	Percentage of Total Emissions	
Source	*Sulfur Oxides*[a]	*Nitrogen Oxides*
Mobile sources[b]	3.8	44.0
Stationary sources[c]	80.0	51.0
Industrial processes	16.0	2.9
Other (agriculture, etc.)	0.2	2.1

[a]SO_2 emissions (tons/year) decreased 33% between 1975 and 1990.
[b]All transportation including recreational vehicles.
[c]Utilities, industrial, institutional, and residential sources.

SOURCE: Council on Environmental Quality, *Twelfth Annual Report,* 1981, and various other sources.

The phenomenon of acid rain was first recognized by Robert Angus Smith in England in 1872, about a century after the beginning of the Industrial Revolution. It was not until 1961, however, that acid rain became a public environmental concern. At that time it was realized that acid rain is corrosive and irritating, kills vegetation, and has a detrimental effect on surface waters. Since then the effects of acid rain have been recognized in Europe (especially in Eastern Europe where so much coal is burned), the eastern United States, and eastern Canada. During the past three decades, acid rain has been getting worse, particularly in Europe and eastern North America (▶ Figure 21-5). During the last decade, the developed countries have made efforts to reduce the impact of acid rain; in the United States the Clean Air Act of 1990 outlined specific steps to reduce the emissions of pollutants that cause acid rain.

The areas most affected by acid rain invariably lie downwind from coal-burning power plants or other industries that emit sulfur gases. Chemical plants and smelters (plants where metal ores are refined) discharge large quantities of sulfur gases and other substances such as heavy metals. The effect of acid rain in these areas may be modified by the local geology (▶ Figure 21-6). If an area is underlain by limestone or alkaline soils, the acid rain tends to be neutralized by the limestone or soil. Areas underlain by granite, on the other hand, are acidic to begin with and have little or no effect on the rain.

The effects of acid rain vary. Small lakes become more acid as they lose the ability to neutralize the acid rainfall. As the lakes increase in acidity, various types of organisms disappear, and, in some cases, all life-forms eventually die. Acid rain also causes increased weathering of limestone and marble (recall that both are soluble in weak acids) and, to a lesser degree, sandstone. Such effects are particularly visible on buildings, monuments, and tombstones; a notable example is Gettysburg National Military Park in Pennsylvania, which lies in an area that receives some of the most acidic rain in the country.

Although industries that emit sulfur gas have clear effects on vegetation in the immediate vicinity, some people have questioned whether acid rain has much effect on forests and crops distant from these sources. But many forests in the eastern United States show signs of stress that cannot be attributed to other causes. In Germany's Black Forest, the needles of firs, spruce, and pines are turning yellow and falling off.

Currently, about 20 million tons of sulfur dioxide are released yearly into the atmosphere in the United States, mostly from coal-burning power plants. Power plants built before 1975 have no emission controls but the problems they pose must be addressed if emissions are to be reduced to an acceptable level. The most effective way to reduce

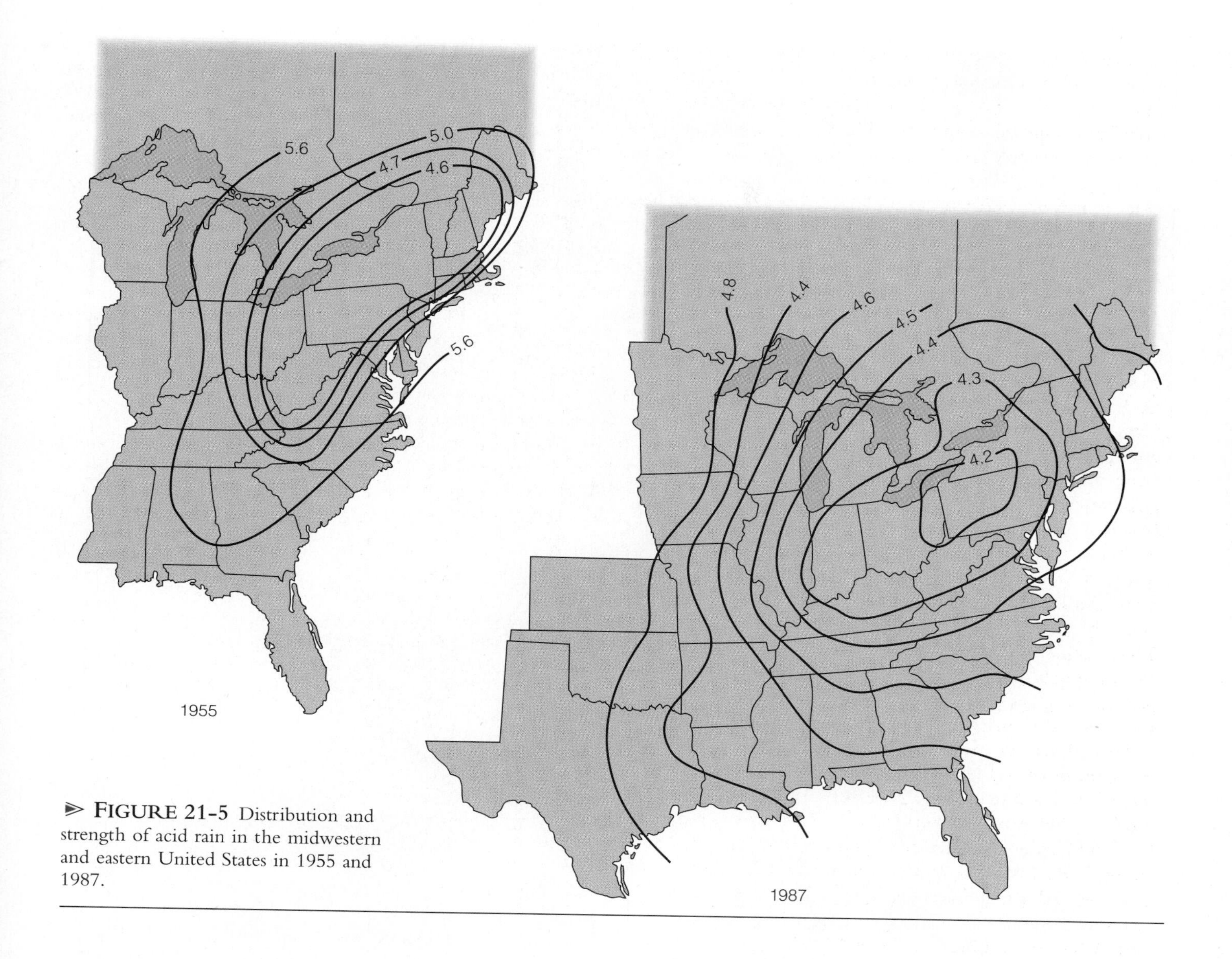

➢ **FIGURE 21-5** Distribution and strength of acid rain in the midwestern and eastern United States in 1955 and 1987.

emissions from these older plants is with flue-gas desulfurization, a process that removes up to 90% of the sulfur dioxide from exhaust gases. Flue-gas desulfurization has some drawbacks. One is that some plants are simply too old to be profitably upgraded; the 85-year-old Phelps Dodge copper smelter in Douglas, Arizona, closed in 1987 for this reason. Other problems with flue-gas desulfurization include disposal of sulfur wastes, the lack of control on nitrogen gas emissions, and reduced efficiency of the power plant, which must burn more coal to make up the difference.

Other ways to control emissions include the conservation of electricity; the less electricity used, the lower the emissions of pollutants. Natural gas contains practically no sulfur, but converting to this alternate energy source would require the installation of expensive new furnaces in existing plants.

Acid rain, like global warming, is a global problem that knows no national boundaries. Wind currents may blow pollutants from the source in one country to another where the effects are felt. Developed nations have the economic resources to reduce emissions, but many underdeveloped nations cannot afford to do so. Furthermore, many nations have access to only high-sulfur coal and cannot afford to install flue-gas desulfurization devices. Nevertheless, acid rain can be controlled only by the cooperation of all nations contributing to the problem.

Nuclear Waste Disposal

Most environmentalists have long opposed the use of nuclear power, primarily because of the potential for catastrophic accidents occurring at the plants and problems with radioactive waste disposal. Nevertheless, because such problems as global warming and acid rain are partially caused by the burning of fossil fuels, nuclear power may experience a resurgence, particularly if acceptable safety features can be incorporated into the plants and the problem of waste

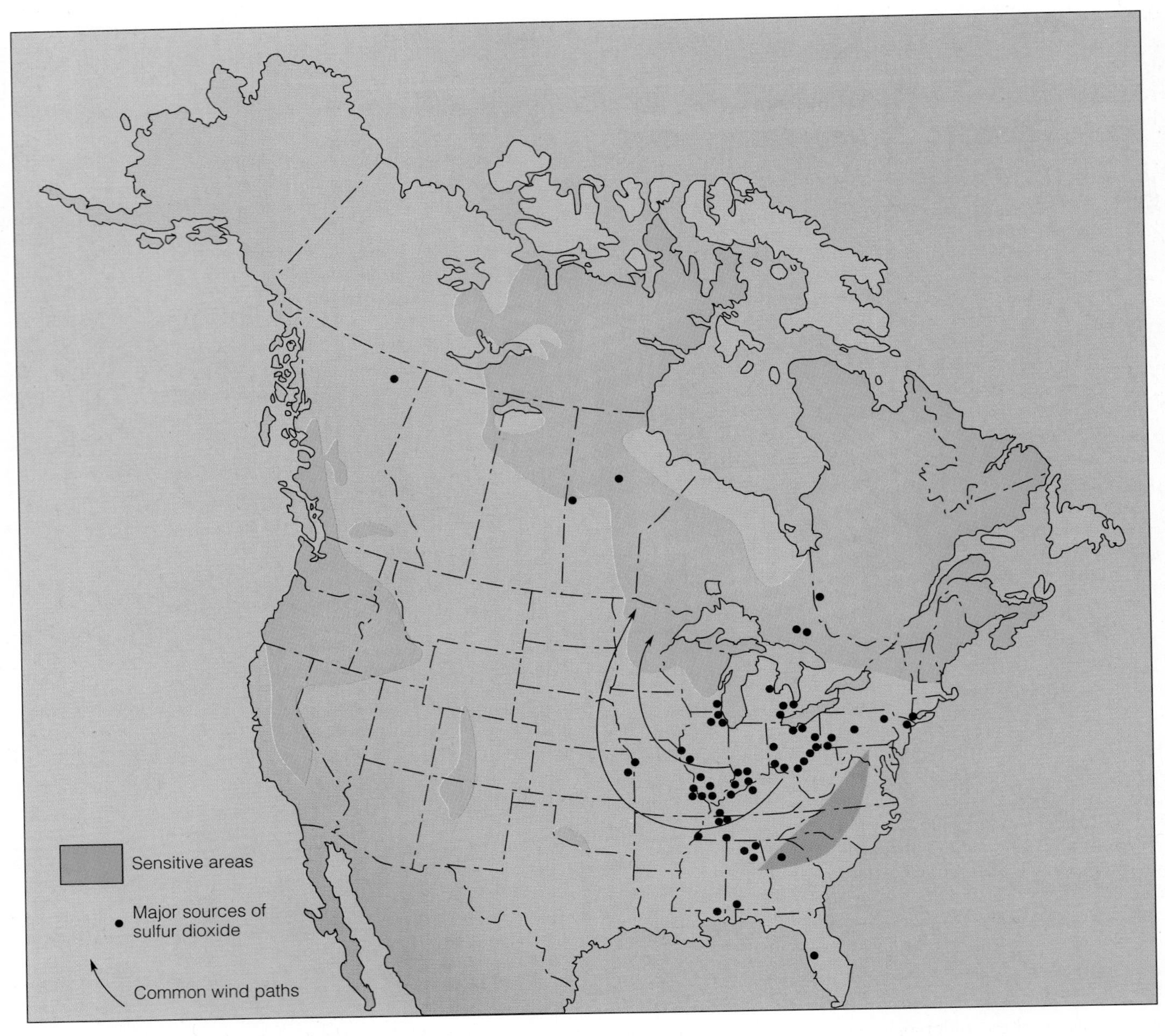

▶ **FIGURE 21-6** Acid-sensitive areas in North America and major acid rain–producing sources.

disposal is satisfactorily resolved. Currently, uranium inventories remain high around the world, and little change in demand is expected during the next few years.

A discussion of the safety features of nuclear power plants is beyond the scope of this book, but an examination of the environmental problems associated with nuclear waste disposal is certainly appropriate. One of the problems of nuclear energy is finding sites suitable for storing radioactive waste products, most of which must be isolated from the environment for thousands of years.

Near the end of 1987, Congress passed the Nuclear Waste Policy Amendments Act, authorizing the U.S Department of Energy's Civilian Radioactive Waste Management Program to evaluate the potential of Nevada's Yucca Mountain to serve as the nation's first high-level radioactive waste dump (see Perspective 16-2). Since then, scientists from many disciplines have been analyzing the site to determine if it can isolate high-level radioactive waste from the environment for 10,000 years, the minimum time such waste will remain dangerous.

In addition to concerns about the geology of the Yucca Mountain area, including possible volcanic activity, the depth of the water table, and the likelihood of a change in climate in the area during the next 10,000 years, an additional factor must be considered, and that is its economic potential. The question is whether the proposed site has a natural resource potential greater than the surrounding areas, or whether it currently has economically extractable resources. One of the reasons for concern about potentially valuable natural resources is the possibility that direct or indirect exploration activities might disturb the nuclear waste and cause environmental damage. Studies done to date indicate a low probability of such natural resources as oil and gas and a low potential for metals.

One of the reasons for concern about the proper disposal of nuclear waste is the environmental catastrophes that are now known to have occurred in uranium mining areas in Eastern Europe. For example, after years of supplying the former Soviet Union with more than 220,000 tons of uranium ore, the environmental, human, and monetary costs of East Germany's uranium industry are finally coming due. For years, East Germany was the world's third largest producer of uranium ore (following the United States and Canada), shipping most of it to the Soviet Union and helping it become a nuclear superpower. With the collapse of the Soviet Union, East Germany's uranium industry also collapsed as there was no longer a market for its low-grade uranium ore.

Within East Germany's uranium mining area, vast amounts of low-grade uranium ore were mined and processed with virtually no regard for the environmental devastation resulting from such operations. Huge open pits as deep as 165 m were dug during mining operations, with the uranium tailings piled up into hills nearby, many as high as 150 m. Winds would carry the radioactive dust from these tailings and deposit it on the surrounding towns and villages. In addition to radioactive dust, many of the mines generated waste highly concentrated in heavy metals, pyrite, and arsenic.

As a result of the mining and processing activities, the soil and groundwater in many areas of East Germany are highly contaminated. The German government has allocated about 1 billion marks (U.S. $682 million) per year during the next 15 years to deal with the massive cleanup of the area. Unfortunately, many officials think that this expenditure will not be enough.

In addition to the environmental cleanup costs, the costs in terms of human suffering are still to be determined. Although detailed health records of the miners were kept, the overall effect on the health of the local residents is not known. To date, more than 6,000 confirmed cases of lung cancer have been reported among mine workers who worked in the uranium mines during the 1950s. Some of the mining areas have high rates of certain cancers, but as yet no effort has been made to see if these are statistically significant and if they can be related to the mining and processing activities.

Another recent example of unregulated nuclear waste concerns a huge waste pile of 1,200 tons of uranium and 4 million tons of uranium mining residue dumped in a pond at a secret Soviet military plant 187 km east of Tallinn, Estonia. The pond covers 82 acres and has radiation levels 500 times greater than accepted limits. There are no safety measures at this location, and there is great concern that the pond could overflow during stormy weather, sending a stream of radioactive water into the Gulf of Finland. During its operation from 1940 to 1991, more than 4 million tons of uranium, most of it imported from Hungary and Czechoslovakia, were processed at the plant.

The four examples just discussed point out the need for qualified geologists and people trained in other sciences to help solve many of our major environmental problems.

Geology and the Search for Natural Resources

The search for natural resources has been the main area of employment for geologists in the past and will continue to be in the future. The industrialized economies will always need energy and minerals to sustain themselves, and finding those resources is the job of geologists. No country is self-sufficient in all of the mineral resources it needs and uses, and therefore must import some of them. The United States, for example, is rich in a variety of mineral resources (◉ Table 21-4) but still must import much of its oil and many of its metals (see Table 2-7). Such a dependence on other nations for certain basic economic needs translates into political policy in protecting those areas or regions. The 1991 Gulf War was fought, in part, to make sure Kuwait's oil was available to the West.

Today, geologists use increasingly sophisticated techniques in their worldwide exploration for natural resources. Remote sensing and computer technology are just two of the ways geologists are applying modern technology in their work. The application of plate tectonic theory to the formation and distribution of natural resources is helping geologists to focus their attention on areas that have a high potential for economic success. In this section we will focus on coal, oil and natural gas, and metals.

Coal

Because of dwindling supplies of oil in the United States and increasing reliance on imported oil, coal has been championed as a major source of energy and one that is free of foreign control. In fact, coal supplies about 24% of the nation's current energy needs, compared to 18% in 1977. It is estimated that the United States has coal reserves (that part of the resource base that can be economically extracted) of 243 billion metric tons, and at the current rate of production of nearly 1 billion metric tons per year, a roughly 243-year supply of coal.

In recent years, however, this estimate of coal reserves has been challenged, and it is possible that the seemingly inexhaustible supply is not as large as it once appeared. In part, the problem has been that the Department of Energy relies on databases compiled by the United States Geological Survey and state geological surveys, which are not uniform in what they consider to be minable coal or in the classification standards used. The Department of Energy has interpreted the information provided and applied it as best it

TABLE 21-4 Value of Selected Mineral Resources Produced in the United States

Mineral Resources	Production Value (Millions of Dollars)	Principal Producing States
Energy resources		
Coal	20,987.0	WY, KY, WV, PA, IL
Natural gas	30,096.0	TX, LA, OK, NM
Petroleum	37,447.4	AK, TX, CA, LA, OK
Uranium	336.8	WY, NM, TX
Industrial minerals		
Bromine	144.0	AR, MI
Cement, Portland	3,575.9	CA, TX, PA, MI
Cement, masonry	243.9	FL, IN, PA, AL
Clays	1,400.8	GA, OH, NC, TX
Gypsum	109.2	OK, IA, MI, TX
Phosphate rock	887.8	FL, NC, ID, UT
Salt	680.2	LA, TX, NY, OH
Sulfur	430.8	TX, LA
Metals		
Copper	3,771.6	AZ, NM, UT, MT
Gold	2,831.3	NV, CA, SD, UT
Iron ore	1,716.7	MN, MI, MO, UT
Lead	315.2	MO, ID, CO, MT
Molybdenum	266.9	AZ, CO, MT, VT
Silver	349.3	NV, ID, MT, AZ
Zinc	324.2	TN, NY, MO, MT

SOURCE: *Statistical Abstract of the United States,* 1991.

could to the national standard to produce an annual summary of the nation's minable coal reserves. This is the figure that is usually quoted when referring to the nation's coal reserves.

One of the problems with this figure is that coals that have neither current nor future economic mining potential are included in it. Other coals are economically minable, but may be unavailable for mining because of land-use restrictions or because they do not meet current emission standards. A joint study conducted by the United States Geological Survey and the Kentucky, Virginia, and West Virginia state geological surveys found that about half of the original coal reserves in the initial 12 study areas are not available for mining. This study has been expanded to other states, and initial results show similarly disturbing trends in the actual amount of coal available for mining. What once seemed like a secure and almost inexhaustible source of energy is no longer the only answer to our energy needs.

Oil and Natural Gas

During the 1850s, the demand for petroleum was increasing in the United States as people sought a cheap alternative to other sources to be used for lighting, as a lubricant for machinery, and as an ingredient in liniments. In 1859, Edwin L. Drake drilled an oil well 21 m deep at Titusville, Pennsylvania, and began pumping 10 to 35 barrels of oil per day (one barrel is equal to 42 gallons). The United States quickly became the world's leading producer, a position it maintained until 1965. In 1992, it was in third place, behind the Commonwealth of Independent States and Saudi Arabia (▶ Figure 21-7).

In little more than one hundred years after Drake drilled his well, the United States had become a net petroleum importer. In 1992, the United States imported more than 6 million barrels of oil a day, which represented 47% of its daily consumption. The cost of importing foreign oil was $50 billion and accounted for almost 50% of the United States' trade deficit. Even with increased conservation and improved energy saving devices, it won't be long before the United States imports more than half of its oil from overseas, with much of it coming from the Persian Gulf region.

Even though petroleum was discovered as early as 1908 in Iran, the Persian Gulf did not become a significant petroleum-producing area until the economic recovery after World War II. Following the war, Western Europe and Japan in particular became dependent on Gulf oil and still rely heavily on this region for most of their supply. The United States is also dependent on imports from the Gulf, but receives significant quantities of petroleum from other

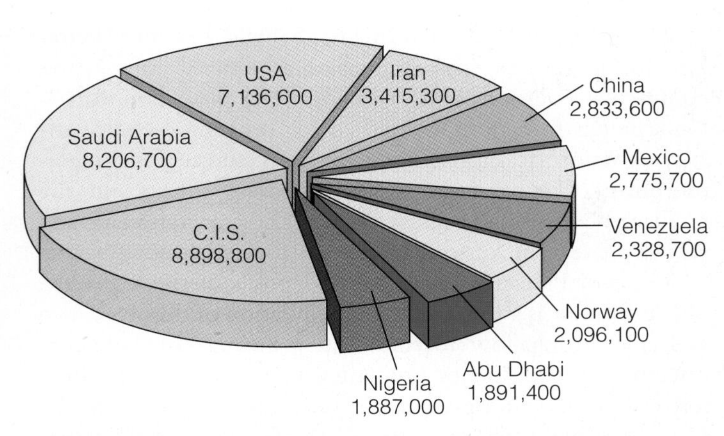

➢ **FIGURE 21-7** The top 10 oil-producing countries for 1992. Numbers indicate barrels of oil produced daily.

sources such as Mexico and Venezuela. Currently, fully 40% of all petroleum imports in the world come from the Gulf countries, and more than 60% of all proven reserves are in that region (➢ Figure 21-8).

Many nations including the United States are heavily dependent on imports of Persian Gulf oil, a dependence that will increase in the future. It is projected that the world's demand for petroleum will increase another 15% by the end of the century. With the decline in production in the United States and the Commonwealth of Independent States, the world will be even more dependent on oil from the Persian Gulf, a fact that will be important in the global political arena.

One of the major problems faced in promoting natural gas as an energy source has been establishing the infrastructure to get it from the gas fields to market. The use of natural gas in the United States has been steadily growing during the past decade, in part because of large reserves in North America (◉ Table 21-5), and is expected to continue to increase during the 1990s.

According to the International Energy Agency, the demand for natural gas in Western Europe is projected to increase by one-third, or an additional 100 billion m^3, by the end of this century. Natural gas has already surpassed oil as the main fuel used by power generating plants in Europe, and its use as a source for heating is also increasing. The European Bank for Reconstruction and Development estimates that natural gas usage will double in Western Europe in the next 20 years.

Waiting to serve that market are Russia's largely untapped natural gas reserves in Siberia. These reserves are the largest in the world and are estimated to represent 40% of the world's total natural gas reserves. Western Europe currently relies upon Russia for about 25% of its natural gas needs and could become even more dependent should Russia be able to develop Siberia's natural gas potential.

The problem lies in getting the gas from Siberia to its markets in Europe. It is estimated that it will cost about U.S. $15 billion to construct a pipeline from western Siberia to the German-Czech border where it can be fed into Europe's natural gas pipeline grid. Such an undertaking presents enormous economic and political risks. The two major ones are Russia's ability to raise the money to build a pipeline,

TABLE 21-5 Natural Gas Reserves in North America

Country	Billions of Cubic Feet	Percentage
Mexico	70,900	21.2
United States	167,062	50.1
Canada	95,734	28.7

SOURCE: *Oil and Gas Journal* 90 (December 28, 1992): 44–45.

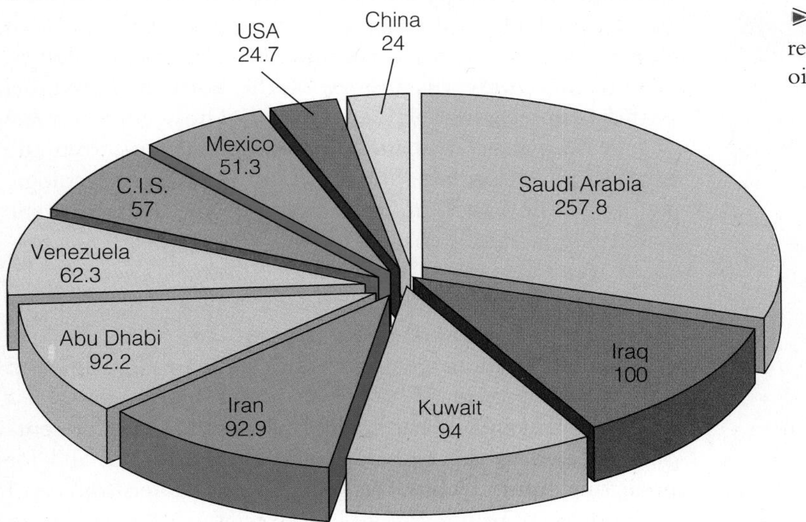

➢ **FIGURE 21-8** The top 10 countries in proven oil reserves in 1992. Numbers indicate billions of barrels of oil.

and whether it would be profitable at the low prices natural gas currently commands.

Even if the pipeline is built, there is no guarantee there will be enough new customers for the gas. Several countries expect to increase production and build pipelines of their own to serve Europe in the near future. For example, Algeria plans to double its natural gas exports to Europe by the end of this century and is expanding its present pipeline capacity to Italy and building a new pipeline to Spain. Iran has 13% of the world's natural gas reserves, second only to Russia, and is planning a pipeline through Turkey to serve Europe.

Currently, oil is still the major source of energy in the world, but as pipeline grids are constructed and natural gas fields are brought on-line, natural gas looms as a major threat to oil's dominance in the world energy market. Though there will always be a need for oil, clean-burning natural gas is an attractive alternative to oil as a source for power and heating needs.

Metals

Metals are an essential part of the world's economy. Recall that the distribution and concentration of many metal ore bodies are controlled by plate tectonics (see Chapter 12), and geologists are increasingly using plate tectonic theory in their search for metals. Because of the importance of plate tectonics to the formation of various metal ores, we will briefly examine its role in relation to Canada's mineral wealth (◉ Table 21-6), specifically in the Canadian Shield (see Chapter 7).

Canada is a leading producer of iron ore, nickel, zinc, lead, gold, and silver, many of which are found in the Canadian Shield. In fact, all of Canada's iron and nickel, nearly 90% of its gold, 50% of its copper and silver, and about 30% of its zinc comes from the Canadian Shield.

TABLE 21-6 1990 Production of Leading Metals in Canada

Metal	Millions of Dollars
Copper	2,500
Zinc	2,480
Gold	2,380
Nickel	2,050
Iron	1,350
Uranium	970
Lead	260
Silver	256
Platinum	230
Molybdenum	140

SOURCE: *Canada Year Book,* 1992.

As the history of the Canadian Shield becomes better known, it appears that it formed as a result of various continental collisions and breakups along plate boundaries. Continental formation is marked by orogenic activity involving deformation, magma generation, and regional metamorphism. Breakup was accompanied by rifting and the emplacement of mafic dikes. During these accretionary and rifting phases, the various metal ores formed.

The vast Precambrian iron ore deposits of the Canadian Shield formed as a result of the precipitation of dissolved iron and silica in shallow marine waters during the quiescent intervals between major orogenies, mainly during the Proterozoic Eon. All of Canada's iron ore is mined in the Canadian Shield, and more than 90% of Canada's gold comes from Archean and Early Proterozoic rocks of the Canadian Shield. These rocks formed during the various orogenic events occurring along convergent plate boundaries. When the metal-rich magmas and fluids that were produced during convergence cooled, they concentrated the gold and other metals that are now mined.

In addition to the metals formed as a result of convergence, the Canadian Shield also has massive volcanic-associated sulfide deposits containing copper and zinc. These deposits are thought to be the result of major volcanic events occurring along mid-oceanic ridges or in back-arc basins corresponding to the beginning of orogenic episodes during the Archean and Proterozoic eons.

Current knowledge about the history of the Canadian Shield indicates that most of its metal resources were formed during three major orogenic events and their intervening quiescent periods. These three orogenic events occurred during the Late Archean, Early Proterozoic, and Middle Proterozoic eons.

 THE EVOLUTION OF THE EARTH

Having examined the role and impact of geology in relation to the environment and the search for natural resources, we can now focus our attention on what lessons we can learn from the study of the Earth's geologic history. One of the fundamental concepts to be learned from the study of geology is that the Earth continuously changes or evolves. One cannot study the history of the Earth and its biota without appreciating the role of time and how imperceptible changes can have dramatic effects when viewed from the perspective of geologic time. In some ways, a geologic history of the Earth can be summed up by saying that the only thing certain is change.

As we discussed in Chapter 1, one of the cornerstones of geology is the principle of uniformitarianism. This principle is based on the premise that present-day processes have operated throughout geologic time, sometimes at different rates than at present. What makes uniformitarianism such a powerful principle is that it allows geologists to use present-day processes as the basis for interpreting the past and for predicting future events. We can use our understanding of the underlying principles and processes of geology to help us predict volcanic eruptions or earthquakes or to avoid potential disasters such as landslides and floods.

Another aspect of uniformitarianism is the cyclical nature of Earth history. In terms of plate tectonics, many geologists now view the movement of plates in terms of a supercontinent cycle—the formation and breakup of continental masses through time (see Perspective 12-1). It also appears that the Earth has undergone cyclic occurrences of glacial and interglacial episodes, and we must keep this perspective of climatic change in mind when debating whether the rise in mean global temperatures during the past century is due to human activity or part of a longer glacial-interglacial cycle. Lastly, the history of the interior region of continents can be studied from the perspective of transgressive and regressive cycles of shallow seas throughout geologic time. Again, the unstated but underlying principle of uniformitarianism plays a pivotal role in deciphering the Earth's history.

When we examine the history of life, we also see the theme of change. The history of life is the result of natural selection operating on random mutations ever since life first evolved sometime about 3.5 billion years ago. Another important implication derived from studying the history of life is the interaction between the living and nonliving components of the global ecosystem and how each part affects the other. Just as plate tectonics plays an important role in the formation and distribution of natural resources, it plays a pivotal part in the evolution of life. The distribution of plants and animals is controlled to a large extent by the distribution of continents and oceans. Atmospheric and oceanic circulation patterns are also controlled by plate movement.

The study of the cause of the mass extinctions at the end of the Cretaceous Period led researchers to realize that a nuclear war could lead to what has popularly been called nuclear winter. The somewhat serendipitous discovery of a 2.5-cm-thick iridium-rich clay layer at the Cretaceous-Tertiary boundary in Italy led to the hypothesis that a collision with a meteorite 66 million years ago was responsible for the mass extinctions observed in the fossil record at that time (see Figure 20-9). It is thought that the tremendous amount of dust released into the atmosphere from the impact of the meteorite, combined with the soot produced by massive forest fires, resulted in a decrease in light reaching the Earth's surface. This caused a reduction in photosynthesis and a lowering of global temperatures that disrupted the global ecosystem, leading to mass extinctions.

Applying the same reasoning to the amount of dust and soot that would be thrown into the atmosphere as the result of a nuclear war, scientists concluded that the same thing could happen again, resulting in major crop failures that would lead to chaos for today's ecosystem. We have already examined the effects of volcanic eruptions on global climates (see Perspective 4-1) and seen how volcanic ash and gases can affect global temperatures. Thus, an event that happened 66 million years ago may have important implications for the continued survival of the human race.

The study of geology is more than learning numerous facts abut the Earth. Geology is an integral part of our lives. As individuals and societies, the standard of living that we enjoy is directly dependent on the consumption of natural resources and interaction with the environment. An appreciation of geology and its relationship to the environment is critical if we, as a species, are to continue to exist on this planet.

Chapter Summary

1. The Earth is a complex, dynamic planet that has continuously changed since it formed 4.6 billion years ago.
2. Geology is part of the human experience, and a basic understanding of it is important for dealing with the many environmental problems and issues facing society today.
3. In addition to their traditional role in finding new deposits of natural resources, geologists are also playing an increasingly important role in environmental issues and regulations.
4. The greatest environmental problem facing the world today is overpopulation. The Earth's carrying capacity is being overwhelmed by an increasingly greater number of humans, resulting in misery and suffering for a large part of the population.
5. The Earth's carrying capacity is limited by three factors: food supply, resource supply, and the ability to assimilate pollution produced by organisms. All three factors can be addressed from a geologic perspective.
6. The contribution by humans to global warming is currently being debated. Though the global mean temperature has risen 0.6°C since the late 1800s, it is not at all clear whether this is due to human activities, especially the burning of fossil fuels, or part of the natural global temperature cycle. The consequences of a continued rise in global temperatures, however, may be catastrophic.
7. Acid rain is another problem brought on by industrialization. The areas most affected by acid rain usually lie downwind from coal-burning plants or other sulfur-emitting industries. The most effective way to minimize the effects of acid rain is to reduce sulfur emissions.
8. Another problem affecting society is nuclear waste disposal. As a result of mining and processing activities, vast areas of Eastern Europe are highly contaminated and polluted. Scientists in the United States are currently studying the feasibility of Nevada's Yucca Mountain as a high-level radioactive waste repository.

9. The search for natural resources has been the main area of employment for geologists in the past and will continue to be in the future.
10. The three main energy sources are coal, oil, and natural gas. Coal has long been touted as a major source of energy in the world and particularly in the United States because of its abundance.
11. Oil continues to be the energy source of choice for most nations because of its abundance and relatively low cost. However, most oil reserves occur in the Persian Gulf region, a fact that will become increasingly important in the global political arena.
12. As oil becomes scarcer, natural gas will make up a larger portion of the energy picture. The major problem in the utilization of natural gas has been in establishing an infrastructure to get it from the gas fields to market.
13. Metals are an essential part of the world's economy. Their distribution and concentration are largely controlled by plate tectonics.
14. One of the fundamental concepts to be learned from the study of geology is that the Earth and its biota are continually evolving.

REVIEW QUESTIONS

1. The greatest environmental problem facing the world today is:
   a. ____ decreasing food supply;
   b. ____ increasing pollution;
   c. ____ overpopulation;
   d. ____ depletion of natural resources;
   e. ____ none of these.
2. Which of the following factors limits the carrying capacity of an ecosystem?
   a. ____ food supply;
   b. ____ resource supply;
   c. ____ its ability to assimilate the pollution produced by organisms;
   d. ____ all of these;
   e. ____ none of these.
3. Some of the world's greatest population densities occur in countries in:
   a. ____ Antarctica;
   b. ____ the Pacific Rim;
   c. ____ Greenland;
   d. ____ Europe;
   e. ____ Africa.
4. One of the by-products of respiration and the burning of organic material and a probable cause of global warming is:
   a. ____ carbon dioxide;
   b. ____ sulfur dioxide;
   c. ____ carbon monoxide;
   d. ____ hydrogen;
   e. ____ nitrogen.
5. The three major contributors to global warming are:
   a. ____ nitrogen, hydrogen, and oxygen;
   b. ____ carbon dioxide, methane, and chlorofluorocarbons;
   c. ____ carbon monoxide, carbon dioxide, and methane;
   d. ____ methane, ethane, and carbon monoxide;
   e. ____ nitrogen, chlorofluorocarbons, and hydrogen.
6. By how many degrees has the mean global temperature increased since the late 1800s?
   a. ____ 0.6°C;
   b. ____ 1.0°C;
   c. ____ 1.6°C;
   d. ____ 2.0°C;
   e. ____ 6.0°C.
7. The major contributor to acid rain is:
   a. ____ carbon dioxide;
   b. ____ carbon monoxide;
   c. ____ nitrogen;
   d. ____ sulfur dioxide;
   e. ____ none of these.
8. Which of the following fossil fuels does the United States have enough reserves of to last for at least another century?
   a. ____ geothermal energy;
   b. ____ natural gas;
   c. ____ oil;
   d. ____ nuclear energy;
   e. ____ coal.
9. What region of the world contains more than 60% of all proven oil reserves?
   a. ____ North America;
   b. ____ Commonwealth of Independent States;
   c. ____ South America;
   d. ____ Persian Gulf;
   e. ____ Australia.
10. Roughly 40% of the world's natural gas reserves occur in what country?
    a. ____ United States;
    b. ____ Canada;
    c. ____ Russia;
    d. ____ Iran;
    e. ____ Iraq.
11. Why is it important that people have a better understanding of science and how it works?
12. Why is the principle of uniformitarianism one of the cornerstones of geology?
13. Discuss the advantages and disadvantages of each of the following as an energy resource: oil, natural gas, coal, and nuclear energy.
14. What is the role of plate tectonics in the formation and distribution of metallic ores?
15. How can geology help determine whether global warming is part of a long-term temperature trend or is primarily the result of human activity during the past 200 years?
16. Discuss the causes of and possible remedies for the acid rain problem.

Points to Ponder

1. Do the industrialized nations, having achieved a relatively high standard of living, at the expense of a polluted environment as some would argue, have the right to impose strict environmental standards or regulations on the undeveloped or underdeveloped nations in an effort to reduce the environmental damage they have caused during their industrialization? What are the arguments for and against this proposition?
2. Considering the overreaction to the so-called asbestos crisis (see Chapter 7) and the problems of radon contamination (see Chapter 8), is it possible that some of the dire predictions concerning global warming may be somewhat exaggerated, especially when viewed from a geologic perspective?
3. According to Stephen Jay Gould, if we wound the tape of life back to its beginning and let it play again, the odds are that the evolutionary history of life would not repeat itself. Can this analogy be applied to the physical history of the Earth, or are the physical and biologic history of the Earth so intimately related, as James Lovelock and others postulate, that they cannot be separated?

Additional Readings

Brookins, D. G. 1990. *Mineral and energy resources: Occurrence, exploitation, and environmental impact.* Columbus, Ohio: Merrill.

Davis, G. R. 1990. Energy for planet Earth. *Scientific American* 263, no. 3: 55–74.

Edmonds, J., and J. M. Reilly. 1985. *Global energy: Assessing the future.* New York: Oxford University Press.

Ellis, W. S. 1990. A Soviet sea lies dying. *National Geographic* 177, no. 2: 73–93.

Holloway, M. 1993. Sustaining the Amazon. *Scientific American* 269, no. 1: 90–100.

Howell, D. G., K. J. Bird, and D. L. Gautier. 1993. Oil: Are we running out? *Earth* 2, no. 2: 26–33.

Howell, D., F. Cole, M. Faneui, and K. Wiese. 1993. Natural gas. *Earth* 2, no. 5: 52–59.

Jones, P. D., and T. M. L. Wigley. 1990. Global warming trends. *Scientific American* 263, no. 2: 84–91.

Lovelock, J. E. 1988. *The ages of Gaia: A biography of our living Earth.* New York: Norton.

Micklin, P. P. 1993. The shrinking Aral Sea. *Geotimes* 38, no. 4: 14–18.

Mohnen, V. A. 1988. The challenge of acid rain. *Scientific American* 259, no. 2: 30–38.

Rowland, F. S. 1989. Chlorofluorocarbons and the depletion of stratospheric ozone. *American Scientist* 77, no. 1: 36–45.

Sawkins, F. J. 1984. *Metal deposits in relation to plate tectonics.* New York: Springer-Verlag.

Skinner, B. J. 1986. *Earth resources.* Englewood Cliffs, N.J.: Prentice-Hall.

World Commission on Environment and Development. 1987. *Our common future.* New York: Oxford University Press.

Answers to Multiple-Choice and Fill-in-the-Blank Review Questions

CHAPTER 1

1. c; 2. c; 3. e; 4. b; 5. c; 6. d; 7. e; 8. a; 9. d; 10. c; 11. b; 12. a; 13. c; 14. d; 15. a; 16. a; 17. e. 18. b.

CHAPTER 2

1. b; 2. e; 3. c; 4. d; 5. b; 6. b; 7. a; 8. c; 9. b; 10. a; 11. a; 12. b; 13. e; 14. c.

CHAPTER 3

1. b; 2. a; 3. d; 4. a; 5. c; 6. d; 7. d; 8. e; 9. b; 10. d; 11. a; 12. a; 13. d.

CHAPTER 4

1. a; 2. c; 3. a; 4. e; 5. b; 6. b; 7. e; 8. b; 9. a; 10. c; 11. a; 12. c; 13. d; 14. a; 15. d.

CHAPTER 5

1. b; 2. e; 3. a; 4. b; 5. c; 6. d; 7. d; 8. a; 9. a; 10. d; 11. e; 12. b; 13. c.

CHAPTER 6

1. c; 2. d; 3. a; 4. e; 5. a; 6. d; 7. b; 8. c; 9. e; 10. c; 11. c; 12. b; 13. d.

CHAPTER 7

1. c; 2. e; 3. a; 4. c; 5. a; 6. c; 7. d; 8. c; 9. d; 10. b; 11. b; 12. d; 13. b; 14. a; 15. d.

CHAPTER 8

1. c; 2. a; 3. e; 4. d; 5. a; 6. d; 7. b; 8. c; 9. e; 10. b.

CHAPTER 9

1. c; 2. a; 3. d; 4. e; 5. a; 6. b; 7. c; 8. d; 9. e; 10. c; 11. b.

CHAPTER 10

1. b; 2. c; 3. c; 4. a; 5. e; 6. c; 7. b; 8. c; 9. b; 10. d; 11. b; 12. c; 13. e; 14. b.

CHAPTER 11

1. b; 2. d; 3. a; 4. e; 5. c; 6. c; 7. d; 8. b; 9. b; 10. d; 11. c; 12. d; 13. e; 14. c; 15. b.

CHAPTER 12

1. d; 2. a; 3. e; 4. c; 5. e; 6. b; 7. c; 8. d; 9. b; 10. c; 11. a; 12. b; 13. c; 14. b; 15. divergent; 16. oceanic-oceanic convergent; 17. transform; 18. oceanic-continental convergent.

CHAPTER 13

1. c; 2. e; 3. d; 4. a; 5. b; 6. c; 7. c; 8. a; 9. a; 10. d; 11. a; 12. b; 13. c.

CHAPTER 14

1. e; 2. e; 3. b; 4. d; 5. c; 6. a; 7. e; 8. e; 9. c; 10. a; 11. e.

CHAPTER 15

1. d; 2. a; 3. c; 4. e; 5. b; 6. a; 7. c; 8. b; 9. a; 10. c; 11. d; 12. c; 13. c.

CHAPTER 16

1. a; 2. c; 3. b; 4. d; 5. e; 6. d; 7. e; 8. b; 9. d; 10. e; 11. e; 12. e; 13. e; 14. b.

CHAPTER 17

1. c; 2. a; 3. b; 4. c; 5. e; 6. b; 7. c; 8. b; 9. e; 10. c; 11. b; 12. a; 13. c.

CHAPTER 18

1. b; 2. a; 3. c; 4. e; 5. d; 6. c; 7. a; 8. b; 9. e; 10. c; 11. b; 12. d; 13. a.

CHAPTER 19

1. e; 2. c; 3. b; 4. d; 5. b; 6. a; 7. c; 8. a; 9. d; 10. e; 11. c; 12. c; 13. b; 14. a.

CHAPTER 20

1. e; 2. c; 3. b; 4. a; 5. c; 6. d; 7. a; 8. c; 9. e; 10. a; 11. a; 12. e; 13. d; 14. b; 15. c; 16. b.

CHAPTER 21

1. c; 2. d; 3. b; 4. a; 5. b; 6. a; 7. d; 8. e; 9. d; 10. c.

PERIODIC TABLE OF THE ELEMENTS

47	— Atomic Number
Ag	— Symbol of Element
silver	— Name of Element
107.9	— Atomic Weight (rounded to four significant figures)

Representative Elements | Transition Elements | Inner-transition Elements | Noble Gases

Period	(1)* I A	(2) II A	(3) III B	(4) IV B	(5) V B	(6) VI B	(7) VII B	(8) VIII B	(9) VIII B
1	1 H hydrogen 1.008								
2	3 Li lithium 6.941	4 Be beryllium 9.012							
3	11 Na sodium 22.99	12 Mg magnesium 24.31							
4	19 K potassium 39.10	20 Ca calcium 40.08	21 Sc scandium 44.96	22 Ti titanium 47.90	23 V vanadium 50.94	24 Cr chromium 52.00	25 Mn manganese 54.94	26 Fe iron 55.85	27 Co cobalt 58.93
5	37 Rb rubidium 85.47	38 Sr strontium 87.62	39 Y yttrium 88.91	40 Zr zirconium 91.22	41 Nb niobium 92.91	42 Mo molybdenum 95.94	43 Tc^x technetium 98.91	44 Ru ruthenium 101.1	45 Rh rhodium 102.9
6	55 Cs cesium 132.9	56 Ba barium 137.3	57 La lanthanum 138.9	72 Hf hafnium 178.5	73 Ta tantalum 180.9	74 W tungsten 183.9	75 Re rhenium 186.2	76 Os osmium 190.2	77 Ir iridium 192.2
7	87 Fr^x francium (223)	88 Ra^x radium 226.0	89 Ac^x actinium (227)	104 Unq^x (261)	105 Unp^x (262)	106 Unh^x (263)	107 Uns^x (262)	108 Uno^x (265)	109 Une^x (266)

Lanthanides	58 Ce cerium 140.1	59 Pr praseodymium 140.9	60 Nd neodymium 144.2	61 Pm^x promethium (147)	62 Sm samarium 150.4
Actinides	90 Th^x thorium 232.0	91 Pa^x protactinium 231.0	92 U^x uranium 238.0	93 Np^x neptunium 237.0	94 Pu^x plutonium (244)

x: All isotopes are radioactive.

( ) Indicates mass number of isotope with longest known half-life.

* Number in ( ) heading each column represents the group designation recommended by the ACS Committee on Nomenclature.

(10)	(11) I B	(12) II B	(13) III A	(14) IV A	(15) V A	(16) VI A	(17) VII A	(18) Noble Gases
								2 **He** helium 4.003
			5 **B** boron 10.81	6 **C** carbon 12.01	7 **N** nitrogen 14.01	8 **O** oxygen 16.00	9 **F** fluorine 19.00	10 **Ne** neon 20.18
			13 **Al** aluminum 26.98	14 **Si** silicon 28.09	15 **P** phosphorus 30.97	16 **S** sulfur 32.06	17 **Cl** chlorine 35.45	18 **Ar** argon 39.95
28 **Ni** nickel 58.71	29 **Cu** copper 63.55	30 **Zn** zinc 65.37	31 **Ga** gallium 69.72	32 **Ge** germanium 72.59	33 **As** arsenic 74.92	34 **Se** selenium 78.96	35 **Br** bromine 79.90	36 **Kr** krypton 83.80
46 **Pd** palladium 106.4	47 **Ag** silver 107.9	48 **Cd** cadmium 112.4	49 **In** indium 114.8	50 **Sn** tin 118.7	51 **Sb** antimony 121.8	52 **Te** tellurium 127.6	53 **I** iodine 126.9	54 **Xe** xenon 131.3
78 **Pt** platinum 195.1	79 **Au** gold 197.0	80 **Hg** mercury 200.6	81 **Tl** thallium 204.4	82 **Pb** lead 207.2	83 **Bi** bismuth 209.0	84 Po^{x} polonium (210)	85 At^{x} astatine (210)	86 Rn^{x} radon (222)

63 **Eu** europium 152.0	64 **Gd** gadolinium 157.3	65 **Tb** terbium 158.9	66 **Dy** dysprosium 162.5	67 **Ho** holmium 164.9	68 **Er** erbium 167.3	69 **Tm** thulium 168.9	70 **Yb** ytterbium 173.0	71 **Lu** lutetium 175.0
95 Am^{x} americium (243)	96 Cm^{x} curium (247)	97 Bk^{x} berkelium (247)	98 Cf^{x} californium (251)	99 Es^{x} einsteinium (254)	100 Fm^{x} fermium (257)	101 Md^{x} mendelevium (258)	102 No^{x} nobelium (255)	103 Lr^{x} lawrencium (256)

GLOSSARY

A

aa A lava flow with a surface of rough, angular blocks and fragments.

abrasion The wearing and scraping of exposed rock surfaces by the impact of solid particles.

absolute dating The process of assigning actual ages in years before the present to geologic events. Various radioactive decay dating techniques yield absolute ages. See also *relative dating.*

abyssal plain Vast, flat area on the sea floor adjacent to the continental rises of passive continental margins.

active continental margin A continental margin characterized by volcanism and seismicity at the leading edge of a continental plate where oceanic lithosphere is subducted. Possesses a narrow continental shelf and a slope that descends directly into an oceanic trench. See also *passive continental margin.*

aftershock An earthquake following a main shock resulting from adjustments along a fault. Aftershocks are common after a large earthquake, but most are smaller than the main shock.

alluvial fan A cone-shaped alluvial deposit formed where a stream flows from mountains onto an adjacent lowland.

alluvium A collective term for all solid materials transported and deposited by streams.

alpha decay A type of radioactive decay involving the emission of a particle consisting of two protons and two neutrons from the nucleus of an atom; decreases the atomic number by two and the atomic mass number by four.

angular unconformity An unconformity below which older strata dip at a different angle (usually steeper) than the overlying strata. See also *disconformity* and *nonconformity.*

anticline An up-arched fold in which the oldest exposed rocks coincide with the fold axis, and all strata dip away from the axis.

aphanitic An igneous rock texture in which individual mineral grains are too small to be seen without magnification; results from rapid cooling and generally indicates an extrusive origin.

aquiclude Any material that prevents the movement of groundwater.

aquifer A permeable layer that allows the movement of groundwater.

arête A narrow, serrated ridge separating two glacial valleys or adjacent cirques.

artesian system A confined groundwater system in which high hydrostatic pressure builds, causing water in the system to rise above the level of the aquifer.

aseismic ridge Long ridge or broad plateaulike feature lacking seismic activity that rises as much as 3 km above the surrounding sea floor.

ash Pyroclastic material measuring less than 2 mm.

assimilation A process in which magma changes composition by reacting with country rock.

asthenosphere The part of the upper mantle over which lithospheric plates move.

atom The smallest unit of matter that retains the characteristics of an element.

atomic mass number The total number of protons and neutrons in the nucleus of an atom.

atomic number The number of protons in the nucleus of an atom.

aureole A zone surrounding a pluton in which contact metamorphism has occurred.

B

bajada A broad alluvial apron formed at the base of a mountain range by overlapping alluvial fans.

barchan dune A crescent-shaped sand dune with the tips of the crescent pointing downwind.

barrier island A long, narrow island composed of sand and separated from the mainland by a lagoon.

basal slip A type of glacial movement in which a glacier slides over its underlying surface.

basalt plateau A plateau built up by numerous lava flows from fissure eruptions.

base level The lowest limit to which a stream can erode.

basin The circular equivalent of a syncline. All strata in a basin dip toward a central point, and the youngest exposed rocks are in the center of the fold.

batholith A discordant, irregularly shaped pluton composed chiefly of granitic rocks.

baymouth bar A spit that has grown until it cuts off a bay from the open ocean.

beach A deposit of unconsolidated sediment extending landward from low tide to a change in topography or where permanent vegetation begins.

bed (bedding) Bed refers to an individual layer of rock, especially sedimentary rock, whereas bedding refers to the layered arrangement of rocks. See also *strata (stratification).*

bed load The part of a stream's sediment load that is transported along its bed; consists of sand and gravel.

berm The nearly horizontal or gently landward-sloping part of a beach consisting of sand deposited by waves.

beta decay A type of radioactive decay during which a fast-moving electron is emitted from a neutron, thereby converting it to a proton; results in an increase of one atomic number, but does not change atomic mass number.

Big Bang A model for the evolution of the universe in which a dense, hot state is followed by expansion, cooling, and a less dense state.

biochemical sedimentary rock A sedimentary rock resulting from the chemical processes of organisms.

bonding The process whereby atoms are joined to other atoms.

Bowen's reaction series A mechanism accounting for the derivation of intermediate and felsic magmas from a mafic magma. It has a discontinuous branch of ferromagnesian minerals that change from one to another over specific temperature ranges and a continuous branch of plagioclase feldspars whose composition changes as the temperature decreases.

braided stream A stream possessing an intricate network of dividing and rejoining channels. Braiding occurs when sand and gravel bars are deposited within channels.

breaker A wave that oversteepens as it enters shallow water until its crest plunges forward.

butte An isolated, steep-sided, pinnacle-like erosional feature found in arid and semiarid regions.

C

caldera A large, steep-sided circular to oval volcanic depression usually formed by summit collapse resulting from partial draining of an underlying magma chamber.

capillary fringe The area where water extends upward from the water table by capillary action.

carbon 14 dating technique An absolute age dating method relying on determining the ratio of C^{14} to C^{12} in a sample; useful back to about 70,000 years ago; can be applied only to organic substances.

carbonate mineral A mineral containing the negatively charged carbonate ion $(CO_3)^{-2}$, e.g., calcite ($CaCO_3$).

carbonate rock A sedimentary rock containing mostly carbonate minerals (e.g., limestone and dolostone).

cave A naturally formed subsurface opening that is generally connected to the surface and is large enough for a person to enter.

cementation The precipitation of minerals as binding material between and around sediment grains, thus converting sediment to sedimentary rock.

chemical sedimentary rock Rock formed of minerals derived from the ions taken into solution during chemical weathering.

chemical weathering The decomposition of rock materials by chemical alteration of parent material.

cinder cone A small steep-sided volcano composed of pyroclastic materials that accumulated around a vent.

circum-Pacific belt A zone of seismic and volcanic activity that nearly encircles the margins of the Pacific Ocean basin.

cirque A steep-walled, bowl-shaped depression formed by erosion at the upper end of a glacial trough.

cleavage The breaking or splitting of mineral crystals along planes of weakness.

column A cave deposit formed when stalagmites and stalactites join.

columnar joint A type of joint in some igneous rocks in which six-sided columns form as a result of cooling.

compaction A method of lithification whereby the pressure exerted by the weight of overlying sediment reduces the amount of pore space and thus the volume of a deposit.

complex movement A combination of different types of mass movements in which one type is not dominant; most complex movements involve sliding and flowing.

composite volcano A volcano composed of pyroclastic layers, lava flows typically of intermediate composition, and mudflows; also called stratovolcano.

compound A substance resulting from the bonding of two or more different elements, e.g., water (H_2O) and quartz (SiO_2).

compression Stress resulting when rocks are squeezed by external forces directed toward one another.

concordant Refers to plutons whose boundaries are parallel to the layering in the country rock. See also *discordant*.

cone of depression The cone-shaped depression in the water table around a well; results from pumping water from an aquifer faster than it can be replenished.

contact metamorphism Metamorphism of country rock adjacent to a pluton.

continental-continental plate boundary A convergent plate boundary along which two continental lithospheric plates collide, e.g., the collision of India with Asia.

continental crust The Earth's crust beneath the continents and continental shelves consisting of a wide variety of igneous, sedimentary, and metamorphic rocks; it has an overall granitic composition and an average density of 2.70 g/cm^3. See also *oceanic crust*.

continental drift The theory that the continents were once joined into a single landmass that broke apart with the various fragments moving with respect to one another.

continental glacier A glacier covering at least 50,000 km^2 and unconfined by topography. Also called an ice sheet.

continental margin The area separating the part of a continent above sea level from the deep-sea floor. Consists of a continental shelf and slope and, in some places, a rise.

continental rise Part of the continental margin; the gently sloping area beyond the base of the continental slope.

continental shelf Part of the continental margin; the area between the shoreline and the continental slope where the sea floor slopes very gently in a seaward direction.

continental slope Part of the continental margin; the relatively steep area between the shelf-slope break and the continental rise or an oceanic trench.

convergent plate boundary The boundary between two plates that are moving toward one another See also *continental-continental plate boundary*, *oceanic-continental plate boundary*, and *oceanic-oceanic plate boundary*.

core The innermost part of the Earth below the mantle at about 2,900 km; divided into an outer liquid core and an inner solid core.

Coriolis effect The deflection of winds to the right of their direction of motion (clockwise) in the Northern Hemisphere and to the left (counterclockwise) in the Southern Hemisphere due to the Earth's rotation.

correlation The demonstration of time equivalency of rock units in different areas.

country rock Any rock that is invaded by and surrounds a pluton.

covalent bond A bond formed by the sharing of electrons between atoms.

crater A circular or oval depression at the summit of a volcano resulting from the extrusion of gases, pyroclastic materials, and lava.

craton The relatively stable part of a continent; consists of a shield and a buried extension of a shield known as a platform; the ancient nucleus of a continent.

creep A type of mass wasting involving slow downslope movement of soil or rock.

crest The highest part of a wave.

cross-bedding Layers in sedimentary rocks that were deposited at an angle to the surface upon which they were accumulating.

crust The outermost part of the Earth overlying the mantle. See also *continental crust and oceanic crust.*

crystal settling The physical separation and concentration of minerals in the lower part of a magma chamber by crystallization and gravitational settling.

crystalline solid A solid in which the constituent atoms are arranged in a regular, three-dimensional framework.

Curie point The temperature at which iron-bearing minerals in a cooling magma attain their magnetism.

D

daughter element An element formed by the radioactive decay of another element, e.g., argon 40 is the daughter element of potassium 40. See also *parent element.*

debris avalanche A complex type of mass wasting common in steep mountains; typically starts as a rockfall.

debris flow A mass wasting process; much like a mudflow but more viscous and containing larger particles.

deflation The removal of loose surface sediment by the wind.

deflation hollow A shallow depression resulting from the differential erosion of surface materials by the wind.

deformation A general term referring to any change in shape or volume, or both, of rocks in response to stress. Deformation involves folding and fracturing.

delta An alluvial deposit formed where a stream discharges into a lake or an ocean.

depositional environment Any area where sediment is deposited such as on a stream's floodplain or on a beach.

desert Any area that receives less than 25 cm of rain per year and has a high evaporation rate.

desert pavement A surface mosaic of close-fitting gravel particles formed by the removal of sand-sized and smaller particles by the wind.

desertification The expansion of deserts into formerly productive lands.

detrital sedimentary rock Sedimentary rock consisting of the solid particles (detritus) of preexisting rocks, e.g., sandstone and conglomerate.

differential pressure Pressure that is not applied equally to all sides of a rock body.

differential weathering Weathering of rock at different rates, producing an uneven surface.

dike A tabular or sheetlike discordant pluton.

dilatancy model A model used to predict earthquakes based on changes occurring in rocks subjected to very high pressure.

dip A measure of the maximum angular deviation of an inclined plane from horizontal.

dip-slip fault A fault on which all movement is parallel with the dip of the fault plane. See also *normal fault* and *reverse fault.*

discharge The volume of water in a stream moving past a particular point in a given period of time.

disconformity An unconformity above and below which the strata are parallel. See also *angular unconformity* and *nonconformity.*

discontinuity A boundary within the Earth across which a marked change in the velocity of seismic waves occurs, indicating a significant change in composition or properties of Earth materials.

discordant Refers to plutons with boundaries cutting across the layering in the country rock. See also *concordant.*

dissolved load The part of a stream's load consisting of ions in solution.

divergent plate boundary The boundary between two plates that are moving apart.

divide A topographically high area that separates adjacent drainage basins.

dome A circular equivalent of an anticline. All strata dip outward from a central point, and the oldest exposed rocks are in the center of the dome.

drainage basin The surface area drained by a stream and its tributaries.

drainage pattern The regional arrangement of stream channels in a drainage system.

dripstone Various cave deposits resulting from the deposition of calcite.

drumlin An elongate hill of till formed by the movement of a continental glacier.

dry climate A climate that occurs in the low and middle latitudes where the potential loss of water by evaporation exceeds the yearly precipitation; divided into arid and semiarid regions.

dune A mound or ridge of wind-deposited sand.

dynamic metamorphism Metamorphism occurring in fault zones where rocks are subjected to high differential pressure.

E

earthflow A type of mass wasting process; involves downslope flowage of water-saturated soil.

earthquake Vibration of the Earth caused by the sudden release of energy, usually as a result of displacement of rocks along faults.

elastic rebound theory A theory that explains the sudden release of energy when rocks are deformed by movement on a fault.

elastic strain A type of strain in which the material returns to its original shape when stress is relaxed.

electron A negatively charged particle of very little mass that orbits the nucleus of an atom.

electron capture A type of radioactive decay in which a proton captures an electron and is thereby converted to a neutron; results in a loss of one atomic number, but no change in atomic mass number.

electron shell Electrons orbit rapidly around the nuclei of atoms at specific distances known as electron shells.

element A substance composed of all the same atoms; it cannot be changed into another element by ordinary chemical means.

emergent coast A coast where the land has risen with respect to sea level.

end moraine A pile of rubble deposited at the terminus of a glacier. See also *recessional moraine* and *terminal moraine*.

epicenter The point on the Earth's surface directly above the focus of an earthquake.

erosion The removal of weathered materials from their source area.

esker A long, sinuous ridge of stratified drift formed by deposition by running water in a tunnel beneath stagnant ice.

evaporite A sedimentary rock formed by inorganic chemical precipitation of minerals from solution e.g., rock salt and rock gypsum.

Exclusive Economic Zone An area extending 371 km seaward from the coast of the United States and its territories in which the United States claims all sovereign rights.

exfoliation The process whereby slabs of rock bounded by sheet joints slip or slide off the host rock.

exfoliation dome A large rounded dome of rock resulting from the process of exfoliation.

F

fault A fracture along which movement has occurred parallel to the fracture surface.

fault plane A fracture surface along which blocks of rock on opposite sides have moved relative to one another.

felsic magma A type of magma containing more than 65% silica and considerable sodium, potassium, and aluminum, but little calcium, iron, and magnesium. See also *intermediate magma* and *mafic magma*.

ferromagnesian silicate A silicate mineral containing iron and magnesium or both. See also *nonferromagnesian silicate*.

fetch The distance the wind blows over a continuous water surface.

fiord A glacial valley below sea level.

firn Granular snow formed by the partial melting and refreezing of snow.

firn limit The elevation to which the snowline recedes during a wastage season.

fission track dating The process of dating samples by counting the number of small linear tracks (fission tracks) that result when a mineral crystal is damaged by rapidly moving alpha particles generated by radioactive decay of uranium.

fissure eruption An eruption in which lava or pyroclastic materials are emitted along a long, narrow fissure or group of fissures.

floodplain A low-lying, relatively flat area adjacent to a stream that is partly or completely covered by water when the stream overflows its banks.

fluid activity An agent of metamorphism in which water and carbon dioxide promote metamorphism by increasing the rate of chemical reactions.

focus The place within the Earth where an earthquake originates and energy is released.

foliated texture A texture of metamorphic rocks in which platy and elongate minerals are arranged in a parallel fashion.

footwall block The block of rock that lies beneath a fault plane.

fossil Remains or traces of prehistoric organisms preserved in rocks of the Earth's crust.

fracture A break in rock resulting from intense applied pressure.

frost action The mechanical weathering process that disaggregates rocks by repeated freezing and thawing of water in cracks and crevices.

frost heaving The process whereby a mass of sediment or soil undergoes freezing, expansion, and actual lifting, followed by thawing, contraction, and lowering of the mass.

frost wedging The opening and widening of cracks by the repeated freezing and thawing of water.

G

geologic time scale A geologic chart with the designation for the earliest interval of geologic time at the bottom, followed upward by designations for more recent intervals of time.

geology The science concerned with the study of the Earth; includes studies of Earth materials (minerals and rocks), surface and internal processes, and Earth history.

geothermal energy Energy that comes from the steam and hot water trapped within the Earth's crust.

geothermal gradient The temperature increases with depth in the Earth; it averages about 25°C/km near the surface.

geyser A hot spring that intermittently ejects hot water and steam.

glacial budget The balance between accumulation and wastage in a glacier.

glacial drift A collective term for all sediment deposited by glacial activity, including till deposited directly by glacial ice and outwash deposited by streams derived from melting ice.

glacial erratic A boulder transported by a glacier from its original source.

glacial ice Ice that has formed from firn.

glacial polish A smooth, glistening bedrock surface formed by the movement of a sediment-laden glacier over it.

glacial striation A straight scratch on a rock caused by the movement of sediment-laden glacial ice.

glacier A mass of ice on land that moves by plastic flow and basal slip.

Glossopteris flora A Late Paleozoic association of plants found only on the Southern Hemisphere continents and India.

Gondwana One of six large Paleozoic continents; composed mostly of present-day South America, Africa, Antarctica, Australia, and India.

graded bedding A type of sedimentary bedding in which an individual bed shows a decrease in grain size from bottom to top.

graded stream A stream possessing an equilibrium profile in which a delicate balance exists between gradient, discharge, flow velocity, channel characteristics, and sediment load such that neither significant erosion nor deposition occurs within the channel.

gradient The slope over which a stream flows; usually expressed in m/km or ft/mi.

gravity anomaly A departure from the expected force of gravity at a particular location; gravity anomalies may be either positive or negative.

greenhouse effect The retention of the Sun's heat in the atmosphere, resulting in an increase in global temperatures.

ground moraine A deposit formed of sediment liberated from melting ice as a glacier's terminus retreats.

groundwater The underground water stored in the pore spaces in rocks, sediment, or soil.

guide fossil Any fossil that can be used to determine the relative geologic ages of rocks and to correlate rocks of the same age in different areas.

guyot A flat-topped seamount of volcanic origin rising more than 1 km above the sea floor.

H

half-life The time required for one-half of the original number of atoms of a radioactive element to decay to a stable daughter product, e.g., the half-life of potassium 40 is 1.3 billion years.

hanging valley A tributary glacial valley whose floor is at a higher level than that of the main glacial valley.

hanging wall block The block of rock that lies above a fault plane.

headland Part of a shoreline projecting as a point of land into a body of water.

heat An agent of metamorphism.

heat flow The flow of heat from the Earth's interior to its surface.

horn A steep-walled, pyramidal peak formed by the headward erosion of cirques.

hot spot A localized zone of melting below the lithosphere.

hot spring A spring in which the water temperature is warmer than the temperature of the human body (37°C).

humus The material in soils derived by bacterial decay of organic matter.

hydraulic action The power of moving water.

hydrologic cycle The continuous recycling of water from the oceans, through the atmosphere, to the continents, and back to the oceans.

hydrolysis The chemical reaction between the hydrogen (H^+) ions and hydroxyl (OH^-) ions of water and a mineral's ions.

hypothesis A provisional explanation for observations; subject to continual testing and modification. If well supported by evidence, hypotheses are then generally called theories.

I

igneous rock Any rock formed by cooling and crystallization of magma or lava, or by the accumulation and consolidation of pyroclastic materials.

incised meander A deep, meandering canyon cut into solid bedrock by a stream.

index mineral A mineral that forms within a specific temperature and pressure range during metamorphism.

infiltration capacity The maximum rate at which sediment or soil can absorb water.

inselberg An isolated steep-sided erosional remnant rising above a desert plain.

intensity The subjective measure of the kind of damage done by an earthquake as well as people's reaction to it.

intermediate magma A magma having a silica content between 53 and 64% and an overall composition intermediate between felsic and mafic magmas. See also *felsic magma* and *mafic magma.*

intrusive igneous rock See *plutonic rock.*

ion An electrically charged atom produced by adding or removing electrons from its outermost electron shell.

ionic bond A bond resulting from the attraction of positively and negatively charged ions.

irons A group of meteorites composed mostly of iron and nickel.

isograd A line on a map connecting the first appearances of a particular index mineral and thus indicating equal metamorphic intensity.

isostatic rebound The phenomenon in which unloading of the Earth's crust causes it to rise upward until equilibrium is again attained. See also *principle of isostasy.*

isotope Two or more forms of an element having the same atomic number and the same chemical properties, but a different atomic mass number, e.g., carbon 12 and carbon 14.

J

joint A fracture along which no movement has occurred parallel with the fracture surface.

Jovian planet Any of the four planets that resemble Jupiter (Jupiter, Saturn, Uranus, and Neptune). All are large and have low mean densities, indicating they are composed mostly of lightweight gases and ices. See also *terrestrial planet.*

K

kame Conical hill of stratified drift originally deposited in a depression on a glacier's surface.

karst topography A topography with numerous caves, springs, sinkholes, solution valleys, and disappearing streams developed by groundwater erosion.

L

laccolith A concordant pluton with a mushroomlike geometry.

lahar A mudflow consisting of volcanic materials such as ash.

lateral moraine The sediment deposited as a long ridge of till along the margin of a valley glacier.

laterite A red soil rich in iron and aluminum or both that forms in the tropics by intense chemical weathering.

Laurasia A Late Paleozoic Northern hemisphere continent consisting of present-day continents of North America, Greenland, Europe, and Asia.

lava Magma at the Earth's surface; the molten rock material that flows from a volcano or a fissure.

lava dome A bulbous, steep-sided mass of very viscous magma forced upward through a volcanic conduit.

lava flow A stream of magma issuing from a volcano or fissure.

leaching The dissolution or removal of soluble minerals from a soil or rock by percolating water.

lithification The process of converting sediment into sedimentary rock.

lithosphere The outer, rigid part of the Earth consisting of the upper mantle, oceanic crust, and continental crust.

lithostatic pressure Pressure exerted on rock resulting from the weight of the overlying rock; it is applied equally in all directions.

loess Windblown silt and clay deposits derived from deserts, Pleistocene glacial outwash, and floodplains of streams in semiarid regions.

longitudinal dune A long ridge of sand aligned generally parallel to the direction of the prevailing wind.

longshore current A current between the breaker zone and the beach flowing parallel to the shoreline and produced by wave refraction.

longshore drift The movement of sediment along a shoreline by longshore currents.

Love wave (L-wave) A surface wave in which the individual particles of the material only move back and forth in a horizontal plane perpendicular to the direction of wave travel.

low-velocity zone The zone within the mantle between 100 and 250 km deep where the velocity of both P- and S-waves decreases markedly; it corresponds closely to the asthenosphere.

M

mafic magma A silica-poor magma containing between 45 and 52% silica and proportionately more calcium, iron, and magnesium than intermediate and felsic magmas. See also *felsic magma* and *intermediate magma.*

magma Molten rock material generated within the Earth.

magma mixing The process of mixing together magmas of different composition, thereby producing a modified version of the parent magmas.

magnetic anomaly Any change, such as a change in average strength, of the Earth's magnetic field.

magnetic declination The angle between lines drawn from a compass position to the north magnetic and geographic poles.

magnetic field The area in which magnetic substances are affected by lines of magnetic force emanating from the Earth.

magnetic inclination The deviation from horizontal of the magnetic lines of force around the Earth.

magnetic reversal The phenomenon in which the north and south magnetic poles are completely reversed.

magnitude The total amount of energy released by an earthquake at its source. See also *Richter Magnitude Scale.*

mantle The thick layer between the Earth's crust and core.

mantle plume A stationary column of magma originating deep within the mantle that slowly rises to the Earth's surface to form volcanoes or flood basalts.

marine regression The withdrawal of the sea from a continent or coastal area resulting in the emergence of the land.

marine terrace A wave-cut platform now elevated above sea level.

marine transgression The invasion of coastal areas or much of a continent by the sea.

mass wasting The downslope movement of rock, sediment, or soil under the influence of gravity.

meandering stream A stream possessing a single, sinuous channel with broadly looping curves.

mechanical weathering The disaggregation of rock materials by physical forces yielding smaller pieces that retain the chemical composition of the parent material.

medial moraine A moraine formed where lateral moraines of two valley glaciers merge.

Mediterranean belt A zone of seismic and volcanic activity extending westerly from Indonesia through the Himalayas, across Iran and Turkey, and through the Mediterranean region of Europe.

mesa A broad, flat-topped erosional remnant bounded on all sides by steep slopes and capped by resistant rock.

metamorphic facies A group of metamorphic rocks characterized by mineral assemblages formed under the same broad temperature-pressure conditions.

metamorphic rock Any rock that has been altered by heat, pressure, or chemical fluids or a combination of these agents of metamorphism.

metamorphic zone The region between isograds.

meteorite A mass of matter of extraterrestrial origin that has fallen to the Earth.

microplate A small lithospheric plate that is clearly of different origin than the rocks of the surrounding area.

Milankovitch theory A theory that explains cyclic variations in climate and the onset of glacial episodes as a consequence of irregularities in the Earth's rotation and orbit.

mineral A naturally occurring, inorganic, crystalline solid having characteristic physical properties and a narrowly defined chemical composition.

Modified Mercalli Intensity Scale A scale having values ranging from I to XII that is used to characterize earthquake intensity based on damage.

Mohorovičić discontinuity (Moho) The boundary between the Earth's crust and mantle.

monocline A simple bend or flexure in otherwise horizontal or uniformly dipping rock layers.

mud crack A sedimentary structure found in clay-rich sediment that has dried out. When drying occurs, the sediment shrinks and intersecting fractures form.

mudflow A mass wasting process; a flow consisting of mostly clay- and silt-sized particles and more than 30% water.

N

native element A mineral composed of a single element, e.g., gold and silver.

natural levee A ridge of sandy alluvium deposited along the margins of a stream channel during floods.

nearshore sediment budget The balance between additions and losses of sediment in the nearshore zone.

neutron An electrically neutral particle found in the nucleus of an atom.

nonconformity An unconformity in which stratified rocks above an erosion surface overlie igneous or metamorphic rocks. See also *angular unconformity* and *disconformity*.

nonferromagnesian silicate A silicate mineral that does not contain iron or magnesium. See also *ferromagnesian silicate*.

nonfoliated texture A metamorphic texture in which there is no discernible preferred orientation of minerals.

normal fault A dip-slip fault on which the hanging wall block has moved down relative to the footwall block. See also *reverse fault*.

normal polarity A record of magnetism in which the direction of the magnetic field is the same as it is at the present. See also *reversed polarity*.

nucleus The central part of an atom consisting of one or more protons and neutrons.

nuée ardente A mobile dense cloud of hot pyroclastic materials and gases ejected from a volcano.

O

oblique-slip fault A fault having both dip-slip and strike-slip movement.

oceanic-continental plate boundary A type of convergent plate boundary along which oceanic lithosphere and continental lithosphere collide; characterized by subduction of the oceanic plate beneath the continental plate and by volcanism and seismicity.

oceanic crust The Earth's crust underlying the ocean basins; it consistis mostly of basalt and gabbro and has an average density of 3.0 g/cm^3.

oceanic-oceanic plate boundary A type of convergent plate boundary along which two oceanic lithospheric plates collide and one is subjected beneath the other; characterized by seismicity, volcanism, and the origin of a volcanic island arc.

oceanic ridge A submarine mountain system found in all of the oceans; it is composed of volcanic rock (mostly basalt) and displays features produced by tension.

oceanic trench A long, narrow depression in the sea floor where subduction occurs.

ooze Deep-sea pelagic sediment composed mostly of shells of marine animals and plants.

ophiolite A sequence of igneous rocks thought to represent a fragment of oceanic lithosphere; composed of peridotite overlain successively by gabbro, sheeted basalt dikes, and pillow lavas.

orogeny The process of forming mountains, especially by folding and thrust faulting; an episode of mountain building.

outgassing The process whereby gases derived from the Earth's interior are released into the atmosphere by volcanic activity.

outwash plain The sediment deposited by the meltwater discharging from the terminus of a continental glacier.

oxbow lake A cutoff meander filled with water.

oxidation The reaction of oxygen with other atoms to form oxides or, if water is present, hydroxides.

P

pahoehoe A type of lava flow with a smooth ropy surface.

paleocurrent The direction of an ancient current as indicated by sedimentary structures such as cross-bedding.

paleomagnetism The study of remanent magnetism in rocks so that the intensity and direction of the Earth's past magnetic field can be determined.

Pangaea The name proposed by Alfred Wegener for a supercontinent consisting of all the Earth's landmasses that existed at the end of the Paleozoic Era.

parabolic dune A crescent-shaped dune in which the tips of the crescent point upwind.

parent element An unstable element that changes by radioactive decay into a stable daughter element. See also *daughter element*.

parent material The material that is mechanically and chemically weathered to yield sediment and soil.

passive continental margin The trailing edge of a continental plate consisting of a broad continental shelf and a continental slope and rise, commonly with an abyssal plain adjacent to the rise. Passive continental margins lack volcanism and intense seismic activity. See also *active continental margin*.

pedalfer A soil formed in humid regions with an organic-rich A horizon and aluminum-rich clays and iron oxides in horizon B.

pediment An erosion surface of low relief gently sloping away from a mountain base.

pedocal A soil of arid and semiarid regions with a thin A horizon and a calcium carbonate–rich B horizon.

pelagic clay Generally brown or reddish deep-sea sediment composed of clay-sized particles derived from the continents and oceanic islands.

perched water table A water table that forms above the main water table where a local aquiclude occurs within a larger aquifer.

permafrost Ground that remains permanently frozen.

permeability A material's capacity for transmitting fluids.

phaneritic A coarse-grained texture in igneous rocks in which the minerals are easily visible without magnification; results from slow cooling and generally indicates an intrusive origin.

pillow lava Bulbous masses of basalt resembling pillows formed when lava is rapidly chilled underwater.

planetesimal An asteroid-sized body that along with other planetesimals accreted to form protoplanets.

plastic flow The flow that occurs in response to pressure and causes permanent deformation.

plastic strain The result of stress in which a material cannot recover its original shape and retains the configuration produced by the stress such as folding of rocks.

plate An individual piece of lithosphere that moves over the asthenosphere.

plate tectonic theory The theory that large segments of the lithosphere move relative to one another.

playa A dry lake bed found in deserts.

playa lake A broad, shallow, temporary lake that forms in an arid region and quickly evaporates to dryness.

plunging fold A fold with an inclined axis.

pluton An intrusive igneous body that forms when magma cools and crystallizes within the Earth's crust, e.g., batholith and sill.

plutonic (intrusive igneous) rock Igneous rock that crystallizes from magma intruded into or formed in place within the Earth's crust.

point bar The sedimentary body deposited on the gently sloping side of a meander loop.

porosity The percentage of a material's total volume that is pore space.

porphyritic An igneous texture with mineral grains of markedly different sizes.

precursor Any change within the Earth that precedes an earthquake.

pressure release A mechanical weathering process in which rocks formed deep within the Earth, due to release of pressure, expand upon being exposed at the surface.

pressure ridge A buckled area on the surface of a lava flow that formed because of pressure on the partly solid crust of a moving flow.

principle of cross-cutting relationships A principle used to determine the relative ages of events; holds that an igneous intrusion or fault must be younger than the rocks that it intrudes or cuts.

principle of fossil succession A principle holding that fossils, and especially assemblages of fossils, succeed one another through time in a regular and determinable order.

principle of inclusions A principle that holds that inclusions, or fragments, in a rock unit are older than the rock unit itself, e.g., granite fragments in a sandstone are older than the sandstone.

principle of isostasy The theoretical concept of the Earth's crust "floating" on a denser underlying layer.

principle of lateral continuity A principle that holds that sediment layers extend outward in all directions until they terminate.

principle of original horizontality A principle that holds that sediment layers are deposited horizontally or very nearly so.

principle of superposition A principle that holds that younger layers of strata are deposited on top of older strata.

principle of uniformitarianism A principle that holds that we can interpret past events by understanding present-day processes; based on the assumption that natural laws have not changed through time.

proton A positively charged particle found in the nucleus of an atom.

P-wave A compressional, or push-pull, wave; the fastest seismic wave and one that can travel through solids, liquids, and gases; also known as a primary wave.

P-wave shadow zone The area between 103° and 143° from an earthquake focus where little P-wave energy is recorded.

pyroclastic material Fragmental material such as ash explosively ejected from a volcano.

pyroclastic (fragmental) texture A fragmental texture found in igneous rocks composed of pyroclastic materials.

pyroclastic sheet deposit Vast, sheet-like deposits of felsic pyroclastic materials erupted from fissures.

Q

quick clay A clay that spontaneously liquefies and flows like water when disturbed.

R

radioactive decay The spontaneous change of an atom to an atom of a different element.

rainshadow desert A desert found on the leeward side of a mountain range; forms because moist marine air moving inland yields precipitation on the windward side of the mountain range and the air descending on the leeward side is much warmer and drier.

rapid mass movement A type of mass movement involving a visible movement of material.

Rayleigh wave (R-wave) A surface wave in which the individual particles of material move in an elliptical path within a vertical plane oriented in the direction of wave movement.

recessional moraine A type of end moraine formed when a glacier's terminus retreats, then stabilizes and deposits till. See also *end moraine* and *terminal moraine*.

recharge The addition of water to the zone of saturation.

reef A moundlike, wave-resistant structure composed of the skeletons of organisms.

reflection The return to the source of some of a seismic wave's energy when it encounters a boundary separating materials of different density or elasticity within the Earth.

refraction The change in direction and velocity of a seismic wave when it travels from one material into another of different density and elasticity.

regional metamorphism Metamorphism that occurs over a large area resulting from high temperature and pressure, and the action of chemical fluids within the Earth's crust.

regolith The layer of unconsolidated rock and mineral fragments and soil that covers much of the Earth's surface.

relative dating The process of determining the age of an event relative to other events; involves placing geologic events in their correct chronological order, but involves no consideration of when the events occurred in terms of number of years ago. See also *absolute dating*.

reserve The part of the resource base that can be extracted economically.

resource A concentration of naturally occurring solid, liquid, or gaseous material in or on the Earth's crust in such form and amount that economic extraction of a commodity from the concentration is currently or potentially feasible.

reverse fault A dip-slip fault in which the hanging wall block moves upward relative to the footwall block. See also *normal fault*.

reversed polarity A record of magnetism in which the direction of the magnetic field is the opposite of its present orientation. See also *normal polarity*.

Richter Magnitude Scale An open-ended scale that measures the amount of energy released during an earthquake.

rill erosion Erosion by running water that scours small channels in the ground.

rip current A narrow surface current that flows out to sea through the breaker zone.

ripple mark Wavelike (undulating) structure produced in granular sediment such as sand by unidirectional wind and water currents, or by oscillating wave currents.

rock A solid aggregate of minerals of one or more kinds, as in granite, and limestone, or a consolidated aggregate of particles of other rocks, as in sandstone and conglomerate; although exceptions to this definition, coal and natural glass are considered rocks.

rock cycle A sequence of processes through which Earth materials may pass as they are transformed from one rock type to another.

rockfall A common type of extremely rapid mass wasting in which rocks fall through the air.

rock-forming mineral A common mineral that comprises a significant portion of a rock.

rock glide A type of rapid mass wasting in which rocks move downslope along a more or less planar surface.

rock varnish A thin, red, brown, or black shiny coating on the surface of many desert rocks; composed of iron and manganese oxides.

rounding The process by which the sharp corners and edges of sedimentary particles are abraded during transport.

runoff The surface flow of streams.

S

salt crystal growth A mechanical weathering process in which rocks are disaggregated by the growth of salt crystals in crevices and pores.

saltwater incursion The displacement of fresh groundwater by saltwater as a result of excessive pumping in coastal areas.

scientific method A logical, orderly approach that involves gathering data, formulating and testing hypotheses, and proposing theories.

sea-floor spreading The theory that the sea floor moves away from spreading ridges and is eventually consumed at subduction zones.

seamount A structure of volcanic origin rising more than 1 km above the sea floor.

sediment Loose aggregate of solids derived from preexisting rocks, or solids precipitated from solution by inorganic chemical processes or extracted from solution by organisms.

sedimentary facies Any aspect of a sedimentary rock unit that makes it recognizably different from adjacent sedimentary rocks of the same, or approximately the same, age.

sedimentary rock Any rock composed of sediment, e.g., sandstone and limestone.

sedimentary structure Any structure in sedimentary rock such as cross-bedding, mud cracks, and animal burrows that formed at the time of deposition or shortly thereafter.

seismic profiling A method in which strong waves generated at an energy source penetrate the layers beneath the sea floor. Some of the energy is reflected from the various layers to the surface, making it possible to determine the nature of the layers.

seismic risk map A map based on the distribution and intensity of past earthquakes. Such maps indicate the potential severity of future earthquakes and are useful in planning.

seismograph An instrument that detects, records, and measures the various vibrations produced by an earthquake.

seismology The study of earthquakes.

shear strength The resisting forces helping to maintain slope stability.

shear stress The result of forces acting parallel to one another but in opposite directions; results in deformation by displacement of adjacent layers along closely spaced planes.

sheet erosion Erosion that is more or less evenly distributed over the surface and removes thin layers of soil.

sheet joint A large fracture more or less parallel to a rock surface resulting from pressure released by expansion of the rock.

shield A vast area of exposed ancient rocks on a continent; the nucleus of a continent around which accretion occurred.

shield volcano A large, dome-shaped volcano with a low rounded profile built up mostly of overlapping basalt lava flows (e.g., Mauna Loa and Kilauea on the island of Hawaii).

shoreline The line of intersection between the sea or a lake and the land.

silica A compound of silicon and oxygen atoms.

silica tetrahedron The basic building block of all silicate minerals. It consists of one silicon atom and four oxygen atoms.

silicate A mineral containing silica (e.g., quartz [SiO_2] and orthoclase [$KAlSi_3O_8$]).

sill A tabular or sheetlike concordant pluton.

sinkhole A depression in the ground that forms in karst regions by the solution of the underlying carbonate rocks or by the collapse of a cave roof.

slide A type of mass movement involving movement of material along one or more surfaces of failure.

slow mass movement Mass movement that advances at an imperceptible rate and is usually only detectable by its effects.

slump A type of mass wasting that occurs along a curved surface of failure and results in the backward rotation of the slump block.

soil Regolith consisting of weathered material, water, air, and humus that can support plants.

soil degradation Any processes leading to a loss of soil productivity; may involve erosion, chemical pollution, and compaction.

soil horizon A distinct soil layer that differs from other soil layers in texture, structure, composition, and color.

solar nebula theory A theory for the evolution of the solar system from a rotating cloud of gas.

solifluction A type of mass wasting involving the slow downslope movement of water-saturated surface materials.

solution A reaction in which the ions of a substance become dissociated in a liquid, and the solid substance dissolves.

sorting A term referring to the degree to which all particles in sediment or sedimentary rock are about the same size.

spatter cone A small, steep-sided cone that forms when gases escaping from a lava flow hurl globs of molten lava into the air that fall back to the surface and adhere to one another.

spheroidal weathering A type of chemical weathering in which corners and sharp edges of angular rocks weather more rapidly than flat surfaces, thus yielding spherical shapes.

spit A continuation of a beach forming a point of land that projects into a body of water, commonly a bay.

spreading ridge A location where plates separate and new oceanic lithosphere is formed.

spring A place where groundwater flows or seeps out of the ground. Springs occur where the water table intersects the ground surface.

stalactite An icicle-shaped structure hanging from a cave ceiling formed of calcium carbonate precipitated from evaporating water.

stalagmite An icicle-shaped structure rising from a cave floor formed of calcium carbonate precipitated from evaporating water.

stock An irregularly-shaped discordant pluton with a surface area less than 100 km^2.

stones A group of meteorites composed of iron and magnesium silicate minerals; comprise about 93% of all meteorites.

stony-irons A group of meteorites composed of nearly equal amounts of iron and nickel and silicate minerals; comprise about 1% of all meteorites.

stoping A process in which rising magma detaches and engulfs pieces of country rock.

strain Deformation caused by stress. See also *elastic strain* and *plastic strain*.

strata (stratification) Strata (singular *stratum*) refers to the layers in sedimentary rocks, whereas stratification refers to the layered aspect of sedimentary rocks. See also *bed (bedding)*.

stratified drift Glacial drift displaying both sorting and stratification.

stream Runoff confined to channels regardless of size.

stream terrace An erosional remnant of a floodplain that formed when a stream was flowing at a higher level.

stress The force per unit area applied to a material such as rock within the Earth's crust.

strike The direction of a line formed by the intersection of a horizontal plane with an inclined plane, such as a rock layer.

strike-slip fault A fault involving horizontal movement so that blocks on opposite sides of a fault plane slide sideways past one another.

subduction The process whereby the leading edge of one plate descends beneath the margin of another plate.

subduction zone A long, narrow zone at a convergent plate boundary where an oceanic plate descends beneath another plate, e.g., the subduction of the Nazca plate beneath the South American plate.

submarine canyon A steep-sided canyon below sea level in the continental shelf and slope.

submarine fan A cone-shaped sedimentary deposit that accumulates on the continental slope and rise.

submergent coast A coast along which sea level rises with respect to the land or the land subsides.

superposed stream A stream that once flowed on a higher surface and eroded downward into resistant rocks, while still maintaining its course.

suspended load The smallest particles carried by a stream, such as silt and clay, which are kept suspended by fluid turbulence.

S-wave A shear wave that moves material perpendicular to the direction of travel, thereby producing shear stresses in the material it moves through; also known as a secondary wave.

S-wave shadow zone Those areas more than 103° from an earthquake focus where no S-waves are recorded.

syncline A down-arched fold in which the youngest exposed rocks coincide with the fold axis, and all strata dip inward toward the axis.

T

talus An accumulation of angular pieces of mechanically weathered rock at the base of a slope.

tension A type of stress in which forces act in opposite directions but along the same line and tend to stretch an object.

terminal moraine A type of end moraine; the outermost moraine marking the greatest extent of a glacier. See also *end moraine* and *recessional moraine*.

terrestrial planet Any of the four innermost planets of the solar system (Mercury, Venus, Earth, and Mars). All are small and have high mean densities, indicating that they are composed of rock and metallic elements. See also *Jovian planet*.

theory An explanation for some natural phenomenon that has a large body of supporting evidence; to be considered scientific, a theory must be testable e.g., plate tectonic theory.

thermal convection cell A type of circulation of material in the asthenosphere during which hot material rises, moves laterally, cools and sinks, and is reheated and reenters the cycle.

thermal expansion and contraction A type of mechanical weathering in which the volume of rock changes in response to heating and cooling.

thrust fault A type of reverse fault with a fault plane dipping less than 45°.

tide The regular fluctuation in the sea's surface in response to the gravitational attraction of the Moon and Sun.

till All sediment deposited directly by glacial ice.

time-distance graph A graph showing the average travel times for P- and S-waves for any specific distance from an earthquake's focus.

tombolo A type of spit that extends out into the sea or a lake and connects an island to the mainland.

transform fault A type of fault that changes one type of motion between plates into another type of motion.

transform plate boundary Plate boundary along which plates slide past one another, and crust is neither produced nor destroyed; on land recognized as a strike-slip fault.

transport The mechanism by which weathered material is moved from one place to another, commonly by running water, wind, or glaciers.

transverse dune A long ridge of sand perpendicular to the prevailing wind direction.

tree-ring dating The process of determining the age of a tree by counting the number of annual growth rings.

trough The lowest point between wave crests.

tsunami A destructive sea wave that is usually produced by an earthquake but can also be caused by submarine landslides or volcanic eruptions.

turbidity current A sediment-water mixture denser than normal seawater that flows downslope to the deep-sea floor.

U

unconformity An erosion surface separating younger strata from older rocks. See also *angular unconformity, disconformity*, and *nonconformity*.

U-shaped glacial trough A valley with very steep or vertical walls and a broad, rather flat floor. Formed by the movement of a glacier through a stream valley.

V

valley glacier A glacier confined to a mountain valley or to an interconnected system of mountain valleys.

valley train A long, narrow deposit of stratified drift confined within a glacial valley.

velocity A measure of the downstream distance water travels per unit of time.

ventifact A stone whose surface has been polished, pitted, grooved, or faceted by wind abrasion.

vesicle A small hole or cavity formed by gas trapped in a cooling lava.

viscosity A fluid's resistance to flow.

volcanic island arc A curved chain of volcanic islands parallel to a deep-sea trench where oceanic lithosphere is subducted causing volcanism and the origin of volcanic islands.

volcanic neck An erosional remnant of the material that solidified in a volcanic pipe.

volcanic pipe The conduit connecting the crater of a volcano with an underlying magma chamber.

volcanic (extrusive igneous) rock Igneous rock formed when magma is extruded onto the Earth's surface where it cools and crystallizes, or when pyroclastic materials become consolidated.

volcanism The process whereby magma and its associated gases rise through the Earth's crust and are extruded onto the surface or into the atmosphere.

volcano A conical mountain formed around a vent as a result of the eruption of lava and pyroclastic materials.

W

water table The surface separating the zone of aeration from the underlying zone of saturation.

water well A well made by digging or drilling into the zone of saturation.

wave base A depth of about one-half wave length, where the diameter of the orbits of water particles in waves is essentially zero; the depth below which water is not affected by surface waves.

wave-cut platform A beveled surface that slopes gently in a seaward direction; formed by erosion and landward retreat of a sea cliff.

wave height The vertical distance from wave trough to wave crest.

wave length The distance between successive wave crests or troughs.

wave period The time required for two successive wave crests (or troughs) to pass a given point.

wave refraction The bending of waves so that they more nearly parallel the shoreline.

weathering The physical breakdown and chemical alteration of rocks and minerals at or near the Earth's surface.

Y

yardang An elongated and streamlined ridge of rock that looks like an overturned ship's hull; formed by wind erosion.

Z

zone of accumulation In soil terminology, another name for horizon B where soluble minerals leached from horizon A accumulate as irregular masses. In glacial terminology, the part of a glacier where additions exceed losses and the glacier's surface is perennially covered by snow.

zone of aeration The zone above the water table that contains both water and air within the pore spaces of the rock, sediment, or soil.

zone of saturation The zone below the water table in which all pore spaces are filled with groundwater.

zone of wastage The part of a glacier where losses from melting, sublimation, and calving of icebergs exceed the rate of accumulation.

INDEX

B

C

D

E

F

G

H

I

M

N

O

P

Q

R

T

U

V

W

Y

Z

Credits

CHAPTER 1

Opener: Photo ©Krafft, K, Explorer le Monde de L'Image
1-1: Krakatau 1883, by Tom Simpkin and Richard S. Fiske, Smithsonian Institution, 1983
1-2: Precision Graphics
1-3: The Geological Society, London
1-4: Photo courtesy of NASA
1-5a: Patricia K. Armstrong/Visuals Unlimited
1-5b: American Association of Petroleum Geologists/IBM
1-6: Collection of the New York Public Library Astor, Lenox, and Tilden Foundations
1-7: British Tourist Authority
1-8: THE FAR SIDE Copyright © 1991 Far Works, Distributed by Universal Press Syndicate. Reprinted with permission. All rights reserved.
1-10: Precision Graphics
1-11: Victor Royer
1-12: Victor Royer
1-13: Carlyn Iverson
1-14: Precision Graphics
1-15: Carlyn Iverson
1-16: Precision Graphics
1-18: Carlyn Iverson
Table 1-2: Modified from R.V. Dietrich and R. Wicander, *Minerals, Rocks, and Fossils* (New York: John Wiley & Sons, Inc., 1983): 160, Table IV-2.
1-19: Precision Graphics. From A.R. Palmer, "The Decade of North American Geology, 1983 Geologic Time Scale." *Geology* (Boulder, Colo.: Geological Society of America, 1983): 504. Reprinted by permission of the Geological Society of America.

CHAPTER 2

Opener: Photo courtesy of Jeff Scovil
2-2: Precision Graphics
2-3: Precision Graphics
2-4: Precision Graphics
2-5: Precision Graphics
2-6: Precision Graphics
2-7: Precision Graphics
2-8: Precision Graphics. From R.V. Dietrich and Brian J. Skinner, *Gems, Granites, and Gravels: Knowing and Using Rocks and Minerals* (New York: Cambridge University Press, 1990): 39, Figure 3.4.
2-9: Precision Graphics
2-10: Precision Graphics
2-15b: Ward's Natural Science Establishment, Inc.
2-16: Precision Graphics
2-17: Precision Graphics
2-19: Photo courtesy of Jerry Jacka Photography
2-20: ©Gemological Institute of America. Photo by Shane McClure and Robert Kane.
Review Question 16: Precision Graphics

CHAPTER 3

3-2: Photo of painting by Herbert Collins, courtesy Devil's Tower National Monument
3-3: Precision Graphics
3-6: Precision Graphics
3-8: Precision Graphics
3-9: Precision Graphics
3-11: Precision Graphics
3-12: Precision Graphics. Modified from R.V. Dietrich, *Geology and Michigan: Fortynine Questions and Answers. 1979.*
Perspective 3-1, Figure 1: Precision Graphics
3-20: Precision Graphics
3-21: Martin G. Miller/Visuals Unlimited
3-22: Precision Graphics
3-25: Precision Graphics
3-26: Precision Graphics
3-27: Precision Graphics

CHAPTER 4

4-1: United States Department of the Interior, U.S. Geological Survey, David A. Johnston Cascades Volcano Observatory, Vancouver, Washington
4-2: Precision Graphics
4-3: United States Department of the Interior, U.S. Geological Survey, David A. Johnston Cascades Volcano Observatory, Vancouver, Washington
4-4: U.S. Geological Survey
4-6a: U.S. Geological Survey
4-6b: John S. Shelton
4-7a: D.W. Peterson, U.S. Geological Survey
4-7b: J.B. Stokes, U.S. Geological Survey
4-8a: T.J. Takahashi, U.S. Geological Survey
4-10: Reproduced by permission of Marie Tharp, 1 Washington Ave., South Nyack, NY 10960.

4-11a: University of Colorado
4-11b: J.P. Lockwood, U.S. Geological Survey
4-12a-d: Precision Graphics. From Howel Williams, *Crater Lake: The Story of Its Origin* (Berkeley, Calif.: University of California Press): Illustrations from page 84. Copyright © 1941 Regents of the University of California, ©renewed 1969 Howel Williams.
4-14a: Precision Graphics
4-14b: Solarfilma/GeoScience Features
Perspective 4-2, Figure 1: Precision Graphics. From R.I. Tilling, U.S. Geological Survey.
4-13: Precision Graphics
4-15a: Precision Graphics
4-15b: Lawrence R. Solkoski, consulting geologist, Vancouver, B.C. Canada
4-15c: D.R. Crandell, U.S. Geological Survey
4-16: Precision Graphics
4-17: I.C. Russell, U.S. Geological Survey
4-18: Ward's Natural Science Establishment, Inc.
4-19: Precision Graphics
Perspective 4-3, Figure 2: Precision Graphics. From R.I. Tilling, U.S. Geological Survey.
Perspective 4-3, Figure 3: Keith Ronnholm
Perspective 4-4, Figure 4: U.S. Geological Survey
4-21: Precision Graphics. Modified from R.I. Tilling, C. Heliker, and T.L. Wright, *Eruptions of Hawaiian Volcanoes: Past, Present, and Future*. 1987. U.S. Geological Survey.
4-22: Carlyn Iverson
4-23: Precision Graphics
4-24: Precision Graphics. From "Hot Spots on the Earth's Surface," copyright © 1976, by Scientific American, Inc., George V. Kelvin, all rights reserved.
4-25: Carlyn Iverson
4-26: Precision Graphics

CHAPTER 5

Opener: Paul Johnson
5-1: From *Tropical Rainforests* by Arnold Newman. Copyright (c) 1990 Arnold Newman. Reprinted with permission of Facts On File, Inc., New York.
5-2: Visuals Unlimited/ © Frank Lembrecht
5-5: Precision Graphics. From A. Cox and R.R. Doell, "Review of Paleomagnetism." *GSA Bulletin* 71 (1960): 758, Figure 33.
5-6: University of Colorado
Perspective 5-1, Figure 2: N.K. Huber, U.S. Geological Survey
5-11: Precision Graphics
5-12: Bill Beatty/Visuals Unlimited
5-13: Precision Graphics
5-14: Precision Graphics
5-15: Precision Graphics
5-16: Precision Graphics
5-18a: John S. Shelton
5-18b: John D. Cunningham/Visuals Unlimited
5-19: Precision Graphics
5-20: Precision Graphics
5-21a: Walt Anderson/Visuals Unlimited
5-22: Precision Graphics
Perspective 5-2, Figure 1: Precision Graphics. Modified from Donald Worster, *Dust Bowl* (New York: Oxford University Press, 1979): 30.
Perspective 5-2, Figure 2: Kansas State Historical Society
5-23: Precision Graphics
5-25: Science VU/Visuals Unlimited
5-26: Precision Graphics

CHAPTER 6

6-2: R.L. Elderkin, U.S, Geological Survey
6-5: Carlyn Iverson
6-6: Precision Graphics
6-12: Precision Graphics
6-16: Precision Graphics. From the U.S. Geological Survey.
6-17: Precision Graphics
6-18: Precision Graphics
Perspective 6-1, Figure 2: Precision Graphics
6-19: Precision Graphics
6-21: Precision Graphics
Perspective 6-2, Figure 1a: Photo courtesy of Mike Johnson
6-27c-e: Precision Graphics
6-29: Alan L. Mayo, GeoPhoto Publishing Company
6-30: Precision Graphics
6-31a-b: Precision Graphics
6-32: Precision Graphics

CHAPTER 7

7-1: Arthur M. Sackler Museum, Harvard University, Cambridge, Massachusetts, John Randolph Coleman Memorial Fund
7-2: Collection of the J. Paul Getty Museum, Malibu, California
7-3: Precision Graphics
Perspective 7-1, Figure 1: Smithsonian Institution
7-3: Precision Graphics
7-4a: Precision Graphics. From C. Gillen, *Metamorphic Geology (London: Chapman & Hall, 1982): 24 and 73, Figures 2.3 and 4.4. Reprinted by permission of Chapman & Hall and C. Gillen.*
7-6: Precision Graphics
Table 7-1: From C. Gillen, *Metamorphic Geology* (London: Chapman & Hall, 1982): 49, Figure 3.2; and Table 4-1, p.70. Reprinted by permission of Chapman & Hall and C. Gillen.
7-9: Precision Graphics. From C. Gillen, *Metamorphic Geology* (London: Chapman & Hall, 1982): 49, Figure 3.2; and Table 4-1, p.70. Reprinted by permission of Chapman & Hall and C. Gillen.

7-10a: Precision Graphics

7-18: Precision Graphics. From H.L. James, *GSA Bulletin* 66 (1955): 1454, Plate 1. Reprinted by permission of the Geological Society of America.

7-19: Precision Graphics. Reprinted with permission from AGI Data Sheet 35.4, *AGI Data Sheets,* 3d ed., 1989, American Geological Institute.

7-20: Carlyn Iverson

7-21a: Precision Graphics. From "Effects of Late Jurassic-Early Tertiary Subduction in California." *Late Mesozoic and Cenozoic Sedimentation and Tectonics in California,* San Joaquin Geological Society Short Course (1977): 66, Figure 5-9. Reprinted by permission of the San Joaquin Geological Society.

7-21b: Carlyn Iverson

CHAPTER 8

8-1: Darwen and Vally Hennings. Modified from *Geologic Time.* 1981. U.S. Geological Survey.

8-2: Precision Graphics. From A.R. Palmer, "The Decade of North American Geology, 1983 Geologic Time Scale." *Geology* (Boulder, Colo.: Geological Society of America, 1983): 504. Reprinted by permission of the Geological Society of America.

8-5: Precision Graphics

8-6: Precision Graphics

8-7: Precision Graphics. From *The Story of the Great Geologists* by Carroll Lane Fenton and Mildred Adams Fenton. Used by permission of Doubleday, a division of Bantam Doubleday Dell Publishing Group, Inc.

8-8: Precision Graphics

8-9a: Precision Graphics

8-10a: Precision Graphics

8-11a: Precision Graphics

8-12: Precision Graphics

8-13: Precision Graphics

Perspective 8-1, Figure 2: Precision Graphics

Perspective 8-1, Figure 3: Precision Graphics

8-14: Precision Graphics. From *History of the Earth: An Introduction to Historical Geology, second* edition, by Bernhard Kummer. © 1970 by W.H. Freeman and Company; reprinted by permission.

8-15: Precision Graphics. From *Geologic Time.* 1981. U.S. Geological Survey.

8-16: Precision Graphics

8-17: Precision Graphics

8-18: Precision Graphics

Perspective 8-2, Figure 1: Precision Graphics

Perspective 8-2, Figure 2: Precision Graphics. Data from Environmental Protection Agency.

8-19: Precision Graphics. Data from S.M. Richardson and H.Y. McSween, Jr., *Geochemistry—Pathways and Processes* (Englewood Cliffs, NJ: Prentice-Hall, 1989).

8-20: Precision Graphics

8-21: Precision Graphics

8-22: Precision Graphics

8-24: Precision Graphics

8-25: Precision Graphics. From E.K. Ralph, H.N. Michael, and M.C. Han, "Radiocarbon Dates and Reality." *MASCA Newsletter* 9 (1973): 5, Figure 8.

8-26: Precision Graphics

8-27: Precision Graphics. From L.W. Mintz, *Historical Geology: The Science of a Dynamic Earth,* 3d ed. (Westerville, Ohio: Charles E. Merrill Publishing Company, 1981): 27, Figure 2.18.

8-28: Precision Graphics

Review Question 25: Precision Graphics

CHAPTER 9

Opener: AP/Wide World Photos/Douglas C. Pizac

9-1a: Precision Graphics

9-1b: Lee Stone/Sygma

9-1c: Ted Soqui/Sygma

9-1d: Don Boomer/Time

9-1e: Photo by Rick Rickman

9-2: Katsuhiko Ishida

9-3a: Precision Graphics

9-3b: U.S. Geological Survey

9-4: Reproduced by the Trustees of the Science Museum, London

9-5a: *Earthquake Information Bulletin 181,* U.S. Geological Survey

9-5b-c Precision Graphics

9-6: Precision Graphics

9-7: Precision Graphics. Data from National Oceanic and Atmospheric Administration.

9-8: Carlyn Iverson

9-9: J.K. Hillers, U.S. Geological Survey

9-10: Precision Graphics. From *Nuclear Explosions and Earthquakes: The Parted Veil* by B.A. Bolt. Copyright © 1976 by W.H. Freeman and Co. Reprinted by permission.

9-11: Precision Graphics

9-12: Precision Graphics. Data from C.F. Richter, *Elementary Seismology.* 1958. W.H. Freeman and Co.

9-13: Precision Graphics

9-14: Precision Graphics

9-15: Precision Graphics. From R.D. Borcherdt, ed., *U.S. Geological Survey Professional Paper 941-A.* 1975.

9-16: Precision Graphics. From *Earthquakes* by Bruce A. Bolt. Copyright © 1978, 1988 by W.H. Freeman and Co. Reprinted by permission.

9-17: Precision Graphics

9-18: National Geophysical Data Center

9-19: Photo courtesy of ChinaStock Photo Agency

9-20: San Francisco Public Library

Perspective 9-1, Figure 1: Precision Graphics

Perspective 9-1, Figure 2: R. Kachadoorian, U.S. Geological Survey
Perspective 9-1, Figure 3: Precision Graphics. From *Earthquakes* by Bruce A. Bolt. Copyright © 1978, 1988 by W.H. Freeman and Co. Reprinted by permission.
Perspective 9-1, Figure 4: M. Celebi, U.S. Geological Survey
9-21: Matin E. Klimek, Marin *Independent Journal*
9-22: Photo courtesy of Bishop Museum
9-23: Precision Graphics. Data from NOAA.
9-25a: Precision Graphics. From M.L. Blair and W.W. Spangle, *U.S. Geological Survey Professional Paper 941-B*. 1979.
9-25b: Precision Graphics. Modified from S.T. Algermissen and D.M. Perkins, "A Probabilistic Estimate of Maximum Acceleration in the Contiguous United States." *U.S. Geological Survey Open-File Report 76-416*. July 1976.
9-26: From *The Loma Prieta Earthquake of October 17, 1989*. 1989. U.S. Geological Survey.
9-27: Precision Graphics. Reprinted with permission from *Predicting Earthquakes,* 1976. Published by National Academy Press, Washington, D.C.
Perspective 9-2, Figure 1: Precision Graphics
9-28: Precision Graphics. Reprinted with permission from *Geotimes* 10 (1966): 17.

CHAPTER 10

Opener: Photo courtesy of Cornelius Gillen
10-2: Victor Royer
10-3: Precision Graphics
10-4: Precision Graphics
10-5: Precision Graphics
10-6: Precision Graphics. From G.C. Brown and A.E. Musset, *The Inaccessible Earth* (London: Chapman & Hall, 1981): 17 and 124, Figures 12.7a and 7.11. Reprinted by permission of Chapman & Hall.
10-7: Precision Graphics
10-8a: Kort-og Matrikelstyrelsen (National Survey and Cadastre—Denmark).
10-8b: Precision Graphics
10-9: Precision Graphics
10-10: Precision Graphics
10-11: Precision Graphics
10-12: Precision Graphics. From G.C. Brown and A.E. Musset, *The Inaccessible Earth* (London: Chapman & Hall, 1981): 17 and 124, Figures 12.7a and 7.11. Reprinted by permission of Chapman & Hall.
10-13: Precision Graphics. Andrew Christie/Copyright © 1987 Discover Publications.
Perspective 10-1, Figure 1: Precision Graphics
Perspective 10-1, Figure 2: Precision Graphics. From Keith G. Cox, "Kimberlite Pipes." Original illustration by Adolph E. Brotman. Copyright © April 1978 by Scientific American, Inc. All rights reserved.
10-14: Precision Graphics
10-15: Precision Graphics
10-16: Precision Graphics
10-17: Precision Graphics
10-18: Precision Graphics
10-19: Precision Graphics
10-20: Precision Graphics
10-21: Precision Graphics
10-22a: Precision Graphics. From Beno Gutenberg, *Physics of the Earth's Interior* (Orlando, Florida: Academic Press, 1959): 194, Figure 9.1. Reprinted by permission of Academic Press.
10-22b: Precision Graphics. From R.F. Flint, *Glacial and Quaternary Geology* (New York: John Wiley & Sons, Inc., 1971): 363, Figure 13-13.
10-23: Fundamental Photographs
10-24: Precision Graphics
10-25: Precision Graphics
10-26: Precision Graphics
10-27: Precision Graphics
10-28: Precision Graphics
10-29: Precision Graphics
10-30: Precision Graphics

CHAPTER 11

Opener: Woods Hole Oceanographic Institution
11-1: Precision Graphics
11-2: Precision Graphics
11-3: Precision Graphics
11-4: Precision Graphics
11-5: World Ocean Floor map by Bruce C. Heezen and Marie Tharp, 1977. Copyright © Marie Tharp 1977. Reproduced by permission of Marie Tharp, 1 Washington Ave., South Nyack, NY 10960.
11-6: Ocean Drilling Program, Texas A & M University
11-7: Precision Graphics. From U.S. Geological Survey.
Perspective 11-1, Figure 1: Precision Graphics. From Phyllis Young Forsyth, *Atlantis: The Making of a Myth* (Montreal: McGill-Queen's University Press): 13, Figure 2.
Perspective 11-1, Figure 2: Precision Graphics
Perspective 11-1, Figure 3: Painting by Lloyd K. Townsend copyright © National Geographic Society
11-8: Precision Graphics
11-9: Precision Graphics
11-10: Precision Graphics
11-11: Precision Graphics. Modified from Bruce C. Heezen and Charles D. Hollister, *The Face of the Deep* (New York: Oxford University Press, 1971): 297, Figure 8.15.
11-12: Precision Graphics
11-13: Precision Graphics. From Alyn and Alison Duxbury, *An Introduction to the World's Oceans*. Copyright © 1984 Addison-Wesley Publishing Company, Inc., Reading, Massachusetts. Reprinted by permission of Wm. C. Brown Publishers, Dubuque, Iowa. All rights reserved.
11-14: From Bruce C. Heezen and Charles D. Hollister, *The Face of the Deep* (New York: Oxford University Press, 1971): 329, Figure 8.48.

11-15: Precision Graphics
11-16: Precision Graphics
11-17: Carlyn Iverson
11-18: Precision Graphics
11-19: Dr. Bruce Heezen, Lamont- Doherty Geologic Observatory, Columbia University, courtesy Scripps Institution of Oceanography, University of California, San Diego.
11-20: Precision Graphics. From T.A. Davies and D.S. Gorsline, "Oceanic Sediments and Sedimentary Processes." In J.P. Riley and R. Chester, eds., *Chemical Oceanography 5,* 2d ed. (Orlando, Florida: Academic Press, 1976): 26, Figure 24.7. Reprinted with permission from Academic Press and T.A. Davies
11-21: Photo courtesy of Carl Roessler
11-22: Precision Graphics
11-23: © Douglas Faulkner, Science Source/Photo Researchers
11-24: Carlyn Iverson
11-25: Precision Graphics. From U.S. Geological Survey.
11-26: Precision Graphics

CHAPTER 12

Opener: Robert Caputo/Aurora
12-1a: Precision Graphics
12-1b: Carlyn Iverson
12-4: Bildarchiv Preussischer Kulturbesitz
12-5: Precision Graphics. From E. Bullard, J.E. Everett, and A.G. Smith, "The Fit of the Continents Around the Atlantic." *Philosophical Transactions of the Royal Society of London* 258 (1965). Reproduced with permission of the Royal Society, J.E. Everett, and A.G. Smith.
12-6: Precision Graphics. Reprinted with permission of Macmillan Publishing Company from R.J. Foster, *General Geology,* 4/e, Fig 20-2, p. 351. Copyright © 1983 by Merrill Publishing Company.
12-7: Precision Graphics
12-8: Precision Graphics
12-9: Precision Graphics. Modified from E.H. Colbert, *Wandering Lands and Animals* (1973): 72, Figure 31.
12-10: Precision Graphics. Reprinted with permission from Cox and Doell, "Review of Paleomagnetism," *GSA,* v. 71, 1960, p. 758 (Fig. 33), Geological Society of America.
12-11: Precision Graphics. From A. Cox, "Geomagnetic Reversals." *Science* 163 (17 January 1969): 240, Figure 4. Copyright 1969 by the AAAS.
12-12: Precision Graphics. From *The Bedrock Geology of the World.* Copyright © 1984 by R.L. Larson and W.C. Pitman III. Reprinted with permission by W.H. Freeman and Company.
Perspective 12-1, Figure 1: Carlyn Iverson
12-13: Precision Graphics
Perspective 12-2, Figure 1: AP/Wide World Photos
12-14: Precision Graphics
12-15: Carlyn Iverson
12-16: Precision Graphics
12-17: Carlyn Iverson
12-18: Carlyn Iverson
12-19: Carlyn Iverson
12-20: Carlyn Iverson
12-21a: Precision Graphics
12-22: Precision Graphics. Data from J.B. Minster and T.H. Jordan, "Present-day Plate Motions." *Journal of Geophysical Research* 83 (1978): 5331-51.
12-23a: Carlyn Iverson
12-23b: Precision Graphics
12-24: Precision Graphics
12-25: Carlyn Iverson
12-26: Precision Graphics
Review Question: Precision Graphics

CHAPTER 13

13-2: Precision Graphics. From *Structural Geology of North America* by A.J. Eardley. Copyright © 1951 by Harper & Row, Publishers, Inc. Copyright © 1951 by A.J. Eardley. Reprinted by permission of HarperCollins Publishers.
13-3: U.S. Geological Survey
13-4: Precision Graphics
13-5: Precision Graphics
13-6: John S. Shelton
13-8: Precision Graphics
13-9a: Precision Graphics
13-11: Precision Graphics
13-12: Martin F. Schmidt, Jr.
13-13: Precision Graphics
13-14: Precision Graphics
13-15a: Precision Graphics
13-15b: Precision Graphics
13-15c: John S. Shelton
13-16: Precision Graphics
Perspective 13-1, Figure 2: S.W. Lohman, U.S. Geological Survey
13-18: Precision Graphics
13-19a: Precision Graphics
13-20: Precision Graphics
13-22: B. Bradley, University of Colorado Geology Department
13-23a: Precision Graphics
13-24: John S. Shelton
13-25: Precision Graphics
Perspective 13-2, Figure 1: Precision Graphics
Perspective 13-2, Figure 2: Precision Graphics
13-26a: Precision Graphics
13-27: Precision Graphics. Reproduced by permission of the Geological Society and A.M. Spencer, ed., *Mesozoic-Cenozoic Orogenic Belts* (Bath: Geological Society Publishing House, 1974).
13-28: Precision Graphics
13-29: Precision Graphics
13-31: Precision Graphics. From Peter Molnar, "The Geologic History and Structure of the Himalayas."

American Scientist 74: 148-149, Figure 4. Reprinted by permission of *American Scientist,* journal of Sigma Xi, The Scientific Research Society.

13-32: Precision Graphics. From Zvi Ben-Avraham, "The Movement of Continents." *American Scientist* 69: 291-299, Figure 9, p.298. Reprinted by permission of *American Scientist,* journal of Sigma Xi, The Scientific Research Society.

13-33: Precision Graphics

13-34: Precision Graphics

13-35: Precision Graphics. Reproduced with permission from P.F. Hoffman, "United Plates of America, The Birth of a Craton: Early Proterozoic Assembly and Growth of Laurentia," Annual Review of Earth and Planetary Sciences, v. 16, p. 544, © 1988 by Annual Reviews, Inc.

13-37a: Precision Graphics

13-38: Precision Graphics

13-39: Precision Graphics. Figure 175 (Page 386) from *Structural Geology of North America,* 2nd Edition by A.J. Eardley. Copyright © 1962 by A.J. Eardley. Copyright renewed. Reprinted by permission of HarperCollins, Publishers, Inc.

13-40a: Precision Graphics. Reprinted with permission from W.R. Dickinson, "Cenozoic Plate Tectonic Setting of the Cordilleran Region in the U.S.," in *Cenozoic Paleogeography of the Western U.S.,* Pacific Coast, Symposium 3, 1979, pp. 10-11 and pp. 2, 4.

CHAPTER 14

Opener: Hong Kong Government, Geotechnical Control Office

14-1a: Precision Graphics

14-1b: George Plafker, U.S. Geological Survey

14-2: U.S. Geological Survey

14-3: Precision Graphics

14-4a-b: Precision Graphics

14-5a-c: Precision Graphics

Perspective 14-1, Figure 1a: Precision Graphics

Perspective 14-1, Figure 1b: T. Spencer/Colorific

14-6: Boris Yaro, *Los Angeles Times*

14-7: Precision Graphics

14-8: Precision Graphics

14-9: W.R. Hansen, U.S. Geological Survey

14-10: Precision Graphics

14-11a: John S. Shelton

14-11b: Precision Graphics

14-12: Precision Graphics

14-13a: Precision Graphics

14-13b: Precision Graphics. From A.C. Waltham, *Catastrophe: The Violent Earth* (New York: Macmillan, 1978): 51.

14-14 (map): Precision Graphics

14-14 (photo): Stephen R. Lower, GeoPhoto Publishing Company

14-15a: Precision Graphics

14-15b: B. Bradley and the University of Colorado's Geology Department to National Geophysical Data Center, NOAA, Boulder, Colorado.

14-17a: Precision Graphics

14-18a: Precision Graphics

14-18b: NGDC

14-19a: Precision Graphics. From O.J. Ferrians, Jr., R. Kachadoorian, and G.W. Greene, *U.S. Geological Survey Professional Paper 678.* 1969.

14-20: O.J. Ferrians, Jr., U.S. Geological Survey.

14-21a: Precision Graphics

14-22: Precision Graphics

Perspective 14-2, Figure 1: Precision Graphics. Reprinted with permission ASCE. From G.A. Kiersch, "Vaiont Reservoir Disaster." *Civil Engineering* 34 (1964).

Perspective 14-2, Figure 2: UPI/Bettmann

Perspective 14-3, Figure 3: Precision Graphics. Reprinted with permission ASCE. From G.A. Kiersch, "Vaiont Reservoir Disaster." *Civil Engineering* 34 (1964).

14-23: Precision Graphics

14-24a: Precision Graphics

14-25: Precision Graphics

14-26a: Precision Graphics

14-26b: John D. Cunningham/Visuals Unlimited

14-27a: Precision Graphics

14-27b: Dell R. Foutz/Visuals Unlimited

CHAPTER 15

Opener: Photo courtesy of the Kentucky Department of Parks

15-1a: Photo courtesy of Michael Lawton

15-1b: Photo courtesy of Michael Lawton

15-2: Precision Graphics

15-3: Martin G. Miller/Visuals Unlimited

15-5: Carlyn Iverson

15-6: Precision Graphics

15-7: Precision Graphics

15-8: Precision Graphics

15-9: Precision Graphics

15-10: Precision Graphics

15-14: Precision Graphics

15-15: Precision Graphics

15-17: John S. Shelton

15-19a: Precision Graphics

15-20: Precision Graphics

15-21: Precision Graphics

15-22: Precision Graphics

Perspective 15-1, Figure 1: Precision Graphics

Perspective 15-1, Figure 2: Precision Graphics

15-23a: Precision Graphics

15-24a-b: Precision Graphics

15-24c: Precision Graphics. From W.L. Fisher et al., *Delta Systems in the Exploration for Oil and Gas—A Research Colloquium* (1969).

15-25a: Precision Graphics
15-26: Precision Graphics
15-27: Precision Graphics
15-28: Precision Graphics
15-29: Precision Graphics
15-30: Alan Smith/Tony Stone Worldwide
15-31: Precision Graphics
15-32: Precision Graphics
15-34: Petley Studios
15-35: Precision Graphics
15-36: Precision Graphics
15-37: Precision Graphics
15-38: J.R. Stacy/U.S. Geological Survey
Perspective 15-1, Figure 2: Precision Graphics. From *Natural Bridges*. National Park Service.
15-39: Precision Graphics
15-40: John S. Shelton

CHAPTER 16

Opener: Used with permission of U.S. Forest Service
16-1a,c: Precision Graphics. From *Trapped* by Robert K. Murray and Roger W. Brucker. Copyright © 1979 by Murray and Brucker, copyright renewed. Used by permission.
16-1b: Brown Brothers
16-2: Precision Graphics
16-3: Precision Graphics
16-4: Precision Graphics
16-5: Photo courtesy of Nassau County Department of Public Works
16-6: Precision Graphics
16-7: Linda D. Mayo, GeoPhoto Publishing Company
16-8: Precision Graphics
16-9: Precision Graphics
16-10: Precision Graphics
16-11: J.R. Stacy, U.S. Geological Survey
16-12: Precision Graphics
Perspective 16-1, Figure 1: Ed Cooper
Perspective 16-1, Figure 2: W.L. McCoy
16-13: Precision Graphics
16-14: Frank Kujawa, University of Central Florida, GeoPhoto Publishing Company.
16-15: Precision Graphics
16-16b: John S. Shelton
16-17: Precision Graphics
16-18: Daniel W. Gotshall/Visuals Unlimited
16-19: Precision Graphics. From J.B. Weeks et al., *U.S. Geological Survey Professional Paper 1400-A*. 1988.
16-20: Precision Graphics
16-21a: Precision Graphics
16-21b: U.S. Geological Survey
16-22: Sarah Stone/Tony Stone Worldwide
16-24: City of Long Beach Department of Oil Properties
Perspective 16-2, Figure 1: Courtesy of United States Department of Energy
Perspective 16-2, Figure 2: Precision Graphics. Modified from *U.S. News & World Report* (18 March 1991): 72-73.
16-25: Precision Graphics
16-27: British Tourist Authority
16-29: Precision Graphics

CHAPTER 17

Opener: Photo courtesy of Michael J. Hambrey
17-1a: Offentliche Kunstammlung, Kupferstichkabinett, Basel
17-1b: Reproduced courtesy of the Board of Directors of the Budapest Museum of Fine Arts
17-2: Precision Graphics
17-3: Precision Graphics
17-4: Engineering Mechanics, Virginia Polytechnic Institute and State University
17-5: Frank Awbrey/Visuals Unlimited
17-6: Precision Graphics
17-7: Precision Graphics
17-8: Precision Graphics
17-9: Precision Graphics
17-10: National Park Service photograph by Ruth and Louis Kirk
17-12: Precision Graphics
17-15: Precision Graphics
17-19: John S. Shelton
17-21: Swiss National Tourist Office
17-22: Alan Kesselheim/Mary Pat Ziter, ©JLM Visuals
17-24: Precision Graphics
17-25a-b: Precision Graphics
17-25c: Engineering Mechanics, Virginia Polytechnic Institute and State University
17-27b: John S. Shelton
17-28: Precision Graphics
17-29b: John S. Shelton
17-30: Photo courtesy Canadian Geological Survey
Perspective 17-1, Figure 1: Precision Graphics
Perspective 17-1, Figure 2: P. Weis, U.S. Geological Survey
Perspective 17-1, Figure 3: P. Weis, U.S. Geological Survey
17-31: Precision Graphics
17-32: Precision Graphics
17-33: Precision Graphics. From H.E. Wright and D.G. Frey, eds., *Quaternary Geology of the United States*. Copyright © 1965 by Princeton University Press. Reprinted by permission of Princeton University Press.
Perspective 17-2, Figure 1: Precision Graphics. From Kelley and Farrand, "The Glacial Lakes Around Michigan." *Geological Survey Bulletin* 4 (1967). Michigan Department of Natural Resources.
Perspective 17-2, Figure 2: Precision Graphics. From V.K. Prest, *Geology and Economic Minerals of Canada: Department of Energy, Mines, and Resources Economic Geology Report 1,*

5/e (1970): 90-91, Figure 7-6. Reproduced with the permission of the Minister of Supply and Services Canada, 1991.

17-34: Precision Graphics. From J.T. Andrews, "Earth Science Symposium on Hudson Bay, Ottawa, 1968." *Canadian Geological Surbey Paper 68-53* (1969): 53. Reproduced with the permission of the Minister of Supply and Services Canada, 1991.

17-35: Precision Graphics

CHAPTER 18

Opener: Photo courtesy of Edward Mann
18-1: Precision Graphics
18-2: Steve McCurry/Magmum
18-3: Precision Graphics
18-4: Martin G. Miller/Visuals Unlimited
18-5a: Precision Graphics
Perspective 18-1, Figure 1: Mary A. Dale-Bannister, Washington University, St. Louis
Perspective 18-1, Figure 2: NASA
Perspective 18-1, Figure 3: NASA
18-7: Martin G. Miller/Visuals Unlimited
18-8a-b: Precision Graphics
18-9: Precision Graphics
18-10: Precision Graphics
18-12a: Precision Graphics
18-12b: John S. Shelton
18-13a: Precision Graphics
18-13b: © 1994 CNES. Provided by SPOT Image Corporation
18-14a: Precision Graphics
18-14b: Alan L. and Linda D. Mayo, GeoPhoto Publishing Company.
18-15: Willard Clay/Tony Stone Worldwide
18-16a: Precision Graphics
18-18: Precision Graphics
18-19: Precision Graphics. Based on F.K. Lutgens and E.J. Tarbuck, *The Atmosphere: An Introduction to Meteorology* (Englewood Cliffs, New Jersey: Prentice-Hall, 1979): 150, Figure 7.3.
18-20: Precision Graphics
18-21: Precision Graphics
Perspective 18-2, Figure 1: Precision Graphics. From C.B. Hunt and D.R. Mabey, *U.S. Geological Survey Professional Paper 494A* (1966): A5, Figure 2.
Perspective 18-2, Figure 3: John D. Cunningham/Visuals Unlimited
Perspective 18-2, Figure 4: United States Borax & Chemical Corporation
18-24a-b: John S. Shelton
18-25: Martin G. Miller/Visuals Unlimited
18-26: Alan L. and Linda D. Mayo, GeoPhoto Publishing Company.
18-27a: Precision Graphics

CHAPTER 19

19-1: Rosenberg Library, Galveston, Texas
19-2: Steve Starr/SABA
19-3: Susan Trossbach, University of Virginia
19-4: Precision Graphics
19-5: Precision Graphics
19-7: John S. Shelton
19-8a: Precision Graphics
19-8b: John S. Shelton
19-9b: Michael Slear
19-10: Precision Graphics
Perspective 19-1, Figure 1: Department of the Interior/USGS. EROS Data Center.
Perspective 19-1, Figure 2: Precision Graphics
19-11a: Precision Graphics
19-11b: John S. Shelton
19-12: Precision Graphics
19-13a: Precision Graphics
19-14a: Precision Graphics
19-15: John S. Shelton
19-17: Precision Graphics. From *U.S. Geological Survey Circular 1075.*
19-18: NASA
19-19: Precision Graphics
19-20: Precision Graphics
Perspective 19-2, Figure 1: Precision Graphics. From *U.S. Geological Survey Circular 1075.*
Perspective 19-2, Figure 2: Precision Graphics
Perspective 19-2, Figure 3: U.S. Army Corps of Engineers
19-21a: Precision Graphics
19-22a: Precision Graphics
19-22b: Nick Harvey
19-22c: Suzanne and Nick Geary, Tony Stone Worldwide
19-23: Precision Graphics
19-25: GEOPIC, Earth Satellite Corporation
19-28: Karl Kuhn
19-29: Precision Graphics

CHAPTER 20

Opener: Photo courtesy of Dana Berry
20-1: JPL/NASA
20-2: NASA
20-3: NASA
20-4a: John and Judy Waller
20-4b: Precision Graphics
Perspective 20-1, Figure 1: Precision Graphics
Perspective 20-1, Figure 2: TASS from Sovfoto
20-6a: Precision Graphics
20-6b: Fotosmith
20-6c: Fotosmith

20-6d:	Meteor Crater, Northern Arizona
20-7:	D.J. Roddy, U.S. Geological Survey
20-8:	Paul Dimare
20-10a:	NASA
20-10b:	Victor Royer
Perspective 20-2, Figure 1:	Carlyn Iverson
20-11a:	Finley Holiday Film
20-11b:	NASA
20-11c:	AP/Wide World Photos
20-11d:	Victor Royer
20-12a:	Victor Royer
20-12b:	JPL/NASA
20-13a:	NASA
20-13b:	Victor Royer
20-14b:	Victor Royer
20-15a-b:	JPL/NASA
20-15c:	Victor Royer
20-16:	Victor Royer
20-17:	Victor Royer
20-19:	Lick Observatory
20-20:	Precision Graphics
20-21:	W. Benz and W. Slattery, Los Alamos National Laboratories

CHAPTER 21

Opener:	Photo courtesy Black Star Publishing. © 1990 David Turnley
21-1a:	Precision Graphics
21-1b:	NASA
21-2:	AP/Wide World Photos
21-3:	Precision Graphics
21-4:	Precision Graphics
21-5:	Precision Graphics
21-6:	Precision Graphics
21-7:	Precision Graphics. Data from *Oil and Gas Journal,* v. 90, pp. 44-45 (Dec. 28, 1992).
21-8:	Precision Graphics. Data from *Oil and Gas Journal, v. 90, pp. 44-45 (Dec. 28, 1992).*